Fundamental Interactions

Orlando Luongo

Fundamental Interactions

A Roadmap to Classical and Quantum Field Theories

Springer

Orlando Luongo
Physics Division
School of Science and Technology
University of Camerino
Camerino (MC), Italy

ISBN 978-3-032-27412-0 ISBN 978-3-032-27413-7 (eBook)
https://doi.org/10.1007/978-3-032-27413-7

This Springer imprint is published by the registered company Springer Nature Switzerland AG
The registered company address is: Gewerbestrasse 11, 6330 Cham, Switzerland

To the loving memory of my father, Benito, who was my teacher, guide and inspiration, especially in physics. To my wife, Maryam, for her patience and for always reminding me what truly matters.

Foreword

Fundamental interactions constitute the grammar of Nature. They shape the structure of matter, control the evolution of the cosmos, and determine which phenomena are allowed. Yet, for the student who first approaches the subject in its modern form, the landscape can appear fragmented: symmetries are taught in one language, relativistic dynamics in another, and quantum fields in yet another, often with abrupt transitions and a heavy reliance on "standard" shortcuts that conceal the inner logic. This is precisely where this work, made by Orlando Luongo, succeeds in an exemplary way.

Orlando Luongo has written a remarkable and beautiful book: ambitious in scope, rigorous in its foundations, and pedagogically attentive without ever diluting the conceptual depth. The distinctive merit of this volume is that it treats the subject as an articulated whole. The reader is guided from the language of symmetries and group theory, which is not at the end of the work, but rather at the very beginning. The route starts from special relativity and relativistic field theory, passing through relativistic quantum mechanics and quantum field theory, in a sequence that is logically compelling.

At each step, the narrative emphasizes not only how the formalism works but also why it takes the form it does.

The book reflects the physical interpretation in every new topic that is presented. In this respect, it turns out to be a *unicum* in the literature, since standard books intend to present, generally speaking, rather than comment on the presented physics.

A particularly valuable aspect of the book is the deliberate bridge between classical and quantum viewpoints. The reader is thus prepared to appreciate the conceptual unity behind diverse tools.

The breadth of topics covered is impressive, but breadth alone does not make a textbook exceptional. What makes this work outstanding is its coherence. The chapters are written with a consistent notation, a careful ordering of concepts, and an evident commitment to intellectual honesty: the author does not hide subtleties under the carpet.

Gauge invariance and its physical content are treated with the seriousness that they deserve, and the discussion naturally extends to modern viewpoints in which the electromagnetic interaction serves as a prototype for more general gauge structures. The inclusion of nonlinear electrodynamics, quasiparticles as well as the interplay between field theory and effective-medium interpretations, reflects an informed and contemporary perspective that will be appreciated by graduate students and researchers alike.

Equally noteworthy is the style of exposition. The book does not read as a compilation of results, but as a carefully designed learning path. The proposed exercises, distributed throughout the text, are not ornamental. They are strategically chosen to consolidate key ideas, to train the reader in standard techniques, and to invite deeper reflection on the conceptual architecture of the theory.

I conclude that this volume is splendid and provides an authoritative and modern roadmap through classical and quantum field theories, suitable for advanced Master students, Ph.D. candidates, and postdoctoral researchers, and it will also serve as a reliable reference for anyone who wishes to revisit the main concepts of fundamental physics with fresh eyes.

I am confident that many readers will return to these pages repeatedly, first to learn, then to refine their understanding and ultimately to draw inspiration for their own work.

March 2026

Prof. Hernando Quevedo
UNAM University
Mexico City, Mexico

Preface

Why writing another book on fields and interactions? The intention grew out of a simple ambition: to present the fundamental interactions as a logically-connected framework, rather than as a collection of techniques that are learned in isolation, typically in more than one book.

The subject is often taught along parallel tracks, symmetry and group theory on one side, relativistic dynamics on another and quantum fields on yet another. *The central message of this volume is that these tracks are a single road.*

Physics and physical interpretation are absolutely central and represent a guiding principle throughout the text: *formal developments must remain tied to their physical meaning.*

The material is based on lecture courses I have taught for several years at the University of Camerino. In particular, the book reflects and systematizes my teaching in *Theoretical Physics, Quantum Field Theory, Advanced Quantum Field Theory and Cosmology*. The intended audience includes advanced master students, doctoral students and postdoctoral researchers in physics, astrophysics, mathematics, and related disciplines. The exposition aims to be self contained at the level required by a graduate curriculum, while preserving the conceptual sharpness that is needed for research oriented study.

The selection of topics and the pacing of the presentation are designed so that readers can progress from classical field concepts to quantum fields and effective descriptions without losing the thread that connects them.

The driving exercises consolidate core derivations, sharpening intuition on subtle points and encouraging readers to move fluently between different viewpoints, just *adopting one single book, instead of multiple.*

In several instances, the exercises are also meant to open a door toward active research directions, where the fundamental interactions are explored beyond their simplest textbook limits. All this justifies the fact that the book appears far from being short as it intends to explain very different concepts into one single amazing path.

Additional information, updates and related material can also be found on the website cosmology.unicam.it, where I will regularly report updates and additional material related to this volume.

As for several books and authors, also this book would not exist without the steady presence, encouragement and support of my wife, Maryam Azizinia. Her confidence and generosity supported me along the editorial effort behind this volume and I am deeply grateful to her for being an essential part of this achievement. Using a bit of mathematical humor: if I were a differential equation, she would be my boundary condition, i.e., the one that makes the solution unique.

I am also very indebted to some of my collaborators.

In primis, I am especially grateful to Alessio Belfiglio, my right-hand collaborator, whose careful feedback and attention to detail were invaluable in helping me reach a stylistically accurate level for the manuscript. I am also very thankful to Anna Chiara Alfano, Youri Carloni and Mattia Dubbini for their great support in fixing errors, typos and in reading the work. They all carefully checked parts of the book and provided valuable feedback throughout the writing process. They were essential for my work on this book and remind me that a scientist should never walk alone. *Walking with the right collaborators is the path to success.* Additionally, I thank Jesus Anaya-Galeana, Francesco Lottatori and Aniello Quaranta for contributing to the conceptual development of this volume and for encouraging comments on it.

I am also very grateful to colleagues, Roberto Giambò, Stefano Mancini, Marco Muccino, Hernando Quevedo and Francesco Russo for stimulating discussions on a variety of topics covered in this volume and for reading entire chapters, giving me precious feedback and comments.

I regard each day as an opportunity to learn and to engage in discussions on new concepts and emerging physical ideas with my colleagues, striving to revisit established notions with renewed perspective, reformulating them where necessary and attempting to open the Pandora's box of discovery. Paraphrasing Einstein, *we stand in front of a vast ocean: knowledge, culture, and flashes of insight constitute the foundation allowing us to advance, step by step, toward uncovering the profound beauty of Nature.*

Finally, this book is dedicated to the memory of my father. During the years in which I studied the very subjects that I now have the privilege to teach, his unwavering support guided and sustained me. We are all stardust and, with quiet nostalgia, I like to think that, somewhere in the vastness of the universe, he continues to accompany my journey.

March 2026

Orlando Luongo
Associate Professor of Theoretical
Physics, Physics Division
University of Camerino
Camerino (MC), Italy

Competing Interests The author has no competing interests to declare that are relevant to the content of this manuscript.

Contents

Acronyms

ADM	Arnowitt—Deser—Misner
ATLAS	A Toroidal LHC ApparatuS
BBMB	Bocharova—Bronnikov—Melnikov—Bekenstein
BRST	Becchi—Rouet—Stora—Tyutin
CKM	Cabibbo—Kobayashi—Maskawa
CMS	Compact Muon Solenoid
CP	Charge—parity
CPT	Charge—parity—time reversal
EFT	Effective field theory
EKG	Einstein—Klein—Gordon
FCNC	Flavor-changing neutral current
GIM	Glashow—Iliopoulos—Maiani
GR	General relativity
IR	Infrared
JNW	Janis—Newman—Winicour
LSZ	Lehmann—Symanzik—Zimmermann
MTZ	Martinez—Troncoso—Zanelli
QCD	Quantum chromodynamics
QED	Quantum electrodynamics
QFT	Quantum field theory
QGP	Quark—gluon plasma
RG	Renormalization group
SM	Standard Model
UV	Ultraviolet
WKB	Wentzel—Kramers—Brillouin

1 The Language of Symmetries

God Always Geometrizes.
—Plato

Abstract

Modern physics is deeply intertwined with geometry and symmetry. While classical physics often relied on intuitive or phenomenological arguments, the formulation of physical laws in a mathematically precise language has become an essential ingredient of contemporary theoretical physics. In this context, group theory provides the natural framework for encoding symmetry principles and for organizing the fundamental interactions of nature. In many textbooks, group-theoretical methods are confined to appendices or presented only in a highly compressed form. In contrast, the present chapter treats group theory as a central component of the theoretical framework, rather than as a technical addendum. The basic concepts of groups, representations, and continuous symmetries are introduced in a systematic and pedagogical manner, with emphasis on their physical interpretation. The discussion begins with general symmetry transformations and abstract group structures, and then proceeds to Lie groups and Lie algebras, highlighting the role of generators, invariants, and global properties. Concrete examples drawn from classical physics are used to illustrate the formalism and to prepare the ground for later applications. In subsequent chapters, these tools will be employed to describe the symmetries underlying modern fundamental interactions.

O. Luongo, *Fundamental Interactions*,
https://doi.org/10.1007/978-3-032-27413-7_1

Fig. 1.1 Portrait of Sophus Lie by L. Szaciński, 1896. Reprinted from Wikimedia. Public domain

A key character: *Sophus Lie (1842–1899), Fig.* 1.1.

Sophus Lie was born into a Lutheran family as the son of a pastor and the sixth of seven children. Lie was unable to pursue a military career because of serious eyesight problems. He therefore enrolled at the University of Christiania (now Oslo), where he attended courses in mathematics, astronomy, mechanics, botany, zoology, and physics, without initially committing to a specific subject for his degree.

Over time, his interests increasingly focused on geometry. He published his first paper on imaginary numbers, which earned him a scholarship and enabled him to travel extensively across Europe.

In 1870 Lie traveled to Paris together with Klein, where he met Gaston Darboux and Camille Jordan. These encounters marked the beginning of his systematic work on the theory of transformation groups.

In 1886 Lie was appointed professor of mathematics at the University of Leipzig, succeeding Klein. Despite his scientific success, his integration into the Prussian academic environment proved difficult and contributed to periods of severe depression. After receiving the Lobachevsky Prize for his mathematical achievements, and already in poor health, he left Leipzig to accept a chair in Christiania. Shortly after returning to Norway, Lie died of pernicious anemia.

Last but not least, up to now the *Sophus Lie Seminars* are famous advanced research seminars[1] dedicated to Lie symmetries and their applications in geometry, general relativity, and quantum field theory (QFT). Moreover, a scientific community associated with Sophus Lie still exists. It is primarily active in Norway and focuses on group theory, geometry, differential equations, and physics. Key hubs include the Sophus Lie Conference Center in Nordfjordeid and the Lie-Størmer Center in Tromsø.

1.1 Symmetries as Transformations

In physics[2], a *symmetry* is a transformation that leaves the physical content of a system unchanged. More precisely, a symmetry is a map

$$\mathcal{T} : \mathcal{S} \longrightarrow \mathcal{S}, \tag{1.1}$$

where $\mathcal{S}$ denotes the space of states (or configurations), such that all physically observable quantities remain invariant under the transformation $\mathcal{T}$.

Invariance does not mean that the system is unchanged at the descriptive level. Rather, it means that different descriptions related by $\mathcal{T}$ correspond to the same physical situation.

▶ **Pause and think** If a transformation changes the mathematical description, in which sense can it still be a symmetry? Try to formulate one example where the description changes but all measurable predictions do not.

[1] See, e.g., www.mat.uniroma1.it/it/sophus-lie-seminar.

[2] The main purpose of this book is to deal genuinely with physics, i.e., from Greek *physis*, namely *Nature*. Hence, for our purposes, the princess of science, pure mathematics, will be used with parsimony. Thus, we will reduce the formal definitions of mathematics, mainly focusing on physical concepts and on the beauty of Nature, formulated in mathematical language.

1.1.1 Symmetries in Classical Mechanics

In classical mechanics the state space is given by generalized coordinates and momenta, $\mathcal{S} = \{q_i(t),\ p_i(t)\}$. A transformation is a symmetry if it leaves the action,

$$S[q] = \int_{t_1}^{t_2} L(q, \dot{q}, t)\, dt, \tag{1.2}$$

invariant.

For non-relativistic conservative systems the Lagrangian, $L(q, \dot{q}, t)$, takes the form

$$L(q, \dot{q}, t) = T - V, \tag{1.3}$$

where T denotes the kinetic energy and V the potential energy. The structure in Eq. (1.3) is not an arbitrary assumption, but follows from the requirement that the equations of motion reproduce *Newtonian dynamics* while preserving the fundamental symmetries of space and time.
Indeed, substituting Eq. (1.3) into the Euler-Lagrange equations,

$$\frac{d}{dt}\left(\frac{\partial L(q, \dot{q}, t)}{\partial \dot{q}_i}\right) - \frac{\partial L(q, \dot{q}, t)}{\partial q_i} = 0, \tag{1.4}$$

one obtains

$$\frac{d}{dt}\left(\frac{\partial T}{\partial \dot{q}_i}\right) = -\frac{\partial V}{\partial q_i}, \tag{1.5}$$

which is equivalent to Newton's second law for forces derived from a scalar potential. **The minus sign in Eq.** (1.5) **ensures that forces are generated as the negative gradient of the potential, in agreement with physical observations**. Alternative choices, such as $L(q, \dot{q}, t) = T + V$, do not reproduce the correct dynamical equations.

Explicitly, a transformation, $q_i \longrightarrow q_i' = q_i + \delta q_i$, $t \longrightarrow t' = t + \delta t$, is a symmetry if

$$S[q'] = S[q]. \tag{1.6}$$

Fundamental examples include:

- time translations: $t \to t + \varepsilon$, with $\varepsilon > 0$,
- spatial translations: $\mathbf{x} \to \mathbf{x} + \mathbf{a}$,
- rotations: $\mathbf{x} \to R\,\mathbf{x}$, with rotation matrix R.

The use of the variational principle is physically motivated by the fact that the action provides a global, not local, description of the system's evolution. For systems whose dynamics is invariant under continuous transformations of space and time, the action represents the quantity that remains well-defined under arbitrary changes of generalized coordinates.
Requiring the action to be stationary does not imply that the system minimizes or maximizes a physical quantity. Instead, the condition

$$\delta S = 0, \tag{1.7}$$

derived from Eq. (1.6), selects those trajectories for which infinitesimal deformations of the entire path do not lead to a first order change and, physically, this expresses the fact that the realized motion is stable under small variations of the history of the system, *once the initial and final configurations are fixed*, that is the reason for which $\delta q_i(t_1) = \delta q_i(t_2) = 0$.

▶ **Pause and think** For each of the three examples above, try to anticipate what should be conserved *before* invoking Noether's theorem.

1.1.2 Symmetries in Quantum Mechanics

In quantum mechanics the space of the states is a complex Hilbert space $\mathcal{H}$. A symmetry is a transformation that preserves transition probabilities,

$$|\langle\psi|\phi\rangle|^2 = |\langle\psi'|\phi'\rangle|^2. \tag{1.8}$$

Wigner's theorem states that any symmetry transformation is represented by either

- a unitary operator, or
- an antiunitary operator.

For continuous symmetries,

$$U(\varepsilon) = \exp\left(-\frac{i}{\hbar}\varepsilon G\right), \tag{1.9}$$

where G is the generator of the transformation. If the Hamiltonian, H, i.e., the Legendre transformation of the Lagrangian, is invariant,

$$HG - GH \equiv [H, G] = 0, \tag{1.10}$$

then the observable associated with G is conserved in time.

Pause and think Knowing how a symmetry acts on states, how would you identify its generator G? What property of $U(\varepsilon)$ makes G observable in quantum mechanics?

1.1.3 Discrete Symmetries

Not all symmetries are continuous. Important discrete symmetries include

$$P : \mathbf{x} \to -\mathbf{x} \quad \text{(parity)}, \tag{1.11}$$

$$T : t \to -t \quad \text{(time reversal)}, \tag{1.12}$$

$$C : \text{particles} \leftrightarrow \text{antiparticles}. \tag{1.13}$$

These transformations do not admit infinitesimal generators and therefore do not lead to Noether currents. Nevertheless, they play a crucial role in the structure of fundamental interactions.

We will discuss their role, later in the book.

1.1.4 Symmetries in Field Theory

In field theories, symmetries are realized as transformations acting on the spacetime coordinates and/or on the fields themselves. For a scalar field $\phi(x)$, a generic symmetry transformation can be written as

$$x \longrightarrow x', \qquad \phi(x) \longrightarrow \phi'(x'). \tag{1.14}$$

A global symmetry is defined as a transformation of the fields such that the action remains invariant. At the level of the Lagrangian density this implies

$$\mathcal{L}(\phi', \partial\phi') = \mathcal{L}(\phi, \partial\phi) + \partial_\mu K^\mu. \tag{1.15}$$

Promoting a global symmetry to a local one requires the introduction of gauge fields, in order to preserve invariance under spacetime-dependent transformations. This is achieved by replacing ordinary derivatives with covariant derivatives,

$$D_\mu = \partial_\mu + igA_\mu, \tag{1.16}$$

where A_μ is the gauge field associated with the symmetry and g a coupling constant.

A precise analogy with gauge symmetry can be drawn from fluid mechanics, where the same physical system admits both an Eulerian and a Lagrangian description. In the Eulerian formulation, the fluid is described in terms of spacetime fields, such as the velocity field $\mathbf{v}(\mathbf{x}, t)$ and the density $\rho(\mathbf{x}, t)$, defined at fixed spatial points. The time evolution of a scalar quantity $\psi(\mathbf{x}, t)$ carried by the fluid is governed by the material (or convective) derivative

$$\frac{D\psi}{Dt} \equiv \partial_t \psi + \mathbf{v} \cdot \nabla \psi, \tag{1.17}$$

which accounts for both explicit time dependence and transport by the flow. In contrast, the Lagrangian description follows individual fluid elements, whose trajectories $\mathbf{X}(\mathbf{a}, t)$ are labeled by initial coordinates $\mathbf{a}$. Physical observables are then functions of these labels and time, and the dynamics is invariant under time-dependent relabelings

$$\mathbf{a} \longrightarrow \mathbf{a}'(\mathbf{a}, t), \tag{1.18}$$

leaving the physical configuration of the fluid unchanged. This invariance reflects a redundancy in the description, rather than a physical symmetry.
The passage from the Lagrangian to the Eulerian formulation requires the introduction of the velocity field

$$\mathbf{v}(\mathbf{x}, t) = \left. \frac{d\mathbf{X}(\mathbf{a}, t)}{dt} \right|_{\mathbf{x}=\mathbf{X}(\mathbf{a},t)}, \tag{1.19}$$

which plays a role analogous to that of a connection. Indeed, the convective derivative can be interpreted as a covariant time derivative ensuring invariance under time-dependent relabelings of fluid elements.

In this sense, interactions emerge as a direct consequence of enforcing local symmetry principles.

▶ **Pause and think** If you replace a constant group parameter by a spacetime function, what term do you expect to appear when differentiating a transformed field? How might a new field compensate it?

1.2 Basics of Group Theory

A *group* G is a set that holds a binary operation $\cdot$ that associates to every ordered pair $(g, h) \in G \times G$ an element $g \cdot h \in G$, guaranteeing the following properties,

- **Closure**: if $g, h \in G$, then $g \cdot h \in G$.
- **Associativity**: for all $f, g, h \in G$, $(f \cdot g) \cdot h = f \cdot (g \cdot h)$.
- **Identity**: there exists $e \in G$ such that $g \cdot e = e \cdot g = g$ for all $g \in G$.
- **Inverse**: for every $g \in G$ there exists $g^{-1} \in G$ such that $g \cdot g^{-1} = g^{-1} \cdot g = e$.

A group is *finite* if it contains finitely many elements and its cardinality is dubbed *group order*.

An immediate example may be given by the cyclic group $\mathbb{Z}_3$, where a finite group of order 3 can be represented by $\mathbb{Z}_3 = \{e, a, b\}$ with a Cayley table,[3]

$\cdot$	e	a	b
e	e	a	b
a	a	b	e
b	b	e	a

This group is ***Abelian***, i.e. $g \cdot h = h \cdot g$ for all $g, h \in G$ and translates the meaning of group whose elements commute.

Abelian groups play a fundamental role in physics. For example, in quantum mechanics, physical symmetries are implemented by (projective) unitary representations of groups acting on a Hilbert space.

As a consequence, generators of Abelian symmetries commute, leading to simultaneously diagonalizable observables. A paradigmatic example is the Abelian group of spatial translations $\mathbb{R}^n$, whose unitary representations yield momentum operators via Stone's theorem,[4]

$$U(a) = e^{-ia \cdot \hat{P}}, \tag{1.21}$$

with commuting components, $[\hat{P}_i, \hat{P}_j] = 0$.

More generally, Abelian groups control conserved charges and Fourier decompositions of fields.

▶ **Pause and think** If two symmetry generators commute, what does that suggest about the possibility of measuring the associated observables simultaneously? How does this connect to choosing a basis of states?

---

[3] A *Cayley table* is a compact way to represent, for instance, a group. More precisely, given a set S with a binary operation $* : S \times S \to S$, the Cayley table lists the result of $a * b$ for all ordered pairs $(a, b) \in S \times S$, typically arranging the elements of S along the rows and columns.

[4] Let $\mathcal{H}$ be a complex Hilbert space and let$\{U(t)\}_{t\in\mathbb{R}}$ be a strongly continuous one-parameter family of unitary operators on $\mathcal{H}$, i.e.

$$U(t+s) = U(t)U(s), \qquad U(0) = \mathbb{I}, \qquad \lim_{t\to 0} U(t)\psi = \psi \ \ \forall \psi \in \mathcal{H}. \tag{1.20}$$

Stone's theorem states that there exists a *unique self-adjoint operator* P on $\mathcal{H}$ such that $U(t) = e^{-itP}$, where the exponential is defined via the spectral theorem. Conversely, every self-adjoint operator P generates a strongly continuous one-parameter unitary group through the above exponential map.

1.2.1 Group Representations

A linear *representation* of a group G is a map D that associates each element $g \in G$ with an invertible linear operator $D(g)$ acting on a vector space V,

$$g \longmapsto D(g), \tag{1.22}$$

such that

- $D(e) = \mathbb{1}$, the identity operator on V,
- $D(g \cdot h) = D(g)\, D(h)$ for all $g, h \in G$.

If $\dim V = n < \infty$, then each $D(g)$ can be represented by an invertible $n \times n$ matrix $[D(g)]_{ij}$, with $i, j = 1, \ldots, n$; this is a *matrix representation* and n is the *dimension* of the representation.

Intuitive examples are given below. A one-dimensional representation on $V = \mathbb{C}$ is

$$D(e) = 1, \qquad D(a) = e^{2\pi i/3}, \qquad D(b) = e^{4\pi i/3}. \tag{1.23}$$

On the other hand, a three-dimensional representation on $V = \mathbb{R}^3$ is given by

$$D(e) = \begin{pmatrix} 1 & 0 & 0 \\ 0 & 1 & 0 \\ 0 & 0 & 1 \end{pmatrix}, \qquad D(a) = \begin{pmatrix} 0 & 0 & 1 \\ 1 & 0 & 0 \\ 0 & 1 & 0 \end{pmatrix}, \qquad D(b) = \begin{pmatrix} 0 & 1 & 0 \\ 0 & 0 & 1 \\ 1 & 0 & 0 \end{pmatrix}. \tag{1.24}$$

This is named *regular representation*, as *its dimension equals the order of the group*.

1.2.2 Equivalence and Unitarity

Two representations D and D' are *equivalent* if there exists an invertible linear map $S : V \to V$ (a change of basis in the carrier space), independent of g, such that

$$D'(g) = S^{-1} D(g) S, \qquad \forall\, g \in G. \tag{1.25}$$

A representation is *unitary* if each $D(g)$ is unitary, i.e. $D(g)^\dagger = D(g)^{-1}$. For finite groups, one can always choose an equivalent representation that is unitary.

1.2.3 Invariant Subspaces and reducibility

A subspace $W \subset V$ is *invariant* if $D(g)w \in W$ for all $g \in G$ and all $w \in W$, with $W \neq \{0\}$ and $W \neq V$. A representation is:

- *reducible* if it admits a non-trivial invariant subspace,
- *irreducible* otherwise.

A representation is *completely reducible* if it is equivalent to one in block-diagonal form

$$D(g) = \begin{pmatrix} D_1(g) & 0 & \cdots \\ 0 & D_2(g) & \cdots \\ \vdots & \vdots & \ddots \end{pmatrix}, \tag{1.26}$$

where each block $D_j(g)$ is irreducible. In that case one writes

$$D \simeq D_1 \oplus D_2 \oplus \cdots . \tag{1.27}$$

▶ **Driving exercise**
For the matrices in Eq. (1.24), consider the similarity transformation

$$S = \frac{1}{3}\begin{pmatrix} 1 & 1 & 1 \\ 1 & \omega^2 & \omega \\ 1 & \omega & \omega^2 \end{pmatrix}, \qquad \omega \equiv e^{2\pi i/3}, \tag{1.28}$$

which yields an equivalent representation with

$$D'(e) = \mathbb{1}, \qquad D'(a) = \begin{pmatrix} 1 & 0 & 0 \\ 0 & \omega & 0 \\ 0 & 0 & \omega^2 \end{pmatrix}, \qquad D'(b) = \begin{pmatrix} 1 & 0 & 0 \\ 0 & \omega^2 & 0 \\ 0 & 0 & \omega \end{pmatrix}, \tag{1.29}$$

so the regular representation is indeed completely reducible. This was expected: all finite-group representations are completely reducible.

Groups share immediately an intriguing property. In physics, often one deals with tensors and irreducible representations.

A tensor usually transforms according to a *reducible representation* of the symmetry group, while *its properly projected components transform irreducibly*.

For this reason, when one speaks of an *irreducible tensor*, one actually refers to a tensorial object belonging to an invariant subspace that carries an irreducible representation of the underlying symmetry group.

Let us see it in detail and take G be a group acting linearly on a finite-dimensional vector space V through a representation $\rho : G \to \mathrm{GL}(V)$.

From this fundamental representation, one can naturally construct higher-rank tensor spaces of the form $V^{\otimes p} \otimes (V^*)^{\otimes q}$, where V^* denotes the dual space of V, giving rise to tensors of type (p, q).

Now, if $g \in G$ acts on V as $\rho(g)$, then its action on a tensor

$$T \in V^{\otimes p} \otimes (V^*)^{\otimes q} \tag{1.30}$$

is given componentwise by

$$T^{i_1\ldots i_p}{}_{j_1\ldots j_q} \longrightarrow \rho(g)^{i_1}{}_{k_1}\cdots\rho(g)^{i_p}{}_{k_p}\left(\rho(g)^{-1}\right)^{\ell_1}{}_{j_1}\cdots\left(\rho(g)^{-1}\right)^{\ell_q}{}_{j_q}\,T^{k_1\ldots k_p}{}_{\ell_1\ldots\ell_q}\,. \tag{1.31}$$

In general, the representation of G carried by a tensor space $V^{\otimes p}\otimes(V^*)^{\otimes q}$ is *not irreducible*. Indeed, such spaces typically contain non-trivial invariant subspaces, meaning that they can be decomposed into a direct sum of smaller subspaces, each of which is stable under the action of the group. These invariant subspaces correspond precisely to irreducible representations of G.

▶ **Pause and think** Given a tensor space that is reducible, what kind of "projection" would you look for to isolate invariant subspaces? Think in terms of symmetrization, antisymmetrization, and trace removal.

▶ **Driving exercise** As a robust example, let us take a rank–2 tensor (as later we will see, related to $SO(3)$ symmetry), T_{ij}, with indices $i, j = 1, 2, 3$. This representation is reducible and admits the following canonical decomposition:

$$T_{ij} = \underbrace{\frac{1}{3}\,\delta_{ij}\,T_{kk}}_{\text{scalar (irreducible)}} + \underbrace{\left(T_{(ij)} - \frac{1}{3}\,\delta_{ij}\,T_{kk}\right)}_{\text{symmetric traceless tensor (irreducible)}} + \underbrace{T_{[ij]}}_{\text{antisymmetric tensor (irreducible)}}\,, \tag{1.32}$$

where

$$T_{(ij)} = \frac{1}{2}(T_{ij} + T_{ji}), \tag{1.33a}$$

$$T_{[ij]} = \frac{1}{2}(T_{ij} - T_{ji}), \tag{1.33b}$$

denote the symmetric and antisymmetric parts of the tensor, respectively. Each term on the right-hand side transforms according to an irreducible representation, namely

- the trace part transforms as a scalar ($\ell = 0$),
- the symmetric traceless part corresponds to a spin-2 representation ($\ell = 2$),
- the antisymmetric part is dual to a vector and corresponds to a spin-1 representation ($\ell = 1$).

The concept of spin will be faced later on in the text, in the framework of special relativity and field theories. As above stated, however, its foundations are classical and may be recovered directly in group theory.

At this stage, the discussion has been entirely algebraic: we have introduced groups and their representations without making any reference to continuity, topology, or geometry. These additional structures become essential as soon as one moves from abstract symmetry principles to physically relevant symmetry groups.

1.3 Lie Groups

An important class of groups is represented by the *Lie groups*. For that, the elements depend on a set of continuous parameters α^a, with $a = 1, \dots, N$.

Hence, a Lie group preserves the aforementioned abstract definition for groups and includes an additional smooth manifold structure compatible with the group operations.

We could generically state: *a Lie group is a group endowed with an additional differentiable structure.*

Writing a generic element as $g(\alpha)$, one typically chooses coordinates so that the identity corresponds to $\alpha^a = 0$, i.e. $g(0) = e$. In a representation D, the matrices $D(g(\alpha))$ inherit a smooth dependence on α and so, expanding at first order, one obtains

$$D(\alpha) = \mathbb{1} + i\alpha^a T_a + O(\alpha^2), \tag{1.34}$$

where the operators

$$T_a \equiv -i \left. \frac{\partial D(\alpha)}{\partial \alpha^a} \right|_{\alpha=0} \tag{1.35}$$

are the *generators* of the group in that representation. The factor of i is chosen so that, in a unitary representation, the generators are Hermitian. One may equivalently write the group elements *near the identity* as

$$D(\alpha) = e^{i\alpha^a T_a}. \tag{1.36}$$

Generators are thus linear and easy to handle, while the entire group may lead to complicate properties. As it will appear clear later, *symmetries are defined by groups, but the underlying physics may be found by investigating the corresponding generators.*

▶ **Pause and think Where does non-commutativity enter?** If the generators all commuted, what would you expect for $\gamma^c(\alpha, \beta)$ beyond first order? Which term must therefore detect non-Abelian structure?

Driving exercise Now, consider two group elements, $g_1 = g(\alpha)$ and $g_2 = g(\beta)$, expanded around the identity as before, then having by hypothesis,

$$D(g_1)D(g_2) = e^{i\alpha^a T_a} e^{i\beta^b T_b} = D(g_1 g_2) = e^{i\gamma^c(\alpha,\beta)\, T_c}, \tag{1.37}$$

for some parameters $\gamma^c(\alpha, \beta)$.
Recall now the Taylor expansion $e^X = \mathbb{I} + X + \frac{1}{2}X^2 + O(X^3)$ with

$$X = i\alpha^a T_a, \qquad Y = i\beta^b T_b.$$

Then

$$\begin{aligned} e^X e^Y &= \left(\mathbb{I} + X + \tfrac{1}{2}X^2 + O(\alpha^3)\right)\left(\mathbb{I} + Y + \tfrac{1}{2}Y^2 + O(\beta^3)\right) \\ &= \mathbb{I} + (X + Y) + \tfrac{1}{2}(X^2 + Y^2) + XY + O(3), \end{aligned} \tag{1.38}$$

where $O(3)$ denotes terms cubic (or higher) in (α, β), e.g. $\alpha^2\beta$, $\alpha\beta^2$, etc.
Afterwards, substitute $X = i\alpha^a T_a$ and $Y = i\beta^b T_b$:

$$X + Y = i(\alpha^a + \beta^a)T_a, \tag{1.39}$$

$$\tfrac{1}{2}X^2 = \tfrac{1}{2}(i\alpha^a T_a)(i\alpha^b T_b) = -\tfrac{1}{2}\alpha^a\alpha^b\, T_a T_b, \tag{1.40}$$

$$\tfrac{1}{2}Y^2 = -\tfrac{1}{2}\beta^a\beta^b\, T_a T_b, \tag{1.41}$$

$$XY = (i\alpha^a T_a)(i\beta^b T_b) = -\,\alpha^a\beta^b\, T_a T_b. \tag{1.42}$$

Hence, up to second order,

$$e^{i\alpha^a T_a}\, e^{i\beta^b T_b} = \mathbb{I} + i(\alpha^a + \beta^a)T_a - \frac{1}{2}\alpha^a\alpha^b T_a T_b - \frac{1}{2}\beta^a\beta^b T_a T_b - \alpha^a\beta^b T_a T_b + O(3). \tag{1.43}$$

Let us expand the right-hand side to second order and similarly one obtains

$$e^{i\gamma^c T_c} = \mathbb{I} + i\gamma^c T_c - \frac{1}{2}\gamma^c\gamma^d\, T_c T_d + O(3). \tag{1.44}$$

At first order, comparing Eqs. (1.43) and (1.44) gives

$$\gamma^a(\alpha, \beta) = \alpha^a + \beta^a + O(2). \tag{1.45}$$

Therefore we may write, to second order,

$$\gamma^a(\alpha, \beta) = \alpha^a + \beta^a + \gamma^a_{(2)}(\alpha, \beta) + O(3), \tag{1.46}$$

where $\gamma^a_{(2)}$ collects all second-order terms, and in particular it must contain terms of the type $\alpha\beta$.
The crucial information comes from the *mixed* terms proportional to $\alpha^a\beta^b$.
Hence, since the mixed term is $-\alpha^a\beta^b\, T_a T_b$, from the right-hand side of Eq. (1.44), using Eq. (1.46), the mixed terms come from two places:

- the linear term $i\gamma^c T_c$ contributes

$$i\,\gamma^c_{(2)}(\alpha,\beta)\,T_c \quad \text{(this can contain } \alpha\beta\text{)},$$

- the quadratic term $-\frac{1}{2}\gamma^c\gamma^d T_c T_d$ with $\gamma = \alpha + \beta + O(2)$ contributes

$$-\frac{1}{2}(\alpha^c + \beta^c)(\alpha^d + \beta^d)\,T_c T_d \supset -\frac{1}{2}\left(\alpha^c\beta^d + \beta^c\alpha^d\right)T_c T_d.$$

Thus the mixed part of the right-hand side is

$$-\frac{1}{2}\left(\alpha^c\beta^d + \beta^c\alpha^d\right)T_c T_d \;+\; i\,\gamma^e_{(2)}(\alpha,\beta)\,T_e. \tag{1.47}$$

Comparing the mixed terms, it is possible to extract a commutator and, in fact, equating mixed terms from left and right yields

$$-\,\alpha^a\beta^b\,T_a T_b = -\frac{1}{2}\left(\alpha^a\beta^b + \beta^a\alpha^b\right)T_a T_b \;+\; i\,\gamma^c_{(2)}(\alpha,\beta)\,T_c. \tag{1.48}$$

Move the symmetric piece to the left.The difference between $T_a T_b$ and $T_b T_a$ appears:

$$\begin{aligned}
-\,\alpha^a\beta^b\,T_a T_b + \frac{1}{2}\left(\alpha^a\beta^b + \beta^a\alpha^b\right)T_a T_b &= -\frac{1}{2}\alpha^a\beta^b\,T_a T_b + \frac{1}{2}\beta^a\alpha^b\,T_a T_b \\
&= \frac{1}{2}\alpha^a\beta^b\,(T_b T_a - T_a T_b) \\
&= -\frac{1}{2}\alpha^a\beta^b\,[T_a, T_b].
\end{aligned} \tag{1.49}$$

Therefore, Eq. (1.48) becomes

$$-\frac{1}{2}\alpha^a\beta^b\,[T_a, T_b] = i\,\gamma^c_{(2)}(\alpha,\beta)\,T_c. \tag{1.50}$$

The commutator should be a linear combination of generators and the left-hand side of Eq. (1.50) is proportional to $\alpha^a\beta^b$ times $[T_a, T_b]$, whereas the right-hand side is proportional to $\alpha^a\beta^b$ times a *linear combination of* T_c. Since Eq. (1.50) must hold for arbitrary small parameters α^a, β^b, it follows that $[T_a, T_b]$ turns out to give

$$[T_a, T_b] = i f_{ab}{}^c\,T_c, \tag{1.51}$$

where the numbers $f_{ab}{}^c$ are called the *structure constants* of the Lie algebra associated with the representation.

By definition of the commutator, $[T_a, T_b] = -([T_b, T_a])$, and, so, using Eq. (1.51) for both $[T_a, T_b]$ and $[T_b, T_a]$ provides

$$i f_{ab}{}^{c} T_c = -i f_{ba}{}^{c} T_c.$$

Assuming the generators $\{T_c\}$ are linearly independent, we conclude $f_{ab}{}^{c} = -f_{ba}{}^{c}$.

The overall result is a second-order computation of the *Baker-Campbell-Hausdorff formula*, by setting $X = i\alpha^a T_a$ and $Y = i\beta^b T_b$.

This viewpoint is particularly useful in physics: commutators encode the deviation from an Abelian composition law and therefore control the structure constants, selection rules, and (in quantum theory) the algebra of conserved charges.

1.3.1 The Jacobi Identity and Adjoint Representation

The generators necessarily obey the Jacobi identity,

$$[T_a, [T_b, T_c]] + [T_b, [T_c, T_a]] + [T_c, [T_a, T_b]] = 0, \tag{1.52}$$

which translates into a cyclic constraint on the structure constants:

$$f_{bc}{}^{d} f_{ad}{}^{e} + f_{ca}{}^{d} f_{bd}{}^{e} + f_{ab}{}^{d} f_{cd}{}^{e} = 0. \tag{1.53}$$

► **Driving exercise**
Moreover, the structure constants define a representation, the *adjoint representation*, whose generator matrices are

$$(T_a)_{bc} \equiv -i f_{ab}{}^{c}. \tag{1.54}$$

Using the Jacobi identity one checks that these matrices satisfy the same algebra,

$$[T_a, T_b] = i f_{ab}{}^{c} T_c, \tag{1.55}$$

so they provide a faithful realization of the Lie algebra.

A Lie group which is compact as a manifold (e.g. spatial rotations) is called a *compact Lie group*. The Lorentz group is instead non-compact.

For compact Lie algebras, one can choose a basis of generators such that

$$\mathrm{Tr}(T_a T_b) = \kappa\, \delta_{ab}, \qquad \kappa > 0. \tag{1.56}$$

In such a basis the structure constants can be written as

$$f_{abc} \propto -i \ \mathrm{Tr}([T_a, T_b]T_c)\,, \tag{1.57}$$

and using cyclicity of the trace one shows that, for compact groups, the f_{abc} are totally antisymmetric:

$$f_{abc} = f_{bca} = f_{cab} = -f_{bac} = -f_{acb} = -f_{cba}. \tag{1.58}$$

1.3.2 Lie Algebras

It appears clear that Lie groups and algebras are intertwined.

Many symmetry groups of physical relevance are continuous and can be parametrized by real variables. Such groups are naturally described within the framework of Lie groups, which provide the appropriate mathematical language for treating continuous symmetries in both classical and quantum physics.
Why is this possible? Easily, a Lie group G is a group in the abstract sense introduced in Sect. 1.3, endowed with the additional structure of a smooth manifold, such that the group multiplication map $G \times G \to G$ and the inversion map $G \to G$ are smooth. This additional structure allows one to define infinitesimal transformations and to relate global group elements to local generators.
Associated with any Lie group G is a **Lie algebra**, $\mathfrak{g}$, defined as the *tangent space* to the group manifold at the identity element. Elements of $\mathfrak{g}$ generate one-parameter subgroups of G and may be represented, in a given representation, by matrices T^a. The Lie algebra is characterized by the commutation relations

$$[T^a, T^b] = i f^{ab}{}_c T^c, \tag{1.59}$$

where the real constants f^{ab}_c encode the local structure of the group.
For matrix Lie groups, the relation between the algebra and the group is made explicit by the exponential map. A group element sufficiently close to the identity can be written as

$$U(\epsilon) = \exp\bigl(i\,\epsilon_a T^a\bigr)\,, \tag{1.60}$$

where the real parameters ϵ_a label the infinitesimal transformation. In general, the exponential map provides a local diffeomorphism between a neighborhood of the origin in $\mathfrak{g}$ and a neighborhood of the identity in G, but it need not be globally surjective.

The commutation relations of the generators determine the local structure of the group, while global properties such as connectedness and simple connectedness depend on the topology of the underlying manifold. As a consequence, distinct Lie groups may share the same Lie algebra while differing in their global properties, a fact that has profound physical implications.

A paradigmatic example is provided by SU(2) and SO(3), which possess isomorphic Lie algebras but differ in their global topology, leading to the existence of spinorial representations in quantum mechanics.

1.3.3 Invariants and Casimir Operators

How can one distinguish groups in an intrinsic way, without referring to the dimension or to a particular choice of basis?

A powerful way to label irreducible representations is through *invariants*. A *Casimir operator* C is built from the generators and commutes with all of them,

$$[C, T_a] = 0, \qquad \forall a. \tag{1.61}$$

Hence, C does not see any group transformation and does not mix states that the group can distinguish.

Accordingly, we can invoke the Schur's lemma, that states: *any linear operator C acting on V and commuting with all generators of the representation must be proportional to the identity*, say $[C, T_a] = 0, \forall a, g \implies C = \lambda \mathbb{I}$, where λ is a complex, or real, in the unitary case, that

- depends only on the representation R,
- is independent of the particular state within the representation.

Last but not least, given two representations D_1 and D_2 on spaces V_1 and V_2, their tensor product acts on $V_1 \otimes V_2$ by the natural factor-wise action $D_{1\otimes 2}(g) = D_1(g) \otimes D_2(g)$. Tensor products can be decomposed into direct sums of irreducible representations; an explicit and important example is provided by SU(2).

▶ **Pause and think** If an operator commutes with all generators, why can it not mix states inside an irreducible representation? What does this suggest about using its eigenvalue to label the representation?

1.3.3.1 Covering Groups and the Relation SU(2) → SO(3)

The groups SO(3) and SU(2) share the *same Lie algebra*,

$$\mathfrak{so}(3) \simeq \mathfrak{su}(2), \tag{1.62}$$

yet they are globally different Lie groups. This distinction is invisible at the level of commutators but becomes decisive for quantum theory.

There exists a smooth surjective homomorphism

$$\pi : \mathrm{SU}(2) \longrightarrow \mathrm{SO}(3), \tag{1.63}$$

whose kernel is the two-element subgroup $\{\pm\mathbb{1}\}$. Therefore,

$$\mathrm{SO}(3) \simeq \mathrm{SU}(2)/\mathbb{Z}_2. \tag{1.64}$$

Equivalently, SU(2) is the universal covering group of SO(3).

> Representations of $\mathfrak{su}(2)$ always exist for spins $j = 0, \frac{1}{2}, 1, \frac{3}{2}, \ldots$, but only those with *integer* j descend to honest representations of SO(3). Half-integer representations are single-valued only on SU(2): this is the group-theoretic origin of spinors and of the familiar 2π versus 4π rotation behavior in quantum mechanics.

1.3.3.2 Curves, Connectedness, and Simple Connectedness

So far, the discussion has been purely local: generators describe infinitesimal transformations and the Lie algebra captures the structure of the group in a neighborhood of the identity. To understand how this local information extends to the full group, one must take into account the global and topological properties of Lie groups.

A Lie group G is not only a group but also a smooth manifold. This allows one to define continuous curves in G, namely maps

$$\gamma : \mathbb{R} \to G, \qquad t \mapsto \gamma(t), \tag{1.65}$$

with $\gamma(t) \in G$ for all t. Two elements $g_1, g_2 \in G$ are said to be *connected* if there exists a curve $\gamma(t)$ such that $\gamma(t_1) = g_1$ and $\gamma(t_2) = g_2$ for some t_1, t_2.

> The group G is called *connected* if every pair of elements can be joined by a continuous curve. This distinction is physically relevant: transformations that belong to disconnected components cannot be reached continuously from the identity.

A connected Lie group is said to be *simply connected* if every closed curve in G can be continuously contracted to a point. Simple connectedness controls whether local (infinitesimal) data are sufficient to reconstruct global properties of the group, a fact that will play a crucial role in the discussion of representations.

The relevance of these global properties becomes apparent when one attempts to parametrize group elements in terms of infinitesimal generators, a procedure that is inherently local and requires additional care at the global level.

▶ **Pause and think** If you know the Lie algebra (i.e. all commutators), what extra information might still be missing to characterize the group globally? Try to think of one physical reason why this could matter for representations.

1.3.3.3 From Lie-Algebra Representations to Group Representations

In physics one often constructs a representation by specifying how the generators act. Concretely, let $\rho : \mathfrak{g} \to \mathrm{End}(V)$ be a Lie-algebra representation,

$$\rho([X, Y]) = [\rho(X), \rho(Y)] \qquad \forall\, X, Y \in \mathfrak{g}. \tag{1.66}$$

A natural question is whether ρ exponentiates to a representation of the *full* Lie group G on the same space V, i.e. whether there exists a homomorphism $D : G \to \mathrm{GL}(V)$ such that, locally,

$$D\left(e^X\right) = e^{\rho(X)} \qquad \text{for } X \text{ near } 0 \text{ in } \mathfrak{g}. \tag{1.67}$$

Independently of global issues, $e^{\rho(X)}$ always defines a representation of the *local* group obtained by restricting to a sufficiently small neighborhood of the identity in G.

The extension to a global representation of G may fail if G is not simply connected. A standard and extremely useful result is:

if G is connected and simply connected, then every finite-dimensional representation of its Lie algebra $\mathfrak{g}$ integrates uniquely to a representation of G.

If G is connected but not simply connected, a Lie-algebra representation may still integrate, but typically only to a representation of the *universal covering group* $\tilde{G}$, with additional constraints required for it to descend to G itself.
This is the conceptual origin of the distinction between vector and spinor representations in rotation symmetry, discussed in the next subsection.

With these conceptual issues clarified, we are now in a position to focus on Lie groups, namely symmetry groups that possess both an algebraic structure and a compatible differentiable manifold structure.

► **Pause and think** If exponentiation always works locally, what kind of obstruction could prevent it from working globally? Try to connect your answer to the notion of closed curves and simple connectedness.

1.3.4 Global Versus Local Symmetries and the Gauge Principle

The group-theoretical concepts introduced so far acquire their sharpest physical meaning once one distinguishes between *global* and *local* symmetry transformations. This distinction is not merely terminological: in relativistic field theory it is precisely the promotion of a global symmetry to a local one that compels the introduction of interaction fields.

► **Pause and think** **What breaks when $\epsilon \to \epsilon(x)$?** Consider a globally invariant kinetic term built with ∂_μ. What extra contribution do you expect when differentiating a locally transformed field?

1.3.4.1 Global Internal Symmetries

Consider, for definiteness, a multiplet of matter fields $\psi(x)$ taking values in a complex vector space V that carries a representation R of a Lie group G. Let $\{T^a\}$ be a basis of the Lie algebra $\mathfrak{g}$ in that representation, so that

$$[T^a, T^b] = i f^{ab}{}_c\, T^c, \tag{1.68}$$

with real structure constants f_c^{ab} (for compact groups one can choose T^a Hermitian). A *global* transformation is defined by a group element $U \in G$ that is independent of spacetime:

$$\psi(x) \longrightarrow \psi'(x) = U\,\psi(x), \qquad U = \exp\big(i\epsilon_a T^a\big), \qquad \epsilon_a = \text{const.} \tag{1.69}$$

In this case, the ordinary derivative transforms covariantly,

$$\partial_\mu \psi(x) \longrightarrow \partial_\mu \psi'(x) = U\, \partial_\mu \psi(x), \tag{1.70}$$

and any Lagrangian built from ψ, $\partial_\mu \psi$, and invariant tensors of the representation may be made invariant under G.

It is in this sense that *representations label charges*: specifying the representation R fixes how the field multiplet transforms under the symmetry, and the generators T^a determine the corresponding conserved currents and charges in the quantum theory. For example, for an Abelian factor $U(1)$ one has $U = \exp(iq\alpha)$ and the single real number q is the charge; for a non-Abelian group, the "charge" is encoded by the weight content of the chosen representation.

1.3.4.2 Why Locality Forces New Fields

The situation changes qualitatively if one attempts to promote the symmetry parameter to a spacetime-dependent function. Consider now

$$\psi(x) \longrightarrow \psi'(x) = U(x)\, \psi(x), \qquad U(x) = \exp\big(i\alpha_a(x)\, T^a\big), \tag{1.71}$$

with $\alpha_a(x)$ arbitrary smooth functions. Taking a derivative produces an additional term:

$$\begin{aligned} \partial_\mu \psi'(x) &= \partial_\mu\big(U(x)\psi(x)\big) \\ &= U(x)\, \partial_\mu \psi(x) + \big(\partial_\mu U(x)\big)\psi(x). \end{aligned} \tag{1.72}$$

The second contribution, proportional to $\partial_\mu U(x)$, prevents $\partial_\mu \psi$ from transforming in the same way as ψ itself. Therefore, a Lagrangian that is invariant under the global group generally ceases to be invariant once the transformation is made local.

The remedy is to introduce a new field $A_\mu(x)$ taking values in the Lie algebra $\mathfrak{g}$ and to replace the ordinary derivative by a *covariant* derivative,

$$D_\mu := \partial_\mu + ig\, A_\mu(x), \qquad A_\mu(x) = A_\mu^a(x)\, T^a, \tag{1.73}$$

where g is the coupling constant. The defining requirement of the gauge principle is that $D_\mu \psi$ transform covariantly,

$$(D_\mu \psi)'(x) = U(x)\, D_\mu \psi(x). \tag{1.74}$$

▶ **Driving exercise** Imposing Eq.(1.74) fixes the transformation law of A_μ uniquely. Indeed, applying D'_μ to ψ' and using Eq.(1.73) yields

$$\begin{aligned} (D_\mu \psi)'(x) &= \Big(\partial_\mu + igA'_\mu(x)\Big)\, U(x)\psi(x) \\ &= \big(\partial_\mu U(x)\big)\psi(x) + U(x)\partial_\mu \psi(x) + igA'_\mu(x)\, U(x)\psi(x). \end{aligned} \tag{1.75}$$

Demanding equality with $U(x)\big(\partial_\mu + igA_\mu(x)\big)\psi(x)$ for arbitrary $\psi(x)$ gives

$$A'_\mu(x) = U(x)\,A_\mu(x)\,U^{-1}(x) - \frac{i}{g}\big(\partial_\mu U(x)\big)U^{-1}(x). \tag{1.76}$$

1.3.4.3 Field Strength and the Meaning of the Coupling

The commutator of covariant derivatives defines the gauge-invariant notion of curvature (field strength),

$$[D_\mu, D_\nu]\psi = ig\,F_{\mu\nu}\psi, \qquad F_{\mu\nu} = \partial_\mu A_\nu - \partial_\nu A_\mu + ig\,[A_\mu, A_\nu], \tag{1.77}$$

which transforms covariantly,

$$F'_{\mu\nu}(x) = U(x)\,F_{\mu\nu}(x)\,U^{-1}(x). \tag{1.78}$$

For an Abelian group (such as $U(1)$) the commutator term vanishes, recovering $F_{\mu\nu} = \partial_\mu A_\nu - \partial_\nu A_\mu$; for non-Abelian groups (such as $\mathrm{SU}(N)$) the term $ig[A_\mu, A_\nu]$ encodes the intrinsic self-interactions of the gauge field.

At this level, the group-theoretic message can be stated succinctly: *matter fields are classified by representations of G, while gauge fields take values in the Lie algebra* $\mathfrak{g}$ *and couple through the covariant derivative.* In subsequent chapters, this framework will reappear as the common language behind electromagnetism ($U(1)$), the weak interaction, and quantum chromodynamics ($\mathrm{SU}(3)$).

The discussion above reveals a unifying principle that will guide the remainder of this book: *fundamental interactions arise from the requirement of local symmetry invariance*. Once a symmetry group is specified, its representations classify matter fields, while the corresponding Lie algebra determines the structure of the interaction through gauge connections and covariant derivatives. In this sense, group theory is not a mathematical embellishment, but the organizing language of modern fundamental physics.

1.4 The Groups $\mathrm{SU}(N)$ and $\mathrm{SU}(2)$

The special unitary group $\mathrm{SU}(N)$ is the group of unitary $N \times N$ matrices with unit determinant:

$$U^\dagger U = UU^\dagger = \mathbb{1}, \qquad \det U = 1. \tag{1.79}$$

Any unitary matrix can be written as the exponential of a Hermitian matrix,

$$U = e^{iH}, \tag{1.80}$$

with $H^\dagger = H$ from unitarity. The determinant constraint implies

$$1 = \det U = \det(e^{iH}) = e^{i\,\mathrm{Tr}\,H} \quad\Longrightarrow\quad \mathrm{Tr}\,H = 0, \tag{1.81}$$

so H is Hermitian and traceless. Such matrices form a real vector space of dimension $N^2 - 1$, hence SU(N) has $N^2 - 1$ generators T_a:

$$H = \sum_{a=1}^{N^2-1} \epsilon_a T_a, \qquad \epsilon_a \in \mathbb{R}, \tag{1.82}$$

and therefore

$$U = \exp\left(i \sum_{a=1}^{N^2-1} \epsilon_a T_a \right). \tag{1.83}$$

Since SU(N) is compact, one may choose the conventional normalization

$$\mathrm{Tr}(T_a T_b) = \frac{1}{2}\,\delta_{ab}, \tag{1.84}$$

so that the corresponding structure constants are totally antisymmetric.

The Lie algebra of SU(2) is generated by three operators J_1, J_2, J_3 obeying

$$[J_a, J_b] = i\,\epsilon_{abc}\,J_c, \qquad a, b, c = 1, 2, 3, \tag{1.85}$$

with $\epsilon_{123} = 1$.

A distinguished invariant (Casimir) is

$$J^2 \equiv J_1^2 + J_2^2 + J_3^2, \tag{1.86}$$

which commutes with all J_a. One introduces a basis that diagonalizes J^2 and J_3 simultaneously,

$$J^2\,|j, j_3\rangle = j(j+1)\,|j, j_3\rangle, \tag{1.87}$$

$$J_3\,|j, j_3\rangle = j_3\,|j, j_3\rangle, \tag{1.88}$$

and the ladder operators

$$J_\pm \equiv J_1 \pm i J_2. \tag{1.89}$$

From Eq. (1.85) it follows that

$$[J_+, J_-] = 2J_3, \qquad [J_3, J_\pm] = \pm J_\pm, \tag{1.90}$$

and consequently

$$J_+|j, j_3\rangle = \sqrt{j(j+1) - j_3(j_3+1)}\,|j, j_3+1\rangle, \tag{1.91}$$

$$J_-|j, j_3\rangle = \sqrt{j(j+1) - j_3(j_3-1)}\,|j, j_3-1\rangle. \tag{1.92}$$

For fixed j, the allowed values of j_3 form a finite ladder:

$$j_3 \in \{-j, -j+1, \ldots, j-1, j\}, \tag{1.93}$$

so the irreducible representation labelled by $j = 0, \frac{1}{2}, 1, \frac{3}{2}, \ldots$ has dimension

$$\dim(j) = 2j + 1. \tag{1.94}$$

In particular:

- $j = 0$: trivial (singlet), dim $= 1$;
- $j = \frac{1}{2}$: fundamental (spinorial), dim $= 2$;
- $j = 1$: adjoint (vector), dim $= 3$.

1.4.1 Spin-$\frac{1}{2}$ Representation and Pauli Matrices

As claimed above, the concept of spin emerges. In the case of spin-$\frac{1}{2}$, typical of Fermions, in particle physics, the fundamental representation is thus realized by

$$J_a = \frac{\sigma_a}{2} \qquad (a = 1, 2, 3), \tag{1.95}$$

where the Pauli matrices are

$$\sigma_1 = \begin{pmatrix} 0 & 1 \\ 1 & 0 \end{pmatrix}, \quad \sigma_2 = \begin{pmatrix} 0 & -i \\ i & 0 \end{pmatrix}, \quad \sigma_3 = \begin{pmatrix} 1 & 0 \\ 0 & -1 \end{pmatrix}, \tag{1.96}$$

satisfying the identity

$$\sigma_i \sigma_j = \delta_{ij}\,\mathbb{1} + i\,\epsilon_{ijk}\,\sigma_k, \tag{1.97}$$

which immediately implies that $\sigma_i/2$ obey the SU(2) commutation relations.

Some properties of Pauli matrices are

$$\begin{cases} \sigma_x^2 = \sigma_y^2 = \sigma_z^2 = 1 \\ \sigma_x\sigma_y = -\sigma_y\sigma_x = i\sigma_z \\ \sigma_y\sigma_z = -\sigma_z\sigma_y = i\sigma_x \\ \sigma_z\sigma_x = -\sigma_x\sigma_z = i\sigma_y \end{cases}, \tag{1.98}$$

it is possible to demonstrate that:

$$(\boldsymbol{\sigma} \cdot \mathbf{a})(\boldsymbol{\sigma} \cdot \mathbf{b}) = (\mathbf{a} \cdot \mathbf{b}) + i\boldsymbol{\sigma} \cdot \mathbf{a} \times \mathbf{b}. \tag{1.99}$$

▶ **Driving exercise** Let us prove it. Consider that:

$$\boldsymbol{\sigma} \cdot \mathbf{a} = a_1\sigma_1 + a_2\sigma_2 + a_3\sigma_3. \tag{1.100}$$

The Eq. (1.99) becomes:

$$\begin{aligned}\Rightarrow (a_1\sigma_1 + a_2\sigma_2 + a_3\sigma_3)(b_1\sigma_1 + b_2\sigma_2 + b_3\sigma_3) = \underline{a_1b_1\sigma_1\sigma_1} + a_1b_2\sigma_1\sigma_2 \\ + a_1b_3\sigma_1\sigma_3 + a_2b_1\sigma_2\sigma_1 + \underline{a_2b_2\sigma_2\sigma_2} + a_2b_3\sigma_2\sigma_3 \\ + a_3b_1\sigma_3\sigma_1 + a_3b_2\sigma_3\sigma_2 + \underline{a_3b_3\sigma_3\sigma_3}.\end{aligned}$$

Noting that the underlined terms correspond exactly to the scalar product $\mathbf{a} \cdot \mathbf{b}$, we have that:

$$\begin{aligned}(\mathbf{a} \cdot \mathbf{b}) + (a_1b_2\sigma_1\sigma_2 + a_1b_3\sigma_1\sigma_3 \\ + a_2b_1\sigma_2\sigma_1 + a_2b_3\sigma_2\sigma_3 + a_3b_1\sigma_3\sigma_1 + a_3b_2\sigma_3\sigma_2).\end{aligned}$$

Using now the above properties, Eq. (1.98), we obtain:

$$\begin{aligned}(\mathbf{a} \cdot \mathbf{b}) + [a_1b_2(i\sigma_3) + a_1b_3(-i\sigma_2) + a_2b_1(-i\sigma_3) + a_2b_3(i\sigma_1) \\ + a_3b_1(i\sigma_2) + a_3b_2(-i\sigma_1)]\end{aligned}$$

$$\Rightarrow (\mathbf{a} \cdot \mathbf{b}) + i[\sigma_3(a_1b_2 - a_2b_1) - \sigma_2(a_1b_3 - a_3b_1) + \sigma_1(a_2b_3 - a_3b_2)]$$

$$\Rightarrow (\mathbf{a} \cdot \mathbf{b}) + i\boldsymbol{\sigma} \cdot \mathbf{a} \times \mathbf{b}. \tag{1.101}$$

A concrete realization of the adjoint (spin-1) representation is provided by the 3×3 matrices

$$J_1 = \frac{1}{\sqrt{2}}\begin{pmatrix}0 & 1 & 0\\ 1 & 0 & 1\\ 0 & 1 & 0\end{pmatrix}, \quad J_2 = \frac{1}{\sqrt{2}}\begin{pmatrix}0 & -i & 0\\ i & 0 & -i\\ 0 & i & 0\end{pmatrix}, \quad J_3 = \begin{pmatrix}1 & 0 & 0\\ 0 & 0 & 0\\ 0 & 0 & -1\end{pmatrix}, \tag{1.102}$$

which satisfy Eq. (1.85).

▶ **Driving exercise**
Let us consider the tensor products in SU(2) that turns out to be a paradigmatic example of two spin-$\frac{1}{2}$ representations:

$$\frac{1}{2} \otimes \frac{1}{2} = 0 \oplus 1, \tag{1.103}$$

i.e. two spinors combine into a singlet (one antisymmetric state) plus a triplet (three symmetric states). In terms of dimensions this is

$$2 \otimes 2 = 1 \oplus 3, \tag{1.104}$$

which preserves the total dimensionality.

1.5 Proposed Exercises

Exercise 1

Let $G = \mathbb{Z}_{16}^*$ be the group of invertible elements in $\mathbb{Z}_{16}$ under multiplication modulo 16.

- Find one cyclic subgroup of order 4 in G.
- Analogously, find a non-cyclic one.

Exercise 2

Consider the following two subsets of $\mathrm{GL}_2(\mathbb{R})$:

$$T = \left\{ \begin{pmatrix} a & b \\ 0 & 1 \end{pmatrix} \,\middle|\, a \in \mathbb{R}^*,\ b \in \mathbb{R} \right\}, \qquad U = \left\{ \begin{pmatrix} 1 & b \\ 0 & 1 \end{pmatrix} \,\middle|\, b \in \mathbb{R} \right\}. \tag{1.105}$$

1. Show that T is a subgroup of $\mathrm{GL}_2(\mathbb{R})$.
2. Show that U is a normal subgroup of T. *Hint.* Notice that U consists of unipotent matrices and examine how such elements transform under conjugation by elements of T.

Exercise 3

Let G be a connected Lie group and $H \subset G$ a closed Lie subgroup. Denote by $\mathfrak{g}$ and $\mathfrak{h}$ their Lie algebras, respectively.

1. Show that the homogeneous space G/H admits a smooth manifold structure such that the natural projection $\pi : G \to G/H$ is a smooth submersion.
2. Prove that there exists a vector space decomposition
$$\mathfrak{g} = \mathfrak{h} \oplus \mathfrak{m}, \tag{1.106}$$
where $\mathfrak{m}$ is invariant under the adjoint action of H if and only if G/H admits a G-invariant affine connection.
3. Using the exponential map, show that a neighborhood of the identity coset in G/H can be identified with a neighborhood of the origin in $\mathfrak{m}$. Explain why this identification fails globally in general.

Essential Bibliography

- Georgi (1999) Lie algebras in particle physics: from isospin to unified theories, 2nd edn. Westview Press.

An introduction to Lie algebras and compact groups from the viewpoint of physics, with clear conventions for SU(N), generators, structure constants, and representation-theory tools used throughout high-energy physics.

- Jones HF (1998) Groups, representations and physics, 2nd ed., CRC Press.
  A concise but systematic account of representations (reducibility, Schur's lemma, tensor products) and the bridge between abstract group concepts and concrete physics applications.
- Hall BC (2015) Lie groups, lie algebras, and representations: an elementary introduction, 2nd edn. Springer.
  A modern, clean treatment of Lie groups as manifolds, exponential map, connectedness/simply-connectedness, and the local–global distinction underlying the $SU(2) \to SO(3)$ covering.
- Bagarello F, Bavuma Y, Russo FG (2021) Topological decompositions of the Pauli group and their influence on dynamical systems. Math Phys Anal Geom 24:16.
  A relevant approach toward the way of explaining Pauli group in reference to physics.

Special Relativity 2

There are only two ways to live your life. One is as though nothing is a miracle. The other is as though everything is a miracle.

—Albert Einstein

Abstract

The physics of the twentieth century marked a radical turning point in the history of Physics. It explicitly revealed a **dissolution of absolutes** that had characterized classical physics, which until then had successfully described natural phenomena. At the close of the nineteenth century, it was widely believed that humankind had attained such a deep understanding of Nature that no further fundamental breakthroughs would be necessary. Quantum mechanics and Einstein's Special Relativity, however, overturned this conviction by introducing revolutionary concepts such as wave–particle duality, energy quantization, the invariance of signal velocity with respect to the state of motion, and the non-absoluteness of space and time. The birth of relativity can be regarded as a "silent revolution", largely because Einstein did not belong to the established international scientific community of the time. This is evident from the fact that his original papers contained no bibliography and were thus not written in the orthodox style. Quite interestingly, it is worth mentioning that relativity identifies the *invariant quantities*, or in a funny nomenclature the **absolute quantities**, namely those that do not change with the change of the inertial frame of reference. Motivated by all of this, the present chapter begins with an introduction to the scientific context of the late nineteenth century, proceeds through the basic principles of relativity, develops its covariant formulation, and concludes with the study of relativistic collisions.

O. Luongo, *Fundamental Interactions*,
https://doi.org/10.1007/978-3-032-27413-7_2

Fig. 2.1 Portrait of Mileva Maric, 1896. Reprinted from Wikimedia. Public domain

A key character: *Mileva Marić (1875–1948), Fig.* 2.1. No theories are known as Maric's models. However, the first Einstein's wife played a role in allowing the husband to develop the special theory of relativity. To her, with no doubt, humanity is in debit. Mileva Marić was born in Titel, then part of the Austro–Hungarian Empire, into a Serbian family that strongly supported her education at a time when advanced scientific training for women was highly uncommon. From an early age she exhibited remarkable intellectual abilities, particularly in mathematics and physics. Her academic talent allowed her to attend prestigious secondary schools in the Habsburg Empire and, later, to enroll at the Swiss Federal Polytechnic in Zürich (now ETH Zürich), one of the very few institutions in Europe that admitted women to scientific studies at the end of the nineteenth century. At the Polytechnic, Marić pursued a curriculum in mathematics and physics and was the only woman in her cohort. Among her classmates was Albert Einstein, with whom she developed both an intellectual partnership and a personal relationship. Contemporary correspondence between Marić and Einstein reveals a shared engagement with the conceptual foundations of theoretical physics, including problems related to electrodynamics, molecular forces, and the emerging crisis of classical physics. Despite her strong performance in several courses, Marić did not complete the diploma program, failing the final examinations. Marić married Einstein in 1903 and supported him during the period in which he produced his *annus mirabilis* papers. While the precise extent of her scientific contribution to Einstein's early work remains a subject of historical debate, it is well established that she played a significant role as an intellectual interlocutor, discussing ideas, calculations, and physical interpretations during their formative years. Following the deterioration of their marriage, Marić and Einstein divorced in 1919. She spent the remainder of her life in Zürich, devoting herself primarily to the care and education of their children, particularly their son Eduard, who suffered from severe mental illness. Today, Mileva Marić is remembered as one of the earliest women to receive advanced training in theoretical physics in Europe and as a figure emblematic of the many talented scientists whose potential contributions were constrained by the social and institutional barriers of their time.

2.1 Physics Before Einstein's Theory

Special relativity did not arise from a single "paradox" but from a sustained tension between mechanics and Maxwellian electrodynamics.

Newtonian dynamics presupposes an absolute time parameter shared by all inertial observers, whereas Maxwell's equations single out a universal propagation speed for electromagnetic waves, suggesting that the naive composition of velocities implicit in Galilean kinematics cannot be the last word.

The crisis was therefore conceptual and structural and based on two main points.

- In Newtonian mechanics space and time are independent arenas. The notion of an *inertial frame* is defined operationally by the validity of Newton's first law, and the principle of relativity is encoded as the statement that the laws of mechanics have the same form in any two inertial frames related by a uniform relative velocity. This is implemented through the *Galilean group*. For two frames S and S' with relative velocity $\mathbf{v} = v\,\hat{\mathbf{x}}$ one writes

$$x' = x - vt, \qquad y' = y, \qquad z' = z, \qquad t' = t, \tag{2.1}$$

and the transformation laws for velocities and accelerations follow immediately,

$$\mathbf{u}' = \mathbf{u} - \mathbf{v}, \qquad \mathbf{a}' = \mathbf{a}. \tag{2.2}$$

Because accelerations are invariant under Eq. (2.1), Newton's second law

$$\mathbf{F} = \frac{d(m\mathbf{v})}{dt}, \tag{2.3}$$

retains its form in all inertial frames, provided forces are defined appropriately[1].

- A second, equally important ingredient was the availability of variational formulations. The principle of stationary action,

$$\delta S = \delta \int_{t_1}^{t_2} L(q_i, \dot{q}_i, t)\, dt = 0, \tag{2.4}$$

with $L = T - V$, produces the Euler–Lagrange equations,

$$\frac{d}{dt}\left(\frac{\partial L}{\partial \dot{q}_i}\right) - \frac{\partial L}{\partial q_i} = 0. \tag{2.5}$$

This point of view will later become essential: special relativity is, at a deep level, a modification of kinematics that preserves the variational structure while replacing the underlying symmetry group.

2.1.1 The Crisis Is Due to Electrodynamics!

In the above subsection, something incomplete has been stated. One needs to wonder which kind of real problem arose to suggest scientists to reconsider the Nature as seen by classical physics.

The answer is into the nineteenth–century electrodynamics culminated in Maxwell's synthesis, which unified disparate empirical laws into a set of field equations,

$$\nabla \cdot \mathbf{E} = \frac{\rho}{\varepsilon_0}, \tag{2.6}$$

$$\nabla \cdot \mathbf{B} = 0, \tag{2.7}$$

$$\nabla \times \mathbf{E} = -\frac{\partial \mathbf{B}}{\partial t}, \tag{2.8}$$

$$\nabla \times \mathbf{B} = \mu_0 \mathbf{J} + \mu_0 \varepsilon_0 \frac{\partial \mathbf{E}}{\partial t}. \tag{2.9}$$

[1] It is interesting to stress that the second law written as $\mathbf{F} = \frac{d(m\mathbf{v})}{dt}$ holds iff the mass is a constant. However, in the attempt to formulate a generalized relativistic version, we will start from Eq. (2.3).

In vacuum ($\rho = 0, \mathbf{J} = 0$) these imply wave equations. Taking the curl of Eq. (2.8) and using Eq. (2.9) yields

$$\nabla^2 \mathbf{E} - \mu_0 \varepsilon_0 \frac{\partial^2 \mathbf{E}}{\partial t^2} = 0, \qquad \nabla^2 \mathbf{B} - \mu_0 \varepsilon_0 \frac{\partial^2 \mathbf{B}}{\partial t^2} = 0, \tag{2.10}$$

so electromagnetic solutions propagate with speed

$$c \equiv \frac{1}{\sqrt{\mu_0 \varepsilon_0}}. \tag{2.11}$$

The striking numerical agreement of c with the measured speed of light suggested that light is electromagnetic in nature. At the same time, it introduced a new conceptual element: c appears as a constant derived from electromagnetic parameters and does not resemble the frame-dependent speeds typical in Newtonian kinematics.

The natural nineteenth–century interpretation was to regard fields as properties of a medium. Thus one postulated the *luminiferous ether*, a subtle medium filling space, providing the support for electromagnetic waves much as air supports sound. The ether hypothesis carries an implicit kinematical commitment: there should exist a preferred rest frame in which Maxwell's equations take their simplest form and in which the speed c is to be interpreted as the propagation speed of waves with respect to that medium.

To make the tension explicit, consider two inertial frames related by a Galilean boost with relative velocity $\mathbf{v}$ and by the chain rule, the spatial derivatives are unchanged, since $\mathbf{x} = \mathbf{x}' + \mathbf{v}t'$, giving rise to

$$\nabla = \nabla', \tag{2.12}$$

while the time derivative at fixed *unprimed* position differs from the time derivative at fixed *primed* position:

$$\frac{\partial}{\partial t} = \left.\frac{\partial}{\partial t'}\right|_{\mathbf{x}'} + \left(\frac{\partial \mathbf{x}'}{\partial t}\right) \cdot \nabla' = \frac{\partial}{\partial t'} - \mathbf{v} \cdot \nabla'. \tag{2.13}$$

If Maxwell's equations were invariant under Galilean transformations, then they might preserve their forms in the primed frame.

▶ **Driving exercise**
Let us prove that electromagnetism is not invariant. Starting from

$$\frac{\partial^2}{\partial t^2} = \left(\frac{\partial}{\partial t'} - \mathbf{v}\cdot\nabla'\right)^2 = \frac{\partial^2}{\partial t'^2} - 2(\mathbf{v}\cdot\nabla')\frac{\partial}{\partial t'} + (\mathbf{v}\cdot\nabla')^2, \tag{2.14}$$

we find

$$\begin{aligned} 0 &= \nabla'^2\mathbf{E} - \mu_0\varepsilon_0\left[\frac{\partial^2\mathbf{E}}{\partial t'^2} - 2(\mathbf{v}\cdot\nabla')\frac{\partial\mathbf{E}}{\partial t'} + (\mathbf{v}\cdot\nabla')^2\mathbf{E}\right] \\ &= \underbrace{\left(\nabla'^2\mathbf{E} - \mu_0\varepsilon_0\frac{\partial^2\mathbf{E}}{\partial t'^2}\right)}_{\text{same form as in the unprimed frame}} + \mu_0\varepsilon_0\left[2(\mathbf{v}\cdot\nabla')\frac{\partial\mathbf{E}}{\partial t'} - (\mathbf{v}\cdot\nabla')^2\mathbf{E}\right]. \end{aligned} \tag{2.15}$$

The last bracketed terms are *new* and do not vanish in general. Therefore the wave equation does not retain its form under a Galilean boost. This already proves that vacuum Maxwell theory cannot be invariant under Galilean transformations.
Equivalently, consider a plane-wave solution in the unprimed frame,

$$\mathbf{E}(\mathbf{x}, t) = \mathbf{E}_0\, e^{i(\mathbf{k}\cdot\mathbf{x}-\omega t)}, \qquad \omega = c|\mathbf{k}|. \tag{2.16}$$

Expressing it in primed variables using $\mathbf{x} = \mathbf{x}' + \mathbf{v}t'$ and $t = t'$ gives

$$\mathbf{E}(\mathbf{x}', t') = \mathbf{E}_0\, e^{i(\mathbf{k}\cdot\mathbf{x}'-(\omega-\mathbf{k}\cdot\mathbf{v})t')}. \tag{2.17}$$

Thus the primed frequency is $\omega' = \omega - \mathbf{k}\cdot\mathbf{v}$. If the same wave equation with the same constant c held in the primed frame, one would require the same dispersion relation $\omega' = c|\mathbf{k}'|$ with $\mathbf{k}' = \mathbf{k}$ (since $\nabla = \nabla'$). But

$$\omega' = c|\mathbf{k}| - \mathbf{k}\cdot\mathbf{v} \neq c|\mathbf{k}|$$

for generic $\mathbf{v}$ and $\mathbf{k}$. Again, the Maxwellian wave law is incompatible with Galilean kinematics.

2.1.2 Why a Crisis?

At the end of the nineteenth century, the real difficulty was not that classical mechanics failed for ordinary velocities, nor that Maxwell's equations failed in the laboratory. The difficulty was that the two theories presupposed different transformation groups.

Three possibilities were therefore licit,

- kinematics was wrong,
- electromagnetism was wrong,
- kinematics and mechanics were wrong.

In Newtonian mechanics the principle of relativity is implemented by the Galilean transformations, Eq. (2.1) and it seemed quite impossible to presume that mechanics was wrong. Under these transformations, time is absolute and the addition of velocities is linear. If light were a material signal, one would expect its speed to transform like any other velocity.

Hence, the solution turned out to be the *existence of the ether*. The ether was postulated as a pervasive medium filling the entire space, providing the mechanical support required for the propagation of electromagnetic waves, in close analogy with the role played by air for sound waves or by elastic media for mechanical vibrations.

The ether hypothesis naturally reconciled Maxwell's theory with Galilean kinematics. If electromagnetic waves propagate with speed c with respect to the ether, then c is no longer required to be the same in all inertial frames. Instead, observers moving with velocity $\mathbf{v}$ relative to the ether would measure different light velocities, in accordance with the Galilean law of velocity addition. In this sense, the ether provided a preferred reference frame, namely the rest frame of the medium, in which Maxwell's equations were assumed to take their simplest and most symmetric form.

From this perspective, the principle of relativity retained its Newtonian meaning only approximately: it applied strictly to mechanical phenomena, while electromagnetic processes were expected to reveal motion with respect to the ether. The experimental detection of such motion, often referred to as *ether drift*, thus became a central objective of late nineteenth century physics. Interferometric experiments, most notably the Michelson-Morley experiment, were specifically designed to measure anisotropies in the speed of light induced by the Earth's motion through the ether.

Thus, if a light beam has $u_x = c$ in the ether frame, it would have $u'_x = c - v$ in a frame moving with speed v along the beam direction. Maxwell's electrodynamics, however, predicts in vacuum a wave speed determined by $\mu_0\varepsilon_0$ with no explicit dependence on the state of motion of the source or observer.

On the other side, the ether hypothesis was not invented to solve a single problem; it was the default language for wave physics.

The Michelson-Morley experiment (1887) was designed to detect an "ether wind" via differences in optical path length along orthogonal arms of an interferometer. In the simplest ether model, the time of flight should depend on the orientation of the apparatus relative to Earth's velocity through the ether, producing a measurable

Table 2.1 Key ether experiments and roles in testing the existence of a preferred inertial frame

Experiment	Physical quantity tested	Experimental strategy	Implication for ether
Michelson-Morley (1887)	Directional dependence of the speed of light ("ether wind")	Interferometric comparison of light travel times along two orthogonal arms of equal length, rotated with respect to the Earth's motion	Excludes anisotropy of *c* due to motion through a preferred frame
Kennedy-Thorndike (1932)	Dependence of the speed of light on the absolute velocity of the laboratory	Modified Michelson interferometer with unequal arm lengths, monitored over time as the Earth's velocity changes	Rules out ether models relying solely on length contraction without time dilation
Trouton-Noble (1901–1903)	Existence of electromagnetic torques induced by motion through the ether	Electrostatically charged capacitor suspended on a torsion fiber, searching for orientation-dependent torques	Excludes detectable dynamical effects of absolute motion in electrodynamics

fringe shift under rotation. The observed null result, within experimental sensitivity, was inconsistent with naive ether drift.

Additional tests were proposed in subsequent years, including the Kennedy–Thorndike and Trouton–Noble experiments. The final outputs are summarized in Table 2.1.

2.2 Lorentz Theory Before Einstein

The theory of ether served as the starting point from which experiments were performed. In this framework, a key role was played by Hendrik A. Lorentz, quite appreciated by Einstein, who still believed in the mainstream idea according to which the ether defined a preferred reference frame.

However, Lorentz proposed a theory that singled out a preferred frame, the dynamics of charges and electromagnetic fields in moving matter were modified in such a way that *experiments performed with material apparatus would be insensitive to uniform motion through the ether.*

A key step was the introduction of *local time*,

$$t' = t - \frac{v}{c^2}x, \tag{2.18}$$

as an auxiliary variable that naturally appears when transforming field equations to a moving coordinate system. In Lorentz's interpretation, Eq. (2.18) did not initially imply a new physical time. It was a mathematical device reflecting how moving clocks would have to behave if Maxwell's equations were to preserve their form approximately.

Independently, FitzGerald and Lorentz proposed a *length contraction* of moving bodies along the direction of motion,

$$L = L_0\sqrt{1 - \frac{v^2}{c^2}}, \tag{2.19}$$

again not as a kinematical principle but as a dynamical response of matter to motion through the ether. The purpose was clear: the contraction could cancel the expected Michelson-Morley fringe shift.

The modern formulation organizes these ideas into a single set of linear transformations between inertial frames:

$$x' = \gamma(x - vt), \tag{2.20a}$$

$$y' = y, \tag{2.20b}$$

$$z' = z, \tag{2.20c}$$

$$t' = \gamma\left(t - \frac{v}{c^2}x\right), \tag{2.20d}$$

where $\gamma = \frac{1}{\sqrt{1-\frac{v^2}{c^2}}}$. Lorentz arrived at these relations as the transformation laws under which Maxwell's equations retain their form. Starting from Eq. (2.20) the wave equation is covariant and the speed c is invariant in the sense that a light ray with $x = ct$ in one frame satisfies $x' = ct'$ in the other.

A compact characterization is provided by the invariance of the quadratic form

$$s^2 = c^2t^2 - x^2 - y^2 - z^2, \tag{2.21}$$

namely $s'^2 = s^2$ under Eqs. (2.20). In Lorentz's pre-Einstein view, the quantity in Eq. (2.21) was not yet interpreted as a fundamental spacetime structure. Rather, it was a mathematical invariant reflecting the symmetry of electrodynamics. That will represent the main difference between the Lorentz and Einstein approaches.

Once Eqs. (2.20) are accepted, the main relativistic effects follow as kinematical statements.

- **Time dilation**. A clock at rest in S' has $\Delta x' = 0$. Using (2.20),

$$\Delta t' = \gamma\left(\Delta t - \frac{v}{c^2}\Delta x\right) = \gamma\Delta t \quad (\Delta x = v\Delta t), \tag{2.22}$$

  so for a clock comoving with S' one obtains

$$\Delta t = \gamma\Delta t', \tag{2.23}$$

  i.e. moving clocks run slower as seen from S.

- **Length contraction**. A rod at rest in S' has endpoints measured simultaneously in S ($\Delta t = 0$). From $x' = \gamma(x - vt)$ one finds $\Delta x' = \gamma \Delta x$, hence

$$L = \Delta x = \frac{\Delta x'}{\gamma} = \frac{L_0}{\gamma}. \tag{2.24}$$

- **Relativity of simultaneity**. Two events simultaneous in S satisfy $\Delta t = 0$, but then

$$\Delta t' = -\gamma \frac{v}{c^2} \Delta x, \tag{2.25}$$

so simultaneity is frame-dependent. Lorentz theory succeeded phenomenologically: it explained the null ether-wind results by endowing matter with precisely the deformation properties required, to conceal uniform motion relative to the ether itself. What it did not provide, in its original interpretation, was a clean conceptual reason *why* the ether should be undetectable and *why* time itself should be affected.

Henri Poincaré emphasized the *principle of relativity* as a fundamental constraint: *the laws of physics should take the same form in all inertial frames, with no experiment (mechanical or electromagnetic) capable of detecting absolute uniform motion. Poincaré also clarified the group properties of the Lorentz transformations and introduced the language of invariants, foreshadowing the geometric interpretation later crystallized by Minkowski.*

In this perspective, the Lorentz group is not an ad hoc artifact of electrodynamics, but rather a candidate symmetry of nature.

Very often, it is claimed that Poincaré could have formulated a special theory of relativity in the actual form and this statement is far from being wrong. Both Lorentz and Poincaré opened the avenues to the unification scheme proposed in one shot by Albert Einstein.

2.3 Special Relativity and Its "Aesthetic" Motivation

Einstein's 1905 breakthrough is often portrayed as a response to the ether-wind experiments, but an equally decisive motivation was the demand for conceptual economy and structural unity. The pre-Einstein landscape contained a patchwork: Newtonian kinematics with Galilean symmetry on one side, and Maxwell's electrodynamics with an apparent invariant speed on the other. Lorentz's solution repaired the mismatch by introducing dynamical deformations of matter, but it did so by preserving a preferred ether frame that no observation could detect. This situation is methodologically unstable: a theoretically unobservable structure is difficult to justify as fundamental.

From the modern viewpoint, the guiding aesthetic principle can be stated precisely: *the laws of physics should be expressed in a form that is manifestly invariant under the symmetry group that relates inertial observers.* If Maxwell's equations are invariant under the Lorentz group, then either electrodynamics is a special theory with exceptional transformation properties, or Lorentz symmetry is the correct kinematics for all fundamental physics. Einstein adopted the second option.

This shift is not merely philosophical, but reorganizes the conceptual foundations outlined below.

- It elevates the transformation laws, Eq. (2.20), from a dynamical trick to a kinematical principle.
- It replaces absolute time with an operational notion of time tied to synchronization procedures and invariant signals.
- It turns the invariant, Eq. (2.21), into a metric statement about spacetime, preparing the stage for Minkowski's geometry.

Finally, the geometric formulation of electrodynamics provides the cleanest expression of this unity. Introducing the four-position $x^{\mu} = (ct, \mathbf{x})$ and the field strength tensor $F^{\mu\nu}$, Maxwell's equations condense into

$$\partial_{\mu} F^{\mu\nu} = \mu_0 J^{\nu}, \qquad \partial_{\lambda} F_{\mu\nu} + \partial_{\mu} F_{\nu\lambda} + \partial_{\nu} F_{\lambda\mu} = 0, \tag{2.26}$$

which are manifestly Lorentz covariant. In that language, the conceptual tension that opened this chapter becomes sharply formulated: the symmetry of electrodynamics dictates the symmetry of spacetime itself. Special relativity is the systematic implementation of that dictum.

▶ **Pause and think**
Propose the improvements required to fully characterize a theory overcoming the Lorentz formulation with no ether.

In the above parts, we described the improvements toward a theory formulation that could incorporate electromagnetism into Galilean group. The incredible solution provided by Einstein was to recognize this limitation and formulate a unified scheme, bypassing the Galilean group and, then, recognizing the need of extending mechanics through a new group, namely the Lorentz group.

Precisely, special relativity is a theory founded on two physical postulates,

1. **Galilean principle of relativity**: The laws of physics take the same form (are invariant) in all inertial reference frames;
2. **Postulate of the constancy of the speed of light** (c): The speed of light in vacuum has the same value in all inertial reference frames.

The first postulate highlights the inadequacy of Galilean transformations. Indeed, if all the laws of physics are invariant under the change from one inertial frame to another, it appears contradictory that this invariance does not hold for Maxwell's equations. The second postulate has the immediate consequence that simultaneity between two events is not absolute. Two events that are simultaneous in one inertial frame are not necessarily simultaneous in another frame, see Eq. (2.25).

2.3.1 Vectors in $\mathbb{R}^3$

In classical mechanics, physical phenomena take place at points of three-dimensional Euclidean space $E = \mathbb{R}^3$. Each point in space is identified by a Cartesian coordinate triple $(x, y, z) \in E$, parametrized by an absolute time.

We define the *Euclidean distance* between two infinitesimally close points (x, y, z) and $(x + dx, y + dy, z + dz)$ as

$$|d\mathbf{x}| = \left(\sum_{i=1}^{3} dx_i^2\right)^{1/2} = \sqrt{dx^2 + dy^2 + dz^2}. \tag{2.27}$$

The previous expression can be rewritten using the Kronecker delta

$$\delta_{ij} = \begin{cases} 0, & \text{if } i \neq j \\ 1, & \text{if } i = j \end{cases} \tag{2.28}$$

as

$$d\mathbf{x}^2 = \sum_{i,j} \delta_{ij} dx_i dx_j = \delta_{ij} dx_i dx_j, \tag{2.29}$$

where we adopt Einstein's summation convention over repeated indices. By adopting this notion, we are indicating that Euclidean space is a Riemannian space with δ_{ij} as the metric tensor.

Having defined the concept of distance in $\mathbb{R}^3$, we now examine transformations between inertial reference frames. A linear transformation of coordinates is defined as

$$x'_i = \sum_{j=1}^{3} R_{ij} x_j = R_{ij} x_j, \tag{2.30}$$

where $R = \{R_{ij}\}$ is a matrix.

A *rotation* in Euclidean space is defined as a linear transformation that preserves distances, i.e., such that $d\mathbf{x}'^2 = d\mathbf{x}^2$. In other words, rotations in Euclidean space are *isometries*.

Combining Eqs. (2.29) and (2.30), we can write

$$d\mathbf{x}'^2 = \delta_{ij} dx'_i dx'_j = \delta_{ij} R_{ik} dx_k R_{jl} dx_l = R_{ik} dx_k R_{il} dx_l = R_{ik} R_{il} dx_k dx_l, \tag{2.31}$$

and noting that $R_{ik} R_{il} = \delta_{kl}$, we rewrite this as

$$d\mathbf{x}'^2 = \delta_{kl} dx_k dx_l = d\mathbf{x}^2. \tag{2.32}$$

We have thus demonstrated that distances remain invariant under rotations. The condition $R_{ik} R_{il} = \delta_{kl}$ can also be written in matrix form as

$$R^T R = \mathbb{I}, \tag{2.33}$$

which defines the *rotation matrix*, which is *orthogonal*.

Rotations in $\mathbb{R}^3$ are specified by three parameters. Indeed, considering the nine elements of R and the six constraints represented by $R_{ik} R_{il} = \delta_{kl}$, we obtain $9 - 6 = 3$ parameters. This can also be understood by considering a rotation angle θ and three parameters representing the components of the rotation axis $\mathbf{n}$ such that $|\mathbf{n}|^2 = n_x^2 + n_y^2 + n_z^2 = 1$.

Another important property of rotation matrices, useful for classifying rotations into *proper* and *improper*, is based on their determinant. Taking the determinant of Eq. (2.33) yields

$$\det(R^T R) = \det R^T \det R = (\det R)^2 = \det \mathbb{I} \rightarrow \det R = \pm 1. \tag{2.34}$$

Thus, we can distinguish two cases:

1. $\det R = 1$: rotations are called ***proper***, i.e., without axis inversion;
2. $\det R = -1$: rotations are called ***improper***, i.e., with axis inversion.

Proper rotations belong to the group of special orthogonal rotations, $SO(3)$, which is a subgroup of $O(3)$, the group of all rotations in $\mathbb{R}^3$.

In Euclidean space, *vectors* $\mathbf{A} = (A_1, A_2, A_3)$ are defined as objects that transform like the position differential $d\mathbf{x}$, i.e.,

$$A'_i = R_{ij} A_j. \tag{2.35}$$

If the components A_j are functions of the coordinates, we can more generally define a *vector field* $\mathbf{A}(\mathbf{x})$, which under the transformation $x' = Rx$ transforms as

$$A'_i(\mathbf{x}') = R_{ij} A_j(\mathbf{x}). \tag{2.36}$$

We now define *tensors* as objects containing 3^n components $T_{i_1,i_2,\ldots,i_n}$ with $i_j = 1, 2, 3$ for $j = 1, \ldots, n$. These objects transform according to

$$T'_{i_1,\ldots,i_n} = R_{i_1 j_1} \ldots R_{i_n j_n} T_{j_1,\ldots,j_n}. \tag{2.37}$$

In particular, we notice that:

$$n = \begin{cases} 1, & T_i \text{ is a vector,} \\ 2, & T_{ij} \text{ is a } 3 \times 3 \text{ matrix.} \end{cases} \tag{2.38}$$

Let us examine the case $n = 2$ in detail. A rank-2 tensor has nine components, since $3^2 = 9$, and transforms as

$$T'_{ij} = R_{ik} R_{jl} T_{kl}. \tag{2.39}$$

From vectors and tensors, we can construct *scalars*, which are invariant under rotations and obtained by contracting all indices of a tensor. The scalar products are defined as

$$\mathbf{A} \cdot \mathbf{B} = \sum_{i=1}^{3} A_i B_i = A_i B_i, \tag{2.40}$$

while the squared norm is

$$\mathbf{A}^2 = \mathbf{A} \cdot \mathbf{A} = A_i^2. \tag{2.41}$$

▶ **Driving exercise**
Let us demonstrate that these quantities are indeed invariant under rotations. For the scalar product:

$$\mathbf{A}' \cdot \mathbf{B}' = A_i' B_i' = R_{ij} A_j R_{ik} B_k = \delta_{jk} A_j B_k = \mathbf{A} \cdot \mathbf{B}, \qquad (2.42)$$

and for the squared norm:

$$\mathbf{A}'^2 = A_i' A_i' = R_{ij} A_j R_{ik} A_k = \delta_{jk} A_j A_k = A_j A_j = \mathbf{A}^2. \qquad (2.43)$$

Since all vectors and tensors transform in the same way under rotations, Newtonian physical laws are said to be *manifestly covariant* because they are vectorial or tensorial and transform consistently. Specifically, a law written as $\mathbf{A} = \mathbf{B}$, $A_{ijk\ldots} = B_{ijk\ldots}$, or as an equation only involving scalar products such as $\mathbf{A} \cdot \mathbf{B}$, retains its form under rotations of the axes.

However, it is important to distinguish between *covariance* and *invariance* under rotations. A law is *covariant* if it retains the same form after a rotation of the axes; this property depends on the tensorial nature of the law rather than the explicit values of its components. In contrast, a law is *invariant under rotations* when the quantities themselves do not change, depending explicitly on their components.

To provide a geometric formulation of relativity, we follow the Euclidean-adapted scheme:

1. identify the transformations in spacetime containing events,
2. define mathematical objects that transform like events,
3. write the physical laws in a covariant form so that they comply with the principle of relativity.

We are now ready to provide a geometric formulation of special relativity following the scheme outlined above.

2.3.2 Vectors in Minkowski Space

The physical universe, where phenomena occur as a succession of events, is described by *Minkowski space*, $\mathbf{M}(3, 1)$, endowed with three spatial dimensions and one temporal dimension. The distinction between spatial and temporal dimensions emphasizes the difference between $\mathbf{M}$ and four-dimensional Euclidean space $\mathbf{R}^4$.

Points in Minkowski space are called *events*. The coordinates of an event occurring at time t at the spatial point (x, y, z) are defined as

$$x^\mu = (x^0, x^1, x^2, x^3) = (ct, x, y, z). \tag{2.44}$$

By definition, the metric in Minkowski space is

$$ds^2 = \sum_{\mu=0}^{3}\sum_{\nu=0}^{3} g_{\mu\nu}dx^\mu dx^\nu \equiv g_{\mu\nu}dx^\mu dx^\nu = (dx^0)^2 - (d\mathbf{x})^2 = c^2dt^2 - d\mathbf{x}^2, \tag{2.45}$$

where $g_{\mu\nu} = \mathrm{diag}(1, -1, -1, -1)$ is the metric tensor, which is symmetric, i.e., $g_{\mu\nu} = g_{\nu\mu}$. Moreover, Eq. (2.45) represents the spacetime interval between events, which can be classified as:

1. $ds^2 > 0$: *timelike*,
2. $ds^2 = 0$: *lightlike*,
3. $ds^2 < 0$: *spacelike*.

The metric tensor $g_{\mu\nu}$ plays the role of raising or lowering indices of objects in $\mathbf{M(3, 1)}$. Before providing a practical example of this property, it is useful to introduce a nomenclature for "upper" and "lower" indices. A coordinate of the type x^μ is called contravariant, while one of the type x_μ is covariant. Using the metric tensor to lower an index, we have

$$x_\mu = g_{\mu\nu}x^\nu = g_{\mu 0}x^0 + g_{\mu 1}x^1 + g_{\mu 2}x^2 + g_{\mu 3}x^3. \tag{2.46}$$

In general, we can write

$$x^\nu = (x^0, x^1, x^2, x^3) \xrightarrow{g_{\mu\nu}} x_\mu = (x_0, x_1, x_2, x_3) = (x^0, -x^1, -x^2, -x^3). \tag{2.47}$$

Thus, within our convention rules, when lowering or raising an index, the sign of the temporal coordinate does not change, while the signs of the spatial coordinates are reversed.

We can also define the metric tensor as $g^{\mu\nu}$ such that $g^{\mu\alpha}g_{\alpha\nu} = \delta^\mu_\nu = g^\mu_\nu$, where δ^μ_ν is the four-dimensional Kronecker delta. The inverse matrix can then be written as $g^{\mu\nu} = (g^{-1})_{\mu\nu}$. In this case, g^μ_ν coincides with the identity matrix, so that its trace is $g^\mu_\mu = 4$.

2.3.3 Lorentz Transformations in Special Relativity

The isometries of Minkowski spacetime take the form

$$x'^{\mu} = \Lambda^{\mu}{}_{\nu} x^{\nu} + a^{\mu}. \tag{2.48}$$

The expression above is called a *Poincaré transformation*. If the pure translation term $a^{\mu} = 0$, the expression reduces to a *homogeneous Lorentz transformation*.

We can consider a special case where $\Lambda^{\mu}{}_{\nu} = \delta^{\mu}{}_{\nu}$, so that Eq. (2.48) becomes $x'^{\mu} = \delta^{\mu}{}_{\nu} x^{\nu} + a^{\mu} = x^{\mu} + a^{\mu}$, thus recovering a pure translation.
Let us focus on homogeneous Lorentz transformations,[2] ' i.e., setting $a^{\mu} = 0$. These transformations leave the Minkowski metric invariant:

$$\begin{aligned} ds'^2 &= g_{\mu\nu} dx'^{\mu} dx'^{\nu} = g_{\mu\nu} \Lambda^{\mu}{}_{\rho} dx^{\rho} \Lambda^{\nu}{}_{\sigma} dx^{\sigma} = \\ &= g_{\mu\nu} \Lambda^{\mu}{}_{\rho} \Lambda^{\nu}{}_{\sigma} dx^{\rho} dx^{\sigma} = g_{\rho\sigma} dx^{\rho} dx^{\sigma} = ds^2, \end{aligned} \tag{2.49}$$

that holds iff the following condition is satisfied:

$$\boxed{g_{\mu\nu} \Lambda^{\mu}{}_{\rho} \Lambda^{\nu}{}_{\sigma} = g_{\rho\sigma}}, \tag{2.50}$$

which can be written in matrix form as

$$\boxed{\Lambda^T g \Lambda = g}, \tag{2.51}$$

known as the *pseudo-orthogonality condition*.

Recall that $\Lambda \equiv \{\Lambda^{\mu}{}_{\nu}\}$ is a 4×4 matrix with components $\Lambda^{\mu}{}_{\nu}$:

$$\Lambda \equiv \{\Lambda^{\mu}{}_{\nu}\} = \begin{pmatrix} \Lambda^0_0 & \Lambda^0{}_1 & \Lambda^0{}_2 & \Lambda^0{}_3 \\ \Lambda^1{}_0 & \Lambda^1{}_1 & \Lambda^1{}_2 & \Lambda^1{}_3 \\ \Lambda^2{}_0 & \Lambda^2{}_1 & \Lambda^2{}_2 & \Lambda^2{}_3 \\ \Lambda^3{}_0 & \Lambda^3{}_1 & \Lambda^3{}_2 & \Lambda^3{}_3 \end{pmatrix}. \tag{2.52}$$

It is important to note that Lorentz transformations depend on six parameters. Considering that $\Lambda^{\mu}{}_{\nu}$ is a 4×4 matrix, there are initially 16 components. The pseudo-orthogonality condition provides 10 independent constraints (since both sides are

[2] In any case, even if translations are included, we would have $dx'^{\mu} = \Lambda^{\mu}{}_{\nu} dx^{\nu}$.

symmetric in indices ρ and σ), leaving $16 - 10 = 6$ real parameters. Adding the four translation constants yields the 10 parameters of the full Poincaré group.

The composition of two Lorentz matrices is

$$(\Lambda\Lambda')^{\mu}{}_{\nu} = \Lambda^{\mu}{}_{\alpha}\Lambda'^{\alpha}{}_{\nu}, \tag{2.53}$$

and the transpose is

$$(\Lambda^T)^{\mu}{}_{\nu} = \Lambda^{\nu}{}_{\mu}. \tag{2.54}$$

$\Lambda^T g\Lambda = g$ reduces to an ordinary orthogonality condition if $g = \mathbb{I} = \text{diag}(1, 1, 1, 1)$.

Now consider Lorentz transformations for covariant coordinates,

$$x'_{\mu} = \Lambda^{\nu}{}_{\mu}x_{\nu}. \tag{2.55}$$

Driving exercise

Let us find $\Lambda^{\nu}{}_{\mu}$. Note that $x'_{\mu} = g_{\mu\rho}x'^{\rho} = g_{\mu\rho}\Lambda^{\rho}{}_{\alpha}x^{\alpha}$, and also $\Lambda^{\nu}{}_{\mu}x_{\nu} = \Lambda^{\nu}{}_{\mu}g_{\nu\sigma}x^{\sigma}$, setting $\sigma = \alpha$:

$$g_{\mu\rho}\Lambda^{\rho}{}_{\alpha}x^{\alpha} = \Lambda^{\nu}{}_{\mu}g_{\nu\alpha}x^{\alpha} \quad \forall x^{\alpha}. \tag{2.56}$$

Thus, $g_{\mu\rho}\Lambda^{\rho}{}_{\alpha} = \Lambda^{\nu}{}_{\mu}g_{\nu\alpha}$. Multiplying both sides by $g^{\alpha\beta}$ gives

$$g^{\alpha\beta}g_{\mu\rho}\Lambda^{\rho}_{\alpha} = \Lambda^{\nu}{}_{\mu}g^{\alpha\beta}g_{\nu\alpha} = \Lambda^{\beta}{}_{\mu}, \tag{2.57}$$

where we used $g^{\alpha\beta}g_{\nu\alpha} = \delta^{\beta}{}_{\nu}$. Equation (2.57) is called the *transformation condition for covariant coordinates.*

Let us now relate $\Lambda^{\beta}{}_{\mu}$ to $\Lambda^{\mu}{}_{\nu}$. First, consider the Lorentz transformation for contravariant coordinates:

$$x'^{\mu} = \Lambda^{\mu}{}_{\nu}x^{\nu}. \tag{2.58}$$

Multiplying $\Lambda^{\mu}{}_{\nu}$ and $\Lambda^{\beta}{}_{\mu}$, and applying Eq. (2.57) on $\Lambda^{\beta}{}_{\mu}$, we obtain

$$\Lambda^{\mu}{}_{\nu}\Lambda^{\beta}{}_{\mu} = \Lambda^{\mu}{}_{\nu}g^{\alpha\beta}g_{\mu\rho}\Lambda^{\rho}{}_{\alpha} = g^{\alpha\beta}\left(g_{\mu\rho}\Lambda^{\mu}{}_{\nu}\Lambda^{\rho}{}_{\alpha}\right) = g^{\alpha\beta}g_{\nu\alpha} = \delta^{\beta}{}_{\nu}. \tag{2.59}$$

Note that in the last step, we used the pseudo-orthogonality condition $g_{\nu\alpha} = g_{\mu\rho}\Lambda^{\mu}{}_{\nu}\Lambda^{\rho}{}_{\alpha}$. Therefore, we can write

$$\Lambda_{\mu}{}^{\beta} = (\Lambda^{-1})^{\beta}{}_{\mu}. \tag{2.60}$$

Hence, covariant coordinates transform as

$$dx'_\mu = \Lambda_\mu{}^\nu dx_\nu = (\Lambda^{-1})^\nu{}_\mu dx_\nu, \tag{2.61}$$

with $\Lambda_\mu{}^\nu = \frac{\partial x^\nu}{\partial x'^\mu} = \frac{\partial x'_\mu}{\partial x_\nu}$.

2.3.3.1 Classification of Homogeneous Lorentz Transformations

A first classification of Lorentz transformations can be made based on the determinant of the corresponding matrices. Taking the determinant on both sides of $\Lambda^T g \Lambda = g$ we obtain

$$\det(\Lambda^T g \Lambda) = \det \Lambda^T \cancel{\det g} \det \Lambda = \cancel{\det g} \Rightarrow (\det \Lambda)^2 = 1 \Rightarrow \det \Lambda = \pm 1. \tag{2.62}$$

According to the determinant, transformations are divided into

1. $\det \Lambda = 1$: Lorentz transformations are called **proper**,
2. $\det \Lambda = -1$: Lorentz transformations are called **improper**.

Another characteristic of the transformations is the sign of Λ^0_0. Indeed, from $g_{\rho\sigma} = g_{\mu\nu}\Lambda^\mu{}_\rho \Lambda^\nu{}_\sigma$, setting $\rho = \sigma = 0$:

$$\begin{aligned} 1 = & g_{00} = g_{\mu\nu}\Lambda^\mu{}_0\Lambda^\nu{}_0 = (\Lambda^0_0)^2 - (\Lambda^i{}_0)^2 \\ & \Rightarrow (\Lambda^0_0)^2 = (\Lambda^i{}_0)^2 + 1 \geq 1 \Rightarrow (\Lambda^0_0)^2 - 1 \geq 0 \\ & \Rightarrow \Lambda^0_0 \geq +1, \quad \text{or} \quad \Lambda^0_0 \leq -1. \end{aligned} \tag{2.63}$$

If Λ^0_0 is positive, the temporal component does not change sign; if Λ^0_0 is negative, we have a time reversal. Thus, according to the sign of Λ^0_0:

1. $\Lambda^0_0 \geq 0$: Lorentz transformations are called *orthochronous*,
2. $\Lambda^0_0 \leq 0$: Lorentz transformations are called *antichronous*.

Considering the above definitions, we can identify four main classes as in Table 2.2.

Table 2.2 Lorentz transformations classification

$\det \Lambda$	Λ^0_0	
+1	≥ 0	$L^\uparrow_+$
	≤ 0	$L^\downarrow_+$
−1	≥ 0	$L^\uparrow_-$
	≤ 0	$L^\downarrow_-$

The class $L_+^\uparrow$ contains Lorentz transformations that are proper and orthochronous, also called *restricted*. Moreover, $L_+^\uparrow$ includes velocity transformations or boosts and proper rotations, $L_-^\uparrow$ includes spatial inversion or parity, $L_-^\downarrow$ includes time reversal, and $L_+^\downarrow$ includes spacetime inversion. Let us examine these transformations in detail.

2.3.4 Velocity Transformations and Boosts

We define velocity transformations, or boosts, as Lorentz transformations connecting two inertial frames moving at constant velocity relative to each other. A boost along the x-axis can be written as

$$\begin{cases} x'^0 = \frac{x^0-\beta x^1}{\sqrt{1-\beta^2}}, \\ x'^1 = \frac{x^1-\beta x^0}{\sqrt{1-\beta^2}}, \\ x'^2 = x^2, \\ x'^3 = x^3, \end{cases} \tag{2.64}$$

where $\beta \equiv v/c$, with v the relative velocity between the two frames and c the speed of light. Introducing $\gamma \equiv 1/\sqrt{1-\beta^2}$, we can rewrite Eq. (2.64) as

$$\begin{cases} x'^0 = \gamma\left(x^0 - \beta x^1\right), \\ x'^1 = \gamma\left(x^1 - \beta x^0\right), \\ x'^2 = x^2, \\ x'^3 = x^3. \end{cases} \tag{2.65}$$

Driving exercise
How do we write the Λ as matrixes?

$$\Lambda_{BOOST} = \begin{pmatrix} \gamma & -\beta\gamma & 0 & 0 \\ -\beta\gamma & \gamma & 0 & 0 \\ 0 & 0 & 1 & 0 \\ 0 & 0 & 0 & 1 \end{pmatrix}. \tag{2.66}$$

Let us now verify that these transformations are indeed proper and orthochronous. For properness, we check that $\det\Lambda = 1$, while for orthochronous behavior, we require $\Lambda^0_0 \geq 0$. Computing the determinant:

$$\det \Lambda_{BOOST} = \gamma^2 - \beta^2\gamma^2 = \frac{1-\beta^2}{1-\beta^2} = 1, \tag{2.67}$$

showing that the transformation is proper. To show it is orthochronous, consider the first matrix element $(\Lambda_{BOOST})^0{}_0 = \gamma > 0$, confirming that it is indeed orthochronous.

We can also express Λ_{BOOST} in terms of the *rapidity* parameter α:

$$\Lambda_{BOOST} = \begin{pmatrix} \cosh\alpha & -\sinh\alpha & 0 & 0 \\ -\sinh\alpha & \cosh\alpha & 0 & 0 \\ 0 & 0 & 1 & 0 \\ 0 & 0 & 0 & 1 \end{pmatrix}, \tag{2.68}$$

where the rapidity α is defined via

$$\frac{\sinh\alpha}{\cosh\alpha} = \frac{\beta\gamma}{\gamma} = \beta \rightarrow \alpha = \tanh^{-1}\left(\frac{v}{c}\right). \tag{2.69}$$

Accordingly, we identify the following special cases.

- **Proper rotation.** In general, a rotation is represented by a matrix of the form

$$\Lambda_R(\theta) = \begin{pmatrix} 1 & 0\ 0\ 0 \\ 0 & \\ 0 & R(\theta) \\ 0 & \end{pmatrix}, \tag{2.70}$$

where $R(\theta)$ is a 3×3 orthogonal matrix.
From the rotation matrix, we have $\Lambda^0_0 = 1 > 0$, and therefore rotations are orthochronous. If $\det R = +1$, then $\det \Lambda_R = +1$ and the rotation is proper; if $\det R = -1$, then $\det \Lambda_R = -1$ and the rotation is improper.
- **Spatial inversion or parity.**
Spatial inversion, or parity transformation, is the Lorentz transformation defined by

$$\begin{cases} x'^0 = x^0, \\ x'^i = -x^i. \end{cases} \tag{2.71}$$

The transformation is represented by the matrix

$$\Lambda_P = \begin{pmatrix} 1 & 0 & 0 & 0 \\ 0 & -1 & 0 & 0 \\ 0 & 0 & -1 & 0 \\ 0 & 0 & 0 & -1 \end{pmatrix}. \tag{2.72}$$

From the first matrix element $(\Lambda_P)^0{}_0 = 1$ we see that the transformation is orthochronous, while from the determinant $\det \Lambda_P = -1$ we see that it is improper.

- **Time reversal.**
  Time reversal is characterized by

$$\begin{cases} x'^0 = -x^0, \\ x'^i = x^i. \end{cases} \tag{2.73}$$

  In matrix form:

$$\Lambda_T = \begin{pmatrix} -1 & 0 & 0 & 0 \\ 0 & 1 & 0 & 0 \\ 0 & 0 & 1 & 0 \\ 0 & 0 & 0 & 1 \end{pmatrix}. \tag{2.74}$$

  From the first matrix element $(\Lambda_T)^0{}_0 = -1$, the time reversal is antichronous, and from the determinant $\det \Lambda_T = -1$, we conclude that it is improper.
- **Space-time inversion.**

  Space-time inversion is characterized by

$$\begin{cases} x'^0 = -x^0, \\ x'^i = -x^i. \end{cases} \tag{2.75}$$

  Its matrix representation is given by the product of the time reversal and spatial inversion matrices:

$$\Lambda_I = \Lambda_T \Lambda_P = \begin{pmatrix} -1 & 0 & 0 & 0 \\ 0 & -1 & 0 & 0 \\ 0 & 0 & -1 & 0 \\ 0 & 0 & 0 & -1 \end{pmatrix}. \tag{2.76}$$

  From the first matrix element $(\Lambda_I)^0{}_0 = -1$, we deduce that the spacetime inversion is antichronous, and from the determinant $\det \Lambda_I = +1$, it is an proper transformation.
  It can be shown that any element of $L_+^\uparrow$ can be written as

$$\Lambda = \Lambda_{BOOST} \Lambda_R(\theta) = \Lambda_R(\theta) \Lambda_{BOOST}. \tag{2.77}$$

  Moreover, starting from an element of $L_+^\uparrow$, one can obtain elements of:

1. $L_-^\uparrow$ by applying Λ_P,
2. $L_-^\downarrow$ by applying Λ_T,
3. $L_+^\downarrow$ by applying Λ_I.

Special Relativity provides a unified spacetime framework in which discrete transformations such as parity, time reversal, and spacetime inversion can be defined geometrically as isometries of Minkowski spacetime. While analogous parity and time-reversal symmetries already exist in classical mechanics, their interpretation in Special Relativity is fundamentally spacetime-based, reflecting the Lorentzian structure rather than separate notions of space and time.

2.4 Covariant and Contravariant Tensors

Pause and think
Manage to understand by yourself how the above definitions can be extended to vectors.

In Minkowski space, it is possible to define objects that transform like events under Lorentz transformations. Restricting ourselves to Lorentz transformations belonging to the group $L_+^\uparrow$, we define a *contravariant four-vector* as

$$A^\mu = (A^0, A^1, A^2, A^3). \tag{2.78}$$

Under a Lorentz transformation, this quantity transforms as

$$A'^\mu = \Lambda^\mu{}_\nu A^\nu. \tag{2.79}$$

In general, the components A^μ may depend on x^μ, i.e., they can be *vector fields*, so that the transformation becomes

$$A'^\mu(x') = \Lambda^\mu{}_\nu A^\nu(x). \tag{2.80}$$

Pause and think
Try to understand how each component of a vector in Minkowski space might transform.

Consider now a Lorentz transformation along a generic direction

$$\begin{cases} A'^0 = \gamma\left(A^0 - \boldsymbol{\beta}\cdot\mathbf{A}\right), \\ \mathbf{A}'_\parallel = \left(\mathbf{A}_\parallel - \boldsymbol{\beta}\mathbf{A^0}\right), \\ \mathbf{A}'_\perp = \mathbf{A}_\perp, \end{cases} \tag{2.81}$$

where $\boldsymbol{\beta} = \mathbf{v}/c$ defines the boost direction and $\mathbf{A}' = \mathbf{A}'_{\parallel} + \mathbf{A}'_{\perp}$. Substituting the expressions for $\mathbf{A}'_{\parallel}$ and $\mathbf{A}'_{\perp}$, we obtain

$$\mathbf{A}' = \mathbf{A}'_{\parallel} + \mathbf{A}'_{\perp} = \left(\mathbf{A}_{\parallel} - \boldsymbol{\beta}\mathbf{A^0}\right) + \mathbf{A}_{\perp} + \left(\mathbf{A}_{\parallel} - \mathbf{A}_{\parallel}\right) \tag{2.82}$$

$$= (\gamma - 1)\mathbf{A}_{\parallel} - \boldsymbol{\beta}\mathbf{A^0} + \mathbf{A}. \tag{2.83}$$

We can rewrite $\gamma - 1$ as $\gamma - 1 = \frac{\gamma^2-1}{\gamma+1} = \frac{\beta^2\gamma^2}{\gamma+1}$. Moreover, by construction $\mathbf{A}_{\perp} \cdot \boldsymbol{\beta} = \mathbf{0}$, so that multiplying both sides of $\mathbf{A} = \mathbf{A}_{\parallel} + \mathbf{A}_{\perp}$ by $\boldsymbol{\beta}$ gives

$$\boldsymbol{\beta} \cdot \mathbf{A} = \boldsymbol{\beta} \cdot \mathbf{A}_{\parallel} \quad \Rightarrow \quad \mathbf{A} = \mathbf{A}_{\parallel}. \tag{2.84}$$

Substituting these relations back, we finally get

$$\begin{cases} A'^0 = \gamma(A^0 - \boldsymbol{\beta} \cdot \mathbf{A}), \\ \mathbf{A}' = \mathbf{A} + \left(\frac{\mathbf{2}}{\mathbf{+1}}(\boldsymbol{\beta} \cdot \mathbf{A}) - \mathbf{A^0}\right)\boldsymbol{\beta}. \end{cases} \tag{2.85}$$

► **Pause and think**
Try to formulate how a covariant vector transforms.

A *covariant four-vector* A_μ is defined as

$$A_\mu = g_{\mu\nu}A^\nu = (A^0, -A^1, -A^2, -A^3) = (A^0, -\mathbf{A}). \tag{2.86}$$

Under a Lorentz transformation, it transforms like an event but with the inverse of $(\Lambda^\mu{}_\nu)$:

$$A'_\mu = \Lambda^\nu{}_\mu A_\nu = (\Lambda^{-1})^\nu{}_\mu A_\nu. \tag{2.87}$$

► **Pause and think**
Before going on, manage to understand by yourself how to extend vectors to tensors using the above definitions.

We can generalize these definitions to *four-tensors*. A (p, q)-type four-tensor of rank n is a collection of 4^n components $T^{\mu_1 \ldots \mu_p}{}_{\nu_1 \ldots \nu_q}$, where p counts contravariant indices and q covariant indices, with $p + q = n$. Under a Lorentz transformation, it transforms as

$$T'^{\mu_1 \ldots \mu_p}{}_{\nu_1 \ldots \nu_q} = (\Lambda^{\mu_1}{}_{\alpha_1} \ldots \Lambda^{\mu_p}{}_{\alpha_p} \Lambda^{\beta_1}{}_{\nu_1} \ldots \Lambda^{\beta_q}{}_{\nu_q}) T^{\alpha_1 \ldots \alpha_p}{}_{\beta_1 \ldots \beta_q}. \tag{2.88}$$

For rank $n = 1$, we have a *four-vector*, e.g., A^μ or A_μ. For rank $n = 2$, we have a *contravariant tensor* $T^{\mu\nu}$, a *covariant tensor* $T_{\mu\nu}$, or a *mixed tensor* $T^\mu{}_\nu$. Rank $n = 0$ corresponds to scalars.

The metric tensor $g_{\mu\nu}$ is a rank-2 tensor, since it satisfies

$$g_{\mu\nu}\Lambda^{\mu}{}_{\rho}\Lambda^{\nu}{}_{\sigma} = g_{\rho\sigma}. \tag{2.89}$$

Multiplying both sides by the inverses of Λ yields

$$g_{\alpha\beta} = g_{\rho\sigma}\Lambda^{\rho}{}_{\alpha}\Lambda^{\sigma}{}_{\beta}, \tag{2.90}$$

which confirms that $g_{\mu\nu}$ transforms as a covariant rank-2 tensor. In contrast, the transformation matrices $\Lambda^{\mu}{}_{\nu}$ themselves **do not** transform as tensors.

2.4.1 Properties of Four-Tensors

When writing a tensorial expression, certain rules must be followed.

- To write covariant laws explicitly, two tensors are set equal:

$$T^{\mu_1\ldots\mu_p}{}_{\nu_1\ldots\nu_q} = S^{\mu_1\ldots\mu_p}{}_{\nu_1\ldots\nu_q}. \tag{2.91}$$

  The *free indices* on both sides of the equation must be identical.
- If an index appears twice in the same term, it must appear once as a contravariant index and once as a covariant index:

$$g_{\alpha\beta} = g_{\rho\sigma}\Lambda^{\rho}{}_{\alpha}\Lambda^{\sigma}{}_{\beta}. \tag{2.92}$$

  Such an index is called a *dummy (contracted) index*, and summation over it is implied.
- A sum or linear combination of tensors of the same type is itself a tensor of that type:

$$\alpha T^{\mu_1\ldots\mu_p}{}_{\nu_1\ldots\nu_q} + \beta S^{\mu_1\ldots\mu_p}{}_{\nu_1\ldots\nu_q} = C^{\mu_1\ldots\mu_p}{}_{\nu_1\ldots\nu_q}. \tag{2.93}$$

- The product of a (p, q) tensor with an (r, s) tensor is a $(p + r, q + s)$ tensor, unless n contravariant indices are contracted with n covariant indices, in which case the resulting tensor has type $(p + r - n, q + s - n)$. Examples include:

$$A^{\mu}B^{\nu} = P^{\mu\nu} \quad \text{contravariant four-tensor of rank 2,} \tag{2.94a}$$

$$T^{\mu\nu\rho}S_{\sigma\tau} = P^{\mu\nu\rho}{}_{\sigma\tau} \quad \text{four-tensor of type (3,2), rank 5,} \tag{2.94b}$$

$$T^{\mu\nu\rho}S_{\rho\tau} = P^{\mu\nu}{}_{\tau} \quad \text{four-tensor of type (2,1), rank 3 (1 contracted index),} \tag{2.94c}$$

$$T^{\mu\nu\rho}S_{\alpha\mu\nu\rho} = P_{\alpha} \quad \text{covariant four-vector (3 contracted indices).} \tag{2.94d}$$

– Differentiating a four-tensor with respect to x^μ (or x_μ) yields another four-tensor with increased rank:

$$\frac{\partial a}{\partial x^\mu} \equiv \partial_\mu a = A_\mu, \tag{2.95a}$$

$$\frac{\partial a}{\partial x_\mu} \equiv \partial^\mu a = A^\mu. \tag{2.95b}$$

If a is a scalar,

$$da = \frac{\partial a}{\partial x^\mu} dx^\mu, \tag{2.96a}$$

$$da = \frac{\partial a}{\partial x_\mu} dx_\mu, \tag{2.96b}$$

i.e., contraction of a covariant and a contravariant four-vector is required for symmetry.

For a covariant derivative, the resulting tensor has type $(p, q+1)$:

$$\partial_\mu T^{\mu_1 \dots \mu_p}{}_{\nu_1 \dots \nu_q} = t^{\mu_1 \dots \mu_p}{}_{\nu_1 \dots \nu_q \mu}. \tag{2.97}$$

For a contravariant derivative, the resulting tensor has type $(p+1, q)$:

$$\partial^\mu T^{\mu_1 \dots \mu_p}{}_{\nu_1 \dots \nu_q} = t^{\mu_1 \dots \mu_p \mu}{}_{\nu_1 \dots \nu_q}. \tag{2.98}$$

– Scalars invariant under Lorentz transformations can always be constructed from four-tensors and four-vectors:

$$A^\mu B_\mu = g_{\mu\nu} A^\mu B^\nu = A^0 B^0 - \mathbf{A} \cdot \mathbf{B}, \tag{2.99a}$$

$$A^\mu A_\mu = (A^0)^2 - \mathbf{A}^2, \tag{2.99b}$$

$$T^{\mu\nu} S_{\mu\nu} = T^{\mu\nu} g_{\mu\rho} g_{\nu\sigma} S^{\rho\sigma}. \tag{2.99c}$$

The trace of a rank-2 four-tensor is also a scalar:

$$T^\mu{}_\mu = T^0{}_0 + T^1{}_1 + T^2{}_2 + T^3{}_3. \tag{2.100}$$

For the metric tensor:

$$g^\mu{}_\mu = g^0{}_0 + g^1{}_1 + g^2{}_2 + g^3{}_3 = \delta^\mu{}_\mu = 4. \tag{2.101}$$

Table 2.3 Main differences between Euclidian four-dimensional and Minkowski spaces

	$\mathbb{R}^4$	M(3,1)
Coordinates	$x^\mu = (x^1, x^2, x^3, x^4 = ict = ix^0)$	$x^\mu = (x^0, x^1, x^2, x^3)$
Metric	$g_{ij} = \text{diag}(1, 1, 1, 1)$	$g_{\mu\nu} = \text{diag}(1, -1, -1, -1)$

Difference between $\mathbb{R}^4$ and Minkowski space M(3,1). The metrics in $\mathbb{R}^4$ (ds_E^2) and M(3,1) (ds^2) are related by

$$\begin{aligned} ds^2 &= (dx^0)^2 - (dx^1)^2 - (dx^2)^2 - (dx^3)^2 \\ &= -(dx^1)^2 - (dx^2)^2 - (dx^3)^2 - (dx^4)^2 = -ds_E^2, \end{aligned} \tag{2.102}$$

where we used the Wick rotation, i.e., $x^0 = -ix^4$ (see Table 2.3).

2.4.2 Symmetric and Antisymmetric Tensors

A rank-2 tensor $T^{\mu\nu}$ is *symmetric* if $T^{\mu\nu} = T^{\nu\mu}$, and is *antisymmetric* if $T^{\mu\nu} = -T^{\nu\mu}$. Any tensor can be decomposed into its symmetric $T^{(\mu\nu)}$ and antisymmetric parts $T^{[\mu\nu]}$ as follows:

$$T^{\mu\nu} = T^{(\mu\nu)} + T^{[\mu\nu]}, \tag{2.103}$$

where

$$T^{(\mu\nu)} = \frac{T^{\mu\nu} + T^{\nu\mu}}{2}, \quad T^{[\mu\nu]} = \frac{T^{\mu\nu} - T^{\nu\mu}}{2}. \tag{2.104}$$

2.4.3 Pseudotensors and the Levi-Civita Symbol

If the Lorentz transformation with respect to which vectors or tensors in **M(3,1)** are defined is no longer restricted to $L_+^\uparrow$ but is any orthochronous transformation $\in L_\pm^\uparrow$ (including proper and improper rotations and spatial inversions), then one deals with *pseudotensors* or *pseudovectors*.

Pseudotensors transform as

$$T'^{\mu_1 \ldots \mu_p}_{\nu_1 \ldots \nu_q} = (\det \Lambda) \Lambda^{\mu_1}_{\alpha_1} \ldots \Lambda^{\mu_p}_{\alpha_p} \Lambda^{\beta_1}_{\nu_1} \ldots \Lambda^{\beta_q}_{\nu_q} T^{\alpha_1 \ldots \alpha_p}_{\beta_1 \ldots \beta_q}. \tag{2.105}$$

Pseudovectors transform as

$$A'^\mu = (\det \Lambda) \Lambda^\mu_\nu A^\nu, \tag{2.106}$$

and pseudoscalars as

$$A'(x') = (\det \Lambda) A(x). \tag{2.107}$$

The distinction between proper tensors and pseudotensors lies in the factor $\det \Lambda$: under spatial inversions, $\det \Lambda = -1$. In such cases, the objects are called *pseudotensors* and transform according to Eqs. (2.105)–(2.107).

Under a parity transformation $P = \text{diag}(1, -1, -1, -1)$, a polar four-vector transforms as

$$A^{\mu} \to (A^0, -\mathbf{A}), \tag{2.108}$$

whereas an axial pseudovector transforms as

$$A^{\mu} \to (-A^0, \mathbf{A}), \tag{2.109}$$

since it acquires an additional factor $\det P = -1$.

It follows that the product of two proper tensors or two pseudotensors yields a proper tensor, while the product of a proper tensor with a pseudotensor yields a pseudotensor.

We now introduce the *Levi-Civita symbol* $\varepsilon^{\mu\nu\rho\sigma}$, which is a completely antisymmetric pseudotensor:

$$\varepsilon^{\mu\nu\rho\sigma} = \begin{cases} 1, & \text{if}(\mu, \nu, \rho, \sigma) \text{ is an even permutation,} \\ -1, & \text{if}(\mu, \nu, \rho, \sigma) \text{ is an odd permutation,} \\ 0, & \text{if any two indices are equal.} \end{cases} \tag{2.110}$$

The Levi-Civita symbol has the following properties:

1. Being a rank-4 tensor, it would have $4^4 = 256$ components, but due to antisymmetry only $4! = 24$ independent components are nonzero:

$$\varepsilon^{\mu\nu\rho\sigma}\varepsilon_{\mu\nu\rho\sigma} = -24; \tag{2.111}$$

2.

$$\varepsilon^{\mu\nu\rho\sigma}\varepsilon_{\mu\nu\alpha\beta} = -2(\delta^{\rho}_{\alpha}\delta^{\sigma}_{\beta} - \delta^{\rho}_{\beta}\delta^{\sigma}_{\alpha}); \tag{2.112}$$

3. From the previous property, we also obtain

$$\varepsilon^{\mu\nu\rho\sigma}\varepsilon_{\lambda\nu\rho\sigma} = -6\delta^{\mu}_{\lambda}, \tag{2.113}$$

which follows by setting $\tau = \nu$ in

$$\varepsilon^{\mu\nu\rho\sigma}\varepsilon_{\lambda\tau\rho\sigma} = -2(\delta^{\mu}_{\lambda}\delta^{\nu}_{\tau} - \delta^{\mu}_{\tau}\delta^{\nu}_{\lambda}) = -2(\delta^{\mu}_{\lambda}\delta^{\tau}_{\tau} - \delta^{\mu}_{\tau}\delta^{\tau}_{\lambda}) = -6\delta^{\mu}_{\lambda}; \tag{2.114}$$

4. The determinant of a general 4×4 matrix M can be expressed as

$$\det M = \varepsilon^{\mu\nu\rho\sigma} M^0_\mu M^1_\nu M^2_\rho M^3_\sigma . \tag{2.115}$$

In particular, for a Lorentz transformation Λ, we have

$$\det \Lambda = \varepsilon^{\mu\nu\rho\sigma} \Lambda^0_\mu \Lambda^1_\nu \Lambda^2_\rho \Lambda^3_\sigma . \tag{2.116}$$

5. The Levi-Civita symbol transforms as a pseudotensor. Indeed, multiplying both sides of $\det \Lambda$ by $(\det \Lambda)\varepsilon^{0123}$ yields

$$\begin{aligned} (\varepsilon^{0123} \det \Lambda) \det \Lambda &= (\det \Lambda)\varepsilon^{\mu\nu\rho\sigma}\varepsilon^{0123}\Lambda^0_\mu \Lambda^1_\nu \Lambda^2_\rho \Lambda^3_\sigma \\ \longrightarrow \varepsilon^{0123} &= (\det \Lambda)\varepsilon^{\mu\nu\rho\sigma}\Lambda^0_\mu \Lambda^1_\nu \Lambda^2_\rho \Lambda^3_\sigma . \end{aligned} \tag{2.117}$$

The property discussed in point 2 above also provides a way to construct a generic antisymmetric matrix. Consider the matrix $M^{(\rho\sigma)}_{\alpha\beta}$ and let us verify its antisymmetry. For instance, take $\alpha = 1$ and $\beta = 0$ with $\rho, \sigma = 0, 1, 2, 3$, then

$$M^{(\rho\sigma)}_{10} = \delta^\rho_1 \delta^\sigma_0 - \delta^\rho_0 \delta^\sigma_1 , \tag{2.118}$$

where the first term indicates a nonzero entry at position $(1, 0)$, and the second term at position $(0, 1)$. In matrix form:

$$M^{(\rho\sigma)}_{10} = \begin{pmatrix} 0&0&0&0 \\ 1&0&0&0 \\ 0&0&0&0 \\ 0&0&0&0 \end{pmatrix} - \begin{pmatrix} 0&1&0&0 \\ 0&0&0&0 \\ 0&0&0&0 \\ 0&0&0&0 \end{pmatrix} = \begin{pmatrix} 0&-1&0&0 \\ 1&0&0&0 \\ 0&0&0&0 \\ 0&0&0&0 \end{pmatrix} , \tag{2.119}$$

which is manifestly antisymmetric.

In general, one can write the transformation of the Levi-Civita symbol as

$$\varepsilon^{\alpha\beta\gamma\delta} = (\det \Lambda)\, \Lambda^\alpha_\mu \Lambda^\beta_\nu \Lambda^\gamma_\rho \Lambda^\delta_\sigma \varepsilon^{\mu\nu\rho\sigma} . \tag{2.120}$$

2.5 Covariant Formulation of Dynamics

We now aim to exploit tensor calculus in the attempt to provide a covariant formulation of relativistic mechanics, such as velocity and momentum.

We start by considering the motion of a particle. Classically, it is described by a trajectory $x = x(t)$, but in relativity, the time t is not invariant and is therefore not a suitable parameter to describe motion. What we need is a *scalar parameter*. The natural choice for this parameter is given by

$$ds^2 = g_{\mu\nu} dx^\mu dx^\nu \begin{cases} > 0 & \text{timelike curve,} \\ = 0 & \text{lightlike curve,} \\ < 0 & \text{spacelike curve.} \end{cases} \tag{2.121}$$

Hence, we may conclude what follows.

- For *massive particles* $ds^2 > 0$, so we can parameterize the particle's motion using the invariant s as $x^\mu = x^\mu(s)$.
- For *massless particles* $ds^2 = 0$, in which case we cannot use s, and instead we introduce an affine parameter ξ, such that $x^\mu = x^\mu(\xi)$.

2.5.1 From Four-Velocity to four-momentum

We start with the definition of *four-velocity* for a particle whose motion is described along $x^\mu(s)$. The derivative of x^μ with respect to the invariant s is a four-vector, which we call the particle's four-velocity:

$$u^\mu = \frac{dx^\mu}{ds}. \tag{2.122}$$

Its components can be computed considering $\mu = 0$ (temporal) and $\mu = i$ (spatial).

For the temporal component, $\mu = 0$, taking into account $ds = \sqrt{c^2dt^2 - d\mathbf{x}^2}$ and $dx^0 = cdt$, we have:

$$u^0 = \frac{cdt}{cdt\sqrt{1 - \frac{d\mathbf{x}^2}{c^2dt^2}}} = \frac{\cancel{cdt}}{\cancel{cdt}\sqrt{1 - \frac{v^2}{c^2}}} = \frac{1}{\sqrt{1 - \frac{v^2}{c^2}}} = \gamma. \tag{2.123}$$

The spatial components, $\mu = i$, are:

$$u^i = \frac{dx^i}{ds} = \frac{\gamma}{c}\frac{dx^i}{dt} = \frac{\gamma}{c}v^i. \tag{2.124}$$

Introducing $\boldsymbol{\beta} = \mathbf{v}/c$, the components of the four-velocity are

$$u^\mu = (\gamma, \gamma\boldsymbol{\beta}). \tag{2.125}$$

Its squared norm is

$$u^\mu u_\mu = \frac{dx^\mu}{ds}\frac{dx_\mu}{ds} = \gamma^2 - \gamma^2\beta^2 = 1. \tag{2.126}$$

If we had used the parametrization $x^\mu = x^\mu(\tau)$, where τ is the proper time and $ds = c d\tau$, we would have:

$$u^\mu = \frac{dx^\mu}{d\tau} = (c\gamma, c\gamma\boldsymbol{\beta}) = (c\gamma, \gamma\mathbf{v}). \tag{2.127}$$

In this case, the norm is

$$u^\mu u_\mu = c^2\gamma^2 - \gamma^2 v^2 = c^2. \tag{2.128}$$

We now define the *four-acceleration* as the derivative of the four-velocity with respect to s:

$$w^\mu = \frac{du^\mu}{ds} = \frac{d^2x^\mu}{ds^2}. \tag{2.129}$$

The temporal component is

$$w^0 = \frac{du^0}{ds} = \frac{\gamma}{c}\frac{d\gamma}{dt} = \frac{\gamma}{c}\frac{d}{dt}\left[(1-\beta^2(t))^{-1/2}\right]. \tag{2.130}$$

Expanding the derivative gives

$$\frac{d\gamma}{dt} = \gamma^3\boldsymbol{\beta}\cdot\frac{d\boldsymbol{\beta}}{dt} = \frac{\gamma^3}{c}\mathbf{v}\cdot\mathbf{a} \quad \Rightarrow \quad w^0 = \frac{\gamma^4}{c^3}\mathbf{v}\cdot\mathbf{a}. \tag{2.131}$$

The spatial components are

$$w^i = \frac{\gamma^2}{c^2}\left[a^i + \frac{\gamma^2}{c^2}(\mathbf{v}\cdot\mathbf{a})v^i\right]. \tag{2.132}$$

Thus, the four-acceleration is

$$w^\mu = \left(\frac{\gamma^4}{c^3}\mathbf{v}\cdot\mathbf{a}, \frac{\gamma^2}{c^2}\left[\mathbf{a} + \frac{\gamma^2}{c^2}(\mathbf{v}\cdot\mathbf{a})\mathbf{v}\right]\right), \tag{2.133}$$

showing that the spatial component has a contribution along $\mathbf{v}$ depending on the relative orientation of $\mathbf{a}$ and $\mathbf{v}$.

The norm satisfies

$$\frac{d}{ds}(u^\mu u_\mu) = 2w^\mu u_\mu = 0, \tag{2.134}$$

so that w^μ and u_μ are orthogonal.

Finally, the *four-momentum* is defined. Classically, $\mathbf{p} = m\mathbf{v}$, so one might naively define $p^\mu \propto u^\mu$, but the proportionality constant must be determined experimentally to preserve the physics. Relativistic experiments show that the spatial component of four-momentum is

$$\mathbf{p} = m\gamma\mathbf{v}. \tag{2.135}$$

In the classical case, the linear relation $\mathbf{p} = m\mathbf{v}$ holds. Performing a Galilean velocity transformation $\mathbf{v}' = \mathbf{v} - \mathbf{V}$ gives $\mathbf{p}' = m\mathbf{v}' = m\mathbf{v} - m\mathbf{V}$. If $m\mathbf{v}$ is conserved, then $m\mathbf{v}'$ is also conserved, ensuring momentum conservation in all reference frames.

In the relativistic case, the transformation law changes, so momentum is no longer automatically conserved. To restore conservation, one should consider a *nonlinear relation* of the form

$$\mathbf{p} = m\, f(\beta^2)\mathbf{v}, \tag{2.136}$$

where f depends on $\mathbf{v}$ through $\beta = v/c$. In the Newtonian limit $v/c \ll 1$, one must have

$$f(0) = 1 \quad \Rightarrow \quad \mathbf{p} = m\mathbf{v}. \tag{2.137}$$

Consider an *elastic collision* between identical particles and impose the conservation of total momentum in three frames: k_1 moving with particle 1, k_2 moving with particle 2, and $k^\star$ moving with the center of mass (see Fig. 2.2).

In $k^\star$, the momenta and velocities of the two particles are equal and opposite in sign,

$$\mathbf{p}_1^\star = -\mathbf{p}_2^\star \quad \text{and} \quad \mathbf{v}_1^\star = -\mathbf{v}_2^\star. \tag{2.138}$$

Indeed, from Eq. (2.136),

$$\mathbf{p}_1^\star = -\mathbf{p}_2^\star \quad \Rightarrow \quad f(\mathbf{v}_1^\star)\mathbf{v}_1^\star = -f(\mathbf{v}_2^\star)\mathbf{v}_2^\star. \tag{2.139}$$

The two frames are completely equivalent, so, choosing k_2, we observe that k_1 is related to k_2 by a velocity boost of $-\mathbf{v}$, and that k_1 is obtained from a boost of $-\mathbf{v}_1^\star$ from $k^\star$, while k_2 is obtained from a boost of $-\mathbf{v}_2^\star = \mathbf{v}_1^\star$ from $k^\star$. We can then write the *momentum conservation law* in k_1 along the y direction (see Fig. 2.3):

$$-mf(v)v_y = -mf(u)u_y - mf(w)w_y. \tag{2.140}$$

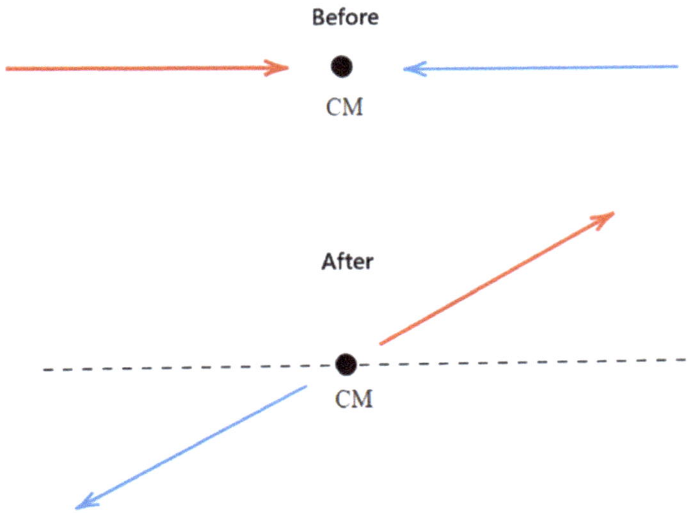

Fig. 2.2 Two-body collision between particle 1 (red) and particle 2 (blue) in the center-of-mass frame

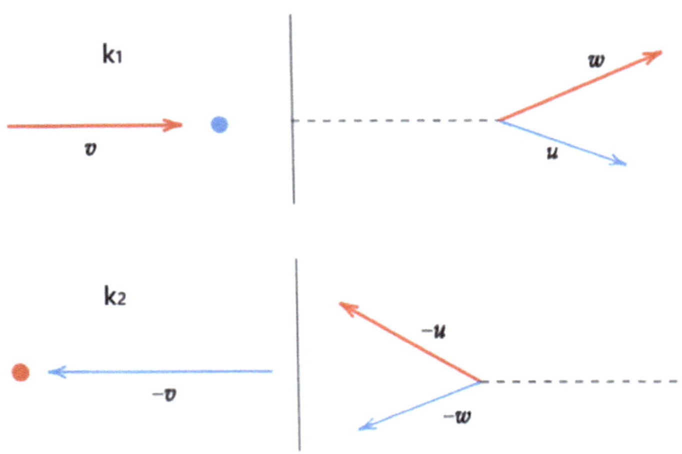

Fig. 2.3 Process seen by k_1 and k_2

Regarding the boost for the y component from k_2 ($\mathbf{u}$) to k_1 ($-\mathbf{w}$) along x:

$$- w_y = \frac{u_y\sqrt{1-(v/c)^2}}{1-\frac{(-v)u_x}{c^2}}. \tag{2.141}$$

The velocity transformation laws are obtained from Lorentz boosts. Specifically:

$$x^{0\prime} = \gamma(x^0 - \beta x^1) \quad \Rightarrow \quad dt' = \gamma\left(dt - \beta\frac{dx}{c}\right) = \gamma dt\left(1 - \frac{vu_x}{c^2}\right), \tag{2.142a}$$

$$x^{1\prime} = \gamma(x^1 - \beta x^0) \quad \Rightarrow \quad dx' = \gamma(dx - \beta c dt) = \gamma dt(u_x - v), \tag{2.142b}$$

$$x^{2\prime} = x^2 \quad \Rightarrow \quad dy' = dy, \tag{2.142c}$$

$$x^{3\prime} = x^3 \quad \Rightarrow \quad dz' = dz. \tag{2.142d}$$

Dividing Eqs. (2.142b)–(2.142d) by Eq. (2.142a) gives the velocity transformations:

$$u^{1\prime} = \frac{u_x - v}{1 - \frac{vu_x}{c^2}}, \tag{2.143a}$$

$$u^{2\prime} = \frac{u_y}{\gamma\left(1 - \frac{vu_x}{c^2}\right)} = \frac{u_y\sqrt{1-(v/c)^2}}{1 - \frac{vu_x}{c^2}}, \tag{2.143b}$$

$$u^{3\prime} = \frac{u_z}{\gamma\left(1 - \frac{vu_x}{c^2}\right)} = \frac{u_z\sqrt{1-(v/c)^2}}{1 - \frac{vu_x}{c^2}}. \tag{2.143c}$$

Using Eqs. (2.140)–(2.141), and noting that $v_y = 0$, rewriting Eq. (2.141) via Eq. (2.143b) gives

$$0 = -mf(u)u_y - mf(w)w_y, \tag{2.144a}$$

$$-w_y = \frac{u_y}{\gamma\left(1 + \frac{\mathbf{v}\cdot\mathbf{u}}{c^2}\right)}, \tag{2.144b}$$

where $-\mathbf{v}\cdot\mathbf{u} = -vu_x$. Substituting Eq. (2.144b) into Eq. (2.144a) yields

$$f(u) = \frac{f(w)}{\gamma\left(1 + \frac{\mathbf{v}\cdot\mathbf{u}}{c^2}\right)}. \tag{2.145}$$

In the particular case where particle 2 is at rest in frame k_2, we have $\mathbf{u} = 0$: Then,

$$1 = f(0) = \frac{f(w=v)}{\gamma} = \frac{f(v)}{\gamma} \quad \Rightarrow \quad f(v) = \gamma. \tag{2.146}$$

Thus, we have found the expression for the relativistic momentum. Substituting this into Eq. (2.136) gives

$$\mathbf{p} = m\gamma\mathbf{v}. \tag{2.147}$$

At this stage, a comparative analysis of the fundamental concepts can be found in Table 2.4.

Table 2.4 Covariant, invariant, and form-invariant quantities under symmetry transformations

Term	Transformation rule	Physical meaning
Covariant	Transforms according to a representation of the symmetry group	The quantity changes under the transformation, but in a well-defined way. Covariant objects ensure that equations retain their structure when written in a symmetry-respecting form
Invariant	Unchanged under the action of the group	The numerical value is preserved by the symmetry. Invariants correspond to physical observables or labels of representations, such as masses, charges, or Casimir eigenvalues
Form-invariant	Functional form preserved, components may transform	The mathematical form of the physical laws is unchanged, even though the individual quantities appearing in the equations transform covariantly

2.6 On-Shell Relation and Negative Energies

To obtain the temporal component of p^μ, we consider the relativistic kinetic energy expression, where the total energy E is the sum of kinetic energy T and rest energy $E_0 = mc^2$:

$$m\gamma c^2 = E = T + E_0. \tag{2.148}$$

One can show that there exists a relation between total energy and four-momentum, called the *mass-shell relation*:

$$\begin{cases} E = m\gamma c^2, \\ \mathbf{p} = m\gamma \mathbf{v}, \end{cases} \quad \Rightarrow \quad \begin{cases} E^2 = m^2\gamma^2 c^4, \\ \mathbf{p}^2 = m^2\gamma^2 \mathbf{v}^2. \end{cases} \tag{2.149}$$

Writing $E^2/c^2 = m^2\gamma^2c^2$ and subtracting the momentum squared gives

$$\frac{E^2}{c^2} - p^2 = m^2\gamma^2c^2\left(1 - \frac{v^2}{c^2}\right) = m^2\cancel{\gamma^2}c^2\frac{1}{\cancel{\gamma^2}} \quad \Rightarrow \quad \frac{E^2}{c^2} - p^2 = m^2c^2. \tag{2.150}$$

Manipulating the previous expression yields

$$E^2 = p^2c^2 + m^2c^4 \quad \Rightarrow \quad E = \pm\sqrt{p^2c^2 + m^2c^4}. \tag{2.151}$$

Note that this relation is analogous to the classical relation between energy and momentum of a free particle, $E = p^2/2m$. The negative sign corresponds to negative-energy states, which are not physically acceptable in classical physics, as there is no continuous process allowing transitions from $E > 0$ to $E < 0$ (Fig. 2.4).

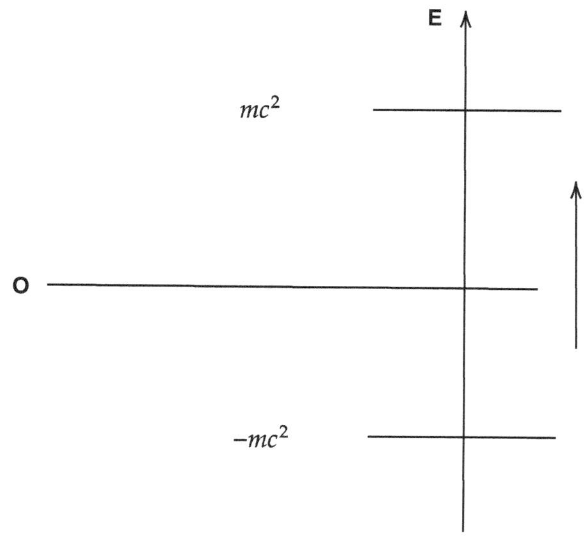

Fig. 2.4 The existence of negative-energy solutions implies a spectral gap, defined as the distance between the minimum of positive energy and the maximum of negative energy, namely $\Delta E \equiv mc^2 - (-mc^2) = 2mc^2$. So, any transition between them necessarily involves a *discrete energy exchange equal to twice the rest energy*. This picture will be reconsidered by the Dirac's hole theory

As clear from Fig. 2.3, between $E > mc^2$ and $E < -mc^2$ there is at least a $2mc^2$ discontinuity, which cannot be traversed by changing inertial frames.

In quantum physics, however, transitions between discrete energy states are allowed. In particular, this issue is resolved in QFT, where only positive energies are considered, and each particle is associated with an antiparticle having the same positive energy but opposite momentum.

We can now check the physical constraints on $p^\mu = mcu^\mu$ noting that:

$$p^0 = mc\gamma = \frac{E}{c}, \tag{2.152a}$$

$$p^i = m\gamma v^i, \tag{2.152b}$$

where p^0 comes from the relativistic total energy and p^i from the experimental relativistic momentum. Thus, we can write

$$p^\mu = \left(\frac{E}{c}, \mathbf{p}\right). \tag{2.153}$$

Note that E and $\mathbf{p}$ form a four-vector and therefore transform like an event. For a boost along x we have

$$\frac{E'}{c} = \frac{\frac{E}{c} - \beta p_x}{\sqrt{1-\beta^2}}, \tag{2.154a}$$

$$p'_x = \frac{p_x - \beta\frac{E}{c}}{\sqrt{1-\beta^2}}, \tag{2.154b}$$

$$p'_y = p_y, \tag{2.154c}$$

$$p'_z = p_z. \tag{2.154d}$$

The invariant square of p^μ is

$$\boxed{p^\mu p_\mu = \frac{E^2}{c^2} - \mathbf{p}^2 = m^2c^2.} \tag{2.155}$$

This represents the mass-shell condition, which is both a geometric and a physical constraint.

If we consider $u^\mu = (\gamma, \gamma\beta)$, then the invariant square is $u^\mu u_\mu = 1$, representing a geometric constraint. Therefore, for $p^\mu = mcu^\mu$ we have $p^\mu p_\mu = m^2c^2$, which is a physical constraint. This a posteriori determines the relation between p^μ and u^μ. If we had used $u^\mu = (\gamma c, \gamma \mathbf{v})$, then $u^\mu u_\mu = c^2$ and, in order to satisfy the mass-shell condition, we would have written $p^\mu = mu^\mu$.

One can also derive the Lorentz transformation laws for velocities, confirming that E and $\mathbf{p}$ form a four-vector.

The mass-shell relation $\frac{E^2}{c^2} - \mathbf{p}^2 = m^2c^2$ holds for massive particles. For massless particles, we have $\frac{E^2}{c^2} - \mathbf{p}^2 = 0$, i.e. $\frac{E^2}{c^2} = \mathbf{p}^2$. In the non-relativistic case, this would lead to a singularity since $E = \frac{p^2}{2m}$. Relativistic theory naturally accommodates such particles. Indeed, with $E = m\gamma c^2$ and $\mathbf{p} = m\gamma\mathbf{v}$, we get

$$v = \frac{pc^2}{E} \to v = \frac{p\cancel{c^2}}{\cancel{pc}} = c. \tag{2.156}$$

Thus, relativistic massless particles move at the speed of light. These are the *photons*.

Moreover, for $m = 0$, both $E = m\gamma c^2$ and $\mathbf{p} = m\gamma\mathbf{v}$, as well as $p^\mu = mcu^\mu$, lose meaning and u^μ cannot be defined because γ diverges. A new definition is introduced: for massless particles, satisfying the mass-shell condition $p^\mu p_\mu = 0$,

$$p^\mu = (|\mathbf{p}|, \mathbf{p}), \tag{2.157}$$

which is a *light-like four-vector*.

Using the relativistic momentum $\mathbf{p}$, one can write a *relativistic equation of motion*

$$\mathbf{F} = \frac{d\mathbf{p}}{dt} = \frac{d}{dt}(m\gamma\mathbf{v}), \tag{2.158}$$

known as *Minkowski's force law*. Explicitly calculating Eq. (2.158) gives

$$\mathbf{F} = m\frac{d\gamma}{dt}\mathbf{v} + m\gamma\frac{d\mathbf{v}}{dt} = m\frac{\gamma^3}{c^2}(\mathbf{v}\cdot\mathbf{a})\mathbf{v} + m\gamma\mathbf{a}, \tag{2.159}$$

where we assumed constant mass, i.e. $\frac{dm}{dt} = 0$, giving

$$\mathbf{F} = m\gamma\mathbf{a} + m\frac{\gamma^3}{c^2}(\mathbf{v}\cdot\mathbf{a})\mathbf{v}. \tag{2.160}$$

Unlike the classical law, the relativistic equation remains valid at high velocities since the $\gamma \to \infty$ factor regulates the growth of $\mathbf{v}$.

Rewriting Eq. (2.160) as $m\gamma\mathbf{a} = \mathbf{F} - m\frac{\gamma^3}{c^2}(\mathbf{v}\cdot\mathbf{a})\mathbf{v}$ and taking the scalar product with $\mathbf{v}$, we get

$$m\gamma(\mathbf{v}\cdot\mathbf{a}) = \mathbf{F}\cdot\mathbf{v} - m\frac{\gamma^3}{c^2}(\mathbf{v}\cdot\mathbf{a})v^2, \tag{2.161}$$

where $\mathbf{F}\cdot\mathbf{v}$ is the power. Rearranging, it gives

$$\mathbf{F}\cdot\mathbf{v} = m\gamma(\mathbf{v}\cdot\mathbf{a}) + m\frac{\gamma^3}{c^2}(\mathbf{v}\cdot\mathbf{a})v^2 = m\gamma(\mathbf{v}\cdot\mathbf{a})\left[1 + \gamma^2\frac{v^2}{c^2}\right] = m\gamma^3(\mathbf{v}\cdot\mathbf{a}), \tag{2.162}$$

yielding the *relativistic power*, which differs from the classical one by the factor γ^3.

We can also write the force as

$$\mathbf{F} = m\gamma\mathbf{a} + m\frac{\gamma^3}{c^2}(\mathbf{v}\cdot\mathbf{a})\mathbf{v} = m\gamma\mathbf{a} + (\mathbf{F}\cdot\mathbf{v})\frac{\mathbf{v}}{c^2} \quad \Rightarrow \quad m\gamma\mathbf{a} = \mathbf{F} - (\mathbf{F}\cdot\mathbf{v})\frac{\mathbf{v}}{c^2}. \tag{2.163}$$

Equations (2.160)–(2.163) are equivalent forms of Minkowski's force law. If $\mathbf{a} \parallel \mathbf{F} \parallel \mathbf{v}$, then

$$m\gamma\mathbf{a} = \mathbf{F} - \mathbf{F}\frac{v^2}{c^2} = \mathbf{F}(1-\beta^2) = \frac{\mathbf{F}}{\gamma^2} \quad \Rightarrow \quad \mathbf{F} = m\gamma^3\mathbf{a}, \tag{2.164}$$

analogous to the classical $\mathbf{F} = m\mathbf{a}$ except for the γ^3 factor.

2.6.1 Relativistic Angular Momentum

The relativistic generalization of the orbital angular momentum $\mathbf{L} = \mathbf{r}\times\mathbf{p}$ is not known *a priori*.

Assuming that all the aforementioned definitions work well, i.e., agree with the experiments, then it is possible to formulate an antisymmetric *angular momentum tensor*, in analogy to the classical definition, by

$$L^{\mu\nu} = x^\mu p^\nu - x^\nu p^\mu. \tag{2.165}$$

The tensor can be separated into

- **spatial parts,**

$$L^i = \frac{1}{2}\varepsilon^{ijk}L^{jk} = (L^{23}, L^{31}, L^{12}), \tag{2.166}$$

that reduces to the *exact* definition of the classical angular momentum;
- **temporal part,**

$$k^i = \frac{L^{i0}}{c} = \frac{x^i p^0 - x^0 p^i}{c} = \frac{x^i m\gamma c - ctm\gamma v^i}{c} = \tag{2.167}$$

$$= \frac{x^i(E/c) - ctp^i}{c} = x^i\frac{E}{c^2} - tp^i = m\gamma(x^i - v^i t). \tag{2.168}$$

The temporal part defines the *center-of-mass motion*:

$$\mathbf{k} = \mathbf{x}\frac{E}{c^2} - t\mathbf{p} = m\gamma(\mathbf{x} - \mathbf{v}t). \tag{2.169}$$

Driving exercise

Let us prove the above statement. Given a system of particles with masses m_a, we define the velocity of the center of mass in the relativistic case as

$$\mathbf{v}_{\text{CM}} = \frac{\sum_a m_a \gamma(v_a) \mathbf{v}_a}{\sum_a m_a \gamma(v_a)} = \frac{\sum_a \mathbf{p}_a c^2}{\sum_a m_a \gamma(v_a) c^2} = \frac{c^2 \mathbf{p}_{\text{tot}}}{E_{\text{tot}}}. \tag{2.170}$$

This expression reduces to the Newtonian case for $\gamma \to 1$.
In this context, the quantity $\mathbf{k}$ is written as

$$\mathbf{k}_{\text{tot}} = \frac{1}{c^2} \sum_a E_a \mathbf{x}_a - t \sum_a \mathbf{p}_a. \tag{2.171}$$

Defining analogously to $\mathbf{v}_{\text{CM}}$ the center-of-mass position as $\mathbf{x}_{\text{CM}} = \frac{\sum_a m_a \gamma(v_a) \mathbf{x}_a}{\sum_a m_a \gamma(v_a)}$ and substituting into Eq. (2.171), we obtain

$$\mathbf{k}_{\text{tot}} = \frac{1}{c^2} \left[\sum_a m_a \gamma(v_a) c^2 \right] \mathbf{x}_{\text{CM}} - t \mathbf{p}_{\text{tot}} = \frac{1}{c^2} E_{\text{tot}} \mathbf{x}_{\text{CM}} - t \mathbf{p}_{\text{tot}}. \tag{2.172}$$

Dividing by E_{tot}/c^2 gives

$$\mathbf{x}_{\text{CM}} - \frac{t c^2}{E_{\text{tot}}} \mathbf{p}_{\text{tot}} = \frac{c^2 \mathbf{k}_{\text{tot}}}{E_{\text{tot}}} \quad \Rightarrow \quad \mathbf{x}_{\text{CM}} = \mathbf{v}_{\text{CM}} t + \frac{c^2 \mathbf{k}_{\text{tot}}}{E_{\text{tot}}}, \tag{2.173}$$

where $c^2 \mathbf{k}_{\text{tot}}/E_{\text{tot}}$ is the *center-of-mass position shift*. The expression for the temporal part has no correspondence in classical physics, indicating that additional contributions to the angular momentum are forced to exist in special relativity.

2.6.2 Covariant Equations of Motion

To define the equations of motion in covariant form, we consider three cases:

1. *free particle with* $m > 0$,
2. *massless particle* $m = 0$,
3. *particle under interaction*.

We first examine the case of a ***free particle*** with $m > 0$.

Classically, $\frac{d\mathbf{p}}{dt} = 0$, while in the relativistic case the covariant equation of motion for a free particle reads

$$\frac{dp^\mu}{ds} = 0 \quad \text{with} \quad p^\mu p_\mu = m^2 c^2. \tag{2.174}$$

Considering $p^\mu = mcu^\mu$ and substituting into Eq. (2.174) yields

$$mc\frac{du^\mu}{ds} = 0 \quad \text{with} \quad u^\mu u_\mu = 1. \tag{2.175}$$

This expression is equivalent to the non-covariant equations

$$\frac{d\gamma}{dt} = 0 \quad \text{and} \quad \frac{d}{dt}(m\gamma\mathbf{v}) = 0. \tag{2.176}$$

Substituting $u^\mu = \frac{dx^\mu}{ds}$ into Eq. (2.175) gives

$$\frac{d^2x^\mu}{ds^2} = 0 \quad \text{with} \quad dx^\mu dx_\mu = ds^2, \tag{2.177}$$

which is equivalent to the non-covariant equation

$$\frac{d^2\mathbf{x}}{dt^2} = 0. \tag{2.178}$$

All these equations are equivalent to three independent non-covariant forms due to the constraints, with the difference that $u^\mu u_\mu = 1$ has a geometric origin from the choice of the parameter s, while $p^\mu p_\mu = m^2c^2$ has a physical origin, representing an ineliminable link between mass, momentum, and energy.

Next, we consider the case of a ***massless particle*** $m = 0$.

The equations of motion are written in terms of an affine parameter ξ since $ds^2 = dx^\mu dx_\mu = 0$:

$$\frac{dp^\mu}{d\xi} = 0 \quad \Rightarrow \quad \frac{d^2x^\mu}{d\xi^2} = 0, \tag{2.179}$$

with $p^\mu p_\mu = 0$ and $dx^\mu dx_\mu = ds^2 = 0$.

Finally, for a ***particle under interaction***, we introduce the concept of *four-force* as a four-vector:

$$\mathcal{F}^\mu = \frac{dp^\mu}{ds}. \tag{2.180}$$

This is called the *covariant (Minkowski) equation of motion*. Its temporal and spatial components are

$$\mathcal{F}^\mu = (\mathcal{F}^0, \boldsymbol{\mathcal{F}}) = \left(\frac{d(E/c)}{ds}, \frac{d\mathbf{p}}{ds}\right) = \left(\frac{\gamma}{c^2}\frac{dE}{dt}, \frac{\gamma}{c}\frac{d\mathbf{p}}{dt}\right), \tag{2.181}$$

using $ds = \frac{c\,dt}{\gamma}$.

Since the non-covariant Minkowski equation is $\mathbf{F} = d\mathbf{p}/dt$ and relativistic power is

$$\frac{dE}{dt} = \frac{d}{dt}(m\gamma c^2) = mc^2\frac{d\gamma}{dt} = m\cancel{c^2}\frac{\gamma^3}{\cancel{c^2}}(\mathbf{v}\cdot\mathbf{a}) = \mathbf{F}\cdot\mathbf{v}, \tag{2.182}$$

we can rewrite Eq. (2.181)as

$$\mathcal{F}^\mu = \left(\frac{\gamma}{c^2}\mathbf{F}\cdot\mathbf{v},\, \frac{\gamma}{c}\mathbf{F}\right), \tag{2.183}$$

which depends explicitly on velocity. As a consequence, an important property of the four-force is

$$\mathcal{F}^\mu u_\mu = mc\frac{du^\mu}{ds}u_\mu = mcw^\mu u_\mu = 0. \tag{2.184}$$

Thus, the four-force is orthogonal to the four-velocity, which leads to the following relations.

1. The components of $\mathcal{F}^\mu$ always depend on velocity, indeed

$$\mathcal{F}^\mu u_\mu = 0 \quad\Rightarrow\quad \mathcal{F}^0 u_0 - \boldsymbol{\mathcal{F}}\cdot\mathbf{u} = \mathcal{F}^0\gamma - \boldsymbol{\mathcal{F}}\cdot\gamma\frac{\mathbf{v}}{c} = 0 \quad\Rightarrow\quad \mathcal{F}^0 = \frac{\boldsymbol{\mathcal{F}}\cdot\mathbf{v}}{c}. \tag{2.185}$$

2. The dependence occurs through an antisymmetric tensor:

$$\mathcal{F}^\mu = f^{\mu\nu}u_\nu \quad\text{with}\quad \mathcal{F}^\mu u_\mu = f^{\mu\nu}u_\mu u_\nu = 0. \tag{2.186}$$

To satisfy this condition, since $u_\mu u_\nu$ is symmetric, $f^{\mu\nu}$ must be antisymmetric.

A prime example is the *Lorentz force* on a charged particle in an electromagnetic field. Classically,

$$\mathbf{F} = q\mathbf{E} + \frac{q}{c}\mathbf{v}\times\mathbf{B}, \tag{2.187}$$

which depends linearly on velocity.
We can write

$$\mathcal{F}^\mu = \left(\frac{\gamma}{c^2}\mathbf{F}\cdot\mathbf{v},\, \frac{\gamma}{c}\mathbf{F}\right). \tag{2.188}$$

Substituting Eq. (2.187)into Eq. (2.188) and using $(\mathbf{v}\times\mathbf{B})\cdot\mathbf{v} = 0$, we obtain

$$\mathcal{F}^\mu = \left(\frac{\gamma}{c^2}q\mathbf{E}\cdot\mathbf{v},\, \frac{\gamma}{c}q\mathbf{E} + \frac{\gamma}{c^2}q\mathbf{v}\times\mathbf{B}\right) = \left(\frac{\gamma}{c}\frac{q}{c}\mathbf{E}\cdot\mathbf{v},\, \frac{q}{c}\gamma\left[\mathbf{E} + \frac{\mathbf{v}\times\mathbf{B}}{c}\right]\right), \tag{2.189}$$

with $\mathcal{F}^\mu = \widetilde{F}^{\mu\nu} u_\nu$, where $\widetilde{F}^{\mu\nu} = \frac{q}{c} F^{\mu\nu}$ is the *electromagnetic field tensor*:

$$\mathcal{F}^\mu = \widetilde{F}^{\mu\nu} u_\nu = \frac{q}{c} F^{\mu\nu} u_\nu, \tag{2.190}$$

with $u_\nu = (\gamma, -\gamma \mathbf{v}/c)$ and

$$F^{\mu\nu} = \begin{pmatrix} 0 & -E_x & -E_y & -E_z \\ E_x & 0 & -B_z & B_y \\ E_y & B_z & 0 & -B_x \\ E_z & -B_y & B_x & 0 \end{pmatrix}, \tag{2.191}$$

which is antisymmetric, containing the electric and magnetic field components. The temporal and spatial components of $\mathcal{F}^\mu$ are

$$\mathcal{F}^0 = \frac{q}{c}\left[-E_x\left(-\gamma\frac{v_x}{c}\right) - E_y\left(-\gamma\frac{v_y}{c}\right) - E_z\left(-\gamma\frac{v_z}{c}\right)\right] = \frac{q}{c}\frac{\gamma}{c}(\mathbf{E}\cdot\mathbf{v}), \tag{2.192a}$$

$$\mathcal{F}^i = \frac{q}{c}(E^i\gamma + \frac{\varepsilon^{ijk}}{c} v^j B^k). \tag{2.192b}$$

For instance, if we consider the spatial component x $(i = 1)$, we have

$$\mathcal{F}^1 = \frac{q}{c}\left(E_x\gamma + \gamma\frac{B_z v_y}{c} - \gamma\frac{B_y v_z}{c}\right) = \frac{q}{c}\gamma\left[E_x + \frac{1}{c}\left(B_z v_y - B_y v_z\right)\right]. \tag{2.193}$$

It should be noted that the force not included in the relativistic treatment is the *gravitational force*, understood as an interaction within Minkowski spacetime.

▶ **Pause and think**
Before going on, propose your generalized torque.

Finally, we can define the *generalized moment of the force*, namely the torque, again in analogy to the classical moment $\mathbf{M} = \mathbf{r} \times \mathbf{F}$:

$$N^{\mu\nu} = x^\mu \mathcal{F}^\nu - x^\nu \mathcal{F}^\mu, \tag{2.194}$$

which satisfies $N^{\mu\nu} = \frac{dL^{\mu\nu}}{ds}$ and is analogous to the fundamental law of dynamics $\mathbf{M} = \frac{d\mathbf{L}}{dt}$.

2.7 Phase Invariance and Doppler Effect

Let us consider a plane wave

$$\Psi = A \exp i(\omega t - \mathbf{k} \cdot \mathbf{x}), \tag{2.195}$$

with $\mathbf{k} = (k_x, k_y, 0) = (k \cos\theta, k \sin\theta, 0)$ the wavevector. Substituting the wavevector into the plane wave expression, we obtain

$$\begin{aligned} \Psi &= A \exp\left[i(\omega t - k\cos\theta \cdot x - k \sin\theta \cdot y)\right] \\ &= A \exp\left[2\pi\nu i \left(t - \frac{kx}{2\pi\nu}\cos\theta - \frac{ky}{2\pi\nu}\sin\theta\right)\right], \end{aligned} \tag{2.196}$$

where the relation between angular frequency and frequency, *i.e.*, $\omega = 2\pi\nu$, has been used. Moreover, from the dispersion relation $\omega = kc$ we have

$$\Psi = A \exp\left[2\pi\nu i \left(t - \frac{x\cos\theta + y\sin\theta}{c}\right)\right] = A\exp(2\pi i\phi), \tag{2.197}$$

where ϕ is the phase of the wave,

$$\phi = \nu\left(t - \frac{x\cos\theta + y\sin\theta}{c}\right) = \nu\left(t - \frac{r}{c}\right), \tag{2.198}$$

and ν represents the number of wave crests passing through a point with coordinates (x, y) per unit time. This point is reached by a crest originating at $(0, 0)$ after a time $\frac{r}{c}$.

▶ **Pause and think**
Manage to understand what is the phase and what it measures.

Therefore, the time $t - \frac{r}{c}$ appearing in Eq. (2.198) represents the time taken by an observer to count $\phi = \nu\left(t - \frac{r}{c}\right)$ crests.

Now consider a reference frame moving with respect to the previous one along the x-axis with velocity v, and suppose that at time $t' - \frac{r'}{c}$ the same number of crests is counted as at time $t - \frac{r}{c}$.

Hence, the phase counts the number of crests and, accordingly, might be relativistic invariant, so we have

$$\phi = \phi' \longrightarrow \nu\left(t - \frac{r}{c}\right) = \nu'\left(t' - \frac{r'}{c}\right). \tag{2.199}$$

Applying the Lorentz transformations given by Eq. (2.65) leads to

$$
\begin{aligned}
\phi' = \nu'\left[\gamma\left(t - \frac{vx}{c^2}\right) - \frac{x'\cos\theta' + y'\sin\theta'}{c}\right] = \\
= \nu'\left[\gamma\left(t - \frac{vx}{c^2}\right) - \frac{\gamma\,(x - vt)\cos\theta' + y\sin\theta'}{c}\right] = \\
= \nu'\left[\gamma t - \gamma\frac{vx}{c^2} - \frac{\gamma x}{c}\cos\theta' + \frac{\gamma v t}{c}\cos\theta' - \frac{y}{c}\sin\theta'\right] = \\
= \nu'\left[t\gamma\left(1 + \beta\cos\theta'\right) - \frac{x\gamma}{c}\left(\beta + \cos\theta'\right) - \frac{y}{c}\sin\theta'\right]
\end{aligned}
\tag{2.200}
$$

and

$$
\phi = \nu\left[t - \frac{x\cos\theta + y\sin\theta}{c}\right] = \nu\left[t - \frac{x\cos\theta}{c} - \frac{y\sin\theta}{c}\right]. \tag{2.201}
$$

Equating the corresponding terms in Eqs. (2.200) and (2.201 gives

$$
\begin{cases}
\nu'\cancel{t}\gamma(1 + \beta\cos\theta') = \nu\cancel{t}, \\
\nu'\cancel{\frac{x}{c}}\gamma(\beta + \cos\theta') = \nu\cancel{\frac{x}{c}}\cos\theta, \\
\nu'\cancel{\frac{y}{c}}\sin\theta' = \nu\cancel{\frac{y}{c}}\sin\theta,
\end{cases}
\longrightarrow
\begin{cases}
\nu = \nu'\gamma(1 + \beta\cos\theta'), \\
\nu\cos\theta = \nu'\gamma(\beta + \cos\theta'), \\
\nu\sin\theta = \nu'\sin\theta'.
\end{cases}
\tag{2.202}
$$

The first expression is known as the *frequency transformation law*, while dividing the last two expressions term by term in Eq. (2.202) yields the *aberration formula* (see Fig. 2.5)

$$
\tan\theta = \frac{\sin\theta'}{\gamma(\beta + \cos\theta')}. \tag{2.203}
$$

Let us consider a light ray emitted with frequency ν propagating in the xy plane, forming an angle θ with the x-axis

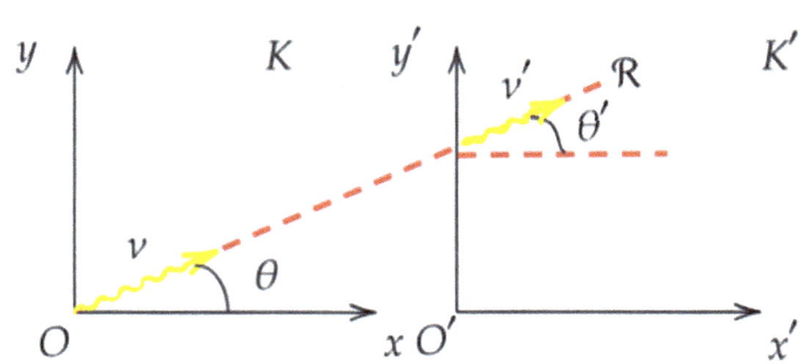

Fig. 2.5 Schematic description of aberration

The frame K' is moving with respect to the frame K with velocity v. The light ray emitted in the reference frame K is detected with a frequency $\nu' \neq \nu$ by a receiver $\mathcal{R}$ located in the reference frame K'. The phenomenon for which $\nu \neq \nu'$ is called the *Doppler effect*, while the phenomenon for which $\theta \neq \theta'$ is called *aberration.*

By solving for ν' in the first equation of Eqs. (2.202 and recalling $\gamma = (1 - \beta^2)^{-1/2}$, we obtain

$$\nu' = \frac{\nu\sqrt{1 - \frac{v^2}{c^2}}}{1 + \frac{v}{c}\cos\theta'}. \tag{2.204}$$

In the non-relativistic limit $\gamma \sim 1$, so that $\nu' \sim \frac{\nu}{1+\frac{v}{c}}$.

Pause and think

Remember that θ' can be any value (in principle). So, what would you expect from fixing it to special ones?

Let us now consider two effects depending on the direction of light propagation: the *longitudinal effect* and the *transverse effect.*

- In the longitudinal effect, light propagates along the direction of the relative motion between the source and the receiver. In this case $\theta' = 0$, yielding

$$\nu' = \frac{\nu\sqrt{1 - \frac{v^2}{c^2}}}{1 + \frac{v}{c}} = \nu\sqrt{\frac{\left(1 - \frac{v}{c}\right)\cancel{\left(1 + \frac{v}{c}\right)}}{\left(1 + \frac{v}{c}\right)^{\cancel{2}}}} = \nu\sqrt{\frac{1 - \frac{v}{c}}{1 + \frac{v}{c}}}. \tag{2.205}$$

In the non-relativistic limit, to first order in β, we obtain

$$\begin{aligned} \nu' &\simeq \nu\sqrt{(1-\beta)(1+\beta)^{-1}} = \nu\sqrt{(1-\beta)(1-\beta)} = \nu(1-\beta) \\ &= \begin{cases} > \nu & \text{if} \quad v < 0, \\ < \nu & \text{if} \quad v > 0. \end{cases} \end{aligned} \tag{2.206}$$

This shows that if the receiver approaches ($v < 0$) the source O, it measures a frequency $\nu' > \nu$, while if it recedes from the source ($v > 0$), it measures a frequency $\nu' < \nu$.

- The second effect, the transverse effect, occurs when light propagates perpendicular to the relative motion between the source and the receiver. In this case $\theta' = \frac{\pi}{2}$, giving

$$\nu' = \nu\sqrt{1 - \frac{v^2}{c^2}} = \nu\sqrt{1 - \beta^2}. \tag{2.207}$$

In the non-relativistic limit we obtain

$$\nu' \sim \nu\left(1 - \frac{\beta^2}{2}\right) < \nu, \tag{2.208}$$

which is smaller than the frequency ν because the term $\frac{\beta^2}{2}$ is always positive. This means that the receiver always measures a lower frequency.

2.8 Relativistic Spin

Spin is one of the fundamental properties of particles and is fully framed within relativistic quantum mechanics. Its introduction is attributed to George Uhlenbeck and Samuel Goudsmit in 1926 to explain the anomalous Zeeman effect. According to Uhlenbeck and Goudsmit's hypothesis, the electron possesses an intrinsic angular momentum, called *spin* σ, and a magnetic moment μ related to the spin by

$$\mu = g\frac{e}{2mc}\sigma = \frac{\not{2}e}{\not{2}mc}\sigma = \frac{e}{mc}\sigma, \tag{2.209}$$

where $g = 2$ is the *gyromagnetic factor* for the electron.

We define the *spin vector* $\mathbf{s}$ by considering $\mathbf{n}$ as the direction along which the expectation value of the spin ($\langle\sigma \cdot \mathbf{n}\rangle$) is maximal, so that

$$\mathbf{s} = \langle\sigma \cdot \mathbf{n}\rangle\mathbf{n}. \tag{2.210}$$

The spin vector can also be treated classically using the *Ehrenfest theorem*, which states that the expectation values of quantum observables obey classical equations of motion.

We now define the *spin four-vector* $S^\mu = (S^0, \mathbf{S})$ for massive particles. To find its form, we consider the spin and angular momentum in the particle's rest frame, so that

$$S^{\mu\star} = (0, \mathbf{s}) \quad \text{and} \quad p^{\mu\star} = (mc, 0, 0, 0). \tag{2.211}$$

Note that from these expressions $S^{\mu\star}p_{\mu\star} = 0$. Since this quantity is a Lorentz scalar and it vanishes in the rest frame, we have the general relation

$$S^\mu p_\mu = 0 \longrightarrow S^0 p_0 - \mathbf{S}\cdot\mathbf{p} = S^0 m\gamma c - m\gamma\mathbf{S}\cdot\mathbf{v} = 0 \longrightarrow S^0 = \frac{\mathbf{S}\cdot\mathbf{v}}{c} = \mathbf{S}\cdot\beta, \tag{2.212}$$

which relates the temporal component to the spatial components of S^μ.

To obtain $(S^0, \mathbf{S})$ in terms of $\mathbf{s}$, we write the Lorentz transformation relating $S^{\mu\star}$ to S^μ along a generic direction, obtaining

$$\begin{cases} S^{0\prime} = \gamma(S^0 + \beta\cdot\mathbf{s}), \\ \mathbf{S}' = \mathbf{s} + \left(\frac{\gamma^2}{\gamma+1}\beta\cdot\mathbf{s} + \gamma S^0\right)\beta, \end{cases} \longrightarrow \begin{cases} S^{0\prime} = \gamma\beta\cdot\mathbf{s}, \\ \mathbf{S}' = \mathbf{s} + \left(\frac{\gamma^2}{\gamma+1}\beta\cdot\mathbf{s}\right)\beta. \end{cases} \tag{2.213}$$

The general expression for S^μ is therefore

$$S^\mu = \left(\gamma\beta\cdot\mathbf{s}, \mathbf{s} + \frac{\gamma^2}{\gamma+1}(\beta\cdot\mathbf{s})\beta\right). \tag{2.214}$$

We verify that $S^0 = \gamma\beta\cdot\mathbf{s}$.

$$\begin{aligned} S^0 = \gamma\beta\cdot\mathbf{s} = \mathbf{S}\cdot\beta = \mathbf{s}\cdot\beta + \frac{\gamma^2}{\gamma+1}(\mathbf{s}\cdot\beta)\beta^2 &= \\ = (\mathbf{s}\cdot\beta)\left[1 + \frac{\beta^2}{1-\beta^2}\frac{1}{\gamma+1}\right] &= \\ = (\mathbf{s}\cdot\beta)\left[\frac{(1-\beta^2)\gamma + 1 - \cancel{\beta^2} + \cancel{\beta^2}}{(1-\beta^2)(\gamma+1)}\right] &= \\ = (\mathbf{s}\cdot\beta)\left[\frac{\cancel{(1-\beta^2)}\gamma}{\cancel{(1-\beta^2)}(\gamma+1)} + \frac{1}{(1-\beta^2)(\gamma+1)}\right] &= \\ = (\mathbf{s}\cdot\beta)\left[\frac{\gamma}{1+\gamma} + \frac{\gamma^2}{\gamma+1}\right] &= \\ = (\mathbf{s}\cdot\beta)\frac{\gamma\cancel{(\gamma+1)}}{\cancel{\gamma+1}} &= \gamma\mathbf{s}\cdot\beta. \end{aligned} \tag{2.215}$$

Moreover, the squared norm is given by

$$\begin{aligned} S^\mu S_\mu &= \gamma^2(\beta\cdot\mathbf{s})^2 - \mathbf{s}^2 - \frac{\gamma^4}{(\gamma+1)^2}(\beta\cdot\mathbf{s})^2 - 2\frac{\gamma^2}{\gamma+1}(\beta\cdot\mathbf{s})^2 = \\ &= -\mathbf{s}^2 + \gamma^2(\beta\cdot\mathbf{s})^2 - \frac{\gamma^4}{(\gamma+1)^2}\beta^2(\beta\cdot\mathbf{s})^2 - \frac{2\gamma^2}{\gamma+1}(\beta\cdot\mathbf{s})^2. \end{aligned} \tag{2.216}$$

Considering that we can write $\gamma^2 = \frac{1}{1-\beta^2}$, manipulating this relation we obtain $\beta^2 = \frac{\gamma^2-1}{\gamma^2}$.

Substituting this expression into Eq. (2.216) we find

$$\begin{aligned} S^\mu S_\mu = -\mathbf{s}^2 + \gamma^2(\beta\cdot\mathbf{s})^2 - (\beta\cdot\mathbf{s})^2\left[\frac{\gamma^4}{(\gamma+1)^2}\frac{(\gamma-1)(\gamma+1)}{\gamma^2} + \frac{2\gamma^2}{\gamma+1}\right] &= \\ = -\mathbf{s}^2 + \gamma^2(\beta\cdot\mathbf{s})^2 - (\beta\cdot\mathbf{s})^2\left[\gamma^2\frac{\gamma-1}{\gamma+1} + \frac{2\gamma^2}{\gamma+1}\right] &= \\ = -\mathbf{s}^2 + \gamma^2(\beta\cdot\mathbf{s})^2 - (\beta\cdot\mathbf{s})^2\left[\frac{\gamma^3 - \gamma^2 + 2\gamma^2}{\gamma+1}\right] &= \\ = -\mathbf{s}^2 + \gamma^2(\beta\cdot\mathbf{s})^2 - (\beta\cdot\mathbf{s})^2\left[\frac{\gamma^2\cancel{(\gamma+1)}}{\cancel{\gamma+1}}\right] &= -\mathbf{s}^2. \end{aligned} \tag{2.217}$$

Hence, the squared norm of the spin four-vector is

$$S^\mu S_\mu = -\mathbf{s}^2. \tag{2.218}$$

In the ultrarelativistic limit

$$S^\mu = \left(\gamma\beta\cdot\mathbf{s}, \mathbf{s} + \frac{\gamma^2}{\gamma+1}(\beta\cdot\mathbf{s})\beta\right) \sim (\gamma\beta\cdot\mathbf{s}, \gamma(\beta\cdot\mathbf{s})\beta) = (\gamma mc, \gamma mc\beta)\frac{\beta\cdot\mathbf{s}}{mc}$$
$$= (p^0, \mathbf{p})\frac{\beta\cdot\mathbf{s}}{mc} = \frac{\beta\cdot\mathbf{s}}{mc}p^\mu. \tag{2.219}$$

We define the ***helicity*** as

$$\lambda \equiv \beta\cdot\mathbf{s} = \frac{\mathbf{v}}{c}\langle\sigma\cdot\mathbf{n}\rangle\mathbf{n} = \langle\sigma\cdot\mathbf{n}\rangle\frac{\mathbf{v}}{c}\mathbf{n} = \langle\sigma\cdot\mathbf{n}\rangle\frac{\mathbf{p}\cdot\mathbf{n}}{m}. \tag{2.220}$$

As we will see, helicity is introduced with the aim of quantifying the physically measurable correlation, and it is the quantity to which relativistic interactions, especially parity violating ones, are sensitive.
Substituting the helicity into Eq. (2.219) yields

$$S^\mu = \frac{\lambda}{mc}p^\mu. \tag{2.221}$$

As we have seen, the orbital angular momentum of a particle is given by $L^{\mu\nu} = x^\mu p^\nu - x^\nu p^\mu$, which is antisymmetric. If the particle has non-zero spin, it possesses intrinsic angular momentum, represented by $S^{\mu\nu}$, which, when combined with the orbital angular momentum, gives the total angular momentum of a particle, expressed as

$$J^{\mu\nu} = L^{\mu\nu} + S^{\mu\nu}. \tag{2.222}$$

Consider the orbital contribution to the so-called **Pauli-Lubanski four-vector**,

$$W^\mu_{\text{orb}} \equiv \frac{1}{2}\varepsilon^{\mu\nu\rho\sigma}L_{\nu\rho}p_\sigma = \frac{1}{2}\varepsilon^{\mu\nu\rho\sigma}(x_\nu p_\rho - x_\rho p_\nu)p_\sigma. \tag{2.223}$$

The two terms cancel identically. Indeed, rename the dummy indices $\nu \leftrightarrow \rho$ in the second term and use the antisymmetry of $\varepsilon^{\mu\nu\rho\sigma}$:

$$
\begin{aligned}
W^{\mu}_{\rm orb} &= \frac{1}{2}\,\varepsilon^{\mu\nu\rho\sigma} x_\nu p_\rho p_\sigma - \frac{1}{2}\,\varepsilon^{\mu\nu\rho\sigma} x_\rho p_\nu p_\sigma \\
&= \frac{1}{2}\,\varepsilon^{\mu\nu\rho\sigma} x_\nu p_\rho p_\sigma - \frac{1}{2}\,\varepsilon^{\mu\rho\nu\sigma} x_\nu p_\rho p_\sigma \\
&= \frac{1}{2}\,\varepsilon^{\mu\nu\rho\sigma} x_\nu p_\rho p_\sigma + \frac{1}{2}\,\varepsilon^{\mu\nu\rho\sigma} x_\nu p_\rho p_\sigma = 0,
\end{aligned}
\tag{2.224}
$$

because $\varepsilon^{\mu\rho\nu\sigma} = -\varepsilon^{\mu\nu\rho\sigma}$. Equivalently, the contraction $\varepsilon^{\mu\nu\rho\sigma} p_\rho p_\sigma$ vanishes since $p_\rho p_\sigma$ is symmetric in (ρ, σ) while $\varepsilon^{\mu\nu\rho\sigma}$ is antisymmetric.

Therefore, the Pauli-Lubanski vector depends only on the intrinsic part of the angular momentum:

$$
W^{\mu} \equiv \frac{1}{2}\,\varepsilon^{\mu\nu\rho\sigma} J_{\nu\rho} p_\sigma = \frac{1}{2}\,\varepsilon^{\mu\nu\rho\sigma} S_{\nu\rho} p_\sigma. \tag{2.225}
$$

A direct consequence is the orthogonality relation

$$
W^{\mu} p_\mu = \frac{1}{2}\,\varepsilon^{\mu\nu\rho\sigma} S_{\nu\rho} p_\sigma p_\mu = 0, \tag{2.226}
$$

again because $p_\sigma p_\mu$ is symmetric while $\varepsilon^{\mu\nu\rho\sigma}$ is antisymmetric.

▶ **Driving exercise**
We can now distinguish two main cases, when mass is zero or non-zero.

- **Massive case.** For a massive particle ($p^2 = m^2c^2 > 0$) one can introduce the rest frame $p^{\mu} = (mc, \mathbf{0})$ and define a covariant spin four-vector S_{μ} orthogonal to the momentum,

$$
S^{\mu} p_\mu = 0, \qquad W^{\mu} = mc\, S^{\mu}, \tag{2.227}
$$

so that in the rest frame $W^{\mu} = (0, mc\,\mathbf{s})$ and $S^{\mu} = (0, \mathbf{s})$. The two Casimir invariants of the Poincaré group can then be written as

$$
p^2 = m^2c^2, \qquad W^2 = W_\mu W^\mu = -m^2c^2\,\mathbf{s}^2, \tag{2.228}
$$

showing that, for $m > 0$, the mass m and the spin magnitude $|\mathbf{s}|$ characterize the irreducible representation.

- **Massless case.** For a massless particle ($p^2 = 0$) no rest frame exists, hence one cannot define S^μ by boosting a rest-frame spin vector. Nevertheless, polarization remains well defined and is still encoded in W^μ. In the massless case the second Casimir reduces to

$$W^2 = 0, \tag{2.229}$$

so W^μ must be lightlike and orthogonal to p^μ. The physically relevant realization is that W^μ becomes collinear with p^μ,

$$W^\mu = \lambda\, p^\mu, \tag{2.230}$$

where the proportionality factor λ is Lorentz invariant. The scalar λ is the *helicity*, i.e. the projection of the spin along the direction of motion (in units of $\hbar$ in quantum theory). Thus, while for $m > 0$ the invariants are $(m, |\mathbf{s}|)$, for $m = 0$ the intrinsic label of the representation is the helicity λ, with the two physically distinct states corresponding to $\lambda = \pm s$ for a given spin s.

2.9 Proposed Exercises

Exercise 1

Consider an accelerated observer in flat Minkowski spacetime, whose coordinates (τ, ρ, y, z) are related to inertial coordinates (t, x, y, z) by

$$t = \rho \sinh(\alpha\tau), \qquad x = \rho \cosh(\alpha\tau), \qquad y = y, \qquad z = z, \tag{2.231}$$

where $\alpha > 0$ is a constant and units $c = 1$ are used.

1. Compute the spacetime metric in the accelerated coordinate system (τ, ρ, y, z) and identify the non-vanishing components of the metric tensor.
2. Consider a massive particle at rest in the inertial frame, with worldline $x = x_0$, $y = z = 0$. Determine its four-velocity components in the accelerated frame.
3. An observer following the trajectory $\rho = \rho_0 = \text{const.}$, $y = z = 0$ measures proper time τ. Compute the proper acceleration of this observer and discuss its dependence on ρ_0.

Exercise 2

Consider the following four-forces acting on a point particle with four-velocity u^μ,

$$\begin{aligned} \mathcal{G}_1^\mu &= C^\mu(x) + D_{\nu\rho}(x)\, u^\nu u^\rho\, u^\mu, \\ \mathcal{G}_2^\mu &= H^{\mu\nu}(x)\, u_\nu, \end{aligned} \tag{2.232}$$

where C^μ is a vector field, $D_{\nu\rho}$ is a rank–2 tensor field, and $H^{\mu\nu}$ is an arbitrary rank–2 tensor.

1. State the conditions under which $\mathcal{G}_1^\mu$ and $\mathcal{G}_2^\mu$ are orthogonal to the four-velocity.
2. Show explicitly if these conditions are independent of the choice of affine parameter along the worldline.
3. By comparing Eqs. (2.232) with the Lorentz force, identify the main structural differences.

Exercise 3

A particle of mass m moves along the x–axis under the action of a position–dependent *three–dimensional* force

$$f(x) = \frac{\beta\, mc^2}{(x_0 - x)^3} + \delta\, mc^2, \tag{2.233}$$

where β and δ are real constants and x_0 is a fixed length scale. Assume the motion is strictly one–dimensional.

(a) Describe qualitatively the motion of the particle and the role of β and δ.
(b) Write the relativistic equation of motion and compute the kinetic energy as a function of x. Determine the instantaneous power delivered by the force.
(c) Suppose that at time $t = t^\star$ the particle is located at $x = x^\star$ and has velocity $v(t^\star) = 0$. Determine the constants β and δ in terms of $x^\star$ and x_0. Compute the velocity of the particle at time $t = t^\star$ in the classical limit and compare it with the relativistic prediction.

Essential Bibliography

- Susskind L, Friedman A (2017) Special relativity and classical field theory. Basic Books.
  A clear treatment of Einstein's special relativity and the classical field formulation in a covariant framework.
- Rindler W (1991) Introduction to special relativity, 2nd edn. Oxford University Press (1991).
  A clear pedagogical treatment of relativistic kinematics, causality, and the geometric structure of Minkowski spacetime.
- Weinberg S (1995) The quantum theory of fields, vol. I Foundations. Cambridge University Press .
  A modern and systematic discussion of Lorentz and Poincaré symmetries, including their role in classifying relativistic states and conserved quantities.

Relativistic Field Theory 3

What we observe as material bodies and forces are nothing but shapes and variations in the structure of fields.

— Richard P. Feynman

Abstract

This chapter introduces the basic structure of relativistic field theory, motivating the field concept from a discrete mechanical model and progressively building the covariant formulation required by special relativity. Starting from a lattice of coupled masses and springs, the continuum limit is taken explicitly, leading to the emergence of a scalar field description, its Lagrangian and Hamiltonian formulations, and the relativistic wave equation. The correspondence between the continuum theory and the underlying lattice dynamics is examined in the long-wavelength regime, and an optional extension shows how a mass term naturally gives rise to the Klein-Gordon equation. The chapter then develops the symmetry framework of relativistic theories by introducing the Lorentz and Poincaré groups and discussing their representations. Relativistic Lagrangians are formulated in a manifestly covariant manner, emphasizing the role of geometric contractions and closure relations.

O. Luongo, *Fundamental Interactions*,
https://doi.org/10.1007/978-3-032-27413-7_3

Fig. 3.1 Portrait of Hermann Minkowski, 1896. Reprinted from Wikimedia. Public domain

A key character: *Hermann Minkowski (1864–1909), Fig.* 3.1.

Hermann Minkowski was born in Aleksotas, then part of the Russian Empire, into a family of Baltic German origin. He received his early education in Germany and quickly distinguished himself as an exceptional mathematician. At the age of eighteen he entered the University of Königsberg, where he studied mathematics under the supervision of Adolf Hurwitz and David Hilbert. His early work already revealed a remarkable ability to combine geometric intuition with analytical rigor.

Minkowski made foundational contributions to number theory and geometry before turning his attention to problems motivated by physics. His most profound and lasting impact on theoretical physics came from his reformulation of Einstein's special theory of relativity in geometric terms. In a celebrated lecture delivered in 1908, Minkowski introduced the concept of a four-dimensional continuum unifying space and time into a single geometric entity, now known as *Minkowski spacetime*. In this framework, physical events are represented as points in spacetime, while physical laws acquire a manifestly covariant form under Lorentz transformations.

The geometric structure introduced by Minkowski provided the natural setting for relativistic field theory. By embedding physical fields in spacetime and formulating their dynamics in terms of invariant intervals, four-vectors, and tensors, Minkowski laid the conceptual groundwork for constructing relativistically invariant Lagrangian densities. His insight that "space by itself, and time by itself, are doomed to fade away into mere shadows" marked a decisive shift in the understanding of relativistic physics, transforming covariance from a formal requirement into a geometric principle.
Minkowski's ideas profoundly influenced the subsequent development of both classical and quantum field theories. The formulation of relativistic wave equations, the classification of fields according to their transformation properties, and the modern use of action principles in spacetime all rely implicitly on the geometric language he introduced. Although his life was cut short at the age of forty-four, Minkowski's contribution permanently reshaped the conceptual foundations of modern theoretical physics.
Today, Hermann Minkowski is recognized not merely as a mathematician who refined Einstein's theory, but as the architect of the spacetime framework within which relativistic field theory is naturally formulated.

3.1 From Field Definition to Relativistic Fields

In classical physics, the concept of a field arose as a convenient way to describe interactions that act at a distance. In Newtonian mechanics, forces such as gravity were initially treated as instantaneous actions between point particles. However, this description is conceptually unsatisfactory when one considers the principle of locality: *physical influences should propagate continuously through space, rather than acting instantaneously over finite distances.*

The introduction of fields allows the description of interactions to be expressed as local quantities defined at every point in space and time. Formally, a basic example of field may be given as a function

$$\Phi(x^{\mu}), \tag{3.1}$$

which assigns a physical quantity, namely scalar, vector, tensor, or spinor, to each event in spacetime. *This concept naturally accommodates the idea of an energy density distributed throughout space, as opposed to being confined solely to point particles.*

In classical electrodynamics, the electromagnetic field $F_{\mu\nu}$ provides a paradigmatic example. Maxwell's equations demonstrate that the dynamics of the electromagnetic field can be fully described in terms of local field quantities, with the electric and magnetic fields $\mathbf{E}(\mathbf{x}, t)$ and $\mathbf{B}(\mathbf{x}, t)$ mediating forces between charged particles.

Importantly, the field propagates at a finite speed c, in contrast to the instantaneous action-at-a-distance of Newtonian gravity. Here, the field serves as the mediator of interactions, and its local character ensures compliance with the principle of causality.

A field is a function that assigns a physical quantity to every point in spacetime. In relativistic physics, fields must transform covariantly under Lorentz transformations to preserve the form of physical laws across all inertial frames.

The advent of special relativity introduced a radical conceptual shift in the formulation of physical laws. The invariance of the speed of light and the unification of space and time into Minkowski spacetime impose stringent constraints on the mathematical and physical structure of fields. Specifically, *any physical field must transform consistently under Lorentz transformations, ensuring that the form of the physical laws remains invariant in all inertial frames.*

This requirement implies that naive non-relativistic field formulations are inadequate: derivatives with respect to time and space must enter in combinations that are covariant under Lorentz transformations, such as $\partial_\mu \Phi$, to guarantee invariance of the action.

Consequently, the relativistic generalization of a field is not merely an extension of its classical counterpart but a profound restructuring. Scalars, vectors, and tensors in Minkowski spacetime are no longer independent objects; their components mix under Lorentz boosts, and the distinction between space and time derivatives is replaced by four-vector or tensor structures.

Moreover, the concept of a relativistic field is essential when considering the creation and annihilation of particles. In non-relativistic mechanics, particle number is conserved, and fields are secondary constructs, useful primarily as mediators of forces. However, in relativistic regimes, energy and mass equivalence, $E = mc^2$, allows processes in which particles can be created or destroyed. *A consistent description of such phenomena requires fields that can encode not only the dynamical evolution of particles but also the possibility of variable particle number. This requirement leads naturally to the formalism of QFT, where fields are promoted to operators acting on a Hilbert space of states, and particles arise as excitations of these underlying fields.*

Relativistic invariance also constrains the permissible interactions between fields.

Relativistic fields are not merely kinematic extensions of classical quantities. Their structure is dictated by symmetry principles, locality, and causality. This framework forms the bridge between classical field theory and modern QFT.

The notion of a relativistic field reconciles the local nature of interactions, the invariance under Lorentz transformations, and the capacity to describe processes involving creation and annihilation of particles, forming the essential conceptual framework for particle physics and QFT.

3.2 From Discrete to Continuous Scenarios

Pause and think
Sketch the reasons toward having fields in Nature. Moreover, try to imagine how many degrees of freedom might be associated with fields.

Let us try to clarify two more points.

- Why do we need fields?
- How many degrees of freedom does a field have?

To do so, consider a chain of identical point masses m located at equilibrium positions

$$x_n = n\,a, \qquad n \in \mathbb{Z}, \tag{3.2}$$

where a is the lattice spacing. Let $y_n(t)$ denote the transverse displacement of the n-th mass from its equilibrium position. Assume nearest-neighbour harmonic couplings with spring constant K. The discrete Lagrangian is taken as

$$L = \sum_n \left[\frac{m}{2}\, \dot{y}_n^2(t) - \frac{K}{2} \Big(y_{n+1}(t) - y_n(t) \Big)^2 \right], \tag{3.3}$$

where $\dot{y}_n \equiv dy_n/dt$. The first term represents kinetic energy, while the second term represents the elastic energy stored in the springs due to discrete gradients.

We now derive the equations of motion by applying the Euler-Lagrange equation to each discrete degree of freedom $y_n(t)$:

$$\frac{d}{dt}\left(\frac{\partial L}{\partial \dot{y}_n}\right) - \frac{\partial L}{\partial y_n} = 0. \tag{3.4}$$

From Eq. (3.3), we immediately obtain $\frac{\partial L}{\partial \dot{y}_n} = m\,\dot{y}_n$.

Driving exercise
For the potential part, we note that y_n appears in the two neighbouring spring terms:

$$\frac{K}{2}(y_{n+1} - y_n)^2, \qquad \frac{K}{2}(y_n - y_{n-1})^2.$$

Hence,

$$\frac{\partial}{\partial y_n}\left[-\frac{K}{2}(y_{n+1}-y_n)^2\right] = -\frac{K}{2}\cdot 2(y_{n+1}-y_n)\cdot(-1) \;=\; K\,(y_{n+1}-y_n), \tag{3.5}$$

$$\frac{\partial}{\partial y_n}\left[-\frac{K}{2}(y_n-y_{n-1})^2\right] = -\frac{K}{2}\cdot 2(y_n-y_{n-1})\cdot(1) \;=\; -K\,(y_n-y_{n-1}). \tag{3.6}$$

Summing the two contributions,

$$\frac{\partial L}{\partial y_n} = K\,(y_{n+1}-y_n) \;-\; K\,(y_n-y_{n-1}) = K\,(y_{n+1}-2y_n+y_{n-1}), \tag{3.7}$$

we end up with

$$m\,\ddot{y}_n \;=\; K\,(y_{n+1}-2y_n+y_{n-1}). \tag{3.8}$$

The right-hand side is the discrete Laplacian along the chain.

Pause and think
Propose a class of solutions and motivate your choice. Then, proceed with imposing the infinite chain case.

To extract the propagation properties of the discrete system, consider

$$y_n(t) = A\,e^{i(\omega t - k x_n)} = A\,e^{i(\omega t - kna)}. \tag{3.9}$$

Then, $\ddot{y}_n = -\omega^2 y_n$, $y_{n\pm1} = A\,e^{i(\omega t - k(n\pm1)a)} = y_n\,e^{\mp ika}$ and so the discrete Laplacian reads

$$y_{n+1}-2y_n+y_{n-1} = y_n\left(e^{-ika}-2+e^{ika}\right) = y_n\,(2\cos(ka)-2) = -4\,y_n\,\sin^2\!\left(\frac{ka}{2}\right). \tag{3.10}$$

Substituting Eqs. (3.9) and (3.10) into Eq. (3.8) yields $m(-\omega^2)y_n = K\left[-4\,y_n\,\sin^2\!\left(\frac{ka}{2}\right)\right]$, and therefore the dispersion relation becomes

$$\omega^2(k) = \frac{4K}{m}\,\sin^2\!\left(\frac{ka}{2}\right), \tag{3.11}$$

giving rise, at small wave number $|ka| \ll 1$, $\sin(ka/2) \simeq ka/2$ and thus

$$\omega^2(k) \simeq \frac{4K}{m}\left(\frac{ka}{2}\right)^2 = \frac{Ka^2}{m}\,k^2, \tag{3.12}$$

namely *a continuum wave equation with speed* $c^2 = Ka^2/m$.

We now construct the continuum field theory by introducing a smooth field $\phi(x,t)$ such that

$$\boxed{\phi(x_n,t) \equiv y_n(t), \qquad x_n = na}. \tag{3.13}$$

Assuming that ϕ varies slowly on the lattice scale, one may expand

$$\phi(x \pm a, t) = \phi(x,t) \pm a\, \partial_x \phi(x,t) + \frac{a^2}{2}\, \partial_x^2 \phi(x,t) + O(a^3). \tag{3.14}$$

From (3.14), the discrete first difference becomes

$$y_{n+1} - y_n = \phi(x+a,t) - \phi(x,t) = a\, \partial_x \phi(x,t) + \frac{a^2}{2}\, \partial_x^2 \phi(x,t) + O(a^3), \tag{3.15}$$

so that, to leading order,

$$(y_{n+1} - y_n)^2 = a^2\, (\partial_x \phi)^2 + O(a^3). \tag{3.16}$$

Similarly, the discrete second difference becomes

$$y_{n+1} - 2y_n + y_{n-1} = \phi(x+a,t) - 2\phi(x,t) + \phi(x-a,t) = a^2\, \partial_x^2 \phi(x,t) + O(a^4). \tag{3.17}$$

Finally, the discrete sum converts into an integral via

$$\sum_n (\cdots) \longrightarrow \int \frac{dx}{a}\, (\cdots), \tag{3.18}$$

since each lattice site occupies a length a.

Using Eq. (3.18) and the identifications (3.13)–(3.16), the discrete Lagrangian of Eq. (3.3) becomes

$$L = \sum_n \left[\frac{m}{2} \dot{y}_n^2 - \frac{K}{2} (y_{n+1} - y_n)^2 \right] \longrightarrow \int \frac{dx}{a} \left[\frac{m}{2}\, (\partial_t \phi)^2 - \frac{K}{2}\, a^2 (\partial_x \phi)^2 \right] + O(a). \tag{3.19}$$

It is convenient to introduce continuum parameters: the linear mass density ρ and the effective tension (stiffness) T,

$$\rho \equiv \frac{m}{a}, \qquad T \equiv Ka. \tag{3.20}$$

Then Eq. (3.19) becomes

$$L \longrightarrow \int dx \left[\frac{\rho}{2}\, (\partial_t \phi)^2 - \frac{T}{2}\, (\partial_x \phi)^2 \right]. \tag{3.21}$$

Hence the action $S = \int dt\, L$ is

$$S = \int dt\, dx\, \mathcal{L}, \qquad \boxed{\mathcal{L} = \frac{\rho}{2}\,(\partial_t \phi)^2 - \frac{T}{2}\,(\partial_x \phi)^2.} \tag{3.22}$$

This is the Lagrangian density of a real massless scalar field in $1+1$ dimensions, up to a normalization constant.

▶ **Pause and think**
Before going on, ask yourself:

- What are the roles of ρ and T?
- How could one reproduce the Euler-Lagrange equations?
- Is the Hamiltonian well-defined?

The continuum equations of motion can be readily derived starting from

$$\partial_\mu \left(\frac{\partial \mathcal{L}}{\partial(\partial_\mu \phi)} \right) - \frac{\partial \mathcal{L}}{\partial \phi} = 0, \qquad \mu \in \{t, x\}, \tag{3.23}$$

which, for constant ρ and T, give

$$\rho\, \partial_t^2 \phi - T\, \partial_x^2 \phi = 0. \tag{3.24}$$

Defining the propagation speed $c^2 \equiv \frac{T}{\rho}$, we immediately recognize the standard wave equation, as expected.

Nevertheless, a plane wave $\phi \propto e^{i(\omega t - kx)}$ yields $\omega^2 = c^2 k^2$. On the other hand, the discrete dispersion in Eq. (3.11) gives, for $|ka| \ll 1$, $\omega^2(k) \simeq \frac{Ka^2}{m}\, k^2 = \left(a^2 \frac{K}{m}\right) k^2$.

Therefore, we find a precise match with the long-wavelength limit, i.e. Eq. (3.12) of the lattice theory.

▶ **Pause and think**
Where is the mass?

In the above picture, mass is not present. Inertia is associated with forces, so that a common generalization is to add an on-site harmonic restoring force, which can model (for instance) a string coupled to an elastic substrate.

This corresponds to adding a term

$$V_{\text{site}} = \sum_n \frac{m\Omega^2}{2}\, y_n^2, \tag{3.25}$$

so that the discrete Lagrangian becomes

$$L = \sum_n \left[\frac{m}{2} \dot{y}_n^2 - \frac{K}{2} (y_{n+1} - y_n)^2 - \frac{m\Omega^2}{2} y_n^2 \right]. \tag{3.26}$$

Repeating the continuum limit exactly as above, the additional term contributes

$$-\sum_n \frac{m\Omega^2}{2} y_n^2 \longrightarrow -\int \frac{dx}{a} \frac{m\Omega^2}{2} \phi^2 = -\int dx \frac{\rho\Omega^2}{2} \phi^2. \tag{3.27}$$

Hence the continuum Lagrangian density becomes

$$\boxed{\mathcal{L} = \frac{\rho}{2} (\partial_t \phi)^2 - \frac{T}{2} (\partial_x \phi)^2 - \frac{\rho\Omega^2}{2} \phi^2.} \tag{3.28}$$

The corresponding Euler-Lagrange equation is

$$\rho\, \partial_t^2 \phi - T\, \partial_x^2 \phi + \rho\Omega^2\, \phi = 0, \tag{3.29}$$

or equivalently

$$\boxed{\partial_t^2 \phi - c^2\, \partial_x^2 \phi + \Omega^2\, \phi = 0,} \tag{3.30}$$

which turns out to be the so-called *Klein-Gordon equation in* $1 + 1$ *dimensions*, where the mass scale is set by Ω. We will discuss the Klein-Gordon equation, in more detail in Sect. 4.2.

3.2.1 Hamiltonian Formulation and Canonical Momentum

For the massless theory, see Eq. (3.22), the canonical momentum conjugate to ϕ is

$$\pi(x, t) \equiv \frac{\partial \mathcal{L}}{\partial(\partial_t \phi)} = \rho\, \partial_t \phi. \tag{3.31}$$

The Hamiltonian density is defined as

$$\mathcal{H} \equiv \pi\, \partial_t \phi - \mathcal{L}, \tag{3.32}$$

namely it involves derivatives w.r.t. the time only.

Accordingly, using Eqs. (3.31) and (3.22),

$$\mathcal{H} = \pi\left(\frac{\pi}{\rho}\right) - \left[\frac{\rho}{2}\left(\frac{\pi}{\rho}\right)^2 - \frac{T}{2}(\partial_x\phi)^2\right] = \frac{\pi^2}{\rho} - \frac{1}{2\rho}\pi^2 + \frac{T}{2}(\partial_x\phi)^2 = \frac{1}{2\rho}\pi^2 + \frac{T}{2}(\partial_x\phi)^2. \tag{3.33}$$

Including mass yields

$$\boxed{\mathcal{H} = \frac{1}{2\rho}\pi^2 + \frac{T}{2}(\partial_x\phi)^2 + \frac{\rho\Omega^2}{2}\phi^2}. \tag{3.34}$$

Since this is a Hamiltonian density, the full Hamiltonian is $H = \int dx\, \mathcal{H}$, *since we are dealing with one dimension.*

The entire above treatment suggests that

The continuum limit does not introduce additional spatial dimensions; rather, it promotes the discrete label of the oscillators to a continuous spatial coordinate, thereby yielding an infinite set of degrees of freedom. This implies

$$\phi : \mathbb{R} \to \mathbb{R}, \qquad x \mapsto \phi(x), \qquad N_{\text{dof}} = \text{card}(\mathbb{R}), \tag{3.35}$$

namely **in field theory the number of degrees of freedom is infinite.**

3.3 The Lorentz Group L

Let us follow here the prescriptions of Chap. 1. The Lorentz group L consists of homogeneous Lorentz transformations, isomorphic to $O(3, 1)$, and represents the group of isometries of $(3 + 1)$-dimensional Minkowski spacetime. It satisfies the following properties.

1. *Closure*: If $\Lambda, \Lambda' \in L$, then also $\Lambda\Lambda' \in L$. Indeed, from the definition of a pseudo-orthogonal group:

$$\Lambda^T g\Lambda = g, \quad \Lambda'^T g\Lambda' = g \implies (\Lambda\Lambda')^T g(\Lambda\Lambda') = g. \tag{3.36}$$

2. *Associativity*: This follows from the associativity of matrix multiplication

$$(\Lambda\Lambda')\Lambda'' = \Lambda(\Lambda'\Lambda''). \tag{3.37}$$

3. *Identity*: There exists $\mathbb{1}$ such that $\Lambda\mathbb{1} = \mathbb{1}\Lambda = \Lambda$, with

$$\mathbb{1} = \begin{pmatrix} 1\,0\,0\,0 \\ 0\,1\,0\,0 \\ 0\,0\,1\,0 \\ 0\,0\,0\,1 \end{pmatrix} \in L. \tag{3.38}$$

4. *Inverse*: There exists Λ^{-1} such that $\Lambda^{-1}\Lambda = \Lambda\Lambda^{-1} = \mathbb{1}$. One can verify that

$$(\Lambda^{-1})^T g \Lambda^{-1} = g \implies \Lambda^{-1} \in L. \tag{3.39}$$

Transformations with $\det \Lambda = 1$ and $\Lambda^0_0 \geq 0$ belong to the *proper Lorentz group* $L^\uparrow_+$. Improper Lorentz transformations do not form a closed group.

$L^\uparrow_+$ is isomorphic to the group of special pseudo-orthogonal matrices $SO(3,1)$ and contains both *boosts* and proper rotations (the latter being isomorphic to $SO(3)$). Altogether, $L^\uparrow_+$ has 6 independent parameters.

As stated in the previous chapter, an *irreducible representation* of $L^\uparrow_+$ is a map $D : G \longrightarrow GL(V)$ that admits no nontrivial invariant subspaces. A *reducible representation*, instead, admits invariant subspaces $W \subset V$, such that $D(g)w \in W$ for every $g \in G$ and $w \in W$.

Finite-dimensional irreducible representations are not unitary, since $L^\uparrow_+$ is noncompact. However, they are fundamental for describing how physical quantities and fields transform. On these representations, one defines the *Casimir operators*, which are invariant under $L^\uparrow_+$.

3.3.1 The Special Lorentz Group

We first derive the number of parameters of $L^\uparrow_+$. We start by considering an infinitesimal representation $\Lambda = \mathbb{1} + \omega$, which in components reads

$$\Lambda^\mu_\nu = \delta^\mu_\nu + \omega^\mu_\nu, \tag{3.40}$$

with $\Lambda^\mu_\nu \in L^\uparrow_+$ and $\omega^\mu_\nu \ll 1$. Considering $g_{\mu\nu}\Lambda^\mu_\rho\Lambda^\nu_\sigma = g_{\rho\sigma}$, we obtain

$$g_{\mu\nu}(\delta^\mu_\rho + \omega^\mu_\rho)(\delta^\nu_\sigma + \omega^\nu_\sigma) = g_{\mu\nu}\delta^\mu_\rho\delta^\nu_\sigma + g_{\mu\nu}\delta^\mu_\rho\omega^\nu_\sigma + g_{\mu\nu}\omega^\mu_\rho\delta^\nu_\sigma = g_{\rho\sigma}, \tag{3.41}$$

to first order. From Eq. (3.41) one finds

$$\cancel{g_{\rho\sigma}} + g_{\rho\nu}\omega^\nu_\sigma + g_{\mu\sigma}\omega^\mu_\rho = \cancel{g_{\rho\sigma}} \longrightarrow \omega_{\rho\sigma} = -\omega_{\sigma\rho}. \tag{3.42}$$

Thus, the parameters of a Lorentz transformation must satisfy the antisymmetry condition, giving 6 independent parameters. We now write $\omega_{\rho\sigma}$ as a linear combination of antisymmetric matrices[1] :

$$\omega_{\rho\sigma} = \frac{a_{\alpha\beta}}{2} A^{(\alpha\beta)}_{\rho\sigma} = i\frac{a_{\alpha\beta}}{2} M^{(\alpha\beta)}_{\rho\sigma}, \tag{3.43}$$

where $A^{(\alpha\beta)}_{\rho\sigma} = \delta^{\alpha}_{\rho}\delta^{\beta}_{\sigma} - \delta^{\alpha}_{\sigma}\delta^{\beta}_{\rho}$ and $M^{(\alpha\beta)}_{\rho\sigma} = -iA^{(\alpha\beta)}_{\rho\sigma}$. Raising and lowering indices, we obtain

$$\omega^{\mu}_{\ \nu} = g^{\mu\rho}\omega_{\rho\nu} = \frac{g^{\mu\rho}}{2} a_{\alpha\beta} A^{(\alpha\beta)}_{\rho\nu} = \frac{g^{\mu\rho}}{2} a_{\alpha\beta}(\delta^{\alpha}_{\rho}\delta^{\beta}_{\nu} - \delta^{\alpha}_{\nu}\delta^{\beta}_{\rho}) = \frac{a_{\alpha\beta}}{2}(g^{\mu\alpha}\delta^{\beta}_{\nu} - g^{\mu\beta}\delta^{\alpha}_{\nu}). \tag{3.44}$$

The equation above can also be written as

$$\omega^{\mu}_{\ \nu} = i\frac{a_{\alpha\beta}}{2} g^{\mu\rho} M^{(\alpha\beta)}_{\rho\nu} = i\frac{a_{\alpha\beta}}{2}(M^{(\alpha\beta)})^{\mu}_{\ \nu}. \tag{3.45}$$

▶ **Pause and think**
Try to find a generic basis for antisymmetric matrices.

Comparing Eqs. (3.44) and (3.45), we obtain the *generic basis for antisymmetric matrices*:

$$(M^{(\alpha\beta)})^{\mu}_{\ \nu} = -i(g^{\mu\alpha} g^{\beta}_{\nu} - g^{\mu\beta} g^{\alpha}_{\nu}). \tag{3.46}$$

Substituting into Eq. (3.40) gives

$$\Lambda^{\mu}_{\ \nu} = \delta^{\mu}_{\nu} + \omega^{\mu}_{\ \nu} = \delta^{\mu}_{\nu} + \frac{i}{2} a_{\alpha\beta}(M^{(\alpha\beta)})^{\mu}_{\ \nu}. \tag{3.47}$$

In matrix form

$$\Lambda = \mathbb{1} + \frac{i}{2} a_{\alpha\beta} M^{(\alpha\beta)}. \tag{3.48}$$

This represents the infinitesimal transformation for $a_{\alpha\beta} \ll 1$, obtained from

$$\boxed{\Lambda = \exp\left(\frac{i}{2} a_{\alpha\beta} M^{(\alpha\beta)}\right)}. \tag{3.49}$$

[1] Note that here the parentheses are introduced to label the matrices.

Driving exercise
Let us now determine the generators. We compute

$$-i\frac{\partial\Lambda(a=0)}{\partial a_{\alpha\beta}}=\frac{i}{2}\frac{\partial a_{\alpha\beta}}{\partial a_{\alpha\beta}}M^{\alpha\beta}=\frac{i}{2}\sum_{\alpha,\beta}\left[\frac{\partial a_{\alpha\beta}}{\partial a_{\alpha\beta}}M^{\alpha\beta}\right]=M^{\alpha\beta}:=J^{\alpha\beta}. \tag{3.50}$$

The $M^{\alpha\beta}$ are a particular linear representation of the special Lorentz group, namely the representation acting on Minkowski space. A generic representation of $L_+^\uparrow$ on a vector space V is a linear invertible operator with $J^{\alpha\beta}$ as generators.

A generic representation of $L_+^\uparrow$ on a vector space V is a linear invertible operator $D(\Lambda)$

$$D: L_+^\uparrow \longrightarrow GL(V) \quad\Rightarrow\quad D(\Lambda)=\exp\left(\frac{i}{2}a_{\alpha\beta}J^{\alpha\beta}\right). \tag{3.51}$$

A representation of $L_+^\uparrow$ on Minkowski space $M(3,1)$ is an operator $D(\Lambda)$ such that

$$D: L_+^\uparrow \longrightarrow GL(M(3,1)) \quad\Rightarrow\quad D(\Lambda)=\exp\left(\frac{i}{2}a_{\alpha\beta}M^{\alpha\beta}\right). \tag{3.52}$$

The commutation relation among the generators $J^{\alpha\beta}$ is given by

$$[J^{\mu\nu},J^{\rho\sigma}]=-i\left[g^{\mu\sigma}J^{\nu\rho}+g^{\nu\rho}J^{\mu\sigma}-g^{\mu\rho}J^{\nu\sigma}-g^{\nu\sigma}J^{\mu\rho}\right]. \tag{3.53}$$

To better understand the structure of $L_+^\uparrow$, it is useful to introduce the six operators $J^{\alpha\beta}$

$$J^i=\frac{i}{2}\varepsilon^{ijk}J^{jk}=(J^{23},J^{31},J^{12}),\qquad K^i=J^{0i}=(J^{01},J^{02},J^{03}), \tag{3.54}$$

with

$$J^1=\frac{\varepsilon^{123}}{2}J^{23}-\frac{\varepsilon^{132}}{2}J^{32}=J^{23}, \tag{3.55a}$$

$$J^2=\frac{\varepsilon^{231}}{2}J^{31}-\frac{\varepsilon^{213}}{2}J^{13}=J^{31}, \tag{3.55b}$$

$$J^3=\frac{\varepsilon^{312}}{2}J^{12}-\frac{\varepsilon^{321}}{2}J^{21}=J^{12}. \tag{3.55c}$$

Thus, we can write

$$J^{\mu\nu}=\begin{pmatrix}J^{00}&J^{01}&J^{02}&J^{03}\\J^{10}&J^{11}&J^{12}&J^{13}\\J^{20}&J^{21}&J^{22}&J^{23}\\J^{30}&J^{31}&J^{32}&J^{33}\end{pmatrix}=\begin{pmatrix}0&K^1&K^2&K^3\\-K^1&0&J^3&-J^2\\-K^2&-J^3&0&J^1\\-K^3&J^2&-J^1&0\end{pmatrix}. \tag{3.56}$$

Exercise
Prove the results below.
The components satisfy the following commutation relations:

$$[J^i, J^j] = i\varepsilon^{ijk} J^k, \tag{3.57a}$$

$$[K^i, K^j] = -i\varepsilon^{ijk} J^k, \tag{3.57b}$$

$$[J^i, K^j] = i\varepsilon^{ijk} K^k. \tag{3.57c}$$

Pause and think
Wonder why it is so important to recognize the existence of commutators in a generic theory.

Generally speaking, commutators reveal the presence of constraints and symmetries. They preserve the non-trivial dynamical content of a theory. At a fundamental level, a physical theory is not merely a collection of quantities, but a framework specifying which quantities can be simultaneously defined, measured, or evolved independently. *Commutators provide the* **algebraic criterion** *that determines this structure.*

We now introduce the parameters

$$\theta^i = \frac{\varepsilon^{ilm}}{2} a^{lm} = (a^{23}, a^{31}, a^{12}), \tag{3.58a}$$

$$\alpha^i = a^{0i} = (a^{01}, a^{02}, a^{03}). \tag{3.58b}$$

It follows that

$$\begin{aligned} \theta^i J^i &= \frac{1}{4}\varepsilon^{ijk} J^{jk} \varepsilon^{ilm} a^{lm} = \frac{1}{4}(\delta^j_l \delta^k_m - \delta^j_m \delta^k_l) J^{jk} a^{lm} = \frac{1}{4}(a^{jk} J^{jk} - a^{kj} J^{jk}) \\ &= \frac{1}{4}(a^{jk} J^{jk} + a^{jk} J^{jk}) = -\frac{a_{jk} J^{jk}}{2}, \end{aligned} \tag{3.59a}$$

$$\alpha^i K^i = a^{0i} J^{0i} = -a_{0i} J^{0i}. \tag{3.59b}$$

Substituting into Eq. (3.51) gives

$$\frac{i}{2} a_{\alpha\beta} J^{\alpha\beta} = \frac{i}{2}[a_{0i} J^{0i} + a_{i0} J^{i0} + a_{jk} J^{jk}], \tag{3.60}$$

and noting that $a_{0i} = -a_{i0}$ and $J^{i0} = -J^{0i}$, we find

$$\frac{i}{2} a_{\alpha\beta} J^{\alpha\beta} = \frac{i}{2}[2a_{0i} J^{0i} + a_{jk} J^{jk}] = i\left(a_{0i} J^{0i} + \frac{a_{jk} J^{jk}}{2}\right) = i[-\alpha^i K^i + \theta^i J^i]. \tag{3.61}$$

In general, we have

$$D(\Lambda) = \exp\left[i(\boldsymbol{\theta} \cdot \boldsymbol{J} - \boldsymbol{\alpha} \cdot \boldsymbol{K})\right], \tag{3.62}$$

where $\exp[i(\boldsymbol{\theta} \cdot \boldsymbol{J})]$ represents a rotation of angle θ about the axis $\mathbf{n} = \boldsymbol{\theta}/\theta$, while $\exp[-i(\boldsymbol{\alpha} \cdot \boldsymbol{K})]$ represents a boost along the axis $\hat{\alpha} = \boldsymbol{\alpha}/\alpha = \hat{v}$ with rapidity $\alpha = \tanh^{-1}(v/c)$. Here, $\boldsymbol{J}$ is the *rotation generator*, representing angular momentum, while $\boldsymbol{K}$ is the *boost generator*, representing the motion of the center of mass.

Exercise
Do boosts alone form a group? Prove that.

Boosts alone do not form a group, so we **cannot define** $\{K^i\}$ as a basis. For this reason, we perform a change of basis by introducing the operators

$$J^i_\pm = \frac{J^i \pm i K^i}{2}, \tag{3.63}$$

which satisfy the following properties:

$$[J^i_+, J^j_+] = i\varepsilon^{ijk} J^k_+, \tag{3.64a}$$

$$[J^i_+, J^j_-] = 0, \tag{3.64b}$$

$$[J^i_-, J^j_-] = i\varepsilon^{ijk} J^k_-. \tag{3.64c}$$

Equations (3.64a)–(3.64c) correspond to two independent $SU(2)$ algebras, the group of 2×2 Hermitian traceless matrices. Therefore, the algebra of $L^\uparrow_+$ is the algebra of $SU(2) \oplus SU(2)$. Note that, although it shares the same algebra as $SO(3, 1)$, the groups themselves are not isomorphic (only locally).

We now classify the irreducible representations. They are labeled by the eigenvalues of the Casimir operators, J^2_+ and J^2_-, which commute with the basis elements of the algebra in the context of irreducible representations. The eigenvalues of J^2_+ and J^2_- are of the form $J_+(J_+ + 1)$ and $J_-(J_- + 1)$ with $J_\pm$ integer or half-integer. We denote by (J_+, J_-) the representation of dimension $(2J_+ + 1)(2J_- + 1)$ associated with these eigenvalues.

The most important finite-dimensional representations of $L^\uparrow_+$ are listed below.

1. *Scalar representation* $(0, 0)$: Here $J_+ = J_- = 0$, generating only the identity $\mathbb{I}$. Fields transforming under this representation remain invariant under Lorentz transformations; these are *scalar fields*:

$$\phi'(x') = \phi(x).$$

2. *Spinor representations* $(0, 1/2)$ *and* $(1/2, 0)$: These are inequivalent but can be related by parity, under which $\mathbf{K}$ changes sign while $\mathbf{J}$ remains unchanged. The generators are 2×2 matrices satisfying the $SU(2)$ commutation relations, and are proportional to the Pauli matrices $\boldsymbol{\sigma} = (\sigma_x, \sigma_y, \sigma_z)$:

$$\sigma_x = \begin{pmatrix} 0 & 1 \\ 1 & 0 \end{pmatrix}, \quad \sigma_y = \begin{pmatrix} 0 & -i \\ i & 0 \end{pmatrix}, \quad \sigma_z = \begin{pmatrix} 1 & 0 \\ 0 & -1 \end{pmatrix}.$$

Specifically,

$$\left(0, \frac{1}{2}\right) \Rightarrow \mathbf{J}_+ = \mathbf{0}, \quad \mathbf{J}_- = \frac{\boldsymbol{\sigma}}{2} \quad \left(\mathbf{J} = \frac{\boldsymbol{\sigma}}{2}, \ \mathbf{K} = \mathbf{i}\frac{\boldsymbol{\sigma}}{2}\right), \tag{3.65a}$$

$$\left(\frac{1}{2}, 0\right) \Rightarrow \mathbf{J}_+ = \frac{\boldsymbol{\sigma}}{2}, \quad \mathbf{J}_- = \mathbf{0} \quad \left(\mathbf{J} = \frac{\boldsymbol{\sigma}}{2}, \ \mathbf{K} = -\mathbf{i}\frac{\boldsymbol{\sigma}}{2}\right), \tag{3.65b}$$

which generate the *spinor representations*:

$$D(\Lambda) = \exp\left[i\left(\boldsymbol{\theta}\frac{\boldsymbol{\sigma}}{2} \pm \boldsymbol{\alpha}\frac{\boldsymbol{\sigma}}{2}\right)\right]. \tag{3.66}$$

Fields transforming under these representations are called *Weyl spinor fields*:

$$\xi'(x') = D(\Lambda)\xi(x), \qquad \text{with} \qquad \xi(x) = \begin{pmatrix} \xi_1(x) \\ \xi_2(x) \end{pmatrix}. \tag{3.67}$$

3. *Vector representation* $(1/2, 1/2)$: The generators are the 4×4 matrices $M^{\alpha\beta}$ with $\mathbf{J}_+ = \mathbf{J}_-$. These matrices generate the representation

$$D(\Lambda) = \exp\left(\frac{i}{2} a_{\alpha\beta} M^{\alpha\beta}\right). \tag{3.68}$$

Fields transforming according to this representation are *four-vector fields*:

$$A'^{\mu} = \Lambda^{\mu}_{\nu} A^{\nu}(x). \tag{3.69}$$

4. *Representation* $(1/2, 0) \oplus (0, 1/2)$: This representation, different from $(1/2, 1/2)$, is reducible. The generators are

$$J = \begin{pmatrix} \frac{1}{2}\boldsymbol{\sigma} & 0 \\ 0 & \frac{1}{2}\boldsymbol{\sigma} \end{pmatrix}, \qquad \mathbf{K} = \begin{pmatrix} -\frac{i}{2}\boldsymbol{\sigma} & 0 \\ 0 & \frac{i}{2}\boldsymbol{\sigma} \end{pmatrix}. \tag{3.70}$$

Fields transforming under this representation are *four-component Dirac spinor fields*

$$\Psi(x) = \begin{pmatrix} \xi_1(x) \\ \xi_2(x) \\ \zeta_1(x) \\ \zeta_2(x) \end{pmatrix}. \tag{3.71}$$

What is a spinor? Conceptually it remains a mysterious quantity, composed by four scalars, since its defining feature is how it transforms, not how it is constructed. We will use them in the Dirac equation, later in Sect. 4.4.

3.3.2 The Poincaré Group $\mathcal{P}$

Recalling the results obtained in Sect. 2.3.3, we reconsider the Poincaré group, i.e., the group of isometries of spacetime, with L as its homogeneous subgroup. We denote an element of the Poincaré group as a combination of a pure translation[2] $\{\mathbb{I}, \mathrm{a}\} \in T$ and a homogeneous Lorentz transformation $\{\Lambda, 0\} \in L$, i.e.,

$$\{\Lambda, a\} = \{\mathbb{I}, \mathrm{a}\}\{\Lambda, 0\}. \tag{3.72}$$

Pure translations have 4 parameters, while homogeneous Lorentz transformations have 6, so the Poincaré group has a total of 10 parameters. Consequently, there are 10 generators as well. Having already established the generators of $L_+^\uparrow$, $J^{\rho\sigma}$, we now find those of T. Consider infinitesimal translations:

$$x'^\mu = x^\mu + \varepsilon^\mu = x^\mu + \varepsilon^\nu \delta^\mu_\nu = x^\mu + \varepsilon^\nu \frac{\partial x^\mu}{\partial x^\nu} = x^\mu + \partial_\nu x^\mu \, \varepsilon^\nu. \tag{3.73}$$

Defining the operators $P_\nu = -i\partial_\nu$, the previous equation becomes

$$x'^\mu = x^\mu + i\varepsilon^\nu P_\nu x^\mu. \tag{3.74}$$

▶ **Pause and think**
Manage to state which are the generators of T.

The P_ν are the generators of T. In matrix form:

$$\{\mathbb{I}, \mathrm{a}\} = \mathbb{I} + \mathrm{i}\mathrm{a}^\nu \mathrm{P}_\nu. \tag{3.75}$$

We can write the generators as $P^\mu = (P^0, \mathbf{P})$, where P^0 is the temporal component generating time translations, called the *Hamiltonian operator*, and $\mathbf{P}$ is the spatial component generating spatial translations, called the *momentum operator*.

---

[2] T is an abelian invariant subgroup of $\mathcal{P}$.

An infinitesimal Poincaré transformation reads:

$$\Pi = 1\!\!1 + \frac{i}{2} a_{\alpha\beta} J^{\alpha\beta} + i\varepsilon_\nu P^\nu. \tag{3.76}$$

The generators $J^{\alpha\beta}$ and P^ν, with 6 and 4 parameters respectively, satisfy the commutation relations:

$$[J^{\mu\nu}, J^{\rho\sigma}] = -i\left(g^{\nu\sigma} J^{\rho\mu} + g^{\nu\rho} J^{\mu\sigma} - g^{\mu\rho} J^{\nu\sigma} - g^{\mu\sigma} J^{\rho\nu}\right), \tag{3.77a}$$

$$[P^\mu, P^\nu] = 0, \tag{3.77b}$$

$$[J^{\mu\nu}, P^\rho] = -i\left(g^{\mu\rho} P^\nu - g^{\nu\rho} P^\mu\right). \tag{3.77c}$$

Separating components, we define $J^i = \frac{1}{2}\varepsilon^{ijk} J^{jk}$, $P^0 = H$, $P^i = P^i$, and $K^i = J^{0i}$. To construct a basis commuting with P^μ, we recall the *Pauli-Lubanski operator*, see Eq. (2.223),

$$W^\mu = \frac{1}{2}\varepsilon^{\mu\nu\rho\sigma} J_{\nu\rho} P_\sigma. \tag{3.78}$$

This operator is orthogonal to P_μ:

$$W^\mu P_\mu = \frac{1}{2}\varepsilon^{\mu\nu\rho\sigma} J_{\nu\rho} P_\sigma P_\mu = 0. \tag{3.79}$$

Similarly, W^μ and P^α commute:

$$[W^\mu, P^\alpha] = \frac{1}{2}\varepsilon^{\mu\nu\rho\sigma}[J_{\nu\rho}, P^\alpha] P_\sigma = 0. \tag{3.80}$$

The *irreducible representations* of the Poincaré group are labeled by the eigenvalues of P^2 and W^2.

1. For *massive particles* ($m > 0$), in the rest frame of the particle the momentum is $P^\mu = (P^0, \mathbf{P}) = (mc, 0, 0, 0)$, with invariant norm

$$P^\mu P_\mu = P^2 = m^2c^2, \tag{3.81}$$

which remains invariant under the Poincaré representation. In this frame, one obtains

$$W^\mu = (0, mc\mathbf{J}) = (0, mc\mathbf{S}), \tag{3.82}$$

where, since $\mathbf{J} = \mathbf{L} + \mathbf{S}$ and the particle has no orbital angular momentum in its rest frame, $\mathbf{L} = 0$, so $\mathbf{J} = \mathbf{S}$. Its norm is

$$W^\mu W_\mu = W^2 = -m^2c^2\, S(S+1). \tag{3.83}$$

Thus, for massive particles, representations are classified by mass m and spin S. This representation, denoted $O(3)$, has an algebra generated by angular momentum and, if relevant, momentum.

2. For *massless particles* ($m = 0$), $P^2 = 0$ and $W^2 = 0$. In this case, $W^\mu = \lambda P^\mu$, with λ being the helicity: $\lambda = \mathbf{J} \cdot \hat{\mathbf{P}}$. For massless particles, it is the helicity that classifies the representations. The generators form a rotation group in the plane orthogonal to $\hat{\mathbf{P}}$, called $O(2)$ due to the preferred direction.[3]

3.4 The Multiple Representations of Relativistic Lagrangian

We now intend to construct the analytical mechanics of relativistic systems. Let us start by defining the *relativistic action* $S = \alpha \int ds$. To derive the equations of motion, we require that the variation of the action vanishes, i.e., $\delta S = 0$. Moreover, these equations must satisfy the principles of relativity, hence we impose that $Ldt = \alpha ds$. Let us first consider the case of a *free particle*. In this context, the action reads

$$S = \int Ldt = \alpha \int ds \quad \text{with} \quad ds = \frac{cdt}{\gamma}. \tag{3.84}$$

This leads to

$$Ldt = \sqrt{1 - \frac{v^2}{c^2}}\alpha cdt \rightarrow L = \alpha c\sqrt{1 - \frac{v^2}{c^2}}. \tag{3.85}$$

Substituting this expression back into the action, we obtain

$$S = \alpha c \int \sqrt{1 - \frac{v^2}{c^2}}dt, \tag{3.86}$$

▶ **Pause and think**
Find the simplest way to fix α.

where α is determined by requiring that the action reduces to the integral of the kinetic energy in the non-relativistic limit, i.e.,

$$L_{NR} = T - V = T = \frac{1}{2}mv^2 \rightarrow S_{NR} = \int \frac{1}{2}mv^2dt + \text{const}. \tag{3.87}$$

[3] No longer $O(3)$, since a privileged direction is considered.

For $c \to \infty$ we must have[4]

$$S = \alpha c \int \sqrt{1 - \frac{v^2}{c^2}} dt \sim \alpha c \int \left(1 - \frac{v^2}{2c^2}\right) dt = \int \frac{m}{2} v^2 dt + \text{const}, \tag{3.88}$$

where the last equality holds only if $\alpha = -mc$.

Therefore, for a free particle, the *relativistic action* and the *relativistic Lagrangian* are given, respectively, by

$$S = -mc \int ds, \tag{3.89a}$$

$$L = -mc^2 \sqrt{1 - \frac{v^2}{c^2}}. \tag{3.89b}$$

Pause and think
Try to vary yourself the Lagrangian using the principle of stationary action.

We can now derive the equations of motion by applying the *principle of stationary action*.

$$\delta \int_A^B (-mcds) = 0 \quad \text{with fixed endpoints} \quad \delta x_A^\mu = \delta x_B^\mu = 0. \tag{3.90}$$

We thus obtain

$$0 = \delta \int_A^B ds = \delta \int_A^B (dx^\mu dx_\mu)^{1/2} = \int_A^B \frac{1}{\cancel{2} ds} (\cancel{2} \delta(dx^\mu) dx_\mu) = \int_A^B \frac{dx^\mu}{ds} d\delta x_\mu, \tag{3.91}$$

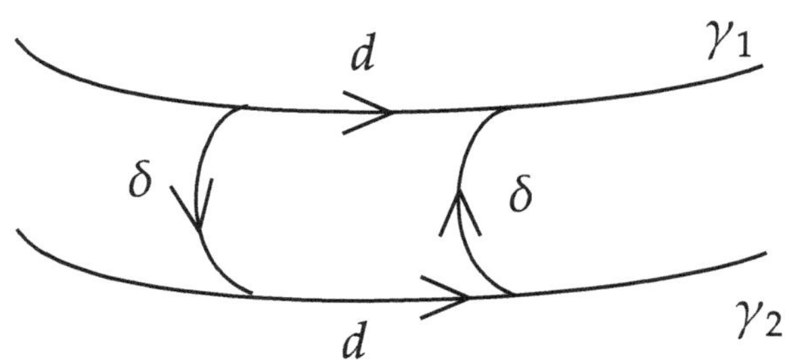

Fig. 3.2 Schematic representation of the two different types of differential, adopting two different curves, γ_1 and γ_2: variational, δ, is between the two curves, whereas differential, d, on each curve

[4] The Newtonian limit of a relativistic theory is obtained by taking the formal limit $c \to \infty$, rather than by assuming that particle velocities satisfy $v \ll c$. The condition $v \ll c$ is a *kinematical regime* describing slow motion, whereas the Newtonian limit is a *structural limit* of the theory itself. More generally, the Newtonian limit corresponds to an expansion in powers of $1/c$, in which relativistic corrections vanish order by order as $c \to \infty$. The assumption $v \ll c$ is only meaningful *after* this limit has been taken, when the resulting theory admits a Galilean kinematics. The condition $v \ll c$ merely selects a low-velocity sector within an already non-relativistic framework.

where we have used the fact that $\delta dx^\mu = d\delta x^\mu$, which can be more generally written as $\delta d - d\delta = [\delta, d] = 0$, see Fig. 3.2. Integrating by parts Eq. (3.91), we get

$$0 = \frac{dx^\mu}{ds}\delta x_\mu\Big|_A^B - \int_A^B \frac{du^\mu}{ds}\delta x_\mu \, ds, \tag{3.92}$$

where the surface term $\frac{dx^\mu}{ds}\delta x_\mu\Big|_A^B = 0$. We then obtain the *equations of motion in covariant form*

$$\frac{du^\mu}{ds} = 0 \longrightarrow \frac{d^2x^\mu}{ds^2} = 0. \tag{3.93}$$

Special relativity seeks what is absolute w.r.t. the Lorentz transformations!

We define the *canonical momentum* as

$$\mathbf{p} = \frac{\partial L}{\partial v} = \frac{\partial L}{\partial \dot{x}} = \frac{m\cancel{c^2}}{2\sqrt{1-\frac{v^2}{c^2}}}\left(\frac{\cancel{2}\mathbf{v}}{\cancel{c^2}}\right) = \frac{m\mathbf{v}}{\sqrt{1-\frac{v^2}{c^2}}} = m\gamma\mathbf{v}, \tag{3.94}$$

which coincides with the relativistic mechanical momentum.

Using the canonical momentum and the Lagrangian, we can define the *Hamiltonian* as

$$\begin{aligned}
&H(\mathbf{x},\mathbf{p}) = \mathbf{p}\cdot\dot{\mathbf{x}} - L(\mathbf{x},\dot{\mathbf{x}}) = \mathbf{p}\cdot\mathbf{v} - L(\mathbf{x},\mathbf{v}) = \frac{p^2}{m\gamma} + \frac{mc^2}{\gamma} = \\
&= \frac{mc^2}{\gamma}\left(1 + \frac{p^2\cancel{\gamma}}{m^2c^2\cancel{\gamma}}\right) = \frac{mc^2}{\gamma}\left(1 + \frac{m^2\gamma^2v^2}{m^2c^2}\right) = \frac{mc^2}{\gamma}\left(1 + \frac{v^2}{c^2}\frac{1}{1-\frac{v^2}{c^2}}\right) = \frac{mc^2}{\gamma}\left[\frac{1-\cancel{\frac{v^2}{c^2}}+\cancel{\frac{v^2}{c^2}}}{1-\frac{v^2}{c^2}}\right] = \\
&= mc^2\frac{\sqrt{1-\frac{v^2}{c^2}}}{1-\frac{v^2}{c^2}}.
\end{aligned} \tag{3.95}$$

Let us now consider the non-covariant Euler-Lagrange equations

$$\frac{d}{dt}\frac{\partial L}{\partial \dot{x}} - \frac{\partial L}{\partial x} = 0. \tag{3.96}$$

Substituting Eq. (3.94) into the previous expression, we obtain the *non-covariant equation of motion*

$$\frac{d}{dt}(m\gamma\mathbf{v}) = 0 \longrightarrow \frac{d\mathbf{p}}{dt} = 0. \tag{3.97}$$

In the case of an interacting particle, the Lagrangian is given by adding a term that accounts for the interaction, denoted L_{int}:

$$L = L_0 + L_{int}, \tag{3.98}$$

with

$$L_0 = -mc^2\sqrt{1 - \frac{v^2}{c^2}} \tag{3.99}$$

being the free particle Lagrangian.

3.4.1 Covariant Lagrangian Formulation of Analytical Mechanics

We will argue that in special relativity we should expect,

- **geometric constraint:** $u^\mu u_\mu = 1$, i.e., that can be reparametrized,
- **physical constraint:** $p^\mu p_\mu = m^2c^2$, i.e., that cannot be reparameterized.

Accordingly, we do not have a counterpart in the non-relativistic case, acting the same way.

Indeed, from the action

$$S = \int L(x^\mu(s), u^\mu(s), s)ds \rightarrow \delta \int Lds = 0, \tag{3.100}$$

it is not possible to directly derive Euler-Lagrange equations because $ds = (dx^\mu dx_\mu)^{1/2}$ represents a constraint between coordinates: ds and x_μ are **NOT** independent. Therefore, one must vary both L and ds, making it impossible to obtain a single equation for the Lagrangian.

The simplest way to circumvent this issue is to introduce a *scalar parameter* ξ for parametrizing the trajectories (see Chap. 2). We define thus

$$x^\mu = x^\mu(\xi) \quad \text{and} \quad v^\mu = \frac{dx^\mu}{d\xi} = \frac{dx^\mu}{ds}\frac{ds}{d\xi} = u^\mu \frac{ds}{d\xi}. \tag{3.101}$$

Note that the velocity just introduced, v^μ, is different from the classical u^μ introduced so far. The squared norm of v^μ is indeed

$$v^\mu v_\mu = \left(\frac{ds}{d\xi}\right)^2 \tag{3.102}$$

and, in general, it is not subject to any constraint. In this way, we can write the action S as

$$S = \int L(x^\mu(\xi), v^\mu(\xi), \xi) d\xi. \tag{3.103}$$

We now impose that L is *invariant under reparametrization*. The consequence of this invariance is that the Lagrangian is a homogeneous function of first order in v^μ. Let us define the reparametrization

$$\xi \longrightarrow \xi' = \xi + \delta\xi \tag{3.104}$$

so that

$$x'^\mu(\xi') = x^\mu(\xi) + \delta' x^\mu(\xi), \tag{3.105}$$

where $\delta' x^\mu(\xi)$ is defined as the *rotated variation*, which can be expressed in terms of the ordinary variation

$$\delta x^\mu(\xi) = x'^\mu(\xi) - x^\mu(\xi). \tag{3.106}$$

Expanding to first order in $\delta\xi$ we have

$$x'^\mu(\xi') \simeq x'^\mu(\xi) + v'^\mu \delta\xi = x^\mu(\xi) + \delta x^\mu(\xi) + v^\mu \delta\xi. \tag{3.107}$$

In general, the total variation and the ordinary variation are related by

$$\delta' = \delta + \frac{d}{d\xi} \cdot \delta\xi. \tag{3.108}$$

Using reparametrization, we can write the action as

$$S = \int L(x(\xi), v(\xi), \xi) d\xi = \int L(x(\xi'), v(\xi'), \xi') d\xi'. \tag{3.109}$$

Varying the action using Eq. (3.108) we obtain

$$\delta' S = 0 \iff 0 = \int \left(\frac{\partial L}{\partial x^\mu} \delta' x^\mu + \frac{\partial L}{\partial v^\mu} \delta' v^\mu + \frac{\partial L}{\partial \xi} \delta' \xi \right) d\xi + \int L\, \delta' d\xi. \tag{3.110}$$

Using

$$\delta' x^\mu = \delta x^\mu + v^\mu \delta\xi, \tag{3.111a}$$
$$\delta' v^\mu = \delta v^\mu + \dot{v}^\mu \delta\xi, \tag{3.111b}$$
$$\delta'\xi = \delta\xi, \tag{3.111c}$$

with $\dot{v}^\mu = \frac{dv^\mu}{d\xi}$, we obtain

$$0 = \int \left(\frac{\partial L}{\partial x^\mu}\delta x^\mu + v^\mu \frac{\partial L}{\partial x^\mu}\delta\xi + \frac{\partial L}{\partial v^\mu}\delta v^\mu + \dot{v}^\mu \frac{\partial L}{\partial v^\mu}\delta\xi + \frac{\partial L}{\partial \xi}\delta\xi \right) d\xi + \int L\delta d\xi =$$
$$= \int \left(\frac{\partial L}{\partial x^\mu}\delta x^\mu + \frac{\partial L}{\partial v^\mu}\delta v^\mu + \frac{dL}{d\xi}\delta\xi \right) d\xi + \int L\delta d\xi, \tag{3.112}$$

where $\frac{dL}{d\xi}\delta\xi = v^\mu \frac{\partial L}{\partial x^\mu}\delta\xi + \dot{v}^\mu \frac{\partial L}{\partial v^\mu}\delta\xi + \frac{\partial L}{\partial \xi}\delta\xi$ is the total derivative.

Using $\delta v^\mu = \delta \frac{dx^\mu}{d\xi} = \frac{d}{d\xi}\delta x^\mu$, we can write

$$\frac{\partial L}{\partial x^\mu}\delta x^\mu + \frac{\partial L}{\partial v^\mu}\delta v^\mu = \frac{\partial L}{\partial x^\mu}\delta x^\mu + \frac{d}{d\xi}\left(\frac{\partial L}{\partial v^\mu}\delta x^\mu \right) - \frac{d}{d\xi}\left(\frac{\partial L}{\partial v^\mu} \right)\delta x^\mu$$
$$= \left(\frac{\partial L}{\partial x^\mu} - \frac{d}{d\xi}\frac{\partial L}{\partial v^\mu} \right)\delta x^\mu + \frac{d}{d\xi}\left(\frac{\partial L}{\partial v^\mu}\delta x^\mu \right). \tag{3.113}$$

Moreover,

$$\int \frac{dL}{d\xi}\delta\xi\, d\xi + \int L\delta d\xi = \int \frac{dL}{d\xi}\delta\xi\, d\xi + \int L\frac{d}{d\xi}(\delta\xi) d\xi$$
$$= \int \left[\frac{dL}{d\xi}\delta\xi + L\frac{d}{d\xi}(\delta\xi) \right] d\xi = \int \frac{d}{d\xi}(L\delta\xi) d\xi. \tag{3.114}$$

In this way, we can write

$$0 = \delta' S = \int \left[\left(\frac{\partial L}{\partial x^\mu} - \frac{d}{d\xi}\frac{\partial L}{\partial v^\mu} \right)\delta x^\mu + \frac{d}{d\xi}\left(\frac{\partial L}{\partial v^\mu}\delta x^\mu + L\delta\xi \right) \right]$$
$$\rightarrow \left(\frac{\partial L}{\partial x^\mu} - \frac{d}{d\xi}\frac{\partial L}{\partial v^\mu} \right)\delta x^\mu + \frac{d}{d\xi}\left(\frac{\partial L}{\partial v^\mu}\delta x^\mu + L\delta\xi \right) = 0. \tag{3.115}$$

In the previous expression, the *covariant equations of motion* arise from setting the first term to zero:

$$\boxed{\frac{d}{d\xi}\left(\frac{\partial L}{\partial v^\mu}\delta x^\mu + L\delta\xi \right) = 0.} \tag{3.116}$$

Pause and think
Think on the fact that the above treatment is valid if one can have $\delta\xi$ and δx^μ. How do we find them?

To obtain δx^μ and $\delta\xi$, we observe that, since $\delta\xi$ is arbitrary, we can denote it as $\varepsilon(\xi)$. Exploiting reparametrization invariance of the event,

$$\delta x^\mu + v^\mu \delta\xi = 0 \quad \Rightarrow \quad \delta x^\mu = -v^\mu \delta\xi = -v^\mu \varepsilon, \tag{3.117}$$

and substituting the previous expression into Eq. (3.116), we obtain

$$\frac{d}{d\xi}\left(-\frac{\partial L}{\partial v^\mu} v^\mu \varepsilon + L\varepsilon\right) = 0 \quad \Rightarrow \quad \frac{d}{d\xi}\left[\left(\frac{\partial L}{\partial v^\mu} v^\mu - L\right)\varepsilon(\xi)\right] = 0. \tag{3.118}$$

Since ε is arbitrary, we can set the quantity inside the derivative to zero, so that

$$\frac{\partial L}{\partial v^\mu} v^\mu - L = 0 \quad \Rightarrow \quad L = \frac{\partial L}{\partial v^\mu} v^\mu. \tag{3.119}$$

Pause and think
Now think on the form of the Hamiltonian. How does it look?

Since $P_\mu = \frac{\partial L}{\partial v^\mu}$, the Hamiltonian $H = P_\mu v^\mu - L$ is equivalently zero, and the Lagrangian in Eq. (3.119) is a first-order homogeneous function of v^μ, i.e., $L(\alpha v^\mu) = \alpha L(v^\mu)$.

To obtain the covariant Euler-Lagrange equations, we need to introduce the parameter ξ because S is subject to the constraint $ds = (dx_\mu dx^\mu)^{1/2}$, i.e.,

$$L = L(x^\mu(\xi), v^\mu(\xi), \xi) \quad \text{and} \quad S = \int L\, d\xi \neq \int L\, dt, \tag{3.120}$$

where $d\xi$ is an invariant while dt is not. Using reparametrization invariance $L = L'$ under $\xi \to \xi'$, we have

$$L = \frac{\partial L}{\partial v^\mu} v^\mu = \text{function of } v^\mu. \tag{3.121}$$

We thus end up with two main consequences.

- In particular, the conjugate momentum p is a function of degree zero in v^μ, so that $P_\mu = \frac{\partial L}{\partial v^\mu}$ cannot be inverted to obtain v^μ.

- Another consequence of $L = L'$ arising from reparametrization is the vanishing Hamiltonian in Eq. (3.119), where, as before, $P = \frac{\partial L}{\partial v}$ cannot be inverted to find v. In this case, the Lagrangian is said to be *singular*, i.e., there exists a constraint $\phi(x, P) = 0$ which can be used to define an effective Hamiltonian $H_{\text{eff}} = f(\xi)\phi(x, P)$ such that Hamilton's equations hold. This allows us to construct an effective Hamiltonian

$$H_{\text{eff}} = f(\xi)\phi(x, P). \tag{3.122}$$

Pause and think
Bearing in mind the above choice for H_{eff}, which are the equations of motion? Which kind of interpretation may you give to them?

The canonical equations of motion are easy to compute,

$$\frac{dx^\mu}{d\xi} = \frac{\partial H}{\partial P_\mu}, \qquad \frac{dP^\mu}{d\xi} = -\frac{\partial H}{\partial x_\mu}. \tag{3.123}$$

Using these equations, the function $f(\xi)$ can be expressed in terms of the dynamical variables.

3.4.2 Relativistic Free Particle

Let us now apply all the above considerations to the case of a *free massive particle* $m > 0$. We start by writing

$$ds = (dx_\mu dx^\mu)^{1/2} = (v_\mu v^\mu)^{1/2} d\xi. \tag{3.124}$$

The action for a free particle may be rewritten as $S_0 = -mc \int ds = -mc \int (v^\mu v_\mu)^{1/2} d\xi$, which is invariant under reparametrization $\xi \to \xi'(\xi)$. The covariant Lagrangian is then $L_0 = -mc(v^\mu v_\mu)^{1/2} = L(x(\xi), v(\xi), \xi)$.

The particle's momentum reads

$$P^\mu = \frac{\partial L_0}{\partial v_\mu} = -\frac{mc}{\cancel{2}(v^\mu v_\mu)^{1/2}}\cancel{2}v^\mu = -mc\frac{dx^\mu/d\xi}{\left(dx^\mu/d\xi\, dx_\mu/d\xi\right)^{1/2}} = -mc\frac{dx^\mu}{ds} = -mcu^\mu = -p^\mu. \tag{3.125}$$

The equations of motion are given by

$$\frac{\partial L_0}{\partial x_\mu} = 0 \quad \Rightarrow \quad \frac{d}{d\xi}\frac{\partial L_0}{\partial v_\mu} = \frac{d}{d\xi}\left[\frac{mcv^\mu}{(v_\mu v^\mu)^{1/2}}\right] = 0, \tag{3.126}$$

with $v^\mu v_\mu = \text{const}$.

Since P^μ cannot be inverted to obtain v^μ, L_0 is singular. Indeed,

$$\begin{aligned} H &= P^\mu v_\mu - L = -mcv^\mu v_\mu + mc(v^\mu v_\mu)^{1/2} = \\ &= -\frac{mcv^\mu v_\mu}{(v^2)^{1/2}} + mc(v^2)^{1/2} = \frac{(-mcv^\mu + mcv^\mu)v_\mu}{(v^2)^{1/2}} = 0. \end{aligned} \tag{3.127}$$

Thus, there exists a constraint of the form

$$\phi(x, P) \approx 0 \quad \Longleftrightarrow \quad p^\mu p_\mu = m^2c^2\frac{v^\mu v_\mu}{v^2} = m^2c^2 \quad \rightarrow \quad p^\mu p_\mu - m^2c^2 = 0, \tag{3.128}$$

which is the *mass-shell condition.*

The effective Hamiltonian can then be constructed to preserve the canonical equations

$$H_{\text{eff}} = f(\xi)(p^2 - m^2c^2). \tag{3.129}$$

In this way,

$$\begin{cases} \frac{dx^\mu}{d\xi} = \frac{\partial H}{\partial p_\mu} = f(\xi)2p^\mu, \\ \frac{dp^\mu}{d\xi} = -\frac{\partial H}{\partial x_\mu} = 0 \end{cases} \Rightarrow v^\mu = 2fp^\mu = 2f\frac{mcv^\mu}{(v^2)^{1/2}} \quad \Rightarrow \quad f = \frac{(v^2)^{1/2}}{2mc}. \tag{3.130}$$

We can thus write the *effective Hamiltonian for a free particle* as

$$\boxed{H_{\text{eff}} = \frac{(v^2)^{1/2}}{2mc}(p^2 - m^2c^2).} \tag{3.131}$$

► **Pause and think**
Try to interpret the above result in view of Eq. (2.155).

Equation (3.131) generates the free relativistic dynamics and its role is analogous to that of a Lagrange multiplier enforcing the mass-shell condition.

The appearance of the invariant factor $\sqrt{v^2}$ reflects the freedom in the parametrization of the worldline. The four-velocity $v^\mu = \dot{x}^\mu$ is defined with respect to an arbitrary evolution parameter, and the action is reparametrization invariant. As a consequence, the Hamiltonian formulation does not single out the coordinate time as a preferred evolution variable, and the canonical Hamiltonian vanishes on-shell.

From a physical point of view, H_{eff} vanishes identically when the equations of motion are satisfied, since $p^2 - m^2c^2 = 0$. **Its dynamical content is therefore not encoded in its numerical value, but in the Hamilton equations it induces.**

In particular, variations of H_{eff} reproduce the free equations of motion,

$$\dot{p}^\mu = 0, \qquad \dot{x}^\mu \propto p^\mu, \tag{3.132}$$

showing that the particle propagates along straight worldlines in Minkowski spacetime.

Additionally, $\phi \equiv p^2 - m^2c^2 \approx 0$ is a *primary first-class constraint* in the Dirac classification. Its origin is the invariance of the relativistic action under reparametrizations of the worldline, which implies that the evolution parameter carries no physical meaning.

From the viewpoint of dynamical systems, the relativistic particle is not described by unconstrained Hamiltonian flow on phase space, but by a flow restricted to the constraint surface $\phi = 0$. Physical trajectories correspond to integral curves lying on this hypersurface, while the Hamiltonian vector field generated by H_{eff} is tangent to it. The vanishing of H_{eff} on shell is therefore a direct manifestation of gauge redundancy rather than a dynamical degeneracy.

The presence of the constraint implies that time evolution is generated by a linear combination of constraints rather than by a genuine Hamiltonian. In this sense, the relativistic free particle provides a paradigmatic example of a reparametrization-invariant dynamical system, closely analogous to other generally covariant systems such as general relativity.

Physically speaking, it is very remarkable to imagine that, while physical states satisfy the mass-shell condition $\phi = 0$, the Hamiltonian formulation naturally allows for configurations with

$$p^2 \neq m^2c^2, \tag{3.133}$$

corresponding to *off-shell* trajectories in phase space. These configurations do not describe observable free particles, but play a crucial role in quantum theory, where intermediate states in perturbative processes are not constrained to lie on the mass shell.

From this perspective, the constraint $\phi = 0$ distinguishes between physical, asymptotic states and unphysical, virtual configurations.

Virtual particles correspond precisely to excursions away from the constraint surface.

The effective Hamiltonian provides a compact classical framework, *anticipating the off-shell structure that becomes central in QFT.*

3.4.3 Relativistic Interacting Particle

Now, let us consider the case of an *interacting particle*. In this context, the Lagrangian contains an additional term representing the interaction, i.e., $L = L_0 + L_{\text{int}}$. If the interaction term does not depend explicitly on ξ, it is, like L_0, a first-order homogeneous function in v.

The Euler-Lagrange equations are written as

$$\frac{\partial L}{\partial x_\mu} = \frac{d}{d\xi}\frac{\partial L}{\partial v_\mu} \quad \Longleftrightarrow \quad \frac{\partial L_0}{\partial x_\mu} + \frac{\partial L_{\text{int}}}{\partial x_\mu} = \frac{d}{d\xi}\frac{\partial L_0}{\partial v_\mu} + \frac{d}{d\xi}\frac{\partial L_{\text{int}}}{\partial v_\mu}. \tag{3.134}$$

For a free particle, $\frac{\partial L_0}{\partial x_\mu} = 0$ and therefore

$$\frac{d}{d\xi}\frac{\partial L_0}{\partial v_\mu} = \frac{\partial L_{int}}{\partial x_\mu} - \frac{d}{d\xi}\frac{\partial L_{int}}{\partial v_\mu}. \tag{3.135}$$

Reparametrizing the previous equation $\left(\frac{d}{d\xi} = \frac{d}{ds}\frac{ds}{d\xi}\right)$, we find

$$\frac{d}{ds}\frac{ds}{d\xi}\frac{\partial L_0}{\partial v_\mu} = \frac{\partial L_{\text{int}}}{\partial x_\mu} - \frac{d}{ds}\frac{ds}{d\xi}\frac{\partial L_{\text{int}}}{\partial v_\mu}. \tag{3.136}$$

Multiplying by $\frac{d\xi}{ds}$, we get

$$\frac{d}{ds}\frac{\partial L_0}{\partial v_\mu} = \frac{d\xi}{ds}\left(\frac{\partial L_{\text{int}}}{\partial x^\mu} - \frac{d}{d\xi}\frac{\partial L_{\text{int}}}{\partial v_\mu}\right), \tag{3.137}$$

where $\frac{\partial L_0}{\partial v_\mu} = P^\mu = p^\mu$. Comparing with the Minkowski equation $\mathcal{F}^\mu = \frac{d}{ds}p^\mu$, we obtain the four-force acting on the particle:

$$\mathcal{F}^\mu = \frac{d\xi}{ds}\left(\frac{\partial L_{\text{int}}}{\partial x_\mu} - \frac{d}{d\xi}\frac{\partial L_{\text{int}}}{\partial v_\mu}\right). \tag{3.138}$$

3.4.4 Reparametrization with the Einbein Action

A free relativistic point particle can be described by several *classically equivalent* actions. This non-uniqueness is only apparent and rather not a physical ambiguity. Indeed, it reflects a gauge redundancy associated with reparametrizations of the worldline.

Different Lagrangians define the same set of physical worldlines once constraints and gauge choices are properly implemented.

Let $x^\mu(\xi)$ be the worldline, parametrized by an arbitrary monotonic parameter ξ, and define

$$v^\mu(\xi) \equiv \frac{dx^\mu}{d\xi}. \tag{3.139}$$

Let us introduce the worldline einbein $e(\xi)$, acting as a Lagrange multiplier, and consider

$$L_e = \frac{1}{2} e^{-1}(\xi)\, v^2 + \frac{1}{2} e(\xi)\, m^2 c^2. \tag{3.140}$$

Under a reparametrization $\xi \mapsto \xi'(\xi)$ with $d\xi'/d\xi > 0$, one has

$$v'^{\mu}(\xi') = \frac{dx^\mu}{d\xi'} = \frac{d\xi}{d\xi'}\, v^\mu(\xi), \qquad e'(\xi') = \frac{d\xi}{d\xi'}\, e(\xi), \tag{3.141}$$

so that $e^{-1}v^2\, d\xi$ and $e\, d\xi$ are invariant, and therefore the action $S = \int d\xi\, L_e$ is reparametrization invariant. This gauge symmetry is the origin of the constraint structure.

▶ **Driving exercise**
Find the Euler-Lagrange equations and the mass-shell constraint.
Since L_e does not depend on $\dot{e}$, the equation of motion for e reduces to

$$\frac{\partial L_e}{\partial e} = 0. \tag{3.142}$$

Computing the derivative explicitly,

$$\frac{\partial}{\partial e}\left(\frac{1}{2} e^{-1} v^2\right) = \frac{1}{2}(-1) e^{-2}\, v^2, \qquad \frac{\partial}{\partial e}\left(\frac{1}{2} e\, m^2 c^2\right) = \frac{1}{2}\, m^2 c^2, \tag{3.143}$$

hence

$$-\frac{1}{2} e^{-2} v^2 + \frac{1}{2} m^2 c^2 = 0 \implies v^2 = e^2\, m^2 c^2. \tag{3.144}$$

For $m > 0$ this implies $v^2 > 0$, i.e. the worldline is timelike. The equation is not a dynamical restriction but a *constraint* reflecting gauge freedom: e acts as a Lagrange multiplier enforcing a relation between the parametrization and the normalization of v^μ.
The equation of motion for x^μ follows from

$$\frac{d}{d\xi}\left(\frac{\partial L_e}{\partial v_\mu}\right) = 0, \qquad \frac{\partial L_e}{\partial v_\mu} = \frac{1}{2} e^{-1} \frac{\partial (v^2)}{\partial v_\mu} = \frac{1}{2} e^{-1}(2 v^\mu) = e^{-1} v^\mu, \tag{3.145}$$

so that

$$\frac{d}{d\xi}\left(e^{-1} v^\mu\right) = 0. \tag{3.146}$$

Therefore $e^{-1} v^\mu$ is constant along the worldline.

Pause and think
Manage to find the Lagrange multiplier if the mass does not vanish and then if $m = 0$.

For $m > 0$, the constraint in Eq. (3.144) can be solved algebraically as

$$e(\xi) = \frac{\sqrt{v^2}}{mc}, \tag{3.147}$$

where the positive branch is chosen for future-directed timelike motion. Inserting Eq. (3.147) into Eq. (3.140) yields

$$\begin{aligned} L_e\Big|_{e=\sqrt{v^2}/(mc)} &= \frac{1}{2}\frac{mc}{\sqrt{v^2}}\,v^2 + \frac{1}{2}\frac{\sqrt{v^2}}{mc}\,m^2c^2 = \frac{1}{2}mc\,\sqrt{v^2} + \frac{1}{2}mc\,\sqrt{v^2} \\ &= mc\,\sqrt{v^2}. \end{aligned} \tag{3.148}$$

Thus the einbein action is classically equivalent to the standard square-root action

$$S_0 = mc \int d\xi\,\sqrt{v^2} = mc \int ds, \tag{3.149}$$

with $ds^2 = dx_\mu dx^\mu$ the Minkowski proper time element.

Equation (3.144) shows that changing $e(\xi)$ corresponds to changing the parametrization of the same geometric worldline.

Two convenient gauges are the following.

- **Proper-time gauge**. Choosing

  $$e(\xi) = \frac{1}{mc} \tag{3.150}$$

  gives, from Eq. (3.144),

  $$v^2 = 1 \implies v^\mu = u^\mu, \qquad d\xi = ds, \tag{3.151}$$

  so ξ coincides with proper time. In this gauge,

  $$L_e = \frac{1}{2}(mc)\,v^2 + \frac{1}{2}mc = \frac{1}{2}mc\,(v^2 + 1), \tag{3.152}$$

  and using $v^2 = 1$ on-shell one recovers $L_e = mc$ (a constant), consistently with the fact that for a free particle the physical content is in the equations of motion, not in the on-shell value of the Lagrangian.

– **Affine gauge (constant** e**).** Taking $e(\xi) = e_0 = \text{const}$ yields from Eq. (3.146)

$$\frac{dv^\mu}{d\xi} = 0 \quad \Longrightarrow \quad x^\mu(\xi) = a^\mu\,\xi + b^\mu, \tag{3.153}$$

i.e. straight worldlines. The constraint in Eq. (3.144) fixes the normalization $a^2 = e_0^2 m^2 c^2$, leaving only the parametrization-dependent scale undetermined, as expected for a gauge system.

Driving exercise
Compute the canonical momenta.
To make the constraint nature explicit, compute the momentum conjugate to x^μ:

$$p_\mu \equiv \frac{\partial L_e}{\partial v^\mu} = e^{-1} v_\mu, \qquad \Longleftrightarrow \qquad v^\mu = e\,p^\mu. \tag{3.154}$$

Then

$$v^2 = e^2 p^2, \tag{3.155}$$

and inserting into Eq. (3.144) gives the familiar mass-shell constraint

$$p^2 = m^2 c^2. \tag{3.156}$$

The canonical Hamiltonian reads

$$\begin{aligned} H_c = p_\mu v^\mu - L_e &= p_\mu (e p^\mu) - \left[\frac{1}{2} e^{-1}(e^2 p^2) + \frac{1}{2} e m^2 c^2\right] \\ &= e p^2 - \left[\frac{1}{2} e p^2 + \frac{1}{2} e m^2 c^2\right] = \frac{e}{2}\,(p^2 - m^2 c^2). \end{aligned} \tag{3.157}$$

Hence e multiplies the constraint, and $H_c \approx 0$ on the constraint surface. This is the standard hallmark of reparametrization-invariant systems: different Lagrangian representatives are related by gauge fixing and elimination of auxiliary fields, but they define the same physical dynamics.

The massless case follows similarly. Setting $m = 0$ in Eq. (3.140) gives

$$L_e = \frac{1}{2} e^{-1} v^2. \tag{3.158}$$

The e-equation yields

$$\frac{\partial L_e}{\partial e} = -\frac{1}{2} e^{-2} v^2 = 0 \quad \Longrightarrow \quad v^2 = 0, \tag{3.159}$$

so the worldline is null. The x^μ-equation is again

$$\frac{d}{d\xi}\left(e^{-1}v^\mu\right) = 0. \tag{3.160}$$

Choosing an affine gauge $e = e_0 = \text{const}$ gives $dv^\mu/d\xi = 0$ and therefore

$$x^\mu(\xi) = k^\mu\,\xi + b^\mu, \qquad k^2 = 0, \tag{3.161}$$

i.e. straight null lines. Unlike the massive case, one cannot impose a proper-time gauge because $ds^2 = 0$ along null curves; the residual gauge freedom reflects the absence of an intrinsic clock along lightlike trajectories.

In phase space, the conjugate momentum is again $p_\mu = e^{-1}v_\mu$, and the only invariant constraint is $p^2 = 0$, which is the massless mass-shell condition.

▶ **Pause and think**
Imagine some alternatives, but equivalent formulations.

It is instructive to display other standard representatives of the same free-particle dynamics.

- **Square-root (Nambu–Goto) form**. For $m > 0$,

$$S_0 = mc \int d\xi\,\sqrt{v^2} \tag{3.162}$$

  is manifestly reparametrization invariant and leads to the geodesic equation in flat spacetime. It becomes ill-defined for $m = 0$ because the overall scale m vanishes and the normalization of v^μ cannot be fixed by ds.
- **Phase-space (first-order) form.** A formulation that treats massive and massless particles uniformly is

$$S_{\text{ph}} = \int d\xi\left[p_\mu\dot{x}^\mu - \frac{e}{2}\left(p^2 - m^2c^2\right)\right], \tag{3.163}$$

  where e enforces the mass-shell constraint. Varying p_μ gives $\dot{x}^\mu = ep^\mu$, and substituting back reproduces Eq. (3.140). Varying e yields Eq. (3.156). This form is particularly useful in quantization and in coupling to backgrounds.

3.5 Proposed Exercises

Exercise 1

Consider Lorentz boosts along the x–axis parametrized by the rapidity ϕ, such that the corresponding velocity is $v = c \tanh\phi$.

1. Show that the composition of two boosts with rapidities ϕ_1 and ϕ_2 is again a Lorentz boost along the x–axis, with rapidity $\phi = \phi_1 + \phi_2$.
2. Consider now the following Lorentz matrix

$$\Lambda = \begin{pmatrix} -\cosh\alpha & \sinh\alpha & 0 & 0 \\ -\sinh\alpha & \cosh\alpha & 0 & 0 \\ 0 & 0 & \frac{1}{2} & -\frac{\sqrt{3}}{2} \\ 0 & 0 & \frac{\sqrt{3}}{2} & \frac{1}{2} \end{pmatrix}, \qquad \alpha > 0. \tag{3.164}$$

   (a) Verify explicitly that Λ is a Lorentz transformation.
   (b) Determine to which connected component of the Lorentz group Λ belongs.
   (c) Construct a Lorentz transformation $\tilde{\Lambda}$ in the proper orthochronous subgroup $L_+^\uparrow$ that is related to Λ by discrete symmetries.

Exercise 2

Consider a relativistic point particle whose dynamics is described by the Lagrangian

$$\mathcal{L} = \sqrt{v^2}\,\big(mc + \beta\,\psi(x)\big), \tag{3.165}$$

where $v^\mu = dx^\mu/d\xi$ is the four–velocity with respect to an arbitrary affine parameter ξ, $\psi(x)$ is a scalar field, and β is a constant.

- Derive the equations of motion.
- Compute the canonical momentum conjugate to x^μ.
- Determine the associated four–force acting on the particle.
- Compute the total energy and identify the kinetic contribution.

Discuss whether the constant β has a direct physical effect. Then assume that $\beta = \beta(\xi)$ depends explicitly on the affine parameter. How do the above results change? Does the physical interpretation of β remain the same?

Exercise 3

Consider a relativistic point particle described by the action

$$S = \int L\, d\xi, \tag{3.166}$$

where ξ is an affine parameter and $v^\mu = dx^\mu/d\xi$.

(a) Let the Lagrangian be

$$L_1 = \left[mc + \phi(x)\right](v^2)^{1/2}, \tag{3.167}$$

where $\phi(x)$ is a scalar field. Derive the canonical momentum and the Euler–Lagrange equations in terms of the proper time s. Identify the corresponding four–force acting on the particle.

(b) Show that the particle energy satisfies the dispersion relation

$$E^2 = \mathbf{p}^2c^2 + (mc + \phi)^2c^2. \tag{3.168}$$

(c) Now consider the modified Lagrangian

$$L_2 = \left[mc - f_{\mu\nu}(x)\frac{v^\mu v^\nu}{(v^2)^{1/2}}\right](v^2)^{1/2}, \tag{3.169}$$

where $f_{\mu\nu}(x)$ is a generic rank–two tensor field. Derive the equations of motion and discuss under which conditions the resulting four–force is orthogonal to the four–velocity.

(d) Compare the physical interpretation of the interactions encoded in $\phi(x)$ and $f_{\mu\nu}(x)$, emphasizing the differences with the electromagnetic coupling.

Essential Bibliography

- Landau LD, Lifshitz EM (1975) The classical theory of fields, 4th edn. Butterworth–Heinemann.
  A compact and rigorous reference for relativistic field theory, covariant Lagrangians, Noether currents, energy–momentum tensors, and the foundations of Lorentz-invariant dynamics.
- Arnold VI (1989) Mathematical methods of classical mechanics, 2nd edn. Springer.
  A geometric formulation of analytical mechanics, essential for understanding reparametrization invariance, constraints, and equivalent Lagrangian descriptions.
- Jackiw R (1995) Diverse topics in theoretical and mathematical physics. World Scientific.
  Insightful discussions on Noether currents, stress–energy tensors, and the Belinfante-Rosenfeld symmetrization from a modern field-theoretic viewpoint.

4 Relativistic Quantum Mechanics

Relativity dictates the kinematics; quantum mechanics dictates the probabilities.

Eugene Wigner

Abstract

Combining relativity and quantum puzzles appears a difficult task unless one bypasses the concept of wave function in favor of fields. The advantage is to consider objects whose independent variable is the event, thus ensuring that they transform consistently under Lorentz transformations. The main purpose of this chapter is to introduce two kinds of particles, namely spin 0 and $1/2$ particles, whose dynamics is associated with the Klein-Gordon and Dirac equations, respectively. These revolutionary concepts have opened new avenues in the context of particle and theoretical physics, unveiling how spacetime symmetries constrain the energy-momentum structure of fields and predict conserved quantities through invariance principles.

O. Luongo, *Fundamental Interactions*,
https://doi.org/10.1007/978-3-032-27413-7_4

Fig. 4.1 Portrait of Paul Adrien Maurice Dirac, 1933. Reprinted from Wikimedia. Public domain

A key character: *Paul Adrien Maurice Dirac (1902–1984), Fig.* 4.1.

Paul Adrien Maurice Dirac was born in Bristol, England, into a family of Swiss and British origin. His father, a Swiss national, was a teacher of French and imposed a rigid and austere upbringing, which strongly shaped Dirac's reserved personality and extreme precision in language. From an early age, Dirac displayed an exceptional aptitude for mathematics, coupled with a tendency toward abstract and highly structured reasoning.

Dirac initially pursued formal training in engineering at the University of Bristol, where he obtained a degree in electrical engineering. Unsatisfied with purely practical applications, he soon turned to mathematics and theoretical physics, earning a doctorate at the University of Cambridge under the supervision of Ralph Fowler. He rapidly emerged as one of the central figures in the development of modern quantum theory and was later appointed to the Lucasian Chair of Mathematics at Cambridge, the same professorship once held by Isaac Newton.

Dirac's most influential contribution to physics lies in the formulation of a relativistically consistent quantum theory of the electron. In 1928, he introduced the equation that bears his name, a first-order relativistic wave equation that naturally incorporates spin and predicts the existence of antiparticles.

Beyond the Dirac equation, his work profoundly shaped the formal structure of quantum mechanics. In his book *The Principles of Quantum Mechanics*, Dirac introduced a powerful and abstract formulation based on linear operators, bras and kets, and transformation theory, which remains the standard language of the field. Within this framework, he also introduced the distribution now known as the Dirac delta, providing an indispensable mathematical tool for the continuous spectra of observables and for the formulation of quantum states in momentum and position space.

A distinctive feature of Dirac's scientific philosophy was his unwavering belief in the guiding role of mathematical beauty. He regarded elegance, simplicity, and internal coherence as essential criteria for fundamental theories, famously asserting that it is more important for equations to be beautiful than for them to agree immediately with experiment.

Dirac's impact extended well beyond relativistic quantum mechanics. His ideas contributed decisively to the early development of QFT, canonical quantization, and the modern understanding of constrained dynamical systems. In recognition of his achievements, he was awarded the Nobel Prize in Physics in 1933, shared with Erwin Schrödinger, for the discovery of new productive forms of atomic theory.

In the later years of his life, Dirac held academic positions in the United States, spending a significant period at Florida State University.

4.1 Continuous Symmetries

As we know, the principle of relativity ensures the complete equivalence for every inertial reference system, allowing each observer to independently describe the field dynamics through an action and a Lagrangian with the same functional dependence on the fields. This is equivalent to requiring that

$$S(\phi') = \int d^4x' \mathcal{L}(\phi'(x'), \partial_\mu \phi'(x'), x') = \int d^4x \mathcal{L}(\phi(x), \partial_\mu \phi(x), x) = S(\phi). \tag{4.1}$$

We can consider this expression as a symmetry of the action concerning transformations in a single reference system

$$x \longrightarrow x', \phi \longrightarrow \phi'. \tag{4.2}$$

Such transformations include two principal groups:

1. internal symmetries: such that $x = x'$ and $\phi \longrightarrow \phi'$, which change the fields but not the coordinates;
2. spacetime symmetries of Minkowski: i.e., transformations of the Poincaré group, including translations and proper, orthochronous Lorentz transformations.

▶ **Pause and think**
Prepare a lecture of a few minutes in which you explain how fields would transform.

The symmetries of Minkowski spacetime are those considered to impose the principle of relativity. Let us consider continuous isometries of spacetime, which can be expressed in infinitesimal form

$$\begin{cases} x'^{\mu} = x^{\mu} + \delta x^{\mu}, \\ \phi'(x') = \phi(x) + \delta_T \phi, \end{cases} \tag{4.3}$$

where $\delta_T \phi$ is the total variation of the field, which, in addition to varying the field, also varies the event. We can decompose it in the following way

$$\delta_T \phi = \phi'(x') - \phi(x) = \phi'(x') - \phi(x) \pm \phi(x') = \phi'(x') - \phi(x') + \phi(x') - \phi(x), \tag{4.4}$$

where $\delta\phi = \phi'(x') - \phi(x')$ and $\phi(x') - \phi(x)$ is the translation, which we can rewrite as, using $x' = x + \delta x$

$$\phi(x') - \phi(x) = \phi(x + \delta x) - \phi(x) \sim \phi(x) + (\partial_\mu \phi)\delta x^{\mu} - \phi(x) = (\partial_\mu \phi)\delta x^{\mu}. \tag{4.5}$$

Substituting the previous expression into Eq. (4.4), we obtain

$$\delta_T \phi = \delta\phi + (\partial_\mu \phi)\delta x^{\mu}. \tag{4.6}$$

For each field, we can write

$$\delta_T = \delta + \delta x^{\mu} \partial_\mu, \tag{4.7}$$

where δ_T is the total variation, δ is the functional variation, and $\delta x^{\mu} \partial_\mu$ represents translations.

Since the volume element in $M(3,1)$ is a Lorentz invariant, i.e., $d^4x' = d^4x$, we need to impose

$$\mathcal{L}(\phi', \partial_\mu\phi', x') = \mathcal{L}(\phi, \partial_\mu\phi, x) \tag{4.8}$$

or equivalently

$$\mathcal{L}(\phi', \partial_\mu\phi', x') - \mathcal{L}(\phi, \partial_\mu\phi, x) = \delta_T\mathcal{L} = 0. \tag{4.9}$$

Let us write

$$\delta_T\mathcal{L} = \delta\mathcal{L} + \delta x^\mu\partial_\mu\mathcal{L} = \frac{\partial\mathcal{L}}{\partial\phi}\delta\phi + \frac{\partial\mathcal{L}}{\partial(\partial_\mu\phi)}\delta(\partial_\mu\phi) + \delta x^\mu\frac{\partial\mathcal{L}}{\partial x^\mu}. \tag{4.10}$$

Let us impose $\delta_T\mathcal{L} = 0$ to get

$$\begin{aligned} 0 = \delta_T\mathcal{L} &= \frac{\partial\mathcal{L}}{\partial\phi}\delta\phi + \partial_\mu\left[\frac{\partial\mathcal{L}}{\partial(\partial_\mu\phi)}\delta\phi\right] - \partial_\mu\left(\frac{\partial\mathcal{L}}{\partial(\partial_\mu\phi)}\right)\delta\phi + \delta x^\mu\frac{\partial\mathcal{L}}{\partial x^\mu} = \\ &= \left[\frac{\partial\mathcal{L}}{\partial\phi} - \partial_\mu\frac{\partial\mathcal{L}}{\partial(\partial_\mu\phi)}\right]\delta\phi + \partial_\mu\left[\frac{\partial\mathcal{L}}{\partial(\partial_\mu\phi)}\delta\phi\right] + \frac{\partial\mathcal{L}}{\partial x^\mu}\delta x^\mu. \end{aligned} \tag{4.11}$$

Within the previous expression, we have that

$$\frac{\partial\mathcal{L}}{\partial\phi} - \partial_\mu\frac{\partial\mathcal{L}}{\partial(\partial_\mu\phi)} = 0, \tag{4.12a}$$

$$\frac{\partial\mathcal{L}}{\partial x^\mu}\delta x^\mu = \frac{\partial}{\partial x^\mu}\left[\mathcal{L}(\phi, \partial_\mu\phi, x)\delta x^\mu\right] - \mathcal{L}\left[\frac{\partial}{\partial x^\mu}\delta x^\mu\right]. \tag{4.12b}$$

The second term in Eq. (4.12b) is proven to be zero using the invariance of d^4x.

Driving exercise
Prove the above statement.
In general, we have

$$d^4x'^\mu = \left\|\frac{\partial x'^\mu}{\partial x^\nu}\right\| d^4x^\nu, \tag{4.13}$$

where $\left\|\frac{\partial x'^\mu}{\partial x^\nu}\right\|$ represents the determinant of the Jacobian matrix of the transformation. Let us write

$$\det\left[\frac{\partial x'^\mu}{\partial x^\nu}\right] = \det\left[\delta^\mu_\nu + \partial_\nu\right]. \tag{4.14}$$

Using the fact that $\det(e^A) = e^{\mathrm{Tr}(A)}$ with A small, we can write $\det(1 + A) \simeq 1 + \mathrm{Tr}(A)$. Therefore,

$$d^4x'^\mu = \det\left[\delta^\mu_\nu + \partial_\nu\delta x^\mu\right] \simeq (1 + \partial_\mu\delta x^\mu)d^4x, \tag{4.15}$$

where $\partial_\mu\delta x^\mu = \mathrm{Tr}(\partial_\nu\delta x^\mu)$. However, $d^4x' = d^4x$, so

$$1 + \partial_\mu\delta x^\mu = 1 \longrightarrow \partial_\mu\delta x^\mu = 0. \tag{4.16}$$

In this way, we obtain a *conservation law* (the continuity equation)

$$\delta_T\mathcal{L}=0 \longrightarrow 0=\partial_\mu\left[\frac{\partial\mathcal{L}}{\partial(\partial_\mu\phi)}\delta\phi\right]+\partial_\mu\left[\mathcal{L}\delta x^\mu\right], \tag{4.17}$$

which can be written more compactly as

$$\partial_\mu\left[\frac{\partial\mathcal{L}}{\partial(\partial_\mu\phi)}\delta\phi+\mathcal{L}\delta x^\mu\right]=0. \tag{4.18}$$

Since we are considering infinitesimal transformations

$$x'^\mu=x^\mu+\delta x^\mu, \tag{4.19a}$$
$$\phi'(x)=\phi(x)+\delta\phi, \tag{4.19b}$$

we can denote them by ε_a, where $a=1,\ldots,n$, as infinitesimal parameters and write each transformation as

$$T\sim \mathbb{1}+\varepsilon_a\left.\frac{\partial T}{\partial\varepsilon_a}\right|_{\varepsilon=0}, \tag{4.20}$$

which acts either on x or on ϕ, and with $\frac{\partial T}{\partial\varepsilon_a}$ being the generators of the transformation, where a can indicate either an index or a set of indices depending on the dimension of the representation. In this way, infinitesimal transformations can be written as

$$\delta x^\mu=\sum_a\varepsilon_a(\Delta^a)^\mu(x), \tag{4.21a}$$
$$\delta\phi=\sum_a\varepsilon_a(\Sigma^a)\phi. \tag{4.21b}$$

Substituting these transformations into Eq. (4.18), we obtain

$$\partial_\mu\left[\frac{\partial\mathcal{L}}{\partial(\partial_\mu\phi)}\sum_a\varepsilon_a(\Sigma^a)\phi+\mathcal{L}\sum_a\varepsilon_a(\Delta^a)^\mu\right]=\sum_a\varepsilon_a\left[\partial_\mu\left(\frac{\partial\mathcal{L}}{\partial(\partial_\mu\phi)}\Sigma^a\phi+\mathcal{L}(\Delta^a)^\mu\right)\right]=0. \tag{4.22}$$

However, considering that the equation must be independent of ε_a, we write

$$\partial_\mu\left[\frac{\partial\mathcal{L}}{\partial(\partial_\mu\phi)}\Sigma^a\phi+\mathcal{L}\Delta^{a\mu}\right]=0. \tag{4.23}$$

These equations are n *continuity equations*. We can define n conserved currents, each associated with a generator

$$(J^a)^\mu=\frac{\partial\mathcal{L}}{\partial(\partial_\mu\phi)}\Sigma^a\phi+\mathcal{L}\Delta^{a\mu}\longrightarrow\partial_\mu(J^a)^\mu=0, \tag{4.24}$$

where $a=1,\ldots,n$. These currents are effectively tensors in Minkowski space.

4.1.1 The Noether Theorem

The result obtained at the end of the previous section is the *Noether theorem.*

> To each symmetry of the system, there are associated n conserved currents, as many as the generators of the continuous symmetry.

In addition, each conserved current determines a *constant of motion*, defined by integrating the continuity equation over a three-dimensional volume parameterized by $x^0 = ct = t$ in Minkowski space $M(3, 1)$

$$0 = \int d^3x \partial_\mu J^{a\mu} = \int d^3x \left(\frac{\partial}{\partial t} J^{a0} + \nabla \cdot \boldsymbol{J}^a \right) = \frac{d}{dt} \int d^3x J^{a0} + \int d^3x \nabla \cdot \boldsymbol{J}^a. \tag{4.25}$$

The *Noether charge* is defined as the integral over space of the temporal component of $J^{a\mu} = (J^{a0}, \boldsymbol{J}^a)$

$$Q^a = \int d^3x J^{a0} \tag{4.26}$$

and it is observed that

$$0 = \int d^3x \partial_\mu J^{a\mu} \longrightarrow \frac{d}{dt} Q^a = -\int_V d^3x \nabla \cdot \boldsymbol{J}^a = -\int_\Sigma d\sigma \boldsymbol{J}^a \cdot \mathbf{n}, \tag{4.27}$$

where $\Sigma = \partial V$. When $\Sigma \to \infty$, the integral goes to zero as the surface extends to infinity because the fields vanish. In this way, we obtain the *conservation of charge*

$$\frac{d}{dt} Q^a = 0. \tag{4.28}$$

> The considered symmetries are also called *global gauge* symmetries and do not involve changes in coordinates in spacetime. In other words, the transformation parameters are not functions of x^μ, i.e., the ε_a **do not** depend on x^μ.

Let us write the continuity equation for $J^\mu = (J^{1\mu}, \ldots, J^{n\mu})$ and integrate over a 4-dimensional volume:

$$\partial_\mu J^\mu = 0 \longrightarrow \int_{V_4} d^4x \partial_\mu J^\mu = 0. \tag{4.29}$$

We can choose a cylindrical volume where the base is parametrized in one system by $\Gamma_0 : t = 0$ and in another by $\Gamma_1 : t' = \text{const}$. Let I denote the contribution from the lateral surface, so we have:

$$\int_{V_4} d^4x \partial_\mu J^\mu = -\int_{\Gamma_0} d\Gamma_0 n_\mu J^\mu + \int_{\Gamma_1} d\Gamma_1 n'_\mu J^\mu + I = 0, \tag{4.30}$$

where $I = 0$ if the lateral surface extends to infinity. The minus sign is due to choosing the outward direction as the positive one. Therefore, we can write:

$$\int_{\Gamma_0} d\Gamma_0 n_\mu J^\mu = \int_{\Gamma_1} d\Gamma_1 n'_\mu J^\mu. \tag{4.31}$$

Now, since there are scalars inside the integral, we can calculate them in the most convenient reference frame. In this case, consider:

$$d\Gamma_0 n_\mu = (d^3x, 0, 0, 0), \tag{4.32a}$$

$$d\Gamma_1 n'_\mu = (d^3x', 0, 0, 0). \tag{4.32b}$$

In this case, Eq. (4.31) becomes:

$$\int d^3x J^0(x) = \int d^3x' J^0(x'). \tag{4.33}$$

Thus, we have concluded that charge is a Lorentz invariant.

We write $J'^\mu = J^\mu + s^\mu$ where

1. if s^μ is conserved for physical reasons, then $\partial_\mu s^\mu = 0$,
2. if s^μ is conserved for algebraic reasons, then we can write it as the divergence of the antisymmetric tensor $G^{\alpha\beta} = -G^{\beta\alpha}$, so that $s^\mu = \partial_\alpha G^{\alpha\mu}$ with $G^{\alpha\mu}$ being a superpotential. In this case, we can write

$$\partial_\mu s^\mu = \partial_\mu \partial_\alpha G^{\alpha\mu} = \partial_\alpha \partial_\mu G^{\mu\alpha} = -\partial_\mu \partial_\alpha G^{\alpha\mu} = 0, \tag{4.34}$$

and then we have

$$\partial_\mu J'^\mu = \partial_\mu J^\mu. \tag{4.35}$$

The currents are defined up to a conserved four-vector, for physical or algebraic reasons. The continuity equation remains unchanged, and the charge is conserved. In fact,

$$Q' = \int d^3x J^0 + s^0 = \int d^3x J^0 + \int d^3x s^0 = \int d^3x J^0 + \int d^3x \partial_\alpha G^{\alpha 0}, \tag{4.36}$$

with $\alpha = 0, 1, 2, 3$. Since G is antisymmetric, we consider only the spatial components, so the above expression becomes

$$Q' = \int d^3x J^0 + \int d^3x \partial_i G^{i0} = \int d^3x J^0 + \int d\sigma_i G^{i0} = \int d^3x J^0 = Q. \quad (4.37)$$

Considering continuous symmetries of $M(3, 1)$, i.e., group transformations P separately, we calculate Noether currents with associated charges, recalling the notion of the generator of an infinitesimal transformation. We write $\delta\phi = \sum_a \varepsilon_a (\Sigma^a)\phi$ and explicitly calculate Q^a so that

$$Q^a = \int d^3x J^{a0} = \int d^3x \left[\frac{\partial \mathcal{L}}{\partial(\partial_t \phi)} \Sigma^a \phi + \mathcal{L}(\nabla^a)^0 \right], \quad (4.38)$$

where $\frac{\partial \mathcal{L}}{\partial(\partial_t \phi)} = \pi$ and $(\nabla^a)^0 = (\nabla^a)(t)$. We find that the variation of Q^a with respect to the conjugate field momentum is equal to the transformation of the field generator Σ^a, i.e.,

$$\frac{\delta Q^a}{\delta \pi} = \{\phi, Q^a\} = \Sigma^a \phi. \quad (4.39)$$

This implies that Q^a is the generator of the a-th component of the infinitesimal symmetry transformation $\delta\phi$ and is different from Σ^a, which, instead, is a matrix and represents the generator related to a particular representation of $\phi'(x) = \phi(x) + \delta\phi$.

4.1.2 The Energy-Momentum Tensor

Let us consider the Poincaré transformation group, which includes translations in spacetime and special Lorentz transformations.

▶ **Pause and think** Manage to do a lecture of translations of spacetime. How does it influence our field scheme?

Consider the coordinate transformations for the event $x'^\mu = x^\mu + a^\mu$ and for the field $\phi'(x') = \phi(x)$. We impose invariance under translations

$$\mathcal{L}(\phi', \phi'_\mu, x') = \mathcal{L}(\phi, \phi'_\mu, x) \longrightarrow \mathcal{L}(\phi', \phi'_\mu, x + a) = \mathcal{L}(\phi, \phi'_\mu, x), \quad (4.40)$$

where, from now on, it is imposed that $\mathcal{L}$ does not explicitly depend on the event but only through ϕ and ϕ_μ, so $\mathcal{L} = \mathcal{L}(\phi, \partial_\mu \phi)$. Now, assuming $\delta x^\mu = a^\mu$ constant, the generators will take the form $\delta x^\mu = \delta^\mu_\nu a^\nu$. We derive the form of the functional variation of the field

$$0 = \delta_T \phi = \delta\phi + (\partial_\mu \phi) a^\mu \longrightarrow \delta\phi = -(\partial_\mu \phi) a^\mu. \tag{4.41}$$

We lower the indices of the parameters

$$\delta x^\mu = \delta^\mu_\nu a^\nu = \delta^\mu_\nu g^{\alpha\nu} a_\alpha = g^{\alpha\mu} a_\alpha, \tag{4.42a}$$

$$\delta\phi = -(\partial_\mu \phi) a^\mu = -(\partial_\mu \phi) g^{\alpha\mu} a_\alpha = -(\partial^\alpha \phi) a_\alpha. \tag{4.42b}$$

We write the conserved currents

$$J^{a\mu} = \frac{\partial \mathcal{L}}{\partial(\partial_\mu \phi)} (-\partial^a \phi) + \mathcal{L} g^{a\mu}. \tag{4.43}$$

Changing sign, we define the *canonical energy-momentum tensor*

$$T^{\mu\nu} = \frac{\partial \mathcal{L}}{\partial(\partial_\mu \phi)} \partial^\nu \phi - \mathcal{L} g^{\mu\nu} \longrightarrow \partial_\mu T^{\mu\nu} = 0. \tag{4.44}$$

Two important components of $T^{\mu\nu}$ are

- $T^{00} = \frac{\partial \mathcal{L}}{\partial(\partial_t \phi)} \partial^t \phi - \mathcal{L} g^{00} = \pi \dot{\phi} - \mathcal{L} = \mathcal{H} \longrightarrow \int d^3x T^{00} = \int d^3x \mathcal{H} = H$ (temporal component of the four-momentum);
- $T^{0i} = \frac{\partial \mathcal{L}}{\partial(\partial_t \phi)} \partial^i \phi = \pi \partial^i \phi \longrightarrow \int d^3x T^{0i} = \int d^3x \pi \partial^i \phi = P^i$ (spatial component).

Thus, the charges are identified with the four-momentum

$$\int d^3x T^{0\nu} = P^\nu = (P^0, \mathbf{P}). \tag{4.45}$$

This is consistent with two facts already derived in non-relativistic settings.

1. The generator of spacetime translations is $P^\mu = (E, \mathbf{P})$, where E generates temporal translations, and $\mathbf{P}$ generates spatial translations.
2. The charge associated with invariance under spacetime translations is precisely the four-momentum. This symmetry leads to the conservation of energy and momentum.

4.1.3 Special Lorentz Transformations

The invariance under Lorentz transformations is conveniently expressed, at the infinitesimal level, by writing

$$\omega^\mu_\nu = g^{\mu\rho}\delta^\sigma_\nu \omega_{\rho\sigma} = \frac{1}{2}\left(g^{\mu\rho}\delta^\sigma_\nu \omega_{\rho\sigma} + g^{\mu\sigma}\delta^\rho_\nu \omega_{\sigma\rho}\right) = \frac{\omega_{\rho\sigma}}{2}\left(g^{\mu\rho}\delta^\sigma_\nu - g^{\mu\sigma}\delta^\rho_\nu\right). \tag{4.46}$$

To avoid calling the parameters $\omega_{\rho\sigma}$ as the transformation matrix and thus create confusion, let us use a notation consistent with translations, defining

$$\omega_{\rho\sigma} \longrightarrow \varepsilon_{\rho\sigma}. \tag{4.47}$$

Let us consider $x^\mu = \Lambda^\mu_\nu x^\nu$ and expand it in the vicinity of the identity $\Lambda \simeq \mathbb{I} + \omega$ obtaining

$$x'^\mu = (\delta^\mu_\nu + \omega^\mu_\nu)x^\nu \quad \text{with} \quad |\omega^\mu_\nu| \ll 1. \tag{4.48}$$

The ω^μ_ν are not 16 independent elements because Λ satisfies $\Lambda^T g \Lambda = g$. Applying the same procedure to find the generators of $L^\uparrow_+$, we obtain

$$g_{\mu\nu}\Lambda^\mu_\rho \Lambda^\nu_\sigma = g_{\rho\sigma} \longrightarrow g_{\mu\nu}(\delta^\mu_\rho + \omega^\mu_\rho)(\delta^\nu_\sigma + \omega^\nu_\sigma) = g_{\rho\sigma}. \tag{4.49}$$

Expanding the previous expression

$$g_{\mu\nu}\delta^\mu_\rho\delta^\nu_\sigma + g_{\mu\nu}\delta^\mu_\rho\omega^\nu_\sigma + g_{\mu\nu}\omega^\mu_\rho\delta^\nu_\sigma + O(\omega^2) = g_{\rho\sigma} \to g_{\rho\sigma} + \delta^\mu_\rho\omega_{\mu\sigma} + \delta^\nu_\sigma\omega_{\nu\rho} = g_{\rho\sigma}. \tag{4.50}$$

In the end, we obtain

$$\omega_{\rho\sigma} + \omega_{\sigma\rho} = 0 \longrightarrow \omega_{\rho\sigma} = -\omega_{\sigma\rho}. \tag{4.51}$$

Now, let us write $\omega_{\rho\sigma}$ and ω^μ_ν as

$$\omega_{\rho\sigma} = i\frac{a_{\alpha\beta}}{2}M^{(\alpha\beta)}_{\rho\sigma} \quad \text{with} \quad M^{(\alpha\beta)}_{\rho\sigma} = -i(\delta^\alpha_\rho\delta^\beta_\sigma - \delta^\alpha_\sigma\delta^\beta_\rho), \tag{4.52a}$$

$$\omega^\mu_\nu = i\frac{a_{\alpha\beta}}{2}M^{(\alpha\beta)\mu}_\nu \quad \text{with} \quad M^{(\alpha\beta)\mu}_\nu = g^{\mu\rho}\delta^\sigma_\nu M^{(\alpha\beta)}_{\rho\sigma} = -i(g^{\mu\alpha}\delta^\beta_\nu - g^{\mu\beta}\delta^\alpha_\nu). \tag{4.52b}$$

Then, the infinitesimal Lorentz transformations read

$$\Lambda^\mu_\nu = \delta^\mu_\nu + i\frac{\varepsilon_{\rho\sigma}}{2}M^{(\rho\sigma)\mu}_\nu \iff \Lambda = \mathbb{I} + i\frac{\varepsilon_{\rho\sigma}}{2}M^{\rho\sigma}. \tag{4.53}$$

Furthermore, let us define the representation of $L_+^\uparrow$ in the Minkowski spacetime $M(3,1)$ as

$$D(\Lambda) = e^{\frac{i}{2}\varepsilon_{\rho\sigma}M^{\rho\sigma}}. \tag{4.54}$$

Now we know the generators of the event transformation, i.e.,

$$\begin{aligned}\delta x^\mu =&\omega^\mu_\nu x^\nu = i\frac{\varepsilon_{\rho\sigma}}{2}M_\nu^{(\rho\sigma)\mu}x^\nu = i\frac{\varepsilon_{\rho\sigma}}{2}(-i)(g^{\mu\rho}\delta^\sigma_\nu - g^{\mu\sigma}\delta^\rho_\nu)x^\nu = \\ =&\frac{\varepsilon_{\rho\sigma}}{2}(g^{\mu\rho}\delta^\sigma_\nu - g^{\mu\sigma}\delta^\rho_\nu)x^\nu.\end{aligned} \tag{4.55}$$

Let us consider a set of fields $\{\phi_1, \ldots, \phi_n\}$ with $L, N = 1, \ldots, n$ labeling the components of the field. The most general transformation one can think of is through a general transformation of $L_+^\uparrow$ represented in any generic space V. We have

1. infinitesimal parameters $\varepsilon_{\rho\sigma}$;
2. antisymmetric generators $F^{\rho\sigma}$ acting on

$$\begin{aligned}\phi'_L(x') = D(\Lambda)_{LN}\phi_N(x) = \left(e^{\frac{i}{2}\varepsilon_{\rho\sigma}F^{\rho\sigma}}\right)_{LN}\phi_N \simeq \\ \simeq \left[\delta_{LN} + \frac{i}{2}\varepsilon_{\rho\sigma}(F^{\rho\sigma})_{LN}\right]\phi_N(x).\end{aligned} \tag{4.56}$$

Obviously, the field does not live in $M(3,1)$ but in a generic V; thus, $D(\Lambda)$ will map the element of $L_+^\uparrow$ to a transformation group in a generic V, and the generators are $F^{\rho\sigma} \neq M^{\rho\sigma}$.

If the field is scalar, it has only one component!

Let us consider the variation in the form of ϕ_L, obtaining

$$\delta\phi_L = \phi'_L(x) - \phi_L(x) = \left[\delta_{LN} + \frac{i\varepsilon_{\rho\sigma}}{2}F^{\rho\sigma}_{LN}\right]\phi_N(x-\delta_x) - \phi_L(x). \tag{4.57}$$

Using $\phi_N(x-\delta x) \sim \phi_N(x) - (\partial_\mu \phi_N)\delta x^\mu$, we can rewrite the above expression as

$$\delta\phi_L = \left[\delta_{LN} + \frac{i}{2}\varepsilon_{\rho\sigma} F^{\rho\sigma}_{LN}\right]\left[\phi_N(x) - \partial_\mu\phi_N\delta x^\mu\right] - \phi_L(x) =$$
$$\delta_{LN}\phi_N(x) - \delta_{LN}(\partial_\mu\phi_N)\delta x^\mu + \frac{i}{2}\varepsilon_{\rho\sigma} F^{\rho\sigma}_{LN}\phi_N(x) +$$
$$-\left(\frac{i}{2}\varepsilon_{\rho\sigma} F^{\rho\sigma}_{LN}\right)(\partial_\mu\phi_N)\delta x^\mu - \phi_L(x). \tag{4.58}$$

Since $\left(\frac{i}{2}\varepsilon_{\rho\sigma} F^{\rho\sigma}_{LN}\right)(\partial_\mu\phi_N)\delta x^\mu$ is of second order, we can ignore it, obtaining

$$\delta\phi_L \simeq \phi_L(x) - (\partial_\mu\phi_L)\delta x^\mu + \frac{i}{2}\varepsilon_{\rho\sigma} F^{\rho\sigma}_{LN}\phi_N(x) - \phi_L(x) \simeq$$
$$\simeq -(\partial_\mu\phi_L)\frac{i\varepsilon_{\rho\sigma}}{2}(M^{\rho\sigma})^\mu_\nu x^\nu + \frac{i}{2}\varepsilon_{\rho\sigma} F^{\rho\sigma}_{LN}\phi_N(x), \tag{4.59}$$

where $\delta x^\mu = i\frac{\varepsilon_{\rho\sigma}}{2}(M^{\rho\sigma})^\mu_\nu x^\nu$. Highlighting the common terms, we obtain

$$\delta\phi_L \simeq i\frac{\varepsilon_{\rho\sigma}}{2}\left[F^{\rho\sigma}_{LN}\phi_N(x) - (\partial_\mu\phi_L)(M^{\rho\sigma})^\mu_\nu x^\nu\right]. \tag{4.60}$$

Now, let us write the conserved currents

$$J^{a\mu} = \frac{\partial\mathcal{L}}{\partial(\partial_\mu\phi_L)}\Sigma^a\phi_L + \mathcal{L}\Delta^a x^\mu \quad \text{with} \quad a = (\rho\sigma). \tag{4.61}$$

We can express the *canonical angular momentum tensor* as

$$M^{\mu(\rho\sigma)} = \frac{\partial\mathcal{L}}{\partial(\partial_\mu\phi_L)}\left[i\left(F^{\rho\sigma}_{LN}\phi_N(x) - (\partial_\alpha\phi_L)(M^{\rho\sigma})^\alpha_\nu x^\nu\right)\right] + \mathcal{L}(g^{\mu\rho}\delta^\sigma_\nu - g^{\mu\sigma}\delta^\rho_\nu)x^\nu = \tag{4.62}$$
$$= -i\frac{\partial\mathcal{L}}{\partial(\partial_\mu\phi_L)}(\partial_\alpha\phi_L)(M^{\rho\sigma})^\alpha_\nu x^\nu + \mathcal{L}(g^{\mu\rho}\delta^\sigma_\nu - g^{\mu\sigma}\delta^\rho_\nu)x^\nu + i\frac{\partial\mathcal{L}}{\partial(\partial_\mu\phi_L)}F^{\rho\sigma}_{LN}\phi_N, \tag{4.63}$$

where we have written the generator of $L^\uparrow_+$ in the representation on $M(3,1)$ as a generic antisymmetric matrix. Now, multiply the deltas by x^ν, obtaining

$$M^{\mu(\rho\sigma)} = -\frac{\partial\mathcal{L}}{\partial(\partial_\mu\phi_L)}(\partial_\alpha\phi_L)(g^{\alpha\rho}x^\sigma - g^{\alpha\sigma}x^\rho) + \mathcal{L}(g^{\mu\rho}x^\sigma - g^{\mu\sigma}x^\rho) + i\frac{\partial\mathcal{L}}{\partial(\partial_\mu\phi_L)}F^{\rho\sigma}_{LN}\phi_N. \tag{4.64}$$

Collecting the common terms, we have

$$M^{\mu(\rho\sigma)} = \left[-\frac{\partial \mathcal{L}}{\partial(\partial_\mu \phi_L)} g^{\alpha\rho} + \mathcal{L} g^{\mu\rho}\right] x^\sigma - \left[-\frac{\partial \mathcal{L}}{\partial(\partial_\mu \phi_L)} (\partial_\alpha \phi_L) g^{\alpha\sigma} + \mathcal{L} g^{\mu\sigma}\right] (-x^\rho)$$
$$+ i \frac{\partial \mathcal{L}}{\partial(\partial_\mu \phi_L)} F^{\rho\sigma}_{LN} \phi_N =$$
$$= \left[-\frac{\partial \mathcal{L}}{\partial(\partial_\mu \phi_L)} (\partial^\rho \phi_L) + \mathcal{L} g^{\mu\rho}\right] x^\sigma - \left[-\frac{\partial \mathcal{L}}{\partial(\partial_\mu \phi_L)} (\partial^\sigma \phi_L) + \mathcal{L} g^{\mu\sigma}\right] x^\rho +$$
$$+ i \frac{\partial \mathcal{L}}{\partial(\partial_\mu \phi_L)} F^{\rho\sigma}_{LN} \phi_N. \tag{4.65}$$

The first two terms inside $M^{\mu(\rho\sigma)}$ correspond to the energy-momentum tensor -$T^{\mu\rho}$ and -$T^{\mu\sigma}$, while the last term is $S^{\mu(\rho\sigma)}$, a generic generator of $L_+^\uparrow$. We can finally write

$$M^{\mu(\rho\sigma)} = T^{\mu\sigma} x^\rho - T^{\mu\rho} x^\sigma + S^{\mu(\rho\sigma)} \longrightarrow \partial_\mu M^{\mu(\rho\sigma)} = 0. \tag{4.66}$$

By analogy with the angular momentum $L^{\mu\nu} = x^\mu p^\nu - x^\nu p^\mu$, we identify the *orbital angular momentum component*

$$T^{\mu\sigma} x^\rho - T^{\mu\rho} x^\sigma = L^{\mu(\rho\sigma)} \longrightarrow M^{\mu(\rho\sigma)} = L^{\mu(\rho\sigma)} + S^{\mu(\rho\sigma)}, \tag{4.67}$$

where $L^{\mu(\rho\sigma)}$ is the orbital part dependent on the spacetime event, while $S^{\mu(\rho\sigma)}$ is the spin part dependent on the morphological properties of the Lagrangian (spin seems to be a relativistic prediction of Noether's theorem). The charges are obtained by integrating the temporal part over space.

$$\int d^3x M^{0(\rho\sigma)} = \int d^3x (x^\rho T^{0\sigma} - x^\sigma T^{0\rho}) = L^{\rho\sigma} = L^{\rho\sigma}, \tag{4.68}$$

with

$$L^{\rho\sigma} \to \begin{cases} K^i = L^{0i} & \text{time part,} \\ J^i = \frac{\varepsilon^{ijk}}{2} L^{jk} & \text{spatial part.} \end{cases} \tag{4.69}$$

The division into L and S as the orbital and intrinsic parts has no physical significance because we can always redefine $T^{\mu\nu}$ in such a way as to nullify the intrinsic term.

From general relativity, the energy-momentum tensor found in Einstein's field equations must be symmetric. However, in general, the canonical tensor is not symmetric. To make it symmetric, we perform the *symmetrization of the energy-momentum tensor*. Note that

$$\partial_\mu M^{\mu(\rho\sigma)} = 0 \longrightarrow \partial_\mu (x^\rho T^{\mu\sigma} - x^\sigma T^{\mu\rho}) + \partial_\mu S^{\mu(\rho\sigma)} = 0. \tag{4.70}$$

Taking into account that $\partial_\mu T^{\mu\alpha} = 0$ and $\partial_\mu x^\alpha = \delta^\alpha_\mu$, we can write the above expression as

$$T^{\mu\sigma}\delta^\rho_\mu - T^{\mu\rho}\delta^\sigma_\mu + \partial_\mu S^{\mu(\rho\sigma)} = T^{\rho\sigma} - T^{\sigma\rho} + \partial_\mu S^{\mu(\rho\sigma)} = 0. \tag{4.71}$$

In general, $\partial_\mu S^{\mu(\rho\sigma)} \neq 0$ because it only cancels if $S^{\mu(\rho\sigma)} = 0$ for a scalar field.

▶ **Pause and think** Manage to show why the scalar field is spin 0 and why the spinor field is spin 1/2.

4.1.4 Belinfante and Rosenfeld Procedure

We will now construct an antisymmetric superpotential with respect to the last two indices. This represents a particular case of the Christoffel symbols for curved spacetime. Let us begin by writing

$$G^{\alpha\rho\sigma} = \frac{1}{2}(S^{\alpha\rho\sigma} + S^{\rho\alpha\sigma} - S^{\sigma\alpha\rho}), \tag{4.72}$$

and it is antisymmetric because, by swapping the last two indices, we obtain

$$\begin{aligned} G^{\alpha\sigma\rho} &= \frac{1}{2}(S^{\alpha\sigma\rho} + S^{\sigma\alpha\rho} - S^{\rho\alpha\sigma}) = \frac{1}{2}(-S^{\alpha\rho\sigma} + S^{\sigma\alpha\rho} - S^{\rho\alpha\sigma}) = \\ &= -\frac{1}{2}(S^{\alpha\rho\sigma} + S^{\rho\alpha\sigma} - S^{\sigma\alpha\rho}) = -G^{\alpha\rho\sigma}. \end{aligned} \tag{4.73}$$

We redefine the tensor

$$\Theta^{\mu\nu} = T^{\mu\nu} - \partial_\alpha G^{\nu\alpha\mu}, \tag{4.74}$$

where $\partial_\alpha G^{\nu\alpha\mu}$ is the divergence of the superpotential. This tensor is *symmetric*, as

$$\begin{aligned} \Theta^{\mu\nu} - \Theta^{\nu\mu} &= T^{\mu\nu} - T^{\nu\mu} - \partial_\alpha G^{\nu\alpha\mu} + \partial_\alpha G^{\mu\alpha\nu} = T^{\mu\nu} - T^{\nu\mu} - \partial_\alpha (G^{\nu\alpha\mu} - G^{\mu\alpha\nu}) \\ &= T^{\mu\nu} - T^{\nu\mu} - \frac{1}{2}\partial_\alpha (S^{\nu\alpha\mu} + S^{\alpha\nu\mu} - S^{\mu\nu\alpha} - S^{\mu\alpha\nu} - S^{\alpha\mu\nu}) \\ &= T^{\mu\nu} - T^{\nu\mu} - \frac{1}{2}\partial_\alpha (S^{\nu\alpha\mu} - S^{\nu\alpha\rho} + S^{\alpha\nu\mu} + S^{\alpha\nu\mu} - S^{\mu}{\alpha\nu} + S^{\mu\alpha\nu}) \\ &= T^{\mu\nu} - T^{\nu\mu} - \frac{1}{2}\partial_\mu (2S^{\alpha\nu\mu}) = -\partial_\alpha S^{\alpha(\mu\nu)} + \partial_\alpha S^{\alpha\mu\nu} = 0, \end{aligned} \tag{4.75}$$

where we used the fact that $T^{\mu\nu} - T^{\nu\mu} = -\partial_\alpha S^{\alpha(\mu\nu)}$.

The tensor $\Theta^{\mu\nu}$ is also *conserved* since

$$\partial_\mu \Theta^{\mu\nu} = \partial_\mu T^{\mu\nu} - \partial_\mu T^{\mu\nu} - \partial_\mu \partial_\alpha G^{\nu\alpha\mu} = 0, \tag{4.76}$$

because $T^{\mu\nu}$ is conserved, and $\partial_\mu \partial_\alpha$ is symmetric in (α, μ) while $G^{\nu\alpha\mu}$ is antisymmetric in (α, μ).

Moreover, the conserved charges remain invariant

$$P'^{\nu} = \int d^3x \Theta^{0\nu} = \int d^3x T^{0\nu} - \int d^3x \partial_\alpha G^{\nu\alpha 0} = P^\nu - \int d^3x \partial_0 G^{\nu 00} - \int d^3x \partial_i G^{\nu i 0} = \tag{4.77}$$

$$= P^\nu - \int d^3x \frac{1}{2} \partial_0 (S^{\nu 00} + S^{0\nu 0} - S^{0\nu 0}) - \int d\sigma_i G^{\nu i 0} = P^\nu, \tag{4.78}$$

where the last integral has been neglected because it is a boundary condition.

We could construct a canonical angular momentum tensor of the form

$$N^{\mu(\rho\sigma)} = x^\rho \Theta^{\mu\sigma} - x^\sigma \Theta^{\mu\rho} \longrightarrow \partial_\rho N^{\mu(\rho\sigma)} = 0. \tag{4.79}$$

Let us prove that this tensor coincides with $\partial_\mu M^{\mu(\rho\sigma)} = 0$. If it is demonstrated that the charges are the same, then this confirms that the separation into orbital and spin parts has no physical reasons. First, let us make $\Theta^{\mu\rho}$ explicit, obtaining

$$\begin{aligned} N^{\mu(\rho\sigma)} &= x^\rho (T^{\mu\sigma} - \partial_\alpha G^{\sigma\alpha\mu}) = x^\sigma (T^{\mu\rho} - \partial_\alpha G^{\rho\alpha\mu}) = x^\rho T^{\mu\rho} - x^\sigma T^{\mu\rho} = \\ &- (x^\rho \partial_\alpha G^{\sigma\alpha\mu} - x^\sigma \partial_\alpha G^{\rho\alpha\mu}) = \\ &= x^\rho T^{\mu\sigma} - x^\sigma T^{\mu\rho} - \partial_\alpha (x^\rho G^{\sigma\alpha\mu} - x^\sigma G^{\rho\alpha\mu}) + G^{\sigma\alpha\mu} \partial_\alpha x^\rho - G^{\rho\alpha\mu} \partial_\alpha x^\sigma = \\ &= x^\rho T^{\mu\sigma} - x^\sigma T^{\mu\rho} - \partial_\alpha (x^\rho G^{\sigma\alpha\mu} - x^\sigma G^{\rho\alpha\mu} + G^{\sigma\alpha\mu} \delta^\rho_\alpha - G^{\rho\alpha\mu} \delta^\sigma_\alpha) \end{aligned} \tag{4.80}$$

Now, let us consider that

$$G^{\sigma\alpha\mu} \delta^\rho_\alpha - G^{\rho\alpha\mu} \delta^\sigma_\alpha = G^{\sigma\rho\mu} - G^{\rho\sigma\mu} = \frac{1}{2} (S^{\sigma\rho\mu} + S^{\rho\sigma\mu} - S^{\mu\sigma\rho} - S^{\rho\sigma\mu} - S^{\sigma\rho\mu} + S^{\mu\rho\sigma}) = S^{\mu\rho\sigma} \tag{4.81}$$

and that $x^\rho T^{\mu\sigma} - x^\sigma T^{\mu\rho} = M^{\mu(\rho\sigma)} - S^{\mu(\rho\sigma)}$.

Substituting inside Eq. (4.80), we obtain

$$N^{\mu(\rho\sigma)} = M^{\mu(\rho\sigma)} - S^{\mu(\rho\sigma)} + S^{\mu(\rho\sigma)} - \partial_\alpha (x^\rho G^{\sigma\alpha\mu} - x^\sigma G^{\rho\alpha\mu}) = M^{\mu(\rho\sigma)} - \partial_\alpha X^{\rho\sigma\alpha\mu}, \tag{4.82}$$

where $X^{\rho\sigma\alpha\mu}$ is antisymmetric for (α, μ). Moreover,

$$\partial_\mu N^{\mu(\rho\sigma)} = \partial_\mu M^{\mu(\rho\sigma)} - \partial_\mu \partial_\alpha X^{\rho\sigma\alpha\mu} = 0, \tag{4.83}$$

as $\partial_\mu \partial_\alpha$ is symmetric in (α, μ) whereas $X^{\rho\sigma\alpha\mu}$ is antisymmetric in (α, μ).

The corresponding conserved charges are

$$N^{\rho\sigma} = \int d^3x N^{0(\rho\sigma)} = \int d^3x M^{0(\rho\sigma)} - \int d^3x \partial_\alpha X^{\rho\sigma\alpha} = M^{\rho\sigma} - \int d^3x \partial_\alpha (x^\rho G^{\sigma\alpha 0} - x^\sigma G^{\rho\alpha 0}). \tag{4.84}$$

Let us now consider the temporal and spatial parts of the last integral

$$\int d^3x\,\partial_0(x^\rho G^{\sigma 00} - x^\sigma G^{\rho 00}) \quad \text{temporal part,} \tag{4.85a}$$

$$\int d^3x\,\partial_i(x^\rho G^{\sigma i0} - x^\sigma G^{\rho i0}) \quad \text{spatial part.} \tag{4.85b}$$

The integral related to the temporal part vanishes because the terms $G^{\sigma 00}$ and $G^{\rho 00}$ cancel by construction, while the integral related to the spatial part cancels out by applying Gauss's theorem. This results in

$$N^{\rho\sigma} = \int d^3x N^{0(\rho\sigma)} = M^{\rho\sigma}. \tag{4.86}$$

Now, without invoking special relativity, let us find the meaning of the components of the nonsymmetric canonical tensor $T^{\mu\nu}$.

- T^{00} represents the energy density $\mathcal{H} = \pi\dot{\phi} - \mathcal{L}$;
- T^{0i} represents the momentum density in the x^i direction $\pi\,\partial_i\phi$;
- T^{i0} represents the energy flux density along x^i;
- T^{ij} represents the flux density of the j-th momentum component along x^i.

The first two contributions provide the conserved charge associated with invariance under spacetime translations, namely the four-momentum

$$P^\nu = \int d^3x T^{0\nu} = (P^0, \mathbf{P}) = \left(\int d^3x T^{00}, \int d^3x T^{0i}\right), \tag{4.87}$$

while the last two can be obtained by considering the integral of the continuity equation over a three-dimensional volume V

$$\int_V d^3x\,\partial_\mu T^{\mu\nu} = 0. \tag{4.88}$$

Now, let us consider two cases in the above expression, when $\mu = 0$ and when $\mu = i$, obtaining

$$\int_V d^3x(\partial_t T^{0\nu} - \partial_i T^{i\nu}) = 0 \longrightarrow \frac{\partial}{\partial t}\int_V d^3x T^{0\nu} = -\int_V d^3x\,\partial_i T^{i\nu} = -\int_{\partial V} d\sigma_i T^{i\nu}, \tag{4.89}$$

where $-\int_{\partial V} d\sigma_i T^{i\nu}$ is the flow of P^ν through the surface ∂V. Therefore,

$$\frac{\partial}{\partial t}P^0 = \frac{\partial E}{\partial t} = -\int_{\partial V} d\sigma_i T^{i0}, \tag{4.90a}$$

$$\frac{\partial}{\partial t}P^j = -\int_{\partial V} d\sigma_i T^{ij}. \tag{4.90b}$$

Furthermore, T^{ij} coincides with the non-relativistic stress tensor.

▶ **Driving exercise** The requirement for $T^{\mu\nu}$ to be symmetric has both a relativistic and Newtonian justification.

1. Relativistic Reasoning: For the temporal part, T^{i0} is the energy flux density, which coincides with the momentum density T^{0i}.
2. Newtonian Reasoning: For the spatial part, consider a cube of side length L and mass M. Apply a force along the direction in which we calculate the momentum.

As sketched in Fig. 4.2, we can compute the angular acceleration and, from it, show that we do expect the energy-momentum tensor to be symmetric.

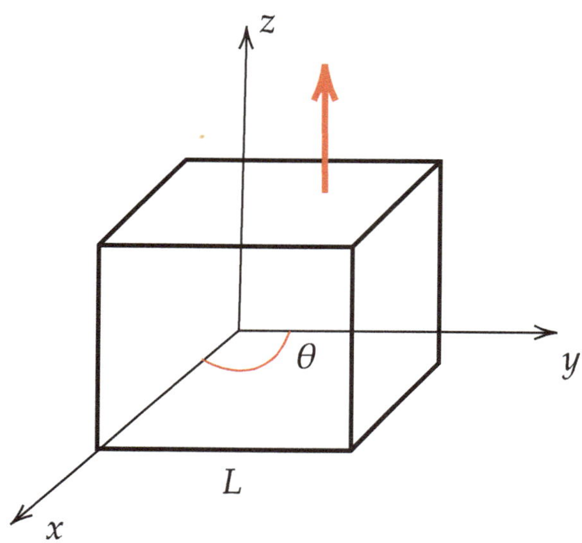

Fig. 4.2 Infinitesimal element on which we apply the stresses

To do so, we readily notice the moment of inertia will be proportional to $ML^2 = (T^{00}L^3)L^2 = T^{00}L^5$. The center of the cube is the origin of the axes. Let us calculate the moment of force along $\hat{z}$

$$(\mathbf{r} \times \mathbf{F}_z) = \begin{vmatrix} \hat{x} & \hat{y} & \hat{z} \\ x & y & z \\ F_x & F_y & F_z \end{vmatrix} = xF_y - yF_x, \tag{4.91}$$

where $F_y = T^{yx}L^2, F_x = T^{xy}L^2, x = y = L/2$. This leads to

$$xF_y - yF_x = L^2T^{yx}\frac{L}{2} - L^2T^{xy}\frac{L}{2} = \frac{L^3}{2}(T^{yx} - T^{xy}) \propto \text{moment of force.} \tag{4.92}$$

This implies that

$$\text{angular acceleration} \propto \frac{\text{moment of force}}{\text{moment of inertia}} = \frac{\frac{L^3}{2}(T^{yx} - T^{xy})}{T^{00}L^5} \sim \frac{(T^{yx} - T^{xy})}{L^2}. \tag{4.93}$$

The last expression tends to infinity as $L \longrightarrow 0$. However, since the acceleration cannot be infinite, it is the numerator that must go to zero, i.e., $T^{ij} = T^{ji}$.

4.2 Spin-0 Particles: The Klein-Gordon Equation

Relativistic quantum mechanics aims to provide a relativistic formulation of the quantum theory.

Incorporating special relativity into quantum mechanics *is not* a simple task.

The objective is to uphold the foundational principles of quantum mechanics, as they define the framework.

1. The state of a physical system is described by a complex-valued function $\psi(\mathbf{x}, t)$, where we consider particles with spin 0, such that ψ does not include spin variables. Moreover, $|\psi|^2$ represents the probability of finding a particle at $\mathbf{x}$ at time t.
2. Observables are treated as Hermitian operators.
3. The evolution of a system is described by the Schrödinger equation, which is an eigenvalue equation $E\psi = \hat{H}\psi$, where E represents the energy and $\hat{H}$ is the Hamiltonian operator

$$i\hbar\frac{\partial}{\partial t}\psi(\mathbf{x}, t) = H\psi(\mathbf{x}, t). \tag{4.94}$$

Furthermore, we aim to satisfy the principle of relativity: the laws of physics (expressed in covariant form according to Lorentz transformations) should be the same in all inertial reference frames.

Let us begin by studying the Schrödinger equation for a free particle with zero spin, differentiating between the non-relativistic and relativistic cases.

In the *non-relativistic* case, we replace the operators in the relativistic expression for energy, i.e., $E = p^2/2m$. This yields $\hat{H} = \frac{\hat{p}^2}{2m}$, and considering $E \rightarrow i\hbar\frac{\partial}{\partial t}$ and $p \rightarrow -i\hbar\nabla$, this gives

$$i\hbar\frac{\partial}{\partial t}\psi(\mathbf{x},t) = -\frac{\hbar^2}{2m}\nabla^2\psi(\mathbf{x},t), \tag{4.95}$$

whose solutions are plane waves, $\psi(\mathbf{x},t) = Ne^{-\frac{i}{\hbar}(Et-\mathbf{p}\cdot\mathbf{x})}$ with N normalization factor. So, substituting into the Schrödinger equation, we have

$$i\hbar\frac{\partial}{\partial t}\psi = i\hbar N\left(\frac{1}{i\hbar}\right)Ee^{-\frac{i}{\hbar}(Et-\mathbf{p}\cdot\mathbf{x})} = E\psi. \tag{4.96}$$

In the relativistic case, we substitute the operators using the energy and momentum representation through the mass shell $E = \sqrt{p^2c^2 + m^2c^4}$, yielding $\hat{H} = \sqrt{\hat{p}^2c^2 + m^2c^4}$ and so

$$i\hbar\frac{\partial}{\partial t}\psi(\mathbf{x},t) = \sqrt{(-\hbar^2\nabla^2c^2 + m^2c^4)}\psi(\mathbf{x},t). \tag{4.97}$$

▶ **Pause and think** Try to think on the latter. Is that covariant? What is the problem with it, in your opinion?

This equation presents two issues: it is not manifestly covariant, and it involves an operator under a square root. We can address these problems by considering the square of the mass shell, i.e., $E^2 = p^2c^2 + m^2c^4$, resulting in

$$\left(i\hbar\frac{\partial}{\partial t}\right)^2\psi(\mathbf{x},t) = (-\hbar^2\nabla^2c^2 + m^2c^4)\psi(\mathbf{x},t). \tag{4.98}$$

Let us now work in natural units, where $\hbar = c = 1$, yielding

$$-\frac{\partial^2}{\partial t^2}\psi(\mathbf{x},t) = (-\nabla^2 + m^2)\psi(\mathbf{x},t) \longrightarrow \left(\frac{\partial^2}{\partial t^2} - \nabla^2 + m^2\right)\psi(\mathbf{x},t) = 0. \tag{4.99}$$

Recalling the definition of four-momentum, $p^\mu = (E,\mathbf{p}) \to \left(i\hbar\frac{\partial}{\partial t}, -i\hbar\nabla\right) = i\hbar\left(\frac{\partial}{\partial t}, -\nabla\right)$, we introduce the contravariant four-divergence as

$$\partial^\mu = \frac{\partial}{\partial x_\mu} = \left(\frac{\partial}{\partial t}, -\nabla\right) \tag{4.100}$$

thus writing $p^\mu \to i\hbar\partial^\mu$. Substituting Eq. (4.100) into Eq. (4.99) yields the *Klein-Gordon equation*

$$\left(\partial_\mu\partial^\mu + m^2\right)\psi = 0 \longrightarrow \left(\Box + m^2\right)\psi = 0, \tag{4.101}$$

where $\partial_\mu\partial^\mu = \frac{\partial^2}{\partial t^2} - \nabla^2 = \Box$ is the d'Alembertian.

For solutions ψ, the eigenvalue equations hold:

$$\hat{E}\psi = i\frac{\partial}{\partial t}\psi = E\psi \longrightarrow \hat{E}^2\psi = -\frac{\partial^2}{\partial t^2}\psi = E^2\psi, \tag{4.102a}$$

$$\hat{p}\psi = -i\nabla\psi = \mathbf{p}\psi \longrightarrow \hat{p}^2\psi = -\nabla^2\psi = \mathbf{p}^2\psi. \tag{4.102b}$$

Thus, the Klein-Gordon equation implies that substituting the eigenvalue solutions into ψ results in the square of the mass shell:

$$0 = \left(\frac{\partial^2}{\partial t^2} - \nabla^2 + m^2\right)\psi = (-E^2 + \mathbf{p}^2 + m^2)\psi \longrightarrow E^2 = \mathbf{p}^2 + m^2. \tag{4.103}$$

4.2.1 Solutions to the Klein-Gordon Equation

To understand which problems are associated with the extractions of solutions from the Klein-Gordon equation, let us compute the corresponding solutions.

Plane wave solutions

The plane wave $\psi = a(p)e^{-i(Et-\mathbf{p}\cdot\mathbf{x})} = a(p)e^{ip^\mu x_\mu}$ is a solution of the Klein-Gordon equation. We then consider a superposition of waves, yielding

$$\psi(x,t) = \psi(x^\mu) = \int \frac{d^4p}{(2\pi)^4}a(p)e^{-i(p^0t-\mathbf{p}\cdot\mathbf{x})}. \tag{4.104}$$

However, we need to select only momenta that satisfy the expression

$$p^2 = p^\mu p_\mu = (p^0)^2 - \mathbf{p}^2 = m^2, \tag{4.105}$$

thus replacing d^4p with $d^4p\delta(p^2 - m^2)$. Additionally, considering the possibility of $p^0 \gtrless 0$, to account for both cases, we separate the integrand into two terms

$$a(p)e^{-i(|p^0|t-\mathbf{p}\cdot\mathbf{x})} + b(p)e^{-i(-|p^0|t-\mathbf{p}\cdot\mathbf{x})}, \tag{4.106}$$

where, for the first term, we integrate over $p = (p^0, \mathbf{p})$ that satisfies $p^0 = +\sqrt{\mathbf{p}^2 + m^2}$, and for the second term, we integrate over $p = (p^0, \mathbf{p})$ that satisfies $p^0 = -\sqrt{\mathbf{p}^2 + m^2} = -|p^0|$.

To express this, we set the first integral to zero for $p^0 > 0$ and the second integral to zero for $p^0 < 0$, resulting in

$$\psi(x^\mu) = \int d^4p\, a(p)e^{-i(p^0t-\mathbf{p}\cdot\mathbf{x})}\delta(p^2 - m^2)\theta(p^0) + \\ + \int d^4p\, b(p)e^{-i(p^0t+\mathbf{p}\cdot\mathbf{x})}\delta(p^2 - m^2)\theta(-p^0), \tag{4.107}$$

with the assumption $p^0 > 0$. Now, changing the sign of the integration variables, we get

$$\psi(x^\mu) = \int d^4p\, a(p) e^{-i(p^0 t - \mathbf{p}\cdot\mathbf{x})} \delta(p^2 - m^2)\theta(p^0) + \\ + \int d^4p'\, b(-p') e^{i(p'^0 t - \mathbf{p}'\cdot\mathbf{x})} \delta(p'^2 - m^2)\theta(p'^0). \quad (4.108)$$

By replacing p' with p, we can combine everything into a single integral, resulting in

$$\psi(x^\mu) = \int d^4p \left[a(p) e^{-i(p^0 t - \mathbf{p}\cdot\mathbf{x})} + b(-p) e^{i(p^0 t - \mathbf{p}\cdot\mathbf{x})}\right] \delta(p^2 - m^2)\theta(p^0), \quad (4.109)$$

where the quantities inside the integral are each other's complex conjugates, provided that $b(-p) = a^\star(p)$:

$$\psi(x^\mu) = \int d^4p\, \theta(p^0)\delta(p^2 - m^2) \left[a(p) e^{-ip^\mu x_\mu} + a^\star(p) e^{ip^\mu x_\mu}\right]. \quad (4.110)$$

This expression represents the superposition of incident and reflected waves with momenta on the shell, considering only the positive roots where $(p^0)^2 = \mathbf{p}^2 + m^2$.

An important property of the δ distribution is

$$\delta\left[g(x)\right] = \sum_i \left(\frac{\delta(x - x_i)}{|g\prime(x_i)|}\right),$$

with x_i being the zeros of the function g.

Using the property in the box for our delta, we obtain

$$\delta(p^2 - m^2) = \delta((p^0)^2 - \mathbf{p}^2 - m^2)). \quad (4.111)$$

Defining $E_p = \sqrt{\mathbf{p}^2 + m^2}$ as the zeros of the function, where $(p^0)^2 - \mathbf{p}^2 - m^2 = 0$ gives $(p^0) = \pm E_p$, and $g'(p^0 = \pm E_p) = \pm 2E_p$, so

$$\delta(p^2 - m^2) = \frac{\delta(p^0 - E_p)}{|2E_p|} + \frac{\delta(p^0 + E_p)}{|-2E_p|} = \frac{\delta(p^0 - E_p) + \delta(p^0 + E_p)}{2E_p} = \frac{\delta(p^0 - E_p)}{2E_p}, \quad (4.112)$$

where $\delta(p^0 + E_p)$ is zero because we assumed $p^0 > 0$. Substituting this into Eq. (4.110), we get

$$\psi(x^\mu) = \int \frac{d^3p}{2E_p}\left[a(\mathbf{p})e^{-i(E_pt-\mathbf{p}\cdot\mathbf{x})} + a^\star(\mathbf{p})e^{i(E_pt-\mathbf{p}\cdot\mathbf{x})}\right]. \tag{4.113}$$

Complete solution

Recalling the Klein-Gordon equation, $(\Box + m^2)\phi = 0$ is a second-order partial differential equation, with $\phi(\mathbf{x}, t)$ a continuous *delocalized field*, we can impose the following boundary conditions,

$$\phi(x, y, z, t) = \phi(x + L, y, z, t), \tag{4.114a}$$
$$\phi(x, y, z, t) = \phi(x, y + L, z, t), \tag{4.114b}$$
$$\phi(x, y, z, t) = \phi(x, y, z + L, t), \tag{4.114c}$$

indicating that the function is the same on opposite faces of a three-dimensional cube with side length L.

So, consider the complete plane wave solution:

$$\phi(x, y, z, t) = N^{-ip^\mu x_\mu} = Ne^{-ik^\mu x_\mu}, \tag{4.115}$$

where in the last equality, we used the fact that $k^\mu = (\omega, \mathbf{k})$ and $E = \omega, \mathbf{p} = \mathbf{k}$, when $c = 1$. Imposing the boundary conditions within this solution, we obtain:

$$Ne^{-i(\omega t-k_x(x+L)-k_yy-k_zz)} = Ne^{-i(\omega t-k_xx-k_yy-k_zz)}, \tag{4.116a}$$
$$Ne^{-i(\omega t-k_xx-k_y(y+L)-k_zz)} = Ne^{-i(\omega t-k_xx-k_yy-k_zz)}, \tag{4.116b}$$
$$Ne^{-i(\omega t-k_xx-k_yy-k_z(z+L))} = Ne^{-i(\omega t-k_x-k_yy-k_zz)}. \tag{4.116c}$$

This leads to:

$$e^{ik_xL} = e^{2\pi in_x}, \tag{4.117a}$$
$$e^{ik_yL} = e^{2\pi in_y}, \tag{4.117b}$$
$$e^{ik_zL} = e^{2\pi in_z}, \tag{4.117c}$$

with $\mathbf{k} = \frac{2\pi}{L}(n_x, n_y, n_z)$ and k^μ satisfying $k^\mu k_\mu = (k^0)^2 - \mathbf{k}^2 = \omega^2 - \mathbf{k}^2 = m^2$, so $\omega = \pm\sqrt{\mathbf{k}^2 + m^2}$. This allows us to write the general solution as a superposition of waves:

$$\begin{aligned}\psi(x) &= \sum_{\mathbf{n}} N(\omega)\left[a(\mathbf{k})e^{-i(|\omega|t-\mathbf{k}\cdot\mathbf{x})} + b(\mathbf{k})e^{i(|\omega|t+\mathbf{k}\cdot\mathbf{x})}\right] \\ &= \sum_{\mathbf{n}} N(\omega)\left[a(\mathbf{k})e^{-i(|\omega|t-\mathbf{k}\cdot\mathbf{x})} + c^\star(\mathbf{k})e^{i(|\omega|t+\mathbf{k}\cdot\mathbf{x})}\right].\end{aligned} \tag{4.118}$$

We change the index twice: first $\mathbf{n} \to -\mathbf{n} = \mathbf{n}'$ in the second summation and then $\mathbf{n}' \to \mathbf{n}$, obtaining:

$$\begin{aligned}\psi(x) &= \sum_{\mathbf{n}} N(\omega) a(\mathbf{k}) e^{-i(|\omega|t - \mathbf{k}\cdot\mathbf{x})} + \sum_{\mathbf{n}'} N(\omega) b(-\mathbf{k}') e^{i(|\omega|t - \mathbf{k}'\cdot\mathbf{x})} \\ &= \sum_{\mathbf{n}} N(\omega) \left[a(\mathbf{k}) e^{-i(|\omega|t - \mathbf{k}\cdot\mathbf{x})} + a^{\star}(\mathbf{k}) e^{i(|\omega|t - \mathbf{k}\cdot\mathbf{x})} \right], \end{aligned} \tag{4.119}$$

with $b(-\mathbf{k}) = a^{\star}(\mathbf{k}) = c^{\star}$ so that ψ is a real scalar field. The $a(\mathbf{k})$ are complex and thus encapsulate two pieces of information: the real and complex parts. This corresponds to the need to have an initial value of the field $\phi(\mathbf{x}, 0)$ and the initial time derivative at $t = 0$, $\partial_t \psi(\mathbf{x}, 0)$. This implies, in the continuous limit, that the Klein-Gordon equation is a second-order differential equation with solutions as in Eq. (4.113).

4.2.2 The Non-relativistic Case

In quantum mechanics, it is easy to construct the continuity equation, by starting from the Schrödinger equation. Let us write

$$\begin{aligned} & i\hbar \left[\psi^{\star} \frac{\partial}{\partial t}\psi + \psi \frac{\partial}{\partial t}\psi^{\star} \right] = -\frac{\hbar^2}{2m} \left[\psi^{\star}\nabla^2\psi - \psi\nabla^2\psi^{\star} \right] \to \\ & \to \frac{\partial}{\partial t}(\psi\psi^{\star}) = -\frac{\hbar}{2im}\nabla \left[\psi^{\star}\nabla\psi - \psi\nabla\psi^{\star} \right]. \end{aligned} \tag{4.120}$$

Recalling $\rho = \psi\psi^{\star} = |\psi|^2$ *probability density* and $\mathbf{J} = -\frac{i\hbar}{2m}\left[\psi^{\star}\nabla\psi - \psi\nabla\psi^{\star}\right]$ the *probability current*, we recast the above relation as

$$\frac{\partial \rho}{\partial t} + \nabla \mathbf{J} = 0. \tag{4.121}$$

▶ **Pause and think**
Prepare a lecture of at least 30 minutes in which you explain the various contexts in which the fluid-representation, provided in Eq. (4.121), has been adopted in Physics.

We now derive the continuity equation in the relativistic case starting from the Klein-Gordon equation. Multiply the Klein-Gordon equation by $\psi^{\star}$ and the complex conjugate of the equation by ψ. Subtracting the two, we obtain:

$$0 = \psi^{\star}(\Box + m^2)\psi - \psi(\Box + m^2)\psi^{\star} = \psi^{\star}\Box\psi + \psi^{\star}m^2\psi - \psi\Box\psi^{\star} - \psi m^2\psi^{\star}. \tag{4.122}$$

Explicitly expressing the d'Alembertian, i.e., $\Box = \partial_\mu \partial^\mu$, and substitute it into the previous expression, we obtain:

$$\psi^\star \partial_\mu \partial^\mu \psi - \psi \partial_\mu \partial^\mu \psi^\star = \partial_\mu \left[\psi^\star \partial^\mu \psi - \psi \partial^\mu \psi^\star \right] = 0. \tag{4.123}$$

Identify the expression in parentheses with the four-current J^μ so that the continuity equation

$$\partial_\mu J^\mu = 0 \tag{4.124}$$

is satisfied. Separating J^μ into $J^\mu = (J^0, \mathbf{J})$ gives the temporal and spatial components of the four-current

$$J^0 = \psi^\star \partial_t \psi - \psi \partial_t \psi^\star, \tag{4.125a}$$

$$\mathbf{J} = -\psi^\star \nabla \psi + \psi \nabla \psi^\star. \tag{4.125b}$$

In this way, Eq. (4.124) becomes

$$\partial_t J^0 + \nabla \mathbf{J} = 0. \tag{4.126}$$

Introducing now

$$\tilde{\rho} = i J^0 = i \left[\psi^\star \partial_t \psi - \psi \partial_t \psi^\star \right], \tag{4.127a}$$

$$\tilde{J} = i \mathbf{J} = -i \left[\psi^\star \nabla \psi - \psi \nabla \psi^\star \right]. \tag{4.127b}$$

In this way, the continuity equation is exactly the same as in the non-relativistic case

$$\partial_t \tilde{\rho} + \nabla \tilde{J} = 0. \tag{4.128}$$

However, the temporal part cannot be interpreted as a true probability density. In fact, substituting the plane wave solution $\psi = N e^{-ip^\mu x_\mu}$ gives

$$\begin{aligned} J^0 &= |N|^2 \left[-iEe^{ip^\mu x_\mu} e^{-ip^\mu x_\mu} - iEe^{-ip^\mu x_\mu} e^{ip^\mu x_\mu} \right] = -2iE|N|^2 \\ &\to \tilde{\rho} = iJ^0 = 2E|N|^2 = \pm 2E_p |N|^2, \end{aligned} \tag{4.129}$$

which can also be negative. The last part of the previous expression is written considering that $E \gtrless 0$ and thus $E = \pm\sqrt{p^2 + m^2} = \pm E_p$. In the next section, we will motivate the existence of negative energy states introducing the Feynman-Stueckelberg interpretation.

4.2.3 The Scalar Field Lagrangian

In classical physics, we can distinguish two entities.

- **Particles**: They constitute matter and are localized in space, e.g., Euclidean E, Minkowskian $M(3, 1)$, etc., with coordinates $x^\mu(s)$.
- **Fields**: They describe interactions, are delocalized, and are described by functions of coordinates $\phi(x)$.

Let us focus on fields. They are crucial when studying processes at small scales, particularly in the quantum realm, where quantizing relativistic particles as classical particles is not feasible.

It will be observed that when dealing with the wave equation for relativistic particles, solutions involving particle-antiparticle pairs emerge, accompanied by other inconsistencies like negative probability density. In other words, the quest for wave equations in the relativistic scenario leads to solutions for particle-antiparticle pairs. To address this, the approach shifts from the interpretation of a wave function to the concept of a quantized field.

Lagrangian field theory introduces the formalism of classical field theory but applies it in a quantum context. The optimal way to frame fields in the Lagrangian formalism without relinquishing the principle of relativity is to start from the action S

$$S = \int_{t_1}^{t_2} L dt = \int_{x_1^0}^{x_2^0} L dx^0 = \int_{V_4} d^4x \mathcal{L}(\phi, \partial_\mu \phi, x), \tag{4.130}$$

where V_4 is the volume in **M(3, 1)** bounded by the hypersurfaces $\Gamma_1 : t = t_1$ and $\Gamma_2 : t = t_2$, while $L = \int d^3x \mathcal{L}(\phi, \partial_\mu \phi, x)$ is the Lagrangian written as an integral over space of a *Lagrangian density* $\mathcal{L}(\phi, \partial_\mu \phi, x)$ with

$$\partial_\mu \phi = \frac{\partial \phi}{\partial x^\mu} \equiv \phi_\mu. \tag{4.131}$$

The dependence on coordinates is explicit only for **NON**-isolated systems.

► **Driving exercise** Let us assume that ϕ is fixed on the boundary of $V_4 : \partial V_4 = \Gamma_1 \cup \Gamma_2$. Note that $\mathcal{L}$ is not unique for a system, but if we subtract the *four-divergence* $\partial_\mu A^\mu$ from the Lagrangian density, the action will change while the equations of motion remain invariant. In fact, if

$$\mathcal{L} \longrightarrow \mathcal{L} - \partial_\mu A^\mu, \quad \text{with} \quad A^\mu = A^\mu(\phi, x) \tag{4.132}$$

then the action will transform into

$$S = \int_{V_4} d^4x \mathcal{L}(\phi, \partial_\mu \phi, x) \longrightarrow S' = \int d^4x (\mathcal{L} - \partial_\mu A^\mu) = \int d^4x \mathcal{L} - \int d^4x \partial_\mu A^\mu. \tag{4.133}$$

Applying Gauss's theorem to the last term in the previous expression, we obtain

$$\int d^4x \partial_\mu A^\mu = \int_{\partial V_4} d\sigma_\mu A^\mu, \tag{4.134}$$

with σ_μ hypersurface, obtained as boundary of V_4.
By varying the previous expression, we get

$$\delta \int_{\partial V_4} d\sigma_\mu A^\mu = \int d\sigma_\mu \left(\frac{\partial A^\mu}{\partial \phi} \delta\phi + \frac{\partial A^\mu}{\partial x_\nu} \delta x^\nu \right), \tag{4.135}$$

since $\delta\phi = 0$ on ∂V_4, we obtain

$$\delta \int_{\partial V_4} d\sigma_\mu A^\mu = \int d\sigma_\mu \left(\frac{\partial A^\mu}{\partial x_\nu} \delta x^\nu \right). \tag{4.136}$$

Furthermore, $\delta S' = 0 \iff \delta S = 0$, meaning that we obtain the same equations of motion through the principle of stationary action. Let us now derive the equations of motion from the principle of least action

$$\delta S = 0 \longrightarrow \delta \int d^3x \mathcal{L}(\phi, \phi_\mu, x) = 0. \tag{4.137}$$

Following $\phi(x) \longrightarrow \phi(x) + \delta\phi(x)$ with $\delta\phi(\mathbf{x}, t_1) = \delta\phi(\mathbf{x}, t_2) = 0$, we obtain

$$0 = \int d^4x \left[\frac{\partial \mathcal{L}}{\partial \phi} \delta\phi + \frac{\partial \mathcal{L}}{\partial(\partial_\mu \phi)} \delta(\partial_\mu \phi) + \frac{\partial \mathcal{L}}{\partial x^\mu} \delta x^\mu \right] = \int d^4x \left[\frac{\partial \mathcal{L}}{\partial \phi} \delta\phi + \frac{\partial \mathcal{L}}{\partial(\partial_\mu \phi)} \partial_\mu(\delta\phi) \right], \tag{4.138}$$

where $\delta x^\mu = 0$ because the coordinates are **NOT** varied. Moreover, we used $\delta(\partial_\mu \phi) = \partial_\mu(\delta\phi)$. Let us write the expression inside the integral as

$$\partial_\mu \left[\frac{\partial \mathcal{L}}{\partial(\partial_\mu \phi)} \delta\phi \right] = \partial_\mu \frac{\partial \mathcal{L}}{\partial(\partial_\mu \phi)} \delta\phi + \frac{\partial \mathcal{L}}{\partial(\partial_\mu \phi)} \partial_\mu(\delta\phi) \tag{4.139}$$

so that the integral becomes

$$\begin{aligned} 0 &= \int d^4x \left[\frac{\partial \mathcal{L}}{\partial \phi} \delta\phi + \partial_\mu \left[\frac{\partial \mathcal{L}}{\partial(\partial_\mu \phi)} \delta\phi \right] - \partial_\mu \frac{\partial \mathcal{L}}{\partial(\partial_\mu \phi)} \delta\phi \right] \\ &= \int d^4x \left[\frac{\partial \mathcal{L}}{\partial \phi} - \partial_\mu \frac{\partial \mathcal{L}}{\partial(\partial_\mu \phi)} \right] \delta\phi + \int d^4x \partial_\mu \left[\frac{\partial \mathcal{L}}{\partial(\partial_\mu \phi)} \delta\phi \right]. \end{aligned} \tag{4.140}$$

The last term inside the integral is zero because $\delta\phi = 0$ on ∂V_4, in fact

$$\int d^4x\,\partial_\mu\left[\frac{\partial\mathcal{L}}{\partial(\partial_\mu\phi)}\delta\phi\right]=\frac{\partial\mathcal{L}}{\partial(\partial_\mu\phi)}\delta\phi\bigg|_{\partial V_4}=0. \tag{4.141}$$

We thus obtain

$$0=\int d^4x\left[\frac{\partial\mathcal{L}}{\partial\phi}-\partial_\mu\frac{\partial\mathcal{L}}{\partial(\partial_\mu\phi)}\right]\delta\phi. \tag{4.142}$$

Due to the arbitrariness of $\delta\phi$, we have $\delta S=0$, and the other term necessarily vanishes, yielding the *Euler-Lagrange equations for the field* ϕ

$$\partial_\mu\frac{\partial\mathcal{L}}{\partial(\partial_\mu\phi)}=\frac{\partial\mathcal{L}}{\partial\phi}. \tag{4.143}$$

For a system of multiple fields $\{\phi_1,\ldots,\phi_N\}$, the Euler-Lagrange equations (for N fields) are given by

$$\partial_\mu\frac{\partial\mathcal{L}}{\partial(\partial_\mu\phi_l)}=\frac{\partial\mathcal{L}}{\partial\phi_l},\quad\text{with } l=1,\ldots,N. \tag{4.144}$$

Analogous to the classical case where the law of motion is specified once the position and velocity are known at a fixed instant of time, so the equations of motion are second-order partial differential equations. This implies that the Euler-Lagrange equations must be second-order in partial derivatives. This further implies that $\mathcal{L}$ cannot contain derivatives higher than the first order but must be quadratic in $\partial_\mu\phi$ so that

$\dfrac{\partial\mathcal{L}}{\partial(\partial_\mu\phi)}$ it turns out to be linear in $\partial_\mu\phi$,

$\partial_\mu\dfrac{\partial\mathcal{L}}{\partial(\partial_\mu\phi)}=\dfrac{\partial\mathcal{L}}{\partial\phi}$ it turns out to be second order.

The action S **MUST** be real because if it were complex, there would be $2N$ equations of motion for N fields, one for the complex part and one for the real part.

The invariance of the action under Lorentz transformations automatically yields equations of motion that respect the principle of relativity. Moreover, since $d^4x' = d^4x$ is a Lorentz invariant, for the action to be an invariant as well, $\mathcal{L}$ must necessarily be a Lorentz invariant.

4.2.4 Field Transformations

To satisfy the principle of relativity, fields must transform appropriately under restricted Lorentz transformations (proper and orthochronous) of spacetime coordinates:

$$x'^{\mu} = \Lambda^{\mu}_{\nu} x^{\nu}. \tag{4.145}$$

The simplest case is that of a scalar field such that:

$$\phi'(x') = \phi(x). \tag{4.146}$$

It is generally assumed that fields change according to a representation D of the Lorentz group $L^{\uparrow}_{+}$. Remembering the classification via the values of Casimirs, we have:

1. *scalar fields:* transform according to the identity ($\dim D = 1$),
2. *Weyl spinor fields:* transform according to the inverse representation of $\dim D = 2$ with $D\left(\frac{1}{2}, 0\right)$ and $D\left(0, \frac{1}{2}\right)$,
3. *Dirac spinor fields:* transform according to reducible representation of $\dim D = 4$,
4. *vector fields:* transform according to irreducible representations of $\dim D = 4$ and are actions of Λ in Minkowski space with $D_{(1,1)} = \exp\left[\frac{1}{2} a_{\alpha\beta} M^{\alpha\beta}\right]$, where $M^{\alpha\beta}$ is a 4×4 antisymmetric matrix.

Let us consider the covariant derivative of the field:

$$\partial'_{\mu}\phi' = \frac{\partial \phi'(x')}{\partial x'^{\mu}} = \frac{\partial \phi(x)}{\partial x'^{\mu}} = \frac{\partial \phi}{\partial x^{\nu}} \frac{\partial x^{\nu}}{\partial x'^{\mu}}, \tag{4.147}$$

where $\frac{\partial x^{\nu}}{\partial x'^{\mu}}$ is the inverse matrix, $\left(\Lambda^{-1}\right)^{\nu}_{\ \mu}$.

Similarly, the contravariant derivative is:

$$\partial'^{\mu}\phi' = \frac{\partial\phi'(x')}{\partial x'_{\mu}} = \frac{\partial\phi(x)}{\partial x'_{\mu}} = \frac{\partial\phi}{\partial x_{\nu}}\frac{\partial x_{\nu}}{\partial x'_{\mu}} = \Lambda^{\mu}_{\nu}\partial^{\nu}\phi, \tag{4.148}$$

where $\frac{\partial x_{\nu}}{\partial x'_{\mu}}$ is obtained from:

$$x'_{\mu} = \Lambda^{\nu}_{\mu}x_{\nu} \to \Lambda^{\mu}_{\nu}x'_{\mu} = \Lambda^{\mu}_{\lambda}\Lambda^{\nu}_{\mu}x_{\nu} = \delta^{\nu}_{\lambda}x_{\nu} \to x_{\lambda} = \Lambda^{\mu}_{\lambda}x'_{\mu} \longrightarrow dx_{\lambda} = \frac{\partial x_{\lambda}}{\partial x'_{\mu}}dx'_{\mu}. \tag{4.149}$$

Thus, we obtain the covariant and contravariant derivatives of a scalar field ϕ:

$$\phi'_{\mu} = \Lambda^{\nu}_{\mu}\phi_{\nu},\ \phi'^{\mu} = \Lambda^{\mu}_{\nu}\phi^{\nu}. \tag{4.150a}$$

Further derivatives can be used to construct tensors. The p times covariant and q times contravariant derivative gives:

$$\phi^{\beta_1,\ldots,\beta_q}\alpha_1,\ldots,\alpha_p(x) = \partial\alpha_1,\ldots,\partial_{\alpha_p}\partial^{\beta_1},\ldots,\partial^{\beta_q}\phi(x). \tag{4.151}$$

The tensor (p, q) is then:

$$tensor(p,q) = \frac{\partial}{\partial x^{\alpha_1}},\ldots,\frac{\partial}{\partial x^{\alpha_p}}\frac{\partial}{\partial x_{\beta_1}},\ldots,\frac{\partial}{\partial x_{\beta_q}}\phi(x). \tag{4.152}$$

By combining tensor fields and derivatives of fields so that all indices are contracted, Lorentz invariants are obtained. We will also speak equivalently of tensors and irreducible representations of $L_+^{\uparrow}$.

4.3 Lagrangian Description

We want to establish general rules for writing Lagrangians in a way that reproduces field equations. Additionally, we aim to be able to write correct mass shell relations through Lagrangians.

In analogy to the discrete case, we will have

$$L = \text{kinetic energy} - \text{potential energy} \longrightarrow L = \frac{1}{2}(\partial_{\mu}\phi)^2 - V(\phi) = L(\partial_{\mu}\phi, \phi) \quad \text{for fields.} \tag{4.153}$$

Since we are dealing with fields and we want the Lagrangian to contain Lorentz invariants instead of x, we will consider ϕ in place of x, and instead of v, we will consider $\partial_{\mu}\phi$. Considering that the Euler-Lagrange equation must be second order, we already know that

$$\frac{\partial \mathcal{L}}{\partial(\partial_\mu \phi)} \propto \partial_\mu \phi \longrightarrow \partial_\mu \frac{\partial \mathcal{L}}{\partial(\partial_\mu \phi)} \propto \partial^2 \phi. \tag{4.154}$$

This implies that the kinetic energy should be

$$\text{kinetic energy} \propto \left(\partial_\mu \phi\right)^2. \tag{4.155}$$

As for the potential $V(\phi)$, we choose it in such a way that

$$\frac{\partial \mathcal{L}}{\partial \phi} = -\alpha \phi. \tag{4.156}$$

To obtain field equations of the form

$$\partial_\mu \frac{\partial \mathcal{L}}{\partial(\partial_\mu \phi)} - \frac{\partial \mathcal{L}}{\partial \phi} = 0 \longrightarrow (\partial^2 + \alpha)\phi = 0. \tag{4.157}$$

This implies that the potential energy must be

$$\text{potential energy} \propto \alpha \phi^2. \tag{4.158}$$

Now, since $i\partial^\mu = (E, \mathbf{p})$ remembering that in the non-relativistic quantum case $\hat{\mathbf{p}} \equiv i\hbar\nabla$ and $\hat{E} \equiv i\hbar\frac{\partial}{\partial t}$ then in the relativistic case, we have

$$p^\mu = (\mathbf{E}, \mathbf{p}) \equiv \left(i\frac{\partial}{\partial t}, -i\nabla\right) = i\partial^\mu. \tag{4.159}$$

Let us check Eq. (4.157) remembering that $\partial^2 = \partial_\mu \partial^\mu = -E^2 + \mathbf{p}^2$

$$(-E^2 + \mathbf{p}^2 + \alpha)\phi = 0 \longrightarrow -E^2 + \mathbf{p}^2 + \alpha = 0. \tag{4.160}$$

If $\alpha = m^2$, the previous expression is the mass shell.

We are now able to write Lagrangians for massive and massless scalar fields. Let us now describe the Lagrangians for various types of fields.

4.3.1 Massive Real Scalar Fields

The Lagrangian associated with a real massive scalar field is

$$\mathcal{L} = \frac{1}{2}(\partial_\mu \phi)^2 - \frac{1}{2}m^2\phi^2, \tag{4.161}$$

so that $\partial_\mu \frac{\partial \mathcal{L}}{\partial(\partial_\mu \phi)} = \frac{\partial \mathcal{L}}{\partial \phi}$ gives the field equations, namely the homogeneous equation

$$(\partial_\mu \partial^\mu + m^2)\phi = 0. \tag{4.162}$$

Driving exercise

Let us now compute the Lagrangian associated with a massless scalar field. It is

$$\mathcal{L} = \frac{1}{2}(\partial_\mu \phi)^2. \tag{4.163}$$

In this way, the field equations are obtained by considering $\frac{\partial \mathcal{L}}{\partial \phi} = 0$, leading to

$$(\partial_\mu \partial^\mu \phi) = 0 \longrightarrow \frac{\partial^2}{\partial t^2}\phi = \nabla^2 \phi. \tag{4.164}$$

The mass shell condition for the case $m = 0$ is given by $(E^2 - \mathbf{p}^2)\phi = 0$, implying $E = |\mathbf{p}|$.

On the other hand, let us focus on a massive real scalar field coupled to an external source.

We define $J(x)\phi(x)$ as the interaction term, which represents the minimal coupling, i.e., the simplest way to write the coupling. In the minimal coupling regime, J plays the role of the source, while ϕ plays the role of the background. The Lagrangian representing this field is

$$\mathcal{L} = \frac{1}{2}(\partial_\mu \phi)^2 - \frac{1}{2}m^2\phi^2 + J\phi. \tag{4.165}$$

In this case, the field equations yield the non-homogeneous equation

$$(\partial_\mu \partial^\mu + m^2)\phi = J. \tag{4.166}$$

4.3.2 Self-interacting Fields

The term ϕ^4 refers to the ϕ^4 theory with spontaneous symmetry breaking potential. The Lagrangian associated with this type of field is

$$\mathcal{L} = \frac{1}{2}(\partial_\mu \phi)^2 - \frac{1}{2}m^2\phi^2 - \frac{\lambda \phi^4}{4!}, \tag{4.167}$$

where λ is the strength of the interaction. The field equation associated with this Lagrangian is

$$(\partial_\mu \partial^\mu + m^2)\phi = -\frac{\lambda}{3!}\phi^3. \tag{4.168}$$

This equation is not solvable, but we can make predictions using perturbation theory.

4.3.3 Coupled Scalar Fields

Consider two coupled scalar fields $\phi_1(x)$ and $\phi_2(x)$. The associated Lagrangian in this case is

$$\mathcal{L} = \frac{1}{2}(\partial_\mu \phi_1)^2 - \frac{1}{2}m^2\phi_1^2 + \frac{1}{2}(\partial_\mu \phi_2)^2 - \frac{1}{2}m^2\phi_2^2 - g(\phi_1^2 + \phi_2^2). \tag{4.169}$$

Suppose we rotate the system, having

$$\begin{pmatrix} \phi_1' \\ \phi_2' \end{pmatrix} = \begin{pmatrix} \cos\theta & -\sin\theta \\ \sin\theta & \cos\theta \end{pmatrix} \begin{pmatrix} \phi_1 \\ \phi_2 \end{pmatrix}. \tag{4.170}$$

Since

$$\phi_1'^2 + \phi_2'^2 = \phi_1^2 \cos^2\theta + \phi_2^2 \sin^2\theta - 2\phi_1\phi_2 \cos\theta \sin\theta \tag{4.171a}$$
$$+ \phi_1^2 \sin^2\theta + \phi_2^2 \cos^2\theta + 2\phi_1\phi_2 \cos\theta \sin\theta = \phi_1^2 + \phi_2^2,$$
$$(\partial_\mu \phi_1')^2 = \partial_\mu \phi_1' \partial^\mu \phi_1' = \partial_\mu (\phi_1 \cos\theta - \phi_2 \sin\theta)\, \partial^\mu (\phi_1 \cos\theta - \phi_2 \sin\theta) \tag{4.171b}$$
$$= (\partial_\mu \phi_1)^2 \cos^2\theta + (\partial_\mu \phi_2)^2 \sin^2\theta - 2(\partial_\mu \phi_1)(\partial^\mu \phi_2) \cos\theta \sin\theta,$$
$$(\partial_\mu \phi_2')^2 = (\partial_\mu \phi_1)^2 \sin^2\theta + (\partial_\mu \phi_2) \cos^2\theta + 2(\partial_\mu \phi_1)(\partial^\mu \phi_2) \cos\theta \sin\theta \tag{4.171c}$$

We obtain

$$(\partial_\mu \phi_1')^2 + (\partial_\mu \phi_2')^2 = (\partial_\mu \phi_1)^2 + (\partial_\mu \phi_2)^2 \longrightarrow \mathcal{L}' = \mathcal{L}, \tag{4.172}$$

so we have invariance under $SO(2)$ symmetries in the $\phi_1 - \phi_2$ plane.

Analogously, the Lagrangian for complex fields coupled to real scalar fields is given by

$$\mathcal{L} = \partial_\mu \psi^\star \partial^\mu \psi - m^2 \psi^\star \psi + \frac{1}{2}(\partial_\mu \phi)^2 - \frac{1}{2}m^2\phi^2 - g\psi^\star \psi \phi, \tag{4.173}$$

where ψ represents the complex field, ϕ the real field and $g\psi^\star\psi\phi$ the coupling term between fields.

▶ **Pause and think** How do scalar fields interact?

4.3.4 The Feynman-Stuckelberg Interpretation

Historically, the problem associated with the Klein-Gordon equation, which involves adding physically unacceptable states with negative energy, leading to the collapse of the interpretation of $\tilde{\rho}$ as a probability and ψ as a wave function, is resolved by temporarily abandoning the Klein-Gordon equation itself. Instead, an equation of first order in ∂_t is sought, allowing for the interpretation of probability density for its solution. This equation is the *Dirac equation*.

However, an interpretation for negative energy solutions of the Klein-Gordon equation was proposed by Feynman and Stueckelberg. There, negative energy states are featured in terms of antiparticles.

To focus on this interpretation, consider the equation of motion for a charge in an electromagnetic field:

$$m\frac{d^2x^\mu}{d\tau^2} = \frac{q}{c}F^\mu_\nu\frac{dx^\nu}{d\tau}. \tag{4.174}$$

Changing $\tau \to -\tau$ in the equation of motion, we find the dynamics of a particle with a change in charge $-q$, which is equivalent to considering a particle traveling backward in time. We thus obtain

$$m\frac{d^2x^\mu}{d(-\tau)^2} = \frac{q}{c}F^\mu_\nu\frac{dx^\nu}{d(-\tau)} \longrightarrow m\frac{d^2x^\mu}{d\tau^2} = -\frac{q}{c}F^\mu_\nu\frac{dx^\nu}{d\tau}. \tag{4.175}$$

Now, leveraging the invariance of the phase of the wave describing the particle's state, which must satisfy $\phi' = \phi$ following $\tau \to \tau' = -\tau$, we obtain

$$\phi' = (E'\tau' - \mathbf{p}' \cdot \mathbf{x}) = -(E'\tau + \mathbf{p}' \cdot \mathbf{x}), \tag{4.176a}$$

$$\phi = (E\tau - \mathbf{p} \cdot \mathbf{x}). \tag{4.176b}$$

Therefore, to have phase invariance, we must accompany the temporal inversion $\tau \to -\tau$ with the inversion of energy $E \to -E$ and momentum $\mathbf{p} \to -\mathbf{p}$. At this point, we can interpret negative energy states as particle states that

1. have positive energy;
2. move backward in time;
3. have momentum with opposite sign.

However, since temporal inversion is equivalent to inverting the charge, the *negative energy particle states* are states of particles that

1. have positive energy;
2. have opposite charge;
3. have opposite momentum.

From these three conditions, we define the notion of *antiparticle*. This is in agreement, for the specific case of the Klein-Gordon equation, with the discussion about the sign of the electromagnetic current density. From the particle current density $J^{\mu} = (J^0, \mathbf{J})$, we write the current density for the opposite charge as

$$\tilde{J} = -i\left[\psi^{\star}\nabla\psi - \psi\nabla\psi^{\star}\right] = -i\left[|N|^2\left(e^{ip^{\mu}x_{\mu}}\nabla e^{-iEt-\mathbf{p}\cdot\mathbf{x}} - e^{-ip^{\mu}x_{\mu}}\nabla e^{i(Et-\mathbf{p}\cdot\mathbf{x})}\right)\right], \quad (4.177)$$

where we have substituted $\psi = Ne^{-ip^{\mu}x_{\mu}} = -i|N|^2(i\mathbf{p}+i\mathbf{p}) = 2\mathbf{p}|N|^2$. We thus obtain

$$J_q^{\mu} = (2Eq|N|^2, 2\mathbf{p}q|N|^2) = 2\,|N|^2 q(E,\mathbf{p}) = 2\,|N|^2 q p^{\mu} = 2\,|N|^2 q(\pm E_p, \mathbf{p}). \quad (4.178)$$

Pause and think Explain how to interpret a particle with the characteristics above described.

Let us consider the two cases

$$J_q^{\mu} = 2\,N|^2 q(E_p, \mathbf{p}), \quad \text{is the current density for positive energy charge,} \quad (4.179a)$$

$$J_q^{\mu} = 2\,N|^2 q(-E_p, \mathbf{p}), \quad \text{is the density for charge } q \text{ with negative energy.} \quad (4.179b)$$

We can also write Eq. (4.179b) as $J_q^{\mu} = 2|N|^2(-q)(E_p, -\mathbf{p})$, interpreting it as a probability density for charge $-q$ with momentum $-\mathbf{p}$ and positive energy.

The discussion made has mathematical reasons but has been experimentally verified in terms of creation and annihilation of particles (and antiparticles) with zero spin.

Let us consider a particle with energy E_0, charge q_0, and momentum $\mathbf{p_0}$ that *absorbs* another particle with energy E, charge q, and momentum $\mathbf{p}$. After absorption, the particle will have energy $E_0 + E$, charge $q_0 + q$, and momentum $\mathbf{p_0} + \mathbf{p}$.

In the case of *emission* of a particle with initial energy E, charge q, and momentum $\mathbf{p}$ by a particle with energy E_0, charge q_0, and momentum $\mathbf{p_0}$, the final energy, charge, and momentum of the system will be $E_0 - E$, $q_0 - q$, and $\mathbf{p_0} - \mathbf{p}$, respectively.

It can be verified that the absorption of a particle with energy $-E$ is equivalent to the emission of an antiparticle with energy E, and the emission of a particle with energy $-E$ is equivalent to the absorption of an antiparticle with energy E.

By not disregarding the negative root of E and interpreting it as a state of antiparticles with positive energy, the general solution of the Klein-Gordon equation is no longer for a single particle, as we must also consider the antiparticle. In summary,

$$\psi(x) = \left| \begin{array}{c} \text{particle} \\ \text{incoming} \\ \text{with } E > 0 \\ \propto e^{-i(Et-\mathbf{p}\cdot\mathbf{x})} \end{array} \right| + \left| \begin{array}{c} \text{antiparticle} \\ \text{outgoing} \\ \text{with } E > 0 \\ \propto e^{i(Et-\mathbf{p}\cdot\mathbf{x})} \end{array} \right| .$$

However, this is because the context of the wave function is not the most suitable for the Klein-Gordon equation. In fact, it is in the dynamics of fields, whose excitations are spinless particles, that the Klein-Gordon equation provides a complete and consistent interpretation.

4.4 Spin-1/2 Particles

The Klein-Gordon equation presents some conceptual problems.

The problems are mainly those following.

- Negative-energy states are allowed, leading to the temporal component of the four-current in the continuity equation not being interpretable as a positively defined probability density:

$$E = \pm\sqrt{\mathbf{p}^2 + m^2} = \pm E_p \longrightarrow \rho = \psi^\star \partial_t \psi - \psi \partial_t \psi^\star = \pm 2E_p |N|^2; \quad (4.180)$$

- The Feynman-Stueckelberg interpretation makes the Klein-Gordon acceptable for describing relativistic particles by predicting the existence of antiparticles. Historically, the Klein-Gordon equation is used to find relativistically covariant equations for a wave function that maintains the interpretation of a positively defined probability density.

Recall that, to achieve this, one started from

$$\begin{aligned} &E^2 = \mathbf{p}^2 + m^2 \longrightarrow (\Box + m^2)\psi = 0, \\ &\hat{E} \longrightarrow i\partial_t \quad \text{and} \quad \hat{p} \longrightarrow -i\nabla. \end{aligned} \quad (4.181)$$

Squaring the mass shell introduces solutions with negative energies.

4.4.1 The Dirac Equation

Dirac noticed that the presence of ∂_t^2 is precisely the problem in the equation of motion for a single particle. This is because this makes the equation second-order, implying that the wave function ψ is determined if ψ and $\partial_t \psi$ are known at $t = 0$, undermining the postulate of quantum mechanics, where knowing $\psi(t, \mathbf{x})$ is sufficient to have all information about the particle's state and its temporal dynamic evolution.

Therefore, to avoid violating the postulate of quantum mechanics, one postulates the Hamiltonian in the equation of motion in such a way that the following requirements are satisfied.

- Linearity in temporal and spatial derivatives so that ∂_t and ∇ can be treated symmetrically, as required by relativistic covariance.
- Covariance under Lorentz transformations.
- Compatibility with the Klein-Gordon equation, meaning that the energy must satisfy the mass shell relation.

By imposing these requirements, the following is obtained:

- ψ is no longer a wave function but a field with N components.
- The equation of motion is matrix-valued.
- It describes a particle with spin 1/2, both massive and massless.

Let us verify the first requirement, namely the linearity of temporal and spatial derivatives. This is satisfied by choosing the following Hamiltonian operator as an ansatz:

$$\hat{H} = (-i\boldsymbol{\alpha} \cdot \nabla + \beta m) = (\boldsymbol{\alpha} \cdot \mathbf{p} + \beta m), \tag{4.182}$$

with the coefficients $\boldsymbol{\alpha}$ and β whose meaning will be clarified later.

Recalling that $i\frac{\partial}{\partial_t}\psi(t, \mathbf{x}) = \hat{H}\psi(t, \mathbf{x})$, substituting the Hamiltonian operator yields the Dirac equation:

$$i\frac{\partial}{\partial_t}\psi(x) = (-i\boldsymbol{\alpha}\nabla + \beta m)\psi(x). \tag{4.183}$$

Now, let us demonstrate that the third requirement, *compatibility with the Klein-Gordon equation*, is satisfied by choosing the ansatz:

$$\hat{H}^2 = \mathbf{p}^2 + m^2, \tag{4.184}$$

where $\mathbf{p}^2 + m^2$ represents a *relativistic energy-momentum closure relation*. Substituting the expression for $\hat{H}$ from Eq. (4.182) into Eq. (4.184), we obtain:

$$(\boldsymbol{\alpha} \cdot \mathbf{p} + \beta m)(\boldsymbol{\alpha} \cdot \mathbf{p} + \beta m) = \mathbf{p}^2 + m^2 \longrightarrow \tag{4.185}$$

$$\longrightarrow \left(\sum_{j=1}^{3} \alpha_j p_j + \beta m\right)\left(\sum_{k=1}^{3} \alpha_k p_k + \beta m\right) = \left(\sum_{jk} p_j p_k \delta_{jk}\right) + m^2. \tag{4.186}$$

Explicitly multiplying out the products in the previous expression, we obtain

$$\sum_{j,k} \left(\alpha_j p_j \alpha_k p_k + \alpha_j p_j \beta m + \beta m \alpha_k p_k\right) + (\beta m)^2 = \sum_{jk} \left(p_j p_k \delta_{jk}\right) + m^2. \tag{4.187}$$

In the end, manipulating the expression again, we obtain

$$\alpha_j \alpha_k p_j p_k + m(\alpha_j p_j \beta + \beta \alpha_j p_j) + \beta^2 m^2 = p_j p_k \delta_{jk} + m^2. \tag{4.188}$$

Setting the terms equal to each other, we get

$$\beta^2 m^2 = m^2 \longrightarrow \beta^2 = 1, \tag{4.189a}$$

$$m(\alpha_j p_j \beta + \beta \alpha_j p_j) = 0 \longrightarrow (\alpha_j \beta + \beta \alpha_j) p_j = 0 \longrightarrow \{\alpha_j, \beta\} = 0, \tag{4.189b}$$

$$\alpha_j \alpha_k p_j p_k = p_j p_k \delta_{jk} \rightarrow \frac{1}{2}\left(\{\alpha_j, \alpha_k\} + [\alpha_j, \alpha_k]\right) p_j p_k = p_j p_k \delta_{jk} \rightarrow$$
$$\frac{1}{2}\{\alpha_j, \alpha_k\} p_j p_k = p_j p_k \delta_{jk} \rightarrow \{\alpha_j, \alpha_k\} = 2\delta_{jk}, \tag{4.189c}$$

where, in Eq. (4.189c), we have considered the antisymmetry of $[\alpha_j, \alpha_k]$ and the symmetry of $p_j p_k$ so that their product is zero. Additionally, the fact that

$$[\alpha_j, \alpha_k] = \alpha_j \alpha_k - \alpha_k \alpha_j, \tag{4.190a}$$

$$\{\alpha_j, \alpha_k\} = \alpha_j \alpha_k + \alpha_k \alpha_j, \tag{4.190b}$$

implies that their sum gives $[\alpha_j, \alpha_k] + \{\alpha_j, \alpha_k\} = 2\alpha_j \alpha_k$.

Taking Eq. (4.189c) into consideration, if $j = k$, we have

$$\{\alpha_j, \alpha_j\} = 2 \rightarrow 2\alpha_j^2 = 2 \longrightarrow \alpha_j^2 = 1. \tag{4.191}$$

Both α_i and β cannot be null simultaneously because one contributes to the kinetic term, while the other to the term $\propto m$. To have a null anticommutator, they must be $N \times N$ matrices.

Therefore, from the previously derived conditions, which we rewrite as

$$\{\alpha_i, \alpha_j\} = 2\delta_{ij}, \tag{4.192a}$$
$$\{\alpha_i, \beta\} = 0, \tag{4.192b}$$
$$\alpha_i^2 = \beta^2 = 1, \tag{4.192c}$$
$$\hat{H}^\dagger = \hat{H} \to \alpha_i^\dagger = \alpha_i \quad \text{and} \quad \beta^\dagger = \beta \tag{4.192d}$$

we obtain that α_i and β are $N \times N$ hermitian matrices because $\hat{H}$ has real eigenvalues equal to ± 1. In this way, we can say that

$$i\frac{\partial}{\partial_t}\psi = (\boldsymbol{\alpha} \cdot \mathbf{p} + \beta m)\psi \tag{4.193}$$

is the matrix equation for an object $\psi = \begin{pmatrix} \psi_1 \\ \vdots \\ \psi_N \end{pmatrix}$ that is **NOT** a wave function and gives rise to N equations for $\psi_1, \ldots, \psi_N$.

Introducing, for convenience, a set of 4 $N \times N$ matrices called *gamma matrices*,

$$\gamma^\mu, \quad \text{with} \quad \mu = 0, 1, 2, 3, \tag{4.194}$$

these matrices can be grouped into a four-vector that transforms like an event, $\gamma^\mu = (\gamma^0, \boldsymbol{\gamma})$ where $\gamma^0 = \beta$ and $\gamma^i = \beta\alpha_i$. Furthermore, $(\gamma^0)^\dagger = \gamma^0$ is hermitian because β is hermitian, while for γ^i,

$$(\gamma^i)^\dagger = (\beta\alpha_i)^\dagger = \alpha_i^\dagger\beta^\dagger = \alpha_i\beta = \beta\beta\alpha_i\beta = \gamma^0\gamma^i\gamma^0. \tag{4.195}$$

In general,

$$(\gamma^\mu)^\dagger = \gamma^0\gamma^\mu\gamma^0, \tag{4.196}$$

with $(\gamma^0)^2 = 1$ and $(\gamma^i)^2 = -1$. Moreover, the gamma matrices satisfy the Clifford algebra, i.e.,

$$\{\gamma^\mu, \gamma^\nu\} = 2g^{\mu\nu}, \tag{4.197}$$

where $\mu = 0$ and $\nu = i$, and $g^{\mu\nu} = (+1, -1, -1, -1)$. Expanding the previous expression, we obtain

$$\gamma^0\gamma^i + \gamma^i\gamma^0 = 2g^{0i} = 0 \longrightarrow \gamma^0\gamma^i = -\gamma^i\gamma^0 \longrightarrow \gamma^0\gamma^i\gamma^0 = -\gamma^i(\gamma^0\gamma^0) = -\gamma^i$$
$$\longrightarrow 2(\gamma^i)^2 = 2g^{ii} \longrightarrow (\gamma^{ii})^2 = -1. \tag{4.198}$$

Using the γ^{μ}, we can derive the Dirac equation. Consider

$$i\frac{\partial}{\partial_t}\psi = (\alpha_j p_j + \beta m)\psi \longrightarrow i\beta\frac{\partial}{\partial_t}\psi = (\beta\alpha_j p_j + \beta^{2m})\psi. \quad (4.199)$$

Subsequently, by using the gamma matrices, we can write

$$i\gamma^0\partial_t\psi = \left[\gamma^j\left(-i\frac{\partial}{\partial x^j}\right) + m\right]\psi \longrightarrow (i\gamma^0\partial_t + i\gamma^j\partial_j - m)\psi = 0. \quad (4.200)$$

Defining $\partial_\mu = (\partial_t, \nabla)$, we obtain the *Dirac equation using the gamma matrices*,

$$(i\gamma^{\mu}\partial_\mu - m)\psi = 0. \quad (4.201)$$

We can also write the above equation using the *Feynman notation* in which the contraction of a covariant four-vector with the gamma matrices is denoted as $\gamma^{\mu}a_\mu = \not{a}$ so that

$$(i\not{\partial} - m)\psi = 0. \quad (4.202)$$

Now, let us use $p^{\mu} = (E, \mathbf{p}) = (i\partial_t, -i\nabla) = i(\partial_t, -\nabla) = i\partial^{\mu}$ to have two equivalent ways of writing the Dirac equation:

$$(\gamma^{\mu}p_\mu - m)\psi = 0 \longrightarrow (\not{p} - m)\psi = 0. \quad (4.203)$$

4.4.2 The Klein-Gordon Equation from the Dirac Picture

We can obtain the Klein-Gordon equation multiplying the Dirac equation (Eq. (4.201)) by $i\gamma_\nu\partial^\nu$:

$$(i\gamma_\nu\partial^\nu i\gamma^{\mu}\partial_\mu - i\gamma_\nu\partial^\nu m)\psi = (-\gamma_\nu\gamma^{\mu}\partial^\nu\partial_\mu - i\gamma_\nu\partial^\nu m)\psi = 0. \quad (4.204)$$

Inside the above equation, $m(i\gamma^\nu\partial_\nu\psi) = m^2\psi$, so

$$(-\gamma_\nu\gamma^{\mu}\partial^\nu\partial_\mu - m^2)\psi = (-\delta^{\mu}_{\nu}\partial^\nu\partial_\mu - m^2)\psi = (-\partial^{\mu}\partial_\mu - m^2)\psi = 0 \longrightarrow (\partial^{\mu}\partial_\mu + m^2)\psi = 0. \quad (4.205)$$

In Eq. (4.205), we used the identity

$$\frac{\gamma_\nu \gamma^\mu + \gamma^\mu \gamma_\nu}{2} = \delta^\mu_\nu . \tag{4.206}$$

This can be shown as follows:

$$\frac{\gamma_\nu \gamma^\mu + \gamma^\mu \gamma_\nu}{2} = g_{\alpha\nu} \frac{\gamma^\alpha \gamma^\mu + \gamma^\mu \gamma^\alpha}{2} = g_{\alpha\nu} \frac{2 g^{\alpha\mu}}{2} = \delta^\mu_\nu . \tag{4.207}$$

We have thus found the Klein-Gordon equation starting from the Dirac equation.

Let us now focus on the inverse process, that is, deriving the Dirac equation starting from the Klein-Gordon equation $(\partial^2 + m^2)\psi = 0$. We try to factorize the second-order operator, namely ∂^2. To do this, we write

$$(\partial^2 + m^2) = (\sqrt{\partial^2} + im)(\sqrt{\partial^2} - im), \tag{4.208}$$

but $\sqrt{\partial^2} = \sqrt{\partial_\mu \partial^\mu}$ is not defined. Therefore, we introduce the object $\gamma^0 = (\gamma^0, \boldsymbol{\gamma})$ satisfying the anticommutation relation of the Clifford algebra $\{\gamma^\mu, \gamma^\nu\} = 2g^{\mu\nu}$ and replace ∂_μ with $\gamma^\mu \partial_\mu$, trying to show that

$$(\partial_\mu \partial^\mu) = (\gamma^\mu \partial_\mu)^2. \tag{4.209}$$

Indeed,

$$\begin{aligned}
(\gamma^\mu \partial_\mu)^2 &= \partial^2 = (\gamma^0 \partial_t + \gamma^1 \partial_1 + \gamma^2 \partial_2 + \gamma^3 \partial_3)^2 = (\gamma_0 \partial_t + \boldsymbol{\gamma} \nabla)^2 \\
&= (\gamma^0)^2 \partial_t^2 + (\gamma^i)^2 (\partial_i)^2 + \gamma^0 \gamma^i \partial_t \partial_i + \gamma^i \gamma^0 \partial_t \partial_i = \\
&= \partial_t^2 - \nabla^2 + \{\gamma^0 \gamma^i + \gamma^i \gamma^0\} \partial_t \partial_i = \partial_t^2 + 2g^{0i} \partial_t \partial_i \\
&= \partial_t^2 - \nabla^2 = \Box = (\partial_\mu \partial^\mu).
\end{aligned} \tag{4.210}$$

At this point, we can factorize, obtaining

$$(\Box + m^2)\psi = 0 \longrightarrow (\partial + im)(\partial - im)\psi = 0. \tag{4.211}$$

Consider that ψ satisfies either $(\partial + im)\psi = 0$ or $(\partial - im)\psi = 0$, so we can rewrite the final form of the equation as:

$$(\partial + im)\psi = 0 \longrightarrow (i\partial - m)\psi = 0 \longrightarrow (i\gamma^\mu \partial_\mu - m)\psi = 0. \tag{4.212}$$

We thus found the Dirac equation by starting from the Klein-Gordon equation.

4.4.3 Weyl Representation for γ

Let us start by writing the γ matrices in 2×2 block form as

$$\gamma^0 = \begin{pmatrix} 0 & \mathbb{I} \\ \mathbb{I} & 0 \end{pmatrix}, \quad \text{with} \quad \mathbb{I} = \begin{pmatrix} 1 & 0 \\ 0 & 1 \end{pmatrix}, \tag{4.213a}$$

$$\gamma^1 = \begin{pmatrix} 0 & \sigma_x \\ -\sigma_x & 0 \end{pmatrix}, \quad \text{with} \quad \sigma_x = \begin{pmatrix} 0 & 1 \\ 1 & 0 \end{pmatrix}, \tag{4.213b}$$

$$\gamma^2 = \begin{pmatrix} 0 & \sigma_y \\ -\sigma_y & 0 \end{pmatrix}, \quad \text{with} \quad \sigma_y = \begin{pmatrix} 0 & -i \\ i & 0 \end{pmatrix}, \tag{4.213c}$$

$$\gamma^3 = \begin{pmatrix} 0 & \sigma_z \\ -\sigma_z & 0 \end{pmatrix}, \quad \text{with} \quad \sigma_z = \begin{pmatrix} 1 & 0 \\ 0 & -1 \end{pmatrix}, \tag{4.213d}$$

where the σ_i are the Pauli matrices, satisfying the properties $(\sigma_i)^2 = 1$, $[\sigma_i, \sigma_j] = 2i\varepsilon^{ijk}$, and $\{\sigma_i, \sigma_j\} = 2\delta_{ij}$. We can express these matrices compactly as

$$\gamma^0 = \begin{pmatrix} 0 & \mathbb{I} \\ \mathbb{I} & 0 \end{pmatrix} \quad \text{and} \quad \boldsymbol{\gamma} = \begin{pmatrix} 0 & \boldsymbol{\sigma} \\ -\boldsymbol{\sigma} & 0 \end{pmatrix} \tag{4.214}$$

or, equivalently, defining $\sigma^\mu = (\mathbb{I}, \boldsymbol{\sigma})$ and $\boldsymbol{\sigma}^\mu = (\mathbb{I}, -\boldsymbol{\sigma})$

$$\gamma^\mu = \begin{pmatrix} 0 & \sigma^\mu \\ \boldsymbol{\sigma}^\mu & 0 \end{pmatrix}. \tag{4.215}$$

The Dirac equation in this representation becomes

$$(\gamma^\mu p_\mu - m)\psi = 0 \longrightarrow (\gamma^0 \hat{p}^0 - \boldsymbol{\gamma} \cdot \hat{\mathbf{p}} - m)\psi(x) = 0, \tag{4.216}$$

with $p_\mu = (p^0, -\mathbf{p})$. Writing explicitly the γ matrices, we have

$$\left[\begin{pmatrix} 0 & \hat{p}^0 \\ \hat{p}^0 & 0 \end{pmatrix} - \begin{pmatrix} 0 & \boldsymbol{\sigma} \cdot \hat{\mathbf{p}} \\ -\boldsymbol{\sigma} \cdot \hat{\mathbf{p}} & \end{pmatrix} - \begin{pmatrix} m & 0 \\ 0 & m \end{pmatrix} \right] \psi(x) = 0. \tag{4.217}$$

Since we write the γ matrices in 2×2 block form, we can split $\psi(x)$ into two components, which we identify as

$$\psi = \begin{pmatrix} \psi_L \\ \psi_R \end{pmatrix}, \tag{4.218}$$

where ψ_L and ψ_R are component fields called *Weyl spinor fields*. In this way, we can write the Dirac equation as:

$$(\hat{p}^0 - \boldsymbol{\sigma} \cdot \hat{\mathbf{p}})\psi_R = m\psi_L, \tag{4.219a}$$

$$(\hat{p}^0 + \boldsymbol{\sigma} \cdot \hat{\mathbf{p}})\psi_L = m\psi_R. \tag{4.219b}$$

Thus, we obtain a system of two coupled linear equations for ψ_L, left-handed particles, and ψ_R, right-handed particles. In the case of massless particles, the equations decouple as follows:

$$(\hat{p}^0 - \boldsymbol{\sigma} \cdot \hat{\mathbf{p}})\psi_R = 0, \tag{4.220a}$$

$$(\hat{p}^0 + \boldsymbol{\sigma} \cdot \hat{\mathbf{p}})\psi_L = 0. \tag{4.220b}$$

ψ_R and ψ_L are eigenstates of the Dirac operator for the case $m = 0$. Furthermore, a left-handed particle can never become a right-handed particle.

Let us define the operator

$$\gamma^5 = i\gamma^0\gamma^1\gamma^2\gamma^3 = i\begin{pmatrix} 0 & \mathbb{1} \\ \mathbb{1} & 0 \end{pmatrix}\begin{pmatrix} 0 & \sigma_x \\ -\sigma_x & 0 \end{pmatrix}\begin{pmatrix} 0 & \sigma_y \\ -\sigma_y & 0 \end{pmatrix}\begin{pmatrix} 0 & \sigma_z \\ -\sigma_z & 0 \end{pmatrix} = i\begin{pmatrix} -\sigma_x & 0 \\ 0 & \sigma_x \end{pmatrix}\begin{pmatrix} -\sigma_y\sigma_z & 0 \\ 0 & -\sigma_y\sigma_z \end{pmatrix} =$$
$$= i\begin{pmatrix} \sigma_x\sigma_y\sigma_z & 0 \\ 0 & -\sigma_x\sigma_y\sigma_z \end{pmatrix} = \begin{pmatrix} -\mathbb{1} & 0 \\ 0 & \mathbb{1} \end{pmatrix}, \tag{4.221}$$

where, for the last step, we used the relations $\sigma_i\sigma_j - \sigma_j\sigma_i = 2i\varepsilon^{ijk}\sigma_k$ and $\sigma_i\sigma_j + \sigma_j\sigma_i = 2\delta_{ij}$, adding them together

$$2\sigma_i\sigma_j = 2(i\varepsilon^{ijk}\sigma_k + \delta_{ij}) \longrightarrow \sigma_i\sigma_j\sigma_k = i\varepsilon^{ijk}(\sigma_k)^2 + \sigma_k \cdot 0 = i\mathbb{1} \longrightarrow \sigma_x\sigma_y\sigma_z = i. \tag{4.222}$$

The operator γ^5 is also called the *chirality operator* because

$$\gamma^5\begin{pmatrix} \psi_L \\ 0 \end{pmatrix} = -\begin{pmatrix} \psi_L \\ 0 \end{pmatrix}, \quad \gamma^5\begin{pmatrix} 0 \\ \psi_R \end{pmatrix} = \begin{pmatrix} 0 \\ \psi_R \end{pmatrix}. \tag{4.223a}$$

In the first case, the entirely left-handed state is an eigenstate of chirality with eigenvalue -1, while in the second case, the entirely right-handed state is an eigenstate with eigenvalue $+1$. Thanks to γ^5, we can define the projectors:

$$\hat{P}_L = \frac{1-\gamma^5}{2} \quad \text{gets the left-handed part} \longrightarrow \hat{P}_L\begin{pmatrix} \psi_L \\ \psi_R \end{pmatrix} = \frac{1}{2}\begin{pmatrix} 2\mathbb{1} & 0 \\ 0 & 0 \end{pmatrix}\begin{pmatrix} \psi_L \\ 0 \end{pmatrix}, \tag{4.224a}$$

$$\hat{P}_R = \frac{1+\gamma^5}{2} \quad \text{gets the right-handed part} \longrightarrow \hat{P}_R\begin{pmatrix} \psi_L \\ \psi_R \end{pmatrix} = \frac{1}{2}\begin{pmatrix} 0 & 0 \\ 0 & 2\mathbb{1} \end{pmatrix}\begin{pmatrix} 0 \\ \psi_R \end{pmatrix}. \tag{4.224b}$$

Again, a particle with left-handed chirality cannot transition into a particle with right-handed chirality.

Let us substitute $\hat{P}^0\psi_a = E_p\psi_a$, where the operator eigenvalues for the massless case are $E = E_p = |\mathbf{p}|$. Substituting for $a = L, R$, we obtain:

$$(|\mathbf{p}| - \boldsymbol{\sigma} \cdot \hat{\mathbf{p}})\psi_R = 0 \longrightarrow \frac{\boldsymbol{\sigma} \cdot \hat{\mathbf{p}}}{|\mathbf{p}|}\psi_R = \psi_R, \qquad (|\mathbf{p}| + \boldsymbol{\sigma} \cdot \hat{\mathbf{p}})\psi_L = 0 \longrightarrow \frac{\boldsymbol{\sigma} \cdot \hat{\mathbf{p}}}{|\mathbf{p}|}\psi_L = -\psi_L. \tag{4.225a}$$

The above equations are eigenvalue equations for an operator $\hat{h} = \frac{\boldsymbol{\sigma}\cdot\hat{\mathbf{p}}}{|\mathbf{p}|}$ called the *helicity operator*. It represents the projection of the particle's spin along the direction of its momentum. In this way, we can write:

$$\gamma^5 (0\ \psi_R) = (0\ \psi_R) \iff \hat{h}\psi_R = \psi_R, \quad \gamma^5 (\psi_L\ 0) = -(\psi_L\ 0) \iff \hat{h}\psi_L = -\psi_L. \tag{4.226a}$$

These equations tell us that left-handed particles have spin and momentum antiparallel with eigenvalue -1, while right-handed particles have them parallel with eigenvalue $+1$. Only in the massless case do $\hat{h}$ and γ^5 have the same eigenvalues and eigenvectors, so helicity and chirality are equivalent.

In general, helicity depends on the reference frame in which it is measured. Therefore, for massive particles $m > 0$, it is possible to switch from a reference frame O where the particle has $h > 0$ to a reference frame O' where the particle has $h < 0$; the sign changes over the momentum but leaves unaltered the spin.

The mass–shell relation is obtained by applying the operator $\hat{p}^0 + \boldsymbol{\sigma}\cdot\hat{\mathbf{p}}$ to

$$(\hat{p}^0 + \boldsymbol{\sigma}\cdot\hat{\mathbf{p}})\left[(\hat{p}^0 - \boldsymbol{\sigma}\cdot\hat{\mathbf{p}})\psi_R\right] = m(\hat{p}^0 + \boldsymbol{\sigma}\cdot\hat{\mathbf{p}})\psi_L = m^2\psi_R, \tag{4.227}$$

that is, ψ_R is an eigenstate of the operator $(\hat{p}^0 + \boldsymbol{\sigma}\cdot\hat{\mathbf{p}})(\hat{p}^0 - \boldsymbol{\sigma}\cdot\hat{\mathbf{p}})$ with eigenvalue m^2. This follows from the assumption that ψ_R and ψ_L are simultaneous eigenstates of the energy operator $\hat{p}^0$ and of the momentum operator $\hat{\mathbf{p}}$. Indeed, recalling that

$$\begin{cases} \hat{p}^0|\psi\rangle = p^0|\psi\rangle = E|\psi\rangle, \\ \boldsymbol{\sigma}\cdot\hat{\mathbf{p}}|\psi\rangle = \boldsymbol{\sigma}\cdot\mathbf{p}\,|\psi\rangle, \end{cases} \tag{4.228}$$

we directly obtain the mass–shell condition

$$\left[(\hat{p}^0)^2 - (\boldsymbol{\sigma}\cdot\hat{\mathbf{p}})^2\right] = m^2 \longrightarrow E^2 - \boldsymbol{\sigma}^2\mathbf{p}^2 = m^2 \longrightarrow E^2 = \mathbf{p}^2 + m^2. \tag{4.229}$$

It should be noted that Dirac does not eliminate the problem of negative energies at this stage, since the mass–shell relation still admits the negative square root. The resolution of this issue is achieved only within the framework of QFT.

Is helicity conserved? One might have

$$(-E_p - \boldsymbol{\sigma}\cdot\hat{\mathbf{p}})\psi_R = 0 \longrightarrow \frac{\boldsymbol{\sigma}\cdot\hat{\mathbf{p}}}{|\mathbf{p}|}\psi_R = -\psi_R, \tag{4.230a}$$

$$(-E_p + \boldsymbol{\sigma}\cdot\hat{\mathbf{p}})\psi_L = 0 \longrightarrow \frac{\boldsymbol{\sigma}\cdot\mathbf{p}}{|\mathbf{p}|}\psi_L = \psi_L, \tag{4.230b}$$

where we substituted $E_p = |\mathbf{p}|$, obtaining the opposite case compared to Eq. (4.225).
So, one concludes that for $m = 0$,

$$E = +E_p > 0 \longrightarrow \hat{h}\psi_R = \psi_R, \hat{h}\psi_L = -\psi_L \quad \text{particles,} \tag{4.231a}$$

$$E = -E_p < 0 \longrightarrow \hat{h}\psi_R = -\psi_R, \hat{h}\psi_L = \psi_L \quad \text{antiparticles.} \tag{4.231b}$$

4.4.4 Properties of the α_i and β Matrices

Both traces of the α_i and β matrices are zero, i.e.,

$$\mathrm{Tr}(\alpha_i) = 0, \tag{4.232a}$$

$$\mathrm{Tr}(\beta) = 0. \tag{4.232b}$$

Let us prove the above equations starting from the Poisson brackets of the matrices

$$\{\alpha_i, \beta\} = \alpha_i\beta + \beta\alpha_i = 0. \tag{4.233}$$

Starting with Eq. (4.232a), we multiply the previous equation by β, obtaining

$$\beta\alpha_i\beta + \beta^2\alpha_i = 0 \longrightarrow \alpha_i = -\beta\alpha_i\beta \longrightarrow \mathrm{Tr}(\alpha_i) = \mathrm{Tr}(-\beta\alpha_i\beta) = -\mathrm{Tr}(\beta\beta\alpha_i) = -\mathrm{Tr}(\alpha_i). \tag{4.234}$$

Since the trace is equal to its opposite, this implies Eq. (4.232a).
To derive Eq. (4.232b), we multiply Eq. (4.233) by α_i, obtaining

$$\alpha_i\alpha_i\beta + \alpha_i\beta\alpha_i = 0 \longrightarrow \beta = -\alpha_i\beta\alpha_i \longrightarrow \mathrm{Tr}(\beta) = \mathrm{Tr}(-\alpha_i\beta\alpha_i) = -\mathrm{Tr}(\alpha_i\alpha_i\beta) = -\mathrm{Tr}(\beta). \tag{4.235}$$

As in the previous case, since the trace is equal to its opposite, this implies Eq. (4.232b).

Recalling that α_i and β are Hermitian, with eigenvalues ± 1, and with zero trace, we conclude that the sum of eigenvalues equal to $+1$ and that of eigenvalues equal to -1 must be zero because the trace is zero.

The only possibility is that $N = 2n$. Let us consider the case where $N = 2$ and try to prove that it cannot describe the particle dynamics associated with the Dirac equation because $\alpha_i \equiv \sigma_i$, i.e., Hermitian matrices with zero trace and eigenvalues ± 1 satisfying $\{\sigma_i, \sigma_j\} = 2\delta_{ij}$.

Let us write a generic matrix using $(\mathbb{1}, \boldsymbol{\sigma})$ as a basis

$$M = M_0\mathbb{1} + M_i\sigma_i. \tag{4.236}$$

Let us consider the anticommutator with the σ_i matrices:

$$\{M, \sigma_i\} = \{M_0\mathrm{I} + M_j\sigma_j, \sigma_i\} = M_0(\mathbb{1}\sigma_i + \sigma_i\mathbb{1}) + M_j(\sigma_j\sigma_i + \sigma_i\sigma_j) = M_0(\mathbb{1}\sigma_i + \sigma_i\mathbb{1}) + 2M_j\delta_{ij}. \tag{4.237}$$

If $i = j$, then $\{M, \sigma_i\} = M_0\sigma_j + 2M_j$; if $i \neq j$, then $\{M, \sigma_i\} = M_0 2\sigma_j$. We want both to be zero, which implies $M_0 = 0$. However, in this case, we would no longer have M independent of σ_i because it would be written as a linear combination of them. Therefore, we can conclude that β cannot exist independently and anticommutative for $N = 2$.

Let us see what happens when $N = 4$. In this case, α_i and β exist with the smaller dimension, i.e., 4×4 matrices which can be written in blocks of 2×2 matrices

$$\alpha_i = \begin{pmatrix} 0 & \sigma_i \\ \sigma_i & 0 \end{pmatrix}, \quad \beta = \begin{pmatrix} \mathrm{I} & 0 \\ 0 & -\mathrm{I} \end{pmatrix}. \tag{4.238}$$

In this representation, the γ matrices are

$$\gamma^0 = \begin{pmatrix} \mathrm{I} & 0 \\ 0 & -\mathrm{I} \end{pmatrix}, \quad \boldsymbol{\gamma} = \beta \cdot \boldsymbol{\alpha} = \begin{pmatrix} 0 & \boldsymbol{\sigma} \\ -\boldsymbol{\sigma} & 0 \end{pmatrix}. \tag{4.239}$$

This nomenclature is called *Dirac or Pauli representation.*

Given γ^μ as one representation and $\tilde{\gamma}^\mu$ as another representation, one can always obtain one from the other through a unitary matrix that is an isometry of spacetime:

$$\tilde{\gamma}^\mu = S^{-1}\gamma^\mu S, \tag{4.240}$$

where $\tilde{\gamma}^\mu$ satisfies the same anticommutation rule as γ^μ:

$$\begin{aligned}\{\tilde{\gamma}^\mu, \tilde{\gamma}^\nu\} &= \tilde{\gamma}^\mu\tilde{\gamma}^\nu + \tilde{\gamma}^\nu\tilde{\gamma}^\mu = S^{-1}\gamma^\mu S S^{-1}\gamma^\nu S + S^{-1}\gamma^\nu S S^{-1}\gamma^\mu S = \\ &= S^{-1}[\gamma^\mu\gamma^\nu + \gamma^\nu\gamma^\mu]S = S^{-1}\{\gamma^\mu, \gamma^\nu\}S = S^{-1}\{2g^{\mu\nu}\}S = 2g^{\mu\nu}.\end{aligned} \tag{4.241}$$

The Hamiltonian $\boldsymbol{\alpha} \cdot \mathbf{p} + \beta m$ does not commute with the angular momentum along $\hat{z}$. For instance, if we consider $L_z = x_1 p_2 - x_2 p_1$,

$$[\hat{H}, L_z] = [\alpha_i p_i + \beta m, x_1 p_2 - x_2 p_1] = -i(\alpha_1 p_2 - \alpha_2 p_1). \tag{4.242}$$

Defining $\Sigma_z = \frac{i}{4}(\gamma_1\gamma_2 - \gamma_2\gamma_1) = -\frac{i}{4}(\alpha_1\alpha_2 - \alpha_2\alpha_1)$, where we have used the fact that $\gamma_1\gamma_2 = \beta\alpha_1\beta\alpha_2 = -\alpha_1\beta\beta\alpha_2 = -\alpha_1\alpha_2$, we notice that

$$\begin{aligned}[\hat{H}, \Sigma_z] &= \frac{i}{4}[\alpha_i p_i + \beta m, \alpha_2\alpha_1 - \alpha_1\alpha_2] = \frac{i}{4}\{[\alpha_i p_i, \alpha_2\alpha_1] - [\alpha_i p_i, \alpha_1\alpha_2]\} = \\ &= \frac{i}{4}\{[\alpha_i, \alpha_2]\alpha_1 p_i + \alpha_2[\alpha_i, \alpha_1]p_i - \alpha_1[\alpha_i, \alpha_2]p_i - [\alpha_i, \alpha_1]\alpha_2 p_i\}.\end{aligned} \tag{4.243}$$

Using the fact that $\{\alpha_i, \alpha_j\} = 2\delta_{ij}$ and, therefore,

$$\alpha_i\alpha_j = 2\delta_{ij} - \alpha_j\alpha_i \longrightarrow \alpha_i\alpha_j - \alpha_j\alpha_i = 2\delta_{ij} - 2\alpha_j\alpha_i$$
$$\longrightarrow [\alpha_i, \alpha_j] = 2(\delta_{ij} - \alpha_j\alpha_i) = \begin{cases} 0, & \text{if}\, i = j, \\ -2\alpha_2\alpha_1, & \text{if}\, i = \alpha_1,\ j = \alpha_2, \end{cases} \tag{4.244}$$

and also

$$\alpha_i\alpha_j + \alpha_j\alpha_i = 2\delta_{ij}, \quad \text{if}\, i \neq j, \tag{4.245a}$$
$$\alpha_i\alpha_j\alpha_i = -\alpha_j. \tag{4.245b}$$

Equation (4.243) becomes

$$[\hat{H}, \Sigma_z] = \frac{i}{4}\{-2\alpha_2\alpha_1^2 p_1 + (-2\alpha_2\alpha_1\alpha_2 p_2) - (-2\alpha_1\alpha_2\alpha_1 p_1) - (-2\alpha_2^2\alpha_1 p_2)\} =$$
$$= \frac{i}{4}(-2\alpha_2 p_1 - 2\alpha_2 p_1 + 2\alpha_1 p_2 + 2\alpha_1 p_2) = -i(\alpha_2 p_1 - \alpha_1 p_2) = -[\hat{H}, L_z]. \tag{4.246}$$

The last expression implies that

$$[\hat{H}, L_z + \Sigma_z] = 0. \tag{4.247}$$

It becomes natural to identify Σ_z with the spin along z, and indeed

$$\Sigma_z = \frac{i}{4}(\gamma_1\gamma_2 - \gamma_2\gamma_1) = \frac{i}{4}\left[\begin{pmatrix} 0 & \sigma_1 \\ -\sigma_1 & 0 \end{pmatrix}\begin{pmatrix} 0 & \sigma_2 \\ -\sigma_2 & 0 \end{pmatrix} - \begin{pmatrix} 0 & \sigma_2 \\ -\sigma_2 & 0 \end{pmatrix}\begin{pmatrix} 0 & \sigma_1 \\ -\sigma_1 & 0 \end{pmatrix}\right] =$$
$$\frac{i}{4}\left[\begin{pmatrix} -\sigma_1\sigma_2 & 0 \\ 0 & -\sigma_1\sigma_2 \end{pmatrix} - \begin{pmatrix} -\sigma_2\sigma_1 & 0 \\ 0 & -\sigma_2\sigma_1 \end{pmatrix}\right] = \frac{i}{4}\left[\begin{pmatrix} \sigma_2\sigma_1 - \sigma_1\sigma_2 & 0 \\ 0 & \sigma_2\sigma_1 - \sigma_1\sigma_2 \end{pmatrix}\right] =$$
$$\frac{i}{4}\left[\begin{pmatrix} -2i\sigma_3 & 0 \\ 0 & -2i\sigma_3 \end{pmatrix}\right] = \frac{1}{2}\begin{pmatrix} \sigma_3 & 0 \\ 0 & \sigma_3 \end{pmatrix} = \Sigma_3. \tag{4.248}$$

It will be seen that the solutions are eigenstates of Σ_3 with eigenvalues $\pm\frac{1}{2}$ if $\mathbf{p} = p\hat{z}$. This can also be concluded by observing that $\hat{H}$ and $\hat{J} = \hat{\mathbf{L}} + \hat{\Sigma}$ commute. Therefore, in the reference frame where the particle has $\hat{\mathbf{L}} = 0$, the eigenstates of $\hat{H}$ are eigenstates of $\hat{\boldsymbol{\sigma}} = \frac{1}{2}\begin{pmatrix} \boldsymbol{\sigma} & 0 \\ 0 & \boldsymbol{\sigma} \end{pmatrix}$, which is diagonal in the Pauli representation. Furthermore, as Σ is the spin operator, the Dirac equation describes particles with spin $1/2$.

Finally, the quantity related to spin that always commutes with $\hat{H}$ is the helicity $\hat{h} = \frac{(\hat{\Sigma}\hat{\mathbf{p}})}{|\mathbf{p}|}$, which is the spin projection along the direction of $\mathbf{p}$. In this case, the state is not an eigenstate of Σ_3 but of $\hat{h}$.

▶ **Pause and think** Report the main differences between the Klein-Gordon and Dirac equations.

4.4.5 Relativistic Invariance of the Dirac Equation

We want to impose that if ψ satisfies the Dirac equation in a reference frame O, then ψ' must satisfy the Dirac equation in the reference frame O'. Recall that the transformation of coordinates in spacetime is given by:

$$x'^{\mu} = \Lambda^{\mu}_{\nu} x^{\nu}, \tag{4.249}$$

so the state might transform linearly as

$$\psi'(x) = S(\Lambda)\psi(x), \tag{4.250}$$

where $S(\Lambda)$ is a representation of the Lorentz group that a priori we do not know.

We know that it will be in the form $S(\Lambda) = e^{\frac{1}{2}\omega_{\rho\sigma}F^{\rho\sigma}}$ and that it preserves the operations $S(\Lambda_1, \Lambda_2) = S(\Lambda_1)S(\Lambda_2)$.

Let us write the Dirac equation in terms of γ in a way that preserves the symmetry of spatial and temporal derivatives. In the reference frame O, we have

$$0 = (i\gamma^{\mu}\partial_{\mu} - m)\psi(x), \tag{4.251}$$

where $\psi(x) = S^{-1}(\Lambda)\psi'(x')$.

Writing $\partial_{\mu} = \frac{\partial}{\partial x^{\mu}} = \frac{\partial}{\partial x'^{\nu}}\frac{\partial x'^{\nu}}{\partial x^{\mu}} = \Lambda^{\nu}_{\mu}\partial'_{\nu}$, we substitute it into the Dirac equation and obtain

$$0 = (i\gamma^{\mu}\Lambda^{\nu}_{\mu}\partial'_{\nu} - m)S^{-1}(\Lambda)\psi'(x') \longrightarrow 0 = (iS(\Lambda)\gamma^{\mu}S^{-1}(\Lambda)\Lambda^{\nu}_{\mu}\partial'_{\nu} - m)\psi'(x), \tag{4.252}$$

where we have multiplied by S. Next, we impose that it must be equal to the Dirac equation in the reference frame O', that is

$$0 = (i\tilde{\gamma}^{\mu}\partial'_{\mu} - m)\psi'(x). \tag{4.253}$$

By imposing equality between Eqs. (4.253) and (4.252), we obtain the transformation law for γ.

$$\tilde{\gamma}^{\mu} = S(\Lambda)\gamma^{\nu}S^{-1}(\Lambda)\Lambda^{\mu}_{\nu}. \tag{4.254}$$

Since $\tilde{\gamma}^{\mu} = \gamma^{\mu}$, this implies

$$\gamma^{\mu} = \Lambda^{\mu}_{\nu}S(\Lambda)\gamma^{\nu}S^{-1}(\Lambda) \longrightarrow S^{-1}(\Lambda)\gamma^{\mu}S(\Lambda) = \Lambda^{\mu}_{\nu}\gamma^{\nu}, \tag{4.255}$$

where we have multiplied on the right and left by S^{-1}.

At this point, we need to find the form of S. First, let us write both S and Λ in infinitesimal form

$$\Lambda^{\mu}_{\nu} = \delta^{\mu}_{\nu} + \omega^{\mu}_{\nu}, \quad \text{with } |\omega^{\mu}_{\nu}| \ll 1 \text{ and } \omega_{\mu\nu} = -\omega_{\nu\mu}, \tag{4.256}$$

derived in 4.1.3.

▶ **Driving exercise** Now write $S(\Lambda)$ and $S^{-1}(\Lambda)$ at first order and substitute your outcomes inside Eq. (4.255). What would you expect?
We have:

$$S(\Lambda) \simeq \mathbb{1} + \frac{1}{2}\omega_{\alpha\beta}F^{\alpha\beta}, \tag{4.257a}$$

$$S^{-1}(\Lambda) \simeq \mathbb{1} - \frac{1}{2}\text{!}_{\text{fi}}\text{F}^{\text{fi}}. \tag{4.257b}$$

Substitute it inside Eq. (4.255) to obtain

$$\left(\mathbb{1} - \frac{1}{2}\omega_{\alpha\beta}F^{\alpha\beta}\right)\gamma^{\mu}\left(\mathbb{1} + \frac{1}{2}\omega_{\alpha\beta}F^{\alpha\beta}\right) = (\delta^{\mu}_{\nu} + \omega^{\mu}_{\nu})\gamma^{\nu}, \tag{4.258}$$

where we write $\omega^{\mu}_{\nu} = g^{\mu\alpha}\delta^{\beta}_{\nu}\omega_{\alpha\beta}$, equivalent to $g^{\mu\beta}\delta^{\alpha}_{\nu}\omega_{\beta\alpha} = -g^{\mu\beta}\delta^{\alpha}_{\nu}\omega_{\alpha\beta}$ due to antisymmetry. Therefore, we write

$$\omega^{\mu}_{\nu} = \frac{1}{2}\omega_{\alpha\beta}(g^{\mu\alpha}\delta^{\beta}_{\nu} - g^{\mu\beta}\delta^{\alpha}_{\nu}). \tag{4.259}$$

This way, ω^{μ}_{ν} is a linear combination of antisymmetric matrices $(A^{\alpha\beta})^{\mu}_{\nu} = g^{\mu\alpha}\delta^{\beta}_{\nu} - g^{\mu\beta}\delta^{\alpha}_{\nu}$. Substituting Eq. (4.259) into Eq. (4.258), we obtain

$$\left(\mathbb{1} - \frac{\omega_{\alpha\beta}}{2}F^{\alpha\beta}\right)\gamma^{\mu}\left(\mathbb{1} + \frac{\omega_{\alpha\beta}}{2}F^{\alpha\beta}\right) = \left[\delta^{\mu}_{\nu} + \frac{\omega_{\alpha\beta}}{2}(g^{\mu\alpha}\delta^{\beta}_{\nu} - g^{\mu\beta}\delta^{\alpha}_{\nu})\right] \tag{4.260}$$

that we rewrite as

$$\gamma^\mu - \frac{\omega_{\alpha\beta}}{2}F^{\alpha\beta}\gamma^\mu + \gamma^\mu\frac{\omega_{\alpha\beta}}{2}F^{\alpha\beta} + O(\omega^2) = \gamma^\mu + \frac{\omega_{\alpha\beta}}{2}(g^{\mu\alpha}\delta^\beta_\nu - g^{\mu\beta}\delta^\alpha_\nu). \tag{4.261}$$

Collecting,

$$\frac{\omega_{\alpha\beta}}{2}\left[\gamma^\mu F^{\alpha\beta} - F^{\alpha\beta}\gamma^\mu\right] = \frac{\omega_{\alpha\beta}}{2}(g^{\mu\alpha}\gamma^\beta - g^{\mu\beta}\gamma^\alpha) \longrightarrow [\gamma^\mu, F^{\alpha\beta}] = g^{\mu\alpha}\gamma^\beta - g^{\mu\beta}\gamma^\alpha. \tag{4.262}$$

Finding S means to obtain the form of $F^{\alpha\beta}$, which are six independent elements due to antisymmetry. Let us try to derive the form of $F^{\alpha\beta}$ with the antisymmetric product of γ given by the commutator $[\gamma^\alpha, \gamma^\beta]$, which appears in the definition of the antisymmetric tensor $\sigma^{\alpha\beta} = \frac{i}{2}[\gamma^\alpha, \gamma^\beta]$. Write

$$F^{\alpha\beta} = \frac{1}{4}[\gamma^\alpha, \gamma^\beta] = -\frac{i}{2}\sigma^{\alpha\beta}. \tag{4.263}$$

Substitute the previous expression into Eq. (4.262)

$$\left[\gamma^\mu, \frac{[\gamma^\alpha, \gamma^\beta]}{4}\right] = \frac{1}{4}[\gamma^\mu, \gamma^\alpha\gamma^\beta - \gamma^\beta\gamma^\alpha] = \frac{1}{4}\{\gamma^\mu\gamma^\alpha\gamma^\beta - \gamma^\mu\gamma^\beta\gamma^\alpha - \gamma^\alpha\gamma^\beta\gamma^\mu + \gamma^\beta\gamma^\alpha\gamma^\mu\}. \tag{4.264}$$

Using $\{\gamma^\mu, \gamma^\alpha\} = 2g^{\mu\alpha}$, we obtain

$$\begin{aligned}
&\frac{1}{4}\{\gamma^\mu\gamma^\alpha\gamma^\beta + \gamma^\beta\gamma^\alpha\gamma^\mu - \gamma^\mu\gamma^\beta\gamma^\alpha - \gamma^\alpha\gamma^\beta\gamma^\mu\} \\
&= \frac{1}{4}\{2g^{\mu\alpha}\gamma^\beta - \gamma^\alpha\gamma^\mu\gamma^\beta + \gamma^\beta\gamma^\alpha\gamma^\mu - 2g^{\mu\beta}\gamma^\alpha - \gamma^\beta\gamma^\mu\gamma^\alpha - \gamma^\alpha\gamma^\beta\gamma^\mu\} \\
&= \frac{1}{4}\{2g^{\mu\alpha}\gamma^\beta + 2g^{\mu\alpha}\gamma^\beta - \gamma^\alpha\gamma^\mu\gamma^\beta - \gamma^\beta\gamma^\mu\gamma^\alpha - 2g^{\mu\beta}\gamma^\alpha + \gamma^\beta\gamma^\mu\gamma^\alpha - 2g^{\mu\beta}\gamma^\alpha + \gamma^\alpha\gamma^\mu\gamma^\beta\} \\
&= \frac{1}{4}\{4g^{\mu\alpha}\gamma^\beta - 4g^{\mu\beta}\gamma^\alpha\} = g^{\mu\alpha}\gamma^\beta - g^{\mu\beta}\gamma^\alpha.
\end{aligned} \tag{4.265}$$

This way, we obtain Eq. (4.263). So the S we are looking for is

$$S = \mathbb{1} + \frac{1}{2}\omega_{\alpha\beta}F^{\alpha\beta} = \mathbb{1} - \frac{i}{4}\omega_{\alpha\beta}\sigma^{\alpha\beta}. \tag{4.266}$$

The last equality is the first-order Taylor expansion of $S(\Lambda) = e^{-\frac{i}{4}\omega_{\alpha\beta}\sigma^{\alpha\beta}}$. This corresponds to the representation of $L_+^\uparrow$ labeled by $(0, 1/2) \oplus (1/2, 0)$, which is reducible and obtained as the direct sum of the inequivalent irreducible representations according to which it transforms two-component spinor fields. Again, $(1/2, 0)$

and (0, 1/2) are the values of (j_+, j_-) eigenvalues of the raising and lowering operators of angular momentum. Moreover,

$$(0, 1/2) \longrightarrow \hat{J}_- = \frac{\boldsymbol{\sigma}}{2}, \hat{J}_+ = 0, \qquad (4.267a)$$

$$(1/2, 0) \longrightarrow \hat{J}_+ = \frac{\boldsymbol{\sigma}}{2}, \hat{J}_- = 0. \qquad (4.267b)$$

Furthermore, through parity, one can switch from one to the other; therefore, the direct sum of the two remains invariant under parity, and according to this, it transforms spinor fields with four components.

In this case, γ^5 is a Casimir because it commutes with all generators. Indeed,

$$\begin{aligned}[\gamma^5, \sigma^{\alpha\beta}] &= \left[\gamma^5, \frac{i}{2}[\gamma^\alpha, \gamma^\beta]\right] \\ &= \frac{i}{2}\{\gamma^5\gamma^\alpha\gamma^\beta - \gamma^5\gamma^\beta\gamma^\alpha - \gamma^\alpha\gamma^\beta\gamma^5 + \gamma^\beta\gamma^\alpha\gamma^5\} \\ &= \frac{i}{2}\{2\gamma^5 g^{\alpha\beta} - \gamma^5\gamma^\beta\gamma^\alpha - 2\gamma^5 g^{\alpha\beta} + \gamma^5\gamma^\alpha\gamma^\beta \\ &- 2g^{\alpha\beta}\gamma^5 + \gamma^\beta\gamma^\alpha\gamma^5 + 2g^{\alpha\beta}\gamma^5 - \gamma^\alpha\gamma^\beta\gamma^5\} \\ &= \frac{i}{2}\{\gamma^\beta\gamma^5\gamma^\alpha - \gamma^\alpha\gamma^5\gamma^\beta - \gamma^\beta\gamma^5\gamma^\alpha + \gamma^\alpha\gamma^5\gamma^\beta\} = 0, \end{aligned} \qquad (4.268)$$

where we used the condition $\{\gamma^5, \gamma^\mu\} = 0$. For the case $\mu = 0$,

$$\begin{aligned} i\{\gamma^0\gamma^1\gamma^2\gamma^3, \gamma^\mu\} &= i[\gamma^0\gamma^1\gamma^2\gamma^3\gamma^\mu + \gamma^\mu\gamma^0\gamma^1\gamma^2\gamma^3] = i[\gamma^0\gamma^1\gamma^2\gamma^3\gamma^0 + (\gamma^0)^2\gamma^1\gamma^2\gamma^3] = \\ &= i[-\gamma^1\gamma^0\gamma^2\gamma^3\gamma^0 + \gamma^1\gamma^2\gamma^3]. \end{aligned} \qquad (4.269)$$

In the case where $\mu \neq \nu$,

$$\{\gamma^\mu, \gamma^\nu\} = i[\gamma^1\gamma^2\gamma^0\gamma^3\gamma^0 + \gamma^1\gamma^2\gamma^3] = i[-\gamma^1\gamma^2\gamma^3(\gamma^0)^2 + \gamma^1\gamma^2\gamma^3] = 0. \qquad (4.270)$$

Now, let us examine the properties of S:

$$\gamma^0 S^\dagger \gamma^0 = \gamma^0(\mathbb{I} + \frac{i}{4}\gamma^{\alpha\beta^\dagger})\gamma^0 = \mathbb{I} + \frac{i}{4}\omega_{\alpha\beta}\gamma^0\sigma^{\alpha\beta^\dagger}\gamma^0. \qquad (4.271)$$

We use the condition

$$\gamma^0\sigma^{\alpha\beta}\gamma^0 = \gamma^0[\gamma^\alpha\gamma^\beta - \gamma^\beta\gamma^\alpha]\gamma^0 = \frac{\gamma^0\gamma^\alpha\gamma^\beta\gamma^0 - \gamma^0\gamma^\beta\gamma^\alpha\gamma^0}{2i} = \frac{\gamma^{\alpha^\dagger}\gamma^{\beta^\dagger} - \gamma^{\beta+}\gamma^{\alpha^\dagger}}{2i}$$
$$= \frac{(\gamma^\beta\gamma^\alpha)^\dagger - (\gamma^\alpha\gamma^\beta)^\dagger}{2i} = -\frac{[\gamma^\alpha, \gamma^\beta]^\dagger}{2i} = \sigma^{\alpha\beta^\dagger}. \tag{4.272}$$

Substituting the above expression into Eq. (4.271), we obtain

$$\gamma^0 S^\dagger \gamma^0 = \mathbb{1} + \frac{i}{4}\omega_{\alpha\beta}\sigma^{\alpha\beta} = S^{-1}. \tag{4.273}$$

This implies that S **is not** unitary, $S^{-1} \neq S^\dagger$.

4.5 Conserved Currents and Dirac Lagrangian

We want to construct scalars, i.e., objects that remain unchanged under $S(\Lambda)$. To do this, we introduce the *adjoint spinor*, which we obtain by considering

$$\psi^\dagger = \begin{pmatrix} \psi_1 \\ \vdots \\ \psi_N \end{pmatrix}^\dagger = \psi_1^\star \dots \psi_N^\star = (\psi^\star)^T, \tag{4.274}$$

where $(\psi^\star)^T$ is complex conjugate transposed. We notice that $\psi^\dagger\gamma^0 = \bar{\psi}$ is the adjonct spinor that transforms as

$$\bar{\psi}'(x') = (\psi'(x'))^\dagger\gamma^0 = (S\psi(x))^\dagger\gamma^0 = \psi^\dagger(x)S^\dagger\gamma^0 = \psi^\dagger(x)\gamma^0\gamma^0S^\dagger\gamma^0 = \psi^\dagger(x)\gamma^0S^{-1} = \bar{\psi}S^{-1}. \tag{4.275}$$

Let us now consider the two forms $\psi^\dagger\psi$ and $\bar{\psi}\psi$

- $\psi^\dagger\psi$ transform as
$$\psi'^\dagger\psi' = (S\psi)^\dagger(S\psi) = \psi^\dagger S^\dagger S\psi, \tag{4.276}$$
where $S^\dagger S \neq 1$, since S is not unitary. Moreover, $\psi^\dagger\psi$ **is not** a scalar;
- $\bar{\psi}\psi$, differently from the previous case, is a scalar and transforms as

$$\bar{\psi}'\psi' = (\psi'^\dagger\gamma^0)(S\psi) = (S\psi)^\dagger\gamma^0(S\psi) = \psi^\dagger S^\dagger\gamma^0 S\psi = \psi^\dagger\gamma^0\gamma^0S^\dagger\gamma^0S\psi =$$
$$= \psi^\dagger\gamma^0(S^{-1}S)\psi = \bar{\psi}\psi. \tag{4.277}$$

4.5.1 The Dirac Lagrangian

Thanks to the bilinear forms, it is possible to construct the Lagrangian density, which must be a scalar.

▶ **Pause and think** How would you create your suitable version of the Lagrangian for the Dirac field?

Moreover, we can provide the Lagrangian as $\psi^\dagger \longrightarrow \bar{\psi}$ because it is thanks to the adjoint spinor that it is possible to construct a scalar. Let us start by writing the Dirac equation as

$$[(i\gamma^\mu\partial_\mu - m)\psi]^\dagger = 0 \longrightarrow \psi^\dagger(-i\gamma^{\mu\dagger}\partial_\mu - m) = 0, \tag{4.278}$$

and recalling that $\gamma^{\mu\dagger} = \gamma^0\gamma^\mu\gamma^0$, we obtain

$$\psi^\dagger(i\gamma^0\gamma^\mu\gamma^0\partial_\mu + m) = \bar{\psi}(i\gamma^\mu\partial_\mu + m) = 0. \tag{4.279}$$

At this point, we write the Lagrangian in such a way that the Euler-Lagrange equations reproduce those of Dirac for ψ and $\bar{\psi}$

$$\mathcal{L} = \bar{\psi}(i\gamma^\mu\partial_\mu - m)\psi. \tag{4.280}$$

The previously written Lagrangian is linear in $\partial\!\!\!/ = \gamma^\mu\partial_\mu$, and $\bar{\psi} = \psi^\dagger\gamma^0 \neq \psi^\dagger$.

We recall that the Lagrangian for a complex field for the Klein-Gordon equation is

$$\mathcal{L} = (\partial_\mu\phi^\star)(\partial^\mu\phi) - m^2\phi^\star\phi. \tag{4.281}$$

It is quadratic in ∂_μ, contains the mass term m^2, and $\phi^\star$ appears.

Writing $\psi = \psi_L + \psi_R = \begin{pmatrix}\psi_L\\0\end{pmatrix} + \begin{pmatrix}0\\\psi_R\end{pmatrix}$, we obtain

$$\mathcal{L} = \bar{\psi}_L i\partial\!\!\!/\psi_L + \bar{\psi}_R i\partial\!\!\!/\psi_R - m(\bar{\psi}_L\psi_R + \bar{\psi}_R\psi_L), \tag{4.282}$$

which is not invariant under the exchange $\psi_R \longleftrightarrow \psi_L$ unless $m = 0$.

4.5.2 Conserved Current

Let us start by multiplying $i\frac{\partial}{\partial t}\psi = (-i\boldsymbol{\alpha}\cdot\nabla + \beta m)\psi$ by $\psi^\dagger$ and the Hermitian conjugate version $-i\frac{\partial}{\partial t}\psi^\dagger = \psi^\dagger(i\boldsymbol{\alpha}^\dagger\cdot\nabla + \beta^\dagger m)$ by ψ. Thus,

$$i\psi^\dagger\partial_t\psi = \psi^\dagger(-i\boldsymbol{\alpha}\cdot\nabla + \beta m)\psi, \tag{4.283a}$$

$$-i\psi\partial_t\psi^\dagger = \psi(i\boldsymbol{\alpha}\cdot\nabla + \beta m)\psi^\dagger. \tag{4.283b}$$

Subtracting these equations from each other, we obtain

$$i[\psi^\dagger\partial_t\psi + \psi\partial_t\psi^\dagger] = \psi^\dagger(-i\boldsymbol{\alpha}\cdot\nabla)\psi - \psi(i\boldsymbol{\alpha}\cdot\nabla)\psi^\dagger \rightarrow \frac{\partial}{\partial_t}(\psi^\dagger\psi) = -[\psi^\dagger\boldsymbol{\alpha}\cdot\nabla\psi + \psi\boldsymbol{\alpha}\cdot\nabla\psi^\dagger]. \tag{4.284}$$

In the end, we get

$$\frac{\partial}{\partial_t}(\psi^\dagger\psi) + \nabla[\psi^\dagger\boldsymbol{\alpha}\psi] = 0. \tag{4.285}$$

Now, let us define $\rho = \psi^\dagger\psi$ and $\mathbf{J} = \psi^\dagger\boldsymbol{\alpha}\psi$. Substituting them into the previous equation gives us the continuity equation. ρ and $\mathbf{J}$ form the spatial and temporal components of a four-vector. In particular,

$$J^\mu = (\psi^\dagger\psi, \psi^\dagger\boldsymbol{\alpha}\psi). \tag{4.286}$$

This can be better understood by writing

$$J^0 = \psi^\dagger(\gamma^0)^2\psi = \bar{\psi}\gamma^0\psi, \tag{4.287a}$$

$$J^i = \psi^\dagger(\gamma^0)^2\alpha_i\psi = \bar{\psi}\gamma^i\psi. \tag{4.287b}$$

Thus, $J^\mu = \bar{\psi}\gamma^\mu\psi$ is the conserved current under the $U(1)$ symmetry. It can be shown that it transforms as an event, turning into

$$J'^\mu(x') = \bar{\psi}'(x')\gamma^\mu\psi'(x') = \bar{\psi}(x)S^{-1}\gamma^\mu S\psi(x) = \bar{\psi}\Lambda^\mu_\nu\gamma^\nu\psi = \Lambda^\mu_\nu J^\nu. \tag{4.288}$$

Moreover, it satisfies the continuity equation $\partial_\mu J^\mu = 0$ written in covariant form.

This is further evidence that $\psi^\dagger\psi$ cannot be a scalar because it is the temporal component of a four-vector.

4.5.3 Bilinear Covariant Forms

Starting from the γ matrices, we can consider antisymmetric products of them, leading to 16 linearly independent matrices, forming the *Dirac algebra*. We reach the total of 16 since γ_μ has four components, $\sigma_{\mu\nu}$ has six, γ_5 has one, and $\gamma_\mu\gamma_5$ has four, making their total sum 16. Compactly writing them as $\{\mathbb{I}, \gamma_\mu, \sigma_{\mu\nu}, \gamma_5, \gamma_\mu\gamma_5\}$, we can express them as

$$\Gamma^S = \mathbb{I}, \quad \Gamma^V_\mu = \gamma_\mu, \quad \Gamma^T_{\mu\nu} = \sigma_{\mu\nu}, \quad \Gamma^P = \gamma_5 = \gamma^5 = i\gamma^0\gamma^1\gamma^2\gamma^3, \quad \Gamma^A_\mu = \gamma_\mu\gamma_5, \tag{4.289}$$

where $\Gamma = \gamma^0\Gamma^\dagger\gamma^0$ holds for all except $\gamma_5 = \gamma_5^\dagger$, which is hermitian.

The bilinear forms obtained transform according to particular objects, allowing us to write any observable as $\bar{\psi}\Gamma\psi$. In our case, we have

1. $\bar{\psi}\Gamma^S\psi = \bar{\psi}\psi$ transforms as a scalar;
2. $\bar{\psi}\Gamma^\mu_\nu\psi = \bar{\psi}\gamma^\mu\psi$ transforms as a vector

$$\bar{\psi}'\gamma^\mu\psi' = \bar{\psi}(S^{-1}\gamma^\mu S)\psi = \bar{\psi}\Lambda^\mu_\nu\gamma^\nu\psi = \Lambda^\mu_\nu\bar{\psi}\gamma^\nu\psi; \tag{4.290}$$

3. $\bar{\psi}\Gamma^{\mu\nu}_T\psi = \bar{\psi}\sigma^{\mu\nu}\psi$ transforms as a tensor

$$\begin{aligned}
&\bar{\psi}'\sigma^{\mu\nu}\psi' = \bar{\psi}(S^{-1}\sigma^{\mu\nu}S)\psi = \bar{\psi}\left(S^{-1}\frac{[\gamma^\mu\gamma^\nu - \gamma^\nu\gamma^\mu]}{2i}S\right)\psi = \\
&= \frac{1}{2i}\bar{\psi}(S^{-1}\gamma^\mu SS{-}1\gamma^\nu S - S^{-1}\gamma^\nu SS^{-1}\gamma^\mu S) = \\
&= \frac{1}{2i}\bar{\psi}(\Lambda^\mu_\alpha\gamma^\alpha\Lambda^\nu_\beta\gamma^\beta - \Lambda^\nu_\beta\gamma^\beta\Lambda^\mu_\alpha\gamma^\alpha) = \bar{\psi}\Lambda^\mu_\alpha\Lambda^\nu_\beta\frac{[\gamma^\alpha, \gamma^\beta]}{2i}\psi = \\
&\Lambda^\mu_\alpha\Lambda^\nu_\beta\bar{\psi}\sigma^{\alpha\beta}\psi.
\end{aligned} \tag{4.291}$$

4. $\bar{\psi}\Lambda^P\psi = \bar{\psi}\gamma_5\psi$ transforms as a pseudoscalar

$$\bar{\psi}'\gamma_t\psi' = \bar{\psi}(S^{-1}\gamma_5 S)\psi = \det(\Lambda)\bar{\psi}\gamma_5\psi, \tag{4.292}$$

where the last equality is provided once considering $\gamma^5 = \frac{i}{4!}\varepsilon_{\mu\nu\alpha\beta}\gamma^\mu\gamma^\nu\gamma^\alpha\gamma^\beta$ so that

$$S^{-1}\gamma_5 S = \frac{i}{4!}\varepsilon_{\mu\nu\alpha\beta}S^{-1}\gamma^\mu\gamma^\nu\gamma^\alpha\gamma^\beta S = \frac{i}{4!}\varepsilon_{\mu\nu\alpha\beta}\Lambda^\mu_\rho\Lambda^\nu_\sigma\Lambda^\alpha_\lambda\Lambda^\beta_\tau\gamma^\rho\gamma^\sigma\gamma^\lambda\gamma^\tau. \tag{4.293}$$

Further, recalling that $\varepsilon_{\mu\nu\alpha\beta} = \det(\Lambda)\Lambda^\rho_\mu\Lambda^\sigma_\nu\Lambda^\lambda_\alpha\Lambda^\tau_\beta\varepsilon_{\rho\sigma\lambda\tau}$ since it is a pseudotensor with all covariant indexes, then it is a covariant tensor if $\det(\Lambda) = 1$ and changes sign under parity, $\det(\Lambda) = -1$. Multiplying for the inverses we obtain

$$\begin{aligned}
\varepsilon_{\mu\nu\alpha\beta}\Lambda^\mu_\xi\Lambda^\nu_\eta\Lambda^\alpha_\gamma\Lambda^\beta_\delta &= \det(\Lambda)\Lambda^\rho_\mu\Lambda^\mu_\xi\Lambda^\sigma_\nu\Lambda^\nu_\eta\Lambda^\lambda_\alpha\Lambda^\alpha_\gamma\Lambda^\tau_\beta\Lambda^\beta_\delta\varepsilon_{\rho\sigma\lambda\tau} = \\
&= \det(\Lambda)\delta^\rho_\xi\delta^\sigma_\eta\delta^\lambda_\gamma\delta^\tau_\delta\varepsilon_{\rho\sigma\lambda\tau} = \det(\Lambda)\varepsilon_{\xi\eta\gamma\delta}.
\end{aligned} \tag{4.294}$$

In the end, we end up with

$$S^{-1}\gamma_5 S = \frac{i}{4!}\det(\Lambda)\varepsilon_{\rho\sigma\lambda\tau}\gamma^\rho\gamma^\sigma\gamma^\lambda\gamma^\tau = \det(\Lambda)\gamma_5. \tag{4.295}$$

5. $\bar{\psi}\Gamma^\mu_A\psi = \bar{\psi}\gamma^\mu\gamma^5\psi$ transforms as a pseudovector or axial vector. It transforms as a vector if $\det(\Lambda) = 1$ so that $A'^\mu = \Lambda^\mu_\nu A^\nu$ while it changes behavior under parity, viz. $A'^\mu = -\Lambda^\mu_\nu A^\nu$.

4.5.4 Solutions of the Dirac Equations

Let us consider the case of a free particle at rest ($\mathbf{p} = 0$). We know that if ψ satisfies the Dirac equation, then each of its components satisfies the Klein-Gordon equation. Therefore, it is natural to write a solution for a free particle in the form of a plane wave:

$$\psi \sim u(\mathbf{p})e^{-ip_\mu x^\mu} \sim u(\mathbf{p})e^{-i(Et-\mathbf{p}\cdot\mathbf{x})}, \tag{4.296}$$

where $u(\mathbf{p})$ is a four-component spinor.

For a particle at rest, the Dirac equation becomes

$$(\gamma^0\hat{p}^0 - \boldsymbol{\gamma}\cdot\hat{\mathbf{p}} - m)\psi = 0 \longrightarrow \gamma^0\hat{p}^0\psi = m\psi \quad \text{or} \quad i\frac{\partial}{\partial_t}\psi = \beta m\psi. \tag{4.297}$$

Substituting $\psi = u(\mathbf{p})e^{-iEt}$ gives an equation for $u(\mathbf{p})$:

$$i\gamma^0(-iE)u(\mathbf{p})e^{-iEt} = mu(\mathbf{p})e^{-iEt} \longrightarrow \gamma^0 Eu(\mathbf{p}) = mu(\mathbf{p}). \tag{4.298}$$

We have obtained an eigenvalue equation for $\hat{E}$. Now, for $E = \pm m$, which are the eigenvalues of the energy for the case $\mathbf{p} = 0$, we get:

$$\pm\gamma^0 u(\mathbf{p}) = u(\mathbf{p}) \longrightarrow \begin{pmatrix} 1 & 0 & 0 & 0 \\ 0 & 1 & 0 & 0 \\ 0 & 0 & -1 & 0 \\ 0 & 0 & 0 & -1 \end{pmatrix}\begin{pmatrix} u_1 \\ u_2 \\ u_3 \\ u_4 \end{pmatrix} = \pm\begin{pmatrix} u_1 \\ u_2 \\ u_3 \\ u_4 \end{pmatrix}, \tag{4.299}$$

with γ^0 in the *Dirac representation* equal to

$$\begin{pmatrix} \mathbb{1} & 0 \\ 0 & -\mathbb{1} \end{pmatrix}. \tag{4.300}$$

Let us separately consider the two cases $E = \pm m$.

1. If $E = m > 0$ then we have that

$$\begin{pmatrix} \mathbb{1} & 0 \\ 0 & -\mathbb{1} \end{pmatrix} \begin{pmatrix} u_1 \\ u_2 \\ u_3 \\ u_4 \end{pmatrix} = \begin{pmatrix} u_1 \\ u_2 \\ u_3 \\ u_4 \end{pmatrix}, \tag{4.301}$$

and clearly $u_3 = u_4 = 0$. We thus select two spinors with eigenstates under the form

$$\begin{pmatrix} 1 \\ 0 \\ 0 \\ 0 \end{pmatrix} \quad \text{and} \quad \begin{pmatrix} 0 \\ 1 \\ 0 \\ 0 \end{pmatrix}. \tag{4.302}$$

These spinors are called *substate spinors*.
2. If $E = -m < 0$ then we obtain that

$$\begin{pmatrix} \mathbb{1} & 0 \\ 0 & -\mathbb{1} \end{pmatrix} \begin{pmatrix} u_1 \\ u_2 \\ u_3 \\ u_4 \end{pmatrix} = - \begin{pmatrix} u_1 \\ u_2 \\ u_3 \\ u_4 \end{pmatrix}. \tag{4.303}$$

Then $u_1 = u_2 = 0$ and as spinors are taken

$$\begin{pmatrix} 0 \\ 0 \\ 1 \\ 0 \end{pmatrix} \quad \text{and} \quad \begin{pmatrix} 0 \\ 0 \\ 0 \\ 1 \end{pmatrix}, \tag{4.304}$$

called *superstate spinors*.

The four independent solutions are

$$\begin{pmatrix} 1 \\ 0 \\ 0 \\ 0 \end{pmatrix} e^{-imt}, \quad \begin{pmatrix} 0 \\ 1 \\ 0 \\ 0 \end{pmatrix} e^{-imt}, \quad \begin{pmatrix} 0 \\ 0 \\ 1 \\ 0 \end{pmatrix} e^{imt}, \quad \begin{pmatrix} 0 \\ 0 \\ 0 \\ 1 \end{pmatrix} e^{imt}. \tag{4.305}$$

The first two are solutions with positive energy and correspond to states of particles with spin $1/2$, while the last two are solutions with negative energy and represent states of antiparticles. It can be shown that the two-component spinors in the upper part are eigenstates of the σ_3 operator with eigenvalues ± 1. This implies that $\begin{pmatrix} 1 \\ 0 \\ 0 \\ 0 \end{pmatrix}$

and $\begin{pmatrix}0\\1\\0\\0\end{pmatrix}$ are eigenstates of $\Sigma_3 = \frac{1}{2}\begin{pmatrix}\sigma_3 & 0\\ 0 & \sigma_3\end{pmatrix}$ with eigenvalues $\pm 1/2$. Let us prove the latter proposition by

$$\sigma_3 \begin{pmatrix}1\\0\end{pmatrix} = \begin{pmatrix}1 & 0\\ 0 & -1\end{pmatrix}\begin{pmatrix}1\\0\end{pmatrix} = \begin{pmatrix}1\\0\end{pmatrix}, \tag{4.306a}$$

$$\sigma_3 \begin{pmatrix}0\\1\end{pmatrix} = \begin{pmatrix}1 & 0\\ 0 & -1\end{pmatrix}\begin{pmatrix}0\\1\end{pmatrix} = \begin{pmatrix}0\\-1\end{pmatrix} = -\begin{pmatrix}0\\1\end{pmatrix}. \tag{4.306b}$$

This implies that for each energy value $E \gtrless 0$, there are two spin states. That's why a 4-component spinor is needed to account for both two energy states and two spin states for a given $\mathbf{p}$.

Now let us examine the *plane wave solutions for* $\mathbf{p} \neq 0$. In this case, it is convenient to split $u(\mathbf{p})$ into two components u_A and u_B, i.e., $u(\mathbf{p}) = \begin{pmatrix}u_A\\u_B\end{pmatrix}$. Inserting $\psi = u(\mathbf{p})e^{-i(Et-\mathbf{p}\cdot\mathbf{x})}$ into the Dirac equation

$$(\gamma^0 \hat{p}^0 - \boldsymbol{\gamma}\hat{\vec{p}} - m)\psi = 0 \longrightarrow (i\gamma^0 \frac{\partial}{\partial_t} + i\boldsymbol{\gamma}\nabla - m)\psi = 0 \tag{4.307}$$

we obtain

$$(i\gamma^0(-iE)u(\mathbf{p}) + i\boldsymbol{\gamma}(i\mathbf{p})u(\mathbf{p})) = mu(\mathbf{p}) \longrightarrow (\gamma^0 E - \boldsymbol{\gamma}\mathbf{p} - m)u(\mathbf{p}) = 0. \tag{4.308}$$

Using the Dirac representation we write

$$\gamma^0 = \begin{pmatrix}\mathbb{1} & 0\\ 0 & -\mathbb{1}\end{pmatrix}, \quad \boldsymbol{\gamma} = \begin{pmatrix}0 & \boldsymbol{\sigma}\\ -\boldsymbol{\sigma} & 0\end{pmatrix} \longrightarrow \begin{pmatrix}E-m & -\boldsymbol{\sigma}\cdot\mathbf{p}\\ \boldsymbol{\sigma}\cdot\mathbf{p} & -E-m\end{pmatrix}\begin{pmatrix}u_A\\u_B\end{pmatrix} = 0, \tag{4.309}$$

with $E = \pm\sqrt{\mathbf{p}^2 + m^2}$. We arrive at two coupled equations

$$(E-m)u_A = (\boldsymbol{\sigma}\cdot\mathbf{p})u_B \longrightarrow u_A = \frac{\boldsymbol{\sigma}\cdot\mathbf{p}}{E-m}u_B, \tag{4.310a}$$

$$(\boldsymbol{\sigma}\cdot\mathbf{p})u_A = (E+m)u_B \longrightarrow u_B = \frac{\boldsymbol{\sigma}\cdot\mathbf{p}}{E+m}u_A. \tag{4.310b}$$

Let us distinguish the two cases for $E \gtrless 0$.

1. In the case where $E > 0$, we make an ansatz for the upper part, assuming that for u_A one of the two 2-component spinors is

$$\chi_1 = \begin{pmatrix}1\\0\end{pmatrix}, \quad \chi_2 = \begin{pmatrix}0\\1\end{pmatrix}. \tag{4.311}$$

We search for u_B by writing

$$\boldsymbol{\sigma}\cdot\mathbf{p}=\sigma_1 p_1+\sigma_2 p_2+\sigma_3 p_3=\begin{pmatrix}0&1\\1&0\end{pmatrix}p_1+\begin{pmatrix}0&-i\\i&0\end{pmatrix}p_2+\begin{pmatrix}1&0\\0&-1\end{pmatrix}p_3=\begin{pmatrix}p_3&p_1-ip_2\\p_1+ip_2&-p_3\end{pmatrix}. \tag{4.312}$$

We obtain two independent solutions

$$u_B^{(1)}=\frac{\boldsymbol{\sigma}\cdot\mathbf{p}}{E+m}\chi_1=\frac{1}{E+m}\begin{pmatrix}p_3&p_1-ip_2\\p_1+ip_2&-p_3\end{pmatrix}\begin{pmatrix}1\\0\end{pmatrix}=\begin{pmatrix}p_3/(E+m)\\(p_1+ip_2)/(E+m)\end{pmatrix}, \tag{4.313a}$$

$$u_B^{(2)}=\frac{\boldsymbol{\sigma}\cdot\mathbf{p}}{E+m}\chi_2=\begin{pmatrix}(p_1-ip_2)/(E+m)\\-p_3/(E+m)\end{pmatrix}. \tag{4.313b}$$

This implies that

$$u_+^{(1)}=N\begin{pmatrix}1\\0\\p_3/(E_p+m)\\(p_1+ip_2)/(E_p+m)\end{pmatrix},\quad u_+^{(2)}=N\begin{pmatrix}0\\1\\(p_1-ip_2)/(E_p+m)\\-p_3/(E_p+m)\end{pmatrix}, \tag{4.314}$$

where N denotes normalization, as usual.

2. In the case where $E<0$, we make an ansatz for the lower part, assuming for u_B one of the two spinors

$$\chi_1=\begin{pmatrix}1\\0\end{pmatrix},\quad \chi_2=\begin{pmatrix}0\\1\end{pmatrix}, \tag{4.315}$$

and we obtain

$$u_A^{(1)}=\frac{\boldsymbol{\sigma}\cdot\mathbf{p}}{E-m}\chi_1=\begin{pmatrix}p_3/(E-m)\\(p_1+ip_2)/(E-m)\end{pmatrix}, \tag{4.316}$$

$$u_A^{(2)}=\frac{\boldsymbol{\sigma}\cdot\mathbf{p}}{E-m}\chi_2=\begin{pmatrix}(p_1-ip_2)/(E-m)\\-p_3/(E-m)\end{pmatrix}. \tag{4.317}$$

We can then write, considering $E=-E_p<0$,

$$u_-^{(1)}=N\begin{pmatrix}-p_3/(E_p+m)\\-(p_1+ip_2)/(E_p+m)\\1\\0\end{pmatrix},\quad u_-^{(2)}=N\begin{pmatrix}-(p_1-ip_2)/(E_p+m)\\p_3/(E_p+m)\\0\\1\end{pmatrix}. \tag{4.318}$$

We can use the compact notation $\chi_r = \chi_1, \chi_2$, finding in both cases

$$u_+^{(r)} = \begin{pmatrix} \chi_r \\ \frac{\boldsymbol{\sigma}\cdot\mathbf{p}}{E_p+m}\chi_r \end{pmatrix}, \quad \text{where } E = +E_p = +\sqrt{\mathbf{p}^2+m^2} > 0, \tag{4.319a}$$

$$u_-^{(r)} = \begin{pmatrix} -\frac{\boldsymbol{\sigma}\cdot\mathbf{p}}{E_p+m}\chi_r \\ \chi_r \end{pmatrix}, \quad \text{where } E = -E_p = -\sqrt{\mathbf{p}^2+m^2} < 0. \tag{4.319b}$$

For $\mathbf{p} \longrightarrow 0$, the above quantities reduce to

$$\begin{pmatrix} 1 \\ 0 \\ 0 \\ 0 \end{pmatrix}, \begin{pmatrix} 0 \\ 1 \\ 0 \\ 0 \end{pmatrix} \quad \text{and} \quad \begin{pmatrix} 0 \\ 0 \\ 1 \\ 0 \end{pmatrix}, \begin{pmatrix} 0 \\ 0 \\ 0 \\ 1 \end{pmatrix}. \tag{4.320}$$

It is no longer true, as in the case $\mathbf{p} = 0$, that $u_\pm^{(r)}$ are eigenstates of Σ_3, but it is true in the particular case where $\mathbf{p} = (0, 0, p)$ because there

$$\boldsymbol{\sigma}\cdot\mathbf{p} = \begin{pmatrix} p & 0 \\ 0 & -p \end{pmatrix}.$$

In general, it is always possible to choose solutions as eigenstates of $\frac{\boldsymbol{\sigma}\cdot\hat{\mathbf{p}}}{|\mathbf{p}|} = \frac{\boldsymbol{\sigma}\cdot\hat{\mathbf{p}}}{|\mathbf{p}|}$ helicity, the spin projection along the momentum, but not along an arbitrary direction. The eigenstates of helicity coincide with the right-handed states u_R and left-handed states u_L related to eigenvalues $\pm 1 (E > 0)$ only if $m = 0$.

We can consider the solution

$$\psi(x) = N \begin{cases} u_r(\mathbf{p})e^{-ipx} & \text{per } E = E_p > 0, \\ v_r(\mathbf{p})e^{ipx} & \text{per } E = -E_p < 0, \end{cases} \tag{4.321}$$

with $u_r(\mathbf{p}) = u_+^{(r)}(\mathbf{p})$ e $v_r(\mathbf{p}) = \{u_-^{(2)}(-\mathbf{p}), -u_-^{(1)}(-\mathbf{p})\}$.

We note that plugging $\psi = u(\mathbf{p})e^{-i(Et-\mathbf{p}\cdot\mathbf{x})}$ into $(i\gamma^\mu\partial_\mu - m)\psi = 0$ we obtain

$$(i\gamma^0(-iE) + i(i\mathbf{p}) - m)u = (\gamma^0 E - \cdot\mathbf{p} - m)u = 0, \tag{4.322}$$

satisfied for $\pm\begin{pmatrix} \frac{\boldsymbol{\sigma}\cdot\mathbf{p}}{E_p+m}\chi_r \\ \chi_r \end{pmatrix}$.

Moreover, plugging $\psi = v(\mathbf{p})e^{i(Et-\mathbf{p}\cdot\mathbf{x})}$ we get

$$(i\gamma^0(iE) + i\boldsymbol{\gamma}(-i\mathbf{p}) - m)v = 0 \longrightarrow (\gamma^0 E - \gamma \cdot \mathbf{p} + m)v = 0, \tag{4.323}$$

satisfied for $\pm \begin{pmatrix} \frac{\sigma\cdot\mathbf{p}}{E_p+m}\chi_r \\ \chi_r \end{pmatrix}$.

Let us assume that $r = 1$ (the case $r = 2$ is analogous). In that case,

$$\begin{pmatrix} E+m & -\sigma\cdot\mathbf{p} \\ \sigma\cdot\mathbf{p} & -E+m \end{pmatrix} \begin{pmatrix} \frac{\sigma\cdot\mathbf{p}}{E_p+m}\chi_2 \\ \chi_2 \end{pmatrix} = 0 \longrightarrow \begin{cases} (E_p+m)\frac{\sigma\cdot\mathbf{p}}{E_p+m}\chi_2 - \sigma\cdot\mathbf{p}\chi_2 = 0, \\ \frac{(\sigma\cdot\mathbf{p})^2}{E_p+m}\chi_2 + (-E_p+m)\chi_2 = 0. \end{cases} \tag{4.324}$$

Now we can conventionally normalize

$$u_r^\dagger(\mathbf{p})u_s(\mathbf{p}) = v_r^\dagger(\mathbf{p})v_s(\mathbf{p}) = \delta_{rs}\frac{E_p}{m}, \tag{4.325}$$

to obtain the expression for N from $\int d^3x \psi^\dagger\psi = 1$, namely

$$\int d^3x |N|^2 u_r^\dagger(\mathbf{p})u_r(\mathbf{p}) = 1 \longrightarrow |N|^2 = \frac{1}{V u_r^\dagger(\mathbf{p})u_r(\mathbf{p})} = \frac{1}{V2E_p}(E_p+m), \tag{4.326}$$

where it has been argued that $\int d^3x = V$ and

$$\begin{aligned} u_r^\dagger(\mathbf{p})u_r(\mathbf{p}) &= \begin{pmatrix} \chi_r & \frac{\sigma\cdot\mathbf{p}}{E_p+m}\chi_r \end{pmatrix}\begin{pmatrix} \chi_r \\ \frac{\sigma\cdot\mathbf{p}}{E_p+m}\chi_r \end{pmatrix} = \chi_r^2 + \left(\frac{\sigma\cdot\mathbf{p}}{E_p+m}\right)^2\chi_r^2 = 1 + \frac{\mathbf{p}^2}{(E_p+m)^2} = \\ &= \frac{E_p^2 + m^2 + 2E_p m + \mathbf{p}^2}{(E_p+m)^2} = \frac{E_p^2 + m^2 + E_p m + E_p m + \mathbf{p}^2}{(E_p+m)^2} = \\ &= \frac{E_p(E_p+m) + m^2 + E_p m + E_p^2 - m^2}{(E_p+m)^2} = \frac{2E_p(E_p+m)}{(E_p+m)^2} = \frac{2E_p}{E_p+m}. \end{aligned} \tag{4.327}$$

Turning back to Eq. (4.326), we find the normalization factor

$$N = \sqrt{\frac{E_p+m}{2E_pV}}. \tag{4.328}$$

Now, let us impose that $u_r^\dagger u_r = \frac{E_p}{m}$ as in Eq. (4.325) so that

$$N = \frac{N'}{(u_r^\dagger u_r)^{\frac{1}{2}}} = N'\sqrt{\frac{m}{E_p}} \longrightarrow N' = \sqrt{\frac{E_p+m}{2E_pV}\frac{E_p}{m}} = \sqrt{\frac{E_p+m}{2mV}}. \tag{4.329}$$

Thus, we obtain

$$\psi(x) = N'\sqrt{\frac{m}{E_p}} \cdot \begin{cases} ue^{-ipx} & E > 0, \\ ve^{ipx} & E < 0, \end{cases} \tag{4.330}$$

with N' given by Eq. (4.329).

The v and u spinors are orthonormal; in particular, they constitute a complete set. However, we first need to find the constant C that normalizes them separately. The relations obtained from $(\gamma^\mu p_\mu - m)u = 0$ and $(\gamma^\mu p_\mu + m)u = 0$ are

$$\bar{u}_r u_s = u_r^\dagger \gamma^0 u_s = C^\dagger C \begin{pmatrix} \chi_r & \frac{\sigma\cdot\mathbf{p}}{E_p+m}\chi_r \end{pmatrix} \begin{pmatrix} \mathbb{I} & 0 \\ 0 & -\mathbb{I} \end{pmatrix} \begin{pmatrix} \chi_s \\ \frac{\sigma\cdot\mathbf{p}}{E_p+m}\chi_s \end{pmatrix} =$$

$$C^\dagger C \begin{pmatrix} \chi_r & \frac{\sigma\cdot\mathbf{p}}{E_p+m}\chi_r \end{pmatrix} \begin{pmatrix} \chi_s \\ -\frac{\sigma\cdot\mathbf{p}}{E_p+m}\chi_s \end{pmatrix} = C^\dagger C \left[\chi_r\chi_s - \left(\frac{\sigma\cdot\mathbf{p}}{E_p+m}\right)^2 \chi_s\chi_r \right] = C^\dagger C \left[\chi_r\chi_s \left(1 - \frac{E_p^2 - m^2}{(E_p+m)^2}\right)\right] = \tag{4.331}$$

$$C^\dagger C \left[\chi_r\chi_s \left(1 - \frac{E_p - m}{E_p+m}\right)\right] = C^\dagger C \left[\chi_r\chi_s \left(\frac{E_p - E_p + m + m}{E_p+m}\right)\right] = C^\dagger C \left[\chi_r\chi_s \frac{2m}{E_p+m}\right], \tag{4.332}$$

where we used $(\boldsymbol{\sigma} \cdot \mathbf{p})^2 = \mathbf{p}^2 = E_p^2 - m^2$. Also, $\chi_r\chi_s = \delta_{rs}$. The above equation must also be equal to the *normalization relation* $u_r^\dagger \gamma^0 u_s = \delta_{rs}$, so

$$C = \sqrt{\frac{E_p + m}{2m}}. \tag{4.333}$$

At this point, we find the *orthogonality relation*

$$\sum_{r=\pm} \int \frac{d^3p}{(2\pi)^3} \frac{1}{2E_p} \left[u_r^\dagger(\mathbf{p})u_r(\mathbf{p}) + v_r^\dagger(\mathbf{p})v_r(\mathbf{p})\right] = \delta_{rs}, \tag{4.334}$$

showing that the solutions are orthonormal. Indeed,

$$\bar{u}_r v_s = \frac{E_p+m}{2m} \begin{pmatrix} \chi_r & \frac{\sigma\cdot\mathbf{p}}{E_p+m}\chi_r \end{pmatrix} \begin{pmatrix} \mathbb{I} & 0 \\ 0 & -\mathbb{I} \end{pmatrix} \begin{pmatrix} \chi_s \frac{\sigma\cdot\mathbf{p}}{E_p+m} \\ \chi_s \end{pmatrix} = \frac{E_p+m}{2m} \left[\chi_r\chi_s \frac{\sigma\cdot\mathbf{p}}{E_p+m} - \frac{\sigma\cdot\mathbf{p}}{E_p+m}\chi_r\chi_s \right] = 0. \tag{4.335}$$

Let us now define the *completeness relations* starting with

$$\sum_r \left[(u_r)_\alpha(\bar{u}_r)_\beta - (v_r)_\alpha(\bar{v}_r)_\beta\right] = \delta_{\alpha\beta}, \tag{4.336}$$

where $u_r = C \begin{pmatrix} \chi_r \\ \frac{\sigma\cdot\mathbf{p}}{E_p+m}\chi_r \end{pmatrix} = \frac{E_p+m}{2m} \begin{pmatrix} \chi_r \\ \frac{\sigma\cdot\mathbf{p}}{E_p+m}\chi_r \end{pmatrix}$.

Let us write it as follows:

$$\sum_r u_r\bar{u}_r = \sum_r u_r u_r^\dagger \gamma^0 = \frac{E_p+m}{2m}\sum_r \begin{pmatrix} \chi_r \\ \frac{\sigma\cdot\mathbf{p}}{E_p+m}\chi_r \end{pmatrix} \begin{pmatrix} \chi_r & \frac{\sigma\cdot\mathbf{p}}{E_p+m}\chi_r \end{pmatrix} \begin{pmatrix} \mathbb{1} & 0 \\ 0 & -\mathbb{1} \end{pmatrix}$$
$$= \frac{E_p+m}{2m}\sum_r \begin{pmatrix} \chi_r \\ \frac{\sigma\cdot\mathbf{p}}{E_p+m}\chi_r \end{pmatrix} \begin{pmatrix} \chi_r & 0 \\ 0 & -\frac{\sigma\cdot\mathbf{p}}{E_p+m}\chi_r \end{pmatrix} =$$
$$= \frac{E_p+m}{2m}\sum_r \begin{pmatrix} \chi_r\chi_r & -\chi_r\chi_r\frac{\sigma\cdot\mathbf{p}}{E_p+m} \\ \frac{\sigma\cdot\mathbf{p}}{E_p+m}\chi_r\chi_r & -\frac{(\sigma\cdot\mathbf{p})^2}{(E_p+m)^2}\chi_r\chi_r \end{pmatrix} =$$
$$= \frac{1}{2m}\begin{pmatrix} E_p+m & -\sigma\cdot\mathbf{p} \\ \sigma\cdot\mathbf{p} & -(E_p+m)\frac{(E_p+m)(E_p-m)}{(E_p+m)^2} \end{pmatrix}$$
$$= \frac{1}{2m}\begin{pmatrix} E_p+m & -\sigma\cdot\mathbf{p} \\ \sigma\cdot\mathbf{p} & -(E_p-m) \end{pmatrix} = \frac{1}{2m}(\not{p}+m). \tag{4.337}$$

Analogously

$$\sum_r v_r\bar{v}_r = \frac{1}{2m}(\not{p}-m). \tag{4.338}$$

Since the completeness relation tells us that the sum of projectors onto eigenstates is the identity, we notice that $\sum_r u_r\bar{u}_r$ and $\sum_r v_r\bar{v}_r$ respectively correspond to the projectors onto eigenspaces $E>0$ and $E<0$, for a given $\mathbf{p}$. We write then

$$\sum_r u_r\bar{u}_r = \Lambda_p^+, \tag{4.339a}$$
$$\sum_r v_r\bar{v}_r = \Lambda_p^-. \tag{4.339b}$$

where $\Lambda_p^+ + \Lambda_p^- = 1$, $(\Lambda_p^\pm)^2 = \Lambda_p^\pm$ and $\Lambda_p^+\Lambda_p^- = 0$.

Exploiting completeness, we can write the *general solution* by summing over $\mathbf{p}$:

$$\psi(x) = \sum_{\mathbf{p}}\sum_r \sqrt{\frac{m}{E_p}}\left[a_r(\mathbf{p})u_r(\mathbf{p})e^{-ipx} + b_r^\star(\mathbf{p})v_r(\mathbf{p})e^{ipx}\right], \tag{4.340}$$

where the first part represents the contribution at positive energies, while the second part represents the contribution at negative energies.

> The solutions could also have been found through a boost from O, where the particle is at rest ($\mathbf{p}=0$), to O' in motion with $\mathbf{v}=$ const.

First of all, $S(\Lambda) = \mathbb{1} - \frac{i}{4}\omega_{\alpha\beta}\sigma^{\alpha\beta}$ is written in the case where $\Lambda^{\mu}_{\ \nu}$ is a boost along $\hat{x}$ with rapidity θ that satisfies $v = \tanh\theta \sim \delta\theta$ for small θ. Given the infinitesimal coordinate transformation $x'^{\mu} = \Lambda^{\mu}_{\ \nu}x^{\nu}$, we can write

$$x'^0 = \frac{x^0 - vx^1}{\sqrt{1-v^2}} = \cosh\theta x^0 - \sinh\theta x^1 = \cosh\theta(x^0 - \tanh\theta x^1) \sim x^0 - \delta\theta x^1, \tag{4.341a}$$

$$x'^1 = \frac{x^1 - vx^0}{\sqrt{1-v^2}} = -\sinh\theta x^0 + \cosh\theta x^1 = \cosh\theta(x^1 - \tanh\theta x^0) \sim x^1 - \delta\theta x^0, \tag{4.341b}$$

where $\delta\theta x^1$ and $\delta\theta x^0$ represent two infinitesimal translations described by a single parameter $\delta\theta = \omega_{10} = -\omega_{01}$.

We define

$$\gamma^2 = \frac{1}{1-v^2} \iff 1 - v^2 = \frac{1}{\gamma^2} \iff v^2 = \frac{\gamma^2 - 1}{\gamma^2}. \tag{4.342}$$

Let us define also the *infinitesimal field transformation*

$$S(\Lambda) = e^{-\delta\boldsymbol{\theta}\cdot\mathbf{k}} \quad \text{with } \delta\boldsymbol{\theta} = (-\omega_{01}, 0, 0) \text{ and } \mathbf{k} = \left(-\frac{i}{2}\sigma^{01}, 0, 0\right). \tag{4.343}$$

Since $\delta\theta$ is infinitesimal, we can expand the above equation in a series, obtaining

$$S(\Lambda) = e^{-i\delta\boldsymbol{\theta}\cdot\mathbf{k}} \sim \mathbb{1} + \frac{i}{2}\delta\theta\sigma^{01} = \mathbb{1} - \frac{\delta\theta}{2}\alpha_1, \tag{4.344}$$

where

$$\sigma^{01} = \frac{i}{2}[\gamma^0, \gamma^1] = \frac{i}{2}(\gamma^0\gamma^1 - \gamma^1\gamma^0) = \frac{i}{2}(\gamma^0\gamma^1 - 2g^{10} + \gamma^0\gamma^1) = i\gamma^0\gamma^1 = i\alpha_1. \tag{4.345}$$

Let us define $\delta\theta = \frac{\theta}{N}$ with N large and write Eq. (4.344) as $S(\Lambda) = e^{-\frac{\theta}{2}\alpha_1}$ which, when expanded, becomes

$$S(\Lambda) = e^{-\frac{\theta}{2}\alpha_1} \sim \cosh\frac{\theta}{2}\mathbb{1} - \hat{\alpha}_1 \sinh\frac{\theta}{2}, \tag{4.346}$$

that can be written in matrix form as

$$S(\Lambda) = \cosh\frac{\theta}{2}\begin{pmatrix}\mathbb{1} & 0\\ 0 & \mathbb{1}\end{pmatrix} - \begin{pmatrix}0 & \sigma_1\\ \sigma_1 & 0\end{pmatrix}\sinh\frac{\theta}{2} = \cosh\frac{\theta}{2}\begin{pmatrix}\mathbb{1} & -\sigma_1\tanh\frac{\theta}{2}\\ -\sigma_1\tanh\frac{\theta}{2} & \mathbb{1}\end{pmatrix}. \tag{4.347}$$

Now, since $\psi'(x') = u'(\mathbf{p}')e^{-ip'x'}$ then $\psi'(x') = S\psi(x) = Su(\mathbf{p})e^{-ipx}$. Since $p'x' = px$, it is enough to transform $u'(\mathbf{p}') = Su(\mathbf{p})$ and $\mathbf{p}$ with

$$p'^{\mu} = \Lambda^{\mu}_{\nu} p^{\nu} \iff \begin{cases} p'^{0} = \gamma(p^0 - vp'), \\ p'^{1} = \gamma(p' - vp^0). \end{cases} \longrightarrow \mathbf{p}' = \pm(m\gamma, -mv\gamma). \quad (4.348)$$

In the reference frame where $\mathbf{p} = 0$, we will have $u_+^{(r)} = \begin{pmatrix} \chi_r \\ 0 \end{pmatrix}$ and $u_-^{(r)} = \begin{pmatrix} 0 \\ \chi_r \end{pmatrix}$ where

$$\chi_r = \begin{cases} \begin{pmatrix} 1 \\ 0 \end{pmatrix} & \text{if } r = 1, \\ \begin{pmatrix} 0 \\ 1 \end{pmatrix} & \text{if } r = 2. \end{cases} \quad (4.349)$$

Before applying the transformation, let us write $\cosh\theta = \gamma = \pm\frac{|p^{0\prime}|}{m}$ using $\cosh\frac{\theta}{2} = \sqrt{\frac{\cosh\theta+1}{2}}$ and $\tanh\frac{\theta}{2} = \frac{\sinh\frac{\theta}{2}}{\cosh\frac{\theta}{2}} = \sqrt{\frac{\cosh\theta-1}{\cosh\theta+1}}$ so that

$$\cosh\frac{\theta}{2} = \sqrt{\frac{\gamma+1}{2}} = \sqrt{\frac{p^{0\prime}+m}{2m}}, \quad (4.350a)$$

$$\tanh\frac{\theta}{2} = \sqrt{\frac{\gamma-1}{\gamma+1}} = \frac{\gamma v}{\gamma+1} = \frac{\mp\frac{|\mathbf{p}'|}{m}}{\frac{p^{0\prime}}{m}+1} = \mp\frac{|\mathbf{p}'|}{p^{0\prime}+m}. \quad (4.350b)$$

We have then

$$u'(\mathbf{p}') = \sqrt{\frac{p^{0\prime}+m}{2m}} \begin{pmatrix} 1\!\mathrm{l} & \pm\frac{\sigma_1 p'}{p^{0\prime}+m} \\ \frac{\pm\sigma_1 p'}{p^{0\prime}+m} & 1\!\mathrm{l} \end{pmatrix} u(\mathbf{p}). \quad (4.351)$$

We can now consider the case of positive and negative energies.

1. For positive energies $E' = p^{0\prime} > 0$ we have

$$u_+^{\prime(r)} = \sqrt{\frac{p^{0\prime}+m}{2m}} \begin{pmatrix} 1\!\mathrm{l} & \frac{\sigma_1 p'}{p^{0\prime}+m} \\ \frac{\sigma_1 p'}{p^{0\prime}+m} & 1\!\mathrm{l} \end{pmatrix} \begin{pmatrix} \chi_r \\ 0 \end{pmatrix} = \sqrt{\frac{p^{0\prime}+m}{2m}} \begin{pmatrix} \chi_r \\ \frac{\sigma_1 p'}{p^{0\prime}+m}\chi_r \end{pmatrix}. \quad (4.352)$$

2. For negative energies $E' = -|p^{0\prime}| < 0$ we have

$$u_-^{\prime(r)} = \sqrt{\frac{p^{0\prime}+m}{2m}} \begin{pmatrix} 1\!\mathrm{l} & \frac{-\sigma_1 p'}{p^{0\prime}+m} \\ \frac{-\sigma_1 p'}{p^{0\prime}+m} & 1\!\mathrm{l} \end{pmatrix} \begin{pmatrix} 0 \\ \chi_r \end{pmatrix} = \sqrt{\frac{p^{0\prime}+m}{2m}} \begin{pmatrix} -\frac{\sigma_1 p'}{p^{0\prime}+m}\chi_r \\ \chi_r \end{pmatrix}. \quad (4.353)$$

▶ **Pause and think** Which are the main differences between the solutions in the Klein-Gordon and Dirac fields?

4.5.5 Discrete Symmetries

We want to investigate the symmetry of the Dirac equation under parity, time reversal, and charge conjugation.

▶ **Pause and think** Which kind of symmetries have you found in classical and quantum physics?

4.5.5.1 Parity or Spatial Inversion

Let us start by defining *parity* or *spatial inversion*. Consider the inversion of spatial coordinates so that

$$\begin{cases} t' = t, \\ \mathbf{x}' = -\mathbf{x}' \end{cases}, \quad \text{and} \quad \Lambda^{\mu}_{\nu} = P^{\mu}_{\nu} = \begin{pmatrix} 1 & 0 & 0 & 0 \\ 0 & -1 & 0 & 0 \\ 0 & 0 & -1 & 0 \\ 0 & 0 & 0 & -1 \end{pmatrix} = g^{\mu}_{\nu}. \tag{4.354}$$

Let us now show the covariance under Lorentz transformations by writing the Dirac equation in the reference frame O.

$$0 = (i\gamma^{\mu}\partial_{\mu} - m)\psi(x), \tag{4.355}$$

where $\psi(x) = P^{-1}\psi'(x')$. Asking for invariance means requiring that Eq. (4.355) is equal to

$$0 = (i\gamma^{\mu}\partial_{\mu} - m)\psi'(x'). \tag{4.356}$$

▶ **Driving exercise**

Compute how P commutes or anticommutes with $\gamma^{0;i}$.

To do so, let us develop Eq. (4.355) by replacing $\psi(x)$ with $P^{-1}\psi'(x')$. Let us write

$$x'^{\mu} = P^{\mu}_{\nu}x^{\nu} = g^{\mu\nu}x^{\nu} \longrightarrow dx'^{\mu} = \frac{\partial x'^{\mu}}{\partial x^{\nu}}dx^{\nu} \quad \text{with} \quad P^{\mu}_{\nu} = \frac{\partial x'^{\mu}}{\partial x^{\nu}}, \tag{4.357a}$$

$$\frac{\partial}{\partial x^{\mu}} = \frac{\partial}{\partial x'^{\nu}}\frac{\partial x'^{\nu}}{\partial x^{\mu}} = \frac{\partial}{\partial x'^{\nu}}P^{\nu}_{\mu}. \tag{4.357b}$$

Equation (4.355) becomes

$$0 = (i\gamma^{\mu} P^{\nu}_{\mu} \partial'_{\nu} - m) P^{-1} \psi'(x') \longrightarrow 0 = (i P\gamma^{\mu} P^{-1} P^{\nu}_{\mu} \partial'_{\nu} - m)\psi'(x'), \tag{4.358}$$

where we have multiplied by P. Imposing the latter being equal to Eq. (4.356), we obtain the relation for P

$$P\gamma^{\mu} P^{-1} P^{\nu}_{\mu} = \gamma^{\nu} \longrightarrow P^{\nu}_{\mu}\gamma^{\mu} = P^{-1}\gamma^{\nu} P, \tag{4.359}$$

where we have multiplied on the right by P and on the left by $(P^{-1})^{\rho}_{\nu}$, then changed $\rho \to \mu$. Expanding P^{μ}_{ν}, we get

$$g^{\nu\mu}\gamma^{\mu} = P^{-1}\gamma^{\nu} P. \tag{4.360}$$

This implies that P commutes with γ^0 while anticommutes with γ^i, indeed

$$P^{-1}\gamma^0 P = \gamma^0 \quad \text{for } \nu = 0, \tag{4.361a}$$

$$P^{-1}\gamma^i P = -\gamma^i \quad \text{for } \nu = i. \tag{4.361b}$$

This condition is satisfied if P is proportional to γ^0 through the constant η. In fact, choosing $P = \eta\gamma^0$, we will have

$$[\gamma^0, \eta\gamma^0] = 0, \tag{4.362a}$$

$$\{\gamma^i, \eta\gamma^0\} = \gamma^i\eta\gamma^0 + \eta\gamma^0\gamma^i = \eta(\gamma^i\gamma^0 - \gamma^i\gamma^0) = 0. \tag{4.362b}$$

In order to find η, we write $P^4 = \mathbb{1}$, which directly follows from the invariance of the probability density $\rho = \psi^{\dagger}\psi$ under parity. Therefore, we find η requiring $(P^{\mu}_{\nu})^4 = \mathbb{1}$

$$\begin{pmatrix} 1 & 0 & 0 & 0 \\ 0 & -1 & 0 & 0 \\ 0 & 0 & -1 & 0 \\ 0 & 0 & 0 & -1 \end{pmatrix}^4 = \begin{pmatrix} 1 & 0 & 0 & 0 \\ 0 & -1 & 0 & 0 \\ 0 & 0 & -1 & 0 \\ 0 & 0 & 0 & -1 \end{pmatrix}\begin{pmatrix} 1 & 0 & 0 & 0 \\ 0 & -1 & 0 & 0 \\ 0 & 0 & -1 & 0 \\ 0 & 0 & 0 & -1 \end{pmatrix}\begin{pmatrix} 1 & 0 & 0 & 0 \\ 0 & 1 & 0 & 0 \\ 0 & 0 & 1 & 0 \\ 0 & 0 & 0 & 1 \end{pmatrix} =$$

$$= \begin{pmatrix} 1 & 0 & 0 & 0 \\ 0 & -1 & 0 & 0 \\ 0 & 0 & -1 & 0 \\ 0 & 0 & 0 & -1 \end{pmatrix}\begin{pmatrix} 1 & 0 & 0 & 0 \\ 0 & -1 & 0 & 0 \\ 0 & 0 & -1 & 0 \\ 0 & 0 & 0 & -1 \end{pmatrix} = \begin{pmatrix} 1 & 0 & 0 & 0 \\ 0 & 1 & 0 & 0 \\ 0 & 0 & 1 & 0 \\ 0 & 0 & 0 & 1 \end{pmatrix} = \mathbb{1} \longrightarrow |\eta|^4 = \mathbb{1}. \tag{4.363}$$

From the probability density, we have

$$\psi'^{+}\psi = (P\psi)^{\dagger}(P\psi) = \psi^{\dagger}P^{\dagger}P\psi, \tag{4.364}$$

since $P^{\dagger}P = \mathbb{1}$, it implies that $|\eta|^2 = 1$. This condition turns out to be more stringent than the one found previously. It tells us that η must be a fourth root of the identity

$$\eta = \pm 1, \quad \eta = \pm i, \tag{4.365}$$

where the last one is to be excluded because we want P to be real. Regarding the condition found with the matrix product, namely $|\eta|^4 = 1$, it is found that $\eta = \pm 1, \pm i$, which are located at the vertices of the square in the complex plane. Having excluded complex solutions, we obtain that

$$\psi'(x') = \pm\gamma^0\psi(x). \tag{4.366}$$

Let us now check how scalars, vectors and tensors change under parity.

$$\Gamma^S : \bar{\psi}'\psi' = \bar{\psi}P^{-1}P\psi = \bar{\psi}\psi, \tag{4.367a}$$
$$\Gamma^{\mu}_{\nu} : \bar{\psi}'\gamma^{\mu}\psi' = \bar{\psi}P^{-1}\gamma^{\mu}P\psi = P^{\mu}_{\nu}\bar{\psi}\gamma^{\nu}\psi, \tag{4.367b}$$
$$\Gamma^P : \bar{\psi}\gamma^5\psi' = \bar{\psi}P^{-1}\gamma^5 P\psi = -\bar{\psi}\gamma^5\psi, \tag{4.367c}$$
$$\Gamma^{\mu}_{A} : \bar{\psi}'\gamma^{\mu}\gamma^5\psi' = \bar{\psi}P^{-1}\gamma^{\mu}\gamma^5 P\psi = -P^{\mu}_{\nu}\bar{\psi}\gamma^{\nu}\gamma^5\psi. \tag{4.367d}$$

The vector Γ^{μ}_{ν} satisfies $\Gamma^{\mu}_{\nu} \longrightarrow P^{\mu}_{\nu}\Gamma^{\nu}_{\nu} = (\Gamma^0_{\nu}, -\boldsymbol{\gamma}_{\nu})$ and is called *polar*. While Γ^{μ}_{A} satisfies $\Gamma^{\mu}_{A} \longrightarrow -P^{\mu}_{\nu}\Gamma^{\nu}_{A} = (-\Gamma^0_{A}, \boldsymbol{\gamma}_{A})$ and is called *axial* transforming in the opposite way with respect to the polar one.

Remembering the division of ψ into right-handed and left-handed components, we observe that we can switch from one component to the other through parity. Consider

$$\begin{pmatrix}\psi_L\\0\end{pmatrix} = \frac{1-\gamma^5}{2}\psi, \quad \begin{pmatrix}0\\\psi_R\end{pmatrix} = \frac{1+\gamma^5}{2}\psi. \tag{4.368}$$

We now apply the parity operator to the left-handed state, having

$$\begin{pmatrix}\psi_L\\0\end{pmatrix} \longrightarrow P^{-1}\frac{(1-\gamma^5)}{2}P\psi = \frac{1-P^{-1}\gamma_5 P}{2}\psi = \frac{1+\gamma_5}{2}\psi = \begin{pmatrix}0\\\psi_R\end{pmatrix}, \tag{4.369}$$

and vice versa, $\psi_R \longrightarrow \psi_L$.

4.5.5.2 The Hole Theory

Before addressing the other discrete symmetries, let us explain why they must be satisfied.

The problem of negative energy solutions, $E < 0$, in Dirac's theory remains unresolved because the Dirac equation alone is not sufficient to interpret these solutions.

For a possible interpretation, Dirac proposed the *hole theory*.[1] As a premise, we start by saying that in quantum mechanics, transitions between discrete energy levels are allowed. However, there is a state called the *ground state*, which represents the lower limit where excited electrons emit radiation. If all electrons are in the E_0 level, then the atom is stable.

The hole theory proposes a set of postulates analogous to electronic transition phenomena.

1. Under normal conditions, negative energy states are occupied, and due to the Pauli exclusion principle, transitions to these states by positively energized electrons are not possible.
2. This sea of infinite electrons at negative energy constitutes the *vacuum state*, which possesses $E = 0$ and $Q = 0$, where electrons have $E = -E_p$ and $Q = -e$.
3. Occasionally, electrons in this sea may absorb a photon with sufficient energy to transition to a positive energy level.

Moreover, it is understood that transitions in the opposite direction are not allowed. If an electron with $E > 0$ were to emit a photon with energy $> 2mc^2$, it would fall to a level with $E < 0$ and then continuously decrease its energy, leading to the instability of the atom. The result of the transition $E < 0 \longrightarrow E > 0$ is the creation of a hole in the Dirac sea.

This hole will have energy and charge given by

$$E(\text{hole}) = E(\text{vacuum}) - (-E_p) = E_p > 0, \tag{4.370a}$$

$$q(\text{hole}) = Q(\text{vacuum}) - q = -(-e) = e > 0. \tag{4.370b}$$

Therefore, the absence of electrons with energies $E = -E_p < 0$ is interpreted as the presence of positrons e^+ with $E = E_p > 0$ and opposite charge to that of electrons. Initially, the predicted antiparticle was not known, leading to the rejection of the hole theory.

---

[1] We highlight that such an interpretation is not entirely satisfying; hole theory attempts not to abandon the interpretation of ψ as the single-particle wave function, requiring the quantization of the Dirac field.

It is not possible to construct a hole theory from the Klein-Gordon equations since they describe particles with zero spin that can occupy a single state simultaneously.

The hole theory predicts the simultaneous existence of positrons and electrons, surpassing the Feynman-Stueckelberg interpretation. It is necessary to abandon the interpretation of ψ as the wave function of a single particle due to the QFT, where ψ satisfying Dirac's equation is a quantized field.

From the hole theory, a particular symmetry emerges: for each particle, there is an antiparticle (charge conjugation). To make this symmetry apparent, we need to formalize the transformation with respect to which the Dirac equation is invariant. If ψ satisfies the Dirac equation in the presence of electromagnetic interaction $-eA_m$, then ψ_C (transformed by C) satisfies the Dirac equation in the presence of the electromagnetic interaction with the sign reversed, $+eA_m$. Testing invariance means finding C in

$$\psi_C(x') = C\psi^\star(x). \tag{4.371}$$

The form of C depends on the representation for the γ matrices. According to the hole theory, ψ_C describes the state with energy $E < 0$, the absence of which appears as the presence of an antiparticle state. In fact, acting with the transformation on the plane wave for a particle with

1. spin down,
2. $E < 0$,
3. momentum $\mathbf{p}$,

we obtain another plane wave that describes a particle with

1. spin up,
2. $E > 0$,
3. momentum $-\mathbf{p}$.

The physical content of the conjugation is that the dynamics of an electron with energy $E > 0$ in $A_\mu(x)$ is equivalent to the behavior of a positron with energy $E < 0$ in $-A_\mu(x)$.

▶ **Pause and think** Emphasize why Dirac is considered the father of antiparticles, although even the Klein-Gordon equation keeps the concept of antiparticle inside it.

4.5.5.3 Time Reversal

We define the *time reversal* by

$$\begin{cases} t' = -t, \\ \mathbf{x}' = \mathbf{x} \end{cases}, \quad \text{and} \quad \Lambda^{\mu}_{\nu} = I^{\mu}_{\nu} = \begin{pmatrix} -1 & 0 & 0 & 0 \\ 0 & 1 & 0 & 0 \\ 0 & 0 & 1 & 0 \\ 0 & 0 & 0 & 1 \end{pmatrix}. \tag{4.372}$$

We want to check the invariance of the Dirac equation by writing it in separate spatial and temporal parts:

$$i\frac{\partial \psi}{\partial t} = H\psi \quad \text{with} \quad H = (-i \cdot \nabla + \beta m). \tag{4.373}$$

Let us write $\psi(\mathbf{x}, t) = J^{-1}\psi'(\mathbf{x}, -t)$ and $\frac{\partial}{\partial t} = \frac{\partial}{\partial t'}\frac{\partial t'}{\partial t} = -\frac{\partial}{\partial t'}$. Therefore, Eq. (4.373) becomes

$$-i\frac{\partial}{\partial t'}J^{-1}\psi' = HJ^{-1}\psi'. \tag{4.374}$$

Multiplying on the right and left by J, we get

$$-J(i)J^{-1}\frac{\partial}{\partial t'}\psi' = JHJ^{-1}\psi' \longrightarrow \begin{cases} H' = JHJ^{-1}, \\ i = -J(i)J^{-1}. \end{cases} \tag{4.375}$$

The first equation does not depend on time, while the second condition tells us that the action of J involves conjugation, i.e., $i \longrightarrow -i$ under inversion. Let us express J as:

$$\psi'(x') = J\psi(x) = T\psi^{\star}(x), \tag{4.376}$$

where T is a matrix. Now, let us rewrite the action on the Hamiltonian to find T:

$$-i \cdot \nabla + \beta m = T(-i \cdot \nabla + \beta m)^{\star}T^{-1} = (+iT^{\star}T^{-1}\nabla + T\beta^{\star}T^{-1}). \tag{4.377}$$

Define $T^{\star}T' = -$ and $T\beta T^{-1} = \beta$. Separate $= (\alpha_1, \alpha_2, \alpha_3)$ where

$$\alpha_1^{\star} = \begin{pmatrix} 0 & \sigma_1^{\star} \\ \sigma_1^{\star} & 0 \end{pmatrix} = \begin{pmatrix} 0 & \sigma_1 \\ \sigma_1 & 0 \end{pmatrix} = \alpha_1, \tag{4.378a}$$

$$\alpha_2^{\star} = \begin{pmatrix} 0 & \sigma_2^{\star} \\ \sigma_2^{\star} & 0 \end{pmatrix} = \begin{pmatrix} 0 & -\sigma_2 \\ -\sigma_2 & 0 \end{pmatrix} = -\alpha_2, \tag{4.378b}$$

$$\alpha_3^{\star} = \begin{pmatrix} 0 & \sigma_3^{\star} \\ \sigma_3^{\star} & 0 \end{pmatrix} = \begin{pmatrix} 0 & \sigma_3 \\ \sigma_3 & 0 \end{pmatrix} = \alpha_3. \tag{4.378c}$$

Using

$$\begin{pmatrix} 0 & 0 & 0 & 1 \\ 0 & 0 & 1 & 0 \\ 0 & 1 & 0 & 0 \\ 1 & 0 & 0 & 0 \end{pmatrix}^{\star} = \begin{pmatrix} 0 & 0 & 0 & 1 \\ 0 & 0 & 1 & 0 \\ 0 & 1 & 0 & 0 \\ 1 & 0 & 0 & 0 \end{pmatrix}, \quad \begin{pmatrix} 0 & 0 & 1 & 0 \\ 0 & 0 & 0 & -1 \\ 1 & 0 & 0 & 0 \\ 0 & -1 & 0 & 0 \end{pmatrix}^{\star} = \begin{pmatrix} 0 & 0 & 1 & 0 \\ 0 & 0 & 0 & -1 \\ 1 & 0 & 0 & 0 \\ 0 & -1 & 0 & 0 \end{pmatrix}, \tag{4.379a}$$

$$\begin{pmatrix} 0 & 0 & 0 & -i \\ 0 & 0 & i & 0 \\ 0 & -i & 0 & 0 \\ i & 0 & 0 & 0 \end{pmatrix}^{\star} = \begin{pmatrix} 0 & 0 & 0 & i \\ 0 & 0 & i & 0 \\ 0 & i & 0 & 0 \\ -i & 0 & 0 & 0 \end{pmatrix} \tag{4.379b}$$

we obtain that T anticommutes with α_1 and α_3 and commutes with α_2 and β.

$$\begin{cases} T\alpha_1 T^{-1} = -\alpha_1, \\ T\alpha_3 T^{-1} = -\alpha_3 \end{cases}, \quad \begin{cases} T\alpha_2 T^{-1} = \alpha_2, \\ T\beta T^{-1} = \beta. \end{cases} \tag{4.380}$$

We consider a product because in the case of a single matrix, it would not anti-commute with itself due to $\{\alpha_i, \alpha_j\} = 2\delta_{ij}$. Thus,

$$T = \eta\alpha_1\alpha_3. \tag{4.381}$$

We can write

$$[\eta\alpha_1\alpha_3, \beta] = \eta[\alpha_1\alpha_3\beta - \beta\alpha_1\alpha_3] = \eta[\alpha_1\alpha_3\beta + \alpha_1\beta\alpha_3] = \eta[\alpha_1\alpha_3\beta - \alpha_1\alpha_3\beta] = 0, \tag{4.382a}$$

$$\{\eta\alpha_1\alpha_3, \alpha_1\} = \eta[\alpha_1\alpha_3\alpha_1 + \alpha_1\alpha_1\alpha_3] = \eta[-\alpha_1\alpha_1\alpha_3 + \alpha_1\alpha_1\alpha_3] = 0, \tag{4.382b}$$

$$[\eta\alpha_1\alpha_3, \alpha_2] = \eta[\alpha_1\alpha_3\alpha_2 - \alpha_2\alpha_1\alpha_3] = \eta[\alpha_1\alpha_3\alpha_2 + \alpha_1\alpha_2\alpha_3] = \eta[\alpha_1\alpha_3\alpha_2 - \alpha_1\alpha_3\alpha_2] = 0, \tag{4.382c}$$

$$\{\eta\alpha_1\alpha_3, \alpha_3\} = \eta[\alpha_1\alpha_3\alpha_3 + \alpha_3\alpha_1\alpha_3] = \eta[-\alpha_3\alpha_1\alpha_3 + \alpha_3\alpha_1\alpha_3] = 0. \tag{4.382d}$$

We obtain η by imposing that

$$T^2 = \begin{pmatrix} -1 & 0 & 0 & 0 \\ 0 & 1 & 0 & 0 \\ 0 & 0 & 1 & 0 \\ 0 & 0 & 0 & 1 \end{pmatrix} \begin{pmatrix} -1 & 0 & 0 & 0 \\ 0 & 1 & 0 & 0 \\ 0 & 0 & 1 & 0 \\ 0 & 0 & 0 & 1 \end{pmatrix} = \mathbb{I} \longrightarrow \eta^2 = 1 \rightarrow \eta = \pm 1, \pm i. \tag{4.383}$$

If $T = \eta\alpha_2\beta$, would it commute with both α_2 and β? No, because $\{\alpha_i, \beta\} = 0$. In fact,

$$[\eta\alpha_2\beta, \alpha_2] = \eta[\alpha_2\beta\alpha_2 - \alpha_2\alpha_2\beta] = 2\eta\alpha_2\beta\alpha_2, \tag{4.384}$$

$$[\eta\alpha_2\beta, \beta] = \eta[\alpha_2\beta\beta - \beta\alpha_2\beta] = 2\eta\alpha_2\beta\beta. \tag{4.385}$$

4.6 Spin-3/2 Particles

A relativistic field describing a particle of spin $s = \frac{3}{2}$ is naturally realized as a *vector–spinor* $\psi_\mu(x)$: it carries a Lorentz index $\mu = 0, 1, 2, 3$ and, simultaneously, transforms as a Dirac spinor. This field contains more components than the physical degrees of freedom of a genuine spin-$\frac{3}{2}$ particle.

4.6.1 The Rarita-Schwinger Equation

The purpose of the Rarita-Schwinger dynamics is precisely to ensure that the unphysical spin-$\frac{1}{2}$ content is removed (consistently with Lorentz invariance and positivity).

Throughout this section we use

$$\bar{\psi}_\mu \equiv \psi_\mu^\dagger \gamma^0, \qquad \partial\!\!\!/ \equiv \gamma^\mu \partial_\mu, \qquad \{\gamma^\mu, \gamma^\nu\} = 2g^{\mu\nu}. \tag{4.386}$$

We also introduce antisymmetrized products of gamma matrices (with unit weight)

$$\gamma^{\mu\nu} \equiv \gamma^{[\mu}\gamma^{\nu]} = \frac{1}{2}(\gamma^\mu\gamma^\nu - \gamma^\nu\gamma^\mu), \tag{4.387}$$

$$\gamma^{\mu\nu\rho} \equiv \gamma^{[\mu}\gamma^\nu\gamma^{\rho]} = \frac{1}{3!}\sum_{\pi\in S_3} \mathrm{sgn}(\pi)\, \gamma^{\pi(\mu)}\gamma^{\pi(\nu)}\gamma^{\pi(\rho)}. \tag{4.388}$$

A useful identity, repeatedly employed below, is

$$\gamma_\mu\gamma^{\mu\nu\rho} = (d-2)\gamma^{\nu\rho} \quad \Rightarrow \quad \gamma_\mu\gamma^{\mu\nu\rho} = 2\,\gamma^{\nu\rho} \quad \text{in } d = 4. \tag{4.389}$$

Another standard identity is

$$\gamma_\mu\gamma^{\mu\nu} = (d-1)\gamma^\nu \quad \Rightarrow \quad \gamma_\mu\gamma^{\mu\nu} = 3\,\gamma^\nu \quad \text{in } d = 4. \tag{4.390}$$

4.6.2 Massless Rarita-Schwinger Case

The massless Rarita-Schwinger Lagrangian density can be written in a compact, manifestly Lorentz-covariant form as

$$\mathcal{L}_0 = -\,\bar{\psi}_\mu\, \gamma^{\mu\nu\rho}\, \partial_\nu \psi_\rho. \tag{4.391}$$

Treating ψ_μ and $\bar{\psi}_\mu$ as independent fields in the variational principle, variation with respect to $\bar{\psi}_\mu$ gives

$$\delta\mathcal{L}_0 = -\,\delta\bar{\psi}_\mu\, \gamma^{\mu\nu\rho}\, \partial_\nu\psi_\rho \quad \Rightarrow \quad \gamma^{\mu\nu\rho}\, \partial_\nu\psi_\rho = 0,$$

so that the equations of motion are

$$\gamma^{\mu\nu\rho}\,\partial_\nu\psi_\rho = 0. \tag{4.392}$$

A key structural feature of the massless theory is the fermionic gauge symmetry

$$\psi_\mu \longrightarrow \psi_\mu + \partial_\mu\epsilon, \tag{4.393}$$

where $\epsilon(x)$ is an arbitrary Dirac spinor parameter. To verify invariance, consider the variation induced by $\delta\psi_\rho = \partial_\rho\epsilon$:

$$\begin{aligned}\delta\mathcal{L}_0 &= -\,\bar{\psi}_\mu\,\gamma^{\mu\nu\rho}\,\partial_\nu(\partial_\rho\epsilon)\\ &= -\,\bar{\psi}_\mu\,\gamma^{\mu\nu\rho}\,\partial_\nu\partial_\rho\epsilon.\end{aligned} \tag{4.394}$$

Since $\partial_\nu\partial_\rho$ is symmetric in (ν,ρ) whereas $\gamma^{\mu\nu\rho}$ is antisymmetric in (ν,ρ), their contraction vanishes:

$$\gamma^{\mu\nu\rho}\,\partial_\nu\partial_\rho\epsilon = 0 \quad\Rightarrow\quad \delta\mathcal{L}_0 = 0. \tag{4.395}$$

Thus $\mathcal{L}_0$ is gauge invariant. This symmetry is essential: it removes unphysical components of ψ_μ and represents the massless analogue of how the Maxwell gauge symmetry eliminates the longitudinal photon polarization.

4.6.3 Massive Rarita-Schwinger Case

A standard Lorentz-covariant massive Rarita-Schwinger Lagrangian is

$$\mathcal{L}_m = -\,\bar{\psi}_\mu\,\gamma^{\mu\nu\rho}\,\partial_\nu\psi_\rho - m\,\bar{\psi}_\mu\,\gamma^{\mu\nu}\,\psi_\nu. \tag{4.396}$$

Varying with respect to $\bar{\psi}_\mu$ yields the massive equations of motion:

$$\gamma^{\mu\nu\rho}\,\partial_\nu\psi_\rho + m\,\gamma^{\mu\nu}\,\psi_\nu = 0. \tag{4.397}$$

The mass term breaks the gauge symmetry, Eq. (4.393). Accordingly, the theory now propagates $2s+1=4$ physical polarizations for a massive spin-$\frac{3}{2}$ particle.

4.6.3.1 γ-Trace Constraint

A crucial step is to show that Eq. (4.397) implies constraints that eliminate the spin-$\frac{1}{2}$ admixture. Take the γ-trace by contracting Eq. (4.397) with γ_μ. Using Eqs.

(4.389)–(4.390), we obtain

$$\begin{aligned} 0 &= \gamma_\mu \gamma^{\mu\nu\rho}\,\partial_\nu \psi_\rho + m\,\gamma_\mu \gamma^{\mu\nu}\psi_\nu \\ &= 2\,\gamma^{\nu\rho}\,\partial_\nu \psi_\rho + 3\,m\,\gamma^\nu \psi_\nu. \end{aligned} \tag{4.398}$$

Now expand $\gamma^{\nu\rho}\partial_\nu\psi_\rho$:

$$\gamma^{\nu\rho}\partial_\nu\psi_\rho = \frac{1}{2}(\gamma^\nu\gamma^\rho - \gamma^\rho\gamma^\nu)\partial_\nu\psi_\rho = \frac{1}{2}\big(\gamma^\nu \not{\partial}\psi_\nu - \gamma^\rho\partial_\rho(\gamma^\nu\psi_\nu)\big),$$

where in the last step we relabeled dummy indices and used $\gamma^\rho\partial_\rho = \not{\partial}$. Hence

$$2\,\gamma^{\nu\rho}\partial_\nu\psi_\rho = \gamma^\nu\not{\partial}\psi_\nu - \not{\partial}(\gamma^\nu\psi_\nu). \tag{4.399}$$

Substituting into Eq. (4.398),

$$\gamma^\nu\not{\partial}\psi_\nu - \not{\partial}(\gamma^\nu\psi_\nu) + 3\,m\,\gamma^\nu\psi_\nu = 0. \tag{4.400}$$

At this point one uses Eq. (4.397) itself to relate $\gamma^\nu\not{\partial}\psi_\nu$ to $\gamma^\nu\psi_\nu$. Contract Eq. (4.397) with ∂_μ:

$$\partial_\mu\gamma^{\mu\nu\rho}\partial_\nu\psi_\rho + m\,\partial_\mu\gamma^{\mu\nu}\psi_\nu = 0. \tag{4.401}$$

Since gamma matrices are constant and $\partial_\mu\partial_\nu$ is symmetric while $\gamma^{\mu\nu\rho}$ is antisymmetric in (μ, ν), the first term vanishes:

$$\gamma^{\mu\nu\rho}\partial_\mu\partial_\nu\psi_\rho = 0,$$

so Eq. (4.401) reduces to

$$m\,\gamma^{\mu\nu}\partial_\mu\psi_\nu = 0 \quad\Rightarrow\quad \gamma^{\mu\nu}\partial_\mu\psi_\nu = 0 \qquad (m \neq 0). \tag{4.402}$$

Using Eq. (4.399) with $\partial_\nu\psi_\rho$ replaced by $\partial_\mu\psi_\nu$ in a dummy relabeling, one recognizes that $\gamma^{\mu\nu}\partial_\mu\psi_\nu = 0$ implies

$$\gamma^\nu\not{\partial}\psi_\nu = \not{\partial}(\gamma^\nu\psi_\nu). \tag{4.403}$$

Then Eq. (4.400) collapses to

$$3\,m\,\gamma^\nu\psi_\nu = 0 \quad\Rightarrow\quad \gamma^\nu\psi_\nu = 0 \qquad (m \neq 0). \tag{4.404}$$

Equation (4.404) is the *γ–tracelessness* condition: it projects out the spin-$\frac{1}{2}$ part contained in the vector–spinor.

4.6.3.2 Transversality (Divergence) Constraint

With $\gamma^\nu\psi_\nu = 0$ established, we now show that $\partial^\nu\psi_\nu = 0$ follows on-shell. Contract Eq. (4.397) with γ_μ again, but now use $\gamma^\nu\psi_\nu = 0$ to simplify the intermediate expressions. A clean route is to take the divergence of Eq. (4.397) and then employ Eq. (4.404). Start from Eq. (4.401) and use Eq. (4.402). Write $\gamma^{\mu\nu}$ explicitly:

$$\gamma^{\mu\nu}\partial_\mu\psi_\nu = \frac{1}{2}(\gamma^\mu\gamma^\nu - \gamma^\nu\gamma^\mu)\partial_\mu\psi_\nu = \frac{1}{2}\left(\partial\!\!\!/(\gamma^\nu\psi_\nu) - \gamma^\nu\partial_\nu(\gamma^\mu\psi_\mu)\right) = \partial\!\!\!/(\gamma^\nu\psi_\nu),$$

where the last equality follows by relabeling indices in the second term. Thus Eq. (4.402) gives

$$\partial\!\!\!/(\gamma^\nu\psi_\nu) = 0. \tag{4.405}$$

Together with Eq. (4.404), this is consistent but does not yet yield $\partial^\nu\psi_\nu = 0$. To obtain transversality, contract Eq. (4.397) with ∂_μ *and* use Eq. (4.404) inside the original equation to reduce it to a Dirac-type form, then take the divergence.

Using $\gamma^\nu\psi_\nu = 0$, the identity

$$\gamma^{\mu\nu\rho}\partial_\nu\psi_\rho = \gamma^\nu\partial_\nu\psi^\mu - \partial^\mu(\gamma^\rho\psi_\rho) + \gamma^\mu\,\partial^\rho\psi_\rho \tag{4.406}$$

simplifies to

$$\gamma^{\mu\nu\rho}\partial_\nu\psi_\rho = \partial\!\!\!/\psi^\mu + \gamma^\mu\,\partial^\rho\psi_\rho, \qquad \text{if } \gamma^\rho\psi_\rho = 0. \tag{4.407}$$

Substituting Eq. (4.407) into Eq. (4.397) gives

$$\partial\!\!\!/\psi^\mu + \gamma^\mu\,\partial^\rho\psi_\rho + m\,\gamma^{\mu\nu}\psi_\nu = 0. \tag{4.408}$$

Now use $\gamma^{\mu\nu}\psi_\nu = \gamma^\mu(\gamma^\nu\psi_\nu) - \psi^\mu = -\psi^\mu$ because $\gamma^\nu\psi_\nu = 0$. Therefore, Eq. (4.408) becomes

$$(\partial\!\!\!/ - m)\psi^\mu = -\gamma^\mu\,\partial^\rho\psi_\rho. \tag{4.409}$$

Apply γ_μ to Eq. (4.409) and use $\gamma_\mu\psi^\mu = 0$:

$$0 = -\gamma_\mu\gamma^\mu\,\partial^\rho\psi_\rho = -4\,\partial^\rho\psi_\rho, \tag{4.410}$$

hence

$$\partial^\mu\psi_\mu = 0. \tag{4.411}$$

With both constraints Eqs. (4.404) and (4.411) in hand, Eq. (4.409) reduces to the free Dirac equation for each physical component:

$$(\partial\!\!\!/ - m)\psi^\mu = 0, \qquad \gamma^\mu\psi_\mu = 0, \qquad \partial^\mu\psi_\mu = 0. \tag{4.412}$$

The last two relations ensure that only the genuine spin-$\frac{3}{2}$ sector propagates.

4.7 Proposed Exercises

Exercise 1

The following (real) double-scalar field (ϕ and ψ) Lagrangian is given,

$$\mathcal{L} = \frac{1}{2}\partial_\mu\varphi\partial^\mu\varphi - V_0\tanh(\beta(\psi)\varphi^{2+2n}),$$

with β analytical in ψ, n an integer and V_0 a free constant.

- Describe the symmetries, if any, of this Lagrangian.
- Consider plausible limiting cases on ψ and feature the physical consequences that you expect on $\mathcal{L}$.
- Use the Noether's theorem to compute the conserved charge.
- Compute at least one component of the shear stress [hint. The energy-momentum tensor is composed by several components, among which shear stresses].

Exercise 2

The following (real) double-scalar field Lagrangian is given,

$$\mathcal{L} = \frac{1}{2}G^{\mu\nu}(\psi)\partial_\mu\varphi\partial_\nu\varphi - V_0\exp^{-\alpha\varphi^2},$$

with ψ and ϕ scalar fields, $G^{\mu\nu}$ a superpotential, α and V_0 free constants.

- Use the Noether's theorem to compute the conserved charge.
- Evaluate the energy momentum tensor. Baptizing ρ and P, as pressure and density, respectively, under which conditions can we express the tensor as a perfect fluid?

Hint: for a perfect fluid, $T_{\mu\nu}$ is diagonal, namely $T_{\mu\nu} \equiv diag(\rho, -P, -P, -P)$.

Exercise 3

The following (real) double-scalar field Lagrangian is given,

$$\mathcal{L} = \frac{1}{2}G(\psi)\partial_\mu\varphi\partial^\mu\varphi - V_0\exp^{-\alpha\varphi^2},$$

with ψ and ϕ scalar fields, $G(\psi)$ a superpotential, α and V_0 free constants.

– If one adds a constant to $\mathcal{L}$, namely V_1, does the result change? Explain why.
– Add a Dirac field and write a minimal and non-minimal interactions between fields.

Essential Bibliography

- Greiner W (2000) Relativistic quantum mechanics: wave equations. Springer.
  A modern and comprehensive reference on relativistic wave equations, covering the Klein-Gordon, Dirac, and higher-spin formalisms with explicit emphasis on Lorentz covariance and conserved currents.
- Srednicki M (2007) Quantum field theory. Cambridge University Press.
  A modern perspective bridging relativistic quantum mechanics and quantum field theory, clarifying the limitations of the single-particle interpretation and the role of relativistic fields.
- Ryder LH (1996) Quantum field theory, 2nd edn. Cambridge University Press.
  A clear and geometrically motivated introduction to relativistic spinors, γ-matrix algebra, and bilinear covariants, closely aligned with the formalism used in this chapter.
- Zee A (2010) Quantum field theory in a Nutshell. Princeton University Press.
  Modern conceptual overview emphasizing symmetry principles, discrete transformations, and the physical interpretation of relativistic quantum equations.

Generalities About Classical and Quantum Scattering

5

Only invariant quantities have an absolute physical meaning.

- Max Von Laue

Abstract

This chapter reports the basic demands of classical and quantum scattering, starting from relativistic kinematics and culminating in the operator formulation of non-relativistic quantum collision theory. After defining Mandelstam variables and counting independent kinematic degrees of freedom through Poincaré invariance, we construct the invariant phase-space measure for n-body final states and derive explicitly the two-body phase space. The second part addresses quantum scattering in potential theory, up to formulate the optical theorem. The overall aim of this chapter is to show how symmetry principles, invariant measures, and operator methods combine to produce a coherent description of collision and decay phenomena, bridging relativistic kinematics and quantum scattering theory within a common conceptual framework.

O. Luongo, *Fundamental Interactions*,
https://doi.org/10.1007/978-3-032-27413-7_5

Fig. 5.1 Portrait of Hendrik Antoon Lorentz, 1902. Reprinted from Wikimedia. Public domain

A key character: *Hendrik Antoon Lorentz (1853–1928), Fig.* 5.1.

Hendrik Antoon Lorentz was born in Arnhem, in the Netherlands, and displayed an early talent for mathematics and the natural sciences. He studied physics and mathematics at the University of Leiden, where he obtained his doctoral degree in 1875 with a dissertation on the theory of reflection and refraction of light. At the remarkably young age of twenty-four, Lorentz was appointed Professor of Theoretical Physics at Leiden, becoming the first holder of a chair explicitly dedicated to theoretical physics in the Netherlands.

Lorentz's scientific work was driven by the attempt to reconcile electrodynamics with the principle of relativity and the experimental evidence concerning the propagation of light. In the context of his electron theory, he introduced the transformations that now bear his name, providing the mathematical structure that leaves Maxwell's equations invariant.

Although Einstein later reformulated relativity in a conceptually more radical way, Lorentz continued to support and promote the new theory, recognizing its physical significance. His influence extended well beyond his own research through his role as a mentor, organizer, and international scientific mediator. In 1902 he was awarded the Nobel Prize in Physics, shared with Pieter Zeeman, for their investigation of the influence of magnetism on radiation phenomena. Lorentz remained scientifically active until the end of his life and was widely regarded by his contemporaries as a model of intellectual integrity and clarity. His work permanently shaped the language of modern theoretical physics, and his name remains inseparably linked to the relativistic description of scattering processes, decays, and fundamental interactions.

5.1 Decays and Scatterings

A *decay* is a process in which a single particle disintegrates into two or more final-state particles,

$$a \to a' + b' + \cdots . \tag{5.1}$$

A *scattering* (or collision) is a process in which two (or more) incoming particles produce a final state containing two (or more) particles, possibly different from the initial ones,

$$a + b + \cdots \to a' + b' + \cdots . \tag{5.2}$$

In practice one typically focuses on $2 \to n$ scatterings, since the probability of having three (or more) particles collide at the same spacetime point is negligible in dilute beams.[1]

At this stage, one can wonder whether invariance principles can reduce the number of independent parameters needed to specify a process. Indeed, for a system with N total particles, counting both initial and final states, the configuration space would naively require $3N$ variables to specify their positions (or, in momentum space, their 3-momenta). Thus, Poincaré invariance provides 10 conserved quantities (corresponding to the 10 generators: 4 translations and 6 Lorentz transformations), so one expects $3N - 10$ independent variables to characterize the process. Some examples follow.

- **Two-body decay** $a \to b + c$: here $N = 3$ and $3N - 10 = -1$. The system is (over-)constrained: in the parent rest frame the final state kinematics are fixed up to trivial overall rotations.
- $2 \to 2$ **scattering** $a + b \to c + d$: here $N = 4$ and $3N - 10 = 2$. The process can be described by two independent Lorentz invariants. A standard choice is to use two Mandelstam variables.

[1] This statement can fail in dense media (liquids, condensates, or strongly correlated systems), where multi-body collisions may become relevant.

For $a(p_1) + b(p_2) \to c(p_3) + d(p_4)$ it is important to define the Mandelstam variables as

$$s = (p_1 + p_2)^2 = (p_3 + p_4)^2, \tag{5.3}$$
$$t = (p_1 - p_3)^2 = (p_2 - p_4)^2 \equiv q^2, \tag{5.4}$$
$$u = (p_1 - p_4)^2 = (p_2 - p_3)^2. \tag{5.5}$$

Here $q \equiv p_1 - p_3$ is the four-momentum transfer in the t-channel. In general u is *not* equal to q^2; rather, s, t, u are related by the on-shell identity

$$s + t + u = m_1^2 + m_2^2 + m_3^2 + m_4^2. \tag{5.6}$$

The three variables are not independent because the degrees of freedom for a two-body scattering (two initial and two final particles) are $3(N = 4) - 10 = 12 - 10 = 2$. This can be verified by summing the three variables:

$$\begin{aligned} s + t + u &= p_1^2 + 2p_1p_2 + p_2^2 + p_1^2 - 2p_1p_3 + p_3^2 + p_1^2 - 2p_1p_4 + p_4^2 \\ &= 3p_1^2 + p_2^2 + p_3^2 + p_4^2 - 2p_1(p_3 + p_4 - p_2) \\ &= 3p_1^2 - 2p_1^2 + p_2^2 + p_3^2 + p_4^2 = m_1^2 + m_2^2 + m_3^2 + m_4^2 = \sum_{i=1}^{4} m_i^2. \end{aligned} \tag{5.7}$$

Hence, there are $3 - 1 = 2$ independent variables.

5.2 Phase Space and Lorentz-Invariant Measures

To formulate relativistic decay rates and cross sections, we need a Lorentz-invariant integration measure over on-shell final-state momenta. We start from the invariant Dirac delta

$$\delta(p^2 - m^2) = \delta(p_\mu p^\mu - m^2) = \delta\big((p^0)^2 - |\mathbf{p}|^2 - m^2\big), \tag{5.8}$$

whose argument is a Lorentz scalar. On the mass shell, $p^2 = m^2$ implies

$$p^0 = \pm\sqrt{|\mathbf{p}|^2 + m^2} \equiv \pm E(\mathbf{p}). \tag{5.9}$$

Using the standard identity

$$\delta\big(g(x)\big) = \sum_i \frac{\delta(x - x_i)}{|g'(x_i)|}, \tag{5.10}$$

(where x_i are the simple roots of g), one finds

$$\delta(p^2 - m^2) = \frac{1}{2E(\mathbf{p})}\Big[\delta\big(p^0 - E(\mathbf{p})\big) + \delta\big(p^0 + E(\mathbf{p})\big)\Big]. \tag{5.11}$$

Introduce now the Heaviside function

$$\theta(p^0) = \begin{cases} 1, & p^0 > 0, \\ 0, & p^0 \leq 0, \end{cases} \tag{5.12}$$

which is invariant under *proper* Lorentz transformations (those preserving time orientation). Since $d^4 p$ is also invariant under proper Lorentz transformations (Jacobian determinant $= 1$), the combination

$$d^4 p \, \delta(p^2 - m^2)\, \theta(p^0) \tag{5.13}$$

is Lorentz invariant. Integrating over p^0 and using $\theta(p^0)$ to select the positive-energy root gives

$$\begin{aligned} \int_{-\infty}^{+\infty} dp^0 \, d^3\mathbf{p} \, \delta(p^2 - m^2)\, \theta(p^0) &= \int_0^{+\infty} dp^0 \, d^3\mathbf{p} \, \frac{\delta(p^0 - E(\mathbf{p}))}{2E(\mathbf{p})} \\ &= \frac{d^3\mathbf{p}}{2E(\mathbf{p})}. \end{aligned} \tag{5.14}$$

Hence, the Lorentz-invariant one-particle measure is conventionally written as

$$\frac{d^3\mathbf{p}}{(2\pi)^3\, 2E(\mathbf{p})}. \tag{5.15}$$

▶ **Pause and think**
Figure out the main differences between phase space and coordinate space.

5.2.1 The n-Body Differential Phase Space

Consider a process where an initial state with total four-momentum P_i^μ goes into a final state with total four-momentum P_f^μ. Four-momentum conservation is imposed by the invariant delta

$$\delta^{(4)}(P_i^\mu - P_f^\mu). \tag{5.16}$$

For n final-state particles with momenta p_k^μ (so that $P_f^\mu = \sum_{k=1}^n p_k^\mu$), the Lorentz-invariant *n-body differential phase space* is defined as

$$d\Phi^{(n)} \equiv (2\pi)^4 \, \delta^{(4)}\left(P_i - \sum_{k=1}^n p_k\right) \prod_{k=1}^n \frac{d^3\mathbf{p}_k}{(2\pi)^3\, 2E_k}. \tag{5.17}$$

With these conventions, each δ contributes a factor (2π) and each momentum integration a factor $(2\pi)^{-1}$, leading to the standard overall normalization.

Dimensional analysis shows that

$$[d\Phi^{(n)}] = \text{mass}^{\,2n-4}. \tag{5.18}$$

5.2.2 Two-Body Phase Space

Let us compute explicitly the case $n = 2$. Let P^μ be the total incoming four-momentum, and p_1^μ, p_2^μ the final momenta, so that (sometimes we indicate the four-momenta with indexes, other times no. This does not change anything about our discussion):

$$P = p_1 + p_2. \tag{5.19}$$

From Eq. (5.17) one has

$$d\Phi^{(2)} = (2\pi)^4\, \delta^{(4)}(P - p_1 - p_2)\, \frac{d^3\mathbf{p}_1}{(2\pi)^3\, 2E_1}\, \frac{d^3\mathbf{p}_2}{(2\pi)^3\, 2E_2}. \tag{5.20}$$

Because $d\Phi^{(2)}$ is invariant, we evaluate it in the center-of-mass (CM) frame, where

$$P = (E_c, \mathbf{0}), \qquad p_1 = (E_1, \mathbf{p}_1), \qquad p_2 = (E_2, \mathbf{p}_2). \tag{5.21}$$

Momentum conservation implies $\mathbf{p}_1 + \mathbf{p}_2 = \mathbf{0}$, so $\mathbf{p}_2 = -\mathbf{p}_1$ and $|\mathbf{p}_1| = |\mathbf{p}_2| \equiv p_c$, while energy conservation gives $E_c = E_1 + E_2$. Then

$$\delta^{(4)}(P - p_1 - p_2) = \delta(E_c - E_1 - E_2)\, \delta^{(3)}(\mathbf{p}_1 + \mathbf{p}_2). \tag{5.22}$$

Using $\delta^{(3)}(\mathbf{p}_1 + \mathbf{p}_2)$ to integrate over $d^3\mathbf{p}_2$ yields

$$d\Phi^{(2)} = \frac{1}{(2\pi)^2}\, \delta(E_c - E_1 - E_2)\, \frac{d^3\mathbf{p}_1}{4E_1E_2}. \tag{5.23}$$

This readily gives

$$d\Phi^{(2)} = \frac{1}{(2\pi)^2}\, \delta(E_c - E_1 - E_2)\, \frac{p_1^2\, dp_1\, d\Omega}{4E_1E_2}. \tag{5.24}$$

To integrate over p_1 it is convenient to convert the energy delta into a momentum delta. Since $E_1 = \sqrt{p_1^2 + m_1^2}$ and $p_1\, dp_1 = E_1\, dE_1$, one finds

$$\begin{aligned} \delta(E_c - E_1 - E_2) &= \delta(p_1 - p_c) \left| \frac{\partial(E_1 + E_2)}{\partial p_1} \right|^{-1}_{p_1 = p_c} \\ &= \delta(p_1 - p_c) \left(\frac{p_c}{E_1} + \frac{p_c}{E_2} \right)^{-1} \\ &= \frac{E_1E_2}{p_c(E_1 + E_2)}\, \delta(p_1 - p_c). \end{aligned} \tag{5.25}$$

Substituting back gives

$$d\Phi^{(2)} = \frac{1}{16\pi^2}\,\frac{p_c}{E_c}\,d\Omega. \tag{5.26}$$

5.2.3 Flux Factor and Invariant Form

We define the Lorentz-invariant *flux factor* as

$$I \equiv \sqrt{(p_1\cdot p_2)^2 - m_1^2 m_2^2}\,, \tag{5.27}$$

where p_1^μ and p_2^μ are the four-momenta of the incoming particles, satisfying the on-shell conditions, $p_1^2 = m_1^2$, $p_2^2 = m_2^2$.

To express I in terms of the Mandelstam invariant s, we start from its definition, $s \equiv (p_1 + p_2)^2 = p_1^2 + p_2^2 + 2\,p_1\cdot p_2$ and, then, using the on-shell relations above, we find

$$s = m_1^2 + m_2^2 + 2\,p_1\cdot p_2, \tag{5.28}$$

which can be inverted to give $p_1\cdot p_2 = \frac{1}{2}\left(s - m_1^2 - m_2^2\right)$.

We now substitute the last relation into the definition in Eq. (5.27), obtaining

$$I^2 = (p_1\cdot p_2)^2 - m_1^2 m_2^2 = \frac{1}{4}\left(s - m_1^2 - m_2^2\right)^2 - m_1^2 m_2^2. \tag{5.29}$$

Expanding the square, $\left(s - m_1^2 - m_2^2\right)^2 = s^2 + m_1^4 + m_2^4 - 2\,sm_1^2 - 2\,sm_2^2 + 2m_1^2m_2^2$, so that

$$I^2 = \frac{1}{4}\Big[s^2 - 2\,s(m_1^2 + m_2^2) + (m_1^2 - m_2^2)^2\Big]. \tag{5.30}$$

The expression in square brackets can be factorized as

$$s^2 - 2\,s(m_1^2 + m_2^2) + (m_1^2 - m_2^2)^2 = \big[s - (m_1 + m_2)^2\big]\big[s - (m_1 - m_2)^2\big]. \tag{5.31}$$

Combining Eqs. (5.30) and (5.31), we finally obtain

$$I^2 = \frac{1}{4}\big[s - (m_1 + m_2)^2\big]\big[s - (m_1 - m_2)^2\big], \tag{5.32}$$

and therefore

$$I = \frac{1}{2}\sqrt{\big[s - (m_1 + m_2)^2\big]\big[s - (m_1 - m_2)^2\big]}. \tag{5.33}$$

Equation (5.33) shows explicitly that the flux factor is a Lorentz-invariant function of the Mandelstam variable s and the particle masses, and it vanishes at the physical threshold $s = (m_1 + m_2)^2$, as expected.

In the CM frame, $E_c = \sqrt{s}$ and one finds

$$I = p_c\, E_c \qquad \text{(CM frame)}. \tag{5.34}$$

In the lab frame where particle 2 is at rest, $p_2 = (m_2, \mathbf{0})$ and

$$I = m_2\, |\mathbf{p}_1| \qquad \text{(lab frame)}. \tag{5.35}$$

Using $I = p_c E_c$ in Eq. (5.26), the two-body phase space can be written as

$$d\Phi^{(2)} = \frac{1}{16\pi^2}\,\frac{I}{s}\,d\Omega = \frac{1}{32\pi^2}\sqrt{1-\frac{(m_1+m_2)^2}{s}}\,\sqrt{1-\frac{(m_1-m_2)^2}{s}}\;d\Omega. \tag{5.36}$$

Two useful special cases are:

– equal masses $m_1 = m_2 = m$:

$$d\Phi^{(2)} = \frac{d\Omega}{32\pi^2}\sqrt{1-\frac{4\,m^2}{s}}; \tag{5.37}$$

– one massless particle $m_1 = m$, $m_2 = 0$:

$$d\Phi^{(2)} = \frac{d\Omega}{32\pi^2}\left(1-\frac{m^2}{s}\right). \tag{5.38}$$

5.3 Lifetime and Decay Width

Consider N_0 unstable particles, each having a constant (time-independent) decay probability per unit time,

$$\frac{dP}{dt} = \frac{1}{\tau}, \tag{5.39}$$

where τ is the (mean) lifetime. If decays are independent, the number of decays $N(t)$ satisfies

$$\frac{dN(t)}{dt} = \frac{1}{\tau}\big(N_0 - N(t)\big), \tag{5.40}$$

with solution

$$N(t) = N_0\big(1 - e^{-t/\tau}\big). \tag{5.41}$$

The number of remaining particles is

$$N_{\text{remaining}}(t) = N_0 - N(t) = N_0\, e^{-t/\tau}. \tag{5.42}$$

The *decay width* is defined as

$$\Gamma \equiv \frac{1}{\tau}, \tag{5.43}$$

which has dimensions of mass.

5.3.1 Differential Decay Probability and Invariant Normalization

Let i denote the initial state, and let α denote the discrete quantum labels of the final state (particle species, spins, internal charges, etc.). For a decay into n final particles with momenta $p_1, \ldots, p_n$, the decay probability per unit time can be written as

$$\frac{dP(i, \alpha, p_1, \ldots, p_n)}{dt} = \widetilde{\Gamma}(i, \alpha, p_1, \ldots, p_n)\, d^3\mathbf{p}_1 \cdots d^3\mathbf{p}_n. \tag{5.44}$$

Energy-momentum conservation implies that $\widetilde{\Gamma}$ must contain a factor $\delta^{(4)}(P_i - P_f)$, and it is natural to trade the non-invariant product $d^3\mathbf{p}_1 \cdots d^3\mathbf{p}_n$ for the invariant phase space $d\Phi^{(n)}$. We therefore rewrite the number of decays per unit time and per unit volume as

$$\frac{dN}{dt\, dV} = \rho\, \widetilde{\Gamma}(i, \alpha, p_1, \ldots, p_n)\, d\Phi^{(n)}, \tag{5.45}$$

where ρ is the number density of decaying particles (so that $\rho\, dV$ is the number of parents inside dV).

Lorentz invariance requires dN to be invariant and also $dV\, dt$ to be a Lorentz invariant 4-volume. Since $d\Phi^{(n)}$ is invariant, the product $\rho\, \widetilde{\Gamma}$ must be invariant as well. The number density transforms as the energy:

$$\rho = \rho_0 \frac{P_i^0}{m}, \tag{5.46}$$

where ρ_0 is the density in the parent rest frame and P_i^0 is the parent energy in the considered frame. Therefore $\widetilde{\Gamma}$ must transform as $1/P_i^0$. One introduces the conventional redefinition

$$\widetilde{\Gamma} \to \frac{1}{2P_i^0}\, \Gamma, \tag{5.47}$$

so that Γ is Lorentz invariant and Eq. (5.45) becomes

$$\frac{dN}{dt\, dV} = \rho\, \frac{1}{2P_i^0}\, \Gamma(i, \alpha, p_1, \ldots, p_n)\, d\Phi^{(n)}. \tag{5.48}$$

Integrating over all final states yields

$$\frac{1}{\tau_i} = \frac{\Gamma(i)}{2P_i^0}, \tag{5.49}$$

where $\Gamma(i)$ is Lorentz invariant and depends only on the quantum numbers of the initial state.

A useful consequence is that the lifetime is invariant under spatial rotations (hence independent of the spin projection of the decaying particle), while under boosts the observed lifetime is dilated in the expected relativistic manner.

5.4 Cross Sections

Consider a scattering experiment where a beam (species 1) with number density ρ_1 and velocity v_1 impinges on a target (species 2) at rest with density ρ_2. The number of scattering events per unit time and per unit volume takes the form

$$\frac{dN}{dV\,dt} = \rho_1 v_1\,\rho_2\,\sigma, \tag{5.50}$$

where $\rho_2\,dV$ is the number of targets in dV and $\rho_1 v_1$ is the incident flux. Thus

$$\sigma = \frac{\#\ \text{events/time}}{(\#\ \text{targets}) \times (\text{incident flux})}. \tag{5.51}$$

The cross section has dimensions of area, $[\sigma] = \text{length}^2 = \text{mass}^{-2}$. Typical scales are

$$\sigma \sim 10^{-20}\,\text{cm}^2\,(\text{atomic}), \qquad \sigma \sim 10^{-25}\,\text{cm}^2\,(\text{nuclear}), \qquad \sigma \sim 10^{-34}\,\text{cm}^2\,(\text{weak}). \tag{5.52}$$

The barn is defined as $1\text{ barn} \equiv 10^{-24}\,\text{cm}^2$, with common submultiples nanobarn (10^{-9} b), picobarn (10^{-12} b), femtobarn (10^{-15} b).

The differential cross section is another interesting quantity. For an initial state i and final-state discrete labels α, with final momenta $p_1, \ldots, p_n$, one writes in the lab frame

$$\frac{dN(i,\alpha,p_1,\ldots,p_n)}{dt\,dV} = \rho_1^{\text{lab}} v_1\,\rho_2^{\text{lab}}\,\Sigma_{\text{lab}}(i,\alpha,p_1,\ldots,p_n)\,d^3\mathbf{p}_1\cdots d^3\mathbf{p}_n, \tag{5.53}$$

where Σ_{lab} is the (lab-frame) differential cross section density. As in the decay case, it is convenient to trade the non-invariant measure and the energy-momentum delta for the invariant phase space:

$$\frac{dN(i,\alpha,p_1,\ldots,p_n)}{dt\,dV} = \rho_1^{\text{lab}} v_1\,\rho_2^{\text{lab}}\,\Sigma_{\text{lab}}(i,\alpha,p_1,\ldots,p_n)\,d\Phi^{(n)}. \tag{5.54}$$

Lorentz invariance implies that dN and $dV\,dt$ are invariant, and since $d\Phi^{(n)}$ is invariant, the quantity $dN/(dt\,dV\,d\Phi^{(n)})$ must be invariant. One thus seeks an invariant that reduces to $\rho_1^{\text{lab}} v_1\, \rho_2^{\text{lab}}$ in the lab frame.

In a generic frame,

$$p_1 = (E_1, \mathbf{p}_1) = E_1(1, \mathbf{v}_1), \qquad p_2 = (E_2, \mathbf{p}_2) = E_2(1, \mathbf{v}_2), \tag{5.55}$$

with $E_i = m_i \gamma(\mathbf{v}_i)$ for $i = 1, 2$. Using the flux factor I of Eq. (5.27), one finds

$$\begin{aligned} I^2 &= (p_1 \cdot p_2)^2 - m_1^2 m_2^2 = E_1^2 E_2^2 \Big[(1 - \mathbf{v}_1 \cdot \mathbf{v}_2)^2 - (1 - \mathbf{v}_1^2)(1 - \mathbf{v}_2^2) \Big] \\ &= E_1^2 E_2^2 \Big[(\mathbf{v}_1 - \mathbf{v}_2)^2 - (\mathbf{v}_1 \times \mathbf{v}_2)^2 \Big], \end{aligned} \tag{5.56}$$

so

$$I = E_1 E_2 \sqrt{(\mathbf{v}_1 - \mathbf{v}_2)^2 - (\mathbf{v}_1 \times \mathbf{v}_2)^2}. \tag{5.57}$$

The *relative velocity* v_{rel} is defined by the invariant relation

$$p_1 \cdot p_2 = \frac{m_1 m_2}{\sqrt{1 - v_{\text{rel}}^2}} \quad \Longrightarrow \quad v_{\text{rel}} = \sqrt{1 - \frac{m_1^2 m_2^2}{(p_1 \cdot p_2)^2}} = \frac{I}{p_1 \cdot p_2} = \frac{\sqrt{(\mathbf{v}_1 - \mathbf{v}_2)^2 - (\mathbf{v}_1 \times \mathbf{v}_2)^2}}{1 - \mathbf{v}_1 \cdot \mathbf{v}_2}. \tag{5.58}$$

It follows that

$$v_{\text{rel}}\, \gamma(v_{\text{rel}}) = \frac{I}{m_1 m_2}. \tag{5.59}$$

The number densities transform as energies:

$$\rho_1 = \rho_1^0 \frac{E_1^0}{m_1}, \qquad \rho_2 = \rho_2^0 \frac{E_2^0}{m_2}, \tag{5.60}$$

where $\rho_{1,2}^0$ are the densities in the respective rest frames. In the lab frame, $\rho_1^{\text{lab}} = \rho_1^0 \gamma(v_1)$ and $\rho_2^{\text{lab}} = \rho_2^0$, hence

$$\rho_1^{\text{lab}} v_1\, \rho_2^{\text{lab}} = \rho_1^0 \rho_2^0\, v_1 \gamma(v_1). \tag{5.61}$$

In a generic frame one may replace this by $\rho_1^0 \rho_2^0\, v_{\text{rel}} \gamma(v_{\text{rel}})$, i.e.

$$\rho_1^0 \rho_2^0\, v_{\text{rel}} \gamma(v_{\text{rel}}) = \rho_1^0 \rho_2^0 \frac{I}{m_1 m_2}, \tag{5.62}$$

which is manifestly invariant (since $\rho_1^0 \rho_2^0$ and I are invariants). Equivalently, using velocities,

$$\rho_1^0 \rho_2^0\, v_{\text{rel}} \gamma(v_{\text{rel}}) = \rho_1 \rho_2 \sqrt{(\mathbf{v}_1 - \mathbf{v}_2)^2 - (\mathbf{v}_1 \times \mathbf{v}_2)^2}. \tag{5.63}$$

Therefore, in a generic frame the differential event rate can be written as

$$\frac{dN(i,\alpha,p_1,\ldots,p_n)}{dt\,dV} = \rho_1\rho_2\sqrt{(\mathbf{v}_1-\mathbf{v}_2)^2-(\mathbf{v}_1\times\mathbf{v}_2)^2}\;\Sigma(i,\alpha,p_1,\ldots,p_n)\,d\Phi^{(n)}, \tag{5.64}$$

where Σ is a Lorentz-invariant differential cross section density containing all the physics of the scattering process.

5.5 Quantum Scattering

We consider a single spinless particle of mass m in $\mathbb{R}^3$, with no internal degrees of freedom. The dynamics is governed by the time-dependent Schrödinger equation

$$i\,\frac{\partial}{\partial t}\,|\psi(t)\rangle = H\,|\psi(t)\rangle\,, \qquad H = H_0 + V\,, \tag{5.65}$$

where

$$H_0 = \frac{\hat{\mathbf{p}}^2}{2m}, \tag{5.66}$$

and $V = V(\hat{\mathbf{x}})$ is a localized interaction (typically short-range). The unitary evolution is described by

$$|\psi(t)\rangle = U(t)\,|\psi\rangle\,, \qquad U(t) = e^{-iHt}, \tag{5.67}$$

with $|\psi\rangle \equiv |\psi(0)\rangle$. The free evolution is

$$|\phi(t)\rangle = U_0(t)\,|\phi\rangle\,, \qquad U_0(t) = e^{-iH_0t}. \tag{5.68}$$

The central idea of scattering theory is that, sufficiently far in the past and in the future, the interaction becomes ineffective and the motion is indistinguishable from free propagation. Accordingly, one introduces two free asymptotic states, $|\psi_{\text{in}}\rangle$ and $|\psi_{\text{out}}\rangle$, such that

$$\lim_{t\to-\infty}\left\|\,U(t)\,|\psi\rangle - U_0(t)\,|\psi_{\text{in}}\rangle\,\right\| = 0, \tag{5.69a}$$

$$\lim_{t\to+\infty}\left\|\,U(t)\,|\psi\rangle - U_0(t)\,|\psi_{\text{out}}\rangle\,\right\| = 0. \tag{5.69b}$$

The state $U(t)\,|\psi\rangle$ is the full (interacting) orbit, whereas $U_0(t)\,\left|\psi_{\text{in/out}}\right\rangle$ are the free in/out asymptotes.

For the existence of the wave operators in the strong sense and for the standard plane-wave asymptotics, it is convenient to assume that $V(\mathbf{x})$ is "well behaved" in the ultraviolet and short range in the infrared. A simple sufficient set of conditions is:

- as $r \to \infty$, $V(r) = O(r^{-3-\epsilon})$ for some $\epsilon > 0$ (faster than $1/r^3$ at infinity);
- as $r \to 0$, $V(r) = O(r^{-3/2+\delta})$ for some $\delta > 0$ (less singular than $1/r^{3/2}$ at the origin);
- V is piecewise continuous with at most finitely many discontinuities.

These assumptions cover a broad class of localized interactions (screened or Yukawa-type potentials, finite-range wells, atomic effective potentials). They intentionally exclude the pure Coulomb potential, which requires Coulomb-distorted asymptotic dynamics rather than the free asymptotes in Eqs. (5.69)–(5.69b).

5.6 Møller Wave Operators and Asymptotic Completeness

Given the asymptotic relations in Eqs. (5.69)–(5.69b), multiply on the left by $U^\dagger(t) = U(-t)$ to obtain

$$|\psi\rangle = \lim_{t\to-\infty} U^\dagger(t)\, U_0(t)\; |\psi_{\mathrm{in}}\rangle\,, \tag{5.70a}$$

$$|\psi\rangle = \lim_{t\to+\infty} U^\dagger(t)\, U_0(t)\; |\psi_{\mathrm{out}}\rangle\,. \tag{5.70b}$$

This motivates the definition of the Møller (wave) operators

$$\Omega_+ \equiv \lim_{t\to-\infty} U^\dagger(t)\, U_0(t), \tag{5.71a}$$

$$\Omega_- \equiv \lim_{t\to+\infty} U^\dagger(t)\, U_0(t), \tag{5.71b}$$

so that, whenever the limits exist in the strong sense on the appropriate domain,

$$|\psi\rangle = \Omega_+\; |\psi_{\mathrm{in}}\rangle\,, \tag{5.72a}$$

$$|\psi\rangle = \Omega_-\; |\psi_{\mathrm{out}}\rangle\,. \tag{5.72b}$$

Physically, Ω_+ maps an incoming free state to the interacting state that, evolved backwards in time, approaches the free profile $|\psi_{\mathrm{in}}\rangle$. Analogously, Ω_- maps an outgoing free state to the interacting orbit whose forward evolution approaches the free profile $|\psi_{\mathrm{out}}\rangle$.

A key structural identity is the intertwining relation

$$H\,\Omega_\pm = \Omega_\pm\, H_0, \tag{5.73}$$

which expresses that $\Omega_\pm$ map free generalized eigenstates into generalized eigenstates of H at the same energy (on the scattering spectrum).

A concise derivation starts from Eq. (5.71). For any fixed τ and for either choice of sign,

$$e^{iH\tau}\Omega_\pm = \lim_{t\to\mp\infty} e^{iH\tau}U^\dagger(t)U_0(t) = \lim_{t\to\mp\infty} U^\dagger(t-\tau)U_0(t) = \lim_{t\to\mp\infty} U^\dagger(t)U_0(t+\tau) = \Omega_\pm e^{iH_0\tau}, \tag{5.74}$$

where we only reparametrize the limit variable. Differentiating with respect to τ at $\tau = 0$ yields Eq. (5.73).

On the absolutely continuous subspace of H_0 relevant to scattering, the wave operators are isometries,

$$\Omega_\pm^\dagger \Omega_\pm = \mathbb{I}. \tag{5.75}$$

In general they are not surjective on the full Hilbert space: if H admits bound states, then $\mathrm{Ran}(\Omega_\pm)$ does not exhaust $\mathcal{H}$. One has instead

$$\Omega_\pm \Omega_\pm^\dagger = P_{R_\pm}, \tag{5.76}$$

the orthogonal projector onto the scattering range $R_\pm \equiv \mathrm{Ran}(\Omega_\pm)$.

Let $B \subset \mathcal{H}$ be the bound-state subspace (pure point spectrum of H) and let

$$R_\pm \equiv \mathrm{Ran}(\Omega_\pm) \tag{5.77}$$

be the set of interacting states possessing in/out free asymptotes. Under standard hypotheses for short-range potentials one has:

- **orthogonality:** $B \perp R_\pm$;
- **asymptotic completeness:**

$$\mathcal{H} = B \oplus R, \qquad R \equiv R_+ = R_- = B^\perp. \tag{5.78}$$

Thus, every state decomposes uniquely into a bound component (which remains localized for all times) and a scattering component (which admits free in/out asymptotes). In particular, the scattering theory is fully encoded by the restriction of the dynamics to R.

Pause and think

A scattering process may be highly nontrivial to describe explicitly. Therefore, it is natural to ask why one should expect the existence of a scattering operator. Explain your reasons.

5.6.1 The Scattering Operator

Since experiments only access free asymptotic data, one aims at relating $|\psi_{\text{out}}\rangle$ to $|\psi_{\text{in}}\rangle$. From Eqs. (5.72a)–(5.72b) one obtains

$$|\psi_{\text{out}}\rangle = \Omega_{-}^{\dagger}\,|\psi\rangle\,, \qquad |\psi\rangle = \Omega_{+}\,|\psi_{\text{in}}\rangle\,. \tag{5.79}$$

Hence,

$$|\psi_{\text{out}}\rangle = S\,|\psi_{\text{in}}\rangle\,, \qquad S \equiv \Omega_{-}^{\dagger}\Omega_{+}. \tag{5.80}$$

The operator S acts on the free Hilbert space of asymptotic states (equivalently, on the absolutely continuous subspace of H_0 relevant to scattering). Using Eq. (5.75), one finds on that space

$$S^{\dagger}S = SS^{\dagger} = \mathbb{I}, \tag{5.81}$$

i.e. S is unitary.

Because S maps free asymptotic orbits to free asymptotic orbits, conservation of energy in scattering is encoded by

$$[S, H_0] = 0. \tag{5.82}$$

This follows directly from the intertwining property in Eq. (5.73):

$$SH_0 = \Omega_{-}^{\dagger}\Omega_{+}H_0 = \Omega_{-}^{\dagger}H\Omega_{+} = H_0\Omega_{-}^{\dagger}\Omega_{+} = H_0 S, \tag{5.83}$$

where we used $H\Omega_{+} = \Omega_{+}H_0$ and its adjoint $\Omega_{-}^{\dagger}H = H_0\Omega_{-}^{\dagger}$.

Consequently, if $|\psi_{\text{out}}\rangle = S\,|\psi_{\text{in}}\rangle$, then for any incoming free wave packet,

$$\langle\psi_{\text{out}}|H_0|\psi_{\text{out}}\rangle = \left\langle\psi_{\text{in}}|S^{\dagger}H_0 S|\psi_{\text{in}}\right\rangle = \langle\psi_{\text{in}}|H_0|\psi_{\text{in}}\rangle\,. \tag{5.84}$$

5.6.2 Momentum Representation and On-Shell Structure

Let $|\mathbf{p}\rangle$ be the generalized eigenstates of H_0,

$$H_0\,|\mathbf{p}\rangle = E_{\mathbf{p}}\,|\mathbf{p}\rangle\,, \qquad E_{\mathbf{p}} = \frac{\mathbf{p}^2}{2m}, \tag{5.85}$$

normalized as

$$\left\langle\mathbf{p}'|\mathbf{p}\right\rangle = \delta^{(3)}(\mathbf{p}' - \mathbf{p}), \tag{5.86}$$

with coordinate representation (one consistent convention is)

$$\langle\mathbf{x}|\mathbf{p}\rangle = \frac{1}{(2\pi)^{3/2}}\,e^{i\mathbf{p}\cdot\mathbf{x}}. \tag{5.87}$$

A generic free state is expanded as

$$|\psi\rangle = \int d^3p\, \psi(\mathbf{p})\, |\mathbf{p}\rangle\,, \qquad \psi(\mathbf{p}) = \langle \mathbf{p}|\psi\rangle\,. \tag{5.88}$$

Because $[S, H_0] = 0$, in the momentum basis one has

$$0 = \langle \mathbf{p}'|\, [H_0, S]\, |\mathbf{p}\rangle = (E_{\mathbf{p}'} - E_{\mathbf{p}})\, \langle \mathbf{p}'|\, S\, |\mathbf{p}\rangle\,. \tag{5.89}$$

Hence $\langle \mathbf{p}'|\, S\, |\mathbf{p}\rangle$ vanishes unless $E_{\mathbf{p}'} = E_{\mathbf{p}}$, i.e. unless the transition is on-shell. Accordingly, one writes the standard decomposition

$$\langle \mathbf{p}'|\, S\, |\mathbf{p}\rangle = \delta^{(3)}(\mathbf{p}' - \mathbf{p}) - 2\pi i\, \delta\big(E_{\mathbf{p}'} - E_{\mathbf{p}}\big)\, \langle \mathbf{p}'|\, T(E_{\mathbf{p}})\, |\mathbf{p}\rangle\,, \tag{5.90}$$

which defines the on-shell T-matrix (often simply denoted by T when no ambiguity arises). The first term is the amplitude for no scattering; the second encodes the scattering event and contains the energy-conserving delta distribution.

It is convenient to introduce the scattering amplitude f via

$$f(\mathbf{p}' \leftarrow \mathbf{p}) \equiv -(2\pi)^2\, m\, \langle \mathbf{p}'|\, T(E_{\mathbf{p}})\, |\mathbf{p}\rangle\,, \tag{5.91}$$

so that Eq. (5.90) becomes

$$\langle \mathbf{p}'|\, S\, |\mathbf{p}\rangle = \delta^{(3)}(\mathbf{p}' - \mathbf{p}) + \frac{i}{2\pi m}\, \delta\big(E_{\mathbf{p}'} - E_{\mathbf{p}}\big)\, f(\mathbf{p}' \leftarrow \mathbf{p}). \tag{5.92}$$

5.7 Lippmann-Schwinger Equation and the T-Matrix

Define the free resolvent

$$G_0(E) \equiv \frac{1}{E - H_0 + i0}, \tag{5.93}$$

where $+i0$ is the limiting prescription encoding outgoing boundary conditions (similarly, $-i0$ corresponds to incoming).

The stationary scattering states $\left|\psi_{\mathbf{p}}^{(\pm)}\right\rangle$ are defined by the Lippmann-Schwinger equation

$$\left|\psi_{\mathbf{p}}^{(\pm)}\right\rangle = |\mathbf{p}\rangle + G_0(E_{\mathbf{p}} \pm i0)\, V\, \left|\psi_{\mathbf{p}}^{(\pm)}\right\rangle. \tag{5.94}$$

These are generalized eigenstates of H at energy $E_{\mathbf{p}}$:

$$H\left|\psi_{\mathbf{p}}^{(\pm)}\right\rangle = E_{\mathbf{p}}\left|\psi_{\mathbf{p}}^{(\pm)}\right\rangle. \tag{5.95}$$

The $\pm$ prescription fixes the large-r asymptotics (outgoing vs incoming spherical waves).

Introduce now the T operator by

$$T(E) \equiv V + V\,G_0(E)\,T(E), \tag{5.96}$$

equivalently,

$$T(E) = V + V\,G_0(E)\,V + V\,G_0(E)\,V\,G_0(E)\,V + \cdots, \tag{5.97}$$

i.e. the Born series (when convergent). By acting with $\langle \mathbf{p}'|$ and $|\mathbf{p}\rangle$ one recovers the integral equation for matrix elements,

$$\langle \mathbf{p}'|\,T(E)\,|\mathbf{p}\rangle = \langle \mathbf{p}'|\,V\,|\mathbf{p}\rangle + \int d^3k\,\frac{\langle \mathbf{p}'|\,V\,|\mathbf{k}\rangle\ \langle \mathbf{k}|\,T(E)\,|\mathbf{p}\rangle}{E - \frac{\mathbf{k}^2}{2m} + i0}. \tag{5.98}$$

For short-range potentials and stationary scattering at energy $E = p^2/(2m)$, one has in coordinate space the asymptotic form

$$\psi_{\mathbf{p}}^{(+)}(\mathbf{x}) \equiv \left\langle \mathbf{x}|\psi_{\mathbf{p}}^{(+)}\right\rangle \sim \frac{1}{(2\pi)^{3/2}}\left[e^{i\mathbf{p}\cdot\mathbf{x}} + \frac{e^{ipr}}{r}\,f(\hat{\mathbf{x}} \leftarrow \hat{\mathbf{p}})\right], \qquad r = |\mathbf{x}| \to \infty, \tag{5.99}$$

where $\hat{\mathbf{x}} = \mathbf{x}/r$ and the scattering amplitude depends on the scattering angle θ via $\cos\theta = \hat{\mathbf{x}} \cdot \hat{\mathbf{p}}$ (for central potentials, f depends only on θ). The f introduced in Eq. (5.91) is precisely the coefficient of the outgoing spherical wave in Eq. (5.99); this identification fixes the conventional constants.

The decomposition in Eq. (5.99) requires that $V(r)$ decays sufficiently fast at infinity, so that the asymptotic dynamics is genuinely free up to an outgoing spherical wave. Long-range interactions, notably the Coulomb potential $V(r) = \alpha/r$, distort the phase already at arbitrarily large distances. In that case the free plane wave must be replaced by a Coulomb-distorted incoming wave, and the asymptotic analysis must be modified accordingly.
Nevertheless, the differential cross section remains operationally well defined. For non-relativistic scattering of a particle of charge $Z_1 e$ off a fixed center of charge $Z_2 e$, one obtains the Rutherford formula

$$\left.\frac{d\sigma}{d\Omega}\right|_{\text{Rutherford}} = \left(\frac{Z_1 Z_2 \alpha}{4E}\right)^2 \frac{1}{\sin^4(\theta/2)}, \tag{5.100}$$

with $E = p^2/(2m)$ and $\alpha = e^2/(4\pi)$ in natural units. Equation (5.100) is a useful benchmark: it exhibits the forward singularity characteristic of long-range forces.

Pause and think
Repeatability is crucial in experimental physics. Explain why and relate your explanations to scattering.

5.8 Cross Sections and Repeatability of Scattering Experiments

A single scattering event only reveals whether the particle was deflected into a given detector. Meaningful information requires repetition, leading to the operational definition of the differential cross section.

For an incident beam characterized by an incident flux j_{inc} (particles per unit time per unit area) and a number of scattered particles per unit time detected in a solid angle element $d\Omega$,

$$\frac{dN_{\text{sc}}}{dt}(d\Omega), \tag{5.101}$$

the differential cross section is defined by

$$\frac{d\sigma}{d\Omega} \equiv \frac{1}{j_{\text{inc}}}\frac{dN_{\text{sc}}}{dt}(d\Omega). \tag{5.102}$$

For a plane-wave normalization, the standard result (consistent with Eq. (5.99)) is

$$\frac{d\sigma}{d\Omega} = \left|f(\mathbf{p}' \leftarrow \mathbf{p})\right|^2, \qquad |\mathbf{p}'| = |\mathbf{p}| = p, \tag{5.103}$$

where $\mathbf{p}' = p\,\hat{\mathbf{x}}$ labels the asymptotic outgoing direction.

A rigorous connection between the abstract S operator and measurable event rates is obtained by considering incident wave packets peaked around $\mathbf{p}_0$, repeated many times with randomized transverse impact parameters. Let $\psi_{\text{in}}(\mathbf{p})$ be a normalized packet peaked at $\mathbf{p}_0$. A spatial translation by $\boldsymbol{\rho}$ in the plane perpendicular to $\mathbf{p}_0$ is implemented by

$$(D_{\boldsymbol{\rho}}\psi)(\mathbf{x}) = \psi(\mathbf{x} - \boldsymbol{\rho}), \tag{5.104}$$

which in momentum space yields the phase factor

$$(D_{\boldsymbol{\rho}}\psi)(\mathbf{p}) = e^{-i\boldsymbol{\rho}\cdot\mathbf{p}}\,\psi(\mathbf{p}). \tag{5.105}$$

Repeating the experiment with a uniform distribution of $\boldsymbol{\rho}$ realizes a translational average that removes dependence on the detailed packet shape and isolates the on-shell scattering information. Under this averaging, the detection probability in $d\Omega$ scales linearly with the illuminated transverse area, and the proportionality constant is precisely the differential cross section in Eq. (5.103).

5.9 The Optical Theorem

Write

$$S = \mathbb{I} + i\mathcal{T}, \tag{5.106}$$

where $\mathcal{T}$ is (up to conventions) the transition operator. Unitarity in Eq. (5.81) implies

$$\mathcal{T} - \mathcal{T}^\dagger = i\mathcal{T}^\dagger\mathcal{T}. \tag{5.107}$$

Taking matrix elements between momentum eigenstates and using the on-shell structure in Eqs. (5.90)–(5.92), one obtains a relation between the imaginary part of the forward scattering amplitude and the total cross section. In the present non-relativistic normalization in Eqs. (5.91)–(5.92), the optical theorem reads

$$\sigma_{\text{tot}}(p) = \int d\Omega \frac{d\sigma}{d\Omega} = \frac{4\pi}{p} \Im f(\theta = 0), \tag{5.108}$$

where $\theta = 0$ corresponds to forward scattering ($\mathbf{p}' = \mathbf{p}$). The content of Eq. (5.108) is that a measurable quantity (the total cross section) is fixed by the imaginary part of a single complex number (the forward amplitude), which cannot be inferred from $|f|^2$ alone.

5.10 Proposed Exercises

Exercise 1

Consider the generic relativistic process

$$1 + 2 \longrightarrow 3 + 4 + \cdots + N, \tag{5.109}$$

with spinless asymptotic particles and on-shell four-momenta $p_i^2 = m_i^2$. Work in natural units $c = 1$.

(a) Starting from the $4N$ components of $\{p_i^\mu\}$, show that the on-shell constraints and total four-momentum conservation reduce the number of independent kinematic degrees of freedom to $3N - 10$.
(b) Define the invariant mass W of the initial state as

$$W^2 \equiv (p_1 + p_2)^2, \tag{5.110}$$

and show that W is Lorentz invariant. Evaluate W explicitly in:

- the center-of-mass (CM) frame of the *two-body initial state*;
- the laboratory (target) frame where particle 2 is at rest, $p_2 = (m_2, \mathbf{0})$.

Derive the standard target-frame expression

$$W^2 = m_1^2 + m_2^2 + 2m_2 E_1, \tag{5.111}$$

and comment on the inequality $W < E_1 + m_2$ in the target frame.

(c) Consider now a decay $1 \to 2 + 3 + \cdots + N$. Explain why it is conceptually misleading to speak of a "target frame" for a decay, and identify the natural frame in which $W = m_1$ is read off immediately. Rewrite Eq. (5.110) accordingly.

Exercise 2

Consider elastic two-body scattering

$$1 + 2 \longrightarrow 1' + 2', \tag{5.112}$$

with $m_1 = m_2 = m$ (equal masses), in natural units $c = 1$. Let the Mandelstam invariants be

$$s = (p_1 + p_2)^2, \qquad t = (p_1 - p_{1'})^2, \qquad u = (p_1 - p_{2'})^2. \tag{5.113}$$

(a) In the CM frame, show that the magnitude of the three-momentum is

$$p^\star = \frac{1}{2}\sqrt{s - 4m^2}, \tag{5.114}$$

and derive the relation between t and the CM scattering angle $\theta^\star$:

$$t = -2\, p^{\star 2}\,(1 - \cos\theta^\star). \tag{5.115}$$

(b) Using four-momentum conservation, prove the identity

$$s + t + u = 4m^2. \tag{5.116}$$

(c) The laboratory (target) frame is defined by $p_2 = (m, \mathbf{0})$ and an incoming beam along $\hat{z}$. Show that the CM and laboratory frames are related by a boost along $\hat{z}$ with $\gamma_{\text{CM}} = E^\star/m$ (where $E^\star = \sqrt{s}/2$), and derive the angle relations

$$\tan\theta = \frac{\sin\theta^\star}{\gamma_{\text{CM}}\,(1 + \cos\theta^\star)}, \qquad \tan\varphi = \frac{\sin\theta^\star}{\gamma_{\text{CM}}\,(1 - \cos\theta^\star)}. \tag{5.117}$$

Deduce the compact constraint

$$\tan\theta\,\tan\varphi = 1 - \beta_{\text{CM}}^2, \tag{5.118}$$

and recover the non-relativistic limit $\theta + \varphi \to \pi/2$ as $\beta_{\text{CM}} \to 0$.

Exercise 3

This exercise collects threshold and decay kinematics in a way that explains how to use the derivations of this Chapter.

(a) (*Production threshold in the laboratory frame*) Consider the reaction

$$1 + 2 \longrightarrow 3 + 4 + \cdots + N, \tag{5.119}$$

with particle 2 at rest in the laboratory frame. Starting from the invariant condition

$$W^2 = (p_1 + p_2)^2 \geq \left(\sum_{i=3}^{N} m_i \right)^2, \tag{5.120}$$

derive the laboratory threshold on the projectile energy,

$$E_{1,\mathrm{thr}} = \frac{\left(\sum_{i=3}^{N} m_i \right)^2 - (m_1^2 + m_2^2)}{2m_2}, \tag{5.121}$$

and rewrite it as a condition on the threshold kinetic energy $T_{1,\mathrm{thr}} = E_{1,\mathrm{thr}} - m_1$.

(b) (*Two-body decay in the parent rest frame*) Consider a two-body decay $1 \rightarrow 2 + 3$ in the parent rest frame $p_1 = (m_1, \mathbf{0})$. Show that the decay-product energies are fixed purely by masses:

$$E_2^\star = \frac{m_1^2 + m_2^2 - m_3^2}{2m_1}, \qquad E_3^\star = \frac{m_1^2 - m_2^2 + m_3^2}{2m_1}. \tag{5.122}$$

Compute the common momentum magnitude $p^\star$ in terms of the Källén function $\lambda(m_1^2, m_2^2, m_3^2)$.

(c) (*Energy endpoints for decays in flight*) Assume particle 1 decays in flight in the laboratory frame with boost parameters β, γ along $\hat{z}$. Show that the lab energy of particle 2 satisfies

$$E_2 = \gamma \left(E_2^\star + \beta p^\star \cos\theta^\star \right), \tag{5.123}$$

and derive the endpoints

$$E_{2,\max} = \gamma(E_2^\star + \beta p^\star), \qquad E_{2,\min} = \gamma(E_2^\star - \beta p^\star), \tag{5.124}$$

corresponding to $\cos\theta^\star = \pm 1$. Explain physically why two distinct lab configurations (forward/backward emission in the parent rest frame) map to different energies measured in the lab.

Essential Bibliography

- Barone V (2006) Lezioni di Relatività, Bollati Boringhieri. A rigorous and elegant introduction to special relativity and relativistic kinematics. Particularly useful for mastering Lorentz invariants, Mandelstam variables, and frame transformations essential to relativistic scattering processes.
- Taylor JR (2006) Scattering theory: the quantum theory of nonrelativistic collisions. Dover Publications, Mineola, NY, reprint of the 1972 Wiley edition. A clear and systematic development of the Lippmann– Schwinger equation, Green functions, and the Born series. It remains one of the most pedagogically effective presentations of stationary scattering theory.
- Newton RG (2002) Scattering theory of waves and particles. Dover Publications (2002), reprint of the 2nd ed., Springer (1982). A rigorous and comprehensive treatment of classical and quantum scattering theory, including partial waves, analytic properties of the S-matrix, phase shifts, resonances, and inverse scattering. It provides the mathematical and conceptual backbone underlying both potential scattering and collision theory.

Quantum Field Theory

6

What we observe is not nature itself, but nature exposed to our method of questioning.

–Werner Heisenberg

Abstract

This Chapter develops the minimal conceptual and computational toolkit of quantum field theory needed to describe fundamental interactions in a relativistically consistent way. Starting from the canonical (Hamiltonian) structure of classical fields, we motivate second quantization through the failure of fixed-particle relativistic quantum mechanics, and we construct Fock space, creation/annihilation operators, and operator-valued fields. We then introduce Pauli-Weisskopf quantization for scalar fields (complex and real), derive the associated Noether charge and the particle-antiparticle interpretation, and emphasize the role of locality and microcausal structure. The second half of the Chapter builds the perturbative framework: propagators as Green functions, the $i\epsilon$ prescription and time ordering, Dyson expansions, and Feynman diagrams as devices for S-matrix elements and correlation functions. The Chapter closes with representative quantum electrodynamics (QED) tree-level applications (spin sums, trace technology, cross sections, and crossing symmetry).

O. Luongo, *Fundamental Interactions*,
https://doi.org/10.1007/978-3-032-27413-7_6

Fig. 6.1 Portrait of Freeman John Dyson by Betsy Devine, 2008. Reprinted from Wikimedia under CC license

A key character: *Freeman John Dyson (1923–2020), Fig.* 6.1.

Freeman Dyson was one of the central figures in shaping the modern understanding of QFT, not by introducing a single new formalism, but by revealing the deep unity hidden behind apparently different approaches. Born in Crowthorne, England, into an intellectually rich family environment: his father was a composer and his mother a lawyer. Dyson showed exceptional mathematical talent from an early age. He studied mathematics at Trinity College, Cambridge, during the final years of World War II, and later moved to the United States, where he became associated with the Institute for Advanced Study in Princeton.

Dyson's decisive contribution came in the late 1940s, when QED was undergoing a profound conceptual and technical crisis. Three distinct formulations of QED had emerged: Schwinger's operator-based formalism, Tomonaga's covariant approach, and Feynman's path-integral and diagrammatic method. These appeared, at first sight, to be radically different theories. Dyson demonstrated that they were, in fact, mathematically equivalent and physically identical. His synthesis clarified the structure of renormalization and showed how Feynman diagrams arise naturally.

Furthermore, Dyson made important contributions to statistical mechanics, condensed matter physics, random matrix theory, and nuclear physics. His name is associated with the Dyson series, Dyson equations, and Dyson's Brownian motion model. Yet he always resisted narrow specialization, preferring to work at the boundaries between disciplines. At the Institute for Advanced Study he cultivated an atmosphere of intellectual openness, interacting not only with physicists and mathematicians but also with biologists, historians, and philosophers.

Dyson was also a remarkable scientific writer and thinker. His essays and books, including *Disturbing the Universe* and *Infinite in All Directions*, combined technical depth with literary clarity, reflecting on science as a human enterprise shaped by creativity, uncertainty, and moral responsibility.

In the context of QFT, his role was that of the architect rather than the builder: showing how the pieces assembled into a single, logically consistent edifice.

6.1 Fock Space

Let us start with states and wave functions in the first-quantized picture.

Hence, the construction starts from the one-particle Hilbert space of states, denoted by $\mathcal{H}_1$. The first essential choice is a complete orthonormal basis $\{\,|\underline{\alpha}\rangle\,\}_{\underline{\alpha}}$ for a single particle, so that

$$\mathcal{H}_1 = \mathrm{Span}\{\,|\underline{\alpha}\rangle\,\}_{\underline{\alpha}}. \tag{6.1}$$

Here $|\underline{\alpha}\rangle$ (with $\underline{\alpha} = 1, 2, \ldots$) denotes a basis vector of the one-particle Hilbert space; the underlining is a notational reminder that the label refers to *one-particle* basis states.[1] Given the basis vector $|\underline{\alpha}\rangle$, the corresponding one-body wave function is

$$\psi_{\underline{\alpha}}(x) = \langle x|\underline{\alpha}\rangle. \tag{6.2}$$

The entire second-quantized construction hinges on this initial one-particle basis choice.

Consider now a system of N identical particles, each described by the same one-particle Hilbert space $\mathcal{H}_1$. If, for the moment, we pretend that the particles are

[1] A basis is an *ordered* set. Fixing a basis in $\mathcal{H}_1$ implicitly fixes an enumeration of one-particle eigenstates. For instance, for a one-dimensional harmonic oscillator one can order the basis by increasing energy: ground state $\underline{\alpha} = 1$, first excited state $\underline{\alpha} = 2$, etc. Another common choice is the momentum basis (plane waves) for a particle in a box with suitable boundary conditions.

distinguishable, the N-particle Hilbert space is the N-fold tensor product

$$\mathcal{H}_N = \underbrace{\mathcal{H}_1 \otimes \cdots \otimes \mathcal{H}_1}_{N \text{ times}}. \tag{6.3}$$

A natural product basis of $\mathcal{H}_N$ is

$$|\alpha_1 \cdots \alpha_N\rangle \equiv |\alpha_1\rangle \otimes \cdots \otimes |\alpha_N\rangle, \tag{6.4}$$

where $|\alpha_i\rangle$ denotes the one-particle basis eigenstate assigned to the i-th particle ($i = 1, \ldots, N$). In this notation the indices are *not* underlined because they label the state of a specific particle in the N-body tensor-product space. The corresponding wave functions factorize:

$$\psi_{\alpha_1 \cdots \alpha_N}(x_1, \ldots, x_N) = \psi_{\alpha_1}(x_1) \cdots \psi_{\alpha_N}(x_N). \tag{6.5}$$

Since $\dim(V \otimes W) = \dim(V)\, \dim(W)$, one has

$$\dim \mathcal{H}_N = \left(\dim \mathcal{H}_1\right)^N. \tag{6.6}$$

Consider N identical particles with only two available one-particle states, e.g. spin up/down: $\mathcal{H}_1 = \mathrm{Span}\{|\uparrow\rangle, |\downarrow\rangle\}$, so $\dim \mathcal{H}_1 = 2$. Then $\dim \mathcal{H}_N = 2^N$. Equivalently, basis vectors of $\mathcal{H}_N$ correspond to binary words of length N.

The physically relevant case for identical quantum particles is *indistinguishability*. Exchanging two particles cannot generate a new physical state; consequently the N-body wave function must be either totally symmetric (bosons) or totally antisymmetric (fermions). Concretely, for any i, j,

$$\psi(x_1, \ldots, x_i, \ldots, x_j, \ldots, x_N) = \zeta\, \psi(x_1, \ldots, x_j, \ldots, x_i, \ldots, x_N), \qquad \zeta = \pm 1, \tag{6.7}$$

with

$$\zeta = +1 \iff \text{bosons} \quad \text{(totally symmetric)}, \tag{6.8}$$

$$\zeta = -1 \iff \text{fermions} \quad \text{(totally antisymmetric)}. \tag{6.9}$$

It is convenient to denote by $\mathcal{H}_N^{(\zeta)}$ the subspace of $\mathcal{H}_N$ with the correct exchange symmetry. Formally,

---

[2] S_N is the group of permutations π of N elements. Any permutation can be decomposed into a (non-unique) product of transpositions, but the parity (even/odd number of transpositions) is well-defined. The sign is $\mathrm{sign}(\pi) = +1$ for even π and $\mathrm{sign}(\pi) = -1$ for odd π (do not confuse π with the canonical momentum).

$$\mathcal{H}_N^{(\zeta)} = \left\{ |\psi\rangle \in \mathcal{H}_N : \pi|\psi\rangle = \zeta^{\mathrm{sign}(\pi)}|\psi\rangle, \ \forall\pi \in S_N \right\}, \tag{6.10}$$

where S_N is the permutation group of N objects.[2] By construction, $\mathcal{H}_N^{(\zeta)}$ is strictly smaller than the tensor-product space $\mathcal{H}_N$.

A basis of $\mathcal{H}_N^{(\zeta)}$ is obtained by symmetrizing (bosons) or antisymmetrizing (fermions) the product basis of $\mathcal{H}_N$.

Fermions: Slater determinants.
Given one-particle orbitals $\psi_{\underline{\alpha}}(x)$, the antisymmetrized N-fermion wave function is

$$\psi_{\alpha_1\cdots\alpha_N}(x_1,\ldots,x_N) = \frac{1}{\sqrt{N!}}\det\begin{pmatrix} \psi_{\alpha_1}(x_1) & \cdots & \psi_{\alpha_1}(x_N) \\ \vdots & \ddots & \vdots \\ \psi_{\alpha_N}(x_1) & \cdots & \psi_{\alpha_N}(x_N) \end{pmatrix}. \tag{6.11}$$

This determinant changes sign under exchange of two coordinates (exchange of two columns), ensuring total antisymmetry. If $\alpha_i = \alpha_j$ for some $i \neq j$, two rows coincide and the determinant vanishes: two fermions cannot occupy the same one-particle state (Pauli exclusion). The prefactor $1/\sqrt{N!}$ ensures normalization.

Bosons: symmetrized products.
For bosons, basis vectors are symmetric sums of products. If $n_{\underline{i}}$ denotes the occupation of the $\underline{i}$-th one-particle eigenstate, the properly normalized symmetric wave function can be written as

$$\psi_{\alpha_1\cdots\alpha_N}(x_1,\ldots,x_N) = \left(\frac{n_{\underline{1}}!\cdots n_{\underline{\alpha}}!}{N!}\right)^{1/2} \sum_{\pi(x_1,\ldots,x_N)} \psi_{\alpha_1}(x_1)\cdots\psi_{\alpha_N}(x_N), \tag{6.12}$$

where the sum runs over permutations of the coordinates that produce distinct terms (if some α's repeat, there are fewer distinct permutations). The prefactor fixes normalization.

A properly symmetrized/antisymmetrized N-particle basis ket will be denoted by

$$|\psi\rangle = |\alpha_1\cdots\alpha_N\rangle_\zeta. \tag{6.13}$$

At this stage, then, the notation still resembles first quantization: a state is specified by listing which one-particle eigenstate is assigned to each particle. For large N this becomes unwieldy, and moreover it carries redundant information when particles are indistinguishable. Finally, this framework is naturally tied to a fixed particle number N, whereas many applications require a Hilbert space that accommodates varying N (grand-canonical settings and, in QFT, genuine particle creation/annihilation).

6.1.1 Occupation-Number Representation

The key simplification is to describe the same N-particle states by *how many* particles occupy each one-particle basis state, rather than which particle occupies which state.

For each one-particle eigenstate label $\underline{i} \geq 1$, define the occupation number $n_{\underline{i}}$ as the number of particles sitting in that eigenstate. Equivalently, one may write $n_{\underline{\alpha}}$ as the occupation of the state $|\underline{\alpha}\rangle$. Thus $n_{\underline{i}} = n$ means: "the $\underline{i}$-th basis state of $\mathcal{H}_1$ is occupied by n particles". In general the one-particle spectrum is infinite (e.g. harmonic oscillator), hence the sequence of occupations is infinite.

The allowed values depend on statistics:

$$n_{\underline{i}} \in \mathbb{N} \quad \text{for bosons,} \qquad n_{\underline{i}} \in \{0, 1\} \quad \text{for fermions.} \tag{6.14}$$

The total number of particles in an N-particle sector is recovered as

$$N = \sum_{\underline{i} \geq 1} n_{\underline{i}}. \tag{6.15}$$

Fixing once and for all the one-particle basis $\{|\underline{\alpha}\rangle\}$, we map the description

$$|\alpha_1 \cdots \alpha_N\rangle \longrightarrow |n_{\underline{1}} n_{\underline{2}} \cdots\rangle. \tag{6.16}$$

Underlining again distinguishes the role of indices: in the first notation, indices keep track of particles (even though that becomes physically redundant after symmetrization/antisymmetrization); in the second, indices enumerate *one-particle basis states.*

The occupation-number sequence is infinite but contains only finitely many non-zero entries for finite N: at most N occupations can be non-zero, while all others vanish. In practice one lists only the non-zero occupations. If $\mathcal{H}_1$ is finite dimensional (e.g. two-level systems), the occupation list is finite.

Example (*three-level system*)

Take $\mathcal{H}_1 = \text{Span}\{|1\rangle, |2\rangle, |3\rangle\}$. For any N, an occupation-number basis state has the form $|n_{\underline{1}}, n_{\underline{2}}, n_{\underline{3}}\rangle$. For $N = 1$, the first-quantized state $|\alpha_1\rangle = |3\rangle$ maps to $|0\,0\,1\rangle$, i.e. one particle in level 3.

For $N = 2$ indistinguishable particles, a first-quantized basis ket can be written as $|\alpha_1\alpha_2\rangle_\zeta = |1, 3\rangle_\zeta$, which maps to $|1\,0\,1\rangle$. This occupation pattern is admissible both for bosons and fermions (no double occupation).

For $N = 4$ (possible only for bosons in a three-level system), a first-quantized label $|1, 1, 2, 3\rangle_\zeta$ maps to $|2\,1\,1\rangle$: two particles occupy level 1, and one particle occupies each of levels 2 and 3. In first-quantized language the underlying product wave function would be $\psi_1(x_1)\psi_1(x_2)\psi_2(x_3)\psi_3(x_4)$, then symmetrized for bosons.

For bosons, the occupation numbers fix the symmetric wave function uniquely. On the other hand, for fermions, antisymmetry implies an overall sign ambiguity. To remove it, one adopts a convention:the many-fermion basis ket is taken with positive sign when the occupied one-particle eigenstates appear in increasing order with respect to the chosen one-particle ordering $\underline{1} < \underline{2} < \underline{3} < \cdots$. For example, one uses $|2\,4\rangle_\zeta$ rather than $|4\,2\rangle_\zeta$.

Fock space.

The Fock space $\mathcal{F}$ is the Hilbert space spanned by all admissible occupation-number basis vectors, for all particle numbers $N \geq 0$. It decomposes as the direct sum

$$\mathcal{F} = \mathcal{F}_0 \oplus \mathcal{F}_1 \oplus \mathcal{F}_2 \oplus \cdots = \bigoplus_{N \geq 0} \mathcal{F}_N, \tag{6.17}$$

where $\mathcal{F}_N$ is the fixed-N sector (isomorphic to $\mathcal{H}_N^{(\zeta)}$ written in occupation numbers).

The sector $\mathcal{F}_0 = \text{Span}\{|0\rangle\}$ is one-dimensional and is generated by the vacuum state $|0\rangle$. The symbol $|0\rangle$ is conventional and *does not* mean the number 0 nor the null vector of the Hilbert space:

$$|0\rangle \neq 0, \qquad |0\rangle \neq |e\rangle, \tag{6.18}$$

where $|e\rangle$ denotes the additive identity (null vector) of the vector space.[3] The sector $\mathcal{F}_1$ coincides with the one-particle Hilbert space in occupation form. For $N \geq 2$, $\mathcal{F}_N$ contains only the properly (anti)symmetrized states; e.g. $\mathcal{F}_2$ is not all of $\mathcal{F}_1 \otimes \mathcal{F}_1$, but only its symmetric (bosons) or antisymmetric (fermions) subspace.

A generic Fock-space vector is a superposition of occupation-number basis kets,

$$|\Psi\rangle = \sum_{n_{\underline{1}}, n_{\underline{2}}, \ldots} c_{n_{\underline{1}} n_{\underline{2}} \cdots} \, |n_{\underline{1}}\, n_{\underline{2}} \cdots\rangle. \tag{6.19}$$

Such superpositions over different particle numbers are physical. They arise, for instance, in superconductivity in the grand-canonical formulation (BCS-type states), and in quantum optics where coherent states superpose different photon-number sectors.

[3] The neutral (null) vector $|e\rangle \in \mathcal{F}$ satisfies $|\psi\rangle + |e\rangle = |\psi\rangle$ for all $|\psi\rangle \in \mathcal{F}$, and has zero norm: $\langle e|e\rangle = 0$ and $\langle \psi|e\rangle = 0$ for all $|\psi\rangle$.

Finally, the occupation-number representation makes sense precisely because particles are indistinguishable: it is meaningful to count how many particles occupy each one-particle state only when exchanging particles has no physical content. This is the structural reason why the map in Eq. (6.16) is viable and why Fock space is the natural arena for many-body and field-theoretic problems.

6.1.2 Creation and Annihilation Operators

We now introduce the basic operators that generate the entire operator algebra on $\mathcal{F}$. Since $\{|n_{\underline{1}}\, n_{\underline{2}} \cdots\rangle\}$ is a basis, an operator on $\mathcal{F}$ is defined once its action on all basis vectors is specified; linearity then extends the definition to all vectors.

6.1.3 Definitions on the Occupation-Number Basis

The creation operator $a^{\dagger}_{\underline{\alpha}}$ adds one particle in the one-particle state $|\underline{\alpha}\rangle$. For bosons it acts as

$$a^{\dagger}_{\underline{\alpha}} \,|\, \cdots\, n_{\underline{\alpha}}\, \cdots\rangle \equiv \sqrt{n_{\underline{\alpha}}+1}\,|\, \cdots\, (n_{\underline{\alpha}}+1)\, \cdots\rangle. \tag{6.20}$$

For fermions it is defined by

$$a^{\dagger}_{\underline{\alpha}} \,|\, \cdots\, n_{\underline{\alpha}}\, \cdots\rangle \equiv \begin{cases} (-1)^{\sum_{\underline{i}<\underline{\alpha}} n_{\underline{i}}}\,|\, \cdots\, 1\, \cdots\rangle, & n_{\underline{\alpha}} = 0, \\ 0, & n_{\underline{\alpha}} = 1, \end{cases} \tag{6.21}$$

so that attempting to create a fermion in an already occupied state annihilates the vector (Pauli principle), while creating into an empty state produces a sign determined by the number of occupied states preceding $\underline{\alpha}$ in the fixed one-particle ordering. This sign is essential for obtaining the canonical fermionic anticommutation relations.

The annihilation operator $a_{\underline{\alpha}}$ removes one particle from the one-particle state $|\underline{\alpha}\rangle$. It is the Hermitian conjugate of the creation operator, $a_{\underline{\alpha}} = (a^{\dagger}_{\underline{\alpha}})^{\dagger}$. Explicitly, for bosons

$$a_{\underline{\alpha}} \,|\, \cdots\, n_{\underline{\alpha}}\, \cdots\rangle \equiv \sqrt{n_{\underline{\alpha}}}\,|\, \cdots\, (n_{\underline{\alpha}}-1)\, \cdots\rangle, \tag{6.22}$$

and for fermions

$$a_{\underline{\alpha}} \,|\, \cdots\, n_{\underline{\alpha}}\, \cdots\rangle \equiv \begin{cases} (-1)^{\sum_{\underline{i}<\underline{\alpha}} n_{\underline{i}}}\,|\, \cdots\, 0\, \cdots\rangle, & n_{\underline{\alpha}} = 1, \\ 0, & n_{\underline{\alpha}} = 0. \end{cases} \tag{6.23}$$

For bosons, the square-root factors can be remembered as the square root of the larger of the initial and final occupation numbers; in particular, $\sqrt{n_{\underline{\alpha}}}$ ensures that annihilation on an empty level returns the null vector rather than producing an unphysical occupation -1. For fermions, the same ordering sign appears as in Eq. (6.21).

At this stage, it is interesting to deal with the vacuum state. This is defined by the requirement that every annihilation operator kills it:

$$a_{\underline{\alpha}} |0\rangle = 0, \qquad \forall \underline{\alpha}. \tag{6.24}$$

This definition applies both to bosons and fermions.[4]

By construction $\mathcal{F}$ contains all particle-number sectors, cf. Eq. (6.17). Creation and annihilation operators map $\mathcal{F}_N \to \mathcal{F}_{N+1}$ and $\mathcal{F}_N \to \mathcal{F}_{N-1}$, respectively, and thus change the total particle number.

Repeated application of creation operators builds the occupation-number basis from the vacuum state $|0\rangle$:

$$|n_{\underline{1}} n_{\underline{2}} \cdots\rangle = \prod_{\underline{\alpha}} \frac{(a^\dagger_{\underline{\alpha}})^{n_{\underline{\alpha}}}}{\sqrt{n_{\underline{\alpha}}!}} |0\rangle. \tag{6.25}$$

We notice that, for fermions, factorials are irrelevant because $n_{\underline{\alpha}} \in \{0, 1\}$, hence $\sqrt{n_{\underline{\alpha}}!} = 1$. In the fermionic case the product is understood as ordered left-to-right according to $\underline{1} < \underline{2} < \underline{3} < \cdots$.

▶ **Pause and Think** Annihilate versus destroy: the two words may appear synonymous. Why, then, do we speak of an *annihilation operator* rather than a *destruction operator*? Reflect on the physical and mathematical meaning of a and $a^\dagger$.

6.1.4 Algebra of Creation and Annihilation Operators

We now introduce the commutation/anticommutation relations associated with ladder operators. For bosons we use the commutator $[\,\cdot\,,\,\cdot\,]$,

$$[A, B] \equiv AB - BA, \tag{6.26}$$

[4] In Eq. (6.24), the symbol 0 on the right-hand side is a shorthand for the null vector $|e\rangle$ of $\mathcal{F}$, since operators map vectors in $\mathcal{F}$ to vectors in $\mathcal{F}$, not to real numbers.

while for fermions we use the anticommutator $\{\cdot, \cdot\}$,

$$\{A, B\} \equiv AB + BA. \tag{6.27}$$

Bosons.
The canonical commutation relations are

$$[a_{\underline{\alpha}}^\dagger, a_{\underline{\beta}}^\dagger] = 0, \qquad [a_{\underline{\alpha}}, a_{\underline{\beta}}] = 0, \qquad [a_{\underline{\alpha}}, a_{\underline{\beta}}^\dagger] = \delta_{\underline{\alpha\beta}}. \tag{6.28}$$

Two creation operators commute, and two annihilation operators commute. A creation and an annihilation operator commute unless they refer to the same one-particle state, in which case their commutator is the Kronecker delta. These relations follow directly from the defining action in Eqs. (6.20)–(6.22) on the Fock basis. For instance, assuming $\underline{\alpha} = \underline{\beta}$ one checks

$$a_{\underline{\alpha}} a_{\underline{\alpha}}^\dagger \,|\cdots n_{\underline{\alpha}} \cdots\rangle = \sqrt{n_{\underline{\alpha}} + 1}\, a_{\underline{\alpha}} \,|\cdots (n_{\underline{\alpha}} + 1) \cdots\rangle = (n_{\underline{\alpha}} + 1)\,|\cdots n_{\underline{\alpha}} \cdots\rangle, \tag{6.29}$$

$$a_{\underline{\alpha}}^\dagger a_{\underline{\alpha}} \,|\cdots n_{\underline{\alpha}} \cdots\rangle = \sqrt{n_{\underline{\alpha}}}\, a_{\underline{\alpha}}^\dagger \,|\cdots (n_{\underline{\alpha}} - 1) \cdots\rangle = n_{\underline{\alpha}}\,|\cdots n_{\underline{\alpha}} \cdots\rangle, \tag{6.30}$$

hence $(a_{\underline{\alpha}} a_{\underline{\alpha}}^\dagger - a_{\underline{\alpha}}^\dagger a_{\underline{\alpha}})$ acts as the identity on all basis kets.

Fermions.
The canonical anticommutation relations are

$$\{a_{\underline{\alpha}}^\dagger, a_{\underline{\beta}}^\dagger\} = 0, \qquad \{a_{\underline{\alpha}}, a_{\underline{\beta}}\} = 0, \qquad \{a_{\underline{\alpha}}, a_{\underline{\beta}}^\dagger\} = \delta_{\underline{\alpha\beta}}. \tag{6.31}$$

Thus two creation operators anticommute, two annihilation operators anticommute, and the only non-vanishing mixed anticommutator is the Kronecker delta. These relations are a direct consequence of the sign convention in Eqs. (6.21)–(6.23). In particular,

$$\{a_{\underline{\alpha}}^\dagger, a_{\underline{\alpha}}^\dagger\} = 2(a_{\underline{\alpha}}^\dagger)^2 = 0 \iff (a_{\underline{\alpha}}^\dagger)^2 = 0, \tag{6.32}$$

expressing again the impossibility of double occupation for fermions.

Number operators.
The occupation-number operator for the level $\underline{\alpha}$ is defined as

$$\hat{n}_{\underline{\alpha}} = a_{\underline{\alpha}}^\dagger a_{\underline{\alpha}}, \tag{6.33}$$

and the total particle-number operator is

$$\hat{N} = \sum_{\underline{\alpha}} \hat{n}_{\underline{\alpha}}. \tag{6.34}$$

On any Fock basis ket one finds, for both bosons and fermions,

$$\hat{n}_{\underline{\alpha}} \,|\, \cdots n_{\underline{\alpha}} \cdots \rangle = n_{\underline{\alpha}} \,|\, \cdots n_{\underline{\alpha}} \cdots \rangle, \tag{6.35}$$

i.e. the basis vectors are simultaneous eigenvectors of all $\hat{n}_{\underline{\alpha}}$ with eigenvalues equal to the corresponding occupations. Accordingly, $\hat{N}$ counts the total number of particles in a fixed-N sector.

6.1.5 Change of One-Particle Basis

The operators $a^\dagger_{\underline{\alpha}}$ and $a_{\underline{\alpha}}$ are defined with respect to the chosen one-particle basis $\{|\underline{\alpha}\rangle\}$. Consider a new one-particle basis $\{|\underline{\tilde{\alpha}}\rangle\}$, related to the old one by

$$|\underline{\tilde{\alpha}}\rangle = \sum_{\underline{\alpha}} \langle\underline{\alpha}|\underline{\tilde{\alpha}}\rangle \; |\underline{\alpha}\rangle, \tag{6.36}$$

where the coefficients are the usual Hilbert-space overlaps.

To relate the corresponding creation operators, note that acting on the vacuum, a creation operator produces a one-particle state:

$$a^\dagger_{\underline{\tilde{\alpha}}}|0\rangle = |0 \cdots 1_{\underline{\tilde{\alpha}}} \cdots 0\rangle = |\underline{\tilde{\alpha}}\rangle = \sum_{\underline{\alpha}} \langle\underline{\alpha}|\underline{\tilde{\alpha}}\rangle \; |\underline{\alpha}\rangle = \sum_{\underline{\alpha}} \langle\underline{\alpha}|\underline{\tilde{\alpha}}\rangle \; a^\dagger_{\underline{\alpha}}|0\rangle. \tag{6.37}$$

Note that the one-particle Hilbert space is naturally isomorphic to the one-particle sector of the Fock space and, then, the equalities above are strict.

Since every Fock vector is obtained from $|0\rangle$ by creation operators, this identity lifts to the operator relation

$$a^\dagger_{\underline{\tilde{\alpha}}} = \sum_{\underline{\alpha}} \langle\underline{\alpha}|\underline{\tilde{\alpha}}\rangle \; a^\dagger_{\underline{\alpha}}. \tag{6.38}$$

Taking the Hermitian conjugate yields

$$a_{\underline{\tilde{\alpha}}} = \sum_{\underline{\alpha}} \langle\underline{\tilde{\alpha}}|\underline{\alpha}\rangle \; a_{\underline{\alpha}}. \tag{6.39}$$

▶ **Driving exercise**
The relations Eqs. (6.28) and (6.31) are invariant under basis changes. Indeed, using Eqs. (6.38)–(6.39),

$$\begin{aligned}[] [\, a_{\underline{\tilde{\alpha}}}, a^\dagger_{\underline{\tilde{\beta}}} \,]_\pm &= \left[\sum_{\underline{\alpha}} \langle\underline{\tilde{\alpha}}|\underline{\alpha}\rangle a_{\underline{\alpha}}, \sum_{\underline{\beta}} \langle\underline{\beta}|\underline{\tilde{\beta}}\rangle a^\dagger_{\underline{\beta}} \right]_\pm = \sum_{\underline{\alpha},\underline{\beta}} \langle\underline{\tilde{\alpha}}|\underline{\alpha}\rangle \langle\underline{\beta}|\underline{\tilde{\beta}}\rangle [\, a_{\underline{\alpha}}, a^\dagger_{\underline{\beta}} \,]_\pm \\ &= \sum_{\underline{\alpha},\underline{\beta}} \langle\underline{\tilde{\alpha}}|\underline{\alpha}\rangle \langle\underline{\beta}|\underline{\tilde{\beta}}\rangle \delta_{\underline{\alpha}\underline{\beta}} = \sum_{\underline{\alpha}} \langle\underline{\tilde{\alpha}}|\underline{\alpha}\rangle \langle\underline{\alpha}|\underline{\tilde{\beta}}\rangle = \langle\underline{\tilde{\alpha}}|\underline{\tilde{\beta}}\rangle = \delta_{\underline{\tilde{\alpha}}\underline{\tilde{\beta}}}, \end{aligned} \tag{6.40}$$

where $[\cdot, \cdot]_\pm$ stands for the commutator (bosons) or anticommutator (fermions), and completeness $\sum_{\underline{\alpha}} |\underline{\alpha}\rangle\langle\underline{\alpha}| = \mathbb{I}$ was used. The vanishing (anti)commutators between two creations or two annihilations follow analogously.

6.2 The Pauli-Weisskopf Quantization

When dealing with the Klein-Gordon equation for a scalar field $\phi(x)$, several problems arise if one naively interprets ϕ as a single-particle wavefunction. For instance, even if the conserved current satisfies $\partial_\mu j^\mu = 0$, yet the corresponding "density" j^0 is not positive definite, as we saw in Chap. 4. This undermines the probabilistic interpretation that is central to single-particle quantum mechanics.

The Pauli-Weisskopf idea is to abandon the single-particle interpretation: *the relativistic wave equation is not the quantum equation of a particle; it is the classical equation of motion of a field that must itself be quantized.*

Conceptually, this implies a physical revolution, as it provides the first example of second quantization.

Hence, *replace "first quantization" (quantize particle trajectories) with "second quantization" (quantize the field degrees of freedom).*

The conceptual revolution consists of quantizing the field, not the particle.

Hence, the Pauli-Weisskopf quantization promotes ϕ and π to operators.

How to pursue the quantization scheme? One has to make up equal-time commutators. Hence, the quantized field $\hat{\phi}(t, \mathbf{x})$ and its conjugate momentum $\hat{\pi}(t, \mathbf{x})$ satisfy

$$[\hat{\phi}(t, \mathbf{x}), \hat{\phi}(t, \mathbf{y})] = 0, \qquad [\hat{\pi}(t, \mathbf{x}), \hat{\pi}(t, \mathbf{y})] = 0, \qquad [\hat{\phi}(t, \mathbf{x}), \hat{\pi}(t, \mathbf{y})] = i\hbar\,\delta^{(3)}(\mathbf{x} - \mathbf{y}). \tag{6.41}$$

The field obeys the Heisenberg equation generated by $\hat{H}$:

$$\frac{\partial}{\partial t}\hat{\phi}(t, \mathbf{x}) = \frac{1}{i\hbar}[\hat{\phi}(t, \mathbf{x}), \hat{H}], \qquad \frac{\partial}{\partial t}\hat{\pi}(t, \mathbf{x}) = \frac{1}{i\hbar}[\hat{\pi}(t, \mathbf{x}), \hat{H}], \tag{6.42}$$

which reproduce the Klein-Gordon equation at the operator level:

$$(\partial_\mu \partial^\mu + m^2)\hat{\phi}(x) = 0. \tag{6.43}$$

Let us focus on how to pursue the above scheme in detail. For a *complex* Klein-Gordon field ϕ (required to describe a conserved charge). Consider

$$\mathcal{L} = \partial_\mu \phi^* \, \partial^\mu \phi - m^2 \phi^* \phi, \tag{6.44}$$

invariant under the global $U(1)$ transformation $\phi \mapsto e^{i\alpha}\phi$. As seen in Chapter 4, Noether's theorem yields the conserved current $j^\mu = i\,(\phi^* \partial^\mu \phi - (\partial^\mu \phi^*)\phi)$, fulfilling $\partial_\mu j^\mu = 0$, and the conserved charge

$$Q = \int d^3x \, j^0. \tag{6.45}$$

If one insists on a single-particle interpretation, j^0 fails to be positive. In the field interpretation, j^0 is *not* a probability density; it is a *charge density*. The operator-valued field admits a mode expansion involving two independent ladder operators:

$$\hat{\phi}(x) = \int \frac{d^3k}{(2\pi)^3} \frac{1}{\sqrt{2\omega_{\mathbf{k}}}} \left(\hat{a}(\mathbf{k})\, e^{-ik\cdot x} + \hat{b}^\dagger(\mathbf{k})\, e^{+ik\cdot x} \right), \tag{6.46}$$

$$\hat{\phi}^\dagger(x) = \int \frac{d^3k}{(2\pi)^3} \frac{1}{\sqrt{2\omega_{\mathbf{k}}}} \left(\hat{a}^\dagger(\mathbf{k})\, e^{+ik\cdot x} + \hat{b}(\mathbf{k})\, e^{-ik\cdot x} \right), \tag{6.47}$$

with commutators

$$[\hat{a}(\mathbf{k}), \hat{a}^\dagger(\mathbf{k}')] = (2\pi)^3 \delta^{(3)}(\mathbf{k} - \mathbf{k}'), \qquad [\hat{b}(\mathbf{k}), \hat{b}^\dagger(\mathbf{k}')] = (2\pi)^3 \delta^{(3)}(\mathbf{k} - \mathbf{k}'), \tag{6.48}$$

and all mixed commutators vanishing. The existence of both $\hat{a}$ and $\hat{b}$ encodes the presence of *particles and antiparticles* as independent excitations.

In terms of normal-ordered operators one finds

$$: \hat{Q} := \int \frac{d^3k}{(2\pi)^3} \left(\hat{a}^\dagger(\mathbf{k})\hat{a}(\mathbf{k}) - \hat{b}^\dagger(\mathbf{k})\hat{b}(\mathbf{k}) \right), \tag{6.49}$$

showing that $\hat{a}^\dagger$ creates quanta with charge $+1$ (in suitable units) and $\hat{b}^\dagger$ creates quanta with charge -1. The energy, conversely, is positive for both sectors:

$$: \hat{H} := \int \frac{d^3k}{(2\pi)^3} \, \hbar\omega_{\mathbf{k}} \left(\hat{a}^\dagger(\mathbf{k})\hat{a}(\mathbf{k}) + \hat{b}^\dagger(\mathbf{k})\hat{b}(\mathbf{k}) \right). \tag{6.50}$$

Thus the Pauli-Weisskopf quantization resolves the single-particle difficulties in a conceptually clean way: the problematic "negative densities" become physically meaningful charge densities, while the Hilbert-space norm remains positive definite.

The aforementioned scheme leads to

- **Advantages**. The Pauli-Weisskopf strategy yields:

  1. a positive-definite Hilbert-space interpretation (the norm is a quantum inner product, not j^0),
  2. a natural place for antiparticles (complex fields require two sets of ladder operators),
  3. a framework compatible with relativistic causality at the level of local fields,
  4. and an immediate route to many-particle sectors and variable particle number.

- **Residual issues**. Even in the free theory, one meets the vacuum energy divergence and must choose an ordering prescription. Moreover, in interacting theories the canonical procedure becomes technically delicate: renormalization, the definition of interacting fields, and the construction of the physical Hilbert space require additional conceptual input. These issues do not invalidate the Pauli-Weisskopf revolution; rather, they reflect the intrinsic richness of QFT as the correct language for relativistic quantum physics.

In this framework, negative-frequency solutions are reinterpreted as antiparticle excitations, the conserved current is understood as a charge current rather than a probability current, and variable particle number becomes unavoidable. Thus, Pauli-Weisskopf quantization is best viewed as the canonical quantization of a complex scalar field, with emphasis on its physical interpretation rather than on new quantization rules.

6.2.1 Complex Massive Scalar Field

Let us start by writing the Lagrangian, the conjugate momentum, the Hamiltonian, and then express them in operator form, namely

$$\mathcal{L} = \partial^\mu \phi^\star \partial_\mu \phi - m^2 \phi^\star \phi, \tag{6.51a}$$

$$\pi_\sigma = \frac{\partial \mathcal{L}}{\partial(\partial_t \phi_\sigma)} = \begin{cases} \partial_t \phi^\star & \text{if } \phi_\sigma = \phi, \\ \partial_t \phi & \text{if } \phi_\sigma = \phi^\star, \end{cases} \tag{6.51b}$$

$$\mathcal{H} = \pi_\sigma \partial_t \phi_\sigma - \mathcal{L} = \partial_t \phi\, \partial_t \phi^\star + \nabla \phi^\star \nabla \phi + m^2 \phi^\star \phi = \pi^\star \pi + \nabla \phi^\star \nabla \phi + m^2 \phi^\star \phi, \tag{6.51c}$$

$$\phi_\sigma(x) \longrightarrow \hat{\phi}_\sigma(x) \quad \text{and} \quad \pi_\sigma(x) \longrightarrow \hat{\pi}_\sigma(x), \tag{6.51d}$$

where σ is the index labeling the two components $\phi_\sigma : \phi^\star, \phi$. The commutation relations for the operators are

$$[\hat{\phi}(t,\mathbf{x}),\hat{\pi}(t,\mathbf{y})]=[\hat{\phi}^\dagger(t,\mathbf{x}),\hat{\pi}^\dagger(t,\mathbf{y})]=i\delta^{(3)}(\mathbf{x}-\mathbf{y}), \tag{6.52a}$$

$$[\hat{\phi}(x),\hat{\phi}(y)]=[\hat{\phi}^\dagger(x),\hat{\phi}^\dagger(y)]=0, \tag{6.52b}$$

$$[\hat{\phi}(x),\hat{\phi}^\dagger(y)]=[\hat{\pi}(x),\hat{\pi}^\dagger(y)]=0, \tag{6.52c}$$

$$[\hat{\pi}(x),\hat{\pi}(y)]=[\hat{\pi}^\dagger(x),\hat{\pi}^\dagger(y)]=0. \tag{6.52d}$$

We now expand $\hat{\phi}$ and $\hat{\phi}^\dagger$ in normal modes. In this general case, there is no reason to require $\hat{\phi}^\dagger=\hat{\phi}$, so

$$\hat{\phi}(x)=\int\frac{d^3p}{(2\pi)^{3/2}(2E_{\mathbf{p}})^{1/2}}[\hat{a}_{\mathbf{p}}e^{-ipx}+\hat{b}^\dagger_{\mathbf{p}}e^{ipx}], \tag{6.53a}$$

$$\hat{\phi}^\dagger(x)=\int\frac{d^3p}{(2\pi)^{3/2}(2E_{\mathbf{p}})^{1/2}}[\hat{a}^\dagger_{\mathbf{p}}e^{ipx}+\hat{b}_{\mathbf{p}}e^{-ipx}]. \tag{6.53b}$$

Accordingly, the operators $\hat{a}_{\mathbf{p}}$ and $\hat{b}_{\mathbf{p}}$ annihilate two different types of particles. By constructing $\hat{H}$ we obtain

$$\hat{H}=\int d^3x\,\left(\hat{\pi}^\star\hat{\pi}+\nabla\hat{\phi}^\dagger\cdot\nabla\hat{\phi}+m^2\hat{\phi}^\dagger\hat{\phi}\right). \tag{6.54}$$

Using $\hat{\phi}^\dagger$ and $\hat{\phi}$ together with

$$\hat{\pi}^\dagger(x)=\int\frac{d^3p}{(2\pi)^{3/2}(2E_{\mathbf{p}})^{1/2}}(-iE_{\mathbf{p}})\left[\hat{a}_{\mathbf{p}}e^{-ipx}-\hat{b}^\dagger_{\mathbf{p}}e^{ipx}\right]=\partial_t\hat{\phi}(x), \tag{6.55a}$$

$$\hat{\pi}(x)=\int\frac{d^3p}{(2\pi)^{3/2}(2E_{\mathbf{p}})^{1/2}}(iE_{\mathbf{p}})\left[\hat{a}^\dagger_{\mathbf{p}}e^{ipx}-\hat{b}_{\mathbf{p}}e^{-ipx}\right]=\partial_t\hat{\phi}^\dagger(x), \tag{6.55b}$$

$$\nabla\hat{\phi}^\dagger(x)=\int\frac{d^3p}{(2\pi)^{3/2}(2E_{\mathbf{p}})^{1/2}}(-i\mathbf{p})\left[\hat{a}^\dagger_{\mathbf{p}}e^{ipx}-\hat{b}_{\mathbf{p}}e^{-ipx}\right], \tag{6.55c}$$

$$\nabla\hat{\phi}(x)=\int\frac{d^3p}{(2\pi)^{3/2}(2E_{\mathbf{p}})^{1/2}}(i\mathbf{p})\left[\hat{a}_{\mathbf{p}}e^{-ipx}-\hat{b}^\dagger_{\mathbf{p}}e^{ipx}\right], \tag{6.55d}$$

we find from Eq. (6.54)

$$\hat{H}=\int\frac{d^3x\,d^3p\,d^3q}{(2\pi)^3(2E_{\mathbf{p}}2E_{\mathbf{q}})^{1/2}}\Big\{[E_{\mathbf{p}}E_{\mathbf{q}}+\mathbf{p}\cdot\mathbf{q}][\hat{a}_{\mathbf{p}}e^{-ipx}-\hat{b}^\dagger_{\mathbf{p}}e^{ipx}][\hat{a}^\dagger_{\mathbf{q}}e^{iqx}-\hat{b}_{\mathbf{q}}e^{-iqx}] \tag{6.56}$$
$$+m^2[\hat{a}_{\mathbf{p}}e^{-ipx}+\hat{b}^\dagger_{\mathbf{p}}e^{ipx}][\hat{a}^\dagger_{\mathbf{q}}e^{iqx}+\hat{b}_{\mathbf{q}}e^{-iqx}]\Big\}.$$

Before proceeding, let us introduce the following notation within the previous expression:

$$\int d^3x\, e^{i(\mathbf{p}\pm\mathbf{q})\cdot\mathbf{x}} = (2\pi)^3\delta(\mathbf{p}\pm\mathbf{q}), \quad \int \frac{d^3p\, d^3q}{(2E_{\mathbf{p}}2E_{\mathbf{q}})^{1/2}} \equiv \int_{p,q}. \tag{6.57}$$

Thus, replacing $\mathbf{p} \to -\mathbf{q}$ and substituting into $\hat{H}$, we obtain

$$\hat{H} = \int \frac{d^3p}{2E_{\mathbf{p}}} \Big\{ \cancel{[-E_{\mathbf{p}}^2 + \mathbf{p}^2 + m^2]}[\hat{a}_{\mathbf{p}}\hat{b}_{-\mathbf{p}}e^{2iE_{\mathbf{p}}t} + \hat{b}^\dagger_{\mathbf{p}}\hat{a}^\dagger_{-\mathbf{p}}e^{2iE_{\mathbf{p}}t}]$$
$$+ [E_{\mathbf{p}}^2 + \mathbf{p}^2 + m^2][\hat{a}_{\mathbf{p}}\hat{a}^\dagger_{\mathbf{p}} + \hat{b}^\dagger_{\mathbf{p}}\hat{b}_{\mathbf{p}}] \Big\}. \tag{6.58}$$

Since $E_{\mathbf{p}}^2 + \mathbf{p}^2 + m^2 = 2E_{\mathbf{p}}^2$, the Hamiltonian reduces to

$$\hat{H} = \int d^3p\, E_{\mathbf{p}}\,(\hat{a}_{\mathbf{p}}\hat{a}^\dagger_{\mathbf{p}} + \hat{b}^\dagger_{\mathbf{p}}\hat{b}_{\mathbf{p}}). \tag{6.59}$$

For the normal ordering one has

$$N[\hat{H}] = \int d^3pE_{\mathbf{p}}(\hat{a}^\dagger_{\mathbf{p}}\hat{a}_{\mathbf{p}} + \hat{b}^\dagger_{\mathbf{p}}\hat{b}_{\mathbf{p}}) = \int d^3pE_{\mathbf{p}}(\hat{n}^{(a)}_{\mathbf{p}} + \hat{n}^{(b)}_{\mathbf{p}}), \tag{6.60}$$

where $\hat{n}^{(a)}_{\mathbf{p}}$ and $\hat{n}^{(b)}_{\mathbf{p}}$ are the number operators for the two types of particles, a and b, counting how many particles of type a and b, respectively, are present in a state with momentum $\mathbf{p}$ and energy $E_{\mathbf{p}}$.

Particles of type a are created by $\hat{a}^\dagger_{\mathbf{p}}$ and annihilated by $\hat{a}_{\mathbf{p}}$, whereas particles of type b are the corresponding antiparticles, created by $\hat{b}^\dagger_{\mathbf{p}}$ and annihilated by $\hat{b}_{\mathbf{p}}$.

6.2.1.1 Internal Symmetry $U(1)$

The complex scalar field exhibits a symmetry under the transformations $\psi \longrightarrow e^{i\alpha}\psi$ and $\psi^\dagger \longrightarrow e^{-i\alpha}\psi^\dagger$, such that the Lagrangian $\mathcal{L}$ remains invariant. By Noether's theorem, there exists a conserved current associated with this symmetry. Let us construct it to identify the *associated conserved quantity* by considering

$$\delta\phi = \phi'(x) - \phi(x) \quad \text{with} \quad \phi'(x) = e^{i\alpha}\phi(x), \tag{6.61a}$$
$$\delta\phi^\star = \phi'^\star(x) - \phi^\star(x) \quad \text{with} \quad \phi'^\star(x) = e^{-i\alpha}\phi^\star(x). \tag{6.61b}$$

Writing the above expressions in infinitesimal form, we obtain

$$\phi'(x) \sim \phi(x) + i\alpha\phi(x) \longrightarrow \delta\phi = i\alpha\phi = \Sigma_a\varepsilon_a\Sigma^a\phi, \tag{6.62a}$$
$$\phi'^\star(x) \sim \phi^\star(x) - i\alpha\phi^\star(x) \longrightarrow \delta\phi^\star = -i\alpha\phi^\star = \Sigma_a\eta_a\Lambda^a\phi^\star. \tag{6.62b}$$

Now consider the current J^μ and use the notation $\varepsilon_a = \eta_a = \alpha$, $\Sigma^a \phi = i\phi$ and $\Lambda^a \phi^\star = -i\phi^\star$:

$$J^\mu = \frac{\partial \mathcal{L}}{\partial(\partial_\mu \phi)}(i\phi) + \frac{\partial \mathcal{L}}{\partial(\partial_\mu \phi^\star)}(-i\phi^\star) \longrightarrow J^\mu = i\left[(\partial^\mu \phi^\star)\phi - (\partial^\mu \phi)\phi^\star\right]. \tag{6.63}$$

Replacing $\phi \longrightarrow \hat{\phi}$ and $\phi^\star \longrightarrow \hat{\phi}^\dagger$, we obtain the operator form of the current $\hat{J}^\mu$. The conserved charge is then obtained by integrating $\hat{J}^0$ over space:

$$\begin{aligned}
\hat{Q} &= \int d^3x\, \hat{J}^0 = i\int d^3x\, [(\partial_t \hat{\phi}^\dagger)\hat{\phi} - (\partial_t \hat{\phi})\hat{\phi}^\dagger] \\
&= i\int d^3x \int \frac{d^3p\, d^3q}{(2\pi)^3 (2E_\mathbf{p})^{1/2}(2E_\mathbf{q})^{1/2}} \{(iE_\mathbf{q})[\hat{a}_\mathbf{p} e^{-ipx} + \hat{b}^\dagger_\mathbf{p} e^{ipx}][\hat{a}^\dagger_\mathbf{q} e^{iqx} - \hat{b}_\mathbf{q} e^{-iqx}]\} \\
&= i\int_{p,q} \{(+iE_\mathbf{q})[\delta(\mathbf{p}+\mathbf{q})(-\hat{a}_\mathbf{p}\hat{b}_\mathbf{q} e^{i(+)t} + \hat{b}_\mathbf{p}\hat{a}^\dagger_\mathbf{q} e^{i(+)t}) + \delta(\mathbf{p}-\mathbf{q})(\hat{a}_\mathbf{p}\hat{a}^\dagger_\mathbf{q} e^{-i(-)t} - \hat{b}^\dagger_\mathbf{p}\hat{b}_\mathbf{q} e^{i(-)t})] + \\
&\quad (iE_\mathbf{q})[\delta(\mathbf{p}+\mathbf{q})(-\hat{a}^\dagger_\mathbf{p}\hat{b}^\dagger_\mathbf{q} e^{i(+)t} + \hat{b}_\mathbf{p}\hat{a}_\mathbf{q} e^{-i(+)t}) + \delta(\mathbf{p}-\mathbf{q})(\hat{a}^\dagger_\mathbf{p}\hat{a}_\mathbf{q} e^{i(-)t} - \hat{b}_\mathbf{p}\hat{b}^\dagger_\mathbf{q} e^{-i(-)t})]\}.
\end{aligned} \tag{6.64}$$

Collecting the δ-functions, we obtain

$$\begin{aligned}
\hat{Q} = \int_{p,q} &\{\delta(\mathbf{p}+\mathbf{q})[e^{-i(+)t}(-\hat{a}_\mathbf{p}\hat{b}_\mathbf{q} + \hat{b}_\mathbf{q}\hat{a}_\mathbf{q})(+iE_\mathbf{q}) + e^{i(+)t}(\hat{b}^\dagger_\mathbf{p}\hat{a}^\dagger_\mathbf{q} - \hat{a}^\dagger_\mathbf{p}\hat{b}^\dagger_\mathbf{q})(+iE_\mathbf{q})] + \\
&+\delta(\mathbf{p}-\mathbf{q})[e^{-i(-)t}(\hat{a}_\mathbf{p}\hat{a}^\dagger_\mathbf{q} - \hat{b}_\mathbf{p}\hat{b}^\dagger_\mathbf{q})(+iE_\mathbf{q}) + e^{i(-)t}(-\hat{b}^\dagger_\mathbf{p}\hat{b}_\mathbf{q} + \hat{a}^\dagger_\mathbf{p}\hat{a}_\mathbf{q})(+iE_\mathbf{q})]\}.
\end{aligned} \tag{6.65}$$

Integrating over $d\mathbf{q}$ and rewriting the first term as $\mathbf{q} \longrightarrow -\mathbf{p}$ and the second as $\mathbf{q} \longrightarrow \mathbf{p}$, we obtain

$$\begin{aligned}
\hat{Q} = i\int \frac{d^3p}{2E_\mathbf{p}} &\{(+iE_\mathbf{p})[e^{-2iE_\mathbf{p}t}(-\hat{a}_\mathbf{p}\hat{b}_{-\mathbf{p}} + \hat{b}_\mathbf{p}\hat{a}_{-\mathbf{p}}) + e^{2iE_\mathbf{p}t}(\hat{b}^\dagger_\mathbf{p}\hat{a}_{-\mathbf{p}} + \hat{b}_\mathbf{p}\hat{a}^\dagger_{-\mathbf{p}}) \\
&+(+iE_\mathbf{p})(\hat{a}_\mathbf{p}\hat{a}^\dagger_\mathbf{p} - \hat{b}_\mathbf{p}\hat{b}^\dagger_\mathbf{p} - \hat{b}^\dagger_\mathbf{p}\hat{b}_\mathbf{p} + \hat{a}^\dagger_\mathbf{p}\hat{a}_\mathbf{p})]\}.
\end{aligned} \tag{6.66}$$

The first term vanishes due to the commutation relations, while $\hat{a}_{-\mathbf{p}}\hat{b}_\mathbf{p} = \hat{a}_\mathbf{p}\hat{b}_{-\mathbf{p}}$, so that

$$\begin{aligned}
\hat{Q} &= -\int \frac{d^3p}{2\cancel{E_\mathbf{p}}} \cancel{E_\mathbf{p}} (\hat{a}^\dagger_\mathbf{p}\hat{a}_\mathbf{p} - \hat{b}^\dagger_\mathbf{p}\hat{b}_\mathbf{p} + \hat{a}_\mathbf{p}\hat{a}^\dagger_\mathbf{p} - \hat{b}_\mathbf{p}\hat{b}^\dagger_\mathbf{p}) \\
&= \frac{1}{2}\int d^3p(-\hat{a}^\dagger_\mathbf{p}\hat{a}_\mathbf{p} + \hat{b}^\dagger_\mathbf{p}\hat{b}_\mathbf{p} - \hat{a}_\mathbf{p}\hat{a}^\dagger_\mathbf{p} + \hat{b}_\mathbf{p}\hat{b}^\dagger_\mathbf{p}).
\end{aligned} \tag{6.67}$$

Note that the annihilation of a particle with momentum $-\mathbf{p}$ and an antiparticle with momentum $\mathbf{p}$ is equivalent to annihilating a particle with momentum $\mathbf{p}$ and an antiparticle with momentum $-\mathbf{p}$. Using normal ordering, we obtain

$$N[\hat{Q}] = \int d^3p\,(\hat{b}^\dagger_{\mathbf{p}}\hat{b}_{\mathbf{p}} - \hat{a}^\dagger_{\mathbf{p}}\hat{a}_{\mathbf{p}}) = N_{\text{antiparticles}} - N_{\text{particles}}. \tag{6.68}$$

The complex scalar field is charged, and the condition for it to be real would set the charge to zero. Under the Pauli-Weisskopf, the Klein-Gordon equation acquires a direct and complete interpretation, overcoming the Feynman-Stuckelberg interpretation.

The $U(1)$ symmetry implies the conservation of the difference between the number of particles and antiparticles. In this sense, the existence of antiparticles arises from the necessity to conserve the Noether charge associated with the symmetry of $\mathcal{L}$ for a complex scalar field. By convention, particles contribute $+q$ to this charge (created by $\hat{a}^\dagger_{\mathbf{p}}$), whereas antiparticles contribute $-q$ (created by $\hat{b}^\dagger_{\mathbf{p}}$). With the standard normalization, the conserved Noether charge associated with the global $U(1)$ symmetry takes the diagonal form

$$\hat{Q} = \int \frac{d^3p}{(2\pi)^3}\left(\hat{a}^\dagger_{\mathbf{p}}\hat{a}_{\mathbf{p}} - \hat{b}^\dagger_{\mathbf{p}}\hat{b}_{\mathbf{p}}\right), \tag{6.69}$$

so that particle quanta created by $\hat{a}^\dagger_{\mathbf{p}}$ carry charge $+q$, while antiparticle quanta created by $\hat{b}^\dagger_{\mathbf{p}}$ carry charge $-q$ (up to the overall unit q fixed by convention). In particular, charge conservation enforces conservation of the difference $N_a - N_b$ in any process.

6.2.2 Massive Real Scalar Field

Let us follow the steps introduced above, starting with the first three.

It appears evident that to come back to real field one can ensure that $\hat{b}_{\mathbf{p}} \to \hat{a}_{\mathbf{p}}$. So immediately it is evident that a real field has $\hat{Q}_C = 0$, see Eq. (6.69).

However, it is instructive to redo the computation, above reported, for the real field too. In this case, the Lagrangian, the momentum, and the Hamiltonian can be reobtained as above, imposing $\phi = \phi^\star$.

At this stage, the field and the momentum are promoted to operators, namely $\phi(x) \longrightarrow \hat{\phi}(x)$ and $\pi(x) \longrightarrow \hat{\pi}(x)$ with $x = (t, \mathbf{x})$. For this reason, one imposes the equal-time commutator

$$[\hat{\phi}(t, \mathbf{x}), \hat{\pi}(t, \mathbf{y})] = i\delta^{(3)}(\mathbf{x} - \mathbf{y}), \tag{6.70a}$$

$$[\hat{\phi}(x), \hat{\phi}(y)] = [\hat{\pi}(x), \hat{\pi}(y)] = 0. \tag{6.70b}$$

We are now ready to move to step 4), namely to expand the fields in terms of creation and annihilation operators, writing

$$\hat{\phi}(x) = \int \frac{d^3p}{(2\pi)^{3/2}(2E_p)^{1/2}}(\hat{a}_{\mathbf{p}}e^{-ipx} + \hat{a}^\dagger_{\mathbf{p}}e^{ipx}). \tag{6.71}$$

We thus obtain a field operator which is a solution of the Klein-Gordon equation, with $\hat{a}^\dagger_{\mathbf{p}}$ and $\hat{a}_{\mathbf{p}}$ being creation and annihilation operators introduced in the framework of the quantum harmonic oscillator. Moreover, $\hat{\phi}$ is expanded over normal modes, which are essentially quantized harmonic oscillators. Recall that ϕ, as a classical field, can be obtained as the continuum limit of the displacement of an infinite chain of coupled oscillators, while the corresponding $\hat{\phi}$ describes quantized oscillators acting on the Fock space. This space contains Fock states that can be represented as shown in Sect. 6.1.

When quantizing a scalar field, since ϕ satisfies the Klein-Gordon equation, it is expressed as a superposition of plane waves. From commutation relations between $\hat{\phi}$ and $\hat{\pi}$ does one obtain that $\hat{a}_{\mathbf{p}}$ and $\hat{a}^\dagger_{\mathbf{p}}$ play the role of annihilation and creation operators, which satisfy

$$[\hat{a}_{\mathbf{p}}, \hat{a}^\dagger_{\mathbf{q}}] = 2E_{\mathbf{p}}(2\pi)^3\delta(\mathbf{p} - \mathbf{q}). \tag{6.72}$$

Note that one first writes the time-independent operator

$$\hat{\phi}(\mathbf{x}) = \int \frac{d^3p}{(2\pi)^{3/2}(2E_{\mathbf{p}})^{1/2}}(\hat{a}_{\mathbf{p}}e^{i\mathbf{p}\cdot\mathbf{x}} + \hat{a}^\dagger_{\mathbf{p}}e^{-i\mathbf{p}\cdot\mathbf{x}}) \tag{6.73}$$

and, in order to arrive at

$$\begin{aligned}\hat{\phi}(x) &= \int \frac{d^3p}{(2\pi)^{3/2}(2E_{\mathbf{p}})^{1/2}}[\hat{a}_{\mathbf{p}}e^{-ipx} + \hat{a}^\dagger_{\mathbf{p}}e^{ipx}] \overset{\partial_t\hat{\phi}}{\longrightarrow} \hat{\pi}(x) \\ &= \int \frac{d^3p}{(2\pi)^{3/2}(2E_{\mathbf{p}})^{1/2}}(-iE_{\mathbf{p}})[\hat{a}_{\mathbf{p}}e^{-ipx} - \hat{a}^\dagger_{\mathbf{p}}e^{ipx}]\end{aligned} \tag{6.74}$$

one constructs the time evolution of

$$\hat{\phi}(\mathbf{x}) \longrightarrow \hat{\phi}(\mathbf{x}, t) = U^\dagger\hat{\phi}(\mathbf{x})U \tag{6.75}$$

according to the Heisenberg representation, with $U = e^{-i\hat{H}t}$. One can show that U acts on $\hat{a}_{\mathbf{p}}$ and $\hat{a}^{\dagger}_{\mathbf{p}}$ as follows:

$$e^{i\hat{H}t}\hat{a}_{\mathbf{p}}e^{-i\hat{H}t} = e^{-iE_{\mathbf{p}}t}\hat{a}_{\mathbf{p}}, \tag{6.76a}$$

$$e^{i\hat{H}t}\hat{a}^{\dagger}_{\mathbf{p}}e^{-i\hat{H}t} = e^{iE_{\mathbf{p}}t}\hat{a}^{\dagger}_{\mathbf{p}}. \tag{6.76b}$$

To demonstrate this, we apply $U^{\dagger}\hat{a}_{\mathbf{p}}U$ to a state $|n_{\mathbf{p}}\rangle$, obtaining

$$\begin{aligned} e^{i\hat{H}t}\hat{a}_{\mathbf{p}}e^{-i\hat{H}t}|n_{\mathbf{p}}\rangle &= e^{i\hat{H}t}\hat{a}_{\mathbf{p}}e^{-iE_{\mathbf{p}}n_{\mathbf{p}}t}|n_{\mathbf{p}}\rangle = e^{i\hat{H}t}\sqrt{n_{\mathbf{p}}}|n_{\mathbf{p}}-1\rangle e^{-iE_{\mathbf{p}}n_{\mathbf{p}}t} \\ &= \sqrt{n_{\mathbf{p}}}\cancel{e^{-iE_{\mathbf{p}}n_{\mathbf{p}}t}}e^{+iE_{\mathbf{p}}(n_{\mathbf{p}}-1)t}|n_{\mathbf{p}}-1\rangle, \end{aligned} \tag{6.77}$$

which is equivalent to applying the operator $\hat{a}_{\mathbf{p}}$ on $|n_{\mathbf{p}}\rangle$ with the addition of a phase factor $e^{-iE_{\mathbf{p}}t}$, that is

$$U^{\dagger}\hat{a}_{\mathbf{p}}U = \hat{a}_{\mathbf{p}}e^{-iE_{p}t}. \tag{6.78}$$

The normalization factor appears because we are integrating over d^4p with $p^{\mu} = (p^0, \mathbf{p})$ subject to the on-shell condition $p^2 = m^2$. To make this explicit in the integral we follow the same procedure as in the solution of the Klein-Gordon equation, considering $d^4p \longrightarrow d^4p\delta(p^2 - m^2)$. Since the integration is performed, it amounts to the sum of two separate contributions with $p_0 = +E_{\mathbf{p}}$ and $p_0 = -E_{\mathbf{p}}$. In this way, the following equality holds:

$$\begin{aligned} &\int d^4p\,\delta(p^2 - m^2)\hat{a}_{\mathbf{p}}e^{-i(p^0t - \mathbf{p}\cdot\mathbf{x})} \\ &\int d^4p\,\delta(p^2 - m^2)[\hat{a}_{\mathbf{p}}e^{-i(p^0t - \mathbf{p}\cdot\mathbf{x})}\theta(p^0) + \hat{a}^{\dagger}_{\mathbf{p}}e^{-i(p^0t - \mathbf{p}\cdot\mathbf{x})}\theta(-p^0)] \\ &= \int d^4p\,\delta(p^2 - m^2)\theta(p^0)[\hat{a}_{\mathbf{p}}e^{-ipx} + \hat{a}^{\dagger}_{\mathbf{p}}e^{ipx}]. \end{aligned} \tag{6.79}$$

Using again $\delta[f(x)] = \sum_i \frac{1}{|f'(x_i)|}\delta(x - x_i)$ with x_i the zeros of f, we obtain

$$\delta((p^0)^2 - \mathbf{p}^2 - m^2) = \delta[f(p^0)] = \frac{\delta(p^0 - E_{\mathbf{p}})}{2E_{\mathbf{p}}} + \frac{\delta(p^0 + E_{\mathbf{p}})}{2E_{\mathbf{p}}}. \tag{6.80}$$

Including the normalization factor $1/(2\pi)^4$ from the integral, we obtain

$$
\begin{aligned}
&\int \frac{d^4p}{2E_{\mathbf{p}}} \frac{\theta(p^0)\delta(p^0 - E_{\mathbf{p}})}{(2\pi)^4} [\hat{a}_{\mathbf{p}} e^{-ipx} + \hat{a}^{\dagger}_{\mathbf{p}} e^{ipx}] \\
&= \frac{1}{(2\pi)^3} \int \frac{d^3p}{2E_{\mathbf{p}}} \int \frac{dp^0}{2\pi} \theta(p^0)\delta(p^0 - E_{\mathbf{p}})[\hat{a}_{\mathbf{p}} e^{-ipx} + \hat{a}^{\dagger}_{\mathbf{p}} e^{ipx}] \\
&= \frac{1}{(2\pi)^3} \int \frac{d^3p}{2E_{\mathbf{p}}} \hat{a}_{\mathbf{p}} e^{-i(E_p t - \mathbf{p}\cdot\mathbf{x})} + \hat{a}^{\dagger}_{\mathbf{p}} e^{i(E_p t - \mathbf{p}\cdot\mathbf{x})}.
\end{aligned} \tag{6.81}
$$

The operators $\hat{a}_{\mathbf{p}}$ and $\hat{a}^{\dagger}_{\mathbf{p}}$ cannot be left in this form. At this point, we impose the normalization on the states $< \mathbf{p}|\mathbf{q} >= \delta^{(3)}(p - q)$ with $\hat{a}^{\dagger}_{\mathbf{p}}|0 >= |\mathbf{p} >$ *(the application of a creation operator must yield a one-particle state at a given momentum)*. We then introduce constants such that $\alpha\hat{a}_{\mathbf{p}}|0 >= |\mathbf{p} >$ and $\alpha\hat{a}^{\dagger}_{\mathbf{q}}|0 >= |\mathbf{q} >$, so that

$$
\begin{aligned}
&< \mathbf{p}|\mathbf{q} >= [\alpha\hat{a}^{\dagger}_{\mathbf{p}}|0 >]^{\dagger}[\alpha\hat{a}^{\dagger}_{\mathbf{q}}|0 >] = \alpha^*\alpha < 0|\hat{a}_{\mathbf{p}}\hat{a}^{\dagger}_{\mathbf{q}}|0 > -\alpha^*\alpha < 0|\hat{a}^{\dagger}_{\mathbf{q}}\hat{a}_{\mathbf{p}}|0 > \\
&= \alpha^*\alpha < 0|[\hat{a}_{\mathbf{p}}, \hat{a}^{\dagger}_{\mathbf{q}}]|0 >= (2\pi)^3(2E_{\mathbf{p}})|\alpha|^2\delta^{(3)}(\mathbf{p} - \mathbf{q}) \longrightarrow \alpha = [(2\pi)^{3/2}(2E_p)^{1/2}]^{-1}.
\end{aligned} \tag{6.82}
$$

6.2.3 Hamiltonian in Second Quantization

First, we integrate the Hamiltonian density written in terms of $\hat{\phi}$ and $\hat{\pi}$ over space, obtaining

$$
\hat{H} = \int d^3x \frac{1}{2}[\hat{\pi}^2 + (\nabla\hat{\pi})^2 + m^2\hat{\phi}^2], \tag{6.83}
$$

where

$$
\hat{\phi}(x) = \int \frac{d^3p}{(2\pi)^{3/2}(2E_{\mathbf{p}})^{1/2}} [\hat{a}_{\mathbf{p}} e^{-ipx} + \hat{a}^{\dagger}_{\mathbf{p}} e^{ipx}], \tag{6.84a}
$$

$$
\hat{\pi}(x) = \int \frac{d^3p}{(2\pi)^{3/2}(2E_{\mathbf{p}})^{1/2}} (-iE_{\mathbf{p}})[\hat{a}_{\mathbf{p}} e^{-ipx} - \hat{a}^{\dagger}_{\mathbf{p}} e^{ipx}]. \tag{6.84b}
$$

▶ **Driving exercise** We can write

$$
\nabla\hat{\phi} = \int \frac{d^3p}{(2\pi)^{3/2}(2E_{\mathbf{p}})^{1/2}} (i\mathbf{p})[\hat{a}_{\mathbf{p}} e^{-ipx} - \hat{a}^{\dagger}_{\mathbf{p}} e^{ipx}], \tag{6.85a}
$$

$$
\hat{\pi}^2 = \int \frac{d^3p\,d^3q}{(2\pi)^3(2E_p 2E_q)^{1/2}} (-iE_{\mathbf{p}})(-iE_{\mathbf{q}})[\hat{a}_{\mathbf{p}} e^{-ipx} - \hat{a}^{\dagger}_{\mathbf{p}} e^{ipx}][\hat{a}_{\mathbf{q}} e^{-iqx} - \hat{a}^{\dagger}_{\mathbf{q}} e^{iqx}]. \tag{6.85b}
$$

Substituting Eqs. (6.84)–(6.85) into (6.83), we obtain

$$\hat{H} = \frac{1}{2}\int \frac{d^3x d^3p d^3q}{(2\pi)^3(2E_{\mathbf{p}})^{1/2}(2E_{\mathbf{q}})^{1/2}}\{(-E_{\mathbf{p}}E_{\mathbf{q}} - \mathbf{p}\cdot\mathbf{q})[\hat{a}_{\mathbf{p}}e^{-ipx} - \hat{a}^{\dagger}_{\mathbf{p}}e^{ipx}]$$
$$[\hat{a}_{\mathbf{q}}e^{-iqx} - \hat{a}^{\dagger}_{\mathbf{q}}e^{iqx}] + m^2[\hat{a}_{\mathbf{p}}e^{-ipx} + \hat{a}^{\dagger}_{\mathbf{p}}e^{ipx}][\hat{a}_{\mathbf{q}}e^{-iqx} + \hat{a}^{\dagger}_{\mathbf{q}}e^{iqx}]\}. \tag{6.86}$$

Since the only dependence on $\mathbf{x}$ appears in the exponentials, we use

$$\int d^3x\, e^{i\mathbf{p}\cdot\mathbf{x}}e^{i\mathbf{q}\cdot\mathbf{x}} = \int d^3x\, e^{i(\mathbf{p}+\mathbf{q})\cdot\mathbf{x}} = (2\pi)^3\delta^{(3)}(\mathbf{p}+\mathbf{q}), \tag{6.87a}$$
$$\int d^3x\, e^{i\mathbf{p}\cdot\mathbf{x}}e^{-i\mathbf{q}\cdot\mathbf{x}} = (2\pi)^3\delta^{(3)}(\mathbf{p}-\mathbf{q}), \tag{6.87b}$$
$$\int d^3x\, e^{-i\mathbf{p}\cdot\mathbf{x}}e^{i\mathbf{q}\cdot\mathbf{x}} = (2\pi)^3\delta^{(3)}(\mathbf{p}-\mathbf{q}), \tag{6.87c}$$
$$\int d^3x\, e^{-i\mathbf{p}\cdot\mathbf{x}}e^{-i\mathbf{q}\cdot\mathbf{x}} = (2\pi)^3\delta^{(3)}(\mathbf{p}+\mathbf{q}). \tag{6.87d}$$

After integrating over d^3x, the Hamiltonian becomes

$$\hat{H} = \frac{1}{2}\int \frac{d^3p d^3q}{[(2E_{\mathbf{p}})(2E_{\mathbf{q}})]^{1/2}}\{(-E_{\mathbf{p}}E_{\mathbf{q}} - \mathbf{p}\cdot\mathbf{q})[\hat{a}_{\mathbf{p}}\hat{a}_{\mathbf{q}}\delta(\mathbf{p}+\mathbf{q})e^{-i(E_{\mathbf{p}}+E_{\mathbf{q}})t}$$
$$+ \hat{a}^{\dagger}_{\mathbf{p}}\hat{a}^{\dagger}_{\mathbf{q}}\delta(\mathbf{p}+\mathbf{q})e^{i(E_{\mathbf{p}}+E_{\mathbf{q}})t} + \quad -\hat{a}_{\mathbf{p}}\hat{a}^{\dagger}_{\mathbf{q}}\delta(\mathbf{p}-\mathbf{q})e^{-i(E_{\mathbf{p}}-E_{\mathbf{q}})t} - \hat{a}^{\dagger}_{\mathbf{p}}\hat{a}_{\mathbf{q}}\delta(\mathbf{p}-\mathbf{q})e^{i(E_{\mathbf{p}}-E_{\mathbf{q}})t}]$$
$$- m^2[\delta(\mathbf{p}+\mathbf{q})(\hat{a}_{\mathbf{p}}\hat{a}_{\mathbf{q}}e^{-i(E_{\mathbf{p}}-E_{\mathbf{q}})t} + \hat{a}^{\dagger}_{\mathbf{p}}\hat{a}^{\dagger}_{\mathbf{q}}e^{i(E_{\mathbf{p}}+E_{\mathbf{q}})t})$$
$$+ \delta(\mathbf{p}-\mathbf{q})(\hat{a}_{\mathbf{p}}\hat{a}^{\dagger}_{\mathbf{q}}e^{-i(E_{\mathbf{p}}-E_{\mathbf{q}})t} + \hat{a}^{\dagger}_{\mathbf{p}}\hat{a}_{\mathbf{q}}e^{i(E_{\mathbf{p}}-E_{\mathbf{q}})t})]\}. \tag{6.88}$$

Performing the d^3q integration, with $\mathbf{q} \to -\mathbf{p}$ in the first term and $\mathbf{q} \to \mathbf{p}$ in the second, we obtain

$$\hat{H} = \frac{1}{2}\int \frac{d^3p}{2E_{\mathbf{p}}}\{[-E^2_{\mathbf{p}} + \mathbf{p}^2 + m^2][\hat{a}_{\mathbf{p}}\hat{a}_{-\mathbf{p}}e^{-i(E_{\mathbf{p}}+E_{-\mathbf{p}})t} + \hat{a}^{\dagger}_{\mathbf{p}}\hat{a}^{\dagger}_{-\mathbf{p}}e^{i(E_{\mathbf{p}}+E_{-\mathbf{p}})t}]$$
$$+ [E^2_{\mathbf{p}} + \mathbf{p}^2 + m^2][\hat{a}_{\mathbf{p}}\hat{a}^{\dagger}_{\mathbf{p}}e^{-i(E_{\mathbf{p}}-E_{\mathbf{p}})t} + \hat{a}^{\dagger}_{\mathbf{p}}\hat{a}_{\mathbf{p}}e^{i(E_{\mathbf{p}}-E_{\mathbf{p}})t}]\}. \tag{6.89}$$

Using $E^2_{\mathbf{p}} = \mathbf{p}^2 + m^2$, the first term vanishes, while in the second term the exponentials cancel, leading to

$$\hat{H} = \frac{1}{2}\int d^3p \frac{2E^2_{\mathbf{p}}}{2E_{\mathbf{p}}}[\hat{a}_{\mathbf{p}}\hat{a}^{\dagger}_{\mathbf{p}} + \hat{a}^{\dagger}_{\mathbf{p}}\hat{a}_{\mathbf{p}}] = \frac{1}{2}\int d^3p E_{\mathbf{p}}[\hat{a}_{\mathbf{p}}\hat{a}^{\dagger}_{\mathbf{p}} + \hat{a}^{\dagger}_{\mathbf{p}}\hat{a}_{\mathbf{p}}]$$
$$= \frac{1}{2}\int d^3p E_{\mathbf{p}}[2\hat{a}^{\dagger}_{\mathbf{p}}\hat{a}_{\mathbf{p}} + \delta(0)] = \int d^3p E_{\mathbf{p}}\left[\hat{a}^{\dagger}_{\mathbf{p}}\hat{a}_{\mathbf{p}} + \frac{\delta(0)}{2}\right]. \tag{6.90}$$

We observe two points: first, the Hamiltonian obtained is structurally similar to that of the quantum harmonic oscillator, while the term $\delta(0)/2$ diverges. **This divergence represents a problem since the integral becomes ill-defined.**

There are two ways to eliminate it.

1. The Hamiltonian can be redefined by noticing that the divergent term

$$E_0 = \langle 0|\hat{H}|0\rangle = \int d^3p \left[\langle 0|\hat{a}^\dagger_{\mathbf{p}}\hat{a}_{\mathbf{p}}|0\rangle E_{\mathbf{p}} + \frac{\delta(0)}{2}\langle 0|0\rangle \right] = \int d^3p \, \frac{\delta(0)}{2} E_{\mathbf{p}} \tag{6.91}$$

coincides with the ground state energy, and it is always possible to consider the redefinition $\hat{H} \longrightarrow \hat{H} - E_0$ through normal ordering. **This procedure justifies how to remove the *infrared divergence***, i.e., the lowest energy divergence appearing in the expression above.
2. The Hamiltonian defines an energy that appears infinite even due to the fact that in the phase-space the volume element diverges as well. Accordingly, one can notice that the right object to measure is the *energy density*, rather than energy. Accordingly, one can ensure that the measurement can be done on ϵ,

$$\epsilon \equiv \frac{\hat{H}}{\mathcal{V}}, \tag{6.92}$$

where the above $\hat{H}$ is $\hat{H} \longrightarrow \hat{H} - E_0$ and $\mathcal{V} \equiv d^3p$. **This technique removes conceptually the other divergence, namely the ultraviolet divergence**, that appears the highest one.

Hence, we may conclude as follows. In non-gravitational contexts, shifting $H \to H + \text{const.}$ does not alter Heisenberg equations of motion and does not affect transition amplitudes between states of fixed particle content. Therefore the vacuum-energy constant can be removed by a convention (normal ordering), and what remains is the physically relevant excitation spectrum.

6.2.4 Normal Ordering

Normal ordering is a prescription that places all creation operators to the left of all annihilation operators. For bosons this is achieved by commuting operators; for fermions one must anticommute operators, producing sign factors.

For any operator-valued expression $O(a, a^\dagger)$ built from creation and annihilation operators, we define its normal-ordered form by

$$: O : \equiv N[O] \;\equiv\; \text{“}O \text{ with all } a^\dagger \text{ moved to the left of all } a\text{”}. \tag{6.93}$$

The defining property is that, for the Fock vacuum $|0\rangle$ satisfying $a_{\mathbf{p}}|0\rangle = 0$, one has

$$\langle 0| : O : |0\rangle = 0 \qquad \text{(for polynomials in free-field operators)}. \tag{6.94}$$

Applying normal ordering to Eq. (6.90) removes the constant term:

$$:H: = \int \frac{d^3p}{(2\pi)^3}\, E_{\mathbf{p}}\, a_{\mathbf{p}}^\dagger a_{\mathbf{p}}, \qquad \Rightarrow \qquad \langle 0| :H: |0\rangle = 0. \tag{6.95}$$

This is not a dynamical approximation; it is a choice of energy zero point, legitimate as long as only energy differences matter (or gravity is neglected).

For fermionic operators, normal ordering must account for anticommutation. If $\{c_i, c_j^\dagger\} = \delta_{ij}$ and one reorders a string of fermionic operators into normal order, each swap introduces a minus sign. Equivalently, the normal-ordered expression differs by $(-1)^P$, where P is the parity of the permutation required to bring the string into normal order.

As we will see, normal ordering will naturally appear when developing Wick's theorem: time-ordered products can be rewritten as normal-ordered products plus sums over contractions, and the contraction is precisely the vacuum two-point function.

6.2.5 The Meaning of the Mode Expansion

The mode expansion is often introduced as an ansatz. Here we clarify *why* this is the correct structure and what the operators $a_{\mathbf{p}}$ and $a_{\mathbf{p}}^\dagger$ actually mean.

(i) Solutions of the classical field equation.
The Klein-Gordon equation,

$$(\partial^2 + m^2)\phi = 0, \tag{6.96}$$

admits plane-wave solutions $e^{-ip\cdot x}$ provided $p^0 = \pm E_{\mathbf{p}}$. Thus the most general real classical solution can be expanded as a superposition of positive- and negative-frequency modes. Quantization promotes the mode coefficients to operators, and the reality condition $\phi^\dagger = \phi$ forces the coefficient of the negative-frequency part to be the adjoint of the positive-frequency coefficient.

(ii) Why the factor $1/\sqrt{2E_{\mathbf{p}}}$?
The normalization is fixed by the canonical commutator $[\phi(t,\mathbf{x}), \pi(t,\mathbf{y})] = i\delta^{(3)}(\mathbf{x}-\mathbf{y})$. Starting from a general expansion

$$\phi(t,\mathbf{x}) = \int \frac{d^3p}{(2\pi)^3} \left(f_{\mathbf{p}}\, a_{\mathbf{p}}\, e^{-i(E_{\mathbf{p}}t - \mathbf{p}\cdot\mathbf{x})} + f_{\mathbf{p}}^\star\, a_{\mathbf{p}}^\dagger\, e^{+i(E_{\mathbf{p}}t - \mathbf{p}\cdot\mathbf{x})} \right), \tag{6.97}$$

one finds that the commutator fixes $|f_{\mathbf{p}}|^2 = 1/(2E_{\mathbf{p}})$. This choice simultaneously ensures that the Hamiltonian diagonalizes to a sum of harmonic-oscillator Hamiltonians labelled by $\mathbf{p}$.

(iii) What do $a_{\mathbf{p}}$ and $a_{\mathbf{p}}^\dagger$ do?
The operator $a_{\mathbf{p}}^\dagger$ creates a one-particle excitation with momentum $\mathbf{p}$:

$$|\mathbf{p}\rangle \equiv a_{\mathbf{p}}^\dagger |0\rangle, \tag{6.98}$$

and using Eq. (6.90) (or its normal-ordered form) one verifies

$$H\,|\mathbf{p}\rangle = E_{\mathbf{p}}\,|\mathbf{p}\rangle \;+\; (\text{vacuum constant})\,|\mathbf{p}\rangle, \tag{6.99}$$

which confirms the interpretation of $E_{\mathbf{p}}$ as the relativistic one-particle energy.

(iv) Completeness and the continuum normalization. Assuming an infinite volume, momentum labels are continuous and the commutator $[a_{\mathbf{p}}, a_{\mathbf{q}}^{\dagger}] = (2\pi)^3\delta^{(3)}(\mathbf{p}-\mathbf{q})$ is the field-theory analogue of $[a, a^{\dagger}] = 1$ for a single oscillator. On the other hand, in a finite box of volume $V = L^3$ with periodic boundary conditions, $\mathbf{p}$ becomes discrete and

$$(2\pi)^3\delta^{(3)}(\mathbf{p}-\mathbf{q}) \;\longrightarrow\; V\,\delta_{\mathbf{p},\mathbf{q}}, \tag{6.100}$$

thus clarifying the relation between the delta function at the origin and the spatial volume.

The mode expansion is therefore not merely a mathematical trick: it expresses the fundamental statement that a free relativistic quantum field is equivalent to an infinite collection of independent harmonic oscillators, one for each momentum mode $\mathbf{p}$. Creation and annihilation operators provide the natural language for counting quanta of these oscillators, and the Hamiltonian becomes a weighted sum over number operators, plus the vacuum-energy constant removed by normal ordering.

6.3 Wick's Theorem

Perturbation theory in the interaction picture produces Dyson series containing time-ordered strings of field operators, whose vacuum expectation values (VEVs) must be computed repeatedly. The central technical difficulty is that

$$\langle 0|\, T[\hat{A}\hat{B}\hat{C}\cdots\hat{Z}]\,|0\rangle \tag{6.101}$$

is, in general, nontrivial, whereas the corresponding normal-ordered VEV is immediate:

$$\langle 0|\, N[\hat{A}\hat{B}\hat{C}\cdots\hat{Z}]\,|0\rangle = 0, \tag{6.102}$$

because normal ordering places annihilation operators to the right and creation operators to the left. In particular, every term in a normal-ordered product contains at least one annihilation operator acting on the vacuum on the right, or one creation operator acting on the vacuum on the left, and therefore vanishes. The goal of Wick's theorem is to convert time-ordered products into normal-ordered products plus c-number corrections.

6.3.1 Creation/Annihilation Split

For a free (interaction-picture) bosonic field operator one can split

$$\hat{\phi}(x) = \hat{\phi}^{-}(x) + \hat{\phi}^{+}(x), \tag{6.103}$$

where $\hat{\phi}^{-}$ contains annihilation operators and $\hat{\phi}^{+}$ contains creation operators. Consequently,

$$\hat{\phi}^{-}(x)|0\rangle = 0, \qquad \langle 0|\hat{\phi}^{+}(x) = 0. \tag{6.104}$$

The same decomposition applies to any operator built linearly from free-field modes; for the present discussion it suffices that $\hat{A} = \hat{A}^{-} + \hat{A}^{+}$ and $\hat{B} = \hat{B}^{-} + \hat{B}^{+}$.

Consider first the ordinary product $\hat{A}\hat{B}$:

$$\begin{aligned} \hat{A}\hat{B} &= (\hat{A}^{+} + \hat{A}^{-})(\hat{B}^{+} + \hat{B}^{-}) \\ &= \hat{A}^{+}\hat{B}^{+} + \hat{A}^{-}\hat{B}^{-} + \hat{A}^{+}\hat{B}^{-} + \hat{A}^{-}\hat{B}^{+}. \end{aligned} \tag{6.105}$$

Normal ordering only affects terms in which a creation operator appears to the right of an annihilation operator. In Eq. (6.105) the only such term is $\hat{A}^{-}\hat{B}^{+}$. Thus

$$N[\hat{A}\hat{B}] = \hat{A}^{+}\hat{B}^{+} + \hat{A}^{-}\hat{B}^{-} + \hat{A}^{+}\hat{B}^{-} + \hat{B}^{+}\hat{A}^{-}. \tag{6.106}$$

Subtracting,

$$\begin{aligned} \hat{A}\hat{B} - N[\hat{A}\hat{B}] &= \hat{A}^{-}\hat{B}^{+} - \hat{B}^{+}\hat{A}^{-} \\ &= [\hat{A}^{-}, \hat{B}^{+}]. \end{aligned} \tag{6.107}$$

For free bosonic fields this commutator is a c-number (it contains no operator content), a fact ultimately traceable to the canonical commutation relations.

6.3.2 Normal Ordering and Contraction

For spacetime-dependent operators, time ordering is defined by

$$T[\hat{A}(x)\hat{B}(y)] = \begin{cases} \hat{A}(x)\hat{B}(y), & x^0 > y^0, \\ \hat{B}(y)\hat{A}(x), & x^0 < y^0. \end{cases} \tag{6.108}$$

Using the same reasoning as above (the difference between $T[\cdot]$ and $N[\cdot]$ is entirely due to permuting creation/annihilation parts), one obtains

$$T[\hat{A}(x)\hat{B}(y)] - N[\hat{A}(x)\hat{B}(y)] = \begin{cases} [\hat{A}^{-}(x), \hat{B}^{+}(y)], \ x^0 > y^0, \\ [\hat{B}^{-}(y), \hat{A}^{+}(x)], \ x^0 < y^0. \end{cases} \tag{6.109}$$

Taking the vacuum expectation value and using $\langle 0|N[\cdots]|0\rangle = 0$ gives

$$\langle 0|\, T[\hat{A}(x)\hat{B}(y)]\, |0\rangle = \begin{cases} \langle 0|\, [\hat{A}^{-}(x), \hat{B}^{+}(y)]\, |0\rangle, & x^0 > y^0, \\ \langle 0|\, [\hat{B}^{-}(y), \hat{A}^{+}(x)]\, |0\rangle, & x^0 < y^0. \end{cases} \tag{6.110}$$

In particular, for $\hat{A} = \hat{B} = \hat{\phi}$ (a free scalar field), the left-hand side is the Feynman propagator.

It is therefore convenient to define the *contraction* of two operators as

$$\overset{\sqcap}{\hat{A}\hat{B}} \equiv T[\hat{A}\hat{B}] - N[\hat{A}\hat{B}]. \tag{6.111}$$

Because the difference between time ordering and normal ordering is a c-number, the contraction is itself a c-number and may equivalently be written as

$$\overset{\sqcap}{\hat{A}\hat{B}} = \langle 0|\, T[\hat{A}\hat{B}]\, |0\rangle. \tag{6.112}$$

Since c-numbers are unaffected by normal ordering, one can package the two-operator identity as

$$T[\hat{A}\hat{B}] = N[\hat{A}\hat{B}] + \overset{\sqcap}{\hat{A}\hat{B}} = N\left[\hat{A}\hat{B} + \overset{\sqcap}{\hat{A}\hat{B}}\right]. \tag{6.113}$$

Equations (6.111)–(6.113) are the elementary building blocks of Wick's theorem.

▶ **Driving Exercise**

Let us now see how to handle Wick's theorem for longer strings.

For a product of many free-field operators, the general statement is:

$$T[\hat{A}\hat{B}\hat{C}\cdots\hat{Z}] = N\Big[\hat{A}\hat{B}\hat{C}\cdots\hat{Z} + \text{(all possible single contractions)} + \text{(all possible double contractions)} + \cdots\Big]. \tag{6.114}$$

Here, each "k-fold contraction" term is obtained by selecting k disjoint pairs from the string and replacing each selected pair by its contraction, leaving all remaining uncontracted operators inside a normal-ordered product. The theorem is valid for *free* fields (as in the interaction picture), where contractions reduce to c-number two-point functions.

As an illustrative example, for four bosonic operators,

$$
\begin{aligned}
T[\hat{A}\hat{B}\hat{C}\hat{D}] = N[\hat{A}\hat{B}\hat{C}\hat{D}]
&+ N\left[\overbracket{\hat{A}\hat{B}}\,\hat{C}\hat{D}\right] + N\left[\overbracket{\hat{A}\hat{C}}\,\hat{B}\hat{D}\right] + N\left[\overbracket{\hat{A}\hat{D}}\,\hat{B}\hat{C}\right] \\
&+ N\left[\overbracket{\hat{B}\hat{C}}\,\hat{A}\hat{D}\right] + N\left[\overbracket{\hat{B}\hat{D}}\,\hat{A}\hat{C}\right] + N\left[\overbracket{\hat{C}\hat{D}}\,\hat{A}\hat{B}\right] \\
&+ N\left[\overbracket{\hat{A}\hat{B}}\,\overbracket{\hat{C}\hat{D}}\right] + N\left[\overbracket{\hat{A}\hat{C}}\,\overbracket{\hat{B}\hat{D}}\right] + N\left[\overbracket{\hat{A}\hat{D}}\,\overbracket{\hat{B}\hat{C}}\right].
\end{aligned}
\tag{6.115}
$$

The last line contains the distinct double-contraction patterns. Notice that, in the fermionic case, swapping operators to achieve the displayed ordering produces additional minus signs; one must account for the parity of the required permutations.

6.3.3 Vacuum Expectation Values

A key corollary, constantly used in perturbation theory, is that the VEV of a time-ordered product reduces to a sum of products of two-point contractions.

Consider any term in the Wick expansion containing at least one uncontracted operator inside the normal ordering, schematically

$$
N\left[\overbracket{\hat{C}\hat{F}}\,\hat{A}\hat{B}\hat{D}\hat{E}\hat{G}\cdots\right]. \tag{6.116}
$$

Because the contracted factor is a c-number it can be pulled out:

$$
N\left[\overbracket{\hat{C}\hat{F}}\,\hat{A}\hat{B}\hat{D}\hat{E}\hat{G}\cdots\right] = \overbracket{\hat{C}\hat{F}}\; N[\hat{A}\hat{B}\hat{D}\hat{E}\hat{G}\cdots]. \tag{6.117}
$$

Taking the vacuum expectation value annihilates such terms,

$$
\langle 0|\, N[\hat{A}\hat{B}\hat{D}\hat{E}\hat{G}\cdots]\,|0\rangle = 0, \tag{6.118}
$$

and therefore *only those Wick terms in which every operator is contracted* can contribute to the VEV. As a consequence, any VEV with an odd number of (bosonic) field operators vanishes, since one cannot pair all operators.

Hence

$$\langle 0|\, T[\hat{A}\hat{B}\hat{C}\cdots\hat{Y}\hat{Z}]\,|0\rangle = \sum_{\text{all complete pairings}} \prod_{\text{pairs } (i,j)} \langle 0|\, T[\hat{O}_i\hat{O}_j]\,|0\rangle, \tag{6.119}$$

i.e. the VEV of a time-ordered string equals the sum over all complete pairwise contractions (pairings), each term being the product of the corresponding two-point functions.

Driving Exercise

As an example, let us consider again four bosonic operators. From Eq. (6.119) one finds

$$\langle 0|\, T[\hat{A}\hat{B}\hat{C}\hat{D}]\,|0\rangle = \langle 0|\, T[\hat{A}\hat{B}]\,|0\rangle\langle 0|\, T[\hat{C}\hat{D}]\,|0\rangle$$
$$+\langle 0|\, T[\hat{A}\hat{C}]\,|0\rangle\langle 0|\, T[\hat{B}\hat{D}]\,|0\rangle + \langle 0|\, T[\hat{A}\hat{D}]\,|0\rangle\langle 0|\, T[\hat{B}\hat{C}]\,|0\rangle. \tag{6.120}$$

For a free scalar field this becomes

$$\begin{aligned}\langle 0|\, T[\hat{\phi}(x_1)\hat{\phi}(x_2)\hat{\phi}(x_3)\hat{\phi}(x_4)]\,|0\rangle &= \Delta_F(x_1-x_2)\Delta_F(x_3-x_4)\\ &+ \Delta_F(x_1-x_3)\Delta_F(x_2-x_4)\\ &+ \Delta_F(x_1-x_4)\Delta_F(x_2-x_3),\end{aligned} \tag{6.121}$$

where $\Delta_F(x-y) = \langle 0|\, T[\hat{\phi}(x)\hat{\phi}(y)]\,|0\rangle$ is the Feynman propagator. This factorization property is the algebraic engine behind the diagrammatic expansion of perturbation theory.

6.4 Field Operators and Second Quantization

Field operators are central in field theory. They are again defined relative to a chosen one-particle basis $\{|\underline{\alpha}\rangle\}$ with wave functions $\psi_{\underline{\alpha}}(x) = \langle x|\underline{\alpha}\rangle$.

The annihilation field operator is defined as the basis expansion

$$\hat{\psi}(x) = \sum_{\underline{\alpha}} \langle x|\underline{\alpha}\rangle\, a_{\underline{\alpha}} = \sum_{\underline{\alpha}} \psi_{\underline{\alpha}}(x)\, a_{\underline{\alpha}}. \tag{6.122}$$

Its Hermitian conjugate is the creation field operator

$$\hat{\psi}^\dagger(x) = \sum_{\underline{\alpha}} \langle\underline{\alpha}|x\rangle\, a^\dagger_{\underline{\alpha}} = \sum_{\underline{\alpha}} \psi^*_{\underline{\alpha}}(x)\, a^\dagger_{\underline{\alpha}}. \tag{6.123}$$

The variable x is continuous (position), so $\hat{\psi}(x)$ and $\hat{\psi}^\dagger(x)$ are operator-valued distributions. Formally, their definition resembles a change of basis from the discrete set $\{|\underline{\alpha}\rangle\}$ to the continuous position basis $\{|x\rangle\}$: one may view $\hat{\psi}(x)$ and $\hat{\psi}^\dagger(x)$ as annihilation and creation operators in the $|x\rangle$ basis. In physical terms, $\hat{\psi}(x)$ annihilates a particle at position x and $\hat{\psi}^\dagger(x)$ creates a particle at position x (with the usual caveat that perfect localization implies completely delocalized momentum).

For bosons (commutators) and fermions (anticommutators) one has

$$[\,\hat{\psi}^\dagger(x), \hat{\psi}^\dagger(x')\,]_\pm = 0, \qquad [\,\hat{\psi}(x), \hat{\psi}(x')\,]_\pm = 0, \qquad [\,\hat{\psi}(x), \hat{\psi}^\dagger(x')\,]_\pm = \delta^{(3)}(x - x'), \tag{6.124}$$

where $\delta^{(3)}$ is the three-dimensional Dirac delta distribution. The mixed relation follows from Eqs. (6.122)–(6.123) and the canonical algebra of $a_{\underline{\alpha}}, a^\dagger_{\underline{\alpha}}$:

$$\begin{aligned}[\,\hat{\psi}(x), \hat{\psi}^\dagger(x')\,]_\pm &= \left[\sum_{\underline{\alpha}} \langle x|\underline{\alpha}\rangle a_{\underline{\alpha}}, \sum_{\underline{\beta}} \langle \underline{\beta}|x'\rangle a^\dagger_{\underline{\beta}}\right]_\pm = \sum_{\underline{\alpha},\underline{\beta}} \langle x|\underline{\alpha}\rangle\langle\underline{\beta}|x'\rangle [\,a_{\underline{\alpha}}, a^\dagger_{\underline{\beta}}\,]_\pm \\ &= \sum_{\underline{\alpha},\underline{\beta}} \langle x|\underline{\alpha}\rangle\langle\underline{\beta}|x'\rangle \delta_{\underline{\alpha}\underline{\beta}} = \sum_{\underline{\alpha}} \langle x|\underline{\alpha}\rangle\langle\underline{\alpha}|x'\rangle = \langle x|\Big(\sum_{\underline{\alpha}} |\underline{\alpha}\rangle\langle\underline{\alpha}|\Big)|x'\rangle \\ &= \langle x|x'\rangle = \delta^{(3)}(x - x').\end{aligned} \tag{6.125}$$

The vanishing relations in Eq. (6.124) follow similarly: they reduce to the vanishing (anti)commutators among two annihilators or two creators in the discrete basis.

6.4.1 Extending Canonical Quantization for Any Field

The Pauli-Weisskopf prescription is well-posed but limited to a complex scalar field. Conceptually, even assuming the limiting case to real fields appears non-conclusive, as it was formulated to provide a direct explanation of the apparent Klein-Gordon inconsistencies.

So, one can wonder whether the second quantization scheme, constructed by means of creation and annihilation operators, could be general, namely valid for *any field*.

We thus generalize, as an ansatz, the Pauli-Weisskopf quantization through the following steps,

1. write the Lagrangian density in terms of the field;
2. compute the momentum density and the Hamiltonian density;

3. treat fields and momenta as operators and impose commutation rules for them;
4. expand the fields in terms of creation and annihilation operators;
5. introduce an interpretation based on normal ordering;

This procedure then applies to scalars, spinors, vectors, etc.

▶ **Pause and Think** Canonical quantization is not a theorem but a prescription, guaranteeing Lorentz invariance, locality, and stability. Explain all the reasons that, in your view, permit to accept it for any field. Then, ask yourself (after reading Chap. 10): is it valid even for gravity?

6.4.2 Massive Vector Fields

Starting from the massive scalar field framework, we can consider a massive extension to the vector sector. Hence, a massive spin-1 field is described by a Lorentz four-vector $A_\mu(x)$ with Lagrangian density

$$\mathcal{L} = -\frac{1}{4} F_{\mu\nu} F^{\mu\nu} + \frac{1}{2} m^2 A_\mu A^\mu, \qquad F_{\mu\nu} \equiv \partial_\mu A_\nu - \partial_\nu A_\mu. \tag{6.126}$$

Varying with respect to A_ν yields the Euler-Lagrange equations

$$\partial_\mu F^{\mu\nu} + m^2 A^\nu = 0, \tag{6.127}$$

known as the *Proca equation*. Taking the divergence and using antisymmetry of $F^{\mu\nu}$,

$$\partial_\nu \partial_\mu F^{\mu\nu} = 0 \quad \Rightarrow \quad m^2 \partial_\nu A^\nu = 0, \tag{6.128}$$

so that for $m \neq 0$ the field satisfies the subsidiary condition

$$\partial_\mu A^\mu = 0. \tag{6.129}$$

Substituting Eq. (6.129) back into Eq. (6.127) gives

$$(\Box + m^2) A^\nu = 0, \tag{6.130}$$

i.e. each component obeys a Klein-Gordon type equation, while the constraint in Eq. (6.129) removes one unphysical degree of freedom, leaving three physical polarizations.

We can then proceed with the same quantization procedure as presented above.

- **Canonical structure and Hamiltonian.** The canonical momentum tensor conjugate to A_ν is

$$\Pi^{\mu\nu} \equiv \frac{\partial \mathcal{L}}{\partial(\partial_\mu A_\nu)} = -F^{\mu\nu}, \tag{6.131}$$

so that the time components satisfy $\Pi^{00} = 0$ and

$$\Pi^{0i} = -F^{0i} \equiv E^i, \tag{6.132}$$

which motivates the notation $\mathbf{E}$ (with $\mathbf{B}$ defined as usual by $B^i = \frac{1}{2}\epsilon^{ijk}F_{jk}$). The Hamiltonian density may be arranged into a manifestly positive form once the non-dynamical component A_0 is eliminated using the equations of motion. A compact derivation proceeds as follows.
Start from

$$\mathcal{H} = \Pi^{0\nu}\dot{A}_\nu - \mathcal{L} = \mathbf{E}\cdot\dot{\mathbf{A}} + \frac{1}{4}F_{\mu\nu}F^{\mu\nu} - \frac{1}{2}m^2 A_\mu A^\mu, \tag{6.133}$$

use

$$\frac{1}{4}F_{\mu\nu}F^{\mu\nu} = \frac{1}{2}(\mathbf{B}^2 - \mathbf{E}^2), \qquad A_\mu A^\mu = A_0^2 - \mathbf{A}^2, \tag{6.134}$$

and integrate by parts the term $\mathbf{E}\cdot\nabla A_0$ generated when rewriting $\mathbf{E}\cdot\dot{\mathbf{A}}$ in terms of $F^{0i} = \dot{A}^i - \partial^i A^0$:

$$\mathbf{E}\cdot\nabla A_0 = \nabla\cdot(A_0\mathbf{E}) - A_0(\nabla\cdot\mathbf{E}). \tag{6.135}$$

The Proca equation with $\nu = 0$ gives a constraint relating A_0 to $\nabla\cdot\mathbf{E}$:

$$\partial_\mu F^{\mu 0} + m^2 A^0 = 0 \quad\Rightarrow\quad \nabla\cdot\mathbf{E} + m^2 A_0 = 0 \quad\Rightarrow\quad A_0 = -\frac{1}{m^2}\nabla\cdot\mathbf{E}. \tag{6.136}$$

Dropping the total derivative $\nabla\cdot(A_0\mathbf{E})$ under the standard assumption that fields vanish sufficiently fast at spatial infinity, one arrives at

$$\mathcal{H} = \frac{1}{2}\left(\mathbf{E}^2 + \mathbf{B}^2 + m^2\mathbf{A}^2 + \frac{1}{m^2}(\nabla\cdot\mathbf{E})^2\right), \tag{6.137}$$

which exhibits positivity and makes transparent why the mass term in Eq. (6.126) must enter with the chosen sign.
Canonical quantization is implemented by imposing equal-time commutators for the dynamical spatial components:

$$\left[\hat{A}^i(t,\mathbf{x}), \hat{E}^j(t,\mathbf{y})\right] = -i\,g^{ij}\,\delta^{(3)}(\mathbf{x}-\mathbf{y}) = i\,\delta^{ij}\,\delta^{(3)}(\mathbf{x}-\mathbf{y}), \tag{6.138}$$

while A_0 is fixed by the constraint in Eq. (6.136).

- **Mode expansion and polarization vectors.**
Plane-wave solutions satisfy the on-shell condition $p^2 = m^2$ and the transversality constraint $p_\mu A^\mu = 0$. Accordingly, one expands the quantum field as

$$\hat{A}^\mu(x) = \int \frac{d^3p}{(2\pi)^3} \frac{1}{\sqrt{2E_{\mathbf{p}}}} \sum_{\lambda=1}^{3} \left(\epsilon_\lambda^\mu(p)\, \hat{a}_{\mathbf{p}\lambda}\, e^{-ip\cdot x} + \epsilon_\lambda^{\mu *}(p)\, \hat{a}_{\mathbf{p}\lambda}^\dagger\, e^{ip\cdot x} \right), \quad (6.139)$$

The polarization vectors are chosen to satisfy

$$p_\mu \epsilon_\lambda^\mu(p) = 0, \qquad \epsilon_\lambda^*(p) \cdot \epsilon_{\lambda'}(p) \equiv g_{\mu\nu} \epsilon_\lambda^{\mu *}(p) \epsilon_{\lambda'}^\nu(p) = -\delta_{\lambda\lambda'}. \quad (6.140)$$

In the rest frame $p^\mu = (m, 0, 0, 0)$ a convenient linear basis is

$$\epsilon_1^\mu(m, \mathbf{0}) = (0, 1, 0, 0), \quad \epsilon_2^\mu(m, \mathbf{0}) = (0, 0, 1, 0), \quad \epsilon_3^\mu(m, \mathbf{0}) = (0, 0, 0, 1), \quad (6.141)$$

and in an arbitrary frame one may obtain $\epsilon_\lambda^\mu(p)$ by boosting with a Lorentz transformation taking $(m, \mathbf{0}) \mapsto p$. For example, for motion along the z-axis, $p^\mu = (E_{\mathbf{p}}, 0, 0, |\mathbf{p}|)$, one finds

$$\epsilon_1^\mu = (0, 1, 0, 0), \quad \epsilon_2^\mu = (0, 0, 1, 0), \quad \epsilon_3^\mu = \left(\frac{|\mathbf{p}|}{m}, 0, 0, \frac{E_{\mathbf{p}}}{m} \right), \quad (6.142)$$

which explicitly displays the third (longitudinal) physical polarization present for massive spin-1 quanta.
Using Eqs. (6.139) in (6.137) and normal ordering yields the diagonal Hamiltonian

$$\hat{H} = \int d^3p\, E_{\mathbf{p}} \sum_{\lambda=1}^{3} \hat{a}_{\mathbf{p}\lambda}^\dagger \hat{a}_{\mathbf{p}\lambda}, \quad (6.143)$$

so the excitations are massive spin-1 particles with three polarizations.

- **Projection tensors and polarization completeness.**
Given a nonzero three-momentum $\mathbf{p}$, the longitudinal and transverse projectors acting on spatial vectors are

$$(P_L)^{ij} = \frac{p^i p^j}{|\mathbf{p}|^2}, \qquad (P_T)^{ij} = \delta^{ij} - \frac{p^i p^j}{|\mathbf{p}|^2}, \quad (6.144)$$

so that for any X^i one has $X_L^i = (P_L)^{ij} X^j$ and $X_T^i = (P_T)^{ij} X^j$. A covariant upgrade is obtained by replacing δ^{ij} with $g^{\mu\nu}$ and $|\mathbf{p}|^2$ with p^2:

$$(P_L)^{\mu\nu} = \frac{p^\mu p^\nu}{p^2}, \qquad (P_T)^{\mu\nu} = g^{\mu\nu} - \frac{p^\mu p^\nu}{p^2}. \quad (6.145)$$

For massive modes $p^2 = m^2$. In a linear polarization basis, the polarization sum satisfies the completeness relation

$$\sum_{\lambda=1}^{3} \epsilon_\lambda^\mu(p)\epsilon_\lambda^\nu(p) = -g^{\mu\nu} + \frac{p^\mu p^\nu}{m^2} = -(P_T)^{\mu\nu}\Big|_{p^2=m^2}, \tag{6.146}$$

which can be verified directly in the rest frame using Eq. (6.141) and then boosted to any frame, or derived covariantly by noting that the left-hand side is a rank-2 tensor built only from $g^{\mu\nu}$ and $p^\mu p^\nu$ and fixing coefficients in the rest frame.

6.4.3 Canonical Quantization of the Electromagnetic Field in Coulomb Gauge

The Proca Lagrangian *is not* a gauge theory. Electromagnetism, experimentally speaking, requires $m \to 0$. The corresponding theory appears immediately a gauge theory.

The electromagnetic potential A_μ is defined up to gauge transformations $A_\mu \to A_\mu + \partial_\mu \alpha$. This redundancy reduces the number of physical degrees of freedom: although A_μ has four components, only two independent transverse polarizations propagate as physical photons. This feature must be handled carefully in canonical quantization.

When a matter field carries a global $U(1)$ symmetry, promoting it to a local symmetry introduces the gauge field and dictates the interaction structure. For a complex scalar ψ, minimal coupling replaces ∂_μ by $D_\mu = \partial_\mu + iqA_\mu$, generating interaction terms linear and quadratic in A_μ. This illustrates the gauge principle: local symmetry fixes the form of the coupling.

We quantize the free electromagnetic field starting from

$$\mathcal{L} = -\frac{1}{4}F_{\mu\nu}F^{\mu\nu}, \qquad F_{\mu\nu} = \partial_\mu A_\nu - \partial_\nu A_\mu, \tag{6.147}$$

and impose the Coulomb gauge conditions

$$A_0 = 0, \qquad \nabla \cdot \mathbf{A} = 0, \tag{6.148}$$

so that only transverse spatial components remain as dynamical variables.

The canonical momentum tensor is

$$\Pi^{\mu\nu} = \frac{\partial \mathcal{L}}{\partial(\partial_\mu A_\nu)} = -F^{\mu\nu}, \tag{6.149}$$

hence $\Pi^{0i} = -F^{0i} \equiv E^i$ and the Hamiltonian density reduces to

$$\mathcal{H} = \frac{1}{2}\left(\mathbf{E}^2 + \mathbf{B}^2\right), \tag{6.150}$$

after discarding a total divergence and using the constraints implied by the chosen gauge.

A naive equal-time commutator $[\hat{A}^i(\mathbf{x}), \hat{E}^j(\mathbf{y})] = i\delta^{ij}\delta^{(3)}(\mathbf{x}-\mathbf{y})$ is incompatible with $\nabla \cdot \hat{\mathbf{A}} = 0$, because taking a divergence would produce a nonzero right-hand side. The correct commutator must project onto transverse components:

$$\left[\hat{A}^i(t,\mathbf{x}), \hat{E}^j(t,\mathbf{y})\right] = i \int \frac{d^3p}{(2\pi)^3}\, e^{i\mathbf{p}\cdot(\mathbf{x}-\mathbf{y})} \left(\delta^{ij} - \frac{p^i p^j}{\mathbf{p}^2}\right) \equiv i\,\delta^{(3)}_{\mathrm{T}ij}(\mathbf{x}-\mathbf{y}), \tag{6.151}$$

where $\delta_{\mathrm{T}}^{(3)ij}$ is the *transverse delta function*. Indeed,

$$\partial_i\, \delta^{(3)}_{\mathrm{T}ij}(\mathbf{x}-\mathbf{y}) = 0, \tag{6.152}$$

so the commutator is consistent with the Coulomb gauge condition.

6.4.3.1 Mode Expansion and Hamiltonian

The field operator only contains the two transverse polarizations:

$$\hat{A}^\mu(x) = \int \frac{d^3p}{(2\pi)^3} \frac{1}{\sqrt{2E_\mathbf{p}}} \sum_{\lambda=1}^{2} \left(\epsilon_\lambda^\mu(p)\, \hat{a}_{\mathbf{p}\lambda}\, e^{-ip\cdot x} + \epsilon_\lambda^{\mu *}(p)\, \hat{a}^\dagger_{\mathbf{p}\lambda}\, e^{ip\cdot x}\right), \qquad E_\mathbf{p} = |\mathbf{p}|. \tag{6.153}$$

The polarization vectors satisfy $p_\mu \epsilon_\lambda^\mu(p) = 0$ and $\epsilon_\lambda^* \cdot \epsilon_{\lambda'} = -\delta_{\lambda\lambda'}$, with no physical longitudinal or time-like modes. Normal ordering gives the diagonal Hamiltonian

$$\hat{H} = \int d^3p \sum_{\lambda=1}^{2} E_\mathbf{p}\, \hat{a}^\dagger_{\mathbf{p}\lambda} \hat{a}_{\mathbf{p}\lambda}, \tag{6.154}$$

so the excitations are massless spin-1 quanta (photons) with two transverse polarizations.

For a photon propagating along the z-axis, $q^\mu = (|q|, 0, 0, |q|)$, a linear basis may be chosen as

$$\epsilon_1^\mu(q) = (0,1,0,0), \qquad \epsilon_2^\mu(q) = (0,0,1,0). \tag{6.155}$$

Equivalently, one may use circular polarizations

$$\epsilon_R^\mu(q) = \frac{1}{\sqrt{2}}(0,1,i,0), \qquad \epsilon_L^\mu(q) = \frac{1}{\sqrt{2}}(0,1,-i,0), \tag{6.156}$$

which are eigenstates of helicity $h = \mathbf{J}\cdot\hat{\mathbf{q}}$ with eigenvalues $h = \pm 1$ for right- and left-handed photons, respectively. The absence of an $h = 0$ mode is the statement that longitudinal polarization is unphysical for the massless gauge field (in contrast to the massive Proca field).

▶ **Driving Exercise**
Alternative quantization for the electromagnetic field starts from the fact that canonical quantization in Coulomb gauge is not manifestly Lorentz covariant.
So, one may retain all four components of A_μ and work with a covariant gauge-fixed Lagrangian (Lorenz gauge)

$$\mathcal{L} = -\frac{1}{4} F_{\mu\nu} F^{\mu\nu} - \frac{1}{2\xi} (\partial_\mu A^\mu)^2, \tag{6.157}$$

where ξ is a gauge parameter. The price is that the state space carries unphysical polarizations. Physical states are selected by imposing the Gupta-Bleuler condition

$$(\partial_\mu \hat{A}^\mu)^{(+)} \, |\text{phys}\rangle = 0, \tag{6.158}$$

where $(+)$ denotes the positive-frequency part. This eliminates negative-norm components from observable matrix elements and ensures that only the two transverse photon polarizations contribute to physical processes.

6.4.4 The Quantized Dirac Field

This section introduces the free Dirac field as a quantum field whose excitations describe spin-$\frac{1}{2}$ fermions and antifermions. We develop the canonical quantization of the free theory, derive the associated global $U(1)$ Noether current, construct the free fermion propagator via the generating functional with Grassmann sources, summarize the momentum space Feynman rules for fermion lines, and illustrate their use with a simple tree-level scattering process. Finally, we promote the global symmetry to a local one and obtain the minimally coupled QED Lagrangian.

The free Dirac Lagrangian density is

$$\mathcal{L} = \bar{\psi} \, (i\gamma^\mu \partial_\mu - m) \, \psi, \qquad \bar{\psi} \equiv \psi^\dagger \gamma^0, \tag{6.159}$$

where ψ is a four-component spinor field. Treating ψ and $\bar{\psi}$ as independent fields in the variational principle, the Euler-Lagrange equation for $\bar{\psi}$ yields

$$\frac{\partial \mathcal{L}}{\partial \bar{\psi}} - \partial_\mu \left(\frac{\partial \mathcal{L}}{\partial(\partial_\mu \bar{\psi})} \right) = 0 \quad \Rightarrow \quad (i\gamma^\mu \partial_\mu - m)\psi = 0, \tag{6.160}$$

i.e. the Dirac equation. Varying with respect to ψ gives the adjoint equation $\partial_\mu \bar{\psi} \, i\gamma^\mu + m\bar{\psi} = 0$.

▶ **Driving Exercise** The canonical momentum conjugate to ψ is defined as

$$\Pi^{\mu}_{\psi} \equiv \frac{\partial \mathcal{L}}{\partial(\partial_{\mu}\psi)}. \tag{6.161}$$

From Eq. (6.159) one has $\mathcal{L} \supset i\bar{\psi}\gamma^{\mu}\partial_{\mu}\psi$, hence

$$\Pi^{\mu}_{\psi} = i\bar{\psi}\gamma^{\mu}, \qquad \Pi^{\mu}_{\bar{\psi}} = \frac{\partial \mathcal{L}}{\partial(\partial_{\mu}\bar{\psi})} = 0. \tag{6.162}$$

In particular,

$$\Pi^{0}_{\psi} = i\bar{\psi}\gamma^{0} = i\psi^{\dagger}, \tag{6.163}$$

since $\bar{\psi}\gamma^{0} = \psi^{\dagger}\gamma^{0}\gamma^{0} = \psi^{\dagger}$.
The canonical Hamiltonian density is

$$\mathcal{H} = \Pi^{0}_{\psi}\,\partial_0\psi - \mathcal{L} = i\psi^{\dagger}\partial_0\psi - \bar{\psi}(i\gamma^{\mu}\partial_{\mu} - m)\psi. \tag{6.164}$$

Separating time and space derivatives in the Lagrangian,

$$\bar{\psi}\,i\gamma^{\mu}\partial_{\mu}\psi = \bar{\psi}\,i\gamma^{0}\partial_0\psi + \bar{\psi}\,i\gamma^{i}\partial_i\psi, \tag{6.165}$$

we obtain

$$\mathcal{H} = i\psi^{\dagger}\partial_0\psi - \bar{\psi}\,i\gamma^{0}\partial_0\psi - \bar{\psi}\,i\gamma^{i}\partial_i\psi + m\bar{\psi}\psi. \tag{6.166}$$

Using $\bar{\psi}\gamma^{0} = \psi^{\dagger}$, the first two terms cancel exactly, and thus

$$\mathcal{H} = -\bar{\psi}\,i\gamma^{i}\partial_i\psi + m\bar{\psi}\psi = \psi^{\dagger}\big(-i\gamma^{0}\gamma^{i}\partial_i + m\gamma^{0}\big)\psi. \tag{6.167}$$

Therefore the Hamiltonian is $H = \int d^3x\,\mathcal{H}$.

It is also useful to note that the Dirac equation (6.160) implies

$$i\gamma^{0}\partial_0\psi = \left(-i\gamma^{i}\partial_i + m\right)\psi \quad \Rightarrow \quad \mathcal{H} = \psi^{\dagger}i\partial_0\psi, \tag{6.168}$$

which is equivalent to Eq. (6.167) on-shell.

At this stage, canonical quantization for fermions requires equal-time *anticommutation* relations (commutators would lead to negative-norm states and an unstable vacuum):

$$\{\hat{\psi}_a(t,\mathbf{x}), \hat{\psi}^{\dagger}_b(t,\mathbf{y})\} = \delta_{ab}\,\delta^{(3)}(\mathbf{x}-\mathbf{y}), \tag{6.169}$$

$$\{\hat{\psi}_a(t,\mathbf{x}), \hat{\psi}_b(t,\mathbf{y})\} = \{\hat{\psi}^{\dagger}_a(t,\mathbf{x}), \hat{\psi}^{\dagger}_b(t,\mathbf{y})\} = 0. \tag{6.170}$$

A standard plane-wave mode expansion consistent with Eqs. (6.169)–(6.170) is

$$\hat{\psi}(x) = \int \frac{d^3p}{(2\pi)^3} \frac{1}{\sqrt{2E_{\mathbf{p}}}} \sum_{s=1}^{2} \left[u_s(\mathbf{p})\, \hat{a}_s(\mathbf{p})\, e^{-ip\cdot x} + v_s(\mathbf{p})\, \hat{b}_s^\dagger(\mathbf{p})\, e^{+ip\cdot x} \right], \quad (6.171)$$

with $p^0 = E_{\mathbf{p}} = \sqrt{\mathbf{p}^2 + m^2}$. The operators satisfy

$$\{\hat{a}_s(\mathbf{p}), \hat{a}_r^\dagger(\mathbf{q})\} = \{\hat{b}_s(\mathbf{p}), \hat{b}_r^\dagger(\mathbf{q})\} = (2\pi)^3 \delta_{sr} \delta^{(3)}(\mathbf{p} - \mathbf{q}), \quad (6.172)$$

with all other anticommutators vanishing.

6.4.4.1 Normal-Ordered Hamiltonian

Substituting Eq. (6.171) into $H = \int d^3x\, \mathcal{H}$ and using orthogonality of spinors and the anticommutation relations, one finds that the normal-ordered Hamiltonian takes the diagonal form

$$:\hat{H} := \int \frac{d^3p}{(2\pi)^3} \sum_{s=1}^{2} E_{\mathbf{p}} \left(\hat{a}_s^\dagger(\mathbf{p}) \hat{a}_s(\mathbf{p}) + \hat{b}_s^\dagger(\mathbf{p}) \hat{b}_s(\mathbf{p}) \right), \quad (6.173)$$

i.e. the energy is the sum of particle and antiparticle energies, as expected for a free theory.

The Lagrangian in Eq. (6.159) is invariant under the global phase rotation

$$\psi \to e^{i\alpha} \psi, \qquad \bar{\psi} \to \bar{\psi}\, e^{-i\alpha}, \quad (6.174)$$

with constant α. For an infinitesimal transformation $\delta\psi = i\alpha\psi$, $\delta\bar{\psi} = -i\alpha\bar{\psi}$, the Noether current is

$$J^\mu = \frac{\partial \mathcal{L}}{\partial(\partial_\mu \psi)} \delta\psi + \frac{\partial \mathcal{L}}{\partial(\partial_\mu \bar{\psi})} \delta\bar{\psi} = \Pi_\psi^\mu\, (i\psi) + 0 = \bar{\psi}\gamma^\mu\psi, \quad (6.175)$$

and it obeys $\partial_\mu J^\mu = 0$ upon using the equations of motion. The conserved charge is

$$\hat{Q} = \int d^3x\, :\hat{\psi}^\dagger \hat{\psi} := \int \frac{d^3p}{(2\pi)^3} \sum_{s=1}^{2} \left(\hat{a}_s^\dagger(\mathbf{p}) \hat{a}_s(\mathbf{p}) - \hat{b}_s^\dagger(\mathbf{p}) \hat{b}_s(\mathbf{p}) \right), \quad (6.176)$$

i.e. particle number minus antiparticle number.

6.4.5 Canonical Quantization of the Rarita-Schwinger Field

The Rarita-Schwinger field $\psi_\mu(x)$ provides the relativistic description of a spin-$\frac{3}{2}$ particle and carries simultaneously a Lorentz vector index μ and a spinor index. As such, it represents a constrained fermionic field theory whose canonical quantization requires particular care, due to the presence of gauge-like redundancies and subsidiary conditions.

The free massive Rarita-Schwinger Lagrangian in flat spacetime can be written as

$$\mathcal{L}_{\mathrm{RS}} = \bar{\psi}_\mu \left(i\gamma^{\mu\nu\rho}\partial_\nu - m\gamma^{\mu\rho}\right)\psi_\rho, \tag{6.177}$$

where

$$\gamma^{\mu\nu\rho} = \frac{1}{2}\left(\gamma^\mu\gamma^\nu\gamma^\rho - \gamma^\rho\gamma^\nu\gamma^\mu\right), \qquad \gamma^{\mu\rho} = \frac{1}{2}[\gamma^\mu, \gamma^\rho].$$

The Euler-Lagrange equations yield the Rarita-Schwinger equation

$$\left(i\gamma^{\mu\nu\rho}\partial_\nu - m\gamma^{\mu\rho}\right)\psi_\rho = 0. \tag{6.178}$$

Consistency implies the subsidiary constraints

$$\gamma^\mu\psi_\mu = 0, \qquad \partial^\mu\psi_\mu = 0, \tag{6.179}$$

which project out the unphysical spin-$\frac{1}{2}$ components, leaving the correct $2s+1=4$ physical degrees of freedom.

▶ **Driving Exercise** Treating ψ_μ as a Grassmann-valued field, the canonical conjugate momenta are defined as

$$\pi^\mu = \frac{\partial \mathcal{L}_{\mathrm{RS}}}{\partial(\partial_0\psi_\mu)}. \tag{6.180}$$

Due to the first-order nature of the Lagrangian, not all components of π^μ are independent, and primary constraints arise in the sense of Dirac. In particular, the time component ψ_0 appears without a time derivative and plays the role of a Lagrange multiplier, enforcing the transversality condition.

The full constraint structure ensures the elimination of non-dynamical and redundant degrees of freedom and is essential for a consistent quantization. In order to proceed with quantization, we then need to get rid of the unphysical components of ψ_μ via the Rarita-Schwinger constraints (or, equivalently, by an appropriate gauge fixing in the massless case).

The field operator can then be expanded in *physical* polarizations as

$$\widehat{\psi}_\mu(x) = \sum_{r=1}^{4}\int \frac{d^3p}{(2\pi)^3}\frac{1}{\sqrt{2E_{\mathbf{p}}}}\left[u_\mu(\mathbf{p},r)\,\widehat{a}(\mathbf{p},r)\,e^{-ip\cdot x} + v_\mu(\mathbf{p},r)\,\widehat{b}^\dagger(\mathbf{p},r)\,e^{+ip\cdot x}\right], \tag{6.181}$$

with $p^0 = E_{\mathbf{p}} = \sqrt{\mathbf{p}^2 + m^2}$. Here u_μ and v_μ are vector-spinors solving the free Rarita-Schwinger equation and satisfying the momentum space version of Eq. (6.179),

$$
\begin{aligned}
(\not{p} - m)u_\mu &= 0, \quad p^\mu u_\mu = 0, \quad \gamma^\mu u_\mu = 0, \\
(\not{p} + m)v_\mu &= 0, \quad p^\mu v_\mu = 0, \quad \gamma^\mu v_\mu = 0,
\end{aligned}
\tag{6.182}
$$

so that Eq. (6.181) contains only the spin-$\frac{3}{2}$ content.

Quantization is implemented by imposing the standard equal-time fermionic algebra on the mode operators,

$$
\begin{aligned}
\{\widehat{a}(\mathbf{p}, r), \widehat{a}^\dagger(\mathbf{p}', r')\} &= (2\pi)^3 \delta^{(3)}(\mathbf{p} - \mathbf{p}')\, \delta_{rr'}, \\
\{\widehat{b}(\mathbf{p}, r), \widehat{b}^\dagger(\mathbf{p}', r')\} &= (2\pi)^3 \delta^{(3)}(\mathbf{p} - \mathbf{p}')\, \delta_{rr'},
\end{aligned}
\tag{6.183}
$$

with all other anticommutators vanishing.

► **Driving exercise** For a free relativistic field the Hamiltonian is the generator of time translations,

$$
\widehat{H} = \int d^3x\, \widehat{\mathcal{H}}(x), \qquad \widehat{\mathcal{H}} = \widehat{T}^{00},
\tag{6.184}
$$

where $T^{\mu\nu}$ can be obtained (up to improvement terms) either canonically or from variation with respect to the metric. When the field is expanded as in Eq. (6.181) and the constraints (6.182) are used to eliminate the unphysical sector, the free Hamiltonian reduces to the expected sum of harmonic-oscillator contributions for each physical polarization:

$$
\widehat{H} = \sum_{r=1}^{4} \int \frac{d^3p}{(2\pi)^3}\, E_{\mathbf{p}} \left[\widehat{a}^\dagger(\mathbf{p}, r)\widehat{a}(\mathbf{p}, r) + \widehat{b}^\dagger(\mathbf{p}, r)\widehat{b}(\mathbf{p}, r)\right] + E_{\text{vac}}.
\tag{6.185}
$$

The constant E_{vac} is the (divergent) vacuum energy and is removed by normal ordering. In normal-ordered form one writes

$$
:\widehat{H}: = \sum_{r=1}^{4} \int \frac{d^3p}{(2\pi)^3}\, E_{\mathbf{p}} \left[\widehat{a}^\dagger(\mathbf{p}, r)\widehat{a}(\mathbf{p}, r) + \widehat{b}^\dagger(\mathbf{p}, r)\widehat{b}(\mathbf{p}, r)\right].
\tag{6.186}
$$

Equations (6.185)–(6.186) are *not* automatic if one quantizes ψ_μ as an unconstrained vector-spinor. The Rarita-Schwinger theory is a constrained fermionic system: the conditions $\gamma^\mu \psi_\mu = 0$ and $\partial^\mu \psi_\mu = 0$ (or their gauge-fixed counterparts) are essential to guarantee that $\widehat{H}$ is bounded below and that only the spin-$\frac{3}{2}$ excitations propagate. Once the constraints are properly implemented, the quantized Hamiltonian takes precisely the free-particle form Eq. (6.186), i.e., a sum of positive energies weighted by number operators for particles and antiparticles.

Table 6.1 Relativistic quantum fields classified by spin, together with their fundamental equations of motion and representative particle examples

Spin	Field type	Relativistic quantum equation	Examples of particles
0	Scalar	Klein-Gordon equation $(\Box + m^2)\,\phi = 0$	Higgs boson, π^0, inflaton (model-dependent)
1/2	Spinor	Dirac equation $(i\gamma^\mu \partial_\mu - m)\,\psi = 0$	Electron, muon, quarks, neutrinos (Dirac case)
1	Vector field	Proca or Maxwell equations $\partial_\mu F^{\mu\nu} + m^2 A^\nu = 0$	Photon, $W^\pm$, Z^0, gluons
3/2	Vector-spinor	Rarita-Schwinger equation $\epsilon^{\mu\nu\rho\sigma}\gamma_5\gamma_\nu\partial_\rho\psi_\sigma + m\,\psi^\mu = 0$	Gravitino (supergravity), spin-3/2 baryon resonances

Canonical quantization of the Rarita-Schwinger field reveals explicitly that spin-$\frac{3}{2}$ particles cannot be treated as unconstrained vector-spinors. The subsidiary conditions are not optional but are required to ensure positivity of the Hamiltonian and the correct spin content. In interacting theories, these constraints become even more restrictive, explaining why consistent couplings of spin-$\frac{3}{2}$ fields are highly nontrivial and naturally arise only within structured frameworks such as supergravity.

As a final suggestion, we summarize our findings in Table 6.1.

Pause and Think Is canonical quantization valid for free and interacting fields or only for free ones? Explain why.

6.5 Gauge Theories

One of the main consequences of field theory is that it describes systems through multiple equivalent formulations. In other words, there exists a *gauge invariance* expressing the equivalence of different formulations describing the same system, which should not be confused with a symmetry of the system itself.

Gauge invariance means that one can describe physics adopting a different language, but leaving unaltered the physical interpretation.

To better see this, we start from the Lagrangian for a complex scalar field, which is invariant ($\phi(x) \longrightarrow \phi(x)e^{i\alpha}$) under a global $U(1)$ transformation:

$$\mathcal{L} = (\partial^\mu \phi)^\dagger (\partial_\mu \phi) - m^2 \phi^\dagger \phi. \tag{6.187}$$

Now consider a *local phase transformation*, i.e. with $\alpha = \alpha(x)$ varying at each spacetime point, and seek a Lagrangian that remains invariant under $\phi(x) \longrightarrow \phi(x)e^{i\alpha(x)}$. First, let us see how the different terms transform:

$$m^2\psi^\dagger\psi \longrightarrow m^2(\psi^\dagger \cancel{e^{-i\alpha(x)}}\psi \cancel{e^{i\alpha(x)}}) = m^2\psi^\dagger\psi, \tag{6.188a}$$

$$\begin{aligned}\partial_\mu\psi \longrightarrow \partial_\mu(\psi e^{i\alpha(x)}) &= e^{i\alpha(x)}\partial_\mu\psi + i(\partial_\mu\alpha(x))\psi e^{i\alpha(x)} \\ &= e^{i\alpha(x)}[\partial_\mu + i(\partial_\mu\alpha(x))]\psi \neq \partial_\mu\psi,\end{aligned} \tag{6.188b}$$

$$(\partial_\mu\psi)^\dagger \longrightarrow e^{-i\alpha(x)}[\partial_\mu - i\partial_\mu\alpha(x)]\psi^\dagger. \tag{6.188c}$$

Thus, under this transformation, the Lagrangian becomes

$$\begin{aligned}\mathcal{L}' &= \cancel{e^{i\alpha(x)}e^{-i\alpha(x)}}[\partial^\mu - i\partial^\mu\alpha(x)]\psi^\dagger[\partial_\mu + i\partial_\mu\alpha(x)]\psi - m^2\psi^\dagger\psi \\ &= (\partial^\mu\psi^\dagger)(\partial_\mu\psi) - i\big[(\partial^\mu\alpha(x))\psi^\dagger(\partial_\mu\psi) - (\partial^\mu\psi^\dagger)(\partial_\mu\alpha(x))\psi\big] + \\ &+ (\partial^\mu\alpha(x))(\partial_\mu\alpha(x))\psi^\dagger\psi - m^2\psi^\dagger\psi \neq \mathcal{L}.\end{aligned} \tag{6.189}$$

Hence, we find that the Lagrangian **is not invariant** under local $U(1)$ transformations.

▶ **Pause and Think** Before going on, manage to make the Lagrangian invariant under local $U(1)$.

How can we make it invariant? We introduce a new field $A^\mu(x)$ which transforms in such a way as to cancel the terms with $\partial_\mu\alpha(x)$, thereby restoring invariance.
We define the covariant derivative:

$$\boxed{D_\mu = \partial_\mu + iqA_\mu(x)}, \tag{6.190}$$

with A_μ transforming as

$$A_\mu \longrightarrow A_\mu - \frac{1}{q}\partial_\mu\alpha(x), \tag{6.191}$$

where q is the *coupling constant* specifying the strength of the interaction of A_μ with other fields.[5]

[5] The introduction of $A^\mu(x)$ is motivated by fluid dynamics, where one can describe motion either from an Eulerian viewpoint (observing the field at fixed points) or a Lagrangian viewpoint (following the particle).

▶ **Driving exercise** Prove how $D_\mu\psi$ transforms.

$$\begin{aligned} D_\mu\psi &\longrightarrow e^{i\alpha(x)}[\partial_\mu + i\partial_\mu\alpha(x)]\psi + iq\left(A_\mu - \frac{1}{q}\partial_\mu\alpha(x)\right)\psi e^{i\alpha(x)} \\ &= e^{i\alpha(x)}[\partial_\mu\psi + \cancel{i\partial_\mu\alpha(x)\psi} + iqA_\mu\psi - \cancel{i\partial_\mu\alpha(x)\psi}] = \\ &= e^{i\alpha(x)}[\partial_\mu + iqA_\mu]\psi = e^{i\alpha(x)}D_\mu\psi, \end{aligned} \tag{6.192a}$$

$$(D_\mu\psi')(D^\mu\psi')^\dagger = (e^{i\alpha(x)}D_\mu\psi)(e^{i\alpha(x)}D^\mu\psi)^\dagger = (D_\mu\psi)(D^\mu\psi)^\dagger. \tag{6.192b}$$

Replacing the ordinary derivatives with covariant derivatives in the Lagrangian, we obtain a Lagrangian invariant under $\psi(x) \longrightarrow \psi(x)e^{i\alpha(x)}$ and $A_\mu(x) \longrightarrow A_\mu(x) - \frac{1}{q}\partial_\mu\alpha(x)$:

$$\mathcal{L} = (\partial^\mu\psi)^\dagger(\partial_\mu\psi) - m^2\psi^\dagger\psi \longrightarrow \mathcal{L} = (D^\mu\psi)^\dagger(D_\mu\psi) - m^2\psi^\dagger\psi. \tag{6.193}$$

A gauge theory is a theory in which a *gauge field* is introduced to ensure invariance under local transformations. The addition of A_μ modifies the dynamics of the Lagrangian, i.e., it changes the field theory itself. The simplest gauge field theory is *electromagnetism*, whose Lagrangian describing the dynamics of the electromagnetic field is

$$\mathcal{L} = -\frac{1}{4}(\partial_\mu A_\nu - \partial_\nu A_\mu)(\partial^\mu A^\nu - \partial^\nu A^\mu) - J^\mu A_\mu, \tag{6.194}$$

where $A^\mu = (V, \mathbf{A})$, with V the scalar potential and $\mathbf{A}$ the vector potential, and $J^\mu = (\rho, \mathbf{J})$ is the electromagnetic current. The antisymmetric tensor $F^{\mu\nu} = \partial^\mu A^\nu - \partial^\nu A^\mu$ is the *electromagnetic field tensor*, whose components represent the electric and magnetic fields. Rewriting Eq. (6.194) in terms of $F^{\mu\nu}$ gives

$$\mathcal{L} = -\frac{1}{4}F_{\mu\nu}F^{\mu\nu} - J^\mu A_\mu, \tag{6.195}$$

where the first term is the free electromagnetic field Lagrangian, and the second term describes the minimal coupling between the external source J^μ and the field A_μ.

From this Lagrangian, the equations of motion follow as

$$\partial_\mu \frac{\partial \mathcal{L}}{\partial(\partial_\mu A_\nu)} = \frac{\partial \mathcal{L}}{\partial A_\nu} \longrightarrow \begin{cases} \frac{\partial \mathcal{L}}{\partial A_\nu} = -J^\nu, \\ \frac{\partial \mathcal{L}}{\partial(\partial_\mu A_\nu)} = -[\partial^\mu A^\nu - \partial^\nu A^\mu] = -F^{\mu\nu}. \end{cases} \tag{6.196}$$

Hence, the equations of motion are Maxwell's equations:

$$\partial_\mu F^{\mu\nu} = J^\nu \iff \Box A^\nu - \partial^\nu(\partial_\mu A^\mu) = J^\nu. \tag{6.197}$$

Electromagnetism is the simplest gauge theory. The Lagrangian in Eq. (6.195) is invariant under transformations of the form

$$A_\mu(x) \longrightarrow A_\mu(x) - \partial_\mu \chi(x) \iff \begin{cases} V \longrightarrow V - \partial_t \chi, \\ \mathbf{A} \longrightarrow \mathbf{A} - \nabla\chi. \end{cases} \tag{6.198}$$

Choosing χ to simplify the description corresponds to a *gauge choice*. Common choices are:

- *Lorenz gauge*: $\partial_\mu A^\mu(x) = 0$;
- *Coulomb gauge*: $\nabla \cdot \mathbf{A} = 0$;
- *Weyl gauge*: $A^0 = 0$;
- *axial gauge*: $A^3 = 0$.

The electromagnetic field in the absence of sources, in the Lorenz gauge, satisfies

$$\partial_\mu A^\mu = 0 \iff \partial_\mu A^\mu - \Box\chi = 0 \iff \Box\chi = \partial_\mu A^\mu = 0. \tag{6.199}$$

The Maxwell equations in vacuum ($J^\nu = 0$) are then given by

$$\Box A^\nu - \partial^\nu(\Box\chi) = 0 \longrightarrow \Box A^\nu = 0. \tag{6.200}$$

The obtained equation is the Klein-Gordon equation for a massless particle. Its solutions are plane waves of the form $A^\mu = a^\mu(p)e^{-ipx}$ with $E_\mathbf{p} = |\mathbf{p}|$. Moreover, A_μ behaves as a vector field; in particular, the condition $\partial_\mu A^\mu = 0$ reduces the degrees of freedom from $4 \rightarrow 3$, so that the number of independent components of A_μ is 3.

Let us consider a further shift of A'_μ:

$$A'_\mu \longrightarrow A'_\mu - \partial_\mu \xi \quad \text{with} \quad \Box \xi = 0. \tag{6.201}$$

This implies that $\partial_\mu A'^\mu - \partial_\mu \partial^\mu \xi = 0 = \partial_\mu A'^\mu$, i.e., both A'_μ and $A'_\mu - \partial_\mu \xi$ satisfy the Lorenz gauge. By imposing a new condition

$$\partial_t \xi = A'_0 \longrightarrow \begin{cases} A'_0 \longrightarrow A'_0 - \partial_t \xi = 0 = A''_0, \\ \mathbf{A}' \longrightarrow \mathbf{A}' - \nabla \xi = \mathbf{A}''. \end{cases} \tag{6.202}$$

This means that the Lorenz gauge condition $\partial_\mu A''^\mu = 0$ reduces to the Coulomb gauge $\nabla \cdot \mathbf{A}'' = 0$, which reduces the number of independent field components to 2. This condition can be expressed as

$$A_\mu = (0, \mathbf{a} e^{-ipx}) \quad \text{with} \quad \nabla \cdot \mathbf{A} = 0 \longrightarrow \mathbf{p} \cdot \mathbf{a} = 0, \tag{6.203}$$

where $\mathbf{p}$ is the propagation direction and $\mathbf{a}$ is the polarization vector, both being perpendicular. There are two possible choices for $\mathbf{a}$ given the direction of $\mathbf{p}$, corresponding to *linear polarization* or *circular polarization*.

6.5.1 Coupling with Matter

We introduce the *minimal coupling prescription*, which consists in replacing the ordinary derivative of the matter field with the covariant derivative. Considering a complex scalar field in an electromagnetic field, one has

$$\mathcal{L} = (\partial^\mu \psi)^+ (\partial_\mu \psi) - m^2 \psi^+ \psi - \frac{1}{4} F_{\mu\nu} F^{\mu\nu}. \tag{6.204}$$

By replacing $\partial_\mu \longrightarrow D_\mu$, the previous Lagrangian can be written as

$$\begin{aligned} \mathcal{L} &= (D^\mu \psi)^+ (D_\mu \psi) - m^2 \psi^+ \psi - \frac{1}{4} F_{\mu\nu} F^{\mu\nu} = \\ &= (\partial^\mu \psi^+ - iq A^\mu \psi^+)(\partial_\mu \psi + iq A_\mu \psi) - m^2 \psi^+ \psi - \frac{1}{4} F_{\mu\nu} F^{\mu\nu}. \end{aligned} \tag{6.205}$$

Combining Eqs. (6.204) with (6.205), we obtain

$$\begin{aligned} \mathcal{L} &= (\partial^\mu \psi)^+ (\partial_\mu \psi) - m^2 \psi^+ \psi - \frac{1}{4} F_{\mu\nu} F^{\mu\nu} \\ &+ \big(- iq A^\mu \psi^+ (\partial_\mu \psi) + iq (\partial^\mu \psi^+) A_\mu \psi + q^2 A^\mu A_\mu \psi^+ \psi \big). \end{aligned} \tag{6.206}$$

Notice that the substitution with the covariant derivative introduces a *minimal coupling term* between the matter field and the electromagnetic field.

6.5.2 Measuring the Gauge Potential: The Aharonov-Bohm Effect

The Aharonov-Bohm effect is a quantum effect in which a particle is influenced by electromagnetic fields in regions where the fields themselves vanish.

The consequence of this effect is that the gauge field will turn out to be measurable.

Consider the double-slit experiment (see Fig. 6.2), very well-known for diffraction effects. Electrons pass through the slits producing an interference pattern with $\psi \propto e^{-i\frac{\mathbf{p}\cdot\mathbf{r}}{\hbar}}$ (in this subsection we retain $\hbar$ as it appears pedagogically useful to show it). Differently from the standard optical experiment, here we can introduce a solenoid carrying a current which produces a magnetic field $\mathbf{B}$ inside the solenoid while $\mathbf{B} = 0$ outside. One then observes a shift in the interference pattern.

▶ **Pause and Think** Before going on, manage to distinguish the classical case from the quantum picture.

We can distinguish two cases.

- Classically, the electron should not be influenced due to the vanishing magnetic field outside the solenoid.
- From a quantum perspective, however, the presence of $\mathbf{A}$ cannot be ignored. The wavefunction ψ and the vector potential $\mathbf{A}$ couple according to the minimal coupling prescription, causing an additional phase along the trajectory of the electron:

$$\psi \longrightarrow e^{i\Delta\alpha}\psi \quad \text{with} \quad \Delta\alpha = -\frac{q}{\hbar}\int \mathbf{A}\cdot d\mathbf{r}. \tag{6.207}$$

It is now instructive to clarify what a *Berry phase* is and how its definition enters the Aharonov-Bohm effect.

The Berry phase is a purely geometric phase acquired by a quantum system when its Hamiltonian depends on a set of external parameters that are varied slowly and cyclically in time. Unlike the dynamical phase, which depends on the energy spectrum and the duration of the evolution, the Berry phase is determined solely by

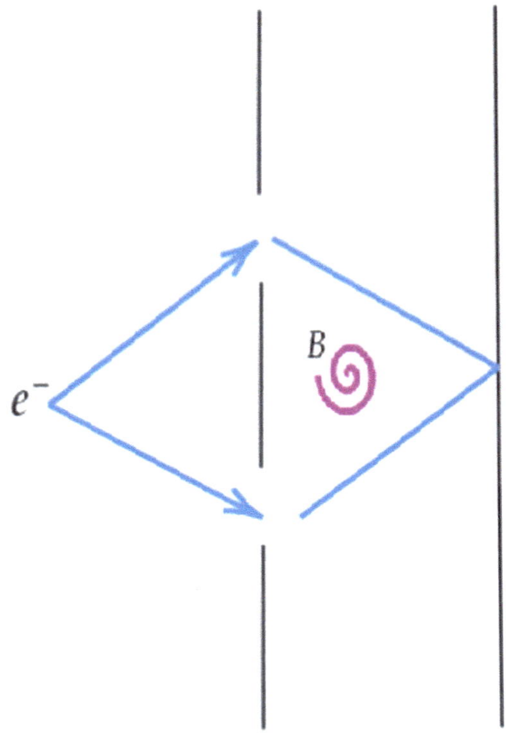

Fig. 6.2 Schematic realization of the Aharonov-Bohm effect, where a solenoid induces a magnetic field **B**. The Aharonov-Bohm effect is topological because the measurable phase shift depends only on the winding number of the particle's path around a confined flux, namely on the global topology of the configuration space, rather than on local electromagnetic fields along the trajectory

the path followed in parameter space and reflects the global geometric structure of the quantum state manifold.

Let us consider an adiabatic formulation. Consider a Hamiltonian $\hat{H}(\boldsymbol{\lambda})$ depending smoothly on a set of real parameters $\boldsymbol{\lambda} = (\lambda_1, \ldots, \lambda_n)$. Let $|n(\boldsymbol{\lambda})\rangle$ be an instantaneous, non-degenerate eigenstate,

$$\hat{H}(\boldsymbol{\lambda})\,|n(\boldsymbol{\lambda})\rangle = E_n(\boldsymbol{\lambda})\,|n(\boldsymbol{\lambda})\rangle\,. \tag{6.208}$$

If the parameters are varied adiabatically along a closed loop C in parameter space, the quantum state evolves as

$$|\psi_n(t)\rangle = e^{i\gamma_n(t)} e^{-\frac{i}{\hbar}\int_0^t E_n(t')\,dt'}\,|n(\boldsymbol{\lambda}(t))\rangle\,, \tag{6.209}$$

where the first exponential contains the Berry phase. After completing the cycle, the geometric contribution is

$$\gamma_n[C] = i\oint_C \langle n(\boldsymbol{\lambda})|\nabla_{\boldsymbol{\lambda}} n(\boldsymbol{\lambda})\rangle \cdot d\boldsymbol{\lambda}. \tag{6.210}$$

The quantity $\mathbf{A}_n(\boldsymbol{\lambda}) = i \langle n(\boldsymbol{\lambda})|\nabla_{\boldsymbol{\lambda}} n(\boldsymbol{\lambda})\rangle$ is known as the *Berry connection* and plays a role formally analogous to a gauge potential in parameter space. Its curl defines the Berry curvature,

$$\boldsymbol{\Omega}_n(\boldsymbol{\lambda}) = \nabla_{\boldsymbol{\lambda}} \times \mathbf{A}_n(\boldsymbol{\lambda}), \tag{6.211}$$

which is a gauge-invariant quantity. By Stokes' theorem, the Berry phase can be written as a surface integral, $\gamma_n[C] = \int_S \boldsymbol{\Omega}_n \cdot d\mathbf{S}$, where S is any surface bounded by C.

The Berry phase has a clear geometric origin: it measures the anholonomy associated with parallel transport of the quantum state along a closed path. Even though the system returns to the same physical configuration, the state vector may acquire a nontrivial phase reflecting the curvature of the projective Hilbert space.

This phase is insensitive to the rate of evolution and to local details of the Hamiltonian, provided the adiabatic condition is satisfied. From a physical perspective, the Berry phase encodes information about global properties of the spectrum, such as degeneracies and level crossings, which act as effective "sources" of Berry curvature. In this sense, it provides a bridge between local quantum dynamics and global, topological features of parameter space.

One might think that the Berry phase in the Aharonov-Bohm effect, dubbed $\Delta\alpha$ in our picture, could be removed through a gauge transformation $\mathbf{A} \longrightarrow \mathbf{A} + \nabla\chi$. The shift in the interference pattern, however, is unavoidable because it is calculated considering the difference of the additional phases along two trajectories arriving at the same point:

$$\Delta\delta = \Delta\alpha_1 - \Delta\alpha_2 = -\frac{q}{\hbar}\left[\int_{a,1}^{b} \mathbf{A} \cdot d\mathbf{r} - \int_{a,2}^{b} \mathbf{A} \cdot d\mathbf{r}\right] = -\frac{q}{\hbar}\oint \mathbf{A} \cdot d\mathbf{r}. \tag{6.212}$$

Using Stokes' theorem, we obtain the *topological effect from gauge invariance*:

$$\Delta\delta = \Delta\alpha_1 - \Delta\alpha_2 = -\frac{q}{\hbar}\int d\boldsymbol{\Sigma}\,(\nabla \times \mathbf{A}) = -\frac{q}{\hbar}\int d\boldsymbol{\Sigma}\,\mathbf{B} = -\frac{q}{\hbar}\Phi \neq 0. \tag{6.213}$$

Furthermore, the expression above measures the magnetic flux **B** and thus provides information about the vector potential **A**.
The effect, as above stated, is topological since the solenoid acts as a hole in the mathematical domain where the electrons span.

The Aharonov-Bohm effect is experimentally accessed through phase-sensitive interference phenomena. Since the effect is purely quantum and topological in nature, its observation relies on interferometric techniques capable of resolving relative phases with high precision.

The most direct and historically significant measurements of the Aharonov-Bohm effect are performed using electron interferometers. In these setups, a coherent electron beam is split into two spatially separated paths that enclose a region containing a confined magnetic flux, typically generated by a long solenoid or a magnetically shielded toroidal coil. Crucially, the electrons propagate exclusively in regions where the magnetic field is negligible, while the vector potential remains non-zero.

After traversing the two paths, the electron wave packets are recombined, producing an interference pattern whose phase depends on the enclosed magnetic flux. The observed phase shift,

$$\Delta\varphi_{\mathrm{AB}} = \frac{q}{\hbar} \oint \mathbf{A} \cdot d\ell = \frac{q}{\hbar} \Phi_B, \qquad (6.214)$$

is extracted from the displacement of the interference fringes as the magnetic flux Φ_B is varied. Modern electron holography techniques, based on transmission electron microscopes (TEM), allow for nanometer-scale control of electron paths and provide a direct real-space visualization of the Aharonov-Bohm phase.

A second major class of experiments exploits phase-coherent electron transport in mesoscopic conductors. Metallic or semiconducting rings with characteristic dimensions smaller than the electronic phase-coherence length are fabricated using nanolithography. When such rings are threaded by a magnetic flux, the conductance exhibits oscillations as a function of Φ_B, with a fundamental period equal to the magnetic flux quantum

$$\Phi_0 = \frac{h}{q}. \qquad (6.215)$$

These conductance oscillations arise from the interference between electronic wave functions propagating along different arms of the ring and constitute a robust manifestation of the Aharonov-Bohm effect in condensed matter systems. Measurements are typically performed at low temperatures to suppress decoherence, using four-terminal transport techniques to accurately resolve the oscillatory signal.

6.6 A First Look at Quantum Electrodynamics

QED is the QFT of electromagnetism and describes the interaction of matter (charged particles) with the electromagnetic field (whose excitations are photons). It is a gauge theory, meaning that it is necessary to introduce a field A_μ in order to write the Lagrangian describing charged particles, resulting in a field-current coupling.

We have studied the free-field Lagrangian (which satisfies the Dirac equation), and now we begin introducing interactions. These are added to the free-field Lagrangian, denoted as $\mathcal{L}_{\text{free}}$, through a trial-and-error procedure: one writes a candidate interaction term and compares it with experimental results until a consistent agreement between theory and observations is found. When predicting interactions, several principles are kept fixed:

1. the classical limit is recovered for large systems;
2. certain symmetries must be preserved.

In physics, interactions are generally divided into three main categories: *electromagnetic*, *weak nuclear*, and *strong nuclear* interactions.

From the Dirac theory Lagrangian, we know that for electrons:

$$\mathcal{L}_D = \bar{\psi}(i\not{\partial} - m)\psi. \tag{6.216}$$

Whereas from the Lagrangian whose equations of motion yield Maxwell's equations, we know that the electromagnetic Lagrangian is

$$\mathcal{L}_{EM} = -\frac{1}{4}F_{\mu\nu}F^{\mu\nu}. \tag{6.217}$$

To study electrons in an electromagnetic field ($q = -e$), we first combine the two Lagrangians above, obtaining

$$\mathcal{L} = -\frac{1}{4}F_{\mu\nu}F^{\mu\nu} + \bar{\psi}(i\not{\partial} - m)\psi. \tag{6.218}$$

We then perform the minimal substitution $\not{\partial} \longrightarrow \not{D} = \gamma^\mu[\partial_\mu + iqA_\mu] = \not{\partial} + iq\not{A} = \not{\partial} - ie\not{A}$, which represents the covariant derivative contracted with γ^μ. This is necessary to ensure gauge invariance of the theory.

We are thus promoting the Dirac theory to a gauge theory by introducing the potential A_μ and replacing derivatives with covariant derivatives. In this way, electromagnetic effects on matter are described. The resulting Lagrangian reads

$$\begin{aligned}\mathcal{L} &= -\frac{1}{4}F_{\mu\nu}F^{\mu\nu} + \bar{\psi}(i\not{D} - m)\psi = -\frac{1}{4}F_{\mu\nu}F^{\mu\nu} + \bar{\psi}(i\not{\partial} + e\not{A} - m)\psi \\ &= -\frac{1}{4}F_{\mu\nu}F^{\mu\nu} + \bar{\psi}(i\not{\partial} - m)\psi + e\bar{\psi}\gamma^\mu A_\mu\psi = \mathcal{L}_{EM} + \mathcal{L}_D + \mathcal{L}_{int}, \end{aligned} \quad (6.219)$$

where $J^\mu = \bar{\psi}\gamma^\mu\psi$ is the conserved electron current. The replacement $\not{\partial} \longrightarrow \not{D}$ gives rise to the interaction term $\mathcal{L}_{int}$, following the minimal coupling prescription.

Driving Exercise
As an example of **non-minimal** coupling, let us consider the *Pauli term*. We start from the total Lagrangian as the sum of the Dirac Lagrangian and the Pauli term:

$$\mathcal{L} = \mathcal{L}_D + \mathcal{L}_P = \mathcal{L}_D - \frac{ek}{4m}\bar{\psi}\sigma_{\mu\nu}\psi F^{\mu\nu}. \quad (6.220)$$

We can study the non-relativistic limit by starting from $\mathcal{L} = \mathcal{L}_D$ and performing the substitution $\partial \longrightarrow D$, yielding

$$\mathcal{L} = \bar{\psi}(i\not{\partial} - m)\psi + eA_\mu\bar{\psi}\gamma^\mu\psi, \quad (6.221)$$

with the corresponding equation of motion given by

$$(i\not{\partial} + e\not{A} - m)\psi = 0. \quad (6.222)$$

Using

$$E = \pm m\sqrt{1 + \frac{\mathbf{p}^2}{m^2}} \sim \pm m\left(1 + \frac{\mathbf{p}^2}{2m^2}\right) \sim \pm m, \quad (6.223)$$

one recovers the Pauli equation in the case of minimal coupling with A_μ, which reproduces $g = 2$ in agreement with the prescription.
If, instead, we add a non-minimal coupling term to $\mathcal{L} = \mathcal{L}_D$, both the equation of motion and the Pauli equation are modified. The Lagrangian becomes

$$\mathcal{L} = \bar{\psi}(i\not{D} - m)\psi + \mathcal{L}_P. \quad (6.224)$$

Considering $\mathbf{E} = 0$ and $\Phi = 0$, and only a magnetic field, the two tensors $\sigma_{\mu\nu}$ and $F^{\mu\nu}$, separated into spatial and temporal components, reduce to

$$\begin{cases} \sigma^{ij} = \varepsilon^{ijk}\sigma^k, \\ F^{ij} = -\varepsilon^{ijk}B^k \end{cases} \implies \bar{\psi}\sigma^{ij}\psi F_{ij} = -\bar{\psi}\varepsilon^{ijk}\sigma^k\psi\varepsilon_{ijl}B_l = -2\bar{\psi}\boldsymbol{\sigma}\cdot\mathbf{B}\psi, \tag{6.225}$$

where the last term represents a *spin-magnetic field coupling term*. The equation of motion then reads

$$(i\partial\!\!\!/ + eA\!\!\!/ - m)\psi = \frac{ek}{4m}(2\boldsymbol{\sigma}\cdot\mathbf{B})\psi = \frac{ek}{2m}\boldsymbol{\sigma}\cdot\mathbf{B}\psi. \tag{6.226}$$

In the Pauli representation, decomposing $\tilde{\psi} = \begin{pmatrix}\psi_A \\ \psi_B\end{pmatrix}$, we obtain the coupled equations for ψ_A and ψ_B:

$$\left(i\frac{\partial}{\partial t} - \frac{ek}{2m}\boldsymbol{\sigma}\cdot\mathbf{B}\right)\psi_A = \boldsymbol{\sigma}\cdot(\hat{\mathbf{p}} + e\hat{\mathbf{A}})\,\psi_B, \tag{6.227a}$$

$$\left(i\frac{\partial}{\partial t} - \frac{ek}{2m}\boldsymbol{\sigma}\cdot\mathbf{B} + 2m\right)\psi_B = \boldsymbol{\sigma}\cdot(\hat{\mathbf{p}} + e\hat{\mathbf{A}})\,\psi_A. \tag{6.227b}$$

Since

$$\left(i\frac{\partial}{\partial t} - \frac{ek}{2m}\boldsymbol{\sigma}\cdot\mathbf{B} + 2m\right) \sim 2m, \tag{6.228}$$

we can approximate

$$\psi_B = \frac{\boldsymbol{\sigma}\cdot(\hat{\mathbf{p}} + e\hat{\mathbf{A}})}{2m}\psi_A \implies i\frac{\partial}{\partial t}\psi_A = \left[\frac{(\hat{\mathbf{p}} + e\hat{\mathbf{A}})^2}{2m} + \frac{e\boldsymbol{\sigma}\cdot\mathbf{B}}{2m} + \frac{ek}{2m}\boldsymbol{\sigma}\cdot\mathbf{B}\right]\psi_A. \tag{6.229}$$

We thus recover the Schrödinger equation for an electron in an electromagnetic field with spin corrections. At this point, the Hamiltonian operator $\hat{H}$ contains two terms: a Pauli term and a spin-magnetic field coupling term.

$$\hat{H} = \frac{(\hat{\mathbf{p}} + e\hat{\mathbf{A}})^2}{2m} + e\frac{\boldsymbol{\sigma}\cdot\mathbf{B}}{2m}(1+k) \longrightarrow \begin{cases} H^{\text{(Pauli)}} = e\frac{\boldsymbol{\sigma}\cdot\mathbf{B}}{2m}(1+k), \\ H^{\text{spin}} = -\hat{\boldsymbol{\mu}}\cdot\mathbf{B} = -\frac{ge}{4m}\boldsymbol{\sigma}\cdot\mathbf{B}. \end{cases} \tag{6.230}$$

Comparing these two terms, we obtain $g = 2(1+k)$, which is no longer $g = 2$ as in the case of minimal coupling, but includes corrections given by $1 + k$ that account for non-minimal interactions at second order in the fields $A_\mu\bar{\psi}\gamma^\mu\psi \longrightarrow (\partial_\mu A_\nu - \partial_\nu A_\mu)\bar{\psi}[\gamma^\mu, \gamma^\nu]$.

For electrons, the Pauli correction is not required, and one can restrict to considering the interaction Lagrangian according to the minimal coupling prescription to reproduce electromagnetic interactions with electrons.
For protons and neutrons, we will see that minimal coupling does not correctly reproduce the interaction, and the Pauli term is necessary.

6.6.1 Quantum Electrodynamics for Leptons

Leptons are fermionic particles, belonging to the lepton family: the electron e, muon μ, tau τ, and neutrinos. In analogy with the Lagrangian describing electromagnetic interactions, we write the Lagrangian for the three leptons listed above as

$$\mathcal{L}_{\text{leptons}} = -\frac{1}{4}F_{\mu\nu}F^{\mu\nu} + \bar{\psi}_e(i\not{D} - m_e)\psi_e + \bar{\psi}_\mu(i\not{D} - m_\mu)\psi_\mu + \bar{\psi}_\tau(i\not{D} - m_\tau)\psi_\tau \tag{6.231}$$

$$= -\frac{1}{4}F_{\mu\nu}F^{\mu\nu} + \bar{\psi}_e(i\not{\partial} - m_e)\psi_e + \bar{\psi}_\mu(i\not{\partial} - m_\mu)\psi_\mu + \bar{\psi}_\tau(i\not{\partial} - m_\tau)\psi_\tau + \mathcal{L}^{(\text{int})}_{\text{leptons}}. \tag{6.232}$$

The latter Lagrangian for the leptons represents the free Lagrangians $\mathcal{L}_0$, describing the leptons according to Dirac dynamics. The Lagrangian can be written as

$$\mathcal{L} = \mathcal{L}_{EM} + \mathcal{L}_{0e} + \mathcal{L}_{0\mu} + \mathcal{L}_{0\tau} + eA_\alpha\left[\bar{\psi}_e\gamma^\alpha\psi_e + \bar{\psi}_\mu\gamma^\alpha\psi_\mu + \bar{\psi}_\tau\gamma^\alpha\psi_\tau\right], \tag{6.233}$$

where $eA_\alpha\bar{\psi}_i\gamma^\alpha\psi_i$ with $i = e, \mu, \tau$ are the conserved lepton currents, i.e.,

$$-J^\alpha_{\text{leptons}} = -J^\alpha_e + (-J^\alpha_\mu) + (-J^\alpha_\tau). \tag{6.234}$$

Equation (6.233) can thus be rewritten as

$$\mathcal{L}_{QED} = \mathcal{L}_{EM} + \mathcal{L}_{0e} + \mathcal{L}_{0\mu} + \mathcal{L}_{0\tau} - eA_\alpha J^\alpha_{\text{leptons}}, \tag{6.235}$$

where e is the coupling constant of the electromagnetic interaction.

6.6.2 Quantum Electrodynamics for Nucleons

We identify protons with $g = 2$ and neutrons with $g = 0$, which are also affected by nuclear interactions. Being half-integer spin particles, their dynamics are described by Dirac equations. However, with minimal coupling, comparing $\mu^{(\text{Dirac})}$ with the nucleon magnetic moment μ_N, we have

$$\mu_N = g_N\mu_p\mathbf{S} = \frac{g_N}{2}\frac{e}{2m_p}\boldsymbol{\sigma}, \quad \text{where} \quad \mu_p = \frac{e}{2m_p} \quad \text{and} \quad \mathbf{S} = \frac{\boldsymbol{\sigma}}{2}. \tag{6.236}$$

Experimentally, the gyromagnetic factor disagrees with the Dirac prediction $\mu^{(\text{Dirac})}$. To correct this, the Pauli term $\frac{eK_N}{4m_p}\bar{\psi}_N\sigma_{\alpha\beta}\psi_N F^{\alpha\beta}$ is added, so that the nucleon Lagrangian becomes

$$\begin{aligned}\mathcal{L}_{\text{nucleons}} &= -\frac{1}{4}F_{\mu\nu}F^{\mu\nu} + \bar{\psi}_p(i\not{\partial} - m_p)\psi_p + \bar{\psi}_n(i\not{\partial} - m_n)\psi_n + \frac{e}{4m_p}[k_p\bar{\psi}_p\sigma_{\alpha\beta}\psi_p \\ &+ k_n\bar{\psi}_n\sigma_{\alpha\beta}\psi_n]F^{\alpha\beta} = \mathcal{L}_{EM} + \mathcal{L}_{0p} + \mathcal{L}_{0n} - eA_\alpha\bar{\psi}_p\gamma^\alpha\psi_p \\ &+ \frac{e}{4m_p}\left[k_p\bar{\psi}_p\sigma_{\alpha\beta}\psi_p + k_n\bar{\psi}_n\sigma_{\alpha\beta}\psi_n\right]F^{\alpha\beta}.\end{aligned} \tag{6.237}$$

Combining the nucleon Lagrangian with the lepton Lagrangian, we obtain the full QED Lagrangian for matter fields:

$$\begin{aligned}\mathcal{L}_{\text{leptons+nucleons}} &= \mathcal{L}_{EM} + \mathcal{L}_{0e} + \mathcal{L}_{0\mu} + \mathcal{L}_{0\tau} - eA_\alpha J^\alpha_{\text{leptons}} + \mathcal{L}_{0p} + \mathcal{L}_{0n} \\ &- eA_\alpha\bar{\psi}_p\gamma^\alpha\psi_p + \frac{e}{4m_p}[k_p\bar{\psi}_p\sigma_{\alpha\beta}\psi_p + k_n\bar{\psi}_n\sigma_{\alpha\beta}\psi_n]F^{\alpha\beta},\end{aligned} \tag{6.238}$$

Here, $-eA_\alpha\bar{\psi}_p\gamma^\alpha\psi_p$ represents the *proton-field coupling* with charge $q = +e$, $\mathcal{L}_0 = \bar{\psi}(i\not{\partial} - m)\psi$ denotes the free Lagrangians of the particles, and $J^\alpha_{\text{leptons}}$ is the conserved lepton current.

Each Pauli interaction term can be rewritten as an additive correction to the corresponding current term. In other words, the effect of non-minimal coupling for nucleons can be incorporated as modifications to the conserved currents, leading to the total interaction Lagrangian

$$\mathcal{L}_{\text{int}} = -eA_\alpha J^\alpha_{\text{leptons}} - eA_\alpha\bar{\psi}_p\gamma^\alpha\psi_p + \frac{e}{4m_p}\left[k_p\bar{\psi}_p\sigma_{\alpha\beta}\psi_p + k_n\bar{\psi}_n\sigma_{\alpha\beta}\psi_n\right]F^{\alpha\beta}. \tag{6.239}$$

$$\frac{ek}{4m_N}\bar{\psi}_N\sigma_{\alpha\beta}\psi_N F^{\alpha\beta} = \frac{ek}{2m_N}\bar{\psi}_N\sigma_{\alpha\beta}\psi_N\partial^\alpha A^\beta, \tag{6.240}$$

where we have substituted $F^{\alpha\beta} = \partial^\alpha A^\beta - \partial^\beta A^\alpha$, so that

$$\sigma_{\alpha\beta}F^{\alpha\beta} = \sigma_{\alpha\beta}(\partial^\alpha A^\beta - \partial^\beta A^\alpha) = 2\sigma_{\alpha\beta}\partial^\alpha A^\beta, \tag{6.241}$$

because $\sigma_{\alpha\beta}$ is antisymmetric. Now, let us consider that

$$\partial^\alpha(\bar{\psi}\sigma_{\alpha\beta}\psi A^\beta) = A^\beta\partial^\alpha(\bar{\psi}\sigma_{\alpha\beta}\psi) + (\partial^\alpha A^\beta)(\bar{\psi}\sigma_{\alpha\beta}\psi),$$

and we substitute this into Eq. (6.240), obtaining

$$- \frac{ek}{2m_N} \partial^\alpha (\bar{\psi}_N \sigma_{\alpha\beta} \psi_N) A^\beta + \text{four-divergence}. \tag{6.242}$$

Hence, we can rewrite the Lagrangian for leptons and nucleons as

$$\begin{aligned} \mathcal{L} = \mathcal{L}^{(1)}_{QED} + \mathcal{L}_{0p} + \mathcal{L}_{0n} - eA_\alpha \bar{\psi}_p \gamma^\alpha \psi_p \\ - \frac{e}{2m_p} [k_p \partial^\alpha (\bar{\psi}_p \sigma_{\alpha\beta} \psi_p) + k_n \partial^\alpha (\bar{\psi}_n \sigma_{\alpha\beta} \psi_n)] A^\beta. \end{aligned} \tag{6.243}$$

Analogously to the lepton current, we can define the nucleon current. We report below both the lepton and nucleon currents

$$J^\alpha_{\text{leptons}} = -[\bar{\psi}_e \gamma^\alpha \psi_e + \bar{\psi}_\mu \gamma^\alpha \psi_\mu + \bar{\psi}_\tau \gamma^\alpha \psi_\tau], \tag{6.244a}$$

$$J^\alpha_{\text{nucleons}} = \bar{\psi}_p \gamma^\alpha \psi_p + \frac{k_p}{2m_p} \partial^\beta (\bar{\psi}_p \sigma_{\alpha\beta} \psi_p) + \frac{k_n}{2m_p} \partial^\beta (\bar{\psi}_n \sigma_{\alpha\beta} \psi_n). \tag{6.244b}$$

The last two terms in the nucleon current represent corrections describing the deviation of the magnetic moment from the value obtained solely with minimal coupling. We then write

$$\mathcal{L} = \mathcal{L}^{(1)}_{QED} + \mathcal{L}_{0p} + \mathcal{L}_{0n} - eA_\alpha J^\alpha_{\text{nucleons}}, \tag{6.245}$$

where $\mathcal{L}^{(1)}_{QED} = \mathcal{L}_{EM} + \mathcal{L}_{0e} + \mathcal{L}_{0\mu} + \mathcal{L}_{0\tau} - eA_\alpha J^\alpha_{\text{leptons}}$.

Since the corrections are expressed in terms of the four-divergence of an anti-symmetric tensor, they do not contribute to the conserved charge, which then remains unchanged

$$Q = \int d^3x J^0_{\text{nucleons}} = \int d^3x \bar{\psi}_p \gamma^0 \psi_p. \tag{6.246}$$

Indeed, one can apply Gauss' theorem to

$$\int \frac{d^3x}{2m_p} [k_p \partial^\beta (\bar{\psi}_p \sigma_{\beta 0} \psi_p) + k_n \partial^\beta (\bar{\psi}_n \sigma_{\beta 0} \psi_n)] = \tag{6.247}$$

$$\int \frac{d\sigma_i}{2m_p} [k_p (\bar{\psi}_p \sigma_{i0} \psi_p) + k_n (\bar{\psi}_n \sigma_{i0} \sigma_0)] = 0. \tag{6.248}$$

In a more compact form, we can write

$$\mathcal{L}_{QED} = \mathcal{L}_{\text{leptons and nucleons}} = \mathcal{L}_{EM} + \mathcal{L}_{0,\text{tot}} + \mathcal{L}_{\text{int,tot}}, \tag{6.249}$$

with

$$\mathcal{L}_{0,\text{tot}} = \sum_{k=e,\mu,\tau,p,n} \bar{\psi}_k(i\not{\partial} - m_k)\psi_k \quad \text{free part gauge-invariant,} \tag{6.250a}$$

$$\mathcal{L}_{EM} = -\frac{1}{4}F_{\mu\nu}F^{\mu\nu} \quad \text{electromagnetic Lagrangian,} \tag{6.250b}$$

$$\mathcal{L}_{\text{int,tot}} = -eA_\alpha(J^\alpha_{\text{leptons}} + J^\alpha_{\text{nucleons}}) \quad \text{minimal coupling terms.} \tag{6.250c}$$

6.7 Scattering in Quantum Field Theory

In order to discuss scattering processes in QFT, we first need to pick the most viable representation.

More precisely, the different representations in quantum mechanics involve the following prescriptions.

- In the Schrödinger picture, operators are time-independent and states evolve though the Schrödinger equation,

$$\partial_t\psi = \mathcal{H}\psi. \tag{6.251}$$

- In the Heisenberg picture, states are time-independent and operators evolve as

$$O_H(t) = U^\dagger(t)\, O_S\, U(t), \tag{6.252}$$

  with equation of motion

$$\frac{dO_H(t)}{dt} = \frac{1}{i}\,[O_H(t), H_H(t)], \tag{6.253}$$

  for time-independent H. These prescriptions have been widely explored in Chap. 5.

Expectation values are invariant

$$\langle\psi(t)|\, O_S\, |\psi(t)\rangle = \langle\psi|\, O_H(t)\, |\psi\rangle\,. \tag{6.254}$$

The interaction picture consists in a convenient and useful compromise between the use of both the above pictures.

Indeed, split $H = H_0 + V$ and define the interaction-picture operators by free evolution:

$$O_I(t) \equiv e^{iH_0t}\, O_S\, e^{-iH_0t}. \tag{6.255}$$

Interaction-picture states are defined by factoring out the free evolution:

$$|\psi_I(t)\rangle \equiv e^{iH_0t}\, |\psi(t)\rangle\,. \tag{6.256}$$

Hence,

$$i\,\frac{d}{dt}\,|\psi_I(t)\rangle = V_I(t)\,|\psi_I(t)\rangle\,, \qquad V_I(t) \equiv e^{iH_0t}\, V e^{-iH_0t}. \tag{6.257}$$

The solution is written in terms of the interaction-picture evolution operator $U_I(t, t_0)$,

$$|\psi_I(t)\rangle = U_I(t, t_0)\,|\psi_I(t_0)\rangle\,, \qquad i\,\frac{\partial}{\partial t} U_I(t, t_0) = V_I(t)\,U_I(t, t_0), \quad U_I(t_0, t_0) = \mathbb{I}. \tag{6.258}$$

Formally,

$$U_I(t, t_0) = \mathcal{T} \exp\!\left(-i \int_{t_0}^{t} dt'\, V_I(t')\right), \tag{6.259}$$

where $\mathcal{T}$ denotes time ordering. Expanding Eq. (6.259) yields the Dyson series.

Hence, since we have seen that in scattering theory one chooses $t_0 \to -\infty$ and $t \to +\infty$, the S operator can be expressed as

$$S = \lim_{\substack{t\to+\infty \\ t_0\to-\infty}} U_I(t, t_0) = \mathcal{T} \exp\!\left(-i \int_{-\infty}^{+\infty} dt\, V_I(t)\right), \tag{6.260}$$

interpreted as a strong limit on the scattering subspace. Equation (6.260) is the natural non-relativistic analog of the scattering operator as an "interaction switched on" between asymptotic times, but here it is purely quantum-mechanical and does not rely on field theory (see Chap. 5).

6.7.1 Scattering Problems

The interaction picture is particularly suited for the description of particle-collision processes. Let us consider a Hamiltonian of the form

$$\hat{H} = \hat{H}_0 + \hat{V}\,.$$

Very long before and very long after the collision, the state of the particle in the interaction representation tends to a constant. Namely,

$$|\psi_{-\infty}\rangle_I = e^{i\hat{H}_0t}\, |\psi\rangle_S, \tag{6.261a}$$

$$|\psi_{+\infty}\rangle_I = e^{i\hat{H}_0t}\, |\psi\rangle_S. \tag{6.261b}$$

These expressions mean that the states $|\psi_{\pm\infty}\rangle_I$ evolve in a way that is indistinguishable from that of a free system governed by $\hat{H}_0$. In the Schrödinger representation, these states coincide with $|\psi_{\text{in}}\rangle_S$ and $|\psi_{\text{out}}\rangle_S$, respectively. One may therefore schematically describe the time evolution of the particle before and after the collision as

- **before the collision:** $|\psi(t)\rangle_I$ evolves from $|\psi_{-\infty}\rangle_I \equiv |\psi_{\text{in}}\rangle_S$,
- **collision**,
- **after the collision**: $|\psi(t)\rangle_I$ tends to $|\psi_{+\infty}\rangle_I \equiv |\psi_{\text{out}}\rangle_S$.

The probability amplitude that the state before the collision is detected in the asymptotic out-state is given by

$$A = (\langle\chi|\hat{S}|\phi\rangle)_S = (\langle\chi(t)|\phi(t)\rangle)_S = (\langle\chi(t)|\phi(t)\rangle)_I. \tag{6.262}$$

We note that states labeled by S belong to the free system with Hamiltonian $\hat{H} = \hat{H}_0$, whereas those labeled by I belong to the system in which the perturbation is taken into account. Let us express χ and ϕ in the interaction representation by writing the probability amplitude as

$$A = (\langle\chi|\hat{S}|\phi\rangle)_S = (\langle\chi(t')|\hat{U}_I^\dagger(t,t')\,\hat{U}_I(t,t'')|\phi(t'')\rangle)_I. \tag{6.263}$$

Using the fact that

$$\hat{U}_I^\dagger(t,t')\,\hat{U}_I(t,t'') = \hat{U}_I(t',t'')$$

is still a unitary operator, together with the limits

$$\lim_{t''\to-\infty} |\phi(t'')\rangle = |\phi_{-\infty}\rangle_I = |\psi_{\text{in}}\rangle_S = |\phi\rangle_S, \tag{6.264a}$$

$$\lim_{t'\to+\infty} |\chi(t')\rangle = |\chi_{+\infty}\rangle_I = |\psi_{\text{out}}\rangle_S = |\chi\rangle_S, \tag{6.264b}$$

one finds that the probability amplitude is

$$A = (\langle\chi|\hat{S}|\phi\rangle)_S = \lim_{t'\,(t'')\to+\infty\,(-\infty)} (\langle\chi(t')|\hat{U}_I(t',t'')|\phi(t'')\rangle)_S = \lim(\langle\chi|\hat{U}_I(t',t'')|\phi\rangle)_S. \tag{6.265}$$

This implies that the scattering matrix $\hat{S}$ can be defined as

$$\hat{S} = \hat{U}_I(+\infty,-\infty). \tag{6.266}$$

6.7.2 The Born Approximation

Any unitary operator in the interaction picture satisfies the differential equation

$$i\frac{d}{dt}\hat{U}_I(t,t_0) = \hat{V}_I(t)\,\hat{U}_I(t,t_0). \tag{6.267}$$

Equivalently, for the interaction-picture state $\psi_I(t)$ one has

$$\begin{aligned}\hat{V}_I(t)\psi_I(t) &= \hat{V}_I(t)\hat{U}_I(t,t_0)\psi_I(t_0)\\ &= i\frac{d}{dt}\psi_I(t) = i\frac{d}{dt}\big(\hat{U}_I(t,t_0)\psi_I(t_0)\big) = i\frac{d}{dt}\hat{U}_I(t,t_0)\psi_I(t_0).\end{aligned} \tag{6.268}$$

A naive attempt at solving this equation would be

$$\int_{t_0}^{t} i\frac{d\hat{U}_I}{\hat{U}_I} = \int_{t_0}^{t}\hat{V}_I(t')\,dt' \quad\Longrightarrow\quad \hat{U}_I(t,t_0) = e^{-i\int_{t_0}^{t}\hat{V}_I(t')\,dt'}. \tag{6.269}$$

However, this expression is in general incorrect, because the interaction Hamiltonians at different times do not necessarily commute,

$$[\hat{V}_I(t_1), \hat{V}_I(t_2)] \neq 0.$$

One therefore introduces the *time-ordered product* $T[\hat{A}_1(t_1)\hat{A}_2(t_2)\ldots\hat{A}_n(t_n)]$, where T denotes Wick's time-ordering symbol and operators are arranged such that those acting at later times appear to the left. With this prescription one can write the *Dyson series* in compact form as

$$\hat{U}_I(t,t_0) = T\left[e^{-i\int_{t_0}^{t}dt'\,\hat{V}_I(t')}\right]. \tag{6.270}$$

It is often more convenient to express this in terms of the explicit Dyson expansion,

$$\begin{aligned}\hat{U}_I(t,t_0) &= T\Big[\mathbb{I} - i\int_{t_0}^{t}dt'\,\hat{V}_I(t') + (-i)^2\int_{t_0}^{t}dt'\int_{t_0}^{t'}dt''\,\hat{V}_I(t')\hat{V}_I(t'') + \cdots\Big]\\ &\simeq \mathbb{I} - i\int_{t_0}^{t}dt'\,\hat{V}_I(t') + (-i)^2\int_{t_0}^{t}dt'\int_{t_0}^{t'}dt''\,\hat{V}_I(t')\hat{V}_I(t''),\end{aligned} \tag{6.271}$$

with $+\infty > t' > t'' > \cdots > -\infty$.

If we consider the matrix element $\langle \mathbf{p}'|\hat{S}|\mathbf{p}\rangle$, it becomes

$$\langle \mathbf{p}'|\hat{S}|\mathbf{p}\rangle = \langle \mathbf{p}'|\mathbf{p}\rangle - i\int_{-\infty}^{+\infty} dt'\,\langle \mathbf{p}'|\hat{V}_I(t')|\mathbf{p}\rangle + \cdots \simeq \delta^{(3)}(\mathbf{p}'-\mathbf{p}) + \langle \mathbf{p}'|S^{(1)}|\mathbf{p}\rangle + \cdots . \tag{6.272}$$

Here $\delta^{(3)}(\mathbf{p}'-\mathbf{p})$ is the probability amplitude for a particle to propagate without being scattered, while $\langle \mathbf{p}'|S^{(1)}|\mathbf{p}\rangle$ is the time integral of the matrix element

$$-\,i\langle \mathbf{p}'|e^{i\hat{H}_0 t'}\hat{V}_S e^{-i\hat{H}_0 t'}|\mathbf{p}\rangle,$$

with $|\mathbf{p}\rangle$ and $|\mathbf{p}'\rangle$ energy eigenstates with energies $E_{\mathbf{p}}$ and $E_{\mathbf{p}'}$. To first order this coincides with the Born approximation for $\langle \mathbf{p}'|\hat{S}|\mathbf{p}\rangle$. The integral represents a process in which the particle in $|\mathbf{p}\rangle$ propagates freely and is then scattered at time t' into $|\mathbf{p}'\rangle$.

At second order one finds

$$\langle \mathbf{p}'|S^{(2)}|\mathbf{p}\rangle = -\int_{-\infty}^{+\infty} dt' \int_{-\infty}^{t'} dt''\,\langle \mathbf{p}'|\hat{V}_I(t')\hat{V}_I(t'')|\mathbf{p}\rangle. \tag{6.273}$$

This term represents free propagation of the particle in $|\mathbf{p}\rangle$, which at time t'' is scattered by $\hat{V}$ into an intermediate state $|\mathbf{k}\rangle$; it then propagates freely between t'' and t', and finally, at time t', is scattered again by $\hat{V}$ into $|\mathbf{p}'\rangle$.

Feynman diagrams therefore provide a graphical representation of the terms in the Born expansion, which describes multiple scattering processes of a particle between the initial state, intermediate free-propagating (virtual) states, and the final state.

6.7.2.1 Interpretation of the T and S Operators

The matrix element can be written as

$$\langle \mathbf{p}'|S|\mathbf{p}\rangle = \delta^{(3)}(\mathbf{p}'-\mathbf{p}) - 2\pi i\,\delta(E_{\mathbf{p}'}-E_{\mathbf{p}})\langle \mathbf{p}'|T|\mathbf{p}\rangle,$$

which in matrix form reads $S = \mathbb{I} - T$. In the Born approximation the operator T is approximately

$$T \simeq V + VG^0V + VG^0VG^0V + \cdots , \tag{6.274}$$

where

$$G^0 = (E - H_0 + i\varepsilon)^{-1}$$

is the operator describing free propagation and is called the *free propagator*.

At the basis of the expression of $\langle \mathbf{p}'|\hat{S}|\mathbf{p}\rangle$ as a Born series and of the interpretation of each term in terms of Feynman diagrams lies the idea that the transition amplitude $\mathbf{p} \to \mathbf{p}'$ can be viewed as a sequence of successive scatterings separated by free propagations through intermediate (virtual) states.

An important remark consists in singling out the kind of expansions. We call *Dyson series* the perturbative expansion of the time-evolution operator in the interaction picture, so

$$\begin{aligned} U_I(t, t_0) &= T \exp\left[-i \int_{t_0}^{t} dt' \, V_I(t')\right] \\ &= \mathbb{I} - i \int dt' \, V_I(t') + (-i)^2 \int dt' dt'' \, T[V_I(t') V_I(t'')] + \cdots . \end{aligned} \tag{6.275}$$

We have seen, then, that in scattering theory one evaluates $S = U_I(+\infty, -\infty) = \mathbb{I} + S^{(1)} + S^{(2)} + \cdots$ and $S^{(1)} = -i \int_{-\infty}^{+\infty} dt \, V_I(t)$.
The *Born approximation* consists in truncating this expansion at lowest order, i.e. keeping only

$$S \simeq \mathbb{I} + S^{(1)}.$$

Thus, the Born approximation turns out to be the Dyson series itself, but at the lowest-order truncation, applied to scattering amplitudes.

► **Pause and Think** Having the scattering matrix looks incomplete. Why, then, propagators?

6.8 Propagators

We begin by reviewing Green functions, which provide the natural prototype for the propagators that will be introduced in quantum mechanics and, later on, in QFT. Consider a linear differential operator $\hat{L}$ acting on an unknown function $x(t)$,

$$\hat{L} \, x(t) = f(t). \tag{6.276}$$

The source $f(t)$ can be represented as a superposition of Dirac delta distributions,

$$f(t) = \int_0^{+\infty} du \, f(u) \, \delta(t - u), \tag{6.277}$$

(where the limits may be adapted to the domain under consideration; in many applications one integrates over the whole real line).

To construct the *Green function* $G(t, u)$ associated with $\hat{L}$, it is sufficient to solve the equation for a single delta source,

$$\hat{L} \, G(t, u) = \delta(t - u). \tag{6.278}$$

Once G is known, the solution to Eq. (6.276) is

$$x(t) = \int_0^{+\infty} du\, G(t,u)\, f(u). \tag{6.279}$$

Indeed, applying $\hat{L}$ to Eq. (6.279) and using Eq. (6.278) yields

$$\hat{L}\, x(t) = \int_0^{+\infty} du\, \hat{L}\, G(t,u)\, f(u) = \int_0^{+\infty} du\, \delta(t-u)\, f(u) = f(t), \tag{6.280}$$

which reproduces Eq. (6.276).

Driving Exercise

Two Green functions frequently recur in scattering theory and field theory. In three spatial dimensions, with $\mathbf{x}, \mathbf{u} \in \mathbb{R}^3$:

1. for the Laplacian ∇^2, the Green function $G(\mathbf{x}, \mathbf{u})$ solving

$$\nabla^2\, G(\mathbf{x}, \mathbf{u}) = -\delta^{(3)}(\mathbf{x} - \mathbf{u}) \tag{6.281}$$

is

$$G(\mathbf{x}, \mathbf{u}) = \frac{1}{4\pi\, |\mathbf{x} - \mathbf{u}|}; \tag{6.282}$$

2. for the Helmholtz operator $\nabla^2 + k^2$, the (outgoing-wave) Green function $G(\mathbf{x}, \mathbf{u})$ solving

$$(\nabla^2 + k^2)\, G(\mathbf{x}, \mathbf{u}) = -\delta^{(3)}(\mathbf{x} - \mathbf{u}) \tag{6.283}$$

is

$$G(\mathbf{x}, \mathbf{u}) = \frac{e^{ik|\mathbf{x}-\mathbf{u}|}}{4\pi\, |\mathbf{x} - \mathbf{u}|}. \tag{6.284}$$

Equations (6.282)–(6.284) correct the common missing factors in informal formulae: the normalization $1/(4\pi)$ is fixed by Gauss' theorem, while the overall sign depends on the convention chosen for the delta source (here we adopted the standard $-\delta^{(3)}$ on the right-hand side). The close analogy between Green functions and quantum propagators will become explicit below.

6.8.1 Quantum-Mechanical Propagator as a Green Function

In quantum mechanics one cannot, in general, assign a definite classical trajectory to a particle. Instead, one introduces transition amplitudes: the probability amplitude that a particle prepared near y at time t_0 is detected near x at time t.

Let $|\psi(t)\rangle$ be a Schrödinger-picture state obeying

$$i\frac{\partial}{\partial t}|\psi(t)\rangle = \hat{H}\,|\psi(t)\rangle. \tag{6.285}$$

Time evolution is generated by the unitary operator $\hat{U}(t, t_0)$,

$$|\psi(t)\rangle = \hat{U}(t, t_0)\,|\psi(t_0)\rangle. \tag{6.286}$$

In the coordinate representation, by inserting a resolution of the identity, $\mathbb{I} = \int dy\,|y\rangle\langle y|$, one obtains

$$\langle x|\psi(t)\rangle = \langle x|\hat{U}(t, t_0)|\psi(t_0)\rangle = \int dy\,\langle x|\hat{U}(t, t_0)|y\rangle\,\langle y|\psi(t_0)\rangle. \tag{6.287}$$

Therefore,

$$\psi(x, t) = \int dy\,\langle x|\hat{U}(t, t_0)|y\rangle\,\psi(y, t_0). \tag{6.288}$$

This motivates the definition of the (retarded) propagator or *transition amplitude*

$$G^{+}(x, t; y, t_0) \equiv \theta(t - t_0)\,\langle x|\hat{U}(t, t_0)|y\rangle. \tag{6.289}$$

The factor $\theta(t - t_0)$ enforces causality in the nonrelativistic setting by ensuring $G^{+} = 0$ for $t < t_0$.

One may also express the kernel as an overlap of time-evolved position eigenstates. Defining

$$|x, t\rangle \equiv \hat{U}(t, t_0)|x\rangle, \tag{6.290}$$

it follows that

$$\langle x|\hat{U}(t, t_0)|y\rangle = \langle x, t|y, t_0\rangle, \tag{6.291}$$

and hence

$$G^{+}(x, t; y, t_0) = \theta(t - t_0)\,\langle x, t|y, t_0\rangle. \tag{6.292}$$

6.8.1.1 Spectral Representation

Let $\{|n\rangle\}$ be a complete set of energy eigenstates,

$$\hat{H}|n\rangle = E_n|n\rangle, \qquad \mathbb{I} = \sum_n |n\rangle\langle n|. \tag{6.293}$$

Then, for time-independent $\hat{H}$,

$$\begin{aligned} G^+(x,t;y,t_0) &= \theta(t-t_0)\,\langle x|e^{-i\hat{H}(t-t_0)}|y\rangle \\ &= \theta(t-t_0)\sum_n \langle x|n\rangle\langle n|y\rangle\, e^{-iE_n(t-t_0)} \\ &= \theta(t-t_0)\sum_n \phi_n(x)\phi_n^*(y)\, e^{-iE_n(t-t_0)}, \end{aligned} \tag{6.294}$$

where $\phi_n(x) \equiv \langle x|n\rangle$.

6.8.1.2 The Schrödinger Green Function Equation

The propagator G^+ is the Green function of the Schrödinger operator. More precisely, it satisfies

$$\left(\hat{H}_x - i\frac{\partial}{\partial t}\right) G^+(x,t;y,t_0) = -i\,\delta(x-y)\,\delta(t-t_0), \tag{6.295}$$

where $\hat{H}_x$ acts on the x-dependence, while y is a fixed parameter.

▶ **Driving Exercise Proof.**
Insert the spectral representation of Eq. (6.294) into the left-hand side:

$$\left(\hat{H}_x - i\frac{\partial}{\partial t}\right)\left[\theta(t-t_0)\sum_n \phi_n(x)\phi_n^*(y)\, e^{-iE_n(t-t_0)}\right]. \tag{6.296}$$

We compute separately the time derivative and the action of $\hat{H}_x$.
First, using $\frac{d}{dt}\theta(t-t_0) = \delta(t-t_0)$,

$$\begin{aligned} -i\frac{\partial}{\partial t}G^+ &= -i\left[\delta(t-t_0)\sum_n \phi_n(x)\phi_n^*(y)\, e^{-iE_n(t-t_0)}\right] \\ &\quad -i\left[\theta(t-t_0)\sum_n (-iE_n)\phi_n(x)\phi_n^*(y)\, e^{-iE_n(t-t_0)}\right] \\ &= -i\,\delta(t-t_0)\sum_n \phi_n(x)\phi_n^*(y)\, e^{-iE_n(t-t_0)} - \theta(t-t_0) \\ &\quad \sum_n E_n\phi_n(x)\phi_n^*(y)\, e^{-iE_n(t-t_0)}. \end{aligned} \tag{6.297}$$

Second, since $\hat{H}_x \phi_n(x) = E_n \phi_n(x)$,

$$\hat{H}_x\, G^+ = \theta(t - t_0) \sum_n E_n \phi_n(x) \phi_n^*(y)\, e^{-iE_n(t-t_0)}. \tag{6.298}$$

Adding Eqs. (6.297) and (6.298) gives the cancellation of the θ-terms:

$$\left(\hat{H}_x - i\frac{\partial}{\partial t} \right) G^+ = -i\,\delta(t - t_0) \sum_n \phi_n(x) \phi_n^*(y)\, e^{-iE_n(t-t_0)}. \tag{6.299}$$

At $t = t_0$ one has $e^{-iE_n(t-t_0)} = 1$, and by completeness,

$$\sum_n \phi_n(x)\phi_n^*(y) = \sum_n \langle x|n\rangle\langle n|y\rangle = \langle x|\mathbb{I}|y\rangle = \delta(x - y). \tag{6.300}$$

Therefore,

$$\left(\hat{H}_x - i\frac{\partial}{\partial t} \right) G^+(x, t; y, t_0) = -i\,\delta(t - t_0)\,\delta(x - y), \tag{6.301}$$

which is precisely Eq. (6.295).

One defines the *retarded* and *advanced* Green functions as

$$G^+(x, t; y, t_0) = \theta(t - t_0)\,\langle x, t|y, t_0\rangle = \begin{cases} 0, & t < t_0, \\ \langle x, t|y, t_0\rangle, & t > t_0, \end{cases} \tag{6.302a}$$

$$G^-(x, t; y, t_0) = \theta(t_0 - t)\,\langle x, t|y, t_0\rangle = \begin{cases} \langle x, t|y, t_0\rangle, & t < t_0, \\ 0, & t > t_0. \end{cases} \tag{6.302b}$$

6.8.1.3 Propagator for a Free Particle

Consider the free Hamiltonian

$$\hat{H}_0 = \frac{\hat{p}^2}{2m}, \qquad E_p = \frac{p^2}{2m}. \tag{6.303}$$

For simplicity we work in one spatial dimension; the extension to three dimensions is straightforward. The retarded propagator reads

$$G_0^+(x, t; y, t_0) = \theta(t - t_0) \int_{-\infty}^{+\infty} \frac{dp}{2\pi}\, e^{ip(x-y)}\, e^{-iE_p(t-t_0)}. \tag{6.304}$$

Here we have replaced the discrete sum over momenta by an integral (e.g. by taking the infinite-volume limit), and used plane-wave eigenfunctions $\phi_p(x) = (2\pi)^{-1/2} e^{ipx}$.

Substituting $E_p = p^2/(2m)$,

$$G_0^+(x,t;y,t_0) = \theta(t-t_0)\int_{-\infty}^{+\infty} \frac{dp}{2\pi}\exp\left\{i\left[p(x-y) - \frac{p^2}{2m}(t-t_0)\right]\right\}. \tag{6.305}$$

We complete the square in the exponent. Let $\Delta t \equiv t - t_0$ and $\Delta x \equiv x - y$. Then

$$\begin{aligned} p\,\Delta x - \frac{p^2}{2m}\Delta t &= -\frac{\Delta t}{2m}\left[p^2 - \frac{2m}{\Delta t}p\,\Delta x\right] \\ &= -\frac{\Delta t}{2m}\left[\left(p - \frac{m\Delta x}{\Delta t}\right)^2 - \left(\frac{m\Delta x}{\Delta t}\right)^2\right] \\ &= -\frac{\Delta t}{2m}\left(p - \frac{m\Delta x}{\Delta t}\right)^2 + \frac{m(\Delta x)^2}{2\Delta t}. \end{aligned} \tag{6.306}$$

Inserting Eqs. (6.306) into (6.305) yields

$$G_0^+(x,t;y,t_0) = \theta(\Delta t)\, e^{i\frac{m(\Delta x)^2}{2\Delta t}} \int_{-\infty}^{+\infty} \frac{dp}{2\pi}\exp\left[-i\frac{\Delta t}{2m}\left(p - \frac{m\Delta x}{\Delta t}\right)^2\right]. \tag{6.307}$$

Define the shifted variable

$$p' \equiv p - \frac{m\Delta x}{\Delta t}, \tag{6.308}$$

so that the Gaussian integral becomes

$$\int_{-\infty}^{+\infty} \frac{dp'}{2\pi}\exp\left(-i\frac{\Delta t}{2m}p'^2\right) = \frac{1}{2\pi}\sqrt{\frac{2\pi m}{i\Delta t}} = \sqrt{\frac{m}{2\pi i\Delta t}}\,\frac{1}{\sqrt{2\pi}}\sqrt{2\pi}. \tag{6.309}$$

Thus, one obtains the standard free-particle kernel

$$G_0^+(x,t;y,t_0) = \theta(t-t_0)\sqrt{\frac{m}{2\pi i(t-t_0)}}\exp\left[i\frac{m(x-y)^2}{2(t-t_0)}\right]. \tag{6.310}$$

Note that Eq. (6.310) corrects a common typographical confusion: the prefactor is a square root in one dimension, while in d spatial dimensions it becomes $(m/2\pi i\Delta t)^{d/2}$.

6.8.1.4 Propagator in the Presence of an Interaction: Born Series and Dyson Equation

Let the Hamiltonian be split as

$$\hat{H} = \hat{H}_0 + \hat{V}, \tag{6.311}$$

where $\hat{H}_0$ is solvable (known eigenvalues and eigenstates) and $\hat{V}$ is treated as a perturbation. It is useful to introduce the *resolvent* (energy-domain Green operator)

$$\hat{G}(E) \equiv \frac{1}{E - \hat{H} + i\varepsilon}, \qquad \hat{G}_0(E) \equiv \frac{1}{E - \hat{H}_0 + i\varepsilon}, \tag{6.312}$$

where the infinitesimal $+i\varepsilon$ selects the causal (retarded) prescription and ensures convergence.

Formally,

$$\hat{G}(E) = \frac{1}{E - \hat{H}_0 - \hat{V} + i\varepsilon}. \tag{6.313}$$

We now derive the Born expansion. Start from the operator identity

$$\frac{1}{A - B} = \frac{1}{A} + \frac{1}{A}B\frac{1}{A} + \frac{1}{A}B\frac{1}{A}B\frac{1}{A} + \cdots, \tag{6.314}$$

which holds whenever the series converges (e.g. for sufficiently small B in an appropriate operator norm). Apply Eq. (6.314) with

$$A = E - \hat{H}_0 + i\varepsilon, \qquad B = \hat{V}, \tag{6.315}$$

to obtain

$$\begin{aligned} \hat{G}(E) &= \hat{G}_0(E) + \hat{G}_0(E)\hat{V}\hat{G}_0(E) + \hat{G}_0(E)\hat{V}\hat{G}_0(E)\hat{V}\hat{G}_0(E) + \cdots \\ &= \hat{G}_0(E)\left[\mathbb{I} + \hat{V}\hat{G}_0(E) + (\hat{V}\hat{G}_0(E))^2 + \cdots\right]. \end{aligned} \tag{6.316}$$

Equation (6.316) is the *Born series*: it expands the full propagator as a sum of free propagation segments $\hat{G}_0$ separated by interaction insertions $\hat{V}$.

Summing the geometric series in the last bracket (again, where it converges) yields the Dyson-type closed form

$$\hat{G}(E) = \hat{G}_0(E)\frac{1}{\mathbb{I} - \hat{V}\hat{G}_0(E)} = \frac{1}{\mathbb{I} - \hat{G}_0(E)\hat{V}}\hat{G}_0(E). \tag{6.317}$$

Equivalently, one may write the Dyson equation in iterative form,

$$\hat{G}(E) = \hat{G}_0(E) + \hat{G}_0(E)\hat{V}\,\hat{G}(E), \tag{6.318}$$

which follows immediately from multiplying Eq. (6.317) by $(\mathbb{I} - \hat{V}\hat{G}_0)$.

6.8.2 Green Functions in Momentum and Energy Space

In momentum space, for the free evolution generated by $\hat{H}_0$, one has

$$G_0(p, t; q, t_0) = \langle p|\hat{U}_0(t, t_0)|q\rangle = \langle p|e^{-i\hat{H}_0(t-t_0)}|q\rangle = \delta(p-q)\, e^{-iE_p(t-t_0)}. \tag{6.319}$$

Fourier transforming with respect to time gives the energy-domain Green function,

$$G_0(p, q; E) = \int_{-\infty}^{+\infty} dt\, e^{iEt}\, G_0(p, t; q, 0) = \frac{i}{E - E_p + i\varepsilon}\, \delta(p-q), \tag{6.320}$$

in agreement with the resolvent definition in Eq. (6.312). Analogously, starting from the coordinate-space kernel $G_0(x, y; t)$ one defines

$$G_0(x, y; E) = \int_{-\infty}^{+\infty} dt\, e^{iEt}\, G_0(x, y; t), \tag{6.321a}$$

$$G_0(p, q; E) = \int_{-\infty}^{+\infty} dt\, e^{iEt}\, G_0(p, q; t), \tag{6.321b}$$

where the replacement $E \to E + i\varepsilon$ (equivalently $E_n \to E_n - i\varepsilon$ in spectral representations) guarantees convergence as $t \to +\infty$ or $t \to -\infty$ depending on the chosen boundary conditions, and selects the causal prescription.

6.8.3 Field Propagators and Feynman Propagators

We now move to QFT, where propagators encode the spacetime evolution of quanta created as excitations of the vacuum. Let

$$\hat{H} = \hat{H}_0 + \hat{V}, \tag{6.322}$$

and distinguish two situations.

Free theory ($\hat{V} = 0$).

In the absence of interactions, the fields are quantized and become *field operators* obeying canonical (anti)commutation relations. The problem can be solved exactly: one expands $\hat{\phi}$ and $\hat{\phi}^\dagger$ in terms of creation and annihilation operators multiplying plane waves. The corresponding Hamiltonian is diagonal in this basis, and the vacuum state $|0\rangle$ (the *free vacuum*) is defined by annihilation by all annihilation operators. Acting with creation operators on $|0\rangle$ generates the free multi-particle Fock space.

Interacting theory ($\hat{V} \neq 0$).

In an interacting system the exact eigenstates of $\hat{H}$ are not known in closed form, and the interacting fields are not, in general, expressible as simple plane-wave expansions in terms of free creation and annihilation operators. One therefore adopts a perturbative strategy, and the relevant vacuum is no longer the free vacuum $|0\rangle$ but the *interacting vacuum* $|\Omega\rangle$.

In QFT, propagators are vacuum expectation values of products of field operators. A typical object has the schematic form

$$\langle\Omega|\,(\text{product of field operators})\,|\Omega\rangle \;=\; G, \tag{6.323}$$

where the product depends on two spacetime events, namely $x^\mu = (t, \mathbf{x})$ and $y^\mu = (t_0, \mathbf{y})$, denoting spacetime points at which quanta are created and annihilated, with $t \neq t_0$.

For an interacting (Heisenberg) scalar field, the *retarded* two-point function is

$$G^+(x, y) = \theta(t - t_0)\,\langle\Omega|\hat{\phi}(x)\hat{\phi}^\dagger(y)|\Omega\rangle, \tag{6.324}$$

with the Heisenberg operators

$$\hat{\phi}(x) = e^{i\hat{H}t}\,\hat{\phi}(\mathbf{x})\,e^{-i\hat{H}t}, \tag{6.325a}$$

$$\hat{\phi}(y) = e^{i\hat{H}t_0}\,\hat{\phi}(\mathbf{y})\,e^{-i\hat{H}t_0}. \tag{6.325b}$$

The interpretation of Eq. (6.324) as a transition amplitude can be made explicit by reading the operator sequence from right to left:

1. $e^{-i\hat{H}t_0}|\Omega\rangle$ is the interacting vacuum evolved to time t_0;
2. $\hat{\phi}^\dagger(y)\,e^{-i\hat{H}t_0}|\Omega\rangle$ creates a quantum at y and time t_0;
3. evolving this state to time t gives $e^{-i\hat{H}(t-t_0)}\hat{\phi}^\dagger(y)\,e^{-i\hat{H}t_0}|\Omega\rangle$;
4. $\hat{\phi}(x)$ annihilates the quantum at spacetime point x and time t;
5. $\langle\Omega|$ projects the final state back onto the interacting vacuum.

Thus, one may view

$$G(x, y) = \langle\Omega|\hat{\phi}(x)\hat{\phi}^\dagger(y)|\Omega\rangle \tag{6.326}$$

as the amplitude that the system returns to the vacuum after the creation, propagation, and annihilation of a quantum.

In the free case ($\hat{V} = 0$), $|\Omega\rangle \to |0\rangle$, and the corresponding correlator reduces to the free two-point function,

$$\Delta(x, y) = \langle 0|\hat{\phi}(x)\hat{\phi}^\dagger(y)|0\rangle. \tag{6.327}$$

As written, this object only captures particle propagation in the sense explained below; to incorporate both particles and antiparticles in a Lorentz-covariant manner one must use time ordering.

For a complex scalar field one may write the mode expansion

$$\hat{\phi}(x) = \int_{\mathbf{p}} \left[\hat{a}_{\mathbf{p}}\, e^{-ip\cdot x} + \hat{b}^\dagger_{\mathbf{p}}\, e^{+ip\cdot x}\right], \tag{6.328a}$$

$$\hat{\phi}^\dagger(x) = \int_{\mathbf{p}} \left[\hat{a}^\dagger_{\mathbf{p}}\, e^{+ip\cdot x} + \hat{b}_{\mathbf{p}}\, e^{-ip\cdot x}\right], \tag{6.328b}$$

Then $\langle 0|\hat{\phi}(x)\hat{\phi}^\dagger(y)|0\rangle$ describes a particle created at y (via $\hat{a}^\dagger_{\mathbf{p}}$ inside $\hat{\phi}^\dagger$) and annihilated at x (via $\hat{a}_{\mathbf{p}}$ inside $\hat{\phi}$).

Wick time ordering.

Let T denote time ordering. For bosonic fields (e.g. a scalar),

$$T\big[\hat{\phi}(x)\hat{\phi}(y)\big] = \begin{cases} \hat{\phi}(x)\hat{\phi}(y), & x^0 > y^0, \\ \hat{\phi}(y)\hat{\phi}(x), & x^0 < y^0, \end{cases} \tag{6.329}$$

whereas for fermionic (spin-1/2) fields the exchange introduces a minus sign,

$$T\big[\hat{\psi}(x)\hat{\psi}(y)\big] = \begin{cases} \hat{\psi}(x)\hat{\psi}(y), & x^0 > y^0, \\ -\hat{\psi}(y)\hat{\psi}(x), & x^0 < y^0. \end{cases} \tag{6.330}$$

This distinction is essential when antiparticle propagation is accounted for.

The *Feynman propagator* is defined by the time-ordered vacuum expectation value

$$G_F(x, y) = \langle\Omega|T\hat{\phi}(x)\hat{\phi}^\dagger(y)|\Omega\rangle = \tag{6.331}$$

$$= \theta(x^0 - y^0)\langle\Omega|\hat{\phi}(x)\hat{\phi}^\dagger(y)|\Omega\rangle + \theta(y^0 - x^0)\langle\Omega|\hat{\phi}(y)\hat{\phi}^\dagger(x)|\Omega\rangle. \tag{6.332}$$

The first term propagates a particle created at y forward to x, where it is annihilated; the second term propagates an antiparticle from x to y (equivalently, a negative-frequency component propagating backward in time).

In the free theory ($\hat{V} = 0$) one has

$$\Delta_F(x, y) = \langle 0|T\hat{\phi}(x)\hat{\phi}^\dagger(y)|0\rangle, \tag{6.333}$$

which describes free propagation between local interaction events. This is a central point in perturbation theory: interactions are localized in spacetime, while propagation between interaction vertices is free.

Driving Exercise

Using Eq. (6.328), one finds

$$\hat{\phi}^\dagger(y)|0\rangle = \int_{\mathbf{p}} \left[\hat{a}^\dagger_{\mathbf{p}} e^{ip\cdot y} + \hat{b}_{\mathbf{p}} e^{-ip\cdot y}\right]|0\rangle = \int_{\mathbf{p}} |\mathbf{p}\rangle\, e^{ip\cdot y}, \tag{6.334a}$$

$$\langle 0|\hat{\phi}(x) = \int_{\mathbf{q}} \left[\langle 0|\hat{a}_{\mathbf{q}} e^{-iq\cdot x} + \langle 0|\hat{b}^\dagger_{\mathbf{q}} e^{iq\cdot x}\right] = \int_{\mathbf{q}} \langle \mathbf{q}|\, e^{-iq\cdot x}. \tag{6.334b}$$

For $x^0 > y^0$ one therefore obtains a particle propagating from y to x,

$$\begin{aligned}\Delta(x, y) = \langle 0|\hat{\phi}(x)\hat{\phi}^\dagger(y)|0\rangle &= \int \frac{d^3p\, d^3q}{(2\pi)^3\sqrt{2E_{\mathbf{p}}2E_{\mathbf{q}}}}\, e^{ip\cdot y - iq\cdot x}\, \langle \mathbf{q}|\mathbf{p}\rangle \\ &= \int \frac{d^3p}{(2\pi)^3(2E_{\mathbf{p}})}\, e^{-ip\cdot(x-y)}.\end{aligned} \tag{6.335}$$

For $x^0 < y^0$ one obtains antiparticle propagation,

$$\Delta(x, y) = \langle 0|\hat{\phi}(y)\hat{\phi}^\dagger(x)|0\rangle = \int \frac{d^3p}{(2\pi)^3(2E_{\mathbf{p}})}\, e^{+ip\cdot(x-y)}. \tag{6.336}$$

Combining the two time orderings,

$$\begin{aligned}\Delta_F(x, y) &= \langle 0|T\hat{\phi}(x)\hat{\phi}^\dagger(y)|0\rangle \\ &= \int \frac{d^3p}{(2\pi)^3(2E_{\mathbf{p}})}\left[\theta(x^0 - y^0)e^{-ip\cdot(x-y)} + \theta(y^0 - x^0)e^{+ip\cdot(x-y)}\right].\end{aligned} \tag{6.337}$$

We have to reformulate the θ-functions to the $i\varepsilon$ prescription. A standard integral representation of the Heaviside function is

$$\theta(\tau) = i\int_{-\infty}^{+\infty} \frac{dz}{2\pi}\, \frac{e^{-iz\tau}}{z + i\varepsilon}. \tag{6.338}$$

Applying Eq. (6.338) to (6.337) (with $\tau = x^0 - y^0$) and combining the two time orderings yields the compact four-dimensional representation

$$\Delta_F(x, y) = i\int \frac{d^4p}{(2\pi)^4}\, \frac{e^{-ip\cdot(x-y)}}{p^2 - m^2 + i\varepsilon}, \tag{6.339}$$

where $p^2 \equiv (p^0)^2 - \mathbf{p}^2$. The corresponding momentum space Feynman propagator is therefore

$$\tilde{\Delta}_F(p) = \frac{i}{p^2 - m^2 + i\varepsilon}. \tag{6.340}$$

(Equation (6.339) corrects the intermediate algebra: once time ordering is implemented, the factor $(2E_{\mathbf{p}})^{-1}$ is absorbed into the covariant d^4p integral, yielding the standard Lorentz-invariant form.)

6.8.4 The Yukawa Potential

The denominator in Eq. (6.340) is generically *off-shell*, as it vanishes for $p^0 = \sqrt{\mathbf{p}^2 + m^2}$. One expresses this by saying that $\tilde{\Delta}_F(p)$ describes the propagation of *virtual quanta*. Virtual particles are off-shell excitations that appear only as internal lines in Feynman diagrams and never as asymptotic in- or out-states. Their four-momentum is not fixed by a mass-shell condition but is instead integrated over all values allowed by energy-momentum conservation at the interaction vertices.
The terminology "particle" in this context is therefore only schematic: virtual quanta do not correspond to detectable objects, nor can they be assigned a classical trajectory or a well-defined lifetime. They are mathematical carriers of interactions within perturbation theory, encoding how fields transmit correlations between spacetime points. The $i\varepsilon$ prescription specifies how the poles of the propagator are to be bypassed in the complex energy plane and guarantees causal propagation, selecting retarded boundary conditions in time-ordered amplitudes.

The statement that virtual particles "exist for a short time" is best understood as a heuristic interpretation of the uncertainty principle rather than as a literal dynamical process. In perturbative calculations, virtual quanta are integrated over all energies and momenta, including values that would be forbidden for real particles. Their apparent violation of the mass-shell condition reflects the fact that they are not physical states but intermediate fluctuations of the quantum field.

To estimate the lifetime, retaining $\hbar$ for pedagogical explanations, one can presume to violate the uncertainty principle, $\Delta t \Delta E \geq \hbar$, for them, namely assuming that

$$\tau \lesssim \frac{\hbar}{\mathcal{E}}, \tag{6.341}$$

where the virtual particle lifetime is assumed to be $\tau \equiv \Delta t$ and its corresponding energy is set to $\mathcal{E} \equiv \Delta E$.

In particular, when the exchanged field has a nonzero mass m, the propagator decays exponentially in coordinate space at large spacelike separations, with a characteristic length scale of order $1/m$.

▶ **Driving Exercise** To see this explicitly, let us start from the free scalar Feynman propagator in momentum space,

$$\tilde{\Delta}_F(p) = \frac{i}{p^2 - m^2 + i\varepsilon}.$$

In coordinate space,

$$\Delta_F(x) = i \int \frac{d^4p}{(2\pi)^4} \frac{e^{-ip\cdot x}}{p^2 - m^2 + i\varepsilon}.$$

Consider a purely spacelike separation, $x^0 = 0, \mathbf{x} \neq 0$, so that $x^2 = -\mathbf{x}^2 < 0$. Performing the p^0 integral by contour methods and then the angular part of the spatial integral, one finds the standard result

$$\Delta_F(0, \mathbf{x}) = \int \frac{d^3p}{(2\pi)^3} \frac{e^{i\mathbf{p}\cdot\mathbf{x}}}{2E_{\mathbf{p}}}, \qquad E_{\mathbf{p}} = \sqrt{\mathbf{p}^2 + m^2}.$$

This integral can be evaluated in spherical coordinates. Writing $r = |\mathbf{x}|$,

$$\Delta_F(0, r) = \frac{1}{4\pi^2 r} \int_0^\infty dp \, \frac{p \sin(pr)}{\sqrt{p^2 + m^2}}.$$

The integral is known and gives a modified Bessel function,

$$\Delta_F(0, r) = \frac{m}{4\pi^2 r} K_1(mr).$$

For large spacelike separations $mr \gg 1$, the asymptotic behavior of the Bessel function is

$$K_1(mr) \sim \sqrt{\frac{\pi}{2mr}} \, e^{-mr}.$$

Therefore,

$$\Delta_F(0, r) \sim \frac{1}{(mr)^{3/2}} e^{-mr}, \qquad r \to \infty.$$

Hence, for a massive field the propagator decays exponentially at large spacelike separations,

$$\Delta_F(x) \propto e^{-m|x|},$$

showing that the characteristic range of the interaction mediated by a particle of mass m is of order $1/m$.

This is the origin of the short-range character of forces mediated by massive fields. By contrast, for massless fields the propagator has a long-range behavior, leading to interactions of infinite range, as in the electromagnetic and gravitational cases.

A paradigmatic example to describe the exponential decay of virtual particles has been therefore presented by Yukawa at the level of potential. The *Yukawa potential*, describing an interaction mediated by the exchange of a massive scalar of mass m, reads

$$U(r) \propto -\frac{e^{-mr}}{r}, \qquad r \equiv |\mathbf{r}|, \tag{6.342}$$

where the characteristic range is set by the Compton length,

$$a = \frac{1}{m}. \tag{6.343}$$

Remarkably, the Yukawa potential is the Green function associated with the static Helmholtz operator. In particular, the function

$$G(\mathbf{r}) = \frac{e^{-mr}}{4\pi r}$$

satisfies $(\nabla^2 - m^2)\, G(\mathbf{r}) = -\delta^{(3)}(\mathbf{r})$.

Returning to the coordinate-space propagator for a free complex scalar field,

$$\Delta_F(x, y) = \theta(x^0 - y^0)\langle 0|\hat{\phi}(x)\hat{\phi}^\dagger(y)|0\rangle + \theta(y^0 - x^0)\langle 0|\hat{\phi}(y)\hat{\phi}^\dagger(x)|0\rangle, \tag{6.344}$$

we see that the first term describes a (virtual) particle created at y and annihilated at x, while the second term describes a (virtual) antiparticle created at x and annihilated at y (see Fig. 6.3).

Thus, the Feynman propagator encodes the exchange of virtual quanta propagating freely between localized interaction events.

Finally, in order to connect back to scattering theory, we observe that that the momentum space factor

$$\tilde{\Delta}_F(p) = \frac{i}{p^2 - m^2 + i\varepsilon} \tag{6.345}$$

is the field-theoretic analogue of the free resolvent in energy space,

$$G_0(E) = \frac{1}{E - \hat{H}_0 + i\varepsilon}, \tag{6.346}$$

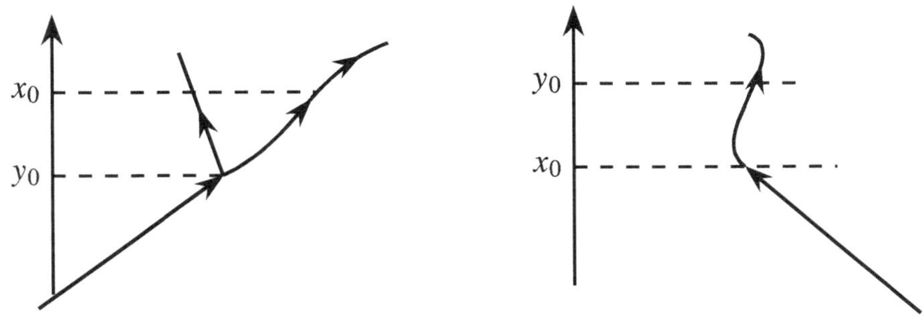

Fig. 6.3 Left: propagation of a virtual particle emitted at y and absorbed at x. Right: propagation of a virtual antiparticle emitted at x and absorbed at y

Fig. 6.4 Free propagation between interaction events: in perturbation theory, internal lines represent free propagation, while vertices represent localized interactions

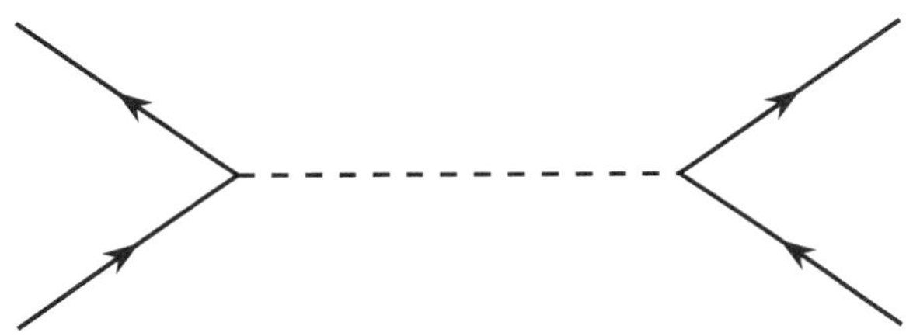

which describes free propagation between asymptotic in/out states (see Fig. 6.4).

In the interacting case one therefore writes the (time-ordered) two-point function as

$$G(x, y) = \langle \Omega | T \hat{\phi}(x) \hat{\phi}^\dagger(y) | \Omega \rangle, \tag{6.347}$$

which is the fundamental building block of perturbative expansions and Feynman-diagrammatics in QFT.

6.9 Feynman Diagrams

Feynman diagrams provide a compact representation of the terms generated by the Dyson expansion of the S-matrix. They are *not* literal particle trajectories, rather they are a graphical representation of algebraic objects: time-ordered products, Wick contractions, propagators, and vertex factors. A diagram becomes meaningful only once a precise *dictionary* maps each graphical element to its corresponding mathematical factor.

Throughout this chapter we adopt the spacetime convention in which *time flows upward*. External lines represent asymptotic on-shell states, while internal lines represent off-shell propagation (virtual quanta). Antiparticles are drawn as particle lines with arrows opposite to the fermion-number flow; in the time-up convention this is consistent with the standard interpretation of negative-frequency components as positive-energy antiparticles.

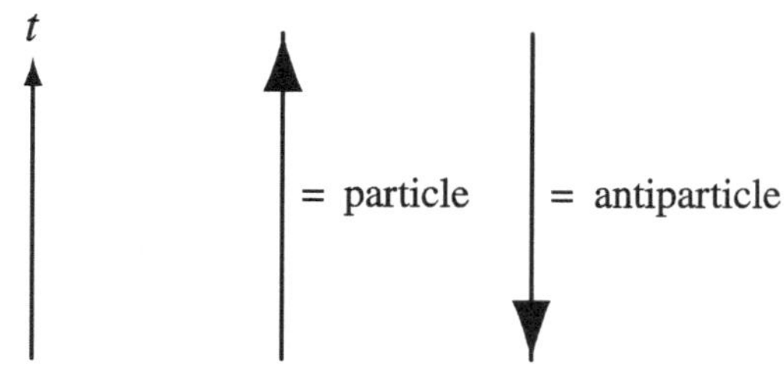

Fig. 6.5 Time orientation of fermion lines. A particle propagates forward in time, while an antiparticle can be interpreted as a particle propagating backward in time

What a diagram really encodes. A Feynman diagram is a graphical encoding of a term in the perturbative expansion of correlation functions and S-matrix elements. Lines stand for propagators (Green functions) of the free theory; vertices stand for local monomials in the interaction Lagrangian/Hamiltonian. The topology of the graph records which fields are contracted with which, and the integration over vertex positions/momenta implements the sum over all interaction events compatible with the chosen external states.

Thus, a key novelty of QFT (as opposed to single-particle quantum mechanics) is that particle number is not conserved: particles and antiparticles can be created and annihilated. Diagrammatically, pair creation at $t = t_0$ is represented by two lines emerging from one spacetime event: a particle line (arrow aligned with fermion-number flow) and an antiparticle line (opposite arrow). Pair annihilation at $t = t_1$ is represented by two lines ending at the same event.

A drawing that appears to show "a particle propagating backward in time" is read physically by following the diagram in time order: one starts with an incoming particle, creates a pair at some intermediate time, and then annihilates the newly created antiparticle with the original particle. The remaining newly created particle continues as the outgoing line. The interpretation does not require any actual backward-time evolution of a physical on-shell object; it is a statement about how antiparticle degrees of freedom enter the field operator.

► **Pause and Think** Starting from Fig. 6.5, explain, using the mode expansion of a complex scalar or Dirac field, why the antiparticle line carries the opposite conserved charge (or fermion number) with respect to the particle line. Where does the arrow convention come from algebraically?

6.10 Dyson Expansion and Interaction Hamiltonians

In the interaction picture the S-operator is

$$\hat{S} = \mathcal{T} \exp\left[-i \int d^4z\, \hat{H}_I(z)\right] = \sum_{n=0}^{\infty} \frac{(-i)^n}{n!} \int d^4z_1 \cdots d^4z_n\, \mathcal{T}\left[\hat{H}_I(z_1) \cdots \hat{H}_I(z_n)\right], \tag{6.348}$$

where $\mathcal{T}$ denotes time ordering and $\hat{H}_I$ is the interaction Hamiltonian density. Each insertion of $\hat{H}_I$ produces one interaction vertex; Wick's theorem then rewrites time-ordered products as sums of normal products plus contractions, and contractions are precisely free-theory propagators. This is the structural origin of the diagrammatic method.

At this stage it is useful to distinguish three layers: (i) the *Lagrangian/Hamiltonian* specifies which vertices exist; (ii) the *free quadratic terms* specify propagators; (iii) the *external states* specify which fields appear as external legs.

Driving Exercise Interaction with an external source.
A real scalar coupled to a classical source,

$$\hat{H}_I(z) = J(z)\,\hat{\phi}(z), \tag{6.349}$$

is drawn as a "blob" emitting/absorbing a scalar line (see Fig. 6.6).
ϕ^4 theory: local scalar self-interaction.
The prototypical scalar interaction,

$$\hat{H}_I(z) = \frac{\lambda}{4!}\,\hat{\phi}^4(z), \tag{6.350}$$

corresponds to a single vertex with four scalar legs (see Fig. 6.7).
A mixed interaction $g\,\psi^\dagger\psi\phi$.
For a complex scalar ψ coupled to a real scalar ϕ,

$$\hat{H}_I(z) = g\,\hat{\psi}^\dagger(z)\hat{\psi}(z)\hat{\phi}(z), \tag{6.351}$$

Fig. 6.6 Source coupling $J\phi$ represented as a point-like insertion

Fig. 6.7 Local ϕ^4 vertex (contact interaction)

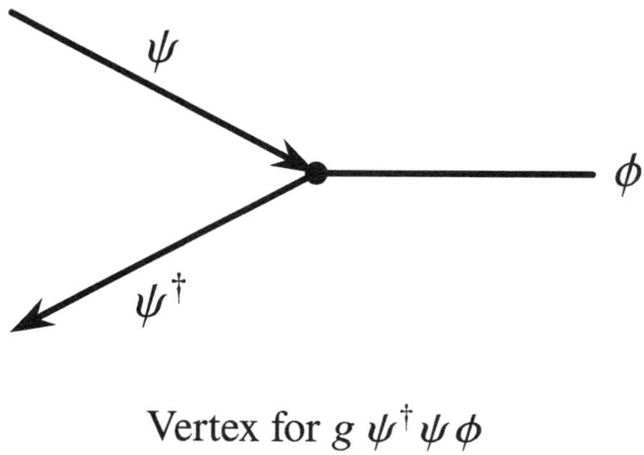

Fig. 6.8 A charge-flow convention for a complex scalar interaction vertex

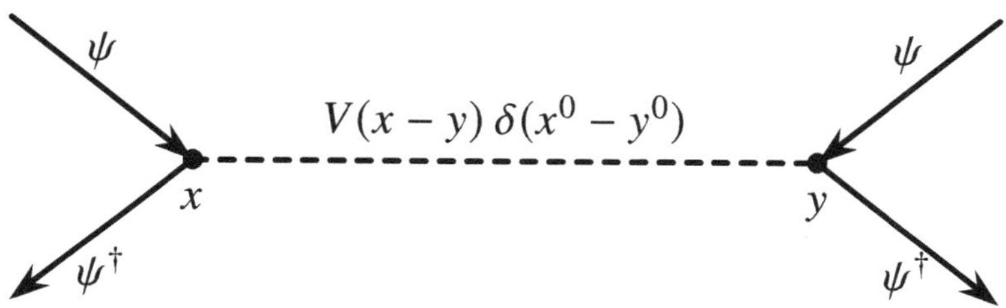

Fig. 6.9 A spatially non-local, time-local effective interaction

we draw arrows on the ψ-lines to track the conserved $U(1)$ charge (see Fig. 6.8).

A non-relativistic instantaneous two-body interaction.
An effective "instantaneous" interaction (see Fig. 6.9),

$$\hat{H}_I(x-y) = \frac{1}{2}\,\hat{\psi}^\dagger(x)\hat{\psi}^\dagger(y)\,V(x-y)\,\delta(x^0-y^0)\,\hat{\psi}(y)\hat{\psi}(x), \tag{6.352}$$

is non-local in space and is better represented by two points connected by an interaction bridge encoding V.

In perturbative QFT the building blocks of diagrams are the free fields and their propagators. The interaction is treated as a perturbation of the free theory, so internal lines always correspond to *free* propagators, while vertices are dictated by the interaction terms.

6.11 From S-Matrix Elements to Observables: The "Golden Rules"

The experimentally accessible quantities are decay widths and scattering cross sections. In relativistic QFT, it is convenient to factor out kinematical normalization and define the invariant amplitude $\mathcal{M}$ by

$$\langle\beta|\hat{S}|\alpha\rangle = \delta_{\beta\alpha} - i(2\pi)^4\delta^{(4)}(p_\beta - p_\alpha)\,\mathcal{M}_{\beta\alpha}\prod_{i\in\alpha,\beta}\frac{1}{\sqrt{(2\pi)^3\,2E_i}}\,. \tag{6.353}$$

In terms of $\mathcal{M}$, the differential decay rate for $1 \to 1' + 2' + \cdots + n'$ reads

$$d\Gamma = \frac{1}{2E_1}\,|\mathcal{M}|^2\,(2\pi)^4\delta^{(4)}\Big(p_1 - \sum_{i=1}^{n} p_i'\Big)\prod_{i=1}^{n}\frac{d^3p_i'}{(2\pi)^3\,2E_i'}\,, \tag{6.354}$$

while the differential cross section for $1 + 2 \to 1' + \cdots + n'$ is

$$d\sigma = \frac{1}{4\sqrt{(p_1\cdot p_2)^2 - m_1^2m_2^2}}\,|\mathcal{M}|^2\,(2\pi)^4\delta^{(4)}\Big(p_1 + p_2 - \sum_{i=1}^{n} p_i'\Big)\prod_{i=1}^{n}\frac{d^3p_i'}{(2\pi)^3\,2E_i'}\,. \tag{6.355}$$

The entire diagrammatic machinery is, in practice, a method to compute $|\mathcal{M}|^2$ efficiently.

▶ **Driving Exercise** Derive the flux factor $4\sqrt{(p_1\cdot p_2)^2 - m_1^2m_2^2}$ by matching the Lorentz-invariant expression for $d\sigma$ to the non-relativistic limit in the laboratory frame. Where does the relative velocity enter?

6.12 Feynman Rules for ϕ^4

The simplest interacting relativistic scalar theory is

$$\mathcal{L} = \frac{1}{2}\,\partial_\mu\phi\,\partial^\mu\phi - \frac{1}{2}m^2\phi^2 - \frac{\lambda}{4!}\phi^4, \qquad \lambda \ll 1. \tag{6.356}$$

The free Euler-Lagrange equation is $(\Box + m^2)\phi = 0$, so in momentum space $(p^2 - m^2)\tilde{\phi}(p) = 0$. The corresponding Feynman propagator is

$$\tilde{\Delta}_F(p) = \frac{i}{p^2 - m^2 + i\epsilon}, \qquad \Delta_F(x-y) = \int\frac{d^4p}{(2\pi)^4}\,e^{-ip\cdot(x-y)}\,\tilde{\Delta}_F(p). \tag{6.357}$$

A term at order n in the Dyson series contains n insertions of $\hat{H}_I$, hence n vertices. Wick contractions between two fields generate internal lines; contractions between fields and external operators generate external legs. The diagram is simply a compact encoding of these pairings, together with the integration over unconstrained vertex positions (or loop momenta).

Position space rules

For an amplitude associated with a diagram in position space:

- each vertex contributes $-i\lambda$ (for ϕ^4; replace accordingly);
- each internal line connecting x and y contributes $\Delta_F(x-y)$;
- integrate over all vertex positions $\int d^4x$;
- divide by the symmetry factor when identical contractions produce the same graph.

Momentum space rules (most practical)

For an amplitude associated with a diagram in momentum space:

- assign a momentum q to each internal line and include the propagator $\frac{i}{q^2-m^2+i\epsilon}$ (or the appropriate spin-dependent propagator);
- enforce momentum conservation at each vertex;
- integrate over each independent loop momentum with $\int \frac{d^4q}{(2\pi)^4}$;
- include the overall $(2\pi)^4\delta^{(4)}(\sum p_{\text{in}} - \sum p_{\text{out}})$.

6.12.1 External Lines and Universal Spin Dependence

A convenient summary of external-line factors (in momentum space) is:

External particle	Factor
Ingoing fermion	$u(p,s)$
Outgoing fermion	$\bar{u}(p,s)$
Ingoing antifermion	$\bar{v}(p,s)$
Outgoing antifermion	$v(p,s)$
Ingoing photon	$\varepsilon_\mu(p,\lambda)$
Outgoing photon	$\varepsilon^*_\mu(p,\lambda)$
Ingoing/outgoing scalar	1

This table is the practical form of the statement that external legs are determined by the free-field mode expansion.

Vertices depend on the *interaction* part of $\mathcal{L}$; propagators and external-line wavefunctions depend only on the *free* part of $\mathcal{L}$.

6.13 A Second Look at QED

The free Maxwell Lagrangian is $\mathcal{L}_{\text{EM}} = -\frac{1}{4}F_{\mu\nu}F^{\mu\nu}$, with $F_{\mu\nu} = \partial_\mu A_\nu - \partial_\nu A_\mu$. In momentum space the equations of motion read

$$\left(k^2 g_{\mu\nu} - k_\mu k_\nu\right)\tilde{A}^\nu(k) = 0, \tag{6.358}$$

which cannot be inverted to obtain a propagator because the operator $k^2 g_{\mu\nu} - k_\mu k_\nu$ has a null direction (a direct manifestation of gauge redundancy). The obstruction is physical: the field A_μ contains unphysical components that do not correspond to propagating degrees of freedom.

To obtain an invertible quadratic form, one adds a covariant gauge-fixing term

$$\mathcal{L}_{\text{gf}} = -\frac{1}{2\xi}\left(\partial_\mu A^\mu\right)^2, \tag{6.359}$$

so that the quadratic Lagrangian becomes $\mathcal{L}_{\text{EM}} + \mathcal{L}_{\text{gf}}$. The momentum space equations of motion are then

$$\left[-k^2 g_{\mu\nu} + k_\mu k_\nu - \frac{1}{\xi}k_\mu k_\nu\right]\tilde{A}^\nu(k) = 0 \quad\Longleftrightarrow\quad M_{\mu\nu}(k)\tilde{A}^\nu(k) = 0, \tag{6.360}$$

with

$$M_{\mu\nu}(k) = -k^2 g_{\mu\nu} + \left(1 - \frac{1}{\xi}\right)k_\mu k_\nu. \tag{6.361}$$

By Lorentz covariance, the inverse must be of the form

$$(M^{-1})^{\mu\nu}(k) = A(k)\, g^{\mu\nu} + B(k)\, k^\mu k^\nu. \tag{6.362}$$

Imposing $M_{\mu\alpha}(M^{-1})^{\alpha\nu} = \delta_\mu^{\ \nu}$ yields

$$\begin{aligned}\delta_\mu^{\ \nu} &= \left[-k^2 g_{\mu\alpha} + \left(1 - \frac{1}{\xi}\right)k_\mu k_\alpha\right]\left[A g^{\alpha\nu} + B k^\alpha k^\nu\right] \\ &= -k^2 A\,\delta_\mu^{\ \nu} + \left[-k^2 B + A\left(1 - \frac{1}{\xi}\right) + B\left(1 - \frac{1}{\xi}\right)k^2\right]k_\mu k^\nu.\end{aligned} \tag{6.363}$$

Matching the $\delta_\mu^{\ \nu}$ term gives $A = -1/k^2$. Matching the coefficient of $k_\mu k^\nu$ then gives $B = (1-\xi)/k^4$. The free photon propagator, defined as $i(M^{-1})^{\mu\nu}$ with the Feynman prescription, is therefore

$$\tilde{D}^{(0)}_{\mu\nu}(k) = -\frac{i}{k^2 + i\epsilon}\left[g_{\mu\nu} - (1-\xi)\frac{k_\mu k_\nu}{k^2}\right]. \tag{6.364}$$

The choice $\xi = 1$ (Feynman gauge) yields the particularly simple form

$$\tilde{D}^{(0)}_{\mu\nu}(k) = -\frac{i g_{\mu\nu}}{k^2 + i\epsilon}. \tag{6.365}$$

Physical amplitudes couple photons to a conserved electromagnetic current. This fact has a decisive consequence: the longitudinal term proportional to $k_\mu k_\nu$ in Eq. (6.364) does not affect gauge-invariant matrix elements. The underlying reason is summarized by the Ward identity discussed in Sect. 6.13.4.

6.13.1 Photon Exchange and the Coulomb Limit

Consider the exchange of a virtual photon between two conserved currents J_a^μ and J_b^ν. The tree-level amplitude is

$$\mathcal{A}(k) = J_a^\mu(k)\, \tilde{D}^{(0)}_{\mu\nu}(k)\, J_b^\nu(-k) \simeq J_a^\mu(k) \left(\frac{-i g_{\mu\nu}}{k^2}\right) J_b^\nu(-k), \tag{6.366}$$

where we used the Feynman-gauge form for simplicity. Choose a frame in which $k^\mu = (k^0, 0, 0, k^3)$. Then

$$\mathcal{A}(k) = -\frac{i}{(k^0)^2 - (k^3)^2}\left(J_a^0 J_b^0 - \mathbf{J}_a \cdot \mathbf{J}_b\right). \tag{6.367}$$

Current conservation $k_\mu J^\mu = 0$ gives $k^0 J^0 - k^3 J^3 = 0$, hence $J^3 = (k^0/k^3) J^0$ (for $k^3 \neq 0$). Substituting,

$$\begin{aligned} J_a^0 J_b^0 - \mathbf{J}_a \cdot \mathbf{J}_b &= J_a^0 J_b^0 - \left(J_a^1 J_b^1 + J_a^2 J_b^2 + J_a^3 J_b^3\right) \\ &= J_a^0 J_b^0 \left(1 - \frac{(k^0)^2}{(k^3)^2}\right) - (J_a^1 J_b^1 + J_a^2 J_b^2), \end{aligned} \tag{6.368}$$

so that

$$\mathcal{A}(k) = \frac{i\, J_a^0 J_b^0}{(k^3)^2} + \frac{i\,(J_a^1 J_b^1 + J_a^2 J_b^2)}{(k^0)^2 - (k^3)^2}. \tag{6.369}$$

The first term involves charge densities and is instantaneous in a non-covariant splitting; it reproduces the Coulomb potential. Indeed, inverse Fourier transforming the factor $1/\mathbf{k}^2$ yields

$$\int \frac{d^3\mathbf{k}}{(2\pi)^3} \frac{e^{i\mathbf{k}\cdot\mathbf{r}}}{\mathbf{k}^2} = \frac{1}{4\pi r}, \tag{6.370}$$

so the interaction energy between static charges is $\sim q^2/(4\pi r)$. The second term describes propagation of the two transverse photon polarizations.

Physical interpretation. In covariant perturbation theory the photon propagator simultaneously contains an instantaneous Coulombic contribution (tied to charge densities) and a genuinely radiative contribution (tied to transverse currents). Current conservation reorganizes these terms so that the final result is consistent with causality and Lorentz covariance once the full amplitude is assembled.

6.13.2 Feynman Rules for QED

QED is defined by Eq. (6.218), namely $\mathcal{L}_{\text{QED}} = \bar{\psi}\,(i\partial\!\!\!/ - m)\psi - e\,\bar{\psi}\gamma^\mu\psi\,A_\mu - \frac{1}{4}\,F_{\mu\nu}F^{\mu\nu}$, $\quad F_{\mu\nu} = \partial_\mu A_\nu - \partial_\nu A_\mu$, where the electric charge is denoted by e. The interaction term fixes the fermion-photon vertex, while the quadratic terms fix the free propagators. In momentum space one uses

$$\tilde{S}_F(p) = \frac{i(p\!\!\!/ + m)}{p^2 - m^2 + i\epsilon}, \qquad \tilde{D}_F^{\mu\nu}(p) = \frac{-ig^{\mu\nu}}{p^2 + i\epsilon} \quad \text{(Feynman gauge)}, \tag{6.371}$$

and the vertex factor is $-ie\gamma^\mu$. These are the universal building blocks for tree-level QED amplitudes.

Equivalently, starting from Eq. (6.218), the interaction Hamiltonian density is $\mathcal{H}_I = +e\,\bar{\psi}\gamma^\mu\psi\,A_\mu$, and perturbation theory in e leads to the standard momentum space rules:

- Internal fermion line (momentum p): $\dfrac{i(p\!\!\!/ + m)}{p^2 - m^2 + i\epsilon}$.
- Internal photon line (momentum k): $\tilde{D}^{(0)}_{\mu\nu}(k)$, e.g. in Feynman gauge $\dfrac{-ig_{\mu\nu}}{k^2 + i\epsilon}$ from Eq. (6.365).
- Fermion-photon vertex: $-ie\gamma^\mu$.
- External photon: incoming $\epsilon^\mu_\lambda(k)$, outgoing $\epsilon^{\mu*}_\lambda(k)$.

6.13.3 Example: $e^+e^- \to \mu^+\mu^-$ at Tree Level

There is a single tree diagram: an s-channel photon exchanged between the electron and muon currents. The invariant amplitude is

$$-i\mathcal{M} = \left[\bar{u}(p_3, s_3)(-ie\gamma^\mu)v(p_4, s_4)\right] \frac{-ig_{\mu\nu}}{(p_1 + p_2)^2 + i\epsilon} \left[\bar{v}(p_2, s_2)(-ie\gamma^\nu)u(p_1, s_1)\right], \tag{6.372}$$

or equivalently

$$\mathcal{M} = \frac{e^2}{s}\left[\bar{u}(p_3, s_3)\gamma^\mu v(p_4, s_4)\right]\left[\bar{v}(p_2, s_2)\gamma_\mu u(p_1, s_1)\right], \qquad s = (p_1 + p_2)^2. \tag{6.373}$$

The compact form in Eq. (6.373) isolates the kinematic invariant s and factorizes the amplitude into a product of a muonic current and an electronic current, linked by the photon propagator. This factorization is typical of abelian gauge theories at tree level.

Kinematics. The Mandelstam invariants are defined as $s = (p_1 + p_2)^2$, $t = (p_1 - p_3)^2$, and $u = (p_1 - p_4)^2$ for the assignment $1 + 2 \to 3 + 4$. They provide a Lorentz-invariant language for expressing both $\overline{|\mathcal{M}|^2}$ and phase-space factors, particularly in the center-of-mass frame.

For unpolarized beams one averages over initial spins and sums over final spins. The standard tool is completeness,

$$\sum_s u(p,s)\bar{u}(p,s) = \not{p} + m, \qquad \sum_s v(p,s)\bar{v}(p,s) = \not{p} - m, \tag{6.374}$$

which turns spinor chains into traces. One obtains the structure

$$\overline{|\mathcal{M}|^2} = \frac{e^4}{4s^2}\,\mathrm{Tr}\big[(\not{p}_1 + m_e)\gamma^\mu(\not{p}_2 - m_e)\gamma^\nu\big]\,\mathrm{Tr}\big[(\not{p}_4 - m_\mu)\gamma_\mu(\not{p}_3 + m_\mu)\gamma_\nu\big]. \tag{6.375}$$

Equation (6.375) is the precise content of the "Casimir trick": the incoherent sum over spin states is converted into an algebraic trace problem. The reduction from spinors to traces is a structural advantage of relativistic QFT, because it makes the result manifestly Lorentz covariant.

At high energy ($m_e, m_\mu \to 0$) this simplifies to the textbook expression

$$\overline{|\mathcal{M}|^2} = \frac{2e^4}{s^2}\,(t^2 + u^2), \tag{6.376}$$

with Mandelstam invariants $t = (p_1 - p_3)^2$, $u = (p_1 - p_4)^2$.

▶ **Pause and Think** Why do we sum/average $|\mathcal{M}|^2$ rather than $\mathcal{M}$ itself for unpolarized beams? Give a careful argument in terms of (in)coherence of initial spin ensembles and what an experiment actually measures.

The invariant phase space and flux normalization were developed in Chap. 5. Here, we use the result to convert $\overline{|\mathcal{M}|^2}$ to $d\sigma/d\Omega$. Hence, for $2 \to 2$ scattering in the CM frame,

$$\frac{d\sigma_{\mathrm{CM}}}{d\Omega} = \frac{1}{64\pi^2 s}\,\frac{|\mathbf{p}_3|}{|\mathbf{p}_1|}\,\overline{|\mathcal{M}|^2}. \tag{6.377}$$

In the ultra-relativistic regime for $e^+e^- \to \mu^+\mu^-$ one has $|\mathbf{p}_3| = |\mathbf{p}_1|$ and $\overline{|\mathcal{M}|^2}$ reduces to Eq. (6.376). The final result is

$$\frac{d\sigma}{d\Omega} = \frac{\alpha^2}{4s}\,(1+\cos^2\theta), \qquad \alpha = \frac{e^2}{4\pi}. \tag{6.378}$$

This result is a paradigmatic illustration of the workflow: draw diagrams $\to$ write $\mathcal{M} \to$ spin sums $\to$ traces $\to$ kinematics $\to$ phase space.

Pedagogically, it is useful to summarize the various steps and procedures behind Feynman's diagrams through a disciplined six-step workflow. Practically, for computations of decay widths and cross sections, one may follow the following guidelines.

1. Draw all diagrams contributing at the chosen order.
2. Write $-i\mathcal{M}$ from the Feynman rules (propagators, vertices, external lines).
3. Compute $|\mathcal{M}|^2$, performing spin sums/averages when needed.
4. Reduce spinor structures using completeness relations and trace identities.
5. Impose kinematics (often CM frame) and express results via s, t, u when convenient.
6. Integrate over Lorentz-invariant phase space to obtain Γ or σ.

Finally, it is convenient to remark that all trace identities follow from the Clifford algebra $\{\gamma^\mu, \gamma^\nu\} = 2g^{\mu\nu}$. The most frequently used ones are:

$$\mathrm{Tr}(\gamma^{\mu_1}\cdots\gamma^{\mu_{2n+1}}) = 0, \tag{6.379}$$

$$\mathrm{Tr}(\gamma^\mu\gamma^\nu) = 4g^{\mu\nu}, \tag{6.380}$$

$$\mathrm{Tr}(\gamma^\mu\gamma^\nu\gamma^\rho\gamma^\sigma) = 4\big(g^{\mu\nu}g^{\rho\sigma} - g^{\mu\rho}g^{\nu\sigma} + g^{\mu\sigma}g^{\nu\rho}\big). \tag{6.381}$$

These are sufficient to reduce most tree-level QED computations to scalar products of external momenta.

6.13.4 Ward Identity

Gauge invariance under local $U(1)$ transformations $\psi \to e^{i\alpha(x)}\psi$, $A_\mu \to A_\mu - \frac{1}{q}\partial_\mu\alpha$ is not merely a redundancy of description, but a structural principle that fixes the form of the interaction and constrains the quantum theory in a highly non-trivial way.

At the quantum level, the content of gauge invariance is deeper. The local symmetry constrains Green functions and scattering amplitudes, leading to relations known as Ward identities. These identities are the quantum counterpart of the continuity equation and guarantee that the symmetry is preserved order by order in perturbation theory.

To understand their origin pedagogically, consider the time-ordered correlator with one insertion of the electromagnetic current,

$$\langle 0|T\{J^{\mu}(x)\psi(x_1)\bar{\psi}(x_2)\cdots\}|0\rangle. \tag{6.382}$$

Taking the divergence ∂_{μ} and using current conservation, one finds that derivatives acting inside the time-ordering symbol generate contact terms proportional to delta functions. After Fourier transforming, these contact terms become algebraic relations between momentum-space Green functions. In particular, for the fermion two-point function and the fermion-photon vertex function one obtains the Ward-Takahashi identity,

$$k_{\mu}\Gamma^{\mu}(p', p) = S^{-1}(p') - S^{-1}(p), \tag{6.383}$$

where

- $\Gamma^{\mu}(p', p)$ is the full one-particle-irreducible vertex function,
- $S(p)$ is the full fermion propagator,
- $k = p' - p$ is the photon momentum entering the vertex.

Equation (6.383) is an exact statement following directly from gauge invariance. It holds to all orders in perturbation theory provided the regularization procedure respects the symmetry.

When external fermions are placed on-shell, one has

$$S^{-1}(p)u(p) = 0, \qquad \bar{u}(p')S^{-1}(p') = 0, \tag{6.384}$$

and Eq. (6.383) reduces to the on-shell Ward identity

$$k_{\mu}\,\mathcal{M}^{\mu}(k, \ldots) = 0, \tag{6.385}$$

where $\mathcal{M}^{\mu}(k, \ldots)$ denotes the amputated amplitude in which a photon of momentum k is attached to a conserved current insertion. Equation (6.385) is the form most frequently used in practical scattering computations and is the algebraic reason why the longitudinal term in Eq. (6.364) cannot contribute to physical S-matrix elements.

Transversality. The Ward identity in Eq. (6.385) states that physical amplitudes are transverse with respect to the photon momentum. Equivalently, a polarization vector proportional to k_{μ} corresponds to a pure-gauge excitation and decouples from all gauge-invariant observables.

The physical implications are immediate and far-reaching.

First, consider the photon propagator in a general covariant gauge, Eq. (6.364). The second term in Eq. (6.364) is explicitly longitudinal and depends on the arbitrary gauge parameter ξ. Inserting this propagator into a scattering amplitude, one encounters contractions of the form

$$\mathcal{M}^{\mu}(k, \ldots)\, D_{\mu\nu}(k)\, \mathcal{M}^{\nu}(k, \ldots). \tag{6.386}$$

The potentially dangerous longitudinal contribution is proportional to

$$\mathcal{M}^{\mu}(k, \ldots)\, \frac{k_{\mu}k_{\nu}}{k^2}\, \mathcal{M}^{\nu}(k, \ldots). \tag{6.387}$$

However, using Eq. (6.385) twice, one finds

$$k_{\mu}\mathcal{M}^{\mu}(k, \ldots) = 0, \qquad k_{\nu}\mathcal{M}^{\nu}(k, \ldots) = 0, \tag{6.388}$$

so that the entire longitudinal contribution vanishes identically. Therefore, only the transverse part proportional to $g_{\mu\nu}$ survives. This proves that S-matrix elements are independent of the gauge parameter ξ.

Second, the Ward identity constrains renormalization. In QED, one finds that the renormalization constants satisfy

$$Z_1 = Z_2, \tag{6.389}$$

where Z_1 renormalizes the vertex and Z_2 the fermion wave function. This equality is not accidental: it is the renormalized version of the structural relation expressed by Eq. (6.383). It guarantees that charge renormalization is controlled consistently and that gauge symmetry is preserved by the renormalized theory.

Third, the Ward identity ensures the decoupling of unphysical photon polarizations. A photon polarization vector proportional to k_{μ} would correspond to a pure gauge excitation. Because of Eq. (6.385), replacing $\epsilon^{\mu}(k)$ with k^{μ} in any physical amplitude gives zero. Hence only the two transverse helicity states propagate as physical degrees of freedom.

For practical computations one may therefore adopt the particularly convenient Feynman-gauge propagator introduced in Eq. (6.365), confident that physical observables remain unaffected by this choice. The freedom to select a gauge is a computational advantage; the Ward identities are the mechanism that makes this freedom compatible with physical predictivity.

6.14 Proposed Exercises

Exercise 1

Consider the bosonic Fock space $\mathcal{F}$ built from a one-particle Hilbert space $\mathcal{H}_1$ with orthonormal basis $\{|\alpha\rangle\}_{\alpha\geq 1}$. Let $a_\alpha^\dagger$ and a_α denote the associated creation and annihilation operators acting on the occupation-number basis, and let the vacuum $|0\rangle$ satisfy $a_\alpha|0\rangle = 0$ for all α.
Assume a change of one-particle basis

$$|\tilde{\alpha}\rangle = \sum_\beta \langle\beta|\tilde{\alpha}\rangle\,|\beta\rangle, \tag{6.390}$$

with $\langle\tilde{\alpha}|\tilde{\beta}\rangle = \delta_{\tilde{\alpha}\tilde{\beta}}$ and completeness $\sum_\beta |\beta\rangle\langle\beta| = \mathbb{I}$.

(a) Show that the corresponding creation operators satisfy

$$a_{\tilde{\alpha}}^\dagger = \sum_\beta \langle\beta|\tilde{\alpha}\rangle\, a_\beta^\dagger, \qquad a_{\tilde{\alpha}} = \sum_\beta \langle\tilde{\alpha}|\beta\rangle\, a_\beta. \tag{6.391}$$

(b) Prove that the canonical commutation relations are invariant under the basis change, i.e.

$$[a_{\tilde{\alpha}}, a_{\tilde{\beta}}^\dagger] = \delta_{\tilde{\alpha}\tilde{\beta}}, \qquad [a_{\tilde{\alpha}}, a_{\tilde{\beta}}] = [a_{\tilde{\alpha}}^\dagger, a_{\tilde{\beta}}^\dagger] = 0. \tag{6.392}$$

(c) Define the number operator $\hat{n}_\alpha = a_\alpha^\dagger a_\alpha$ and the total number operator $\hat{N} = \sum_\alpha \hat{n}_\alpha$. Show that $\hat{N}$ is basis independent, i.e.

$$\hat{N} = \sum_\alpha a_\alpha^\dagger a_\alpha = \sum_{\tilde{\alpha}} a_{\tilde{\alpha}}^\dagger a_{\tilde{\alpha}}. \tag{6.393}$$

Exercise 2

Consider a *real* Klein-Gordon field ϕ of mass m in Minkowski spacetime, quantized canonically in the Heisenberg picture. The field operator admits the standard mode expansion

$$\hat{\phi}(x) = \int \frac{d^3p}{(2\pi)^{3/2}} \frac{1}{\sqrt{2E_p}} \left(\hat{a}_{\mathbf{p}} e^{-ipx} + \hat{a}_{\mathbf{p}}^\dagger e^{ipx}\right), \qquad E_p = \sqrt{\mathbf{p}^2 + m^2}, \tag{6.394}$$

with the only non-vanishing commutator

$$[\hat{a}_{\mathbf{p}}, \hat{a}_{\mathbf{q}}^\dagger] = (2\pi)^3 \delta^{(3)}(\mathbf{p} - \mathbf{q}). \tag{6.395}$$

(a) Using Eq. (6.394), compute the equal-time commutator $[\hat{\phi}(t, \mathbf{x}), \hat{\phi}(t, \mathbf{y})]$ and show that it vanishes.

(b) Define the conjugate momentum operator $\hat{\pi}(x) = \partial_0 \hat{\phi}(x)$. Show that the canonical equal-time commutator

$$[\hat{\phi}(t, \mathbf{x}), \hat{\pi}(t, \mathbf{y})] = i\,\delta^{(3)}(\mathbf{x} - \mathbf{y}) \tag{6.396}$$

fixes uniquely the normalization factor $1/\sqrt{2E_p}$ in Eq. (6.394).

(c) Define the Pauli-Jordan commutator function

$$\Delta(x - y) = \frac{1}{i}\,[\hat{\phi}(x), \hat{\phi}(y)]. \tag{6.397}$$

Show that it can be written as

$$\Delta(x - y) = \int \frac{d^3p}{(2\pi)^3} \frac{1}{2E_p} \left(e^{-ip(x-y)} - e^{ip(x-y)}\right), \tag{6.398}$$

and verify that $\Delta(x - y)$ is Lorentz invariant.

(d) From the Feynman propagator, $\Delta_F(x - y) = \langle 0\, T\{\hat{\phi}(x)\hat{\phi}(y)\}|0\rangle$, and using Eq. (6.398), show that Δ_F satisfies the inhomogeneous Klein-Gordon equation

$$(\Box_x + m^2)\Delta_F(x - y) = -i\,\delta^{(4)}(x - y), \tag{6.399}$$

and explain the physical meaning of the delta source term in Eq. (6.399).

Exercise 3

In covariant gauges, the photon propagator can be written as

$$D_{\mu\nu}(k) = \frac{-i}{k^2 + i\epsilon}\left(g_{\mu\nu} - (1 - \xi)\frac{k_\mu k_\nu}{k^2}\right), \tag{6.400}$$

where ξ is an arbitrary gauge parameter. Let $\mathcal{M}^\mu(k, \ldots)$ denote an amputated contribution in which a photon of momentum k couples to a conserved current insertion, and assume the on-shell Ward identity

$$k_\mu \mathcal{M}^\mu(k, \ldots) = 0. \tag{6.401}$$

(a) Consider a generic exchange amplitude of the form

$$\mathcal{A} = \mathcal{M}^\mu(k, \ldots)\, D_{\mu\nu}(k)\, \mathcal{N}^\nu(k, \ldots), \tag{6.402}$$

where $\mathcal{N}^\nu$ is another conserved-current amplitude satisfying $k_\nu \mathcal{N}^\nu(k, \ldots) = 0$. Show explicitly that $\mathcal{A}$ is independent of ξ.

(b) In a physical amplitude with an external photon polarization vector $\varepsilon_\mu(k)$, consider the replacement $\varepsilon_\mu(k) \to k_\mu$. Use Eq. (6.401) to explain why this replacement gives a vanishing contribution and interpret the result in terms of unphysical (longitudinal) polarizations.

(c) The Ward-Takahashi identity for the full fermion-photon vertex reads

$$k_\mu \Gamma^\mu(p', p) = S^{-1}(p') - S^{-1}(p), \tag{6.403}$$

with $k = p' - p$. Assuming on-shell external spinors, explain how Eq. (6.403) reduces to Eq. (6.401).

Essential Bibliography

- Peskin ME, Schroeder DV (1995) An introduction to quantum field theory. Addison-Wesley. *A modern classic: perturbation theory, Feynman diagrams, renormalization, and phenomenology, with an operational focus matching the spirit of this chapter.*
- Weinberg S (1995) the quantum theory of fields, vol. I: foundations. Cambridge University Press. *The most systematic modern treatment of QFT foundations: symmetries, locality, fields as operators, and the construction of the S-matrix.*
- Ryder LH (1996) Quantum field theory, 2nd edn. Cambridge University Press. *Pedagogical bridge between relativistic quantum mechanics and field theory, especially useful for understanding the motivation of second quantization.*
- Itzykson C, Zuber J-B (1980) Quantum field theory. McGraw-Hill. *Still unmatched in technical depth on propagators, perturbation theory, and diagrammatic methods; a reference work for serious calculations.*
- Schwartz MD (2014) Quantum field theory and the standard model. Cambridge University Press. *Modern, calculation-oriented approach, with strong emphasis on amplitudes, cross sections, and phenomenological applications.*

7 Quantum Field Theory in Action Beyond Its Foundations

Symmetry, as wide or as narrow as you may define its meaning, is one idea by which man through the ages has tried to comprehend and create order, beauty, and perfection.

Hermann Weyl

Abstract

This chapter equips the reader with the conceptual and technical tools that render quantum field theory an operative framework for physical applications. It begins with discrete symmetries, parity, charge conjugation and time reversal, constructing their operator realization and deriving the CPT theorem together with its immediate consequences for particle and antiparticle properties. The path integral formulation is then introduced as a unifying description of quantum dynamics, from single-particle systems to relativistic fields. Functional measures, generating functionals, propagators, Wick rotation, and the Euclidean formulation are developed explicitly, establishing the connection with statistical mechanics. Finite-temperature field theory is formulated in imaginary time, leading to the Matsubara formalism and its implementation in perturbative analyses. Functional methods are extended to interacting theories, including systematic perturbation theory and the Hubbard–Stratonovich transformation as a tool to expose composite channels. Fermionic functional integrals are constructed using Grassmann variables, yielding propagators and Feynman rules directly from the generating functional. The chapter concludes with symmetry breaking in relativistic and many-body systems, from the Landau paradigm to Goldstone modes, the Higgs and Stueckelberg mechanisms, and Bogoliubov transformations.

O. Luongo, *Fundamental Interactions*,
https://doi.org/10.1007/978-3-032-27413-7_7

Fig. 7.1 Portrait of Richard Phillips Feynman, 1959. Reprinted from Wikimedia. Public domain

A key character: *Richard Phillips Feynman (1918–1988), Fig.* 7.1.

Richard Feynman was one of the most influential theoretical physicists of the twentieth century, whose work reshaped the language and practice of QFT. Born in New York into a family that strongly encouraged scientific curiosity, Feynman showed an early talent for mathematics and physics, developing a problem-oriented and highly intuitive style that later became his scientific trademark.

He studied at the Massachusetts Institute of Technology and then at Princeton University under John A. Wheeler, where he contributed to the absorber theory of radiation. During World War II he played a central role in the Manhattan Project at Los Alamos, where his exceptional ability to solve technical problems rapidly and creatively became widely recognized.

In the late 1940s Feynman introduced the path integral formulation of quantum mechanics, interpreting quantum amplitudes as weighted sums over histories. In parallel he developed the diagrammatic technique now known as Feynman diagrams, which provides a graphical encoding of perturbative expansions in quantum field theory. Together with J. Schwinger and S. Tomonaga, this work led to the modern renormalized formulation of QED, for which the three shared the Nobel Prize in Physics in 1965.

Feynman's influence goes far beyond specific technical results. He transformed the way physicists think about quantum theory, insisting on physical intuition, operational meaning, and calculational effectiveness. His lectures, in particular the *Feynman Lectures on Physics*, educated generations of physicists and established a style in which deep principles such as symmetry, locality, and causality are constantly tested against explicit calculations.

Within the perspective of this chapter, Feynman represents the conceptual bridge between abstract QFT and its concrete applications: discrete symmetries, path integrals, functional methods, symmetry breaking, and quasiparticle descriptions are all naturally organized in a language that owes much to his way of thinking about quantum physics as a theory of processes, amplitudes, and histories.

7.1 From Discrete Symmetries to CPT Theorem: Consequences in Particle Physics

We start this chapter through the notion of discrete symmetries, as anticipated in Chap. 1, applied to the relativistic quantum scenarios, as prompted in Chap. 4. Let us first review, then, the three fundamental discrete transformations in relativistic quantum theory, namely parity (P), charge conjugation (C), and time reversal (T), up to arrive at their combined action leading to the so-called *CPT theorem*. We emphasize the operator properties, their action on Dirac spinors and bilinears, and the most direct physical consequences in particle physics, including parity violation in weak interactions.

7.1.1 Parity

Parity is synonym of spatial inversion and leads to the discrete transformation that reverses spatial coordinates while leaving time unchanged,

$$\begin{cases} t \longrightarrow t, \\ \mathbf{x} \longrightarrow -\mathbf{x}. \end{cases} \tag{7.1}$$

We denote the corresponding operator by P. It is unitary and Hermitian,

$$P^{\dagger} = P^{-1}, \qquad P^2 = \mathbb{1}. \tag{7.2}$$

The geometric meaning of parity is particularly interesting since it leads to the concept of *reflection versus inversion.*

A mirror reflection reverses only the coordinate perpendicular to the mirror plane, leaving the other two unchanged. By contrast, *inversion* reverses all three spatial coordinates, $\mathbf{x} \to -\mathbf{x}$, and cannot be achieved by a pure proper rotation. A convenient way to visualize inversion is as the composition of a mirror reflection with a π-rotation about an axis perpendicular to the mirror: the reflection flips one coordinate, the rotation flips the remaining two, yielding $\mathbf{x} \to -\mathbf{x}$.

Action on Basic Operators.
On the basic quantum-mechanical operators, parity reverses both position and momentum,

$$P^{-1}\,\hat{x}\,P = -\hat{x}, \tag{7.3a}$$

$$P^{-1}\,\hat{p}\,P = -\hat{p}. \tag{7.3b}$$

In particular, parity preserves the canonical commutation relation $[\hat{x}, \hat{p}] = i\hbar$. Indeed,

$$\begin{aligned} P^{-1}[\hat{x}, \hat{p}]P &= P^{-1}(\hat{x}\hat{p} - \hat{p}\hat{x})P \\ &= (P^{-1}\hat{x}P)(P^{-1}\hat{p}P) - (P^{-1}\hat{p}P)(P^{-1}\hat{x}P) \\ &= (-\hat{x})(-\hat{p}) - (-\hat{p})(-\hat{x}) = \hat{x}\hat{p} - \hat{p}\hat{x} = i\hbar. \end{aligned} \tag{7.4}$$

Action on Vectors.
Parity acts on classical fields according to their tensorial character. For a scalar $a(\mathbf{x})$, a four-vector $a^{\mu}(\mathbf{x}, t)$, a rank-two tensor $a^{\mu\nu}(\mathbf{x}, t)$, a pseudoscalar $\tilde{a}(\mathbf{x})$, and a pseudovector $\tilde{a}^{\mu}(\mathbf{x}, t)$, one has respectively

$$P^{-1}a(\mathbf{x})P = a(-\mathbf{x}), \tag{7.5a}$$

$$P^{-1}a^{\mu}(\mathbf{x}, t)P = g^{\mu\mu}\, a^{\mu}(-\mathbf{x}, t), \tag{7.5b}$$

$$P^{-1}a^{\mu\nu}(\mathbf{x}, t)P = g^{\mu\mu}g^{\nu\nu}\, a^{\mu\nu}(-\mathbf{x}, t), \tag{7.5c}$$

$$P^{-1}\tilde{a}(\mathbf{x})P = -\tilde{a}(-\mathbf{x}), \tag{7.5d}$$

$$P^{-1}\tilde{a}^{\mu}(\mathbf{x}, t)P = -g^{\mu\mu}\, \tilde{a}^{\mu}(-\mathbf{x}, t). \tag{7.5e}$$

It is worth stressing the physical distinction between *polar vectors* (ordinary vectors) and *axial vectors* (pseudovectors). If $\mathbf{a}$ and $\mathbf{b}$ are polar vectors, then $\mathbf{c} = \mathbf{a} \times \mathbf{b}$ is an axial vector and does *not* change sign under parity. Likewise, a pseudoscalar can be formed from a scalar triple product and flips sign. In summary,

$$P(\text{scalar}) = +(\text{scalar}), \tag{7.6}$$
$$P(\text{pseudoscalar}) = -(\text{pseudoscalar}), \tag{7.7}$$
$$P(\text{polar vector}) = -(\text{polar vector}), \tag{7.8}$$
$$P(\text{axial vector}) = +(\text{axial vector}). \tag{7.9}$$

Typical examples are the electric field **E** (polar) and the magnetic field **B** and angular momentum **L** (axial):

$$\mathbf{E} \xrightarrow{P} -\mathbf{E}, \qquad \mathbf{B} \xrightarrow{P} +\mathbf{B}, \qquad \mathbf{L} \xrightarrow{P} +\mathbf{L}. \tag{7.10}$$

Parity as a Lorentz Transformation and Its Group Properties.
Equivalently, parity can be represented as a Lorentz transformation

$$P^{\mu}{}_{\nu} = g^{\mu\nu} = \mathrm{diag}(1, -1, -1, -1), \tag{7.11}$$

acting on events as $x^{\mu\prime} = P^{\mu}{}_{\nu} x^{\nu} = (x^0, -\mathbf{x})$. Since $P^2 = \mathbb{1}$, the set $\{\mathbb{1}, P\}$ forms a two-element group under multiplication, isomorphic to $\mathbb{Z}_2$. Consequently, the eigenvalues of P are ± 1. One may therefore speak of *parity-even* and *parity-odd* objects: scalars and pseudovectors carry eigenvalue $+1$, while polar vectors and pseudoscalars carry eigenvalue -1.

Intrinsic Parity.
In relativistic field theory it is natural to attribute to particle species an *intrinsic parity* $\eta_P = \pm 1$, encoding how one-particle states transform under parity (up to the reversal of momentum),

$$P\,|\mathbf{p}, \lambda\rangle = \eta_P\,|-\mathbf{p}, \lambda\rangle, \qquad \eta_P = \pm 1. \tag{7.12}$$

For instance, the photon is an excitation of a massless vector field and has intrinsic parity -1, while the pion is described by a pseudoscalar field and also has intrinsic parity -1. (Later, when fermionic spinor fields are introduced, one finds that the parity of a fermion is opposite to that of its antifermion.)

▶ **Driving Exercise**
Let us now focus on parity and mode operators, through the immediate example of a scalar-field.
To connect the geometrical action $\mathbf{x} \to -\mathbf{x}$ with quantum fields, it is instructive to consider a real scalar field expanded in modes. In the Heisenberg picture one may write schematically

$$\hat{\phi}(t, \mathbf{x}) = \int \frac{d^3p}{(2\pi)^3} \frac{1}{\sqrt{2E_{\mathbf{p}}}} \Big(\hat{a}_{\mathbf{p}}\, e^{-i(E_{\mathbf{p}}t - \mathbf{p}\cdot\mathbf{x})} + \hat{a}^{\dagger}_{\mathbf{p}}\, e^{+i(E_{\mathbf{p}}t - \mathbf{p}\cdot\mathbf{x})} \Big), \tag{7.13}$$

and impose the parity transformation rule appropriate to a scalar,

$$P^{-1}\hat{\phi}(t,\mathbf{x})P = \hat{\phi}(t,-\mathbf{x}). \tag{7.14}$$

Using the mode expansion on the right-hand side, one sees that the equality holds provided the creation and annihilation operators transform by reversing momentum,

$$P^{-1}\hat{a}_{\mathbf{p}}P = \hat{a}_{-\mathbf{p}}, \qquad P^{-1}\hat{a}^{\dagger}_{\mathbf{p}}P = \hat{a}^{\dagger}_{-\mathbf{p}}. \tag{7.15}$$

Thus, for scalar fields, parity acts on the field by $\mathbf{x} \to -\mathbf{x}$ and on the quanta by $\mathbf{p} \to -\mathbf{p}$.

▶ **Driving exercise**

As a natural further step, it is interesting to explore the action of parity on Dirac spinors. On spacetime indices, parity acts with $(+1,-1,-1,-1)$; its action in spinor space is fixed by requiring that the Dirac vector current $\bar{\psi}\gamma^{\mu}\psi$ transform as a four-vector:

$$\bar{\psi}'\gamma^{\mu}\psi' = P^{\mu}{}_{\nu}\,\bar{\psi}\gamma^{\nu}\psi = g^{\mu\mu}\,\bar{\psi}\gamma^{\mu}\psi. \tag{7.16}$$

Here $\psi' = \psi(-\mathbf{x},t)$ and $\psi = \psi(\mathbf{x},t)$. Writing $\psi' = P\psi$ and substituting,

$$(P\psi)^{\dagger}\gamma^{0}\gamma^{\mu}(P\psi) = g^{\mu\mu}\,\bar{\psi}\gamma^{\mu}\psi \quad \Longrightarrow \quad \psi^{\dagger}P^{\dagger}\gamma^{0}\gamma^{\mu}P\psi = g^{\mu\mu}\,\psi^{\dagger}\gamma^{0}\gamma^{\mu}\psi. \tag{7.17}$$

Multiplying appropriately by γ^0 one arrives at the conditions

$$P^{\dagger}\gamma^{0}\gamma^{\mu}P = g^{\mu\mu}\gamma^{0}\gamma^{\mu} \quad \Rightarrow \quad \begin{cases} P^{\dagger}P = \mathbb{1}, \\ P^{\dagger}\alpha_i P = -\alpha_i, \end{cases} \tag{7.18}$$

which leads to

$$P = \pm\gamma^{0}. \tag{7.19}$$

7.1.2 Charge Conjugation

Charge conjugation exchanges particles with antiparticles, leaving spacetime coordinates unchanged. At the level of the mode expansion (schematically for a Dirac field),

$$\hat{\psi}(x) \propto \int d^3p\,\left(\hat{a}_{\mathbf{p}}\,e^{-ipx} + \hat{b}^{\dagger}_{\mathbf{p}}\,e^{+ipx}\right), \tag{7.20a}$$

$$\hat{\psi}^{\dagger}(x) \propto \int d^3p\,\left(\hat{a}^{\dagger}_{\mathbf{p}}\,e^{+ipx} + \hat{b}_{\mathbf{p}}\,e^{-ipx}\right), \tag{7.20b}$$

so that particle creation is associated with $\hat{a}^\dagger$ and antiparticle creation with $\hat{b}^\dagger$.

Charge conjugation is implemented by a *unitary* operator C acting on the Hilbert space,

$$C^\dagger = C^{-1}. \tag{7.21}$$

Its defining action is to exchange particle and antiparticle operators (up to conventional phase choices that can be absorbed in the definition of one-particle states),

$$C^{-1}\hat{a}_{\mathbf{p}}C = \eta_a\,\hat{b}_{\mathbf{p}}, \qquad C^{-1}\hat{b}_{\mathbf{p}}C = \eta_b\,\hat{a}_{\mathbf{p}}, \tag{7.22}$$

with $|\eta_a| = |\eta_b| = 1$. In many conventions one chooses $\eta_a = \eta_b = 1$ and $C^2 = \mathbb{1}$, but neither assumption is essential for the algebraic statements below: only unitarity and the exchange property in Eq. (7.22) matter.

Two different "C" objects. The symbol C is commonly used for two distinct entities:

- a *unitary Hilbert-space operator* implementing charge conjugation on fields, as in Eq. (7.21) and Eq. (7.22);
- a *numerical* 4×4 *matrix* (the *charge-conjugation matrix*) acting on spinor indices, introduced below to relate ψ and its conjugate.

To avoid confusion one may denote the matrix as C, but we keep the standard notation and rely on context.

Charge conjugation acts on the Dirac field by conjugation with the unitary operator,

$$\hat{\psi}(x) \xrightarrow{C} \hat{\psi}_C(x) \equiv C^{-1}\hat{\psi}(x)\,C. \tag{7.23}$$

The field $\hat{\psi}_C$ must transform as a Dirac spinor under Lorentz transformations and must satisfy the Dirac equation with the *opposite* gauge charge. These requirements uniquely fix its relation to $\hat{\psi}$ at the level of spinor indices. The standard definition is

$$\hat{\psi}_C(x) \equiv C\,\overline{\hat{\psi}}(x)^T, \qquad \overline{\hat{\psi}} \equiv \hat{\psi}^\dagger\gamma^0, \tag{7.24}$$

where T denotes transpose in spinor space and the 4×4 matrix C acts on Dirac indices only. Using Eq. (7.24), one also obtains

$$\overline{\hat{\psi}_C}(x) = -\,\hat{\psi}(x)^T C^{-1}. \tag{7.25}$$

Equations (7.24) and (7.25) are the appropriate starting point for deriving the transformation properties of bilinears.

Physical content. The definition in Eq. (7.24) is designed so that, in the mode expansion, particle and antiparticle operators are exchanged in the sense of Eq. (7.22). In particular, $\hat{\psi}_C$ creates antiparticles where $\hat{\psi}$ creates particles, consistently with the interpretation of charge conjugation as a map between sectors of opposite $U(1)$ charge.

Driving Exercise
To determine the matrix C acting in spinor space, one imposes covariance of the Dirac equation under charge conjugation. A clean way is to start from the Dirac equation for ψ,

$$(i\partial\!\!\!/ - m)\psi = 0, \tag{7.26}$$

take the complex conjugate, and then use Eq. (7.24) to rewrite the result as an equation for ψ_C.
Show that Eq. (7.26) implies

$$\left(i\gamma^{\mu\star}\partial_\mu - m\right)\psi^\star = 0, \tag{7.27}$$

and that requiring ψ_C to satisfy again the Dirac equation,

$$(i\partial\!\!\!/ - m)\psi_C = 0, \tag{7.28}$$

forces the defining condition

$$C^{-1}\gamma^\mu C = -\gamma^{\mu T}. \tag{7.29}$$

Verify explicitly that in the standard Dirac representation one may choose

$$C = i\gamma^2\gamma^0, \tag{7.30}$$

up to an irrelevant overall phase, and that $C^T = -C$ and $C^\dagger C = \mathbb{1}$ hold.

Driving Exercise
The action of charge conjugation on bilinear covariants is particularly interesting, because it provides an efficient classification of interaction terms by their C-parity.
Using Eq. (7.24) and Eq. (7.25), show that for a generic Dirac matrix Γ one has the identity

$$\overline{\psi_C}\,\Gamma\,\psi_C = \bar{\psi}\left(C^{-1}\Gamma^T C\right)\psi. \tag{7.31}$$

Equivalently, the bilinear $\bar{\psi}\,\Gamma\,\psi$ transforms under charge conjugation according to

$$\bar{\psi}\,\Gamma\,\psi \xrightarrow{C} \bar{\psi}\,(C^{-1}\Gamma^T C)\,\psi. \tag{7.32}$$

In other words, charge conjugation induces the mapping

$$\begin{cases} \psi \longrightarrow \psi_C, \\ \Gamma \longrightarrow C^{-1}\Gamma^T C. \end{cases} \tag{7.33}$$

Use Eqs. (7.29) and (7.33) to determine the C-parity of the standard bilinears. Show that

$$\bar{\psi}\psi \xrightarrow{C} +\bar{\psi}\psi, \tag{7.34}$$

$$\bar{\psi}\gamma^\mu\psi \xrightarrow{C} -\bar{\psi}\gamma^\mu\psi, \tag{7.35}$$

$$\bar{\psi}\gamma^\mu\gamma^5\psi \xrightarrow{C} +\bar{\psi}\gamma^\mu\gamma^5\psi, \tag{7.36}$$

$$\bar{\psi}\gamma^5\psi \xrightarrow{C} +\bar{\psi}\gamma^5\psi, \tag{7.37}$$

$$\bar{\psi}\sigma^{\mu\nu}\psi \xrightarrow{C} -\bar{\psi}\sigma^{\mu\nu}\psi, \tag{7.38}$$

where $\sigma^{\mu\nu} \equiv \frac{i}{2}[\gamma^\mu, \gamma^\nu]$.

▶ **Pause and think**
Try to explain, from the above, the emerging concept of *intrinsic charge conjugation.*

For neutral fields, one may define an intrinsic charge-conjugation parity $\eta_C = \pm 1$ via

$$C\,\phi(x)\,C^{-1} = \eta_C\,\phi(x). \tag{7.39}$$

This quantum number is meaningful only for self-conjugate fields and plays an important role in selection rules and in the classification of mesonic states.

As it will be clear while studying weak and strong interactions, Chaps. 11 and 12, in the quark model, a neutral meson is described as a bound state of a quark and its corresponding antiquark, $|q\bar{q}\rangle$. Under charge conjugation one has $C\,|q\bar{q}\rangle = |\bar{q}q\rangle$, so that only those flavor combinations that are symmetric or antisymmetric under the interchange $q \leftrightarrow \bar{q}$ can be eigenstates of C. For such states the eigenvalue is fixed by $\eta_C = (-1)^{L+S}$, where L is the orbital angular momentum of the $q\bar{q}$ pair and S their total spin. This relation explains, for example, why the pseudoscalar mesons ($L = 0$, $S = 0$) such as π^0 and η have $\eta_C = +1$, while the vector mesons ($L = 0$, $S = 1$) such as ρ^0, ω and J/ψ have $\eta_C = -1$.

The concept of intrinsic C-parity is therefore not an abstract quantum number, but a concrete tool for organizing the meson spectrum, identifying allowed and forbidden decay channels, and formulating selection rules for electromagnetic and strong processes, which conserve charge conjugation to a very good approximation.

7.1.3 Time Reversal

Define time reversal by

$$\begin{cases} t \longrightarrow -t, \\ \mathbf{x} \longrightarrow \mathbf{x}. \end{cases} \tag{7.40}$$

(We adopt here the standard convention: time is reversed while space is not. If one instead combines time reversal with parity, one would flip both t and $\mathbf{x}$.)

The time-reversal operator $\hat{T}$ commutes with $\hat{x}$ but anticommutes with $\hat{p}$, reflecting the reversal of motion:

$$\hat{T}^{-1}\hat{x}\,\hat{T} = \hat{x}, \tag{7.41a}$$

$$\hat{T}^{-1}\hat{p}\,\hat{T} = -\hat{p}. \tag{7.41b}$$

For the canonical commutator $[\hat{x}, \hat{p}] = i\hbar$ to be preserved, $\hat{T}$ must send $i \to -i$, i.e. it must involve complex conjugation and hence be antiunitary. Indeed,

$$\begin{aligned} \hat{T}^{-1}[\hat{x}, \hat{p}]\hat{T} &= \hat{T}^{-1}\hat{x}\hat{T}\,\hat{T}^{-1}\hat{p}\hat{T} - \hat{T}^{-1}\hat{p}\hat{T}\,\hat{T}^{-1}\hat{x}\hat{T} \\ &= \hat{x}(-\hat{p}) - (-\hat{p})\hat{x} = -(\hat{x}\hat{p} - \hat{p}\hat{x}) = -[\hat{x}, \hat{p}], \end{aligned} \tag{7.42}$$

which is consistent provided

$$\hat{T}^{-1}i\,\hat{T} = -i, \qquad \Rightarrow \qquad \hat{T}^{-1}[\hat{x}, \hat{p}]\hat{T} = \hat{T}^{-1}i\hat{T} = -i = -[\hat{x}, \hat{p}]. \tag{7.43}$$

The time-reversal operator is not a real linear operator: it is antiunitary and necessarily involves complex conjugation.

Driving Exercise

The action of time reversal on Dirac spinors, analogously for parity, can be immediately computed. We can determine the spinor action of T by requiring the proper transformation of a vector current, i.e., let

$$\bar{\psi}'\gamma^{\mu}\psi' = T^{\mu}{}_{\nu}\,\bar{\psi}\gamma^{\nu}\psi, \tag{7.44}$$

with $T^{\mu}{}_{\nu} = \mathrm{diag}(-1, 1, 1, 1) = -g^{\mu\nu}$. Write

$$\psi' = \psi(\mathbf{x}, -t) = \mathcal{T}\,\psi^{\star}(\mathbf{x}, t), \tag{7.45}$$

where we emphasize antiunitarity by separating complex conjugation from a unitary matrix (heuristically $\mathcal{T} = k\,U$ with k the conjugation part). Imposing the current transformation gives

$$(\mathcal{T}\psi^{\star})^{\dagger}\gamma^{0}\gamma^{\mu}(\mathcal{T}\psi^{\star}) = -g^{\mu\mu}\,\bar{\psi}\gamma^{\mu}\psi \quad \Longrightarrow \quad \psi^{T}\mathcal{T}^{\dagger}\gamma^{0}\gamma^{\mu}\mathcal{T}\psi^{\star} = -g^{\mu\mu}\,\bar{\psi}\gamma^{\mu}\psi. \tag{7.46}$$

Then,

$$\bar{\psi}\gamma^{\mu}\psi = \psi^{\dagger}\gamma^{0}\gamma^{\mu}\psi = -\psi^{T}(\gamma^{\mu})^{T}\gamma^{0}\psi^{\star}, \tag{7.47}$$

and matching both sides yields

$$g^{\mu\mu}\,(\gamma^{\mu})^{T}\gamma^{0} = \mathcal{T}^{\dagger}\gamma^{0}\gamma^{\mu}\mathcal{T} \quad \Longrightarrow \quad g^{\mu\mu}\gamma^{0}\gamma^{\mu\star} = \mathcal{T}^{\dagger}\gamma^{0}\gamma^{\mu}\mathcal{T}. \tag{7.48}$$

This implies

$$\begin{cases} (\gamma^{0})^{2} = \mathcal{T}^{\dagger}\mathcal{T} \Rightarrow \mathcal{T}^{-1}\gamma^{0} = \gamma^{0}\mathcal{T}^{\dagger}, \\ -\alpha_{i}^{\star} = \mathcal{T}^{\dagger}\alpha_{i}\mathcal{T} \Rightarrow \alpha_{2} = \mathcal{T}^{\dagger}\alpha_{2}\mathcal{T}, \quad -\alpha_{1,3} = \mathcal{T}^{\dagger}\alpha_{1,3}\mathcal{T}. \end{cases} \tag{7.49}$$

A consistent choice is

$$\mathcal{T} = \pm i\gamma^{1}\gamma^{3} = \sigma_{2}, \tag{7.50}$$

i.e. a matrix that flips the spatial gamma matrices appropriately while leaving γ^{0} invariant.[1]

Now, the action on states, i.e., spinors and mode operators, can be investigated as well.

Collecting the transformations on Dirac spinors that will be used to relate particle and antiparticle states,

$$P^{-1}\psi(\mathbf{x}, t)P = \gamma^{0}\psi(-\mathbf{x}, t), \tag{7.51a}$$
$$C^{-1}\psi(\mathbf{x}, t)C = i\gamma^{2}\psi^{\dagger}(\mathbf{x}, t), \tag{7.51b}$$
$$T^{-1}\psi(\mathbf{x}, t)T = \sigma_{2}\psi^{\star}(\mathbf{x}, t). \tag{7.51c}$$

Expanding ψ in normal modes and equating terms on both sides, one obtains the induced transformations of the mode amplitudes,

$$P^{-1}a_{\mathbf{p}}P = a_{-\mathbf{p}}, \qquad P^{-1}b_{\mathbf{p}}P = -b_{-\mathbf{p}}, \tag{7.52a}$$
$$C^{-1}a_{\mathbf{p}}C = i\,b_{\mathbf{p}}, \qquad C^{-1}b_{\mathbf{p}}C = -i\,a_{\mathbf{p}}, \tag{7.52b}$$
$$T^{-1}a_{\mathbf{p}}T = a_{\mathbf{p}}^{\star}, \qquad T^{-1}b_{\mathbf{p}}T = b_{\mathbf{p}}^{\star}. \tag{7.52c}$$

[1] In what follows, it should be clear from the context whether σ_2 denotes the 4×4 matrix defined in Eq. (7.50) or the standard Pauli matrix.

Driving Exercise
In the same way, we can study the action of time-reversal on a complex scalar field. Starting from

$$\hat{\phi}(\mathbf{x}, t) = \int \frac{d^3p}{(2\pi)^{3/2}(2E_\mathbf{p})^{1/2}} \left[\hat{a}_\mathbf{p} e^{-ipx} + \hat{b}^\dagger_\mathbf{p} e^{+ipx}\right]. \tag{7.53}$$

Interestingly, let us now examine how $\hat{P}, \hat{C}$, and $\hat{T}$ act.
Applying parity,

$$\hat{P}^{-1}\hat{\phi}(\mathbf{x}, t)\hat{P} = \int \frac{d^3p}{(2\pi)^{3/2}(2E_\mathbf{p})^{1/2}} \left[\hat{P}^{-1}\hat{a}_\mathbf{p}\hat{P}\, e^{-ipx} + \hat{P}^{-1}\hat{b}^\dagger_\mathbf{p}\hat{P}\, e^{+ipx}\right]. \tag{7.54}$$

After the change of variables $\mathbf{p} \to -\mathbf{p}$ in the first term, equality with $\hat{\phi}(-\mathbf{x}, t)$ holds provided

$$\hat{P}^{-1}\hat{a}_\mathbf{p}\hat{P} = \hat{a}_{-\mathbf{p}}, \quad \hat{P}^{-1}\hat{b}^\dagger_\mathbf{p}\hat{P} = \hat{b}^\dagger_{-\mathbf{p}}. \tag{7.55}$$

Charge conjugation exchanges particle and antiparticle operators. Requiring $\hat{C}^{-1}\hat{\phi}\,\hat{C} = \hat{\phi}^\dagger$ gives

$$\hat{C}^{-1}\hat{a}_\mathbf{p}\hat{C} = \hat{b}_\mathbf{p}, \quad \hat{C}^{-1}\hat{b}^\dagger_\mathbf{p}\hat{C} = \hat{a}^\dagger_\mathbf{p}. \tag{7.56}$$

For a real field, particle and antiparticle coincide: $\hat{a}_\mathbf{p} = \hat{b}_\mathbf{p}$ and $\hat{a}^\dagger_\mathbf{p} = \hat{b}^\dagger_\mathbf{p}$.
Under time reversal one has $t \to -t$, and in practice this corresponds to reversing the sign in the exponential factors. One may also note that combining parity and time reversal implies $\mathbf{p} \to -\mathbf{p}$ when comparing plane-wave phases. In this setting, creation and annihilation operators can remain invariant under PT:

$$(\hat{P}\hat{T})^{-1}\hat{a}_\mathbf{p}(\hat{P}\hat{T}) = \hat{a}_\mathbf{p}, \quad (\hat{P}\hat{T})^{-1}\hat{b}^\dagger_\mathbf{p}(\hat{P}\hat{T}) = \hat{b}^\dagger_\mathbf{p}. \tag{7.57}$$

Furthermore, comparing

$$\begin{aligned} \hat{T}^{-1}\hat{\phi}(\mathbf{x}, t)\hat{T} &= \int \frac{d^3p}{(2\pi)^{3/2}(2E_\mathbf{p})^{1/2}} \left[\hat{T}^{-1}\hat{a}_\mathbf{p}\hat{T}\, e^{-ipx} + \hat{T}^{-1}\hat{b}^\dagger_\mathbf{p}\hat{T}\, e^{+ipx}\right] \\ &= \hat{\phi}^\dagger(\mathbf{x}, -t) = \int \frac{d^3p}{(2\pi)^{3/2}(2E_\mathbf{p})^{1/2}} \left[\hat{a}^\dagger_\mathbf{p} e^{-ipx} + \hat{b}_\mathbf{p} e^{+ipx}\right], \end{aligned} \tag{7.58}$$

one infers

$$\hat{T}^{-1}\hat{a}_\mathbf{p}\hat{T} = \hat{a}^\dagger_\mathbf{p}, \quad \hat{T}^{-1}\hat{b}^\dagger_\mathbf{p}\hat{T} = \hat{b}_\mathbf{p}. \tag{7.59}$$

We have now all the ingredients to go through the so called *CPT theorem*, i.e., the ordered combination of the three above-presented discrete symmetries.

7.1.4 The CPT Theorem and Its Consequences

Let $S_{ab} = \bar{\psi}_a(\mathbf{x}, t)\Gamma\psi_b(\mathbf{x}, t)$ be a generic bilinear built from two fields ψ_a and ψ_b.

The actions of P, C, and T on such bilinears can be summarized as

$$P^{-1}S_{ab}(\mathbf{x}, t)P = \eta_P(S)\, S_{ab}(-\mathbf{x}, t), \tag{7.60a}$$

$$C^{-1}S_{ab}(\mathbf{x}, t)C = \eta_C(S)\, S_{ba}(\mathbf{x}, t), \tag{7.60b}$$

$$T^{-1}S_{ab}(\mathbf{x}, t)T = \eta_T(S)\, S_{ab}(\mathbf{x}, -t), \tag{7.60c}$$

where $\eta_{P,C,T}(S) = \pm 1$ depend both on the transformation and on the specific bilinear structure.

Define $\theta = CPT$. The signs for standard bilinears (scalar S, pseudoscalar P, vector V, axial vector A, and tensor T) may be organized as in Table 7.1. The CPT action can be represented as an antilinear operator

$$\theta\big[g_S\, S_{ab}(\mathbf{x})\big]\theta^\dagger = (-1)^N\, g_S^\star\, \big(S_{ab}(-\mathbf{x})\big)^\dagger, \tag{7.61}$$

where g_S is a complex coefficient and N is the number of Lorentz indices carried by the bilinear $\bar{\psi}\Gamma\psi$ (so that index-dependent sign factors are tracked).

To see explicitly the action on a spinor field, apply T, then P, then C:

$$\begin{aligned}(CPT)\,\psi(\mathbf{x})\,(CPT)^\dagger &= C\,P\big(\sigma_2\psi^\star(\mathbf{x}, -t)\big)P\,C \\ &= C\big(\sigma_2\gamma^0\psi^\star(-\mathbf{x}, t)\big)C = \sigma_2\gamma^0(i\gamma^2)\,\psi^T(-\mathbf{x}).\end{aligned} \tag{7.62}$$

Defining $\theta = CPT$, this can be written in the compact form

$$\theta\,\psi(\mathbf{x})\,\theta^\dagger = i\gamma^5\,\psi^\dagger(-\mathbf{x}), \tag{7.63}$$

Table 7.1 Sign factors for the standard Dirac bilinears under parity P, charge conjugation C, time reversal T, and their product CPT

	S	P	V	A	T
η_P	$+1$	-1	$g^{\mu\mu}$	$-g^{\mu\mu}$	$g^{\mu\mu}g^{\nu\nu}$
η_C	$+1$	$+1$	-1	$+1$	-1
η_T	$+1$	-1	$-g^{\mu\mu}$	$g^{\mu\mu}$	$-g^{\mu\mu}g^{\nu\nu}$
η_{CPT}	$+1$	$+1$	-1	-1	$+1$

where the matrix identity used is

$$\begin{aligned}\sigma_2\gamma^0(i\gamma^2) &= (i\gamma^1\gamma^3)\gamma^0(i\gamma^2) = -(\gamma^1\gamma^3\gamma^0\gamma^2)\\ &= -(\gamma^1(-\gamma^0\gamma^3)\gamma^2) = -(-\gamma^0\gamma^1)(-\gamma^2\gamma^3)\\ &= -\gamma^0\gamma^1\gamma^2\gamma^3 = -(-i\gamma^5) = i\gamma^5. \end{aligned}\tag{7.64}$$

Hence, CPT maps a state into (up to a phase) its chirality operator times Hermitian conjugation and inversion of coordinates.

Now apply θ to a bilinear $S_{ab} = \bar{\psi}_a\Gamma\psi_b$:

$$\begin{aligned}\theta S_{ab}\theta^\dagger &= \theta\bar{\psi}_a\Gamma\psi_b\theta^\dagger = \theta\psi_a^\dagger\gamma^0\Gamma\psi_b\theta^\dagger\\ &= \theta\psi_a^\dagger\theta^\dagger\ \theta(\gamma^0\Gamma)\theta^\dagger\ \theta\psi_b\theta^\dagger\\ &= -\theta\psi_a^\dagger\theta^\dagger\ (\gamma^0\Gamma)^\star\,\theta\psi_b\theta^\dagger\\ &= -\left[i\gamma^5\psi_a(-\mathbf{x})\right]\ (\gamma^0\Gamma)^\dagger\ \left[i\gamma^5\psi_b^\dagger(-\mathbf{x})\right]\\ &= \left[\gamma^5\psi_a(-\mathbf{x})\right]\ (\gamma^0\Gamma)^\star\ \left[\gamma^5\psi_b^\dagger(-\mathbf{x})\right], \end{aligned}\tag{7.65}$$

where we used that, under T and P, time and space indices pick up opposite signs so that the $g^{\mu\mu}$ factors cancel, and C contributes complex conjugation. Swapping transpose pairs and using anticommutation rules when required, we obtain

$$\theta\,\bar{\psi}_a\Gamma\psi_b\,\theta^\dagger = -\psi_b^\dagger(-\mathbf{x})\gamma^{5T}\left((\gamma^0\Gamma)^\star\right)^T\gamma^5\psi_a(-\mathbf{x}) = -\psi_b^\dagger(-\mathbf{x})\gamma^5(\gamma^0\Gamma)^\dagger\gamma^5\psi_a(-\mathbf{x}). \tag{7.66}$$

Since $\{\gamma^0, \gamma^5\} = 0$, this becomes

$$\theta\,\bar{\psi}_a\Gamma\psi_b\,\theta^\dagger = \bar{\psi}_b(-\mathbf{x})\,\gamma^0\gamma^5\Gamma^\dagger\gamma^5\gamma^0\,\psi_a(-\mathbf{x}). \tag{7.67}$$

Furthermore,

$$\gamma^0\gamma^5\Gamma^\dagger\gamma^5\gamma^0 = \gamma^0(\gamma^5\Gamma\gamma^5)^\dagger\gamma^0 = (-1)^N\gamma^0\Gamma^\dagger\gamma^0 = (-1)^N\Gamma, \tag{7.68}$$

so inserting this into Eq. (7.67) gives

$$\theta\,\bar{\psi}_a\Gamma\psi_b\,\theta^\dagger = (-1)^N\,\bar{\psi}_b(-\mathbf{x})\Gamma\psi_a(-\mathbf{x}) = (-1)^N\Big[\bar{\psi}_a(-\mathbf{x})\Gamma\psi_b(-\mathbf{x})\Big]^\dagger. \tag{7.69}$$

Thus θ implements Hermitian conjugation together with a complete inversion of the coordinates. It is a proper isometry of Euclidean $\mathbb{R}^4$, though not a proper orthochronous Lorentz transformation in Minkowski spacetime (since it would require $\Lambda^0_{\ 0} > 0$).

7.1.5 Pauli-Lüders Formulation of the CPT theorem

The CPT transformation $\theta = CPT$ is a symmetry of any relativistic QFT. Equivalently, the Lagrangian density must be invariant under total inversion and Hermitian conjugation in order to describe a consistent relativistic QFT.

Driving Exercise

To outline the theorem, consider a relativistic QFT Lagrangian $\mathcal{L}(\mathbf{x})$ satisfying:

1. $\mathcal{L}^\dagger = \mathcal{L}$ (Hermiticity);
2. locality: $\mathcal{L} = \mathcal{L}(\phi, \partial_\mu\phi, \ldots)$ with a finite number of derivatives;
3. fermions appear in bilinears $\bar{\psi}\Gamma\psi$ (hence an even number of ψ's);
4. Lorentz invariance;
5. normal ordering.

One must show

$$\theta\,\mathcal{L}(\mathbf{x})\,\theta^\dagger = \mathcal{L}(-\mathbf{x}), \tag{7.70}$$

i.e. invariance up to full inversion. Consequently, the action is invariant:

$$\theta S\theta^\dagger = \int d^4x\,\theta\mathcal{L}\theta^\dagger = \int d^4x\,\mathcal{L}(-\mathbf{x}) = \int d^4(-\mathbf{x})\,\mathcal{L}(-\mathbf{x}) = S. \tag{7.71}$$

Thus the dynamics is unchanged and CPT is a symmetry.
To see how Eq. (7.70) arises, write the Lagrangian as

$$\mathcal{L}(\mathbf{x}) = \sum_i c_i\, O_i(\mathbf{x}), \tag{7.72}$$

with complex coefficients c_i, where each O_i is a normal-ordered local operator built from fields and derivatives, schematically

$$O_i(\mathbf{x}) =: \cdots A^\mu(\mathbf{x}) \cdots (\bar{\psi}_a\Gamma\psi_b) \cdots \partial^\nu \cdots :, \tag{7.73}$$

for instance interaction terms such as $A_\mu J^\mu = A_\mu\bar{\psi}_a\gamma^\mu\psi_b$, mass terms $\bar{\psi}_a m\psi_b$, and derivative bilinears $\bar{\psi}_a\partial_\mu\partial^\mu\psi_b$, with all indices fully contracted if $\mathcal{L}$ is a Lorentz scalar.
Applying θ yields

$$\theta\,\mathcal{L}(\mathbf{x})\,\theta^\dagger = \sum_i (c_i)^\star(-1)^{N_{\text{tot}}} : \cdots A^\mu(-\mathbf{x}) \cdots \left[(\bar{\psi}_a\Gamma\psi_b)(-\mathbf{x})\right]^\dagger \cdots \partial^\nu \cdots :, \tag{7.74}$$

where N_{tot} is the total number of Lorentz indices appearing in O_i. Normal ordering is compatible with conjugation here, e.g.

$$:\left(\bar{\psi}_a\Gamma\psi_b\right)^\dagger :=: \psi_b^\dagger\Gamma^\dagger\gamma^0\psi_a := \left(\bar{\psi}_a\Gamma\psi_b\right)^\dagger. \tag{7.75}$$

Therefore

$$\theta\,\mathcal{L}(\mathbf{x})\,\theta^\dagger = \sum_i (-1)^{N_{\rm tot}}\big[c_i\,O_i(-\mathbf{x})\big]^\dagger = \big(\mathcal{L}(-\mathbf{x})\big)^\dagger = \mathcal{L}(-\mathbf{x}), \qquad (7.76)$$

where we used Lorentz invariance (all indices contracted in a scalar Lagrangian) and Hermiticity. This implies in particular that $(-1)^{N_{\rm tot}} = +1$, i.e. the total number of involved indices is even.

7.1.5.1 Consequences of the CPT Theorem

In any relativistic QFT, CPT symmetry implies a complete correspondence between particle and antiparticle states. From $\theta\mathcal{L}(\mathbf{x})\theta^\dagger = \mathcal{L}(-\mathbf{x})$ one infers $\theta\mathcal{H}(\mathbf{x})\theta^\dagger = \mathcal{H}(-\mathbf{x})$ and hence invariance of the Hamiltonian:

$$\theta H\theta^\dagger = \theta\int d^3x\,\mathcal{H}(\mathbf{x})\theta^\dagger = \int d^3(-\mathbf{x})\,\mathcal{H}(-\mathbf{x}) = H. \qquad (7.77)$$

Further consequences include:

- $\theta\hat{\mathbf{P}}\theta^\dagger = \hat{\mathbf{P}}$, since momentum is a polar vector and flips under total inversion;
- $\theta\,\hat{Q}\theta^\dagger = -\hat{Q}$, because P and T do not change charge while C does. Indeed, acting on a charge eigenstate $|\psi\rangle$ with $\hat{Q}|\psi\rangle = q|\psi\rangle$,

$$\begin{cases} \hat{C}\,\hat{Q}|\psi\rangle = \hat{C}\,(q|\psi\rangle) = q|\bar{\psi}\rangle, \\ \hat{Q}\,\hat{C}|\psi\rangle = \hat{Q}|\bar{\psi}\rangle = -q|\bar{\psi}\rangle, \end{cases} \qquad (7.78)$$

  so $\hat{C}^{-1}\hat{Q}\hat{C} = -\hat{Q}$, and therefore also $\theta\,\hat{Q}\theta^\dagger = -\hat{Q}$;
- $\theta\hat{\mathbf{J}}\theta^\dagger = -\hat{\mathbf{J}}$, since angular momentum is an axial vector (a cross product of polar vectors), unchanged by parity and therefore also unchanged under CPT.

A key physical implication is that in a CPT-invariant theory, antiparticles have the same mass as their corresponding particles. For a massive particle consider the rest frame $\mathbf{p} = 0$, where $\mathbf{J} = \mathbf{S}$. A state is characterized by mass m, momentum $\mathbf{p}$, spin s, and charge q. Given

$$|A\rangle = |m, \mathbf{p} = 0, s, q\rangle, \qquad (7.79)$$

its CPT-conjugate must exist,

$$\theta|A\rangle = |m, \mathbf{p} = 0, s, -q\rangle, \qquad (7.80)$$

with the same m but opposite charge. Two cases follow:

1. if $q = 0$, then $|A\rangle = \theta|A\rangle$ and the particle is its own antiparticle. Examples include neutral mesons such as π^0 and neutral spin-1 mesons such as ρ^0 and ω^0;
2. if $q \neq 0$, then $\theta|A\rangle$ is a distinct antiparticle state with opposite charge but identical mass. Electrons provide the prototypical example.

For massless particles, one characterizes states by helicity. Choose a frame with momentum along z, and define helicity

$$\lambda = \mathbf{S} \cdot \hat{\mathbf{P}} = \frac{\mathbf{S} \cdot \mathbf{p}}{|\mathbf{p}|} = \pm 1. \tag{7.81}$$

For a state $|A\rangle = |p_z, \lambda, q\rangle$, CPT invariance requires the existence of

$$\theta|A\rangle = |-p_z, -\lambda, -q\rangle. \tag{7.82}$$

Again:

1. if $q = 0$, a neutral massless particle must exist in two helicity states. This is the photon, with two polarizations;
2. if $q \neq 0$, the CPT-conjugate is the antiparticle with opposite charge and opposite helicity.

Finally, experimental evidence supports additional CPT-related equalities such as identical lifetimes for particles and antiparticles, and magnetic moments equal in magnitude but opposite in sign.

7.2 Feynman's Path Integrals

The path integral formulation reorganizes quantum dynamics as a weighted sum over histories, see Fig. 7.2. In quantum mechanics (QM) the histories are trajectories $q(t)$ in configuration space, whereas in QFT they are field configurations $\phi(x)$ on spacetime. In both cases, amplitudes are obtained by summing contributions of the form $\exp\left(\frac{i}{\hbar}S\right)$, with S the classical action functional associated with the history (it is convenient to retain $\hbar$ in this Section, where needful, for pedagogical reasons).

Two conceptual consequences are central for applications.

1. The classical limit emerges from a stationary-phase mechanism: contributions localize around the classical path(s) of least action, with quantum corrections supplied by fluctuations about them.
2. The same formalism admits a Euclidean (imaginary-time) continuation in which oscillatory weights become exponentially damped. This maps quantum ampli-

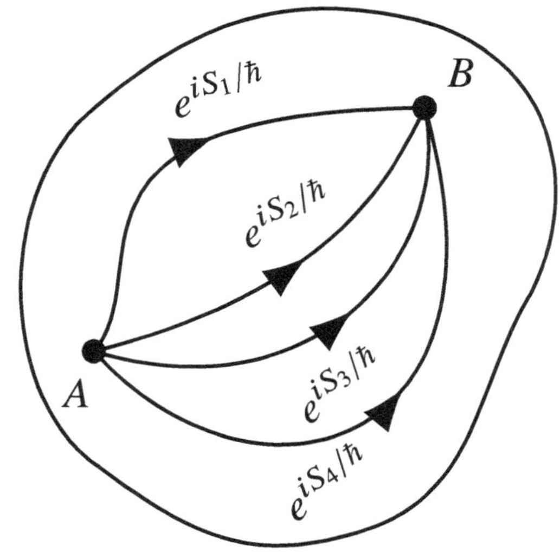

Fig. 7.2 Feynman's revolutionary insight was reformulating quantum dynamics as a sum over histories: instead of a single trajectory, every conceivable path between the initial and final configurations contributes to the quantum amplitude. Each path is weighted by a phase determined by the action. Classical trajectories appear only as the paths for which the phase varies least, surviving constructive interference

tudes onto statistical weights, thereby linking QFT correlators to partition functions of statistical field theory.

7.2.1 Quantum Mechanics as a Sum over Trajectories and Time Slicing

Consider the propagator (kernel) for a single nonrelativistic particle in one dimension,

$$G(q_b, t_b; q_a, t_a) \equiv \left\langle q_b | e^{-\frac{i}{\hbar}\hat{H}(t_b - t_a)} | q_a \right\rangle, \qquad \hat{H} = \frac{\hat{p}^2}{2m} + V(\hat{q}), \tag{7.83}$$

which is the probability amplitude for a particle to start at q_a at time t_a and be found at q_b at time t_b.

Introduce a time slicing: $t_b - t_a \equiv T = N\Delta t$, with intermediate times $t_n = t_a + n\Delta t$ and intermediate positions q_n, where $q_0 = q_a$ and $q_N = q_b$, see Fig. 7.3. Insert $N - 1$ resolutions of the identity,

$$\mathbf{1} = \int_{-\infty}^{+\infty} dq_n \, |q_n\rangle \langle q_n| \,, \tag{7.84}$$

to rewrite the propagator as a product of *mini-propagators*,

$$G = \int \prod_{n=1}^{N-1} dq_n \prod_{n=0}^{N-1} \left\langle q_{n+1} | e^{-\frac{i}{\hbar}\hat{H}\Delta t} | q_n \right\rangle. \tag{7.85}$$

For sufficiently small Δt one uses the Trotter product approximation,

$$e^{-\frac{i}{\hbar}\hat{H}\Delta t} \simeq e^{-\frac{i}{\hbar}\frac{\hat{p}^2}{2m}\Delta t} \, e^{-\frac{i}{\hbar}V(\hat{q})\Delta t} \quad (\Delta t \to 0), \tag{7.86}$$

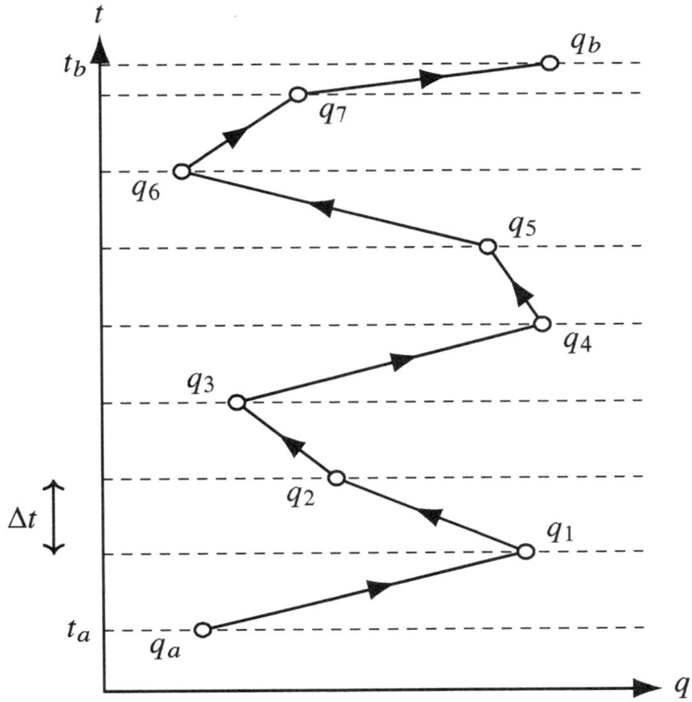

Fig. 7.3 Discretizing a trajectory into time slices from (t_a, q_a) to (t_b, q_b)

with corrections of order $O((\Delta t)^2)$. Insert momentum completeness at each slice,

$$\mathbf{1} = \int \frac{dp_n}{2\pi\hbar}\, |p_n\rangle\langle p_n|\,, \qquad \langle q|p\rangle = \frac{1}{\sqrt{2\pi\hbar}} e^{\frac{i}{\hbar}pq}. \tag{7.87}$$

One obtains (suppressing $\hbar$ factors temporarily for readability)

$$\left\langle q_{n+1}|e^{-\frac{i}{\hbar}\hat{H}\Delta t}|q_n\right\rangle \simeq \int \frac{dp_n}{2\pi\hbar}\exp\left\{\frac{i}{\hbar}p_n(q_{n+1}-q_n) - \frac{i}{\hbar}\frac{p_n^2}{2m}\Delta t - \frac{i}{\hbar}V(q_n)\Delta t\right\}. \tag{7.88}$$

The p_n-integral is Gaussian and can be done exactly, yielding

$$\left\langle q_{n+1}|e^{-\frac{i}{\hbar}\hat{H}\Delta t}|q_n\right\rangle = \left(\frac{m}{2\pi i\hbar\Delta t}\right)^{\frac{1}{2}}\exp\left\{\frac{i}{\hbar}\left[\frac{m}{2}\left(\frac{q_{n+1}-q_n}{\Delta t}\right)^2 - V(q_n)\right]\Delta t\right\}. \tag{7.89}$$

Thus,

$$G(q_b, t_b; q_a, t_a) = \lim_{N\to\infty}\left(\frac{m}{2\pi i\hbar\Delta t}\right)^{\frac{N}{2}}\int\prod_{n=1}^{N-1}dq_n\,\exp\left\{\frac{i}{\hbar}\sum_{n=0}^{N-1}L\left(\frac{q_{n+1}-q_n}{\Delta t}, q_n\right)\Delta t\right\}. \tag{7.90}$$

where $L(\dot{q}, q) = \frac{1}{2}m\dot{q}^2 - V(q)$ is the classical Lagrangian.

In the continuum limit, $\Delta t \to 0$ and $N \to \infty$ at fixed T, the discrete sum becomes an action functional,

$$\sum_{n=0}^{N-1} L\left(\frac{q_{n+1} - q_n}{\Delta t}, q_n\right) \Delta t \longrightarrow \int_{t_a}^{t_b} dt\, L(\dot{q}(t), q(t)) \equiv S[q], \tag{7.91}$$

and the multiple integral defines the functional measure,

$$\int \mathcal{D}q \equiv \lim_{N\to\infty} \left(\frac{m}{2\pi i\hbar\Delta t}\right)^{\frac{N}{2}} \int \prod_{n=1}^{N-1} dq_n. \tag{7.92}$$

Hence the Feynman path integral:

$$G(q_b, t_b; q_a, t_a) = \int_{q(t_a)=q_a}^{q(t_b)=q_b} \mathcal{D}q(t)\, \exp\left(\frac{i}{\hbar} S[q]\right). \tag{7.93}$$

Driving Exercise

Computing Gaussian integrals and quadratic actions appear crucial to evaluate path integrals.

The practical computability of many path integrals rests on Gaussian integration. One starts from

$$\int_{-\infty}^{+\infty} dx\, e^{-x^2} = \sqrt{\pi}, \tag{7.94}$$

and generalizes to

$$\int_{-\infty}^{+\infty} dx\, \exp\left(-\frac{a}{2}x^2 + bx\right) = \sqrt{\frac{2\pi}{a}}\, \exp\left(\frac{b^2}{2a}\right), \qquad \Re(a) > 0, \tag{7.95}$$

which is proved by completing the square. The multi-dimensional analog is

$$\int_{\mathbb{R}^N} d^N q\, \exp\left(-\frac{1}{2}q^T A q + b^T q\right) = \sqrt{\frac{(2\pi)^N}{\det A}}\, \exp\left(\frac{1}{2}b^T A^{-1} b\right), \tag{7.96}$$

for symmetric positive-definite A; oscillatory Minkowski versions follow by analytic continuation $A \mapsto -iA$ with the usual $i\epsilon$ prescription.

For a quadratic action, the functional integral becomes (formally) a determinant:

$$\int \mathcal{D}q \, \exp\left\{\frac{i}{\hbar}\left[\frac{1}{2}\int dt\, dt'\, q(t)\, A(t,t')\, q(t') + \int dt\, b(t) q(t)\right]\right\} \propto \tag{7.97}$$
$$[\det A]^{-1/2} \exp\left\{-\frac{i}{2\hbar}\int dt\, dt'\, b(t)\, A^{-1}(t,t')\, b(t')\right\}.$$

Semiclassical limit: stationary phase and fluctuations.
Let $q_{\rm cl}(t)$ be the classical path solving the Euler-Lagrange equation with fixed endpoints. Expand $q(t) = q_{\rm cl}(t) + \eta(t)$, with $\eta(t_a) = \eta(t_b) = 0$. Then

$$S[q] = S[q_{\rm cl}] + \frac{1}{2}\int dt\, dt'\, \eta(t) \left.\frac{\delta^2 S}{\delta q(t)\delta q(t')}\right|_{q_{\rm cl}} \eta(t') + \cdots, \tag{7.98}$$

since the linear term vanishes on-shell. Truncating at quadratic order yields

$$G \simeq \mathcal{N}\, e^{\frac{i}{\hbar}S[q_{\rm cl}]} \left[\det(\delta^2 S)\right]^{-1/2}, \tag{7.99}$$

which makes explicit that quantum corrections are dominated by fluctuations near the classical saddle.

7.2.2 Bridging Statistical Mechanics and Quantum Field Theory Through Euclidean Continuation

Let us introduce imaginary time[2] $\tau = it$. The oscillatory factor $e^{\frac{i}{\hbar}S}$ becomes $e^{-\frac{1}{\hbar}S_E}$ with a Euclidean action S_E that is typically bounded below for stable systems. This transforms quantum amplitudes into objects closely resembling statistical weights $e^{-\beta H}$ and enables a direct route to partition functions and thermodynamic correlation functions.

At the QM level, the Euclidean path integral can be used to compute traces such as $Z = \mathrm{Tr}(e^{-\beta \hat{H}})$ by integrating over periodic trajectories in imaginary time of extent β. In QFT the same structure generalizes to periodic/antiperiodic boundary conditions for bosonic/fermionic fields and yields Matsubara formalism (see Sect. 7.7.1).

[2] The notion of imaginary time is formally obtained by performing a Wick rotation in the complex time plane, as we will discuss in Sect. 7.7.

7.3 From Particle Paths to Field Configurations

In QFT, the dynamical variable is a field $\phi(x)$ rather than a coordinate $q(t)$. The corresponding action is the spacetime integral of the Lagrangian density $\mathcal{L}[\phi(x)]$:

$$S[\phi] = \int d^4x \; \mathcal{L}[\phi(x)]. \tag{7.100}$$

A functional integral over fields is defined by discretizing spacetime into a lattice with N sites and spacing ℓ, turning $\phi(x)$ into an N-component vector q_j. Differential operators become matrices, and integrals become sums.

For the class of quadratic functionals,

$$S[\phi] = \frac{1}{2} \int d^4x \, d^4y \; \phi(x) \, A(x, y) \, \phi(y) + \int d^4x \; b(x)\phi(x), \tag{7.101}$$

one obtains the computable field integral

$$\begin{aligned} &\int \mathcal{D}\phi \; \exp\left\{\frac{i}{2} \int d^4x \, d^4y \; \phi(x)A(x, y)\phi(y) + i \int d^4x \; b(x)\phi(x)\right\} \\ &= \mathcal{B}\,[\det A]^{-1/2} \exp\left\{-\frac{i}{2} \int d^4x \, d^4y \; b(x)A^{-1}(x, y)b(y)\right\}, \end{aligned} \tag{7.102}$$

where $\mathcal{B}$ absorbs the divergent normalization factor associated with the continuum limit.

7.3.1 Generating Functionals and Green's Functions

The central object of the formalism is the (generally unnormalized) generating functional with a classical source $J(x)$:

$$Z[J] \equiv \int \mathcal{D}\phi \; \exp\left\{i \int d^4x \, [\mathcal{L}(\phi) + J(x)\phi(x)]\right\}. \tag{7.103}$$

Time-ordered vacuum correlators are recovered by functional differentiation:

$$\langle \Omega | T\hat{\phi}(x_1) \cdots \hat{\phi}(x_n) | \Omega \rangle = \frac{1}{i^n} \frac{1}{Z[0]} \left. \frac{\delta^n Z[J]}{\delta J(x_1) \cdots \delta J(x_n)} \right|_{J=0}. \tag{7.104}$$

The normalized generating functional is $\mathcal{Z}[J] \equiv Z[J]/Z[0]$, or briefly $\mathcal{Z}$.

7.4 Free Scalar Field

For a real scalar field with free Lagrangian density, the free generating functional is

$$Z_0[J] = \int \mathcal{D}\phi \, \exp\left\{\frac{i}{2}\int d^4x \, \left[(\partial_\mu\phi)^2 - m^2\phi^2\right] + i\int d^4x \, J\phi\right\}. \tag{7.105}$$

Integrate by parts (discarding boundary terms) to write

$$\frac{1}{2}\int d^4x \, \left[(\partial_\mu\phi)^2 - m^2\phi^2\right] = -\frac{1}{2}\int d^4x \, \phi(x)\,(\Box + m^2)\,\phi(x). \tag{7.106}$$

Thus the quadratic kernel is $A = (\Box + m^2)$, and its inverse is the Green's function Δ defined by

$$(\Box_x + m^2)\Delta(x-y) = \delta^{(4)}(x-y), \tag{7.107}$$

with the Feynman prescription $\Delta \to \Delta_F$ obtained by $m^2 \to m^2 - i\epsilon$ (equivalently $p^2 - m^2 \to p^2 - m^2 + i\epsilon$ in momentum space).

Applying the general Gaussian identity in Eq. (7.102) yields

$$Z_0[J] = Z_0[0]\exp\left\{-\frac{i}{2}\int d^4x\, d^4y \; J(x)\Delta_F(x-y)J(y)\right\}. \tag{7.108}$$

In normalized form,

$$\mathcal{Z}_0[J] \equiv \frac{Z_0[J]}{Z_0[0]} = \exp\left\{-\frac{i}{2}\int d^4x\, d^4y \; J(x)\Delta_F(x-y)J(y)\right\}. \tag{7.109}$$

▶ **Driving Exercise**
Recover the free propagator. Differentiate Eq. (7.108) twice:

$$G_0(x,y) \equiv \langle\Omega|T\hat{\phi}(x)\hat{\phi}(y)|\Omega\rangle = \frac{1}{i^2}\frac{1}{Z_0[0]}\frac{\delta^2 Z_0[J]}{\delta J(x)\delta J(y)}\bigg|_{J=0} = \Delta_F(x-y). \tag{7.110}$$

Equivalently, using the path integral definition,

$$G_0(x,y) = \frac{\int \mathcal{D}\phi \; \phi(x)\phi(y)\exp\left(i\int d^4x\, \mathcal{L}_0\right)}{\int \mathcal{D}\phi \; \exp\left(i\int d^4x\, \mathcal{L}_0\right)}, \tag{7.111}$$

showing the close analogy with statistical averages weighted by $e^{-\beta H}$.

7.5 Interacting Scalar Field: Perturbation Theory from Functional Methods

For an interacting theory with $\mathcal{L} = \mathcal{L}_0 + \mathcal{L}_I$, the generating functional reads

$$Z[J] = \int \mathcal{D}\phi \ \exp\left\{ i \int d^4x \, [\mathcal{L}_0 + \mathcal{L}_I + J\phi] \right\}. \tag{7.112}$$

A standard strategy is to treat $\mathcal{L}_I$ as an operator acting by functional differentiation on the free functional $Z_0[J]$. For example, in ϕ^4 theory,

$$\mathcal{L}_I = -\frac{\lambda}{4!}\phi^4, \tag{7.113}$$

one may use $\phi(x) \to \frac{1}{i}\frac{\delta}{\delta J(x)}$ under the functional integral to write

$$Z[J] = \exp\left\{ -i \int d^4x \, \frac{\lambda}{4!} \left(\frac{1}{i} \frac{\delta}{\delta J(x)} \right)^4 \right\} Z_0[J]. \tag{7.114}$$

Expanding the exponential generates the perturbation series and, after differentiating to obtain correlators, reproduces Wick contractions and Feynman rules. The explicit multi-derivative structure and the appearance of terms of order J^0, J^2, J^4, ... in the first correction can be verified by direct differentiation of $Z_0[J]$.

7.5.1 Hubbard-Stratonovich Transformation

In many contexts it is advantageous to linearize quartic interactions by introducing an auxiliary field. For instance, starting from

$$\mathcal{L} = \frac{1}{2}(\partial_\mu \phi)^2 - \frac{m^2}{2}\phi^2 - \frac{g}{8}\phi^4, \tag{7.115}$$

one may add

$$\Delta\mathcal{L} = \frac{1}{2g}\left(\sigma - \frac{g}{2}\phi^2 \right)^2, \tag{7.116}$$

which does not change the dynamics after integrating out σ but reorganizes perturbation theory by replacing a ϕ^4 vertex with $\phi^2\sigma$ couplings.

To see this explicitly, consider the functional integral formulation. The generating functional for the original theory reads

$$Z = \int \mathcal{D}\phi \ \exp\left[i \int d^4x \left(\frac{1}{2}(\partial_\mu \phi)^2 - \frac{m^2}{2}\phi^2 - \frac{g}{8}\phi^4 \right) \right]. \tag{7.117}$$

Introducing the auxiliary field σ one may rewrite the same functional as

$$Z = \int \mathcal{D}\phi\, \mathcal{D}\sigma\, \exp\left[i \int d^4x \left(\frac{1}{2}(\partial_\mu \phi)^2 - \frac{m^2}{2}\phi^2 + \frac{1}{2g}\sigma^2 - \frac{1}{2}\sigma\phi^2 \right) \right], \tag{7.118}$$

where the ϕ^4 interaction has been traded for a trilinear vertex $\sigma\phi^2$ and a quadratic term for σ. Performing the Gaussian path integral over σ reproduces exactly the original ϕ^4 theory, showing that the two formulations are mathematically equivalent.

From a physical point of view, the field σ does not represent a new fundamental degree of freedom: it is an auxiliary or composite field, whose equation of motion gives

$$\sigma = \frac{g}{2}\phi^2. \tag{7.119}$$

Thus σ encodes the local amplitude of the composite operator ϕ^2. In this sense the Hubbard-Stratonovich transformation promotes a composite operator of the original theory to an independent field variable, which is particularly useful when the dynamics is dominated by collective or correlated degrees of freedom.

This reformulation has several structural advantages. First, perturbation theory is reorganized: instead of a single quartic vertex, one now has two trilinear vertices, which often simplifies diagrammatic expansions and resummations. Second, in large-N generalizations, where ϕ carries an internal index ϕ^a with $a = 1, \ldots, N$, the auxiliary field σ becomes naturally of order N^0 and its propagator captures the leading collective fluctuations of the composite operator $\phi^a\phi^a$. In this framework, σ can be interpreted as mediating an effective interaction between the ϕ quanta, much like an emergent or effective boson. In condensed-matter and statistical field theory, the same mechanism underlies the introduction of order-parameter fields. For example, in theories of superconductivity or magnetism, the Hubbard-Stratonovich field represents the local pairing amplitude or magnetization, and its expectation value signals the onset of an ordered phase.

The Hubbard-Stratonovich transformation therefore reorganizes the theory in terms of variables that are often closer to the physically relevant excitations.

7.6 The Free Massive Vector Field

Functional methods extend to multi-component fields. Consider a massive vector field A_μ with (Proca-type) Lagrangian density

$$\mathcal{L} = -\frac{1}{4} F_{\mu\nu} F^{\mu\nu} + \frac{1}{2} m^2 A_\mu A^\mu, \qquad F_{\mu\nu} = \partial_\mu A_\nu - \partial_\nu A_\mu. \tag{7.120}$$

The free generating functional with source J_μ has the Gaussian structure

$$Z_0[J] = \frac{\int \mathcal{D}A \ \exp\left\{\frac{i}{2} \int d^4x \ A_\mu \hat{K}^{\mu\nu} A_\nu + i \int d^4x \ J_\mu A^\mu\right\}}{\int \mathcal{D}A \ \exp\left\{\frac{i}{2} \int d^4x \ A_\mu \hat{K}^{\mu\nu} A_\nu\right\}} = \\ = \exp\left\{-\frac{1}{2} \int d^4x \, d^4y \ J_\mu(x) \, [i\hat{K}^{-1}]^{\mu\nu}(x, y) \, J_\nu(y)\right\}. \tag{7.121}$$

After integrating by parts, the kernel operator can be cast as

$$\hat{K}^{\mu\nu} = (\partial^2 + m^2) g^{\mu\nu} - \partial^\mu \partial^\nu. \tag{7.122}$$

The propagator is $G_0^{\mu\nu} = i(\hat{K}^{-1})^{\mu\nu}$; in momentum space one finds

$$\widetilde{G}_0^{\nu\lambda}(p) = \frac{-i}{p^2 - m^2 + i\epsilon} \left(g^{\nu\lambda} - \frac{p^\nu p^\lambda}{m^2}\right), \tag{7.123}$$

exhibiting the transverse projector characteristic of a massive spin-1 field.

7.7 Wick Rotation and Euclidean Quantum Field Theory

Imaginary time is introduced via a Wick rotation, i.e. a rotation of the time coordinate in the complex plane by angle $-\pi/2$:

$$x^0 = -i\tau. \tag{7.124}$$

This simultaneously requires rotating the energy component of momentum, $p^0 = i\omega$, so that Fourier factors remain well-behaved. The net effect is the passage from Minkowski signature $(+, -, -, -)$ to Euclidean signature $(+, +, +, +)$ (up to an overall sign).

A useful collection of Euclidean rules is

1. $x^2 = x_0^2 - \mathbf{x}^2 = -(\tau^2 + \mathbf{x}^2) \equiv -x_E^2$, $\quad d^4x = -i\, d^4x_E = -i\, d\tau\, d^3x$.
2. $p^2 = p_0^2 - \mathbf{p}^2 = -(\omega^2 + \mathbf{p}^2) \equiv -p_E^2$, $\quad d^4p = i\, d^4p_E = i\, d\omega\, d^3p$.
3. $p\cdot x = p_0 x_0 - \mathbf{p}\cdot\mathbf{x} = \omega\tau - \mathbf{p}\cdot\mathbf{x}$, $\quad \partial_0 = i\partial_\tau$, $\quad \partial^2 = -(\partial_\tau^2 + \nabla^2) \equiv -\partial_E^2$.
4. $(\partial_\mu\phi)^2 = -\left[(\partial_\tau\phi)^2 + (\nabla\phi)^2\right] \equiv -(\partial_\mu^E\phi)^2$.

Now, consider the scalar theory with source J:

$$iS = i\int d^4x\left[\frac{1}{2}(\partial_\mu\phi)^2 - \frac{m^2}{2}\phi^2 - \frac{\lambda}{4!}\phi^4 + J\phi\right]. \tag{7.125}$$

Under Wick rotation this becomes

$$iS = -S_E, \qquad S_E = \int d^4x_E\left[\frac{1}{2}(\partial_\tau\phi)^2 + \frac{1}{2}(\nabla\phi)^2 + \frac{m^2}{2}\phi^2 + \frac{\lambda}{4!}\phi^4 - J\phi\right]. \tag{7.126}$$

Hence, the Euclidean functional integral is

$$Z[J] = \int \mathcal{D}\phi\; e^{-S_E[\phi;J]}. \tag{7.127}$$

Time rotation removes the purely oscillatory character of the Minkowski weight and clarifies the role of the $i\epsilon$ prescription: in Euclidean space the Gaussian is damped for stable quadratic forms, making the functional integral mathematically more tractable.

7.7.1 Statistical Field Theory and Finite Temperature Quantum Field Theory

In QFT the Green's functions are built from time-evolution operators (schematically $e^{-iEt/\hbar}$), whereas in statistical mechanics equilibrium weights are governed by the Boltzmann factor (schematically e^{-E/k_BT}). The mapping between the two is realized by identifying imaginary time with inverse temperature (we retain $\hbar$ to stress its importance here),

$$\frac{it}{\hbar} \longleftrightarrow \frac{1}{k_BT} \equiv \beta, \tag{7.128}$$

so that β behaves as the extent of the Euclidean time direction.

▶ **Pause and think**
Before going on, manage to infer the partition function as a Euclidean functional integral.

For a quantum system of fields with Hamiltonian $\hat{H}$, the partition function turns out to be

$$Z = \mathrm{Tr}\left(e^{-\beta \hat{H}}\right). \tag{7.129}$$

In the path integral representation this trace becomes an integral over field configurations defined on $\tau \in [0, \beta]$ with periodic boundary conditions (PBC) for bosons:

$$Z = \int_{\mathrm{PBC}} \mathcal{D}\phi \, \exp\left[-\int_0^\beta d\tau \int d^3x \, \mathcal{L}_E(\phi)\right], \qquad \phi(0, \mathbf{x}) = \phi(\beta, \mathbf{x}), \tag{7.130}$$

while for fermions one must impose antiperiodic boundary conditions $\psi(0, \mathbf{x}) = -\psi(\beta, \mathbf{x})$.

7.7.2 Matsubara Frequencies

Place the system in a finite spatial volume V (a box), which discretizes spatial momenta. Periodicity in imaginary time quantizes the Euclidean frequencies. For a bosonic real scalar field one expands

$$\phi(\tau, \mathbf{x}) = \sqrt{\frac{\beta}{V}} \sum_{n,\mathbf{p}} e^{-i\omega_n \tau + i\mathbf{p}\cdot\mathbf{x}} \, \tilde{\phi}_n(\mathbf{p}), \qquad \omega_n = \frac{2\pi n}{\beta}, \quad n \in \mathbb{Z}, \tag{7.131}$$

where ω_n are Matsubara frequencies. For fermions, antiperiodicity shifts the frequencies to $\omega_n = \frac{2\pi(n+1)}{\beta}$.

In this formalism, integrals over continuous energy variables are replaced by discrete sums:

$$\int \frac{d\omega}{2\pi} \longrightarrow \frac{1}{\beta} \sum_{n\in\mathbb{Z}}, \tag{7.132}$$

and finite-temperature propagators become functions of $(\omega_n, \mathbf{p})$ in Euclidean momentum space.

In view of what presented above, a practical set of prescriptions is what follows.

1. Construct $S_E[\phi]$ (including sources) from the Minkowski action by Wick rotation.
2. Compute $\mathcal{Z}_E[J] = Z_E[J]/Z_E[0]$ by Gaussian evaluation (free theory) or by perturbative expansion (interactions), using $\phi \to \delta/\delta J$ to generate Wick contractions.

3. Obtain Euclidean n-point functions by functional differentiation with respect to J.
4. Continue $\omega_n \to -ip^0$ (and appropriate prescriptions) to recover Minkowski correlators when needed.

7.8 From Grassmann Numbers to Fermion Path Integral

The path integral formulation integrates over *c-number* field configurations. For bosonic fields this is consistent with the canonical operator algebra because bosonic operators commute. Fermionic fields, however, satisfy anticommutation relations, and a faithful path integral representation therefore requires an algebra of anticommuting *numbers*. These are the Grassmann variables. In this section we introduce the Grassmann calculus, construct coherent states for a single fermionic mode, and derive the standard generating functional for free fermions with Grassmann-valued sources.

A Grassmann variable η is defined by anticommutation with any other Grassmann variable ζ,

$$\eta\,\zeta = -\,\zeta\,\eta. \tag{7.133}$$

In particular, setting $\zeta = \eta$ implies *nilpotency*,

$$\eta^2 = 0. \tag{7.134}$$

Consequently, any function of a single Grassmann variable truncates after the linear term:

$$f(\eta) = a + b\,\eta, \tag{7.135}$$

where a, b are ordinary (commuting) numbers. For two Grassmann variables η, ζ the most general polynomial truncates at bilinear order,

$$g(\eta, \zeta) = a + b\eta + c\zeta + d\,\eta\zeta. \tag{7.136}$$

Differentiation

Grassmann differentiation is defined by

$$\frac{\partial}{\partial\eta}\,\eta = 1, \qquad \frac{\partial}{\partial\eta}\,a = 0, \tag{7.137}$$

and extended linearly. Since $\partial/\partial\eta$ is itself Grassmann-odd, it anticommutes with Grassmann variables. For example, if ζ is Grassmann-odd,

$$\frac{\partial}{\partial\eta}\,(\zeta\,\eta) = -\,\zeta\,\frac{\partial}{\partial\eta}\eta = -\,\zeta. \tag{7.138}$$

Applying Eq. (7.137) to Eq. (7.135) yields

$$\frac{\partial f}{\partial \eta} = b. \tag{7.139}$$

For Eq. (7.136), one finds

$$\frac{\partial g}{\partial \eta} = b + d\,\zeta, \qquad \frac{\partial g}{\partial \zeta} = c - d\,\eta, \tag{7.140}$$

where the relative sign in the second derivative follows from reordering $\eta\zeta$ into $\zeta\eta$ before differentiating.

Integration

Grassmann integration is defined by the axioms

$$\int d\eta\; \eta = 1, \qquad \int d\eta\; 1 = 0. \tag{7.141}$$

Using Eq. (7.135), one obtains

$$\int d\eta\; f(\eta) = \int d\eta\,(a + b\eta) = b. \tag{7.142}$$

Thus, for Grassmann variables, differentiation and integration act identically on polynomials:

$$\int d\eta\;(\cdots)\;\equiv\;\frac{\partial}{\partial \eta}(\cdots) \qquad \text{(on Grassmann polynomials).} \tag{7.143}$$

For two variables, repeated integration gives

$$\int d\eta\, d\zeta\;g(\eta,\zeta) = \int d\eta\, d\zeta\;\big(a + b\eta + c\zeta + d\,\eta\zeta\big) = d, \tag{7.144}$$

up to the chosen ordering convention for the measure.

7.8.1 Fermionic Coherent States

Consider a single fermionic mode with annihilation and creation operators $\hat{c}$ and $\hat{c}^\dagger$ obeying $\{\hat{c}, \hat{c}^\dagger\} = 1$ and $\hat{c}|0\rangle = 0$, $|1\rangle = \hat{c}^\dagger|0\rangle$. Let η be a Grassmann number that anticommutes with fermionic operators,

$$\{\eta, \hat{c}\} = 0, \qquad \{\eta, \hat{c}^\dagger\} = 0. \tag{7.145}$$

A fermionic coherent state is defined as

$$|\eta\rangle \equiv e^{-\eta\hat{c}^\dagger}\,|0\rangle. \tag{7.146}$$

Because $(\eta\hat{c}^\dagger)^2 = 0$ (Grassmann nilpotency and $(\hat{c}^\dagger)^2 = 0$), the exponential truncates:

$$|\eta\rangle = (1 - \eta\hat{c}^\dagger)|0\rangle = |0\rangle - \eta|1\rangle. \tag{7.147}$$

This state is an eigenstate of $\hat{c}$ with Grassmann eigenvalue η:

$$\hat{c}|\eta\rangle = \hat{c}|0\rangle - \eta\,\hat{c}|1\rangle = 0 - \eta\,|0\rangle = \eta\,(|0\rangle - \eta|1\rangle) = \eta|\eta\rangle, \tag{7.148}$$

where we used $\hat{c}|1\rangle = |0\rangle$ and $\eta^2 = 0$.

One also introduces left coherent states. A convenient parametrization is

$$\langle\bar{\eta}| \equiv \langle 0|\,e^{-\hat{c}\,\bar{\eta}} = \langle 0| + \bar{\eta}\langle 1|, \tag{7.149}$$

where $\bar{\eta}$ is an independent Grassmann variable (not the complex conjugate of η). With these definitions, the overlap evaluates to

$$\langle\bar{\zeta}|\eta\rangle = 1 + \bar{\zeta}\eta = e^{\bar{\zeta}\eta}, \tag{7.150}$$

again using nilpotency to truncate the exponential.

The coherent states provide a resolution of the identity. With the standard convention for the measure ordering, one may write

$$\int d\eta\,d\bar{\eta}\;e^{-\bar{\eta}\eta}\;|\eta\rangle\langle\bar{\eta}| = 1, \tag{7.151}$$

which can be verified by acting on the basis $\{|0\rangle, |1\rangle\}$ and using Eq. (7.141).

7.8.2 Grassmann Gaussian Integrals and determinants

A basic Gaussian identity for Grassmann variables is

$$\int d\eta\,d\bar{\eta}\;\exp\big(-\bar{\eta}\,a\,\eta\big) = a, \tag{7.152}$$

for commuting a. To see this explicitly, expand the exponential:

$$e^{-\bar{\eta}a\eta} = 1 - \bar{\eta}a\eta, \tag{7.153}$$

hence

$$\int d\eta\,d\bar{\eta}\,(1 - \bar{\eta}a\eta) = -a\int d\eta\,d\bar{\eta}\,\bar{\eta}\eta = a, \tag{7.154}$$

where the last equality follows from the adopted convention $\int d\eta\,d\bar{\eta}\,\bar{\eta}\eta = -1$ (equivalently, $\int d\bar{\eta}\,d\eta\,\bar{\eta}\eta = +1$). The sign is convention-dependent, but the physically relevant multi-mode result is unambiguous once a measure ordering is fixed.

Pause and think
Before going on, work out the Grassmann Gaussian integral for N modes.

Let η and $\bar{\eta}$ be N-component Grassmann vectors and let A be an $N \times N$ matrix with commuting entries. Then

$$\int \prod_{i=1}^{N} d\eta_i \, d\bar{\eta}_i \ \exp\big(- \bar{\eta}\, A\, \eta\big) = \det A. \tag{7.155}$$

This identity follows by expanding the exponential and noting that only the term containing the product $\prod_i \bar{\eta}_i \eta_i$ survives the Grassmann integration; its coefficient is precisely $\det A$.

7.8.3 The Fermionic Generating Functional

Let $\psi, \bar{\psi}$ be Grassmann variables, and let K be an invertible matrix (or, in the functional case, an invertible linear operator). Consider the integral with Grassmann sources $\eta, \bar{\eta}$:

$$\mathcal{I}[\bar{\eta}, \eta] \equiv \int d\psi \, d\bar{\psi} \ \exp\big(- \bar{\psi} K \psi + \bar{\eta}\psi + \bar{\psi}\eta\big). \tag{7.156}$$

Define shifted variables

$$\psi' \equiv \psi - K^{-1}\eta, \qquad \bar{\psi}' \equiv \bar{\psi} - \bar{\eta}K^{-1}, \tag{7.157}$$

so that

$$\begin{aligned} - \bar{\psi} K \psi + \bar{\eta}\psi + \bar{\psi}\eta &= -(\bar{\psi}' + \bar{\eta}K^{-1})K(\psi' + K^{-1}\eta) + \bar{\eta}(\psi' + K^{-1}\eta) + (\bar{\psi}' + \bar{\eta}K^{-1})\eta \\ &= -\bar{\psi}' K \psi' + \bar{\eta}K^{-1}\eta, \end{aligned} \tag{7.158}$$

where the cross terms cancel. Therefore,

$$\mathcal{I}[\bar{\eta}, \eta] = \left(\int d\psi' \, d\bar{\psi}' \, e^{-\bar{\psi}' K \psi'} \right) \exp\big(\bar{\eta}K^{-1}\eta\big) \equiv \mathcal{N} \exp\big(\bar{\eta}K^{-1}\eta\big). \tag{7.159}$$

The prefactor $\mathcal{N}$ depends on $\det K$ (cf. Eq. (7.155)) and is removed by normalization in the field-theory generating functional.

In the path integral formulation for free fermions one introduces Grassmann-valued sources $\eta(x)$ and $\bar{\eta}(x)$ and defines the generating functional

$$Z[\bar{\eta}, \eta] \equiv \frac{1}{Z[0, 0]} \int \mathcal{D}\psi \, \mathcal{D}\bar{\psi} \ \exp\Big\{ i \int d^4x \Big[\bar{\psi}(x)\, K\, \psi(x) + \bar{\eta}(x)\psi(x) + \bar{\psi}(x)\eta(x) \Big] \Big\}, \tag{7.160}$$

where K is the Dirac operator for the free theory (e.g. $K = i\partial\!\!\!/ - m$, possibly with an $i\epsilon$ prescription). Applying the functional version of Eq. (7.159) yields the normalized result

$$Z[\bar{\eta}, \eta] = \exp\left\{-i \int d^4x\, d^4y\, \bar{\eta}(x)\, K^{-1}(x, y)\, \eta(y)\right\}. \tag{7.161}$$

Functional differentiation then reproduces the Feynman propagator:

$$S_F(x - y) = \langle 0\, T\{\psi(x)\bar{\psi}(y)\}|0\rangle = i\, K^{-1}(x, y), \tag{7.162}$$

which completes the construction of the fermionic path integral.

▶ **Pause and think**
Before going on, manage to compute the free fermion propagator from the generating functional using Grassmann numbers.

To derive the free propagator, since the action is quadratic in the fields, the functional integral is Gaussian (in Grassmann variables) and can be evaluated by completing the square. One obtains, up to an irrelevant normalization,

$$Z[\bar{\eta}, \eta] = \exp\left\{-i \int d^4x\, d^4y\, \bar{\eta}(x)\, (i\partial\!\!\!/ - m)^{-1}\eta(y)\right\}. \tag{7.163}$$

It is customary to normalize such that $Z[0, 0] = 1$.

7.8.4 Momentum-Space Representation

The Feynman propagator is

$$S_F(x - y) \equiv \langle 0\, T\{\hat{\psi}(x)\hat{\bar{\psi}}(y)\}|0\rangle, \tag{7.164}$$

and it is the Green's function of the Dirac operator:

$$(i\partial\!\!\!/ - m)\, S_F(x) = \delta^{(4)}(x)\, \mathbb{I}_{4\times 4}. \tag{7.165}$$

Fourier transforming,

$$S_F(x) = \int \frac{d^4p}{(2\pi)^4}\, e^{-ip\cdot x}\, \tilde{S}_F(p), \tag{7.166}$$

equation (7.165) becomes

$$(\not{p} - m)\, \tilde{S}_F(p) = \mathbb{I}. \tag{7.167}$$

The inverse is obtained by multiplying numerator and denominator by $(\not{p} + m)$:

$$\tilde{S}_F(p) = \frac{i(\not{p} + m)}{p^2 - m^2 + i\epsilon}, \tag{7.168}$$

which makes explicit that the propagator is a 4×4 matrix in spinor space.

At this stage, we thus recall that, when dealing with perturbation theory in momentum space, internal fermion lines contribute the factor

$$\frac{i(\not{p} + m)}{p^2 - m^2 + i\epsilon} \qquad \text{(equivalently } \frac{i}{\not{p} - m + i\epsilon}\text{)}. \tag{7.169}$$

External legs are associated with spinors: incoming fermion $u_s(p)$, outgoing fermion $\bar{u}_s(p)$, incoming antifermion $\bar{v}_s(p)$, outgoing antifermion $v_s(p)$. For each closed fermion loop one takes the trace over spinor indices and includes an overall factor (-1) arising from fermionic statistics.

Driving Exercise

Let us study the Yukawa-type coupling to a scalar mediator.

Consider a model with a real scalar field ϕ coupled to the Dirac field via

$$\mathcal{L}_I = -g\, \bar{\psi}\psi\, \phi. \tag{7.170}$$

At tree level, fermion-fermion scattering $\psi\psi \to \psi\psi$ proceeds via scalar exchange. For distinguishable fermions, the t-channel amplitude is

$$i\mathcal{M}_t = (-ig)^2 \left[\bar{u}_{s'}(p')u_s(p)\right] \frac{i}{(p' - p)^2 - m_\phi^2 + i\epsilon} \left[\bar{u}_{r'}(k')u_r(k)\right]. \tag{7.171}$$

For identical fermions, there is also a u-channel contribution obtained by exchanging the external legs, and it enters with a relative minus sign from operator interchange:

$$i\mathcal{M}_u = -(-ig)^2 \left[\bar{u}_{r'}(k')u_s(p)\right] \frac{i}{(p' - k)^2 - m_\phi^2 + i\epsilon} \left[\bar{u}_{s'}(p')u_r(k)\right]. \tag{7.172}$$

In the non-relativistic limit $|\mathbf{p}| \ll m$, the positive-energy spinors satisfy $u_s(p) \simeq \sqrt{m}\,(\xi_s, \xi_s)^{\mathrm{T}}$ (up to representation-dependent conventions), so that

$$\bar{u}_{s'}(p)\, u_s(p) \simeq 2\, m\, \delta_{s's}, \tag{7.173}$$

showing that spin is conserved at leading order. For distinguishable particles (where only the t-channel contributes), one then finds

$$i\mathcal{M} \simeq \frac{4ig^2 m^2}{(\mathbf{p}-\mathbf{p}')^2 + m_\phi^2}\,\delta_{s's}\,\delta_{r'r}, \tag{7.174}$$

which is the momentum-space form associated with an attractive Yukawa potential.

7.9 Broken Symmetry

Here we discuss the central idea of *broken symmetry*, a concept that permeates nearly every aspect of physics, focusing on the mechanism of spontaneous symmetry breaking.

As a direct prelude, let us focus on what we mean with this strange name.

- **Why breaking?** Symmetry breaking refers to the situation in which the fundamental equations of motion (or the Lagrangian) of a physical system are invariant under a given symmetry transformation, while the physically realized states of the system are not. In other words, the symmetry is a property of the dynamical laws, but not of the ground state around which the theory is expanded.
- **Which symmetry?** In spontaneous symmetry breaking one considers a *continuous global or local symmetry* of the action, typically associated with an internal transformation group. The symmetry is exact at the level of the Lagrangian, but the space of degenerate minima of the potential does not remain invariant under the full symmetry group.
- **Why spontaneous?** The breaking is termed spontaneous because it is not induced by explicit symmetry-violating terms in the Lagrangian. Instead, it arises from the structure of the potential and from the dynamical selection of a particular ground state among several energetically equivalent ones. The symmetry of the theory remains intact, but hidden by the choice of vacuum.

7.9.1 Classical Broken Symmetry: The Landau Theory

Consider a magnet whose spins can point up or down. At sufficiently high temperature each spin is equally likely to be up or down, so the net magnetization (the spatial average of the magnetic moment) is zero. In this regime the system is symmetric under a global 180° rotation that flips all spins: the microscopic configuration changes, but the macroscopic magnetization remains $M = 0$.

Upon cooling below a critical temperature T_c, the system undergoes a phase transition: spins align and the magnetization becomes nonzero. Two equivalent ordered states exist, corresponding to alignment "up" ($M = +M_0$) or "down" ($M = -M_0$). Flipping all spins maps one ordered state into the other, so the *Hamiltonian* keeps the symmetry, but the *ground state* does not: the symmetry is spontaneously broken.

A mechanical analogy is the Euler strut: above a critical load, the strut buckles either left or right. The underlying equations do not choose a direction; tiny perturbations select one outcome in practice.

A simple Hamiltonian exhibiting the up/down symmetry is

$$\hat{H} = -J \sum_i \hat{S}_i^z \hat{S}_{i+1}^z, \tag{7.175}$$

which has no bias favoring $S^z > 0$ or $S^z < 0$.

Landau's approach is to write the free energy as a function of an *order parameter*. For a magnet, the order parameter is the magnetization field M. The equilibrium state minimizes the free energy $F = U - TS$. Landau expands

$$F(M) = F_0 + aM^2 + bM^4 + \cdots, \tag{7.176}$$

where a and b are independent of M (but may depend on T). The symmetry $M \to -M$ is manifest. For $a > 0$, the minimum is at $M = 0$ (disordered phase). For $a < 0$, the potential develops two minima at $M = \pm M_0$ (ordered phase), and $M = 0$ becomes metastable.

A common parametrization is $a = a_0(T - T_c)$ (with a_0 constant) and $b > 0$ approximately temperature-independent, so that a changes sign at $T = T_c$.

The stationary points satisfy

$$\frac{\partial F}{\partial M} = 2aM + 4bM^3 = 0. \tag{7.177}$$

Besides the maximum $M = 0$, the minima for $a < 0$ occur at

$$M_0^2 = -\frac{a}{2b}, \tag{7.178}$$

so the ordered phase has two degenerate ground states $M = \pm M_0$.

Below, we discuss a different scenario, where time evolution plays a fundamental role.

7.9.2 Time-Dependent Classical Symmetry Breaking: Bifurcations

A conceptually different mechanism of classical symmetry breaking arises when the symmetric configuration is not energetically disfavored, but dynamically unstable. In this case the symmetry is lost not by minimizing a static potential, as in the Landau picture, but through the time evolution of the system itself.

A paradigmatic example is provided by an effective equation for a real amplitude $a(t)$,

$$\dot{a} = \mu\, a - a^3, \tag{7.179}$$

where μ is a control parameter. This equation is invariant under the discrete symmetry $a \to -a$ and therefore admits the symmetric solution, $a(t) = 0$.

For $\mu < 0$ the linearized equation $\dot{a} = \mu a$ implies exponential decay of any small perturbation, $a(t) \sim e^{\mu t}$, so that $a = 0$ is dynamically stable. In this regime the system relaxes toward the symmetric configuration and no symmetry breaking occurs.

The situation changes qualitatively when μ crosses zero. For $\mu > 0$ the solution $a = 0$ becomes dynamically unstable: an arbitrarily small perturbation grows as $a(t) \sim e^{\mu t}$ at early times. At the same time, new stationary solutions appear, obtained by setting $\dot{a} = 0$ in Eq. (7.179),

$$a_{\pm} = \pm\sqrt{\mu}. \tag{7.180}$$

These two solutions are dynamically stable for $\mu > 0$ and are exchanged by the symmetry $a \to -a$. The phase portrait therefore exhibits a pitchfork bifurcation: the single symmetric fixed point at $a = 0$ splits into two non-symmetric stable branches.

If the system is prepared near $a = 0$ with an arbitrarily small perturbation, the subsequent time evolution inevitably drives it toward one of the two stable branches a_+ or a_-. Which branch is selected is determined by initial fluctuations, but once the system has relaxed, the final state is no longer invariant under $a \to -a$. The symmetry is therefore broken by the dynamics itself: the equations of motion remain symmetric, but their long-time solutions are not.

This mechanism differs in spirit from the Landau picture of symmetry breaking. There one inspects a static potential and identifies degenerate minima among which the system must choose. Here, by contrast, the decisive ingredient is the instability of the symmetric solution under time evolution. Symmetry breaking is thus a dynamical process, encoded in the flow of solutions in phase space rather than in the static structure of an energy functional.

In both Landau and bifurcation theories, **the symmetry of the physically realized state is strictly smaller than the symmetry of the underlying equations of motion.**

7.9.3 Relativistic Broken Symmetry: Scalar Field Lagrangian

We now translate this logic into the language of (classical) field theory. The order parameter becomes the ground-state value of a field $\phi(x)$, or in quantum language the vacuum expectation value $\langle\Omega|\hat{\phi}(x)|\Omega\rangle$.

Consider a real scalar field with Lagrangian

$$\mathcal{L} = \frac{1}{2}(\partial_\mu \phi)^2 - U(\phi), \tag{7.181}$$

where $U(\phi)$ is the potential energy density. For ϕ^4 theory, $U(\phi) = \frac{\mu^2}{2}\phi^2 + \frac{\lambda}{4!}\phi^4$. This has a discrete global symmetry $\phi \to -\phi$. For $\mu^2 > 0$, the potential is minimized at $\phi = 0$, corresponding to a vacuum with vanishing expectation value. Excitations about this vacuum are scalar quanta with mass $m = \mu$.

Now reverse the sign of the quadratic term[3]:

$$U(\phi) = -\frac{\mu^2}{2}\phi^2 + \frac{\lambda}{4!}\phi^4. \tag{7.182}$$

The stationary points satisfy $0 = \frac{\partial U}{\partial \phi} = -\mu^2\phi + \frac{\lambda}{3!}\phi^3$, giving $\phi = 0$ and

$$\phi_0 = \pm\left(\frac{6\mu^2}{\lambda}\right)^{1/2}. \tag{7.183}$$

The point $\phi = 0$ is unstable, while $\phi = \pm\phi_0$ are the true minima. The system chooses one vacuum, thereby breaking the symmetry of the ground state.

Choose the vacuum $\phi = +\phi_0$ and define the shifted field, $\phi' = \phi - \phi_0$. Expanding the potential around ϕ_0,

$$\begin{aligned} U(\phi) &= U(\phi_0) + \left.\frac{\partial U}{\partial \phi}\right|_{\phi_0}(\phi - \phi_0) + \frac{1}{2}\left.\frac{\partial^2 U}{\partial \phi^2}\right|_{\phi_0}(\phi - \phi_0)^2 + \cdots \\ &= U(\phi_0) + \mu^2(\phi - \phi_0)^2 + \cdots, \end{aligned} \tag{7.184}$$

[3] One can wonder *how* the sign of m^2 can be reversed. Generally speaking, imagine to have the simplest case, $m^2 = m_b^2\left(1 - \frac{\psi^2}{\psi_c^2}\right)$, where ψ, a scalar *environment* or a *spectator field* is introduced to fuel the background. In this case, the sign of mass can be switched by invoking the cases $\psi > \psi_c$ or $\psi < \psi_c$. This distinguishes the *bare mass*, m_b, namely the real mass of a phion, that cannot change, from the effective mass that may change under the effects of external fields.

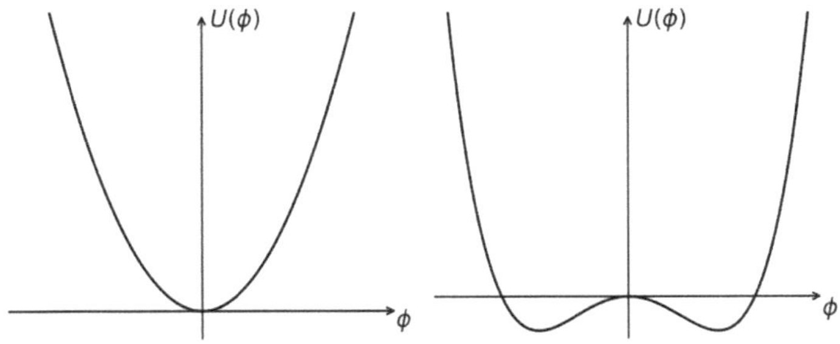

Fig. 7.4 Minima of $U(\phi)$ before (left figure) and after (right figure) symmetry breaking

where the linear term vanishes at the minimum. Dropping the constant $U(\phi_0)$, the Lagrangian becomes

$$\mathcal{L} = \frac{1}{2}(\partial\phi')^2 - \mu^2\phi'^2 + O(\phi'^3). \tag{7.185}$$

Comparing with the standard quadratic form, the excitation mass in the broken phase is

$$m = \sqrt{2}\,\mu. \tag{7.186}$$

Crucially, the *Lagrangian* remains symmetric under $\phi \to -\phi$. Symmetry breaking is a property of the *vacuum* (ground state), i.e., in fact note that the initial Lagrangian is symmetric, while Eq. (7.185) is not, since it contains terms $\propto \phi'$, differently from the original Lagrangian (Fig. 7.4).

7.9.4 Breaking a Continuous Symmetry: Goldstone Modes

More dramatic consequences follow when a *continuous* global symmetry is broken. We know from Chap. 1 that $SO(2)$ is isomorphic to $U(1)$ for a complex scalar field. Hence, let us focus on $SO(2)$ and consider a two-component real field with Lagrangian

$$\mathcal{L} = \frac{1}{2}\Big[(\partial_\mu\phi_1)^2 + (\partial_\mu\phi_2)^2\Big] + \frac{\mu^2}{2}(\phi_1^2 + \phi_2^2) - \frac{\lambda}{4!}(\phi_1^2 + \phi_2^2)^2. \tag{7.187}$$

This theory has a global $SO(2)$ symmetry (rotations in the internal ϕ_1-ϕ_2 plane). The set of minima forms a circle:

$$\phi_1^2 + \phi_2^2 = \frac{6\mu^2}{\lambda}. \tag{7.188}$$

Choosing one point on this circle as the vacuum breaks the continuous symmetry.

Take the vacuum at $(\phi_1, \phi_2) = (\sqrt{6\mu^2/\lambda}, 0)$ and define $\phi_1' = \phi_1 - \sqrt{6\mu^2/\lambda}$, $\phi_2' = \phi_2$. Expanding to quadratic order yields

$$\mathcal{L} = \frac{1}{2}\Big[(\partial\phi_1')^2 + (\partial\phi_2')^2\Big] - \mu^2(\phi_1')^2 + O(\phi'^3). \tag{7.189}$$

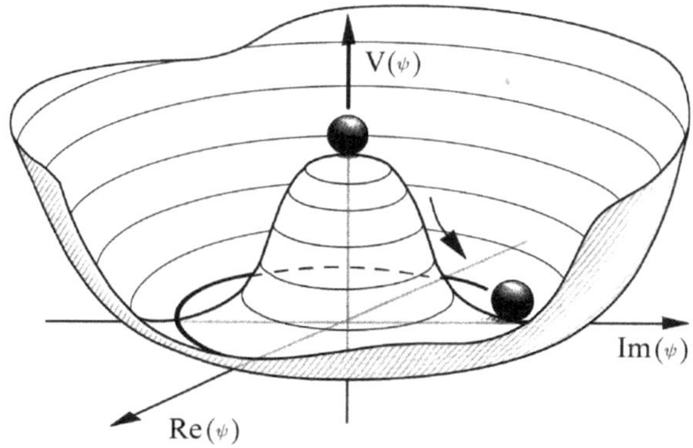

Fig. 7.5 Mexican hat potential in the (Re ψ, Im ψ) plane. The symmetric point at the origin is unstable, while the true vacuum lies along a circle of degenerate minima. After spontaneous symmetry breaking the field settles at one point on this ring. Motion along the valley, illustrated by the particle circulating around the minimum, corresponds to a massless excitation, providing the graphical interpretation of Goldstone's theorem

Thus the ϕ_1' excitation is massive with $m = \sqrt{2}\mu$, while the ϕ_2' field has no quadratic potential term and is therefore massless. Geometrically, small motion along the trough of minima costs no potential energy.

Goldstone's theorem: spontaneous breaking of a continuous global symmetry implies a massless excitation (a Goldstone mode), corresponding to a Goldstone boson.

For a complex field ψ, consider $V(\psi) \equiv -\mu^2\psi^\dagger\psi + \lambda(\psi^\dagger\psi)^2$ and

$$\mathcal{L} = (\partial_\mu\psi)^\dagger(\partial^\mu\psi) + \mu^2\psi^\dagger\psi - \lambda(\psi^\dagger\psi)^2, \tag{7.190}$$

invariant under the global transformation $\psi \to e^{i\alpha}\psi$. Write $\psi(x) = \varrho(x)e^{i\theta(x)}$. Then

$$\mathcal{L} = (\partial_\mu\varrho)^2 + \varrho^2(\partial_\mu\theta)^2 + \mu^2\varrho^2 - \lambda\varrho^4, \tag{7.191}$$

with symmetry $\varrho \to \varrho$, $\theta \to \theta + \alpha$. The minima lie at $\varrho_0 = \sqrt{\mu^2/(2\lambda)}$. Choosing (for convenience) $\theta_0 = 0$ and expanding around $\varrho' = \varrho - \varrho_0$, $\theta' = \theta - \theta_0$ yields a massive radial mode and a massless angular mode (no θ'^2 term), again reflecting motion along the trough (see Fig. 7.5).

7.10 Broken Symmetry in a Gauge Theory: The Higgs Mechanism

In a gauge theory, symmetry breaking leads to a qualitatively different outcome (the Higgs mechanism). Consider the gauged complex scalar theory

$$\mathcal{L} = (\partial_\mu \psi^\dagger - iqA_\mu \psi^\dagger)(\partial^\mu \psi + iqA^\mu \psi) + \mu^2 \psi^\dagger \psi - \lambda(\psi^\dagger \psi)^2 - \frac{1}{4} F_{\mu\nu} F^{\mu\nu}, \tag{7.192}$$

invariant under the local $U(1)$ transformation $\psi \to e^{i\alpha(x)}\psi$ provided $A_\mu \to A_\mu - \frac{1}{q}\partial_\mu \alpha(x)$.

In the symmetric phase, the spectrum includes charged massive scalars and massless photons. After symmetry breaking, the ground state picks a definite phase, thereby breaking the global symmetry and (in effect) the local symmetry as well.

► **Pause and think**
Try to understand what is the role of Goldstone boson here.

► **Driving Exercise**
The "eaten" Goldstone mode. In polar variables, one finds

$$\partial_\mu \psi + iqA_\mu \psi = (\partial_\mu \varrho)e^{i\theta} + i(\partial_\mu \theta + qA_\mu)\varrho e^{i\theta}, \tag{7.193}$$

so A_μ appears only in the gauge-invariant combination

$$C_\mu \equiv A_\mu + \frac{1}{q}\partial_\mu \theta. \tag{7.194}$$

Because $F_{\mu\nu}$ is unchanged by this shift, we can rewrite the theory as

$$\mathcal{L} = (\partial_\mu \varrho)^2 + q^2 \varrho^2 C_\mu C^\mu + \mu^2 \varrho^2 - \lambda \varrho^4 - \frac{1}{4} F_{\mu\nu} F^{\mu\nu}. \tag{7.195}$$

To make the role of the new variables explicit, start from the original complex scalar $\psi = \varrho e^{i\theta}$ coupled to the gauge field through $D_\mu \psi = (\partial_\mu + iqA_\mu)\psi$. Inserting the polar decomposition and expanding,

$$\begin{aligned} D_\mu \psi &= (\partial_\mu \varrho)e^{i\theta} + i\varrho(\partial_\mu \theta)e^{i\theta} + iqA_\mu \varrho e^{i\theta} \\ &= \Big[(\partial_\mu \varrho) + i\varrho(\partial_\mu \theta + qA_\mu)\Big]e^{i\theta}. \end{aligned} \tag{7.196}$$

Hence the kinetic term becomes

$$|D_\mu \psi|^2 = (\partial_\mu \varrho)^2 + \varrho^2(\partial_\mu \theta + qA_\mu)^2, \tag{7.197}$$

which depends on θ and A_μ only through the combination $\partial_\mu \theta + qA_\mu = q\,C_\mu$. This motivates the definition in Eq. (7.194), under which the field strength is invariant, $F_{\mu\nu}(A) = F_{\mu\nu}(C)$, since the shift by a gradient leaves the curl unchanged.

Breaking the symmetry at $\varrho_0 = \sqrt{\mu^2/(2\lambda)}$ and expanding (using a real field χ for radial fluctuations) gives, to leading orders,

$$\varrho(x) = \varrho_0 + \chi(x), \tag{7.198}$$

$$\varrho^2 = \varrho_0^2 + 2\varrho_0\chi + \chi^2. \tag{7.199}$$

Inserting this into Eq. (7.195) and using $\varrho_0^2 = \mu^2/(2\lambda)$ one finds

$$(\partial_\mu\varrho)^2 = (\partial_\mu\chi)^2, \tag{7.200}$$

$$q^2\varrho^2 C_\mu C^\mu = q^2\varrho_0^2 C_\mu C^\mu + 2q^2\varrho_0\chi\, C_\mu C^\mu + q^2\chi^2 C_\mu C^\mu, \tag{7.201}$$

$$\mu^2\varrho^2 - \lambda\varrho^4 = \mu^2(\varrho_0 + \chi)^2 - \lambda(\varrho_0 + \chi)^4. \tag{7.202}$$

Expanding the potential term and using the condition that ϱ_0 is a minimum, $\partial V/\partial\varrho|_{\varrho_0} = 0$, all linear terms in χ cancel. The remaining terms give

$$\mu^2\varrho^2 - \lambda\varrho^4 = -\mu^2\chi^2 - \sqrt{\lambda\mu}\,\chi^3 - \frac{\lambda}{4}\chi^4 + \text{const.} \tag{7.203}$$

Collecting all contributions and dropping irrelevant constants yields

$$\mathcal{L} = \frac{1}{2}(\partial_\mu\chi)^2 - \mu^2\chi^2 - \sqrt{\lambda\mu}\,\chi^3 - \frac{\lambda}{4}\chi^4 - \frac{1}{4}F_{\mu\nu}F^{\mu\nu} + \frac{M^2}{2}C_\mu C^\mu + \cdots, \tag{7.204}$$

with

$$M = q\sqrt{\frac{\mu^2}{\lambda}}. \tag{7.205}$$

The dots denote interaction terms such as $\chi C_\mu C^\mu$ and $\chi^2 C_\mu C^\mu$ that are irrelevant for identifying the particle spectrum at leading order.

Thus the gauge field becomes massive (three polarizations), while the would-be Goldstone field disappears from the physical spectrum: it is removed by a gauge transformation, thus representing a pure gauge contribution. The particle content is rearranged, but degrees of freedom are conserved.

A massive vector field in four dimensions carries three physical polarizations. In the rest frame of the massive gauge boson, they can be characterized as

- two transverse modes, whose polarization vectors are orthogonal both to the momentum and to each other; in the massless limit they reduce to the usual helicity ± 1 photon states;

- one longitudinal mode, whose polarization vector is parallel to the spatial momentum; this mode originates from the phase field θ (the would-be Goldstone boson) and is "eaten" by the gauge field through the Higgs mechanism.

The appearance of the longitudinal polarization is the concrete manifestation of how the gauge field acquires mass without violating gauge invariance: the phase degree of freedom of the scalar field is not lost, but is reshuffled into the physical spectrum as the third polarization of the massive vector boson.

7.10.1 Why Not a Higgs Fermion?

The Higgs mechanism relies on a vacuum order parameter that can be chosen to be *Poincaré invariant.* In a relativistic QFT with unbroken spacetime symmetries, the vacuum is assumed to be invariant under the proper orthochronous Poincaré group. In particular, Lorentz invariance of the vacuum implies that only vacuum expectation values (VEVs) of Lorentz *scalars* (or, more generally, quantities transforming in the trivial representation of the Lorentz group) can be non-vanishing. This immediately explains why the Higgs field may develop a non-zero VEV, whereas a fundamental spinor field cannot play the same role without changing the nature of the ground state.

Lorentz invariance forbids a spinor VEV.
Let $\psi(x)$ be a relativistic spinor field (Dirac, Weyl, or Majorana). Under a proper Lorentz transformation Λ, it transforms as

$$U(\Lambda)\,\psi(x)\,U^{-1}(\Lambda) = S(\Lambda)\,\psi(\Lambda^{-1}x), \tag{7.206}$$

where $S(\Lambda)$ is the spinor representation. If the vacuum is Lorentz invariant,

$$U(\Lambda)\,|0\rangle = |0\rangle, \tag{7.207}$$

then taking the vacuum expectation value at a fixed point (e.g. $x = 0$) yields

$$\langle 0|\psi(0)|0\rangle = \langle 0|U^{-1}(\Lambda)\,U(\Lambda)\psi(0)U^{-1}(\Lambda)\,U(\Lambda)|0\rangle = S(\Lambda)\,\langle 0|\psi(0)|0\rangle. \tag{7.208}$$

Therefore the spinor VEV $v_\psi \equiv \langle 0|\psi(0)|0\rangle$ must satisfy

$$v_\psi = S(\Lambda)\,v_\psi \qquad \forall\,\Lambda \in SO^+(1,3). \tag{7.209}$$

However, there is no non-zero spinor left invariant by *all* Lorentz transformations: the only solution of Eq. (7.209) compatible with Lorentz invariance is

$$\langle 0|\psi(x)|0\rangle = 0. \tag{7.210}$$

In physical terms, a non-vanishing spinor VEV would single out a preferred direction in spinor space, thus inducing a preferred frame and breaking Lorentz invariance [4].

This is the basic obstruction to the notion of a "Higgs fermion": the Higgs mechanism requires an order parameter that can be constant in the vacuum without spoiling Lorentz invariance, a property that fundamental spinors do not possess.

7.10.2 Fermionic Theories Do Break Symmetries

The above conclusion does *not* mean that fermionic theories cannot exhibit spontaneous symmetry breaking. Rather, it implies that the corresponding order parameters must be Lorentz scalars (or, if Lorentz symmetry is allowed to break, Lorentz tensors) constructed from fermions. The simplest examples are fermion bilinears, that we studied before, cfr Chap. 4.

A Lorentz-invariant vacuum may support scalar or pseudoscalar condensates, such as

$$\langle\bar{\psi}\psi\rangle \neq 0, \qquad \langle\bar{\psi}i\gamma^5\psi\rangle \neq 0, \tag{7.211}$$

which can break *internal* symmetries (for instance chiral symmetries) while preserving Poincaré invariance. This mechanism is realized, in different guises, in quantum chromodynamics (QCD) and in effective models such as the Nambu-Jona-Lasinio theory, see Chap. 12. The order parameter is bosonic, even though it is made of fermions.

By contrast, condensates carrying Lorentz indices generally break spacetime symmetries. For example,

$$\langle\bar{\psi}\gamma^\mu\psi\rangle \neq 0 \tag{7.212}$$

selects a preferred timelike or spacelike direction and thus corresponds to a phase with spontaneous Lorentz breaking (as may occur in media, at finite density, or in explicitly Lorentz-violating effective theories). Such phases are conceptually distinct from the standard Higgs phase in vacuum.

► **Pause and think**
Manage to distinguish the scalar and fermionic cases.

In the electroweak theory the Higgs field Φ is a Lorentz scalar, so a constant vacuum configuration $\langle\Phi\rangle = v/\sqrt{2}$ does not distinguish any direction in spacetime. The non-zero VEV breaks only the *internal* gauge symmetry (more precisely, it places the theory in the Higgs phase), while Lorentz invariance remains intact. The Yukawa sector then turns this scalar order parameter into fermion masses (see Chap. 4),

$$\mathcal{L}_Y = -y_f\, \bar{\psi}_L \Phi\, \psi_R + \text{h.c.} \quad \Longrightarrow \quad m_f = \frac{y_f\, v}{\sqrt{2}}. \tag{7.213}$$

Attempting to mimic the same logic with a fermionic order parameter would require $\langle \psi \rangle \neq 0$, which is incompatible with a Lorentz invariant vacuum. Hence, in relativistic particle physics the Higgs mechanism is naturally formulated with scalar fields (elementary or effective), whereas fermions can participate in symmetry breaking only through Lorentz-covariant composite operators.

7.11 Inverse Symmetry Breaking and Symmetry Non-restoration

As we have seen, symmetries which are broken at low temperatures are typically restored as the temperature increases. This expectation is deeply rooted in both statistical mechanics and QFT: thermal fluctuations tend to wash out order.

Remarkably, this intuitive picture is not universally correct. There exist well-defined quantum and statistical field theories in which a symmetry broken at zero temperature remains broken at arbitrarily high temperature, or is even broken *only* at high temperature. These phenomena are known respectively as *symmetry non-restoration* (SNR) and *inverse symmetry breaking* (ISB).

▶ **Driving Exercise**
Consider a scalar field theory with a $\mathbb{Z}_2$ symmetry,

$$\mathcal{L} = \frac{1}{2}(\partial_\mu \phi)^2 + \frac{1}{2}m^2\phi^2 - \frac{\lambda}{4!}\phi^4, \tag{7.214}$$

with $m^2 > 0$. At zero temperature the symmetry is spontaneously broken. At finite temperature, the one-loop effective potential acquires a thermal mass correction,

$$m^2_{\text{eff}}(T) = -m^2 + \frac{\lambda}{24}T^2, \tag{7.215}$$

which eventually becomes positive at sufficiently high T, restoring the symmetry.This behavior underlies the standard picture of phase transitions in the early Universe, including the electroweak phase transition.
Inverse symmetry breaking refers to the opposite behavior: a symmetry that is unbroken at zero temperature becomes spontaneously broken as the temperature increases.
To illustrate this mechanism, consider a theory with two real scalar fields, ϕ and χ, invariant under $\phi \to -\phi, \quad \chi \to -\chi$. The most general renormalizable potential is

$$V(\phi, \chi) = \frac{1}{2}m_\phi^2\phi^2 + \frac{1}{2}m_\chi^2\chi^2 + \frac{\lambda_\phi}{4}\phi^4 + \frac{\lambda_\chi}{4}\chi^4 + \frac{\lambda_{\phi\chi}}{2}\phi^2\chi^2. \tag{7.216}$$

At finite temperature, thermal corrections modify the effective masses:

$$m_{\phi,\mathrm{eff}}^2(T) = m_\phi^2 + \frac{T^2}{12}\left(3\lambda_\phi + \lambda_{\phi\chi}\right), \quad m_{\chi,\mathrm{eff}}^2(T) = m_\chi^2 + \frac{T^2}{12}\left(3\lambda_\chi + \lambda_{\phi\chi}\right). \tag{7.217}$$

If the mixed coupling $\lambda_{\phi\chi}$ is *negative* and sufficiently large in magnitude, the thermal correction to $m_{\phi,\mathrm{eff}}^2$ can become negative at high temperature, even if $m_\phi^2 > 0$ at $T = 0$. As a result, the symmetry associated with ϕ becomes broken only at high temperature. This is the essence of inverse symmetry breaking.

7.11.1 Symmetry Non-restoration

An even more striking phenomenon occurs when a symmetry that is broken at zero temperature remains broken for all temperatures. This happens when thermal corrections fail to restore the positive curvature at the origin of field space.

In multi-field theories, negative mixed couplings can prevent the thermal mass from ever becoming positive:

$$m_{\phi,\mathrm{eff}}^2(T) < 0 \qquad \forall T. \tag{7.218}$$

The symmetry is said to exhibit *non-restoration*.

The key lesson is that thermal fluctuations do not necessarily favor symmetry restoration. In systems with competing order parameters, entropy can actually favor an ordered phase at high temperature. From a statistical point of view, the system gains entropy by populating field configurations that break symmetry.

This mechanism has analogues in condensed matter physics, where re-entrant phases and order-by-disorder phenomena are well known.

7.11.2 Goldstone Modes and Inverse Symmetry Breaking

When a continuous symmetry is broken via ISB, Goldstone's theorem continues to apply: massless excitations appear in the broken phase. However, their thermal population can significantly affect the effective potential, requiring careful resummation techniques (e.g. daisy diagrams[5]) for consistent calculations.

Inverse symmetry breaking and symmetry non-restoration have important consequences in cosmology, still object of current research. In particular:

- they can prevent the formation of unwanted topological defects (e.g. monopoles) by keeping symmetries broken at all temperatures;
- they modify the thermal history of the early Universe, altering phase transition patterns;
- they offer alternative mechanisms for baryogenesis and dark sector dynamics.

In particular, symmetry non-restoration was proposed as a solution to the monopole problem without invoking cosmological inflation, namely a phase of strong acceleration induced by the presence of an extra-standard model scalar field.

7.12 The Stueckelberg Mechanism

Gauge invariance severely constrains the form of admissible mass terms for vector fields. In particular, a naive Proca mass term for an Abelian gauge field,

$$\mathcal{L}_{\text{mass}} = \frac{1}{2} m^2 A_\mu A^\mu, \tag{7.219}$$

explicitly breaks gauge invariance under

$$A_\mu \longrightarrow A_\mu + \partial_\mu \alpha. \tag{7.220}$$

As a consequence, gauge symmetry, renormalizability, and unitarity appear to be in tension with the presence of massive vector bosons.

The Stueckelberg mechanism provides a resolution of this tension by allowing a gauge field to acquire a mass *without breaking gauge invariance*, at the price of introducing an additional scalar degree of freedom. Unlike the Higgs mechanism, no symmetry-breaking potential is required, and the gauge symmetry remains exact at the Lagrangian level.

[5] A daisy (or ring) diagram denotes the infinite class of Feynman diagrams formed by repeated loop insertions on the same propagator, whose resummation yields a self-consistent thermal mass (or self-energy) and is essential for controlling infrared effects at finite temperature and near phase transitions.

7.12.1 Abelian Stueckelberg Theory

Consider a $U(1)$ gauge field A_μ coupled to a real scalar field ϕ. The Stueckelberg Lagrangian is defined as

$$\mathcal{L} = -\frac{1}{4} F_{\mu\nu} F^{\mu\nu} + \frac{1}{2} \left(\partial_\mu \phi - m A_\mu\right) \left(\partial^\mu \phi - m A^\mu\right), \tag{7.221}$$

where $F_{\mu\nu} = \partial_\mu A_\nu - \partial_\nu A_\mu$ and m is a mass parameter.

This Lagrangian is invariant under the *Stueckelberg gauge transformations*

$$\begin{cases} A_\mu \longrightarrow A_\mu + \partial_\mu \alpha, \\ \phi \longrightarrow \phi + m\,\alpha. \end{cases} \tag{7.222}$$

Gauge invariance is thus preserved by construction, even though the vector field effectively becomes massive.

► **Driving Exercise**

Let us compute the physical degrees of freedom associated with the Stueckelberg Lagrangian.

The combination

$$B_\mu \equiv A_\mu - \frac{1}{m} \partial_\mu \phi \tag{7.223}$$

is gauge invariant. In terms of B_μ, the Lagrangian in Eq. (7.221) can be rewritten as

$$\mathcal{L} = -\frac{1}{4} F_{\mu\nu}(B) F^{\mu\nu}(B) + \frac{1}{2} m^2 B_\mu B^\mu, \tag{7.224}$$

which is formally equivalent to the Proca theory for a massive vector field. However, the crucial difference lies in the gauge redundancy: the Stueckelberg formulation contains one additional scalar field and one gauge symmetry, so the total number of physical degrees of freedom is $(4 + 1) - 2 = 3$, corresponding to the three polarizations of a massive spin-1 particle. Gauge invariance ensures that no unphysical longitudinal mode propagates.

7.12.2 Gauge Fixing and Equivalence to Proca Theory

By choosing the unitary gauge

$$\phi = 0, \tag{7.225}$$

the Lagrangian directly reduces to the Proca form,

$$\mathcal{L} = -\frac{1}{4} F_{\mu\nu} F^{\mu\nu} + \frac{1}{2} m^2 A_\mu A^\mu. \tag{7.226}$$

In Eq. (7.226), gauge invariance is no longer manifest, but the physical spectrum is unchanged.

Conversely, in covariant gauges (such as R_ξ gauges) the scalar field ϕ remains explicit and plays a crucial role in preserving good ultraviolet behavior. This is one of the reasons why the Stueckelberg formulation is often preferred in quantum calculations.

It is instructive to contrast the Stueckelberg and Higgs mechanisms.

- In the Higgs mechanism, gauge symmetry is spontaneously broken, and the mass of the vector boson arises from the vacuum expectation value of a scalar field.
- In the Stueckelberg mechanism, gauge symmetry is exact, and the mass term is introduced through a derivative coupling to a scalar.
- No scalar potential or symmetry-breaking vacuum is required in the Stueckelberg case.
- The Stueckelberg scalar does not correspond to a physical Higgs particle.

In the Abelian case, the Stueckelberg mechanism can be viewed as the limit of a Higgs model in which the radial Higgs mode is infinitely heavy and decouples, leaving only the phase degree of freedom.

7.12.3 Non-abelian Generalizations

Extending the Stueckelberg mechanism to non-Abelian gauge groups (see Chap. 1) is conceptually possible but technically subtle. A naive generalization,

$$\partial_\mu \phi^a - m A^a_\mu, \tag{7.227}$$

does not transform covariantly under non-Abelian gauge transformations.

Consistent non-Abelian Stueckelberg theories require the introduction of group-valued fields,

$$U(x) = \exp\left(\frac{i}{m}\phi^a(x)T^a\right), \tag{7.228}$$

and lead to nonlinear sigma models coupled to gauge fields. Such theories are typically non-renormalizable and must be interpreted as effective field theories, valid below a cutoff scale.

Last but not least, the Stueckelberg mechanism appears naturally in several modern contexts:

- effective field theories with massive Abelian gauge bosons,
- string-inspired models with anomalous $U(1)$ symmetries,
- hidden-sector gauge fields and dark photon scenarios,
- theories with kinetic mixing between multiple $U(1)$ factors.

In these settings, the Stueckelberg field often originates from higher-dimensional forms or axion-like degrees of freedom and provides a natural, gauge-invariant description of vector boson masses.

7.13 Bogoliubov Transformations

Bogoliubov transformations are the natural language for *quadratic* (or bilinear) dynamics. These transformations encode:

1. the canonical reparametrizations that preserve the symplectic structure in classical mechanics,
2. the linear maps that preserve commutation relations in quantum mechanics, and
3. the change of mode decompositions in QFT, where the notion of "particle" becomes representation-dependent.

In all cases, the characteristic feature is the mixing between "positive-frequency" and "negative-frequency" components, i.e., between a variable and its complex conjugate (or, quantum mechanically, between annihilation and creation operators).

In this section, we begin with the formulation in terms of complex canonical variables and then develop the corresponding quantum and QFT generalizations.

7.13.1 Bogoliubov Transformations in Classical Physics

Consider a set of complex canonical variables $\{a_j, a_j^*\}$ and a Hamiltonian $H(a, a^*)$. The equations of motion can be written in terms of Poisson brackets as

$$i\,\dot{a}_j = \frac{\partial H}{\partial a_j^*}, \qquad -i\,\dot{a}_j^* = \frac{\partial H}{\partial a_j}, \tag{7.229}$$

which is equivalent to adopting the complex-variable Poisson bracket

$$\{A, B\} = -i \sum_j \left(\frac{\partial A}{\partial a_j}\frac{\partial B}{\partial a_j^*} - \frac{\partial A}{\partial a_j^*}\frac{\partial B}{\partial a_j} \right). \tag{7.230}$$

Driving Exercise
Let us perform a change of variables

$$b_s = b_s(\{a_j, a_j^*\}), \qquad b_s^* = b_s^*(\{a_j, a_j^*\}), \tag{7.231}$$

and ask how the equations of motion look with respect to the new variables. Using Eq. (7.230),

$$i\,\dot{b}_s = i\{H, b_s\} = \sum_j \left[\frac{\partial b_s}{\partial a_j}\frac{\partial H}{\partial a_j^*} - \frac{\partial b_s}{\partial a_j^*}\frac{\partial H}{\partial a_j}\right]. \tag{7.232}$$

The crucial question is whether one can choose $\{b_s\}$ so that, *for any Hamiltonian H*,

$$i\,\dot{b}_s = \frac{\partial H}{\partial b_s^*}. \tag{7.233}$$

This is precisely the definition of a canonical transformation in this complex formulation.

Hence, we end up with necessary and sufficient conditions for canonicity.

To enforce Eq. (7.233) for arbitrary H, we compare differential operators. Starting from Eq. (7.232) and demanding equivalence with Eq. (7.233) for any H yields the operator identity

$$\sum_j \left[\frac{\partial b_s}{\partial a_j}\frac{\partial}{\partial a_j^*} - \frac{\partial b_s}{\partial a_j^*}\frac{\partial}{\partial a_j}\right] \equiv \frac{\partial}{\partial b_s^*}. \tag{7.234}$$

Now expand $\partial/\partial b_s^*$ by the chain rule:

$$\frac{\partial}{\partial b_s^*} = \sum_j \left(\frac{\partial a_j^*}{\partial b_s^*}\frac{\partial}{\partial a_j^*} + \frac{\partial a_j}{\partial b_s^*}\frac{\partial}{\partial a_j}\right). \tag{7.235}$$

Since $\{\partial/\partial a_j, \partial/\partial a_j^*\}$ are linearly independent, matching the coefficients of $\partial/\partial a_j^*$ and $\partial/\partial a_j$ in Eqs. (7.234)–(7.235) gives the necessary and sufficient conditions

$$\boxed{\frac{\partial b_s}{\partial a_j} = \frac{\partial a_j^*}{\partial b_s^*}, \qquad \frac{\partial b_s}{\partial a_j^*} = -\frac{\partial a_j}{\partial b_s^*}.} \tag{7.236}$$

These are the complex-variable analog of symplectic (canonical) conditions.

Driving Exercise
Some elementary examples are immediate.

- *Shift:* $a_j \mapsto a_j + c_j$ with $c_j \in \mathbb{C}$. One has $\partial b/\partial a = 1$ and $\partial b/\partial a^* = 0$, and likewise $\partial a^*/\partial b^* = 1$, $\partial a/\partial b^* = 0$, hence Eq. (7.236) holds.
- *Phase rotation:* $a_j \mapsto e^{i\varphi_j} a_j$ with $\varphi_j \in \mathbb{R}$. Then $\partial b/\partial a = e^{i\varphi_j}$ and $\partial a^*/\partial b^* = e^{i\varphi_j}$, while the mixed derivatives vanish; again Eq. (7.236) holds.

Another intriguing example is given by linear transformations without mixing.
Consider $b_s = \sum_j u_{sj} a_j$ with constant matrix u. A direct computation using the fundamental brackets implied by Eq. (7.230) gives

$$\{b_s, b_r^*\} = -i \sum_j u_{sj} u_{rj}^*. \tag{7.237}$$

Canonicity requires $\{b_s, b_r^*\} = -i\delta_{sr}$ and $\{b_s, b_r\} = 0$, hence

$$\sum_j u_{sj} u_{rj}^* = \delta_{sr}, \qquad \text{i.e.} \qquad u\,u^\dagger = \mathbb{I}, \tag{7.238}$$

so u must be unitary.

7.13.1.1 Invariance of Poisson Brackets

A remarkable consequence of Eq. (7.236) is that Poisson brackets are invariant under canonical transformations:

$$\{A, B\}_{(a,a^*)} = \{A, B\}_{(b,b^*)}. \tag{7.239}$$

Conversely, bracket invariance is *sufficient* for canonicity. One way to see this is to use, for any observable A,

$$\frac{\partial A}{\partial a_j^*} = i\{A, a_j\}, \qquad \frac{\partial A}{\partial a_j} = i\{a_j^*, A\}, \tag{7.240}$$

which follow directly from Eq. (7.230). If Poisson brackets are invariant under the change of variables, then

$$\frac{\partial b_s}{\partial a_j} = i\{a_j^*, b_s\} = \frac{\partial a_j^*}{\partial b_s^*}, \tag{7.241}$$

$$\frac{\partial b_s}{\partial a_j^*} = i\{b_s, a_j\} = -i\{a_j, b_s\} = -\frac{\partial a_j}{\partial b_s^*}, \tag{7.242}$$

which reproduces Eq. (7.236).

7.13.1.2 Quadratic Hamiltonians and the Classical Bogoliubov Map

Near a (local) equilibrium point one expands H to second order in deviations. Up to a canonical shift that moves the equilibrium to $a = 0$, the most general quadratic Hamiltonian reads

$$H = \sum_{ij} \left(A_{ij} a_i^* a_j + \frac{1}{2} B_{ij} a_i a_j + \frac{1}{2} B_{ij}^* a_i^* a_j^* \right), \tag{7.243}$$

with Hermiticity/symmetry constraints

$$A_{ij} = A_{ji}^*, \qquad B_{ij} = B_{ji}. \tag{7.244}$$

If $a = 0$ is a local minimum, H can be diagonalized by a *linear canonical* transformation that mixes a and a^*:

$$b_s = \sum_j \left(u_{sj} a_j + v_{sj} a_j^* \right). \tag{7.245}$$

Diagonalization means

$$H = \sum_s E_s \, b_s^* b_s, \tag{7.246}$$

i.e. a decoupled set of harmonic oscillators (normal modes).

Quite importantly, we can now evaluate the canonical constraints on (u, v). Imposing $\{b_s, b_r^*\} = -i\delta_{sr}$ and $\{b_s, b_r\} = 0$ yields

$$\sum_j \left(u_{sj} u_{rj}^* - v_{sj} v_{rj}^* \right) = \delta_{sr}, \tag{7.247}$$

$$\sum_j \left(u_{sj} v_{rj} - v_{sj} u_{rj} \right) = 0. \tag{7.248}$$

These are the classical Bogoliubov (symplectic) conditions.

Proof of inverse transformation.
Assume Eqs. (7.247)–(7.248). Define

$$a_j = \sum_s \left(\tilde{u}_{js} b_s + \tilde{v}_{js} b_s^* \right), \qquad \tilde{u}_{js} = u_{sj}^*, \qquad \tilde{v}_{js} = -v_{sj}. \tag{7.249}$$

Substituting Eq. (7.249) into Eq. (7.245) gives

$$\begin{aligned} b_r &= \sum_j \left(u_{rj}a_j + v_{rj}a_j^*\right) \\ &= \sum_j u_{rj} \sum_s \left(\tilde{u}_{js}b_s + \tilde{v}_{js}b_s^*\right) + \sum_j v_{rj} \sum_s \left(\tilde{u}_{js}^* b_s^* + \tilde{v}_{js}^* b_s\right) \\ &= \sum_s \Bigg[\underbrace{\sum_j \left(u_{rj}\tilde{u}_{js} + v_{rj}\tilde{v}_{js}^*\right)}_{C_{rs}} b_s + \underbrace{\sum_j \left(u_{rj}\tilde{v}_{js} + v_{rj}\tilde{u}_{js}^*\right)}_{\mathcal{D}_{rs}} b_s^* \Bigg]. \end{aligned} \tag{7.250}$$

Using $\tilde{u}_{js} = u_{sj}^*$ and $\tilde{v}_{js}^* = -v_{sj}^*$,

$$C_{rs} = \sum_j \left(u_{rj}u_{sj}^* - v_{rj}v_{sj}^*\right) = \delta_{rs} \quad \text{by Eq. (7.247)}, \tag{7.251}$$

while $\tilde{v}_{js} = -v_{sj}$ and $\tilde{u}_{js}^* = u_{sj}$ give

$$\mathcal{D}_{rs} = \sum_j \left(-u_{rj}v_{sj} + v_{rj}u_{sj}\right) = 0 \quad \text{by Eq. (7.248)}. \tag{7.252}$$

Hence $b_r = \sum_s (\delta_{rs}b_s + 0 \cdot b_s^*) = b_r$, proving that Eq. (7.249) is indeed the inverse.
From Eq. (7.246) one has

$$E_s b_s = \frac{\partial H}{\partial b_s^*}. \tag{7.253}$$

Differentiate Eq. (7.253) with respect to a_j and a_j^* and use

$$\frac{\partial b_s}{\partial a_j} = u_{sj}, \qquad \frac{\partial b_s}{\partial a_j^*} = v_{sj}. \tag{7.254}$$

Because the transformation is linear, derivatives commute, and we obtain

$$E_s u_{sj} = \frac{\partial}{\partial a_j}\left(\frac{\partial H}{\partial b_s^*}\right) = \frac{\partial}{\partial b_s^*}\left(\frac{\partial H}{\partial a_j}\right), \tag{7.255}$$

$$E_s v_{sj} = \frac{\partial}{\partial a_j^*}\left(\frac{\partial H}{\partial b_s^*}\right) = \frac{\partial}{\partial b_s^*}\left(\frac{\partial H}{\partial a_j^*}\right). \tag{7.256}$$

From Eq. (7.243),

$$\frac{\partial H}{\partial a_j} = \sum_i \left(A_{ij}a_i^* + B_{ij}a_i\right), \tag{7.257}$$

$$\frac{\partial H}{\partial a_j^*} = \sum_i \left(A_{ji}a_i + B_{ij}^* a_i^*\right). \tag{7.258}$$

Next use the inverse map in Eq. (7.249) to evaluate

$$\frac{\partial a_i}{\partial b_s^*} = \tilde{v}_{is} = -v_{si}, \qquad \frac{\partial a_i^*}{\partial b_s^*} = \tilde{u}_{is}^* = u_{si}. \tag{7.259}$$

Substituting Eqs. (7.257)–(7.259) into Eqs. (7.255)–(7.256) gives the coupled system

$$\sum_i \left(A_{ij}u_{si} - B_{ji}v_{si}\right) = E_s u_{sj}, \tag{7.260}$$

$$\sum_i \left(B_{ji}^* u_{si} - A_{ji}v_{si}\right) = E_s v_{sj}. \tag{7.261}$$

Introduce vectors $|u_s\rangle = (u_{s1}, u_{s2}, \dots)^T$ and $|v_s\rangle = (v_{s1}, v_{s2}, \dots)^T$. Using Eq. (7.244), Eqs. (7.260)–(7.261) become

$$A^*|u_s\rangle - B|v_s\rangle = E_s|u_s\rangle, \tag{7.262}$$

$$B^*|u_s\rangle - A|v_s\rangle = E_s|v_s\rangle. \tag{7.263}$$

Equations (7.262)–(7.263) are invariant under

$$|u_s\rangle \mapsto |v_s^*\rangle, \qquad |v_s\rangle \mapsto |u_s^*\rangle, \qquad E_s \mapsto -E_s, \tag{7.264}$$

so eigenvalues come in $\pm E$ pairs. Only one member is compatible with the normalization implied by Eq. (7.247):

$$\langle u_s|u_s\rangle - \langle v_s|v_s\rangle > 0, \tag{7.265}$$

since the opposite sign cannot be rescaled to satisfy $\langle u_s|u_s\rangle - \langle v_s|v_s\rangle = 1$.

▶ **Driving Exercise**
Consider the one-mode quadratic Hamiltonian

$$H = a^*a + \frac{1}{2}\left(\gamma\, aa + \gamma^* a^*a^*\right). \tag{7.266}$$

(i) Removing the phase of γ.
Let $\gamma = |\gamma|e^{i\theta}$. Perform the canonical phase shift $a \mapsto e^{-i\theta/2}a$. Then $aa \mapsto e^{-i\theta}aa$ and $a^*a^* \mapsto e^{+i\theta}a^*a^*$, so Eq. (7.266) becomes

$$H = a^*a + \frac{1}{2}\left(|\gamma|\, aa + |\gamma|\, a^*a^*\right), \tag{7.267}$$

i.e. we may take $\gamma > 0$ without loss of generality.
(ii) Bogoliubov diagonalization for $0 \le \gamma < 1$.

Try $b = u\,a + v\,a^*$ with $u, v \in \mathbb{C}$ constants. Canonicity in Eq. (7.247) gives

$$|u|^2 - |v|^2 = 1. \tag{7.268}$$

Choose u, v real (consistent with $\gamma > 0$) and parameterize

$$u = \cosh r, \qquad v = \sinh r, \tag{7.269}$$

so that Eq. (7.268) is automatic. Invert: $a = u\,b - v\,b^*$. Substitute into Eq. (7.266). A convenient route is to rewrite H in Nambu form

$$H = \frac{1}{2}\begin{pmatrix} a^* & a \end{pmatrix}\begin{pmatrix} 1 & \gamma \\ \gamma & 1 \end{pmatrix}\begin{pmatrix} a \\ a^* \end{pmatrix} - \frac{1}{2}, \tag{7.270}$$

where the constant $-1/2$ accounts for symmetrization (classically it is irrelevant; quantum mechanically it becomes the zero-point shift). Under the squeeze of Eq. (7.269) the off-diagonal terms vanish if

$$\tanh 2r = \gamma. \tag{7.271}$$

This requires $|\gamma| < 1$. Then one finds the diagonal form

$$H = E\,b^*b\; +\; \text{const}, \qquad E = \sqrt{1-\gamma^2}. \tag{7.272}$$

Thus $0 \le \gamma < 1$ corresponds to a stable harmonic mode of frequency E.
(iii) Direct dynamics for $\gamma > 1$ (unstable regime).
From $i\dot{a} = \partial H/\partial a^*$ one gets

$$i\dot{a} = a + \gamma\,a^*, \qquad -i\dot{a}^* = a^* + \gamma\,a. \tag{7.273}$$

Differentiate the first equation with respect to t and use the second:

$$i\ddot{a} = \dot{a} + \gamma\,\dot{a}^*, \tag{7.274}$$

$$\dot{a} = -i(a + \gamma a^*), \qquad \dot{a}^* = i(a^* + \gamma a), \tag{7.275}$$

hence

$$i\ddot{a} = -i(a + \gamma a^*) + \gamma\,i(a^* + \gamma a) = -i(1-\gamma^2)a. \tag{7.276}$$

Therefore

$$\ddot{a} = -\,(1-\gamma^2)a. \tag{7.277}$$

For $\gamma > 1$ the coefficient is positive and solutions grow/decay exponentially:

$$a(t) = c_+\,e^{\sqrt{\gamma^2-1}\,t} + c_-\,e^{-\sqrt{\gamma^2-1}\,t}, \tag{7.278}$$

signaling dynamical instability of the quadratic equilibrium, consistent with the fact that Eq. (7.271) has no real solution.

7.13.1.3 Time Evolution as a Bogoliubov (canonical) Transformation

Define new variables b_j as the value of a_j after time t: $b_j = a_j(t)$.
For infinitesimal t,

$$b_j = a_j - it\,\frac{\partial H}{\partial a_j^*} + O(t^2). \tag{7.279}$$

Then

$$\frac{\partial b_j}{\partial a_k} = \delta_{jk} - it\,\frac{\partial^2 H}{\partial a_k \partial a_j^*} + O(t^2), \tag{7.280}$$

$$\frac{\partial b_j}{\partial a_k^*} = -it\,\frac{\partial^2 H}{\partial a_k^* \partial a_j^*} + O(t^2). \tag{7.281}$$

Moreover, to leading order $\partial/\partial b_j = \partial/\partial a_j + O(t)$. Substituting these into Eq. (7.236) shows that evolution is canonical up to $O(t^2)$; chaining infinitesimal steps yields canonicity for finite t.

If the Hamiltonian is quadratic, the evolution is linear, and in vector notation $|a\rangle = (a_1, a_2, \ldots)^T$ one can write

$$|a(t)\rangle = U(t)|a(0)\rangle + V(t)|a^*(0)\rangle, \tag{7.282}$$

which is precisely a *time-dependent Bogoliubov transformation.*

▶ **Driving Exercise**

Explicit $U(t)$, $V(t)$ in the diagonal basis. If $b_s = \sum_j (u_{sj} a_j + v_{sj} a_j^*)$ diagonalizes H as in Eq. (7.246), then

$$i\dot{b}_s = \frac{\partial H}{\partial b_s^*} = E_s b_s \quad \Rightarrow \quad b_s(t) = e^{-iE_s t}\, b_s(0). \tag{7.283}$$

Using the inverse $a_j = \sum_s (u_{sj}^* b_s - v_{sj} b_s^*)$, we obtain

$$\begin{aligned} a_j(t) &= \sum_s \left[u_{sj}^* e^{-iE_s t} b_s(0) - v_{sj} e^{+iE_s t} b_s^*(0)\right] \\ &= \sum_{s,k} \left[u_{sj}^* e^{-iE_s t} u_{sk} - v_{sj} e^{+iE_s t} v_{sk}^*\right] a_k(0) \\ &\quad + \sum_{s,k} \left[u_{sj}^* e^{-iE_s t} v_{sk} - v_{sj} e^{+iE_s t} u_{sk}^*\right] a_k^*(0), \end{aligned} \tag{7.284}$$

so that Eq. (7.282) holds with

$$U_{jk}(t) = \sum_s \left(u_{sj}^* u_{sk}\, e^{-iE_s t} - v_{sj} v_{sk}^*\, e^{+iE_s t}\right), \tag{7.285}$$

$$V_{jk}(t) = \sum_s \left(u_{sj}^* v_{sk}\, e^{-iE_s t} - v_{sj} u_{sk}^*\, e^{+iE_s t}\right). \tag{7.286}$$

7.13.1.4 Stability from the Linearized Spectrum

Let $a^{(0)}$ be an equilibrium point:

$$\left.\frac{\partial H}{\partial a_s}\right|_{a^{(0)}} = 0, \qquad \left.\frac{\partial H}{\partial a_s^*}\right|_{a^{(0)}} = 0. \tag{7.287}$$

Expanding around $a^{(0)}$ yields a quadratic Hamiltonian as in Eq. (7.243), with

$$A_{ij} = \left.\frac{\partial^2 H}{\partial a_i\, \partial a_j^*}\right|_{a^{(0)}}, \qquad B_{ij} = \left.\frac{\partial^2 H}{\partial a_i\, \partial a_j}\right|_{a^{(0)}}. \tag{7.288}$$

The linearized equation of motion is

$$i\frac{\partial}{\partial t}|a\rangle = A|a\rangle + B^*|a^*\rangle. \tag{7.289}$$

Seek elementary solutions of the form $|a(t)\rangle = e^{-i\omega t}|f\rangle + e^{+i\omega^* t}|g\rangle$, which leads, upon substitution into Eq. (7.289) and separating the independent exponentials, to

$$A|f\rangle + B^*|g^*\rangle = \omega|f\rangle, \tag{7.290}$$

$$B|f\rangle + A^*|g^*\rangle = -\omega|g^*\rangle. \tag{7.291}$$

This system matches the diagonalization problem in Eqs. (7.262)–(7.263): for real ω the solutions describe oscillatory normal modes, while complex ω implies exponential growth/decay, i.e., dynamical instability.

7.13.2 Bogoliubov Transformations in Quantum Physics

In quantum mechanics, the classical canonical variables are replaced by operators. For a set of bosonic modes $\{a_j, a_j^\dagger\}$,

$$[a_j, a_k^\dagger] = \delta_{jk}, \qquad [a_j, a_k] = [a_j^\dagger, a_k^\dagger] = 0. \tag{7.292}$$

A (bosonic) Bogoliubov transformation is a linear map

$$b_s = \sum_j \left(u_{sj}a_j + v_{sj}a_j^\dagger\right), \qquad b_s^\dagger = \sum_j \left(u_{sj}^*a_j^\dagger + v_{sj}^*a_j\right), \tag{7.293}$$

such that the new operators satisfy the same commutation relations:

$$[b_s, b_r^\dagger] = \delta_{sr}, \qquad [b_s, b_r] = 0. \tag{7.294}$$

Using Eq. (7.292), one computes

$$[b_s, b_r^\dagger] = \sum_{jk} \left(u_{sj} u_{rk}^* [a_j, a_k^\dagger] + v_{sj} v_{rk}^* [a_j^\dagger, a_k] \right) = \sum_j \left(u_{sj} u_{rj}^* - v_{sj} v_{rj}^* \right), \tag{7.295}$$

$$[b_s, b_r] = \sum_j \left(u_{sj} v_{rj} - v_{sj} u_{rj} \right), \tag{7.296}$$

so Eq. (7.294) is equivalent to the same algebraic constraints as in the classical case, namely Eqs. (7.247)–(7.248). In matrix form, with $U_{sj} = u_{sj}$ and $V_{sj} = v_{sj}$,

$$UU^\dagger - VV^\dagger = \mathbb{I}, \qquad UV^T - VU^T = 0. \tag{7.297}$$

These conditions define the bosonic symplectic group in the complex (annihilation/creation) representation.

7.13.2.1 Unitary Implementability and Squeezing

A commutator-preserving linear map does not automatically guarantee the existence of a unitary operator $\mathcal{U}$ such that $b_s = \mathcal{U}\, a_s\, \mathcal{U}^\dagger$. For a finite number of modes this is always possible: every solution of Eq. (7.297) is unitarily implementable, and the corresponding $\mathcal{U}$ is generated by a quadratic operator (a metaplectic representation).

For one mode, $b = ua + va^\dagger$, $|u|^2 - |v|^2 = 1$. Writing $u = e^{i\phi} \cosh r$ and $v = e^{i\psi} \sinh r$, one can realize the transformation via the squeeze operator $S(\zeta) = \exp\left[\frac{1}{2}\left(\zeta^* a^2 - \zeta (a^\dagger)^2\right)\right]$, $\zeta = r\, e^{i(\psi - \phi)}$, for which

$$S(\zeta)\, a\, S(\zeta)^\dagger = a \cosh r - a^\dagger e^{i(\psi - \phi)} \sinh r. \tag{7.298}$$

Accordingly, Bogoliubov transformations provide the operator-theoretic implementation of gaussian unitaries and squeezed states.

7.13.2.2 Diagonalizing Quadratic Hamiltonians

The quantum counterpart of the classical quadratic Hamiltonian in Eq. (7.243) is

$$\hat{H} = \sum_{ij} \left(A_{ij}\, a_i^\dagger a_j + \frac{1}{2} B_{ij}\, a_i a_j + \frac{1}{2} B_{ij}^*\, a_i^\dagger a_j^\dagger \right) + \text{const}, \tag{7.299}$$

with the same constraints, Eq. (7.244). A Bogoliubov transformation, Eq. (7.293), that satisfies Eq. (7.297) can be chosen to diagonalize $\hat{H}$:

$$\hat{H} = \sum_s E_s\, b_s^\dagger b_s + E_0, \tag{7.300}$$

where E_0 is the (renormalized) ground-state energy. Technically, the same eigenvalue problem, as in Eqs. (7.262)–(7.263), arises, now as the diagonalization of a quadratic form on the bosonic Fock space. The paired $\pm E$ structure remains, and one keeps the positive branch to ensure a stable vacuum.

Vacua and quasiparticles.
If $|0_a\rangle$ is the Fock vacuum annihilated by all a_j, the new vacuum $|0_b\rangle$ is defined by $b_s|0_b\rangle = 0$. When $V \neq 0$, $|0_b\rangle$ is not empty of a-quanta:

$$\langle 0_b|\, a_j^\dagger a_j \,|0_b\rangle = \sum_s |v_{sj}|^2, \tag{7.301}$$

a quantitative statement of "pair creation" in the original basis.

Pause and think
Why quasiparticles appear? And what they are?

The operators $a_j^\dagger$ are naturally interpreted as particle creation operators only as long as the Hamiltonian is diagonal in the corresponding number operators $a_j^\dagger a_j$. When pairing terms of the form $a_i a_j$ and $a_i^\dagger a_j^\dagger$ are present, the number of a-quanta is no longer conserved and states with a definite number of a-particles are not stationary. In this situation the physically meaningful excitations are those that diagonalize the quadratic form of the Hamiltonian. The Bogoliubov transformation constructs such excitations by mixing creation and annihilation operators, defining new operators b_s whose quanta are eigenmodes of the energy operator.

The quanta created by $b_s^\dagger$ are referred to as quasiparticles: they behave as free particles with well-defined energies E_s with respect to the diagonal Hamiltonian, but their internal structure reflects the interactions and correlations encoded in the original quadratic form.

Now, we ended up with something new: the existence of quasiparticles. Very naively in the Dirac *hole theory* we have had some hints about something that was not exactly a particle, but behaving like it (the hole!).

We will formalize better the concept of quasiparticles in Chap. 8, whereas below we will turn into the role of Bogoliubov transformations in QFT.

7.13.3 Bogoliubov Transformations in Quantum Field Theory

In QFT the canonical variables become field operators. For definiteness, consider a real scalar field $\hat{\phi}(x)$ on a globally hyperbolic spacetime. Choose a complete set of complex solutions $\{u_k(x)\}$ of the classical field equation, normalized with respect to the Klein-Gordon inner product

$$(u, v)_{\mathrm{KG}} = i \int_{\Sigma} d\Sigma^{\mu} \left(u^* \nabla_{\mu} v - v \nabla_{\mu} u^*\right), \tag{7.302}$$

on a Cauchy surface Σ. A choice of *positive-frequency* modes defines the expansion

$$\hat{\phi}(x) = \sum_k \left(a_k\, u_k(x) + a_k^{\dagger}\, u_k^*(x)\right), \tag{7.303}$$

where the coefficients are operators satisfying Eq. (7.292). The vacuum is defined by $a_k|0\rangle = 0$.

Now consider another complete normalized set $\{v_k(x)\}$, corresponding to a different notion of positive frequency (e.g. different time coordinate, different asymptotic region, or different observer). Completeness implies a linear relation at the level of classical mode functions:

$$v_k(x) = \sum_j \left(\alpha_{kj}\, u_j(x) + \beta_{kj}\, u_j^*(x)\right). \tag{7.304}$$

Substituting Eq. (7.304) into the field expansion and matching coefficients yields the *QFT Bogoliubov transformation*

$$b_k = \sum_j \left(\alpha_{kj}^* a_j - \beta_{kj}^* a_j^{\dagger}\right), \qquad b_k^{\dagger} = \sum_j \left(\alpha_{kj} a_j^{\dagger} - \beta_{kj} a_j\right), \tag{7.305}$$

where b_k annihilates the vacuum associated with the v-modes.

7.13.3.1 Bogoliubov Identities from the Klein-Gordon Product

The normalization and orthogonality of the mode sets with respect to Eq. (7.302) imply constraints on (α, β). Using $(u_i, u_j)_{\mathrm{KG}} = \delta_{ij}$, $(u_i^*, u_j^*)_{\mathrm{KG}} = -\delta_{ij}$, and $(u_i, u_j^*)_{\mathrm{KG}} = 0$, one finds

$$\begin{aligned}(v_k, v_\ell)_{\mathrm{KG}} &= \sum_{ij} \left(\alpha_{ki}\alpha_{\ell j}^* (u_i, u_j)_{\mathrm{KG}} + \beta_{ki}\beta_{\ell j}^* (u_i^*, u_j^*)_{\mathrm{KG}}\right) \\ &= \sum_i \left(\alpha_{ki}\alpha_{\ell i}^* - \beta_{ki}\beta_{\ell i}^*\right) = \delta_{k\ell},\end{aligned} \tag{7.306}$$

$$(v_k, v_\ell^*)_{\mathrm{KG}} = \sum_i \left(\alpha_{ki}\beta_{\ell i} - \beta_{ki}\alpha_{\ell i}\right) = 0. \tag{7.307}$$

These are the direct analog of Eqs. (7.247)–(7.248) and guarantee that Eq. (7.305) preserves the canonical commutators:

$$[b_k, b_\ell^\dagger] = \delta_{k\ell}, \qquad [b_k, b_\ell] = 0. \tag{7.308}$$

Particle production and number expectation values
If $|0_a\rangle$ is the vacuum annihilated by all a_j, the expected number of b-quanta in $|0_a\rangle$ is

$$\langle 0_a| \, b_k^\dagger b_k \, |0_a\rangle = \sum_j |\beta_{kj}|^2. \tag{7.309}$$

Hence the β-coefficients quantify particle production between the two splittings of the field into positive/negative frequencies.
Two paradigmatic contexts where $\beta \neq 0$ are:

- *time-dependent backgrounds* (e.g. expanding cosmologies), where early-time and late-time adiabatic mode bases differ;
- *observer dependence* (e.g. accelerated vs inertial observers), where different timelike Killing vectors lead to non-equivalent frequency notions.

A standard sufficient condition for unitary implementability (bosons) is the Hilbert-Schmidt property of β:

$$\sum_{kj} |\beta_{kj}|^2 < \infty, \tag{7.310}$$

which ensures that the transformed vacuum has finite total particle number with respect to the original Fock space. When this sum diverges, the two vacua live in inequivalent Hilbert-space representations, and the Bogoliubov map is realized only as an algebraic (not Hilbert-space unitary) isomorphism.

7.14 Proposed Exercises

Exercise 1

Consider a real scalar field ϕ at finite temperature $T = \beta^{-1}$ in the imaginary-time (Euclidean) formulation. Let $\tau \in [0, \beta]$ denote Euclidean time and impose periodicity $\phi(\tau + \beta, \mathbf{x}) = \phi(\tau, \mathbf{x})$.

(a) Starting from the Euclidean action

$$S_E[\phi] = \int_0^\beta d\tau \int d^3x \left[\frac{1}{2}(\partial_\tau\phi)^2 + \frac{1}{2}(\nabla\phi)^2 + \frac{1}{2}m^2\phi^2\right], \tag{7.311}$$

derive the Fourier expansion compatible with periodicity and show that the allowed frequencies are $\omega_n = 2\pi n/\beta$.

(b) Define the Euclidean two-point function

$$G_E(\tau, \mathbf{x}) \equiv \langle\phi(\tau, \mathbf{x})\phi(0, \mathbf{0})\rangle_\beta, \tag{7.312}$$

and show that in momentum space it admits the representation

$$G_E(\tau, \mathbf{x}) = \frac{1}{\beta}\sum_{n\in\mathbb{Z}}\int \frac{d^3p}{(2\pi)^3}\,\frac{e^{i\mathbf{p}\cdot\mathbf{x}}\,e^{i\omega_n\tau}}{\omega_n^2 + \mathbf{p}^2 + m^2}. \tag{7.313}$$

(c) Perform the Matsubara sum in Eq. (7.313) for $0 < \tau < \beta$ and show that the result can be written in the form

$$G_E(\tau, \mathbf{p}) \equiv \int d^3x\, e^{-i\mathbf{p}\cdot\mathbf{x}}\, G_E(\tau, \mathbf{x}) = \frac{1}{2E_\mathbf{p}}\left[(1 + n_B(E_\mathbf{p}))e^{-E_\mathbf{p}\tau} + n_B(E_\mathbf{p})\, e^{+E_\mathbf{p}(\tau-\beta)}\right], \tag{7.314}$$

with $E_\mathbf{p} = \sqrt{\mathbf{p}^2 + m^2}$ and $n_B(E) = (e^{\beta E} - 1)^{-1}$.

Exercise 2

Consider in Minkowski signature a real scalar with quartic self-interaction,

$$\mathcal{L} = \frac{1}{2}(\partial_\mu\phi)^2 - \frac{m^2}{2}\phi^2 - \frac{g}{8}\phi^4, \qquad g > 0. \tag{7.315}$$

Introduce an auxiliary real field σ through

$$\Delta\mathcal{L} = \frac{1}{2g}\left(\sigma - \frac{g}{2}\phi^2\right)^2. \tag{7.316}$$

(a) Expand Eq. (7.316) and show that the Lagrangian $\mathcal{L} + \Delta\mathcal{L}$ is equivalent to

$$\mathcal{L}_{\text{HS}} = \frac{1}{2}(\partial_\mu\phi)^2 - \frac{m^2}{2}\phi^2 + \frac{1}{2g}\sigma^2 - \frac{1}{2}\sigma\phi^2, \tag{7.317}$$

up to an irrelevant field-independent constant.

(b) Consider the generating functional (no sources)

$$Z = \int \mathcal{D}\phi \, e^{i \int d^4x \, \mathcal{L}} \quad \text{and} \quad Z_{\text{HS}} = \int \mathcal{D}\phi \, \mathcal{D}\sigma \, e^{i \int d^4x \, \mathcal{L}_{\text{HS}}}. \tag{7.318}$$

Show explicitly, by performing the Gaussian functional integral over σ, that $Z_{\text{HS}} = Z$ (up to an overall normalization independent of m, g).

(c) Derive the classical equation of motion for σ from Eq. (7.317) and interpret it as a local constitutive relation between σ and the composite operator ϕ^2.

(d) Working at tree level, compute the effective potential $V_{\text{tree}}(\phi, \sigma)$ associated with Eq. (7.317).

Exercise 3

Let $\eta_i, \bar{\eta}_i$ $(i = 1, \ldots, N)$ be independent Grassmann variables and let A be an $N \times N$ matrix with commuting entries.

(a) Prove the multi-mode Grassmann Gaussian identity

$$\int \prod_{i=1}^{N} d\eta_i \, d\bar{\eta}_i \, \exp\left(- \sum_{i,j=1}^{N} \eta_i A_{ij} \bar{\eta}_j \right) = \det A, \tag{7.319}$$

assuming a fixed ordering convention for the measure.

(b) Consider a set of Grassmann variables $\psi, \bar{\psi}$ and an invertible matrix (or operator) K. Evaluate

$$I[\eta, \bar{\eta}] \equiv \int d\psi \, d\bar{\psi} \, \exp\left(-\bar{\psi} K \psi + \bar{\eta}\psi + \bar{\psi}\eta\right), \tag{7.320}$$

by completing the square, and show that it can be written as

$$I[\eta, \bar{\eta}] = (\det K) \, \exp\left(\bar{\eta} K^{-1} \eta\right), \tag{7.321}$$

up to a convention-dependent overall sign that is fixed once the integration measure ordering is fixed.

(c) Using Eq. (7.321), derive the standard identity for two-point functions generated by Grassmann sources,

$$\frac{\delta}{\delta \bar{\eta}_a} \frac{\delta}{\delta \eta_b} \exp\left(\bar{\eta} K^{-1} \eta\right) \bigg|_{\eta = \bar{\eta} = 0} = (K^{-1})_{ba}, \tag{7.322}$$

and explain why the order of functional derivatives matters.

(d) Specialize to the free Dirac operator in momentum space, $K(p) = \not{p} - m + i\epsilon$, and show that the resulting propagator is $\tilde{S}_F(p) = \frac{i(\not{p}+m)}{p^2 - m^2 + i\epsilon}$.

Essential Bibliography

- Feynman R, Hibbs A (1965) Quantum mechanics and path integrals, McGraw-Hill; and Chaichian M, Demichev A (2001) Path integrals in physics, vols I-II, Institute of Physics Publishing, Bristol. Canonical sources for the sum-over-histories viewpoint and its extension from mechanics to fields, including measure issues and semiclassical expansions.
- Zinn-Justin J (2002) Quantum field theory and critical phenomena, 4th edn. Oxford University Press. A technically sophisticated treatment of functional methods, renormalization, and the Euclidean formulation of quantum field theory, with a deep analysis of analytic continuation and the structural relation between quantum and statistical field theory.
- Le Bellac M (1996) Thermal field theory. Cambridge University Press; and Kapusta JI, Gale C (2006) Finite-temperature field theory, 2nd edn. Cambridge University Press. Standard treatments of the Matsubara formalism, thermal correlators, and the Euclidean-time approach to statistical field theory, directly complementing the finite-temperature sections.

8 Quasiparticles

"More is different." - **Philip Warren Anderson**

Abstract

This chapter develops a unifying viewpoint: in a wide class of physical systems, the *observed* moving entity is not a "bare" degree of freedom, but rather a *dressed* effective object whose inertia and dynamics include the response of the surrounding medium. In quantum many-body physics this is the operational definition of a quasiparticle; in classical continuum physics, the same structure appears as added-mass and memory forces. In nonlinear wave physics and field theory, stable localized energy lumps (solitons) behave as effective particles, again with emergent inertial and interaction properties. These themes are here reported, with the ambitious purpose of providing general definitions and formalizations in classical, quantum and field theories.

O. Luongo, *Fundamental Interactions*,
https://doi.org/10.1007/978-3-032-27413-7_8

Fig. 8.1 Portrait of Lev Davidovich Landau by Nobel foundation, 1962. Reprinted from Wikimedia. Public domain

A key character: *Lev Davidovich Landau (1908–1968), Fig.* 8.1.

Lev D. Landau was one of the central architects of twentieth-century theoretical physics and the principal figure behind the modern concept of quasiparticles. Born in Baku, he displayed extraordinary talent at an early age and was trained in the most advanced European schools of theoretical physics, working with Bohr, Pauli, Heisenberg, and Dirac before establishing his own scientific school in the Soviet Union.

Landau's most influential contribution to the physics of many-body systems is the theory of quantum liquids, in which he introduced the idea that the low-energy properties of an interacting system can be described in terms of weakly interacting elementary excitations, now called quasiparticles. In his theory of Fermi liquids, interacting fermions are mapped to long-lived excitations that carry the same quantum numbers as free particles but whose masses, lifetimes, and interactions are renormalized by the medium. This idea provided a unifying framework for the understanding of electrons in metals, liquid helium, nuclear matter, and a wide range of condensed-matter and nuclear systems.

Landau also introduced the concept of collective excitations, such as phonons and rotons in superfluid helium, showing that the relevant physical degrees of freedom are not the microscopic particles themselves but the normal modes of the interacting system. This perspective established a new hierarchy of description: microscopic Hamiltonians define the system, but its observable properties are governed by the spectrum and dynamics of its quasiparticles.

Beyond quasiparticle theory, Landau made foundational contributions to superfluidity, phase transitions, plasma physics, diamagnetism, and quantum electrodynamics. He was awarded the Nobel Prize in Physics in 1962 for his pioneering theories of condensed matter, especially liquid helium. His scientific style, emphasizing the identification of the correct effective degrees of freedom, remains a guiding principle in modern field theory, many-body physics, and the theory of emergent phenomena.

8.1 Classical Quasiparticles: The Key Is Non-Linearity

Nonlinear terms control interactions and mode coupling, but their presence alone does not decide whether an excitation qualifies as a quasiparticle. The decisive criterion that we will study in this chapter is the existence of a long-lived mode, signaled by a well-isolated pole of the retarded Green function with width much smaller than its energy.

We can immediately provide a summary toward this point as follows.

- Nonlinearity + dispersion/diffraction can produce solitons, because dispersive spreading can be exactly balanced by nonlinear self-focusing (or self-defocusing, depending on the model). However, true solitons in the strict sense (shape-preserving waves with elastic collisions) typically require integrability or at least special structural properties (hidden conservation laws, Lax pair, inverse scattering, etc.). Many non-integrable nonlinear systems still support solitary waves or long-lived localized pulses, but collisions radiate, profiles deform, and the "soliton" character is approximate.
- Nonlinearity + sufficient dimensionality/freedom can produce chaos, meaning sensitive dependence on initial conditions, broadband spectra, and positive Lyapunov exponents. But chaos usually needs more than mere nonlinearity: you need a mechanism that prevents the system from being reducible to a small set of conserved actions.

8.1.1 Non-Relativistic Quasiparticles

Consider a distinguished coordinate $x(t)$ (a body position, a collective mode, a charged particle trajectory, *etc.*) interacting with a large set of environmental degrees of freedom, schematically denoted by $Y(t)$. A very general classical action may be written as

$$S[x, Y] = \int dt \left[\frac{m}{2}\dot{x}^2 - U(x)\right] + S_{\mathrm{env}}[Y] + S_{\mathrm{int}}[x, Y]. \tag{8.1}$$

If the environment is not directly monitored (as is almost always the case), one eliminates Y in favor of an *effective* equation for x. At the level of the action, integrating out Y yields

$$S_{\mathrm{eff}}[x] = \int dt \left[\frac{m}{2}\dot{x}^2 - U(x)\right] + \Delta S[x], \tag{8.2}$$

where ΔS summarizes the backreaction of the medium.

Expanding ΔS for slow motion and weak departures from equilibrium, one obtains two generic structures.

1. **Local-in-time inertial renormalization:** a correction quadratic in $\dot{x}$, hence an *effective mass*:

$$\Delta S[x] \supset \frac{1}{2}\int dt\ \delta m\, \dot{x}^2 \qquad \Longrightarrow \qquad m_{\mathrm{eff}} = m + \delta m. \tag{8.3}$$

2. **Nonlocal-in-time memory (history) terms:** a functional dependence on the entire past trajectory:

$$\Delta S[x] \supset -\frac{1}{2}\int dt \int d\tau\ \dot{x}(t)\, K(t-\tau)\, \dot{x}(\tau), \tag{8.4}$$

   which leads, upon variation, to a generalized equation of motion of the form

$$m_{\mathrm{eff}}\, \ddot{x}(t) + \frac{dU}{dx}(x(t)) + \int_0^t d\tau\ \gamma(t-\tau)\, \dot{x}(\tau) = F_{\mathrm{ext}}(t), \tag{8.5}$$

   with a memory kernel γ related to K by differentiation.

Equation (8.5) is the classical blueprint behind a large class of effective theories: generalized Langevin dynamics, viscoelastic response, radiation reaction in classical electrodynamics, added-mass in hydrodynamics, and the "self-energy" description of quasiparticles in quantum theory.

8.1.2 A Paradigmatic Classical Dressing: Electromagnetic Mass

A first, purely classical, example of a quasiparticle mechanism is provided by a charged body, showing that inertia is not necessarily an intrinsic property of a bare mechanical core, but may partly originate from the surrounding field that is dragged along when the body moves. The essential idea is that an accelerated charge must accelerate not only its material support, but also the electromagnetic field configuration attached to it. The energy and momentum stored in this field then contribute to the total inertia.

Consider, for simplicity, a point charge regularized by introducing a short-distance cutoff r_0. The electrostatic field of a charge q at rest is $\mathbf{E}(\mathbf{r}) = \frac{q}{4\pi r^2}\,\hat{\mathbf{r}}$, so that the energy density is $u = \frac{1}{2}\mathbf{E}^2$ (in Heaviside-Lorentz units). The field energy outside the sphere of radius r_0 is then

$$E_{\text{field}} = \int_{r_0}^{\infty} d^3x\,\frac{1}{2}\mathbf{E}^2 = \int_{r_0}^{\infty} 4\pi r^2 dr\,\frac{1}{2}\frac{q^2}{16\pi^2 r^4} \sim \frac{q^2}{8\pi r_0}, \tag{8.6}$$

which diverges as $r_0 \to 0$. Even at the classical level, a charged particle thus carries an infinite amount of field energy unless some short-distance structure is assumed.

If the charge is set into slow motion with velocity $\mathbf{v}$, the electromagnetic field carries momentum. Restoring c, to leading order in v/c, this momentum is proportional to the velocity,

$$\mathbf{p}_{\text{field}} \sim \frac{E_{\text{field}}}{c^2}\,\mathbf{v}. \tag{8.7}$$

This relation suggests interpreting the field momentum as that of an effective mass associated with the electromagnetic dressing,

$$\mathbf{p}_{\text{field}} = m_{\text{EM}}\mathbf{v}, \qquad m_{\text{EM}} \sim \frac{E_{\text{field}}}{c^2} \sim \frac{q^2}{8\pi r_0 c^2}. \tag{8.8}$$

The observed inertia of the charge is therefore not purely mechanical, but receives a contribution from the field that accompanies it. The physical object is not a bare charge plus an independent field, but a composite entity consisting of the charge together with its electromagnetic cloud. In this sense, already at the classical level, the moving charge behaves as a quasiparticle: its effective mass is not a microscopic parameter but a property generated by its interaction with the field.

This mechanism anticipates the general structure of mass renormalization that we will face in this Chapter. One starts from a "bare" object, introduces a short-distance regularization, and finds that the interaction with a surrounding medium or field dresses the object, shifting its inertial parameters. The measured particle is thus a dressed excitation rather than a bare constituent.

An historically important realization of this idea is provided by Abraham's model of the classical electron. There, the electron is modeled as a rigid charged sphere of radius R, whose electrostatic self-energy

$$U = \frac{3}{5}\frac{e^2}{4\pi\varepsilon_0 R} \tag{8.9}$$

is interpreted as contributing to its inertia. For slow motion, the electromagnetic field carries momentum

$$\mathbf{p}_{\mathrm{em}} = \frac{4}{3}\frac{U}{c^2}\mathbf{v}, \tag{8.10}$$

leading to an electromagnetic mass $m_{\mathrm{em}} = (4/3)U/c^2$. Although the appearance of the factor $4/3$ reveals a conceptual inconsistency that required additional cohesive stresses, the physical message remains clear: part of what is observed as inertial mass is in fact field inertia.

From the viewpoint adopted here, the Abraham electron may be regarded as a classical quasiparticle: a charged core together with its self-generated electromagnetic environment. The same structural mechanism reappears in many other cases: in hydrodynamics as added mass, in radiation reaction as history-dependent inertia, and in QFT as mass renormalization and dressing of elementary excitations.

8.1.3 The Added Mass in Fluid Mechanics

Consider a rigid body of characteristic size R and volume V immersed in a fluid of density ρ. When the body accelerates, it cannot do so without accelerating some portion of the surrounding fluid.

Hence the kinetic energy of the *fluid motion* contributes to the total kinetic energy of the body-fluid system. This contribution can be written as if the body had an additional inertia m_{added}:

$$m_{\mathrm{eff}} = m_{\mathrm{bare}} + m_{\mathrm{added}}. \tag{8.11}$$

In the simplest setting (incompressible, non-viscous, irrotational outside the body), one can compute m_{added} exactly by solving a potential-flow boundary value problem.

Driving exercise Assume an incompressible fluid with constant density ρ and negligible viscosity. Outside the body we assume irrotational flow, hence a velocity potential ϕ exists:

$$\mathbf{v} = \nabla\phi, \qquad \nabla\cdot\mathbf{v} = 0 \implies \nabla^2\phi = 0. \tag{8.12}$$

Consider a rigid sphere of radius R translating with instantaneous velocity $\mathbf{U}(t)$ in an otherwise quiescent fluid. In the body frame (or equivalently for the disturbance flow), the far-field condition is $\nabla\phi \to 0$ as $r \to \infty$, while the no-penetration condition on the sphere reads

$$\left.\frac{\partial\phi}{\partial n}\right|_{r=R} = \mathbf{U}\cdot\mathbf{n} \qquad (\mathbf{n} = \hat{\mathbf{r}}). \tag{8.13}$$

For translation along the z axis, $\mathbf{U} = U\,\hat{\mathbf{z}}$, one writes the general axisymmetric harmonic function (decaying at infinity) as

$$\phi(r,\theta) = \frac{A}{r^2}\cos\theta, \tag{8.14}$$

where (r,θ) are spherical coordinates. Compute the radial derivative:

$$\left.\frac{\partial\phi}{\partial r}\right|_{r=R} = -\frac{2A}{r^3}\cos\theta\bigg|_{r=R} = -\frac{2A}{R^3}\cos\theta. \tag{8.15}$$

The boundary condition in Eq. (8.13) becomes

$$-\frac{2A}{R^3}\cos\theta = U\cos\theta \qquad \implies \qquad A = -\frac{1}{2}UR^3. \tag{8.16}$$

Thus the disturbance potential is

$$\phi(r,\theta) = -\frac{1}{2}UR^3\frac{\cos\theta}{r^2}. \tag{8.17}$$

The kinetic energy stored in the fluid motion is

$$T_{\text{fluid}} = \frac{\rho}{2}\int_{\mathbb{R}^3\setminus V_{\text{sphere}}} d^3x\,|\mathbf{v}|^2 = \frac{\rho}{2}\int d^3x\,|\nabla\phi|^2. \tag{8.18}$$

A standard identity for harmonic potentials (Green's first identity) gives, for $\nabla^2\phi = 0$,

$$\int_\Omega d^3x\,|\nabla\phi|^2 = \int_{\partial\Omega} dS\,\phi\,\frac{\partial\phi}{\partial n}, \tag{8.19}$$

where Ω is the fluid domain and $\partial\Omega$ its boundary. Taking Ω as the exterior of the sphere truncated at a large radius and sending the truncation to infinity (the far contribution vanishes by the $1/r^2$ decay), one obtains

$$T_{\text{fluid}} = \frac{\rho}{2} \int_{r=R} dS\, \phi(R,\theta) \left. \frac{\partial \phi}{\partial n} \right|_{r=R}. \tag{8.20}$$

Using Eq. (8.17) at $r = R$,

$$\phi(R,\theta) = -\frac{1}{2} U R \cos\theta, \qquad \left. \frac{\partial \phi}{\partial n} \right|_{r=R} = \left. \frac{\partial \phi}{\partial r} \right|_{r=R} = U \cos\theta, \tag{8.21}$$

so that

$$T_{\text{fluid}} = \frac{\rho}{2} \int_{r=R} dS \left(-\frac{1}{2} U R \cos\theta \right) (U \cos\theta) = -\frac{\rho U^2 R}{4} \int_{r=R} dS\, \cos^2\theta. \tag{8.22}$$

The sign is cured by tracking the outward normal for the exterior domain: for the fluid outside the sphere, the outward normal at $r = R$ points *into* the sphere, giving an extra minus sign. Equivalently, one may take the absolute value after enforcing positivity of energy. The integral is

$$\int_{r=R} dS\, \cos^2\theta = R^2 \int_0^{2\pi} d\varphi \int_0^{\pi} d\theta\, \sin\theta \cos^2\theta = 2\pi R^2 \int_0^{\pi} d\theta\, \sin\theta \cos^2\theta = \frac{4\pi}{3} R^2. \tag{8.23}$$

Hence

$$T_{\text{fluid}} = \frac{\rho U^2 R}{4} \cdot \frac{4\pi}{3} R^2 = \frac{1}{2} \left(\frac{1}{2} \rho \frac{4\pi R^3}{3} \right) U^2. \tag{8.24}$$

This identifies the added mass of a sphere as

$$m_{\text{added}} = \frac{1}{2} \rho \frac{4\pi R^3}{3}. \tag{8.25}$$

Since the total kinetic energy is

$$T_{\text{tot}} = \frac{1}{2} m_{\text{bare}} U^2 + T_{\text{fluid}} = \frac{1}{2} (m_{\text{bare}} + m_{\text{added}}) U^2, \tag{8.26}$$

the translational dynamics (in the simplest model where added mass is constant) is described by a renormalized Newton law

$$(m_{\text{bare}} + m_{\text{added}})\, \dot{\mathbf{U}} = \mathbf{F}_{\text{ext}}. \tag{8.27}$$

The operational statement is: *even if the rigid body is unchanged*, the system responds to forces as if the inertia were larger. In the quasiparticle language, the bare body has been dressed by the kinetic degrees of freedom of the surrounding fluid.

For bodies other than a sphere, m_{added} becomes a tensor (added-mass matrix), because acceleration in different directions sets different fluid volumes into motion. In many applications one summarizes this dependence as

$$m_{\text{added}} \sim C\,\rho\,V_{\text{eff}}, \tag{8.28}$$

where C is a dimensionless geometric factor and V_{eff} is an effective volume of fluid participating in the motion.

8.1.4 Forces on Histories

Added mass is an inertial renormalization that is local in time (within the idealized assumptions above). Viscous fluids introduce an additional and deeper phenomenon: the force on an accelerating body depends on *how the body arrived* at its current velocity. The reason is physical and robust: vorticity production and boundary-layer diffusion possess their own dynamics and relaxation times, so the fluid response is not instantaneous.

We can summarize this by stating that vortices and boundary layers "remember" the past motion and yield a nonlocal-in-time term in the force.

A compact way to encode this is through a convolution integral

$$\mathbf{F}_{\text{history}}(t) = -C \int_0^t d\tau\, \frac{\dot{\mathbf{v}}(\tau)}{\sqrt{t-\tau}}, \tag{8.29}$$

where $\mathbf{v}(t)$ is the *relative* velocity between the body and the undisturbed fluid and the kernel $(t-\tau)^{-1/2}$ is characteristic of diffusion-driven memory (the square-root time dependence is the hallmark of viscous boundary-layer growth).

8.1.5 The Basset-Boussinesq Term for a Sphere

For a sphere of radius R moving in an incompressible viscous fluid with kinematic viscosity ν, the classical Basset-Boussinesq memory force is

$$\mathbf{F}_{\text{Basset}}(t) = -6\pi R^2 \rho\nu \int_0^t d\tau\, \frac{\dot{\mathbf{v}}(\tau)}{\sqrt{\pi\nu\,(t-\tau)}}. \tag{8.30}$$

The kernel $\sim (t-\tau)^{-1/2}$ arises because the vorticity generated at the surface diffuses into the fluid with diffusion constant ν. For an impulsive change of the boundary velocity, the vorticity profile satisfies a diffusion equation in the normal direction with characteristic penetration length $\ell(t) \sim \sqrt{\nu t}$. Since shear stress scales as $\mu\, \partial u/\partial n$ and gradients scale as $u/\ell(t)$, one expects a stress contribution scaling as $\sim u/\sqrt{t}$. The convolution integral is the causal superposition of such responses to a time-dependent acceleration history. This is the precise sense in which the force is *non-Markovian*: it cannot be written as a function of $\mathbf{v}(t)$ and finitely many time derivatives at time t alone.

Combining external forcing, added mass, and memory, the effective dynamics of a single translational degree of freedom $x(t)$ can be cast as

$$m_{\mathrm{eff}}\, \ddot{x}(t) + \int_0^t d\tau\; K(t-\tau)\, \dot{x}(\tau) = F_{\mathrm{ext}}(t). \tag{8.31}$$

In the frequency domain, convolution becomes multiplication:

$$\big[-\omega^2 m_{\mathrm{eff}} - i\omega\, \widetilde{K}(\omega)\big]\tilde{x}(\omega) = \tilde{F}_{\mathrm{ext}}(\omega). \tag{8.32}$$

Thus the memory kernel acts as a frequency-dependent correction to inertia and dissipation. This is the classical analogue of a quantum self-energy:

$$\omega^2 m_{\mathrm{bare}} \longrightarrow \omega^2 m_{\mathrm{bare}} + \Pi(\omega), \tag{8.33}$$

where Π is a response function; in QFT one writes instead $G^{-1} = G_0^{-1} - \Sigma$, and $\Sigma(\omega, \mathbf{k})$ plays the same role as a dynamical dressing term.

8.1.6 Prototypes of Classical Quasiparticles: Solitons

A soliton is a localized, stable solution of a nonlinear evolution equation that propagates without changing shape and, in integrable systems, undergoes collisions after which it reemerges unchanged up to phase shifts. The stability mechanism is an exact balance between dispersion (or diffraction), which tends to spread waves, and nonlinearity, which tends to focus or localize them.

From the standpoint of this chapter, the crucial point is the following.

A soliton is an emergent particle-like degree of freedom of a nonlinear field.

Its "mass" is not a bare parameter but an integral of field energy density, and its effective dynamics in a weakly perturbed environment may acquire memory/friction terms, paralleling the history force structure.

Table 8.1 Comparison between particles, quasiparticles, and solitons. Despite their common particle-like behavior, their physical origin, stability mechanisms, and dynamical interpretation are fundamentally different

Property	Particles	Quasiparticles	Solitons
Origin	Fundamental degrees of freedom of the theory	Emergent collective excitations in an interacting medium	Localized nonlinear solutions of field equations
Existence in vacuum	Yes	No, only within the hosting medium	Yes (the field itself provides the substrate)
Description	Poles of fundamental propagators and asymptotic states	Renormalized propagator poles with residue Z and damping rate Γ	Solutions of nonlinear PDEs (e.g. KdV, NLSE, sine-Gordon)
Mass/inertia	Bare mass (possibly renormalized)	Effective mass m^* induced by medium dressing	Not pointlike; inertia encoded in localized field energy
Nature	Quantum	Quantum, but emergent from many-body interactions	Classical or semiclassical; often topological
Interactions	Fixed by the fundamental Lagrangian and symmetries	Effective and medium-dependent	Elastic in integrable systems; radiative otherwise
Localization	Idealized pointlike object	Localized over a finite coherence length	Spatially localized and shape-preserving
Stability	Dynamical or quantum-mechanical	Limited by the existence and properties of the medium	Topological and/or integrability-protected
Examples	Electrons, photons, quarks	Phonons, polarons, magnons	ϕ^4 kinks, sine-Gordon solitons, NLSE solitons, skyrmions

So, why the solitons are not exactly quasiparticles or particles? See Table 8.1 for a brief explanation.

▶ **Driving exercise** Three paradigmatic integrable equations and their fundamental solitons can be derived as a pedagogical set of examples.

Korteweg-de Vries (KdV)
The KdV equation,

$$u_t + 6u\, u_x + u_{xxx} = 0, \tag{8.34}$$

admits a traveling-wave soliton $u(x,t) = f(\xi)$ with $\xi = x - vt$. Then $u_t = -vf', u_x = f', u_{xxx} = f'''$, and Eq. (8.34) becomes

$$-vf' + 6ff' + f''' = 0. \tag{8.35}$$

Integrate once in ξ (assuming $f \to 0$ and $f' \to 0$ as $|\xi| \to \infty$):

$$-vf + 3f^2 + f'' = 0. \tag{8.36}$$

Multiply by f' and integrate again:

$$-\frac{v}{2}f^2 + f^3 + \frac{1}{2}(f')^2 = 0 \quad \Longrightarrow \quad (f')^2 = vf^2 - 2f^3 = f^2\,(v - 2f). \tag{8.37}$$

The nontrivial localized solution is obtained by setting the peak amplitude $A = v/2$ and integrating, leading to

$$u(x,t) = A\,sech^2\left(\sqrt{\frac{A}{2}}\,(x - 2At - x_0)\right). \tag{8.38}$$

Focusing nonlinear Schrödinger equation (NLSE)
The focusing NLSE,

$$i\psi_t + \psi_{xx} + 2|\psi|^2\psi = 0, \tag{8.39}$$

admits a one-soliton envelope solution. One may verify by direct substitution that

$$\psi(x,t) = \eta\,sech\big(\eta(x - 2\xi t - x_0)\big)\,\exp\Big(i\big[\xi x - (\xi^2 - \eta^2)t + \varphi_0\big]\Big), \tag{8.40}$$

solves Eq. (8.39). Here η controls the amplitude and inverse width, ξ controls the group velocity, and φ_0 is an arbitrary phase.
The same equation also supports localized-in-spacetime "breathers", e.g. the *peregrine soliton*:

$$\psi(x,t) = e^{2it}\left[1 - \frac{4(1 + 4it)}{1 + 4x^2 + 16t^2}\right]. \tag{8.41}$$

Sine-Gordon
The Sine-Gordon equation,

$$\phi_{tt} - \phi_{xx} + \sin\phi = 0, \tag{8.42}$$

admits a traveling kink. Introducing the variable $\xi = (x - vt)/\sqrt{1 - v^2}$, one reduces the PDE to an ODE and finds the kink solution

$$\phi(x,t) = 4\arctan\left[\exp\left(\frac{x - vt - x_0}{\sqrt{1 - v^2}}\right)\right], \tag{8.43}$$

which interpolates between adjacent minima $\phi = 0$ and $\phi = 2\pi$ and is therefore topological.

Sine-Gordon also admits breather solutions describing bound kink-antikink pairs. The explicit form is

$$\phi(x,t) = 4\arctan\left[\frac{\sqrt{1-\omega^2}}{\omega}\,\frac{\sin(\omega t)}{\cosh\big(\sqrt{1-\omega^2}\,x\big)}\right]. \tag{8.44}$$

Particle-like interpretation.
Once a static soliton profile $\phi_{\text{sol}}(x)$ is known, a moving soliton is obtained by Lorentz boosting (in relativistic theories) or by a Galilean transformation (in appropriate nonrelativistic settings). The soliton then carries a localized energy density T_{00} and a momentum density T_{0x}; one defines a soliton "mass" by

$$M_{\text{sol}} = \int dx\; T_{00}(x), \tag{8.45}$$

which is a field-theoretic quantity: the soliton's inertia is *entirely* emergent from the field configuration.

8.2 Topological Quasiparticles

▶ **Pause and think** Try to recall the main differences between topology and geometry.

First of all, we may wonder: what is topology? To understand it, let us focus on two objects that turn out to be *topologically equivalent* when one can be continuously deformed into the other by stretching and bending, without tearing, gluing, puncturing, or sealing holes.

A classic example is the equivalence between a coffee mug and a torus: the handle is a hole, so one can reshape the mug into a doughnut without changing the number of holes.

Topological spaces are commonly denoted as follows: the real line is $\mathbb{R}$; the plane is $\mathbb{R}^2$; and in general $\mathbb{R}^n$. One can form product spaces: $\mathbb{R}\times\mathbb{R}=\mathbb{R}^2$. A line segment with endpoints identified produces the circle S^1. The sphere S^2 denotes the *surface* of a ball (a two-dimensional manifold), which cannot be embedded in $\mathbb{R}^2$ without a cut but can be embedded in $\mathbb{R}^3$. A torus T^2 can be constructed as a product of circles,

$$T^2 = S^1 \times S^1. \tag{8.46}$$

Given a topological space, a *path* is a continuous map from an interval $[a,b]\subset\mathbb{R}$ into the space. If the endpoints coincide, $f(a)=f(b)$, the path is a *loop*. Loops can be continuously deformed into one another, and in any given space the set of loops splits into equivalence classes under such deformations. These classes form a group, the *fundamental group* π_1.

In $\mathbb{R}^n$ every loop is contractible to a point, so π_1 is trivial. When the space contains a hole, loops may wind around it, and different winding numbers define distinct classes. In that case π_1 is isomorphic to $\mathbb{Z}$. In particular,

$$\pi_1(S^1) = \mathbb{Z}, \qquad \pi_1(T^2) = \mathbb{Z} \times \mathbb{Z}, \tag{8.47}$$

reflecting the two independent winding numbers of loops on a torus.

The power of topological reasoning is that it depends only on deformability classes, not on metric geometry: it captures the "knottiness" of the configuration space.

The connection with the earlier sections is conceptual and very useful: in both cases the effective object is not a bare point degree of freedom.

- For solitons, the dressing is the nonlinear field profile itself.
- For quasiparticles, the dressing is the medium polarization cloud.
- For added mass, the dressing is the co-accelerated fluid volume.

Accordingly, consider a real scalar field in $1 + 1$ dimensions with Lagrangian density

$$\mathcal{L} = \frac{1}{2}\partial_\mu \phi \, \partial^\mu \phi - V(\phi), \tag{8.48}$$

leading to the nonlinear Klein-Gordon equation

$$\partial_\mu \partial^\mu \phi + V'(\phi) = 0. \tag{8.49}$$

For static configurations $\phi = \phi(x)$,

$$\partial_t \phi = 0, \qquad \partial_\mu \partial^\mu \phi = -\frac{d^2\phi}{dx^2}, \tag{8.50}$$

and Eq. (8.49) becomes the ODE

$$\frac{d^2\phi}{dx^2} = V'(\phi). \tag{8.51}$$

The energy functional of a static configuration is

$$E[\phi] = \int_{-\infty}^{+\infty} dx \left[\frac{1}{2}\left(\frac{d\phi}{dx}\right)^2 + V(\phi) \right]. \tag{8.52}$$

Finite energy requires $\phi(x) \to \phi_\pm$ as $x \to \pm\infty$, with $V(\phi_\pm) = 0$ (or more generally the same vacuum energy, which can be shifted to zero). These boundary conditions are what allow localized interpolating solutions (kinks) to exist.

8.2.1 Kinks and Antikinks

We may discuss the standard double-well potential

$$V(\phi) = \frac{\lambda}{4}\left(\phi^2 - v^2\right)^2, \tag{8.53}$$

which has two degenerate vacua $\phi = \pm v$, with $V(\pm v) = 0$.

The static equation (8.51) reads

$$\frac{d^2\phi}{dx^2} = V'(\phi) = \lambda\phi(\phi^2 - v^2). \tag{8.54}$$

Multiply Eq. (8.54) by $d\phi/dx$:

$$\frac{d^2\phi}{dx^2}\frac{d\phi}{dx} = \lambda\phi(\phi^2 - v^2)\frac{d\phi}{dx}. \tag{8.55}$$

Observe that

$$\frac{d^2\phi}{dx^2}\frac{d\phi}{dx} = \frac{1}{2}\frac{d}{dx}\left(\frac{d\phi}{dx}\right)^2, \qquad \lambda\phi(\phi^2 - v^2)\frac{d\phi}{dx} = \lambda\frac{d}{dx}\left(\frac{\phi^4}{4} - \frac{v^2\phi^2}{2}\right). \tag{8.56}$$

Integrating once in x yields

$$\frac{1}{2}\left(\frac{d\phi}{dx}\right)^2 = \lambda\left(\frac{\phi^4}{4} - \frac{v^2\phi^2}{2}\right) + C. \tag{8.57}$$

Impose kink boundary conditions: as $x \to \pm\infty$, $\phi \to \pm v$ and $d\phi/dx \to 0$. Substituting $\phi = \pm v$ gives

$$0 = \lambda\left(\frac{v^4}{4} - \frac{v^4}{2}\right) + C = -\frac{\lambda v^4}{4} + C \quad \Longrightarrow \quad C = \frac{\lambda v^4}{4}. \tag{8.58}$$

Therefore

$$\frac{1}{2}\left(\frac{d\phi}{dx}\right)^2 = \frac{\lambda}{4}\left(\phi^2 - v^2\right)^2 = V(\phi). \tag{8.59}$$

Equation (8.59) is the first-order reduction emphasized in the notes. It implies

$$\frac{d\phi}{dx} = \pm\sqrt{2V(\phi)} = \pm\sqrt{\frac{\lambda}{2}}\,|\phi^2 - v^2|. \tag{8.60}$$

For a kink interpolating from $-v$ to $+v$, ϕ increases with x, and in the interval $\phi \in [-v, v]$ one has $|\phi^2 - v^2| = v^2 - \phi^2$. Hence

$$\frac{d\phi}{dx} = \sqrt{\frac{\lambda}{2}}\,(v^2 - \phi^2). \tag{8.61}$$

Explicit kink and antikink

Separating variables:

$$\frac{d\phi}{v^2-\phi^2}=\sqrt{\frac{\lambda}{2}}\,dx, \tag{8.62}$$

and integrating

$$\int \frac{d\phi}{v^2-\phi^2}=\frac{1}{2v}\ln\left|\frac{v+\phi}{v-\phi}\right|=\sqrt{\frac{\lambda}{2}}\,(x-x_0), \tag{8.63}$$

so

$$\frac{v+\phi}{v-\phi}=\exp\left(2v\sqrt{\frac{\lambda}{2}}\,(x-x_0)\right). \tag{8.64}$$

Solving for ϕ yields the kink profile

$$\phi_{\text{kink}}(x)=v\,\tanh\big(\mu(x-x_0)\big),\qquad \mu\equiv v\sqrt{\frac{\lambda}{2}}, \tag{8.65}$$

and the antikink is obtained by $\phi\mapsto-\phi$:

$$\phi_{\text{antikink}}(x)=-v\,\tanh\big(\mu(x-x_0)\big). \tag{8.66}$$

Kink energy

Using Eq. (8.59), the energy in Eq. (8.52) becomes

$$E_{\text{kink}}=\int_{-\infty}^{+\infty}dx\,\left[\frac{1}{2}\phi'^2+V(\phi)\right]=\int_{-\infty}^{+\infty}dx\,\phi'^2. \tag{8.67}$$

Change variables from x to ϕ:

$$E_{\text{kink}}=\int_{\phi(-\infty)}^{\phi(+\infty)}d\phi\,\phi'=\int_{-v}^{+v}d\phi\,\sqrt{2V(\phi)}=\sqrt{\frac{\lambda}{2}}\int_{-v}^{+v}d\phi\,(v^2-\phi^2). \tag{8.68}$$

Compute the integral:

$$\int_{-v}^{+v}(v^2-\phi^2)\,d\phi=\left[v^2\phi-\frac{\phi^3}{3}\right]_{-v}^{+v}=\frac{4}{3}v^3, \tag{8.69}$$

hence

$$E_{\text{kink}}=\frac{4}{3}v^3\sqrt{\frac{\lambda}{2}}, \tag{8.70}$$

that can be put in relation with the phion mass of the system. Indeed, expanding about either vacuum yields conventional particle excitations ("phions") of mass

$$m = \sqrt{2\lambda v^2}, \tag{8.71}$$

we then end up with

$$E = \frac{1}{\sqrt{2}} \frac{4m^3}{3\lambda}. \tag{8.72}$$

The energy is therefore finite, so kinks are admissible classical solutions. A notable feature is that the kink energy scales inversely with the coupling, signalling that such objects are non-perturbative and would not appear in a naive expansion in powers of λ.

The kink energy is localized within a finite spatial extent ℓ. One can estimate ℓ from a balance between gradient energy and potential energy: the gradient contribution scales as $\int dx\, \frac{1}{2}(\partial_x \phi)^2 \sim \ell\, (v/\ell)^2$, favoring a broad profile, while the potential term scales as $\int dx\, U \sim \lambda v^4 \ell$, favoring a narrow transition region. Minimization gives

$$\ell \sim (\lambda v^2)^{-1/2} = \frac{1}{m}. \tag{8.73}$$

The theory is translationally invariant, so the kink center can be located anywhere. It is also Lorentz invariant, so the kink can be boosted: it has a particle-like behaviour.

A kink is stable in isolation: to remove it one would have to change the field at infinity, forcing (semi)infinite regions to move from one vacuum to the other, which costs infinite energy. Equivalently, the kink is not continuously deformable into a trivial vacuum configuration while keeping its boundary conditions fixed.

▶ **Pause and think** Before going on, manage to understand why antikinks might exist, physically speaking.

An *antikink* is the opposite interpolation:

$$\phi(-\infty) = +v, \qquad \phi(+\infty) = -v, \tag{8.74}$$

i.e. the field crosses the axis in the opposite direction. A kink-antikink pair is deformable into a vacuum configuration, so such pairs can annihilate.

Kink stability is encoded by a conserved *topological charge*. Define a topological current (not a Noether current)

$$J_T^\mu = \frac{1}{2v}\, \epsilon^{\mu\nu} \partial_\nu \phi, \tag{8.75}$$

with $\epsilon^{01} = 1$. The associated charge is

$$Q_T = \int_{-\infty}^{+\infty} dx\, J_T^0 = \frac{1}{2v} \int_{-\infty}^{+\infty} dx\, \frac{\partial \phi}{\partial x} = \frac{1}{2v}\Big[\phi(+\infty) - \phi(-\infty)\Big] = 1 \quad (8.76)$$

for a kink, while an antikink has $Q_T = -1$. Ordinary particle excitations have $Q_T = 0$ because they do not change the vacuum at infinity.

The charge is called "topological" because its definition does not rely on the metric structure: it involves contraction with the antisymmetric symbol $\epsilon^{\mu\nu}$ rather than the metric $g_{\mu\nu}$. This reliance on ϵ (rather than g) is a recurring hallmark of topological sectors.

8.2.2 Kinks in the Sine-Gordon Model

Take the periodic potential

$$V(\phi) = 1 - \cos\phi, \quad (8.77)$$

so that the field equation is Eq. (8.42). For static solutions $\phi = \phi(x)$,

$$\frac{d^2\phi}{dx^2} = \sin\phi. \quad (8.78)$$

Multiply by ϕ':

$$\phi''\phi' = \sin\phi\,\phi'. \quad (8.79)$$

Using $\phi''\phi' = \frac{1}{2}\frac{d}{dx}(\phi'^2)$ and $\sin\phi\,\phi' = \frac{d}{dx}(1-\cos\phi)$, integrate once:

$$\frac{1}{2}\phi'^2 = 1 - \cos\phi + C. \quad (8.80)$$

Finite-energy boundary conditions require $\phi' \to 0$ as $\phi \to 2\pi n$ and $V(\phi) = 0$ at $\phi = 2\pi n$, so $C = 0$, yielding

$$\phi'^2 = 2(1-\cos\phi) = 4\sin^2\!\left(\frac{\phi}{2}\right). \quad (8.81)$$

Choosing the kink branch $\phi' > 0$ gives

$$\frac{d\phi}{dx} = 2\sin\!\left(\frac{\phi}{2}\right). \quad (8.82)$$

Separate variables:

$$\frac{d\phi}{2\sin(\phi/2)} = dx, \tag{8.83}$$

let $u = \phi/2$ so $d\phi = 2\,du$, and obtain

$$\int \frac{du}{\sin u} = x - x_0 \quad \Longrightarrow \quad \ln\left|\tan\frac{u}{2}\right| = x - x_0. \tag{8.84}$$

Therefore

$$\tan\frac{\phi(x)}{4} = e^{x-x_0}, \qquad \phi_{\text{kink}}(x) = 4\arctan\big(e^{x-x_0}\big), \tag{8.85}$$

which interpolates between $\phi(-\infty) = 0$ and $\phi(+\infty) = 2\pi$. Again, the existence of distinct asymptotic vacua endows the kink with a topological charge defined by the winding between adjacent minima.

▶ **Driving exercise** Let us now compute the energy and the topological charge. The static energy functional reads

$$E[\phi] = \int_{-\infty}^{+\infty} dx\left[\frac{1}{2}\left(\frac{d\phi}{dx}\right)^2 + 1 - \cos\phi\right]. \tag{8.86}$$

Using the first integral obtained above for finite-energy solutions,

$$\frac{1}{2}\phi'^2 = 1 - \cos\phi, \tag{8.87}$$

one immediately gets for the kink profile

$$\mathcal{E}(x) = \frac{1}{2}\phi'^2 + 1 - \cos\phi = 2\big(1 - \cos\phi\big) = \phi'^2, \tag{8.88}$$

hence the total kink energy is

$$E_{\text{kink}} = \int_{-\infty}^{+\infty} dx\,\phi'^2 = \int_{-\infty}^{+\infty} dx\,\phi'\,\frac{d\phi}{dx} = \int_{\phi(-\infty)}^{\phi(+\infty)} d\phi\,\phi'. \tag{8.89}$$

On the kink branch one has $\phi' = 2\sin(\phi/2)$, so

$$E_{\text{kink}} = \int_0^{2\pi} d\phi\, 2\sin\left(\frac{\phi}{2}\right). \tag{8.90}$$

With $u = \phi/2$ ($d\phi = 2\,du$) this becomes

$$E_{\text{kink}} = 4\int_0^{\pi} \sin u\,du = 4\Big[-\cos u\Big]_0^{\pi} = 4\,(2) = 8, \tag{8.91}$$

showing that the sine-Gordon kink has a *finite* classical rest energy $E_{\rm kink} = 8$ in these units. The corresponding topological charge is defined purely by the asymptotics,

$$Q_{\rm top} = \frac{1}{2\pi}\int_{-\infty}^{+\infty} dx\,\frac{d\phi}{dx} = \frac{1}{2\pi}\Big[\phi(+\infty) - \phi(-\infty)\Big], \tag{8.92}$$

so that for $\phi(-\infty) = 0$ and $\phi(+\infty) = 2\pi$ one finds

$$Q_{\rm kink} = +1, \qquad Q_{\rm antikink} = -1, \tag{8.93}$$

which is invariant under smooth deformations that keep the boundary conditions fixed, and therefore classifies the disconnected sectors of the configuration space.

Winding number and topological sectors.
When the potential admits a discrete set of degenerate minima,

$$\mathcal{M} = \{\phi_n \in \mathbb{R} \mid V'(\phi_n) = 0,\ V(\phi_n) = 0\}, \tag{8.94}$$

finite-energy configurations must approach these vacua at spatial infinity,

$$\lim_{x\to\pm\infty} \phi(x) = \phi_{n_\pm} \in \mathcal{M}, \qquad \lim_{x\to\pm\infty} \frac{d\phi}{dx} = 0, \tag{8.95}$$

so that the spatial line is effectively compactified to a circle by adding the point at infinity. A static field configuration therefore defines a continuous map

$$\phi:\ \mathbb{R}\cup\{\infty\} \simeq S^1_{\rm space} \longrightarrow \mathcal{M} \simeq S^1_{\rm vac}, \tag{8.96}$$

where in the sine-Gordon case the vacuum manifold is the circle $\mathcal{M} \equiv \{\phi = 2\pi n,\ n \in \mathbb{Z}\}$, periodically identified.
The integer that classifies the homotopy class of this map is the *winding number*. It measures how many times the field configuration wraps around the vacuum manifold as space is traversed once. In one spatial dimension it is given explicitly by the boundary data,

$$N_{\rm wind} = \frac{1}{2\pi}\int_{-\infty}^{+\infty} dx\,\frac{d\phi}{dx} = \frac{1}{2\pi}\big[\phi(+\infty) - \phi(-\infty)\big] \in \mathbb{Z}. \tag{8.97}$$

Configurations with different $N_{\rm wind}$ cannot be continuously deformed into each other without violating the finite-energy boundary conditions, since any such deformation would require the field to leave the vacuum manifold at infinity. The configuration space therefore decomposes into disconnected *topological sectors* labelled by $N_{\rm wind}$.

For the sine-Gordon model the elementary kink interpolates between two neighbouring vacua, $\phi(-\infty) = 0$ and $\phi(+\infty) = 2\pi$, and thus carries

$$N_{\text{wind}} = +1, \tag{8.98}$$

while the antikink, interpolating in the opposite direction, has $N_{\text{wind}} = -1$. More general multi-kink configurations carry $|N_{\text{wind}}| > 1$ and may be interpreted as bound or widely separated superpositions of elementary kinks. The winding number is conserved under the classical time evolution because it depends only on the asymptotic behaviour of the field and is therefore insensitive to any local, smooth dynamics in the bulk.

8.2.3 Vortices

We now move to $(2+1)$-dimensional spacetime and consider an analogous symmetry-broken theory whose vacuum manifold is a circle in internal (ϕ_1, ϕ_2) space (the "Mexican hat" picture). It is convenient to use a complex field

$$\phi(x) = \phi_1(x) + i\phi_2(x) \equiv r(x)e^{i\theta(x)}, \tag{8.99}$$

with internal polar coordinates (r, θ).

For kinks, different vacua were realized at spatial infinities in one dimension. In two spatial dimensions, the analogue is to let the field take different vacuum orientations as one goes around the circle at spatial infinity. Examples are swirling configurations characterized by a *winding number* n. A vortex is defined by an asymptotic boundary condition

$$\phi(x) = K\, e^{i\,[n\,\theta(x)+\varphi]} \qquad (|x| \to \infty), \tag{8.100}$$

where $\theta(x) = \tan^{-1}(x^2/x^1)$ is the polar angle in real space, φ is an arbitrary constant phase, and $n \in \mathbb{Z}$ counts how many times the field winds in internal space when one traverses spatial infinity. Changing φ corresponds to a global rotation of the asymptotic field direction and can be absorbed into K.

For a static configuration the Hamiltonian density is

$$\mathcal{H} = \frac{1}{2}\nabla\phi^\dagger \cdot \nabla\phi + U(\phi), \tag{8.101}$$

and take a potential whose minima satisfy $|\phi| = |K|$, e.g. $U(\phi) = \left(|K|^2 - \phi^\dagger\phi\right)^2$, so that $U = 0$ at infinity.

Using cylindrical coordinates in real space, the dominant asymptotic gradient is

$$\nabla\phi = \frac{1}{r}\left(inKe^{in\theta}\right)\hat{\mathbf{e}}_\theta. \tag{8.102}$$

The core may be singular in naive asymptotics, but we assume it is regulated by the full solution and denote the core energy by $E_{\rm core}(a)$, where a is the core size. For $r > a$ the energy becomes

$$E = E_{\rm core} + \int_a^\infty dr\, d\theta\, r\, \mathcal{H} = E_{\rm core} + \pi n^2 |K|^2 \int_a^\infty \frac{dr}{r}. \tag{8.103}$$

This diverges logarithmically, implying that a single vortex in the ungauged theory is not stable.

This instability can be intuitively understood: even at very large radii the configuration remains "swirly", so gradients do not die fast enough. More broadly,

Derrick's theorem implies that time-independent localized topological solitons of this type are not supported in more than one spatial dimension for the simple scalar theory.

▶ **Driving exercise** To stabilize the vortex we modify the theory by introducing a gauge field. Gauging is implemented by replacing ordinary derivatives with covariant derivatives,

$$D_\mu \phi = \partial_\mu \phi + iqA_\mu \phi. \tag{8.104}$$

The goal is to choose a gauge field that cancels the problematic gradient contribution at infinity, i.e. to arrange that $D_\mu \phi \to 0$ as $|x| \to \infty$. A suitable asymptotic choice is

$$\mathbf{A}(r,\theta) \xrightarrow{r\to\infty} \frac{1}{q}\nabla(n\theta). \tag{8.105}$$

At infinity the components behave as (with the sign convention $A_i = -A^i$)

$$A_r \to 0, \qquad A_\theta \to -\frac{n}{qr} \qquad (r \to \infty). \tag{8.106}$$

Then the covariant derivatives satisfy

$$D_r\phi \to 0, \qquad D_\theta \phi = \frac{1}{r}\frac{\partial \phi}{\partial \theta} + iqA_\theta \phi \to 0 \qquad (r \to \infty), \tag{8.107}$$

so the kinetic (gradient) energy that diverged in the ungauged theory is canceled.

One might worry about an additional gauge-field term $-\frac{1}{4}F_{\mu\nu}F^{\mu\nu}$, but the asymptotic gauge field here is *pure gauge*. Indeed at infinity $\mathbf{A} = \nabla\chi$ with $\chi = \frac{n\theta}{q}$, hence $F_{\mu\nu} = 0$ there, so no divergent field-strength energy is generated.

8.2.4 Flux Quantization

▶ **Pause and think** What does the gauge field contribute physically?

Consider the integral around a large circle at infinity:

$$\Phi = \oint \mathbf{A} \cdot d\boldsymbol{\ell}. \tag{8.108}$$

By Stokes' theorem this equals the magnetic flux through the circle. With the asymptotic form above,

$$\Phi = \oint \mathbf{A} \cdot d\boldsymbol{\ell} = \oint A_\theta \, r \, d\theta = \int_0^{2\pi} d\theta \, \frac{n}{q} = \frac{2\pi n}{q}. \tag{8.109}$$

Therefore vortices carry *quantized magnetic flux*.

The natural step beyond the above treatment would be to explore topological objects in (3 + 1) dimensions. For now, however, it is convenient to investigate non-topological quasiparticles first and then the consequences of having quasiparticles in QFT.

8.3 Non-Topological Quasiparticles

Non-topological solitons are localized finite-energy configurations whose stability does *not* rely on a topological invariant.

A (naive but standard) classification suggests that non-topological objects do not have a topological charge, while topological ones show topological charges.

For non-topological quasiparticles, stability is typically ensured by one of the features that follow.

- **A conserved Noether charge** (global symmetry).
- **Integrability** and exact conservation laws (in some models).
- **Dynamical/metastable** trapping (long-lived "oscillons").

8.3.1 Q-Balls

Consider a complex scalar field Φ with a global $U(1)$ symmetry,

$$\Phi \to e^{i\alpha}\Phi, \tag{8.110}$$

and Lagrangian

$$\mathcal{L} = \partial_\mu \Phi^* \, \partial^\mu \Phi - U(|\Phi|), \tag{8.111}$$

where U is a nonnegative potential with $U(0) = 0$. The conserved Noether current is

$$j^\mu = i\big(\Phi^* \partial^\mu \Phi - \Phi\, \partial^\mu \Phi^*\big), \qquad Q = \int d^3x \; j^0. \tag{8.112}$$

A Q-ball is a localized solution that minimizes the energy at fixed charge Q.

▶ **Driving exercise** Using the stationary ansatz

$$\Phi(t, \mathbf{x}) = e^{i\omega t}\,\varphi(\mathbf{x}), \tag{8.113}$$

with real $\varphi \geq 0$ and constant frequency ω, then

$$|\partial_t \Phi|^2 = \omega^2 \varphi^2, \qquad |\nabla \Phi|^2 = |\nabla \varphi|^2, \tag{8.114}$$

and the energy functional becomes

$$E[\varphi; \omega] = \int d^3x \, \Big(|\nabla \varphi|^2 + \omega^2 \varphi^2 + U(\varphi)\Big). \tag{8.115}$$

The charge is

$$Q[\varphi; \omega] = \int d^3x \; 2\omega\, \varphi^2. \tag{8.116}$$

To enforce fixed Q, minimize E with a Lagrange multiplier ω, which is equivalent to minimize the functional

$$\mathcal{E}_\omega[\varphi] \equiv E - \omega Q = \int d^3x \, \Big(|\nabla \varphi|^2 + U(\varphi) - \omega^2 \varphi^2\Big). \tag{8.117}$$

Varying with respect to φ yields the field equation

$$\nabla^2 \varphi = \frac{1}{2} \frac{d}{d\varphi} \Big(U(\varphi) - \omega^2 \varphi^2\Big), \tag{8.118}$$

with boundary conditions $\varphi \to 0$ at spatial infinity to ensure finite energy. A localized solution requires that the effective potential

$$U_\omega(\varphi) \equiv U(\varphi) - \omega^2\varphi^2 \tag{8.119}$$

is negative in some interval $\varphi \in (0, \varphi_0)$ so that the "particle in a potential" analogy allows a nontrivial trajectory returning to $\varphi = 0$. Equivalently, there must exist $\varphi \neq 0$ such that

$$\frac{U(\varphi)}{\varphi^2} < \omega^2, \tag{8.120}$$

while stability against decay into Q free quanta of mass m demands $\omega < m$ (so that energy per unit charge is lower than the free-particle threshold). These are the standard non-topological soliton conditions that replace topological protection by charge conservation.

8.3.2 Oscillons

Even in the absence of an exact conserved charge, nonlinear scalar field theories can support long-lived, spatially localized, oscillatory configurations known as *oscillons*. They are not protected by topology or by a Noether charge and are therefore not strictly stable solitons. Nevertheless, their lifetime can be parametrically large, because energy loss through radiation is strongly suppressed.

Consider, for definiteness, a real scalar field in 3 + 1 dimensions with Lagrangian

$$\mathcal{L} = \frac{1}{2}(\partial_\mu\phi)^2 - \frac{1}{2}m^2\phi^2 - V_{\rm nl}(\phi), \tag{8.121}$$

where $V_{\rm nl}$ contains nonlinear self-interactions, for example

$$V_{\rm nl}(\phi) = \frac{\lambda}{4}\phi^4 + \frac{g}{6}\phi^6 + \cdots. \tag{8.122}$$

The field equation is

$$\ddot{\phi} - \nabla^2\phi + m^2\phi + V'_{\rm nl}(\phi) = 0. \tag{8.123}$$

An oscillon is a solution of the form

$$\phi(t, \mathbf{x}) \simeq \Phi(r)\cos(\omega t) + \text{higher harmonics}, \qquad r = |\mathbf{x}|, \tag{8.124}$$

with $\omega < m$. The inequality $\omega < m$ is crucial: free radiation modes satisfy $\omega_k^2 = k^2 + m^2$, so any mode with $\omega < m$ is evanescent and cannot propagate to infinity.

Substituting the leading ansatz into Eq. (8.123) and keeping only the fundamental harmonic gives, at lowest order,

$$\left[-\nabla^2 + (m^2 - \omega^2)\right] \Phi(r) + \frac{3\lambda}{4}\Phi^3(r) + \cdots = 0. \tag{8.125}$$

This is a nonlinear eigenvalue problem for the profile $\Phi(r)$ and the frequency ω. Localized solutions exist only for special ranges of parameters and only when nonlinear terms balance dispersion.

Because the field oscillates periodically, nonlinearities generate higher harmonics at frequencies $n\omega$ with $n \geq 2$. When $n\omega > m$, those harmonics lie above the mass threshold and correspond to propagating modes. They therefore radiate energy to infinity. The energy of the configuration,

$$E = \int d^3x \left[\frac{1}{2}\dot{\phi}^2 + \frac{1}{2}(\nabla\phi)^2 + \frac{1}{2}m^2\phi^2 + V_{\mathrm{nl}}(\phi)\right], \tag{8.126}$$

is not exactly conserved for the localized core: it slowly decreases due to this weak emission of high-frequency radiation.

A scaling estimate shows why the lifetime can nevertheless be very large. For small amplitude oscillons one finds

$$\omega = m\left(1 - \epsilon^2\right), \qquad 0 < \epsilon \ll 1, \tag{8.127}$$

so that the fundamental mode is deeply below threshold. The first radiating harmonic appears only at order $n \sim m/\omega \sim 1 + O(\epsilon^2)$. The amplitude of these harmonics is suppressed by powers of ϵ, leading to a radiated power of the form

$$\frac{dE}{dt} \sim -A\,\epsilon^p, \tag{8.128}$$

with $p \gg 1$ and A a model-dependent constant. Consequently, the lifetime scales as

$$\tau_{\mathrm{osc}} \sim \epsilon^{-p}, \tag{8.129}$$

which can be many orders of magnitude larger than the natural microscopic timescale m^{-1}. This is the sense in which oscillons are long-lived but not exactly stable.

Oscillon "particle" character is only approximate. *Unlike true solitons, their existence is not guaranteed by topology or by a conserved charge.* **Their longevity is dynamical, resulting from the strong suppression of radiation due to the subthreshold fundamental frequency.**
In the quasiparticle language, an oscillon is a collective excitation of the scalar field: a coherent, many-mode configuration that behaves as a single effective object over long timescales. It illustrates a general warning: **particle-like behavior does not automatically imply exact stability.**

8.4 Quasiparticles in Quantum Field Theory: Renormalization

Renormalization was developed to organize perturbation theory in quantum electrodynamics and to render observable predictions finite. Its modern interpretation is simple and physical: the parameters appearing in the Lagrangian are *bare* inputs, while experiments measure *dressed* quantities after interactions at all scales have been taken into account. Divergences in naive calculations therefore signal an incomplete separation between long-distance observables and short-distance degrees of freedom, not an intrinsic inconsistency of the framework.

In QFT, particles are not isolated: vacuum fluctuations and interactions continuously modify their masses, charges, and wave-function normalizations. Perturbation theory must be reorganized so that such dressing is absorbed into redefined fields and parameters, leaving finite residual corrections. In this sense, renormalized particles are quasiparticles in a precise field-theoretic meaning: they are identified with poles of full propagators, with residues and widths controlled by self-energies and by available decay channels.

In the following, we motivate renormalization from this quasiparticle viewpoint, outline the classification of quantum field theories by their renormalization properties, and introduce the renormalization group as the natural framework for understanding scale dependence and universality.

8.4.1 Regularization Versus Renormalization

Renormalization is therefore not a simple trick or a mathematical tool to hide divergencies arising in most QFTs, but rather a fundamental process without which we wouldn't be able to understand nearly anything about modern physics. In practice, however, one must distinguish *two logically different steps* that are often conflated in informal discussions: introducing a *regularization* to make intermediate expressions well-defined, and performing a *renormalization* to identify which quantities are physical and how they depend on the chosen normalization conditions.

▶ **Pause and think** Figure out the main differences between renormalizing or regularizing a field.

- **Regularization** represents any prescription that renders divergent objects finite by introducing an auxiliary parameter (a cutoff, an analytic continuation, an unphysical mass scale, ...). Regularization *does not* by itself carry physical content: the regulator is not an observable, and different regulators may be used as long as they respect the relevant symmetries and lead to the same renormalized predictions.
- **Renormalization** is the procedure by which divergent (or regulator-dependent) pieces are absorbed into a redefinition of parameters and/or local counterterms already allowed by the symmetries (masses, charges, couplings, vacuum energy, boundary terms, ...), together with a *choice of renormalization conditions* that specify what is meant by "measured" parameters. This step fixes which finite

quantity is to be compared with experiment and explains the scale dependence of couplings (running).

- **Casimir effect as the cleanest separation of roles.** In the Casimir problem one meets divergent zero-point energies already at the level of free fields, $E_0 = \frac{1}{2}\sum_{\mathbf{k}} \hbar\omega_{\mathbf{k}}$, which is meaningless without a regulator. A *regularization* (e.g. an exponential damping $e^{-\epsilon\omega}$ or a zeta-function continuation) makes both the constrained and unconstrained vacuum energies finite. The physically relevant quantity is not either energy separately, but their *difference* at fixed geometry,

$$E_{\text{Cas}}(L) = \Big[E_0(L) - E_0(\infty)\Big]_{\text{reg}}, \qquad (8.130)$$

which removes the bulk divergence. Yet this subtraction is not merely algebraic: it is the simplest instance of *renormalization of the vacuum energy*, i.e. fixing the reference level of the cosmological-constant-like term by a normalization condition (here, that no force is present as $L \to \infty$). In more general geometries and realistic materials, further local counterterms associated with surfaces and curvature may be required; again, these belong to renormalization, not to regularization.
- **Regulator dependence versus scheme dependence.** After renormalization, observable quantities (such as the Casimir force $F(L) = -\partial E_{\text{Cas}}/\partial L$) must be independent of the regulator. Residual dependence on an arbitrary scale (e.g. μ in dimensional regularization) reflects the *renormalization scheme* and is compensated by the running of renormalized parameters. Regularization changes intermediate expressions; renormalization changes the *parametrization* of physics.

8.4.2 Dressed Particles

With the Casimir effect, we saw how to correct the simplest free field, without the need to talk about particles explicitly. Now we turn to interacting theories. In such theories, the standard approach is to describe interactions as small perturbations of the free Hamiltonian. We can imagine this as an analogy to statistical mechanics with a function $\lambda(T)$ that gradually turns on the interactions (see Fig. 8.2), such as

$$\hat{H} = \hat{H}_0 + \lambda(T)\hat{H}'. \qquad (8.131)$$

The new eigenstates of the full Hamiltonian $\hat{H}$, labeled by $|\mathbf{p}_\lambda\rangle$, are different from those of the free Hamiltonian $\hat{H}_0$, called $|\mathbf{p}\rangle$. The ground state is also different, since we now have an interacting ground state $|\Omega\rangle$ instead of the non interacting one $|0\rangle$. Therefore, we cannot really use creation and annihilation to operate on single particles anymore, since $|\Omega\rangle$ now has a a priori unknown non-vacuum configuration. The state $\hat{a}^\dagger_{\mathbf{p}}|\Omega\rangle$ generally will not contain a single-particle excitation, but most likely produces several excitations all at once. Is it possible to find a single-particle excitation in an interacting system? The answer is yes.

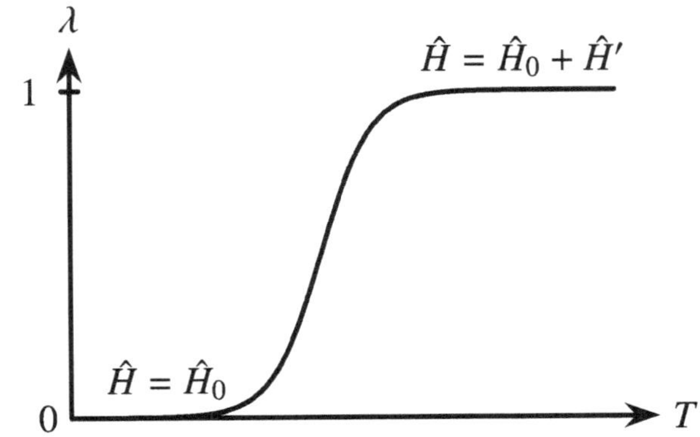

Fig. 8.2 Adiabatic turning on of the interactions. This picture shows what happens to particles in an interacting theory

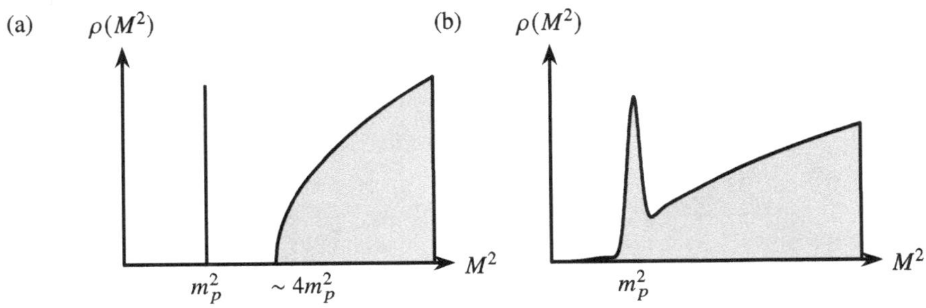

Fig. 8.3 **a** The spectral function showing a single-particle peak. **b** The spectral function showing a quasiparticle peak with lifetime $(2\Gamma_p)^{-1}$

The produced state might have a part we can distinguish that resembles the state $|\mathbf{p}_\lambda\rangle$ we intend to make, the single-particle state:

$$\hat{a}^\dagger_{\mathbf{p}} |\Omega\rangle = |\mathbf{p}_\lambda\rangle \left\langle \mathbf{p}_\lambda | \hat{a}^\dagger_{\mathbf{p}} | \Omega \right\rangle + \sum (\text{multiparticle parts}). \tag{8.132}$$

In the above state, we can recognize the single-particle excitation state as separated from the rest of the particles in other superpositions. But this first part actually only resembles a single-particle state, since it carries a weight $\left\langle \mathbf{p}_\lambda | \hat{a}^\dagger_{\mathbf{p}} | \Omega \right\rangle$. What we are creating is actually a narrow wave packet of width $\Gamma_{\mathbf{p}}$, a phenomenon of resonance, in which the narrower the packet, the more it resembles such a state. However, its energy will be described as complex, $E_{\mathbf{p}} + i\Gamma_{\mathbf{p}}$, which causes the packet to evolve as a function $e^{iE_{\mathbf{p}}t - \Gamma_{\mathbf{p}}t}$, meaning it is unstable, decaying with a lifetime $(2\Gamma_{\mathbf{p}})^{-1}$.

What we are describing is again the concept of **quasiparticle**, this time in field theory, i.e., such a particle-resembling object will give rise to something that resembles a free propagator, which is now considering interactions in a natural way.

In this sense, we say that we are considering a quasiparticle as a particle dressed with its interactions. This will adjust the parameters of the theory, starting from the physical mass of the particle, to return the correct ones, even managing to tame divergences in the process.

The process of renormalization is therefore embodied in the equation

$$\text{quasiparticle} = \text{real particle} + \text{interactions}. \tag{8.133}$$

We therefore define the quasiparticle weight as $Z_{\mathbf{p}}^{1/2} = \left\langle \mathbf{p}_\lambda | \hat{a}_{\mathbf{p}}^\dagger | \Omega \right\rangle$, so that we have a quasiparticle if $Z_{\mathbf{p}}^{1/2} \neq 0$. For a quasiparticle to be physically meaningful, its energy must exceed the decay rate, namely $E_{\mathbf{p}} > \Gamma_{\mathbf{p}}$, otherwise it would decay too rapidly to be observed.

Having momentum change in $\left|\mathbf{p}_\lambda\right\rangle$, so that the energy changes accordingly, it is natural to expect a change in the mass of the particle too, in light of the dispersion relation. The single-particle exctitation mass therefore changes from the **bare mass** m to the **physical mass** m_P, which is the correct mass found in experiments. In fact we now have an energy $(\mathbf{p}^2 + m_P^2)^{1/2}$.

8.4.3 The Propagator for a Quasiparticle

Now, we want to work out the propagator for a quasiparticle, recalling that a Green's function in the interacting picture is written as

$$G(x, y) = \left\langle \Omega | T \hat{\phi}_H(x) \hat{\phi}_H^\dagger(y) | \Omega \right\rangle. \tag{8.134}$$

The completeness relation for one-particle states is

$$1 = \int \frac{d^3 p}{(2\pi)^3} \frac{1}{2E_{\mathbf{p}}} |\mathbf{p}\rangle \langle \mathbf{p}| , \tag{8.135}$$

then, we can write an analogous completeness relation for the entire Hilbert space as

$$\mathbf{1} = |\Omega\rangle \langle\Omega| + \sum_\lambda \int \frac{d^3 p}{(2\pi)^3} \frac{1}{2E_{\mathbf{p}}(\lambda)} \left|\mathbf{p}_\lambda\right\rangle \left\langle\mathbf{p}_\lambda\right| . \tag{8.136}$$

Inserting $\left\langle \Omega|\hat{\phi}_H(x)\hat{\phi}_H^\dagger(y)|\Omega\right\rangle$ between the operators and dropping the uninteresting constant produced by the $|\Omega\rangle\langle\Omega|$ term, we arrive at

$$\left\langle \Omega|\hat{\phi}_H(x)\hat{\phi}_H^\dagger(y)|\Omega\right\rangle = \sum_\lambda \int \frac{d^3p}{(2\pi)^3}\frac{1}{2E_{\mathbf{p}}(\lambda)}\left\langle \Omega|\hat{\phi}_H(x)|\mathbf{p}_\lambda\right\rangle\left\langle \mathbf{p}_\lambda|\hat{\phi}_H(y)|\Omega\right\rangle. \tag{8.137}$$

Recalling that a particle now time evolves according to $e^{iE_{\mathbf{p}}t}e^{-\Gamma_{\mathbf{p}}t}$ and the definition of the weight $Z_{\mathbf{p}}$, we can manipulate as follows

$$\begin{aligned}\left\langle \mathbf{p}_\lambda|\hat{\phi}_H(x)|\Omega\right\rangle &= \left\langle \mathbf{p}_\lambda|e^{ipx}\hat{\phi}(0)e^{ipx}|\Omega\right\rangle e^{-\Gamma_{\mathbf{p}}t} = \\ &\left\langle \mathbf{p}_\lambda|\hat{a}_{\mathbf{p}}^\dagger|\Omega\right\rangle e^{ipx}e^{-\Gamma_{\mathbf{p}}t} = \\ &Z_{\mathbf{p}}^{1/2}e^{ipx}e^{-\Gamma_{\mathbf{p}}t},\end{aligned} \tag{8.138}$$

and analogously for the other bra-ket. Introducing an integration over p^0 and substituting the infinitesimal factor $i\epsilon$ of the Feynman propagator with the decay constant $i\Gamma_{\mathbf{p}}$, the result is

$$\left\langle \Omega|\hat{\phi}_H(x)\hat{\phi}_H^\dagger(y)|\Omega\right\rangle = \sum_\lambda \int \frac{d^4p}{(2\pi)^4}\frac{i}{p^2-m^2+i\Gamma_{\mathbf{p}}}e^{-ip(x-y)}Z_{\mathbf{p}}. \tag{8.139}$$

This goes analogously for both $x^0 < y^0$ and $y^0 > x^0$, means that we can time order the ϕ operators with a time ordering operator T and get the implicit form for the propagator $G(x, y) = \left\langle \Omega|T\hat{\phi}_H(x)\hat{\phi}_H^\dagger(y)|\Omega\right\rangle$. But this still includes the superposition of states produced by $\hat{a}_{\mathbf{p}}^\dagger|\Omega\rangle$, so we do not know how to simplify it for a propagator in general. A way to calculate it is to consider the spectral function as in (b) in Fig. 8.3, where we recognized the quasiparticle as the narrow wave packet resembling a single particle peak, and the rest of the spectrum, the multiparticle states, as higher orders of perturbations, just like for the bare particle. Then we can use the spectral density function $\rho(M^2)$ and modify the free propagator-like term inside (8.139) as $\tilde{\Delta}(p, M^2) = \dfrac{i}{p^2-M^2+i\Gamma_{\mathbf{p}}}$, where M is the mass of the whole spectrum. This produces a propagator written as

$$\tilde{G}(p) = \int_0^\infty \frac{dM^2}{2\pi}\rho(M^2)\tilde{\Delta}(p, M^2). \tag{8.140}$$

The weight $Z_{\mathbf{p}}$ was included in the density spectral function, written as

$$\rho(M^2) = 2\pi\delta(M^2 - m_P^2)Z_{\mathbf{p}} + (\text{multiparticle parts}), \tag{8.141}$$

which allows us to separate the terms in the propagator as the one for the state resembling the single particle and the multiparticles one that starts around $2m_P$ (m_P

being the mass of this wave packet only, what we called the physical mass, as we see just now), as the higher order perturbations. This is also consistent with the required condition $E_{\mathbf{p}} > \Gamma_{\mathbf{p}}$, since otherwise the quasiparticle would be indistinguishable from the background, so unstable and short-lived. We can therefore write the full propagator as

$$\tilde{G}(p) = \frac{iZ_{\mathbf{p}}}{p^2 - m_P^2 + i\Gamma_{\mathbf{p}}} + \text{(multiparticle parts)}. \tag{8.142}$$

Here, we read the physical mass energy of the excited states m_P from the position of the pole, just like in the free propagator, while the residue at that pole gives us the weight $Z_{\mathbf{p}}$, and the lifetime comes from $\Gamma_{\mathbf{p}}$.

Having computed a propagator, we expect to be able to calculate actual physical amplitudes from this. In fact, nowadays a wide range of experiments provide evidence for the existence of quasiparticles, which should not surprise us at all, since we can now see them as the natural description of particles after renormalization.

A subject field where these objects come particularly handy is condensed matter physics. There, a system usually made up of very strongly interacting real particles (such as electrons in a metal) is modified through renormalization into a system of a small number of weakly (or non) interacting particle-like excitations. For a crystal, for instance, the ground state is full of excitations, but in this picture we regard it as a vacuum. Raising the temperature then produces an excited state of the whole system described as a few of these excitations, often called elementary excitations. These can be classified as

- collective excitations: particle-like excitations that occur through the motion of all the components of the system. An example is phonons, seen as excitations in a crystalline structure passing through a chain of coupled masses;
- quasiparticles: single-particle excitations from the non-interacting system dressed in interactions. An example is an electron gas, which we describe with quasielectrons and quasiholes.

This distinction is quite loose and not always agreed upon, but gives an insight into the subject fields that deal with excitations like this on a daily basis.

8.4.4 Procedures of Renormalization

Now that we understand the fundamental reason behind renormalization, we can see actual examples and procedures from QFT. We will see how to move from the bare mass to the physical mass to account for quasiparticles, and also how all other parameters must be shifted from their unphysical (bare) values to the corresponding physical ones.

Driving exercise Two-particle scattering expansion.
A simple and useful example is that of two-particle scattering for a self-interacting scalar field. The system is described by a Lagrangian

$$\mathcal{L} = \frac{1}{2}(\partial_\mu \phi)^2 - \frac{m^2}{2}\phi^2 - \frac{\lambda}{4!}\phi^4. \tag{8.143}$$

The expanded $\hat{S}$-operator up to second order is

$$\begin{aligned}\hat{S} &= T\left[\exp\left(-i\int d^4z \hat{\mathcal{H}}_I(z)\right)\right] \\ &= T\left[1 - \frac{i\lambda}{4!}\int d^4z\hat{\phi}(z)^4 + \frac{(-i)^2}{2!}\left(\frac{\lambda}{4!}\right)^2\int d^4y d^4w\hat{\phi}(y)^4\hat{\phi}(w)^4 + \ldots\right].\end{aligned} \tag{8.144}$$

Then, the amplitude for two-particle scattering process is

$$\left\langle q_1 q_2|\hat{S}|p_2 p_1\right\rangle = (2\pi)^6(16E_{\mathbf{p}_1}E_{\mathbf{p}_2}E_{\mathbf{q}_1}E_{\mathbf{q}_2})^{\frac{1}{2}}\left\langle 0|\hat{a}_{\mathbf{q}_1}\hat{a}_{\mathbf{q}_2}\hat{S}a^\dagger_{\mathbf{p}_2}\hat{a}^\dagger_{\mathbf{p}_1}|0\right\rangle \tag{8.145}$$

and ignoring the zeroth-order diagrams where no interaction occurs, we get the following Feynmnan diagrams, which correspond to all connected, amputated diagrams with four external legs:
The first order expansion gives the single-vertex diagram (a), whose amplitude is simply a coupling constant,

$$i\mathcal{M}_a = -i\lambda. \tag{8.146}$$

For the subsequent diagrams, we can easily figure out through the Feynman rules for ϕ^4 theory the integral needed as

$$\int_0^\Lambda \frac{d^4q}{(2\pi)^4}\frac{i}{q^2-m^2}\frac{i}{(p-q)^2-m^2} = -4ia\ln\left(\frac{\Lambda}{p}\right), \tag{8.147}$$

where a includes all the unimportant numerical constants, 4 comes from the symmetry in ϕ^4 theory, and the logarithm is obtained through standard techniques. We have also introduced a cutoff Λ instead of integrating up to infinity. In fact, this amplitude is in principle divergent, and we must send $\Lambda \to \infty$. By counting powers in the integrals we see that it is $\sim \int \frac{d^4q}{q^4}$, telling us that the ultraviolet behavior of this theory is to diverge logarithmically. Restoring p as the internal momenta of (b), (c) and (d) diagrams, we obtain the amplitudes (see Fig. 8.4)

$$i\mathcal{M}_b = ia\lambda^2\left(\ln\Lambda^2 - \ln s\right), \qquad s = (p_1+p_2)^2, \tag{8.148}$$

$$i\mathcal{M}_c = ia\lambda^2\left(\ln\Lambda^2 - \ln t\right), \qquad t = (p_1-p_3)^2, \tag{8.149}$$

$$i\mathcal{M}_d = ia\lambda^2\left(\ln\Lambda^2 - \ln u\right), \qquad u = (p_1-p_4)^2. \tag{8.150}$$

The total amplitude is therefore

$$\begin{aligned} i\mathcal{M} &= i\left(\mathcal{M}_a + \mathcal{M}_b + \mathcal{M}_c + \mathcal{M}_d\right) \\ &= -i\lambda + ia\lambda^2\left(6\ln\Lambda - \ln s - \ln t - \ln u\right), \end{aligned} \tag{8.151}$$

where the divergence is expressed in the term $6ia\lambda^2 \ln \Lambda$, proportional to the strength λ of the interaction. This means that changing the bare mass into the physical one, while surely being the right way to proceed, won't cancel out the divergence.

We can think in a similar way to how we did for the Casimir effect: the divergence comes from the fact that the integral takes up all frequencies between 0 and $\Lambda \to \infty$, so we can introduce a finite cutoff $\Lambda \neq \infty$, justifying it by saying that we deliberately ignore any fine-structure details in the fields below a length scale $\sim 1/\Lambda$. In fact, divergences in quantum field theories often arise in the UV regime, for ultra-high frequencies and very small scales (infinities can also come up at the infrared scale, although more rarely, an example is the $e^+e^- \to \mu^+\mu^-(+\gamma)$ scattering). Condensed matter physicists actually often perform this cutoff, since they are not interested in any details below the scale of an atom. But if we are talking about fundamental particles, it's not clear why there should be a smallest scale, we might say it represents some graininess in spacetime, but this is quite an assumption for a QFT.

In this picture, we are basically trying to say that some super high-frequency modes somehow do not contribute to the scattering, but we cannot really find a convincing enough argument as the one for the Casimir force.

The answer is still the same: we are not modeling interactions right. But this time the reason is more subtle. When writing a Lagrangian, the coupling terms must always be accompanied by a constant: we know this from simple special relativity, otherwise they would be allowed to have a completely arbitrary strength and therefore even be punctual. This is strictly forbidden by our field theory, that models interactions as virtual particles traveling between the on-shell ones. This is why the coupling constant λ is necessary. However, simply introducing λ is not enough: without further specification, it has no intrinsic physical meaning or bounds, and in principle could still describe unphysical processes.

8.4.5 Renormalization Through Counterterms

At this stage it is clear that the parameter λ appearing in the bare Lagrangian cannot be identified with the quantity one would infer from a scattering experiment: the latter is a physical (renormalized) coupling, which we denote by λ_P.

So, the most direct way to incorporate renormalization in perturbation theory is to introduce **counterterms**. One supplements the Lagrangian with additional local

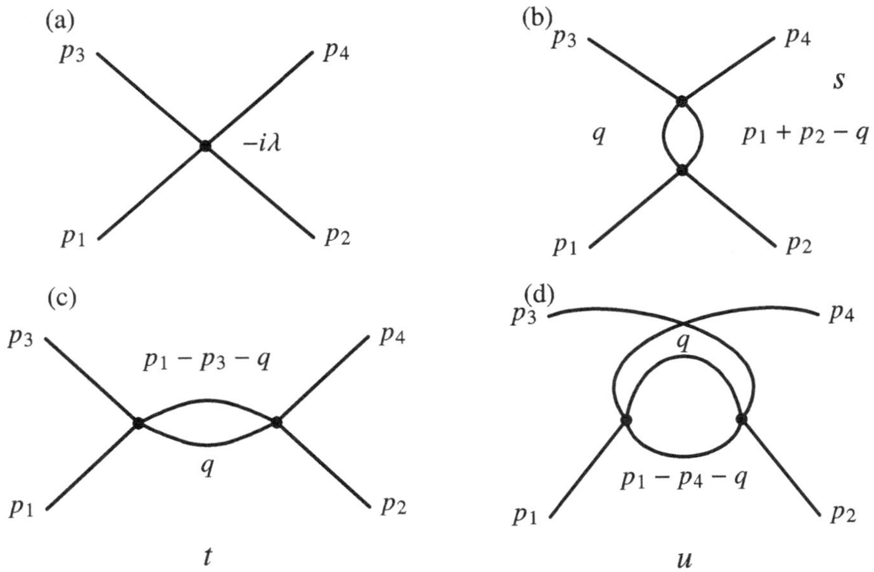

Fig. 8.4 The expansion of four-point diagrams for two-particle scattering up to second order in the interaction strength. Diagrams **b–d** define the processes labelled s, t and u, respectively

operators, with the same field dependence as the original interactions, whose coefficients are chosen so as to cancel the cutoff dependence generated by loop integrals. In doing so, the bare parameters are shifted into physical ones. In principle one should also trade the bare mass m for the physical (pole) mass m_P, which is the parameter naturally associated with quasiparticle excitations; for the moment, however, we postpone this step to keep the notation light.

▶ **Driving exercise** In ϕ^4 theory the only coupling constant is λ, multiplying the operator ϕ^4. At one loop, the divergent part of the amputated four-point amplitude is proportional to $-(6ia\lambda^2 \ln \Lambda)$, so we modify the Lagrangian as

$$\mathcal{L} = \frac{1}{2}(\partial_\mu \phi)^2 - \frac{m^2}{2}\phi^2 - \frac{\lambda}{4!}\phi^4 + \frac{C^{(2)}}{4!}\phi^4, \tag{8.152}$$

where the counterterm $C^{(2)}$ reminds us that we are adjusting the coupling consistently up to second order in perturbation theory. In particular, we choose

$$C^{(2)} = -6a\lambda^2 \ln \Lambda. \tag{8.153}$$

Proceeding as usual, i.e., canonical quantization, Dyson expansion, identification of the modified Feynman rules, and evaluation of the additional diagrams, one finds that the only novelty is the appearance of the counterterm insertion in the Dyson series for the $\hat{S}$-operator,

$$\hat{S} = T\left[\exp\left(-i\int d^4x \frac{1}{4!}\left(\lambda - C^{(2)}\right)\hat{\phi}^4\right)\right]. \tag{8.154}$$

Accordingly, the effective interaction vertex carries the factor

$$-i\lambda_P = -i(\lambda - C^{(2)}) = -(i\lambda - iC^{(2)}). \tag{8.155}$$

Up to second order, the corrected two-particle scattering amplitude becomes

$$i\mathcal{M}^{(2)} = -i\lambda + ia\lambda^2\left[6\ln\Lambda - \ln s - \ln t - \ln u\right] - 6ia\lambda^2\ln\Lambda, \tag{8.156}$$

and one sees explicitly that the divergent contribution proportional to $\ln\Lambda$ cancels against the counterterm (which is itself divergent, by construction). One is left with the finite expression

$$i\mathcal{M}^{(2)} = -i\lambda - ia\lambda^2\left(\ln s + \ln t + \ln u\right), \tag{8.157}$$

which no longer depends on the UV cutoff. In other words, once the coupling has been shifted appropriately, the amplitude is finite and can, in principle, be compared with experiment. This is the practical sense in which the renormalized theory becomes predictive: order by order, unphysical regulator-dependence is absorbed into a redefinition of parameters.

Counterterms as parameter shifts. In renormalized perturbation theory, one does not "remove" loop effects; rather, one reorganizes the expansion so that measurable quantities are expressed in terms of physical parameters. Counterterms implement this reorganization locally in the Lagrangian: they cancel cutoff-dependent pieces and convert (m, λ) into (m_P, λ_P), also rescaling the field in some cases.

► **Pause and think** Think of a scattering experiment: which quantity would you call "the coupling"? Discuss why the identification $\lambda \equiv \lambda_P$ is inconsistent once loop corrections are included, and how counterterms repair this mismatch.

8.4.6 Renormalizable Theories and Power Counting

A conceptual caveat is in order. So far we have only renormalized the coupling to second order: can we truly claim that we have reached the physical value? In principle, higher orders could generate new divergent structures. For instance, at third order one might encounter divergent contributions requiring not only a new counterterm $C^{(3)}\phi^4$, but perhaps also a genuinely new operator such as $D^{(3)}\phi^6$ to absorb additional divergences. If this pattern continued indefinitely, the theory would require an ever-growing (possibly infinite) set of independent counterterms, and predictivity would be lost.

A theory for which *only finitely many* counterterms are needed to absorb all divergences (to all orders) is called **renormalizable**. Many theories of physical interest enjoy this property. From the viewpoint adopted here, this is not miraculous: if the parameters appearing in the theory are to be identified with physical (measurable) ones, then the perturbative expansion must be reorganizable in terms of a finite set of such parameters. Still, the existence of non-renormalizable theories may appear to undermine this logic; we will return to this point in the next chapter.

► **Driving exercise** For the time being, let us complete the renormalization discussion in ϕ^4 theory by examining the so-called Saturn diagram, shown in Fig. 8.5.

Written out explicitly, its amplitude reads

$$i\mathcal{M} = \frac{(-i\lambda)^2}{6} \int_0^\Lambda \frac{d^4q}{(2\pi)^4} \frac{d^4k}{(2\pi)^4} \frac{i}{q^2 - m^2} \frac{i}{k^2 - m^2} \frac{i}{(p - q - k)^2 - m^2}. \tag{8.158}$$

This integral is already substantially more involved than the one-loop integrals considered earlier, and the complexity only increases at higher orders. A simple and remarkably robust diagnostic of the UV behavior is provided by **power counting**. The key point is that UV divergences are controlled by the large loop momenta, whereas external momenta are comparatively small. One may therefore expand the integrand as a Taylor series in the external momentum p. Denoting the loop integral (after extracting overall constants) by I, one writes schematically

$$I = \alpha + \beta p^2 + \gamma p^4 + \dots, \tag{8.159}$$

where odd powers are absent because the theory is invariant under $\phi \to -\phi$, and the corresponding amplitudes respect the associated parity selection rules.

One can now infer which counterterms are required by inspecting the superficial degree of divergence of each term in the expansion.

- The integral I has eight powers of momentum in the measure and six powers in the denominators, hence it is superficially *quadratically* divergent. Setting $p = 0$ isolates the leading term, $I|_{p=0} = \alpha$, and one concludes that α carries a quadratic divergence. This divergence is cancelled by a mass-type counterterm, which we parameterize as $\frac{A}{2}\phi^2$.

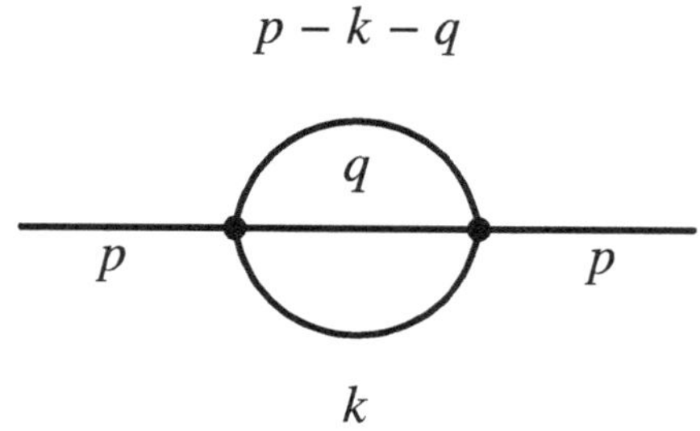

Fig. 8.5 The Saturn diagram

- Differentiating I once with respect to p and setting $p = 0$ gives zero because odd terms are absent. Differentiating twice isolates $I''|_{p=0} = 2\beta$. At this stage the superficial degree of divergence is reduced: one finds a *logarithmic* divergence, consistent with the appearance of a p^2 structure. Since in momentum space $(\partial_\mu \phi(x))^2 \to \tilde{\phi}(-p)\, p^2\, \tilde{\phi}(p)$, the corresponding counterterm has the form $p^2 \frac{B}{2} \to \frac{B}{2}(\partial_\mu \phi)^2$.
- Differentiating two additional times makes the integral convergent, so no further divergences arise from this diagram beyond those already absorbed (including the interaction counterterm discussed earlier in the two-particle scattering amplitude).

Collecting the counterterms, the Lagrangian can be written as

$$\mathcal{L} = \frac{1}{2}(\partial_\mu \phi)^2 - \frac{m^2}{2}\phi^2 - \frac{\lambda}{4!}\phi^4 + \frac{A}{2}\phi^2 + \frac{B}{2}(\partial_\mu \phi)^2 - \frac{C}{4!}\phi^4. \tag{8.160}$$

These counterterms are nothing but parameter shifts: renormalization affects not only explicit interaction vertices but also the normalization of the field and the location of the pole (mass) of the propagator. In practice, the A and B counterterms are often represented compactly by a single two-point insertion carrying the factor $i(B^{(n)} p^2 + A^{(n)})$ at order n.

The Feynman rules for renormalized ϕ^4 theory can be summarized as follows:

- a factor $\frac{i}{p^2 - m^2}$ for each propagator,
- a factor $-i\lambda$ for each interaction,
- add sufficient counterterm diagrams to cancel all infinities,
- a factor $i(B^{(n)} p^2 + A^{(n)})$ for each counterterm propagator, where n is the order of the diagram,
- a factor $iC^{(n)}$ for each interaction counterterm, where n is the order of the diagram,
- all other rules regarding integrating, symmetry factors and overall energy momentum conserving delta functions are identical as for the case of unrenormalized perturbation theory.

Power counting in one line. The UV behavior of a diagram is governed by the large-momentum scaling of its loop integral. Expanding in external momenta identifies which local operators can absorb the divergences: constant terms correspond to ϕ^2 counterterms, p^2 terms to $(\partial\phi)^2$, and so on.

▶ **Pause and think** Why does the external-momentum expansion make sense? Give a clear argument (in words and with a scaling estimate) explaining why UV divergences are controlled by loop momenta and why p can be treated as "small" for the purpose of diagnosing divergences.

8.4.7 The Full Counterterms Picture

We can now reinsert the quasiparticle language introduced previously. Conceptually, this amounts to two steps: (i) trade the bare mass m for the physical (pole) mass m_P, and (ii) account for the fact that the interacting field creates a *dressed* one-particle excitation only with some weight. The latter is encoded in a field-strength renormalization, implemented by the rescaling

$$\phi(x) = \sqrt{Z}\,\phi_r(x), \tag{8.161}$$

where Z is the quasiparticle weight (the residue of the propagator at the pole), previously denoted by $Z_{\mathbf{p}}$. In terms of renormalized fields, the Lagrangian becomes

$$\mathcal{L} = \frac{1}{2}(\partial_\mu\phi_r)^2 - \frac{m_P^2}{2}\phi_r^2 - \frac{\lambda_P}{4!}\phi_r^4 + \frac{A}{2}\phi_r^2 + \frac{B}{2}(\partial_\mu\phi_r)^2 - \frac{C}{4!}\phi_r^4. \tag{8.162}$$

To display the renormalization as an explicit shift of parameters, we collect coefficients:

$$\mathcal{L} = \frac{1+B}{2}(\partial_\mu\phi_r)^2 - \frac{(m_P^2 - A)}{2}\phi_r^2 - \frac{(\lambda_P - C)}{4!}\phi_r^4. \tag{8.163}$$

Defining $-A = \delta m^2$, $-C = \delta\lambda$ and $B = \delta Z$, we can rewrite this as

$$\mathcal{L} = \frac{1+\delta Z}{2Z}(\partial_\mu\phi)^2 - \frac{(m_P^2 + \delta m^2)}{2Z}\phi^2 - \frac{(\lambda_P + \delta\lambda)}{4!Z^2}\phi^4, \tag{8.164}$$

where we have restored the original (unrenormalized) field ϕ. From this expression one reads the relations connecting bare and renormalized quantities:

$$\phi = \sqrt{Z}\phi_r, \quad Z = 1 + \delta Z, \quad m^2 = \frac{m_P^2 + \delta m^2}{Z}, \quad \lambda = \frac{\lambda_P + \delta\lambda}{Z^2}. \tag{8.165}$$

Equivalently, one may run the logic in reverse: first rescale the fields by the quasiparticle weight, then implement parameter shifts,

$$Z = 1 + \delta Z, \quad Zm^2 = m_P^2 + \delta m^2, \quad Z^2\lambda = \lambda_P + \delta\lambda, \tag{8.166}$$

to obtain the standard "renormalized + counterterms" decomposition,

$$\mathcal{L} = \frac{1}{2}(\partial_\mu \phi_r)^2 - \frac{m_P^2}{2}\phi_r^2 - \frac{\lambda_P}{4!}\phi_r^4 + \frac{\delta Z}{2}(\partial_\mu \phi_r)^2 - \frac{\delta m^2}{2}\phi_r^2 - \frac{\delta\lambda}{4!}\phi_r^4. \tag{8.167}$$

This closes the renormalization analysis of ϕ^4 theory via counterterms. The core lesson is that naive perturbation theory expands around the wrong parameters: it treats m and λ as if they were the physical mass and coupling, and loop corrections then punish this choice with regulator-dependent infinities. Renormalization reorganizes the perturbative expansion so as to describe scattering and propagation directly in terms of the dressed (quasiparticle) excitations and their physical parameters.

Quasiparticles meet renormalization. Field-strength renormalization is not an optional technicality: it encodes the fact that the interacting field does not create a pure one-particle state with unit probability. The residue Z is precisely the weight of the quasiparticle peak in the spectral function, and it enters amplitudes through $\phi = \sqrt{Z}\phi_r$.

► **Pause and think** Make a short list of which quantities are, in principle, measurable (m_P, λ_P, Z) and which are scheme-dependent objects (m, λ, δm^2, $\delta\lambda$, δZ). Explain why the latter can be infinite while the former are finite.

8.4.8 On-Shell Perturbation Theory

A complementary viewpoint is to expand directly in terms of physical parameters from the outset. This approach is often referred to as **on-shell perturbation theory**. Rather than introducing counterterms explicitly, one defines renormalized parameters by imposing renormalization conditions on amplitudes evaluated at a chosen kinematical point.

To illustrate the idea, let us recall the example of amplitude, before explored, retaining only the tree-level term and the s-channel one-loop contribution for simplicity. We write it as

$$\mathcal{M}(s) = -\lambda - 3a\lambda^2 \ln \frac{s}{\Lambda^2}. \tag{8.168}$$

Here s plays the role of a center-of-mass energy scale. Fixing a specific value $s = s_0$ is tantamount to selecting a reference scale at which the coupling is *defined*. Instead of adding a counterterm to the Lagrangian, we may therefore introduce the renormalized coupling λ_P as the value of the amplitude at s_0:

$$\lambda_P = -\mathcal{M}(s_0) = \lambda + 3a\lambda^2 \ln \frac{s_0}{\Lambda^2} + \dots . \tag{8.169}$$

Since λ_P is defined as an observable (finite) quantity, the bare λ must contain the compensating cutoff dependence. Next, we solve perturbatively for λ in terms of λ_P by setting

$$\lambda = \lambda_P + c\lambda_P^2 + \dots \tag{8.170}$$

and determining the coefficient c. Substituting into the previous relation gives

$$\lambda = \lambda_P + c\lambda_P^2 + 3a\lambda_P^2 \ln \frac{s_0}{\Lambda^2} + \dots , \tag{8.171}$$

so that $c = -3a \ln \frac{s_0}{\Lambda^2}$, and therefore

$$\lambda = \lambda_P - 3a\lambda_P^2 \ln \frac{s_0}{\Lambda^2} + \dots . \tag{8.172}$$

Now suppose we measure the scattering at a different center-of-mass energy s. Then, to the same perturbative order,

$$\begin{aligned} \mathcal{M}(s) &= -\lambda - 3a\lambda^2 \ln \frac{s}{\Lambda^2} = -\left(\lambda_P - 3a\lambda_P^2 \ln \frac{s_0}{\Lambda^2}\right) - 3a\lambda_P^2 \ln \frac{s}{\Lambda^2} + \dots \\ &= -\lambda_P - 3a\lambda_P^2 \ln \frac{s}{s_0} + \dots . \end{aligned} \tag{8.173}$$

This expression is finite order-by-order, and, more importantly, it is a genuine physical prediction: it relates one observable (the amplitude at scale s) to another observable (the amplitude at scale s_0) through a logarithmic dependence. In this sense, on-shell renormalization is equivalent to the counterterm method discussed earlier, although the latter makes parameter shifts more explicit.

Renormalization point. Defining λ_P at $s = s_0$ does not change the physics, but it changes the *scheme*: different choices of s_0 correspond to different parameterizations of the same theory. Physical predictions are obtained by consistently expressing all amplitudes in terms of the chosen renormalized parameters.

8.5 Renormalization in Action: Propagators and Feynman Diagrams

Having obtained an explicit renormalized Lagrangian, we now examine how renormalization reshapes propagators and Feynman diagrams. Physically, dressing a particle by interactions is encoded in two central objects. The first is the *self-energy*, which governs how interactions shift the pole of the propagator (hence the physical mass). The second is the *vertex function*, which accounts for how virtual fluctuations screen (or, more generally, renormalize) the interaction strength. Before focusing on these quantities, it is useful to sharpen the link between divergences and diagrammatics.

8.5.1 Loops and Divergences

Both the amputated diagrams in ϕ^4 theory and the Saturn diagram in Fig. 8.5 contain loops. Each loop is built out of internal propagators, and these loop integrals precisely generate UV divergences, largely independently of the number of internal lines, vertices, or external legs. Internal lines represent the propagation of virtual quanta, capturing both vacuum fluctuations and the exchange picture of interactions between real particles.

If one attempts to represent scattering within a purely classical diagrammatic language, interactions would be point-like and no meaningful analogue of a loop would arise: there would be no notion of intermediate virtual propagation that can reconnect back onto itself. Loops are therefore not a mistake; they are the natural byproduct of describing interactions quantum mechanically. The pathology is not the existence of loops, but the naive identification of bare parameters with physical ones. Renormalization teaches us that loop effects can be consistently absorbed into redefinitions of parameters and fields: in a renormalized description, the net effect of UV-sensitive virtual processes is packaged into dressed vertices and propagators.

In particular, loops are the diagrammatic origin of ultraviolet divergences. It is in the UV regime that short-distance fluctuations become important and, without renormalization, they contaminate amplitudes with regulator dependence. A useful practical signature is that loop effects generically produce logarithms (and sometimes powers) of kinematical invariants and cutoffs. Tree-level graphs, by contrast, yield only rational functions of momenta and couplings: they do not generate logarithms.

To correct diagrams and propagators systematically, we need a classification of loop subgraphs. The relevant notion is the following, illustrated in Fig. 8.6.

We call a **1-particle irreducible diagram (1PI diagram)** any connected diagram which cannot be disconnected by cutting a single internal propagator line. Examples of 1PI diagrams in ϕ^4 theory are shown in Fig. 8.6(a), while Fig. 8.6(b) displays connected diagrams that are *not* 1PI: they contain a cutline, and a single scissor cut along that line splits them into disconnected pieces. 1PI graphs are the minimal nontrivial building blocks entering the self-energy and vertex corrections; heuristically, they form the “irreducible core” of the dressing.

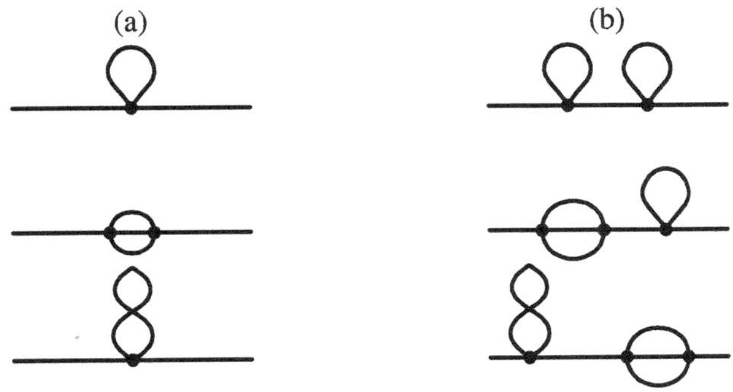

Fig. 8.6 **a** Some 1PI diagrams from ϕ^4 theory. They cannot be turned into two meaningful diagrams by cutting one propagator line. **b** Diagrams that are not 1PI. All can be turned into legitimate 1PI diagrams by cutting at some point along the horizontal line

8.5.2 Renormalization Conditions and pole mass

The sum of all 1PI diagrams with two external legs defines the **1PI self-energy**:

$$-i\tilde{\Sigma}(p) = \sum(\text{all amputated 1PI diagrams with two external lines}). \quad (8.174)$$

Our goal is to express the full propagator $\tilde{G}(p)$ in terms of $\tilde{\Sigma}(p)$. This connection is naturally formulated within statistical field theory. In position space, the propagator can be written as the connected two-point correlation function

$$G_c(x, y) = \left\langle \hat{\phi}(x)\hat{\phi}(y) \right\rangle_t - \left\langle \hat{\phi}(x) \right\rangle_t \left\langle \hat{\phi}(y) \right\rangle_t, \quad (8.175)$$

where $\left\langle \hat{\phi} \right\rangle_t$ denotes the thermal average of the field at temperature t. In the path-integral formulation of QFT, after a Wick rotation to Euclidean space, the generating functional reads

$$Z[J] = \int \mathcal{D}\phi e^{-S_E} = \int \mathcal{D}\phi e^{-\int d^4x_E \{\mathcal{L}_E[\phi] - J(x)\phi(x)\}}. \quad (8.176)$$

Moreover, in statistical physics one has

$$Z[J = 0] = e^{\sum(\text{connected vacuum diagrams})}, \quad (8.177)$$

so that, when $\left\langle \hat{\phi}(x) \right\rangle_t \neq 0$,

$$\begin{aligned} G_c(x, y) &= \frac{\delta^2}{\delta J(x)\delta J(y)} \ln Z[J]|_{J=0} \\ &= \sum(\text{all connected diagrams with two external legs}) \end{aligned} \quad (8.178)$$

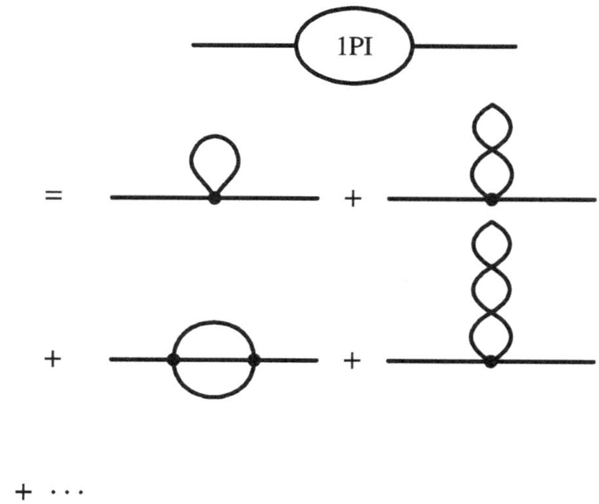

Fig. 8.7 Some contributions to the 1PI self-energy in ϕ^4 theory

and this defines the Green's function $\tilde{G}(p)$ in momentum space (see Fig. 8.7).

The self-energy $-i\tilde{\Sigma}(p)$ collects precisely the interaction insertions that dress the propagator. One uses the standard perturbative resummation trick: by the very definitions of $\tilde{G}$ and $\tilde{\Sigma}$, the full propagator can be expanded as the free propagator plus the free propagator with one self-energy insertion, plus the same with two insertions, and so on. Diagrammatically, this is the iterative dressing of the line. Algebraically,

$$\tilde{G}(p) = \frac{i}{p^2 - m^2} + \frac{i}{p^2 - m^2}\left[-i\tilde{\Sigma}(p)\right]\frac{i}{p^2 - m^2} + \ldots. \tag{8.179}$$

Treating this as a geometric series and summing it yields

$$\begin{aligned}\tilde{G}(p) &= \frac{i}{p^2 - m^2}\left\{1 + \left[-i\tilde{\Sigma}(p)\right]\frac{i}{p^2 - m^2} + \ldots\right\} \\ &= \frac{i}{p^2 - m^2}\left\{\frac{1}{1 - \dfrac{\tilde{\Sigma}(p)}{p^2 - m^2}}\right\} \\ &= \frac{i}{p^2 - m^2 - \tilde{\Sigma}(p)}.\end{aligned} \tag{8.180}$$

This resummation is represented diagrammatically in Fig. 8.8.

The physical mass m_P is determined by the location of the pole of the propagator. In particular, the pole condition reads

$$p^2 - m^2 - \mathrm{Re}\left(\tilde{\Sigma}(p)\right) = 0, \tag{8.181}$$

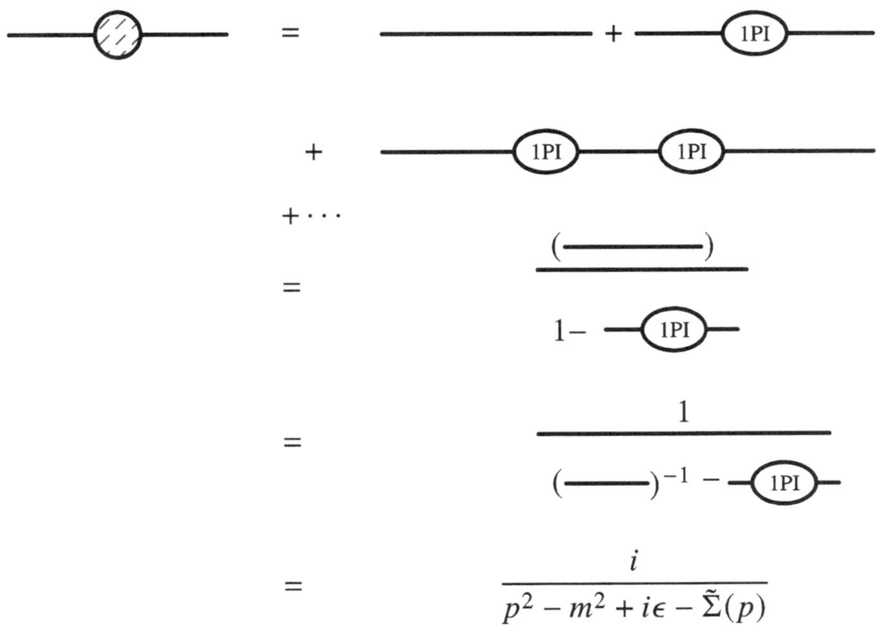

Fig. 8.8 Diagrammatic version of the Dyson resummation for the propagator

evaluated at $p^2 = m_P^2$. Hence,

$$m_P^2 = m^2 + \mathrm{Re}\left(\tilde{\Sigma}(p^2 = m_P^2)\right), \tag{8.182}$$

so that the real part of the self-energy encodes the interaction-induced shift of the mass. This is a prototypical example of a **renormalization condition**.

Up to this point, we have ignored that $\tilde{\Sigma}$ is itself divergent and that counterterms must be included. One may either (i) interpret m as a bare quantity containing the cutoff dependence needed to render m_P finite (the on-shell viewpoint), or (ii) introduce counterterms from the beginning and impose a renormalization condition directly on the finite, renormalized self-energy. In the latter case, the pole condition is implemented by requiring

$$\mathrm{Re}\left(\tilde{\Sigma}(p^2 = m_P^2)\right) = 0, \tag{8.183}$$

since the pole has already been placed at m_P by construction. These two procedures are equivalent when carried out consistently order by order.

8.5.3 The Vertex Function

The second central object is the **vertex function**. Virtual fluctuations screen the interaction between particles, so the effective strength of an interaction depends on

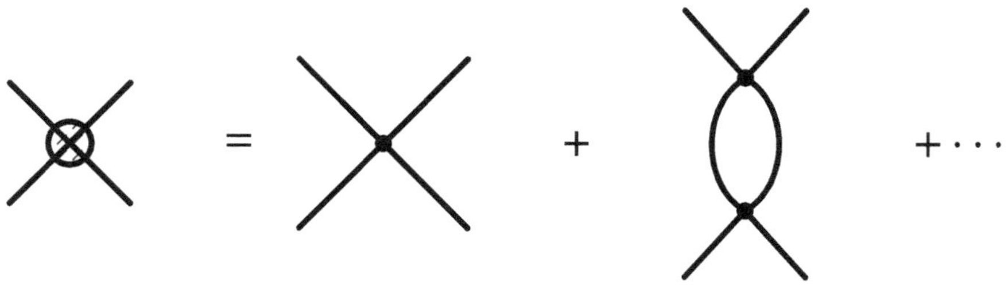

Fig. 8.9 Some contributions to the vertex function in ϕ^4 theory

how the vertex is probed, i.e., on the external kinematics. Renormalization amounts to separating the unphysical (cutoff-dependent) part of this screening from the finite, physically meaningful momentum dependence.

In ϕ^4 theory, the elementary interaction vertex carries the Feynman rule

$$(\text{vertex}) = (2\pi)^4\delta^{(4)}(p_4 + p_3 - p_2 - p_1)(-i\lambda), \tag{8.184}$$

where each vertex contributes $-i\lambda$ and the δ distribution enforces momentum conservation. The vertex function is encoded in the four-point Green's function (e.g. through the Gell-Mann-Low theorem) $\dfrac{\left\langle 0|T\hat{\phi}^4\hat{S}|0\right\rangle}{\left\langle 0|\hat{S}|0\right\rangle}$, and can be represented as

$$\tilde{G}(p_1)\tilde{G}(p_2)\left[-i\tilde{\Gamma}(p_1, p_2, p_3)\right]\tilde{G}(p_3)\tilde{G}(p_4), \tag{8.185}$$

appropriate to two-particle scattering. It follows that, for ϕ^4 theory,

$$-i\tilde{\Gamma} = \sum(\text{all connected four-point diagrams with external legs amputated}), \tag{8.186}$$

as illustrated in Fig. 8.9 below.

To second order the relevant contributions coincide with those entering the amplitude in Eq. (3.13), hence

$$-i\tilde{\Gamma}(p_1, p_2, p_3) = -i\lambda - ia\lambda^2\left(\ln s + \ln t + \ln u\right). \tag{8.187}$$

This explicitly shows that the effective interaction depends on the external kinematics: $\tilde{\Gamma}$ is a function of the independent external momenta (one momentum being redundant due to momentum conservation at the vertex). Consequently, there is no unique, kinematics-independent notion of "the" coupling constant beyond tree level. One is free to *define* the renormalized coupling by fixing $\tilde{\Gamma}$ at a chosen scale. In compact form,

$$-i\lambda_P = -i\tilde{\Gamma}(p_1, p_2, p_3) = \sum\left(\begin{array}{c}\text{amputated, connected diagrams with}\\ \text{four external legs and momenta } p_1, p_2, p_3\end{array}\right). \tag{8.188}$$

The specific choice of (p_1, p_2, p_3) (or, equivalently, of (s, t, u)) is called a **renormalization point**.

In analogy with the previous discussion, we choose a point (s_0, t_0, u_0) and, to second order, we write

$$-i\lambda_P = -i\tilde{\Gamma}^{(2)}(s_0, t_0, u_0) = -i\lambda + ia\lambda^2 \left(3\ln\Lambda^2 - \ln s_0 - \ln t_0 - \ln u_0\right). \tag{8.189}$$

Following the on-shell perturbation logic, we invert this relation order by order and obtain

$$-i\lambda = -i\lambda_P + ia\lambda_P^2 \left(3\ln\Lambda^2 - \ln s_0 - \ln t_0 - \ln u_0\right) + O(\lambda^3), \tag{8.190}$$

which, when substituted into the amplitude $i\mathcal{M}$ in Eq. (3.9), absorbs the cutoff-dependent piece and yields the finite result

$$i\mathcal{M} = -i\lambda_P + ia\lambda_P^2 \left(\ln\frac{s}{s_0} + \ln\frac{t}{t_0} + \ln\frac{u}{u_0}\right) + O(\lambda^3). \tag{8.191}$$

Vertex renormalization in practice. The renormalization point fixes what one means by λ_P beyond tree level. Different choices correspond to different schemes, but physical predictions are scheme-independent once expressed in terms of renormalized parameters consistently defined at the chosen point.

8.6 Renormalizability

We have already seen that ϕ^4 theory is *renormalizable*, in the sense that a *finite* set of counterterms suffices to absorb all ultraviolet divergences that appear in perturbation theory. At the same time, it is worth stressing a point that is often blurred by textbook terminology: the *cancellation of divergences* is not the conceptual content of renormalization, but rather a technical symptom of having reorganized the theory in terms of the appropriate physical parameters. In particular, the Wilsonian viewpoint, discussed below, clarifies that renormalizability is ultimately a statement about how a theory behaves under changes of scale, not merely about the existence of counterterms.

There is also a second subtlety: it is not always obvious, *a priori*, which finite quantities should be regarded as genuine observables. A renormalization prescription can certainly produce finite parameters, but finiteness alone does not guarantee direct measurability. A pragmatic first classification, therefore, is to quantify how "efficiently" ultraviolet infinities can be absorbed into a controlled set of operators.

In perturbation theory, UV divergences organize themselves according to the local operators allowed by the symmetries of the theory. A theory is renormalizable if, once these operators are included with independent coefficients, *no new structures* are required at higher orders. The Wilsonian viewpoint will refine this into a statement about which operators are relevant, marginal, or irrelevant under coarse graining.

8.6.1 Diagrammatical Analysis

To generalize the power-counting argument we used for the Saturn diagram, it is convenient to introduce the **superficial degree of divergence** D associated with the momentum integral arising from a given Feynman diagram:

$$D = \begin{pmatrix} \text{powers of momentum} \\ \text{in numerator} \end{pmatrix} - \begin{pmatrix} \text{powers of momentum} \\ \text{in denominator} \end{pmatrix}. \tag{8.192}$$

We call this degree *superficial* because the true ultraviolet behavior can be altered by gauge cancellations, special kinematics, or other subtleties that are not captured by a naive counting of powers. Nonetheless, as a first diagnostic one expects:

- $D > 0$: power divergence,
- $D = 0$: logarithmic divergence,
- $D < 0$: superficial convergence.

The origin of this criterion is easily understood from the typical structure of a loop amplitude,

$$\mathcal{A} \sim \int \frac{d^4k_1 d^4k_2 \dots d^4k_L}{(k_i^2 - m^2)\dots(k_j^2)\dots(k_n^2)}, \tag{8.193}$$

where L is the number of loops. Each loop contributes an integration over a four-momentum, potentially generating a UV divergence, whereas each propagator improves the UV behavior by supplying powers of momentum in the denominator. Roughly speaking, divergences arise when the denominator does not provide enough momentum suppression to compensate the loop integrations.

This naive perspective can fail for several reasons. First, UV-divergent *subdiagrams* can enhance the effective divergence of a full diagram beyond what a global D would suggest. Second, symmetries (especially gauge symmetries) can enforce cancellations that reduce the actual divergence. Third, D does not distinguish trivial tree graphs: a diagram with no loop integration may have $D = 0$ by the above definition, yet it is obviously finite. Despite these caveats, D is particularly reliable for *1PI* graphs, which serve as the genuinely irreducible loop building blocks. For such graphs, D often coincides with the true UV degree of divergence.

Let us compute D explicitly in ϕ^4 theory. Besides the number of loops L, we introduce V as the number of vertices and B_I as the number of internal lines. Each loop integral carries four powers of momentum, while each internal propagator behaves as $1/k^2$ and contributes -2 powers of momentum. Therefore,

$$D = 4L - 2B_I. \tag{8.194}$$

To relate $L, V,$ and B_I, recall that L counts the number of independent loop momenta. Each internal line introduces a momentum variable, but momentum-conserving delta functions at vertices remove integrations. There is one delta function per vertex, but one overall delta function expresses the conservation of total momentum for the entire diagram and does not remove an independent loop integration. Hence,

$$L = B_I - (V - 1). \tag{8.195}$$

Now, label by B_E the number of external lines. In ϕ^4 theory, four lines meet at each vertex. Each external line attaches to exactly one vertex, whereas each internal line connects two vertices. Consequently,

$$4V = B_E + 2B_I. \tag{8.196}$$

Substituting these relations into the expression for D, one finds the celebrated result

$$D = 4 - B_E. \tag{8.197}$$

Thus, in ϕ^4 theory the superficial degree of divergence depends *only* on the number of external legs. This neatly corroborates the renormalization pattern discussed earlier: only diagrams with $B_E \leq 4$ can diverge. Moreover, diagrams with one or three external legs are forbidden by the $\phi \to -\phi$ symmetry, and we have already accounted for the divergences in the two-point and four-point functions. No further counterterms are required.

8.6.2 Which Theories Are Renormalizable?

In perturbation theory, renormalizable theories are those for which a *finite* set of counterterms cancels all divergences to all orders. Not all theories enjoy this property. A classic example is Fermi's four-fermion model of weak interactions, which treats the interaction as a pointlike contact between four fermionic fields. Its Lagrangian can be written as

$$\mathcal{L} = \bar{\psi}(\hat{\not{p}} - m)\psi + G(\bar{\psi}\psi)^2, \tag{8.198}$$

where ψ is a fermion field. The superficial degree of divergence in this theory takes the form

$$D = 4 - \frac{3}{2}F_E + 2V, \tag{8.199}$$

where F_E is the number of external fermion lines. Here, the problematic feature is the explicit dependence on V. For fermion-fermion scattering one has $F_E = 4$, and already for $V > 1$ there exist divergent contributions; moreover, the superficial divergence worsens as V increases. In this sense, new counterterms are required at each order in perturbation theory.

It is therefore useful to distinguish:

- **super-renormalizable theories**: only a finite number of diagrams are superficially divergent;
- **renormalizable theories**: only a finite number of *amplitudes* are superficially divergent, but divergences occur at all orders in perturbation theory;
- **non-renormalizable theories**: at sufficiently high order, divergences proliferate in such a way that infinitely many independent counterterms are needed.

As a comprehensive illustration, consider a scalar ϕ^n theory in d spacetime dimensions:

$$\mathcal{L} = \frac{1}{2}(\partial_\mu \phi)^2 - \frac{1}{2}m^2\phi^2 - \frac{\lambda}{n!}\phi^n. \tag{8.200}$$

The number of loops is

$$L = B_I - V + 1, \tag{8.201}$$

and because n lines meet at each vertex, one has

$$nV = B_E + 2B_I. \tag{8.202}$$

Hence the superficial degree of divergence becomes

$$D = dL - 2B_I = d + \left[n\left(\frac{d-2}{2}\right) - d\right]V - \left(\frac{d-2}{2}\right)B_E. \tag{8.203}$$

Several familiar conclusions follow immediately. In $d = 4$, the ϕ^4 interaction is renormalizable, while ϕ^n interactions with $n > 4$ are non-renormalizable. In $d = 3$, the ϕ^6 interaction becomes renormalizable, whereas ϕ^4 is super-renormalizable. In $d = 2$, any ϕ^n interaction is super-renormalizable.

▶ **Pause and think** Using the formula for D above, classify the interactions ϕ^3 in $d = 6$, ϕ^4 in $d = 4$, and ϕ^6 in $d = 3$ as super-renormalizable, renormalizable, or non-renormalizable. Justify your answer in one paragraph for each case.

8.7 Renormalization Group

In renormalizable theories, ultraviolet divergences can be absorbed into a small set of parameters: the bare mass and couplings (or, equivalently, the counterterms in renormalized perturbation theory). Beyond these parameter shifts, virtual quanta with extremely large momenta do not explicitly affect low-energy predictions. The cancellation of UV divergences is essential for obtaining quantitative results, in fact: how can quantum fluctuations associated with arbitrarily short distances leave their imprint only through a handful of parameters?

In a renormalizable theory, loop integrals that probe virtual momenta are effectively dominated by scales comparable to the external kinematics. Understanding *why* this happens requires a change of viewpoint: one must track how the description of physics changes as the scale at which we probe the system is varied. Kenneth Wilson pioneered this approach by formulating renormalization as a scale-dependent evolution of couplings. The resulting differential relations are the **renormalization group (RG) equations**.

The RG does not merely "remove infinities": it explains how a theory reorganizes itself when short-distance degrees of freedom are eliminated. Couplings become scale-dependent, and the flow of these couplings under coarse graining encodes both universality and the effective nature of low-energy physics.

8.7.1 Wilson's Approach to Renormalization Theory

In Wilson's view, a QFT is defined together with an explicit UV cutoff Λ (which, in practice, can be chosen arbitrarily within the range where the effective description is meaningful). The conceptual issue raised at the beginning of our discussion then reappears in a sharpened form: changing Λ changes what is counted as "short-distance" physics, hence it reshuffles the parameters in the Lagrangian and, unless this reshuffling is controlled, it may appear to change predictions for long-distance observables.

The RG determines how parameters must depend on scale so that long-distance observables remain invariant under changes of the UV cutoff.

A theory is characterized by a set of parameters $\{g_i\}$ (including masses), which we regard as functions of the cutoff, $g_i = g_i(\Lambda)$. To probe different length scales,

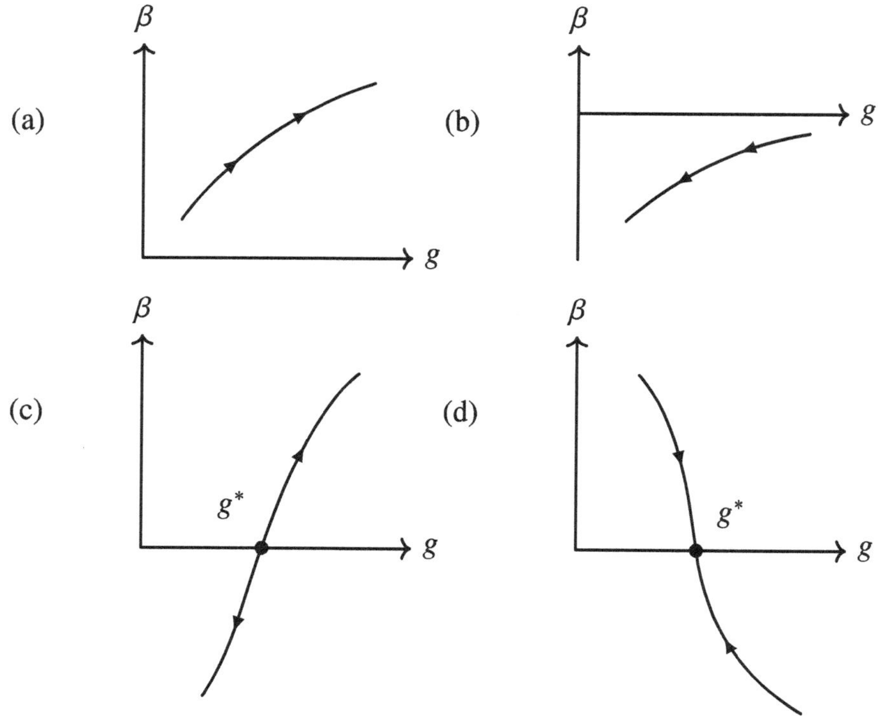

Fig. 8.10 RG flows in a one-coupling theory, represented by plotting $\beta(g) = dg/d\ln b$; arrows indicate the direction of coarse graining towards longer length scales (smaller momentum cutoff).

we lower the cutoff by a factor $b > 1$ and track how the parameters transform,

$$g_i(\Lambda) \to g_i(\Lambda/b). \tag{8.204}$$

For $b > 1$, the new cutoff Λ/b is smaller. This is a *coarse graining*: one integrates out modes in the momentum shell $\Lambda/b \leq |k| < \Lambda$, thereby discarding short-distance fluctuations while keeping the long-wavelength sector $|k| < \Lambda/b$ explicit. The set $\{g_i\}$ may be viewed as coordinates in a (typically infinite-dimensional) space of theories. Under the transformation in Eq. (8.204), the representative point moves along a trajectory called a **renormalization group flow**, schematically illustrated in Fig. 8.10.

A **fixed point** is a point in coupling space at which the theory is invariant under coarse graining, i.e. the couplings no longer change with scale. Fixed points correspond to scale-invariant theories and play a central role in both quantum field theory and critical phenomena.

For a single coupling g, one introduces the beta function

$$\beta(g) = \frac{dg}{d\ln b}, \tag{8.205}$$

which is a convenient parametrization since multiplicative rescalings of the cutoff translate into additive shifts of $\ln b$. Figure 8.10 illustrates typical behaviors:

- if $\beta(g) > 0$ throughout a domain, then g increases monotonically along the flow, as in Fig. 8.10(a); if $\beta(g) < 0$ throughout, g is driven towards smaller values, as in Fig. 8.10(b);
- if there exists $g = g^\star$ such that $\beta(g^\star) = 0$, then $g^\star$ is a fixed point;
- if the flow moves away from $g^\star$ under small perturbations, then $g^\star$ is *repulsive* (unstable), as in Fig. 8.10(c);
- if the flow is attracted towards $g^\star$ from both sides, then $g^\star$ is *attractive* (stable), as in Fig. 8.10(d).

▶ **Pause and think** Explain why a fixed point of the RG flow corresponds to a scale-invariant theory. Then give an explicit example (conceptual, not computational) of a physical system where scale invariance is expected to emerge.

8.7.2 Fast and Slow Components

Consider Euclidean ϕ^4 theory with UV cutoff Λ and Lagrangian density

$$\mathcal{L} = \frac{1}{2}(\partial_\mu \phi)^2 + \frac{m^2}{2}\phi^2 + \frac{\lambda}{4!}\phi^4. \tag{8.206}$$

(The Euclidean sign convention is convenient for the Wilsonian construction; the coarse-graining logic is unchanged.) To implement coarse graining, introduce a factor $b > 1$ and decompose the field into *slow* modes with $0 \le |k| < \Lambda/b$ and *fast* modes with $\Lambda/b \le |k| < \Lambda$, which we integrate out.

Formally, the partition function can be written as

$$Z(\Lambda) = \int_{|k|<\Lambda/b} \mathcal{D}\phi_s \int_{\Lambda/b\le|k|<\Lambda} \mathcal{D}\phi_f \, \exp\Big(-\int d^d x \, \mathcal{L}[\phi_s + \phi_f]\Big). \tag{8.207}$$

To make the decomposition explicit, define the fast-mode field in momentum space by

$$\tilde{\phi}(k) = \begin{cases} \phi(k), & \text{for } \Lambda/b \le |k| < \Lambda, \\ 0, & \text{otherwise,} \end{cases} \tag{8.208}$$

and correspondingly define the slow-mode field (still denoted by $\phi(k)$ for notational simplicity) by taking it to coincide with the original $\phi(k)$ for $|k| < \Lambda/b$ and to vanish for $|k| \geq \Lambda/b$. With this convention, the original field is decomposed as $\phi \to \phi + \tilde{\phi}$, and Eq. (8.207) becomes

$$Z = \int \mathcal{D}\phi \int \mathcal{D}\tilde{\phi}\, \exp\Big[-\int d^dx \left(\frac{1}{2}(\partial_\mu\phi + \partial_\mu\tilde{\phi})^2 + \frac{1}{2}m^2(\phi+\tilde{\phi})^2 + \frac{\lambda}{4!}(\phi+\tilde{\phi})^4\right)\Big]$$
$$= \int \mathcal{D}\phi\, e^{-\int d^dx\, \mathcal{L}[\phi]} \int \mathcal{D}\tilde{\phi}\, \exp\Big[-\int d^dx \left(\frac{1}{2}(\partial_\mu\tilde{\phi})^2 + \frac{1}{2}m^2\tilde{\phi}^2 + \lambda\left(\frac{1}{6}\phi^3\tilde{\phi} + \frac{1}{4}\phi^2\tilde{\phi}^2 + \frac{1}{6}\phi\tilde{\phi}^3 + \frac{1}{4!}\tilde{\phi}^4\right)\right)\Big]. \tag{8.209}$$

The mixed quadratic term $\phi\tilde{\phi}$ does not appear: Fourier components with non-overlapping momentum support are orthogonal, so the quadratic form diagonalizes between slow and fast sectors. We now treat the interactions involving $\tilde{\phi}$ as perturbations and organize them as corrections to an effective Lagrangian for the slow field. For transparency, we focus on the regime $m^2 \ll \Lambda^2$, where the UV shell is effectively massless.

The perturbative contributions are computed using Wick's theorem with the fast-mode propagator (in momentum space)

$$\overline{\tilde{\phi}(k)\tilde{\phi}(p)} = \frac{(2\pi)^d\delta^{(d)}(k+p)}{k^2+m^2}\,\Theta_{\text{shell}}(k), \qquad \Theta_{\text{shell}}(k) = \begin{cases} 1, & \text{if } \Lambda/b \leq |k| < \Lambda, \\ 0, & \text{otherwise.} \end{cases} \tag{8.210}$$

As a concrete example, expand to first order in the $\phi^2\tilde{\phi}^2$ interaction. Contracting the fast fields yields a correction to the slow-field mass term,

$$-\int d^dx\, \frac{\lambda}{4}\,\phi^2\,\overline{\tilde{\phi}\tilde{\phi}} = -\frac{1}{2}\int \frac{d^dk'}{(2\pi)^d}\,\mu\,\phi(k')\phi(-k'), \tag{8.211}$$

where the coefficient μ is obtained by integrating over the momentum shell:

$$\mu = \frac{\lambda}{2}\int_{\Lambda/b\leq|k|<\Lambda} \frac{d^dk}{(2\pi)^d}\,\frac{1}{k^2+m^2} \simeq \frac{\lambda}{2}\int_{\Lambda/b\leq|k|<\Lambda} \frac{d^dk}{(2\pi)^d}\,\frac{1}{k^2}. \tag{8.212}$$

Evaluating the shell integral in d dimensions gives

$$\mu = \frac{\lambda}{(4\pi)^{d/2}\Gamma(d/2)}\,\frac{1-b^{-(d-2)}}{d-2}\,\Lambda^{d-2}. \tag{8.213}$$

In diagrammatic notation, Eq. (8.211) corresponds to the one-loop tadpole built from a fast-mode propagator, as shown in Fig. 8.11.

Higher orders in λ can be treated similarly: each additional order introduces further fast-mode propagators and interaction vertices. Their net effect is to generate corrections to couplings already present in the bare Lagrangian and, generically, to induce higher-dimension operators such as ϕ^6 and operators with more derivatives. Only *connected* contributions survive in the effective action, so one arrives at a Wilsonian effective Lagrangian of the form

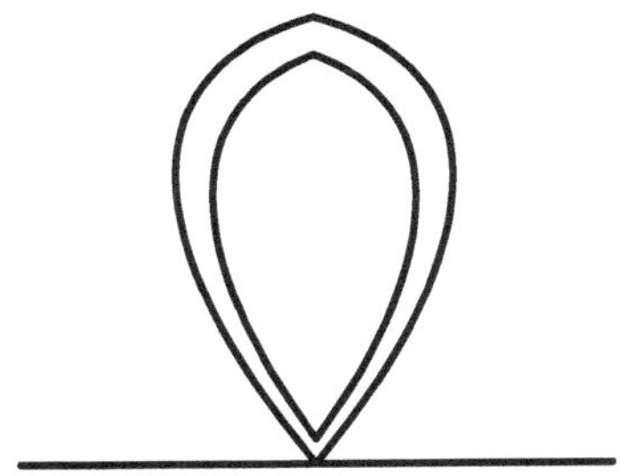

Fig. 8.11 Tadpole correction generated by contracting the two fast fields in the $\phi^2\tilde{\phi}^2$ interaction, yielding the mass shift term in Eq. (8.211)

$$\mathcal{L}_{\text{eff}} = \frac{1}{2}(\partial_\mu\phi)^2 + \frac{m^2}{2}\phi^2 + \frac{\lambda}{4!}\phi^4 + (\text{sum of connected diagrams built from fast modes}). \quad (8.214)$$

Equivalently, one may write schematically

$$\int d^dx\,\mathcal{L}_{\text{eff}} = \int d^dx\left[\frac{1}{2}(1+\delta Z)(\partial_\mu\phi)^2 + \frac{1}{2}(m^2+\delta m^2)\phi^2 + \frac{1}{4!}(\lambda+\delta\lambda)\phi^4 + \delta C(\partial_\mu\phi)^4 + \delta D\phi^6 + \dots\right]. \quad (8.215)$$

In this sense, we are indeed working with effective theories: the cutoff Λ is not merely a regulator, but a physical reminder that the description is not intended to be extrapolated to arbitrarily high energies without additional degrees of freedom and operators.

8.7.3 Momenta and Fields Relabelling

We must now compare the partially integrated functional integral to the original one. After integrating out the fast modes, the remaining slow modes only occupy momenta $|k| < \Lambda/b$ (equivalently, $|k| < b\Lambda$ if one adopts the alternative convention $b < 1$). To restore the original cutoff range $|k'| < \Lambda$, we rescale momenta and coordinates as indicated in Fig. 8.12.

▶ **Driving exercise** We perform the transformations

$$k' = \frac{k}{b}, \qquad x' = b\,x, \quad (8.216)$$

so that $|k| < \Lambda/b$ maps into $|k'| < \Lambda$. In terms of the rescaled coordinate x', the effective action of Eq. (8.215) (from the previous subsection) becomes

$$\int d^dx\,\mathcal{L}_{\text{eff}} = \int d^dx'\,b^{-d}\left[\frac{1}{2}(1+\delta Z)\,b^2\,(\partial'_\mu\phi)^2 + \frac{1}{2}(m^2+\delta m^2)\,\phi^2 + \frac{1}{4!}(\lambda+\delta\lambda)\,\phi^4 + \delta C\,b^4(\partial'_\mu\phi)^4 + \delta D\,\phi^6 + \dots\right]. \quad (8.217)$$

Here $\partial'_\mu \equiv \partial/\partial x'^\mu$, and we used $\partial_\mu = b\,\partial'_\mu$ together with $d^dx = b^{-d}d^dx'$. The factorial in the quartic term is restored for consistency with the normalization in Eq. (8.206).

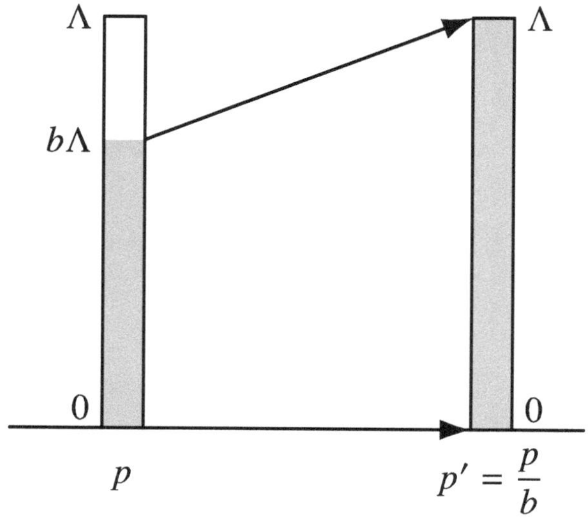

Fig. 8.12 Rescaling of momenta after integrating out a momentum shell

To bring the kinetic term back to canonical normalization (and thus reproduce the same propagator normalization as in Eq. (8.210) for the slow field), we rescale the field as

$$\phi' = \left[b^{2-d}(1+\delta Z)\right]^{1/2}\phi. \tag{8.218}$$

One then obtains an action of the original form,

$$\int d^d x\, \mathcal{L}_{\text{eff}} = \int d^d x' \left[\frac{1}{2}(\partial'_\mu \phi')^2 + \frac{1}{2}m'^2\phi'^2 + \frac{1}{4!}\lambda'\phi'^4 + C'(\partial'_\mu\phi')^4 + D'\phi'^6 + \dots\right], \tag{8.219}$$

but with *renormalized* parameters. Keeping the explicit factors generated by Eqs. (8.216) and (8.218), one finds

$$m'^2 = (m^2+\delta m^2)\,(1+\delta Z)^{-1}\,b^{-2}, \quad \lambda' = (\lambda+\delta\lambda)\,(1+\delta Z)^{-2}\,b^{d-4},$$
$$C' = (C+\delta C)\,(1+\delta Z)^{-2}\,b^{d}, \quad D' = (D+\delta D)\,(1+\delta Z)^{-3}\,b^{2d-6}, \tag{8.220}$$

and similarly for higher operators.

Of course, the bare Lagrangian in Eq. (8.206) had $C = D = 0$, but within the Wilsonian logic this only means that one started from an effective description in which higher-dimension operators were neglected. Even if these operators vanish at the starting scale, coarse graining typically generates them with small coefficients. Provided perturbation theory is justified, such corrections remain subleading compared to the leading relevant and marginal operators.

The next step is to iterate the procedure: integrate over another thin momentum shell, then rescale, and so forth. If one chooses b very close to 1, the shells become infinitesimally thin and the transformation becomes continuous. The net result is that coarse graining defines a trajectory in the space of couplings. Equivalently, one can parametrize the flow by a length scale l (towards the infrared for $l \to \infty$ and towards the ultraviolet for $l \to 0$), obtaining flow equations of the form

$$\frac{dg_i}{d \ln l} = \beta_i(g_1, g_2, \ldots, g_N), \tag{8.221}$$

i.e., a *generalized Gell-Mann-Low* class of equations, determining how parameters evolve with scale.

▶ **Pause and think** Using the scaling factors appearing in Eq. (8.220), argue (without computing loop corrections) which operators become more important and which become less important as one flows towards the infrared.

8.8 Renormalization in QED

This section collects two flagship consequences of renormalization in quantum electrodynamics (QED), both directly tied to loop effects and to the necessity of controlling ultraviolet (UV) divergences:

1. the electromagnetic coupling is *scale dependent* ("running charge"),
2. the electron magnetic moment deviates from the Dirac value $g = 2$ (Schwinger's $g - 2$ at one loop).

The common conceptual origin is the same: virtual processes dress the elementary propagators and vertices, and renormalization provides a precise prescription to separate unphysical short-distance divergences from measurable long-distance parameters (masses, charges, and form factors). In practice, one fixes a renormalization scheme (here we use on-shell conditions) and then interprets finite loop remnants as genuine predictions.

Two complementary viewpoints. Vacuum polarization may be read as a quantum-medium effect: the vacuum behaves as a polarizable dielectric due to virtual e^+e^- pairs, which screen electric charge. The anomalous magnetic moment, instead, probes the short-distance structure of the electron-photon vertex: radiative corrections induce a Pauli term $\bar{\psi}\sigma^{\mu\nu}\psi\, F_{\mu\nu}$, which is absent at tree level for a point Dirac particle.

8.8.1 Renormalization Data and the Photon Dressing

Renormalization in QED can be organized in terms of three one-particle-irreducible (1PI) Green's functions:

$$\begin{aligned} &-i\widetilde{\Sigma}(\not{p}) \quad \text{(electron 1PI self-energy)},\\ &i\widetilde{\Pi}_{\mu\nu}(q) \quad \text{(photon 1PI self-energy)},\\ &-iQ|e|\,\widetilde{\Gamma}_{\mu}(p,p') \quad \text{(1PI vertex)}. \end{aligned} \tag{8.222}$$

Renormalizability implies that the UV divergences appearing in these functions can be absorbed into a *finite* set of local counterterms, corresponding to field-strength renormalizations and to shifts of the mass and charge. Equivalently, one may impose renormalization conditions that *define* the physical mass and charge as the parameters entering the perturbative expansion.

A convenient on-shell set of conditions is:

$$\begin{aligned} &\widetilde{\Sigma}(\not{p}=m)=0 \quad \text{(physical electron mass)},\\ &\widetilde{\Pi}_{\mu\nu}(q=0)=0 \quad \text{(photon remains massless)},\\ &\widetilde{\Gamma}_{\mu}(q=0)=\gamma_\mu \quad \text{(physical charge)}. \end{aligned} \tag{8.223}$$

The second condition ensures that the renormalized photon propagator has its pole at $q^2=0$ with no mass shift; the third fixes the coupling in the Thomson limit, i.e. at vanishing momentum transfer, where the notion of "electric charge" is operationally defined by low-energy scattering.

We now implement these statements in the photon two-point function, from which running charge follows almost immediately.

Dyson resummation with Lorentz indices

Let $\widetilde{D}^{(0)}_{\mu\nu}(q)$ denote the free photon propagator (Feynman gauge),

$$\widetilde{D}^{(0)}_{\mu\nu}(q)=\frac{-ig_{\mu\nu}}{q^2+i\epsilon}. \tag{8.224}$$

Insertions of the photon 1PI self-energy $i\widetilde{\Pi}_{\mu\nu}(q)$ generate the Dyson series

$$\widetilde{D}_{\mu\nu}(q)=\widetilde{D}^{(0)}_{\mu\nu}(q)+\widetilde{D}^{(0)}_{\mu\lambda}(q)\,i\widetilde{\Pi}^{\lambda\eta}(q)\,\widetilde{D}^{(0)}_{\eta\nu}(q)+\widetilde{D}^{(0)}_{\mu\lambda}(q)\,i\widetilde{\Pi}^{\lambda\eta}(q)\,\widetilde{D}^{(0)}_{\eta\rho}(q)\,i\widetilde{\Pi}^{\rho\sigma}(q)\,\widetilde{D}^{(0)}_{\sigma\nu}(q)+\cdots. \tag{8.225}$$

This is a geometric series in Lorentz-index space,

$$\widetilde{D}=\widetilde{D}^{(0)}+\widetilde{D}^{(0)}(i\widetilde{\Pi})\widetilde{D}^{(0)}+\widetilde{D}^{(0)}(i\widetilde{\Pi})\widetilde{D}^{(0)}(i\widetilde{\Pi})\widetilde{D}^{(0)}+\cdots, \tag{8.226}$$

hence the Dyson equation

$$\widetilde{D}_{\mu\nu}(q)=\widetilde{D}^{(0)}_{\mu\nu}(q)+\widetilde{D}^{(0)}_{\mu\lambda}(q)\,i\widetilde{\Pi}^{\lambda\eta}(q)\,\widetilde{D}_{\eta\nu}(q). \tag{8.227}$$

Multiplying Eq. (8.227) on the left by $(\widetilde{D}^{(0)})^{-1}$ yields the inverted form

$$\left[(\widetilde{D}^{(0)})^{-1} - i\widetilde{\Pi}\right]_{\mu}{}^{\eta}\, \widetilde{D}_{\eta\nu} = \delta_{\mu\nu}. \tag{8.228}$$

For Eq. (8.224),

$$\left[(\widetilde{D}^{(0)})^{-1}\right]_{\mu\nu} = i(q^2 + i\epsilon)\, g_{\mu\nu}, \tag{8.229}$$

up to the standard distributional prescription.

Ward identity and transverse structure
Gauge invariance constrains the vacuum polarization to be transverse,

$$q^{\nu}\widetilde{\Pi}_{\mu\nu}(q) = 0. \tag{8.230}$$

Lorentz covariance then forces the decomposition

$$\widetilde{\Pi}_{\mu\nu}(q) = \left(q^2 g_{\mu\nu} - q_\mu q_\nu\right)\widetilde{\Pi}(q^2), \tag{8.231}$$

where $\widetilde{\Pi}(q^2)$ is a scalar function encoding loop effects.
Substituting Eq. (8.231) into Eq. (8.228) and solving for $\widetilde{D}_{\mu\nu}$ gives

$$\widetilde{D}_{\mu\nu}(q) = \frac{-ig_{\mu\nu}}{q^2\left[1 - \widetilde{\Pi}(q^2)\right]} + \frac{iq_\mu q_\nu}{q^4}\,\frac{\widetilde{\Pi}(q^2)}{1 - \widetilde{\Pi}(q^2)}. \tag{8.232}$$

In physical amplitudes the photon propagator is contracted with conserved currents, so $q_\mu J^\mu_{\text{em}} = 0$ and the longitudinal term proportional to $q_\mu q_\nu$ never contributes. Hence the operationally relevant renormalized propagator is

$$\widetilde{D}_{\mu\nu}(q) = \frac{-ig_{\mu\nu}}{q^2\left[1 - \widetilde{\Pi}(q^2)\right]}. \tag{8.233}$$

▶ **Driving exercise** To derive Eq. (8.232) starting from Eq. (8.228) and the decomposition (8.231), write the inverse propagator as

$$\left[\widetilde{D}^{-1}\right]_{\mu\nu} = i(q^2 + i\epsilon)g_{\mu\nu} - i(q^2 g_{\mu\nu} - q_\mu q_\nu)\widetilde{\Pi}(q^2). \tag{8.234}$$

Factor iq^2 and separate transverse and longitudinal projectors,

$$P^T_{\mu\nu} \equiv g_{\mu\nu} - \frac{q_\mu q_\nu}{q^2}, \qquad P^L_{\mu\nu} \equiv \frac{q_\mu q_\nu}{q^2}, \qquad P^T + P^L = g. \tag{8.235}$$

Then

$$\left[\widetilde{D}^{-1}\right]_{\mu\nu} = iq^2\left[(1 - \widetilde{\Pi})P^T_{\mu\nu} + P^L_{\mu\nu}\right], \tag{8.236}$$

since $(q^2 g_{\mu\nu} - q_\mu q_\nu) = q^2 P^T_{\mu\nu}$. The inverse is immediate because P^T and P^L are orthogonal projectors:

$$\widetilde{D}_{\mu\nu} = \frac{-i}{q^2}\Big[\frac{1}{1-\widetilde{\Pi}} P^T_{\mu\nu} + P^L_{\mu\nu}\Big]. \tag{8.237}$$

Finally expand $P^T_{\mu\nu} = g_{\mu\nu} - q_\mu q_\nu / q^2$ to obtain

$$\widetilde{D}_{\mu\nu} = \frac{-i g_{\mu\nu}}{q^2(1-\widetilde{\Pi})} + \frac{i q_\mu q_\nu}{q^4}\frac{\widetilde{\Pi}}{1-\widetilde{\Pi}}, \tag{8.238}$$

which is Eq. (8.232).

On-shell subtraction at $q^2 = 0$

The on-shell renormalization condition for the photon requires

$$\widetilde{\Pi}(0) = 0, \tag{8.239}$$

so that the photon pole remains at $q^2 = 0$ and no mass term is generated. At fixed loop order, one enforces this by subtracting the value at $q^2 = 0$:

$$\widetilde{\Pi}(q^2) \longrightarrow \widetilde{\Pi}(q^2) - \widetilde{\Pi}(0). \tag{8.240}$$

This subtraction removes the UV divergence in the one-loop vacuum polarization (equivalently, it corresponds to a local counterterm compatible with Eq. (8.239)). Physically, it expresses the fact that we *define* the renormalized charge through a low-energy observable, so only *changes* of $\widetilde{\Pi}(q^2)$ with momentum transfer are meaningful predictions.

Subtracting at $q^2 = 0$ is natural. In the Thomson limit $q \to 0$, the photon probes charge only at very long distances. Fixing the vertex at $q^2 = 0$ and requiring $\widetilde{\Pi}(0) = 0$ ensures that the renormalized charge is the one measured in low-energy Coulomb scattering, while deviations at finite q^2 encode the gradual "unscreening" of the charge at shorter distances.

8.8.2 One-Loop Vacuum Polarization and running charge

At one loop (electron-positron bubble), the 1PI insertion reads

$$i\widetilde{\Pi}_{\mu\nu}(q) = (-1)(-ie_0)^2 \int \frac{d^4 p}{(2\pi)^4} \mathrm{Tr}\Big[\gamma_\mu \frac{i(\not p + m)}{p^2 - m^2 + i\epsilon}\gamma_\nu \frac{i(\not p + \not q + m)}{(p+q)^2 - m^2 + i\epsilon}\Big], \tag{8.241}$$

where the initial minus sign is the fermion loop. The tensor reduction implied by Eq. (8.231) is standard; here we focus on the renormalized scalar combination $\widetilde{\Pi}(q^2) - \widetilde{\Pi}(0)$, since it is finite and directly connected to measurable momentum dependence.

Combine denominators using a Feynman parameter x:

$$\frac{1}{(p^2-m^2)\left((p+q)^2-m^2\right)} = \int_0^1 dx\, \frac{1}{\left[(p+xq)^2-\Delta\right]^2}, \qquad \Delta \equiv m^2 - x(1-x)q^2. \quad (8.242)$$

Shift the integration variable $p \to \ell = p + xq$ so that the denominator becomes $(\ell^2 - \Delta)^2$. After performing the Dirac trace, one indeed finds the transverse structure, Eq. (8.231). The remaining scalar loop integral contains a logarithmic divergence which is independent of q^2 (a consequence of gauge invariance and power counting). Therefore the subtraction at $q^2 = 0$ removes the divergence and yields the finite representation

$$\widetilde{\Pi}(q^2) - \widetilde{\Pi}(0) = -\frac{e_0^2}{2\pi^2}\int_0^1 dx\; x(1-x)\, \ln\!\left(\frac{m^2}{m^2 - x(1-x)q^2}\right). \qquad (8.243)$$

This form is convenient for extracting both the infrared ($q^2 \to 0$) and ultraviolet ($|q^2| \gg m^2$) behavior, and for emphasizing that it is the momentum dependence, not the overall divergent constant, that carries physics.

▶ **Driving exercise** Let us show explicitly why the divergence cancels in $\widetilde{\Pi}(q^2) - \widetilde{\Pi}(0)$.
After Feynman parametrization and the shift $p \to \ell$, the scalar integral has the generic form

$$\widetilde{\Pi}(q^2) \sim \int_0^1 dx\, x(1-x) \int \frac{d^4\ell}{(2\pi)^4}\,\frac{1}{(\ell^2-\Delta)^2}, \qquad \Delta = m^2 - x(1-x)q^2,$$

up to finite tensor-reduction factors. The UV divergence resides entirely in the large-ℓ region, where $(\ell^2-\Delta)^2 \simeq \ell^4$ and Δ is negligible. Hence the divergent part is independent of Δ, and therefore independent of q^2. Subtracting the same integral evaluated at $q^2 = 0$ (i.e. at $\Delta = m^2$) removes precisely this q^2-independent divergent constant, leaving the finite logarithm in Eq. (8.243).

For electrostatics one needs the static (space-like) momentum transfer, effectively $q^2 \to -\mathbf{q}^2$ with $\mathbf{q}^2 > 0$. Then

$$\widetilde{\Pi}(-\mathbf{q}^2) - \widetilde{\Pi}(0) = -\frac{e_0^2}{2\pi^2}\int_0^1 dx\; x(1-x)\, \ln\!\left(\frac{m^2}{m^2 + x(1-x)\mathbf{q}^2}\right). \qquad (8.244)$$

For $\mathbf{q}^2 \ll m^2$, expand the logarithm:

$$\ln\left(\frac{m^2}{m^2 + x(1-x)\mathbf{q}^2}\right) = -\ln\left(1 + \frac{x(1-x)\mathbf{q}^2}{m^2}\right) \approx -\frac{x(1-x)\mathbf{q}^2}{m^2} + O\left(\frac{\mathbf{q}^4}{m^4}\right).$$

Substituting into Eq. (8.244) gives

$$\widetilde{\Pi}(-\mathbf{q}^2) - \widetilde{\Pi}(0) \approx \frac{e_0^2}{2\pi^2}\frac{\mathbf{q}^2}{m^2}\int_0^1 dx\, x^2(1-x)^2 = \frac{e_0^2}{60\pi^2}\frac{\mathbf{q}^2}{m^2} + O\left(\frac{\mathbf{q}^4}{m^4}\right), \tag{8.245}$$

since $\int_0^1 dx\, x^2(1-x)^2 = 1/30$. Using $\alpha \equiv e_0^2/(4\pi)$, this becomes

$$\widetilde{\Pi}(-\mathbf{q}^2) - \widetilde{\Pi}(0) = \frac{\alpha}{15\pi}\frac{\mathbf{q}^2}{m^2} + O\left(\frac{\mathbf{q}^4}{m^4}\right). \tag{8.246}$$

To one loop, the effective photon exchange can be expanded as

$$\widetilde{D}_{\mu\nu}(q) \simeq \frac{-ig_{\mu\nu}}{q^2}\left[1 + \widetilde{\Pi}(q^2) - \widetilde{\Pi}(0)\right], \tag{8.247}$$

where we used $(1-x)^{-1} \simeq 1 + x$ to linear order in the loop correction. In the static limit, the correction to $\widetilde{D}_{00}$ modifies the Fourier representation of the electrostatic potential. Using

$$\int \frac{d^3\mathbf{q}}{(2\pi)^3}\frac{e^{i\mathbf{q}\cdot\mathbf{r}}}{\mathbf{q}^2} = \frac{1}{4\pi r}, \qquad \int \frac{d^3\mathbf{q}}{(2\pi)^3}\, e^{i\mathbf{q}\cdot\mathbf{r}} = \delta^{(3)}(\mathbf{r}), \tag{8.248}$$

and Eq. (8.246), one finds that the leading low-$\mathbf{q}$ vacuum-polarization correction is proportional to $\mathbf{q}^2/\mathbf{q}^2 \sim 1$ in momentum space, hence to a contact term in position space:

$$V(\mathbf{r}) = -\frac{\alpha}{r} - \frac{4\alpha^2}{15\, m^2}\,\delta^{(3)}(\mathbf{r}) + \cdots. \tag{8.249}$$

The first term is Coulomb's law. The second encodes the fact that the vacuum polarizes: virtual e^+e^- pairs partially screen the charge at short distances, producing a local correction at the origin when expanded at small momentum.

▶ **Pause and think** Where does the $\delta^{(3)}(\mathbf{r})$ actually matter? Explain why the correction in Eq. (8.249) vanishes for hydrogenic states with orbital angular momentum $\ell \neq 0$. Then estimate the parametric scaling of the energy shift for an ns state in terms of α and m by dimensional analysis.

In processes mediated by a virtual photon, factors of the coupling naturally appear multiplying the propagator. It is therefore convenient to define an effective, momentum-dependent charge by grouping the photon dressing with e_0^2:

$$\frac{e_0^2}{q^2} \frac{1}{1 - \left[\widetilde{\Pi}(q^2) - \widetilde{\Pi}(0)\right]} \equiv \frac{e^2(q)}{q^2}, \tag{8.250}$$

so that (to one loop)

$$e^2(q) \simeq e_0^2 \left[1 + \widetilde{\Pi}(q^2) - \widetilde{\Pi}(0)\right]. \tag{8.251}$$

This makes explicit that the observed coupling depends on the momentum transfer: probing at shorter distances (larger $|q|$) partially penetrates the screening cloud and reveals a larger effective charge.

Screening vs antiscreening. In QED, vacuum polarization screens charge and the coupling grows logarithmically with energy (positive β-function). In non-abelian gauge theories, gauge-boson self-interactions contribute with the opposite sign, leading to antiscreening and asymptotic freedom. The sign of β is a dynamical fingerprint of the field content and the symmetry.

Define an energy scale μ and regard $e(\mu)$ as the renormalized coupling at that scale. From Eq. (8.251), a convenient expansion is

$$e(\mu) \simeq e_0 \left[1 + \frac{1}{2}\left(\widetilde{\Pi}(\mu^2) - \widetilde{\Pi}(0)\right)\right], \tag{8.252}$$

so that, the β-function $\beta(e) \equiv \mu \frac{de}{d\mu}$ becomes

$$\beta(e) \simeq \frac{1}{2}\, \mu e_0 \frac{d}{d\mu} \widetilde{\Pi}(\mu^2), \tag{8.253}$$

where the subtraction term $\widetilde{\Pi}(0)$ drops upon differentiation.

Using Eq. (8.243) with $q^2 = \mu^2$ and considering $\mu \gg m$,

$$\ln\left(\frac{m^2}{m^2 - x(1-x)\mu^2}\right) = \ln\left(\frac{m^2}{\mu^2}\right) - \ln(x(1-x)) + O\left(\frac{m^2}{\mu^2}\right),$$

so only the $-2\ln\mu$ term contributes to $d/d\mu$. Differentiating,

$$\frac{d}{d\mu}\widetilde{\Pi}(\mu^2) \simeq \frac{e_0^2}{\pi^2} \frac{1}{\mu} \int_0^1 dx\, x(1-x) = \frac{e_0^2}{6\pi^2} \frac{1}{\mu}, \tag{8.254}$$

since $\int_0^1 dx\, x(1-x) = 1/6$. Substituting into Eq. (8.253) gives

$$\beta(e) = \mu \frac{de}{d\mu} \simeq \frac{e^3}{12\pi^2}, \tag{8.255}$$

where at this order one may replace $e_0 \to e$ on the right-hand side. The positive sign implies that the QED coupling increases with increasing energy scale.

▶ **Driving exercise** Starting from Eq. (8.243), we can reproduce the one-loop β-function, Eq. (8.255), and justify precisely where the approximation $\mu \gg m$ is used.
Thus, insert $q^2 = \mu^2$ into Eq. (8.243). For $\mu \gg m$, expand

$$\ln\left(\frac{m^2}{m^2 - x(1-x)\mu^2}\right) = \ln\left(\frac{m^2}{\mu^2}\right) - \ln(x(1-x)) + O(m^2/\mu^2),$$

so the only μ dependence at leading order is $-2\ln\mu$. Differentiate under the integral, evaluate $\int_0^1 dx\, x(1-x) = 1/6$, and insert into Eq. (8.253), replacing $e_0 \to e$ at this order. This yields $\beta(e) = e^3/(12\pi^2)$.

8.8.3 Vertex Corrections and the Electron Magnetic Moment

The electron magnetic moment operator is conventionally written as

$$\hat{\mu} = g\left(\frac{Q|e|}{2m}\right)\hat{\mathbf{S}}, \tag{8.256}$$

with $Q = -1$ for the electron and $\hat{\mathbf{S}}$ the spin operator. The Dirac equation predicts $g = 2$ at tree level: a pointlike spin-1/2 particle minimally coupled to electromagnetism has precisely that gyromagnetic ratio. Radiative QED corrections modify the effective electron-photon interaction and shift g away from 2. The deviation is the *anomalous magnetic moment*.

Let $q^\mu \equiv p'^\mu - p^\mu$ be the photon momentum flowing into the vertex. For on-shell external spinors ($p^2 = p'^2 = m^2$), Lorentz covariance, current conservation, and discrete symmetries imply that the renormalized vertex may be parametrized as

$$\widetilde{\Gamma}^\mu(p, p') = \gamma^\mu F_1(q^2) + \frac{i\sigma^{\mu\nu}q_\nu}{2m} F_2(q^2), \qquad \sigma^{\mu\nu} \equiv \frac{i}{2}[\gamma^\mu, \gamma^\nu]. \tag{8.257}$$

Here F_1 is the Dirac form factor and F_2 is the Pauli form factor. The charge renormalization condition fixes

$$F_1(0) = 1, \tag{8.258}$$

while at tree level $F_2(0) = 0$.

To extract the magnetic moment, couple the vertex to a classical background field $A^{\mathrm{cl}}_\mu(x)$ with Fourier transform $\widetilde{A}^{\mathrm{cl}}_\mu(q)$. The scattering amplitude is

$$i\mathcal{M} = -iQ|e|\,\bar{u}(p')\,\widetilde{\Gamma}^\mu(p,p')\,u(p)\,\widetilde{A}^{\mathrm{cl}}_\mu(q). \tag{8.259}$$

Substituting Eq. (8.257) gives

$$i\mathcal{M} = -iQ|e|\,\bar{u}(p')\left[\gamma^\mu F_1(q^2) + \frac{i\sigma^{\mu\nu}q_\nu}{2m}F_2(q^2)\right]u(p)\,\widetilde{A}^{\mathrm{cl}}_\mu(q). \tag{8.260}$$

The Gordon identity (valid for on-shell spinors) reads

$$\bar{u}(p')\gamma^\mu u(p) = \bar{u}(p')\left[\frac{(p'+p)^\mu}{2m} + \frac{i\sigma^{\mu\nu}q_\nu}{2m}\right]u(p). \tag{8.261}$$

Using Eq. (8.261) inside the F_1 term of Eq. (8.260), the amplitude splits into a convection-current piece and a spin (magnetization) piece:

$$\begin{aligned} i\mathcal{M} = &-iQ|e|\,\bar{u}(p')\left[\frac{(p'+p)^\mu}{2m}\right]u(p)\,\widetilde{A}^{\mathrm{cl}}_\mu(q)\,F_1(q^2) \\ &+ Q|e|\,\frac{1}{2m}\,\bar{u}(p')\left[\sigma^{\mu\nu}q_\nu\right]u(p)\,\widetilde{A}^{\mathrm{cl}}_\mu(q)\,\left[F_1(q^2)+F_2(q^2)\right]. \end{aligned} \tag{8.262}$$

The first line is spin independent; the second is spin dependent and contains the information relevant for the magnetic moment. The magnetic moment depends only on the combination $F_1 + F_2$ evaluated at $q^2 \to 0$.

Non-relativistic reduction and identification of g

Assume a purely magnetic external field, so that in the electron rest frame we may take

$$A^{\mathrm{cl}}_\mu(x) = (0, \mathbf{A}^{\mathrm{cl}}(\mathbf{x})), \qquad \widetilde{A}^{\mathrm{cl}}_0(q) = 0. \tag{8.263}$$

In the non-relativistic limit, the Dirac spinor reduces to

$$u(p) \approx \sqrt{2m}\begin{pmatrix}\xi\\0\end{pmatrix}, \qquad \bar{u}(p') \approx \sqrt{2m}\left(\xi'^\dagger,\ 0\right), \tag{8.264}$$

where ξ, ξ' are two-component Pauli spinors. One also needs

$$\sigma^{ij} = \begin{pmatrix}\epsilon^{ijk}\sigma^k & 0\\ 0 & \epsilon^{ijk}\sigma^k\end{pmatrix}, \tag{8.265}$$

so that

$$\bar{u}(p')\,\sigma^{ij}\,u(p) \approx 2m\,\epsilon^{ijk}\,\xi'^\dagger\sigma^k\xi. \tag{8.266}$$

Keeping only the spin-dependent piece in Eq. (8.262) and taking $q \to 0$ so that $F_1(0) = 1$, one finds

$$i\mathcal{M}_{\text{spin}} \approx Q|e|\,\epsilon^{ijk}\,(\xi'^{\dagger}\sigma^k\xi)\,q_j\,\widetilde{A}_i^{\text{cl}}(q)\;[1 + F_2(0)]\,. \tag{8.267}$$

In momentum space the magnetic field is

$$\widetilde{B}^k(q) = i\,\epsilon^{kij}\,q_i\,\widetilde{A}_j^{\text{cl}}(q), \tag{8.268}$$

hence $\epsilon^{ijk}q_j\widetilde{A}_i = (1/i)\widetilde{B}^k$ and

$$i\mathcal{M}_{\text{spin}} \approx i\,Q|e|\,\left(\xi'^{\dagger}\sigma^k\xi\right)\,\widetilde{B}^k(q)\;[1 + F_2(0)]\,. \tag{8.269}$$

Introduce the expectation value of the spin operator

$$\langle\hat{\mathbf{S}}\rangle = \frac{1}{2}\,\xi'^{\dagger}\boldsymbol{\sigma}\,\xi\,. \tag{8.270}$$

Matching to the defining interaction $V = -\langle\hat{\boldsymbol{\mu}}\rangle \cdot \mathbf{B}$ with Eq. (8.256) yields

$$g = 2\,[1 + F_2(0)]\,. \tag{8.271}$$

▶ **Driving exercise** We now fill in the intermediate index manipulations that convert Eq. (8.267) into Eq. (8.269) using Eq. (8.268). Starting from Eq. (8.267),

$$i\mathcal{M}_{\text{spin}} \approx Q|e|\,\epsilon^{ijk}\,(\xi'^{\dagger}\sigma^k\xi)\,q_j\,\widetilde{A}_i\,[1 + F_2(0)].$$

Use antisymmetry $\epsilon^{ijk} = -\epsilon^{kji}$ and relabel dummy indices to obtain $\epsilon^{ijk}q_j\widetilde{A}_i = -\epsilon^{kij}q_i\widetilde{A}_j = (1/i)\widetilde{B}^k$, where the last equality follows from Eq. (8.268). Substitution yields Eq. (8.269).

At one loop, the Pauli form factor at $q^2 = 0$ is finite after including the appropriate counterterm and satisfies

$$F_2(0) = \frac{e^2}{8\pi^2} = \frac{\alpha}{2\pi}\,. \tag{8.272}$$

Substituting into Eq. (8.271) gives Schwinger's result

$$g = 2\left(1 + \frac{\alpha}{2\pi} + O(\alpha^2)\right), \qquad a_e \equiv \frac{g-2}{2} = \frac{\alpha}{2\pi} + O(\alpha^2). \tag{8.273}$$

A physical reading of $F_2(0)$. The term proportional to F_2 corresponds, in an effective low-energy description, to a Pauli operator $\sim \bar{\psi}\sigma^{\mu\nu}\psi\, F_{\mu\nu}$ with a coefficient generated radiatively. Once mass and charge are fixed by the on-shell conditions, the remaining finite vertex structure becomes a prediction.

▶ **Pause and think** Argue why $F_2(0)$ (hence $g-2$) is a physical observable and therefore independent of the renormalization scheme, even though intermediate quantities such as Z_1, Z_2, Z_3 and $\widetilde{\Pi}(q^2)$ are scheme dependent. Your answer should identify what is kept fixed when one changes scheme.

8.9 Quasiparticles from Topological Field Theory

Let us come back to topology again, but this time from the perspective of quantum fields. We already faced before new kind of excitations named *topological objects*, i.e., nontrivial field configurations that can exist in systems exhibiting spontaneous symmetry breaking.

Up to now we have studied topological *objects* (defects) that can appear in scalar field theories with broken symmetry. We now go further and construct a field theory that is itself intrinsically *topological*. This will be special to $(2+1)$-dimensional spacetime, where "flatland" physics admits remarkable phenomena whose origin is deeply tied to topology.

We begin with particle statistics in two spatial dimensions and then introduce a topological Chern-Simons theory, leading to an unconventional "electrodynamics".

8.9.1 Anyons

For identical particles in three dimensions, exchanging labels produces

$$\psi(x_1, x_2) = \pm\psi(x_2, x_1), \tag{8.274}$$

with $+$ for bosons and $-$ for fermions.

In two spatial dimensions, "exchange" must be understood as a continuous process in which particles move around each other. One should distinguish two broad classes of processes for two identical particles initially at x_1 and x_2 in the following way.

- Type A: the particles return to their original positions $x_1 \to x_1$ and $x_2 \to x_2$, possibly looping around each other some number of times.
- Type B: the particles swap positions $x_1 \to x_2$ and $x_2 \to x_1$, again possibly with additional windings.

The relevant topological datum is the net angle ϕ by which one particle winds around the other. Topology enters because smooth deformations of the paths cannot change the integer winding number without introducing a singularity. Type A processes correspond to

$$\phi = 2\pi p, \qquad p \in \mathbb{Z}, \tag{8.275}$$

while type B processes correspond to

$$\phi = \pi(2p+1). \tag{8.276}$$

Different p define distinct topological classes.

Assume that such winding processes contribute a multiplicative pure phase $\Phi(\phi)$ to the quantum amplitude. Composing processes should add angles and multiply phases, so

$$\Phi(\phi_1 + \phi_2) = \Phi(\phi_1)\Phi(\phi_2) \quad \Rightarrow \quad \Phi(\phi) = e^{i\eta\phi}, \tag{8.277}$$

where η is a real parameter that need not be an integer.
Comparing this "physical" exchange with the abstract exchange rule, a basic swap corresponds to $\phi = \pi$ and yields a factor $e^{i\eta\pi}$. The boson/fermion cases are recovered only if η is an even/odd integer, respectively. But in two dimensions nothing forces η to be integer: *continuous families of statistics are possible*. Particles realizing such intermediate statistics are called **anyons**, a term associated with Wilczek's work clarifying fractional statistics.

The restriction to two spatial dimensions is essential: in three dimensions, the extra dimension allows worldlines to pass "around" each other, and all type A processes are deformable to trivial motion, while all type B processes are deformable to a simple swap. Thus only bosons and fermions remain.

8.9.2 Chern-Simons Theory

We now construct a genuinely topological Lagrangian. Ordinary field theories rely on the metric $g_{\mu\nu}$ to contract indices and encode the geometry of rulers and clocks. In a topological theory (in $(2+1)$ dimensions), index contractions instead employ the antisymmetric symbol $\epsilon^{\mu\nu\lambda}$.

A direct consequence emerges when one defines the energy-momentum tensor by varying the action with respect to the metric. For an action S_{top} obtained from a purely topological Lagrangian,

$$T^{\mu\nu} = -\frac{2}{\sqrt{-\det g}} \frac{\delta S_{\text{top}}}{\delta g_{\mu\nu}}. \tag{8.278}$$

Since a topological action does not depend on $g_{\mu\nu}$, one gets $T^{\mu\nu} = 0$. In particular, the Hamiltonian density T^{00} vanishes, implying $H = 0$. All states have energy $E = 0$, implying a large degeneracy whose precise counting depends on manifold topology.

A prototypical topological term in (2 + 1) dimensions is the Chern-Simons Lagrangian for a $U(1)$ gauge field a_μ:

$$\mathcal{L}_{\text{CS}} = -\frac{1}{2}\kappa\,\epsilon^{\mu\nu\lambda}\,a_\mu\,\partial_\nu a_\lambda, \tag{8.279}$$

with constant κ. Such terms are recognizable by the schematic structure $\epsilon\,a\,\partial a$. This theory exists naturally in (2 + 1) dimensions; attempting the same structure in (3 + 1) dimensions fails by index counting.

Under a gauge transformation $a_\mu \to a_\mu + \partial_\mu\chi$ one finds

$$\epsilon^{\mu\nu\lambda}a_\mu\partial_\nu a_\lambda \;\to\; \epsilon^{\mu\nu\lambda}a_\mu\partial_\nu a_\lambda + \epsilon^{\mu\nu\lambda}(\partial_\mu\chi)\,\partial_\nu a_\lambda, \tag{8.280}$$

where $\epsilon^{\mu\nu\lambda}\partial_\mu\partial_\nu\chi = 0$ by antisymmetry. The change in the action can be written as a total derivative,

$$\delta S = \int d^3x\,\epsilon^{\mu\nu\lambda}\,\partial_\mu\big(\chi\,\partial_\nu a_\lambda\big), \tag{8.281}$$

which vanishes if boundary terms can be neglected. Hence the action is gauge invariant modulo boundary contributions.

The equations of motion are

$$\kappa\,\epsilon^{\mu\nu\lambda}\,\partial_\nu a_\lambda = 0. \tag{8.282}$$

Defining $f_{\mu\nu} = \partial_\mu a_\nu - \partial_\nu a_\mu$, this can be rewritten as

$$\frac{1}{2}\kappa\,\epsilon^{\mu\nu\lambda}f_{\nu\lambda} = 0 \quad\Rightarrow\quad f_{\mu\nu} = 0. \tag{8.283}$$

This seems to imply no propagating dynamics (unlike Maxwell theory $\partial_\mu F^{\mu\nu} = 0$ which supports waves). Nevertheless, Chern-Simons theory becomes physically rich once coupled to other fields.

Couple a_μ to a conserved source current J^μ:

$$\mathcal{L} = -\frac{1}{2}\kappa\,\epsilon^{\mu\nu\lambda}a_\mu\partial_\nu a_\lambda + a_\mu J^\mu. \tag{8.284}$$

The equations of motion now impose the constraint

$$J^\mu = \kappa\,\epsilon^{\mu\nu\lambda}\partial_\nu a_\lambda = \frac{1}{2}\kappa\,\epsilon^{\mu\nu\lambda}f_{\nu\lambda}. \tag{8.285}$$

This is not a Noether identity; it is enforced by the chosen coupling.

To interpret these relations, note a special feature of (2 + 1) dimensions: the cross product yields a pseudoscalar rather than a pseudovector. Hence the magnetic field in (2 + 1) dimensions is a pseudoscalar, $B = \epsilon^{ij}\partial_i A_j$.

Driving exercise In three dimensions, $(\mathbf{A}\times\mathbf{B})^i = \epsilon^{ijk}A_j B_k$. In two dimensions the antisymmetric symbol has only two indices ϵ^{ij}, so one forms $S = \epsilon^{ij}A_i B_j$; there is no free index left, hence the result is a pseudoscalar. Writing $J^\mu = (\rho, \mathbf{J})$, Eq. (8.285) implies

$$\rho = \kappa(\partial_1 a_2 - \partial_2 a_1), \quad J_1 = \kappa(-\partial_0 a_2 + \partial_2 a_0), \quad J_2 = \kappa(\partial_0 a_1 - \partial_1 a_0). \tag{8.286}$$

Using electromagnetic language (though a_μ is not the physical EM field), define

$$b = \partial_2 a_1 - \partial_1 a_2, \qquad e_i = \partial_0 a_i - \partial_i a_0, \tag{8.287}$$

so that

$$\rho = -\kappa b, \qquad J_i = -\kappa\, \epsilon_{ij} e_j, \tag{8.288}$$

with $\epsilon_{12} = 1$. The second relation means an a_μ-electric field in (say) the y direction induces a current in the x direction. Integrating the charge density over space gives a relation between total charge Q and the a_μ magnetic flux:

$$Q = -\kappa \int d^2x\, b, \tag{8.289}$$

i.e.

$$\Big(\text{source charge}\Big) \propto \Big(a_\mu\text{-flux}\Big). \tag{8.290}$$

Thus: *Chern-Simons theory attaches gauge flux to source charge.*

Fractional statistics from Chern-Simons theory
Attaching flux to charge changes exchange statistics. In $(2+1)$ dimensions, taking one particle fully around another contributes a statistical phase $e^{2\pi i\eta}$ (with η integer-valued for bosons/fermions). Separately, by the Aharonov-Bohm effect, moving a charge q around flux Φ yields a phase $e^{iq\Phi}$.
In Chern-Simons theory, the charge and flux are tied: using the relation $q = -\kappa\Phi$ (equivalently $\Phi = -q/\kappa$), the Aharonov-Bohm phase becomes $e^{-iq^2/\kappa}$ for a full 2π winding. Particle exchange corresponds to a half-winding (π), so interpreting a full loop as two exchanges, one obtains the statistical relation

$$2\pi\eta = -\frac{q^2}{\kappa}. \tag{8.291}$$

Therefore $\eta = -\frac{q^2}{2\pi\kappa}$, which is generically not an integer: the excitations can exhibit *fractional statistics*.

8.10 Majorana Fermions

For a Dirac field $\Psi(x)$, charge conjugation is implemented by an antilinear map,

$$\Psi^C(x) \equiv C\,\bar{\Psi}^T(x) = C_0\,\Psi^*(x), \tag{8.292}$$

where $\bar{\Psi} = \Psi^\dagger\gamma^0$, T denotes transpose, and C_0 is a (representation-dependent) matrix satisfying

$$C_0\,\gamma^\mu\,C_0^{-1} = -(\gamma^\mu)^*, \qquad C_0^T = -C_0. \tag{8.293}$$

A *Majorana spinor* is defined by imposing the *Majorana condition*

$$\Psi(x) = \Psi^C(x). \tag{8.294}$$

This constraint is consistent only when the Dirac equation admits a real form. Equivalently, there exists a representation (the *Majorana representation*) in which the matrices γ^μ are purely imaginary, so that the differential operator $i\gamma^\mu\partial_\mu$ is real. In that basis, solutions can be chosen real in the sense appropriate to Grassmann-valued fields.

Majorana condition as a reduction of degrees of freedom. A Dirac spinor describes independent particle and antiparticle excitations. The constraint $\Psi = \Psi^C$ identifies them and therefore halves the number of independent degrees of freedom. This is the precise field-theoretic meaning of "a Majorana fermion is its own antiparticle".

A practical construction starts from a single two-component Weyl spinor. Let ψ_L be a left-handed field transforming in the $(\frac{1}{2}, 0)$ representation. Define its charge-conjugate right-handed partner (in two-component language) as

$$\psi_{L,C} \equiv i\sigma^2\psi_L^*, \tag{8.295}$$

which indeed transforms as a right-handed spinor. One may then assemble the four-component object

$$\nu(x) \equiv \begin{pmatrix} \psi_L(x) \\ \psi_{L,C}(x) \end{pmatrix} = \begin{pmatrix} \psi_L(x) \\ i\sigma^2\psi_L^*(x) \end{pmatrix}, \tag{8.296}$$

which, by construction, satisfies Eq. (8.294).

The corresponding Lorentz-invariant Lagrangian density, written in terms of ψ_L, is

$$\mathcal{L} = i\,\psi_L^\dagger\,\bar{\sigma}^\mu\partial_\mu\psi_L - \frac{m_L}{2}\left(\psi_L^T\,i\sigma^2\,\psi_L + \psi_L^\dagger\,(-i\sigma^2)\,\psi_L^*\right), \tag{8.297}$$

where $\bar{\sigma}^{\mu} = (\mathbb{I}, -\boldsymbol{\sigma})$. Varying with respect to $\psi_L^{\dagger}$ gives the two-component Majorana equation,

$$i\bar{\sigma}^{\mu}\partial_{\mu}\psi_L - m_L\, i\sigma^2\psi_L^* = 0. \tag{8.298}$$

It is often convenient to rewrite it in a Dirac-like form for the four-component field ν,

$$\left(i\partial\!\!\!/ - m_L\right)\nu(x) = 0, \tag{8.299}$$

with the essential caveat that ν is constrained by Eq. (8.294) and therefore does *not* describe an independent particle-antiparticle pair.

The Majorana mass term in Eq. (8.297) also illustrates an important structural point. The bilinear $\psi_L^T i\sigma^2\psi_L$ is antisymmetric in spinor indices. If ψ_L were an ordinary commuting (c-number) spinor, then

$$\psi_L^T i\sigma^2\psi_L = -\psi_L^T i\sigma^2\psi_L \quad\Rightarrow\quad \psi_L^T i\sigma^2\psi_L = 0, \tag{8.300}$$

so the mass term would vanish identically. In QFT, however, ψ_L is Grassmann-valued: exchanging two fermionic fields introduces an additional minus sign, which compensates the antisymmetry of $i\sigma^2$ and allows a non-zero Majorana mass term. Thus, the fermionic nature of the fields is not a technical detail but a consistency requirement.

▸ **Driving exercise** Let us show explicitly (with indices) why the Majorana bilinear vanishes for commuting spinors but not for Grassmann fields. To do so, we write the bilinear as

$$\psi_L^T i\sigma^2\psi_L = \psi_{L,\alpha}\,(i\sigma^2)_{\alpha\beta}\,\psi_{L,\beta}, \qquad \alpha,\beta = 1,2, \tag{8.301}$$

and use $(i\sigma^2)_{\alpha\beta} = -(i\sigma^2)_{\beta\alpha}$. For commuting components, interchange $\alpha \leftrightarrow \beta$:

$$\psi_{L,\alpha}(i\sigma^2)_{\alpha\beta}\psi_{L,\beta} = \psi_{L,\beta}(i\sigma^2)_{\beta\alpha}\psi_{L,\alpha} = -\psi_{L,\beta}(i\sigma^2)_{\alpha\beta}\psi_{L,\alpha} = -\psi_{L,\alpha}(i\sigma^2)_{\alpha\beta}\psi_{L,\beta}, \tag{8.302}$$

where in the last step we relabel dummy indices and use commutativity to reorder $\psi_{L,\beta}\psi_{L,\alpha} = \psi_{L,\alpha}\psi_{L,\beta}$. Hence the expression equals minus itself and must vanish.

For Grassmann components, the same manipulation yields an *extra* minus sign from exchanging the fields, $\psi_{L,\beta}\psi_{L,\alpha} = -\psi_{L,\alpha}\psi_{L,\beta}$, which cancels the antisymmetry of $(i\sigma^2)_{\alpha\beta}$. Therefore the bilinear is not forced to vanish in QFT.

8.10.1 Field Operators and Mode Expansions

Recall the standard mode expansion for a Dirac field,

$$\hat{\Psi}(x) = \int \frac{d^3p}{(2\pi)^{3/2}} \frac{1}{\sqrt{2E_{\mathbf{p}}}} \sum_s \left(u_s(\mathbf{p})\, \hat{a}_s(\mathbf{p})\, e^{-ip\cdot x} + v_s(\mathbf{p})\, \hat{b}_s^\dagger(\mathbf{p})\, e^{+ip\cdot x} \right), \tag{8.303}$$

with $p^0 = E_{\mathbf{p}} = \sqrt{\mathbf{p}^2 + m^2}$. Using Eq. (8.292), one finds

$$\begin{aligned} \hat{\Psi}^C(x) &= C_0\, \hat{\Psi}^*(x) \\ &= \int \frac{d^3p}{(2\pi)^{3/2}} \frac{1}{\sqrt{2E_{\mathbf{p}}}} \sum_s \left(C_0 u_s^*(\mathbf{p})\, \hat{a}_s^\dagger(\mathbf{p})\, e^{+ip\cdot x} + C_0 v_s^*(\mathbf{p})\, \hat{b}_s(\mathbf{p})\, e^{-ip\cdot x} \right). \end{aligned} \tag{8.304}$$

In an ordinary Dirac theory, charge conjugation exchanges particle and antiparticle operators, $\hat{a} \leftrightarrow \hat{b}$ (up to phases). The Majorana constraint, instead, identifies them.

Imposing $\hat{\nu}(x) = \hat{\nu}^C(x)$ is achieved by identifying $\hat{b}$ with $\hat{a}$ in the mode expansion, together with corresponding spinor relations (in a convenient phase convention) such as

$$C_0 u_s^*(\mathbf{p}) = v_s(\mathbf{p}), \qquad C_0 v_s^*(\mathbf{p}) = u_s(\mathbf{p}). \tag{8.305}$$

The Majorana field operator may then be written as

$$\hat{\nu}(x) = \int \frac{d^3p}{(2\pi)^{3/2}} \frac{1}{\sqrt{2E_{\mathbf{p}}}} \sum_s \left(u_s(\mathbf{p})\, \hat{a}_s(\mathbf{p})\, e^{-ip\cdot x} + C_0 u_s^*(\mathbf{p})\, \hat{a}_s^\dagger(\mathbf{p})\, e^{+ip\cdot x} \right), \tag{8.306}$$

and one verifies explicitly that $C_0 \hat{\nu}^*(x) = \hat{\nu}(x)$.

A particularly transparent "one-handed" realization arises in the chiral basis. Starting from a left-handed momentum-space spinor $u_{L,s}(\mathbf{p})$, choose the four-spinor

$$u_s(\mathbf{p}) \equiv \begin{pmatrix} u_{L,s}(\mathbf{p}) \\ 0 \end{pmatrix}, \qquad C_0 u_s^*(\mathbf{p}) = \begin{pmatrix} 0 \\ i\sigma^2 u_{L,s}^*(\mathbf{p}) \end{pmatrix}, \tag{8.307}$$

so that Eq. (8.306) becomes a manifestly one-handed construction. Conceptually, this is the key point: unlike a massive Dirac field, which requires independent left and right chiral components, a Majorana field can be built from a single Weyl degree of freedom together with its conjugate.

8.10.2 Mass Terms and Conserved Charges

For a Dirac field $\Psi = \Psi_L + \Psi_R$, the mass term can be written as

$$\mathcal{L}_{m,D} = -m_D \left(\bar{\Psi}_L \Psi_R + \bar{\Psi}_R \Psi_L \right). \tag{8.308}$$

It is Lorentz invariant because it couples inequivalent Weyl representations into a scalar. The bilinears $\bar{\Psi}_L \Psi_L$ and $\bar{\Psi}_R \Psi_R$ do not form Lorentz scalars without additional structure and therefore do not appear as mass terms in a strictly Lorentz-invariant Dirac theory. The Majorana term in Eq. (8.297) circumvents this by using the antisymmetric invariant tensor $i\sigma^2$ to build a Lorentz scalar directly from two left-handed fields.

A Majorana field cannot carry a strictly conserved $U(1)$ charge. A direct way to see the obstruction is to attempt a global phase rotation of the left-handed component,

$$\psi_L \to e^{i\alpha} \psi_L. \tag{8.309}$$

Since $\psi_{L,C} = i\sigma^2 \psi_L^*$, it transforms oppositely,

$$\psi_{L,C} \to e^{-i\alpha} \psi_{L,C}. \tag{8.310}$$

Hence the four-component spinor ν in Eq. (8.296) cannot acquire a single common phase $e^{i\alpha}$ in both chiral slots. In particular, the Majorana equation cannot be made invariant under a local $U(1)$ gauge symmetry, and the corresponding Noether charge cannot exist. Physically, a fermion with an exactly conserved electromagnetic charge cannot be Majorana.

Where Majorana quasiparticles may emerge. In superconductors, charge is not a sharp quantum number for Bogoliubov quasiparticles: excitations are coherent superpositions of electron and hole degrees of freedom. This is precisely the setting in which effective Majorana modes can be engineered, for instance as bound states at vortices, defects, or boundaries, where electron-hole mixing is protected by a spectral gap and by topological constraints. In particle physics, Majorana masses are natural for neutral fermions, most notably neutrinos if lepton number is violated, and for neutral fermions in supersymmetric models (gauginos, higgsinos), which are not obstructed by electromagnetic $U(1)$ charge conservation.

8.11 Magnetic Monopoles

Magnetic monopoles provide a paradigmatic example of how topology may enforce the existence of particle-like objects in relativistic field theory. They behave as isolated sources of magnetic flux, yet in consistent field-theoretic realizations they are not inserted by hand as point singularities but emerge from non-trivial global structure. We begin with Dirac's construction, which is kinematically consistent within abelian electrodynamics but requires singular gauge patching. We then turn to the smooth 't Hooft-Polyakov monopole, which arises dynamically in a spontaneously broken non-abelian gauge theory.

8.11.1 Dirac Monopole and Quantization

Allowing magnetic charge density ρ_m and magnetic current $\mathbf{J}_m$, Maxwell's equations admit a fully symmetric form (Heaviside-Lorentz units),

$$\nabla \cdot \mathbf{E} = \rho_e, \qquad \nabla \cdot \mathbf{B} = \rho_m,$$
$$\nabla \times \mathbf{E} = -\mathbf{J}_m - \frac{\partial \mathbf{B}}{\partial t}, \qquad \nabla \times \mathbf{B} = \mathbf{J}_e + \frac{\partial \mathbf{E}}{\partial t}. \tag{8.311}$$

Lorentz covariance then suggests the generalized Lorentz force acting on a particle carrying electric charge q and magnetic charge g,

$$\mathbf{F} = q(\mathbf{E} + \mathbf{v} \times \mathbf{B}) + g(\mathbf{B} - \mathbf{v} \times \mathbf{E}). \tag{8.312}$$

For a point electric charge at the origin one recovers

$$\nabla \cdot \mathbf{E} = q\,\delta^{(3)}(\mathbf{r}) \quad \Rightarrow \quad \mathbf{E}(\mathbf{r}) = \frac{q}{4\pi}\frac{\mathbf{r}}{r^3} = -\nabla\left(\frac{q}{4\pi r}\right), \tag{8.313}$$

with flux $\Phi_E = q$. Analogously, a magnetic point source would require

$$\nabla \cdot \mathbf{B} = g\,\delta^{(3)}(\mathbf{r}) \quad \Rightarrow \quad \mathbf{B}(\mathbf{r}) = \frac{g}{4\pi}\frac{\mathbf{r}}{r^3}, \tag{8.314}$$

with magnetic flux $\Phi_M = g$.

However, writing $\mathbf{B} = \nabla \times \mathbf{A}$ implies $\nabla \cdot \mathbf{B} = 0$ identically. Therefore the field in Eq. (8.314) cannot arise from a globally defined smooth vector potential. Dirac's resolution is to define $\mathbf{A}$ only patchwise, allowing singular structure along an unphysical line, the Dirac string.

A standard choice employs two potentials on overlapping domains:

$$\mathbf{A}_N = \frac{g}{4\pi r}\frac{1-\cos\theta}{\sin\theta}\,\hat{\boldsymbol{\phi}}, \tag{8.315}$$
$$\mathbf{A}_S = -\frac{g}{4\pi r}\frac{1+\cos\theta}{\sin\theta}\,\hat{\boldsymbol{\phi}}, \tag{8.316}$$

each singular only on one polar axis. On their overlap,

$$\mathbf{A}_N - \mathbf{A}_S = \nabla\left(\frac{g\phi}{2\pi}\right). \tag{8.317}$$

A charged wavefunction transforms under gauge transformations as $\psi \to e^{-iq\Lambda/\hbar}\psi$. Hence (restoring $\hbar$),

$$\psi_S = \psi_N \exp\left[-\frac{iq}{\hbar}\left(\frac{g\phi}{2\pi}\right)\right]. \tag{8.318}$$

Single-valuedness under $\phi \to \phi + 2\pi$ requires

$$\exp\left(-\frac{iqg}{\hbar}\right) = 1 \quad \Rightarrow \quad qg = 2\pi n\hbar, \qquad n \in \mathbb{Z}, \tag{8.319}$$

which is Dirac's quantization condition.

The Dirac monopole is kinematically consistent but intrinsically singular. The quantization condition is not imposed dynamically; it follows from global consistency of the quantum wavefunction under gauge patching.

8.11.2 't Hooft-Polyakov Monopole

Unlike Dirac's construction, the 't Hooft-Polyakov monopole is a smooth, finite-energy classical solution of a spontaneously broken non-abelian gauge theory. Consider an SO(3) $\simeq$ SU(2) gauge theory with scalar triplet Φ^a in the adjoint representation:

$$\mathcal{L} = -\frac{1}{4}G^a_{\mu\nu}G^{a\mu\nu} + \frac{1}{2}(D_\mu\Phi)^a(D^\mu\Phi)^a - \frac{\lambda}{4}(\Phi^a\Phi^a - v^2)^2, \tag{8.320}$$

where $(D_\mu\Phi)^a = \partial_\mu\Phi^a + q\,\epsilon^{abc}W^b_\mu\Phi^c$.

Finite energy requires $\Phi^a\Phi^a \to v^2$ at spatial infinity. The hedgehog boundary condition

$$\Phi^a(\mathbf{r}) \xrightarrow{r\to\infty} v\,\frac{r^a}{r} \tag{8.321}$$

defines a non-trivial map $S^2_\infty \to S^2_{\text{vac}}$ of unit winding number. Topological stability follows from the non-trivial homotopy group $\pi_2(S^2) = \mathbb{Z}$.

A pure scalar configuration would produce divergent gradient energy $(\partial_i\Phi^a)^2 \sim v^2/r^2$. The gauge field compensates this via

$$W^a_i \xrightarrow{r\to\infty} -\epsilon^{aib}\frac{r^b}{qr^2}, \tag{8.322}$$

ensuring $D_i\Phi \to 0$ sufficiently fast. The effective abelian field strength is obtained by projection along $\Phi^a/|\Phi|$:

$$F_{\mu\nu} = \frac{\mathbf{\Phi}}{|\mathbf{\Phi}|}\cdot\mathbf{G}_{\mu\nu} - \frac{\mathbf{\Phi}}{q|\mathbf{\Phi}|^3}\cdot\left(D_\mu\mathbf{\Phi}\times D_\nu\mathbf{\Phi}\right), \tag{8.323}$$

which reduces to the ordinary abelian form in the vacuum. For the hedgehog solution one finds

$$\mathbf{B}(\mathbf{r}) = \frac{1}{q}\frac{\mathbf{r}}{r^3}, \qquad \Phi_M = \frac{4\pi}{q}. \tag{8.324}$$

Thus the magnetic charge is $g = \frac{4\pi}{q}$, which automatically satisfies Dirac quantization.

The decisive conceptual difference is that the 't Hooft-Polyakov monopole is non-singular and dynamically generated. Its mass scale is set by v and the gauge coupling q, and its magnetic charge is fixed by topology rather than by boundary conditions imposed at a point.

8.12 Proposed Exercises

Exercise 1

A rigid sphere of radius R translates in an incompressible fluid of density ρ. Assume that, outside the sphere, the flow is irrotational, so that a velocity potential ϕ exists with $\mathbf{v} = \nabla\phi$ and $\nabla^2\phi = 0$.

(a) Consider a translation along the z axis with instantaneous speed $U(t)$. Assuming the far-field condition $\nabla\phi \to 0$ as $r \to \infty$ and the no-penetration condition $\partial_n\phi|_{r=R} = U\cos\theta$, show that an admissible potential is

$$\phi(r,\theta) = -\frac{1}{2}\frac{UR^3}{r^2}\cos\theta. \tag{8.325}$$

(b) Using Green's identity for harmonic potentials in the exterior domain,

$$\int_\Omega d^3x\, |\nabla\phi|^2 = \int_{\partial\Omega} dS\, \phi\, \frac{\partial\phi}{\partial n}, \tag{8.326}$$

derive the kinetic energy stored in the fluid motion in the form $T_{\text{fluid}} = \frac{1}{2}\, m_{\text{added}}\, U^2$, and compute m_{added} explicitly.

(c) Now assume the fluid is weakly viscous (kinematic viscosity ν). A common effective description adds a causal memory contribution to the force,

$$F_{\text{hist}}(t) = -\int_0^t d\tau\, \gamma(t-\tau)\, \dot{U}(\tau), \qquad \gamma(t) \propto t^{-1/2}. \tag{8.327}$$

Explain qualitatively why a $t^{-1/2}$ kernel is expected from vorticity diffusion and why Eq. (8.327) is non-Markovian.

(d) Consider the Fourier transform convention $U(t) = \int \frac{d\omega}{2\pi} e^{-i\omega t} \tilde{U}(\omega)$. Show that the equation of motion in frequency space can be written as

$$\left[-\omega^2 (m + m_{\text{added}}) - i\omega\, \tilde{\gamma}(\omega)\right] \tilde{U}(\omega) = \tilde{F}_{\text{ext}}(\omega). \tag{8.328}$$

Exercise 2

Consider a real scalar field in $1+1$ dimensions with Lagrangian density

$$\mathcal{L} = \frac{1}{2}\, \partial_\mu \phi\, \partial^\mu \phi - \frac{\lambda}{4}\, (\phi^2 - v^2)^2, \qquad \lambda > 0, \tag{8.329}$$

and focus on static finite-energy configurations.

(a) Show that finite energy requires $\phi(x) \to \phi_\pm$ as $x \to \pm\infty$ with $\phi_\pm = \pm v$ and $d\phi/dx \to 0$. Explain why this yields disconnected sectors of configuration space.
(b) Starting from the static equation of motion, derive the first-integral relation

$$\frac{1}{2}\left(\frac{d\phi}{dx}\right)^2 = V(\phi), \qquad V(\phi) = \frac{\lambda}{4}(\phi^2 - v^2)^2, \tag{8.330}$$

for kink boundary conditions. Then solve Eq. (8.330) and obtain $\phi_{\text{kink}}(x)$.
(c) From the topological current $J_T^\mu = \frac{1}{2v}\, \epsilon^{\mu\nu} \partial_\nu \phi$ show that $\partial_\mu J_T^\mu = 0$ identically and compute the associated charge for the kink and for the antikink.
(d) Using Eq. (8.330), show that the kink energy can be written as

$$E_{\text{kink}} = \int_{-\infty}^{+\infty} dx \left(\frac{d\phi}{dx}\right)^2 = \int_{-v}^{+v} d\phi\, \sqrt{2V(\phi)}. \tag{8.331}$$

Evaluate the integral and comment on the coupling dependence of E_{kink}.

Exercise 3

Consider a (scalar) excitation whose full retarded propagator in a homogeneous medium can be written as

$$G_R(\omega, \mathbf{k}) = \frac{1}{\omega^2 - \omega_{\mathbf{k},0}^2 - \Sigma_R(\omega, \mathbf{k})}, \tag{8.332}$$

where $\omega_{\mathbf{k},0}^2 = \mathbf{k}^2 + m^2$ and Σ_R is the retarded self-energy.

(a) Assume that, near a well-defined quasiparticle pole, one may expand $\Sigma_R(\omega, \mathbf{k})$ around $\omega = \omega_\mathbf{k}$ as $\Sigma_R(\omega, \mathbf{k}) \simeq \Sigma_R(\omega_\mathbf{k}, \mathbf{k}) + (\omega - \omega_\mathbf{k})\, \partial_\omega \Sigma_R(\omega_\mathbf{k}, \mathbf{k})$. Define $\omega_\mathbf{k}$ as the solution of the dispersion relation obtained by neglecting the imaginary part of Σ_R. Derive the approximate pole form

$$G_R(\omega, \mathbf{k}) \simeq \frac{Z_\mathbf{k}}{\omega - \omega_\mathbf{k} + i\Gamma_\mathbf{k}/2} \quad \text{(near the pole)}, \tag{8.333}$$

and give expressions for the residue $Z_\mathbf{k}$ and the width $\Gamma_\mathbf{k}$ in terms of Σ_R and its derivative.

(b) The spectral function is defined by $A(\omega, \mathbf{k}) = -2\,\mathrm{Im}\, G_R(\omega, \mathbf{k})$. Using Eq. (8.333), show that $A(\omega, \mathbf{k})$ is approximately a Breit-Wigner peak and identify its normalization and full width at half maximum.

(c) Provide a precise criterion for "long-lived" quasiparticles in terms of $\Gamma_\mathbf{k}$ and $\omega_\mathbf{k}$.

(d) In the time domain, argue that the pole form implies a decay of the form $\exp[-(\Gamma_\mathbf{k}/2)t]$ for the corresponding mode amplitude. Explain the physical interpretation of $\Gamma_\mathbf{k}^{-1}$ as a lifetime.

Essential Bibliography

- Lancaster T, Blundell SJ (2014) Quantum field theory for the Gifted Amateur, Cambridge University Press. A pedagogically refined introduction to QFT that bridges conceptual foundations and practical computations.
- Mahan GD (2000) Many-particle physics, 3rd edn. Kluwer/Plenum. A comprehensive technical reference on Green functions, spectral functions, self-energies, and response theory.
- Sachdev S (2011) Quantum phase transitions, 2nd edn. Cambridge Univ. Press. It introduces regimes where quasiparticles fail or become ill-defined, especially near quantum critical points, and provides the modern language of scaling and universality used to discuss their breakdown.
- Kamenev A (2011) Field theory of non-equilibrium systems. Cambridge Univ. Press. It provides the formal framework for treating quasiparticles out of equilibrium, including their lifetimes, dressing and memory effects in real time.

Electromagnetism 9

The field is the only reality.
—Julian Schwinger

Abstract

Electromagnetism is experimentally ubiquitous, mathematically rigid, and conceptually decisive in shaping the modern notion of symmetry and relativity. In this chapter we develop Maxwell's theory in a fully covariant language, starting from the familiar three-vector formulation and progressing toward the four-potential and the field-strength tensor as the natural objects in Minkowski spacetime. The covariant formulation is then reorganized around gauge invariance: we analyze the redundancy of the potentials, discuss standard gauge choices (Lorenz and Coulomb) together with residual freedom, and clarify why only two independent dynamical degrees of freedom correspond to physical radiation. After establishing the Lagrangian formulation, we use the variational principle to obtain Maxwell's equations, construct the energy-momentum tensor, and interpret energy and momentum flow through the Poynting vector in a relativistically consistent way. The chapter closes by emphasizing a uniquely quantum message already encoded in classical covariance: the physical content is captured by gauge-invariant observables, yet potentials carry indispensable structural information. This prepares the ground for the deeper quantum-field-theoretic treatment developed in the upcoming chapters, while keeping the present one self-contained as the definitive covariant account of Maxwell theory.

O. Luongo, *Fundamental Interactions*,
https://doi.org/10.1007/978-3-032-27413-7_9

Fig. 9.1 Portrait of Hermann Weyl, 1933. Reprinted from Wikimedia. Public domain

A key character: *Hermann Weyl (1885–1955), Fig.* 9.1.
Hermann Weyl was one of the most profound mathematical physicists of the twentieth century, a scholar whose work repeatedly reshaped the boundary between geometry and fundamental interactions. Born in Elmshorn, near Hamburg, Weyl studied mathematics in Munich and Göttingen at a time when Hilbert, Klein, and Noether were transforming the foundations of the subject. His early contributions rapidly established him as a leading figure in analysis and differential geometry.
Weyl's scientific identity is inseparable from the idea that *geometry is not merely a language for physics, but a dynamical principle.* In the wake of Einstein's general relativity, Weyl sought to push the geometrization program further. In 1918 he proposed a unified framework in which electromagnetism and gravitation arise from a generalized geometric structure, introducing what is now called *Weyl geometry*. Although the original physical proposal did not survive empirical scrutiny, the conceptual move was decisive: Weyl introduced the notion of a local change of scale (a local "calibration") and recognized that physical laws may be organized by local symmetry principles. In modern language, this was among the earliest clear manifestations of the gauge idea.

Weyl's influence on electromagnetism is therefore not limited to any single formula or model, but to a structural perspective that still governs contemporary field theory: potentials are not mere computational conveniences; they are geometric objects, defined up to gauge transformations, and the invariants built from them are the true observables.
Beyond gauge ideas, Weyl contributed to foundational results across physics: his development of representation theory and his systematic analysis of continuous groups became essential for quantum mechanics and QFT; his work on spinors and the *Weyl equation* clarified the structure of relativistic fermions; and his book *Space-Time-Matter* shaped generations of physicists by presenting relativity as a geometrical theory of nature. After appointments in Zürich and Göttingen, Weyl moved in 1933 to the Institute for Advanced Study in Princeton, where he remained an intellectual pillar of the institute's early decades.

9.1 Maxwell Equations

The starting point is introducing the Maxwell equations in vacuum, which can be naturally separated into *constraint equations* and *dynamical equations*. This distinction reflects the different logical roles played by the four equations.

▶ **Pause and think**
Manage to figure out the main differences among the Maxwell equations.

Constraint equations. The divergence equations do not involve time derivatives of the fields and therefore act as instantaneous constraints on the admissible configurations:

$$\nabla \cdot \mathbf{E} = \frac{\rho}{\varepsilon_0}, \tag{9.1}$$

$$\nabla \cdot \mathbf{B} = 0. \tag{9.2}$$

Equation (9.1) fixes the electric flux through any closed surface in terms of the enclosed electric charge. Once the charge distribution is given, it restricts the allowed initial configurations of $\mathbf{E}$. Equation (9.2) states that magnetic field lines have no sources or sinks: they always form closed loops. It constrains $\mathbf{B}$ at any given time but does not prescribe its time evolution.

Dynamical equations. The curl equations contain first-order time derivatives and therefore govern the temporal evolution of the fields:

$$\nabla \times \mathbf{E} = -\frac{\partial \mathbf{B}}{\partial t}, \tag{9.3}$$

$$\nabla \times \mathbf{B} = \mu_0 \mathbf{j} + \varepsilon_0 \mu_0 \frac{\partial \mathbf{E}}{\partial t}. \tag{9.4}$$

Equation (9.3) shows that a time-varying magnetic field necessarily induces a circulating electric field. Equation (9.4) states that magnetic circulation is produced both by electric currents and by time variations of the electric field. Given the fields and sources at an initial time, these two equations determine their subsequent evolution.

Hence, the constraint equations reveal a fundamental asymmetry: electric fields possess sources, magnetic fields do not.

Electric field lines can begin and end on charges, as expressed by Gauss' law, Eq. (9.1). In contrast, Gauss' law for magnetism, Eq. (9.2), enforces $\nabla \cdot \mathbf{B} = 0$, implying that magnetic field lines are always closed.

If magnetic charges (monopoles) existed with density ρ_m, the second constraint would be replaced by

$$\nabla \cdot \mathbf{B} = \rho_m, \tag{9.5}$$

and Faraday's law would acquire an additional source term,

$$\nabla \times \mathbf{E} = -\frac{\partial \mathbf{B}}{\partial t} - \mathbf{j}_m, \tag{9.6}$$

making the system fully symmetric under the exchange $\mathbf{E} \leftrightarrow \mathbf{B}$ and $\rho \leftrightarrow \rho_m, \mathbf{j} \leftrightarrow \mathbf{j}_m$.

The observed absence of magnetic charges therefore explains the structural asymmetry of the Maxwell equations: electric fields are sourced by matter, while magnetic fields are generated only indirectly, through electric currents and through the time variation of electric fields. This empirical fact, rather than any formal principle, is responsible for the different roles played by **E** and **B** in the classical theory.

9.1.1 Electric Charge as Universal Coupling

In special relativity, the electromagnetic interaction enters through the source four-current $j^{\mu} = (c\rho, \mathbf{j})$ and through a universal coupling constant e.

Experimentally, *electric charge* is the attribute of matter that sources the gauge field and determines the response to it.

Two structural facts elevate charge from a mere parameter to a fundamental datum of Nature:

- charge is conserved with extreme accuracy, and
- the coupling to the electromagnetic field is universal.

The interaction between matter and the electromagnetic field is encoded by the minimal coupling prescription (made precise later in this Chapter and in Chaps. 6 and 7),

$$p_{\mu} \longrightarrow p_{\mu} - \frac{e}{c} A_{\mu}, \qquad \partial_{\mu} \longrightarrow D_{\mu} \equiv \partial_{\mu} + \frac{ie}{\hbar c} A_{\mu}. \tag{9.7}$$

A single constant e governs both sourcing and response: the same charge that appears in j^{μ} also multiplies A_{μ} in the covariant derivative.

9.1.2 Longitudinal and Transverse Fields

The electric field $\mathbf{E}$ and the magnetic field $\mathbf{B}$, like all sufficiently regular vector fields, can be separated into a *longitudinal* component (with zero curl) and a *transverse* component (with zero divergence). This decomposition, often referred to as the Helmholtz decomposition, allows us to highlight the true dynamical variables of electromagnetism. We will therefore write

$$\mathbf{E} = \mathbf{E}_L + \mathbf{E}_T \quad , \quad \mathbf{B} = \mathbf{B}_L + \mathbf{B}_T \, . \tag{9.8}$$

Here the longitudinal part is determined instant by instant by the sources, while the transverse part contains the radiative degrees of freedom. More precisely, one requires

$$\nabla \times \mathbf{E}_L = 0 \quad , \quad \nabla \cdot \mathbf{E}_T = 0 \, , \tag{9.9a}$$

$$\nabla \times \mathbf{B}_L = 0 \quad , \quad \nabla \cdot \mathbf{B}_T = 0 \, . \tag{9.9b}$$

In practice, under the usual boundary conditions (fields decaying sufficiently fast at spatial infinity), this decomposition is unique. In particular, $\mathbf{E}_L$ and $\mathbf{B}_L$ may be represented as gradients of scalar potentials, whereas $\mathbf{E}_T$ and $\mathbf{B}_T$ may be expressed as curls of vector potentials.

Using $\nabla \cdot \mathbf{E}_T = 0$ in Gauss's law, one finds

$$\nabla \cdot \mathbf{E}_L = 4\pi\rho \,. \tag{9.10}$$

Accordingly, it follows immediately that $\mathbf{B}$ is purely transverse, i.e.

$$\mathbf{B}_L = 0 \,, \tag{9.11}$$

because Gauss' law for magnetism enforces $\nabla \cdot \mathbf{B} = 0$ and, in the absence of magnetic charges, there is no longitudinal (gradient) contribution that is sourced instantaneously by matter.

This leads to the Faraday-Neumann-Lenz law for the transverse electric field, by virtue of Eq. (9.11) and $\nabla \times \mathbf{E}_L = 0$,

$$\nabla \times \mathbf{E}_T + \frac{1}{c}\frac{\partial \mathbf{B}_T}{\partial t} = 0 \,. \tag{9.12}$$

Equation (9.12) already contains the essential dynamical content: a time-varying transverse magnetic field induces a circulating transverse electric field, which is the mechanism behind electromagnetic wave propagation.

If we decompose the current density $\mathbf{j}$ into its longitudinal part $\mathbf{j}_L$ and its transverse part $\mathbf{j}_T$, the continuity equation ($\nabla \cdot \mathbf{j} + \partial\rho/\partial t = 0$) is written as

$$\nabla \cdot \mathbf{j}_L + \frac{\partial \rho}{\partial t} = 0 \,. \tag{9.13}$$

Here $\mathbf{j}_L$ is precisely the part of the current responsible for local charge redistribution (it is the only part with nonzero divergence), whereas $\mathbf{j}_T$ describes solenoidal current patterns that do not change the local charge density.

Combining Eq. (9.13) with the time derivative of Gauss' law Eq. (9.10), one finds

$$\nabla \cdot \left(\frac{\partial \mathbf{E}_L}{\partial t} + 4\pi \mathbf{j}_L \right) = 0 \,. \tag{9.14}$$

Exploiting now the defining property of the longitudinal field, $\nabla \times \mathbf{E}_L = 0$, the vector field

$$\frac{\partial \mathbf{E}_L}{\partial t} + 4\pi \mathbf{j}_L$$

is simultaneously curl-free (as a time derivative of a curl-free field) and divergence-free (by Eq. (9.14)). Under the standard boundary condition of sufficiently fast decay at infinity, the only such vector field is the trivial one. Hence Eq. (9.14) implies

$$\frac{\partial \mathbf{E}_L}{\partial t} = -4\pi \mathbf{j}_L \,. \tag{9.15}$$

Equation (9.15) is not an independent dynamical law: it is a consistency condition expressing how the instantaneous Coulomb field must adjust when charges move so as to preserve Gauss' law at all times.

This constraint allows us to eliminate from the Ampère-Maxwell law the longitudinal components. We can therefore write

$$\nabla \times \mathbf{B}_T - \frac{1}{c}\frac{\partial \mathbf{E}_T}{\partial t} = \frac{4\pi}{c}\mathbf{j}_T \,. \tag{9.16}$$

Equations (9.12) and (9.16) constitute a closed set for the transverse fields once the transverse current $\mathbf{j}_T$ is specified. In particular, in vacuum with $\rho = 0$ and $\mathbf{j} = 0$ one recovers a pair of wave equations for $\mathbf{E}_T$ and $\mathbf{B}_T$, showing explicitly that only the transverse sector carries propagating degrees of freedom.

We have thus separated the Maxwell equations into two groups: the equations for the transverse fields, namely Eqs. (9.12) and (9.16), which are true equations of motion since they contain time derivatives, and the equations for the longitudinal fields, namely Eqs. (9.10) and (9.11), which are simple constraints.[1] The longitudinal components of the fields $\mathbf{E}_L$, $\mathbf{B}_L$ are therefore not true dynamical variables, being fixed instant by instant by the conditions Eqs. (9.10) and (9.11). The dynamical degrees of freedom are the transverse fields $\mathbf{E}_T$ and $\mathbf{B}_T$.

9.1.3 The Lorentz Force

The equations of motion of a particle of mass m, charge e, and law of motion $\mathbf{z}(t)$, subject to an electromagnetic field, are

$$\frac{d}{dt}\left(m\gamma\mathbf{v}(t)\right) = e\mathbf{E}\left(\mathbf{z}(t), t\right) + \frac{e}{c}\mathbf{v}(t) \times \mathbf{B}\left(\mathbf{z}(t), t\right) \,, \tag{9.17}$$

where $\mathbf{v}(t) = d\mathbf{z}/dt$ is the velocity of the particle, and on the right-hand side appears the *Lorentz force*

$$\mathbf{F}_L = e\mathbf{E} + \frac{e}{c}\mathbf{v} \times \mathbf{B} \,. \tag{9.18}$$

The fields $\mathbf{E}$ and $\mathbf{B}$ are evaluated at the point $z(t)$ where the particle is located. We call Eq. (9.17) the *Minkowski-Lorentz equation*. The momentum appearing on the left-hand side of Eq. (9.17)

$$\mathbf{p} = m\gamma\mathbf{v} \,, \tag{9.19}$$

is the *kinetic momentum*. This momentum, as we will see, does not coincide with the *canonical momentum*, i.e. with the momentum canonically conjugate to the coordinates (which we will call $\mathbf{P}$). The distinction becomes essential in the Lagrangian

[1] Note that Eq. (9.15) is not an independent equation, being a consequence of Eq. (9.10) and the continuity equation.

and Hamiltonian formulations, where **P** differs from **p** by a term proportional to the vector potential.

In general, within electrodynamics, the Maxwell equations and the Minkowski-Lorentz equation are *coupled equations*.

The sources appearing in the Maxwell equations are given by

$$\rho(\mathbf{x}, t) = \sum_a e_a \delta^3(\mathbf{x} - \mathbf{z}(t)), \tag{9.20a}$$

$$\mathbf{j}(\mathbf{x}, t) = \sum_a e_a \mathbf{v}_a(t) \delta^3(\mathbf{x} - \mathbf{z}_a(t)), \tag{9.20b}$$

containing the coordinates and velocities of the particles, which satisfy the Minkowski-Lorentz equation. On the other hand, the dynamics of the charges depends, according to Eq. (9.17), on the fields, which satisfy Maxwell's equations. There is therefore an interdependence between the sources and the fields, as can be seen by writing all the equations explicitly

$$\nabla \cdot \mathbf{E}(\mathbf{x}, t) = 4\pi \sum_a e_a \delta^3(\mathbf{x} - \mathbf{z}_a(t)), \tag{9.21a}$$

$$\nabla \cdot \mathbf{B}(\mathbf{x}, t) = 0, \tag{9.21b}$$

$$\nabla \times \mathbf{E}(\mathbf{x}, t) + \frac{1}{c}\frac{\partial \mathbf{B}(\mathbf{x}, t)}{\partial t} = 0, \tag{9.21c}$$

$$\nabla \times \mathbf{B}(\mathbf{x}, t) - \frac{1}{c}\frac{\partial \mathbf{E}(\mathbf{x}, t)}{\partial t} = \frac{4\pi}{c} \sum_a e_a \mathbf{v}_a(t) \delta^3(\mathbf{x} - \mathbf{z}_a(t)), \tag{9.21d}$$

$$\frac{d}{dt}(m_a \gamma \mathbf{v}_a(t)) = e_a \mathbf{E}_a(\mathbf{z}_a(t), t) + \frac{e_a}{c} \mathbf{v}_a(t) \times \mathbf{B}_a(\mathbf{z}_a(t), t). \tag{9.21e}$$

Solving the coupled equations Eq. (9.21) is a difficult, and often prohibitive, task. In particular, the fields entering Eq. (9.21e) are typically *total* fields, including both externally applied fields and the fields generated by the charges themselves. This is precisely where conceptual and technical subtleties arise (self-fields, regularization, and radiation reaction).

Fortunately, in most situations of practical relevance this is not necessary, because either the sources or the fields are *fixed*, and it is therefore sufficient to solve the Maxwell equations, in the first case, or the Minkowski-Lorentz equation, in the second. There is, however, an important effect to take into consideration. A charged particle, when accelerating, emits electromagnetic radiation which reacts back on it, influencing its motion. This is the so-called *radiation reaction*, which can be accounted for by modifying the equations of motion of the charge.

9.1.4 The Electromagnetic Potentials

The Maxwell equations can be rewritten in a more manageable form by introducing the so-called *electromagnetic potentials*. The strategy is to define the potentials so that they satisfy the homogeneous Maxwell equations identically, and then to substitute them into the inhomogeneous equations in order to determine their dynamics. From the Maxwell equations, the magnetic field is solenoidal, namely $\nabla \cdot \mathbf{B} = 0$. Under standard regularity assumptions (in a simply connected region and away from singularities), this implies that there exists a vector field $\mathbf{A}(\mathbf{x}, t)$, called the *vector potential* (or *magnetic potential*), such that

$$\mathbf{B} = \nabla \times \mathbf{A}\,. \tag{9.22}$$

Substituting into the Maxwell equations, one finds

$$\nabla \times \mathbf{E} + \frac{1}{c}\frac{\partial}{\partial t}(\nabla \times \mathbf{A}) = 0\,, \Rightarrow \quad \nabla \times \left(\mathbf{E} + \frac{1}{c}\frac{\partial \mathbf{A}}{\partial t}\right) = 0\,. \tag{9.23}$$

Since the quantity $\mathbf{E} + \frac{1}{c}\frac{\partial \mathbf{A}}{\partial t}$ is irrotational, it can be expressed as the gradient of a scalar field. Therefore, there exists a function $\varphi(\mathbf{x}, t)$, called the *scalar potential* (or *electric potential*), such that

$$\mathbf{E} + \frac{1}{c}\frac{\partial \mathbf{A}}{\partial t} = -\nabla\varphi\,. \tag{9.24}$$

In the static case, Eq. (9.24) reduces to $\mathbf{E} = -\nabla\varphi$, as expected for electrostatics.

We have introduced φ and $\mathbf{A}$ by exploiting the homogeneous Maxwell equations. By construction, once the potentials are defined through Eqs. (9.22) and (9.24), they automatically satisfy the homogeneous sector. Determining φ and $\mathbf{A}$ therefore reduces to solving their equations of motion, obtained by substituting Eqs. (9.22) and (9.24) into the inhomogeneous Maxwell equations.

How does a *magnetic potential* exist?
The existence of $\mathbf{A}$ in Eq. (9.22) is a geometric statement: a divergence-free vector field can be written locally as a curl. This does *not* mean that $\mathbf{B}$ itself is conservative. In fact, $\mathbf{B}$ is conservative only when it is also irrotational,

$\nabla \times \mathbf{B} = 0$, which by the Ampère-Maxwell law holds in regions free of electric currents and displacement currents.
In the strictly magnetostatic case, $\partial_t \mathbf{E} = 0$, the inhomogeneous equation reduces to $\nabla \times \mathbf{B} = \frac{4\pi}{c}\mathbf{j}$ (Gaussian units). Hence in current-free regions ($\mathbf{j} = 0$) one has $\nabla \times \mathbf{B} = 0$, and therefore $\mathbf{B}$ can be written as the gradient of a scalar magnetic potential,

$$\mathbf{B} = -\nabla \psi_m \,, \tag{9.25}$$

at least locally. This scalar magnetic potential is useful, for instance, in describing magnetostatic fields in simply connected regions without currents (e.g. outside localized current loops), where ψ_m satisfies Laplace's equation. Globally, however, one must be careful: if the region is not simply connected (as happens, for example, in the exterior of a solenoid), $\mathbf{B}$ may have vanishing curl and still fail to admit a single-valued scalar potential. In such situations the vector potential $\mathbf{A}$ remains the natural object: it exists locally in general, and it encodes the global/topological information that cannot be captured by a scalar potential. In other words, $\mathbf{A}$ is not introduced because $\mathbf{B}$ is conservative; it is introduced because $\mathbf{B}$ is solenoidal, and because a curl representation is the correct local primitive for such fields.

▶ Driving exercise

Equations for the potentials. Starting from Eq. (9.24), easily one obtains

$$\nabla^2 \varphi + \frac{1}{c}\nabla \cdot \frac{\partial \mathbf{A}}{\partial t} = -4\pi\rho \,. \tag{9.26}$$

Substituting now Eqs. (9.22) and (9.24), one finds

$$\nabla \times (\nabla \times \mathbf{A}) + \frac{1}{c}\frac{\partial}{\partial t}\nabla\varphi + \frac{1}{c^2}\frac{\partial^2 \mathbf{A}}{\partial t^2} = \frac{4\pi}{c}\mathbf{j} \,, \tag{9.27}$$

and, using the identity $\nabla \times (\nabla \times \mathbf{A}) = \nabla(\nabla \cdot \mathbf{A}) - \nabla^2\mathbf{A}$, one finally arrives at

$$\nabla^2\mathbf{A} - \frac{1}{c^2}\frac{\partial^2 \mathbf{A}}{\partial t^2} - \nabla\left(\nabla \cdot \mathbf{A} + \frac{1}{c}\frac{\partial \varphi}{\partial t}\right) = -\frac{4\pi}{c}\mathbf{j} \,. \tag{9.28}$$

Solving the Maxwell equations is therefore equivalent to solving the potential equations Eqs. (9.26) and (9.28). The fields $\mathbf{E}$ and $\mathbf{B}$ are then reconstructed from the potentials φ and $\mathbf{A}$ through Eqs. (9.22) and (9.24), respectively.

We will now derive, heuristically, the continuity equations that mathematically express the conservation laws of energy, linear momentum and angular momentum of the electromagnetic field. In doing so we will identify the expressions for the energy density, linear momentum density and angular momentum density, in terms of **E**, **B**.

9.1.5 Conservation of Energy

Consider a system of fields and particles contained in a volume V. We assume that particles cannot leave V nor enter it. The law of conservation of energy can be expressed as follows: if the sum of the electromagnetic energy stored in the fields and the mechanical energy of the particles inside V increases, then there must be a net inflow of electromagnetic energy through the boundary surface $S = \partial V$. Conversely, if the total energy (electromagnetic plus mechanical) in V decreases, there is a net outflow of electromagnetic energy. In mathematical terms, this balance is expressed by an equation of the form

$$\frac{dU_{em}}{dt} = -\Phi_{em} - \frac{dW}{dt} \,. \tag{9.29}$$

Here U_{em} is the energy stored in the electromagnetic field inside V, Φ_{em} is the flux of electromagnetic energy through the surface S enclosing V, and W is the work done by the field on the particles contained in V. Upon introducing the *energy density* of the electromagnetic field w_{em} and the *energy flux density* (energy flow per unit area) $\mathcal{P}$, one has

$$U_{em} = \int_V w_{em}\, d^3\mathbf{x}\,, \tag{9.30a}$$

$$\Phi_{em} = \int_S \mathcal{P} \cdot d\boldsymbol{\sigma}\,. \tag{9.30b}$$

Equation (9.29) then takes the form

$$\frac{d}{dt}\int_V w_{em}\, d^3\mathbf{x} = -\int_S \mathcal{P}\cdot d\boldsymbol{\sigma} - \frac{dW}{dt}\,. \tag{9.31}$$

Let us start by determining W. The Lorentz force in Eq. (9.18) acts on the particles (which occupy positions z_a and have velocities $\mathbf{v}_a$) and does work per unit time

$$\frac{dW}{dt} = \sum_a \mathbf{F}_a \cdot \mathbf{v}_a = \sum_a e_a\, \mathbf{E}(\mathbf{z}_a, t)\cdot \mathbf{v}_a\,. \tag{9.32}$$

Recalling that the current density of the particles is $\mathbf{j} = \sum_a e_a \mathbf{v}_a\, \delta^3(\mathbf{x} - \mathbf{z}_a)$, and exploiting the Dirac delta, Eq. (9.32) can be rewritten as

$$\frac{dW}{dt} = \int_V \mathbf{j}\cdot \mathbf{E}\, d^3\mathbf{x}\,, \tag{9.33}$$

so that Eq. (9.31) becomes

$$\frac{d}{dt}\left(\int_V w_{em}\, d^3\mathbf{x}\right) = -\int_S \mathcal{P}\cdot d\sigma - \int_V \mathbf{j}\cdot\mathbf{E}\, d^3\mathbf{x}\,. \tag{9.34}$$

Gauss's theorem allows us to re-express the surface integral as a volume integral of the divergence, thus obtaining

$$\begin{aligned} &\frac{d}{dt}\int_V w_{em}\, d^3\mathbf{x} = -\int_V \nabla\cdot\mathcal{P}\, d^3\mathbf{x} - \int_V \mathbf{j}\cdot\mathbf{E}\, d^3\mathbf{x}\,, \\ &\Rightarrow \int_V \left(\frac{\partial w_{em}}{\partial t} + \nabla\cdot\mathcal{P} + \mathbf{j}\cdot\mathbf{E}\right) d^3\mathbf{x} = 0\,, \\ &\Rightarrow \frac{\partial w_{em}}{\partial t} + \nabla\cdot\mathcal{P} = -\mathbf{j}\cdot\mathbf{E}\,. \end{aligned} \tag{9.35}$$

In the absence of charges (free electromagnetic field) Eq. (9.35) reduces to

$$\frac{\partial w_{em}}{\partial t} + \nabla\cdot\mathcal{P} = 0\,, \tag{9.36}$$

which is the *continuity equation for electromagnetic energy* (in differential form). The source term $-\mathbf{j}\cdot\mathbf{E}$ in Eq. (9.35) encodes the local power density transferred from the field to matter: when $\mathbf{j}\cdot\mathbf{E} > 0$ the field loses energy to the charges, while when $\mathbf{j}\cdot\mathbf{E} < 0$ the charges feed energy back into the field.

The energy-balance equation

$$\frac{\partial w_{em}}{\partial t} + \nabla\cdot\mathcal{P} = -\mathbf{j}\cdot\mathbf{E} \tag{9.37}$$

is the local form of energy conservation for coupled fields and charges. The term $\mathcal{P}$ is the energy flux density, and $\mathbf{j}\cdot\mathbf{E}$ is the power per unit volume delivered to matter by the electromagnetic field.

Let us now derive the expressions for w_{em} and $\mathcal{P}$ in terms of the fields. The starting point employs, as usual, the Maxwell equations, which we rewrite here for convenience

$$\nabla\times\mathbf{B} - \frac{1}{c}\frac{\partial\mathbf{E}}{\partial t} = \frac{4\pi}{c}\mathbf{j}\,, \tag{9.38a}$$

$$\nabla\times\mathbf{E} + \frac{1}{c}\frac{\partial\mathbf{B}}{\partial t} = 0\,. \tag{9.38b}$$

Multiplying the first equation scalarly by **E**, and the second by **B**, and subtracting member by member, we obtain

$$\frac{1}{c}\left(\mathbf{E}\cdot\frac{\partial\mathbf{E}}{\partial t}+\mathbf{B}\cdot\frac{\partial\mathbf{B}}{\partial t}\right)=-\left(\mathbf{B}\cdot\nabla\times\mathbf{E}-\mathbf{E}\cdot\nabla\times\mathbf{B}\right)-\frac{4\pi}{c}\mathbf{j}\cdot\mathbf{E}\,. \tag{9.39}$$

Using the vector identity $\mathbf{B}\cdot\nabla\times\mathbf{E}-\mathbf{E}\cdot\nabla\times\mathbf{B}=\nabla\cdot(\mathbf{E}\times\mathbf{B})$, Eq. (9.39) becomes

$$\frac{1}{2c}\frac{\partial}{\partial t}(\mathbf{E}^2+\mathbf{B}^2)=-\nabla\cdot(\mathbf{E}\times\mathbf{B})-\frac{4\pi}{c}\mathbf{j}\cdot\mathbf{E}\,. \tag{9.40}$$

Comparison with Eq. (9.35) leads to the conclusion that the electromagnetic energy density w_{em} and the energy flux density $\mathcal{P}$ are given by

$$w_{em}=\frac{1}{8\pi}(\mathbf{E}^2+\mathbf{B}^2)\,, \tag{9.41a}$$

$$\mathcal{P}=\frac{c}{4\pi}(\mathbf{E}\times\mathbf{B})\,. \tag{9.41b}$$

The vector $\mathcal{P}$ is called the *Poynting vector*. In particular, $\mathcal{P}\cdot d\boldsymbol{\sigma}$ represents the electromagnetic energy per unit time crossing the oriented surface element $d\boldsymbol{\sigma}$.

9.1.6 Conservation of Momentum

Let us now study the law of conservation of momentum. This law states that, if the total momentum of the system in V (i.e. the sum of the mechanical momentum of the particles and the momentum associated with the fields) varies in time, then there must be a flow of electromagnetic momentum through the surface S surrounding V. We now translate this statement into a local balance equation. Starting from the inhomogeneous Maxwell equations

$$\nabla\cdot\mathbf{E}=4\pi\rho\,, \tag{9.42a}$$

$$\nabla\times\mathbf{B}-\frac{1}{c}\frac{\partial\mathbf{E}}{\partial t}=\frac{4\pi}{c}\mathbf{j}\,, \tag{9.42b}$$

from Eq. (9.42a) we get $\mathbf{E}(\nabla\cdot\mathbf{E})=4\pi\rho\mathbf{E}$. Multiplying vectorially on the left Eq. (9.42b) by **B** we have

$$\mathbf{B}\times(\nabla\times\mathbf{B})-\frac{1}{c}\mathbf{B}\times\frac{\partial\mathbf{E}}{\partial t}=-\frac{4\pi}{c}\mathbf{j}\times\mathbf{B}\,.$$

Subtracting member by member, we obtain

$$\frac{1}{4\pi}\left[\mathbf{E}(\nabla\cdot\mathbf{E})+\frac{1}{c}\mathbf{B}\times\frac{\partial\mathbf{E}}{\partial t}-\mathbf{B}\times(\nabla\times\mathbf{B})\right]=\rho\mathbf{E}+\frac{1}{c}\mathbf{j}\times\mathbf{B}\,. \tag{9.43}$$

The second term in square brackets can be rewritten as

$$\mathbf{B} \times \frac{\partial \mathbf{E}}{\partial t} = -\frac{\partial}{\partial t}(\mathbf{E} \times \mathbf{B}) + \mathbf{E} \times \frac{\partial \mathbf{B}}{\partial t} = -\frac{\partial}{\partial t}(\mathbf{E} \times \mathbf{B}) - c\,\mathbf{E} \times (\nabla \times \mathbf{E}), \quad (9.44)$$

where, in the second step, we used the Faraday-Neumann-Lenz law.

Let us now substitute Eq. (9.44) into Eq. (9.43) and add to the left-hand side the null quantity $\mathbf{B}(\nabla \cdot \mathbf{B})$,

$$\begin{aligned} &\frac{1}{4\pi c}\frac{\partial}{\partial t}(\mathbf{E} \times \mathbf{B}) + \rho \mathbf{E} + \frac{1}{c}\mathbf{j} \times \mathbf{B} \\ &\quad = \frac{1}{4\pi}\Big[\mathbf{E}(\nabla \cdot \mathbf{E}) - \mathbf{E} \times (\nabla \times \mathbf{E}) + \mathbf{B}(\nabla \cdot \mathbf{B}) - \mathbf{B} \times (\nabla \times \mathbf{B})\Big]. \end{aligned} \quad (9.45)$$

The vector identity $-\mathbf{E} \times (\nabla \times \mathbf{E}) = \frac{1}{2}\nabla \mathbf{E}^2 - (\mathbf{E} \cdot \nabla)\mathbf{E}$ allows us to write

$$E_i(\nabla \cdot \mathbf{E}) - [\mathbf{E} \times (\nabla \times \mathbf{E})]_i = \partial_j \left(E_i E_j - \frac{1}{2}\delta_{ij}\mathbf{E}^2 \right). \quad (9.46)$$

Similarly,

$$B_i(\nabla \cdot \mathbf{B}) - [\mathbf{B} \times (\nabla \times \mathbf{B})]_i = \partial_j \left(B_i B_j - \frac{1}{2}\delta_{ij}\mathbf{B}^2 \right), \quad (9.47)$$

and the quantity in square brackets in Eq. (9.45) is therefore the divergence of a tensor:

$$\frac{1}{4\pi}\left[\mathbf{E}(\nabla \cdot \mathbf{E}) - \mathbf{E} \times (\nabla \times \mathbf{E}) + \mathbf{B}(\nabla \cdot \mathbf{B}) - \mathbf{B} \times (\nabla \times \mathbf{B})\right]_i = -\partial_j \tau_{ij}. \quad (9.48)$$

Here

$$\tau_{ij} = \frac{1}{4\pi}\left[\frac{1}{2}\delta_{ij}(\mathbf{E}^2 + \mathbf{B}^2) - E_i E_j - B_i B_j\right] \quad (9.49)$$

is the so-called *Maxwell stress tensor* (a symmetric tensor of rank 2 in three-dimensional Euclidean space). From Eq. (9.45), we therefore derive

$$\frac{1}{4\pi c}\frac{\partial}{\partial t}(\mathbf{E} \times \mathbf{B})_i + \rho E_i + \frac{1}{c}(\mathbf{j} \times \mathbf{B})_i = -\partial_j \tau_{ij}. \quad (9.50)$$

Equation (9.50) is the local momentum-balance law for the coupled system of fields and charges. The term $(4\pi c)^{-1}\partial_t(\mathbf{E} \times \mathbf{B})$ is the time variation of electromagnetic momentum density, $\partial_j \tau_{ij}$ is the divergence of the momentum flux tensor, and $\rho \mathbf{E} + \frac{1}{c}\mathbf{j} \times \mathbf{B}$ is the Lorentz force density transferred to matter.

Let us denote by $\mathbf{P}_{mecc}$ the total mechanical momentum of the particles contained in V, i.e. the sum of the kinetic momenta of the individual particles. The variation of $\mathbf{P}_{mecc}$ with time, under the assumption that the particles cannot leave V nor enter it, is due to the Lorentz force exerted by the fields. We have (the sum is over the particles)

$$\frac{d\mathbf{P}_{mecc}}{dt} = \sum_a e_a \left[\mathbf{E}(\mathbf{z}_a, t) + \frac{1}{c}\mathbf{v}_a \times \mathbf{B}(\mathbf{z}_a, t)\right] = \int_V \left(\rho\mathbf{E} + \frac{1}{c}\mathbf{j} \times \mathbf{B}\right) d^3\mathbf{x} . \tag{9.51}$$

If we integrate Eq. (9.50) over the volume V and identify the quantity

$$\mathbf{P}_{em} = \frac{1}{4\pi c}\int_V (\mathbf{E} \times \mathbf{B})\, d^3\mathbf{x} \tag{9.52}$$

with the total momentum of the electromagnetic field in V, we obtain

$$\frac{d}{dt}(\mathbf{P}_{mecc} + \mathbf{P}_{em})_i = -\int_V \partial_j \tau_{ij}\, d^3\mathbf{x} ,$$

and applying Gauss's theorem to the right-hand side,

$$\frac{d}{dt}(\mathbf{P}_{mecc} + \mathbf{P}_{em})_i = -\int_S \tau_{ij}\, d\sigma_j . \tag{9.53}$$

Equation (9.53) is the conservation law of total momentum (mechanical plus electromagnetic). The integrand in Eq. (9.52) is the *electromagnetic momentum density*

$$\mathbf{\Pi}_{em} = \frac{1}{4\pi c}(\mathbf{E} \times \mathbf{B}) , \tag{9.54}$$

while the Maxwell tensor τ_{ij} represents the *momentum flux density* (in particular, τ_{ij} is the flux density of the i component of momentum in the j direction). Note that $\mathbf{\Pi}_{em}$ is proportional to the Poynting vector, which represents the energy flux density of the electromagnetic field:

$$\mathbf{\Pi}_{em} = \frac{1}{c^2}\boldsymbol{\mathcal{P}} . \tag{9.55}$$

Equation (9.50) can be written as

$$\frac{\partial \Pi_{em,i}}{\partial t} + \partial_j \tau_{ij} = -f_i , \tag{9.56}$$

where $\mathbf{f}$ is the density of the Lorentz force

$$\mathbf{f} = \rho\mathbf{E} + \frac{1}{c}\mathbf{j} \times \mathbf{B} . \tag{9.57}$$

In the free case, Eq. (9.56) reduces to

$$\frac{\partial \Pi_{em,i}}{\partial t} + \partial_j \tau_{ij} = 0\,, \tag{9.58}$$

which is the continuity equation for electromagnetic momentum.

Driving exercise
Radiation Pressure.
The momentum carried by an electromagnetic field manifests itself macroscopically through the phenomenon of *radiation pressure*. This effect was demonstrated experimentally at the beginning of the twentieth century by Nichols and Hull in the United States and, independently, by Lebedev in Russia. The existence of radiation pressure follows directly from the fact that electromagnetic waves transport not only energy but also linear momentum.
For an electromagnetic field in vacuum, the momentum density $\mathbf{\Pi}$ is related to the Poynting vector $\boldsymbol{\mathcal{P}}$ by

$$\mathbf{\Pi} = \frac{1}{c^2}\boldsymbol{\mathcal{P}}, \tag{9.59}$$

so that the momentum flux is controlled by the energy flux. Consider a plane wave normally incident on a flat surface. If the surface is perfectly absorbing, the wave transfers all of its incident momentum to the material. The force per unit area exerted on the surface is therefore equal to the rate at which momentum crosses the surface:

$$\text{pressure} = \frac{\text{force}}{\text{area}} = \frac{\text{momentum transferred per unit time}}{\text{area}} = \Pi\, c, \tag{9.60}$$

where $\mathcal{P}$ denotes the momentum density of the wave. Using $\mathbf{\Pi} = \boldsymbol{\mathcal{P}}/c^2$, one finds

$$\text{pressure} = \mathcal{P}c, \tag{9.61}$$

with $\mathcal{P} = |\boldsymbol{\mathcal{P}}|$ the magnitude of the Poynting vector. Since for a plane wave in vacuum $\mathcal{P} = c\,u$, where u is the electromagnetic energy density, this can also be written as

$$\text{pressure} = u\,. \tag{9.62}$$

If the surface is perfectly reflecting rather than absorbing, the momentum of the wave is reversed upon reflection. The change in momentum is therefore twice as large, and the radiation pressure becomes

$$\text{pressure}_{\text{refl}} = \frac{2\mathcal{P}}{c} = 2u\,. \tag{9.63}$$

Intermediate values occur for partially reflecting or transmitting materials, where the pressure depends on the reflection and absorption coefficients.

Radiation pressure plays a negligible role in most laboratory situations, but it becomes relevant in several physical contexts. It is essential in precision optical experiments, in optical trapping and manipulation of small particles, and in astrophysics, where it competes with gravity in luminous stars and drives the dynamics of dust and gas in intense radiation fields. A particularly striking application is the concept of light sails for space propulsion, where the pressure of sunlight or of intense laser beams is exploited to accelerate macroscopic objects.

9.1.7 Conservation of Angular Momentum

In light of Eq. (9.53) we can introduce the *angular momentum density* of the electromagnetic field,

$$\mathbf{l} = \mathbf{x} \times \mathbf{\Pi}_{em} = \frac{1}{c^2}\mathbf{x} \times \mathcal{P}\,. \tag{9.64}$$

When integrated over a volume V, this density gives the angular momentum of the field $\mathbf{J}_{em}$ contained in V,

$$\mathbf{J}_{em} = \int_V \mathbf{l}_{em}\, d^3\mathbf{x} = \frac{1}{4\pi c}\int_V [\mathbf{x} \times (\mathbf{E} \times \mathbf{B})]\, d^3\mathbf{x}\,. \tag{9.65}$$

Differentiating Eq. (9.64) with respect to time, we find

$$\frac{\partial \mathbf{l}_{em}}{\partial t} = \mathbf{x} \times \frac{\partial \mathbf{\Pi}_{em}}{\partial t} = \frac{1}{c^2}\mathbf{x} \times \frac{\partial \mathcal{P}}{\partial t}\,, \tag{9.66}$$

and, combining this relation with Eq. (9.56),

$$\frac{\partial l_{em,i}}{\partial t} + \varepsilon_{ijk} x_j \partial_l \tau_{kl} = -\varepsilon_{ijk} x_j f_k\,. \tag{9.67}$$

The term containing the stress tensor can be rewritten as

$$\begin{aligned} \varepsilon_{ijk} x_j \partial_l \tau_{kl} &= \varepsilon_{ijk}\partial_l (x_j \tau_{kl}) - \varepsilon_{ijk}(\partial_l x_j)\tau_{kl} \\ &= \varepsilon_{ijk}\partial_l (x_j \tau_{kl}) - \varepsilon_{ijk}\delta_{lj}\tau_{kl} \\ &= \varepsilon_{ijk}\partial_l (x_j \tau_{kl}) - \varepsilon_{ijk}\tau_{kj} \\ &= \varepsilon_{ijk}\partial_l (x_j \tau_{kl})\,, \end{aligned} \tag{9.68}$$

where the last step follows from the symmetry of τ_{ij}. Equation (9.67) therefore becomes

$$\frac{\partial l_{em,i}}{\partial t} + \partial_l(\varepsilon_{ijk} x_j \tau_{kl}) = -\varepsilon_{ijk} x_j f_k\,. \tag{9.69}$$

This is the local balance equation expressing the conservation of angular momentum. The tensor

$$\lambda_{il} = \varepsilon_{ijk} x_j \tau_{kl} \tag{9.70}$$

represents the *angular momentum flux density*. On the right-hand side of Eq. (9.69) we recognize the density of the moment (torque density) of the Lorentz force, $\mathbf{m} = \mathbf{x} \times \mathbf{f}$. Integrating over a volume V and applying Gauss's theorem, we obtain

$$\frac{dJ_{em,i}}{dt} = -\int_S \lambda_{il}\, d\sigma_l - M_i\,, \tag{9.71}$$

where $\mathbf{J}_{em}$ is the angular momentum of the electromagnetic field inside V and $\mathbf{M}$ is the moment of the Lorentz force. Equation (9.71) shows that the variation of the field angular momentum in V is controlled by two mechanisms: the flux of angular momentum through the boundary surface S, and the torque exerted by the field on the charges inside V. In the absence of charges (free case), Eq. (9.69) reduces to

$$\frac{\partial l_{em,i}}{\partial t} + \partial_l \lambda_{il} = 0\,, \tag{9.72}$$

which is the *continuity equation for angular momentum.*

The chain

$$\frac{\partial \Pi_{em,i}}{\partial t} + \partial_j \tau_{ij} = -f_i \implies \frac{\partial l_{em,i}}{\partial t} + \partial_l \lambda_{il} = -(\mathbf{x} \times \mathbf{f})_i \tag{9.73}$$

shows that angular momentum conservation is not an independent statement: it follows from momentum conservation once the stress tensor is symmetric. The symmetry of τ_{ij} is therefore the key structural input that guarantees the absence of an intrinsic torque term in the field sector.

A useful rewriting of Eq. (9.69) is

$$\frac{1}{c^2}\frac{\partial}{\partial t}\left(x_j \mathcal{P}_k - x_k \mathcal{P}_j\right) + \partial_l \left(x_j \tau_{kl} - x_k \tau_{jl}\right) = -\left(x_j f_k - x_k f_j\right)\,, \tag{9.74}$$

which is obtained using Eq. (9.66) and the antisymmetry of the symbol ε_{ijk}.

Let us now return to the expression for the angular momentum of the electromagnetic field, Eq. (9.65). Setting $\mathbf{B} = \nabla \times \mathbf{A}$, the vector product $\mathbf{E} \times \mathbf{B}$ becomes

$$\begin{aligned}(\mathbf{E} \times \mathbf{B})_i &= [\mathbf{E} \times (\nabla \times \mathbf{A})]_i = \varepsilon_{ijk} E_j (\nabla \times \mathbf{A})_k = \varepsilon_{ijk}\varepsilon_{klm} E_j \partial_l A_m \\ &= (\delta_{il}\delta_{jm} - \delta_{im}\delta_{jl}) E_j \partial_l A_m = E_j \partial_i A_j - E_j \partial_j A_i\,,\end{aligned} \tag{9.75}$$

namely $\mathbf{E} \times \mathbf{B} = E_j \nabla A_j - (\mathbf{E}\cdot\nabla)\mathbf{A}$. The integrand of Eq. (9.65) is therefore $\mathbf{x} \times (\mathbf{E} \times \mathbf{B}) = \mathbf{x} \times (E_j \nabla A_j) + ((\mathbf{E}\cdot\nabla)\mathbf{A}) \times \mathbf{x}$. Let us rewrite the last term as

$$E_i(\partial_i \mathbf{A}) \times \mathbf{x} = \partial_i (E_i \mathbf{A} \times \mathbf{x}) - (\mathbf{A} \times \mathbf{x})\partial_i E_i + E_i(\partial_i \mathbf{x}) \times \mathbf{A}\,. \tag{9.76}$$

A further manipulation gives

$$[E_i(\partial_i \mathbf{x}) \times \mathbf{A}]_j = E_i \varepsilon_{jkl}(\partial_i x_k) A_l = E_i \varepsilon_{jkl} \delta_{ik} A_l = \varepsilon_{jil} E_i A_l = (\mathbf{E} \times \mathbf{A})_j \,. \tag{9.77}$$

In the free case ($\partial_i E_i = 0$),

$$\mathbf{x} \times (\mathbf{E} \times \mathbf{B}) = \mathbf{x} \times (E_j \nabla A_j) + \partial_i (E_i \mathbf{A} \times \mathbf{x}) + \mathbf{E} \times \mathbf{A} \,. \tag{9.78}$$

Substituting this result into Eq. (9.65) we find

$$\mathbf{J}_{em} = \frac{1}{4\pi c} \int_V \left[\mathbf{x} \times (E_j \nabla A_j)\right] d^3\mathbf{x} + \frac{1}{4\pi c} \int_V (\mathbf{E} \times \mathbf{A}) d^3\mathbf{x} + \frac{1}{4\pi c} \int_V \partial_i (E_i \mathbf{A} \times \mathbf{x}) d^3\mathbf{x} \,. \tag{9.79}$$

Let us now suppose that the integrations are extended to all space. By means of Gauss's theorem we can transform the last volume integral in Eq. (9.79) into a surface integral, which is zero if the fields go to zero at infinity sufficiently rapidly. The angular momentum of the field therefore results composed of two terms

$$\mathbf{J}_{em} = \mathbf{L}_{em} + \mathbf{S}_{em} \,, \tag{9.80a}$$

$$\mathbf{L}_{em} = \frac{1}{4\pi c} \int \left[\mathbf{x} \times (E_j \nabla A_j)\right] d^3\mathbf{x} \,, \tag{9.80b}$$

$$\mathbf{S}_{em} = \frac{1}{4\pi c} \int (\mathbf{E} \times \mathbf{A}) d^3\mathbf{x} \,. \tag{9.80c}$$

Here $\mathbf{L}_{em}$ is the *orbital angular momentum*, while $\mathbf{S}_{em}$ is the so-called *spin angular momentum*. Note that $\mathbf{J}_{em}$ depends only on the fields, while $\mathbf{L}_{em}$ and $\mathbf{S}_{em}$ also depend on the potentials. The orbital and spin angular momenta are therefore not separately gauge invariant. This is not a contradiction: only the total conserved quantity $\mathbf{J}_{em}$ is observable and gauge invariant, whereas the splitting into "orbital" and "spin" parts depends on the chosen representative of the gauge equivalence class.

The decomposition $\mathbf{J}_{em} = \mathbf{L}_{em} + \mathbf{S}_{em}$ is algebraically useful and becomes particularly suggestive in optics (e.g. structured beams), but the two contributions are not individually gauge invariant. The gauge-invariant, conserved quantity is the total field angular momentum $\mathbf{J}_{em}$.

9.2 Interaction of a Charged Particle with the Electromagnetic Field

Let us now show that the equation of motion of a charged particle in an electromagnetic field, Eq. (9.17), can be derived from the Lagrangian (here $\mathbf{z}(t)$ is the trajectory of the particle and $\mathbf{v} = d\mathbf{z}/dt$ its velocity),

$$L = L_o - U(\mathbf{z}, \mathbf{v}) \,, \tag{9.81a}$$

$$L_0 = -mc^2\sqrt{1 - \frac{v^2}{c^2}} \,, \tag{9.81b}$$

$$U(\mathbf{z}, \mathbf{v}) = e\varphi(\mathbf{z}, t) - \frac{e}{c}\mathbf{v} \cdot \mathbf{A}(\mathbf{z}, t) \,. \tag{9.81c}$$

Here L_0 is the relativistic Lagrangian of a free particle, and U is the generalized potential encoding the interaction with the electromagnetic potentials. The Euler-Lagrange equation reads

$$\begin{aligned} &\frac{d}{dt}\frac{\partial L}{\partial \mathbf{v}} - \frac{\partial L}{\partial \mathbf{z}} = 0 \,, \\ &\Rightarrow \frac{d}{dt}\frac{\partial L_0}{\partial \mathbf{v}} = -\frac{\partial U}{\partial \mathbf{z}} + \frac{d}{dt}\frac{\partial U}{\partial \mathbf{v}} \,. \end{aligned} \tag{9.82}$$

Since

$$\frac{d}{dt}\frac{\partial L_0}{\partial \mathbf{v}} = \frac{d}{dt}(m\gamma\mathbf{v}) \,, \tag{9.83a}$$

$$\frac{\partial U}{\partial \mathbf{z}} + \frac{d}{dt}\frac{\partial U}{\partial \mathbf{v}} = -e\nabla\varphi + \frac{e}{c}\nabla(\mathbf{v} \cdot \mathbf{A}) - \frac{e}{c}\frac{d\mathbf{A}}{dt} \,, \tag{9.83b}$$

using $d\mathbf{A}/dt = \partial\mathbf{A}/\partial t + (\mathbf{v} \cdot \nabla)\mathbf{A}$ and the identity $\mathbf{v} \times (\nabla \times \mathbf{A}) = \nabla(\mathbf{v} \cdot \mathbf{A}) - (\mathbf{v} \cdot \nabla)\mathbf{A}$, Eq. (9.83b) becomes

$$-\frac{\partial U}{\partial \mathbf{z}} + \frac{d}{dt}\frac{\partial U}{\partial \mathbf{v}} = -e\nabla\varphi + \frac{e}{c}\left[\mathbf{v} \times (\nabla \times \mathbf{A})\right] - \frac{e}{c}\frac{\partial \mathbf{A}}{\partial t} \,. \tag{9.84}$$

Equating Eqs. (9.83a) and (9.84), and exploiting Eqs. (9.22) and (9.24), we finally find the Minkowski-Lorentz equation, Eq. (9.17). From the relativistic Lagrangian in Eq. (9.81a), which we rewrite explicitly for convenience,

$$L = -mc^2\sqrt{1 - \frac{v^2}{c^2}} - e\varphi + \frac{e}{c}\mathbf{v} \cdot \mathbf{A} \,, \tag{9.85}$$

we immediately obtain the *canonical momentum*,

$$\mathbf{P} \equiv \frac{\partial L}{\partial \mathbf{v}} = \frac{m\mathbf{v}}{\sqrt{1 - \frac{v^2}{c^2}}} + \frac{e}{c}\mathbf{A} \,. \tag{9.86}$$

The first term of Eq. (9.86) is the *kinetic momentum*, i.e. the momentum of a free particle $\mathbf{p} = m\gamma\mathbf{v}$. The relation between $\mathbf{P}$ and $\mathbf{p}$ is therefore $\mathbf{P} = \mathbf{p} + \frac{e}{c}\mathbf{A}$. Accordingly, the two momenta differ by a term proportional to the vector potential $\mathbf{A}$.

The Hamiltonian associated with the Lagrangian in Eq. (9.85) is

$$\begin{aligned} H &= \mathbf{P}\cdot\mathbf{v} - L\,,\\ &= \frac{mv^2}{\sqrt{1-\frac{v^2}{c^2}}} + \frac{e}{c}\mathbf{v}\cdot\mathbf{A} + mc^2\sqrt{1-\frac{v^2}{c^2}} + e\varphi - \frac{e}{c}\mathbf{v}\cdot\mathbf{A}\,,\\ &= \frac{mc^2}{\sqrt{1-\frac{v^2}{c^2}}} + e\varphi \equiv E\,. \end{aligned} \tag{9.87}$$

Equation (9.87) then displays the sum of the free particle energy $m\gamma c^2$ and the electric potential term $e\varphi$.

To express H as a function of $\mathbf{P}$, we observe that

$$H - e\varphi = \frac{mc^2}{\sqrt{1-\frac{v^2}{c^2}}}\,, \tag{9.88a}$$

$$\mathbf{P} - \frac{e}{c}\mathbf{A} = \frac{m\mathbf{v}}{\sqrt{1-\frac{v^2}{c^2}}}\,. \tag{9.88b}$$

Squaring and subtracting member by member we find

$$\begin{aligned} &\left(\frac{H-e\varphi}{c}\right)^2 - \left(\mathbf{P}-\frac{e}{c}\mathbf{A}\right)^2 = m^2c^2\,,\\ &\Rightarrow H = \left[\left(\mathbf{P}-\frac{e}{c}\mathbf{A}\right)^2 c^2 + m^2c^4\right]^{1/2} + e\varphi\,. \end{aligned} \tag{9.89}$$

Equation (9.89) is formally obtained from the Hamiltonian of a free particle by the minimal-coupling substitution $\mathbf{p} \to \mathbf{P} - \frac{e}{c}\mathbf{A}$ and by adding the electric contribution $e\varphi$.

Finally, let us verify the gauge invariance of the system. The presence of the electromagnetic potentials in the Lagrangian in Eq. (9.85) might suggest that such invariance does not hold, since evidently L changes under a gauge transformation. On the other hand, the equations of motion, Eq. (9.17), contain only the fields and therefore manifestly display gauge invariance. The point is that, performing the gauge transformations, the Lagrangian becomes

$$\begin{aligned} L \to L' &= -mc^2\sqrt{1-\frac{v^2}{c^2}} - e\varphi' + \frac{e}{c}\mathbf{v}\cdot\mathbf{A}'\,,\\ &= L + \frac{e}{c}\frac{\partial\chi}{\partial t} + \frac{e}{c}\mathbf{v}\cdot\nabla\chi\,,\\ &= L + \frac{e}{c}\frac{d\chi}{dt}\,. \end{aligned} \tag{9.90}$$

As can be seen, L' differs from L by a total time derivative. Therefore, although different, L' and L generate the same equations of motion: the particle-field system is thus invariant under gauge transformations.

▶ **Pause and think**
How would you handle the presence of dissipation?

Poynting's theorem.
The dissipated energy per unit of time and per unit volume is

$$\mathbf{j}\cdot\mathbf{E} = \frac{1}{\mu_0}(\nabla\times\mathbf{B})\cdot\mathbf{E} - \varepsilon_0\frac{\partial\mathbf{E}}{\partial t}\cdot\mathbf{E} \tag{9.91}$$

$$= -\nabla\cdot\left(\frac{\mathbf{E}\times\mathbf{B}}{\mu_0}\right) - \frac{\partial}{\partial t}\left(\frac{\varepsilon_0 E^2}{2} + \frac{B^2}{2\mu_0}\right). \tag{9.92}$$

This can be rewritten as:

$$\mathbf{j}\cdot\mathbf{E} = -\nabla\cdot\mathcal{P} - \frac{\partial u}{\partial t}, \tag{9.93}$$

where $\mathcal{P} = \frac{\mathbf{E}\times\mathbf{B}}{\mu_0}$, while $u = \frac{\varepsilon_0 E^2}{2} + \frac{B^2}{2\mu_0}$ is the energy per unit volume. If the Joule dissipation term $\mathbf{j}\cdot\mathbf{E}$ is vanishing, then the previous equation represents a continuity equation by interpreting $\mathcal{P}$ as the flow of energy per unit of area.

9.2.1 The Stress Tensor and Poynting Vector

▶ **Driving exercise**
The force per unit volume can be written starting from the Lorentz's force

$$\begin{aligned}
\mathbf{F} &= \rho\mathbf{E} + \mathbf{j}\times\mathbf{B} \\
&= \varepsilon_0\mathbf{E}(\nabla\cdot\mathbf{E}) + \frac{1}{\mu_0}(\nabla\times\mathbf{B})\times\mathbf{B} - \varepsilon_0\frac{\partial\mathbf{E}}{\partial t}\times\mathbf{B} \\
&= \varepsilon_0\mathbf{E}(\nabla\cdot\mathbf{E}) + \frac{1}{\mu_0}(\nabla\times\mathbf{B})\times\mathbf{B} - \varepsilon_0\frac{\partial(\mathbf{E}\times\mathbf{B})}{\partial t} - \varepsilon_0\mathbf{E}\times(\nabla\times\mathbf{E}) \\
&= \varepsilon_0\left[\mathbf{E}(\nabla\cdot\mathbf{E}) + (\mathbf{E}\cdot\nabla)\mathbf{E} - \frac{1}{2}\nabla(E^2)\right] \\
&\quad + \frac{1}{\mu_0}\left[\mathbf{B}(\nabla\cdot\mathbf{B}) + (\mathbf{B}\cdot\nabla)\mathbf{B} - \frac{1}{2}\nabla(B^2)\right] - \frac{1}{c^2}\frac{\partial}{\partial t}\left(\frac{\mathbf{E}\times\mathbf{B}}{\mu_0}\right).
\end{aligned} \tag{9.94}$$

The last equality can be obtained by taking into account the following formula

$$\mathbf{A} \times (\nabla \times \mathbf{A}) = -(\mathbf{A} \cdot \nabla)\mathbf{A} + \frac{1}{2}\nabla(A^2). \tag{9.95}$$

By writing Eq. (9.94) with the components

$$F_j = \sum_l \frac{\partial}{\partial x_l}\left[\varepsilon_0\left(E_j E_l - \frac{1}{2}\delta_{jl}E^2\right) + \frac{1}{\mu_0}\left(B_j B_l - \frac{1}{2}\delta_{jl}B^2\right)\right] - \frac{1}{c^2}\frac{\partial S_j}{\partial t}. \tag{9.96}$$

This expression can be interpreted as if the field had an impulse $\frac{1}{c^2}S_j$ for unit volume and a stress tensor T_{jl} (such as in the continuum medium mechanics)

$$F_j = \sum_l \frac{\partial T_{jl}}{\partial x_l} - \frac{1}{c^2}\frac{\partial S_j}{\partial t}, \tag{9.97}$$

where

$$T_{jl} = \varepsilon_0 E_j E_l + \frac{1}{\mu_0}B_j B_l - \frac{1}{2}u\delta_{jl} \tag{9.98}$$

is the stress-energy tensor in a non-covariant form.
Therefore the mechanical impulse change is the opposite of the change of the field impulse plus the stress tensor flow.

▶ **Driving exercise**
Poisson's differential equations.

$$\nabla^2\Phi(\mathbf{r}, t) = -4\pi f(\mathbf{r}, t). \tag{9.99}$$

If f and the Φ functions admit Fourier transform with respect to $\mathbf{r}$, then

$$-k^2\hat{\Phi}(\mathbf{k}, t) = -4\pi\hat{f}(\mathbf{k}, t) \Rightarrow \hat{\Phi}(\mathbf{k}, t) = \frac{4\pi\hat{f}(\mathbf{k}, t)}{k^2}. \tag{9.100}$$

The right-hand side of the last equation is then written as the product of the functions $\hat{f}(\mathbf{k}, t)$ and $\frac{4\pi}{k^2}$. Therefore, inverse Fourier transform $\Phi(\mathbf{r}, t)$ is the convolution product of the inverse transforms of $\hat{f}(\mathbf{k},t)$ and $\hat{G}(\mathbf{k}) = \frac{4\pi}{k^2}$, which are respectively $f(\mathbf{r},t)$ and

$$\begin{aligned} G(\mathbf{r}) &= \frac{1}{(2\pi)^3}\int \frac{4\pi}{k^2}e^{i\mathbf{k}\cdot\mathbf{r}}\,d\mathbf{k} \\ &= \frac{1}{2\pi^2}\int_0^\infty k^2\,dk\int_0^\pi e^{ikr\cos\theta}\sin\theta\,d\theta\int_0^{2\pi}d\phi = \frac{1}{r}. \end{aligned} \tag{9.101}$$

Finally, we obtain

$$\Phi(\mathbf{r},t) = \int \frac{f(\mathbf{r}',t)}{|\mathbf{r}-\mathbf{r}'|}\,d\mathbf{r}'. \tag{9.102}$$

▶ **Driving exercise**
The D'Alembert's equation and the retarded potential method. The D'Alembert differential equation is:

$$\nabla^2\Phi(\mathbf{r},t) - \frac{1}{c^2}\frac{\partial^2\Phi(\mathbf{r},t)}{\partial t^2} = -4\pi f(\mathbf{r},t). \tag{9.103}$$

This equation can be rewritten (by Fourier transforming) as:

$$\nabla^2\hat{\Phi}(\mathbf{r},\omega) + \frac{\omega^2}{c^2}\hat{\Phi}(\mathbf{r},\omega) = -4\pi\hat{f}(\mathbf{r},\omega). \tag{9.104}$$

Readily, we get:

$$-k^2\tilde{\Phi}(\mathbf{k},\omega) + \frac{\omega^2}{c^2}\tilde{\Phi}(\mathbf{k},\omega) = -4\pi\tilde{f}(\mathbf{k},\omega) \tag{9.105}$$

and then

$$\tilde{\Phi}(\mathbf{k},\omega) = \frac{4\pi\tilde{f}(\mathbf{k},\omega)}{k^2-\frac{\omega^2}{c^2}} = \tilde{G}(\mathbf{k},\omega)\tilde{f}(\mathbf{k},\omega), \tag{9.106}$$

$$\hat{\Phi}(\mathbf{r},\omega) = \int d^3r'\,\hat{G}(\mathbf{r}-\mathbf{r}',\omega)\hat{f}(\mathbf{r}',\omega), \tag{9.107}$$

$$\hat{G}(\mathbf{r},\omega) = \frac{1}{\pi}\int_0^\infty k^2\,dk\int_0^\pi \sin\theta\,d\theta\,e^{ikr\cos\theta}\frac{1}{k^2-\frac{\omega^2}{c^2}}. \tag{9.108}$$

These integrals can be evaluated by introducing a small imaginary contribution in ω. Depending on whether the imaginary part is positive or negative one obtains two different results:

$$\hat{G}_{\pm}(\mathbf{r},\omega) = \lim_{\varepsilon\to 0}\frac{1}{\pi}\int_0^\infty k^2\,dk\int_0^\pi \sin\theta\,d\theta\,e^{ikr\cos\theta}\frac{1}{(k-\frac{\omega\pm i\varepsilon}{c})(k+\frac{\omega\pm i\varepsilon}{c})}$$
$$= \lim_{\varepsilon\to 0}\frac{1}{i\pi r}\int_0^\infty \frac{e^{ikr}k\,dk}{(k-\frac{\omega\pm i\varepsilon}{c})(k+\frac{\omega\pm i\varepsilon}{c})} = \frac{e^{\pm i\frac{\omega}{c}r}}{r}, \tag{9.109}$$

and therefore:

$$\hat{\Phi}(\mathbf{r},\omega) = \int \frac{e^{\pm i\frac{\omega}{c}|\mathbf{r}-\mathbf{r}'|}}{|\mathbf{r}-\mathbf{r}'|}\hat{f}(\mathbf{r}',\omega)\,d^3r'. \tag{9.110}$$

By inverse transforming $\hat{\Phi}(\mathbf{r},\omega)$, we obtain

$$\Phi(\mathbf{r},t) = \frac{1}{2\pi}\int_{-\infty}^{\infty}\int \frac{e^{-i\omega t\pm i\frac{\omega}{c}|\mathbf{r}-\mathbf{r}'|}}{|\mathbf{r}-\mathbf{r}'|}\hat{f}(\mathbf{r}',\omega)\,d^3r' = \int \frac{f\left(\mathbf{r}',t\mp\frac{|\mathbf{r}-\mathbf{r}'|}{c}\right)}{|\mathbf{r}-\mathbf{r}'|}\,d\mathbf{r}', \tag{9.111}$$

which is the well known formula for the retarded and advanced potentials. By taking into account only the retarded potentials, we have:

$$\Phi(\mathbf{r},t) = \frac{1}{4\pi\varepsilon_0}\int \frac{\rho\left(\mathbf{r}',t-\frac{|\mathbf{r}-\mathbf{r}'|}{c}\right)}{|\mathbf{r}-\mathbf{r}'|}\,d^3r', \tag{9.112a}$$

$$\mathbf{A}(\mathbf{r},t) = \frac{\mu_0}{4\pi}\int \frac{\mathbf{j}\left(\mathbf{r}',t-\frac{|\mathbf{r}-\mathbf{r}'|}{c}\right)}{|\mathbf{r}-\mathbf{r}'|}\,d^3r'. \tag{9.112b}$$

Besides retarded solutions, we then see that advanced ones can also exist, which, however, do not typically satisfy boundary conditions. These retarded solutions seem to indicate that the potential depends on the charge density and on the current at previous times, as if the fields propagate at the finite speed c.

9.3 Radiation of Moving Charges

Starting from equations Eqs. (9.112), it is possible to study "classical" radiation. Any reasonably well-behaved function of time can be expanded in a Fourier integral. Therefore, we will study the radiation generated by a charge and current oscillating with a given angular frequency ω:

$$\rho(\mathbf{r}, t) = \rho(\mathbf{r}) \exp(-i\omega t), \tag{9.113a}$$

$$\mathbf{j}(\mathbf{r}, t) = \mathbf{j}(\mathbf{r}) \exp(-i\omega t), \tag{9.113b}$$

$$\Phi(\mathbf{r}, t) = \Phi(\mathbf{r}) \exp(-i\omega t), \tag{9.113c}$$

$$\mathbf{A}(\mathbf{r}, t) = \mathbf{A}(\mathbf{r}) \exp(-i\omega t), \tag{9.113d}$$

where

$$\Phi(\mathbf{r}) = \frac{1}{4\pi\varepsilon_0} \int \frac{\rho(\mathbf{r}') \exp\left(i\omega \frac{|\mathbf{r}-\mathbf{r}'|}{c}\right)}{|\mathbf{r}-\mathbf{r}'|} d^3r', \tag{9.114}$$

$$\mathbf{A}(\mathbf{r}) = \frac{\mu_0}{4\pi} \int \frac{\mathbf{j}(\mathbf{r}') \exp\left(i\omega \frac{|\mathbf{r}-\mathbf{r}'|}{c}\right)}{|\mathbf{r}-\mathbf{r}'|} d^3r'. \tag{9.115}$$

We can find approximate solutions when $|\mathbf{r}| \gg |\mathbf{r}'|$ and $\frac{\omega}{c} r' \ll 1$. In particular, we consider only two cases, dubbed "near" and "far", with respect to the dimensionless parameter $kr = \omega r / c$. The near zone is dominated by quasi-static contributions (fields not purely transverse), whereas the far zone isolates the radiative $1/r$ terms and yields transverse fields with $\mathbf{B} = \hat{\mathbf{n}} \times \mathbf{E}/c$, having $\frac{\omega}{c}|\mathbf{r}| \ll 1$ and $\frac{\omega}{c}|\mathbf{r}| \gg 1$.

Radiation in the Near Zone.

In this case, $\frac{\omega}{c}|\mathbf{r}| \ll 1$, and therefore we can approximate $\exp\left(i\omega \frac{|\mathbf{r}-\mathbf{r}'|}{c}\right) \approx 1$. Expanding $\frac{1}{|\mathbf{r}-\mathbf{r}'|}$ in a power series:

$$\frac{1}{|\mathbf{r}-\mathbf{r}'|} \approx \frac{1}{r} + \frac{\hat{\mathbf{n}} \cdot \mathbf{r}'}{r^2}, \quad \text{where} \quad \hat{\mathbf{n}} = \frac{\mathbf{r}}{r}, \tag{9.116}$$

$$\Phi(\mathbf{r}) \approx \frac{1}{4\pi\varepsilon_0 r} \int \rho(\mathbf{r}') d^3r' + \frac{\hat{\mathbf{n}}}{4\pi\varepsilon_0 r^2} \cdot \int \mathbf{r}' \rho(\mathbf{r}') d^3r', \tag{9.117}$$

$$\mathbf{A}(\mathbf{r}) \approx \frac{\mu_0}{4\pi r} \int \mathbf{j}(\mathbf{r}') d^3r' + \frac{\mu_0}{4\pi r^2} \int (\hat{\mathbf{n}} \cdot \mathbf{r}') \mathbf{j}(\mathbf{r}') d^3r'. \tag{9.118}$$

The first integrals give the total charge and the dipole moment of the radiating system:

$$Q = \int \rho(\mathbf{r}) d^3r, \tag{9.119}$$

$$\mathbf{P} = \int \mathbf{r}' \rho(\mathbf{r}') d^3r'. \tag{9.120}$$

In the second integral, the first term can be evaluated by $\sum_l \frac{\partial}{\partial x_l}(x_m j_l) = j_m + x_m \sum_l \frac{\partial j_l}{\partial x_l}$. Then, we get

$$\int \sum_l \frac{\partial}{\partial x_l}(x_m j_l) d^3 r = \int j_m d^3 r + \int x_m \sum_l \frac{\partial j_l}{\partial x_l} d^3 r. \tag{9.121}$$

The left-hand side can be transformed by means of Green's theorem into a surface integral, which is zero if the currents are bounded. Accordingly, we get

$$\int \mathbf{j} d^3 r = -\int \mathbf{r}(\nabla \cdot \mathbf{j}) d^3 r = \int \mathbf{r} \frac{\partial \rho}{\partial t} d^3 r = \frac{d\mathbf{P}}{dt}, \tag{9.122}$$

with **P** the dipole moment of the charge distribution. In the near zone, the electric field is then dipolar and, in general, not perpendicular to the radial direction.

Radiation in the Far Zone.

In this case $\frac{\omega}{c}|\mathbf{r}| \gg 1$, so one can make the approximation $\exp\left(i\omega \frac{|\mathbf{r}-\mathbf{r}'|}{c}\right) \approx e^{ikr}$. Using $\frac{1}{|\mathbf{r}-\mathbf{r}'|} \approx \frac{1}{r}$, Eq. (9.112) becomes:

$$\Phi(\mathbf{r}) \approx \frac{1}{4\pi\varepsilon_0} \frac{e^{ikr}}{r} \int \rho(\mathbf{r}') d^3 r' = \frac{1}{4\pi\varepsilon_0} \frac{e^{ikr}}{r} Q, \tag{9.123a}$$

$$\mathbf{A}(\mathbf{r}) \approx \frac{\mu_0}{4\pi} \frac{e^{ikr}}{r} \int \mathbf{j}(\mathbf{r}') d^3 r' = -i\omega \frac{\mu_0}{4\pi} \frac{e^{ikr}}{r} \mathbf{P}. \tag{9.123b}$$

If the total charge Q of the system is zero, more terms must be included in the expansion of the potential $\Phi(\mathbf{r})$. On the other hand, a total oscillating charge violates the conservation law of charge; the total charge must be constant. We can now approximate $|\mathbf{r} - \mathbf{r}'| \approx r - \hat{\mathbf{n}} \cdot \mathbf{r}'$ where $\hat{\mathbf{n}} = \frac{\mathbf{r}}{r}$. Then, $\exp\left(i\omega \frac{|\mathbf{r}-\mathbf{r}'|}{c}\right) \approx e^{ikr}(1 - ik\hat{\mathbf{n}} \cdot \mathbf{r}')$, giving

$$\begin{aligned}\Phi(\mathbf{r}) \approx \frac{1}{4\pi\varepsilon_0} \frac{e^{ikr}}{r} \int \rho(\mathbf{r}')(1 - ik\hat{\mathbf{n}} \cdot \mathbf{r}') d^3 r' &= -\frac{1}{4\pi\varepsilon_0} \frac{e^{ikr}}{r} ik\hat{\mathbf{n}} \cdot \int \rho(\mathbf{r}')\mathbf{r}' d^3 r' \\ &= -\frac{i}{4\pi\varepsilon_0} \frac{e^{ikr}}{r} k\hat{\mathbf{n}} \cdot \mathbf{P}. \end{aligned} \tag{9.124}$$

The **E** and **B** fields can be calculated from the potentials. Neglecting terms that decay faster than $\frac{1}{r}$:

$$\mathbf{B} = \nabla \times \mathbf{A} \approx \left(\frac{\omega^2 \mu_0}{4\pi c} \hat{\mathbf{n}} \times \mathbf{P} \right) \frac{e^{ikr}}{r}, \tag{9.125}$$

$$\mathbf{E} = -\nabla\Phi + i\omega\mathbf{A} \approx \left(-\frac{\omega^2 \mu_0}{4\pi} \mathbf{P} + \hat{\mathbf{n}} \frac{k^2}{4\pi\varepsilon_0} \hat{\mathbf{n}} \cdot \mathbf{P} \right) \frac{e^{ikr}}{r}. \tag{9.126}$$

Finally, recalling that $\frac{1}{\varepsilon_0 \mu_0} = c^2$ and $\mathbf{P} - \hat{\mathbf{n}}(\hat{\mathbf{n}} \cdot \mathbf{P}) = (\hat{\mathbf{n}} \times \mathbf{P}) \times \hat{\mathbf{n}}$, we get:

$$\mathbf{E} = -\nabla\Phi + i\omega\mathbf{A} \approx \left[\frac{k^2}{4\pi\varepsilon_0}(\hat{\mathbf{n}} \times \mathbf{P}) \times \hat{\mathbf{n}}\right]\frac{e^{ikr}}{r}. \tag{9.127}$$

Multipole Expansion.

We can decompose Eqs. (9.112) by expanding $\frac{\exp\left(i\omega\frac{|\mathbf{r}-\mathbf{r}'|}{c}\right)}{|\mathbf{r}-\mathbf{r}'|}$ in a multipole series:

$$\frac{\exp\left(i\omega\frac{|\mathbf{r}-\mathbf{r}'|}{c}\right)}{|\mathbf{r}-\mathbf{r}'|} = 4\pi ik\sum_{l=0}^{\infty} J_l(kr_<)H_l(kr_>)\sum_{m=-l}^{l} Y^*_{lm}(\theta',\varphi')Y_{lm}(\theta,\varphi), \tag{9.128}$$

where $r_< = \min(r, r')$ and $r_> = \max(r, r')$. In our case, $r \ll r'$ and therefore $r_< = r$ and $r_> = r'$.

So, the original integrals lead to the following relations

$$\Phi(r,\theta,\varphi) = \frac{ik}{\varepsilon_0}\sum_{l=0}^{\infty} H_l(kr)\sum_{m=-l}^{l} Y_{lm}(\theta,\varphi)$$
$$\int_0^r J_l(kr')Y^*_{lm}(\theta',\varphi')\rho(r',\theta',\varphi')r'^2\sin\theta' dr'd\theta'd\varphi', \tag{9.129}$$

$$\mathbf{A}(r,\theta,\varphi) = ik\mu_0\sum_{l=0}^{\infty} H_l(kr)\sum_{m=-l}^{l} Y_{lm}(\theta,\varphi)$$
$$\int_0^r J_l(kr')Y^*_{lm}(\theta',\varphi')\mathbf{j}(r',\theta',\varphi')r'^2\sin\theta' dr'd\theta'd\varphi, \tag{9.130}$$

where J_l are the spherical Bessel functions and H_l are the Hankel functions of the first kind.

Note that the spherical harmonics $Y_{lm}(\theta, \varphi)$ are irreducible tensors of order l and the vector potential **A** can also be represented as an irreducible tensor of order 1: $A_1 = A_x + iA_y$, $A_0 = A_z$, $A_{-1} = A_x - iA_y$.

► **Pause and think**
Try to compute by yourself the multipole expansion.

9.3.1 Multipole Radiation

In the radiation equations, Eqs. (9.123)–(9.124), the dipolar terms may vanish. In this case, the introduction of higher-order terms in the expansion of $e^{ik\hat{\mathbf{n}}\cdot\mathbf{r}'}$ is necessary. The next terms (after the dipole term) are the so-called magnetic dipole and electric quadrupole contributions, which arise from the next non-vanishing orders in the long-wavelength expansion:

$$\Phi(\mathbf{r}) \approx -\frac{1}{8\pi\varepsilon_0}\frac{e^{ikr}}{r}\int \rho(\mathbf{r}')k^2(\hat{\mathbf{n}}\cdot\mathbf{r}')^2 d^3r', \tag{9.131}$$

$$\mathbf{A}(\mathbf{r}) \approx -i\frac{\mu_0}{4\pi}\frac{e^{ikr}}{r}\int k(\hat{\mathbf{n}}\cdot\mathbf{r}')\mathbf{j}(\mathbf{r}')d^3r'. \tag{9.132}$$

We can rewrite the first as

$$\Phi(\mathbf{r}) \approx -\frac{k^2}{8\pi\varepsilon_0}\frac{e^{ikr}}{r}\sum_{lm} n_l n_m \int \rho(\mathbf{r}')x_l' x_m' d^3r'. \tag{9.133}$$

Then, we can apply the following relation

$$\frac{\partial(j_s x_l x_m)}{\partial x_s} = j_l x_m + j_m x_l + x_l x_m \sum_s \frac{\partial j_s}{\partial x_s} = j_l x_m + j_m x_l + i\omega x_l x_m \rho, \tag{9.134}$$

where, in the last step, the continuity equation in the frequency domain has been used, $\nabla\cdot\mathbf{j} = -\partial\rho/\partial t = i\omega\rho$ for the assumed time dependence $e^{-i\omega t}$.

Moreover, by taking into account that:

$$\begin{aligned}\sum_m n_m x_m j_l &= \sum_m n_m\left[\frac{1}{2}(x_m j_l + x_l j_m) + \frac{1}{2}(x_m j_l - x_l j_m)\right] \\ &= \sum_m \frac{n_m}{2}(x_m j_l + x_l j_m) + \frac{1}{2}\left[\hat{\mathbf{n}}\times(\mathbf{j}\times\mathbf{r})\right]_m,\end{aligned} \tag{9.135}$$

one separates a symmetric part, which is naturally associated with the electric quadrupole structure, from an antisymmetric part, which can be recast in terms of a cross product, being associated with the magnetic dipole structure.

Since $\mathbf{j}$ vanishes on the surface, one gets:

$$A_l \approx -\frac{\mu_0\omega k}{8\pi}\frac{e^{ikr}}{r}\sum_m n_m \int \rho(\mathbf{r}')x_l' x_m' d^3r' - i\frac{\mu_0}{8\pi}\frac{e^{ikr}}{r}\left[\mathbf{n}\times\int(\mathbf{j}\times\mathbf{r}')d^3r'\right]_l. \tag{9.136}$$

These two terms represent the (a) electric quadrupole radiation and the (b) magnetic dipole radiation, respectively.

The first non-dipolar radiative contributions can be recast into an electric-quadrupole-like term, proportional to $\sum_m n_m \int \rho(\mathbf{r}')x_l'x_m' d^3r'$, and a magnetic-dipole-like term, proportional to $\mathbf{n} \times \int (\mathbf{j} \times \mathbf{r}')\, d^3r'$, as displayed in the expression for A_l above.

9.3.1.1 The Intensity of the Electric Dipole Radiation

The time-averaged Poynting vector is given by

$$\mathcal{P} = \frac{\mathbf{E} \times \mathbf{B}^*}{2\mu_0} = \frac{ck^4}{32\pi^2\varepsilon_0 r^2}\left[P^2 - (\hat{\mathbf{n}} \cdot \mathbf{P})^2\right]\hat{\mathbf{n}} = \frac{ck^4}{32\pi^2\varepsilon_0 r^2} P^2 \sin^2\theta\, \hat{\mathbf{n}}. \tag{9.137}$$

Therefore, the radiated intensity per unit solid angle is $\frac{d\wp}{d\Omega} = \frac{ck^4}{32\pi^2\varepsilon_0}|\mathbf{P}|^2 \sin^2\theta$ and the total intensity is

$$\wp = \frac{ck^4}{12\pi\varepsilon_0}|\mathbf{P}|^2. \tag{9.138}$$

► **Driving exercise**

Let us now calculate the radiation emitted by an antenna and propagation in a medium. To calculate the power emitted by an antenna with respect to the length and the current, we start from defining the cross-section S and $I = I_0 \cos\left(\frac{\pi}{l}z\right) e^{-i\omega t}$ as the current. The current density is given by $j_z = \frac{I(z)}{A}$ and $j_x = j_y = 0$, so we obtain

$$\nabla \cdot \mathbf{j} = -\frac{\partial \rho}{\partial t} = i\omega\rho = \frac{\partial j_z}{\partial z} = -\frac{I_0}{A}\frac{\pi}{l} \sin\left(\frac{\pi}{l}z\right) e^{-i\omega t}. \tag{9.139}$$

Such charge distribution has dipole moment along the z axis:

$$P_z = \int_{-l/2}^{l/2} z\rho dz = \frac{i\pi}{\omega l} I_0 \int_{-l/2}^{l/2} z \sin\left(\frac{\pi}{l}z\right) e^{-i\omega t} dz = \frac{i2I_0 l}{\omega\pi} e^{-i\omega t}, \tag{9.140}$$

while $P_x = P_y = 0$. The radiating power per unit solid angle is:

$$\frac{d\wp}{d\Omega} = \frac{ck^4}{32\pi^2\varepsilon_0}\frac{4I_0^2 l^2}{\omega^2\pi^2} \sin^2\theta. \tag{9.141}$$

9.3.2 The Liénard-Wiechert Potentials

The method of retarded potentials can also be applied to the study of the potential generated by a charged point-like particle. If the particle has charge q and position $\mathbf{R}(t)$, then the charge density and current will be:

$$\rho(\mathbf{r},t)=q\delta(\mathbf{r}-\mathbf{R}(t)),\quad \mathbf{j}(\mathbf{r},t)=q\frac{d\mathbf{R}}{dt}\delta(\mathbf{r}-\mathbf{R}(t)), \tag{9.142}$$

and the $\mathbf{A}$ and Φ potentials will be:

$$\Phi(\mathbf{r},t)=\frac{1}{4\pi\varepsilon_0}\int\frac{q\delta\left[\mathbf{r}'-\mathbf{R}\left(t-\frac{|\mathbf{r}-\mathbf{r}'|}{c}\right)\right]}{|\mathbf{r}-\mathbf{r}'|}d^3r', \tag{9.143a}$$

$$\mathbf{A}(\mathbf{r},t)=\frac{\mu_0}{4\pi}\int\frac{q\frac{d\mathbf{R}}{dt}\delta\left[\mathbf{r}'-\mathbf{R}\left(t-\frac{|\mathbf{r}-\mathbf{r}'|}{c}\right)\right]}{|\mathbf{r}-\mathbf{r}'|}d^3r'. \tag{9.143b}$$

Driving exercise

The Potentials of Point-like Charges. In our case $\mathbf{r}'$ is the variable and $\mathbf{r}$ is only a parameter. $\mathbf{F}(\mathbf{r}')=\mathbf{r}'-\mathbf{R}\left(t-\frac{|\mathbf{r}-\mathbf{r}'|}{c}\right)$, then the Jacobian matrix is:

$$\begin{pmatrix}\frac{\partial F_x}{\partial x'} & \frac{\partial F_x}{\partial y'} & \frac{\partial F_x}{\partial z'}\\ \frac{\partial F_y}{\partial x'} & \frac{\partial F_y}{\partial y'} & \frac{\partial F_y}{\partial z'}\\ \frac{\partial F_z}{\partial x'} & \frac{\partial F_z}{\partial y'} & \frac{\partial F_z}{\partial z'}\end{pmatrix}=\begin{pmatrix}1-\frac{dX}{dt}\frac{n_x}{c} & -\frac{dX}{dt}\frac{n_y}{c} & -\frac{dX}{dt}\frac{n_z}{c}\\ -\frac{dY}{dt}\frac{n_x}{c} & 1-\frac{dY}{dt}\frac{n_y}{c} & -\frac{dY}{dt}\frac{n_z}{c}\\ -\frac{dZ}{dt}\frac{n_x}{c} & -\frac{dZ}{dt}\frac{n_y}{c} & 1-\frac{dZ}{dt}\frac{n_z}{c}\end{pmatrix}, \tag{9.144}$$

with $\hat{\mathbf{n}}=\frac{\mathbf{r}-\mathbf{r}'}{|\mathbf{r}-\mathbf{r}'|}$. The determinant of this matrix is given by $\left|\frac{\partial\mathbf{F}}{\partial\mathbf{r}'}\right|=1-\frac{\hat{\mathbf{n}}}{c}\cdot\frac{d\mathbf{R}}{dt}$. Therefore:

$$\Phi(\mathbf{r},t)=\frac{q}{4\pi\varepsilon_0|\mathbf{r}-\mathbf{R}(\bar{\tau})|}\frac{1}{\left|1-\frac{\hat{\mathbf{n}}}{c}\cdot\frac{d\mathbf{R}}{dt}\right|}, \tag{9.145a}$$

$$\mathbf{A}(\mathbf{r},t)=\frac{\mu_0 q\frac{d\mathbf{R}}{dt}}{4\pi|\mathbf{r}-\mathbf{R}(\bar{\tau})|}\frac{1}{\left|1-\frac{\hat{\mathbf{n}}}{c}\cdot\frac{d\mathbf{R}}{dt}\right|}, \tag{9.145b}$$

where the position $\mathbf{R}$ must be calculated at the retarded time $\bar{\tau}$, for which $t-\frac{|\mathbf{r}-\mathbf{R}(\bar{\tau})|}{c}=\bar{\tau}$. Naturally $\bar{\tau}$ depends on $\mathbf{r}$ and t, so:

$$\nabla\bar{\tau}=-\frac{\hat{\mathbf{n}}}{c}\left(1-\frac{\hat{\mathbf{n}}}{c}\cdot\frac{d\mathbf{R}}{dt}\right)^{-1}\quad\text{and}\quad\frac{d\bar{\tau}}{dt}=\left(1-\frac{\hat{\mathbf{n}}}{c}\cdot\frac{d\mathbf{R}}{dt}\right)^{-1}. \tag{9.146}$$

For example:

$$\nabla\times\mathbf{R}(\bar{\tau})=\frac{d\mathbf{R}}{dt}\times\nabla\bar{\tau}, \tag{9.147a}$$

$$\nabla\times\frac{d\mathbf{R}}{dt}=\frac{d^2\mathbf{R}}{dt^2}\times\nabla\bar{\tau}, \tag{9.147b}$$

$$\nabla|\mathbf{r}-\mathbf{R}(\bar{\tau})|=\hat{\mathbf{n}}-\hat{\mathbf{n}}\cdot\frac{d\mathbf{R}}{dt}\nabla\bar{\tau}=\hat{\mathbf{n}}\left(1-\frac{\hat{\mathbf{n}}}{c}\cdot\frac{d\mathbf{R}}{dt}\right), \tag{9.147c}$$

$$\nabla\left(1-\frac{\hat{\mathbf{n}}}{c}\cdot\frac{d\mathbf{R}}{dt}\right)=-\frac{1}{c}\nabla\left(\frac{\mathbf{r}-\mathbf{R}(\bar{\tau})}{|\mathbf{r}-\mathbf{R}(\bar{\tau})|}\cdot\frac{d\mathbf{R}}{dt}\right)$$
$$=-\frac{\frac{d\mathbf{R}}{dt}}{c|\mathbf{r}-\mathbf{R}(\bar{\tau})|}+\frac{\frac{d\mathbf{R}}{dt}\cdot\frac{d\mathbf{R}}{dt}}{c|\mathbf{r}-\mathbf{R}(\bar{\tau})|}\nabla\bar{\tau}$$
$$+\frac{\hat{\mathbf{n}}(\hat{\mathbf{n}}\cdot\frac{d\mathbf{R}}{dt})}{c|\mathbf{r}-\mathbf{R}(\bar{\tau})|^2}\left(1-\frac{\hat{\mathbf{n}}}{c}\cdot\frac{d\mathbf{R}}{dt}\right)-\frac{\hat{\mathbf{n}}}{c}\cdot\frac{d^2\mathbf{R}}{dt^2}\nabla\bar{\tau}. \tag{9.148}$$

Given the difficulty of performing these calculations directly, we can resort to the Jefimenko's equations:

$$\mathbf{E}(\mathbf{r},t)=\frac{1}{4\pi\varepsilon_0}\int\left[\frac{\rho\left(\mathbf{r}',t-\frac{|\mathbf{r}-\mathbf{r}'|}{c}\right)(\mathbf{r}-\mathbf{r}')}{|\mathbf{r}-\mathbf{r}'|^3}+\frac{1}{|\mathbf{r}-\mathbf{r}'|^2c}\frac{\partial\rho\left(\mathbf{r}',t-\frac{|\mathbf{r}-\mathbf{r}'|}{c}\right)(\mathbf{r}-\mathbf{r}')}{\partial t}\right.$$
$$\left.-\frac{1}{|\mathbf{r}-\mathbf{r}'|^2c}\frac{\partial\mathbf{J}\left(\mathbf{r}',t-\frac{|\mathbf{r}-\mathbf{r}'|}{c}\right)}{\partial t}\right]d^3r', \tag{9.149}$$

$$\mathbf{B}(\mathbf{r},t)=\frac{\mu_0}{4\pi}\int\left[\frac{\mathbf{J}\left(\mathbf{r}',t-\frac{|\mathbf{r}-\mathbf{r}'|}{c}\right)\times(\mathbf{r}-\mathbf{r}')}{|\mathbf{r}-\mathbf{r}'|^3}\right.$$
$$\left.+\frac{1}{|\mathbf{r}-\mathbf{r}'|^2c}\frac{\partial\mathbf{J}\left(\mathbf{r}',t-\frac{|\mathbf{r}-\mathbf{r}'|}{c}\right)\times(\mathbf{r}-\mathbf{r}')}{\partial t}\right]d^3r'. \tag{9.150}$$

▶ Driving exercise

The Electric and Magnetic Fields of a Point-like Charge. The electric and magnetic fields of a point-like charge are given by:

$$\mathbf{E}=\frac{q}{4\pi\varepsilon_0}\left[\frac{\hat{\mathbf{n}}-\boldsymbol{\beta}}{\gamma^2(1-\boldsymbol{\beta}\cdot\hat{\mathbf{n}})^3|\mathbf{r}-\mathbf{R}(\bar{\tau})|^2}\right]+\frac{q}{4\pi\varepsilon_0 c}\left[\frac{\hat{\mathbf{n}}\times\{(\hat{\mathbf{n}}-\boldsymbol{\beta})\times\frac{d\boldsymbol{\beta}}{dt}\}}{(1-\hat{\mathbf{n}}\cdot\boldsymbol{\beta})^3|\mathbf{r}-\mathbf{R}(\bar{\tau})|}\right] \tag{9.151}$$

and $\mathbf{B}=\frac{\hat{\mathbf{n}}\times\mathbf{E}}{c}$, where $\boldsymbol{\beta}=\frac{1}{c}\frac{d\mathbf{R}}{dt}$ and $\gamma=\frac{1}{\sqrt{1-\beta^2}}$. The two contributions to the electric and magnetic fields can be analyzed: a field that at large distance decreases as $\frac{1}{r^2}$, like the electrostatic field, and a field that decreases as $\frac{1}{r}$ and behaves like a radiation field. The latter is present only if there is acceleration. If the speed is small compared to that of light $\beta \ll 1$, the radiation fields will be given by:

$$\mathbf{E} \approx \frac{q}{4\pi\varepsilon_0 c}\left[\frac{\hat{\mathbf{n}}\times\left(\hat{\mathbf{n}}\times\frac{d\boldsymbol{\beta}}{dt}\right)}{|\mathbf{r}-\mathbf{R}(\bar{\tau})|}\right], \tag{9.152}$$

$$\mathbf{B} \approx \frac{\mu_0 q}{4\pi}\hat{\mathbf{n}}\times\left[\frac{\hat{\mathbf{n}}\times\left(\hat{\mathbf{n}}\times\frac{d\boldsymbol{\beta}}{dt}\right)}{|\mathbf{r}-\mathbf{R}(\bar{\tau})|}\right] = -\frac{\mu_0 q}{4\pi}\left[\frac{\left(\hat{\mathbf{n}}\times\frac{d\boldsymbol{\beta}}{dt}\right)}{|\mathbf{r}-\mathbf{R}(\bar{\tau})|}\right]. \tag{9.153}$$

From this one can see that the electromagnetic field is that of a dipole with acceleration $-\omega^2\mathbf{r}$. The Poynting vector in such a case is given by:

$$\mathcal{P} = \frac{q^2}{(4\pi)^2\varepsilon_0 c}\left[\left|\frac{d\boldsymbol{\beta}}{dt}\right|^2 - \left(\hat{\mathbf{n}}\cdot\frac{d\boldsymbol{\beta}}{dt}\right)^2\right]\hat{\mathbf{n}}. \tag{9.154}$$

9.3.3 Bremsstrahlung Radiation

The Bremsstrahlung radiation is characterized by the fact that $\frac{d\boldsymbol{\beta}}{dt} \parallel \boldsymbol{\beta}$. Therefore the radiation field is given by:

$$\mathbf{E} = \frac{q}{4\pi\varepsilon_0 c}\left[\frac{\hat{\mathbf{n}}\times\left\{(\hat{\mathbf{n}}-\boldsymbol{\beta})\times\frac{d\boldsymbol{\beta}}{dt}\right\}}{(1-\hat{\mathbf{n}}\cdot\boldsymbol{\beta})^3|\mathbf{r}-\mathbf{R}(\bar{\tau})|}\right] = \frac{q}{4\pi\varepsilon_0 c}\left[\frac{\hat{\mathbf{n}}\times\left(\hat{\mathbf{n}}\times\frac{d\boldsymbol{\beta}}{dt}\right)}{(1-\beta\cos\theta)^3|\mathbf{r}-\mathbf{R}(\bar{\tau})|}\right]. \tag{9.155}$$

Because $\mathbf{B} = \frac{\hat{\mathbf{n}}\times\mathbf{E}}{c}$, $\mathcal{P} = \frac{\mathbf{E}\times\mathbf{B}}{\mu_0} = \frac{E^2\hat{\mathbf{n}}-(\mathbf{E}\cdot\hat{\mathbf{n}})\mathbf{E}}{\mu_0 c}$, the radiating power per unit solid angle is:

$$\frac{dP}{d\Omega} = \hat{\mathbf{n}}\cdot\mathcal{P}r^2 = \frac{E^2-(\hat{\mathbf{n}}\cdot\mathbf{E})^2}{\mu_0 c}r^2. \tag{9.156}$$

The electric field is orthogonal to $\hat{\mathbf{n}}$ and in the plane containing $\hat{\mathbf{n}}$ and $\frac{d\boldsymbol{\beta}}{dt}$. The magnitude of the field is:

$$E = \frac{q}{4\pi\varepsilon_0 c}\frac{\left|\frac{d\boldsymbol{\beta}}{dt}\right|\sin\theta}{(1-\beta\cos\theta)^3 r}, \tag{9.157}$$

and therefore:

$$\frac{dP}{d\Omega} = \frac{q^2\left|\frac{d\boldsymbol{\beta}}{dt}\right|^2\sin^2\theta}{16\pi^2\varepsilon_0 c(1-\beta\cos\theta)^6}. \tag{9.158}$$

The emission per unit solid angle vanishes for $\theta = 0$ and $\theta = \pi$. The maximum occurs when $\cos\theta(1-\beta\cos\theta) - 3\beta(1-\cos^2\theta) = 0$, i.e., when $\cos\theta = \frac{-1\pm\sqrt{1+24\beta^2}}{4\beta}$.

9.3.4 Synchrotron Radiation

The synchrotron radiation is characterized by the fact that $\frac{d\boldsymbol{\beta}}{dt} \perp \boldsymbol{\beta}$. Then the radiation field is given by:

$$\mathbf{E} = \frac{q}{4\pi\varepsilon_0 c}\left[\frac{\hat{\mathbf{n}} \times \left\{(\hat{\mathbf{n}} - \boldsymbol{\beta}) \times \frac{d\boldsymbol{\beta}}{dt}\right\}}{(1 - \hat{\mathbf{n}}\cdot\boldsymbol{\beta})^3|\mathbf{r} - \mathbf{R}(\bar{\tau})|}\right], \tag{9.159}$$

$$\mathbf{B} = \frac{q}{4\pi\varepsilon_0 c^2}\left[\frac{(\boldsymbol{\beta} - \hat{\mathbf{n}}) \times \frac{d\boldsymbol{\beta}}{dt} - \left\{\hat{\mathbf{n}}\cdot\left(\boldsymbol{\beta} \times \frac{d\boldsymbol{\beta}}{dt}\right)\right\}\hat{\mathbf{n}}}{(1 - \hat{\mathbf{n}}\cdot\boldsymbol{\beta})^3|\mathbf{r} - \mathbf{R}(\bar{\tau})|}\right], \tag{9.160}$$

$$\mathcal{P} = \frac{q^2}{16\pi^2\varepsilon_0 c}\left[\frac{|\hat{\mathbf{n}} - \boldsymbol{\beta}|^2\left|\frac{d\boldsymbol{\beta}}{dt}\right|^2 - \left\{(\hat{\mathbf{n}} - \boldsymbol{\beta})\cdot\frac{d\boldsymbol{\beta}}{dt}\right\}^2 - \left\{\hat{\mathbf{n}}\cdot\left(\boldsymbol{\beta} \times \frac{d\boldsymbol{\beta}}{dt}\right)\right\}^2}{(1 - \hat{\mathbf{n}}\cdot\boldsymbol{\beta})^6|\mathbf{r} - \mathbf{R}(\bar{\tau})|^2}\right]\hat{\mathbf{n}}. \tag{9.161}$$

If we suppose that $\boldsymbol{\beta}$ is along the z-axis, $\frac{d\boldsymbol{\beta}}{dt}$ is along the x-axis and the origin is such that $\mathbf{R}(\bar{\tau}) = 0$ and, in the xz plane, i.e., $\varphi = 0$:

$$\frac{dP}{d\Omega} \propto \frac{\cos^2\theta - 2\beta\cos\theta + \beta^2}{(1 - \beta\cos\theta)^6} = \frac{(\beta - \cos\theta)^2}{(1 - \beta\cos\theta)^6}, \tag{9.162}$$

which is zero for $\cos\theta = \beta$ and has two maxima for $\theta = 0$ and $\theta = \pi$. In the yz plane, i.e., $\varphi = \pi/2$, we have $\frac{dP}{d\Omega} \propto \frac{1}{(1-\beta\cos\theta)^4}$, which has a maximum for $\theta = 0$ and a minimum for $\theta = \pi$.

9.4 Radiation Reaction and the Self-force

A charged particle that accelerates radiates, and the radiated energy-momentum must be balanced by a recoil on the particle. This back-reaction is the radiation-reaction (self-)force. The key point is that the Lorentz force law with an externally prescribed field is incomplete once the particle's own field is included.

9.4.1 From Larmor Power to a Damping Force

In the instantaneous rest frame of a nonrelativistic charge, the radiated power is (Larmor formula)

$$P = \frac{2e^2}{3c^3}\,\mathbf{a}^2, \qquad \mathbf{a} = \dot{\mathbf{v}}. \tag{9.163}$$

Assume a self-force of the form $\mathbf{F}_{\rm rr} = m\tau\,\dot{\mathbf{a}}$ with

$$\tau \equiv \frac{2e^2}{3mc^3}. \tag{9.164}$$

Then the mechanical power lost to this force is $\mathbf{F}_{\rm rr} \cdot \mathbf{v} = m\tau\,\dot{\mathbf{a}} \cdot \mathbf{v} = m\tau\,\frac{d}{dt}(\mathbf{a} \cdot \mathbf{v}) - m\tau\,\mathbf{a}^2$. Up to a total time derivative (the Schott term), the power loss is $m\tau\,\mathbf{a}^2$. Using Eq. (9.164), this equals the radiated power in Eq. (9.163):

$$m\tau\,\mathbf{a}^2 = \frac{2e^2}{3c^3}\,\mathbf{a}^2 = P. \tag{9.165}$$

This identifies the nonrelativistic Abraham-Lorentz radiation-reaction force

$$\mathbf{F}_{\rm rr} = \frac{2e^2}{3c^3}\,\dot{\mathbf{a}}. \tag{9.166}$$

9.4.2 Covariant Abraham-Lorentz-Dirac Equation

Let $u^\mu = dx^\mu/d\tau$ be the four-velocity and $a^\mu = du^\mu/d\tau$ the four-acceleration. The covariant self-force consistent with orthogonality $u_\mu a^\mu = 0$ takes the Dirac form

$$f_{\rm rr}^\mu = \frac{2e^2}{3c^3}\left(\frac{da^\mu}{d\tau} + \frac{a_\nu a^\nu}{c^2}\,u^\mu\right). \tag{9.167}$$

The equation of motion becomes

$$ma^\mu = \frac{e}{c}F^{\mu\nu}u_\nu + f_{\rm rr}^\mu. \tag{9.168}$$

Equations (9.166)–(9.168) reveal a conceptual limit of classical electrodynamics: higher derivatives introduce runaway and pre-acceleration solutions unless one adopts a reduction-of-order scheme (Landau-Lifshitz approximation), effectively treating radiation reaction as a perturbative correction.

9.5 Wave Propagation in a Material Medium

Until now we have treated the electromagnetic field generated by charges and currents in vacuum. This approach is very general, but difficult to use in material media where the electric and magnetic fields are significantly variable at the microscopic level. In these cases, it is better to make a spatial average of the electric and magnetic fields over a small, but macroscopic, volume.

If $\mathbf{e}$ is the microscopic electric field, the macroscopic one $\mathbf{E}$ will be given by

$$\mathbf{E}(\mathbf{r}, t) = \int f(\mathbf{r} - \mathbf{r}')\mathbf{e}(\mathbf{r}')d^3r', \tag{9.169}$$

where $f(\mathbf{r})$ is a function different from zero only inside a small macroscopic volume V such that $\int f(\mathbf{r})d^3r = 1$. Analogously if $\mathbf{b}$ is the magnetic microscopic induction, the macroscopic one $\mathbf{B}$ will be given by

$$\mathbf{B}(\mathbf{r}, t) = \int f(\mathbf{r} - \mathbf{r}')\mathbf{b}(\mathbf{r}')d^3r'. \tag{9.170}$$

The Maxwell equations can be rewritten in terms of the fields $\mathbf{E}$ and $\mathbf{B}$. In many cases, two new vector fields can be introduced: $\mathbf{D} = \varepsilon\mathbf{E}$ and $\mathbf{H} = \frac{\mathbf{B}}{\mu}$. In such cases, the Poynting vector can be written as $\mathcal{P} = \mathbf{E} \times \mathbf{H}$. The momentum of the field is then given by $\mathbf{g} = \frac{\mathcal{P}}{c^2}$, while the stress tensor reads

$$T_{jl} = \left[E_j D_l + H_j B_l - \frac{1}{2}(\mathbf{E} \cdot \mathbf{D} + \mathbf{B} \cdot \mathbf{H})\delta_{jl}\right]. \tag{9.171}$$

9.5.1 The Kramers-Kronig Relations

If we consider a physical device for which the outgoing signal $h(t)$ is linearly dependent on the incoming signal $f(t)$, we have the general expression

$$h(t) = \int_{-\infty}^{+\infty} G(t - t')f(t')dt'. \tag{9.172}$$

If the response $h(t)$ of the system to an external perturbation $f(t)$ is causal, then $G(\tau) = 0$ for $\tau < 0$. Therefore:

$$h(t) = \int_{-\infty}^{t} G(t - t')f(t')dt'. \tag{9.173}$$

If we perform a Fourier transform, we can write $\hat{h}(\omega) = \hat{G}(\omega)\hat{f}(\omega)$, where $\hat{G}(\omega)$ is analytic in the upper complex half-plane. If the $\hat{G}(\omega)$ function converges uniformly to zero in the complex semicircle in the upper half-plane:

$$\hat{G}(\omega) = \lim_{\varepsilon \to 0} \hat{G}(\omega + i\varepsilon) = \frac{1}{2\pi i}\int_{-\infty}^{+\infty} \frac{\hat{G}(\omega')}{\omega' - \omega - i\varepsilon}d\omega', \tag{9.174}$$

then

$$\pi i\hat{G}(\omega) = \mathcal{P}\int_{-\infty}^{+\infty} \frac{\hat{G}(\omega')}{\omega' - \omega} d\omega'. \tag{9.175}$$

The Dispersion and the Group Velocity.
These relations can be developed if $G(\tau)$ is real. In such a case $\hat{G}(\omega) = \hat{G}(-\omega)^*$:

$$\Re\hat{G}(\omega) = \frac{2}{\pi}\mathcal{P}\int_{0}^{+\infty} \frac{\omega'\Im\hat{G}(\omega')}{\omega'^2 - \omega^2} d\omega', \tag{9.176}$$

$$\Im\hat{G}(\omega) = -\frac{2\omega}{\pi}\int_{0}^{+\infty} \frac{\Re\hat{G}(\omega')}{\omega'^2 - \omega^2} d\omega', \tag{9.177}$$

which are called the Kramers-Kronig relations.
The Dielectric Constants in Dispersive Media. Suppose we can write a dielectric constant ε dependent on the frequency ω, such that:

$$\mathbf{D}(\mathbf{r}, \omega) = \varepsilon(\omega)\mathbf{E}(\mathbf{r}, \omega), \tag{9.178}$$

$$\mathbf{B}(\mathbf{r}, \omega) = \mu(\omega)\mathbf{H}(\mathbf{r}, \omega). \tag{9.179}$$

At fixed frequency, the Maxwell equations become

$$\nabla^2\mathbf{E} + \varepsilon(\omega)\mu(\omega)\omega^2\mathbf{E} = 0, \tag{9.180}$$

$$\nabla^2\mathbf{B} + \varepsilon(\omega)\mu(\omega)\omega^2\mathbf{B} = 0. \tag{9.181}$$

9.5.2 The Group Velocity

We analyze fields by introducing Fourier transforms with respect to space, $\mathbf{E}(\mathbf{r}, t) = \frac{1}{(2\pi)^3}\int \hat{\mathbf{E}}(\mathbf{k})e^{i(\mathbf{k}\cdot\mathbf{r}-\omega(\mathbf{k})t)}d^3k$, and $\mathbf{B}(\mathbf{r}, t) = \frac{1}{(2\pi)^3}\int \hat{\mathbf{B}}(\mathbf{k})e^{i(\mathbf{k}\cdot\mathbf{r}-\omega(\mathbf{k})t)}d^3k$, where $\omega(\mathbf{k})$ is defined by means of the implicit relation $\omega^2\varepsilon(\omega)\mu(\omega) - k^2 = 0$.

The mean value of the position of the wave packet will be

$$\langle\mathbf{r}\rangle = \frac{\int \mathbf{r}\left\{\mathbf{E}(\mathbf{r}, t)\cdot\mathbf{D}(\mathbf{r}, t) + \mathbf{B}\cdot\mathbf{H}(\mathbf{r}, t)\right\} d^3r}{\int \left\{\mathbf{E}(\mathbf{r}, t)\cdot\mathbf{D}(\mathbf{r}, t) + \mathbf{B}\cdot\mathbf{H}(\mathbf{r}, t)\right\} d^3r}, \tag{9.182}$$

where $\int \left\{\mathbf{E}(\mathbf{r}, t)\cdot\mathbf{D}(\mathbf{r}, t) + \mathbf{B}\cdot\mathbf{H}(\mathbf{r}, t)\right\} d^3r$ is constant with respect to time.

On the other hand:

$$r_j E_l(\mathbf{r},t) = \frac{1}{(2\pi)^3}\int\left[i\frac{\partial}{\partial k_j}\hat{E}_l(\mathbf{k}) + \frac{\partial\omega}{\partial k_j}t\right]e^{i(\mathbf{k}\cdot\mathbf{r}-\omega(\mathbf{k})t)}d^3k, \tag{9.183}$$

$$r_j B_l(\mathbf{r},t) = \frac{1}{(2\pi)^3}\int\left[i\frac{\partial}{\partial k_j}\hat{B}_l(\mathbf{k}) + \frac{\partial\omega}{\partial k_j}t\right]e^{i(\mathbf{k}\cdot\mathbf{r}-\omega(\mathbf{k})t)}d^3k. \tag{9.184}$$

So:

$$\begin{aligned}\langle r_j\rangle &= \frac{i\int\sum_j\left\{\hat{E}_l^*(\mathbf{k})\frac{\partial\hat{D}_l(\mathbf{k})}{\partial k_j} + \hat{B}_l^*(\mathbf{k})\frac{\partial\hat{H}_l(\mathbf{k})}{\partial k_j}\right\}d^3k}{\int\sum_j\left\{\hat{E}_l^*(\mathbf{k})\hat{D}_l(\mathbf{k}) + \hat{B}_l^*(\mathbf{k})\hat{H}_l(\mathbf{k})\right\}d^3k} \\ &\quad + t\frac{\int\frac{\partial\omega}{\partial k_j}\sum_j\left\{\hat{E}_l^*(\mathbf{k})\hat{D}_l(\mathbf{k}) + \hat{B}_l^*(\mathbf{k})\hat{H}_l(\mathbf{k})\right\}d^3k}{\int\sum_j\left\{\hat{E}_l^*(\mathbf{k})\hat{D}_l(\mathbf{k}) + \hat{B}_l^*(\mathbf{k})\hat{H}_l(\mathbf{k})\right\}d^3k} \\ &= r_{0,j} + v_{0,j}t.\end{aligned} \tag{9.185}$$

Therefore, if $\hat{\mathbf{E}}(\mathbf{k})$ and $\hat{\mathbf{B}}(\mathbf{k})$ are different from zero in a small region around $\mathbf{k}_0$:

$$\mathbf{v}_0 \approx \nabla_k\omega|_{\mathbf{k}=\mathbf{k}_0}. \tag{9.186}$$

9.6 The Electromagnetic Field Tensor

The fields **E** and **B** are the components of a rank-2 tensor, the *electromagnetic field tensor* $F^{\mu\nu}$

$$F^{\mu\nu} = \partial^\mu A^\nu - \partial^\nu A^\mu. \tag{9.187}$$

By definition, $F^{\mu\nu}$ is antisymmetric:

$$F^{\mu\nu} = \partial^\mu A^\nu - \partial^\nu A^\mu = -\left(\partial^\nu A^\mu - \partial^\mu A^\nu\right) = -F^{\nu\mu}. \tag{9.188}$$

Consequently, the components on the diagonal are zero ($F^{ii} = 0$), while those off the diagonal are related by $F^{ij} = -F^{ji}$. $F^{\mu\nu}$ therefore has only *six independent components*, which correspond to the three components of **E** and the three components of **B**. We will determine the $F^{\mu\nu}$ using the relations linking the potentials φ and **A** to the fields **E** and **B**:

$$\mathbf{E} = -\nabla\varphi - \frac{1}{c}\frac{\partial\mathbf{A}}{\partial t}, \tag{9.189a}$$

$$\mathbf{B} = \nabla\times\mathbf{A}. \tag{9.189b}$$

For the components F^{i0} we have

$$F^{i0} = \partial^i A^0 - \partial^0 A^i = -\partial_i \varphi - \frac{1}{c}\frac{\partial A^i}{\partial t} = E^i, \tag{9.190a}$$

$$F^{10} = E_x, \qquad F^{20} = E_y, \qquad F^{30} = E_z. \tag{9.190b}$$

The components with spatial indices $F^{ij} = \partial^i A^j - \partial^j A^i$, are

$$F^{12} = \partial^1 A^2 - \partial^2 A^1 = -\left(\partial_x A_y - \partial_y A_x\right) = -B_z, \tag{9.191a}$$

$$F^{23} = \partial^2 A^3 - \partial^3 A^2 = -\left(\partial_y A_z - \partial_z A_y\right) = -B_x, \tag{9.191b}$$

$$F^{31} = \partial^3 A^1 - \partial^1 A^3 = -\left(\partial_z A_x - \partial_x A_z\right) = B_y, \tag{9.191c}$$

$$F^{ij} = -\varepsilon^{ijk} B^k. \tag{9.191d}$$

The tensor $F^{\mu\nu}$ therefore has the structure

$$F^{\mu\nu} = \begin{pmatrix} 0 & E_x & E_y & E_z \\ -E_x & 0 & -B_z & B_y \\ -E_y & B_z & 0 & -B_x \\ -E_z & -B_y & B_x & 0 \end{pmatrix}. \tag{9.192}$$

In particular, $F^{\mu\nu}$ is invariant under gauge transformations, which can be easily verified.

Then, by using

$$A'^{\mu}(x) = A^{\mu}(x) - \partial^{\mu}\chi(x), \tag{9.193}$$

where χ is a scalar function of x^{μ}, we obtain

$$\begin{aligned} F^{\mu\nu} \to F'^{\mu\nu} &= \partial^{\mu} A'^{\nu} - \partial^{\nu} A'^{\mu} \\ &= \partial^{\mu}\left(A^{\nu} - \partial^{\nu}\chi\right) - \partial^{\nu}\left(A^{\mu} - \partial^{\mu}\chi\right) \\ &= \partial^{\mu} A^{\nu} - \partial^{\nu} A^{\mu} - \partial^{\mu}\partial^{\nu}\chi + \partial^{\nu}\partial^{\mu}\chi \\ &= \partial^{\mu} A^{\nu} - \partial^{\nu} A^{\mu} = F^{\mu\nu}. \end{aligned} \tag{9.194}$$

By means of the tensors $F^{\mu\nu}$ and $\widetilde{F}^{\mu\nu}$ we can construct the invariants $F^{\mu\nu}F_{\mu\nu}$, $\widetilde{F}^{\mu\nu}F_{\mu\nu}$, defined by

$$F^{\mu\nu}F_{\mu\nu} = -2\left(\mathbf{E}^2 - \mathbf{B}^2\right), \tag{9.195a}$$

$$\widetilde{F}^{\mu\nu}F_{\mu\nu} = -4\mathbf{E}\cdot\mathbf{B}. \tag{9.195b}$$

It is then seen that, if **E** and **B** have the same magnitude and/or are orthogonal in a certain inertial reference frame (i.e., if $|\mathbf{E}| = |\mathbf{B}|$ and/or $\mathbf{E}\cdot\mathbf{B} = 0$), they have the same magnitude and/or are orthogonal in all inertial reference frames.

9.7 The Maxwell Equations in Covariant Form

Electromagnetic theory is intrinsically consistent with the principle of relativity. The simplest and most elegant way to appreciate this fact is to rewrite the laws of electromagnetism using the covariant formalism.

9.7.1 The Action of Electrodynamics and the Four-Potential

The Lagrangian of a charged particle in an electromagnetic field is given by Eq. (9.85). The corresponding action $S = \int L dt$ is made up of two terms $S = S_0 + S_{int}$. The first is the free particle action, $S_0 = -mc \int ds$; the second is the interaction term $S_{int} = -\frac{e}{c} \int (\varphi c dt - \mathbf{A} \cdot d\mathbf{z})$. The principle of relativity states that the equations of motion obtained from S must have the same form in all inertial reference frames, and therefore S_{int}, like S_0, must be a Lorentz invariant. This is verified if φ and $\mathbf{A}$ are the components of a four-vector

$$A^{\mu} = (\varphi, \mathbf{A}). \tag{9.196}$$

Then

$$S_{int} = -\frac{e}{c} \int A_{\mu}(z) dz^{\mu} = -\frac{e}{c} \int A_{\mu}(z) \frac{dz^{\mu}}{ds} ds \tag{9.197}$$

turns out to be indeed a Lorentz scalar. The action S is therefore

$$S = -mc \int ds - \frac{e}{c} \int A_{\mu}(z) dz^{\mu}. \tag{9.198}$$

The four-vector $A^{\mu}(x)$ is the *electromagnetic four-potential*, its temporal component A^0 is the scalar potential φ, while its spatial components are the components of the vector potential $\mathbf{A}$. The fact that φ and $\mathbf{A}$ constitute a four-vector and that S_{int} is invariant is, as we said, a consequence of the principle of relativity, and therefore, ultimately, an empirical fact (since the principle of relativity, as we never tire of repeating, is based on experiment). The *gauge transformations*

$$\varphi' = \varphi - \frac{1}{c} \frac{\partial \chi}{\partial t}, \tag{9.199a}$$

$$\mathbf{A}' = \mathbf{A} + \nabla \chi, \tag{9.199b}$$

rewritten for A^{μ} we obtain Eq. (9.193).

It is easy to verify that, under a gauge transformation, the interaction action S_{int} acquires an additive constant term which has no effect on the equations of motion: this is the *gauge invariance* of electrodynamics that we have already discussed.

Substituting A^{μ} with A'^{μ}, Eq. (9.197) becomes indeed (we explicitly indicate the integration volume Ω)

$$S_{int} \to S'_{int} = -\frac{e}{c}\int_{\Omega} A_{\mu}(z)dz^{\mu} + \frac{e}{c}\int_{\Omega} \frac{\partial \chi}{\partial z^{\mu}} dz^{\mu} = S_{int} + \frac{e}{c}\chi\,[\partial\Omega]\,, \tag{9.200}$$

where $\chi\,[\partial\Omega]$ is the function χ evaluated on the boundary surface $\partial\Omega$ of the volume Ω. We then have $\delta S' = \delta S$, and the equations of motion remain invariant.

9.7.2 A Deeper Look at the Four-Current Density

In order to complete the covariant description of electromagnetism, we need a further ingredient: a four-vector containing the sources, i.e., the charge density ρ and the current density $\mathbf{j}$. Let us consider, for simplicity, a point source (a particle of charge e). The charge density and current density are then given by Eqs. (9.20).

The *four-current density* (or simply *four-current*) J^{μ} is defined as

$$J^{\mu}(x) = (\rho c, \rho \mathbf{v}) = e\frac{dz^{\mu}}{dt}\delta^{3}(\mathbf{x} - \mathbf{z}(t)) = \rho\frac{dz^{\mu}}{dt}, \tag{9.201}$$

with $x^{\mu} = (ct, \mathbf{x})$, $z^{\mu}(t) = (ct, \mathbf{z}(t))$. Although not directly apparent from Eq. (9.201), J^{μ} is, as we will now verify, a four-vector. Due to the invariance of electric charge, the charge density ρ in the reference frame $\mathcal{H}$ where the particle has velocity $\mathbf{v}$ satisfies

$$e = \int_{V} \rho_0 d^3\mathbf{x}' = \int_{V} \rho d^3\mathbf{x}, \tag{9.202}$$

where ρ_0 is the charge density in the rest frame $\mathcal{H}'$ of the particle and V is the volume containing the particle. Taking into account volume contraction, namely $\int_V \rho_0 d^3\mathbf{x}' = \int_V \rho_0\gamma(v)d^3\mathbf{x}$, we readily find

$$\rho = \rho_0\gamma(v) = \rho_0 c\frac{dt}{ds}. \tag{9.203}$$

We can thus rewrite J^{μ} as

$$\begin{aligned} J^{\mu}(x) &= e\frac{dz^{\mu}}{dt}\delta^{3}(\mathbf{x} - \mathbf{z}(t)) \\ &= e\int_{-\infty}^{+\infty} dt'\frac{dz^{\mu}}{dt'}\delta(t - t')\delta^{3}(\mathbf{x} - \mathbf{z}(t')) \\ &= ec\int_{-\infty}^{+\infty} dt'\frac{dz^{\mu}}{dt'}\delta^{4}(x - z(t')), \end{aligned} \tag{9.204}$$

with $z^{\mu}(t') = (ct', z(t'))$. Moving to the invariant variables, Eq. (9.204) becomes

$$J^{\mu}(x) = ec\int_{-\infty}^{+\infty} ds\frac{dz^{\mu}}{ds}\delta^{4}(x - z(s)), \tag{9.205}$$

which is the covariant form of the four-current density associated with a particle having as worldline $z^\mu(s) = (z^0, \mathbf{z}(s))$. In general, the charge density ρ is, up to a factor c, the temporal component of the four-current J^μ, while the current density $\mathbf{j}$ desribes the spatial components of J^μ, i.e. $J^\mu = (J^0, \mathbf{J}) = (c\rho, \mathbf{j})$. The *continuity equation* $\frac{\partial \rho}{\partial t} + \nabla \cdot \mathbf{j} = 0$ can be expressed in a covariant way by $\partial_\mu J^\mu(x) = 0$.

This is, as is obvious, a covariant law, since the conservation of electric charge must hold in all reference frames. It is verified by direct substitution that Eq. (9.205) satisfies the continuity equation. We have indeed

$$\begin{aligned}
\partial_\mu J^\mu(x) &= ec \int_{-\infty}^{+\infty} ds \frac{dz^\mu}{ds} \frac{\partial}{\partial x^\mu} \delta^4(x - z(s)) \\
&= -ec \int_{-\infty}^{+\infty} ds \frac{dz^\mu}{ds} \frac{\partial}{\partial z^\mu} \delta^4(x - z(s)) \\
&= -ec \int_{-\infty}^{+\infty} ds \frac{d}{ds} \delta^4(x - z(s)) \\
&= -ec \left[\delta^4(x - z(s))\right]_{s=-\infty}^{s=+\infty} \\
&= -ec \left[\delta(x^0 - z^0(s)) \delta^3(\mathbf{x} - \mathbf{z}(s))\right]_{s=-\infty}^{s=+\infty},
\end{aligned} \tag{9.206}$$

and, since for $s = \pm\infty$ the delta functions have no support, it follows that $\partial_\mu J^\mu(x) = 0$. By means of J^μ we can rewrite the interaction action Eq. (9.197) between a particle and the electromagnetic field in the form

$$\begin{aligned}
S_{int} &= -\frac{e}{c} \int ds A_\mu(z) \frac{dz^\mu}{ds} = -\frac{e}{c} \int ds \int d^4x A_\mu(x) \delta^4(x - z(s)) \frac{dz^\mu}{ds} \\
&= -\frac{1}{c^2} \int d^4x A_\mu(x) J^\mu(x).
\end{aligned} \tag{9.207}$$

9.7.3 The Dual Tensor

Let us now introduce the *dual tensor* of the electromagnetic field $\widetilde{F}^{\mu\nu}$, defined as

$$\widetilde{F}^{\mu\nu} = \frac{1}{2} \varepsilon^{\mu\nu\rho\sigma} F_{\rho\sigma}. \tag{9.208}$$

$\widetilde{F}^{\mu\nu}$ is also an antisymmetric tensor. Its components are

$$\begin{aligned}
&\widetilde{F}^{i0} = B^i, \qquad \widetilde{F}^{ij} = \varepsilon^{ijk} E^k, \\
&\widetilde{F}^{\mu\nu} = \begin{pmatrix} 0 & B_x & B_y & B_z \\ -B_x & 0 & E_z & -E_y \\ -B_y & -E_z & 0 & E_x \\ -B_z & E_y & -E_x & 0 \end{pmatrix}.
\end{aligned} \tag{9.209}$$

As can be noted, Eq. (9.209) is obtained from Eq. (9.192) with the substitutions $\mathbf{E} \to \mathbf{B}$ and $\mathbf{B} \to -\mathbf{E}$.

Dual Maxwell equations.
The covariant Maxwell equations take the form

$$\partial_\mu F^{\mu\nu} = \frac{4\pi}{c}\, j^\nu, \tag{9.210}$$

$$\partial_\mu \widetilde{F}^{\mu\nu} = 0. \tag{9.211}$$

Equation (9.211) represents the set of *dual Maxwell equations.*

9.7.4 Homogeneous Equations

The homogeneous Maxwell equations in covariant form are

$$\partial_\mu \widetilde{F}^{\mu\nu} = 0. \tag{9.212}$$

For $\nu = 0$ we have indeed $\partial_i \widetilde{F}^{i0} = 0$ which, recalling Eq. (9.209) is equivalent to $\partial_i B^i = 0$, i.e. $\nabla \cdot \mathbf{B} = 0$.

On the other hand, for $\nu = i$ we find $\partial_0 \widetilde{F}^{0i} + \partial_j \widetilde{F}^{ji} = 0$, that is, by virtue of Eq. (9.209), $\varepsilon^{ijk}\partial_j E^k + \partial_0 B^i = 0$, which is the Faraday-Neumann-Lenz law $\nabla \times \mathbf{E} + (1/c)(\partial \mathbf{B}/\partial t) = 0$. In terms of the tensor $F^{\mu\nu}$, the homogeneous Maxwell equations take the form

$$\varepsilon^{\mu\nu\rho\sigma}\partial_\mu F_{\rho\sigma} = 0 \Leftrightarrow \partial_\mu F_{\rho\sigma} + \partial_\rho F_{\sigma\mu} + \partial_\sigma F_{\mu\rho} = 0. \tag{9.213}$$

We have indeed

$$\begin{aligned} 0 &= \varepsilon^{\mu\nu\rho\sigma}\left(\partial_\mu F_{\rho\sigma} + \partial_\rho F_{\sigma\mu} + \partial_\sigma F_{\mu\rho}\right) = \left(\varepsilon^{\mu\nu\rho\sigma} + \varepsilon^{\sigma\nu\mu\rho} + \varepsilon^{\rho\nu\sigma\mu}\right)\partial_\mu F_{\rho\sigma} \\ &= 3\varepsilon^{\mu\nu\rho\sigma}\partial_\mu F_{\rho\sigma} = 6\partial_\mu \widetilde{F}^{\mu\nu}, \end{aligned} \tag{9.214}$$

where, in the second equality, we changed the names of the dummy indices of the Levi-Civita symbol, and, in the last step, we used the properties of $\varepsilon^{\mu\nu\rho\sigma}$.

9.7.5 Inhomogeneous Equations

We are now able to write the Maxwell equations in covariant form. The covariant expression of the inhomogeneous Maxwell equations is

$$\partial_\mu F^{\mu\nu} = \frac{4\pi}{c} J^\nu. \tag{9.215}$$

We can verify that, for $\nu = 0$, we recover Gauss' law, while $\nu = i$ gives the Ampère-Maxwell law.

Setting $\nu = 0$, the left-hand side of Eq. (9.215) becomes $\partial_\mu F^{\mu 0} = \partial_i F^{i0} = \partial_i E^i = \nabla \cdot \mathbf{E}$, while the right-hand side is $(4\pi/c)J^0 = 4\pi\rho$, so that we indeed derive $\nabla \cdot \mathbf{E} = 4\pi\rho$.

For $\nu = i$ we find on the left-hand side $\partial_\mu F^{\mu i} = \partial_0 F^{0i} + \partial_j F^{ji} = -\partial_0 E^i + \varepsilon^{ijk}\partial_j B^k = -(1/c)(\partial E^i/\partial t) + (\nabla \times \mathbf{B})^i$, and therefore $-(1/c)(\partial E^i/\partial t) + (\nabla \times \mathbf{B})^i = (4\pi/c)j^i$ i.e. $\nabla \times \mathbf{B} - (1/c)(\partial \mathbf{E}/\partial t) = (4\pi/c)\mathbf{j}$.

Substituting into the inhomogeneous Maxwell equations the expression Eq. (9.187) for the field tensor, we obtain the covariant equation of motion for the four-potential. On the left-hand side we have

$$\partial_\mu F^{\mu\nu} = \partial_\mu \left(\partial^\mu A^\nu - \partial^\nu A^\mu\right) = \Box A^\nu - \partial^\nu(\partial_\mu A^\mu), \tag{9.216}$$

and therefore the equation for A^μ is

$$\Box A^\nu - \partial^\nu \left(\partial_\mu A^\mu\right) = \frac{4\pi}{c} J^\nu. \tag{9.217}$$

If we adopt the Lorenz gauge, $\partial_\mu A^\mu = 0$, Eq. (9.217) gives the inhomogeneous d'Alembert equation

$$\Box A^\nu = \frac{4\pi}{c} J^\nu. \tag{9.218}$$

9.8 Gauge Invariance and Potentials

The relations in Eqs. (9.22), (9.24) do not uniquely determine the potentials φ and $\mathbf{A}$. Given the fields $\mathbf{E}$ and $\mathbf{B}$, indeed, there correspond infinitely many electromagnetic potentials. Take for example Eq. (9.22). The magnetic field $\mathbf{B}$ does not change if the vector potential undergoes a transformation of the type

$$\mathbf{A}(\mathbf{x}, t) \to \mathbf{A}'(\mathbf{x}, t) = \mathbf{A}(\mathbf{x}, t) + \nabla\chi(\mathbf{x}, t), \tag{9.219}$$

where $\chi(\mathbf{x}, t)$ is an arbitrary scalar function. The curl of a gradient is indeed zero, $\nabla \times \nabla\chi = 0$, and we therefore have

$$\mathbf{B} \to \mathbf{B}' = \nabla \times \mathbf{A}' = \nabla \times \mathbf{A} + \nabla \times \nabla\chi = \nabla \times \mathbf{A} = \mathbf{B}. \tag{9.220}$$

Recalling Eq. (9.24), we see that, following the transformation Eq. (9.219), the electric field $\mathbf{E}$ changes as

$$\mathbf{E} \to \mathbf{E}' = -\nabla\varphi - \frac{1}{c}\frac{\partial \mathbf{A}'}{\partial t} = -\nabla\varphi - \frac{1}{c}\frac{\partial \mathbf{A}}{\partial t} - \frac{1}{c}\frac{\partial}{\partial t}\nabla\chi = \mathbf{E} - \frac{1}{c}\frac{\partial}{\partial t}\nabla\chi. \tag{9.221}$$

If then, simultaneously with Eq. (9.219), we perform the transformation

$$\varphi(\mathbf{x}, t) \rightarrow \varphi'(\mathbf{x}, t) = \varphi(\mathbf{x}, t) - \frac{1}{c}\frac{\partial \chi(\mathbf{x}, t)}{\partial t}, \tag{9.222}$$

$\mathbf{E}$ does not change. Indeed:

$$\mathbf{E} \rightarrow \mathbf{E}' = -\nabla\varphi' - \frac{1}{c}\frac{\partial \mathbf{A}'}{\partial t} = -\nabla\varphi + \frac{1}{c}\nabla\frac{\partial \chi}{\partial t} - \frac{1}{c}\frac{\partial \mathbf{A}}{\partial t} - \frac{1}{c}\frac{\partial}{\partial t}\nabla\chi = -\nabla\varphi - \frac{1}{c}\frac{\partial \mathbf{A}}{\partial t} = \mathbf{E}. \tag{9.223}$$

The transformations Eqs. (9.219)–(9.222) of the potentials, called *gauge transformations*, therefore leave the fields invariant and, consequently, the Maxwell equations. They thus represent a *symmetry* of electromagnetism. Gauge invariance gives us the freedom to choose the potentials in the most convenient manner. That is, we can impose on φ and $\mathbf{A}$ conditions (the *gauge conditions*) that allow us to simplify the description of the system. Among the most common gauge choices are the *Lorenz gauge*, the *Coulomb gauge*, the *temporal gauge*, the *radiation gauge*. Before describing these gauges, let us separate the vector potential $\mathbf{A}$ into its longitudinal and transverse components $\mathbf{A} = \mathbf{A}_L + \mathbf{A}_T$. The gauge transformation Eq. (9.219) is equivalent to (recall that $\nabla\chi$ is purely longitudinal)

$$\mathbf{A'}_T = \mathbf{A}_T \quad , \quad \mathbf{A'}_L = \mathbf{A}_L + \nabla\chi. \tag{9.224}$$

We thus see that the transverse vector potential $\mathbf{A}_T$ is gauge invariant, unlike $\mathbf{A}_L$ and φ which are instead gauge-dependent. While it is possible to eliminate $\mathbf{A}_L$ and/or φ through gauge transformations, the components of $\mathbf{A}_T$ are dynamical degrees of freedom and therefore cannot be eliminated.

9.8.1 Lorenz Gauge

The Lorenz gauge condition explicitly reads

$$\nabla \cdot \mathbf{A} + \frac{1}{c}\frac{\partial \varphi}{\partial t} = 0. \tag{9.225}$$

This relation greatly simplifies Eq. (9.28) for $\mathbf{A}$, which becomes the inhomogeneous d'Alembert equation,

$$\nabla^2\mathbf{A} - \frac{1}{c^2}\frac{\partial^2 \mathbf{A}}{\partial t^2} = -\frac{4\pi}{c}\mathbf{j}. \tag{9.226}$$

Furthermore, substituting Eq. (9.225) into Eq. (9.26), we find

$$\nabla^2\varphi - \frac{1}{c^2}\frac{\partial^2 \varphi}{\partial t^2} = -4\pi\rho, \tag{9.227}$$

which has the same form as Eq. (9.226). In the absence of sources, Eqs. (9.226)–(9.227) give the *wave equations* (homogeneous d'Alembert equations)

$$\nabla^2 \mathbf{A} - \frac{1}{c^2}\frac{\partial^2 \mathbf{A}}{\partial t^2} = 0, \tag{9.228a}$$

$$\nabla^2 \varphi - \frac{1}{c^2}\frac{\partial^2 \varphi}{\partial t^2} = 0. \tag{9.228b}$$

The main advantage of the Lorenz gauge relies in its *relativistic covariance*. The left-hand side of Eq. (9.225) is indeed a Lorentz invariant. Thus, if the condition Eq. (9.225) is imposed in a certain inertial system, it holds in any other inertial system. The Lorenz gauge, however, has the drawback of not removing the non-dynamical variables φ and $\mathbf{A}_L$. It is important to note that Eq. (9.225) does not fix the potentials uniquely. A certain gauge freedom remains, albeit more restricted. Given a pair of potentials φ and $\mathbf{A}$ that satisfy Eq. (9.225), it is possible, by means of a gauge transformation, to obtain new potentials φ' and $\mathbf{A}'$ for which the Lorenz condition still holds

$$\nabla \cdot \mathbf{A}' + \frac{1}{c}\frac{\partial \varphi'}{\partial t} = 0. \tag{9.229}$$

For this to happen, it is necessary that the function χ generating this gauge transformation satisfies the wave equation

$$\nabla^2 \chi - \frac{1}{c^2}\frac{\partial^2 \chi}{\partial t^2} = 0. \tag{9.230}$$

Substituting Eqs. (9.219) and (9.222) into (9.229), and exploiting Eq. (9.225), one finds indeed Eq. (9.230). We notice that Eq. (9.229) allows us to write one of the four quantities φ, A_x, A_y, A_z in terms of the others. By means of a gauge transformation generated by a given $\chi(\mathbf{x}, t)$ satisfying Eq. (9.230), it is then possible to eliminate a further variable. Accordingly, the overall independent dynamical variables are two, as we already know.

9.8.2 Coulomb Gauge

The *Coulomb gauge* consists of setting

$$\nabla \cdot \mathbf{A} = 0, \tag{9.231}$$

i.e., in eliminating the longitudinal component of the vector potential $\mathbf{A}_L = 0$, so that $\mathbf{A} = \mathbf{A}_T$. This gauge, unlike the Lorenz gauge, is not covariant: Eq. (9.231) can only hold in a particular inertial reference frame. Using the condition (9.231) in Eq. (9.26), we find that φ satisfies *Poisson's equation*

$$\nabla^2 \varphi = -4\pi\rho. \tag{9.232}$$

In the Coulomb gauge, therefore, φ is the instantaneous Coulomb potential

$$\varphi(\mathbf{x}, t) = \int \frac{\rho(\mathbf{x}', t)}{|\mathbf{x} - \mathbf{x}'|} d^3\mathbf{x}', \tag{9.233}$$

which, more precisely, is the solution of Eq. (9.232) in the absence of boundary surfaces at finite distances. The equation of motion in Eq. (9.28) for $\mathbf{A}$ becomes, in the Coulomb gauge

$$\nabla^2\mathbf{A} - \frac{1}{c^2}\frac{\partial^2\mathbf{A}}{\partial t^2} = -\frac{4\pi}{c}\mathbf{j} + \frac{1}{c}\nabla\frac{\partial\varphi}{\partial t}. \tag{9.234}$$

Differentiating Eq. (9.232) with respect to time and combining it with the charge continuity equation in the form Eq. (9.13), we obtain

$$\nabla \cdot \left(\nabla\frac{\partial\varphi}{\partial t} - 4\pi\mathbf{j}_L\right) = 0. \tag{9.235}$$

Since the quantity in parentheses has zero divergence and curl (and therefore zero Laplacian), under the hypothesis that φ and $\mathbf{j}_L$ go to zero at infinity sufficiently rapidly, Eq. (9.235) implies

$$\nabla\frac{\partial\varphi}{\partial t} - 4\pi\mathbf{j}_L = 0. \tag{9.236}$$

Using this relation in Eq. (9.234), we finally find

$$\nabla^2\mathbf{A} - \frac{1}{c^2}\frac{\partial^2\mathbf{A}}{\partial t^2} = -\frac{4\pi}{c}\mathbf{j}_T. \tag{9.237}$$

In the Coulomb gauge, the relation (9.24) between the electric field and the potentials separates into two relations, one for the longitudinal component and the other for the transverse component:

$$\mathbf{E}_L = -\nabla\varphi \quad , \quad \mathbf{E}_T = -\frac{1}{c}\frac{\partial\mathbf{A}}{\partial t}. \tag{9.238}$$

It is immediately verified that, given a potential $\mathbf{A}$ that satisfies Eq. (9.231), a gauge transformation generated by a function $\chi(\mathbf{x}, t)$ with $\nabla^2\chi = 0$ produces a potential $\mathbf{A}'$ that is still in the Coulomb gauge. If we require that the transformation $\mathbf{A} \to \mathbf{A}'$ is everywhere regular, the only class of solutions with zero Laplacian is $\chi =$ constant, and we thus have $\mathbf{A} = \mathbf{A}'$: in this case, the Coulomb gauge is unique.

9.8.3 Temporal Gauge

In the temporal gauge, the scalar potential is set to zero

$$\varphi = 0. \tag{9.239}$$

In this gauge, which is not covariant, the potential equations become

$$\frac{\partial}{\partial t}\nabla \cdot \mathbf{A} = -4\pi c\rho, \tag{9.240a}$$

$$\nabla^2 \mathbf{A} - \frac{1}{c^2}\frac{\partial^2 \mathbf{A}}{\partial t^2} - \nabla(\nabla \cdot \mathbf{A}) = -\frac{4\pi}{c}\mathbf{j}. \tag{9.240b}$$

9.8.4 Radiation Gauge

In the absence of sources, it is possible to require that both Eqs. (9.231) and (9.239) hold, i.e.

$$\nabla \cdot \mathbf{A} = 0 \quad , \quad \varphi = 0. \tag{9.241}$$

This is known as *radiation gauge*, in which Maxwell's equation for $\mathbf{A}$ is the wave equation

$$\nabla^2 \mathbf{A} - \frac{1}{c^2}\frac{\partial^2 \mathbf{A}}{\partial t^2} = 0. \tag{9.242}$$

We will now show that the two conditions in Eq. (9.241) are compatible if $\rho = 0$. Given generic potentials φ and $\mathbf{A}$, we perform a gauge transformation generated by the function $\chi(\mathbf{x}, t) = c\int_{t_0}^{t} \varphi(\mathbf{x}, t')dt'$, which gives the temporal gauge condition. By means of a further gauge transformation

$$\varphi'' = \varphi' - \frac{1}{c}\frac{\partial \chi'}{\partial t} \quad , \quad \mathbf{A}'' = \mathbf{A}' - \nabla\chi', \tag{9.243}$$

we find

$$\nabla \cdot \mathbf{A}'' = \nabla \cdot \mathbf{A}' - \nabla^2\chi' = 0. \tag{9.244}$$

A solution of Poisson's equation Eq. (9.244) for χ' is

$$\chi'(\mathbf{x}', t) = \frac{1}{4\pi}\int \frac{\nabla' \cdot \mathbf{A}'(\mathbf{x}', t)}{|\mathbf{x}' - \mathbf{x}|} d^3\mathbf{x}', \tag{9.245}$$

where $\nabla' = \partial/\partial\mathbf{x}'$. If $\rho = 0$, Eq. (9.240a), valid for $\mathbf{A}'$ in the temporal gauge, becomes $\partial/\partial t(\nabla \cdot \mathbf{A}') = 0$ and consequently, from Eq. (9.245) $\partial\chi'/\partial t = 0$. Then Eq. (9.243) gives $\varphi'' = \varphi' = 0$. We have thus shown that, if $\rho = 0$, it is possible to choose the potentials in such a way to satisfy the radiation gauge.

9.9 Electromagnetism as the Prototype Interaction

Electromagnetism is the simplest nontrivial relativistic field theory: a massless spin-1 gauge field coupled to a conserved current. Three structural properties make it the prototype of modern interaction theory: long-range interactions, gauge redundancy, and local symmetry as a constructive principle.

► Driving exercise

Range from the mediator mass. At the level of potentials, the distinction between long- and short-range forces is already visible. Consider a static source and a scalar potential $\phi(\mathbf{x})$. A massive mediator generates a screened Poisson equation (Helmholtz equation),

$$\left(\nabla^2 - \mu^2\right)\phi(\mathbf{x}) = -4\pi\,\rho(\mathbf{x}), \tag{9.246}$$

while electromagnetism corresponds to $\mu = 0$. For a point charge $\rho(\mathbf{x}) = q\,\delta^{(3)}(\mathbf{x})$, solve Eq. (9.246) by Fourier transform:

$$\phi(\mathbf{x}) = \int \frac{d^3k}{(2\pi)^3}\,\tilde{\phi}(\mathbf{k})\,e^{i\mathbf{k}\cdot\mathbf{x}}, \tag{9.247}$$

$$\left(-k^2 - \mu^2\right)\tilde{\phi}(\mathbf{k}) = -4\pi q, \qquad \Rightarrow \qquad \tilde{\phi}(\mathbf{k}) = \frac{4\pi q}{k^2 + \mu^2}. \tag{9.248}$$

Thus

$$\phi(\mathbf{x}) = q\int \frac{d^3k}{(2\pi)^3}\,\frac{4\pi}{k^2 + \mu^2}\,e^{i\mathbf{k}\cdot\mathbf{x}} = q\,\frac{e^{-\mu r}}{r}, \qquad r = |\mathbf{x}|. \tag{9.249}$$

Electromagnetism is recovered at $\mu = 0, \phi_{\text{Coulomb}}(r) = \frac{q}{r}$.

Hence, the observed long range of the electromagnetic interaction is equivalent to the masslessness of its mediator. In a quantum-field description this mediator is the photon. In the present classical language, it corresponds to the statement that the gauge field has only transverse radiative degrees of freedom and no mass term compatible with gauge invariance.

9.10 The Minkowski-Lorentz Equation in Covariant Form

The covariant Lagrangian of a charged particle subject to an electromagnetic field is

$$L = -mc(-v_\nu v^\nu)^{1/2} + \frac{e}{c}A^\mu(z)v_\mu, \tag{9.250}$$

where the worldline $z^\mu(\xi)$ of the particle is parametrized as a function of an arbitrary scalar ξ and the four-velocity v^μ is defined as $v^\mu = dz^\mu/d\xi$. The Euler-Lagrange equations have the general form $(d/d\xi)(\partial L/\partial v_\mu) - (\partial L/\partial z_\mu) = 0$. In the specific case of the Lagrangian in Eq. (9.250), the *canonical four-momentum* is

$$P^\mu = \frac{\partial L}{\partial v_\mu} = \frac{mcv^\mu}{(-v^2)^{1/2}} + \frac{e}{c}A^\mu. \tag{9.251}$$

The equations of motion become

$$\frac{d}{d\xi}\left[\frac{mcv^\mu}{(-v^2)^{1/2}}\right] + \frac{e}{c}\frac{\partial A^\mu}{\partial z_\nu}v_\nu - \frac{e}{c}\frac{\partial A^\nu}{\partial z_\mu}v_\nu = 0 \Rightarrow \frac{d}{d\xi}\left[\frac{mcv^\mu}{(-v^2)^{1/2}}\right] = \frac{e}{c}F^{\mu\nu}(z)v_\nu. \tag{9.252}$$

Choosing as the parameter ξ the proper length s, we get

$$mc\frac{du^\mu}{ds} = \frac{e}{c}F^{\mu\nu}(z)u_\nu, \quad \mathcal{F}^\mu = \frac{e}{c}F^{\mu\nu}u_\nu. \tag{9.253a}$$

This is the *Minkowski-Lorentz equation* written in covariant form. $\mathcal{F}^\mu$ is the *Lorentz four-force*, as confirmed in Chap. 2.

9.11 Lorentz Transformations of Potentials and Fields

Having recognized that φ and $\mathbf{A}$ are, respectively, the temporal component and the spatial components of the four-potential A^μ, it is easy to derive the transformation formulas for φ, $\mathbf{A}$ for a change of inertial system. Considering, for example, a boost along the x axis: the Lorentz transformations of φ and $\mathbf{A}$ are

$$\varphi' = \frac{\varphi - \beta A_x}{\sqrt{1-\beta^2}}, \tag{9.254a}$$

$$A'_x = \frac{A_x - \beta\varphi}{\sqrt{1-\beta^2}}, \tag{9.254b}$$

$$A'_y = A_y, \tag{9.254c}$$

$$A'_z = A_z. \tag{9.254d}$$

The Lorentz transformations of the fields $\mathbf{E}$ and $\mathbf{B}$ are less immediate to derive, because $\mathbf{E}$ and $\mathbf{B}$ are the components of the four-tensor $F^{\mu\nu}$, and four-tensors transform in a more complicated way, namely $F'^{\mu\nu} = \Lambda^\mu{}_\rho \Lambda^\nu{}_\sigma F^{\rho\sigma}$.

Let us consider again a boost along the x axis whose Lorentz matrix is

$$\Lambda = \begin{pmatrix} \gamma & -\beta\gamma & 0 & 0 \\ -\beta\gamma & \gamma & 0 & 0 \\ 0 & 0 & 1 & 0 \\ 0 & 0 & 0 & 1 \end{pmatrix}. \tag{9.255}$$

The transformation formulas for the electric field $F^{i0} = E^i$ are $F'^{i0} = \Lambda^i{}_\rho \Lambda^0{}_\sigma F^{\rho\sigma}$. In particular, for $F^{10} = E^1 = E_x$ we have

$$\begin{aligned} F'^{10} &= \Lambda^1{}_\rho \Lambda^0{}_\sigma F^{\rho\sigma} = \Lambda^1{}_0 \Lambda^0{}_\sigma F^{0\sigma} + \Lambda^1{}_1 \Lambda^0{}_\sigma F^{1\sigma} \\ &= \Lambda^1{}_0 \Lambda^0{}_1 F^{01} + \Lambda^1{}_1 \Lambda^0{}_0 F^{10} = \beta^2\gamma^2(-E_x) + \gamma^2 E_x = \gamma^2(1-\beta^2)E_x = E_x, \end{aligned} \tag{9.256}$$

i.e. $E'_x = E_x$. For $F^{20} = E^2 = E_y$ we have

$$\begin{aligned} F'^{20} &= \Lambda^2{}_\rho \Lambda^0{}_\sigma F^{\rho\sigma} = \Lambda^2{}_2 \Lambda^0{}_\sigma F^{2\sigma} = \Lambda^2{}_2 \left(\Lambda^0{}_0 F^{20} + \Lambda^0{}_1 F^{21}\right) \\ &= \gamma F^{20} + \beta\gamma F^{21} = \gamma(E_y - \beta B_z), \end{aligned} \tag{9.257}$$

i.e. $E'_y = \gamma(E_y - \beta B_z)$. Proceeding in an analogous manner we find $E'_z = \gamma(E_z + \beta B_y)$. As for the spatial components of $F^{\mu\nu}$, i.e., the components of the magnetic field **B**, the transformation formulas are $F'^{ij} = \Lambda^i{}_\rho \Lambda^j{}_\sigma F^{\rho\sigma}$. As an example, let us derive the Lorentz transformation of $F^{12} = -B_z$ for a boost along x:

$$\begin{aligned} F'^{12} &= \Lambda^1{}_\rho \Lambda^2{}_\sigma F^{\rho\sigma} = \Lambda^1{}_\rho \Lambda^2{}_2 F^{\rho 2} = \Lambda^2{}_2 \left(\Lambda^1{}_0 F^{02} + \Lambda^1{}_1 F^{12}\right) \\ &= -\beta\gamma(-E_y) + \gamma(-B_z) = \gamma(-\beta E_y - B_z), \end{aligned} \tag{9.258}$$

i.e. $B'_z = \gamma(B_z + \beta E_y)$.

Overall, the transformations of the fields **E**, **B** for a boost along the x axis are

$$E'_x = E_x, \tag{9.259a}$$
$$E'_y = \gamma(E_y - \beta B_z), \tag{9.259b}$$
$$E'_z = \gamma(E_z + \beta B_y), \tag{9.259c}$$
$$B'_x = B_x, \tag{9.259d}$$
$$B'_y = \gamma(B_y + \beta E_z), \tag{9.259e}$$
$$B'_z = \gamma(B_z - \beta E_y). \tag{9.259f}$$

As can be seen, the components of **E** and **B** in the direction of the boost do not change, while the Lorentz transformations of the other components mix the electric and magnetic fields. If we separate in **E**, **B** the components parallel and perpendicular to the direction of the boost as in Eq. (9.8) with $\mathbf{E}_L \equiv \mathbf{E}_{||}$ and $\mathbf{E}_T \equiv \mathbf{E}_\perp$, the transformations of $\mathbf{E}_{||}$, $\mathbf{E}_\perp$,$\mathbf{B}_{||}$, $\mathbf{B}_\perp$ are

$$\mathbf{E}'_{||} = \mathbf{E}_{||}, \tag{9.260a}$$
$$\mathbf{E}'_\perp = \gamma\left(\mathbf{E}_\perp + \boldsymbol{\beta} \times \mathbf{B}\right), \tag{9.260b}$$
$$\mathbf{B}'_{||} = \mathbf{B}_{||}, \tag{9.260c}$$
$$\mathbf{B}'_\perp = \gamma\left(\mathbf{B}_\perp - \boldsymbol{\beta} \times \mathbf{E}\right). \tag{9.260d}$$

The last equations show that, if in the inertial reference frame $\mathcal{K}$ the electric field is zero $\mathbf{E} = 0$, in another inertial reference frame $\mathcal{K}'$ moving with respect to $\mathcal{K}$ with velocity $\mathbf{v}$ the electric field $\mathbf{E}'$ is perpendicular to the magnetic field $\mathbf{B}'$ and to $\mathbf{v}$, and is $\mathbf{E}' = (\mathbf{v}/c) \times \mathbf{B}'$. Similarly, if in $\mathcal{K}$ the magnetic field is zero, $\mathbf{B} = 0$, in $\mathcal{K}'$ it is $\mathbf{B}' = -(\mathbf{v}/c) \times \mathbf{E}'$.

Conversely, when in $\mathcal{K}$ the electric and magnetic fields are orthogonal ($\mathbf{E} \cdot \mathbf{B} = 0$), there exists another reference frame $\mathcal{K}'$ in which if $|\mathbf{E}| < |\mathbf{B}|$, the electric field $\mathbf{E}'$ is zero, while if $|\mathbf{E}| > |\mathbf{B}|$ the magnetic field $\mathbf{B}'$ is zero. If in $\mathcal{K}$, in addition to $\mathbf{E} \cdot \mathbf{B} = 0$ we have $|\mathbf{E}| = |\mathbf{B}|$, then in another reference frame $\mathcal{K}'$ the fields $\mathbf{E}'$, $\mathbf{B}'$ will be orthogonal $\mathbf{E}' \cdot \mathbf{B}' = 0$, and equal in magnitude $|\mathbf{E}'| = |\mathbf{B}'|$ (recall that $\mathbf{E} \cdot \mathbf{B}$ and $\mathbf{E}^2 - \mathbf{B}^2$ are invariant quantities). The parallel and perpendicular components of the fields are extracted from $\mathbf{E}$ and $\mathbf{B}$ by

$$\mathbf{E}_{||} = \frac{(\mathbf{E} \cdot \boldsymbol{\beta})}{\beta^2} \boldsymbol{\beta}, \tag{9.261a}$$

$$\mathbf{E}_{\perp} = \mathbf{E} - \mathbf{E}_{||}, \tag{9.261b}$$

and analogous relations for the magnetic field. Using these projections it is possible to reconstruct the transformations of the fields $\mathbf{E}$ and $\mathbf{B}$ for a generic boost:

$$\mathbf{E}' = \gamma(\mathbf{E} + \boldsymbol{\beta} \times \mathbf{B}) - \frac{\gamma - 1}{\beta^2} (\mathbf{E} \cdot \boldsymbol{\beta}) \boldsymbol{\beta}, \tag{9.262a}$$

$$\mathbf{B}' = \gamma(\mathbf{B} - \boldsymbol{\beta} \times \mathbf{E}) - \frac{\gamma - 1}{\beta^2} (\mathbf{B} \cdot \boldsymbol{\beta}) \boldsymbol{\beta}. \tag{9.262b}$$

9.12 Fields Generated by a Uniformly Moving Charge

Let us consider a particle of charge e moving in a reference frame $\mathcal{K}$ with constant velocity $\mathbf{v}$ along the x axis and suppose that it is at $x = 0$ when $t = 0$. In the reference frame $\mathcal{K}'$ comoving with the particle and centered on it, the potentials are ($r' \equiv |\mathbf{x}'|$)

$$\varphi' = \frac{e}{r'}, \tag{9.263a}$$

$$\mathbf{A}' = 0. \tag{9.263b}$$

Based on Eq. (9.254), the potentials in $\mathcal{K}$ are

$$\varphi = \frac{\varphi' + \beta A'_x}{\sqrt{1 - \frac{v^2}{c^2}}} = \frac{\varphi'}{\sqrt{1 - \frac{v^2}{c^2}}}, \tag{9.264a}$$

$$\mathbf{A} = \left(\frac{A'_x + \beta \varphi'}{\sqrt{1 - \frac{v^2}{c^2}}}, A'_y, A'_z \right) = \left(\frac{\beta \varphi'}{\sqrt{1 - \frac{v^2}{c^2}}}, 0, 0 \right) = \varphi \frac{\mathbf{v}}{c}. \tag{9.264b}$$

Substituting the potential φ' and using $r' = \sqrt{x'^2 + y'^2 + z'^2} = \sqrt{\gamma^2(x - vt)^2 + y^2 + z^2}$, we find:

$$\varphi = \frac{e\gamma}{\sqrt{\gamma^2(x - vt)^2 + y^2 + z^2}}, \tag{9.265a}$$

$$\mathbf{A} = \frac{e\gamma\mathbf{v}}{c\sqrt{\gamma^2(x - vt)^2 + y^2 + z^2}}. \tag{9.265b}$$

Let us now study the electric and magnetic fields generated by the charge. In $\mathcal{K}'$, the electric field $\mathbf{E}'$ is the Coulomb field, while the magnetic field $\mathbf{B}'$ is zero

$$\mathbf{E}' = \frac{e\mathbf{x}'}{r'^3}, \quad \mathbf{B}' = 0. \tag{9.266}$$

The components of the electric field $\mathbf{E}$ in the system $\mathcal{K}$ in which the particle is moving are obtained from Eq. (9.259), and are

$$E_x = E'_x = \frac{ex'}{r'^3} = \frac{e\gamma(x - vt)}{(\gamma^2(x - vt)^2 + y^2 + z^2)^{3/2}}, \tag{9.267a}$$

$$E_y = \gamma E'_y = \gamma\frac{ey'}{r'^3} = \frac{e\gamma y}{(\gamma^2(x - vt)^2 + y^2 + z^2)^{3/2}}, \tag{9.267b}$$

$$E_z = \gamma E'_z = \gamma\frac{ez'}{r'^3} = \frac{e\gamma z}{(\gamma^2(x - vt)^2 + y^2 + z^2)^{3/2}}. \tag{9.267c}$$

The longitudinal component of the field E_x changes sign when the charge passes through the origin (at time $t = 0$) and has zero average value. The transverse component E_y, on the other hand, always has the same sign and is significantly different from zero only for a time interval $\Delta t \approx b/\gamma v$; this impulsive field is responsible for the ionization phenomena of atoms in a material traversed by charged particles. Let us introduce the relative position vector $\mathbf{R}$ of the observation point $\mathbf{x}$ with respect to the charge, which is located at $z(t) = (vt, 0, 0)$:

$$\mathbf{R}(t) = \mathbf{x} - \mathbf{z}(t) = (x - vt, y, z), \tag{9.268a}$$

$$R^2 = (x - vt)^2 + y^2 + z^2 \tag{9.268b}$$

$$\Rightarrow r'^2 = \gamma^2(x - vt)^2 + y^2 + z^2 = \gamma^2 R^2 - (\gamma^2 - 1)(y^2 + z^2). \tag{9.268c}$$

Letting θ be the angle between $\mathbf{R}$ and $\mathbf{v}$, we will have

$$y^2 + z^2 = R^2 \sin^2\theta \Rightarrow r'^2 = \gamma^2 R^2(1 - \beta^2 \sin^2\theta). \tag{9.269}$$

This relation and Eq. (9.268a) allow us to re-express Eq. (9.267) as

$$\mathbf{E} = \frac{e\mathbf{R}}{\gamma^2 R^3(1 - \beta^2 \sin^2\theta)^{3/2}}. \tag{9.270}$$

In agreement with the previous discussion, since the magnetic field is zero in $\mathcal{K}'$, in $\mathcal{K}$ it is related to the electric field by $\mathbf{B} = (\mathbf{v}/c) \times \mathbf{E}$. From Eq. (9.270), we see that the electric field $\mathbf{E}$ is radial but not isotropic. In particular, for $\theta = 0$ we find $|\mathbf{E}_{||}| = e/\gamma^2 R^2$, and for $\theta = \pi/2$, $|\mathbf{E}_{\perp}| = e\gamma/R^2$. The field $|\mathbf{E}_{||}|$ in the direction of motion of the charge ($\theta = 0$) is therefore contracted compared to the field $|\mathbf{E}_{\perp}|$ in the transverse direction ($\theta = \pi/2$). Furthermore, as the velocity increases, $|\mathbf{E}_{\perp}|$ grows, while $|\mathbf{E}_{||}|$ decreases.

In the non-relativistic limit, Eq. (9.270) becomes

$$\mathbf{E} \simeq \frac{e\mathbf{R}}{R^3}, \tag{9.271}$$

which is the Coulomb expression. The magnetic field is given by

$$\mathbf{B} = \frac{\mathbf{v}}{c} \times \mathbf{E} \simeq \frac{e\mathbf{v} \times \mathbf{R}}{cR^3}. \tag{9.272}$$

This is the Biot-Savart law: in relativstic scenarios, it is a consequence of Coulomb's law, in Eq. (9.271), and the Lorentz transformations of the fields.

9.13 Nonlinear Electrodynamics

In covariant form, classical electrodynamics is most efficiently organized through an action principle. This approach makes gauge invariance manifest, yields Maxwell's equations as Euler-Lagrange equations, and provides a systematic route to the conserved energy-momentum tensor. It also clarifies what modifications are allowed once one departs from linear Maxwell theory. Hence, let us focus on the domain of validity of electrodynamics and, then, shift to variational principle through which we will formulate examples of nonlinear electrodynamics (NLED).

9.13.1 Domain of Validity of Classical Electromagnetism

Classical electrodynamics is extraordinarily accurate at macroscopic scales, yet its status as an effective theory becomes clear in regimes where quantum structure of matter and the vacuum cannot be neglected.

A classical field description of a charged particle breaks down once variations occur on scales comparable to the Compton wavelength

$$\lambda_C \equiv \frac{\hbar}{mc}, \tag{9.273}$$

since particle creation and relativistic quantum fluctuations become unavoidable. Similarly, the classical point-charge idealization leads to divergent self-energy and to the radiation-reaction pathologies discussed in Sect. 9.4.

Even in vacuum, sufficiently strong electric fields destabilize the ground state through electron-positron pair creation. Dimensional analysis already identifies the critical field by requiring that the work done over a Compton length equals $2mc^2$:

$$eE\lambda_C \sim 2mc^2 \quad \Rightarrow \quad E \sim \frac{2mc^2}{e}\frac{mc}{\hbar} \sim \frac{m^2c^3}{e\hbar} \equiv E_S. \tag{9.274}$$

The scale E_S is the Schwinger critical field. Below it, classical vacuum is stable; near it, the effective description must be quantum (as developed in Sec. 1.26.6 and following).

Quantum fluctuations generate corrections to the Maxwell action, leading to an effective NLED (Euler-Heisenberg type). Hence even the "vacuum" acquires polarizability at high fields/frequencies, and the linear constitutive relations underlying classical Maxwell theory must be regarded as low-energy approximations.

These limitations do not diminish classical electromagnetism; they locate it precisely within the hierarchy of effective theories that constitute the modern understanding of fundamental interactions.

9.13.2 Variational Derivation of the Maxwell Equations

Let A_μ be the four-potential and

$$F_{\mu\nu} \equiv \partial_\mu A_\nu - \partial_\nu A_\mu \tag{9.275}$$

the field-strength tensor. Consider the action (Gaussian units)

$$S[A] = \int d^4x\, \mathcal{L} \quad \text{with} \quad \mathcal{L} = -\frac{1}{16\pi} F_{\mu\nu}F^{\mu\nu} - \frac{1}{c}\, j_\mu A^\mu. \tag{9.276}$$

Gauge invariance holds provided $\partial_\mu j^\mu = 0$, since under $A_\mu \mapsto A_\mu + \partial_\mu \chi$ one has

$$\Delta S_{\text{int}} = -\frac{1}{c}\int d^4x\, j_\mu \partial^\mu \chi = \frac{1}{c}\int d^4x\, (\partial_\mu j^\mu)\chi - \frac{1}{c}\int d^4x\, \partial_\mu(j^\mu \chi), \tag{9.277}$$

which vanishes for conserved current up to a boundary term.

To derive the field equations, vary the action with respect to A_μ. From Eq. (9.275),

$$\delta F_{\mu\nu} = \partial_\mu \delta A_\nu - \partial_\nu \delta A_\mu. \tag{9.278}$$

The variation of the Maxwell term is

$$\delta S_{\text{EM}} = -\frac{1}{16\pi}\int d^4x\ \delta\left(F_{\mu\nu}F^{\mu\nu}\right) = -\frac{1}{8\pi}\int d^4x\ F^{\mu\nu}\delta F_{\mu\nu}$$
$$= -\frac{1}{8\pi}\int d^4x\ F^{\mu\nu}\left(\partial_\mu\delta A_\nu - \partial_\nu\delta A_\mu\right). \tag{9.279}$$

Using antisymmetry $F^{\mu\nu} = -F^{\nu\mu}$, the two terms in Eq. (9.279) are equal

$$F^{\mu\nu}\left(\partial_\mu\delta A_\nu - \partial_\nu\delta A_\mu\right) = 2F^{\mu\nu}\partial_\mu\delta A_\nu, \tag{9.280}$$

hence

$$\delta S_{\text{EM}} = -\frac{1}{4\pi}\int d^4x\ F^{\mu\nu}\partial_\mu\delta A_\nu. \tag{9.281}$$

Integrating by parts (and discarding boundary terms for variations vanishing at infinity):

$$\delta S_{\text{EM}} = \frac{1}{4\pi}\int d^4x\ (\partial_\mu F^{\mu\nu})\,\delta A_\nu. \tag{9.282}$$

The source term varies as

$$\delta S_{\text{int}} = -\frac{1}{c}\int d^4x\ j_\nu\,\delta A^\nu. \tag{9.283}$$

From Eqs. (9.282) and (9.283), we obtain

$$\delta S = \int d^4x\left[\frac{1}{4\pi}\partial_\mu F^{\mu\nu} - \frac{1}{c}j^\nu\right]\delta A_\nu. \tag{9.284}$$

Since δA_ν is arbitrary, stationarity $\delta S = 0$ yields the inhomogeneous Maxwell equations:

$$\partial_\mu F^{\mu\nu} = \frac{4\pi}{c}\,j^\nu. \tag{9.285}$$

The homogeneous equations are not Euler-Lagrange equations: they are identities,

$$\partial_{[\lambda}F_{\mu\nu]} = 0, \tag{9.286}$$

following directly from $F = dA$ (or from commuting partial derivatives).

9.13.2.1 Energy-Momentum Tensor of the Electromagnetic Field

In Minkowski spacetime, the energy-momentum tensor follows from invariance under spacetime translations. For clarity, we first consider the electromagnetic field Lagrangian density

$$\mathcal{L}_{\rm EM} = -\frac{1}{16\pi}\, F_{\mu\nu}F^{\mu\nu}, \tag{9.287}$$

and we treat the interaction term as part of the matter sector. Under an infinitesimal translation $x^\mu \mapsto x^\mu + \epsilon^\mu$, the field transforms as

$$\delta A_\lambda(x) = -\epsilon^\nu \partial_\nu A_\lambda(x), \tag{9.288}$$

hence, using the standard Noether construction, the canonical energy-momentum tensor is

$$T^\mu{}_\nu = \frac{\partial \mathcal{L}_{\rm EM}}{\partial(\partial_\mu A_\lambda)}\, \partial_\nu A_\lambda - \delta^\mu{}_\nu\, \mathcal{L}_{\rm EM}. \tag{9.289}$$

▶ **Driving exercise**

Compute the energy momentum tensor.

The only nontrivial step is the computation of $\partial\mathcal{L}_{\rm EM}/\partial(\partial_\mu A_\lambda)$. Since $F_{\alpha\beta} = \partial_\alpha A_\beta - \partial_\beta A_\alpha$, one has

$$\frac{\partial F_{\alpha\beta}}{\partial(\partial_\mu A_\lambda)} = \delta^\mu{}_\alpha \delta^\lambda{}_\beta - \delta^\mu{}_\beta \delta^\lambda{}_\alpha. \tag{9.290}$$

Therefore,

$$\begin{aligned}\frac{\partial\mathcal{L}_{\rm EM}}{\partial(\partial_\mu A_\lambda)} &= -\frac{1}{16\pi}\,\frac{\partial}{\partial(\partial_\mu A_\lambda)}\left(F_{\alpha\beta}F^{\alpha\beta}\right) = -\frac{1}{16\pi}\,2F^{\alpha\beta}\frac{\partial F_{\alpha\beta}}{\partial(\partial_\mu A_\lambda)}\\ &= -\frac{1}{8\pi}\,F^{\alpha\beta}\left(\delta^\mu{}_\alpha\delta^\lambda{}_\beta - \delta^\mu{}_\beta\delta^\lambda{}_\alpha\right) = -\frac{1}{8\pi}\left(F^{\mu\lambda} - F^{\lambda\mu}\right) = -\frac{1}{4\pi}\,F^{\mu\lambda}.\end{aligned} \tag{9.291}$$

Inserting Eq. (9.291) into Eq. (9.289) gives the canonical tensor

$$T^\mu{}_\nu = -\frac{1}{4\pi}F^{\mu\lambda}\partial_\nu A_\lambda + \delta^\mu{}_\nu\,\frac{1}{16\pi}F_{\alpha\beta}F^{\alpha\beta}. \tag{9.292}$$

This object is conserved on-shell for the free field, but it is neither symmetric nor manifestly gauge invariant, since it contains $\partial_\nu A_\lambda$ rather than $F_{\nu\lambda}$. A standard improvement consists in adding the divergence of a superpotential,

$$\Theta^\mu{}_\nu = T^\mu{}_\nu + \partial_\lambda X^{\lambda\mu}{}_\nu, \qquad X^{\lambda\mu}{}_\nu \equiv \frac{1}{4\pi}F^{\lambda\mu}A_\nu, \tag{9.293}$$

which does not alter the conserved four-momentum because $\partial_\mu\partial_\lambda X^{\lambda\mu}{}_\nu = 0$ by antisymmetry $X^{\lambda\mu}{}_\nu = -X^{\mu\lambda}{}_\nu$.

Compute $\Theta^{\mu}{}_{\nu}$ explicitly. From Eq. (9.293),

$$\partial_\lambda X^{\lambda\mu}{}_{\nu} = \frac{1}{4\pi}\left[(\partial_\lambda F^{\lambda\mu})A_\nu + F^{\lambda\mu}\partial_\lambda A_\nu\right]. \tag{9.294}$$

Adding Eqs. (9.294)–(9.292) yields

$$\begin{aligned}\Theta^{\mu}{}_{\nu} &= -\frac{1}{4\pi}F^{\mu\lambda}\partial_\nu A_\lambda + \frac{1}{16\pi}\delta^{\mu}{}_{\nu}F_{\alpha\beta}F^{\alpha\beta} + \frac{1}{4\pi}(\partial_\lambda F^{\lambda\mu})A_\nu + \frac{1}{4\pi}F^{\lambda\mu}\partial_\lambda A_\nu \\ &= \frac{1}{4\pi}\left(F^{\lambda\mu}\partial_\lambda A_\nu - F^{\mu\lambda}\partial_\nu A_\lambda\right) + \frac{1}{16\pi}\delta^{\mu}{}_{\nu}F_{\alpha\beta}F^{\alpha\beta} + \frac{1}{4\pi}(\partial_\lambda F^{\lambda\mu})A_\nu.\end{aligned} \tag{9.295}$$

The first bracket can be rewritten solely in terms of $F_{\lambda\nu}$:

$$\begin{aligned}F^{\lambda\mu}\partial_\lambda A_\nu - F^{\mu\lambda}\partial_\nu A_\lambda &= F^{\lambda\mu}\partial_\lambda A_\nu + F^{\lambda\mu}\partial_\nu A_\lambda \\ &= F^{\lambda\mu}\left(\partial_\lambda A_\nu - \partial_\nu A_\lambda\right) = F^{\lambda\mu}F_{\lambda\nu}.\end{aligned} \tag{9.296}$$

Hence

$$\Theta^{\mu}{}_{\nu} = \frac{1}{4\pi}F^{\lambda\mu}F_{\lambda\nu} + \frac{1}{16\pi}\delta^{\mu}{}_{\nu}F_{\alpha\beta}F^{\alpha\beta} + \frac{1}{4\pi}(\partial_\lambda F^{\lambda\mu})A_\nu. \tag{9.297}$$

When sources are present, the Maxwell equations in Eq. (9.285) imply

$$(\partial_\lambda F^{\lambda\mu})A_\nu = \frac{4\pi}{c}j^{\mu}A_\nu, \tag{9.298}$$

which belongs naturally to the matter-field interaction bookkeeping. The purely field stress-energy tensor is obtained by dropping this term (equivalently, working in the source-free region), namely

$$T^{\mu}{}_{\nu}{}^{\text{(field)}} = \frac{1}{4\pi}F^{\lambda\mu}F_{\lambda\nu} + \frac{1}{16\pi}\delta^{\mu}{}_{\nu}F_{\alpha\beta}F^{\alpha\beta}. \tag{9.299}$$

Raising the second index with the Minkowski metric, one arrives at the standard symmetric, gauge-invariant expression

$$T^{\mu\nu\text{(field)}} = \frac{1}{4\pi}\left(F^{\mu}{}_{\lambda}F^{\nu\lambda} - \frac{1}{4}\eta^{\mu\nu}F_{\alpha\beta}F^{\alpha\beta}\right). \tag{9.300}$$

Pause and think

Apply the divergence to the energy-momentum tensor and interpret physically what you will obtain.

▶ **Driving exercise**
The divergence of the energy-momentum tensor here reproduces the Lorentz four-force density. Starting from Eq. (9.300) and applying the product rule,

$$\partial_\mu T^{\mu\nu} = \frac{1}{4\pi}\left[(\partial_\mu F^{\mu}{}_{\lambda})F^{\nu\lambda} + F^{\mu}{}_{\lambda}\partial_\mu F^{\nu\lambda} - \frac{1}{4}\partial^\nu(F_{\alpha\beta}F^{\alpha\beta})\right]. \tag{9.301}$$

Using Eq. (9.285), the first term becomes

$$(\partial_\mu F^{\mu}{}_{\lambda})F^{\nu\lambda} = \frac{4\pi}{c}j_\lambda F^{\nu\lambda}. \tag{9.302}$$

The remaining two terms cancel by virtue of the Bianchi identity, Eq. (9.286). One convenient way to see this is to contract Eq. (9.286) with $F^{\mu\lambda}$:

$$F^{\mu\lambda}\left(\partial_\mu F^{\nu}{}_{\lambda} + \partial^\nu F_{\lambda\mu} + \partial_\lambda F_{\mu}{}^{\nu}\right) = 0, \tag{9.303}$$

and use antisymmetry to rewrite

$$F^{\mu\lambda}\partial^\nu F_{\lambda\mu} = \frac{1}{2}\partial^\nu(F^{\mu\lambda}F_{\mu\lambda}), \qquad F^{\mu\lambda}\partial_\lambda F_{\mu}{}^{\nu} = -F^{\mu\lambda}\partial_\mu F^{\nu}{}_{\lambda}, \tag{9.304}$$

which implies

$$F^{\mu}{}_{\lambda}\partial_\mu F^{\nu\lambda} = \frac{1}{4}\partial^\nu(F_{\alpha\beta}F^{\alpha\beta}). \tag{9.305}$$

Substituting Eqs. (9.302) and (9.305) into Eq. (9.301) yields

$$\partial_\mu T^{\mu\nu} = \frac{1}{c}j_\lambda F^{\nu\lambda}, \tag{9.306}$$

showing that the field energy-momentum is locally exchanged with matter through the Lorentz four-force density.

9.13.3 Born-Infeld Electrodynamics

A classic nonlinear deformation of Maxwell theory that preserves gauge invariance and removes the divergence of the electrostatic self-energy of a point charge is the Born-Infeld (BI) theory. It is defined by the Lagrangian density

$$\mathcal{L}_{\text{BI}} = b^2\left(1 - \sqrt{1 + \frac{1}{2b^2}F_{\mu\nu}F^{\mu\nu} - \frac{1}{16b^4}\left(F_{\mu\nu}{}^{\star}F^{\mu\nu}\right)^2}\right) - \frac{1}{c}j_\mu A^\mu, \tag{9.307}$$

where b is a field-strength scale and ${}^{\star}F^{\mu\nu}$ is the dual tensor. Introduce the two Lorentz invariants

$$\mathcal{I}_1 \equiv \frac{1}{2} F_{\mu\nu} F^{\mu\nu}, \qquad \mathcal{I}_2 \equiv \frac{1}{4} F_{\mu\nu} {}^{\star}F^{\mu\nu}, \tag{9.308}$$

so that the square root in Eq. (9.307) is $\sqrt{1 + \mathcal{I}_1/b^2 - \mathcal{I}_2^2/b^4}$.

Driving exercise
Maxwell limit.
For fields small compared with b,

$$\sqrt{1 + \frac{\mathcal{I}_1}{b^2} - \frac{\mathcal{I}_2^2}{b^4}} = 1 + \frac{1}{2}\frac{\mathcal{I}_1}{b^2} - \frac{1}{2}\frac{\mathcal{I}_2^2}{b^4} - \frac{1}{8}\frac{\mathcal{I}_1^2}{b^4} + O(b^{-6}), \tag{9.309}$$

hence

$$\mathcal{L}_{\mathrm{BI}} = -\frac{1}{2}\mathcal{I}_1 + \frac{1}{2b^2}\mathcal{I}_2^2 + \frac{1}{8b^2}\mathcal{I}_1^2 + O(b^{-4}) - \frac{1}{c} j_\mu A^\mu. \tag{9.310}$$

Recalling $\mathcal{I}_1 = \frac{1}{2} F_{\mu\nu} F^{\mu\nu}$, the leading term is precisely the Maxwell Lagrangian $-(1/16\pi) F_{\mu\nu} F^{\mu\nu}$ up to unit conventions.

Field equations.
The Euler-Lagrange equations retain the divergence form

$$\partial_\mu G^{\mu\nu} = \frac{4\pi}{c}\, j^\nu, \tag{9.311}$$

but the constitutive relation becomes nonlinear:

$$G^{\mu\nu} \equiv -4\pi \frac{\partial \mathcal{L}_{\mathrm{BI}}}{\partial F_{\mu\nu}} = \frac{F^{\mu\nu} - \frac{1}{4b^2}\left(F_{\alpha\beta}{}^{\star}F^{\alpha\beta}\right) {}^{\star}F^{\mu\nu}}{\sqrt{1 + \frac{1}{2b^2} F_{\rho\sigma} F^{\rho\sigma} - \frac{1}{16b^4}\left(F_{\rho\sigma}{}^{\star}F^{\rho\sigma}\right)^2}}. \tag{9.312}$$

The homogeneous equations remain $\partial_{[\lambda} F_{\mu\nu]} = 0$. Born-Infeld electrodynamics thus provides a controlled nonlinear completion of Maxwell theory, useful both as an effective model of strong-field regimes and as a theoretically consistent arena where field self-energies are softened without breaking Lorentz and gauge invariance.

Pause and think
Compare the energy-momentum tensor in the Born-Infeld electrodynamics with the standard case.

Driving exercise
To construct the stress-energy tensor directly in Minkowski spacetime, one may proceed exactly as in the Maxwell case, exploiting translational invariance and then performing the standard gauge-invariant improvement. Here it is convenient to work with the field part of the Born-Infeld Lagrangian,

$$\mathcal{L}_{\rm BI}^{(\rm field)} = b^2\left(1-\sqrt{1+\frac{1}{2b^2}F_{\mu\nu}F^{\mu\nu}-\frac{1}{16b^4}\left(F_{\mu\nu}{}^{\star}F^{\mu\nu}\right)^2}\right), \tag{9.313}$$

and to keep the interaction term $-(1/c)j_\mu A^\mu$ as part of the matter sector. So, define for brevity,

$$\mathcal{D}\equiv 1+\frac{\mathcal{I}_1}{b^2}-\frac{\mathcal{I}_2^2}{b^4}, \qquad \text{so that} \qquad \mathcal{L}_{\rm BI}^{(\rm field)} = b^2\left(1-\sqrt{\mathcal{D}}\right), \tag{9.314}$$

since $\mathcal{L}_{\rm BI}^{(\rm field)}$ depends on A_μ only through $F_{\mu\nu}$, the improved, gauge-invariant stress-energy tensor can be written in flat spacetime in the compact form

$$T^{\mu\nu} = -2\,\frac{\partial\mathcal{L}_{\rm BI}^{(\rm field)}}{\partial F_{\mu\lambda}}\,F^{\nu}{}_{\lambda}+\eta^{\mu\nu}\mathcal{L}_{\rm BI}^{(\rm field)}, \tag{9.315}$$

which reduces to the standard Maxwell expression when $\mathcal{L}=-(16\pi)^{-1}F_{\alpha\beta}F^{\alpha\beta}$. It remains to compute $\partial\mathcal{L}_{\rm BI}^{(\rm field)}/\partial F_{\mu\lambda}$ explicitly. From $\mathcal{L}_{\rm BI}^{(\rm field)} = b^2(1-\sqrt{\mathcal{D}})$ one has

$$\frac{\partial\mathcal{L}_{\rm BI}^{(\rm field)}}{\partial F_{\mu\lambda}} = -\frac{b^2}{2\sqrt{\mathcal{D}}}\frac{\partial\mathcal{D}}{\partial F_{\mu\lambda}}. \tag{9.316}$$

Moreover,

$$\frac{\partial\mathcal{D}}{\partial F_{\mu\lambda}} = \frac{1}{b^2}\frac{\partial\mathcal{I}_1}{\partial F_{\mu\lambda}}-\frac{2\mathcal{I}_2}{b^4}\frac{\partial\mathcal{I}_2}{\partial F_{\mu\lambda}}. \tag{9.317}$$

The derivatives of the invariants are obtained by direct variation. For $\mathcal{I}_1 = \frac{1}{2}F_{\rho\sigma}F^{\rho\sigma}$,

$$\delta\mathcal{I}_1 = F^{\rho\sigma}\delta F_{\rho\sigma} \qquad \Longrightarrow \qquad \frac{\partial\mathcal{I}_1}{\partial F_{\mu\lambda}} = F^{\mu\lambda}. \tag{9.318}$$

For $\mathcal{I}_2 = \frac{1}{4}F_{\rho\sigma}{}^{\star}F^{\rho\sigma}$, using ${}^{\star}F^{\rho\sigma} = \frac{1}{2}\epsilon^{\rho\sigma\alpha\beta}F_{\alpha\beta}$ one finds

$$\delta\mathcal{I}_2 = \frac{1}{2}\,{}^{\star}F^{\rho\sigma}\,\delta F_{\rho\sigma} \qquad \Longrightarrow \qquad \frac{\partial\mathcal{I}_2}{\partial F_{\mu\lambda}} = \frac{1}{2}\,{}^{\star}F^{\mu\lambda}. \tag{9.319}$$

Substituting these results into $\partial\mathcal{D}/\partial F_{\mu\lambda}$ gives

$$\frac{\partial\mathcal{D}}{\partial F_{\mu\lambda}} = \frac{1}{b^2}F^{\mu\lambda} - \frac{2\mathcal{I}_2}{b^4}\left(\frac{1}{2}\,{}^{\star}F^{\mu\lambda}\right) = \frac{1}{b^2}F^{\mu\lambda} - \frac{\mathcal{I}_2}{b^4}\,{}^{\star}F^{\mu\lambda}. \tag{9.320}$$

Therefore,

$$\frac{\partial\mathcal{L}_{\mathrm{BI}}^{(\mathrm{field})}}{\partial F_{\mu\lambda}} = -\frac{b^2}{2\sqrt{\mathcal{D}}}\left(\frac{1}{b^2}F^{\mu\lambda} - \frac{\mathcal{I}_2}{b^4}\,{}^{\star}F^{\mu\lambda}\right) = -\frac{1}{2\sqrt{\mathcal{D}}}\left(F^{\mu\lambda} - \frac{\mathcal{I}_2}{b^2}\,{}^{\star}F^{\mu\lambda}\right). \tag{9.321}$$

Inserting Eq. (9.321) into the general flat-spacetime formula, Eq. (9.315), one obtains the explicit Born-Infeld energy-momentum tensor

$$T^{\mu\nu} = \frac{1}{\sqrt{\mathcal{D}}}\left(F^{\mu\lambda} - \frac{\mathcal{I}_2}{b^2}\,{}^{\star}F^{\mu\lambda}\right)F^{\nu}{}_{\lambda} + \eta^{\mu\nu}b^2\left(1-\sqrt{\mathcal{D}}\right), \qquad \mathcal{D} = 1 + \frac{\mathcal{I}_1}{b^2} - \frac{\mathcal{I}_2^2}{b^4}. \tag{9.322}$$

This expression is manifestly Lorentz covariant and depends on the gauge potential only through the field strength. In the weak-field regime $|F_{\mu\nu}| \ll b$, one has $\sqrt{\mathcal{D}} = 1 + \mathcal{I}_1/(2b^2) + O(b^{-4})$, so that Eq. (9.322) reduces, at leading order, to the Maxwell stress-energy tensor plus corrections suppressed by powers of b^{-2}.

9.13.4 Beyond Born-Infeld Electrodynamics

Born-Infeld theory is only one representative of a large class of NLED models, in which the Maxwell Lagrangian is replaced by a general function of the two Lorentz invariants, $\mathcal{I}_1$ and $\mathcal{I}_2$.

The most general local, Lorentz- and gauge-invariant Lagrangian without derivatives of $F_{\mu\nu}$ can be written as

$$\mathcal{L} = \mathcal{L}(\mathcal{I}_1, \mathcal{I}_2) - \frac{1}{c}j_\mu A^\mu. \tag{9.323}$$

The field equations follow from the action principle and retain the divergence form

$$\partial_\mu G^{\mu\nu} = \frac{4\pi}{c}j^\nu, \qquad \partial_{[\lambda}F_{\mu\nu]} = 0, \tag{9.324}$$

with the nonlinear constitutive relation

$$G^{\mu\nu} \equiv -4\pi \frac{\partial \mathcal{L}}{\partial F_{\mu\nu}} = -4\pi \left(\frac{\partial \mathcal{L}}{\partial \mathcal{I}_1} F^{\mu\nu} + \frac{\partial \mathcal{L}}{\partial \mathcal{I}_2} {}^{\star}F^{\mu\nu} \right). \tag{9.325}$$

Thus NLED may be viewed as vacuum behaving as a nonlinear medium whose response depends on the local field invariants.

Examples of pedagogical value are briefly summarized below.

- **Euler-Heisenberg effective theory**.
  At energies well below the electron mass, QED vacuum polarization induces an effective NLED. To leading order in $\alpha = e^2/(\hbar c)$, the Euler-Heisenberg Lagrangian reads

$$\mathcal{L}_{\text{EH}} = -\frac{1}{2}\mathcal{I}_1 + \frac{2\alpha^2}{45m^4}\left(4\mathcal{I}_1^2 + 7\mathcal{I}_2^2\right) - \frac{1}{c} j_\mu A^\mu, \tag{9.326}$$

  where m is the electron mass (Gaussian units). The quadratic corrections encode photon-photon scattering and vacuum birefringence. Unlike Born-Infeld theory, Euler-Heisenberg electrodynamics is not fundamental: it is an effective low-energy limit of QED.
- **Logarithmic electrodynamics**.
  A simple toy model that regularizes the Coulomb divergence while preserving gauge and Lorentz invariance is defined by

$$\mathcal{L}_{\log} = -b^2 \ln\left(1 + \frac{\mathcal{I}_1}{b^2}\right) - \frac{1}{c} j_\mu A^\mu, \tag{9.327}$$

  where b sets the nonlinearity scale. Expanding for $|\mathcal{I}_1| \ll b^2$,

$$\mathcal{L}_{\log} = -\frac{1}{2}\mathcal{I}_1 + \frac{1}{2b^2}\mathcal{I}_1^2 - \frac{1}{3b^4}\mathcal{I}_1^3 + \cdots, \tag{9.328}$$

  so Maxwell theory is recovered at weak field. For a point charge, the electric field remains finite at the origin, analogously to the Born-Infeld case.
- **Power-law models**.
  Another useful class of toy models is

$$\mathcal{L}_p = -\frac{1}{2}\mathcal{I}_1 + \lambda\,|\mathcal{I}_1|^p - \frac{1}{c} j_\mu A^\mu, \qquad p > 1, \tag{9.329}$$

  with coupling λ. For $p = 2$ this reduces to the simplest quartic deformation

$$\mathcal{L} = -\frac{1}{2}\mathcal{I}_1 + \lambda\,\mathcal{I}_1^2, \tag{9.330}$$

often used as a toy model for nonlinear wave propagation and shock formation. Different values of p control how strongly large fields are suppressed.

9.14 Differential form Formulation

The covariant tensor formulation of electromagnetism admits an even more compact and conceptually transparent rewriting in the language of differential forms.

> This is not a change of physics, but a change of *geometric packaging*: the dynamical content is unaltered, but the underlying structure becomes more visible. In particular, the form language separates kinematics from dynamics, isolates gauge redundancy as a geometric freedom, exhibits the homogeneous Maxwell equations as a purely kinematic identity, and generalizes to curved spacetimes with essentially no additional effort.

The price to pay is a certain level of abstraction. The reward is that many relations that appear accidental in component notation emerge as direct consequences of simple geometric principles.

9.14.1 Differential Forms as Geometric Objects

On a four-dimensional spacetime manifold $(\mathcal{M}, g)$, a *p-form* is a completely antisymmetric covariant tensor field of rank p. It is designed to be integrated naturally over oriented p-dimensional submanifolds. A 1-form reads

$$\alpha = \alpha_\mu \, dx^\mu, \tag{9.331}$$

and assigns a covector to each point. A 2-form reads

$$\beta = \frac{1}{2} \beta_{\mu\nu} \, dx^\mu \wedge dx^\nu, \qquad \beta_{\mu\nu} = -\beta_{\nu\mu}, \tag{9.332}$$

and represents an oriented area density. The wedge product $\wedge$ is the natural antisymmetric product of forms, e.g.

$$dx^\mu \wedge dx^\nu = -dx^\nu \wedge dx^\mu, \tag{9.333}$$

ensuring that exchanging the orientation of an infinitesimal surface changes the sign of the integral.

The exterior derivative d is the unique linear operator that raises the degree of a form by one, is compatible with the wedge product, and reduces to the ordinary gradient on scalars. For a 0-form (scalar) f,

$$df = \partial_\mu f \, dx^\mu, \tag{9.334}$$

while for a 1-form $\alpha = \alpha_\mu dx^\mu$,

$$d\alpha = \frac{1}{2}(\partial_\mu \alpha_\nu - \partial_\nu \alpha_\mu) \, dx^\mu \wedge dx^\nu. \tag{9.335}$$

The crucial property is

$$d^2 = 0, \tag{9.336}$$

which is the abstract form of the identities "the curl of a gradient vanishes" and "the divergence of a curl vanishes." In the language of forms, both statements are unified as a single geometric fact.

9.14.2 Electromagnetic Potential as a 1-Form

The electromagnetic four-potential is naturally represented by a 1-form

$$A = A_\mu \, dx^\mu. \tag{9.337}$$

This is not merely a notational choice: a 1-form is precisely the object whose integral along a curve is coordinate independent, and the line integral of A is what appears in phase factors, work integrals, and minimal coupling.

The field strength is then defined as the exterior derivative of A,

$$F \equiv dA. \tag{9.338}$$

Using Eq. (9.335), one immediately finds

$$F = \frac{1}{2} F_{\mu\nu} \, dx^\mu \wedge dx^\nu, \qquad F_{\mu\nu} = \partial_\mu A_\nu - \partial_\nu A_\mu, \tag{9.339}$$

which is exactly the standard electromagnetic field tensor.

Gauge transformations acquire an extremely simple meaning:

$$A \mapsto A + d\chi, \tag{9.340}$$

with χ a scalar (0-form). Since $d^2 = 0$, one has

$$F \mapsto d(A + d\chi) = dA + d^2\chi = F, \tag{9.341}$$

so gauge invariance becomes an immediate consequence of the nilpotency of the exterior derivative, rather than a mysterious redundancy imposed by hand.

Pause and think

Before going on, manage to write the homogeneous and inhomogeneous Maxwell equations in terms of differential forms.

Driving exercise

Homogeneous Maxwell equations as a kinematic identity.

The homogeneous Maxwell equations are encoded in the single geometric statement

$$dF = 0. \tag{9.342}$$

If F is defined as in Eq. (9.338), then $dF = d(dA) = 0$ identically, without using any field equation. Thus the homogeneous sector of Maxwell theory is purely *kinematic*: it expresses the fact that F is built from a potential.
In components, Eq. (9.342) reads

$$\partial_{[\lambda} F_{\mu\nu]} = 0, \tag{9.343}$$

which reproduces $\nabla \cdot \mathbf{B} = 0$ and $\nabla \times \mathbf{E} + \frac{1}{c}\partial_t \mathbf{B} = 0$ in the usual $3+1$ decomposition. What appears as two separate vector identities is in fact a single four-dimensional geometric statement.

Inhomogeneous equations from Hodge dual.

In order to write the inhomogeneous Maxwell equations compactly, one typically introduces the Hodge star operator $\star$, defined by the metric g and the spacetime orientation. For a 2-form F, the dual is another 2-form $\star F$ with components

$$(\star F)_{\mu\nu} = \frac{1}{2}\, \varepsilon_{\mu\nu\rho\sigma} F^{\rho\sigma}, \tag{9.344}$$

where indices are raised with $g^{\mu\nu}$ and $\varepsilon_{\mu\nu\rho\sigma}$ is the Levi-Civita tensor density. The Hodge star is what allows one to convert a flux through surfaces into a circulation around their boundaries, encoding the metric structure into the differential calculus.
Introduce the electric four-current 1-form[2]

$$j = j_\mu\, dx^\mu, \qquad j^\mu = (c\rho, \mathbf{j}). \tag{9.345}$$

Then Maxwell's inhomogeneous equations take the compact form

$$d\star F = \frac{4\pi}{c} \star j. \tag{9.346}$$

[2] Equivalently one may use the current 3-form; conventions differ by a Hodge dual. The present choice keeps the link with j^μ transparent.

This single equation contains both Gauss' law for the electric field and the Ampère-Maxwell law. Taking d of both sides and using $d^2 = 0$ gives

$$d{\star}j = 0, \tag{9.347}$$

which is precisely charge conservation written in geometric language.

Pause and think
Explain physically why this formulation can be so powerful.

Equations (9.338)–(9.346) show that electromagnetism is governed by two universal geometric principles:

$$F \text{ is (locally) exact,} \qquad dF = 0, \qquad d{\star}F = \frac{4\pi}{c}\,{\star}j. \tag{9.348}$$

No coordinates, no Christoffel symbols, and no special spacetime structure are needed beyond a metric and an orientation. For this reason, the same equations hold verbatim on curved spacetimes: one simply interprets d and $\star$ as the exterior calculus associated with $(\mathcal{M}, g)$.
Moreover, the form language makes global issues precise. If $dF = 0$ but F is not globally exact, then no single-valued potential A exists on all of $\mathcal{M}$. This is not a technical annoyance: it encodes real physical phenomena. The Aharonov-Bohm effect is a manifestation of the fact that A can be locally written but globally obstructed. Similarly, magnetic monopoles correspond to configurations for which $dF \neq 0$ or, more subtly, for which F is closed but not exact in the presence of nontrivial topology (see Chap. 6). What looks like "exotic physics" in components becomes a transparent statement about the global geometry of differential forms.

9.15 A Third Look at QED

In early relativistic quantum theory, the Dirac equation produced both positive- and negative-energy solutions. A consistent many-body interpretation required a redefinition of the vacuum and of what is meant by "particles" and "antiparticles". One historically important route (associated with Furry and Oppenheimer) was to *modify the quantization scheme of the Dirac field* so that negative-energy modes are not interpreted as physical electrons with negative energy, but are reinterpreted as *positrons* and absorbed into the definition of the vacuum. In modern language this is accomplished by expanding the quantized Dirac field in a complete set of mode solutions and by assigning creation/annihilation operators to positive- and negative-frequency parts in a way compatible with causality and Lorentz covariance.

A second, equally important idea (the *Furry picture* or *bound interaction representation*) is tailored to problems where a strong *classical external field* is present (e.g. the Coulomb field of a heavy nucleus). In this framework, the external field is treated exactly, while the quantization and perturbative expansion are applied only to the interaction with the *radiation* field. This representation provides the natural setting for bound-state QED and for processes in strong electromagnetic fields.

In what follows, we present a self-contained and pedagogical formulation that (i) clarifies the *particle/antiparticle* reinterpretation at the level of field quantization and (ii) constructs the Furry picture in a way directly usable in calculations.

9.15.1 Furry-Oppenheimer Quantization and the Furry Picture

The conserved Noether current of the global $U(1)$ symmetry is

$$j^\mu(x) = -|e| \, : \bar{\psi}(x)\gamma^\mu\psi(x) :, \tag{9.349}$$

and the charge operator $Q = \int d^3x \, j^0$ becomes

$$Q = -|e| \int \frac{d^3\mathbf{p}}{(2\pi)^3} \sum_s \left[b_s^\dagger(\mathbf{p})b_s(\mathbf{p}) - d_s^\dagger(\mathbf{p})d_s(\mathbf{p})\right]. \tag{9.350}$$

Thus, excitations created by $b^\dagger$ carry charge $-|e|$ (electrons), while those created by $d^\dagger$ carry charge $+|e|$ (positrons). This is the operational content of the early "Furry-Oppenheimer" reinterpretation at the level of quantum fields: negative-frequency components are not discarded but reinterpreted as antiparticle creation operators, and the vacuum is defined accordingly.

► **Pause and think**
Before going on, manage to understand how to recover classical fields from QED: coherent-state limit.

At this stage, the statement that "classical electromagnetism is the $\hbar \to 0$ limit of QED" is made concrete by coherent states of the quantized radiation field. A coherent state has a well-defined complex amplitude and, in the large occupation limit, it reproduces a classical Maxwell solution in expectation value.

Hence, for the free electromagnetic field in Coulomb gauge one may write schematically

$$\mathbf{A}(\mathbf{x},t) = \sum_{\lambda=1,2} \int \frac{d^3k}{(2\pi)^{3/2}} \sqrt{\frac{2\pi\hbar c^2}{\omega_k}} \left[\boldsymbol{\epsilon}_\lambda(\mathbf{k})\, a_{\mathbf{k}\lambda}\, e^{i(\mathbf{k}\cdot\mathbf{x}-\omega_k t)} + \boldsymbol{\epsilon}_\lambda^*(\mathbf{k})\, a_{\mathbf{k}\lambda}^\dagger\, e^{-i(\mathbf{k}\cdot\mathbf{x}-\omega_k t)}\right], \tag{9.351}$$

with $\omega_k = c|\mathbf{k}|$. A coherent state $|\alpha\rangle$ is defined by

$$a_{\mathbf{k}\lambda}|\alpha\rangle = \alpha_{\mathbf{k}\lambda}|\alpha\rangle. \tag{9.352}$$

Taking the expectation value,

$$\langle\alpha|\mathbf{A}(\mathbf{x},t)|\alpha\rangle = \sum_\lambda \int \frac{d^3k}{(2\pi)^{3/2}} \sqrt{\frac{2\pi\hbar c^2}{\omega_k}} \left[\epsilon_\lambda(\mathbf{k})\,\alpha_{\mathbf{k}\lambda}\,e^{i(\mathbf{k}\cdot\mathbf{x}-\omega_k t)} + \text{c.c.}\right]. \tag{9.353}$$

Equation (9.353) is precisely a classical superposition of plane waves.

▶ **Driving exercise**
Maxwell equations as expectation-value dynamics.
In Heisenberg picture the operator field satisfies

$$\partial_\mu \hat{F}^{\mu\nu} = \frac{4\pi}{c}\,\hat{j}^\nu, \tag{9.354}$$

where $\hat{j}^\nu$ is the matter current operator. In a semiclassical regime (coherent photons + matter in a state with sharply peaked current),

$$F^{\mu\nu}_{\text{cl}} \equiv \langle\hat{F}^{\mu\nu}\rangle, \qquad j^\nu_{\text{cl}} \equiv \langle\hat{j}^\nu\rangle, \tag{9.355}$$

one obtains, by taking expectation values,

$$\partial_\mu F^{\mu\nu}_{\text{cl}} = \frac{4\pi}{c}\,j^\nu_{\text{cl}}, \tag{9.356}$$

i.e. the classical Maxwell equations for the mean field. The classical limit is thus not an abstract $\hbar \to 0$ slogan but a statement about states: coherent states minimize field fluctuations and make the electromagnetic field behave classically.

9.15.2 External-Field QED and the Furry Picture

Many physically relevant systems involve a strong classical background field, such as the Coulomb field of a nucleus. In such cases, it is advantageous to incorporate the external field nonperturbatively from the outset. One starts from QED in an external four-potential $A^{\text{ext}}_\mu(x)$ and a quantized radiation field $A_\mu(x)$:

$$\mathcal{L} = \bar{\psi}\left(i\gamma^\mu D^{\text{ext}}_\mu - m\right)\psi - \frac{1}{4}F_{\mu\nu}F^{\mu\nu} - |e|\,\bar{\psi}\gamma^\mu A_\mu\psi, \qquad D^{\text{ext}}_\mu \equiv \partial_\mu + i|e|A^{\text{ext}}_\mu. \tag{9.357}$$

The Furry picture is defined by *choosing the unperturbed Hamiltonian* to include the external field exactly:

$$H_0^{(\mathrm{F})} \equiv H_{\mathrm{Dirac}}[A^{\mathrm{ext}}] + H_{\mathrm{EM}}^{\mathrm{free}}, \qquad H_I^{(\mathrm{F})} \equiv -\int d^3x\, |e| \, : \bar{\psi}\gamma^\mu A_\mu \psi :, \tag{9.358}$$

so that perturbation theory is performed only in the interaction with the quantized photon field, while the bound-state structure generated by A_μ^{ext} is built into the "free" theory. This is the essence of the bound interaction (Furry) representation.

Driving exercise
Mode expansion in the external field.
The "free" fermion field in the Furry picture solves the Dirac equation in the external background:

$$\left(i\gamma^\mu D_\mu^{\mathrm{ext}} - m\right) \psi^{(\mathrm{F})}(x) = 0. \tag{9.359}$$

Let $\{U_n(x)\}$ and $\{V_n(x)\}$ be complete sets of positive- and negative-frequency solutions (here n is a collective label: discrete bound-state indices and/or continuous scattering quantum numbers). One can normalize them using the conserved Dirac inner product on a constant-time hypersurface:

$$(U_n, U_{n'}) \equiv \int d^3x\, U_n^\dagger(\mathbf{x}, t)\, U_{n'}(\mathbf{x}, t) = \delta_{nn'}, \qquad (V_n, V_{n'}) = \delta_{nn'}, \qquad (U_n, V_{n'}) = 0. \tag{9.360}$$

Accordingly, the field operator is expanded as

$$\psi^{(\mathrm{F})}(x) = \sum_n \left[b_n\, U_n(x) + d_n^\dagger\, V_n(x)\right], \tag{9.361}$$

with canonical anticommutation relations

$$\{b_n, b_{n'}^\dagger\} = \delta_{nn'}, \qquad \{d_n, d_{n'}^\dagger\} = \delta_{nn'}, \tag{9.362}$$

and the *Furry vacuum* defined by

$$b_n\, |0\rangle_{\mathrm{F}} = 0, \qquad d_n\, |0\rangle_{\mathrm{F}} = 0, \qquad \forall\, n. \tag{9.363}$$

This construction parallels the free-field case, but single-particle wavefunctions are now *exact eigenmodes in the external field*. For a static Coulomb potential, U_n includes the Dirac bound states of the hydrogenic spectrum and the continuum scattering states.

The external field is not quantized and is not treated as a perturbation; it defines the "background" with respect to which particles and antiparticles are defined. This is indispensable in bound-state QED, where the dominant dynamics is Coulombic and radiative corrections are comparatively small.

9.15.3 Furry-Picture Propagator and Wick Contractions

The fundamental contraction in perturbation theory is the time-ordered two-point function in the external background:

$$S_{\mathrm{F}}^{\mathrm{ext}}(x, y) \equiv {}_{\mathrm{F}}\langle 0|\, T\left\{\psi^{(\mathrm{F})}(x)\bar{\psi}^{(\mathrm{F})}(y)\right\} |0\rangle_{\mathrm{F}}. \tag{9.364}$$

In terms of the external-field mode functions, one obtains explicitly

$$S_{\mathrm{F}}^{\mathrm{ext}}(x, y) = \theta(x^0 - y^0) \sum_n U_n(x)\bar{U}_n(y) - \theta(y^0 - x^0) \sum_n V_n(x)\bar{V}_n(y). \tag{9.365}$$

Equivalently, $S_{\mathrm{F}}^{\mathrm{ext}}$ is the Green's function of the Dirac operator in the background:

$$\left(i\gamma^\mu D_\mu^{\mathrm{ext}} - m\right) S_{\mathrm{F}}^{\mathrm{ext}}(x, y) = \delta^{(4)}(x - y). \tag{9.366}$$

In practical computations, Eqs. (9.364)–(9.366) replace the free propagator. Diagrammatically, internal electron lines correspond to $S_{\mathrm{F}}^{\mathrm{ext}}$, while photon lines remain those of the quantized field (usually the free photon propagator, unless one also includes radiative corrections).

9.15.4 S-Matrix in the Furry Picture

The interaction Hamiltonian in the Furry picture is

$$H_I^{(\mathrm{F})}(t) = \int d^3x\, |e| \,:\bar{\psi}^{(\mathrm{F})}(\mathbf{x}, t)\gamma^\mu \psi^{(\mathrm{F})}(\mathbf{x}, t)\, A_\mu^{(\mathrm{F})}(\mathbf{x}, t):, \tag{9.367}$$

where $A_\mu^{(\mathrm{F})}$ is the quantized photon field evolved with $H_0^{(\mathrm{F})}$ (in practice, this is the usual free photon field, because the external field does not directly modify the photon sector at the level of the unperturbed Hamiltonian). The time-evolution operator is then given by the standard Dyson series:

$$S = T \exp\left[-i \int_{-\infty}^{+\infty} dt\, H_I^{(\mathrm{F})}(t)\right]. \tag{9.368}$$

All perturbative amplitudes follow from Eq. (9.368) by Wick contraction, with the replacement of the free fermion propagator by S_F^{ext}.

Furry theorem. In ordinary QED, any fermion loop with an odd number of external photon legs vanishes (charge conjugation symmetry). In external-field QED, analogous selection rules exist but must be formulated with care because the background may break some discrete symmetries. Nevertheless, for many common backgrounds (notably a static Coulomb field) one retains a useful loop-odd/loop-even structure that simplifies calculations; this is often referred to in the literature in the context of "Furry's theorem" and its external-field generalizations.

▶ Driving exercise

A pedagogical example: bound-state radiative correction as a Furry-picture perturbation.

As a model example, consider a hydrogenic bound state $|a\rangle$ defined by the external-field Dirac Hamiltonian $H_{\text{Dirac}}[A^{\text{ext}}]$:

$$H_{\text{Dirac}}[A^{\text{ext}}]\,|a\rangle = E_a\,|a\rangle. \tag{9.369}$$

Radiative effects (self-energy, vacuum polarization, vertex corrections) are computed as perturbative corrections due to $H_I^{(F)}$ in Eq. (9.367). Schematically, the first nontrivial energy shift arises at order e^2:

$$\Delta E_a^{(2)} \sim {}_{\text{F}}\langle a|\, T\left\{\frac{(-i)^2}{2!}\int d^4x\, d^4y\; \mathcal{H}_I^{(F)}(x)\,\mathcal{H}_I^{(F)}(y)\right\}|a\rangle_{\text{F}}, \tag{9.370}$$

where $\mathcal{H}_I^{(F)}(x) = |e| : \bar{\psi}^{(F)}(x)\gamma^\mu\psi^{(F)}(x)A_\mu(x) :$. The Wick contractions generate the familiar loop topologies, but with the crucial difference that internal electron lines are built from the external-field propagator S_F^{ext} rather than the free one. This is precisely what makes the formalism adapted to atomic and strong-field problems.

The Furry-Oppenheimer viewpoint can be summarized in two complementary statements.

1. **Vacuum and antiparticles.** The Dirac field is quantized by splitting it into positive- and negative-frequency modes and associating the latter with antiparticle creation operators. This yields a vacuum with no physical excitations and a Hamiltonian bounded from below.

2. **External fields treated exactly (Furry picture).** When a classical background is present, one defines the unperturbed theory using the Dirac equation in that background, expands the field in its exact eigenmodes, and performs perturbation theory only in the coupling to the quantized radiation field. The resulting propagator is the Green's function of the Dirac operator in the background, and all diagrammatics proceed with this replacement.

9.15.5 The Schwinger Effect and Pair Production

The *Schwinger effect* is the paradigmatic example of a genuinely *nonperturbative* phenomenon in QED: in a sufficiently strong, slowly varying electric field, the QED vacuum becomes unstable and produces real electron-positron pairs. In contrast to ordinary radiative corrections (which are organized as power series in the fine-structure constant α), the pair production rate is controlled by an essential singularity of the form

$$\exp\left(-\frac{\pi m^2}{|e|E}\right),$$

which cannot be recovered by any finite order in perturbation theory.

A useful intuition comes from energy balance. To create an e^-e^+ pair one must supply at least $2mc^2$ (we set $c = 1$ hereafter). In a uniform electric field E, a charge $|e|$ gains work $|e|E\,\ell$ over a distance ℓ. Equating $|e|E\,\ell \sim 2m$ suggests a characteristic length scale

$$\ell_\star \sim \frac{2m}{|e|E}. \tag{9.371}$$

When $\ell_\star$ becomes comparable to the Compton wavelength $\lambda_C \sim 1/m$, quantum fluctuations can be promoted to real pairs. This heuristic criterion yields the celebrated *critical field*

$$E_c \equiv \frac{m^2}{|e|}, \tag{9.372}$$

which sets the natural scale for strong-field QED.

The modern viewpoint is that the Schwinger effect is a manifestation of *vacuum decay* in an external background: the probability for the vacuum to remain the vacuum in the presence of E decreases exponentially with spacetime volume, and the pair production rate is related to the imaginary part of an effective action.

9.15.6 Vacuum Persistence Amplitude and the Effective Action

Consider QED in a prescribed classical electromagnetic background A_μ^{ext}. The vacuum-to-vacuum transition amplitude in this background is

$$\langle 0_{\text{out}} \mid 0_{\text{in}} \rangle_{A^{\text{ext}}} \equiv e^{iW[A^{\text{ext}}]}, \tag{9.373}$$

where $W[A^{\text{ext}}]$ is the (in-out) *effective action*. Its imaginary part controls the vacuum persistence probability:

$$\mathcal{P}_0 \equiv |\langle 0_{\text{out}} \mid 0_{\text{in}} \rangle_{A^{\text{ext}}}|^2 = e^{-2\Im W[A^{\text{ext}}]}. \tag{9.374}$$

If $\Im W > 0$, the probability $\mathcal{P}_0$ is suppressed, signaling that the vacuum is unstable and decays by producing particles.

For a static, spatially uniform electric field occupying a large four-volume VT (spatial volume V and duration T), one defines the pair-production rate per unit volume as the quantity Γ satisfying

$$\mathcal{P}_0 = e^{-\Gamma VT} \quad \Longrightarrow \quad \Gamma = \frac{2}{VT}\,\Im W. \tag{9.375}$$

Thus the computation reduces to extracting $\Im W$ in a constant electric field.

9.15.7 One-Loop Effective Action in a Constant Field

At one loop, the effective action is obtained by integrating out the fermions in the background field. In functional form,

$$e^{iW[A^{\text{ext}}]} = \int \mathcal{D}\bar{\psi}\,\mathcal{D}\psi\, \exp\left\{i\int d^4x\, \bar{\psi}\big(i\not{D} - m\big)\psi\right\} \propto \det\big(i\not{D} - m\big), \tag{9.376}$$

where $D_\mu = \partial_\mu + i|e|A_\mu^{\text{ext}}$ is the background-covariant derivative. Up to an irrelevant constant,

$$W[A^{\text{ext}}] = -i\,\ln\det\big(i\not{D} - m\big). \tag{9.377}$$

A standard trick is to square the Dirac operator to remove gamma matrices in the determinant:

$$\ln\det(i\not{D} - m) = \frac{1}{2}\,\ln\det\Big((i\not{D})^2 - m^2\Big), \tag{9.378}$$

since $\det(i\not{D} - m) = \det(-i\not{D} - m)$ and the product gives the squared operator. Using

$$(i\not{D})^2 = -D^2 - \frac{|e|}{2}\sigma^{\mu\nu}F_{\mu\nu}, \qquad \sigma^{\mu\nu} \equiv \frac{i}{2}[\gamma^\mu, \gamma^\nu], \tag{9.379}$$

one obtains

$$W[A^{\text{ext}}] = \frac{i}{2}\,\mathrm{Tr}\,\ln\!\left(-D^2 - \frac{|e|}{2}\sigma^{\mu\nu}F_{\mu\nu} - m^2\right). \tag{9.380}$$

Schwinger introduced the *proper-time* representation

$$\ln\frac{a}{b} = \int_0^\infty \frac{ds}{s}\left(e^{-isb} - e^{-isa}\right), \tag{9.381}$$

which, applied to the operator logarithm, yields

$$W[A^{\text{ext}}] = -\frac{i}{2}\int_0^\infty \frac{ds}{s}\, e^{-is(m^2 - i\epsilon)}\,\mathrm{Tr}\,\exp\!\left\{-is\left(-D^2 - \frac{|e|}{2}\sigma^{\mu\nu}F_{\mu\nu}\right)\right\} + (\text{subtractions}). \tag{9.382}$$

The "subtractions" implement renormalization (vacuum energy and charge renormalization) and do not affect the *imaginary* part responsible for pair production.

9.15.8 The Origin of the Imaginary Part

For a constant, uniform electric field E (say along the z-direction), the trace in Eq. (9.382) can be computed exactly. The result is the celebrated Euler-Heisenberg-Schwinger effective Lagrangian $\mathcal{L}_{\text{eff}}(E)$ such that $W = \int d^4x\, \mathcal{L}_{\text{eff}}$. The key physical point is that, for an electric field, $\mathcal{L}_{\text{eff}}$ develops an *imaginary part*. The mathematical origin is the appearance of poles on the integration contour in proper time s, whose residues generate $\Im\mathcal{L}_{\text{eff}}$.

For spinor QED one finds

$$\Im\mathcal{L}_{\text{eff}}(E) = \frac{(|e|E)^2}{8\pi^3}\sum_{n=1}^{\infty}\frac{1}{n^2}\exp\!\left(-\frac{n\pi m^2}{|e|E}\right). \tag{9.383}$$

Using Eq. (9.375) and $W = \int d^4x\, \mathcal{L}_{\text{eff}}$, the pair-production rate per unit volume is therefore

$$\Gamma(E) = 2\,\Im\mathcal{L}_{\text{eff}}(E) = \frac{(|e|E)^2}{4\pi^3}\sum_{n=1}^{\infty}\frac{1}{n^2}\exp\!\left(-\frac{n\pi m^2}{|e|E}\right). \tag{9.384}$$

The leading term ($n = 1$) already displays the nonperturbative character:

$$\Gamma(E) \sim \frac{(|e|E)^2}{4\pi^3}\exp\!\left(-\frac{\pi m^2}{|e|E}\right), \qquad (E \ll E_c). \tag{9.385}$$

The sum over n admits a semiclassical interpretation in terms of repeated instanton-like trajectories in Euclidean proper time, or equivalently as multiple tunnelling events. The exponential suppression is controlled by the ratio $m^2/(|e|E)$: for $E \ll E_c$ the effect is extraordinarily small, while for $E \gtrsim E_c$ the suppression weakens and the vacuum decays efficiently.

9.15.9 A Complementary Viewpoint: Tunnelling Picture

Although the full derivation uses the effective action, it is pedagogically useful to complement it with a tunnelling argument. In a constant electric field, the potential energy for an electron along the field direction is

$$U(z) = -|e|Ez, \tag{9.386}$$

while a positron experiences the opposite slope. Pair production may be viewed as tunneling between negative- and positive-energy continua across an energy gap $2m$ over a characteristic distance $\ell_\star$ given in Eq. (9.371). A WKB estimate yields a tunnelling probability of the form

$$\mathcal{P} \sim \exp\left(-2\int_{z_1}^{z_2} dz\,\sqrt{m^2 + p_\perp^2 - (\omega + |e|Ez)^2}\right), \tag{9.387}$$

which, upon optimizing over the transverse momentum $p_\perp$ and energy ω, reproduces the characteristic Schwinger exponent $\exp(-\pi m^2/|e|E)$. This makes explicit that the effect is essentially a relativistic tunnelling process driven by the external field.

Pause and think
Ask yourself about the limit of validity of the above treatment.

Equation (9.384) is exact at one loop for a strictly constant field. For realistic backgrounds, it remains a reliable approximation when the field varies slowly on the Compton scale,

$$\frac{|\partial_\mu E|}{E} \ll m, \tag{9.388}$$

so that locally the field can be treated as approximately constant (locally-constant-field approximation). Beyond this regime, one must address nonlocal and time-dependent effects explicitly; nevertheless, the exponential dependence on $m^2/(|e|E)$ remains the defining signature of Schwinger pair creation.

The same framework extends to:

- **scalar QED**, where the prefactor and alternating signs differ, but the exponent remains $\exp(-\pi m^2/|e|E)$;
- **combined electric and magnetic fields**, where $\Im\mathcal{L}_{\text{eff}}$ depends on the two invariants $F_{\mu\nu}F^{\mu\nu}$ and $F_{\mu\nu}\widetilde{F}^{\mu\nu}$;
- **curved spacetime and horizons**, where the notion of vacuum is observer dependent and pair creation phenomena acquire a geometric interpretation (see Chap. 10).

9.16 Proposed Exercises

Exercise 1

Consider Maxwell theory in Minkowski spacetime (Gaussian units) in terms of the four-potential $A^\mu = (\phi, \mathbf{A})$ and field strength $F_{\mu\nu} = \partial_\mu A_\nu - \partial_\nu A_\mu$. Assume smooth fields vanishing sufficiently fast at spatial infinity.

(a) Starting from the gauge transformation

$$A_\mu \mapsto A'_\mu = A_\mu + \partial_\mu \chi, \tag{9.389}$$

show that $F_{\mu\nu}$ is gauge invariant. Explain why Eq. (9.389) implies that not all components of A^μ can correspond to physical degrees of freedom.

(b) Impose the Lorenz gauge condition

$$\partial_\mu A^\mu = 0. \tag{9.390}$$

Show that in vacuum the Maxwell equations reduce to

$$\Box A^\mu = 0, \tag{9.391}$$

where $\Box \equiv \partial_\mu \partial^\mu$. Determine the residual gauge freedom compatible with Eq. (9.390) and Eq. (9.391).

(c) Consider a plane-wave ansatz $A^\mu(x) = \epsilon^\mu e^{-ik\cdot x}$ in vacuum. Using Eqs. (9.390) and (9.391), show that

$$k^2 = 0, \qquad k_\mu \epsilon^\mu = 0. \tag{9.392}$$

Discuss how the residual gauge freedom can be used to eliminate unphysical polarizations and argue that only two independent polarization states remain.

(d) In Coulomb gauge $\nabla \cdot \mathbf{A} = 0$, decompose the electric field as $\mathbf{E} = \mathbf{E}_L + \mathbf{E}_T$ with $\nabla \times \mathbf{E}_L = 0$ and $\nabla \cdot \mathbf{E}_T = 0$. Show that, for given sources $(\rho, \mathbf{j})$, the longitudinal sector is fixed instantaneously by Gauss' law, while the transverse sector carries the propagating degrees of freedom. State the physical content of this separation in one concise paragraph.

Exercise 2

Consider an isotropic, homogeneous, linear medium characterized by a complex refractive index $n(\omega) = n'(\omega) + in''(\omega)$ for $\omega > 0$, with $n''(\omega) \geq 0$ (passive medium). Assume the standard Fourier convention $f(t) = \int \frac{d\omega}{2\pi} e^{-i\omega t} \tilde{f}(\omega)$.

(a) A monochromatic plane wave propagating along z has the form

$$E(z,t) = E_0\, e^{ik(\omega)z - i\omega t}, \qquad k(\omega) = \frac{n(\omega)\,\omega}{c}. \tag{9.393}$$

Show that $n''(\omega) > 0$ implies attenuation in z and identify the absorption coefficient in terms of $n''(\omega)$ and ω.

(b) Define phase and group velocities by

$$v_{\text{ph}}(\omega) = \frac{\omega}{\operatorname{Re} k(\omega)}, \qquad v_g(\omega) = \left(\frac{d \operatorname{Re} k(\omega)}{d\omega}\right)^{-1}. \tag{9.394}$$

Express v_{ph} and v_g in terms of $n'(\omega)$ and $dn'/d\omega$.

(c) Explain why causality of the linear response implies analyticity of $n(\omega)$ in the upper half complex ω-plane. Using this, motivate the Kramers-Kronig relations in the schematic form

$$n'(\omega) - 1 = \frac{2}{\pi} \mathcal{P} \int_0^\infty d\omega' \, \frac{\omega' n''(\omega')}{\omega'^2 - \omega^2}, \qquad n''(\omega) = -\frac{2\omega}{\pi} \mathcal{P} \int_0^\infty d\omega' \, \frac{n'(\omega') - 1}{\omega'^2 - \omega^2}. \tag{9.395}$$

State clearly what the symbol $\mathcal{P}$ means.

(d) Superluminal group velocity ($v_g > c$) can occur near absorption lines without violating relativistic causality. Provide a precise explanation distinguishing v_g from the signal velocity, and indicate which property of the response function is ultimately responsible for causal propagation.

Exercise 3

Consider a general nonlinear electrodynamics model whose Lagrangian density depends only on the Lorentz invariants

$$I_1 \equiv \frac{1}{2} F_{\mu\nu} F^{\mu\nu}, \qquad I_2 \equiv \frac{1}{2} F_{\mu\nu} {}^{\star}F^{\mu\nu}, \tag{9.396}$$

and on the external current through $-j_\mu A^\mu/c$:

$$\mathcal{L} = \mathcal{L}(I_1, I_2) - \frac{1}{c}\, j_\mu A^\mu. \tag{9.397}$$

(a) Vary the action with respect to A_μ and show that the field equations can be written as

$$\partial_\mu G^{\mu\nu} = \frac{4\pi}{c}\, j^\nu, \qquad \partial_{[\lambda} F_{\mu\nu]} = 0, \tag{9.398}$$

where $G^{\mu\nu}$ is defined by the constitutive relation

$$G^{\mu\nu} \equiv -4\pi\, \frac{\partial \mathcal{L}}{\partial F_{\mu\nu}} = -4\pi \left(\frac{\partial \mathcal{L}}{\partial I_1} F^{\mu\nu} + \frac{\partial \mathcal{L}}{\partial I_2}\, {}^\star F^{\mu\nu} \right). \tag{9.399}$$

Comment on the interpretation of Eq. (9.399) as "vacuum behaving as a non-linear medium".

(b) Specialize to the quartic toy model (weak-field effective theory)

$$\mathcal{L}_{\text{eff}}(I_1, I_2) = -\frac{1}{16\pi}\, F_{\mu\nu} F^{\mu\nu} + \frac{\lambda}{2}\, I_1^2 + \frac{\eta}{2}\, I_2^2, \tag{9.400}$$

with constants λ and η. Compute $G^{\mu\nu}$ explicitly and show that Maxwell theory is recovered for $\lambda = \eta = 0$.

(c) Consider a purely electrostatic configuration ($\mathbf{B} = 0$) sourced by a point charge q at the origin. Using spherical symmetry, argue that the field equation reduces to an algebraic relation between $E(r)$ and r. Derive this relation for the Lagrangian in Eq. (9.400) and obtain the leading correction to Coulomb's law for small λ.

(d) In nonlinear theories, small electromagnetic disturbances on top of a strong background field can propagate with modified characteristics (effective light cones). Provide a qualitative but precise explanation of why birefringence may occur when both invariants I_1 and I_2 enter $\mathcal{L}(I_1, I_2)$, and state one physical consequence that could be searched for in strong-field environments.

Essential Bibliography

- Jackson JD (1999) Classical electrodynamics, 3rd ed. Wiley. Chosen as the unavoidable reference for advanced classical electromagnetism. It is not the most modern in style, but remains unmatched in depth on radiation, multipoles, boundary-value problems, and relativistic electrodynamics.
- Griffiths DJ, Schroeter DF (2018) Introduction to electrodynamics, 4th ed. Cambridge University Press. Selected for clarity and physical intuition. It is ideal when one wants to check basic concepts, boundary problems, and wave propagation without drowning in formalism.

- Hehl FW, Obukhov YN (2003) Foundations of classical electrodynamics. Birkhäuser. Chosen for its geometric and covariant formulation of electromagnetism. It is the best modern reference for differential forms, constitutive relations, media, and the conceptual structure of Maxwell theory.
- Zangwill A (2013) Modern electrodynamics. Cambridge University Press. Chosen as the most modern large-scale textbook on classical electrodynamics: strong on variational principles, stress-energy tensor, radiation, and applications to contemporary physics.
- Tong D Lectures on electromagnetism. University of Cambridge, DAMTP Lecture Notes (available online). Included for its clean, symmetry-oriented presentation of Maxwell theory, gauge invariance, and relativistic formulation, useful as a conceptual complement to heavier textbooks.

Gravity

10

> To myself I seem to have been only like a boy playing on the seashore.
>
> –Isaac Newton

Abstract

Gravity occupies a singular position among the fundamental interactions: while it admits an interpretation as the exchange of a massless spin-2 mediator, its consistent nonlinear completion forces a geometric description in which spacetime itself becomes dynamical. After introducing the kinematics and dynamics of curved spacetimes, the Einstein's field equations are derived and analyzed, together with their physical content and observational consequences. The discussion is then extended to alternative geometrical formulations based on torsion and non-metricity, to modified theories such as $f(R)$, $f(T)$ and $f(Q)$ gravity, and to the problem of frame equivalence. A central part of the Chapter is devoted to black holes (BHs): their formation, classical structure, coupling to matter fields, and thermodynamic behavior, including entropy, temperature, and the information paradox. Gravity is further examined from an emergent and thermodynamic viewpoint, where field equations can be interpreted as equations of state. Finally, the Chapter addresses the quantization of gravity, from canonical approaches and effective field theory to linearized gravitons, and culminates with the cosmological constant problem and QFT in curved backgrounds. Gravity thus appears not only as the fourth interaction, but as a bridge between geometry, quantum theory, and the deep structure of spacetime itself. Differently from the other Chapters, where pedagogically useful, the Lorentz signature can be $(-, +, +, +)$.

O. Luongo, *Fundamental Interactions*,
https://doi.org/10.1007/978-3-032-27413-7_10

Fig. 10.1 Portrait of Albert Einstein by Herman Mishkin, 1921. Reprinted from Wikimedia. Public domain

A key character: Albert Einstein (1879–1955), Fig. 10.1.

Albert Einstein was born in Ulm, in the Kingdom of Württemberg, and received his early education in Germany and Switzerland. After graduating from the Swiss Federal Polytechnic in Zürich, he spent several years working at the Swiss Patent Office in Bern, a period during which he developed many of the ideas that would profoundly transform modern physics. In 1905, his *annus mirabilis*, Einstein published a series of papers that reshaped the foundations of mechanics, electrodynamics, and statistical physics.

Einstein's most enduring contribution to the theory of gravity is the formulation of the *general theory of relativity* (1915), in which gravitation is no longer described as a force acting in space, but as a manifestation of the curvature of spacetime itself. By elevating the principle of relativity to a general covariance principle and identifying the spacetime metric as the dynamical field encoding gravity, Einstein established a geometric theory in which matter and energy determine spacetime curvature through the field equations $G_{\mu\nu} = 8\pi G\, T_{\mu\nu}$. This framework unified gravitation with the structure of spacetime and replaced Newtonian gravity with a fundamentally relativistic and field-theoretic description. Einstein's theory successfully accounted for previously unexplained phenomena, such as the perihelion precession of Mercury and the deflection of light by massive bodies, and predicted entirely new effects, including gravitational waves and BHs. Beyond its empirical successes, general relativity (GR) introduced a radically new conceptual understanding of gravity, in which geometry and dynamics are inseparably linked. Although Einstein later devoted much of his life to the search for a unified field theory, his formulation of GR remains the cornerstone of modern gravitational physics. Its influence extends from classical gravitational phenomena to cosmology and high-energy theoretical physics, where it continues to guide contemporary research into the structure of spacetime and the fundamental interactions of nature.

10.1 Going Beyond Special Relativity

GR is the relativistic theory in which gravitation is identified with the geometry of spacetime. While special relativity can be formulated in arbitrary (including non-inertial) coordinate systems, it is a theory on a fixed, flat background. GR departs from this by promoting the spacetime metric $g_{\mu\nu}$ to a dynamical field and by encoding gravitational effects in curvature.

The conceptual bridge between inertial motion and gravitation is provided by the **equivalence principle**, which can be articulated in three increasingly strong forms.

Weak equivalence principle (WEP). *The trajectories of freely falling test bodies are independent of their internal composition and structure. Operationally, this means that the gravitational acceleration measured in a given external field is universal for all sufficiently small, non-self-gravitating bodies.*

Einstein equivalence principle (EEP). In any sufficiently small neighborhood of any event, the outcome of any local *non-gravitational* experiment in a freely falling laboratory is independent of the laboratory velocity and position in spacetime. Mathematically, EEP leads to the existence of local inertial frames in which, at a point p,

$$g_{\mu\nu}(p) = \eta_{\mu\nu}, \qquad \partial_\alpha g_{\mu\nu}(p) = 0. \tag{10.1}$$

This local inertial frame is realized using so-called normal coordinates at the aforementioned point p.

Strong equivalence principle (SEP). EEP is extended to include gravitational experiments and self-gravitating bodies. SEP is more restrictive: it holds in GR, but fails in many modified gravity theories. In this book we mainly rely on EEP and the metric nature of gravity.

A key consequence is that, in a sufficiently small neighborhood of any event, one can always choose a freely falling frame in which the effects of gravity are *locally* absent, so that the connection coefficients vanish at the chosen point. However, this does *not* imply that gravity is a coordinate artifact: tidal effects, which depend on spacetime curvature, cannot be removed by any choice of coordinates and remain measurable through relative accelerations of nearby freely falling bodies.

► **Pause and think**
If gravity turns out to be local as working hypothesis, how can you write the metric?

Accordingly, the gravitational field is described by the spacetime metric, viewed as a smooth tensor field on the manifold,

$$g_{\mu\nu} = g_{\mu\nu}(x), \qquad x \in \mathcal{M}. \tag{10.2}$$

In regimes where gravity is weak and velocities are non-relativistic, it is useful to regard the metric as a small perturbation of Minkowski spacetime,

$$g_{\mu\nu}(x) = \eta_{\mu\nu} + h_{\mu\nu}(x), \qquad |h_{\mu\nu}| \ll 1, \tag{10.3}$$

with $\eta_{\mu\nu}$ the Minkowski metric.

10.1.1 Dynamics and Kinematics on Curved Spacetime

On a curved spacetime $(\mathcal{M}, g_{\mu\nu})$, partial derivatives do not transform covariantly. The covariant derivative ∇_μ is defined by a connection $\Gamma^\rho_{\mu\nu}$,

$$\nabla_\mu V^\rho = \partial_\mu V^\rho + \Gamma^\rho_{\mu\alpha} V^\alpha, \qquad \nabla_\mu \omega_\nu = \partial_\mu \omega_\nu - \Gamma^\alpha_{\mu\nu} \omega_\alpha. \tag{10.4}$$

In GR one assumes (i) metric compatibility $\nabla_\rho g_{\mu\nu} = 0$ and (ii) vanishing torsion $\Gamma^\rho_{\mu\nu} = \Gamma^\rho_{\nu\mu}$, which uniquely select the Levi-Civita connection,

$$\Gamma^\rho_{\mu\nu} = \frac{1}{2} g^{\rho\sigma} \Big(\partial_\mu g_{\nu\sigma} + \partial_\nu g_{\mu\sigma} - \partial_\sigma g_{\mu\nu} \Big). \tag{10.5}$$

The motion of freely falling test particles is described by timelike geodesics. A convenient derivation starts from the proper-time action

$$S_{\rm pp} = -m \int d\tau = -m \int d\lambda \sqrt{-g_{\mu\nu}\frac{dx^\mu}{d\lambda}\frac{dx^\nu}{d\lambda}}, \tag{10.6}$$

where λ is any monotonic parameter along the worldline. Varying $x^\mu(\lambda)$ yields the geodesic equation

$$\frac{d^2x^\mu}{d\tau^2} + \Gamma^\mu_{\alpha\beta}\frac{dx^\alpha}{d\tau}\frac{dx^\beta}{d\tau} = 0. \tag{10.7}$$

Null geodesics ($ds^2 = 0$) describe the propagation of light rays in the geometric-optics limit.

If a spacetime admits a Killing vector field ξ^μ satisfying

$$\nabla_{(\mu}\xi_{\nu)} = 0, \tag{10.8}$$

then along any geodesic with tangent $u^\mu = dx^\mu/d\tau$ the quantity

$$E_\xi \equiv -\xi_\mu u^\mu \tag{10.9}$$

is conserved. In stationary spacetimes, $\xi^\mu = \partial_t$ yields a conserved energy, while in axisymmetric spacetimes $\xi^\mu = \partial_\varphi$ yields a conserved angular momentum.

To transport a vector parallel to itself along a curve, one may require that it keeps a constant angle with the local geodesic direction along which it is carried (geodesics are *autoparallel* curves). In a curved space, parallel transport along a closed loop does *not* in general return the vector to its original orientation. A standard illustration is provided by a path on the surface of a sphere made of three great-circle arcs that bound a spherical octant (see Fig. 10.2). After completing the loop, the transported vector is rotated with respect to its initial direction.

This immediately suggests a practical definition: a space in which parallel transport along any closed curve is unambiguous (i.e., brings vectors back to themselves) is said to be *flat*; otherwise it is *curved*.

Let B_β be a covector field. Parallel transport along an infinitesimal displacement dx^μ is expressed by the vanishing of its covariant differential,

$$DB_\beta = B_{\beta;\mu}\, dx^\mu = 0, \tag{10.10}$$

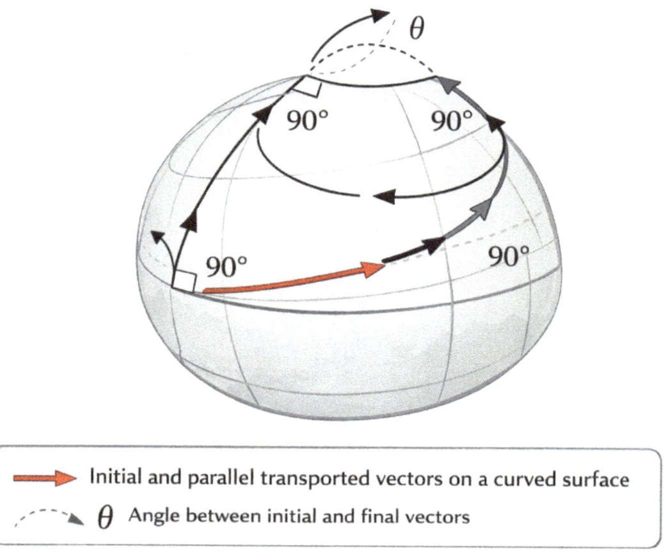

Fig. 10.2 Closed-loop parallel transport on a curved two-dimensional domain: the mismatch after one circuit is encoded in the net rotation θ (holonomy)

so that, along a curve γ, the component variation satisfies

$$dB_\beta = B_{\beta,\mu}\,dx^\mu = \Gamma^\alpha_{\beta\mu}\,B_\alpha\,dx^\mu, \tag{10.11}$$

where we used $B_{\beta;\mu} = B_{\beta,\mu} - \Gamma^\alpha_{\beta\mu}B_\alpha$.

After transport along a *closed* curve C, the covector returns to itself if

$$\oint_C dB_\beta = 0, \rightleftharpoons \oint_C \Gamma^\alpha_{\beta\mu}\,B_\alpha\,dx^\mu = 0. \tag{10.12}$$

To relate the line integral to a local integrability condition, consider the 1-form $\omega_\mu \equiv \Gamma^\alpha_{\beta\mu}B_\alpha$ (with the index β regarded as a fixed label). If the loop C bounds an oriented surface S, Stokes' theorem gives

$$\oint_C \omega_\mu\,dx^\mu = \int_S \left(\omega_{\nu,\mu} - \omega_{\mu,\nu}\right)\,dS^{\mu\nu}, \tag{10.13}$$

so a sufficient condition for the vanishing of the circulation for *arbitrary* infinitesimal loops is the vanishing of the curl:

$$\left[\Gamma^\alpha_{\beta\nu}B_\alpha\right]_{,\mu} = \left[\Gamma^\alpha_{\beta\mu}B_\alpha\right]_{,\nu}. \tag{10.14}$$

Expanding,

$$\Gamma^\alpha_{\beta\nu,\mu}B_\alpha + \Gamma^\alpha_{\beta\nu}B_{\alpha,\mu} = \Gamma^\alpha_{\beta\mu,\nu}B_\alpha + \Gamma^\alpha_{\beta\mu}B_{\alpha,\nu},$$

hence

$$B_\alpha \left(\Gamma^\alpha_{\beta\nu,\mu} - \Gamma^\alpha_{\beta\mu,\nu} \right) + \Gamma^\sigma_{\beta\nu} B_{\sigma,\mu} - \Gamma^\sigma_{\beta\mu} B_{\sigma,\nu} = 0. \tag{10.15}$$

From the definition of covariant derivative and the parallel-transport condition,

$$B_{\sigma;\nu} \equiv B_{\sigma,\nu} - \Gamma^\alpha_{\sigma\nu} B_\alpha = 0, \tag{10.16}$$

so that $B_{\sigma,\nu} = \Gamma^\alpha_{\sigma\nu} B_\alpha$, $B_{\sigma,\mu} = \Gamma^\alpha_{\sigma\mu} B_\alpha$.

Substituting into the previous relation yields

$$B_\alpha \left(\Gamma^\alpha_{\beta\nu,\mu} - \Gamma^\alpha_{\beta\mu,\nu} \right) + \Gamma^\sigma_{\beta\nu} \Gamma^\alpha_{\sigma\mu} B_\alpha - \Gamma^\sigma_{\beta\mu} \Gamma^\alpha_{\sigma\nu} B_\alpha = 0,$$

namely

$$B_\alpha \left(\Gamma^\alpha_{\beta\nu,\mu} - \Gamma^\alpha_{\beta\mu,\nu} + \Gamma^\sigma_{\beta\nu} \Gamma^\alpha_{\sigma\mu} - \Gamma^\sigma_{\beta\mu} \Gamma^\alpha_{\sigma\nu} \right) = 0. \tag{10.17}$$

We therefore define the *Riemann tensor*,

$$R^\alpha{}_{\beta\mu\nu} = \Gamma^\alpha_{\beta\nu,\mu} - \Gamma^\alpha_{\beta\mu,\nu} + \Gamma^\sigma_{\beta\nu} \Gamma^\alpha_{\sigma\mu} - \Gamma^\sigma_{\beta\mu} \Gamma^\alpha_{\sigma\nu}, \tag{10.18}$$

so that the closed-loop condition becomes

$$B_\alpha \, R^\alpha{}_{\beta\mu\nu} = 0. \tag{10.19}$$

Thus, requiring $\oint_C dB_\beta = 0$ for arbitrary infinitesimal loops is equivalent to the vanishing of the Riemann tensor. In particular, a necessary and sufficient condition for flatness (in the sense of path-independence of parallel transport) is

$$R^\alpha{}_{\beta\mu\nu} = 0. \tag{10.20}$$

For instance, in Minkowski spacetime with metric $\eta_{\alpha\beta} = \mathrm{diag}(+1, -1, -1, -1)$, all first and second derivatives of the metric vanish in inertial coordinates; therefore all Christoffel symbols vanish and so does $R^\alpha{}_{\beta\mu\nu}$.

In a compact "determinant" notation, one may write

$$R^\alpha{}_{\beta\mu\nu} = \begin{vmatrix} \frac{\partial}{\partial x^\mu} & \frac{\partial}{\partial x^\nu} \\ \Gamma^\alpha_{\beta\mu} & \Gamma^\alpha_{\beta\nu} \end{vmatrix} + \begin{vmatrix} \Gamma^\alpha_{\sigma\mu} & \Gamma^\alpha_{\sigma\nu} \\ \Gamma^\sigma_{\beta\mu} & \Gamma^\sigma_{\beta\nu} \end{vmatrix}. \tag{10.21}$$

Exactly the same conclusion is obtained transporting a contravariant vector. A contravariant vector A^α is parallel-transported along dx^μ if

$$DA^\alpha = A^\alpha{}_{;\mu} \, dx^\mu = 0.$$

It returns to itself after transport along a closed curve if

$$\oint_C dA^\alpha = 0,$$

and repeating the Stokes argument one finds

$$A^{\beta} R^{\alpha}{}_{\beta\mu\nu} = 0.$$

The discussion above is best regarded as geometric motivation. A fully tensorial and coordinate-independent characterization of curvature is obtained from the commutator of covariant derivatives, which defines the Riemann tensor.

Curvature can also be probed by studying *congruences* of geodesics (see Sect. 10.1.4): depending on the spacetime curvature, neighboring geodesics tend to converge or diverge. This leads to the notion of *geodesic deviation*.

Consider two nearby geodesics. Let one be

$$x^{\mu} = x^{\mu}(s), \tag{10.22}$$

and the other

$$x'^{\mu} = x^{\mu}(s) + \xi^{\mu}(s), \tag{10.23}$$

where s is an affine parameter and ξ^{μ} is the separation vector connecting points with the same value of s on the two curves. The vector ξ^{μ} is often called the *connecting vector*.

The geodesic equation for $x^{\mu}(s)$ is

$$\frac{d^2x^{\mu}}{ds^2} + \Gamma^{\mu}_{\lambda\nu}(x^{\sigma})\frac{dx^{\lambda}}{ds}\frac{dx^{\nu}}{ds} = 0, \tag{10.24}$$

while for $x'^{\mu}(s)$ it reads $\frac{d^2x'^{\mu}}{ds^2} + \Gamma^{\mu}_{\lambda\nu}(x^{\sigma} + \xi^{\sigma})\frac{dx'^{\lambda}}{ds}\frac{dx'^{\nu}}{ds} = 0$. Using $x'^{\mu} = x^{\mu} + \xi^{\mu}$ and keeping only terms linear in ξ^{μ}, we expand the connection:

$$\Gamma^{\mu}_{\lambda\nu}(x^{\sigma} + \xi^{\sigma}) \simeq \Gamma^{\mu}_{\lambda\nu}(x^{\sigma}) + \frac{\partial\Gamma^{\mu}_{\lambda\nu}}{\partial x^{\sigma}}\,\xi^{\sigma}. \tag{10.25}$$

Moreover, $\frac{dx'^{\mu}}{ds} = \frac{dx^{\mu}}{ds} + \frac{d\xi^{\mu}}{ds}$ and $\frac{d^2x'^{\mu}}{ds^2} = \frac{d^2x^{\mu}}{ds^2} + \frac{d^2\xi^{\mu}}{ds^2}$. Substituting into the primed geodesic equation and retaining only linear terms in ξ^{μ} and $d\xi^{\mu}/ds$, one gets

$$\frac{d^2x^{\mu}}{ds^2} + \frac{d^2\xi^{\mu}}{ds^2} + \Gamma^{\mu}_{\lambda\nu}\frac{dx^{\lambda}}{ds}\frac{dx^{\nu}}{ds} + 2\Gamma^{\mu}_{\lambda\nu}\frac{d\xi^{\lambda}}{ds}\frac{dx^{\nu}}{ds} + \Gamma^{\mu}_{\lambda\nu,\sigma}\frac{dx^{\lambda}}{ds}\frac{dx^{\nu}}{ds}\xi^{\sigma} = 0, \tag{10.26}$$

and subtracting the unprimed geodesic equation yields

$$\frac{d^2\xi^{\mu}}{ds^2} + 2\Gamma^{\mu}_{\lambda\nu}\frac{dx^{\lambda}}{ds}\frac{d\xi^{\nu}}{ds} + \Gamma^{\mu}_{\lambda\nu,\sigma}\frac{dx^{\lambda}}{ds}\frac{dx^{\nu}}{ds}\xi^{\sigma} = 0. \tag{10.27}$$

It is now convenient to express this in covariant form. The covariant derivative of ξ^{μ} along the reference geodesic $x^{\mu}(s)$ is

$$\frac{D\xi^{\mu}}{ds} \equiv \frac{d\xi^{\mu}}{ds} + \Gamma^{\mu}_{\nu\lambda}\frac{dx^{\nu}}{ds}\xi^{\lambda}, \tag{10.28}$$

and therefore

$$\frac{D^2\xi^\mu}{ds^2} = \frac{d}{ds}\left(\frac{d\xi^\mu}{ds} + \Gamma^\mu_{\nu\lambda}\frac{dx^\nu}{ds}\xi^\lambda\right) + \Gamma^\mu_{\rho\sigma}\frac{dx^\rho}{ds}\left(\frac{d\xi^\sigma}{ds} + \Gamma^\sigma_{\nu\lambda}\frac{dx^\lambda}{ds}\xi^\nu\right). \quad (10.29)$$

Expanding and eliminating $d^2\xi^\mu/ds^2$ via the linearized equation above, one finds that

$$\frac{D^2\xi^\mu}{ds^2} = \left(\Gamma^\mu_{\nu\lambda,\rho} - \Gamma^\mu_{\lambda\rho,\nu} + \Gamma^\mu_{\rho\sigma}\Gamma^\sigma_{\lambda\nu} - \Gamma^\mu_{\nu\sigma}\Gamma^\sigma_{\rho\lambda}\right)\frac{dx^\rho}{ds}\frac{dx^\lambda}{ds}\xi^\nu. \quad (10.30)$$

Recognizing the bracket as the Riemann tensor components, as above defined, we obtain the *geodesic deviation equation*,

$$\frac{D^2\xi^\mu}{ds^2} = R^\mu{}_{\lambda\nu\rho}\frac{dx^\rho}{ds}\frac{dx^\lambda}{ds}\xi^\nu. \quad (10.31)$$

This equation provides a direct operational measure of curvature: the second covariant derivative of the separation vector along the geodesic equals the tidal term generated by $R^\mu{}_{\lambda\nu\rho}$.

10.1.2 The Concept of Curvature

Curvature is encoded in the Riemann tensor, defined through the commutator of covariant derivatives:

$$[\nabla_\mu, \nabla_\nu]V^\rho = R^\rho{}_{\sigma\mu\nu}V^\sigma. \quad (10.32)$$

In terms of the connection,

$$R^\rho{}_{\sigma\mu\nu} = \partial_\mu\Gamma^\rho_{\nu\sigma} - \partial_\nu\Gamma^\rho_{\mu\sigma} + \Gamma^\rho_{\mu\alpha}\Gamma^\alpha_{\nu\sigma} - \Gamma^\rho_{\nu\alpha}\Gamma^\alpha_{\mu\sigma}. \quad (10.33)$$

A direct physical meaning of $R^\rho{}_{\sigma\mu\nu}$ appears in the geodesic deviation equation. In addition, some properties follow.

- For the Levi-Civita connection (torsion-free and metric compatible), the Riemann tensor obeys the following algebraic symmetries:

$$R_{\alpha\beta\mu\nu} = -R_{\beta\alpha\mu\nu}, \qquad R_{\alpha\beta\mu\nu} = -R_{\alpha\beta\nu\mu}, \qquad R_{\alpha\beta\mu\nu} = R_{\mu\nu\alpha\beta}. \quad (10.34)$$

  In addition, it satisfies the *first (algebraic) Bianchi identity*

$$R_{\alpha[\beta\mu\nu]} = 0 \quad \Longleftrightarrow \quad R_{\alpha\beta\mu\nu} + R_{\alpha\mu\nu\beta} + R_{\alpha\nu\beta\mu} = 0. \quad (10.35)$$

  The corresponding *differential Bianchi identity* reads

$$\nabla_{[\lambda}R_{\alpha\beta]\mu\nu} = 0 \quad \Longleftrightarrow \quad R_{\alpha\beta\mu\nu;\lambda} + R_{\alpha\beta\nu\lambda;\mu} + R_{\alpha\beta\lambda\mu;\nu} = 0, \quad (10.36)$$

and will play a central role in deriving a divergence-free curvature tensor in Sect. 10.2.1.

- The algebraic identities above reduce the number of independent components far below the naive count N^4 in N dimensions, which would be 256 in four dimensions. In four dimensions, this readily results in 20 **independent components**.

10.1.3 Physical Meaning of the Christoffel Symbols

The Christoffel symbols $\Gamma^{\mu}{}_{\alpha\beta}$ are often introduced as auxiliary quantities built from the metric, and then quickly relegated to the role of "connection coefficients" in covariant derivatives. This is formally correct but physically incomplete: the $\Gamma^{\mu}{}_{\alpha\beta}$ encode the *inertial structure* of the chosen coordinate system (and, in a local frame, the inertial structure of the chosen congruence of observers). Their most direct physical manifestation is that they represent the *effective gravitational field* that can be removed at a point by going to a freely falling frame, while their *spacetime variation* (and specific combinations of derivatives and quadratic terms) produces the truly invariant content of gravity, namely tidal effects.

- $\Gamma^{\mu}{}_{\alpha\beta}$ are *not* tensors. One may set them to zero at a given point by an appropriate local choice of coordinates (or local orthonormal frame). Hence, the values of $\Gamma^{\mu}{}_{\alpha\beta}$ at a point are not, by themselves, invariant observables.
- Tidal effects are invariant and are controlled by curvature. In practice, they are governed by derivatives (and specific combinations) of $\Gamma^{\mu}{}_{\alpha\beta}$, i.e., by the Riemann tensor $R^{\mu}{}_{\nu\alpha\beta}$.

The most transparent entry point is the geodesic equation, Eq. (10.24), which has the same structure as "free motion" plus a term that looks like a force. Indeed, for slow motion in a weak, quasi-static field one finds that the dominant component is $\Gamma^{i}{}_{00}$, and Eq. (10.24) reduces to

$$\frac{d^2x^i}{dt^2} \simeq -\Gamma^{i}{}_{00}. \tag{10.37}$$

In the Newtonian limit $g_{00} \simeq -(1+2\Phi)$, one has

$$\Gamma^{i}{}_{00} \simeq \partial^i\Phi, \tag{10.38}$$

so that

$$\frac{d^2x^i}{dt^2} \simeq -\partial^i\Phi, \tag{10.39}$$

recovering the usual gravitational acceleration, reinterpreting Φ as the Newtonian potential. In this regime, therefore, $\Gamma^i{}_{00}$ plays the role of the *local gravitational field* felt by slow test bodies.

However, the decisive conceptual point is that the "field" in Eq. (10.39) is *not* what makes gravity intrinsically different from inertial effects. A uniform gravitational field is locally indistinguishable from a uniformly accelerated frame; correspondingly, one can always choose local coordinates at a point such that $\Gamma^\mu{}_{\alpha\beta} = 0$ there. The genuinely gravitational content is the *relative* acceleration of nearby freely falling particles, i.e. tidal forces.

Consider a congruence of timelike geodesics with unit tangent u^μ and separation vector ξ^μ between neighboring worldlines. The invariant tidal acceleration is governed by the geodesic deviation equation

$$\frac{D^2\xi^\mu}{D\tau^2} = -R^\mu{}_{\nu\alpha\beta}\, u^\nu u^\alpha\, \xi^\beta. \tag{10.40}$$

Thus, the *curvature* (not the connection) is what a gravity gradiometer measures. In a local inertial frame comoving with a reference geodesic at a point, one may impose $\Gamma^\mu{}_{\alpha\beta} = 0$ at that point, but the tidal term remains because it depends on derivatives and quadratic combinations:

$$R^\mu{}_{\nu\alpha\beta} = \partial_\alpha \Gamma^\mu{}_{\nu\beta} - \partial_\beta \Gamma^\mu{}_{\nu\alpha} + \Gamma^\mu{}_{\sigma\alpha}\Gamma^\sigma{}_{\nu\beta} - \Gamma^\mu{}_{\sigma\beta}\Gamma^\sigma{}_{\nu\alpha}.$$

Hence, **tidal effects are the spacetime "curl" of the inertial structure encoded by Γ.**

In the Newtonian limit, Eq. (10.40) reduces to the familiar tidal equation: let $u^\mu \simeq (1, \mathbf{0})$ and $\xi^\mu \simeq (0, \boldsymbol{\xi})$; then

$$\frac{d^2\xi^i}{dt^2} \simeq -R^i{}_{0j0}\,\xi^j, \qquad R^i{}_{0j0} \simeq \partial^i\partial_j\Phi, \tag{10.41}$$

so that

$$\frac{d^2\xi^i}{dt^2} \simeq -(\partial^i\partial_j\Phi)\,\xi^j. \tag{10.42}$$

The Hessian $\partial_i\partial_j\Phi$ is the gravitational gradient responsible for stretching and squeezing: two nearby freely falling particles accelerate relative to one another precisely because $\Gamma^i{}_{00} \simeq \partial^i\Phi$ varies in space.

A useful physical interpretation.

One may regard $\Gamma^{\mu}{}_{\alpha\beta}$ as the *gauge potential* of gravity, in the same sense that A_μ is the gauge potential of electromagnetism:

- A_μ can be changed by a gauge transformation; only $F_{\mu\nu} = \partial_\mu A_\nu - \partial_\nu A_\mu$ is gauge-invariant and directly observable.
- $\Gamma^{\mu}{}_{\alpha\beta}$ can be changed (and even set to zero at a point) by coordinate/frame choice; only curvature, schematically $\partial\Gamma + \Gamma\Gamma$, is invariant and directly measurable.

In this analogy, the equivalence principle is the statement that the gravitational "potential" can be transformed away locally, while Eq. (10.40) shows that the field strength, viz. curvature, cannot.

Finally, there is an operational viewpoint that is often pedagogically effective. A physical observer is described by a worldline and a tetrad (a local orthonormal frame) carried along it. The measured "gravitational acceleration" of nearby free-fall particles is not the value of Γ in an arbitrary coordinate chart, but the *relative acceleration* obtained from Eq. (10.40). In other words, $\Gamma^{\mu}{}_{\alpha\beta}$ is best interpreted as describing how the chosen coordinates (or frame) are *non-inertial*, while curvature quantifies how spacetime itself induces tidal distortions that no coordinate choice can eliminate.

10.1.4 The Raychaudhuri Equation

The Raychaudhuri equation is a purely geometric identity controlling the evolution of the local cross-sectional deformation of a **congruence of curves**.

Consider a smooth congruence of timelike worldlines with unit tangent field u^μ normalized as

$$u^\mu u_\mu = -1, \qquad u^\nu \nabla_\nu u^\mu = a^\mu, \tag{10.43}$$

where a^μ is the four-acceleration (vanishing for geodesics). Introduce the spatial projector orthogonal to u^μ,

$$h_{\mu\nu} \equiv g_{\mu\nu} + u_\mu u_\nu, \qquad h_{\mu\nu} u^\nu = 0, \qquad h^\mu{}_\mu = 3, \tag{10.44}$$

and define the *deformation tensor*

$$B_{\mu\nu} \equiv \nabla_\nu u_\mu. \tag{10.45}$$

The tensor $B_{\mu\nu}$ encodes how the congruence changes: given a separation vector η^μ connecting neighboring worldlines, its evolution along the flow satisfies $\dot{\eta}^\mu = B^\mu{}_\nu \eta^\nu$ (up to the choice of how η^μ is transported), so the eigenstructure of $B^\mu{}_\nu$ determines local expansion/contraction and distortion.
The scalar

$$\theta \equiv \nabla_\mu u^\mu = B^\mu{}_\mu \tag{10.46}$$

is called the *expansion*. It measures the fractional rate of change of an infinitesimal comoving 3-volume element V:

$$\frac{1}{V}\frac{dV}{d\tau} = \theta, \qquad \dot{(\;)} \equiv u^\mu \nabla_\mu. \tag{10.47}$$

We can now proceed with the irreducible decomposition, as prompted in Chap. 1.

Thus, decompose the spatial part of $B_{\mu\nu}$ into irreducible pieces under the local $SO(3)$ acting on the hypersurface orthogonal to u^μ:

$$B_{\mu\nu} = \frac{1}{3}\theta\, h_{\mu\nu} + \sigma_{\mu\nu} + \omega_{\mu\nu} - a_\mu u_\nu, \tag{10.48}$$

where

$$\sigma_{\mu\nu} \equiv B_{(\mu\nu)} - \frac{1}{3}\theta\, h_{\mu\nu} \quad \Rightarrow \quad \sigma_{\mu\nu} = \sigma_{\nu\mu},\ \sigma^\mu{}_\mu = 0,\ \sigma_{\mu\nu}u^\nu = 0, \tag{10.49}$$

$$\omega_{\mu\nu} \equiv B_{[\mu\nu]} \quad \Rightarrow \quad \omega_{\mu\nu} = -\omega_{\nu\mu},\ \omega_{\mu\nu}u^\nu = 0, \tag{10.50}$$

and $u_\mu a_\nu$ is the purely non-geodesic contribution required by $u^\mu u_\mu = -1$.

- θ controls the local *volume* change of the congruence (isotropic dilation or contraction).
- $\sigma_{\mu\nu}$ controls *shape distortion at fixed volume* (a sphere of nearby geodesics becomes an ellipsoid).
- $\omega_{\mu\nu}$ measures *local rotation* (failure of hypersurface orthogonality); nonzero $\omega_{\mu\nu}$ tends to prevent focusing.

Driving Exercise
Let us find the Raychaudhuri equation in a practical and useful form.

- **Step 1: compute $\dot{\theta}$ from $\theta = \nabla_\mu u^\mu$.** We start from

$$\dot{\theta} = u^\nu \nabla_\nu (\nabla_\mu u^\mu) = \nabla_\mu (u^\nu \nabla_\nu u^\mu) - (\nabla_\mu u^\nu)(\nabla_\nu u^\mu), \tag{10.51}$$

where the second equality is the Leibniz rule applied as $\nabla_\mu (u^\nu \nabla_\nu u^\mu) = (\nabla_\mu u^\nu)(\nabla_\nu u^\mu) + u^\nu \nabla_\mu \nabla_\nu u^\mu$, followed by rearrangement.

The first term is simply the divergence of the acceleration:

$$\nabla_\mu (u^\nu \nabla_\nu u^\mu) = \nabla_\mu a^\mu. \tag{10.52}$$

To rewrite the remaining term, use the commutator of covariant derivatives acting on u^μ:

$$(\nabla_\mu \nabla_\nu - \nabla_\nu \nabla_\mu) u^\mu = R^\mu{}_{\lambda\mu\nu}\, u^\lambda = R_{\lambda\nu}\, u^\lambda. \tag{10.53}$$

Contracting with u^ν gives

$$u^\nu \nabla_\mu \nabla_\nu u^\mu = u^\nu \nabla_\nu \nabla_\mu u^\mu + R_{\lambda\nu} u^\lambda u^\nu = \dot{\theta} + R_{\mu\nu} u^\mu u^\nu. \tag{10.54}$$

Equivalently,

$$u^\nu \nabla_\nu \nabla_\mu u^\mu = u^\nu \nabla_\mu \nabla_\nu u^\mu - R_{\mu\nu} u^\mu u^\nu. \tag{10.55}$$

Inserting this relation back into Eq. (10.51) yields the standard intermediate form

$$\dot{\theta} = \nabla_\mu a^\mu - (\nabla_\mu u_\nu)(\nabla^\nu u^\mu) - R_{\mu\nu} u^\mu u^\nu. \tag{10.56}$$

This is already the Raychaudhuri identity, expressed in terms of $B_{\mu\nu}$ and curvature.

- **Step 2: evaluate the quadratic term in irreducible components.** Using $B_{\mu\nu} = \nabla_\nu u_\mu$, the quadratic invariant appearing in Eq. (10.56) is

$$(\nabla_\mu u_\nu)(\nabla^\nu u^\mu) = B_{\nu\mu} B^{\mu\nu}. \tag{10.57}$$

Decompose $B_{\mu\nu}$ as in Eq. (10.48). The acceleration term drops from the spatial contractions because $u^\mu \sigma_{\mu\nu} = u^\mu \omega_{\mu\nu} = 0$ and $u^\mu h_{\mu\nu} = 0$. A short but systematic algebra then gives

$$B_{\nu\mu} B^{\mu\nu} = \frac{1}{3}\theta^2 + \sigma_{\mu\nu}\sigma^{\mu\nu} - \omega_{\mu\nu}\omega^{\mu\nu}. \tag{10.58}$$

The sign difference between shear and vorticity follows from the symmetry/antisymmetry: $\sigma_{\mu\nu}\sigma^{\mu\nu} \geq 0$ contributes positively to $B_{\nu\mu} B^{\mu\nu}$, whereas $\omega_{\nu\mu}\omega^{\mu\nu} = -\omega_{\mu\nu}\omega^{\mu\nu}$ flips sign.

Substituting Eqs. (10.58) into (10.56) yields the Raychaudhuri equation for a timelike congruence:

$$\boxed{\dot{\theta} = -\frac{1}{3}\theta^2 - \sigma_{\mu\nu}\sigma^{\mu\nu} + \omega_{\mu\nu}\omega^{\mu\nu} - R_{\mu\nu}u^\mu u^\nu + \nabla_\mu a^\mu.} \tag{10.59}$$

For *geodesic* flow ($a^\mu = 0$) this reduces to

$$\dot{\theta} = -\frac{1}{3}\theta^2 - \sigma_{\mu\nu}\sigma^{\mu\nu} + \omega_{\mu\nu}\omega^{\mu\nu} - R_{\mu\nu}u^\mu u^\nu. \tag{10.60}$$

Null congruence (for horizon applications). For completeness, consider a congruence of null generators with tangent k^μ, affinely parametrized:

$$k^\mu k_\mu = 0, \qquad k^\nu \nabla_\nu k^\mu = 0. \tag{10.61}$$

One projects onto the two-dimensional screen space orthogonal to k^μ (modulo an auxiliary null vector) and defines the optical scalars $\theta, \sigma_{\mu\nu}, \omega_{\mu\nu}$ analogously. The Raychaudhuri equation becomes

$$\boxed{\frac{d\theta}{d\lambda} = -\frac{1}{2}\theta^2 - \sigma_{\mu\nu}\sigma^{\mu\nu} + \omega_{\mu\nu}\omega^{\mu\nu} - R_{\mu\nu}k^\mu k^\nu,} \tag{10.62}$$

where λ is the affine parameter. In particular, for null hypersurface-orthogonal congruences ($\omega_{\mu\nu} = 0$), positive $R_{\mu\nu}k^\mu k^\nu$ forces focusing of light rays, and underlies the area theorem and the classical second law of BH mechanics (see Sect. 10.5.3).

10.2 General Relativity

The energy-momentum tensor contains the information about mass and energy of a given system. Presuming the validity of the equivalence principle, it turns out that its overall content can gravitate. Accordingly, we need objects that are associated with curvature, but with two indexes, instead of four.

Then, contracting the first and third indices of the Riemann tensor defines the *Ricci tensor*,

$$R_{\alpha\beta} \equiv R^\mu{}_{\alpha\mu\beta} = \frac{\partial}{\partial x^\mu}\Gamma^\mu_{\alpha\beta} - \frac{\partial}{\partial x^\beta}\Gamma^\mu_{\alpha\mu} + \Gamma^\tau_{\alpha\beta}\Gamma^\mu_{\tau\mu} - \Gamma^\sigma_{\alpha\tau}\Gamma^\tau_{\beta\sigma}. \tag{10.63}$$

Using the fact that $\Gamma^\mu_{\alpha\mu} = \frac{\partial}{\partial x^\alpha}\ln\sqrt{-g}$, we can rewrite the previous expression as

$$R_{\alpha\beta} = \frac{\partial}{\partial x^\mu}\Gamma^\mu_{\alpha\beta} - \frac{\partial^2}{\partial x^\alpha\,\partial x^\beta}\ln\sqrt{-g} + \Gamma^\tau_{\alpha\beta}\frac{\partial}{\partial x^\tau}\ln\sqrt{-g} - \Gamma^\sigma_{\alpha\tau}\Gamma^\tau_{\beta\sigma}. \tag{10.64}$$

The following main properties may be therefore inferred.

- The Ricci tensor is symmetric,

$$R_{\alpha\beta} = R_{\beta\alpha}. \tag{10.65}$$

  This symmetry follows from the pair symmetry $R_{\alpha\beta\mu\nu} = R_{\mu\nu\alpha\beta}$ together with the contraction.
- A key consequence of the differential Bianchi identities is the existence of a divergence-free rank-2 tensor constructed from curvature. Starting from

$$R^{\alpha}{}_{\beta\lambda\mu;\nu} + R^{\alpha}{}_{\beta\mu\nu;\lambda} + R^{\alpha}{}_{\beta\nu\lambda;\mu} = 0,$$

  contract α with λ:

$$R_{\beta\mu;\nu} + R^{\lambda}{}_{\beta\mu\nu;\lambda} - R_{\beta\nu;\mu} = 0.$$

  Raise the index β:

$$R^{\beta}{}_{\mu;\nu} + R^{\lambda\beta}{}_{\mu\nu;\lambda} - R^{\beta}{}_{\nu;\mu} = 0.$$

  Finally, contract β with μ:

$$R_{;\nu} - R^{\lambda}{}_{\nu;\lambda} - R^{\mu}{}_{\nu;\mu} = 0 \qquad \Longrightarrow \qquad R_{;\nu} - 2R^{\mu}{}_{\nu;\mu} = 0.$$

  Multiplying by $g_{\mu\rho}$, dividing by 2 and rearranging signs,

$$g_{\mu\rho} R^{\mu}{}_{\nu;\mu} - \frac{1}{2} g_{\mu\rho} R_{;\nu} = 0. \tag{10.66}$$

- Introducing the *Ricci scalar*

$$R = R^{\mu}{}_{\mu} = g^{\mu\nu} R_{\mu\nu}, \tag{10.67}$$

  we can rewrite the previous result in the compact form

$$\left(R_{\mu\rho} - \frac{1}{2} g_{\mu\rho} R \right)_{;\mu} = 0, \tag{10.68}$$

  where we set $(\nu = \mu)$ in the contracted expression.

The combination

$$G_{\mu\nu} \equiv R_{\mu\nu} - \frac{1}{2} g_{\mu\nu} R \tag{10.69}$$

is called the *Einstein tensor*. Owing to the contracted Bianchi identity, it satisfies the geometric identity

$$\nabla^{\mu} G_{\mu\nu} = 0, \tag{10.70}$$

independently of the gravitational field equations. This fact guarantees the consistency of Einstein's equations with the covariant conservation of the stress-energy tensor.

At this point we have the essential mathematical tools required to derive the gravitational field equations.

10.2.1 Einstein's Field Equations

A first guess might be to relate curvature directly to the stress-energy tensor, for instance by positing $R^{\mu}{}_{\nu} \propto T^{\mu}{}_{\nu}$. However, such an ansatz is generically inconsistent with $\nabla^{\mu} T_{\mu\nu} = 0$, because the Ricci tensor does not satisfy a corresponding divergence identity. This motivates the search for a rank-2 curvature tensor with identically vanishing covariant divergence.

Nevertheless, the Riemann tensor obeys the differential Bianchi identity, which implies

$$\nabla^{\mu} G_{\mu\nu} = 0, \qquad G_{\mu\nu} \equiv R_{\mu\nu} - \frac{1}{2} R g_{\mu\nu}. \tag{10.71}$$

This identity is crucial for the consistency of Einstein's equations with energy-momentum conservation.

Accordingly, one can make the following statement.

Contracted Bianchi identity. For the Levi-Civita connection one has $\nabla^{\mu} G_{\mu\nu} = 0$ identically. Therefore, if the field equations are taken to be $G_{\mu\nu} + \Lambda g_{\mu\nu} = \kappa\, T_{\mu\nu}$, they imply $\nabla^{\mu} T_{\mu\nu} = 0$, i.e., covariant energy-momentum conservation.

Einstein's equations are fixed by combining three requirements: (i) the field equations must be *generally covariant*; (ii) in the weak-field, slow-motion limit they must reproduce Newtonian gravity; (iii) the source must be the stress-energy tensor $T_{\mu\nu}$ and the left-hand side must be divergence-free to ensure consistency with $\nabla^{\mu} T_{\mu\nu} = 0$.

The simplest rank-2 tensor built locally from the metric and at most its second derivatives is the Einstein tensor $G_{\mu\nu}$. By the contracted Bianchi identity it satisfies $\nabla^{\mu} G_{\mu\nu} = 0$. Moreover, one may add a term proportional to the metric without spoiling this property, since $\nabla^{\mu}(g_{\mu\nu}) = 0$. One is therefore led to the most general form

$$G_{\mu\nu} + \Lambda g_{\mu\nu} = \kappa\, T_{\mu\nu}, \tag{10.72}$$

where κ is a coupling constant and Λ is the cosmological constant. In units $G = c = 1$ one has $\kappa = 8\pi$, (with units restored, $\kappa = 8\pi G/c^4$).

Taking the trace of Eq. (10.72) in $D = 4$ yields

$$-R + 4\Lambda = \kappa T, \qquad T \equiv T^{\mu}{}_{\mu}, \tag{10.73}$$

so that in vacuum ($T_{\mu\nu} = 0$) one obtains $R = 4\Lambda$ and

$$R_{\mu\nu} = \Lambda g_{\mu\nu}. \tag{10.74}$$

In particular, for $\Lambda = 0$ the vacuum equations reduce to $R_{\mu\nu} = 0$.

Equation (10.72) can be summarized by *Wheeler's dictum.*

Spacetime tells matter how to move; matter tells spacetime how to curve.

A fundamental theory of gravity may be invoked by means of variational principle. The first who disclosed this was Hilbert, proposing for geometry and Λ,

$$S_{\mathrm{HE}}[g] = \frac{1}{16\pi}\int d^4x\,\sqrt{-g}\,(R - 2\Lambda) + S_{\mathrm{bdy}}, \tag{10.75}$$

where S_{bdy} is the boundary term required for a well-posed variational problem (the Gibbons-Hawking-York term in standard setups).

Matter fields are described by an additional action $S_{\mathrm{m}}[g, \Psi]$. Varying the total action $S_{\mathrm{HE}} + S_{\mathrm{m}}$ with respect to the metric gives

$$G_{\mu\nu} + \Lambda g_{\mu\nu} = 8\pi T_{\mu\nu}, \tag{10.76}$$

where the stress-energy tensor is defined as

$$T_{\mu\nu} \equiv -\frac{2}{\sqrt{-g}}\frac{\delta S_{\mathrm{m}}}{\delta g^{\mu\nu}}. \tag{10.77}$$

Equation (10.71) implies $\nabla^{\mu} T_{\mu\nu} = 0$, expressing local energy-momentum conservation in a covariant form.

10.3 Alternatives to Einstein's Gravity

GR provides a remarkably successful geometric description of gravitational phenomena across a wide range of scales, from laboratory experiments to astrophysical and cosmological observations. Its conceptual core, encapsulated by the Einstein's field equations, identifies gravity with the dynamics of spacetime geometry and has passed all classical and weak-field tests to extraordinary precision. Nevertheless, both theoretical and observational considerations motivate the exploration of frameworks that extend or depart from Einstein's theory.

From a theoretical standpoint, GR is a classical field theory and resists a straightforward quantization, signaling its expected incompleteness at very high energies.

While any viable alternative must reproduce the successes of GR in the appropriate regimes, such theories offer a valuable theoretical laboratory to probe the limits of classical gravity and to search for signatures of new physics beyond Einstein's framework.

10.3.1 The Geometrical "Trinity" of Gravity

A conceptually important (and, historically, somewhat underemphasized) structural fact about Einstein gravity is that its classical dynamics can be encoded in *three* inequivalent-looking geometrical languages which, when assembled appropriately, become dynamically equivalent. This is the so-called *geometrical trinity of gravity*: a curvature formulation (standard GR), a torsional formulation (teleparallel gravity), and a non-metric formulation (symmetric teleparallel gravity). The three descriptions differ in which geometric object is taken as fundamental, curvature, torsion, or non-metricity, and, accordingly, in which affine connection is used to define parallel transport. Nevertheless, when the corresponding scalar invariants are chosen in the correct way, they reproduce the same Einstein's field equations, up to boundary terms.

On a differentiable manifold $\mathcal{M}$, an affine connection $\Gamma^{\alpha}{}_{\mu\nu}$ determines covariant differentiation and parallel transport. Its "field content" may be decomposed into three independent geometric tensors:

- curvature $R^{\alpha}{}_{\beta\mu\nu}(\Gamma)$, measuring the non-commutativity of covariant derivatives and the holonomy of parallel transport;
- torsion $T^{\alpha}{}_{\mu\nu} \equiv 2\,\Gamma^{\alpha}{}_{[\mu\nu]}$, measuring the non-closure of infinitesimal parallelograms;
- non-metricity $Q_{\alpha\mu\nu} \equiv \nabla_{\alpha} g_{\mu\nu}$, measuring the failure of the metric to be preserved by parallel transport.

GR selects the Levi-Civita connection, so that gravity is encoded purely in curvature. The other two formulations trade curvature for torsion or for non-metricity by choosing different (flat) connections.

In the standard (curvature) formulation one assumes metric-compatibility and vanishing torsion,

$$Q_{\alpha\mu\nu} = 0, \qquad T^{\alpha}{}_{\mu\nu} = 0, \tag{10.78}$$

so that $\Gamma^{\alpha}{}_{\mu\nu} = \left\{ {\alpha \atop \mu\nu} \right\}$ and the gravitational action is the Hilbert-Einstein functional

$$S_{\text{H-E}} = \frac{1}{2\kappa^2} \int d^4x \sqrt{-g}\, R(\Gamma), \qquad \kappa^2 \equiv 8\pi G. \tag{10.79}$$

The dynamics is then expressed in terms of the Riemann tensor, its contractions, and the Einstein tensor, with matter sourcing geometry through $T_{\mu\nu}$.

In the *teleparallel* formulation one employs instead a curvature-free connection,

$$R^{\alpha}{}_{\beta\mu\nu}(\Gamma) = 0, \tag{10.80}$$

but allows for non-vanishing torsion $T^{\alpha}{}_{\mu\nu} \neq 0$ while imposing metric-compatibility $Q_{\alpha\mu\nu} = 0$. Gravity is not interpreted as spacetime curvature but as torsional distortion of parallel transport. The Teleparallel Equivalent of GR (TEGR) is obtained by choosing the torsion scalar T built from quadratic contractions of $T^{\alpha}{}_{\mu\nu}$ (with the appropriate relative coefficients), and considering the action

$$S_{\text{TEGR}} = \frac{1}{2\kappa^2} \int d^4x \sqrt{-g}\, T. \tag{10.81}$$

The equivalence with Einstein gravity is encoded in an identity of the schematic form

$$R(\Gamma) = -\,T + (\text{total divergence})\,, \tag{10.82}$$

so that S_{HE} and S_{TEGR} differ by a boundary term and yield the same field equations.

A third formulation, say *symmetric teleparallel gravity*, uses a connection which is curvature-free and torsion-free,

$$R^{\alpha}{}_{\beta\mu\nu}(\Gamma) = 0, \qquad T^{\alpha}{}_{\mu\nu} = 0, \tag{10.83}$$

but not metric-compatible: $Q_{\alpha\mu\nu} \neq 0$. In this case, gravitational effects are attributed entirely to non-metricity. One constructs a scalar Q from quadratic contractions of $Q_{\alpha\mu\nu}$ (again with specific relative weights), and defines the action

$$S_{\text{STGR}} = \frac{1}{2\kappa^2} \int d^4x \sqrt{-g}\, Q. \tag{10.84}$$

As in Eq. (10.82), there exists an identity of the form

$$R(\Gamma) = -\,Q + (\text{total divergence})\,, \tag{10.85}$$

which guarantees dynamical equivalence with GR at the level of field equations.

10.3.2 Curvature Versus Torsion and Non-metricity

Curvature is operationally defined by comparing a vector to itself after parallel transport along a *closed* curve. This comparison is meaningful because the initial and final vectors live in the same tangent space. The outcome is governed by the Riemann tensor: a nontrivial holonomy signals nonzero curvature. It is then natural to ask whether curvature exhausts the possible obstructions to "Euclidean" transport. Two additional and logically independent properties are relevant: (i) the closure of infinitesimal parallelograms (torsion), and (ii) the preservation of lengths and angles (non-metricity).

- **Torsion**. In Euclidean geometry, the sum of two vectors at a point $p \in \mathcal{M}$ can be visualized as the diagonal of a parallelogram. Let $u_p, v_p \in T_p\mathcal{M}$ be linearly independent. Construct the parallelogram by moving v_p along u_p and, alternatively, u_p along v_p. In flat Euclidean space these two operations end at the same point, so the parallelogram closes.

  In a general affine geometry, parallel transport is defined by a connection $\Gamma^\alpha{}_{\mu\nu}$ and the parallelogram may fail to close. To quantify this at leading (infinitesimal) order, set

$$\delta u^\mu_p \equiv \epsilon\, u^\mu_p, \qquad \delta v^\mu_p \equiv \lambda\, v^\mu_p, \qquad \epsilon \ll 1,\ \lambda \ll 1, \tag{10.86}$$

  and assume δu_p and δv_p are linearly independent. Using the infinitesimal parallel transport law,

$$\delta v^\alpha_q = \delta v^\alpha_p - \Gamma^\alpha{}_{\mu\nu}(p)\,\delta u^\mu_p\,\delta v^\nu_p, \tag{10.87}$$

  the displacement obtained by first moving along δu_p and then along the transported δv is

$$\delta w^\alpha_p = \delta u^\alpha_p + \delta v^\alpha_p - \Gamma^\alpha{}_{\mu\nu}(p)\,\delta u^\mu_p\,\delta v^\nu_p. \tag{10.88}$$

  If we reverse the order, transporting δu_p along δv_p, we similarly get

$$\delta w'^\alpha_p = \delta v^\alpha_p + \delta u^\alpha_p - \Gamma^\alpha{}_{\nu\mu}(p)\,\delta v^\nu_p\,\delta u^\mu_p. \tag{10.89}$$

  The infinitesimal closure failure is therefore

$$\delta w'^\alpha_p - \delta w^\alpha_p = \left(\Gamma^\alpha{}_{\mu\nu} - \Gamma^\alpha{}_{\nu\mu}\right)(p)\,\delta u^\mu_p\,\delta v^\nu_p \equiv T^\alpha{}_{\mu\nu}(p)\,\delta u^\mu_p\,\delta v^\nu_p, \tag{10.90}$$

  where we introduced the torsion tensor

$$\boxed{T^\alpha{}_{\mu\nu} \;\equiv\; 2\,\Gamma^\alpha{}_{[\mu\nu]} \;=\; \Gamma^\alpha{}_{\mu\nu} - \Gamma^\alpha{}_{\nu\mu}.} \tag{10.91}$$

  The parallelogram closes if and only if $T^\alpha{}_{\mu\nu} = 0$. Moreover, since $T^\alpha{}_{\mu\nu}$ is antisymmetric in (μ, ν), the right-hand side of Eq. (10.90) vanishes whenever δu_p and

δv_p are linearly dependent, in agreement with the fact that two collinear vectors do not span a parallelogram.
A technical point worth emphasizing is that, although $\Gamma^\alpha{}_{\mu\nu}$ is not a tensor (it has an inhomogeneous transformation law), the antisymmetric combination in Eqs. (10.91) is tensorial: the inhomogeneous terms cancel, so $T^\alpha{}_{\mu\nu}$ is a well-defined geometric field.
A connection symmetric in its lower indices, $\Gamma^\alpha{}_{\mu\nu} = \Gamma^\alpha{}_{\nu\mu}$, is called **torsion-free**. Independently of the existence of a metric, one may also define the (unique) trace obtained by contracting the upper index with the second lower one,

$$\boxed{T_\mu \;\equiv\; T^\alpha{}_{\mu\alpha}.} \tag{10.92}$$

One must keep track of index ordering: antisymmetry implies

$$T^\alpha{}_{\alpha\mu} = -T^\alpha{}_{\mu\alpha} = -T_\mu . \tag{10.93}$$

If a metric is present, the contraction $g^{\mu\nu} T^\alpha{}_{\mu\nu}$ vanishes identically because $g^{\mu\nu}$ is symmetric while $T^\alpha{}_{\mu\nu}$ is antisymmetric.

– **Non-metricity**. The third property concerns the behavior of *lengths* (and, more generally, of all metric-dependent quantities) under parallel transport. Let $v \in T\mathcal{M}$ be a vector field, with components v^μ in a chart, and let $g_{\mu\nu}$ be a metric. Its squared norm is

$$\|v\|^2 \;\equiv\; g(v, v) = g_{\mu\nu} v^\mu v^\nu , \tag{10.94}$$

which, in Lorentzian signature, may be positive, negative, or zero even for $v \neq 0$. Let γ be a curve with tangent $u = u^\alpha \partial_\alpha$, and assume that v is parallel transported along γ,

$$u^\alpha \nabla_\alpha v^\mu = 0, \tag{10.95}$$

where ∇ is the covariant derivative induced by $\Gamma^\alpha{}_{\mu\nu}$. Then, by Leibniz's rule,

$$\begin{aligned} u^\alpha \nabla_\alpha \|v\|^2 &= u^\alpha \nabla_\alpha \big(g_{\mu\nu} v^\mu v^\nu\big) \\ &= \big(u^\alpha \nabla_\alpha g_{\mu\nu}\big)\, v^\mu v^\nu + 2 g_{\mu\nu} \big(u^\alpha \nabla_\alpha v^\mu\big)\, v^\nu \\ &= \big(u^\alpha \nabla_\alpha g_{\mu\nu}\big)\, v^\mu v^\nu . \end{aligned} \tag{10.96}$$

This motivates the definition of the **non-metricity tensor**,

$$\boxed{Q_{\alpha\mu\nu} \;\equiv\; \nabla_\alpha g_{\mu\nu}.} \tag{10.97}$$

When $Q_{\alpha\mu\nu} \neq 0$, the metric is not preserved by parallel transport and the length (and angle) of vectors is not constant along parallel-transported congruences. The condition

$$\nabla_\alpha g_{\mu\nu} = 0 \tag{10.98}$$

is called the **metricity condition**, and connections satisfying it are said to be **metric-compatible**.
A common source of confusion is the relation between $Q_{\alpha\mu\nu}$ and the covariant derivative of the inverse metric. Starting from $g_{\mu\lambda}g^{\lambda\nu} = \delta^{\nu}_{\mu}$ and taking a covariant derivative, one finds

$$0 = \nabla_\alpha\left(g_{\mu\lambda}g^{\lambda\nu}\right) = (\nabla_\alpha g_{\mu\lambda})g^{\lambda\nu} + g_{\mu\lambda}(\nabla_\alpha g^{\lambda\nu}), \tag{10.99}$$

hence

$$\boxed{\nabla_\alpha g^{\mu\nu} = -\,Q_\alpha{}^{\mu\nu}, \qquad Q_\alpha{}^{\mu\nu} \equiv g^{\mu\rho}g^{\nu\sigma}\,Q_{\alpha\rho\sigma}.} \tag{10.100}$$

Non-metricity admits further operational interpretations beyond the change of norms. For instance, it controls the transport of metric-dependent scalars such as the volume element. In n dimensions,

$$\mathrm{Vol}(\Omega) \equiv \int_\Omega \sqrt{|g|}\,d^n x, \tag{10.101}$$

and one may compute

$$u^\alpha\nabla_\alpha\sqrt{|g|} = \frac{1}{2}\sqrt{|g|}\,g^{\mu\nu}\nabla_\alpha g_{\mu\nu}\,u^\alpha = \frac{1}{2}\sqrt{|g|}\,g^{\mu\nu}\,Q_{\alpha\mu\nu}\,u^\alpha. \tag{10.102}$$

Therefore, the trace $g^{\mu\nu}Q_{\alpha\mu\nu}$ plays a distinguished role. Since $Q_{\alpha\mu\nu}$ is symmetric in its last two indices, there are two independent vector traces, conveniently defined as

$$\boxed{Q_\alpha \equiv g^{\mu\nu}Q_{\alpha\mu\nu} = Q_\alpha{}^\mu{}_\mu, \qquad \bar{Q}_\alpha \equiv g^{\mu\nu}Q_{\mu\nu\alpha} = Q^\mu{}_{\mu\alpha}.} \tag{10.103}$$

10.3.3 Unimodular or Trace-Free General Relativity

What is, physically, the role of Λ? In the Einstein's field equations it appears as a covariantly constant contribution to the gravitational source, *formally indistinguishable from a spacetime-filling vacuum energy* with equation of state $p = -\rho$. This is far more than a mere integration constant: it controls the late-time acceleration of the Universe and, at the level of the field equations, it is the unique local term compatible with general covariance and second-order metric dynamics.

Where, in microscopic physics, does one encounter an approximately constant energy density? The simplest arena is QFT, where each mode carries a zero-point energy.

For a massive field, a hard ultraviolet cutoff M yields (up to the usual regularization dependence) the vacuum energy density

$$
\begin{aligned}
\langle\rho\rangle &= \frac{1}{4\pi^2}\int_0^M \mathrm{d}k\, k^2\sqrt{k^2+m^2} \\
&= \frac{M^4}{16\pi^2}\left[\sqrt{1+\frac{m^2}{M^2}}\left(1+\frac{1}{2}\frac{m^2}{M^2}\right)-\frac{1}{2}\frac{m^4}{M^4}\ln\left(\frac{M}{m}+\frac{M}{m}\sqrt{1+\frac{m^2}{M^2}}\right)\right] \\
&= \frac{M^4}{16\pi^2}\left(1+\frac{m^2}{M^2}+\cdots\right), \qquad (10.104)
\end{aligned}
$$

where the last line is the asymptotic expansion for $m/M \ll 1$. The leading divergence is quartic in the cutoff scale, a characteristic UV sensitivity of the zero-point sector.

Vacuum energy as an effective cosmological term.
In the gravitational field equations, any contribution proportional to the metric, $T^{(\mathrm{vac})}_{\mu\nu} = -\rho_{\mathrm{vac}}\, g_{\mu\nu}$, is equivalent to shifting the cosmological constant,

$$
G_{\mu\nu} + \Lambda g_{\mu\nu} = 8\pi G\, T_{\mu\nu} \quad \Longrightarrow \quad G_{\mu\nu} + \Lambda_{\mathrm{eff}}\, g_{\mu\nu} = 8\pi G\, T^{(\mathrm{m})}_{\mu\nu}, \qquad (10.105)
$$

with

$$
\Lambda_{\mathrm{eff}} = \Lambda + 8\pi G\,\rho_{\mathrm{vac}}. \qquad (10.106)
$$

Thus, at the level of GR, Eq. (10.104) acts as a source for an effective Λ.

If one naively identifies the vacuum scale with an ultraviolet cutoff, a standard dimensional estimate gives $\rho_{\mathrm{vac}} \sim M^4$, and hence an induced contribution to the Einstein equations that can be parametrized as $\Lambda_{\mathrm{vac}} \sim 8\pi G\, M^4$. Interpreted literally, the observed cosmological constant would then have to arise as a minute residual obtained by cancelling much larger contributions, i.e., an extreme fine-tuning of the renormalized vacuum sector.

Observationally, the late-time acceleration fixes Λ to be of order of the present Hubble scale,

$$
\Lambda \approx H_0^2 = \left(2.13\, h \times 10^{-42}\,\mathrm{GeV}\right)^2, \qquad (10.107)
$$

corresponding to an energy density

$$
\rho_\Lambda = \frac{\Lambda\, m_{\mathrm{pl}}^2}{8\pi} \approx 10^{-47}\,\mathrm{GeV}^4, \qquad (10.108)
$$

where $m_{\rm pl} = 1.22 \times 10^{19}$ GeV is the Planck mass. By contrast, the cutoff estimate Eq. (10.104) grows as M^4. If one assumes that the effective field theory description can be extrapolated up to the vicinity of the Planck scale and sets $M = m_{\rm pl}$, one obtains the characteristic scale

$$\rho_{\rm vac} \approx 10^{74}\,{\rm GeV}^4, \tag{10.109}$$

which overshoots Eq. (10.108) by roughly 10^{121} in relative magnitude. Even adopting a much lower ultraviolet scale, for instance taking M at a hadronic (QCD) scale, yields the parametric estimate

$$\rho_{\rm vac} \approx 10^{-3}\,{\rm GeV}^4, \tag{10.110}$$

still enormously larger than ρ_Λ.

The impasse of not having a cosmological constant cutoff well-defined has motivated a broad literature seeking dynamical alternatives to a rigid cosmological constant. We will turn to the cosmological constant problem in Sect. 10.9.

Trace-free dynamics as a conservative deformation of the problem.

A particularly economical modification of Einstein gravity consists in removing the trace part of the field equations. In its simplest implementation this is equivalent to *unimodular gravity*, where the metric determinant is fixed,

$$\sqrt{-g} = \varepsilon_0 = \text{const}, \tag{10.111}$$

so that admissible variations satisfy $g^{\mu\nu}\delta g_{\mu\nu} = 0$. An equivalent formulation takes as fundamental equations the *trace-free Einstein equations* (TFE).

Start from the Einstein equations with a cosmological term,

$$G_{\mu\nu} + \Lambda g_{\mu\nu} = 8\pi G\, T_{\mu\nu}. \tag{10.112}$$

Taking the trace yields

$$-R + 4\Lambda = 8\pi G\, T, \qquad T \equiv T^{\mu}{}_{\mu}. \tag{10.113}$$

Subtracting one quarter of Eq. (10.113) times $g_{\mu\nu}$ from Eq. (10.112) eliminates Λ identically and gives the trace-free system

$$R_{\mu\nu} - \frac{1}{4} R\, g_{\mu\nu} = 8\pi G \left(T_{\mu\nu} - \frac{1}{4} T\, g_{\mu\nu} \right). \tag{10.114}$$

A key property of Eq. (10.114) is its invariance under vacuum shifts

$$T_{\mu\nu} \longrightarrow T_{\mu\nu} + \rho_{\text{vac}}\, g_{\mu\nu}, \tag{10.115}$$

since $\rho_{\text{vac}} g_{\mu\nu}$ is pure trace and cancels out of the right-hand side. In this precise sense, vacuum energy does not *directly* enter the trace-free gravitational dynamics.

The system in Eqs. (10.114) is not yet closed: the trace information must be reconstructed from an additional condition. At this stage, one imposes the standard local conservation of matter,

$$\nabla^{\mu} T_{\mu\nu} = 0, \tag{10.116}$$

which is independently motivated by diffeomorphism invariance of the matter action and by the canonical definition of $T_{\mu\nu}$. Taking the covariant divergence of Eqs. (10.114) and using the contracted Bianchi identity,

$$\nabla^{\mu} G_{\mu\nu} = 0, \tag{10.117}$$

one obtains

$$\nabla_{\nu} R = -8\pi G\, \nabla_{\nu} T. \tag{10.118}$$

Therefore,

$$R + 8\pi G\, T = 4\Lambda_{\text{int}}, \tag{10.119}$$

where Λ_{int} is a spacetime constant of integration. Combining Eqs. (10.114) with (10.119) reconstructs the full Einstein equations in the usual form.

Thus trace-free (or unimodular) gravity is classically equivalent to GR, but the cosmological constant emerges as an *integration constant* fixed by boundary or initial data, rather than as a coupling parameter in the action.

From an effective field theory viewpoint, additive renormalizations of the matter Lagrangian act primarily as shifts in ρ_{vac}, hence in the trace sector. Because Eqs. (10.114) are insensitive to the shift in Eqs. (10.115), vacuum-energy renormalization does not feed directly into the local trace-free dynamics. This does not, by itself, explain why the observed Λ_{int} is small; rather, it softens the most immediate form of the "vacuum catastrophe" by relocating sensitivity to an integration constant determined globally (by initial/boundary conditions) rather than locally by an arbitrary constant in the matter sector.

10.3.4 Extensions of Einstein's Gravity: $f(R)$ Theories

GR is a classical gauge theory of spacetime geometry whose dynamical variable is the metric $g_{\mu\nu}$ (or, equivalently, a tetrad field). As a QFT treated by straightforward perturbation theory around a fixed background, Einstein gravity is not perturbatively renormalizable: the dimensionful coupling G forces an infinite tower of counterterms. The most conservative reading is therefore effective-field-theoretic: the Hilbert-Einstein action captures the leading operators in a derivative expansion valid below a cutoff scale, while higher-curvature (and possibly nonlocal) operators encode ultraviolet sensitivity.

At low energies one expects an expansion of the schematic form

$$S_{\text{eff}} = \frac{1}{2\kappa}\int d^4x\,\sqrt{-g}\left(R - 2\Lambda + \alpha_1 R^2 + \alpha_2 R_{\mu\nu}R^{\mu\nu} + \alpha_3 R_{\mu\nu\rho\sigma}R^{\mu\nu\rho\sigma} + \cdots\right) + S_{\text{m}}, \tag{10.120}$$

where $\kappa = 8\pi G$ and the coefficients α_i carry inverse powers of the UV scale. A large class of phenomenological modifications corresponds to resumming selected subsets of these operators into functions such as $f(R)$, or to rewriting the theory in equivalent (or inequivalent) geometric languages based on torsion T or non-metricity Q.

A particularly simple and widely used deformation is obtained by replacing the Ricci scalar in the Lagrangian with a generic function $f(R)$. This modification is minimal in the sense that it preserves diffeomorphism invariance and does not introduce additional tensor fields *by hand*; nevertheless, it generically adds a propagating scalar degree of freedom.

► **Driving Exercise**
$f(R)$ **gravity**. The defining action is

$$S = \frac{1}{2\kappa}\int d^4x\,\sqrt{-g}\,f(R) + S_{\text{m}}[g_{\mu\nu},\psi], \tag{10.121}$$

where S_{m} denotes the matter action and ψ collectively denotes the matter fields.
Two inequivalent variational prescriptions are commonly employed.

- **Metric formalism**: the connection is fixed to be Levi-Civita $\Gamma^{\rho}_{\mu\nu} = \{^{\rho}_{\mu\nu}\}$ and one takes variations only with respect to $g_{\mu\nu}$.
- **Palatini formalism**: metric and connection are treated as independent variables, and one varies with respect to both.

In GR the two procedures coincide, whereas in $f(R)$ they define distinct dynamical theories.

In the metric formalism (torsionless and metric-compatible connection), the variation of Eqs. (10.121) with respect to $g^{\mu\nu}$ yields

$$f'(R)R_{\mu\nu} - \frac{1}{2}f(R)g_{\mu\nu} - \nabla_\mu\nabla_\nu f'(R) + g_{\mu\nu}\Box f'(R) = \kappa T_{\mu\nu}, \qquad (10.122)$$

where $f'(R) \equiv df/dR$, $\Box \equiv g^{\alpha\beta}\nabla_\alpha\nabla_\beta$, and $T_{\mu\nu}$ is the matter stress-energy tensor.

Equation (10.122) is generically fourth order in metric derivatives, reflecting additional dynamics beyond Einstein gravity. Taking the trace gives

$$f'(R)\,R - 2f(R) + 3\Box f'(R) = \kappa T, \qquad T \equiv T^\mu{}_\mu, \qquad (10.123)$$

which makes explicit that the Ricci scalar becomes a propagating quantity. In vacuum ($T = 0$), GR enforces $R = 4\Lambda$ (or $R = 0$ when $\Lambda = 0$), while in metric $f(R)$ one has a genuine scalar mode governed by Eq. (10.123).

10.3.5 Scalar-Tensor Representation

The dynamical content of metric $f(R)$ gravity is rendered manifest by a Legendre-type rewriting. Introduce an auxiliary scalar χ and consider

$$S = \frac{1}{2\kappa}\int d^4x\,\sqrt{-g}\left[f(\chi) + f'(\chi)(R-\chi)\right] + S_{\rm m}. \qquad (10.124)$$

Variation with respect to χ gives

$$f''(\chi)\,(R-\chi) = 0, \qquad (10.125)$$

so that, for $f''(\chi) \neq 0$, one recovers $\chi = R$ and hence the original action.
Define $\phi \equiv f'(\chi)$, $\quad V(\phi) \equiv \phi\,\chi - f(\chi)$, with $\chi = \chi(\phi)$ determined implicitly by $\phi = f'(\chi)$. Then Eq. (10.124) becomes

$$S = \frac{1}{2\kappa}\int d^4x\,\sqrt{-g}\,\Big(\phi R - V(\phi)\Big) + S_{\rm m}, \qquad (10.126)$$

which is a Brans-Dicke-type scalar-tensor theory with vanishing Brans-Dicke parameter, $\omega_{\rm BD} = 0$, in the Jordan frame. The additional propagating degree of freedom of metric $f(R)$ is precisely the scalar ϕ.

10.3.6 Modifications of Einstein's Gravity: $f(T)$ and $f(Q)$ Theories

A complementary route to modified gravity exploits the fact that Einsteinian dynamics admits inequivalent geometric reformulations. Instead of modifying the curvature scalar in the metric Levi-Civita formulation, one may (i) trade curvature for torsion (teleparallelism) or (ii) trade curvature for non-metricity (symmetric teleparallelism). These reformulations are not mere changes of language once one promotes the corresponding scalar invariants to nonlinear functions, because the boundary-term equivalences that relate GR to its teleparallel avatars are then lost.

In teleparallel gravity, the fundamental gravitational variable is a tetrad $e^A{}_\mu$, which encodes the metric as

$$g_{\mu\nu} = \eta_{AB}\, e^A{}_\mu\, e^B{}_\nu, \tag{10.127}$$

with $\eta_{AB} = \mathrm{diag}(-1, 1, 1, 1)$. One employs the curvature-free Weitzenböck connection,

$$\Gamma^\rho{}_{\mu\nu} = e_A{}^\rho\, \partial_\mu e^A{}_\mu, \tag{10.128}$$

whose defining feature is vanishing curvature but non-vanishing torsion. From the torsion tensor $T^\rho{}_{\mu\nu}$ one constructs the torsion scalar

$$T = S_\rho{}^{\mu\nu}\, T^\rho{}_{\mu\nu}, \tag{10.129}$$

where $S_\rho{}^{\mu\nu}$ is the standard superpotential, i.e., a specific contraction of torsion and contorsion chosen so that the teleparallel action reproduces Einstein gravity up to a boundary term. The TEGR is

$$S_{\mathrm{TEGR}} = \frac{1}{2\kappa}\int d^4x\, e\, T + S_{\mathrm{m}}, \qquad e \equiv \det(e^A{}_\mu) = \sqrt{-g}, \tag{10.130}$$

and satisfies the identity

$$R(e) = -T + 2\,\nabla_\mu(T^\mu), \qquad T^\mu \equiv T^\nu{}_\nu{}^\mu, \tag{10.131}$$

showing that TEGR and the Hilbert-Einstein action differ by a total divergence and are therefore classically equivalent.

Promoting T to a generic function $f(T)$ defines

$$S = \frac{1}{2\kappa}\int d^4x\, e\, f(T) + S_{\mathrm{m}}. \tag{10.132}$$

Unlike metric $f(R)$, the $f(T)$ field equations remain second order in derivatives of the tetrad because T contains only first derivatives of $e^A{}_\mu$. However, a crucial structural difference emerges: in generic formulations, $f(T)$ models are not invariant under local Lorentz transformations of the tetrad (unless one adopts a fully covariant formulation including an inertial spin connection). Operationally this implies subtleties in the choice of variables and, in many presentations, the appearance of preferred tetrad frames.

Although TEGR and GR are dynamically equivalent, the replacements $R \mapsto f(R)$ and $T \mapsto f(T)$ break the classical equivalence. Metric $f(R)$ typically propagates an extra scalar mode through higher derivatives, whereas $f(T)$ retains second-order equations but trades this for Lorentz-frame subtleties at the level of tetrads. For this reason the two frameworks provide distinct, complementary laboratories for exploring extensions of Einsteinian gravity and for testing how "geometry" is encoded in the gravitational interaction.

A conceptually parallel, but technically distinct, construction is provided by *symmetric teleparallel gravity*, where the connection is chosen to be both torsion-free and curvature-free, so that the only non-Riemannian ingredient is *non-metricity*. Concretely, one introduces an affine connection $\Gamma^{\rho}{}_{\mu\nu}$ constrained by

$$T^{\rho}{}_{\mu\nu} = 0, \qquad R^{\rho}{}_{\sigma\mu\nu}(\Gamma) = 0, \tag{10.133}$$

but allowing $\nabla_{\rho} g_{\mu\nu} \neq 0$. The non-metricity tensor is defined by

$$Q_{\rho\mu\nu} \equiv \nabla_{\rho} g_{\mu\nu}. \tag{10.134}$$

Two independent traces play a special role,

$$Q_{\rho} \equiv Q_{\rho}{}^{\mu}{}_{\mu}, \qquad \tilde{Q}_{\rho} \equiv Q^{\mu}{}_{\rho\mu}, \tag{10.135}$$

and, in complete analogy with teleparallelism, one constructs a particular quadratic scalar Q built from $Q_{\rho\mu\nu}$ such that the resulting action is equivalent to GR up to a boundary term. A convenient definition is

$$Q \equiv -\frac{1}{4}\, Q_{\rho\mu\nu} Q^{\rho\mu\nu} + \frac{1}{2}\, Q_{\rho\mu\nu} Q^{\nu\mu\rho} + \frac{1}{4}\, Q_{\rho} Q^{\rho} - \frac{1}{2}\, Q_{\rho} \tilde{Q}^{\rho}. \tag{10.136}$$

This combination can be compactly written as

$$Q = -\, Q_{\rho\mu\nu}\, P^{\rho\mu\nu}, \tag{10.137}$$

where $P^{\rho\mu\nu}$ is the non-metricity "superpotential" (also called the conjugate of $Q_{\rho\mu\nu}$), defined by the same linear combination appearing in Eqs. (10.136). The symmetric-teleparallel equivalent of GR (STEGR) is then

$$S_{\text{STEGR}} = \frac{1}{2\kappa} \int d^4x \sqrt{-g}\, Q + S_{\text{m}}. \tag{10.138}$$

The equivalence with GR follows from the identity relating the Levi-Civita Ricci scalar to Q up to a total divergence,

$$R(\{\}) = -\,Q + \nabla_\mu\Big(Q^\mu - \tilde{Q}^\mu\Big), \tag{10.139}$$

where $R(\{\})$ denotes the Ricci scalar of the Levi-Civita connection of $g_{\mu\nu}$. Thus, in complete analogy with Eqs. (10.131), the Hilbert-Einstein action and Eq. (10.138) differ by a boundary term and are classically equivalent.

Promoting Q to a nonlinear function defines the $f(Q)$ class,

$$S = \frac{1}{2\kappa}\int d^4x\,\sqrt{-g}\;f(Q) + S_{\rm m}. \tag{10.140}$$

As in $f(T)$, the resulting field equations can be kept second order in the fundamental fields because Q depends on first derivatives of the metric (and the connection variables in a suitable formulation). A standard variational route treats $g_{\mu\nu}$ and $\Gamma^\rho{}_{\mu\nu}$ as independent, with the constraints $T^\rho{}_{\mu\nu} = 0$ and $R^\rho{}_{\sigma\mu\nu} = 0$ enforced by Lagrange multipliers; an equivalent and technically economical gauge choice is the *coincident gauge*, in which the inertial connection is set to zero, $\Gamma^\rho{}_{\mu\nu} = 0$, so that $\nabla_\rho \to \partial_\rho$ and

$$Q_{\rho\mu\nu} = \partial_\rho g_{\mu\nu} \qquad \text{(coincident gauge)}. \tag{10.141}$$

In this gauge, the entire gravitational dynamics is carried by the metric, while the affine structure is purely inertial. Varying Eq. (10.140) yields field equations that can be written in the compact form

$$\frac{2}{\sqrt{-g}}\nabla_\rho\big(\sqrt{-g}\,f_Q\,P^\rho{}_{\mu\nu}\big) + \frac{1}{2}\,g_{\mu\nu}\,f(Q) + f_Q\Big(P_{\mu\rho\sigma}\,Q_\nu{}^{\rho\sigma} - 2Q_{\rho\sigma\mu}\,P^{\rho\sigma}{}_\nu\Big) = -\,\kappa\,T_{\mu\nu}, \tag{10.142}$$

where $f_Q \equiv df/dQ$ and $T_{\mu\nu}$ is the matter stress tensor (defined as usual by variation of $S_{\rm m}$ with respect to $g^{\mu\nu}$). The connection equation (or, equivalently, the Lagrange-multiplier sector) enforces the flat, torsionless character of $\Gamma^\rho{}_{\mu\nu}$ and guarantees the existence of a gauge in which Eqs. (10.141) hold.

10.3.7 Jordan and Einstein Frames

Consider a scalar field non-minimally coupled to curvature,

$$S = \int d^4x\,\sqrt{-g}\left[\frac{1}{2\chi}\Big(1 - \xi\chi\phi^2\Big)R - \frac{1}{2}\,\partial^\mu\phi\,\partial_\mu\phi - V(\phi)\right], \tag{10.143}$$

where ξ is a dimensionless coupling, $V(\phi)$ is a generic potential for the field and $\chi \equiv 8\pi G$. The term $\frac{1}{2}\xi R\phi^2$ is the non-minimal interaction introduced previously: it modifies the gravitational sector because it effectively makes the coefficient of R field-dependent.

From Eq. (10.143) one reads off an effective gravitational coupling,

$$G_{\mathrm{eff}}(\phi) = \frac{G}{1 - \xi\chi\phi^2}, \tag{10.144}$$

which is well-defined only if

$$1 - \xi\chi\phi^2 > 0 \quad\Longrightarrow\quad \phi^2 < \frac{1}{\chi\xi}. \tag{10.145}$$

The action in Eq. (10.143) admits two customary representations: (i) the *Jordan frame*, where matter (here the scalar) is non-minimally coupled to curvature, and (ii) the *Einstein frame*, where a conformal rescaling removes the non-minimal coupling at the price of a nontrivial scalar-field redefinition and a rescaled potential. At the classical level, the two formulations are related by an invertible field redefinition only as long as the conformal factor is smooth and non-vanishing. Hence, Jordan and Einstein frames are not automatically interchangeable as *physical* frames, even if they are algebraically related as *field variables*.

Operationally, the dynamics may be studied in two different frames:

- **Jordan frame**: the interaction term $\propto \xi R\phi^2$ is retained, leading to modified Einstein's field equations and a modified scalar equation of motion;
- **Einstein frame**: a conformal transformation is used to recover a canonical Hilbert-Einstein term, at the expense of nontrivial kinetic mixing and a field-dependent rescaling of the potential.

We now characterize both representations.

Einstein-frame representation. Introduce a conformal transformation with conformal factor $\Omega(\phi)$,

$$g^{(E)}_{\mu\nu} = \Omega^2(\phi)\, g_{\mu\nu}, \qquad \sqrt{-g^{(E)}} = \Omega^4(\phi)\,\sqrt{-g}, \tag{10.146}$$

$$R = \Omega^{-2}(\phi)\left[R_E - 6\,\Box_E \ln\Omega(\phi) - 6\, g_E^{\mu\nu}\partial_\mu \ln\Omega(\phi)\,\partial_\nu \ln\Omega(\phi)\right], \tag{10.147}$$

where $\Box_E \equiv g_E^{\mu\nu}\nabla^{(E)}_\mu\nabla^{(E)}_\nu$. Substituting Eqs. (10.146)–(10.147) into (10.143) and dropping the total divergence $\Box_E \ln\Omega$ (which does not contribute to the field equations), one obtains an equivalent action written in terms of $g^{(E)}_{\mu\nu}$ and ϕ (for notational economy we now drop the label (E) on the metric)

$$S_E = \int d^4x \sqrt{-g}\left[\frac{1}{2\chi}\,\frac{1-\xi\chi\phi^2}{\Omega^2}\,R - \frac{1}{2}\,(\partial\phi)^2\,\frac{\Omega^2 + 6\frac{1-\xi\chi\phi^2}{\chi}\left(\frac{\partial\Omega}{\partial\phi}\right)^2}{\Omega^4} - \frac{V(\phi)}{\Omega^4}\right]. \tag{10.148}$$

To recover a canonical Hilbert-Einstein term one fixes

$$\Omega^2(\phi) = 1 - \xi\chi\phi^2. \tag{10.149}$$

Then the action takes the standard Einstein form once we introduce a canonically normalized scalar h, defined by the Jacobian

$$\frac{\partial h}{\partial\phi} = \frac{\sqrt{\Omega^2 + 6\frac{1-\xi\chi\phi^2}{\chi}\left(\frac{\partial\Omega}{\partial\phi}\right)^2}}{\Omega^2}, \tag{10.150}$$

$$V_E(\phi) = \frac{V(\phi)}{\Omega^4(\phi)}. \tag{10.151}$$

One arrives at

$$S = \int d^4x \sqrt{-g}\left[\frac{R}{2\chi} - \frac{1}{2}\,\partial^\mu h\,\partial_\mu h - V_E(h)\right]. \tag{10.152}$$

The explicit field redefinition follows from Eqs. (10.149) and (10.150). First,

$$\frac{\partial\Omega}{\partial\phi} = -\frac{\xi\chi\phi}{\sqrt{1-\xi\chi\phi^2}}, \qquad \frac{\partial h}{\partial\phi} = \frac{\sqrt{1-\xi(1-6\xi)\chi\phi^2}}{1-\xi\chi\phi^2}. \tag{10.153}$$

Integrating yields

$$h(\phi) = \frac{\sqrt{6\xi}\,\tanh^{-1}\left(\frac{\sqrt{6\chi}\,\xi\,\phi}{\sqrt{1+\phi^2\chi\xi(-1+6\xi)}}\right) - \sqrt{-1+6\xi}\,\sinh^{-1}\left(\sqrt{\chi\xi(-1+6\xi)}\,\phi\right)}{\sqrt{\chi\xi}}. \tag{10.154}$$

Because the map $\phi \mapsto h$ is generally not algebraically invertible in closed form, it is often convenient to work in limiting regimes. In particular, for the conformal coupling and for weak non-minimal coupling one finds, respectively,

$$\phi = \frac{1}{\sqrt{\chi\xi}}\tanh\left(\sqrt{\chi\xi}\,h\right), \qquad \xi = \frac{1}{6}, \tag{10.155}$$

$$\phi = \frac{1}{\sqrt{\chi\xi}}\sin\left(\sqrt{\chi\xi}\,h\right), \qquad \xi \ll 1. \tag{10.156}$$

For $\xi \gg 1$, the field-space stretching implied by Eq. (10.154) becomes extreme and h grows rapidly; in that regime the Einstein-frame description is often used precisely because it makes the large-ξ dynamics technically tractable.

The conformal factor in Eq. (10.149) vanishes at $\phi^2 = 1/(\chi\xi)$, precisely where G_{eff} in Eq. (10.144) diverges. At such points the field redefinition becomes singular, the transformation ceases to be invertible, and the two "frames" are no longer related by a legitimate change of variables. Even when $\Omega^2 > 0$ globally on the field domain of interest, the physical meanings of "time", "mass scales", and "minimal coupling" differ between frames; therefore, comparing observables requires a careful and explicit choice of which metric defines the operational rods and clocks.

Jordan-frame representation. In the Jordan frame the non-minimal coupling is kept explicit, and one derives modified Einstein equations by varying Eq. (10.143) with respect to $g^{\mu\nu}$. The variation of the interaction term reads

$$\begin{aligned}\delta(\xi R\phi^2\sqrt{-g}) &= \xi\,\delta R_{\mu\nu}\,g^{\mu\nu}\phi^2\sqrt{-g} + \xi\,R_{\mu\nu}\,\delta g^{\mu\nu}\,\phi^2\sqrt{-g} + \xi\,R_{\mu\nu}g^{\mu\nu}\phi^2\,\delta\sqrt{-g}\\ &= \xi\,\delta R_{\mu\nu}\,g^{\mu\nu}\phi^2\sqrt{-g} + \xi\,G_{\mu\nu}\phi^2\sqrt{-g}\,\delta g^{\mu\nu},\end{aligned} \tag{10.157}$$

and one uses the standard identity

$$g^{\mu\nu}\delta R_{\mu\nu} = \left(g_{\mu\nu}\partial_\alpha\partial^\alpha - \partial_\mu\partial_\nu\right)\delta g^{\mu\nu}. \tag{10.158}$$

At this stage, it is convenient to find out a compact relationship for the energy-momentum tensor. So, for a generic matter sector with action $S_{\text{m}}[g,\psi]$, where ψ collectively denotes the matter fields, consistency with the variational principle requires that the variation of the total action with respect to $g^{\mu\nu}$ yields Einstein's equations, $G_{\mu\nu} = \kappa\,T_{\mu\nu}$.
This fixes the variation of the matter action to show up the form

$$\delta S_{\text{m}} = -\frac{1}{2}\int d^4x\sqrt{-g}\,T_{\mu\nu}\,\delta g^{\mu\nu},$$

and therefore defines

$$T_{\mu\nu} \equiv -\frac{2}{\sqrt{-g}}\frac{\delta S_{\text{m}}}{\delta g^{\mu\nu}}. \tag{10.159}$$

This definition is also singled out by symmetry. The matter action is invariant under infinitesimal diffeomorphisms $x^\mu \to x^\mu + \xi^\mu(x)$, under which $\delta g^{\mu\nu} = \nabla^\mu \xi^\nu + \nabla^\nu \xi^\mu$. Using the above definition, the induced variation of the action is

$$\delta_\xi S_{\rm m} = -\int d^4x \sqrt{-g}\, (\nabla_\mu T^{\mu\nu})\, \xi_\nu,$$

up to terms proportional to the matter equations of motion. Invariance of $S_{\rm m}$ for arbitrary ξ^μ then implies the covariant conservation law, $\nabla_\mu T^{\mu\nu} = 0$. Hence Eq. (10.159) *is not a convention, but the unique definition that simultaneously ensures*:

1. compatibility of the variational principle with Einstein's equations, and
2. covariant conservation of energy-momentum as a consequence of diffeomorphism invariance.

From the definition of the energy-momentum tensor, Eq. (10.159), one obtains

$$T^{\rm int}_{\mu\nu} = \xi \left(g_{\mu\nu} \partial_\alpha \partial^\alpha - \partial_\mu \partial_\nu \right) \phi^2 + \xi \phi^2 G_{\mu\nu}. \tag{10.160}$$

As a result, the modified Einstein equations can be written as

$$\begin{aligned} \frac{1 - \chi\xi\phi^2}{\chi} G_{\mu\nu} &= (1 - 2\xi)\, \partial_\mu \phi\, \partial_\nu \phi - \left(\frac{1}{2} - 2\xi \right) g_{\mu\nu}\, \partial_\alpha \phi\, \partial^\alpha \phi \\ &\quad - g_{\mu\nu} V(\phi) + 2\xi\, \phi \left(g_{\mu\nu} \Box - \partial_\mu \partial_\nu \right) \phi. \end{aligned} \tag{10.161}$$

For a homogeneous field, the spatial gradients vanish, and the scalar curvature $R = -g^{\mu\nu} G_{\mu\nu}$ becomes

$$\begin{aligned} R &= \frac{\chi}{1 - \chi\xi\phi^2} \left[(1 - 6\xi) \partial_\alpha \phi\, \partial^\alpha \phi + 4V(\phi) - 6\xi \phi\, \Box \phi \right] \\ &= \frac{\chi}{1 - \chi\xi\phi^2} \left[-(1 - 6\xi) \dot{\phi}^2 + 4V(\phi) + 6\xi \phi\, \ddot{\phi} \right]. \end{aligned} \tag{10.162}$$

10.4 Black Holes: A First Journey

Solving Einstein's field equations means determining the spacetime metric $g_{\mu\nu}$ from a specified matter content (or in vacuum), together with appropriate boundary/regularity conditions. In practice one rarely attacks the fully general PDE system directly. Two standard strategies are:

- impose symmetry *a priori* so that the field equations reduce to a manageable set of ODEs (static/stationary, spherical/axial symmetry);
- infer an ansatz for the line element from first principles (causal structure, asymptotics, conserved charges) and determine the remaining functions by the field equations.

Textbooks typically begin with the simplest static and spherically symmetric vacuum geometry (Schwarzschild), which is asymptotically Minkowskian at large radii. Here we adopt a more physical entry point: the operational meaning of a BH is best understood through *gravitational collapse* and the emergence of trapped regions.

In 1939 Oppenheimer and Snyder, and independently Datt, constructed the first fully relativistic model of complete collapse: an exact solution describing a homogeneous spherically symmetric cloud of collisionless matter ("dust") that contracts under its own gravity. Intuitively, collapse occurs when pressure forces, which act against compression, are insufficient to counterbalance gravitational attraction.

Penrose's analysis shows that, within classical GR, curvature singularities (more precisely, geodesic incompleteness) arise as the generic endstate once the following hypotheses are simultaneously satisfied:

1. validity of GR throughout the collapse regime;
2. validity of suitable energy conditions for the matter sector;
3. global hyperbolicity (no "causal pathology" allowing singularity avoidance by acausal leakage);
4. formation of trapped surfaces during the dynamical evolution.

For sufficiently compact and massive configurations, the known non-gravitational interactions cannot halt collapse. If a repulsive mechanism prevents the classical singular endstate, it must originate from **new physics**, becoming dominant at large curvature. This may enter as: (i) a modification of the gravitational dynamics, invalidating hypothesis (1), and/or (ii) a modification of the effective matter sector (or its averaged description), invalidating hypothesis (2).

10.4.1 General Relativity Coupled to a Vector Field

A technically simpler route than the fully dynamical collapse problem is to focus on *static* (and later stationary) solutions in the vacuum exterior of a compact source, where the field equations reduce to a boundary-value problem. The minimal coupling of gravity to an Abelian gauge field yields Einstein-Maxwell theory,

$$S = \frac{1}{16\pi} \int d^4x \sqrt{-g} \left(R - \frac{1}{4} F_{\mu\nu} F^{\mu\nu} \right), \tag{10.163}$$

with $F_{\mu\nu} = \partial_\mu A_\nu - \partial_\nu A_\mu$. Variation with respect to $g^{\mu\nu}$ and A_μ yields

$$G_{\mu\nu} = 8\pi\, T^{(\mathrm{EM})}_{\mu\nu}, \qquad \nabla_\mu F^{\mu\nu} = 0, \qquad \nabla_{[\mu} F_{\nu\rho]} = 0. \tag{10.164}$$

Driving Exercise

Recalling Eq. (10.159), we derive the stress-energy tensor of the electromagnetic field in a generic curved spacetime by direct metric variation of the Maxwell action. The field strength is the usual $F_{\mu\nu} \equiv \nabla_\mu A_\nu - \nabla_\nu A_\mu = \partial_\mu A_\nu - \partial_\nu A_\mu$, where the last equality holds because the Levi-Civita connection is torsionless and $\Gamma^\rho{}_{\mu\nu} = \Gamma^\rho{}_{\nu\mu}$, hence $\nabla_\mu A_\nu - \nabla_\nu A_\mu$ cancels the connection terms.

In the minimal Maxwell theory, the metric enters the electromagnetic action only through

1. the volume element $\sqrt{-g}$;
2. index raising in $F^{\mu\nu} = g^{\mu\alpha} g^{\nu\beta} F_{\alpha\beta}$.

Crucially, $F_{\mu\nu}$ itself is independent of the metric (for a fixed potential A_μ). This is why the metric variation can be performed cleanly and yields a symmetric, gauge-invariant stress tensor.

We readily find

$$T_{\mu\nu} \equiv -\frac{2}{\sqrt{-g}} \frac{\delta S_{\mathrm{EM}}}{\delta g^{\mu\nu}} = -2 \frac{\partial \mathcal{L}_{\mathrm{EM}}}{\partial g^{\mu\nu}} + g_{\mu\nu} \mathcal{L}_{\mathrm{EM}}, \tag{10.165}$$

where in the second equality we used the fact that $\mathcal{L}_{\mathrm{EM}}$ depends on the metric algebraically (no metric derivatives appear). Hence, we may pursue the following steps.

1. The standard identity for the metric variation of the volume element is

$$\delta\sqrt{-g} = -\frac{1}{2}\sqrt{-g}\, g_{\mu\nu}\, \delta g^{\mu\nu}. \tag{10.166}$$

2. Since $F_{\mu\nu}$ does not depend on the metric, the only metric dependence sits in $F^{\mu\nu} = g^{\mu\alpha} g^{\nu\beta} F_{\alpha\beta}$. Hence

$$\delta\big(F_{\mu\nu} F^{\mu\nu}\big) = F_{\mu\nu}\, \delta F^{\mu\nu}. \tag{10.167}$$

Now compute $\delta F^{\mu\nu}$:

$$\begin{aligned} F^{\mu\nu} &= g^{\mu\alpha} g^{\nu\beta} F_{\alpha\beta}, \\ \delta F^{\mu\nu} &= \delta g^{\mu\alpha}\, g^{\nu\beta} F_{\alpha\beta} + g^{\mu\alpha}\, \delta g^{\nu\beta} F_{\alpha\beta}. \end{aligned} \tag{10.168}$$

Insert Eqs. (10.168) into (10.167) and relabel dummy indices to express the result in terms of $\delta g^{\mu\nu}$:

$$\begin{aligned}\delta\big(F_{\mu\nu}F^{\mu\nu}\big) &= F_{\mu\nu}\Big(\delta g^{\mu\alpha}\, g^{\nu\beta}\, F_{\alpha\beta} + g^{\mu\alpha}\,\delta g^{\nu\beta}\, F_{\alpha\beta}\Big)\\ &= \delta g^{\mu\alpha}\, F_{\mu\nu} F_{\alpha}{}^{\nu} + \delta g^{\nu\beta}\, F_{\mu\nu} F^{\mu}{}_{\beta}\\ &= 2\,\delta g^{\mu\nu}\, F_{\mu\alpha} F_{\nu}{}^{\alpha}.\end{aligned} \tag{10.169}$$

In the last step we used the symmetry of $\delta g^{\mu\nu}$ and the fact that the two terms are identical upon exchange of dummy indices.
Therefore,

$$\delta\mathcal{L}_{\mathrm{EM}} = -\frac{1}{16\pi}\,\delta\big(F_{\mu\nu}F^{\mu\nu}\big) = -\frac{1}{8\pi}\, F_{\mu\alpha} F_{\nu}{}^{\alpha}\,\delta g^{\mu\nu}. \tag{10.170}$$

Equivalently, the partial derivative needed in Eq. (10.165) is

$$\frac{\partial\mathcal{L}_{\mathrm{EM}}}{\partial g^{\mu\nu}} = -\frac{1}{8\pi}\, F_{\mu\alpha} F_{\nu}{}^{\alpha}. \tag{10.171}$$

3. Varying the action gives

$$\begin{aligned}\delta S_{\mathrm{EM}} &= \int d^4x\,\Big(\delta\sqrt{-g}\,\mathcal{L}_{\mathrm{EM}} + \sqrt{-g}\,\delta\mathcal{L}_{\mathrm{EM}}\Big)\\ &= \int d^4x\,\sqrt{-g}\left[-\frac{1}{2}\, g_{\mu\nu}\,\mathcal{L}_{\mathrm{EM}}\,\delta g^{\mu\nu} - \frac{1}{8\pi}\, F_{\mu\alpha} F_{\nu}{}^{\alpha}\,\delta g^{\mu\nu}\right]\\ &= -\frac{1}{2}\int d^4x\,\sqrt{-g}\left[\frac{1}{4\pi} F_{\mu\alpha} F_{\nu}{}^{\alpha} + g_{\mu\nu}\,\mathcal{L}_{\mathrm{EM}}\right]\delta g^{\mu\nu}.\end{aligned} \tag{10.172}$$

By comparison with the defining relation

$$\delta S_{\mathrm{EM}} = -\frac{1}{2}\int d^4x\,\sqrt{-g}\, T_{\mu\nu}\,\delta g^{\mu\nu}, \tag{10.173}$$

we read off the stress-energy tensor

$$T_{\mu\nu} = \frac{1}{4\pi}\, F_{\mu\alpha} F_{\nu}{}^{\alpha} + g_{\mu\nu}\,\mathcal{L}_{\mathrm{EM}}. \tag{10.174}$$

Substituting $\mathcal{L}_{\mathrm{EM}} = -(16\pi)^{-1} F_{\alpha\beta} F^{\alpha\beta}$ yields the standard compact expression

$$\boxed{T_{\mu\nu} = \frac{1}{4\pi}\left(F_{\mu\alpha} F_{\nu}{}^{\alpha} - \frac{1}{4}\, g_{\mu\nu}\, F_{\alpha\beta} F^{\alpha\beta}\right).} \tag{10.175}$$

Assume static spherical symmetry and adopt

$$ds^2 = -e^{2\Phi(r)}dt^2 + e^{2\Lambda(r)}dr^2 + r^2 d\Omega^2, \qquad d\Omega^2 = d\theta^2 + \sin^2\theta\, d\varphi^2. \quad (10.176)$$

For an electrostatic configuration compatible with these symmetries, one may choose the gauge potential as

$$A_\mu dx^\mu = A_t(r)\, dt, \qquad F_{tr} = -A_t'(r), \qquad (10.177)$$

(prime denotes d/dr). The Bianchi identity $\nabla_{[\mu} F_{\nu\rho]} = 0$ is automatically satisfied.

The only independent Maxwell equation is $\nabla_\mu F^{\mu t} = 0$. Using $\sqrt{-g} = e^{\Phi+\Lambda} r^2 \sin\theta$ and $F^{rt} = g^{rr} g^{tt} F_{tr} = e^{-2(\Phi+\Lambda)} A_t'$, one obtains

$$0 = \nabla_\mu F^{\mu t} = \frac{1}{\sqrt{-g}} \partial_r \left(\sqrt{-g}\, F^{rt}\right) = \frac{1}{e^{\Phi+\Lambda} r^2}\, \partial_r \left(r^2 e^{-(\Phi+\Lambda)} A_t'\right), \qquad (10.178)$$

hence

$$r^2 e^{-(\Phi+\Lambda)} A_t' = Q, \qquad \Rightarrow \qquad A_t'(r) = \frac{Q}{r^2}\, e^{\Phi+\Lambda}, \qquad (10.179)$$

where Q is the conserved electric charge.

In order to integrate the Einstein equations, it is convenient to introduce the mass function $m(r)$ via

$$e^{-2\Lambda(r)} \equiv 1 - \frac{2\, m(r)}{r}. \qquad (10.180)$$

Then the (t, t) Einstein equation can be cast as

$$m'(r) = 4\pi r^2\, \rho(r), \qquad \rho(r) \equiv -T^t{}_t, \qquad (10.181)$$

with ρ the energy density as seen by static observers.

For a purely radial electric field,

$$\rho(r) = \frac{1}{8\pi}\, E(r)^2, \qquad E(r)^2 \equiv -F_{t\alpha} F^{t\alpha} = -F_{tr} F^{tr}. \qquad (10.182)$$

Using $F_{tr} = -A_t'$ and $F^{tr} = g^{tt} g^{rr} F_{tr} = -e^{-2(\Phi+\Lambda)} A_t'$ gives $E^2 = e^{-2(\Phi+\Lambda)} (A_t')^2$. With Eq. (10.179) one finds the Coulomb law in invariant form,

$$E(r)^2 = e^{-2(\Phi+\Lambda)} \left(\frac{Q}{r^2} e^{\Phi+\Lambda}\right)^2 = \frac{Q^2}{r^4}, \qquad (10.183)$$

hence

$$m'(r) = 4\pi r^2 \cdot \frac{1}{8\pi}\frac{Q^2}{r^4} = \frac{Q^2}{2r^2} \qquad \Rightarrow \qquad m(r) = M - \frac{Q^2}{2r}, \tag{10.184}$$

where M coincides with the gravitational mass measured at infinity, dubbed the Arnowitt-Deser-Misner mass, compactly the **ADM mass**.

Substituting Eqs. (10.184) into (10.180) yields

$$e^{-2\Lambda(r)} = 1 - \frac{2M}{r} + \frac{Q^2}{r^2}. \tag{10.185}$$

A second independent Einstein combination implies (in electro-vacuum)

$$\Phi'(r) + \Lambda'(r) = 0 \qquad \Rightarrow \qquad \Phi(r) = -\Lambda(r) + \text{const.} \tag{10.186}$$

The additive constant is removed by a rescaling of t, so $\Phi = -\Lambda$ and $e^{2\Phi} = e^{-2\Lambda}$. One then arrives at the Reissner-Nordström solution,

$$ds^2 = -f(r)\,dt^2 + f(r)^{-1}dr^2 + r^2 d\Omega^2, \qquad f(r) = 1 - \frac{2M}{r} + \frac{Q^2}{r^2}, \tag{10.187}$$

with potential (up to a constant)

$$A_t(r) = -\frac{Q}{r}, \qquad F_{tr} = \frac{Q}{r^2}. \tag{10.188}$$

The horizon radii are the real roots of $f(r) = 0$,

$$r_\pm = M \pm \sqrt{M^2 - Q^2}. \tag{10.189}$$

For $|Q| < M$ the geometry has an outer and an inner horizon. For $|Q| = M$ it is extremal with a degenerate horizon; for $|Q| > M$ no horizon exists and the curvature singularity at $r = 0$ is naked.

The simplest axisymmetric spacetime, describing an uncharged rotating BH, is offered by the *Kerr solution* that, in Boyer-Lindquist coordinates, reads

$$\begin{aligned} ds^2 = &-\left(1 - \frac{2Mr}{\Sigma}\right)dt^2 - \frac{4Mar\sin^2\theta}{\Sigma}\,dt\,d\varphi + \frac{\Sigma}{\Delta}\,dr^2 + \\ &+ \Sigma\,d\theta^2 + \left(r^2 + a^2 + \frac{2Ma^2 r\sin^2\theta}{\Sigma}\right)\sin^2\theta\,d\varphi^2, \end{aligned} \tag{10.190}$$

where

$$\Sigma \equiv r^2 + a^2\cos^2\theta, \qquad \Delta \equiv r^2 - 2Mr + a^2, \qquad a \equiv \frac{J}{M}. \tag{10.191}$$

The event horizons lie at the roots of $\Delta = 0$,

$$r_{\pm} = M \pm \sqrt{M^2 - a^2}, \tag{10.192}$$

so a Kerr BH exists for $|a| \leq M$; $|a| = M$ is extremal.

Rotation produces an ergoregion, bounded by the *static-limit* surface $g_{tt} = 0$,

$$r_{\text{ergo}}(\theta) = M + \sqrt{M^2 - a^2 \cos^2\theta}, \tag{10.193}$$

inside which the stationary Killing vector ∂_t is spacelike and states with negative conserved energy (as measured at infinity) can exist, enabling energy extraction mechanisms such as Penrose's process.

The most general stationary electro-vacuum BH is Kerr-Newman, obtained by replacing Δ in Eq. (10.190) with

$$\Delta(r) = r^2 - 2Mr + a^2 + Q^2, \tag{10.194}$$

together with the corresponding electromagnetic potential; the horizon radii are again given by $\Delta(r) = 0$.

Schwarzschild limit and the meaning of M.

In the limit $a \to 0$ and $Q \to 0$ one recovers Schwarzschild,

$$ds^2 = -\left(1 - \frac{2M}{r}\right)dt^2 + \left(1 - \frac{2M}{r}\right)^{-1} dr^2 + r^2 d\Omega^2. \tag{10.195}$$

In the weak-field, slow-motion regime $g_{\mu\nu} = \eta_{\mu\nu} + h_{\mu\nu}$ with $|h_{\mu\nu}| \ll 1$, one has $g_{00} \simeq -(1 + 2\Phi)$, where Φ is the Newtonian potential. For Eq. (10.195), $\Phi \simeq -M/r$, so M coincides with the ADM mass.

The loci $r = 2M$ and $r = 0$ in Eq. (10.195) have radically different geometric status. The surface $r = 2M$ is a coordinate singularity: it is removed by adopting coordinates regular on the future horizon, e.g. ingoing Eddington-Finkelstein,

$$v = t + r_*, \qquad r_* = r + 2M \ln\left|\frac{r}{2M} - 1\right|, \tag{10.196}$$

which yields a line element manifestly regular at $r = 2M$. By contrast, $r = 0$ is an intrinsic curvature singularity, diagnosed by invariants such as the Kretschmann scalar[1]

$$K \equiv R_{\mu\nu\alpha\beta} R^{\mu\nu\alpha\beta} = \frac{48M^2}{r^6}, \tag{10.197}$$

which diverges as $r \to 0$ in any coordinate chart.

A physically transparent characterization of the Schwarzschild horizon comes from redshift. For a static emitter at radius r and a static receiver at infinity,

$$1 + z = \frac{d\tau_\infty}{d\tau_r} = \frac{1}{\sqrt{-g_{00}(r)}} = \left(1 - \frac{2M}{r}\right)^{-1/2}, \tag{10.198}$$

which diverges as $r \to 2M^+$. The hypersurface $r = 2M$ is null and defines a causal boundary: outgoing null rays initiated from $r \leq 2M$ do not reach future null infinity. This surface is the (future) event horizon.

10.4.2 General Relativity Coupled with Scalar Fields

Analogously to vector fields, we can now analyze what happens when gravity is coupled with scalar fields (see Table 10.1).

Let us consider the action (Jordan frame)

$$S[g, \phi] = \frac{1}{16\pi} \int d^4x \sqrt{-g} \Big[(1 - 16\pi\xi\phi^2)\, R - 16\pi\, (\nabla\phi)^2 - 32\pi\, V(\phi) \Big], \tag{10.199}$$

where $(\nabla\phi)^2 \equiv g^{\mu\nu}\nabla_\mu\phi\nabla_\nu\phi$, ξ is the non-minimal curvature coupling (with $\xi = \frac{1}{6}$ the conformal value in 4D), and $V(\phi)$ is the scalar potential.

Varying Eq. (10.199) with respect to ϕ yields the generalized Klein-Gordon equation,

$$\Box\phi - \xi R\, \phi - \frac{dV}{d\phi} = 0, \tag{10.200}$$

where $\Box \equiv \nabla_\mu \nabla^\mu$. Varying with respect to $g^{\mu\nu}$ gives

$$(1 - 16\pi\xi\phi^2)\, G_{\mu\nu} = 8\pi\, T^{(\phi)}_{\mu\nu} + 16\pi\xi \Big(\nabla_\mu \nabla_\nu - g_{\mu\nu}\Box\Big)\phi^2 - 16\pi\, g_{\mu\nu}\, V(\phi), \tag{10.201}$$

[1] Second-order curvature scalars are defined as diffeomorphism-invariant quantities quadratic in the curvature. A convenient basis is provided by R^2, $R_{\mu\nu}R^{\mu\nu}$, and $R_{\mu\nu\rho\sigma}R^{\mu\nu\rho\sigma}$. Other notable linear combinations include the Gauss-Bonnet density $\mathcal{G} = R_{\mu\nu\rho\sigma}R^{\mu\nu\rho\sigma} - 4R_{\mu\nu}R^{\mu\nu} + R^2$, which is topological in four dimensions, and the Weyl-squared invariant $C_{\mu\nu\rho\sigma}C^{\mu\nu\rho\sigma}$ (with $C_{\mu\nu\rho\sigma}$ the so-called Weyl tensor), measuring the conformal curvature. A parity-odd pseudoscalar invariant is given by the Pontryagin density $\mathcal{P} = R_{\mu\nu\rho\sigma}\tilde{R}^{\mu\nu\rho\sigma}$.

Table 10.1 Exact and representative solutions of GR coupled to a scalar field, classified by the non-minimal coupling ξ and by the scalar potential $V(\phi)$

Solution	ξ	$V(\phi)$	Asymptotic	Horizon	Main properties
Fisher	0	0	Flat	N	First scalar-vacuum solution; naked singularity; Schwarzschild recovered for $\phi = \text{const}$
JNW	0	0	Flat	N	Massless minimally coupled scalar; naked singularity; continuous deformation of Schwarzschild
Wyman	0	0	Flat	N	Equivalent to Fisher/JNW by reparametrization; same invariant content
BBMB	$\frac{1}{6}$	0	Flat	Y	Conformally coupled scalar; scalar diverges at horizon; extremal BH; evades standard no-hair theorems
Conformal	$\frac{1}{6}$	0	Flat	Y	Alternative parametrizations of BBMB-type solutions; often discussed in the Jordan frame
EB WH	0 (phantom)	0	Flat	–	Traversable wormhole supported by a phantom scalar; no horizon; regular throat
MTZ	$\frac{1}{6}$	$\lambda\phi^4$	AdS	Y	Exact hairy AdS BH; scalar regular on the horizon; conformal symmetry crucial
EKG	0	$\neq 0$	AdS	Y	Massive/self-interacting scalars; typically numerical families; AdS boundary conditions essential
Scalar-tensor	$\neq 0$	0	Frame-dependent	Y/N	Generated via conformal transformations; include BHs, naked singularities, and wormholes

with the "minimal" part of the scalar stress tensor

$$T^{(\phi)}_{\mu\nu} = \nabla_\mu\phi\nabla_\nu\phi - \frac{1}{2}g_{\mu\nu}(\nabla\phi)^2. \tag{10.202}$$

For later use, it is convenient to display the trace of Eq. (10.201). Contracting with $g^{\mu\nu}$ and using $G^\mu{}_\mu = -R$ and $T^{(\phi)} \equiv T^{(\phi)\,\mu}{}_\mu = -(\nabla\phi)^2$ gives

$$-(1 - 16\pi\xi\phi^2)\,R = -8\pi(\nabla\phi)^2 - 48\pi\xi\,\Box(\phi^2) - 64\pi\,V(\phi), \tag{10.203}$$

which makes explicit how ξ introduces higher-derivative terms through $\Box(\phi^2)$. Equation (10.201) is naturally interpreted as Einstein gravity with an effective coupling, see Eq. (10.144), so that the scalar field modulates the normalization of the Einstein tensor.

We now restrict to static, spherically symmetric configurations,

$$ds^2 = -e^{2\Phi(r)}dt^2 + e^{2\Lambda(r)}dr^2 + r^2d\Omega^2, \qquad \phi = \phi(r), \tag{10.204}$$

with $d\Omega^2 = d\theta^2 + \sin^2\theta\, d\varphi^2$. Substituting Eqs. (10.204) into (10.200)–(10.201) produces a closed system of ordinary differential equations. In practice, integrability is highly gauge-sensitive: different closed-form solutions are most transparently obtained in different coordinate choices. Accordingly, rather than insisting on a single universal gauge, we select for each classical solution a parametrization adapted to the relevant first integrals and to the regularity analysis.

Driving Exercise

Minimally coupled massless scalar: Fisher-JNW-Wyman family.

The baseline model is obtained by setting

$$\xi = 0, \qquad V(\phi) = 0. \tag{10.205}$$

The field equations reduce to the Einstein-massless-scalar system

$$G_{\mu\nu} = 8\pi T^{(\phi)}_{\mu\nu}, \qquad \Box\phi = 0, \tag{10.206}$$

with $T^{(\phi)}_{\mu\nu}$ given by Eq. (10.202).

A convenient integrable parametrization is $ds^2 = -A(r)\, dt^2 + \frac{dr^2}{A(r)} + R(r)^2\, d\Omega^2$, where the coordinate r is *not* assumed to be the areal radius; the latter is $R(r)$.

For a purely radial scalar, $\phi = \phi(r)$, the wave equation $\Box\phi = 0$ becomes

$$\partial_r\left(A\, R^2\, \phi'\right) = 0 \quad \Longrightarrow \quad A\, R^2\, \phi' = C, \tag{10.207}$$

where C is an integration constant and primes denote d/dr. The Einstein equations can be rearranged into three independent combinations. The difference between the (t^t) and (r^r) components is particularly diagnostic:

$$G^t{}_t - G^r{}_r = -\frac{2A}{R} R'', \qquad T^t{}_t - T^r{}_r = -A(\phi')^2, \tag{10.208}$$

so that

$$-\frac{2A}{R} R'' = 8\pi\left(T^t{}_t - T^r{}_r\right) = -8\pi\, A(\phi')^2 \quad \Longrightarrow \quad \frac{R''}{R} = 4\pi(\phi')^2. \tag{10.209}$$

Eliminating ϕ' by Eq. (10.207) yields

$$\frac{R''}{R} = 4\pi \frac{C^2}{A^2 R^4}. \tag{10.210}$$

The remaining Einstein equations determine A in terms of R and integration constants. The system is integrable and produces a two-parameter family, conventionally interpreted as ADM mass plus a scalar charge.

A standard closed-form representation is obtained by fixing the areal radius directly. Introducing a radial coordinate ρ for which the angular sector is manifestly $\rho^2 d\Omega^2$ up to a conformal factor, one may write

$$ds^2 = -f(\rho)^{\nu}\, dt^2 + f(\rho)^{-\nu}\, d\rho^2 + f(\rho)^{1-\nu}\, \rho^2\, d\Omega^2, \qquad f(\rho) = 1 - \frac{b}{\rho}, \tag{10.211}$$

with scalar profile

$$\phi(\rho) = \phi_0 \pm \sqrt{\frac{1-\nu^2}{16\pi}}\, \ln f(\rho), \qquad 0 < \nu \leq 1. \tag{10.212}$$

Direct substitution of Eqs. (10.211)–(10.212) into (10.206) confirms that the solution is consistent provided the integration constants are related to the ADM mass M and scalar charge q by

$$\nu = \frac{2M}{b}, \qquad q^2 = \frac{b^2}{4}\,(1-\nu^2). \tag{10.213}$$

The parameter range $0 < \nu \leq 1$ is precisely the statement that the scalar is canonical: the coefficient of $\ln f$ in Eq. (10.212) is real only for $1 - \nu^2 \geq 0$.

Interpretation of the JNW parameters. The geometry carries two independent conserved data at infinity: the ADM mass M and the scalar charge q. Equation (10.213) packages these into (b, ν), where b sets the location of the would-be Schwarzschild radius in the chosen coordinates and ν quantifies the deformation away from the vacuum BH. The limit $\nu \to 1$ suppresses the scalar logarithm and collapses the family to Schwarzschild.

The fundamental properties are as follows.

- **Schwarzschild limit.** Taking $\nu \to 1$ at fixed $b = 2M$ yields

$$ds^2 = -(1 - 2M/\rho)\, dt^2 + (1 - 2M/\rho)^{-1} d\rho^2 + \rho^2 d\Omega^2, \qquad \phi = \phi_0, \tag{10.214}$$

so the scalar hair decouples and vacuum is recovered.
- **Absence of a regular horizon for** $\nu < 1$. The locus $\rho = b$ is not a smooth Killing horizon. Indeed, the scalar in Eq. (10.212) diverges logarithmically as $\rho \to b^+$ and curvature invariants blow up, so $\rho = b$ is a genuine curvature singularity. The generic Fisher/JNW/Wyman solution is therefore a *naked singularity* rather than a BH.

- **Energy conditions and "no hair" obstruction.** For a canonical scalar one has $T_{\mu\nu}k^\mu k^\nu = (k^\mu \nabla_\mu \phi)^2 \geq 0$ for all null vectors k^μ, so the null energy condition is satisfied. In static, asymptotically flat settings this positivity is precisely what obstructs the formation of a regular horizon supporting nontrivial scalar hair, in agreement with standard no hair logic.

Historically the same invariant solution is attributed to Fisher and was later re-derived by Wyman. In practice, one encounters different coordinate choices and parameterizations.

- **Fisher parametrization.** One often introduces a radial coordinate in which the metric functions are simple powers of a harmonic function, making the asymptotic expansion straightforward.
- **Wyman parametrization.** A common gauge choice makes ϕ linear in the chosen coordinate, so that the Einstein equations integrate by quadratures and the metric functions follow algebraically.

All such representations are related by smooth coordinate changes on the domain where the solution is regular and by redefinitions of the two essential parameters (mass and scalar charge). The physical content, and in particular the presence of a naked singularity for nontrivial canonical hair, is invariant.

▶ Driving Exercise
Conformal coupling without potential: BBMB BH.
We now set

$$\xi = \frac{1}{6}, \qquad V(\phi) = 0. \tag{10.215}$$

For $V = 0$ and $\xi = \frac{1}{6}$ the scalar sector is conformally invariant in four dimensions, and this exceptional structure is precisely what permits an asymptotically flat BH geometry supported by a scalar profile.
We seek a static, spherically symmetric configuration in areal gauge, $ds^2 = -F(r)\,dt^2 + \frac{dr^2}{F(r)} + r^2 d\Omega^2, \qquad \phi = \phi(r)$.
The conformal coupling term in Eqs. (10.201) introduces second derivatives of ϕ^2 through $\nabla_\mu \nabla_\nu(\phi^2)$ and $\Box(\phi^2)$. For $\xi = \frac{1}{6}$, a consistent closed-form branch is obtained when the metric function is chosen to coincide with the extremal Reissner-Nordström profile,

$$F(r) = \left(1 - \frac{M}{r}\right)^2, \tag{10.216}$$

without introducing any Maxwell field. The scalar itself provides the effective stress tensor required to support Eq. (10.216).

For $\xi = \frac{1}{6}$ and $V = 0$, the scalar equation, Eq. (10.200), reads

$$\Box\phi - \frac{1}{6} R\,\phi = 0. \tag{10.217}$$

For the metric with Eq. (10.216), the Ricci scalar vanishes identically in the exterior region $r > M$ (as in extremal Reissner-Nordström), so Eq. (10.217) reduces to $\Box\phi = 0$ away from the would-be horizon. For a radial function, $F(r)$,

$$\Box\phi = \frac{1}{r^2}\partial_r\Big(r^2\, F(r)\phi'(r)\Big), \tag{10.218}$$

and hence

$$\partial_r\Big(r^2\, F\,\phi'\Big) = 0 \quad \Longrightarrow \quad r^2\, F(r)\phi'(r) = C, \tag{10.219}$$

for some constant C. Using Eq. (10.216) yields

$$\phi'(r) = \frac{C}{r^2(1 - M/r)^2} = \frac{C}{(r - M)^2}, \tag{10.220}$$

so that

$$\phi(r) = \phi_\infty - \frac{C}{r - M}. \tag{10.221}$$

At this stage the scalar amplitude is still free. Imposing the full Einstein's equations (10.201) (with $\xi = \frac{1}{6}$, $V = 0$, and $R = 0$) fixes C uniquely (up to an overall sign), leading to the BBMB scalar profile

$$\phi(r) = \pm\sqrt{\frac{3}{4\pi}}\,\frac{M}{r - M}, \tag{10.222}$$

together with Eq. (10.216). The $\pm$ sign is immaterial in a Z_2-symmetric theory.

Why BBMB evades the minimally coupled obstruction. In the canonical minimally coupled case, regular horizons are incompatible with nontrivial static hair under asymptotic flatness. The BBMB branch bypasses this rigidity because the conformal term reshapes the effective stress tensor by adding $(\nabla_\mu \nabla_\nu - g_{\mu\nu}\Box)\phi^2$ contributions. These terms can emulate the traceless structure needed to support the extremal geometry, Eq. (10.216), even in the absence of gauge fields. The price is the borderline behavior of the scalar at the horizon.

The main properties are:

- **Horizon.** $r = M$ is a degenerate Killing horizon (vanishing surface gravity), as in extremal Reissner-Nordström.
- **Scalar divergence.** The scalar in Eq. (10.222) diverges as $(r - M)^{-1}$ at the horizon. This is the distinguishing peculiarity of the BBMB solution and is central to discussions of its physical admissibility.
- **Asymptotics.** As $r \to \infty$, one has $\phi(r) \to 0$ and the metric is asymptotically flat with ADM mass M.
- **Effective source.** The geometry in Eq. (10.216) is supported by a scalar-induced effective stress tensor whose algebraic structure mimics that of a traceless source (as would an electromagnetic field in RN), but the supporting degree of freedom is purely scalar.

▶ Driving Exercise

Phantom scalar and the Ellis-Bronnikov wormhole.

A traversable wormhole requires violation of the null energy condition. The most economical realization within a single scalar degree of freedom is to reverse the sign of the kinetic term. Consider the minimally coupled *phantom* action

$$S = \frac{1}{16\pi} \int d^4x \sqrt{-g} \left[R + 16\pi \, (\nabla\phi)^2 \right], \tag{10.223}$$

so that, relative to Eq. (10.202), the stress tensor is sign-flipped in the gradient sector.

A convenient wormhole-adapted ansatz is the Morris-Thorne form

$$ds^2 = -e^{2\Phi(r)} dt^2 + \frac{dr^2}{1 - \frac{b(r)}{r}} + r^2 d\Omega^2, \tag{10.224}$$

where $\Phi(r)$ is the redshift function and $b(r)$ is the shape function.

Without entering deeply the physical meaning of wormholes, we may summarize how, in GR, and in particular from the above form of the metric, one may expect a wormhole to be traversable.

- **Throat location and definition**: the areal radius attains a minimum at the throat $r = r_0 > 0$, with

$$b(r_0) = r_0.$$

- **No horizon (finite redshift)**: $\Phi(r)$ must be finite everywhere (in particular at r_0), so that $e^{2\Phi(r)} \neq 0$ and no infinite-redshift surface forms.
- **Metric signature/absence of coordinate pathology**: for $r \geq r_0$ one must have

$$1 - \frac{b(r)}{r} \geq 0, \qquad \text{with equality only at } r = r_0,$$

so that the radial proper distance

$$\ell(r) = \pm \int_{r_0}^{r} \frac{dr'}{\sqrt{1 - b(r')/r'}}$$

is real and regular.
- **Flaring-out condition (geometric)**: the throat must "flare out", equivalently

$$b'(r_0) < 1,$$

and typically also $b(r) < r$ for $r > r_0$.
- **Asymptotic flatness (for a standard two-end wormhole)**: as $r \to \infty$ (on each side) one imposes

$$\Phi(r) \to 0, \qquad \frac{b(r)}{r} \to 0,$$

so that the geometry approaches Minkowski space in both asymptotic regions.
- **Null energy condition violation at/near the throat**: for radial null vectors k^μ, traversability requires

$$T_{\mu\nu}k^\mu k^\nu < 0 \quad \text{at least at } r = r_0,$$

and, using Einstein equations, this follows from the flaring-out condition. In particular one finds at the throat

$$\rho + p_r = \frac{1}{8\pi}\left(\frac{b'(r_0) - 1}{r_0^2}\right) < 0,$$

where ρ is the energy density and p_r the radial pressure in the static orthonormal frame.

Hence, for the simplest Ellis branch one may set $\Phi(r) = 0$ (no redshift), and a particularly clean derivation is obtained by switching to the proper radial coordinate ℓ, in terms of which the metric takes the manifestly regular form

$$ds^2 = -dt^2 + d\ell^2 + \left(\ell^2 + a^2\right) d\Omega^2, \tag{10.225}$$

with throat radius $a > 0$ and $\ell \in (-\infty, \infty)$.

For Eq. (10.225), the Einstein tensor components are nonvanishing and scale as $a^2/(\ell^2 + a^2)^2$, consistent with a localized source sustaining the throat. The phantom stress tensor for a radial scalar $\phi(\ell)$ reads

$$T^{(\phi)}_{\mu\nu} = -\nabla_\mu\phi\nabla_\nu\phi + \frac{1}{2}g_{\mu\nu}(\nabla\phi)^2, \tag{10.226}$$

and the scalar equation remains $\Box\phi = 0$. One verifies that the profile

$$\phi(\ell) = \phi_0 \pm \sqrt{\frac{1}{4\pi}}\, \arctan\left(\frac{\ell}{a}\right) \tag{10.227}$$

solves $\Box\phi = 0$ on Eq. (10.225) and that Einstein's equations are satisfied with this choice.

The fundamental properties are:

- **Two asymptotic regions**. Both ends $\ell \to \pm\infty$ are asymptotically flat, so the spacetime connects two asymptotically Minkowski domains.
- **Regular throat**. At $\ell = 0$ the areal radius is minimal, $r_{\text{th}} = a$, and curvature invariants remain finite.
- **NEC violation**. The phantom sign implies $T_{\mu\nu}k^\mu k^\nu < 0$ for suitable null vectors k^μ (e.g. radial null directions), which is the local condition enabling traversability.

► Driving Exercise

Conformal coupling with quartic potential in AdS: MTZ BH.

A canonical prototype of an exact *hairy* AdS BH is provided by a conformally coupled scalar with a quartic self-interaction. The model is specified by

$$\xi = \frac{1}{6}, \qquad V(\phi) = \frac{\lambda}{4}\phi^4, \qquad \Lambda < 0, \tag{10.228}$$

where Λ is the cosmological constant. To include Λ in Eq. (10.199) one replaces $R \to R - 2\Lambda$. Exact BHs become possible for two intertwined reasons: AdS boundary conditions weaken the standard asymptotically flat no hair obstructions, and the conformal structure makes it possible to choose a scalar profile that remains regular at the horizon while still solving the nonlinear coupled system.

In the MTZ construction one typically adopts $ds^2 = -F(r)\,dt^2 + \frac{dr^2}{F(r)} + r^2\,d\Sigma_k^2$, where $d\Sigma_k^2$ is the metric on a two-dimensional space of constant curvature and $k = \{+1, 0, -1\}$ labels spherical, planar, or hyperbolic horizon topology (depending on the branch). The scalar is taken radial, $\phi = \phi(r)$. Then, using the first relation in Eqs. (10.147), i.e., $\tilde{g}_{\mu\nu} = \Omega(\phi)^2\, g_{\mu\nu}$, which can map a minimally coupled scalar in an Einstein frame into a non-minimally coupled scalar in the Jordan frame. This viewpoint organizes broad scalar-tensor families (often discussed within Bronnikov-type classifications) into branches distinguished by the behavior of $\Omega(\phi)$.

- **Regular branch**: $\Omega^2 > 0$ throughout the domain of interest, so the map is nonsingular and the causal structure is preserved.
- **Singular/transition branch**: $\Omega^2 \to 0$ or changes sign, signalling an effective strong-coupling locus and potentially generating exotic causal structures or frame-dependent pathologies.

In practice, Fisher/JNW-type seed configurations can generate non-minimally coupled solutions by this route, but the analysis is only trustworthy when the conformal factor remains nondegenerate on the physical domain.

10.5 Black Holes: Thermodynamics

BH thermodynamics studies horizon-scale, macroscopic properties of black holes in close analogy with ordinary thermodynamics. The analogy becomes compelling because stationary BHs behave as equilibrium states labeled by a small set of macroscopic quantities (in electro-vacuum: M, J, Q), and because their horizon dynamics obeys a set of laws that mirror the four laws of thermodynamics.

10.5.1 Bekenstein's Entropy Proposal

Bekenstein was the first to argue that BHs should be endowed with a genuine thermodynamic entropy. His reasoning was motivated by the apparent violation of the second law of thermodynamics when matter carrying entropy falls into a BH.

Classically, BHs are characterized by an event horizon whose area A satisfies Hawking's area theorem:

$$\delta A \geq 0, \tag{10.229}$$

for any classical physical process. Bekenstein interpreted this monotonic behavior as the analogue of entropy increase.

He therefore postulated that a BH carries an entropy $S_{\rm BH}$ proportional to the horizon area:

$$S_{\rm BH} \propto A. \tag{10.230}$$

Dimensional analysis implies

$$S_{\rm BH} = \eta \frac{A}{\hbar G}, \tag{10.231}$$

where η is a dimensionless constant of order unity. At this stage, however, BHs were still assumed to have zero temperature, and the proportionality constant η could not be determined.

Pause and think
Imagine that a body falls inside a BH: try to figure out how the mass, and accordingly entropy, would be modified.

Why does BH entropy scale with the horizon area?
Bekenstein's result was therefore conceptual rather than dynamical: entropy was introduced to preserve the validity of the second law, but without a microscopic or quantum field derivation. Moreover, Bekenstein's reasoning starts from a tension between two classical facts.

- In ordinary physics, entropy S measures the amount of information hidden from an external observer.
- In GR, the event horizon hides information irreversibly from the exterior.

He considers a thought experiment in which a system with energy E, size R, and entropy $S_{\rm matter}$ is slowly lowered toward a BH and then dropped in. If the black hole had no entropy, the total entropy outside would decrease by $S_{\rm matter}$, in violation of the second law. Therefore, the BH itself must carry an entropy $S_{\rm BH}$ that increases when matter falls in.
For a Schwarzschild BH of mass M and radius $r_s = 2GM$, the horizon area is $A = 4\pi r_s^2 = 16\pi G^2 M^2$. Dropping in a body of energy E changes the mass by $\delta M = E$, hence the area by

$$\delta A = \frac{dA}{dM}\,\delta M = 32\pi G^2 M\, E. \tag{10.232}$$

To minimize the increase of area, Bekenstein lowers the body adiabatically until its center is a proper distance of order its own size R from the horizon. Gravitational redshift then reduces the energy actually delivered to the BH to

$$E_{\rm eff} \sim E\sqrt{1-\frac{2GM}{r}} \sim \frac{E\,R}{2GM},$$

so that the minimal area increase is $(\delta A)_{\min} \sim 32\pi G^2 M \frac{ER}{2GM} = 16\pi G\, ER$ If the generalized second law, $\Delta S_{\rm tot} = \Delta S_{\rm matter} + \Delta S_{\rm BH} \geq 0$, is to hold even in this optimal process, the BH entropy must satisfy

$$\delta S_{\rm BH} \geq S_{\rm matter}.$$

Since $\delta S_{\rm BH}$ is induced by the minimal area increase, consistency requires $S_{\rm matter} \leq \eta\, ER$, with some universal constant η, and simultaneously the validity of Eq. (10.230).

10.5.2 Hawking Radiation and Temperature

Hawking's breakthrough consisted in quantizing a quantum field on a fixed BH background. For simplicity, let us consider a Schwarzschild BH. Hawking's semiclassical analysis of quantum fields on a BH background showed that BHs emit thermal radiation with a Planckian spectrum. A Euclidean derivation via periodicity can be then pursued. Thus, introduce the Euclidean time $\tau = it$, yielding the Euclidean metric

$$ds_E^2 = \left(1 - \frac{2GM}{r}\right) d\tau^2 + \left(1 - \frac{2GM}{r}\right)^{-1} dr^2 + r^2 d\Omega^2. \tag{10.233}$$

Near the horizon $r = r_s = 2GM$, define $r = r_s + \frac{\rho^2}{4r_s}$, so that, to leading order, $1 - \frac{2GM}{r} \simeq \frac{\rho^2}{4r_s^2}$. The (τ, r) part of the metric becomes

$$ds_E^2 \simeq d\rho^2 + \rho^2 \left(\frac{d\tau}{2r_s}\right)^2. \tag{10.234}$$

This is the flat metric in polar coordinates provided that τ is periodic with period, $\beta = 4\pi r_s = 8\pi GM$ (see Chap. 7).

The absence of a conical singularity at $\rho = 0$ therefore fixes the inverse temperature, $T_H = \beta^{-1} = \frac{1}{8\pi GM}$, that may be recast by restoring $\hbar$ to read[2]

$$\boxed{T_H = \frac{\hbar}{8\pi GM}}. \tag{10.235}$$

[2] The BH temperature is engraved on Stephen Hawking's memorial stone in Westminster Abbey.

In a generic spacetime, this procedure is still valid. Indeed, consider a static, spherically symmetric line element of the form, $ds^2 = -f(r)\,dt^2 + \frac{dr^2}{f(r)} + r^2 d\Omega_2^2$, and perform the Wick rotation $t \to -i\tau$, obtaining the Euclidean metric, $ds_E^2 = f(r)\,d\tau^2 + \frac{dr^2}{f(r)} + r^2 d\Omega_2^2$.
Near a (simple) horizon $r = r_h$ with $f(r_h) = 0$ and $f'(r_h) \neq 0$,

$$f(r) \simeq f'(r_h)\,(r - r_h) \equiv 2\kappa\,(r - r_h), \tag{10.236}$$

where $\kappa \equiv \frac{1}{2} f'(r_h)$ is the surface gravity. Define a proper radial coordinate ρ by

$$d\rho^2 = \frac{dr^2}{f(r)} \simeq \frac{dr^2}{2\kappa(r - r_h)} \quad \Rightarrow \quad r - r_h \simeq \frac{\kappa}{2}\rho^2, \tag{10.237}$$

so that Eq. (10.236) gives $f(r) \simeq \kappa^2 \rho^2$. The Euclidean metric near the horizon becomes

$$ds_E^2 \simeq \rho^2 \kappa^2\,d\tau^2 + d\rho^2 + r_h^2 d\Omega_2^2. \tag{10.238}$$

Regularity at $\rho = 0$ requires τ to be periodic with the same period, as above. This implies $T = \beta^{-1} = \kappa/(2\pi)$, giving the *Unruh temperature*, that in the precise case of Eq. (10.235) corresponds to the Hawking temperature.

▶ **Pause and think**
Try to understand how Hawking derived the entropy from his temperature definition and try to obtain it by yourself.

▶ **Driving Exercise**
Derivation of the Entropy.

The first law of BH mechanics for a Schwarzschild BH reads

$$dM = \frac{\kappa}{8\pi G} dA, \tag{10.239}$$

where the surface gravity is $\kappa = \frac{1}{4GM}$. Identifying $E = M$ and comparing with the thermodynamic relation $dE = T\,dS$, one finds $T_H dS = \frac{\kappa}{8\pi G} dA$. Substituting $T_H = \hbar\kappa/(2\pi)$,

$$\frac{\hbar\kappa}{2\pi} dS = \frac{\kappa}{8\pi G} dA. \tag{10.240}$$

Canceling κ and integrating yields

$$dS = \frac{1}{4\hbar G} dA \quad \Rightarrow \quad \boxed{S_{\text{BH}} = \frac{A}{4\hbar G}.} \tag{10.241}$$

10.5.3 The Irreducible Mass and Area Law

A BH may be treated as a localized physical object whose external fields are entirely specified by the quantities that are related to the horizon area, namely by (M, J, Q). This will be clarified better in Sect. 10.5.6 while dealing with the concept of *no-hair*. By allowing test particles to fall into the hole, one can induce infinitesimal changes

$$\delta M = E, \qquad \delta J = p_\varphi, \qquad \delta Q = e, \tag{10.242}$$

where E, p_φ, and e are the conserved energy, angular momentum, and charge of the particle, respectively.

Analysis of particle absorption reveals the inequality

$$\delta M \geq \frac{a\,\delta J + r_+ Q\,\delta Q}{r_+^2 + a^2}, \tag{10.243}$$

which distinguishes reversible processes (saturating the inequality) from irreversible ones.

Integrating the reversible transformations leads to the Christodoulou-Ruffini mass formula

$$M^2 = \left(M_{\rm irr} + \frac{Q^2}{4M_{\rm irr}}\right)^2 + \frac{J^2}{4M_{\rm irr}^2}, \tag{10.244}$$

where $M_{\rm irr}$ is the irreducible mass. General processes satisfy

$$\delta M_{\rm irr} \geq 0. \tag{10.245}$$

For Kerr–Newman BHs, the horizon area is

$$A = 4\pi(r_+^2 + a^2) = 16\pi M_{\rm irr}^2, \tag{10.246}$$

so that the increase of $M_{\rm irr}$ is equivalent to the increase of the horizon area.

10.5.4 Black Hole Thermodynamic Laws

The classical theory of stationary BHs exhibits a remarkably rigid set of integral identities and monotonicity statements which, once interpreted through the Hawking temperature and the Bekenstein-Hawking entropy, acquire the precise logical form of a thermodynamics. In the stationary regime, the relevant object is a Killing horizon

$\mathcal{H}$ generated by a horizon Killing field χ^μ, which in the Kerr-Newman family can be written as

$$\chi^\mu = t^\mu + \Omega_H \, \varphi^\mu, \tag{10.247}$$

with t^μ asymptotically timelike and φ^μ axial, and where Ω_H is the angular velocity of the horizon. The associated surface gravity κ is defined intrinsically by

$$\chi^\nu \nabla_\nu \chi^\mu = \kappa \, \chi^\mu \qquad \text{on } \mathcal{H}, \tag{10.248}$$

or equivalently by $\nabla_\mu(\chi^2) = -2\kappa \, \chi_\mu$ on $\mathcal{H}$. In semiclassical gravity, one identifies

$$T_H = \frac{\hbar \kappa}{2\pi k_B c}, \qquad S_{\mathrm{BH}} = \frac{k_B c^3}{4\hbar G} \, A, \tag{10.249}$$

so that the horizon area A plays the role of entropy and κ that of temperature. The thermodynamic laws are then most conveniently viewed as geometrized statements about κ, A, and the conserved charges (M, J, Q) of the stationary exterior.

A concise synopsis of the four laws is as follows.

10.5.4.1 Zeroth Law: Constancy of the Surface Gravity

The zeroth law is the structural statement that stationary horizons are in a genuine equilibrium state: the quantity controlling the near-horizon redshift and the Hawking temperature is uniform across the horizon. Let $\mathcal{H}$ be a Killing horizon generated by χ^μ with $\chi^2 = 0$ on $\mathcal{H}$. The surface gravity κ is defined by Eq. (10.248). Since χ^μ is tangent to the null generators, Eq. (10.248) states that the non-affinity of the horizon generators is constant along each generator if and only if κ is constant.

A useful way to see the role of κ is to compare χ^μ with an affinely parametrized null generator k^μ of the same congruence. Writing $k^\mu = \alpha \, \chi^\mu$ on $\mathcal{H}$, one finds

$$k^\nu \nabla_\nu k^\mu = 0 \quad \Longleftrightarrow \quad \chi^\nu \nabla_\nu \chi^\mu = \kappa \, \chi^\mu, \tag{10.250}$$

with $\kappa = k^\nu \nabla_\nu (\ln \alpha)$ on the horizon. Thus κ measures precisely the failure of the Killing parameter along χ^μ to be affine. In turn, the same quantity controls the leading near-horizon redshift: in the stationary exterior, the norm $-\chi^2$ acts as a gravitational potential, and the relation

$$\nabla_\mu(\chi^2) = -2\kappa \, \chi_\mu \qquad \text{on } \mathcal{H} \tag{10.251}$$

shows that κ fixes the first nontrivial gradient of χ^2 transverse to the horizon.

The nontrivial content of the zeroth law is that κ is not merely constant *along* each generator (which follows from Killing symmetry), but constant *across* the horizon cross-sections. In the Einstein theory this can be established under standard hypotheses by combining (i) the Killing property $\nabla_{(\mu}\chi_{\nu)} = 0$, (ii) the field equations, and

(iii) suitable energy/regularity conditions that imply the relevant components of the Ricci tensor along $\mathcal{H}$ do not source transverse variations of κ. The final result is

$$\nabla_a \kappa = 0 \quad \text{on each connected component of } \mathcal{H}, \tag{10.252}$$

where ∇_a denotes the intrinsic derivative on a horizon cross-section.

10.5.4.2 First Law of BH Mechanics

The first law relates infinitesimal changes of a stationary BH to the variation of its horizon area and conserved charges:

$$dM = \frac{\kappa}{8\pi}\, dA + \Omega_H\, dJ + \Phi_H\, dQ. \tag{10.253}$$

The structure is thermodynamic in form: $\Omega_H dJ$ and $\Phi_H dQ$ are "work terms" associated with rotation and electromagnetism, while the $\kappa\, dA$ term will be identified with $T\, dS$ once the entropy and temperature assignments are fixed.

Driving Exercise

Let us compute the horizon area of a Kerr-Newman configuration.
The induced two-metric (at fixed t and $r = r_+$) gives an area element

$$d\Sigma = \sqrt{\det \bar{g}}\, d\theta\, d\phi = (r_+^2 + a^2)\, \sin\theta\, d\theta\, d\phi, \tag{10.254}$$

hence

$$A = \int_0^{2\pi} \int_0^{\pi} (r_+^2 + a^2) \sin\theta\, d\theta\, d\phi = 4\pi (r_+^2 + a^2). \tag{10.255}$$

The horizon angular velocity is consequently,

$$\Omega_H = \frac{a}{r_+^2 + a^2}, \tag{10.256}$$

and the electrostatic potential (defined as the constant value of $-\chi^\mu A_\mu$ on the horizon in a gauge regular there) is

$$\Phi_H = \frac{Q r_+}{r_+^2 + a^2}. \tag{10.257}$$

10.5.4.3 Second Law: Area Theorem

The second law is a genuinely dynamical statement. It concerns the evolution of event horizons under arbitrary classical processes, not just transitions between stationary states.

Let $\mathcal{H}$ be the future event horizon generated by null geodesics with tangent k^μ. The expansion θ of the congruence measures the rate of change of the cross-sectional area element $\mathcal{A}$ along the generators,

$$\theta = \frac{1}{\mathcal{A}} \frac{d\mathcal{A}}{d\lambda}, \tag{10.258}$$

where λ is an affine parameter. In the Raychaudhuri equation, Eq. (10.62), for horizon generators one has vanishing twist, $\omega_{\mu\nu} = 0$. If the null energy condition holds,

$$T_{\mu\nu} k^\mu k^\nu \geq 0,$$

then by Einstein equations $R_{\mu\nu} k^\mu k^\nu \geq 0$. We argue that θ cannot become negative if it is initially non-negative, since otherwise it would diverge to $-\infty$ in finite affine parameter, producing caustics and violating the definition of an event horizon.

Hence the expansion of the horizon generators satisfies $\theta \geq 0 \quad \Rightarrow \quad \frac{dA}{d\lambda} \geq 0$. Therefore, in any classical process obeying the null energy condition,

$$A_{\text{final}} \geq A_{\text{initial}}. \tag{10.259}$$

10.5.4.4 Third Law: Unattainability of Zero Surface Gravity

The third law states that a BH with vanishing surface gravity,

$$\kappa = 0,$$

cannot be produced from a non-extremal BH by any finite physical process.

In stationary solutions, $\kappa = 0$ characterizes extremal horizons (e.g. extremal Reissner–Nordström or extremal Kerr), for which the inner and outer horizons coincide. Such solutions form a boundary of the space of stationary BHs.

The dynamical content of the third law is that any process attempting to drive a BH toward extremality necessarily slows down and asymptotically stalls. Typical arguments combine:

- the behavior of test-particle or field absorption near extremality,
- backreaction constraints imposed by cosmic censorship,
- the structure of horizon generators as $\kappa \to 0$.

As extremality is approached, the horizon redshift becomes infinite and the response of the BH to further perturbations becomes increasingly suppressed, preventing the attainment of $\kappa = 0$ in finite advanced time.

Formally, if a family of stationary solutions is labeled by a parameter ϵ such that $\kappa(\epsilon) \to 0$ as $\epsilon \to 0$, then any physical evolution satisfies

$$\lim_{v\to\infty} \kappa(v) = 0, \qquad \kappa(v) > 0 \text{ for all finite } v, \tag{10.260}$$

where v is an advanced-time parameter along the horizon.

The third law mirrors the unattainability form of Nernst's theorem: absolute zero cannot be reached by a finite sequence of operations. For BHs, zero temperature corresponds to $\kappa = 0$, i.e., extremality. Extremal BHs are therefore limiting, non-generic states rather than endpoints of finite physical processes.

10.5.5 The Black Hole Information Paradox

The *BH information paradox* is the sharpest formulation of the ensuing conceptual tension: the most conservative set of assumptions that successfully describes low-energy physics appears to be mutually inconsistent when applied to complete BH evaporation.

The paradox is traditionally isolated into three ingredients.

1. *Semiclassical validity outside the horizon*: quantum fields propagate on a classical geometry, and the effective field theory description is trustworthy in the low-curvature exterior region.
2. *Thermality of Hawking quanta*: the outgoing radiation is (to leading order in $\hbar$) thermal, with a temperature fixed by the surface gravity and only mild frequency-dependent corrections from scattering.
3. *Unitarity*: the time evolution of an isolated quantum system is generated by a unitary operator, so that a pure initial state cannot evolve into a mixed final state.

Taken together, these statements imply that information carried by a pure collapsing state is irretrievably lost into a thermal flux, an outcome incompatible with ordinary quantum mechanics.

What is meant by "information" in this context?

Here "information" is not a classical label but the fine-grained quantum correlations that distinguish a pure state from a thermal ensemble. The relevant diagnostic is the von Neumann entropy of the radiation subsystem, which vanishes for a globally pure state but is positive for a mixed state. The paradox is thus a statement about the fate of entanglement under evaporation.

10.5.6 Black Holes Have No Hair

A cornerstone of classical BH physics is the *uniqueness* (often informally called "no hair") picture: within GR, the end-state of gravitational collapse to a *stationary* BH is expected to be described by a very small set of parameters.

Historically, the first rigorous steps were taken by Israel for static, asymptotically flat vacuum and electrovacuum BHs, whereas the modern formulation is that, under appropriate assumptions, stationary, asymptotically flat, four-dimensional *electrovacuum* BHs with a regular connected horizon are members of the Kerr-Newman family, and are completely characterized by three main quantities,

$$(M, J, Q), \tag{10.261}$$

namely the ADM mass M, the total angular momentum J, and the electric charge Q (equivalently $a \equiv J/M$).

What the above statement *does not say*! The slogan "BHs have no hair" is convenient but potentially misleading. The rigorous uniqueness results require the following hypotheses:

- four spacetime dimensions, classical GR, and usually asymptotic flatness;
- electrovacuum outside the horizon (no additional long-range fields);
- stationarity (an equilibrium endpoint, not a dynamical transient);
- regularity and appropriate global conditions (connected horizon, analyticity assumptions in some proofs).

If one relaxes these assumptions, additional "hair" may exist. For instance, non-minimal matter couplings, extra scalar degrees of freedom, or modified gravity theories can support stationary BHs with extra parameters. Hence, it appears necessary to explore solutions that depart from standard BHs such as regular BHs and/or, more broadly, mimickers.

10.6 Regular Black Holes

The essential (curvature) singularity of the Schwarzschild solution is not a logical necessity of Einstein gravity; rather, it follows from the hypotheses entering the classical singularity theorems. In particular, Penrose-type results require suitable global assumptions together with an energy condition that, in the simplest implementations, reduces to the strong energy condition for the effective matter sector sourcing the geometry. If the strong energy condition is violated in a compact region around the center, as is natural for many effective descriptions of ultraviolet (UV) completion, vacuum polarization, or nonlinear gauge sectors, then a trapped region may coexist with a *regular* core. The simplest geometric realization is a de Sitter[3] (dS) interior, which replaces the $r = 0$ singularity by a finite-curvature "vacuum" core.

Regularity mechanism (spherical symmetry). A sufficient local condition to remove the central curvature blow-up is that the Misner-Sharp mass behaves as $m(r) = O(r^3)$ as $r \to 0$. In that case,

$$f(r) = 1 - \frac{2\,m(r)}{r} = 1 - \frac{\Lambda_{\rm eff}}{3}\,r^2 + O(r^3),$$

so the center is locally dS with $\Lambda_{\rm eff} > 0$, and the curvature invariants remain finite. As a matter of completeness, if $\Lambda_{\rm eff} < 0$ the solution is called anti-dS.

Consider a static, spherically symmetric line element in (t, r, θ, ϕ) coordinates, $ds^2 = -f(r)\,dt^2 + \frac{dr^2}{f(r)} + r^2\left(d\theta^2 + \sin^2\theta\,d\phi^2\right)$; a convenient parametrization is obtained by introducing a mass function $m(r)$ via

$$f(r) = 1 - \frac{2\,m(r)}{r}. \tag{10.262}$$

This representation isolates the physical content of the solution: $m(r)$ measures the gravitational mass enclosed within radius r (in the Misner-Sharp sense) and directly controls both horizon structure and the small-r regularity properties.

A broad one-parameter class of regular profiles is captured by the following ansatz for $m(r)$:

$$m(r) = \frac{M_0}{\left[1 + \left(\frac{r_0}{r}\right)^q\right]^{\frac{p}{q}}}, \tag{10.263}$$

where M_0 and r_0 are respectively a mass and a length scale, while p and q are positive integers. The parameter M_0 coincides with the ADM mass, as follows from the asymptotic limit

$$\lim_{r\to\infty} m(r) = M_0, \tag{10.264}$$

[3] The dS solution is particularly interesting, and in general a naive prototype of regular BH. In general, the line element reads $ds^2 = \left(1 - \frac{\Lambda}{3}r^2\right)dt^2 + \left(1 - \frac{\Lambda}{3}r^2\right)^{-1}dr^2 + r^2 d\Omega^2$, introducing a *cosmological constant* term, sometimes interpreted as a vacuum energy effective contribution. Indeed, its energy-momentum tensor is identically constant, with moduli of density and (isotropic) pressures equal to Λ and equation of state $w = -1$.

so that the metric function approaches Schwarzschild at large radius. In fact, for $r \gg r_0$ one may expand

$$m(r) = M_0 \left[1 - \frac{p}{q} \left(\frac{r_0}{r} \right)^q + O\left(\left(\frac{r_0}{r} \right)^{2q} \right) \right], \tag{10.265}$$

and therefore

$$f(r) = 1 - \frac{2M_0}{r} + \frac{2M_0}{r} \frac{p}{q} \left(\frac{r_0}{r} \right)^q + \cdots, \tag{10.266}$$

making explicit that the leading exterior behavior is the Schwarzschild potential plus suppressed corrections.

At small scales, r_0 plays the role of a microscopic length controlling the regular core. Indeed, setting $r_0 = 0$ collapses Eq. (10.263) to $m(r) = M_0$ at all radii and the metric function reduces exactly to the Schwarzschild solution. For $r \ll r_0$ one finds

$$m(r) = M_0 \left(\frac{r}{r_0} \right)^p \left[1 + O\left(\left(\frac{r}{r_0} \right)^q \right) \right], \tag{10.267}$$

so that regularity requires $p \geq 3$, and the dS core is obtained precisely for $p = 3$. In that case

$$m(r) \sim \frac{M_0}{r_0^3} r^3 \quad \Longrightarrow \quad f(r) \sim 1 - \frac{2M_0}{r_0^3} r^2 \equiv 1 - \frac{\Lambda_{\text{eff}}}{3} r^2, \tag{10.268}$$

with an effective cosmological constant $\Lambda_{\text{eff}} = 6M_0/r_0^3 > 0$.

The most commonly discussed representatives are obtained by fixing (p, q):

- **Bardeen-type choice**: $p = 3, q = 2$;
- **Hayward-type choice**: $p = q = 3$.

For these cases the equation $f(r) = 0$ admits at most two horizons, an inner Cauchy horizon r_- and an outer event horizon r_+. A simple sufficient condition for the existence of horizons in this class is that the core scale is smaller than the mass scale, $r_0 < M_0$, in the appropriate units.[4]

The basic properties from Eq. (10.263) can be summarized as follows.

1. **dS core at small r (regularity)**. For $p = 3$, one has $m(r) \sim r^3$, which forces $f(r) \sim 1 - Cr^2$ with $C > 0$, i.e., a locally dS center and bounded curvature invariants. This singles out $p = 3$ as the appropriate value for constructing regular BHs within this ansatz.

[4] More precisely, the relative magnitude of r_0/M_0 controls whether f has two simple zeros, one double zero (extremal case), or no real zeros.

2. **Schwarzschild recovery at large** r. For $r \gtrsim r_+$, the mass profile is approximately constant, $m(r) \approx M_0$, so the exterior geometry is well approximated by Schwarzschild. In particular, for $r_0 = 0$ Schwarzschild is recovered exactly at all radii.
3. **Non-vacuum character**. Metrics generated by Eq. (10.263) are *not* vacuum solutions of Einstein equations: the effective stress tensor satisfies $T_{\mu\nu} \neq 0$. Nevertheless, the source is typically localized and becomes negligible for $r > r_+$, where the geometry is approximately vacuum.

As a criterion of "good" regularity, the class generated by Eq. (10.263) yields bounded scalar invariants. In particular, all BHs of the form Eq. (10.263) are free of curvature blow-ups at the origin. For the studied class, the Kretschmann scalar satisfies

$$\lim_{r\to 0} K(r) = \lim_{r\to 0} R_{\mu\nu\alpha\beta} R^{\mu\nu\alpha\beta} = 96 \left(\frac{M_0}{r_0^3} \right)^2 , \tag{10.269}$$

so the curvature is finite at the center. Notably, this limit is independent of q, reflecting the fact that the parameter q controls the interpolation to the Schwarzschild exterior, while regularity is governed by the leading small-r scaling fixed by $p = 3$.

Regular BHs as effective-matter solutions. In spherically symmetric scenarios, writing $f(r) = 1 - 2m(r)/r$ makes explicit that regularity demands $m(r) = O(r^3)$ near the origin, which in turn enforces an effective dS core. The price is that the geometry cannot be vacuum: an effective stress tensor must violate the strong energy condition in the core region.

A concrete and widely used effective-matter realization is provided by Einstein gravity coupled to NLED, see Chap. 9. The fundamental representations of the Hayward and Bardeen geometries can be obtained from the action

$$S = \int d^4x \sqrt{-g} \left[\frac{R}{16\pi} - \mathcal{L}(\mathcal{F}) \right], \qquad \mathcal{F} \equiv \frac{1}{4} F_{\mu\nu} F^{\mu\nu}, \tag{10.270}$$

where $\mathcal{L}(\mathcal{F})$ is an arbitrary function that may be chosen to reduce to Maxwell theory in a suitable weak-field regime.

Varying with respect to $g^{\mu\nu}$ yields the Einstein equations

$$G_{\mu\nu} = 8\pi\, T_{\mu\nu}, \qquad T_{\mu\nu} = 4\mathcal{L}_{\mathcal{F}}\, F_{\mu\alpha} F_{\nu}{}^{\alpha} - g_{\mu\nu}\, \mathcal{L}, \qquad \mathcal{L}_{\mathcal{F}} \equiv \frac{d\mathcal{L}}{d\mathcal{F}}, \tag{10.271}$$

while variation with respect to the gauge potential gives the NLED field equation

$$\nabla_\mu \left(\mathcal{L}_{\mathcal{F}}\, F^{\mu\nu} \right) = 0. \tag{10.272}$$

Adopt the static, spherically symmetric line element

$$ds^2 = -A(r)\,dt^2 + \frac{dr^2}{A(r)} + r^2 d\Omega^2, \qquad A(r) = 1 - \frac{2\,m(r)}{r}, \tag{10.273}$$

with $m(r)$ the mass function. In the purely magnetic monopole sector (see Sect. 9.14 of Chap. 9),

$$F_{\theta\phi} = g\ \sin\theta, \qquad F_{\mu\nu}dx^{\mu} \wedge dx^{\nu} = g\ \sin\theta\, d\theta \wedge d\phi, \tag{10.274}$$

the invariant is

$$\mathcal{F}(r) = \frac{g^2}{2r^4}. \tag{10.275}$$

For this magnetic ansatz (with constant magnetic charge g), Eqs. (10.272) are automatically satisfied.

From Eqs. (10.271), one finds $T^t{}_t = T^r{}_r = -\mathcal{L}$. Using the standard identity for Eq. (10.273),

$$G^t{}_t = -\frac{2\,m'(r)}{r^2}, \tag{10.276}$$

the tt-equation becomes a single radial quadrature:

$$-\frac{2\,m'(r)}{r^2} = 8\pi\ T^t{}_t = -8\pi\ \mathcal{L}(r) \quad \Longrightarrow \quad m'(r) = 4\pi r^2\ \mathcal{L}(r). \tag{10.277}$$

Therefore, once a target geometry is prescribed (equivalently, once $m(r)$ is specified), the on-shell NLED Lagrangian is fixed as

$$\mathcal{L}(r) = \frac{m'(r)}{4\pi r^2}, \qquad \mathcal{L}(\mathcal{F}) = \left.\frac{m'(r)}{4\pi r^2}\right|_{r=(g^2/2\mathcal{F})^{1/4}}. \tag{10.278}$$

This is the magnetic-sector "reverse engineering" procedure: the effective matter model is reconstructed *a posteriori* from the desired mass profile.

Pause and think

Find the Lagrangian densities for the Bardeen and Hayward solutions.

Driving Exercise

The Bardeen geometry is obtained by choosing Eq. (10.273) with

$$A_{\mathrm{B}}(r) = 1 - \frac{2Mr^2}{(r^2+g^2)^{3/2}} \qquad \Longleftrightarrow \qquad m_{\mathrm{B}}(r) = \frac{Mr^3}{(r^2+g^2)^{3/2}}, \tag{10.279}$$

where M is the ADM mass and g is interpreted as a magnetic monopole charge.
Differentiating $m_{\mathrm{B}}(r)$ yields

$$m'_{\mathrm{B}}(r) = \frac{3Mg^2r^2}{(r^2+g^2)^{5/2}}. \tag{10.280}$$

Then Eq. (10.277) gives the required NLED Lagrangian as a function of r:

$$\mathcal{L}_{\mathrm{B}}(r) = \frac{m'_{\mathrm{B}}(r)}{4\pi r^2} = \frac{3Mg^2}{4\pi\,(r^2+g^2)^{5/2}}. \tag{10.281}$$

Finally, expressing it as a function of $\mathcal{F}$ through Eq. (10.275), namely $r = (g^2/2\mathcal{F})^{1/4}$, one obtains

$$\boxed{\mathcal{L}_{\mathrm{B}}(\mathcal{F}) = \left.\frac{3Mg^2}{4\pi\,(r^2+g^2)^{5/2}}\right|_{r=(g^2/2\mathcal{F})^{1/4}}}, \tag{10.282}$$

which is the magnetic NLED source that reproduces the Bardeen metric through Einstein equations.
As $r \to 0$ one has $m_{\mathrm{B}}(r) \sim (M/g^3)\, r^3$ and therefore

$$A_{\mathrm{B}}(r) \sim 1 - \left(\frac{2M}{g^3}\right) r^2, \tag{10.283}$$

exhibiting explicitly the dS core and hence finite curvature invariants.
The (neutral) Hayward metric is commonly written as

$$A_{\mathrm{H}}(r) = 1 - \frac{2Mr^2}{r^3+2M\ell^2} \qquad \Longleftrightarrow \qquad m_{\mathrm{H}}(r) = \frac{Mr^3}{r^3+2M\ell^2}, \tag{10.284}$$

where ℓ is a length scale controlling the regular core.
Applying the same reconstruction logic, differentiate:

$$m'_{\mathrm{H}}(r) = \frac{6M^2\ell^2\,r^2}{(r^3+2M\ell^2)^2}. \tag{10.285}$$

Equation (10.277) then fixes the on-shell NLED Lagrangian,

$$\mathcal{L}_{\mathrm{H}}(r) = \frac{m'_{\mathrm{H}}(r)}{4\pi r^2} = \frac{6M^2\ell^2}{4\pi\,(r^3+2M\ell^2)^2}. \tag{10.286}$$

Using again $\mathcal{F} = g^2/(2r^4)$ (purely magnetic sector), the associated $\mathcal{L}_{\mathrm{H}}(\mathcal{F})$ follows from the substitution $r = (g^2/2\mathcal{F})^{1/4}$:

$$\boxed{\mathcal{L}_{\mathrm{H}}(\mathcal{F}) = \left.\frac{6M^2\ell^2}{4\pi\,(r^3+2M\ell^2)^2}\right|_{r=(g^2/2\mathcal{F})^{1/4}}}. \tag{10.287}$$

Therefore, *there exists* a magnetic Einstein-NLED model whose stress tensor reproduces the Hayward mass profile exactly, in the precise on-shell sense implied by the reconstruction in Eq. (10.278).

10.6.1 Beyond Regular Black Holes: Mimickers

In astrophysical practice, the expression "BH" is often used in an operational sense: a dark and compact object whose exterior gravitational field is accurately modeled by the Kerr family. From the standpoint of fundamental physics, however, the decisive ingredient is not merely compactness, but the presence of an *infinite-redshift surface*, i.e., an event horizon in the classical stationary regime, together with the associated causal structure that forbids signals from returning to the exterior. However, classical GR predicts spacetime singularities, while semiclassical gravity raises the entropy and information-loss puzzles. These motivations naturally lead to the study of *exotic compact objects* (ECOs), namely horizonless alternatives that are regular (or at least nonsingular in a UV-complete sense) and yet may be as dark and as compact as the observed candidates. A particularly useful taxonomy is the following.

An *ultracompact object* (UCO) is an ECO whose exterior admits a photon sphere (light ring), see Table 10.2. In spherical symmetry, whenever the exterior is well approximated by vacuum GR, the photon sphere lies near $r \simeq 3M$. A necessary condition to host such a structure is that the object be sufficiently compact, in the sense that

$$C \equiv \frac{GM}{Rc^2} \gtrsim \frac{1}{3}, \tag{10.288}$$

where R is the proper circumferential radius.

A *BH mimicker* is a more demanding notion: it is an ECO whose redshift is so large that, on observational timescales, the phenomenology normally attributed to the horizon is effectively reproduced. In particular, the influence of the interior on photon-sphere dynamics must be suppressed to the point of practical invisibility.

Minimal operational criterion for mimicking. In a horizonless spacetime, the key discriminator is not the logical absence of a horizon, but whether signals that probe the interior can return within the timescales accessible to observations. In this sense, "horizon-like" behavior is controlled by *time delay* and *effective absorptivity* as much as by the exterior compactness.

- **Clue I: maximum compactness versus maximum redshift.** A first obstruction to genuine mimicking is purely relativistic: how close can a *stable, physically admissible* object get to the BH compactness $C_{\mathrm{BH}} = 1/2$? In GR, Buchdahl's theorem bounds isotropic perfect-fluid stars as

Table 10.2 Representative BH mimickers (ECOs) in the literature and their main diagnostics.

Class	Support mechanism	Key properties/Observational handles
Gravastars	dS-like core matched to an exterior Schwarzschild region through a thin shell (often anisotropic)	Horizonless ultracompact configurations; photon sphere may exist; nonzero effective surface reflectivity implies possible late-time echoes; tidal response generally differs from Kerr (Love numbers, tidal heating)
Wormholes (thin-shell or traversable)	Nontrivial topology sustained by exotic stresses (NEC violation in GR) or by modified-gravity effective sources	Can reproduce some lensing/shadow features at finite resolution; typically two-sided causal structure; characteristic QNM spectrum and potential echo-like signatures; geodesic structure may differ from BH even when compact
Boson stars (scalar condensates)	Self-gravitating complex scalar field; compactness controlled by mass and self-interactions	Regular and horizonless; in the most compact branches a photon sphere can occur; generically nonzero tidal Love numbers; ringdown differs from Kerr at late times (no perfect absorption)
Proca stars (vector condensates)	Self-gravitating massive vector (Proca) field configurations	Analogues of boson stars with additional vector degrees of freedom; can achieve high compactness in some branches; distinctive QNM spectrum; typically nonzero Love numbers
Anisotropic/elastic ultracompact stars	Pressure anisotropy or elastic stresses engineered to exceed perfect-fluid compactness bounds	May approach $R \simeq 2M(1+\epsilon)$ with $\epsilon \ll 1$ while remaining horizonless; large redshift and long light-travel delay can effectively decouple the interior on observational timescales; departures from Kerr encoded in reflectivity $\mathcal{R}(\omega)$ and tidal observables
Topological stars/solitons	Higher-dimensional Einstein-Maxwell(-like) solitons; effective 4D ECO after dimensional reduction	Provide controlled families interpolating from star-like to near-BH objects; can display Kerr-like early ringdown plus UCO modes and echoes; useful testbeds for stability/formation analyses
Fuzzballs/microstate geometries	String-theory microstates supported by fluxes, topology and extra fields; no classical horizon	Horizon-scale structure replaces the would-be horizon; effective absorption/reflection differs from Kerr; conceptually addresses singularity and information puzzles; phenomenology often captured by near-horizon boundary conditions rather than a classical surface
Phenomenological "quantum" ECOs	Effective near-horizon modification parameterized by a surface location $R = 2M(1+\epsilon)$ and frequency-dependent reflectivity $\mathcal{R}(\omega)$	Model-agnostic framework: echoes, modified tidal heating and non-Kerr QNMs emerge from boundary conditions; provides a compact way to confront data without committing to a specific microphysical realization

$$C \leq \frac{4}{9}, \tag{10.289}$$

and causality (e.g. subluminal sound speed together with reasonable microphysics assumptions) strengthens this to an approximate upper limit $C \lesssim 0.364$ for broad classes of equations of state. Consequently, causal perfect fluids can at best be marginal UCOs, but they are not expected to be true BH mimickers. To exceed the causal bound and approach $C \to 1/2$, one typically requires anisotropic stresses, exotic matter sectors, or modifications of GR; even then, additional requirements such as radial stability and avoidance of gradient instabilities substantially reduce the attainable compactness in concrete models.

Boson stars in GR provide an instructive illustration. Increasing self-interactions tends to drive the macroscopic behavior toward an effective fluid description, so that their maximum compactness is again constrained by the same causal considerations.

Compactness, however, is not the entire story. What controls the appearance of "horizon-like" behavior is often the *delay time* for radiation or perturbations to probe the interior and return. An object whose exterior compactness does not reach $C = 1/2$ may still develop an extremely large *internal redshift*, so that the interior effectively decouples from photon-sphere dynamics for very long times. A clean diagnostic is provided by the gravitational redshift factor for a static metric $ds^2 = -e^{2\Phi(r)}dt^2 + \cdots$, for which the local-to-asymptotic frequency ratio is

$$1 + z(r) = e^{-\Phi(r)}. \tag{10.290}$$

In the (unphysical but conceptually clarifying) incompressible-fluid limit saturating $C \to 4/9$, the central redshift can diverge. Correspondingly, the coordinate time needed for an exterior observer to receive a signal that penetrates to the deep interior can become arbitrarily large.

- **Clue II: reflectivity as a proxy for horizons**. A classical event horizon is a perfect absorber: it has *zero* reflectivity for any linear wave at any frequency. Any nonzero reflectivity is therefore, in principle, a direct indicator of the absence of a classical horizon. For linear perturbations on a fixed background one can schematically reduce the dynamics to a one-dimensional master equation of Schrödinger type,

$$\frac{d^2\Psi}{dx^2} + \left(\omega^2 - V(x)\right)\Psi = 0, \tag{10.291}$$

where x is a tortoise-like coordinate and $V(x)$ is an effective potential. If the ECO is sufficiently compact, then in a neighborhood of the surface $r = R$ the potential becomes negligible and the local solution is a superposition of ingoing and outgoing plane waves,

$$\Psi \sim e^{-i\omega(x-x_0)} + \mathcal{R}(\omega)\, e^{+i\omega(x-x_0)}, \qquad x_0 = x(R). \tag{10.292}$$

Here $\mathcal{R}(\omega)$ is the complex, frequency-dependent reflectivity. BHs correspond to the purely ingoing boundary condition $\mathcal{R} = 0$, whereas generic horizonless objects satisfy $\mathcal{R} \neq 0$.
Since gravitational waves couple only weakly to ordinary matter, near-perfect reflection, $|\mathcal{R}| \approx 1$, is often the natural expectation at sufficiently low frequencies. In that regime one may expand

$$\mathcal{R}(\omega) = 1 + i\omega\mathcal{R}_1 + O(\omega^2)\,, \tag{10.293}$$

where $\mathcal{R}_1$ is model dependent and encodes dispersive and dissipative microphysics. Importantly, a large gravitational redshift inside the object can manifest itself as a large *phase* of $\mathcal{R}(\omega)$: even a perfectly reflecting surface can behave "almost" as a horizon for a finite time, simply because the round-trip time between the photon-sphere barrier and the inner boundary becomes enormous. This sharpens a key conceptual point: the strict horizon boundary condition $\mathcal{R} = 0$ is less relevant observationally than the *effective causal decoupling* produced by large redshift and time delay.
- **Clue III: dynamics and stability (light rings and ergoregions)**. Any viable mimicker must be sufficiently long-lived on astrophysical timescales and must possess a controlled dynamical response under perturbations. The absence of a horizon implies partial reflectivity, and for UCOs this generically leads to low-frequency, long-lived modes that are quasi-trapped by the effective potential barrier. Such modes can seed instabilities.
In rotating configurations a further threat arises. Sufficiently compact, spinning horizonless objects can develop an ergoregion and become subject to the *ergoregion instability* already at the linear level: superradiant amplification combined with reflection can lead to exponential growth. Perfectly reflecting, rapidly spinning UCOs are therefore strongly constrained. A crucial nuance is that even a small amount of absorption (e.g. viscosity, dissipation into internal degrees of freedom, excitation of internal modes) can quench the ergoregion instability altogether. Thus, assessing the effective absorptivity is an integral part of realistic model building.
Beyond linear theory, it has been argued that the mere presence of a *stable* light ring, i.e., a feature allowed in horizonless ultracompact geometries, may induce a nonlinear instability, possibly on extremely long timescales. The current evidence is not definitive, and both the degree of universality and the characteristic timescale remain open. This motivates constructing well-posed ECO models that can be evolved fully nonlinearly. An instructive arena is provided by *topological stars*: regular, horizonless solitons arising in higher-dimensional Einstein-Maxwell constructions, which can interpolate continuously from less compact ECOs to configurations arbitrarily close to BHs.
- **Clue IV: scale invariance as a global consistency requirement.** A frequently overlooked property of vacuum BHs is *scale invariance*: in GR, the Kerr family has no intrinsic mass scale, and the mass simply labels the solution. Replacing *all* astrophysical BH candidates (from stellar-mass to supermassive) with horizonless

alternatives is therefore nontrivial, because coupling to matter almost inevitably introduces additional scales (particle masses, interaction scales, equation-of-state scales, *etc.*). Within GR, scale-invariant mimickers appear to require special anisotropic (elastic) matter models; beyond GR, scale-free solitons may arise in high-energy-motivated settings, including supergravity microstate constructions.

10.7 Is Gravity a Fundamental Force?

At the outset of this Chapter, we pointed out some hints toward the anomalous nature of gravity within fundamental interactions. On the one hand, it admits a fully consistent low-energy description as a long-range interaction universally coupled to stress-energy, with a propagating mediator in the weak-field limit. On the other hand, unlike gauge interactions formulated on a fixed causal background, gravity *is the dynamical agent that determines causal structure, clocks and rulers.*

This background-independence sharply differentiates gravity from the other forces and suggests that the Einstein's field equations may play a role analogous to a macroscopic constitutive relation.

▶ **Pause and think**
Ask yourself: If gravity is geometry, then is it fundamental or *emergent*?

Emergent gravity?
The term *emergent gravity* should not be read as "gravity is not an interaction". Rather, it asserts a hierarchy of descriptions:

- at macroscopic scales, gravity is encoded in geometry and obeys Einsteinian dynamics;
- at microscopic scales, the relevant degrees of freedom need not be the metric itself;
- the classical field equations may arise as thermodynamic/entanglement relations for suitable coarse-grained variables.

In this viewpoint, the central question is not whether the gravitational force exists, but *what microphysical variables reproduce spacetime geometry and its dynamics upon coarse graining.*

Two complementary lines of argument motivate a thermodynamic reading of gravity.

1. Horizons possess entropy and temperature, and their laws strongly resemble the laws of thermodynamics.
2. In appropriate quasi-static settings, Newtonian gravity can be re-expressed as an *entropic force*.

This entire Section, therefore, is developed as a plausible conjecture toward a revision of the gravity origin. Research toward this direction is very active, and the status of gravity as a fundamental interaction will be further clarified in the upcoming years.

10.7.1 Gravity from Entropy?

At least two proposals can be distinguished:

1. an *entropic-force* derivation of Newtonian inertia and Newtonian gravity based on thermodynamic and holographic hypotheses;
2. an *emergent-gravity* scenario in a dS (dark-energy dominated) universe, predicting an additional "dark" gravitational component sourced by baryons in certain quasi-static regimes.

Consider a system whose macroscopic state depends on a generalized coordinate x (e.g. the position of a probe relative to a coarse-grained environment). Under a quasi-static displacement δx, the first law reads

$$\delta E = \delta Q - \delta W, \qquad \delta W = F\,\delta x, \tag{10.294}$$

where F is the generalized force conjugate to x. In an isothermal quasi-static process at temperature T,

$$\delta Q = T\,\delta S, \tag{10.295}$$

if the mechanical response is dominated by changes in entropy at fixed T (i.e., when changes in the internal energy δE are negligible compared to the work extracted), then

$$F\,\delta x = T\,\delta S \quad \Longrightarrow \quad F = T\,\frac{\partial S}{\partial x}, \tag{10.296}$$

leading to *an entropic force*.

The entropic force representation is not the gradient of a fundamental microscopic potential, but a macroscopic bias toward increasing entropy.

Driving Exercise
In view of the above, can we argue the Newton's law from entropy?
To do so, let us consider three inputs.

- **Assumption A (linear entropy shift under displacement).** A particle of mass m at proper distance x from the screen produces, under a small displacement Δx toward the screen, an entropy variation

$$\Delta S = 2\pi k_B \frac{mc}{\hbar}\,\Delta x, \qquad \Rightarrow \qquad \frac{\partial S}{\partial x} = 2\pi k_B \frac{mc}{\hbar}. \tag{10.297}$$

  This is a Bekenstein-type ansatz inspired by horizon thermodynamics, here taken as a postulate.
- **Assumption B (Unruh temperature for acceleration).** A uniformly accelerated observer with proper acceleration a experiences a thermal bath at

$$T_U = \frac{\hbar a}{2\pi c k_B}. \tag{10.298}$$

  This is a standard result of QFT in flat spacetime in Rindler coordinates and is used as an input.
- **Assumption C (area scaling of microscopic degrees of freedom).** A screen of area A carries a number of microscopic degrees of freedom ("bits") proportional to its area,

$$N = \frac{A\,c^3}{G\,\hbar}. \tag{10.299}$$

Given Assumptions A–C, the subsequent derivations are purely algebraic. The physical content resides in the validity (or otherwise) of the postulates, not in the manipulations.

Derivation no. 1. Insert Eqs. (10.297) and (10.298) into the entropic-force identity, Eq. (10.296):

$$\boxed{F = T_U \frac{\partial S}{\partial x} = \left(\frac{\hbar a}{2\pi c k_B}\right)\left(2\pi k_B \frac{mc}{\hbar}\right) = ma.} \tag{10.300}$$

Derivation no. 2. Consider a spherical screen of radius r surrounding baryonic mass M. Its area is $A = 4\pi r^2$, so

$$N(r) = \frac{4\pi r^2 c^3}{G\hbar}. \tag{10.301}$$

Introduce one further hypothesis.

- **Assumption D (equipartition).** The energy stored in the microscopic degrees of freedom at temperature T obeys

$$E = \frac{1}{2} N k_B T. \tag{10.302}$$

Identifying this with the rest energy $E = Mc^2$ gives

$$T(r) = \frac{2Mc^2}{N(r)\, k_B} = \frac{GM\hbar}{2\pi r^2 c\, k_B}. \tag{10.303}$$

Now place a test mass m just outside the screen and displace it radially by Δx. Using Eq. (10.296),

$$F(r) = T(r)\, \frac{\partial S}{\partial x}, \tag{10.304}$$

and inserting Eqs. (10.297) and (10.303), one finds

$$F(r) = \left(\frac{GM\hbar}{2\pi r^2 c\, k_B} \right) \left(2\pi k_B \frac{mc}{\hbar} \right) = \frac{GMm}{r^2}. \tag{10.305}$$

Therefore,

$$\boxed{F(r) = \frac{GMm}{r^2}.} \tag{10.306}$$

In this construction, the inverse-square law is not attributed to a microscopic spin-2 exchange, but emerges from

1. area scaling of information and
2. a thermodynamic equipartition rule on the screen. Accordingly, geometry enters only through the screen area.

From inverse-square law to the Poisson equation (Gauss law). Define $\mathbf{g} = -\nabla\Phi$. For a point mass M,

$$\mathbf{g}(\mathbf{r}) = -\frac{GM}{r^2}\, \hat{\mathbf{r}}. \tag{10.307}$$

The flux through a sphere S_r is

$$\oint_{S_r} \mathbf{g} \cdot d\mathbf{A} = \oint_{S_r} \left(-\frac{GM}{r^2} \hat{\mathbf{r}} \right) \cdot \left(\hat{\mathbf{r}}\, r^2\, d\Omega \right) = -GM \int d\Omega = -4\pi GM. \tag{10.308}$$

By Gauss' theorem,

$$\oint_{S_r} \mathbf{g} \cdot d\mathbf{A} = \int_{V_r} (\nabla \cdot \mathbf{g})\, d^3x = -\int_{V_r} (\nabla^2 \Phi)\, d^3x. \tag{10.309}$$

Consistency with superposition then yields

$$\nabla \cdot \mathbf{g} = -4\pi G\, \rho \qquad \Leftrightarrow \qquad \nabla^2 \Phi = 4\pi G\, \rho. \tag{10.310}$$

The preceding steps are elementary but rely on a nontrivial chain of assumptions. Critiques emphasize that certain postulates can be viewed as re-encodings of Newtonian relations rather than independent microscopic inputs.

10.7.2 Einstein Equations as an Equation of State

A controlled realization of "gravity from thermodynamics" proceeds by applying the Clausius relation to local Rindler horizons. The key idea is that, at any spacetime point p, one may associate to each null direction a local causal horizon, and demand that the horizon satisfies a local thermodynamic balance law. When this condition is required for all local horizons through p, the Einstein equations follow as a consistency requirement.

Local Rindler horizon and heat flux. Fix p and consider an accelerated observer passing through a neighborhood of p. In locally inertial coordinates around p one may construct approximate Rindler coordinates for the accelerated frame. The observer perceives a local horizon patch $\mathcal{H}$ generated by null vectors k^a. The energy flux crossing $\mathcal{H}$ is

$$\delta Q = \int_{\mathcal{H}} T_{ab}\, \chi^a\, d\Sigma^b, \tag{10.311}$$

where χ^a is the approximate boost Killing vector field which becomes null on the horizon and $d\Sigma^b$ is the horizon surface element.

Temperature (Unruh effect). The accelerated observer with acceleration parameter (surface gravity) κ associates to the Hawking temperature and Bekenstein-Hawking entropy, implying that a deformation reads

$$\delta S = \frac{k_B c^3}{4G\hbar}\, \delta A. \tag{10.312}$$

► **Driving Exercise**
To find the area change, let λ be an affine parameter along the null generators and θ the expansion. The area element evolves according to

$$\frac{d}{d\lambda}\ln(dA) = \theta. \tag{10.313}$$

The Raychaudhuri equation for a null congruence has been presented in Eq. (10.62) and for null hypersurface-orthogonal congruences, namely $\omega_{\mu\nu} = 0$, reads

$$\frac{d\theta}{d\lambda} = -\frac{1}{2}\theta^2 - \sigma_{ab}\sigma^{ab} - R_{ab}\,k^a k^b, \tag{10.314}$$

with shear σ_{ab}. Choose generators such that at p one has $\theta|_p = 0$ and $\sigma_{ab}|_p = 0$ (local equilibrium). To leading order near p,

$$\theta(\lambda) \simeq -\lambda\, R_{ab}\, k^a k^b, \tag{10.315}$$

and integrating Eq. (10.313) yields the area change

$$\delta A \propto \int \theta\, d\lambda\, dA \simeq -\int \lambda\, R_{ab}\, k^a k^b\, d\lambda\, dA, \tag{10.316}$$

where the proportionality depends on the normalization of χ^a and k^a.

Hence, one can state the main ansatz **Clausius relation ⇒ Einstein equation.** Impose the Clausius relation for all local horizon patches through p,

$$\delta Q = T\,\delta S. \tag{10.317}$$

Using Eq. (10.311), (10.312) and (10.316) yield a local relation that must hold for all null vectors k^a:

$$T_{ab}\, k^a k^b \;\propto\; R_{ab}\, k^a k^b. \tag{10.318}$$

Since Eq. (10.318) holds for arbitrary null k^a, one infers (up to an undetermined cosmological term) the Einstein equations. This establishes, at a technically sharp level, the interpretation of Einstein gravity as a macroscopic equation of state in which horizon entropy and local energy flux play the role of thermodynamic variables.

10.8 Quantizing Gravity: Why and How

The challenge of quantizing gravity is fundamentally different from the same issue within the other fundamental interactions studied in this book.

There, quantization proceeds by promoting classical fields, defined on a fixed spacetime background, to quantum operators acting on a Hilbert space. The background geometry provides an external notion of time, causality, and energy, which remains unaffected by the quantum dynamics of the fields. None of these assumptions hold in GR.

In Einstein's theory, *the gravitational field is identified with the spacetime metric itself.*

As a consequence, quantizing gravity is not merely the problem of quantizing a field *on* spacetime, but rather the problem of quantizing spacetime geometry. This immediately raises a conceptual tension: standard quantum theory presupposes a fixed causal structure and a preferred notion of time evolution, while GR makes both notions dynamical and observer-dependent. Any attempt to combine the two frameworks must therefore address the absence of a background time parameter at a fundamental level.

From a canonical perspective, this tension manifests itself through the constrained nature of GR The Hamiltonian formulation reveals that time evolution is generated by constraints associated with diffeomorphism invariance, rather than by a genuine Hamiltonian driving evolution with respect to an external time variable.

Upon quantization, these constraints become operator conditions on physical states, leading to the **Wheeler-DeWitt equation**.

In this framework, physical states appear to be "frozen", with no explicit time dependence. This feature, often referred to as the *problem of time*, is not a technical defect but a direct consequence of background independence.

Beyond the conceptual issues, quantizing gravity also poses severe difficulties at the perturbative level. When treated as a QFT around a fixed background, GR is non-renormalizable by power counting. This does not imply that the theory is inconsistent, but rather that it cannot be regarded as a fundamental ultraviolet-complete theory in the same sense as renormalizable gauge theories. Instead, gravity admits a natural interpretation as an *effective field theory*, i.e., valid at energy scales well below the Planck scale, where quantum corrections can be systematically organized in an expansion in powers of energy over the Planck mass.

This fact also signals that new degrees of freedom or new principles are expected to emerge at sufficiently high energies. Whether such a completion requires a radically new formulation of spacetime, additional symmetries, or entirely new fundamental objects *remains an open question*.

In this sense, quantizing gravity should not be understood as a single, uniquely defined procedure, but rather as a collection of complementary approaches, each illuminating different aspects of the problem.

Let us explore the main consequences of applying the *canonical quantization* procedure to gravity.

10.8.1 $3+1$ Decomposition and Arnowitt-Deser-Misner Variables

In order to formulate the Wheeler-deWitt equation, let us consider first the $3+1$ decomposition.

The Hilbert-Einstein action (with the appropriate boundary term) can be written as

$$S_{\rm HE} = \frac{1}{16\pi G_N}\int_V d^4x\,\sqrt{-g}\,R + \frac{1}{8\pi G_N}\int_{\partial V} d^3x\,\sqrt{h}\,K, \tag{10.319}$$

where h_{ij} is the induced metric on the boundary hypersurface, $h \equiv \det(h_{ij})$, and $K = h^{ij}K_{ij}$ is the trace of the extrinsic curvature K_{ij}. The boundary term is chosen so that the variational problem is well-posed when fixing the induced metric on ∂V.

To construct a canonical formulation one introduces a foliation of spacetime by spacelike hypersurfaces Σ_t labelled by a parameter t, with topology $V \simeq \mathbb{R}\times M^3$. In adapted coordinates (t, x^i), the spacetime line element is written in ADM form

$$ds^2 = -\left(N^2 - N^i N_i\right)dt^2 + 2N_i\,dx^i dt + h_{ij}\,dx^i dx^j, \tag{10.320}$$

where $N(t,x)$ is the *lapse* and $N^i(t,x)$ the *shift*. Equivalently,

$$g_{\mu\nu} = \begin{pmatrix} -(N^2 - N^i N_i) & N_j \\ N_i & h_{ij} \end{pmatrix}. \tag{10.321}$$

The unit normal n^μ to Σ_t satisfies $g_{\mu\nu}n^\mu n^\nu = -1$ and $g_{\mu\nu}n^\mu X^\nu_i = 0$, where $X^\mu_i = \partial X^\mu/\partial x^i$ are tangent vectors on Σ_t. The deformation vector that connects points with the same spatial coordinates on neighbouring slices decomposes as

$$\dot{X}^\mu = N n^\mu + N^i X^\mu_i. \tag{10.322}$$

The extrinsic curvature of Σ_t is

$$K_{ij} \equiv \frac{1}{2N}\left(D_i N_j + D_j N_i - \dot{h}_{ij}\right), \tag{10.323}$$

with D_i the covariant derivative associated with h_{ij} and a dot denoting ∂_t.

Using the Gauss-Codazzi relation, the action can be cast (up to surface terms) in the form

$$S = \int dt\, d^3x\, \mathcal{L}, \qquad \mathcal{L} = \frac{\sqrt{h}\, N}{16\pi G_N}\left(K_{ij}K^{ij} - K^2 + {}^{(3)}R\right), \tag{10.324}$$

where ${}^{(3)}R$ is the scalar curvature of (Σ_t, h_{ij}).

A crucial point is that $\mathcal{L}$ contains *no* time derivatives of N or N^i. Therefore their conjugate momenta vanish,

$$\pi_N \equiv \frac{\partial \mathcal{L}}{\partial \dot{N}} = 0, \qquad \pi_i \equiv \frac{\partial \mathcal{L}}{\partial \dot{N}^i} = 0, \tag{10.325}$$

which are primary constraints. The conjugate momentum to h_{ij} is instead

$$\pi^{ij} \equiv \frac{\partial \mathcal{L}}{\partial \dot{h}_{ij}} = \frac{\sqrt{h}}{16\pi G_N}\left(K^{ij} - h^{ij}K\right). \tag{10.326}$$

This relation can be inverted to express K_{ij} in terms of π^{ij}, which is needed to pass to the Hamiltonian formulation.

The Legendre transform yields the canonical action

$$S = \int dt\, d^3x \left(\pi^{ij}\dot{h}_{ij} - N\mathcal{H}_0 - N^i\mathcal{H}_i\right), \tag{10.327}$$

with the *Hamiltonian constraint* (also called *super-Hamiltonian*)

$$\mathcal{H}_0 = 16\pi G_N\, G_{ijkl}\, \pi^{ij}\pi^{kl} - \frac{\sqrt{h}}{16\pi G_N}\, {}^{(3)}R, \tag{10.328}$$

and the *momentum constraints* (also called *super-momenta*)

$$\mathcal{H}_i = -2\, D_j \pi^j{}_i. \tag{10.329}$$

Here,

$$G_{ijkl} \equiv \frac{1}{2\sqrt{h}}\left(h_{ik}h_{jl} + h_{il}h_{jk} - h_{ij}h_{kl}\right) \tag{10.330}$$

is the (covariant) DeWitt supermetric on the space of 3-metrics.

Since N and N^i play the role of Lagrange multipliers, varying the action with respect to them gives the secondary constraints (see Chap. 3)

$$\mathcal{H}_0 = 0, \qquad \mathcal{H}_i = 0. \tag{10.331}$$

At the classical level these are equivalent to the $G_{00} = 0$ and $G_{0i} = 0$ components of Einstein's equations (in vacuum), while the remaining components follow from the Hamilton equations

$$\dot{h}_{ij} = \frac{\delta H}{\delta \pi^{ij}}, \qquad \dot{\pi}^{ij} = -\frac{\delta H}{\delta h_{ij}}, \tag{10.332}$$

where the canonical Hamiltonian is

$$H = \int d^3x \left(N\mathcal{H}_0 + N^i \mathcal{H}_i \right), \tag{10.333}$$

possibly supplemented by surface terms under non-compact boundary conditions (a point we will not pursue here, since we keep the discussion local and structural).

A useful way to read these results is: *the dynamical content of GR is encoded in constraints.* The canonical evolution is a gauge flow generated by $\mathcal{H}_0$ and $\mathcal{H}_i$, reflecting spacetime diffeomorphism invariance.

10.8.2 Wheeler-DeWitt Equation and Superspace

Canonical quantization proceeds by promoting the phase-space variables to operators obeying equal-time commutation relations. In the Schrödinger (metric) representation, one postulates

$$\left[\hat{h}_{ij}(\mathbf{x}), \hat{\pi}^{kl}(\mathbf{x}')\right] = i\, \delta_{ij}^{kl}\, \delta^{(3)}(\mathbf{x} - \mathbf{x}'), \qquad \delta_{ij}^{kl} \equiv \frac{1}{2}\left(\delta_i^k \delta_j^l + \delta_i^l \delta_j^k\right), \tag{10.334}$$

with

$$\hat{h}_{ij}(\mathbf{x})\, \Psi[h] = h_{ij}(\mathbf{x})\, \Psi[h], \qquad \hat{\pi}^{ij}(\mathbf{x})\, \Psi[h] = -i \frac{\delta}{\delta h_{ij}(\mathbf{x})} \Psi[h]. \tag{10.335}$$

The quantum state $\Psi[h]$ is a functional on the configuration space of 3-metrics (more precisely, on equivalence classes under spatial diffeomorphisms; see below).

The classical constraints are then promoted to operator conditions selecting physical states:

$$\hat{\mathcal{H}}_0\, \Psi[h] = 0, \qquad \hat{\mathcal{H}}_i\, \Psi[h] = 0. \tag{10.336}$$

The second equation enforces invariance under spatial diffeomorphisms; the first is the Wheeler-DeWitt equation.

Formally, the Wheeler-DeWitt equation reads

$$\boxed{\left(16\pi G_N\,\hat{\nabla}^2 + \frac{\sqrt{h}}{16\pi G_N}\,{}^{(3)}R\right)\Psi[h] = 0,} \tag{10.337}$$

where the kinetic operator is built from the DeWitt supermetric.

Conceptual issues: time, inner product, and interpretation.
Equation (10.337) contains no external time parameter. This is not a bug introduced by quantization; it is the quantum imprint of a classical fact: in generally covariant systems, the Hamiltonian is a constraint rather than a generator of evolution with respect to an absolute time. In the canonical language, "evolution in t" is gauge, because different foliations describe the same spacetime geometry.
A pragmatic resolution is to use an *internal* or *relational* time: one attempts to identify a functional of the canonical variables that behaves monotonically along a relevant class of configurations and then expresses the dynamics of the remaining degrees of freedom *with respect to that variable*. This is not always globally possible and, when it is, the choice can be region-dependent in configuration space.
Because the kinetic operator is built from an indefinite supermetric, the Wheeler-DeWitt equation naturally admits a conserved current (a functional analogue of the Klein-Gordon current). This suggests a conserved bilinear form, but not a positive-definite norm in general. In other words, the passage from "solution space" to a conventional Hilbert space is subtle and may require additional structure: a choice of time variable, a restriction to a sector (e.g. positive-frequency with respect to an internal time), or a more refined quantization scheme.
Rather than committing to a single interpretational framework, we here emphasize a structural point: the canonical constraints define the *physical* state space, and probability assignments typically become meaningful only after selecting a semiclassical regime where an internal time and an approximate Schrödinger evolution emerge.

10.8.3 General Relativity as an Effective Field Theory

To quantize metric fluctuations one expands[5] around a smooth background $\bar{g}_{\mu\nu}$,

$$g_{\mu\nu} = \bar{g}_{\mu\nu} + \kappa h_{\mu\nu}, \qquad g^{\mu\nu} = \bar{g}^{\mu\nu} - \kappa h^{\mu\nu} + \kappa^2 h^{\mu}{}_{\lambda} h^{\lambda\nu} + \cdots, \tag{10.338}$$

with indices moved by the background metric $\bar{g}_{\mu\nu}$. The field $h_{\mu\nu}$ has a gauge redundancy (linearized diffeomorphisms), so one adds a background-covariant de Donder gauge-fixing term

$$\mathcal{L}_{\text{gf}} = \sqrt{-\bar{g}}\left(h_{\mu\nu}{}^{;\nu} - \frac{1}{2}h_{;\mu}\right)\left(h^{\mu\lambda}{}_{;\lambda} - \frac{1}{2}h^{;\mu}\right), \qquad h \equiv h^{\lambda}{}_{\lambda}, \tag{10.339}$$

and the corresponding Faddeev-Popov ghost Lagrangian

$$\mathcal{L}_{\text{ghost}} = \sqrt{-\bar{g}}\,\eta^{*\mu}\left(\eta_{\mu}{}^{;\lambda}{}_{;\lambda} - \bar{R}_{\mu\nu}\eta^{\nu}\right), \tag{10.340}$$

where η_μ is a complex ghost field.

Quadratic action for one-loop quantization. Expanding the gravitational Lagrangian about $\bar{g}_{\mu\nu}$ gives

$$S_{\text{grav}} = \int d^4x\,\sqrt{-\bar{g}}\left[\frac{2\bar{R}}{\kappa^2} + \mathcal{L}_g^{(1)} + \mathcal{L}_g^{(2)} + \cdots\right], \tag{10.341}$$

with the linear and quadratic pieces

$$\mathcal{L}_g^{(1)} = \frac{1}{\kappa}\,h_{\mu\nu}\left(\bar{g}^{\mu\nu}\bar{R} - 2\bar{R}^{\mu\nu}\right), \tag{10.342}$$

$$\begin{aligned}\mathcal{L}_g^{(2)} &= \frac{1}{2}h_{\mu\nu;\alpha}h^{\mu\nu;\alpha} - \frac{1}{2}h_{;\alpha}h^{;\alpha} + h_{;\alpha}h^{\alpha\beta}{}_{;\beta} - h_{\mu\beta;\alpha}h^{\mu\alpha;\beta} \\ &\quad + \bar{R}\left(\frac{1}{4}h^2 - \frac{1}{2}h_{\mu\nu}h^{\mu\nu}\right) + \left(2h^{\lambda}{}_{\mu}h_{\nu\lambda} - h\,h_{\mu\nu}\right)\bar{R}^{\mu\nu}.\end{aligned} \tag{10.343}$$

Similarly,

$$S_{\text{matter}} = \int d^4x\,\sqrt{-\bar{g}}\left(\mathcal{L}_m^{(0)} + \mathcal{L}_m^{(1)} + \mathcal{L}_m^{(2)} + \cdots\right), \tag{10.344}$$

[5] We previously said that the metric is the field. Accordingly, passing to Minkowski's metric, and handling gravity as a perturbation over spacetime, are the key to formulate a quantum version of GR, at least in weak-energy regimes.

where

$$\mathcal{L}_m^{(0)} = \frac{1}{2}\Big(\partial_\mu\phi\,\partial^\mu\phi - m^2\phi^2\Big), \tag{10.345}$$

$$\mathcal{L}_m^{(1)} = -\frac{\kappa}{2}\,h_{\mu\nu}T^{\mu\nu}, \tag{10.346}$$

$$T_{\mu\nu} = \partial_\mu\phi\,\partial_\nu\phi - \frac{1}{2}\bar{g}_{\mu\nu}\Big(\partial_\lambda\phi\,\partial^\lambda\phi - m^2\phi^2\Big), \tag{10.347}$$

$$\mathcal{L}_m^{(2)} = \kappa^2\Big(\frac{1}{2}h^{\mu\nu}h_\nu{}^\lambda - \frac{1}{4}h\,h^{\mu\nu}\Big)\partial_\mu\phi\,\partial_\nu\phi - \frac{\kappa^2}{8}\Big(h^{\lambda\sigma}h_{\lambda\sigma} - \frac{1}{2}h^2\Big)\Big(\partial_\mu\phi\,\partial^\mu\phi - m^2\phi^2\Big). \tag{10.348}$$

If the background satisfies Einstein's equation with source $T^{\mu\nu}$,

$$\bar{R}^{\mu\nu} - \frac{1}{2}\bar{g}^{\mu\nu}\bar{R} = -\frac{\kappa^2}{4}\,T^{\mu\nu}, \tag{10.349}$$

then the tadpoles cancel and the linear combination $\mathcal{L}_g^{(1)} + \mathcal{L}_m^{(1)}$ vanishes. The one-loop problem is therefore governed by the quadratic action

$$\begin{aligned} S_0 &= \int d^4x\,\sqrt{-\bar{g}}\left(\frac{2\bar{R}}{\kappa^2} + \mathcal{L}_m^{(0)}\right), \\ S_{\text{quad}} &= \int d^4x\,\sqrt{-\bar{g}}\left(\mathcal{L}_g^{(2)} + \mathcal{L}_{\text{gf}} + \mathcal{L}_{\text{ghost}} + \mathcal{L}_m^{(2)}\right), \end{aligned} \tag{10.350}$$

and the corresponding functional integral

$$W[\phi] = e^{iZ[\phi]} = \int [dh_{\mu\nu}]\,[d\eta_\mu]\,\exp\Big(i\big(S_0 + S_{\text{quad}}\big)\Big). \tag{10.351}$$

In general $Z[\phi]$ is nonlocal and divergent. This is not conceptually unusual: the construction parallels Yang-Mills quantization for small fluctuations. However, gravity differs by the presence of the dimensionful coupling κ and by its intrinsic nonlinearity, which make perturbation theory unreliable at very high energies.

Why is an EFT description the correct low-energy quantum theory?

Since $[\kappa] = -1$, perturbative amplitudes organize in powers of $\kappa^2 q^2 \sim q^2/M_{\text{Pl}}^2$. Hence at $E \ll M_{\text{Pl}}$ quantum fluctuations are weakly coupled and Eq. (10.351) admits a controlled loop expansion, while at $E \gtrsim M_{\text{Pl}}$ the coupling grows and the ultraviolet (UV) completion becomes indispensable. From the effective-field-theory (EFT) perspective, one retains the universal low-energy content (massless gravitons and the Einstein vertex) and parametrizes UV ignorance by local higher-derivative operators consistent with general covariance.

Gravity as an EFT: general covariant operator basis and power counting. General covariance allows infinitely many invariants, so the most general gravitational Lagrangian is a derivative expansion,

$$\mathcal{L}_{\text{grav}}^{\text{EFT}} = \Lambda + \frac{2}{\kappa^2} R + c_1 R^2 + c_2 R_{\mu\nu} R^{\mu\nu} + \cdots , \tag{10.352}$$

up to more curvature and/or derivative terms. The counting follows the Christoffel and Ricci symbols: $\Gamma \sim \partial g$ and $R \sim \partial^2 g$, so $R = O(p^2)$ while $(R^2,\ R_{\mu\nu}R^{\mu\nu}) = O(p^4)$.

Hence, at sufficiently low momenta, the Einstein term dominates and the c_i corrections are suppressed.

In the matter sector one analogously writes the most general covariant terms. For a massive scalar, it is convenient to display the first curvature couplings,

$$\begin{aligned}\mathcal{L}_{\text{matter}}^{\text{EFT}} &= \frac{1}{2}\Big(g^{\mu\nu}\partial_\mu\phi\,\partial_\nu\phi - m^2\phi^2\Big) + d_1 R^{\mu\nu}\partial_\mu\phi\,\partial_\nu\phi \\ &\quad + R\Big(d_2\,\partial_\mu\phi\,\partial^\mu\phi + d_3 m^2\phi^2\Big) + \cdots .\end{aligned} \tag{10.353}$$

The EFT ordering counts derivatives acting on light fields (gravitons, and any other massless modes); in particular, derivatives acting on a heavy field need not be small in the same sense.

Renormalization: locality of divergences and predictivity of IR physics. Loop calculations based on Eq. (10.351) generate UV divergences. In an EFT this is expected: divergences are local and can be absorbed into renormalized values of the Wilson coefficients in Eqs. (10.352)–(10.353). By dimensional analysis, graviton loops built from the $O(p^2)$ Einstein vertices produce contributions of $O(p^4)$, matching the operator class multiplying c_1, c_2 in Eq. (10.352); higher loops similarly feed higher-derivative operators.

A crucial distinction is between heavy and massless virtual particles. Heavy fields of mass M_H have range $\Delta r \sim 1/M_H$ and can be integrated out, as encoded by the low-momentum expansion

$$\frac{1}{q^2 - M_H^2} = -\frac{1}{M_H^2} - \frac{q^2}{M_H^4} - \frac{q^4}{M_H^6} + \cdots , \tag{10.354}$$

which translates into local derivative operators. Massless propagators $1/q^2$ cannot be expanded about $q^2 = 0$ and give genuinely long-range (nonlocal) effects.

Leading quantum corrections: nonanalyticity and universality.
Local EFT operators generate analytic momentum expansions, e.g.

$$\mathcal{M}_{\text{local}} = A\, q^2\Big(1 + \alpha\, \kappa^2 q^2 + \cdots\Big). \tag{10.355}$$

Loops of massless particles (gravitons, photons, and possibly neutrinos) generate nonanalytic structures that cannot be reproduced by any local Lagrangian. For instance

$$\mathcal{M} = A\, q^2\Big(1 + \alpha\, \kappa^2 q^2 + \beta\, \kappa^2 q^2 \ln(-q^2) + \cdots\Big), \tag{10.356}$$

and, with massive matter in loops, also terms $\propto m/\sqrt{-q^2}$. These contributions develop imaginary parts for timelike $q^2 > 0$, reflecting on-shell intermediate states and the unitarity of the S-matrix. Since they only depend on the light spectrum and on the lowest-order couplings, they are parameter-free predictions of the low-energy theory, in contrast to the analytic corrections controlled by unknown coefficients such as c_1, c_2, d_i.

10.8.4 Linearizing Gravity: The Fierz-Pauli Theory

In the above treatment, GR has been treated as an EFT and the starting point was to linearize it. Let us restart from this approach, focusing on the weak-field regime, where the spacetime metric deviates only slightly from flat Minkowski spacetime. One writes

$$g_{\mu\nu} = \eta_{\mu\nu} + h_{\mu\nu}, \qquad |h_{\mu\nu}| \ll 1, \tag{10.357}$$

where again $h_{\mu\nu}$ is a small perturbation.

Indices are raised and lowered using $\eta_{\mu\nu}$, and all expressions are expanded to first order in $h_{\mu\nu}$.

▶ **Driving Exercise**
Let us now linerize connection, curvature and Einstein's tensor.
The Christoffel symbols at linear order are

$$\Gamma^{\rho}_{\mu\nu} = \frac{1}{2}\eta^{\rho\sigma}\left(\partial_\mu h_{\nu\sigma} + \partial_\nu h_{\mu\sigma} - \partial_\sigma h_{\mu\nu}\right). \tag{10.358}$$

The linearized Riemann tensor is

$$R_{\mu\nu\rho\sigma} = \frac{1}{2}\left(\partial_\rho \partial_\nu h_{\mu\sigma} + \partial_\sigma \partial_\mu h_{\nu\rho} - \partial_\sigma \partial_\nu h_{\mu\rho} - \partial_\rho \partial_\mu h_{\nu\sigma}\right). \tag{10.359}$$

From this one obtains the Ricci tensor

$$R_{\mu\nu} = \frac{1}{2}\left(\partial_\rho \partial_\mu h^\rho{}_\nu + \partial_\rho \partial_\nu h^\rho{}_\mu - \Box h_{\mu\nu} - \partial_\mu \partial_\nu h\right), \quad (10.360)$$

and the Ricci scalar

$$R = \partial_\mu \partial_\nu h^{\mu\nu} - \Box h, \quad (10.361)$$

where $h = h^\mu{}_\mu$ and $\Box = \eta^{\mu\nu}\partial_\mu\partial_\nu$.

The Einstein tensor at first order is $G^{(1)}_{\mu\nu} = R_{\mu\nu} - \frac{1}{2}\eta_{\mu\nu}R$, giving then the linearized Einstein's field equations,

$$G^{(1)}_{\mu\nu} = 8\pi G\, T_{\mu\nu}. \quad (10.362)$$

Pause and think
Try to argue some advantages of linearizing gravity.

Linearized gravity is invariant under infinitesimal diffeomorphisms

$$x^\mu \to x^\mu + \xi^\mu(x), \quad (10.363)$$

which induce the gauge transformation

$$h_{\mu\nu} \to h_{\mu\nu} + \partial_\mu \xi_\nu + \partial_\nu \xi_\mu. \quad (10.364)$$

This gauge symmetry removes unphysical degrees of freedom and reflects the underlying coordinate invariance of GR.

Define the trace-reversed field $\bar{h}_{\mu\nu} = h_{\mu\nu} - \frac{1}{2}\eta_{\mu\nu}h$.
Imposing the Lorenz gauge condition $\partial^\mu \bar{h}_{\mu\nu} = 0$, the field equations reduce to

$$\Box \bar{h}_{\mu\nu} = -16\pi G\, T_{\mu\nu}. \quad (10.365)$$

In vacuum, $\Box \bar{h}_{\mu\nu} = 0$, whose solutions are plane waves:

$$\bar{h}_{\mu\nu}(x) = \epsilon_{\mu\nu} e^{ik_\rho x^\rho}, \qquad k^2 = 0. \quad (10.366)$$

Using residual gauge freedom, one can impose the transverse-traceless (TT) gauge:

$$h^{\mathrm{TT}}_{0\mu} = 0, \qquad \partial^i h^{\mathrm{TT}}_{ij} = 0, \qquad h^{\mathrm{TT}} = 0. \quad (10.367)$$

Only two physical polarizations remain, corresponding to helicity ± 2.

For non-relativistic sources, the dominant component is $h_{00} = -2\Phi$, where Φ is the Newtonian gravitational potential.

The 00 component of the field equations reduces to Poisson's equation, $\nabla^2\Phi = 4\pi G\rho$.

Driving Exercise

Spin-2 gravitons from the Fierz-Pauli action. The dynamics of linearized gravity is encoded in the (quadratic) Fierz-Pauli action for a symmetric tensor field $h_{\mu\nu}$ on Minkowski spacetime,

$$S_{\mathrm{FP}} = \frac{1}{2}\int d^4x \Big[\partial_\lambda h_{\mu\nu}\partial^\lambda h^{\mu\nu} - 2\partial_\mu h^{\mu\nu}\partial^\lambda h_{\lambda\nu} + 2\partial_\mu h^{\mu\nu}\partial_\nu h - \partial_\lambda h \partial^\lambda h\Big], \tag{10.368}$$

where $h \equiv \eta^{\mu\nu}h_{\mu\nu}$. Equation (10.368) is the unique Lorentz-invariant, ghost-free quadratic action describing a *massless* spin-2 field: any deformation of the relative coefficients reintroduces a propagating scalar ghost already at the linear level.

Varying Eq. (10.368) with respect to $h_{\mu\nu}$ yields the vacuum field equations

$$G^{(1)}_{\mu\nu}(h) = 0, \tag{10.369}$$

where $G^{(1)}_{\mu\nu}$ is the Einstein tensor linearized around $\eta_{\mu\nu}$. Coupling to matter proceeds universally through the conserved stress-energy tensor. We thus introduce

$$S_{\mathrm{int}} = \frac{1}{2}\int d^4x\, h_{\mu\nu}T^{\mu\nu}, \tag{10.370}$$

so that Eq. (10.369) becomes the sourced linearized Einstein equations. Gauge invariance is the linearized remnant of diffeomorphisms,

$$h_{\mu\nu} \longrightarrow h_{\mu\nu} + \partial_\mu\xi_\nu + \partial_\nu\xi_\mu, \tag{10.371}$$

and it enforces stress-energy conservation,

$$\partial_\mu T^{\mu\nu} = 0, \tag{10.372}$$

as a consistency condition for Eq. (10.370). The gauge symmetry in Eq. (10.371) eliminates unphysical polarizations and leaves *exactly two* propagating degrees of freedom, which one may identify with the two helicity states $\lambda = \pm 2$ of a massless graviton.

Finally, the iterative self-coupling of $h_{\mu\nu}$ to its own stress-energy tensor reconstructs, in a unique manner, the full nonlinear Einstein equations: in this sense, Eq. (10.368) is not merely an approximation of GR, but its interaction-theoretic seed.

Pauli-Fierz versus standard gravitational waves.

Linearized GR predicts gravitational waves (GWs) as *solutions* of Eq. (10.369): they are ripples of spacetime geometry propagating at the invariant speed c (in vacuum) and carrying two tensor polarizations. The same statement can be expressed in the particle language: the radiative sector of Eq. (10.368) is a massless spin-2 field with helicities ± 2.
By contrast, the phrase "Pauli-Fierz" is often used to emphasize the *field-theoretic* (and, when desired, *massive*) spin-2 perspective.

- *Massless case*: Eq. (10.368) is equivalent to linearized GR. Its wave solutions are the standard GWs with two transverse tensor modes ($+$ and $\times$).
- *Massive case*: adding the Pauli-Fierz mass deformation (see Eq. (10.373) below) describes a *massive* spin-2 particle and generically predicts up to five polarizations (two tensor, two vector, one scalar/longitudinal). These additional modes modify detector responses and, crucially, the wave dispersion relation.

Standard GWs have now been *directly measured* by ground-based interferometers (LIGO/Virgo/KAGRA), through transient compact-binary coalescences and persistent stochastic/continuous searches. The observed signals are consistent, within experimental accuracy, with propagation essentially luminal and dominated by the tensorial radiative sector expected from Eq. (10.368).

Massive spin-2: Pauli-Fierz mass term and Stückelberg fields. A Lorentz-invariant mass deformation for $h_{\mu\nu}$ that preserves ghost-freedom at quadratic order is uniquely fixed to be the Pauli-Fierz combination,

$$S_{\rm m} \;=\; -\frac{m_{\rm g}^2}{4}\int d^4x\,\Big(h_{\mu\nu}h^{\mu\nu} - h^2\Big), \tag{10.373}$$

so that the full quadratic action is $S_{\rm FP}^{(m)} = S_{\rm FP} + S_{\rm m}$, with $S_{\rm FP}$ given by Eq. (10.368). The special relative coefficient in Eq. (10.373) is not aesthetic: it is precisely what removes the would-be sixth (ghost) polarization at the linear level.

The mass term in Eq. (10.373) breaks the gauge invariance in Eq. (10.371). A technically powerful way to keep track of the degrees of freedom, and to make the symmetry breaking manifestly controlled, is the Stückelberg completion. One introduces auxiliary fields (A_μ, ϕ) and performs the replacement

$$h_{\mu\nu} \;\longrightarrow\; \hat{h}_{\mu\nu} \equiv h_{\mu\nu} + \partial_\mu A_\nu + \partial_\nu A_\mu + 2\,\partial_\mu\partial_\nu\phi. \tag{10.374}$$

Inserting Eq. (10.374) into the mass term Eq. (10.373) yields an action that is invariant under the enlarged gauge symmetry

$$\delta h_{\mu\nu} = \partial_\mu \xi_\nu + \partial_\nu \xi_\mu, \quad \delta A_\mu = -\xi_\mu + \partial_\mu \lambda, \quad \delta\phi = -\lambda, \tag{10.375}$$

where ξ_μ and λ are independent gauge parameters. The Stückelberg formulation thus restores a gauge redundancy that makes transparent how the five physical polarizations of a massive spin-2 field are assembled: the helicity-± 2 sector is carried by $h_{\mu\nu}$, the helicity-± 1 sector is encoded in A_μ, and the helicity-0 sector resides in ϕ (in appropriate gauges and after canonical normalization).

Wave modes and propagation.

For standard (massless) GWs in GR, Eq. (10.369) plus the gauge freedom in Eq. (10.371) imply that radiative solutions can be brought to the transverse-traceless (TT) form, leaving two tensor polarizations and the dispersion $\omega^2 = \mathbf{k}^2$.

For massive Pauli-Fierz waves, Eq. (10.373) implies instead the massive dispersion

$$\omega^2 = \mathbf{k}^2 + m_g^2, \tag{10.376}$$

so the group velocity is subluminal, $v_g = \partial\omega/\partial|\mathbf{k}| < 1$ for finite m_g. Moreover, the radiative content can involve up to five polarizations, whose imprint on interferometric detectors differs from the TT tensor pattern: in particular, the helicity-0 sector produces a characteristic longitudinal response (in addition to breathing-type effects, depending on the basis used). These features make gravitational-wave observations a direct probe of the graviton mass and of possible departures from the strictly massless spin-2 dynamics of Eq. (10.368).

A decisive advantage of the "interaction-first" viewpoint is that the Newtonian limit can be recovered as a bona fide low-energy prediction of a universal, long-range spin-2 interaction. In this approach the metric is expanded about Minkowski space,

$$g_{\mu\nu} = \eta_{\mu\nu} + \kappa\, h_{\mu\nu}, \qquad \kappa \equiv \sqrt{32\pi G}, \tag{10.377}$$

and the leading coupling to matter is fixed by locality and Lorentz invariance,

$$\mathcal{L}_{\text{int}} = -\frac{\kappa}{2}\, h_{\mu\nu} T^{\mu\nu}. \tag{10.378}$$

At tree level, the static potential between two heavy sources follows from the non-relativistic limit of the $2 \to 2$ scattering amplitude mediated by a single graviton.

Minimal ingredients.

- The matter-graviton vertex is dictated by Eq. (10.378). For a heavy scalar (or, more generally, any nonrelativistic source) the on-shell matrix element of the stress tensor reduces to $T^{00} \simeq 2m^2$ (up to standard normalization conventions), while spatial components are velocity-suppressed.
- A convenient gauge fixing for the graviton propagator is the de Donder (harmonic) gauge, $\partial^\mu \bar{h}_{\mu\nu} = 0$, with $\bar{h}_{\mu\nu} \equiv h_{\mu\nu} - \frac{1}{2}\eta_{\mu\nu}h$.
- In momentum space, the graviton propagator reads

$$D_{\mu\nu,\alpha\beta}(q) = \frac{i}{q^2 + i\epsilon} P_{\mu\nu,\alpha\beta}, \quad P_{\mu\nu,\alpha\beta} \equiv \frac{1}{2}\Big(\eta_{\mu\alpha}\eta_{\nu\beta} + \eta_{\mu\beta}\eta_{\nu\alpha} - \eta_{\mu\nu}\eta_{\alpha\beta}\Big).$$

The Newtonian regime corresponds to small momentum transfer $q^\mu = (0, \mathbf{q})$ and $|\mathbf{q}| \ll m_{1,2}$.

Let particles 1 and 2 be heavy, slowly moving, and consider the t-channel exchange. The tree-level amplitude has the schematic form

$$i\mathcal{M} = \left(-\frac{i\kappa}{2}\right) T_1^{\mu\nu}(p_1, p_1')\, D_{\mu\nu,\alpha\beta}(q) \left(-\frac{i\kappa}{2}\right) T_2^{\alpha\beta}(p_2, p_2'), \tag{10.379}$$

with $q = p_1 - p_1' = p_2' - p_2$. In the nonrelativistic limit, the dominant contributions are the 00 components,

$$T_1^{\mu\nu} \simeq 2m_1^2\, \delta_0^\mu \delta_0^\nu, \qquad T_2^{\alpha\beta} \simeq 2m_2^2\, \delta_0^\alpha \delta_0^\beta, \tag{10.380}$$

so that Eq. (10.379) becomes

$$i\mathcal{M} \simeq \left(\frac{i\kappa^2}{4}\right) (2m_1^2)(2m_2^2) \frac{i}{q^2 + i\epsilon} P_{00,00}. \tag{10.381}$$

Since $P_{00,00} = \frac{1}{2}$ and $q^2 \simeq -\mathbf{q}^2$, one finds

$$\mathcal{M}_{\rm NR}(\mathbf{q}) \simeq -\frac{\kappa^2}{8} \frac{m_1^2 m_2^2}{\mathbf{q}^2}. \tag{10.382}$$

To extract the potential, one matches the Born approximation for nonrelativistic scattering,

$$V(\mathbf{r}) = -\int \frac{d^3\mathbf{q}}{(2\pi)^3}\, e^{i\mathbf{q}\cdot\mathbf{r}}\, \frac{\mathcal{M}_{\rm NR}(\mathbf{q})}{4m_1 m_2}, \tag{10.383}$$

where the factor $4m_1m_2$ accounts for the relativistic normalization of external states. Inserting Eqs. (10.382) into (10.383) gives

$$V(\mathbf{r}) = -\frac{\kappa^2}{32} m_1 m_2 \int \frac{d^3\mathbf{q}}{(2\pi)^3} \frac{e^{i\mathbf{q}\cdot\mathbf{r}}}{\mathbf{q}^2}. \tag{10.384}$$

Using the standard Fourier transform

$$\int \frac{d^3\mathbf{q}}{(2\pi)^3} \frac{e^{i\mathbf{q}\cdot\mathbf{r}}}{\mathbf{q}^2} = \frac{1}{4\pi r}, \tag{10.385}$$

one obtains the Newton potential,

$$V(r) = -\frac{\kappa^2}{128\pi} \frac{m_1 m_2}{r} = -G \frac{m_1 m_2}{r}, \tag{10.386}$$

where the final equality follows from $\kappa^2 = 32\pi G$ in Eq. (10.377).

Equation (10.386) shows that, in the low-energy regime, Newton's law is the static limit of a single-quantum exchange governed by the universal vertex in Eq. (10.378) and the massless spin-2 propagator. The *attractiveness* of the force is not imposed by hand: it is fixed by the spin-2 projector $P_{\mu\nu,\alpha\beta}$ together with the positivity of T^{00} for ordinary matter. Equivalently, the "gravitational charge" is energy-momentum, so like charges cannot repel without violating basic stability requirements.
In the massive Pauli-Fierz deformation, the same exchange picture continues to hold but the propagator acquires a mass scale, leading (in the static limit) to a Yukawa-type suppression and to characteristic effects tied to the extra polarizations. This is precisely why gravitational-wave observations, which probe the radiative propagating sector directly, are among the cleanest laboratories to constrain (or reveal) deviations from the strictly massless spin-2 dynamics of Eq. (10.368).

Beyond the static limit, the same framework predicts velocity-dependent corrections and, once loop effects are included, the universal long-distance quantum corrections captured by the gravitational effective field theory expansion. In this sense, the spin-2 exchange picture is not an alternative to geometry but its infrared, interaction-theoretic avatar.

10.9 The Cosmological Constant Problem

Above, we discussed the existence of the cosmological constant Λ and the corresponding fine-tuning problem. One may conclude, very naively, that such an issue is related to the magnitude of Λ and that finding out the right cut-off scale would be enough to define its value. This is only partially true.

The cosmological constant problem is intrinsically related to the fact that there is no way to remove such a contribution from our calculations, unless its value is not *exactly* zero.

The main issue is summarized in the so-called Weinberg's *no-go theorem.*

In order to formulate it, let us notice that the cosmological constant can be computed either in flat or in curved spacetimes. As we have learned, gravity requires a curved spacetime description, yet one can show that calculations produce identical results in both spacetimes, as consequence of the equivalence principle. The above discussion implies the *Sakharov's hypothesis*, as we show below.

10.9.1 Sakharov's Hypothesis

As first emphasized by Sakharov, the expectation value of the stress-energy tensor of a quantum field in its vacuum state must take the form, $\langle 0|T_{\mu\nu}|0\rangle = -\rho_{\rm vac} g_{\mu\nu}$, where $\rho_{\rm vac}$ denotes the vacuum energy density. This relation holds for all quantum fields present in the Universe, as will be shown below.

Since the vacuum state must be identical for all inertial observers, Lorentz invariance requires that its stress-energy tensor expectation value be proportional to $\eta_{\mu\nu}$, namely

$$\langle T_{\mu\nu}\rangle \propto \eta_{\mu\nu}.$$

When this statement is extended to curved space-time, general covariance implies that

$$\langle T_{\mu\nu}\rangle = -\rho_{\rm vac}(t, \mathbf{x})\, g_{\mu\nu}. \tag{10.387}$$

At this stage, the proportionality factor may depend on space-time position. However, the stress-energy tensor must satisfy covariant conservation,

$$\nabla^{\mu}\langle T_{\mu\nu}\rangle = 0.$$

Substituting Eq. (10.387) into this condition immediately yields $\partial_{\nu}\rho_{\rm vac} = 0$, and therefore $\rho_{\rm vac}$ must be a constant. This establishes the general form written in the first equation.

10.9.2 Classical Cosmological Constant Problem

A complementary derivation of the previous result can be obtained by specializing to a concrete dynamical system, for instance a real scalar field Φ.

In the classical ground state, the energy is minimized when the kinetic term vanishes and the field sits at the absolute minimum of its potential. In this configuration, the stress-energy tensor reduces to

$$\langle T_{\mu\nu}\rangle = -V(\Phi_{\text{min}})\, g_{\mu\nu}, \tag{10.388}$$

which reproduces the general vacuum form with the identification $\rho_{\text{vac}} = V(\Phi_{\text{min}})$.

The next conceptual step is to assume that the equivalence principle extends to zero-point contributions, including both the classical and quantum pieces discussed previously: every form of energy gravitates. Accordingly, the Einstein equations in the presence of quantum fields should be written as

$$R_{\mu\nu} - \frac{1}{2}Rg_{\mu\nu} + \Lambda_{\text{B}}g_{\mu\nu} = \kappa T_{\mu\nu}^{\text{matter}} + \kappa\langle T_{\mu\nu}\rangle, \tag{10.389}$$

where the first term on the right-hand side describes ordinary matter, while the second encodes the total vacuum contribution, summed over all fields within a given theory. Using the vacuum form of the stress-energy tensor, one immediately finds

$$R_{\mu\nu} - \frac{1}{2}Rg_{\mu\nu} + \Lambda_{\text{eff}}\, g_{\mu\nu} = \kappa T_{\mu\nu}^{\text{matter}}, \tag{10.390}$$

with

$$\Lambda_{\text{eff}} = \Lambda_{\text{B}} + \kappa\rho_{\text{vac}}. \tag{10.391}$$

Hence, the cosmological constant that governs spacetime dynamics is not the bare parameter Λ_{B} but the effective quantity Λ_{eff}, obtained by adding the vacuum contribution.

The observable quantity constrained by cosmological and local tests of Einstein gravity is Λ_{eff}, not the bare parameter Λ_{B}.
The cosmological constant problem is therefore intrinsically multifaceted: Λ_{eff} receives contributions of very different origin. As already seen for a scalar field, a purely classical contribution arises from the value of the potential at its minimum. If this value is not exactly zero, it directly feeds into Λ_{eff}.

The role of classical contributions becomes particularly transparent in the presence of phase transitions. Consider a scalar field Φ interacting with another scalar Ψ through the potential

$$V(\Phi, \Psi) = V(\Phi) + \frac{\bar{g}}{2}\Phi^2\Psi^2, \tag{10.392}$$

where $\bar{g}$ is dimensionless and $V(\Phi) = V_0 + \frac{\lambda}{4}\left(\Phi^2 - v^2\right)^2$. Here λ controls the self-interaction of Φ, v is the location of the minimum in the absence of Ψ, and V_0 is a classical offset.

If Ψ is in thermal equilibrium, one may replace Ψ^2 by its thermal average $\langle\Psi^2\rangle_T$. Since $\langle\Psi^2\rangle_T \propto T^2$, the effective potential becomes

$$V_{\text{eff}}(\Phi) = V_0 + \frac{\lambda}{4}\left(\Phi^2 - v^2\right)^2 + \frac{\bar{g}}{2}T^2\Phi^2, \tag{10.393}$$

where $\bar{g}$ has been rescaled to absorb numerical factors. This can be rewritten as

$$V_{\text{eff}}(\Phi) = V_0 + \frac{\lambda v^4}{4} + \frac{\lambda v^2}{2}\left(\frac{T^2}{T_{\text{cri}}^2} - 1\right)\Phi^2 + \frac{\lambda}{4}\Phi^4, \tag{10.394}$$

with $T_{\text{cri}} = v\sqrt{\lambda/\bar{g}}$. The interaction with Ψ therefore induces a temperature-dependent effective mass, $m_{\text{eff}}^2(T) \equiv \lambda v^2\left(\frac{T^2}{T_{\text{cri}}^2} - 1\right)$. The phase transition associated with $m_{\text{eff}}^2(T)$ changing sign is precisely the usual mechanism of symmetry breaking (see Chap. 7).

▶ **Pause and think**
Try to locate by yourself where the problem arises when one attempts to eliminate the cosmological constant.

The tension now becomes explicit, and two mutually exclusive choices emerge.

- If one enforces vanishing vacuum energy before the transition, one must choose $V_0 = -\lambda v^4/4$. After the transition, however, the vacuum energy is no longer zero but becomes $\rho_{\text{vac}} = -\lambda v^4/4$.
- If instead one demands $\rho_{\text{vac}} = 0$ after the transition, one must set $V_0 = 0$. Then the vacuum energy before the transition is $\rho_{\text{vac}} = \lambda v^4/4$.

The second option may appear less problematic observationally, since we have no direct constraints on the vacuum energy before Big Bang Nucleosynthesis.

10.9.3 Quantum Cosmological Constant Problem

We now quantize the scalar field introduced above by adopting canonical quantization. In this framework, one can explicitly compute the vacuum expectation value $\langle 0|T_{\mu\nu}|0\rangle$ and track, step by step, how the ultraviolet contribution enters the gravitational setting.

A standard mode expansion in Minkowski space-time yields, after straightforward manipulations,

$$\langle 0|\dot{\Phi}^2|0\rangle = \frac{1}{(2\pi)^3}\int \frac{d^3\boldsymbol{k}}{2\omega(k)}\,\omega^2(k), \tag{10.395}$$

$$\langle 0|\delta^{ij}\partial_i\Phi\partial_j\Phi|0\rangle = \frac{1}{(2\pi)^3}\int \frac{d^3\boldsymbol{k}}{2\omega(k)}\,\boldsymbol{k}^2, \tag{10.396}$$

$$\langle 0|\Phi^2|0\rangle = \frac{1}{(2\pi)^3}\int \frac{d^3\boldsymbol{k}}{2\omega(k)}. \tag{10.397}$$

The energy density is governed by the Hamiltonian density,

$$T_{00} = \mathcal{H} = \frac{1}{2}\dot{\Phi}^2 + \frac{1}{2}\delta^{ij}\partial_i\Phi\partial_j\Phi + \frac{1}{2}m^2\Phi^2, \tag{10.398}$$

so that the energy density $\rho = u^\mu u^\nu T_{\mu\nu}$ measured by a fundamental observer $u^\mu = (1, \mathbf{0})$ reads

$$\langle\rho\rangle = \langle 0|u^0u^0T_{00}|0\rangle = \frac{1}{(2\pi)^3}\frac{1}{2}\int d^3\boldsymbol{k}\,\omega(k). \tag{10.399}$$

Proceeding analogously, the isotropic pressure can be obtained by projecting $T_{\mu\nu}$ orthogonally to u^μ using $\perp^{\mu\nu} \equiv g^{\mu\nu} + u^\mu u^\nu$, namely

$$\langle p\rangle = \left\langle 0\left|\frac{1}{3}\perp^{\mu\nu}T_{\mu\nu}\right|0\right\rangle = \left\langle 0\left|\frac{1}{2}\dot{\Phi}^2 - \frac{1}{6}\delta^{ij}\partial_i\Phi\partial_j\Phi - \frac{1}{2}m^2\Phi^2\right|0\right\rangle. \tag{10.400}$$

Using Eqs. (10.395)–(10.397), (10.400) reduces to

$$\langle p\rangle = \frac{1}{(2\pi)^3}\frac{1}{6}\int d^3\boldsymbol{k}\,\frac{k^2}{\omega(k)}. \tag{10.401}$$

Moreover, $\langle 0|T_{0i}|0\rangle = -\frac{1}{(2\pi)^3}\frac{1}{2}\int d^3\boldsymbol{k}\,k_i = 0$, so there is no momentum flow in the vacuum and one may consistently restrict attention to the diagonal components.

At this point, combining Eqs. (10.391) with (10.399), one obtains

$$\Lambda_{\text{eff}} = \Lambda_{\text{B}} + \frac{\kappa}{(2\pi)^3}\int d^3\boldsymbol{k}\,\frac{1}{2}\omega(k). \tag{10.402}$$

The immediate obstruction is that the integrals in Eqs. (10.399) and (10.401) show ultraviolet divergences, and therefore Eq. (10.402) is formally ill-defined. So a consistent treatment requires a regulator and a renormalization prescription.

A widespread approach is to impose a momentum cut-off $k < M$, where M is interpreted as the scale at which the low-energy description breaks down. Depending on the ultraviolet completion, M might be of order the Planck scale, the string scale, or the supersymmetry-breaking scale. **The precise value will not matter for the conceptual point: the problem persists for any reasonable M.**

We already saw, in Eq. (10.104), that

$$\langle \rho \rangle \simeq \frac{M^4}{16\pi^2}\left(1 + \frac{m^2}{M^2} + \cdots\right), \tag{10.403}$$

plus, it is immediate to see that the corresponding pressure is

$$\langle p \rangle = \frac{1}{3}\frac{1}{4\pi^2}\int_0^M \mathrm{d}k \frac{k^4}{\sqrt{k^2+m^2}} \tag{10.404}$$

$$= \frac{1}{3}\frac{M^4}{16\pi^2}\left[\sqrt{1+\frac{m^2}{M^2}}\left(1-\frac{3}{2}\frac{m^2}{M^2}\right) + \frac{3}{2}\frac{m^4}{M^4}\ln\left(\frac{M}{m}+\frac{M}{m}\sqrt{1+\frac{m^2}{M^2}}\right)\right]$$

$$= \frac{1}{3}\frac{M^4}{16\pi^2}\left(1 - \frac{m^2}{M^2} + \cdots\right). \tag{10.405}$$

A striking feature is that $\langle p \rangle / \langle \rho \rangle \neq -1$. In the massless limit $m \to 0$, we end up with $\langle p \rangle / \langle \rho \rangle = 1/3$, as if the vacuum fluctuations behaved like radiation rather than as a cosmological constant.[6] This contradicts the general covariance argument leading to $\langle T_{\mu\nu} \rangle \propto g_{\mu\nu}$ for a Lorentz invariant vacuum.

The mismatch indicates that the calculation has failed at a structural level: the regularization procedure has spoiled the symmetry that should protect the vacuum equation of state.

Pause and think
Focus on the log-correction only. What do you see?

Interestingly, if one isolates the logarithmic contributions,

$$\langle \rho \rangle^{\log} = -\frac{m^4}{32\pi^2}\ln\left(\frac{M}{m}+\frac{M}{m}\sqrt{1+\frac{m^2}{M^2}}\right), \tag{10.406}$$

$$\langle p \rangle^{\log} = \frac{m^4}{32\pi^2}\ln\left(\frac{M}{m}+\frac{M}{m}\sqrt{1+\frac{m^2}{M^2}}\right), \tag{10.407}$$

then one recovers the correct equation of state $\langle p \rangle^{\log} = -\langle \rho \rangle^{\log}$. As will become clear below, this is not accidental: the logarithms capture contributions consistent

[6] The cosmological constant has always an equation of state, i.e., the ratio between pressure and density, identically equal to -1.

with Lorentz invariance, whereas the power-law terms reflect regulator artifacts when the symmetry is broken.

The problem is not specific to a sharp cut-off. Replacing it by a smooth exponential regulator still breaks Lorentz invariance if it is implemented only on spatial momenta. For $m = 0$ one finds

$$\langle \rho \rangle = \frac{1}{4\pi^2} \int_0^\infty dk\, k^3\, e^{-\alpha k} = \frac{1}{4\pi^2} \frac{\Gamma(4)}{\alpha^4} = \frac{1}{4\pi^2} 6M^4, \tag{10.408}$$

where $\alpha = 1/M$ and $\Gamma(z)$ is the Euler integral of the second kind. One again recovers quartic sensitivity, and one again finds $\langle p \rangle / \langle \rho \rangle = 1/3$. The underlying reason is the same: *imposing a regulator on* **k** *but not on the full four-momentum breaks Lorentz invariance, and therefore contaminates the tensor structure of* $\langle T_{\mu\nu} \rangle$.

The vacuum must be Lorentz invariant. Any regularization that treats time and space differently (e.g. a cut-off on $|\boldsymbol{k}|$ only) breaks Lorentz invariance and can yield an unphysical equation of state for vacuum fluctuations. This is entirely analogous to gauge theories: regulators that violate gauge symmetry typically produce spurious results until symmetry-restoring counterterms are imposed.

Driving Exercise

To fix the problem, let us use the dimensional regularization (see Chap. 8). The Klein-Gordon equation retains its form, with indices now running from 0 to $d-1$. The Fourier expansion of the field becomes

$$\Phi(t, \boldsymbol{x}) = \frac{1}{(2\pi)^{(d-1)/2}} \int \frac{d^{d-1}\boldsymbol{k}}{\sqrt{2\omega(k)}} \left(c_{\boldsymbol{k}} e^{-i\omega t + i\boldsymbol{k}\cdot\boldsymbol{x}} + c^\dagger_{\boldsymbol{k}} e^{i\omega t - i\boldsymbol{k}\cdot\boldsymbol{x}} \right) \tag{10.409}$$

which leads to the d-dimensional energy density

$$\langle \rho \rangle = \frac{\mu^{4-d}}{(2\pi)^{(d-1)}} \frac{1}{2} \int d^{d-1}\boldsymbol{k}\, \omega(k) = \frac{\mu^{4-d}}{(2\pi)^{(d-1)}} \frac{1}{2} \int_0^\infty dk\, k^{d-2}\, d^{d-2}\Omega\, \omega(k), \tag{10.410}$$

where the scale μ is introduced to keep dimensions fixed. Using $\int d^{d-2}\Omega = 2\pi^{(d-1)/2}/\Gamma(d/2 - 1/2)$ one obtains

$$\langle \rho \rangle = \frac{\mu^4}{2\,(4\pi)^{(d-1)/2}} \frac{\Gamma(-d/2)}{\Gamma(-1/2)} \left(\frac{m}{\mu} \right)^d. \tag{10.411}$$

The pressure follows from the same Lorentz-covariant computation,

$$\langle p\rangle = \frac{\mu^{4-d}}{(2\pi)^{(d-1)}}\frac{1}{2(d-1)}\int \mathrm{d}^{d-1}\boldsymbol{k}\,\frac{k^2}{\omega(k)} = \frac{\mu^4}{4\,(4\pi)^{(d-1)/2}}\frac{\Gamma(-d/2)}{\Gamma(1/2)}\left(\frac{m}{\mu}\right)^d. \tag{10.412}$$

Since $\Gamma(-1/2) = -2\Gamma(1/2)$, Eqs. (10.411) and (10.412) imply

$$\langle p\rangle = -\langle\rho\rangle, \tag{10.413}$$

which is the expected vacuum equation of state. In other words, once Lorentz symmetry is preserved by the regulator, the tensor structure of $\langle T_{\mu\nu}\rangle$ is automatically consistent with $\langle T_{\mu\nu}\rangle \propto g_{\mu\nu}$.

The internal consistency can be cross-checked directly from Eqs. (10.410) and (10.412), which satisfy

$$\langle\rho\rangle - 2\,m^2\frac{\mathrm{d}\langle\rho\rangle}{\mathrm{d}m^2} = (d-1)\langle p\rangle. \tag{10.414}$$

Using Eqs. (10.413) in (10.414) one finds

$$m^2\frac{\mathrm{d}\langle\rho\rangle}{\mathrm{d}m^2} = \frac{d}{2}\langle\rho\rangle, \tag{10.415}$$

whose solution is $\langle\rho\rangle \propto m^d$, in agreement with Eq. (10.411).

To exhibit the divergence structure, set $d = 4 - \epsilon$ and expand $\langle\rho\rangle$ for small ϵ. Using

$$\Gamma\left(-2+\frac{\epsilon}{2}\right) = \frac{1}{-2+\epsilon/2}\frac{1}{-1+\epsilon/2}\frac{1}{\epsilon/2}\Gamma\left(1+\frac{\epsilon}{2}\right), \tag{10.416}$$

together with $\Gamma(1+\epsilon/2) \simeq \Gamma(1) + \epsilon\Gamma'(1)/2$, $\Gamma'(1) = \Psi(1)\Gamma(1)$, $\Psi(1) = -\gamma$, and the expansions

$$(4\pi)^{-3/2+\epsilon/2} \simeq \frac{1}{(4\pi)^{3/2}}\left[1+\frac{\epsilon}{2}\ln(4\pi)\right], \quad \left(\frac{m}{\mu}\right)^{4-\epsilon} \simeq \left(\frac{m}{\mu}\right)^4\left(1-\epsilon\frac{m}{\mu}\right),$$

one obtains

$$\langle\rho\rangle \simeq -\frac{m^4}{64\pi^2}\left[\frac{2}{\epsilon}+\frac{3}{2}-\gamma-\ln\left(\frac{m^2}{4\pi\,\mu^2}\right)\right]+\cdots. \tag{10.417}$$

In a minimal subtraction renormalization prescription (here with the convention that the term $3/2$ is subtracted as well), the finite vacuum energy density becomes

$$\langle\rho\rangle = \frac{m^4}{64\pi^2}\ln\left(\frac{m^2}{\mu^2}\right). \tag{10.418}$$

This outcome is qualitatively different from the naive cut-off result. Dimensional regularization removes the power-law terms and retains the logarithmic structure already visible in Eqs. (10.406) and (10.407), precisely the part that respects the vacuum equation of state. The dependence on m^4 is also dimensionally consistent and immediately implies, for example, that strictly massless fields do not contribute to $\langle \rho \rangle$ in this scheme, contrary to what a spatial cut-off would suggest.

► **Pause and think**
Do you think that we solved or alleviated the cosmological constant problem?

Problems persist.

- First of all, Eq. (10.418) is finite and Lorentz invariant, but it does not eliminate the cosmological constant problem: the quantity $\langle \rho \rangle$ remains of order m^4 for each massive species, and the gravitational field is sensitive to the absolute normalization through Eq. (10.391).
- Interactions and renormalization-group effects modify the detailed bookkeeping, yet they do not remove the need to control the net value of Λ_{eff}.
- Finally, note that the computation above has been carried out for free fields. A realistic assessment must incorporate interactions, which feed vacuum diagrams and shift the renormalized parameters.

10.9.4 Weinberg's No-Go Theorem

Is there any dynamical adjustment of the cosmological constant?

The cosmological constant problem can be phrased, in its sharpest field-theoretic form, as a naturalness tension between (i) the existence of a diffeomorphism-invariant low-energy description of gravity and matter, and (ii) the observed smallness of the effective vacuum curvature.

A natural idea is to promote the cancellation in Eq. (10.391) to a *dynamical* adjustment: introduce additional fields (often scalar) whose vacuum expectation values self-tune so that the net vacuum curvature vanishes or becomes parametrically small.

Weinberg's no-go theorem states that such self-adjustment mechanisms in local field theory do *not* solve the problem.

Consider a generally covariant, local four-dimensional theory of the metric $g_{\mu\nu}$ coupled to a set of fields ψ^A (scalars, spinors, vectors, *etc.*), with action

$$S = \int d^4x \sqrt{-g}\, \mathcal{L}\big(g_{\mu\nu}, \psi^A, \nabla\psi^A, \ldots\big). \tag{10.419}$$

Assume that the vacuum is Poincaré invariant (or, more weakly, that it is translationally invariant so that all fields are constant and all derivatives vanish in the vacuum). Then, unless the theory possesses a special symmetry that enforces $\Lambda_{\text{eff}} = 0$ or introduces a genuine degeneracy of vacua, the existence of a Minkowski solution requires a parameter tuning equivalent to setting the vacuum energy to zero by hand. In particular, adding finitely many adjusting fields does not, by itself, eliminate the need to tune at least one combination of couplings so that the stationary-point conditions admit $\Lambda_{\text{eff}} = 0$.

Derivation in the vacuum sector. Under the hypothesis of a translationally invariant vacuum, the fields take constant values

$$g_{\mu\nu} = \eta_{\mu\nu}, \qquad \psi^A = \psi_0^A, \qquad \nabla_\alpha \psi^A = 0, \tag{10.420}$$

and the Lagrangian density reduces to a function of constant parameters and constant field values,

$$\mathcal{L}\Big|_{\text{vac}} = \mathcal{L}_0(\eta_{\mu\nu}, \psi_0^A) \equiv -V(\psi_0), \tag{10.421}$$

where the last equality defines the (possibly multi-field) vacuum energy function V. In such a vacuum, the stress tensor is proportional to the metric,

$$T^{\text{vac}}_{\mu\nu} = -\frac{2}{\sqrt{-g}} \frac{\delta S}{\delta g^{\mu\nu}}\bigg|_{\text{vac}} = \mathcal{L}_0\, g_{\mu\nu}, \tag{10.422}$$

so that a Minkowski solution requires

$$\mathcal{L}_0(\eta, \psi_0) = 0, \tag{10.423}$$

equivalently $V(\psi_0) = 0$ in the convention of Eq. (10.421). At the same time, the vacuum values of the matter fields must satisfy the stationary-point conditions

$$\frac{\partial \mathcal{L}_0}{\partial \psi^A}\bigg|_{\psi=\psi_0} = 0 \qquad \Longleftrightarrow \qquad \frac{\partial V}{\partial \psi^A}\bigg|_{\psi=\psi_0} = 0. \tag{10.424}$$

Equations (10.423) and (10.424) exhibit the core obstruction. For N constant vacuum fields ψ_0^A one has N equations from Eq. (10.424), which may determine (some of) the ψ_0^A in terms of the couplings of the Lagrangian. However Eq. (10.423) is an *additional* scalar constraint: it requires that the *value* of the vacuum Lagrangian at the stationary point vanish. In a generic theory, the stationary value $\mathcal{L}_0(\psi_0)$ is not forced to be zero; it is a number determined by couplings. Thus, achieving Eq. (10.423) typically fixes one extra combination of parameters, i.e. it is a fine tuning.

Pause and think
Try to find by yourself why adding an "adjuster" field does not help to solve the problem!

Driving Exercise
Suppose one introduces an extra scalar χ with the intention of self-adjusting the vacuum energy. In the translationally invariant vacuum, the conditions become

$$\frac{\partial V}{\partial \psi^A}(\psi_0, \chi_0) = 0, \qquad \frac{\partial V}{\partial \chi}(\psi_0, \chi_0) = 0, \tag{10.425}$$

which determine (ψ_0, χ_0) as extrema of V. Yet the requirement of Minkowski space is still

$$V(\psi_0, \chi_0) = 0. \tag{10.426}$$

The value of a generic potential at its extrema is not constrained: it is an output fixed by the parameters of V. Hence Eq. (10.426) does not follow from Eq. (10.425) unless the theory has an additional structural reason enforcing it. Put differently, the adjuster field can move the location of the extremum, but it does not generically force the extremal value to vanish.
This can be made particularly transparent by considering a separable ansatz,

$$V(\psi, \chi) = V_1(\psi) + V_2(\chi), \tag{10.427}$$

for which Eq. (10.425) yields $V_1'(\psi_0) = 0$ and $V_2'(\chi_0) = 0$, while Eq. (10.426) becomes $V_1(\psi_0) + V_2(\chi_0) = 0$. The latter is a relation between two *numbers* (the extremal values) and therefore constrains at least one parameter in V_1 or V_2. More general couplings do not change the logic: without a symmetry or degeneracy, V at an extremum is a parameter-dependent constant which must be tuned to zero.

In a translationally invariant vacuum, the equations of motion for the matter fields supply stationarity conditions, Eq. (10.424). Independently, the metric equation of motion imposes the vanishing of the vacuum stress tensor, Eq. (10.423). The latter is not implied by stationarity: it is a constraint on the *value* of the action density at the stationary point. In a generic local theory, this value is not symmetry-protected, so a Minkowski solution exists only after tuning one combination of couplings.

10.10 Quantum Fields in Curved Background

The formalism of QFT admits a consistent extension when the background geometry is no longer Minkowskian but is described by a generic metric tensor $g_{\mu\nu}$ instead of $\eta_{\mu\nu} = \text{diag}(-1, 1, 1, 1)$. In this setting, special relativistic notions are replaced by generally covariant ones, while locality and causality are preserved.

A convenient starting point is a real scalar field φ with non-minimal coupling to curvature, whose Lagrangian density reads

$$\mathcal{L}_\varphi = -\frac{1}{2} g_{\mu\nu} \nabla^\mu \varphi \nabla^\nu \varphi - \frac{m^2}{2} \varphi^2 - \frac{1}{2} \xi R \varphi^2, \tag{10.428}$$

where ξ is a dimensionless coupling constant and ∇_μ denotes the Levi-Civita covariant derivative associated with $g_{\mu\nu}$. Varying Eq. (10.428) with respect to φ yields the generally covariant Klein-Gordon equation

$$(\Box - m^2 - \xi R)\varphi = 0, \tag{10.429}$$

with $\Box \equiv g^{\mu\nu} \nabla_\mu \nabla_\nu = \frac{1}{\sqrt{-g}} \partial_\mu (\sqrt{-g} g^{\mu\nu} \partial_\nu)$.

The field φ is quantized by promoting it to an operator $\hat{\varphi}$ and expanding it in a complete set of solutions of Eq. (10.429),

$$\hat{\varphi}(x) = \int d\mathbf{k} \left(\hat{a}_\mathbf{k} u_\mathbf{k}(x) + \hat{a}^\dagger_\mathbf{k} u^*_\mathbf{k}(x) \right), \tag{10.430}$$

where the modes $u_\mathbf{k}$ are required to be orthonormal with respect to the Klein-Gordon inner product

$$(u_\mathbf{k}, u_{\mathbf{k}'})_{\text{KG}} = -i \int_\Sigma d\Sigma^\mu \left(u_\mathbf{k} \nabla_\mu u^*_{\mathbf{k}'} - u^*_{\mathbf{k}'} \nabla_\mu u_\mathbf{k} \right), \tag{10.431}$$

with $d\Sigma^\mu = d\Sigma\, n^\mu$, n^μ being the future-directed unit normal to the spacelike Cauchy surface Σ.

Particle localization via smeared field operators can also be introduced in curved spacetime. However, the interpretation of the smearing function becomes observer-dependent, since different foliations of spacetime correspond to inequivalent notions of simultaneity. Moreover, the label $\mathbf{k}$ no longer has the universal meaning of spatial momentum.

10.10.1 Bogoliubov Transformation of the Field Normal Modes

The scalar product in Eq. (10.431) already reveals a central difficulty: in generic curved spacetimes there is no preferred notion of positive frequency. In Minkowski spacetime, the existence of a global timelike Killing vector ∂_t provides an invariant splitting between positive and negative frequency modes. In curved spacetime, such a vector field may not exist globally, and therefore different observers, associated with different time foliations, may decompose the same field into inequivalent sets of modes.

Consider two observers, A and B, with proper times τ_A and τ_B. Each observer defines positive frequency modes by choosing $\tau_A = \text{const}$ or $\tau_B = \text{const}$ as Cauchy surfaces in Eq. (10.431). The corresponding sets of modes are denoted by $\{u_{\mathbf{k}}\}$ and $\{v_{\mathbf{k}}\}$, respectively. The field operator may then be expanded as

$$\hat{\varphi}(x) = \int d\mathbf{k} \left(\hat{b}_{\mathbf{k}} v_{\mathbf{k}}(x) + \hat{b}^{\dagger}_{\mathbf{k}} v^{*}_{\mathbf{k}}(x) \right). \tag{10.432}$$

Completeness implies that the two bases are related by a Bogoliubov transformation,

$$v_{\mathbf{k}} = \int d\mathbf{k}' \left(\alpha_{\mathbf{k}\mathbf{k}'} u_{\mathbf{k}'} + \beta_{\mathbf{k}\mathbf{k}'} u^{*}_{\mathbf{k}'} \right), \tag{10.433}$$

with coefficients $\alpha_{\mathbf{k}\mathbf{k}'} = (v_{\mathbf{k}}, u_{\mathbf{k}'})_{\text{KG}}$ and $\beta_{\mathbf{k}\mathbf{k}'} = (v_{\mathbf{k}}, u^{*}_{\mathbf{k}'})_{\text{KG}}$. Insertion of Eqs. (10.433) into (10.432) yields

$$\hat{b}_{\mathbf{k}} = \int d\mathbf{k}' \left(\alpha^{*}_{\mathbf{k}\mathbf{k}'} \hat{a}_{\mathbf{k}'} - \beta^{*}_{\mathbf{k}\mathbf{k}'} \hat{a}^{\dagger}_{\mathbf{k}'} \right). \tag{10.434}$$

Canonical commutation relations require the Bogoliubov coefficients to satisfy

$$\int d\mathbf{k}'' \left(\alpha_{\mathbf{k}\mathbf{k}''} \alpha^{*}_{\mathbf{k}'\mathbf{k}''} - \beta_{\mathbf{k}\mathbf{k}''} \beta^{*}_{\mathbf{k}'\mathbf{k}''} \right) = \delta^{3}(\mathbf{k} - \mathbf{k}'), \tag{10.435a}$$

$$\int d\mathbf{k}'' \left(\alpha_{\mathbf{k}\mathbf{k}''} \beta_{\mathbf{k}'\mathbf{k}''} - \beta_{\mathbf{k}\mathbf{k}''} \alpha_{\mathbf{k}'\mathbf{k}''} \right) = 0. \tag{10.435b}$$

The vacuum defined by $\hat{a}_{\mathbf{k}} \left| 0_{\text{A}} \right\rangle = 0$ is not the vacuum for observer B. Indeed,

$$N_{\mathbf{k}} = \langle 0_{\text{A}} | \, \hat{b}^{\dagger}_{\mathbf{k}} \hat{b}_{\mathbf{k}} \, | 0_{\text{A}} \rangle = \int d\mathbf{k}' \, |\beta_{\mathbf{k}\mathbf{k}'}|^{2}, \tag{10.436}$$

so that observer B detects particles in the vacuum of A whenever $\beta_{\mathbf{k}\mathbf{k}'} \neq 0$.

The notion of "particle" in curved spacetime is not absolute but observer-dependent. Particle production is a manifestation of the fact that different time foliations induce inequivalent splittings between positive and negative frequency modes.

Driving Exercise

The field stress energy tensor.

From the matter Lagrangian density Eq. (10.428), the stress-energy tensor follows via the standard variational definition. For the quantized scalar field it reads

$$\hat{T}^{\varphi}_{\mu\nu} = (1-2\xi)\nabla_\mu\hat{\varphi}\nabla_\nu\hat{\varphi} - \left(\frac{1}{2}-2\xi\right) g_{\mu\nu}\nabla^\rho\hat{\varphi}\nabla_\rho\hat{\varphi} - 2\xi\,\hat{\varphi}\,\nabla_\mu\nabla_\nu\hat{\varphi} + 2\xi\, g_{\mu\nu}\hat{\varphi}\Box\hat{\varphi} - \frac{1}{2}m^2 g_{\mu\nu}\hat{\varphi}^2 + \xi G_{\mu\nu}\hat{\varphi}^2. \tag{10.437}$$

Even in flat spacetime, $\langle 0|\,\hat{T}^{\varphi}_{\mu\nu}\,|0\rangle$ diverges and must be renormalized. In curved spacetime, different observers define different vacua, $|0_\mathrm{A}\rangle$ and $|0_\mathrm{B}\rangle$, and therefore

$$\langle 0_\mathrm{A}|\,\hat{T}^{\varphi}_{\mu\nu}\,|0_\mathrm{A}\rangle \neq \langle 0_\mathrm{B}|\,\hat{T}^{\varphi}_{\mu\nu}\,|0_\mathrm{B}\rangle\,. \tag{10.438}$$

The difference

$$T^{\varphi}_{\mu\nu}(\mathrm{A}\to\mathrm{B}) = \langle 0_\mathrm{B}|\,\hat{T}^{\varphi}_{\mu\nu}\,|0_\mathrm{B}\rangle - \langle 0_\mathrm{A}|\,\hat{T}^{\varphi}_{\mu\nu}\,|0_\mathrm{A}\rangle \tag{10.439}$$

represents the stress-energy tensor associated with the particles created by the change of observer or by the evolution of the background.

In most applications of QFT in curved spacetime, the backreaction of the created particles on the geometry is neglected. The metric is treated as a fixed classical background, and $\hat{T}_{\mu\nu}$ is used only as a diagnostic of particle creation.

10.10.2 Black Holes and Particle Production

The collapse of a massive object into a BH represents a highly non-stationary gravitational background. At late times the geometry tends to the Schwarzschild metric, as described above.

A Bogoliubov transformation relates the in-modes at past infinity to the out-modes at future infinity. The coefficients are

$$\alpha_{\omega\omega'} = -\frac{1}{\sqrt{4\pi^2\omega\omega'}}e^{2\pi M\omega'}(4M\omega)^{4iM\omega'}4iM\omega'\Gamma(-4iM\omega), \tag{10.440a}$$

$$\beta_{\omega\omega'} = \frac{1}{\sqrt{4\pi^2\omega\omega'}}e^{-2\pi M\omega'}(4M\omega)^{-4iM\omega'}4iM\omega'\Gamma(4iM\omega). \tag{10.440b}$$

From these,

$$|\beta_{\omega\omega'}|^2 = \frac{2M}{\pi\omega'}\frac{1}{e^{8\pi M\omega}-1}, \qquad N_\omega = \frac{1}{e^{8\pi M\omega}-1}, \tag{10.441}$$

which is a thermal spectrum with temperature $T_H = (8\pi M)^{-1}$. This is Hawking radiation.

Hawking radiation is not caused by time dependence of the final Schwarzschild geometry, but by the presence of an event horizon. Horizons generically lead to observer-dependent notions of vacuum and hence to thermal particle spectra.

10.10.3 The Information Paradox from Quantum Field Theory in Curved Spacetime

The semiclassical derivation of Hawking radiation is, at its core, an application of quantum field theory on a fixed but nontrivial causal background. The information puzzle becomes sharp precisely because the QFT-in-curved-spacetime computation is performed in a region where local curvature is small and the effective field theory should be reliable.

We therefore reframe the paradox in the language already developed in Sect. 10.10: the apparent loss of purity is a statement about *state assignment* and *Hilbert-space factorization* in the presence of horizons.

Hence, consider again a real scalar field φ on $(\mathcal{M}, g_{\mu\nu})$ with dynamics given by Eq. (10.429) and quantization based on the Klein-Gordon product in Eq. (10.431). The key structural input is that, in a generic curved spacetime, there is no canonical global notion of positive frequency. As discussed around Eqs. (10.433)–(10.436), two observers (or two asymptotic time notions) define non-equivalent splittings of the same solution space.

In the BH collapse geometry, the natural split is between an "in" basis defined at $\mathscr{I}^-$ and an "out" basis defined at $\mathscr{I}^+$. Denoting by $\{u_\mathbf{k}\}$ the modes adapted to the past and by $\{v_\mathbf{k}\}$ those adapted to the future, completeness yields the Bogoliubov map

$$v_\mathbf{k} = \int d\mathbf{k}' \left(\alpha_{\mathbf{k}\mathbf{k}'} u_{\mathbf{k}'} + \beta_{\mathbf{k}\mathbf{k}'} u^*_{\mathbf{k}'}\right), \qquad \hat{b}_\mathbf{k} = \int d\mathbf{k}' \left(\alpha^*_{\mathbf{k}\mathbf{k}'} \hat{a}_{\mathbf{k}'} - \beta^*_{\mathbf{k}\mathbf{k}'} \hat{a}^\dagger_{\mathbf{k}'}\right), \quad (10.442)$$

which is the operator-level version of Eqs. (10.433) and (10.434). Whenever $\beta \neq 0$, the "in" vacuum is populated by "out" quanta:

$$N_\mathbf{k} = \langle 0_\mathrm{A} | \hat{b}^\dagger_\mathbf{k} \hat{b}_\mathbf{k} | 0_\mathrm{A} \rangle = \int d\mathbf{k}' \, |\beta_{\mathbf{k}\mathbf{k}'}|^2, \qquad (10.443)$$

recovering Eq. (10.436).

For an asymptotically Schwarzschild late-time geometry, the explicit β-coefficients in Eqs. (10.441) lead to the Planckian occupation number

$$N_\omega = \frac{1}{e^{\omega/T_\mathrm{H}} - 1}, \qquad T_\mathrm{H} = \frac{1}{8\pi M}, \tag{10.444}$$

in agreement with Eq. (10.441).

Two notions of "vacuum" are the whole story.

The Hawking effect is not a mysterious dynamical emission mechanism in the final stationary metric: it is the statement that the state defined as "no particles" on $\mathscr{I}^-$ is not the state that looks particle-free on $\mathscr{I}^+$. The mixing coefficients β encode precisely the failure of a global positive-frequency split and are controlled, at late times, by the universal near-horizon redshift.

The Planckian spectrum is a statement about a *reduced state*. The semiclassical computation implicitly factorizes degrees of freedom into an exterior algebra accessible to observers at $\mathscr{I}^+$ and an interior algebra supported behind the horizon. In that factorization, the relevant state is not an eigenstate of the exterior number operator but a highly entangled state between outgoing modes and their partners.

A convenient schematic representation is the near-horizon two-mode squeezing structure. For each frequency band (and angular sector) one may write the late-time state as

$$|\Psi\rangle \propto \prod_\omega \sum_{n=0}^{\infty} e^{-n\omega/(2T_\mathrm{H})} \, |n_\omega\rangle_\mathrm{out} \otimes |n_\omega\rangle_\mathrm{in}\,, \tag{10.445}$$

where the labels "out" and "in" refer to the exterior and interior partners. Tracing over the interior factors produces a thermal density matrix for the exterior:

$$\rho_\mathrm{out} = \mathrm{Tr}_\mathrm{in}\, |\Psi\rangle\langle\Psi| \propto \prod_\omega \left(1 - e^{-\omega/T_\mathrm{H}}\right) \sum_{n=0}^{\infty} e^{-n\omega/T_\mathrm{H}}\, |n_\omega\rangle\langle n_\omega|\,. \tag{10.446}$$

The mixedness of ρ_out is therefore not an additional assumption: it is the direct consequence of (i) horizon-induced mode pairing and (ii) the operational restriction to exterior observables.

The information paradox arises when one insists that the semiclassical factorization behind Eqs. (10.445)–(10.446) remains valid *throughout the entire evaporation process*. If evaporation completes with no remnant and no additional degrees of freedom, the final state on $\mathscr{I}^+$ should be pure, yet Eq. (10.446) suggests a permanently mixed radiation state.

Driving Exercise

In QFT on curved spacetime, particle creation is diagnosed through expectation values such as Eq. (10.439). In particular, the Hawking flux carries positive energy to infinity, implying a decreasing mass parameter $M(u)$ and a shrinking horizon area. At the level of semiclassical gravity one writes schematically

$$G_{\mu\nu} = 8\pi G \left\langle \hat{T}_{\mu\nu} \right\rangle, \tag{10.447}$$

where the right-hand side must be renormalized and is state-dependent. In most practical treatments the backreaction is incorporated only at the level of an adiabatically evolving background, consistent with the quasi-stationarity assumption for macroscopic BHs.

However, the information question is not decided by energy flux alone: it is decided by the *fine-grained entropy* of the radiation. So, from the Von Neumann definition, applied to the reduced density matrix $\rho_{\rm rad}(t)$ of the degrees of freedom collected at $\mathscr{I}^+$:

$$S_{\rm rad}(t) = -\mathrm{Tr}\big[\rho_{\rm rad}(t) \log \rho_{\rm rad}(t)\big]. \tag{10.448}$$

If semiclassical factorization remains valid, the entanglement created in Eq. (10.445) accumulates monotonically, hence $S_{\rm rad}(t)$ grows until the end of evaporation. This monotonic behavior is the semiclassical prediction and is in tension with unitarity, which requires $S_{\rm rad}(t \to \infty) = 0$ for complete evaporation from a pure initial state.

10.11 Proposed Exercises

Exercise 1

Consider the metric (see, R. Giambò and O. Luongo, Class. Quant. Grav., **41**, (2024) 12, 125005)

$$ds^2 = -\left(1 - \frac{\Lambda}{3} r^2\right) dt^2 + \left[\frac{\left(1 - \frac{2}{3} r^2\right)}{\left(1 - \frac{\Lambda}{3} r^2\right)\left(1 + \kappa r^2\right)}\right] dr^2 + r^2 d\Omega^2, \tag{10.449}$$

with $d\Omega^2 \equiv d\theta^2 + \sin^2\theta d\phi^2$. Compute in both cases $\Lambda > 0$ and $\Lambda < 0$,

- the Christoffel symbols and the Riemann tensor,
- the Einstein tensor, finding explicitly ρ, P_r and P_t. What can you say about the pressure?

After that, explain the main properties of this metric w.r.t. the de Sitter space. Moreover,

– in the case $\Lambda > 0$, compute the entropy and temperature for this metric. Does the result change w.r.t. the de Sitter metric?

Exercise 2

Using the gravity–thermodynamic conjecture and the following modified version of entropy,

$$S = \frac{A}{4}\left(\alpha + \beta \tanh A\right), \tag{10.450}$$

– compute the modified Friedmann equations,
– explain the physical meaning of the parameters α and β, and their mutual role,
– sketch the physics associated with this new entropy.

Afterwards, compare the result with the Barrow entropy,

$$S_B = \frac{1}{4} A^{1+\sigma}, \tag{10.451}$$

with σ a constant. Are those versions of entropy extensive or non-extensive? Comment why.

Exercise 3

Bearing the following metric in mind,

$$ds^2 = -\left[1 - \frac{2M}{r}\left(1 - e^{-r^3/r_0^3}\right)\right] dt^2 + \left[1 - \frac{2M}{r}\left(1 - e^{-r^3/r_0^3}\right)\right]^{-1} dr^2 + r^2 d\Omega^2, \tag{10.452}$$

– explain its main properties and report under which conditions the metric is regular by computing the Kretschmann scalar,
– compare the metric w.r.t. to the Bardeen and Hayward spacetimes, evaluating the geodesic equations and the Raychaudhuri equation.

Finally, invoking the irreducible representation of the Riemann tensor and $R_{abcd} \rightarrow \mathbf{R}_{AB}$ $ab \rightarrow A$ (6x6 representation) with $01 \rightarrow 1$, $02 \rightarrow 2$, $03 \rightarrow 3$, $23 \rightarrow 4$, $31 \rightarrow 5$, $12 \rightarrow 6$, writing $\mathbf{R}_{AB} = \mathbf{R}_{BA}$ 6x6 Riemann matrix with the property $R_{a[bcd]} = 0 \rightarrow \mathbf{R}_{14} + \mathbf{R}_{25} + \mathbf{R}_{36} = 0$, compute

$$\mathbf{R}_{AB} - \lambda \mathbf{I}_{AB} = 0 \qquad \Longrightarrow \lambda_k \quad k = 1, 2, ..., 6 \tag{10.453}$$

and find when the eigenvalues change signs (see, O. Luongo and H. Quevedo, Phys. Rev. D, **90** (2014) 8, 084032). What does physically mean for the above metric?

Essential Bibliography

- Misner CW, Thorne KS, Wheeler JA (1973) Gravitation. W. H. Freeman
  A comprehensive reference for the geometric foundations of GR, classical solutions, and the operational meaning of curvature, causal structure, and gravitational radiation. Particularly useful as a long-form encyclopedic companion when building a unified narrative of the gravitational interaction.
- Weinberg S (1972) Gravitation and cosmology: principles and applications of the general theory of relativity. Wiley (1972)
  Excellent for connecting Einstein dynamics to the Friedmann framework and for framing cosmological tests as constraints on gravitational physics.
- Birrell ND, Davies PCW (1982) Quantum fields in curved space. Cambridge University Press
  The standard entry point for semiclassical gravity.
- Kiefer C (2012) Quantum gravity, 3rd edn. Oxford University Press
  A broad and conceptually focused account of the quantum-gravity problem and semiclassical approximations. Useful for structuring the logic behind "why and how" gravity is to be quantized.

Weak Interactions 11

Nature is simple, but not simplistic.

–Abdus Salam

Abstract

Weak interactions entered physics first noticing that beta decays and related processes are slow compared with electromagnetic transitions and with strong rearrangements, indicating a coupling of fundamentally different character. Starting from this phenomenological evidence, the present Chapter develops the weak interaction starting from the pre-gauge era, then deriving the Fermi theory as a universal low-energy current-current interaction, emphasizing how the neutrino hypothesis resolves the continuous beta spectrum and how the mature $V-A$ structure encodes maximal parity violation and the chiral nature of charged currents. Pion leptonic decay is used as a paradigmatic probe of chirality through helicity suppression, while strange hadrons motivate the quark reorganization of hadronic currents and the Cabibbo mixing picture. From there the discussion naturally progresses to flavor physics: the need to suppress flavor-changing neutral currents leads to the Glashow-Iliopoulos-Maiani mechanism and to the role of unitarity cancellations and mass splittings beyond tree level. The extension to three generations introduces the Kobayashi-Maskawa structure, the Cabibbo-Kobayashi-Maskawa (CKM) matrix, and CP violation, together with invariant diagnostics such as the Jarlskog invariant and the geometric interpretation in terms of unitarity triangles. Having identified the conceptual limitations of the four-fermion theory, we embed the weak interaction into a spontaneously broken non-abelian gauge theory, the electroweak $SU(2)_L \otimes U(1)_Y$ model, and show how the mediator picture of $W^{\pm}$ and Z^0 exchange reproduces Fermi's coupling in the infrared while restoring predictivity at high energies. Finally, we move beyond tree level and discuss the logic of electroweak radiative corrections and precision tests, where loop effects become a quantitative probe of the gauge structure itself. The chapter closes by focusing on neutrinos as the most sensitive messengers of weak dynamics.

O. Luongo, *Fundamental Interactions*,
https://doi.org/10.1007/978-3-032-27413-7_11

Fig. 11.1 Portrait of Enrico Fermi by Nobel foundation, 1938. Reprinted from Wikimedia. Public domain

A key character: *Enrico Fermi (1901–1954), Fig.* 11.1.
Enrico Fermi stands as the founder of the modern theory of weak interactions. Born in Rome into a middle-class family, Fermi revealed extraordinary talent at a very early age, mastering advanced mathematics and theoretical physics largely by self-study. After graduating from the Scuola Normale Superiore in Pisa, he rapidly emerged as one of the leading physicists of his generation. In the late 1920s he made foundational contributions to quantum statistics (Fermi-Dirac statistics) and to the theory of radiation, before turning his attention to nuclear physics and radioactive processes. In 1933–34 he formulated the first quantitative theory of β decay, introducing a universal interaction between four fermionic currents that explained nuclear transmutations and the continuous electron spectrum through Pauli's neutrino hypothesis. His theory, now known as Fermi theory, provided a unified description of nuclear beta decay, muon decay, and related processes, and established the constant G_F as the fundamental scale of weak interactions at low energy. Although non-renormalizable and destined to fail at high energies, the Fermi interaction correctly described all known weak processes at accessible energies and became the reference framework for several decades. Its later refinement into the $V-A$ structure after the discovery of parity violation preserved Fermi's basic current-current idea while embedding it in a chiral form. In 1938, following the introduction of racial laws in fascist Italy that directly threatened his Jewish wife Laura

Capon, Fermi emigrated to the United States. There he played a central role in the development of nuclear physics and in the Manhattan Project, leading the construction of the first controlled nuclear chain reaction in Chicago in 1942. After the war, he continued to shape American theoretical and experimental physics through his research and through an exceptional school of students. Despite his later fame in nuclear and particle physics, Fermi always regarded the theory of beta decay as one of his most elegant achievements. It represents the first coherent theoretical structure of weak interactions and the conceptual ancestor of the electroweak theory. Every modern description of weak processes, from W-boson exchange to precision electroweak tests and neutrino physics, can be traced back to the simple but profound idea that Fermi introduced: that weak interactions arise from universal couplings between fermionic currents, governed by a new fundamental scale of Nature.

11.1 The Fermi Theory of Weak Interactions

Weak interactions manifest themselves through processes characterized by remarkably long lifetimes when compared with electromagnetic or strong decays, typically in the range 10^{-13}–10^{-8} s. From an experimental viewpoint, three classes of reactions play a paradigmatic role in identifying the basic features of the weak force:

- nuclear beta decay,
- purely leptonic muon decay,
- leptonic decays of pseudoscalar mesons, in particular the pion.

Although these processes involve very different initial and final states, they share a common underlying mechanism. At energies well below the electroweak scale, all of them can be described in terms of a charged-current interaction mediated by the exchange of a heavy vector boson. In this low-energy regime, the interaction reduces to an effective contact coupling between fermionic currents.

What distinguishes the three cases is the nature of the charged current involved. Muon decay probes a purely leptonic current and therefore provides the cleanest environment to study the weak interaction, free from strong-interaction effects. Nuclear beta decay involves a hadronic charged current, whose matrix elements must be evaluated between bound nucleon or nuclear states. Finally, pion leptonic decay probes the axial component of the weak current and reveals distinctive chiral effects, most notably helicity suppression.

Despite these differences, the remarkable fact is that a single coupling constant, the Fermi constant G_F, governs all low-energy weak processes. This empirical universality is one of the strongest clues that weak interactions are mediated by a single fundamental force with a well-defined microscopic origin.

11.1.1 The Fermi Model of Beta Decay

Let us consider an electronic atomic transition,

$$A^\star \longrightarrow A + \gamma, \tag{11.1}$$

where an excited atom decays by emitting a photon.

This process appears similar to the one originally hypothesized at the dawn of beta-decay studies,

$$n \longrightarrow p + e^-. \tag{11.2}$$

If this decay were a two-body process, energy conservation would impose

$$m_n c^2 = \sqrt{\mathbf{p}_p^2 c^2 + m_p^2 c^4} + \sqrt{\mathbf{p}_e^2 c^2 + m_e^2 c^4}, \tag{11.3}$$

which uniquely fixes the kinetic energy of the emitted electron.

Driving exercise

Why a two-body beta decay would yield a discrete electron energy. Consider the hypothetical process

$$n \to p + e^-, \tag{11.4}$$

with the parent neutron taken at rest in the laboratory frame. Four-momentum conservation implies

$$p_n^\mu = p_p^\mu + p_e^\mu, \qquad p_n^\mu = (m_n c, \mathbf{0}). \tag{11.5}$$

Squaring Eq. (11.5) and using $p^2 = m^2c^2$ gives

$$m_n^2 c^2 = (p_p + p_e)^2 = m_p^2 c^2 + m_e^2 c^2 + 2\, p_p \cdot p_e. \tag{11.6}$$

In the neutron rest frame, the final-state 3-momenta are back-to-back, $\mathbf{p}_p = -\mathbf{p}_e$, and their magnitudes coincide, $|\mathbf{p}_p| = |\mathbf{p}_e| \equiv p$. The corresponding energies are

$$E_p = \sqrt{p^2c^2 + m_p^2 c^4}, \qquad E_e = \sqrt{p^2c^2 + m_e^2 c^4}. \tag{11.7}$$

Energy conservation reads

$$m_n c^2 = E_p + E_e. \tag{11.8}$$

Equations (11.7) and (11.8) contain a single unknown p, which is therefore *uniquely* determined by the masses. Equivalently, one may solve directly for the electron energy. Using the standard two-body kinematics (invariant-mass method), one finds

$$E_e^{(2)} = \frac{m_n^2 + m_e^2 - m_p^2}{2m_n} c^2, \qquad T_e^{(2)} = E_e^{(2)} - m_e c^2. \tag{11.9}$$

Thus, in a genuine two-body decay, the emitted electron would be mono-energetic: all events would populate the single value in Eq. (11.9), producing a line spectrum (schematically shown in Fig. 11.2).

Where can a continuum come from? A continuous spectrum requires at least one additional final-state degree of freedom that can carry a variable share of the available energy and momentum. In beta decay this role is played by the (anti)neutrino. In

$$n \to p + e^- + \bar{\nu}_e, \tag{11.10}$$

the electron kinetic energy T_e is no longer fixed by kinematics alone: it is distributed over an interval $0 \le T_e \le T_0$, where the endpoint T_0 is set by the Q-value (up to recoil corrections), yielding the observed continuous spectrum (schematically shown in Fig. 11.3).

In practice, the detailed spectral shape encodes both phase space and the dynamical structure of the weak current. Nevertheless, the *existence* of a continuum is already a kinematical statement: the two-body hypothesis in Eq. (11.4) is incompatible with the observed distribution of electron energies.

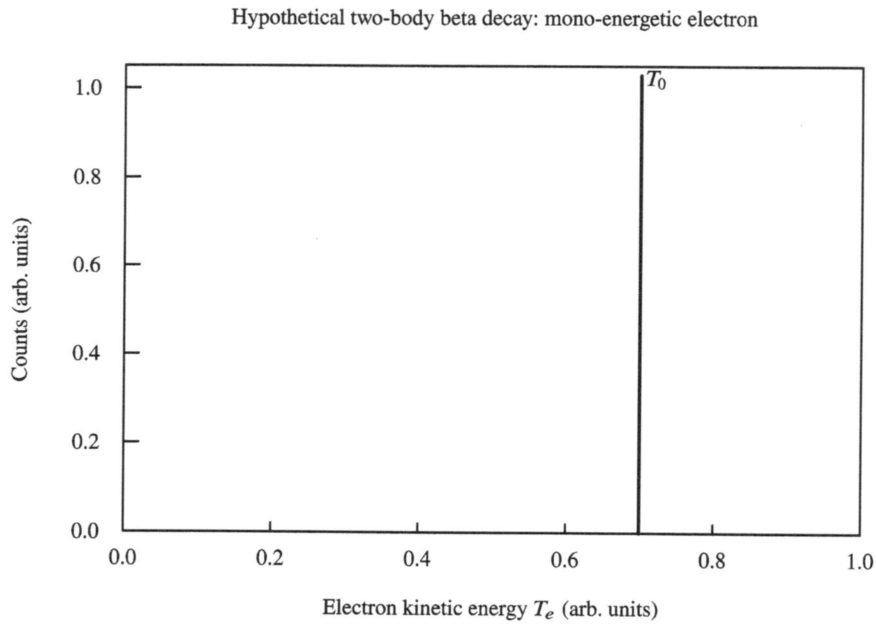

Fig. 11.2 Idealized electron spectrum for a hypothetical two-body decay $n \to p + e^-$. Energy-momentum conservation fixes a unique electron kinetic energy T_0 (up to tiny recoil corrections), producing a mono-energetic line

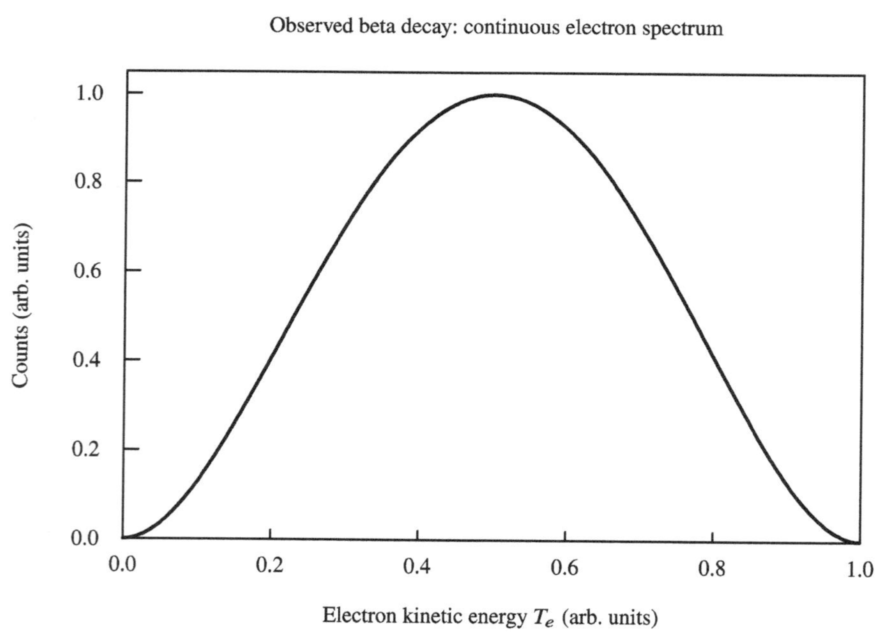

Fig. 11.3 Schematic continuous electron spectrum observed in nuclear beta decay. The endpoint T_0 is set by the available Q-value, while the continuum reflects the three-body kinematics of $n \to p + e^- + \bar{\nu}_e$ and the phase space available to the unobserved neutrino

11.1.2 Pauli's Neutrino Hypothesis

In 1930, Pauli proposed a solution to this puzzle by postulating the existence of a new neutral, very weakly interacting fermion with a small (possibly zero) mass, later named the *neutrino* by Fermi.[1] This hypothesis simultaneously resolved two fundamental problems:

- the continuous beta-decay energy spectrum;
- the so-called *wrong statistics problem*, in which a fermionic initial state appeared to decay into a final bosonic configuration.

With the introduction of the neutrino, the neutron beta decay becomes a three-body process, as shown in Eq. (11.10), allowing the electron energy to vary continuously.

At the nuclear level, beta decay is described by

$$N(A, Z) \longrightarrow N(A, Z+1) + e^- + \bar{\nu}_e, \tag{11.11}$$

which gives Eq. (11.10) at the subnuclear level.

Fermi further recognized that the interaction responsible for this process is neither electromagnetic nor strong, but rather a new, much weaker interaction acting at the nuclear scale. This interaction is now known as the *weak interaction*, intimately connected with neutrino emission.

Motivated by this analogy, Fermi constructed an interaction Lagrangian modeled after QED, involving a product of two currents: one responsible for inducing the transition of the initial state, and one responsible for the creation of the emitted particles. Thus one writes

$$\mathcal{L}_\gamma = -e\, J^\alpha_{\text{em}} A_\alpha, \qquad \text{(electromagnetic interaction)}, \tag{11.12a}$$

$$\mathcal{L}_F = -G_F\big[\bar{\psi}_p \gamma^\mu \psi_n\big]\big[\bar{\psi}_e \gamma_\mu \psi_\nu\big], \qquad \text{(Fermi theory of beta decay)}. \tag{11.12b}$$

Here G_F is the *Fermi constant*, and the interaction is written as a contact term involving a nucleonic current and a leptonic current. If the neutrino is taken to be a particle rather than an antiparticle, it must appear on the left-hand side of the bilinear in order to preserve all quantum numbers.

Once the neutrino was accepted as a necessary participant in beta decay, the weak interaction ceased to be interpreted as an idiosyncratic nuclear effect and started to be treated as a *fundamental* interaction acting on elementary fermions. The first quantitatively successful description was the Fermi theory, formulated as a local four-fermion interaction in which leptonic and hadronic currents couple at a point. In a modern normalization one writes a current-current Hamiltonian density

$$\mathcal{H}_{\text{wk}} = \frac{G_F}{\sqrt{2}} \mathcal{J}^\alpha \mathcal{J}^\dagger_\alpha, \tag{11.13}$$

[1] Fermi called it *neutrino*, since the word *neutron* (in Italian, *neutrone*) was already in use.

where G_F is the Fermi constant and the (charged) weak current has a leptonic and a hadronic part,

$$\mathcal{J}_\alpha = \bar{\psi}_e \gamma_\alpha (1 - \gamma_5) \psi_{\nu_e} + J_\alpha, \qquad J_\alpha = V_\alpha - A_\alpha. \tag{11.14}$$

Equation (11.14) encapsulates two fundamental facts extracted from experiments: the charged weak interaction is *chiral* (it couples to left-handed fermions), and the hadronic charged current decomposes into vector and axial-vector pieces. The corresponding selection rules in nuclear beta decay, as well as the observed pattern of muon decay and pion decay, are then captured by the $V - A$ structure.

Parity violation and the "odd man out" among forces. In the mid-1950s, the assumption that all fundamental interactions respect mirror symmetry (parity) was overturned in the weak sector. The τ-θ puzzle, where two apparently distinct particles of comparable mass ($\sim$ 500 MeV), charge and zero spin, decay into two- and three-pion final states of opposite parity, forced the conclusion that either the identification of the states was wrong or parity is not a good symmetry of weak interactions. The resolution is that there is a single particle (identified as the charged kaon $K^{\pm}$) and that weak interactions violate parity, a conclusion decisively confirmed by the asymmetric emission of electrons from polarized ^{60}Co nuclei.

This episode had consequences far beyond a single decay channel. It implied that "symmetry" could no longer be assumed as a universal property of Nature, but had to be diagnosed interaction by interaction, and sometimes current by current. In particular, the axial current in Eq. (11.14) is parity-odd and therefore cannot be the generator of an exact strong symmetry in any naive sense.

11.1.3 The V-A Generalization of the Fermi Theory

The neutron beta decay is described by the interaction

$$\mathcal{L}_F = -\frac{G_F}{\sqrt{2}} \left[\bar{\psi}_p \gamma^\mu \left(1 + \frac{g_A}{g_V} \gamma_5 \right) \psi_n \right] \left[\bar{\psi}_e \gamma_\mu (1 - \gamma_5) \psi_\nu \right], \tag{11.15}$$

where $G_F \simeq 1.166 \times 10^{-5}$ GeV^{-2}. Here, g_V and g_A are the vector and axial-vector weak coupling constants of the nucleon charged current. In particular, the nucleonic current $H^\mu = \bar{\psi}_p \gamma^\mu (1 + g_A/g_V \, \gamma_5) \psi_n$ describes the $n \to p$ transition, while $L_\mu = \bar{\psi}_e \gamma_\mu (1 - \gamma_5) \psi_\nu$ is the leptonic current.

Several fundamental features of weak interactions are encoded in this expression.

1. **V-A structure**: the weak interaction involves both vector ($\bar{\psi}\gamma^{\mu}\psi$) and axial-vector ($\bar{\psi}\gamma^{\mu}\gamma^{5}\psi$) currents.
2. **Parity violation**: the mixture of polar and axial currents implies that the interaction is not invariant under spatial inversion.
3. **Negative helicity of neutrinos**: the factor $(1-\gamma_5)$ projects onto left-handed components, reflecting the absence of right-handed neutrinos in observed weak processes.

Consider now muon beta decay,

$$\mu^{-} \longrightarrow \nu_{\mu} + e^{-} + \bar{\nu}_{e}, \tag{11.16}$$

which is described by

$$\mathcal{L}_{\mu} = -\frac{G^{(\mu)}}{\sqrt{2}}\left[\bar{\psi}_{\nu_{\mu}}\gamma^{\alpha}(1-\gamma_5)\psi_{\mu}\right]\left[\bar{\psi}_{e}\gamma_{\alpha}(1-\gamma_5)\psi_{\nu_e}\right] + \text{h.c.} \tag{11.17}$$

with $G^{(\mu)} \simeq G_F$.

Neglecting the electron mass and radiative corrections, the total decay rate is

$$\Gamma_{\mu} = \frac{G_F^2 m_{\mu}^5}{192\pi^3}, \qquad \tau_{\mu} = \Gamma_{\mu}^{-1}. \tag{11.18}$$

Two points are worth emphasizing.

- First, the characteristic scaling $\Gamma_{\mu} \propto m_{\mu}^5$ follows from dimensional analysis in a four-fermion theory: G_F has dimension $[\text{mass}]^{-2}$ and the three-body phase space contributes m_{μ}^5.
- Second, the chiral projector $(1-\gamma^5)$ implies maximal parity violation and fixes the angular-energy distributions (Michel spectrum) once muon polarization is included.

Pause and think

What do you argue from the aforementioned equality?

This remarkable equality suggests the *universality* of weak interactions: a single coupling constant governs all weak processes. One may therefore define a universal weak current

$$J_N^{\alpha} = g_V\, \bar{\psi}_p \gamma^{\alpha}\left(1 - \frac{g_A}{g_V}\gamma_5\right)\psi_n, \tag{11.19a}$$

$$J_{\mu}^{\alpha} = \bar{\psi}_{\nu_{\mu}}\gamma^{\alpha}(1-\gamma_5)\psi_{\mu}, \tag{11.19b}$$

$$J_e^{\alpha(+)} = \bar{\psi}_{e}\gamma^{\alpha}(1-\gamma_5)\psi_{\nu_e}, \tag{11.19c}$$

and write the universal current-current interaction as

$$\boxed{\mathcal{L}_{\text{weak}} = -\frac{G}{\sqrt{2}} J_W^{\alpha} J_{\alpha W}^{\dagger},} \tag{11.20}$$

with $J_W^{\alpha} = J_N^{\alpha} + J_{\mu}^{\alpha} + \cdots$.

This framework explains both weak nuclear processes and neutrino-matter interactions, and implies the conservation of separate lepton numbers,

$$N_e = N(e^-) + N(\nu_e) - N(e^+) - N(\bar{\nu}_e), \tag{11.21a}$$

$$N_{\mu} = N(\mu^-) + N(\nu_{\mu}) - N(\mu^+) - N(\bar{\nu}_{\mu}). \tag{11.21b}$$

Fermi's theory of nuclear β decay was the conceptual template for the pre-gauge description of weak interactions. In its mature form-the $V-A$ theory-one assumes that weak amplitudes at low momentum transfer are generated by *local* current-current operators that couple a leptonic charged current to a hadronic (or, more generally, fermionic) charged current. A paradigmatic example is muon decay, $\mu^- \to \nu_{\mu} + e^- + \bar{\nu}_e$, described by the effective Hamiltonian

$$\mathcal{H}_{\text{eff}}^{\mu\text{ decay}} = -\frac{G_{\mu}}{\sqrt{2}} \left[J_{\nu e}^{\rho(+)\dagger} J_{\rho,\nu\mu}^{(+)} + \text{h.c.} \right], \tag{11.22}$$

where the charged current is taken to be purely left-chiral,

$$J_{12}^{\rho(+)} = \bar{\psi}_1 \gamma^{\rho} (1-\gamma_5) \psi_2 \equiv J_{12}^{\rho\,\text{CC}}. \tag{11.23}$$

In complete analogy, neutron β decay is encoded by

$$\mathcal{H}_{\text{eff}}^{\beta\text{ decay}} = -\frac{G_F}{\sqrt{2}} \left[J_{e\nu}^{\mu(+)\dagger} J_{\mu,pn}^{(+)} + \text{h.c.} \right], \tag{11.24}$$

with the hadronic charged current parametrized as

$$J_{pn}^{\mu+} = \bar{\psi}_p \,(1 - 1.26\,\gamma_5)\, \gamma^{\mu} \psi_n, \tag{11.25}$$

so that the departure from a strictly conserved axial current is absorbed, at the hadronic level, into the phenomenological factor 1.26 multiplying γ_5.

Written in terms of the quark constituents of mesons and baryons, the empirical content of weak decays is captured by a *universal* four-fermion operator of the current-current type: one may regard the left panel of Fig. 11.4 as the generic transition $1 \to \bar{2} + 3 + 4$, while the right panel corresponds to the microscopic subprocess underlying neutron decay, $n(udd) \to p(uud) + e^- + \bar{\nu}_e$. The defining postulate of the $V-A$ theory is that only left-handed fermions participate in the charged weak interaction. Accordingly, the effective Hamiltonian can be cast in the compact form

$$\begin{aligned}\mathcal{H}_{\text{eff}}^{4\text{ fermion}} &= -\frac{G_\mu}{\sqrt{2}}\left[J_{24}^{\mu(+)\dagger}\, J_{\mu,31}^{(+)} + \text{h.c.}\right] \\ &= -4\,\frac{G_\mu}{\sqrt{2}}\left[\left(\bar{\psi}_{3L}\gamma^\mu\psi_{1L}\right)\left(\bar{\psi}_{4L}\gamma_\mu\psi_{2L}\right) + \text{h.c.}\right],\end{aligned} \tag{11.26}$$

where

$$\psi_L \equiv \frac{1}{2}(1-\gamma_5)\psi \tag{11.27}$$

is the left-chiral projection. The explicit appearance of ψ_L in Eq. (11.26) is not a notational convenience: it is the structural input that ultimately fixes the way fermion fields must be assigned to representations of the weak isospin group $SU(2)_L$ once the interaction is embedded into a renormalizable gauge theory.

It is instructive to contrast this with electromagnetism, where the interaction is organized in terms of the electromagnetic current

$$J_\mu^{\text{em}} = \bar{\psi}_L\gamma_\mu\psi_L + \bar{\psi}_R\gamma_\mu\psi_R, \tag{11.28}$$

coupled to the gauge field A_μ, with the corresponding vertex shown in the left panel of Fig. 11.5. Equation (11.26) suggests an analogous picture for the weak interaction: the charged current J_μ^+ may be viewed as coupled to a charged vector mediator W_μ^+, producing the elementary transition depicted in the right panel of Fig. 11.5. In this vertex a fermion f is converted into a fermion f' whose electric charge differs by one unit; when f is a strange quark, the same transition also carries $\Delta S = 1$. In that case the charged current induces a process with

$$\Delta S = \Delta Q = 1, \tag{11.29}$$

where S and Q denote strangeness and electric charge, respectively. Thus, neutron decay involves the quark current $J_{ud}^{\mu(+)}$, whereas the decay of a hyperon such as Λ is controlled by $J_{us}^{\mu(+)}$.

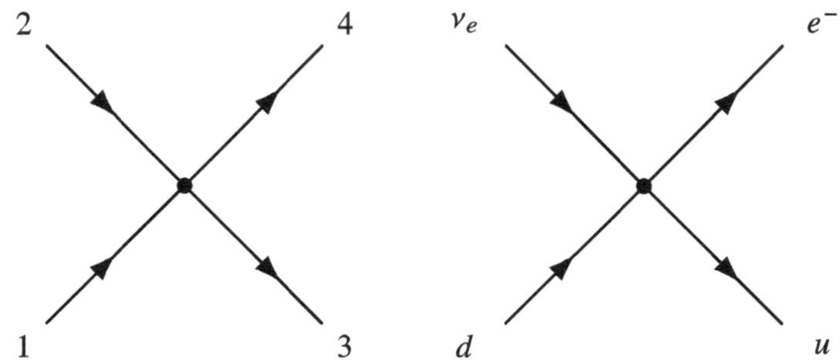

Fig. 11.4 Generic four-fermion interaction responsible for weak processes (left) and the basic diagram describing β decay (right)

In the four-fermion language, the overall strength of the interaction is encoded in the coefficients multiplying the current-current operators, which can be thought of as arising from a single underlying weak coupling, schematically denoted by g_{weak} in Fig. 11.5, once a mediator picture is adopted. Experimentally, the couplings extracted from muon decay and neutron β decay are close but not identical:

$$G_F \simeq 0.98\, G_\mu, \tag{11.30}$$

and the effective coefficient inferred from strangeness-changing hyperon decays is smaller still; for instance, one may summarize the Λ channel by

$$G_\Lambda \simeq 0.20\, G_\mu. \tag{11.31}$$

The near equality in Eq. (11.30) is historically significant: it suggested that the vector part of the hadronic current in Eqs. (11.24)–(11.25) is essentially unaffected by strong interactions and matches the vector current in purely leptonic transitions. This observation crystallized in the Conserved Vector Current (CVC) hypothesis. In discussing the mixing angle at the hadron level, one is effectively referring to the coefficient of this conserved vector component.

11.1.4 Pion Leptonic Decay and Helicity Suppression

Charged pion decay provides a complementary window on the weak interaction because the pion is a pseudoscalar meson, $\pi^+ = u\bar{d}$, and its leptonic decays probe the axial current in a particularly direct way:

$$\pi^+ \to \ell^+ + \nu_\ell \qquad (\ell = e, \mu)\,. \tag{11.32}$$

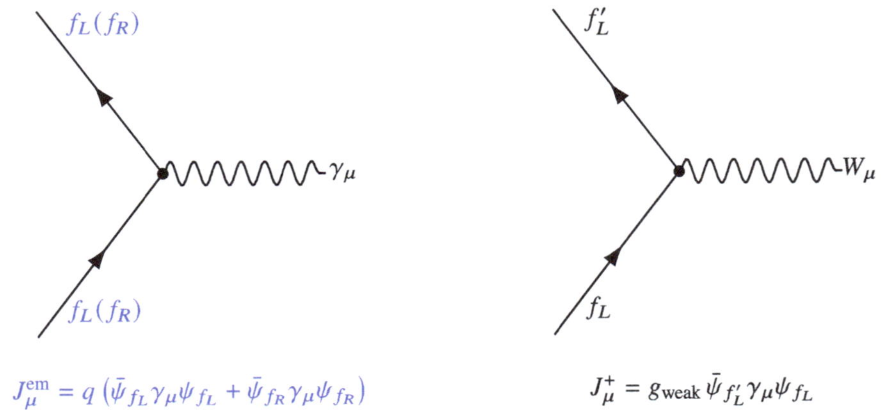

Fig. 11.5 The left panel shows the usual QED vertex depicting the interaction with the QED gauge boson γ_μ, and the right panel shows the generic vertex describing the universal weak interaction among quarks and leptons

The non-perturbative QCD dynamics is encapsulated in a single hadronic parameter, the pion decay constant f_π, defined by the axial current matrix element

$$\langle 0|\bar d\gamma^\mu\gamma^5 u|\pi^+(p)\rangle \;=\; i f_\pi\, p^\mu\,. \tag{11.33}$$

It is immediate to see that the amplitude is proportional to $\bar u_{\nu_\ell}(1-\gamma^5)v_\ell$ and therefore strongly constrained by angular momentum and chirality. Since the pion has spin zero, the final-state lepton and neutrino must carry opposite helicities. However, the weak interaction couples only to left-chiral neutrinos, which for ultrarelativistic particles essentially fixes the neutrino helicity. The charged lepton must then emerge predominantly with the "wrong" helicity, which is only possible through a chirality flip proportional to m_ℓ. This yields the celebrated *helicity suppression*

$$\Gamma(\pi\to\ell\nu)\;\propto\; m_\ell^2\,, \tag{11.34}$$

and explains why $\pi\to\mu\nu$ dominates over $\pi\to e\nu$ despite the larger phase space available to the electron channel. More explicitly, neglecting radiative corrections,

$$\Gamma(\pi\to\ell\nu)\;=\;\frac{G_F^2|V_{ud}|^2 f_\pi^2}{8\pi}\, m_\pi m_\ell^2\left(1-\frac{m_\ell^2}{m_\pi^2}\right)^2. \tag{11.35}$$

The ratio

$$\frac{\Gamma(\pi\to e\nu)}{\Gamma(\pi\to\mu\nu)}\;\simeq\;\left(\frac{m_e}{m_\mu}\right)^2\frac{\left(1-\frac{m_e^2}{m_\pi^2}\right)^2}{\left(1-\frac{m_\mu^2}{m_\pi^2}\right)^2} \tag{11.36}$$

is therefore an exceptionally clean test of the chiral structure of the weak interaction and of lepton universality.

11.1.5 Strange Hadrons and Quarks

A further empirical regularity that emerged in the study of weak interactions is related to processes involving strange hadrons: transitions with $\Delta S = 1$ are systematically suppressed relative to analogous non-strange processes with similar kinematics. In a purely hadronic language this fact was first accommodated by allowing the charged weak current to involve a rotation between strange and non-strange components. At the level of baryonic fields, one may parametrize this mixing via the Cabibbo current (we will discuss about that in Sect. 11.2):

$$J_\alpha \;\propto\; \bar{p}\gamma_\alpha(1-\gamma_5)\big(n\cos\theta + \Lambda\sin\theta\big), \tag{11.37}$$

where θ is the Cabibbo angle, numerically $\theta \simeq 13°$. Equation (11.37) summarizes in one line a wide class of data: strangeness-changing amplitudes are not governed by a different interaction; rather, they arise from the same charged current acting on a rotated hadronic state. The same rotation successfully accounts for kaon decays, turning what appeared as an ad hoc suppression factor into a symmetry-guided mixing rule.

This phenomenology immediately raises a structural question. Strong interactions produce strange particles copiously, often in associated pairs, yet their decays are slow and obey selection rules that single out a different interaction channel. Within a purely hadronic language, this can be encoded in current algebra and flavor quantum numbers, but it becomes conceptually transparent only once the hadronic spectrum is reinterpreted in terms of more elementary degrees of freedom. The quark hypothesis is therefore not an aesthetic simplification: it is a dynamical reorganization that explains at once the pattern of hadron multiplets, the algebraic structure of hadronic currents, and the very meaning of "strangeness-changing" transitions.

The minimal quark content required by the light-hadron spectrum is the $SU(3)$-flavor triplet

$$q = \begin{pmatrix} u \\ d \\ s \end{pmatrix}, \tag{11.38}$$

with electric charges $Q_u = +2/3$ and $Q_d = Q_s = -1/3$. Mesons arise as $q\bar{q}$ states and baryons as qqq states; in particular,

$$p \equiv uud, \qquad n \equiv udd, \tag{11.39}$$

while the strange baryons introduced earlier are those containing one or more s quarks. In this language "strangeness" is no longer a bookkeeping device: it is the statement that certain hadrons contain an s constituent.

From hadronic selection rules to microscopic flavor. Once hadrons are described as bound states of quarks, the suppression of $\Delta S = 1$ weak decays is no longer a separate dynamical assumption. It is the statement that the charged weak current mixes flavors, so that strange and non-strange components enter with different geometric weights.

At low energies, charged-current weak processes are described by the effective current-current interaction in Eq. (11.13). The hadronic charged current can be written schematically as

$$J_\mu^H = V_\mu - A_\mu, \tag{11.40}$$

and its physical content is encoded in matrix elements of J_μ^H between hadronic states. In the quark picture, these currents become quark bilinears with definite flavor structure. In the two-flavor limit, one has

$$V_\mu^{(ud)} = \bar{u}\gamma_\mu d, \qquad A_\mu^{(ud)} = \bar{u}\gamma_\mu\gamma_5 d, \tag{11.41}$$

so that the charged current takes the canonical chiral form

$$J_\mu^{(ud)} = \bar{u}\gamma_\mu(1-\gamma_5)d. \tag{11.42}$$

Equation (11.42) makes explicit what is only implicit in a hadronic description: the charged weak interaction changes flavor by raising a down-type quark to an up-type quark, in direct analogy with the leptonic transition $\nu_e \leftrightarrow e^-$. Because the interaction is purely left-chiral, it is natural to organize left-handed fermions into weak isospin doublets.

The novelty introduced by strange hadrons is the existence of weak decays with $\Delta S = 1$. In quark language, these correspond to the transition $s \to u$, described by

$$J_\mu^{(us)} = \bar{u}\gamma_\mu(1-\gamma_5)s. \tag{11.43}$$

If Eqs. (11.42) and (11.43) entered the weak Hamiltonian with equal weight, $\Delta S = 1$ and $\Delta S = 0$ rates would be comparable after phase-space corrections, contrary to observation. The empirical suppression is naturally implemented by allowing the charged current to couple to a rotated down-type state

$$d' = d\cos\theta_C + s\sin\theta_C, \tag{11.44}$$

so that the universal charged current becomes

$$J_\mu^H = \bar{u}\gamma_\mu(1-\gamma_5)d' = \cos\theta_C\,\bar{u}\gamma_\mu(1-\gamma_5)d + \sin\theta_C\,\bar{u}\gamma_\mu(1-\gamma_5)s. \tag{11.45}$$

Equation (11.45) is the quark-level counterpart of Eq. (11.37). A single coupling multiplies a single charged current, but the current involves a non-trivial rotation in flavor space.

Universality and suppression in one line. From Eq. (11.45) one finds schematically $\mathcal{A}_{\Delta S=0} \propto \cos\theta_C$ and $\mathcal{A}_{\Delta S=1} \propto \sin\theta_C$, so that at fixed kinematics

$$\frac{\Gamma_{\Delta S=1}}{\Gamma_{\Delta S=0}} \sim \tan^2\theta_C,$$

up to hadronic matrix elements and phase-space factors.

Inserting Eq. (11.45) into the low-energy Hamiltonian

$$\mathcal{H}_{\rm wk} = \frac{G_F}{\sqrt{2}}(\bar{\psi}_\ell \gamma^\mu (1-\gamma_5)\psi_{\nu_\ell}) J_\mu^H + \text{h.c.}, \tag{11.46}$$

one sees that all semileptonic hadron decays are governed by the same G_F and the same chiral structure; flavor dependence resides entirely in the current through the mixing angle θ_C. The remaining complication is dynamical: matrix elements of quark bilinears between hadronic states are non-perturbative, but the quark picture makes clear what must be computed and which symmetries can be exploited.

The quark description also provides a microscopic foundation for current algebra. Defining, for any flavor generator T^a, the vector and axial currents

$$V_\mu^a = \bar{q}\gamma_\mu T^a q, \qquad A_\mu^a = \bar{q}\gamma_\mu\gamma_5 T^a q, \tag{11.47}$$

one finds that these currents close under equal-time commutators, explaining why symmetry reasoning at the hadronic level was so successful even before the correct strong dynamics was known.

By the late 1960s the current-current theory in Eq. (11.13) was phenomenologically triumphant but conceptually incomplete. Being of dimension six, the operator in Eq. (11.13) renders the theory non-renormalizable, forcing it to break down above some scale Λ. Loop effects in strangeness-changing processes are extremely sensitive to Λ and require it to be only a few GeV to remain compatible with data. This pointed to new degrees of freedom not far above the hadronic scale.

The charm quark as a consistency requirement. Quark-lepton parallelism in weak interactions suggested that, besides (u, d, s), a fourth quark should exist to match the appearance of a second neutrino. Its deeper role is to enforce cancellations in loop-induced strangeness-changing amplitudes, the Glashow-Iliopoulos-Maiani (GIM) mechanism, preventing unacceptably large flavor-changing effects.

The discovery of charm in 1974 completed the minimal quark content required for this cancellation mechanism and pointed directly toward a unified, renormalizable gauge theory of weak interactions. The effective interaction in Eq. (11.13) is therefore not fundamental: it arises from integrating out heavy charged gauge bosons. For $q^2 \ll M_W^2$ one has

$$\frac{G_F}{\sqrt{2}} \sim \frac{g^2}{8M_W^2}, \tag{11.48}$$

so the apparent "weakness" of low-energy processes is a consequence of the heaviness of the mediators and of the resulting short range of the interaction, while it is not related to a small gauge coupling.

11.1.6 Selection Rules from Conserved Quantum Numbers

The observed pattern of allowed and forbidden processes in particle physics is largely dictated by the conservation of quantum numbers. These conservation laws act as *selection rules*: before any dynamical calculation is performed, one can often decide whether a transition is possible simply by checking which quantum numbers are conserved, violated, or only approximately preserved.

In practice, conservation laws fall into three broad classes:

- exact conservation laws, valid in all known interactions,
- approximate conservation laws, valid only in certain interaction regimes,
- accidentally conserved quantities, not protected by a fundamental symmetry of the Lagrangian.

The most fundamental selection rules follow from exact local or global symmetries of the underlying theory.

- **Electric charge** Q. Gauge invariance under $U(1)_{\text{em}}$ implies exact conservation of electric charge in all physical processes:

$$\sum Q_{\text{initial}} = \sum Q_{\text{final}}. \tag{11.49}$$

No decay or scattering violating this rule has ever been observed.

– **Baryon number** B. In all experimentally observed low-energy processes, baryon number is conserved:

$$\sum B_{\text{initial}} = \sum B_{\text{final}}, \tag{11.50}$$

with $B = 1$ for baryons, $B = -1$ for antibaryons, and $B = 0$ for leptons and gauge bosons. For example, neutron beta decay in Eq. (11.10) satisfies

$$1 \longrightarrow 1 + 0 + 0.$$

Although many extensions of the SM allow baryon-number violation, no such process (e.g. proton decay) has been experimentally confirmed.

– **Lepton numbers.** Each lepton family carries its own lepton number:

$$L_e, \quad L_\mu, \quad L_\tau.$$

In weak interactions at accessible energies, these numbers are conserved to extremely high accuracy. For instance, muon beta decay

$$\mu^- \longrightarrow e^- + \bar{\nu}_e + \nu_\mu \tag{11.51}$$

obeys separate conservation of electron and muon lepton numbers:

$$L_e : \ 0 \longrightarrow +1 + (-1) + 0, \qquad L_\mu : \ +1 \longrightarrow 0 + 0 + (+1).$$

Neutrino oscillations show that individual family numbers are not exact symmetries of Nature, but total lepton number $L = L_e + L_\mu + L_\tau$ remains conserved in all laboratory processes so far observed.

Other quantum numbers generate selection rules that are only approximate, being exact for strong and electromagnetic interactions but violated by weak processes.

– **Strangeness** S**, charm** C**, beauty** B'**, topness** T**.** These flavor quantum numbers are conserved in strong and electromagnetic interactions, but violated by charged weak currents. Thus, decays such as

$$K^+ \to \pi^0 + e^+ + \nu_e, \qquad \Lambda \to p + \pi^-$$

are allowed only because weak interactions permit $\Delta S = \pm 1$, while purely strong transitions always satisfy $\Delta S = 0$.

– **Isospin** I **and hypercharge** Y**.** These arise from approximate flavor symmetries of the strong interaction. They generate useful selection rules for hadronic reactions, but are broken by quark mass differences and by electromagnetic effects.

Discrete symmetries also impose powerful selection rules.

- **Parity** P. Conserved in strong and electromagnetic interactions, but maximally violated in weak interactions.
- **Charge conjugation** C. Also conserved by strong and electromagnetic forces, but violated by weak interactions.
- **Time reversal** T **and CP.** CP is violated in weak interactions involving strange and heavy quarks. Through the CPT theorem, this implies violation of T as well.

Finally, one fundamental principle transcends all specific models: the CPT theorem. Under very general assumptions of locality, Lorentz invariance and unitarity, any QFT must be invariant under the combined transformation CPT. As a consequence, particles and antiparticles have exactly equal masses and lifetimes, and any violation of C, P or T must conspire so that CPT remains exact.

11.2 The Cabibbo Theory of Quark Mixing

In the above sections, we introduced the concept of strange hadrons and the need for a further force, the strong force, that leads to introducing quarks and will represent the main subject of Chap. 12.

▶ **Pause and think**
Do you think that quarks may be or may not be affected by weak force too?

In Fermi's theory, the weak interaction is represented as a local four-fermion interaction, whose dynamics is entirely encoded in bilinear fermionic currents coupled through a universal constant.

Since quarks are fermionic fields with spin $1/2$, on the same footing as electrons and neutrinos, nothing in the structure of Fermi's theory prevents them from entering the weak currents. The theory does not distinguish between "leptonic" and "hadronic" matter at a fundamental level: it only requires the presence of fermionic degrees of freedom capable of forming Lorentz-covariant currents. Phrasing it differently, within Fermi's framework, quarks are subject to the weak force due to their fermionic nature, according to which they can participate in the construction of weak currents and hence in weak interactions, exactly as leptons do.

Hence, it is natural to study now some properties of quarks within the scheme of weak interactions, postponing a detailed treatment to Chap. 12.

In particular, let us start with quark mixing, i.e., the concept for which the eigenstates which diagonalize the charged-current weak interaction are not, in general, the same as the quark mass eigenstates. Historically, this idea emerged from

- the observed pattern of strangeness-changing weak decays and
- the empirical suppression of flavor-changing neutral currents (FCNC). In the SM, mixing originates from the diagonalisation of Yukawa mass matrices and is encoded in a unitary matrix, the CKM matrix.

Let us see this in detail. In the low-energy four-fermion description, semileptonic charged-current processes are governed by an interaction of the schematic form

$$\mathcal{L}_{\text{eff}} = -\frac{G_F}{\sqrt{2}} \left(\bar{\ell}\gamma_\mu(1-\gamma_5)\nu_\ell\right)\left(\bar{u}\gamma^\mu(1-\gamma_5)d'\right) + \text{h.c.}, \tag{11.52}$$

where d' denotes the down-type quark combination that couples to the up quark in the charged current. Phenomenology indicates that weak transitions changing strangeness (e.g. $s \to u$) are suppressed relative to strangeness-conserving ones (e.g. $d \to u$), but the interaction remains universal in the sense that a single G_F controls leptonic and semileptonic decays.

Cabibbo's proposal is that the hadronic charged current involves a rotated down-type field,

$$d' = d\cos\theta_C + s\sin\theta_C, \qquad s' = -d\sin\theta_C + s\cos\theta_C, \tag{11.53}$$

with a single angle θ_C (the Cabibbo angle). Then $\Delta S = 0$ transitions are proportional to $\cos\theta_C$, whereas $|\Delta S| = 1$ transitions are proportional to $\sin\theta_C$.

▶ Driving exercise

Charged current in the two-generation (Cabibbo) form.

Let u be the up-type field and d, s the down-type fields. In the Cabibbo theory, the charged current is

$$J^{\mu(+)} = \bar{u}_L\gamma^\mu d'_L = \bar{u}_L\gamma^\mu\left(d_L\cos\theta_C + s_L\sin\theta_C\right). \tag{11.54}$$

Equivalently, in matrix notation for the down-sector doublet $D_L \equiv (d_L, s_L)^T$,

$$d'_L = (V_C D_L)_1, \qquad V_C = \begin{pmatrix} \cos\theta_C & \sin\theta_C \\ -\sin\theta_C & \cos\theta_C \end{pmatrix}, \tag{11.55}$$

with V_C a real orthogonal matrix, hence unitary.

An immediate consequence is the relative strength of the $\Delta S = 0$ and $|\Delta S| = 1$ transitions.
At the amplitude level, replacing $d_L \to d'_L$ implies

$$\mathcal{A}(u \to d) \propto \cos\theta_C, \qquad \mathcal{A}(u \to s) \propto \sin\theta_C, \tag{11.56}$$

so rates scale as

$$\Gamma(u \to d) \propto \cos^2\theta_C, \qquad \Gamma(u \to s) \propto \sin^2\theta_C. \tag{11.57}$$

This compactly explains why strangeness-changing semileptonic decays are systematically suppressed while preserving the universality of G_F.

It is important to notice that unitarity here matters. Indeed, the rotation in Eq. (11.53) is unitary. This is not a cosmetic choice: unitarity ensures that the total charged-current strength in the down sector is conserved,

$$|d'|^2 + |s'|^2 = |d|^2 + |s|^2, \tag{11.58}$$

and guarantees that the weak interaction remains properly normalised when expressed in the rotated basis.

11.2.1 The Glashow-Iliopoulos-Maiani Mechanism

If one attempts to extend the charged-current picture beyond the lowest-energy Fermi theory without care, one generically predicts large FCNC processes such as:

$$K^0 \to \mu^+ + \mu^-, \qquad K^0 - \bar{K}^0 \text{ mixing}, \qquad s \to d + \ell^+ + \ell^-. \tag{11.59}$$

These processes are in conflict with experimental suppressions. In order to solve this issue, one can deal with the GIM mechanism that states

FCNC are absent at tree level and are strongly suppressed at loop level by a combination of unitarity and mass degeneracy breaking.

▶ **Pause and think**
How do you think to realize this suppression?

The idea is to introduce a new quark, dubbed *charm quark*. It allows to restore the internal consistency of weak interactions involving hadrons. Indeed, in the original formulation with only three quarks (u, d, s), the structure of weak currents led to the appearance of unobserved strangeness-changing neutral currents, in contradiction with experiment.

We saw, in fact, that in Fermi's weak interactions, hadronic currents are built from quark bilinears, but if only u, d, s are present, the charged weak current necessarily mixes d and s in a way that also generates neutral currents changing strangeness. As we claimed above, such processes are strongly suppressed in Nature.

The introduction of a fourth quark, the charm quark c, allows the quarks to be arranged into two weak isospin doublets, (u, d) and (c, s). *The weak charged current then couples only members of the same doublet, while the structure of the neutral current becomes diagonal in flavor.*

▶ **Driving exercise**
Two quark doublets and the structure of the neutral current.
Introduce a second up-type quark c (charm) and assemble left-handed fields into weak doublets

$$Q_{1L} = \begin{pmatrix} u \\ d' \end{pmatrix}_L , \qquad Q_{2L} = \begin{pmatrix} c \\ s' \end{pmatrix}_L , \tag{11.60}$$

where (d', s') are the Cabibbo-rotated fields, Eqs. (11.53). The left-handed neutral current associated with the third generator of $SU(2)_L$ is (up to the coupling factor)

$$J_3^\mu = \sum_{i=1}^{2} \bar{Q}_{iL} \gamma^\mu \frac{\tau^3}{2} Q_{iL} = \frac{1}{2} \left(\bar{u}_L \gamma^\mu u_L - \bar{d}'_L \gamma^\mu d'_L + \bar{c}_L \gamma^\mu c_L - \bar{s}'_L \gamma^\mu s'_L \right). \tag{11.61}$$

Now insert the inverse Cabibbo rotation

$$\begin{pmatrix} d \\ s \end{pmatrix}_L = \begin{pmatrix} \cos\theta_C & -\sin\theta_C \\ \sin\theta_C & \cos\theta_C \end{pmatrix} \begin{pmatrix} d' \\ s' \end{pmatrix}_L , \tag{11.62}$$

and compute the bilinears. Because the rotation is unitary, one has the exact identity

$$\bar{d}'_L \gamma^\mu d'_L + \bar{s}'_L \gamma^\mu s'_L = \bar{d}_L \gamma^\mu d_L + \bar{s}_L \gamma^\mu s_L . \tag{11.63}$$

Importantly, the cross terms $\bar{d}_L\gamma^\mu s_L$ cancel between $\bar{d}'\gamma^\mu d'$ and $\bar{s}'\gamma^\mu s'$. Hence, expressed in the mass basis (d, s), the neutral current remains flavor diagonal:

$$J_3^\mu = \frac{1}{2}\left(\bar{u}_L\gamma^\mu u_L + \bar{c}_L\gamma^\mu c_L - \bar{d}_L\gamma^\mu d_L - \bar{s}_L\gamma^\mu s_L\right). \tag{11.64}$$

This is the tree-level core of the GIM mechanism:

> no flavor-changing neutral current arises from the $SU(2)_L$ neutral current when the quarks form complete weak doublets and the mixing matrix is unitary.

11.2.2 Loop-Level Suppression: Unitarity Cancellation and Mass Splittings

FCNC do occur beyond tree level through loop diagrams such as box and penguin topologies. The key point is that the amplitude involves a sum over internal up-type quarks weighted by CKM factors. In the two-generation case, a typical structure for an $s \to d$ FCNC amplitude is

$$\mathcal{A}_{s\to d}^{\text{FCNC}} \propto \sum_{i=u,c} V_{is}^* V_{id}\, F(m_i^2), \tag{11.65}$$

where $F(m_i^2)$ is a loop function depending on the internal-quark mass. Using the Cabibbo matrix, Eq. (11.55), one finds explicitly

$$V_{us}^* V_{ud} = (\sin\theta_C)(\cos\theta_C), \qquad V_{cs}^* V_{cd} = (\cos\theta_C)(-\sin\theta_C), \tag{11.66}$$

so

$$V_{us}^* V_{ud} + V_{cs}^* V_{cd} = 0. \tag{11.67}$$

Substituting this into Eq. (11.65):

$$\begin{aligned}\sum_{i=u,c} V_{is}^* V_{id} F(m_i^2) &= V_{us}^* V_{ud} F(m_u^2) + V_{cs}^* V_{cd} F(m_c^2) \\ &= V_{us}^* V_{ud}\left[F(m_u^2) - F(m_c^2)\right].\end{aligned} \tag{11.68}$$

Thus the amplitude vanishes if the loop function is mass-independent, and it is suppressed by the *difference* of loop functions when masses are unequal. In practice $m_c \gg m_u$, hence $F(m_u^2)$ can often be neglected relative to $F(m_c^2)$, but the crucial feature remains: the amplitude is proportional to a unitarity factor times a mass-splitting effect. This provides a natural suppression of FCNC amplitudes with respect to naive expectations.

11.3 CP Violation and the Kobayashi-Maskawa Theory

With n generations, the most general charged current for quarks is

$$\mathcal{L}_{\text{CC}} = -\frac{g}{\sqrt{2}} \left(\bar{u}_{Li}\gamma^\mu V_{ij} d_{Lj}\right) W^+_\mu + \text{h.c.}, \tag{11.69}$$

where $u_{Li} = (u_L, c_L, t_L, \dots)$, $d_{Lj} = (d_L, s_L, b_L, \dots)$, and V is a unitary matrix. In the SM, V arises from diagonalising the Yukawa mass matrices. Briefly, the quark Yukawa Lagrangian is

$$\mathcal{L}_Y = -\bar{Q}_L Y_u \tilde{\Phi}\, u_R - \bar{Q}_L Y_d \Phi\, d_R + \text{h.c.}, \tag{11.70}$$

and after electroweak symmetry breaking $\Phi \to (0, v/\sqrt{2})^T$ yields mass matrices

$$M_u = \frac{v}{\sqrt{2}} Y_u, \qquad M_d = \frac{v}{\sqrt{2}} Y_d. \tag{11.71}$$

These are diagonalised via bi-unitary transformations:

$$U^\dagger_{uL} M_u U_{uR} = \text{diag}(m_u, m_c, m_t), \qquad U^\dagger_{dL} M_d U_{dR} = \text{diag}(m_d, m_s, m_b), \tag{11.72}$$

and the CKM matrix is

$$V = U^\dagger_{uL} U_{dL}. \tag{11.73}$$

▶ **Driving exercise**

Counting physical parameters and why three generations are special.
An $n \times n$ unitary matrix contains n^2 real parameters. It can be written as $n(n-1)/2$ mixing angles and $n(n+1)/2$ phases. However, not all phases are physical because one may rephase quark fields without changing physics. Under

$$u_{Li} \to e^{i\alpha_i} u_{Li}, \qquad d_{Lj} \to e^{i\beta_j} d_{Lj}, \tag{11.74}$$

the matrix elements transform as

$$V_{ij} \to e^{-i\alpha_i}\, V_{ij}\, e^{i\beta_j}. \tag{11.75}$$

There are $2n$ arbitrary phases α_i, β_j, but a common overall phase corresponds to baryon-number conservation and leaves V invariant. Hence $2n-1$ phases can be removed. The number of *physical* phases is

$$\#(\text{phases})_{\text{phys}} = \frac{n(n+1)}{2} - (2n-1) = \frac{(n-1)(n-2)}{2}. \tag{11.76}$$

For $n = 2$, $\#(\text{phases})_{\text{phys}} = 0$: the mixing matrix can be chosen real (Cabibbo case), so there is no CP violation from mixing. For $n = 3$, $\#(\text{phases})_{\text{phys}} = 1$: one irreducible complex phase remains, which can generate CP violation. This is the Kobayashi-Maskawa observation.

A standard and convenient parameterisation of the CKM matrix for $n = 3$ uses three angles $\theta_{12}, \theta_{23}, \theta_{13}$ and one phase δ:

$$V = \begin{pmatrix} c_{12}c_{13} & s_{12}c_{13} & s_{13}e^{-i\delta} \\ -s_{12}c_{23} - c_{12}s_{23}s_{13}e^{i\delta} & c_{12}c_{23} - s_{12}s_{23}s_{13}e^{i\delta} & s_{23}c_{13} \\ s_{12}s_{23} - c_{12}c_{23}s_{13}e^{i\delta} & -c_{12}s_{23} - s_{12}c_{23}s_{13}e^{i\delta} & c_{23}c_{13} \end{pmatrix}, \quad (11.77)$$

with $c_{ij} \equiv \cos\theta_{ij}, s_{ij} \equiv \sin\theta_{ij}$. In the limit $s_{13} \to 0, \delta$ drops out and the matrix reduces to a real two-angle rotation, illustrating that all CP-violating effects vanish when any one mixing angle is zero.

11.3.1 The Jarlskog Invariant

Physical CP violation requires a quantity invariant under field rephasings. The simplest such invariant is an imaginary part of the product of V components, defined by

$$J \equiv \Im\Big(V_{ij} V_{kl} V_{il}^* V_{kj}^*\Big), \qquad (i \neq k,\ j \neq l), \quad (11.78)$$

which is independent of the chosen distinct indices (up to sign) when V is 3×3 and unitary. In the standard parameterisation (11.77), one finds explicitly

$$J = c_{12}c_{23}c_{13}^2\, s_{12}s_{23}s_{13}\, \sin\delta. \quad (11.79)$$

This formula makes the necessary conditions for CP violation manifest: $\sin\delta \neq 0$ and all three angles nonzero.

► **Driving exercise**

Unitarity triangles and the geometric interpretation.

Unitarity of V implies orthogonality relations between different rows and columns. For instance, orthogonality of the first and third columns gives

$$V_{ud} V_{ub}^* + V_{cd} V_{cb}^* + V_{td} V_{tb}^* = 0. \quad (11.80)$$

Each term is a complex number. Equation (11.80) is a triangle closure condition in the complex plane. The area of any unitarity triangle equals $|J|/2$. Indeed, for complex vectors a, b, c with $a + b + c = 0$, the triangle area is

$$\mathcal{A} = \frac{1}{2}\, |\Im(ab^*)|, \quad (11.81)$$

and choosing $a = V_{ud} V_{ub}^*$, $b = V_{cd} V_{cb}^*$ yields $\mathcal{A} = |J|/2$ by Eq.(11.78). A nonzero triangle area is thus equivalent to CP violation.

11.3.2 GIM with Three Generations: The General FCNC Sum Rule

The GIM cancellation generalises straightforwardly to three generations. For down-type FCNC $d_j \to d_k$ with $j \neq k$, the relevant loop structure involves

$$\sum_{i=u,c,t} V_{ij}^* V_{ik} \, F(m_i^2). \tag{11.82}$$

By unitarity,

$$\sum_{i=u,c,t} V_{ij}^* V_{ik} = 0 \qquad (j \neq k), \tag{11.83}$$

so one may rewrite Eq. (11.82) by subtracting an arbitrary reference value, e.g. $F(m_u^2)$,

$$\sum_{i=u,c,t} V_{ij}^* V_{ik} \, F(m_i^2) = \sum_{i=u,c,t} V_{ij}^* V_{ik} \left[F(m_i^2) - F(m_u^2) \right], \tag{11.84}$$

making it explicit that only mass splittings contribute. This form is often the most transparent starting point for estimating FCNC amplitudes.

Summing up:

- Cabibbo mixing, Eq. (11.53), explains the systematic suppression of $|\Delta S| = 1$ charged-current transitions with a single parameter while preserving universality.
- The GIM mechanism in Eq. (11.64) shows that, once quarks are arranged into complete weak doublets and the mixing is unitary, neutral currents remain flavor-diagonal at tree level and FCNC are loop-suppressed via unitarity cancellations (11.68).
- For three generations, a single irreducible phase remains after field rephasings, Eq. (11.76), yielding CP violation quantified by the invariant J in Eq. (11.79) and geometrically by the area of unitarity triangles.

11.3.3 Violating CP in Action: Neutral Kaon Oscillations

Neutral-kaon oscillations provide the archetypal realization of flavor mixing in a *decaying* two-level system, and constitute the dynamical backbone of the classic discovery of CP violation. The strong interaction eigenstates

$$K^0 \equiv |d\bar{s}\rangle \qquad \text{and} \qquad \bar{K}^0 \equiv |\bar{d}s\rangle \tag{11.85}$$

carry strangeness $S = \pm 1$, which is conserved by QCD and QED but violated by the charged-current weak interaction. Therefore, once weak radiative effects are included, the flavor basis $\{|K^0\rangle, |\bar{K}^0\rangle\}$ is not a basis of propagation eigenstates.

At energies well below the electroweak scale, the $\Delta S = 2$ transition responsible for $K^0 \leftrightarrow \bar{K}^0$ mixing is described by an effective local interaction generated at second order in the weak couplings (box topologies). Operationally, the time evolution of the unstable two-state system is encoded in a non-Hermitian effective Hamiltonian acting on the flavor doublet,

$$i\,\frac{d}{dt}\begin{pmatrix}|K^0(t)\rangle\\ |\bar{K}^0(t)\rangle\end{pmatrix} = \hat{H}_{\rm eff}\begin{pmatrix}|K^0(t)\rangle\\ |\bar{K}^0(t)\rangle\end{pmatrix}, \qquad \hat{H}_{\rm eff} \equiv \hat{M} - \frac{i}{2}\hat{\Gamma}, \tag{11.86}$$

where $\hat{M}$ and $\hat{\Gamma}$ are 2×2 Hermitian matrices. In the flavor basis,

$$\hat{H}_{\rm eff} = \begin{pmatrix} M_{11} - \frac{i}{2}\Gamma_{11} & M_{12} - \frac{i}{2}\Gamma_{12} \\ M_{12}^* - \frac{i}{2}\Gamma_{12}^* & M_{22} - \frac{i}{2}\Gamma_{22} \end{pmatrix}. \tag{11.87}$$

The off-diagonal elements M_{12} and Γ_{12} quantify, respectively, the dispersive and absorptive parts of the $\Delta S = 2$ transition amplitude; they are precisely the quantities responsible for oscillations.

Oscillations in an open quantum system. Unlike neutrino oscillations in vacuum, the neutral-kaon system is an *open* system: the time evolution is not unitary because kaons decay. Equation (11.86) is the minimal effective description consistent with causality and with the observed exponential decay law, and it naturally predicts both a mass splitting and a lifetime splitting between propagation eigenstates.

The physical states K_S and K_L are defined as right-eigenvectors of $\hat{H}_{\rm eff}$,

$$\hat{H}_{\rm eff}\,|K_{S,L}\rangle = \lambda_{S,L}\,|K_{S,L}\rangle, \qquad \lambda_{S,L} \equiv m_{S,L} - \frac{i}{2}\Gamma_{S,L}, \tag{11.88}$$

with $m_{S,L}$ and $\Gamma_{S,L}$ denoting their masses and decay widths. It is convenient to parameterize the eigenstates in the flavor basis as

$$|K_S\rangle = p\,|K^0\rangle + q\,|\bar{K}^0\rangle, \tag{11.89a}$$

$$|K_L\rangle = p\,|K^0\rangle - q\,|\bar{K}^0\rangle, \tag{11.89b}$$

with normalization $|p|^2 + |q|^2 = 1$. The ratio q/p is fixed by the eigenvalue problem associated with Eq. (11.87) and reads

$$\left(\frac{q}{p}\right)^2 = \frac{M_{12}^* - \frac{i}{2}\Gamma_{12}^*}{M_{12} - \frac{i}{2}\Gamma_{12}}. \tag{11.90}$$

In the limit of exact CP invariance one has $q/p = 1$ (up to a convention), so that K_S and K_L become states of definite CP. As discussed previously, the observed deviation $q/p \neq 1$ is the clean dynamical imprint of CP violation in mixing.

Given Eq. (11.88), the time evolution of the propagation eigenstates is simply

$$|K_{S,L}(t)\rangle = e^{-i\lambda_{S,L}t}\,|K_{S,L}\rangle. \tag{11.91}$$

A state produced as a pure flavor eigenstate is then a coherent superposition of K_S and K_L. Using Eqs. (11.89) and (11.91), one obtains

$$|K^0(t)\rangle = g_+(t)\,|K^0\rangle + \frac{q}{p}\,g_-(t)\,|\bar{K}^0\rangle, \tag{11.92a}$$

$$|\bar{K}^0(t)\rangle = g_+(t)\,|\bar{K}^0\rangle + \frac{p}{q}\,g_-(t)\,|K^0\rangle, \tag{11.92b}$$

where

$$g_\pm(t) \equiv \frac{1}{2}\left(e^{-i\lambda_S t} \pm e^{-i\lambda_L t}\right). \tag{11.93}$$

Equations (11.92) already display the key physical points:

- (i) flavor conversion is controlled by $g_-(t)$ and therefore by the interference between the two eigenmodes;
- (ii) CP violation in mixing enters only through q/p and p/q.

In order to show the oscillatory behaviour explicitly, let us introduce the average and difference parameters

$$m \equiv \frac{m_L + m_S}{2}, \quad \Gamma \equiv \frac{\Gamma_L + \Gamma_S}{2}, \quad \Delta m \equiv m_L - m_S, \quad \Delta\Gamma \equiv \Gamma_L - \Gamma_S, \tag{11.94}$$

so that

$$g_\pm(t) = \frac{1}{2}e^{-imt}\,e^{-\Gamma t/2}\left[e^{-\frac{\Delta\Gamma}{4}t}\,e^{-i\frac{\Delta m}{2}t} \pm e^{+\frac{\Delta\Gamma}{4}t}\,e^{+i\frac{\Delta m}{2}t}\right]. \tag{11.95}$$

If one initially neglects CP violation in mixing ($q/p \simeq 1$), the transition probabilities extracted from Eq. (11.92) take the compact form

$$P\big(K^0 \to K^0; t\big) = |g_+(t)|^2 = \frac{1}{4}\big(e^{-\Gamma_S t} + e^{-\Gamma_L t} + 2e^{-\Gamma t}\cos(\Delta m\, t)\big), \quad (11.96a)$$

$$P\big(K^0 \to \bar{K}^0; t\big) = |g_-(t)|^2 = \frac{1}{4}\big(e^{-\Gamma_S t} + e^{-\Gamma_L t} - 2e^{-\Gamma t}\cos(\Delta m\, t)\big), \quad (11.96b)$$

where the interference term oscillates with angular frequency Δm. The exponential factors encode the fact that coherence is progressively lost, since K_S decays rapidly while K_L survives on much longer time scales. Retaining CP violation in mixing simply multiplies the conversion probabilities by $|q/p|^2$ or $|p/q|^2$, as shown by Eq. (11.92).

For later reference, it is useful to keep in mind the hierarchy

$$\Gamma_S \gg \Gamma_L, \quad (11.97)$$

which implies that kaons produced at $t = 0$ rapidly evolve into a long-lived beam dominated by K_L. It is precisely in this regime that the famous forbidden decay $K_L \to \pi\pi$ becomes experimentally accessible and provides a clean probe of CP violation: K_L is predominantly CP-odd, so a nonzero amplitude to a CP-even two-pion final state requires CP-breaking components in the propagating state and/or in the decay amplitude.

11.3.4 Origin of Quark Mass and Mixing

In the electroweak theory, quarks appear in three generations with definite weak isospin assignments,

$$Q_i = \begin{pmatrix} U_i \\ D_i \end{pmatrix}_L, \qquad U_{R,i} \in \{u_R, c_R, t_R\}, \qquad D_{R,i} \in \{d_R, s_R, b_R\}, \qquad i = 1, 2, 3, \quad (11.98)$$

where Q_i are left-handed $SU(2)_L$ doublets, while $U_{R,i}$ and $D_{R,i}$ are right-handed $SU(2)_L$ singlets. The colour index is understood (each field is a colour triplet), and will be suppressed throughout.

Yukawa invariants

Gauge-invariant fermion mass terms cannot be written at the level of the unbroken $SU(2)_L \otimes U(1)_Y$ symmetry; instead, masses originate from Yukawa interactions with the Higgs doublet ϕ^a ($a = 1, 2$). For fixed family labels i, j, one can form two independent invariants (one for down-type and one for up-type quarks),

$$\mathcal{L}_{ij} = g^D_{ij}\, \bar{Q}_i\, \phi\, D_{R,j} \; + \; g^U_{ij}\, \varepsilon_{ab}\, \bar{U}_{R,i}\, Q^a_j\, \phi^b \; + \; \text{h.c.}, \qquad i, j = 1, 2, 3, \quad (11.99)$$

and the full quark-Higgs Yukawa Lagrangian is

$$\mathcal{L}_{qH} = \sum_{i,j} \mathcal{L}_{ij}. \tag{11.100}$$

Here ε_{ab} is the antisymmetric tensor ($\varepsilon_{12} = +1$), which implements the contraction of $SU(2)_L$ indices in the up-type invariant. The couplings $g_{ij}^{U,D}$ are *a priori* arbitrary complex numbers; hermiticity is ensured only after adding the h.c. term, and complex phases in $g_{ij}^{U,D}$ provide a source of CP (and T) violation, while CPT remains exact.

Spontaneous symmetry breaking and mass matrices

In the unitary gauge, the Higgs doublet can be written as

$$\phi(x) = \frac{1}{\sqrt{2}} \begin{pmatrix} 0 \\ v + h(x) \end{pmatrix}, \qquad \langle \phi \rangle = \frac{1}{\sqrt{2}} \begin{pmatrix} 0 \\ v \end{pmatrix}, \tag{11.101}$$

so that the Yukawa interactions generate Dirac mass terms when ϕ is replaced by its vacuum expectation value. Concretely,

$$g_{ij}^D\, \bar{Q}_i\, \phi\, D_{R,j} \;\longrightarrow\; g_{ij}^D\, \bar{D}_{L,i}\, \frac{v}{\sqrt{2}}\, D_{R,j} + \cdots, \tag{11.102}$$

$$g_{ij}^U\, \varepsilon_{ab}\, \bar{U}_{R,i}\, Q_j^a\, \phi^b \;\longrightarrow\; g_{ij}^U\, \bar{U}_{R,i}\, \frac{v}{\sqrt{2}}\, U_{L,j} + \cdots, \tag{11.103}$$

where the ellipses denote Higgs-fermion interactions proportional to $h(x)$. It is convenient (and equivalent to the notation in the text) to define

$$\eta \equiv \frac{v}{\sqrt{2}}, \tag{11.104}$$

so that the quark mass Lagrangian becomes

$$\mathcal{L}_{qm} = \bar{D}_L\, M_d\, D_R + \bar{U}_R\, M_u\, U_L + \text{h.c.}, \qquad (M_d)_{ij} = g_{ij}^D\, \eta, \qquad (M_u)_{ij} = g_{ij}^U\, \eta. \tag{11.105}$$

In general, M_u and M_d are complex, non-diagonal 3×3 matrices in family space. This is the origin of the mismatch between the *electroweak basis* (fields with definite weak isospin) and the *physical basis* (fields with definite masses).

Bi-unitary diagonalisation

Any complex matrix M admits a singular-value decomposition (bi-unitary diagonalisation): there exist unitary matrices L and R such that

$$M = L^\dagger\, m\, R, \qquad m = \text{diag}(m_1, m_2, m_3), \qquad m_i \geq 0. \tag{11.106}$$

Applying this to the quark mass matrices, one writes

$$M_u = W^\dagger\, m_u\, Z, \qquad M_d = U^\dagger\, m_d\, V, \tag{11.107}$$

with $m_u = \text{diag}(m_u, m_c, m_t)$ and $m_d = \text{diag}(m_d, m_s, m_b)$. Substituting Eq. (11.107) into Eq. (11.105) yields

$$\mathcal{L}_{qm} = \bar{D}_L\, U^\dagger m_d V\, D_R + \bar{U}_R\, W^\dagger m_u Z\, U_L + \text{h.c.} \tag{11.108}$$

– The transformations

$$U_R \;\to\; W\, U_R, \qquad D_R \;\to\; V\, D_R, \tag{11.109}$$

are mere redefinitions of *right-handed singlets*. Since U_R and D_R do not participate in $SU(2)_L$ doublets, these rotations do not disturb the $SU(2)_L \otimes U(1)_Y$ gauge structure.

– One may also redefine the *left-handed doublets* by a common unitary rotation in family space,

$$Q = \begin{pmatrix} U_L \\ D_L \end{pmatrix} \;\to\; Z\, Q, \qquad \text{i.e.} \qquad U_L \to Z\, U_L, \quad D_L \to Z\, D_L, \tag{11.110}$$

which again preserves the electroweak gauge couplings because it acts identically on the two components of each $SU(2)_L$ doublet.

Implementing Eqs. (11.109)–(11.110) in Eq. (11.108), the up-type mass term becomes diagonal,

$$\bar{U}_R\, W^\dagger m_u Z\, U_L \;\longrightarrow\; \bar{U}_R\, m_u\, U_L, \tag{11.111}$$

whereas the down-type term retains a residual unitary matrix,

$$\bar{D}_L\, U^\dagger m_d V\, D_R \;\longrightarrow\; \bar{D}_L\, \left(Z U^\dagger\right) m_d\, D_R. \tag{11.112}$$

It is therefore natural to define the *misalignment matrix*

$$U_{\text{CKM}} \equiv Z U^\dagger, \tag{11.113}$$

so that, in the rotated electroweak basis,

$$\mathcal{L}_{qm} = \bar{D}_L\, U_{\text{CKM}}\, m_d\, D_R + \bar{U}_R\, m_u\, U_L + \text{h.c.} \tag{11.114}$$

This is exactly the statement that one cannot simultaneously diagonalise M_u and M_d by a single unitary rotation on the left-handed doublets: diagonalising M_u fixes the U_L basis, leaving a non-trivial rotation in the D_L sector.

Physical quark fields

Define the *physical* down-type fields by

$$(D_{\text{ph}})_L \equiv U^\dagger_{\text{CKM}}\, D_L, \qquad \Longleftrightarrow \qquad D_L = U_{\text{CKM}}\, (D_{\text{ph}})_L, \tag{11.115}$$

and the corresponding Dirac fields

$$D_{\text{ph}} = (D_{\text{ph}})_L + D_R, \qquad U_{\text{ph}} = U_L + U_R. \tag{11.116}$$

Using Eq. (11.115) in Eq. (11.114), one obtains a fully diagonal mass Lagrangian in terms of physical fields,

$$\mathcal{L}_{qm} = \bar{D}_{\text{ph}}\, m_d\, D_{\text{ph}} + \bar{U}_{\text{ph}}\, m_u\, U_{\text{ph}}, \tag{11.117}$$

which identifies the mass eigenstates with the observed quarks d, s, b and u, c, t.

Where mixing becomes observable: charged currents

The key point is that weak interactions are written in the *electroweak* basis, where Q_i are $SU(2)_L$ doublets. The charged current is

$$J_\mu^{(+)} \equiv J_\mu^1 + i J_\mu^2 = \bar{U}_L \gamma_\mu D_L. \tag{11.118}$$

Expressing U_L and D_L in terms of physical fields and using Eq. (11.115), one finds

$$J_\mu^{(+)} = (\bar{U}_{\text{ph}})_L\, \gamma_\mu\, U_{\text{CKM}}\, (D_{\text{ph}})_L, \tag{11.119}$$

or, componentwise,

$$J_\mu^{(+)} = \frac{1}{2}\,(\bar{u}\ \bar{c}\ \bar{t})\, \gamma_\mu (1 - \gamma_5)\, U_{\text{CKM}} \begin{pmatrix} d \\ s \\ b \end{pmatrix}, \tag{11.120}$$

where, from this point on, it is customary to drop the "ph" subscript and understand u, c, t, d, s, b as physical fields. The dominance of the diagonal entries of U_{CKM} yields the familiar intra-doublet charged transitions ($d \to u, s \to c, b \to t$), whereas the off-diagonal entries mediate inter-family flavor change and therefore encode violations of quark-family conservation in charged-current processes.

Neutral currents remain flavor-diagonal at tree level

The neutral weak current relevant for the Z coupling is

$$J_\mu^{(Z)} = J_\mu^3 - \sin^2\theta_W\, J_\mu^{\text{e.m.}}, \tag{11.121}$$

with $J_\mu^{\text{e.m.}}$ manifestly diagonal in flavor. In the compact notation $Q = (U_L, D_L)^T$, one may write the charged-current structure with a block matrix built from U_{CKM}, and the corresponding third component current is proportional to a commutator-like block combination. The crucial step is unitarity:

$$U_{\text{CKM}}\, U_{\text{CKM}}^\dagger = \mathbb{I}, \tag{11.122}$$

which implies that the left-handed neutral current in the quark sector is *diagonal* in the mass eigenstate basis. Equivalently, the same rotation that introduces $U_{\rm CKM}$ in Eq. (11.119) cancels out in the neutral current, because it always appears as $U_{\rm CKM}U^\dagger_{\rm CKM}$. Therefore, at lowest order in the electroweak interactions, neutral-current processes conserve quark flavor, in sharp contrast to charged-current processes.

The misalignment encoded by $U_{\rm CKM}$ is a direct consequence of electroweak symmetry breaking in the presence of *two* independent Yukawa matrices, g^U and g^D. Once M_u is diagonalised by an allowed basis choice, the down-type sector generically remains rotated with respect to the weak isospin lowering/raising operators. In particular, applying the isospin lowering operator to an up-type mass eigenstate produces a *superposition* of down-type mass eigenstates, i.e. a state not associated with a definite mass. This is precisely why flavor mixing is observable in $W^\pm$-mediated processes and why it is parametrised, in the Standard Model, by the CKM matrix $U_{\rm CKM}$.

11.3.5 The Cabibbo-Kobayashi-Maskawa Matrix

A particularly efficient way to encode quark mixing is provided by the Wolfenstein-type expansion of the CKM matrix. The empirical input behind this choice is simple: flavor-conserving amplitudes are dominated by diagonal entries of order unity, while inter-generation transitions are hierarchically suppressed. One therefore organises the matrix as a power series in a small parameter

$$\lambda \equiv \sin\theta_C, \tag{11.123}$$

θ_C being the Cabibbo angle, and introduces two additional real parameters to set the scale of second- and third-generation mixing. In this language, the "elementary" suppressions are taken to be: λ for first$\leftrightarrow$second generation, $A\lambda^2$ for $b\to c$-type mixing, and $A\lambda^3(\rho - i\eta)$ for $b\to u$-type mixing. The single physical CP phase is carried by $\eta \neq 0$.

Let us start from the usual unitary 3×3 mixing matrix written in terms of three real angles $\theta_{12}, \theta_{23}, \theta_{13}$ and one CP phase δ,

$$U_{\rm CKM} = \begin{pmatrix} c_{12}c_{13} & s_{12}c_{13} & s_{13}e^{-i\delta} \\ -s_{12}c_{23} - c_{12}s_{23}s_{13}e^{i\delta} & c_{12}c_{23} - s_{12}s_{23}s_{13}e^{i\delta} & s_{23}c_{13} \\ s_{12}s_{23} - c_{12}c_{23}s_{13}e^{i\delta} & -c_{12}s_{23} - s_{12}c_{23}s_{13}e^{i\delta} & c_{23}c_{13} \end{pmatrix}, \tag{11.124}$$

with $s_{ij} \equiv \sin\theta_{ij}$ and $c_{ij} \equiv \cos\theta_{ij}$. The Wolfenstein scaling is then implemented by the identifications

$$s_{12} = \lambda, \qquad s_{23} = A\lambda^2, \qquad s_{13}e^{-i\delta} = A\lambda^3(\rho - i\eta), \tag{11.125}$$

so that

$$c_{12} = \sqrt{1-\lambda^2}, \qquad c_{23} = \sqrt{1-A^2\lambda^4}, \qquad c_{13} = \sqrt{1-A^2\lambda^6(\rho^2+\eta^2)}. \tag{11.126}$$

To any fixed order in λ one expands the square roots. Up to $O(\lambda^5)$ one needs

$$c_{12} = 1-\frac{1}{2}\lambda^2-\frac{1}{8}\lambda^4+O(\lambda^6), \qquad c_{23} = 1-\frac{1}{2}A^2\lambda^4+O(\lambda^8), \qquad c_{13} = 1+O(\lambda^6). \tag{11.127}$$

Substituting Eqs. (11.125)–(11.127) into Eq. (11.124) and keeping terms through λ^5 gives an explicitly unitary parameterisation at that accuracy. Carrying out the multiplication entry by entry yields:

$$U_{\rm CKM} = \begin{pmatrix} 1-\frac{1}{2}\lambda^2-\frac{1}{8}\lambda^4 & \lambda & A\lambda^3(\rho-i\eta) \\ -\lambda+\frac{1}{2}A^2\lambda^5[1-2(\rho+i\eta)] & 1-\frac{1}{2}\lambda^2-\frac{1}{8}\lambda^4(1+4A^2) & A\lambda^2 \\ A\lambda^3\left[1-(1-\frac{1}{2}\lambda^2)(\rho+i\eta)\right] & -A\lambda^2+\frac{1}{2}A\lambda^4[1-2(\rho+i\eta)] & 1-\frac{1}{2}A^2\lambda^4 \end{pmatrix} + O(\lambda^6). \tag{11.128}$$

This is the (Buras-Lautenbacher-Ostermaier type) improvement of the original Wolfenstein expansion, where unitarity is enforced consistently at the stated order.

In many applications it is sufficient to truncate at $O(\lambda^3)$. By keeping only the leading non-trivial terms in Eq. (11.128), one finds the familiar approximate form

$$U_{\rm CKM} = \begin{pmatrix} 1-\frac{1}{2}\lambda^2 & \lambda & A\lambda^3(\rho-i\eta) \\ -\lambda & 1-\frac{1}{2}\lambda^2 & A\lambda^2 \\ A\lambda^3[1-(\rho+i\eta)] & -A\lambda^2 & 1 \end{pmatrix} + O(\lambda^4). \tag{11.129}$$

When employing this truncation, tiny imaginary parts in some second-order quantities are, by construction, outside the retained accuracy; this is why more refined estimates may require reverting to Eq. (11.128).

The four Wolfenstein parameters are extracted by a global fit to a variety of flavor observables (semi-leptonic decays, neutral meson mixing, CP-violating asymmetries, and so on). A representative set of best-fit values is

$$\lambda = 0.2253 \pm 0.0007, \qquad A = 0.808^{+0.022}_{-0.015}, \qquad \bar\rho = 0.132^{+0.022}_{-0.014}, \qquad \bar\eta = 0.341 \pm 0.013. \tag{11.130}$$

Here, $\bar\rho$ and $\bar\eta$ are parameters closely related to ρ and η, but arranged so that the point $\bar\rho + i\bar\eta$ is independent of phase conventions and, moreover, the matrix expressed in terms of $\bar\rho$, $\bar\eta$ remains exactly unitary to all orders in λ. The relation between (ρ, η) and $(\bar\rho, \bar\eta)$ can be written as follows:

$$\rho + i\eta = \frac{(\bar\rho+i\bar\eta)\sqrt{1-A^2\lambda^4}}{\sqrt{1-\lambda^2}\left[1-A^2\lambda^4(\bar\rho+i\bar\eta)\right]}. \tag{11.131}$$

With this definition one has, in particular,

$$\bar{\rho} + i\bar{\eta} = -\frac{U_{ud}\, U^*_{ub}}{U_{cd}\, U^*_{cb}}, \tag{11.132}$$

which is explicitly rephasing-invariant. The global fit can be visualised as the intersection of constraint regions in the $(\bar{\rho}, \bar{\eta})$ plane (unitarity triangle apex), coming from different observables such as ϵ_K, Δm_{B_d}, Δm_{B_s}, CP asymmetries, and so on.

11.4 The Particle Standard Model as an Electroweak Unification

From Fermi's theory to the Weinberg-Salam standard model. Fermi's original description of β decay models weak interactions through a local four-fermion operator, which

- correctly reproduces low-energy data and,
- after the discovery of parity violation, incorporates the chiral $V - A$ structure.

However,

- G_F has mass dimension -2, so the corresponding Lagrangian is a non-renormalizable effective interaction: scattering amplitudes grow with energy (schematically $\mathcal{A} \sim G_F s$), signaling a breakdown of perturbative unitarity at sufficiently high $\sqrt{s}$ and demanding a UV completion;
- the theory is punctual, contradicting the Feynman puzzle in which interactions are mediated by virtual particles.

We need a gauge-invariant new model!

The SM of particle physics represents one of the most successful theoretical frameworks ever constructed in fundamental physics. It provides a unified and quantitatively accurate description of three of the four known fundamental interactions-electromagnetic, weak, and strong-and accounts for the properties and interactions of all experimentally observed elementary particles, with the notable exception of gravity.

At its conceptual core, the SM is built upon the principle of *local gauge invariance*. This principle, first introduced in the context of electromagnetism and later generalized by Yang and Mills to non-Abelian symmetry groups, elevates symmetry from a mere classificatory tool to a fundamental organizing principle of Nature. In modern

QFT, interactions are not introduced ad hoc, but emerge as a direct consequence of enforcing invariance under local symmetry transformations.

Historically, the gauge principle arose from the attempt to generalize the global phase symmetry of quantum mechanics to a local one. In electromagnetism, this requirement uniquely leads to the introduction of the electromagnetic four-potential and to Maxwell's equations. The extension of this idea to non-Abelian symmetry groups, pioneered by Yang and Mills in 1954, revealed an entirely new class of gauge theories, characterized by self-interacting gauge fields and a much richer mathematical structure.

However, the application of gauge invariance to the weak interaction posed a deep conceptual challenge. Gauge theories naturally predict massless gauge bosons, as explicit mass terms for vector fields violate local gauge invariance. While this feature is perfectly consistent with electromagnetism, where the photon is experimentally observed to be massless, it is in striking conflict with the phenomenology of weak interactions, which are mediated by massive vector bosons with short-range forces.

This tension between the mathematical rigidity of gauge invariance and the experimental requirement of massive weak bosons was one of the central problems of particle physics throughout the 1950s and early 1960s. Any viable theory of weak interactions had to reconcile these seemingly incompatible demands: exact local gauge symmetry and massive force carriers.

The resolution of this problem came with the discovery of the mechanism of spontaneous symmetry breaking in gauge theories, now universally known as the *Higgs mechanism*. This mechanism allows gauge bosons to acquire mass dynamically, through their interaction with a scalar field whose vacuum state does not share the full symmetry of the underlying Lagrangian, while preserving local gauge invariance at the fundamental level.

The electroweak theory, independently formulated by Weinberg and Salam, implements this idea within a gauge framework based on the group $SU(2)_L \times U(1)_Y$. Through spontaneous symmetry breaking, this theory predicts the existence of massive charged and neutral weak bosons ($W^{\pm}$ and Z^0), a massless photon, and a scalar excitation associated with the symmetry-breaking field: the Higgs boson.

These predictions have been spectacularly confirmed by experiments. The masses of the weak gauge bosons are measured to be

$$M_W \simeq 80.4 \text{ GeV}, \qquad M_Z \simeq 91.2 \text{ GeV}, \tag{11.133}$$

while the weak mixing angle satisfies

$$\sin^2\theta_W \simeq 0.23. \tag{11.134}$$

Most notably, in 2012 the ATLAS and CMS collaborations at the Large Hadron Collider (LHC) announced the discovery of a scalar particle with mass $M_h \simeq 125$ GeV, whose properties are consistent with those predicted for the Higgs boson of the SM.

► **Pause and think**
How would you propose a non-Abelian theory?

11.4.1 Yang-Mills Theory

In 1954, Chen-Ning Yang and Robert Mills extended the concept of local symmetry to more complex non-Abelian transformation groups, which do not satisfy the commutative property.

These groups are represented by the unitary group $SU(2) \times U(1) \times SU(3)$. Here, $SU(2) \times U(1)$ corresponds to the symmetry underlying the electromagnetic and weak interactions, unified in the electroweak theory, while $SU(3)$ corresponds to the symmetry underlying the strong interaction.

At this point, it is necessary to examine the consequences of imposing local symmetry.

Consider the Dirac Lagrangian for two sort of fermions, each with mass m:

$$\mathcal{L} = \bar{f}(i\gamma^{\mu}\partial_{\mu} - m)f + \bar{g}(i\gamma^{\mu}\partial_{\mu} - m)g = \bar{\Psi}(i\gamma^{\mu}\partial_{\mu} - m)\Psi, \tag{11.135}$$

where

$$\Psi = \begin{pmatrix} f \\ g \end{pmatrix}, \quad \bar{\Psi} = \left(\bar{f}\ \bar{g}\right). \tag{11.136}$$

We know that it is invariant under a global $SU(2)$ transformation. Studying now local infinitesimal $SU(2)$ transformations

$$\Psi \to (1 + \frac{i}{2}\boldsymbol{\tau} \cdot \boldsymbol{\alpha}(x))\Psi, \tag{11.137}$$

$$\bar{\Psi} \to \bar{\Psi}(1 - \frac{i}{2}\boldsymbol{\tau} \cdot \boldsymbol{\alpha}(x)), \tag{11.138}$$

where $\boldsymbol{\tau} = (\tau_x, \tau_2, \tau_3)$ are the Pauli matrices,[2] we discover that the dependence of $\boldsymbol{\alpha}$ on x breaks the symmetry. The problem is not the mass term, which is still invariant, but rather the derivative term. To fix it, we need to introduce a new gauge field $\boldsymbol{W}_{\mu}(x) = (W^1_{\mu}, W^2_{\mu}, W^3_{\mu})$ and a new covariant derivative:

$$D_{\mu} = \partial_{\mu} - \frac{i}{2}g\boldsymbol{\tau} \cdot \boldsymbol{W}_{\mu}(x), \tag{11.139}$$

where g is the charge of the theory.

Driving exercise
We now examine how the field $\boldsymbol{W}_{\mu}$ transforms. We ask

$$\Psi \to \Psi' = U\Psi. \tag{11.140}$$

[2] We called these matrices σ_i along the book, as in the standard notation. Since here we are focusing on a non-Abelian theory, we introduce a different nomenclature τ, keeping in mind that $\tau_i \equiv \sigma_i$.

Consequently,

$$(D_\mu \Psi)' = U(D_\mu \Psi), \tag{11.141}$$

also called covariance condition. We also know that:

$$(D_\mu \Psi)' = D'_\mu \Psi'. \tag{11.142}$$

So, Eq. (11.141) becomes:

$$D'_\mu \Psi' = U(D_\mu \Psi). \tag{11.143}$$

Let us now compute the two terms.
Remembering that $U(x)$ is the infinitesimal transformation:

$$U(x) = 1 + \frac{i}{2}\boldsymbol{\tau} \cdot \boldsymbol{\alpha}(x), \tag{11.144}$$

then

$$\begin{aligned}(D_\mu \Psi)' &= U(D_\mu \Psi) = (1 + \frac{i}{2}\boldsymbol{\tau} \cdot \boldsymbol{\alpha}(x))(D_\mu \Psi) = \\ &= (1 + \frac{i}{2}\boldsymbol{\tau} \cdot \boldsymbol{\alpha}(x))(\partial_\mu \Psi - \frac{i}{2} g \boldsymbol{\tau} \cdot \boldsymbol{W}_\mu(x)\Psi) \\ &= \partial_\mu \Psi - \frac{i}{2} g \boldsymbol{\tau} \cdot \boldsymbol{W}_\mu(x)\Psi + \frac{i}{2}\boldsymbol{\tau} \cdot \alpha(x)\partial_\mu \Psi \\ &\quad - \left(\frac{i}{2}\right)^2 g[\boldsymbol{\tau} \cdot \boldsymbol{\alpha}(x)][\boldsymbol{\tau} \cdot \boldsymbol{W}_\mu(x)]\Psi. \end{aligned} \tag{11.145}$$

Suppose now that the covariant derivative transforms as

$$D_\mu \;\rightarrow\; D'_\mu = \partial_\mu - \frac{i}{2} g \boldsymbol{\tau} \cdot \boldsymbol{W}_\mu(x) - \frac{i}{2} g \boldsymbol{\tau} \cdot \delta \boldsymbol{W}_\mu(x), \tag{11.146}$$

where $\delta \boldsymbol{W}_\mu(x)$ is the transformation of the gauge field $\boldsymbol{W}_\mu(x)$; the term $D'_\mu \Psi'$ becomes:

$$\begin{aligned} D'_\mu \Psi' =& (\partial_\mu - \frac{i}{2} g \boldsymbol{\tau} \cdot \boldsymbol{W}_\mu(x) - \frac{i}{2} g \boldsymbol{\tau} \cdot \delta \boldsymbol{W}_\mu(x))(1 + \frac{i}{2}\boldsymbol{\tau} \cdot \boldsymbol{\alpha}(x))\Psi \\ =& \partial_\mu \Psi + \frac{i}{2}\boldsymbol{\tau} \cdot \boldsymbol{\alpha}(x)\partial_\mu \Psi + \frac{i}{2}\boldsymbol{\tau} \cdot \partial_\mu \boldsymbol{\alpha}(x)\Psi - \frac{i}{2} g \boldsymbol{\tau} \cdot \boldsymbol{W}_\mu(x)\Psi \\ &+ \left(\frac{i}{2}\right)^2 g[\boldsymbol{\tau} \cdot \boldsymbol{W}_\mu(x)][\boldsymbol{\tau} \cdot \boldsymbol{\alpha}(x)]\Psi - \frac{i}{2} g \boldsymbol{\tau} \cdot \delta \boldsymbol{W}_\mu(x)\Psi \\ &- \cancel{\left(\frac{i}{2}\right)^2 g[\boldsymbol{\tau} \cdot \delta \boldsymbol{W}_\mu(x)][\boldsymbol{\tau} \cdot \boldsymbol{\alpha}(x)]\Psi}, \end{aligned} \tag{11.147}$$

where the last term has been simplified since it is of higher order.

Considering Eq. (11.143), Eqs. (11.145) and (11.147) are equal, and we obtain:

$$\begin{aligned}
&\partial_\mu \Psi - \frac{i}{2} g\boldsymbol{\tau} \cdot \boldsymbol{W}_\mu(x)\Psi + \frac{i}{2} g\boldsymbol{\tau} \cdot \boldsymbol{W}_\mu(x)\partial_\mu \Psi - \left(\frac{i}{2}\right)^2 g[\boldsymbol{\tau} \cdot \boldsymbol{\alpha}(x)][\boldsymbol{\tau} \cdot \boldsymbol{W}_\mu(x)]\Psi \\
&= \partial_\mu \Psi + \frac{i}{2}\boldsymbol{\tau} \cdot \boldsymbol{\alpha}(x)\partial_\mu \Psi + \frac{i}{2}\boldsymbol{\tau} \cdot \partial_\mu \boldsymbol{\alpha}(x)\Psi - \frac{i}{2} g\boldsymbol{\tau} \cdot \boldsymbol{W}_\mu(x)\Psi - \frac{i}{2} g\boldsymbol{\tau} \cdot \delta\boldsymbol{W}_\mu(x)\Psi \\
&- \left(\frac{i}{2}\right)^2 g[\boldsymbol{\tau} \cdot \boldsymbol{W}_\mu(x)][\boldsymbol{\tau} \cdot \boldsymbol{\alpha}(x)]\Psi. \qquad (11.148)
\end{aligned}$$

Hence,

$$\begin{aligned}
\left(\frac{i}{2}\right)^2 g[\boldsymbol{\tau} \cdot \boldsymbol{\alpha}(x)][\boldsymbol{\tau} \cdot \boldsymbol{W}_\mu(x)]\Psi =& \frac{i}{2}\boldsymbol{\tau} \cdot \partial_\mu \boldsymbol{\alpha}(x)\Psi - \frac{i}{2} g\boldsymbol{\tau} \cdot \delta\boldsymbol{W}_\mu(x)\Psi \\
&- \left(\frac{i}{2}\right)^2 g[\boldsymbol{\tau} \cdot \boldsymbol{W}_\mu(x)][\boldsymbol{\tau} \cdot \boldsymbol{\alpha}(x)]\Psi,
\end{aligned} \qquad (11.149)$$

which readily gives

$$\begin{aligned}
\Rightarrow \left(\frac{i}{2}\right) g[\boldsymbol{\tau} \cdot \boldsymbol{\alpha}(x)][\boldsymbol{\tau} \cdot \boldsymbol{W}_\mu(x)]\Psi =& \boldsymbol{\tau} \cdot \partial_\mu \boldsymbol{\alpha}(x)\Psi - g\boldsymbol{\tau} \cdot \delta\boldsymbol{W}_\mu(x)\Psi \\
&- \left(\frac{i}{2}\right) g[\boldsymbol{\tau} \cdot \boldsymbol{W}_\mu(x)][\boldsymbol{\tau} \cdot \boldsymbol{\alpha}(x)]\Psi
\end{aligned} \qquad (11.150)$$

Finally, we get:

$$\Rightarrow \boldsymbol{\tau} \cdot \delta\boldsymbol{W}_\mu(x) = \frac{1}{g}\boldsymbol{\tau} \cdot \partial_\mu \boldsymbol{\alpha}(x) + \frac{i}{2}\{[\boldsymbol{\tau} \cdot \boldsymbol{\alpha}(x)][\boldsymbol{\tau} \cdot \boldsymbol{W}_\mu(x)] - [\boldsymbol{\tau} \cdot \boldsymbol{W}_\mu(x)][\boldsymbol{\tau} \cdot \boldsymbol{\alpha}(x)]\}. \qquad (11.151)$$

Using the property

$$(\text{œ} \cdot \mathbf{a})(\text{œ} \cdot \mathbf{b}) = (\mathbf{a} \cdot \mathbf{b}) + i\text{œ} \cdot \mathbf{a} \times \mathbf{b}, \qquad (11.152)$$

then Eq. (11.151) becomes:

$$
\begin{aligned}
\boldsymbol{\tau}\cdot\delta\boldsymbol{W}_\mu(x) =& \frac{1}{g}\boldsymbol{\tau}\cdot\partial_\mu\boldsymbol{\alpha}(x)+\frac{i}{2}\{\boldsymbol{\alpha}(x)\cdot\boldsymbol{W}_\mu(x)+i\boldsymbol{\tau}\cdot\boldsymbol{\alpha}(x)\times\boldsymbol{W}_\mu(x)\\
&-\boldsymbol{W}_\mu(x)\cdot\boldsymbol{\alpha}(x)-i\boldsymbol{\tau}\cdot\boldsymbol{W}_\mu(x)\times\boldsymbol{\alpha}(x)\}\\
=& \frac{1}{g}\boldsymbol{\tau}\cdot\partial_\mu\boldsymbol{\alpha}(x)+\frac{i}{2}\{\boldsymbol{\alpha}(x)\cdot\boldsymbol{W}_\mu(x)+i\boldsymbol{\tau}\cdot\boldsymbol{\alpha}(x)\times\boldsymbol{W}_\mu(x)\\
&-\boldsymbol{W}_\mu(x)\cdot\boldsymbol{\alpha}(x)+i\boldsymbol{\tau}\cdot\boldsymbol{\alpha}(x)\times\boldsymbol{W}_\mu(x)\}\\
=& \frac{1}{g}\boldsymbol{\tau}\cdot\partial_\mu\boldsymbol{\alpha}(x)+i(i\boldsymbol{\tau}\cdot\boldsymbol{\alpha}(x)\times\boldsymbol{W}_\mu(x))\\
=& \frac{1}{g}\boldsymbol{\tau}\cdot\partial_\mu\boldsymbol{\alpha}(x)-\boldsymbol{\tau}\cdot\boldsymbol{\alpha}(x)\times\boldsymbol{W}_\mu(x).
\end{aligned}
\tag{11.153}
$$

The transformation of the gauge field is given by:

$$
\begin{aligned}
\boldsymbol{\tau}\cdot\boldsymbol{W}_\mu(x) \to& \boldsymbol{\tau}\cdot\boldsymbol{W}_\mu(x)+\boldsymbol{\tau}\cdot\delta\boldsymbol{W}_\mu(x)\\
=& \boldsymbol{\tau}\cdot\boldsymbol{W}_\mu(x)+\frac{1}{g}\boldsymbol{\tau}\cdot\partial_\mu\boldsymbol{\alpha}(x)-\boldsymbol{\tau}\cdot\boldsymbol{\alpha}(x)\times\boldsymbol{W}_\mu(x).
\end{aligned}
\tag{11.154}
$$

We are now ready to construct the invariant Lagrangian under local $SU(2)$ symmetry.

▶ **Pause and think**
Try to guess how!

In the $U(1)$ theory, we note that a very important contribution is given to the Lagrangian by the gauge field A_μ, whose dynamics gives us the electromagnetism through the $-\frac{1}{4}F^{\mu\nu}F_{\mu\nu}$ term. Similarly, for the gauge field $\boldsymbol{W}_\mu(x)$ the additional term corresponds, as expected, to $-\frac{1}{4}\boldsymbol{G}^{\mu\nu}\boldsymbol{G}_{\mu\nu}$, with

$$
\boldsymbol{G}^{\mu\nu}=\partial^\mu\boldsymbol{W}^\nu-\partial^\nu\boldsymbol{W}^\mu+g(\boldsymbol{W}^\mu\times\boldsymbol{W}^\nu).
\tag{11.155}
$$

Our final expression for the Lagrangian of a locally invariant $SU(2)$ theory is then given by:

$$
\mathcal{L}=\bar{\Psi}(i\gamma^\mu D_\mu-m)\Psi-\frac{1}{4}\boldsymbol{G}^{\mu\nu}\boldsymbol{G}_{\mu\nu},
\tag{11.156}
$$

under the transformation

$$
\begin{aligned}
\boldsymbol{\tau}\cdot\boldsymbol{W}_\mu(x) &\to \boldsymbol{\tau}\cdot\boldsymbol{W}_\mu(x)+\frac{1}{g}\boldsymbol{\tau}\cdot\partial_\mu\boldsymbol{\alpha}(x)-\boldsymbol{\tau}\cdot\boldsymbol{\alpha}(x)\times\boldsymbol{W}_\mu(x)\\
\Psi &\to (1+\frac{i}{2}\boldsymbol{\tau}\cdot\boldsymbol{\alpha}(x))\Psi\\
\bar{\Psi} &\to \bar{\Psi}(1-\frac{i}{2}\boldsymbol{\tau}\cdot\boldsymbol{\alpha}(x)).
\end{aligned}
\tag{11.157}
$$

▶ **Pause and think**
Before going on, do you see any analogy with vector fields?

Finally, in analogy with the $U(1)$ theory, Eq. (11.156) shows that, after quantization, there will be excitations in the $\boldsymbol{W}_\mu$ gauge field which mediate the interactions between the fermions of the theory. Explicitly, we expect massless W_1, W_2 and W_3 photon-like particles, each with two possible transverse polarizations. The particles couple to the isospin of the field with a strength set by the charge g.

11.4.2 Gluon Propagator

We will later be interested in the massless spin-1 particles arising in QCD, **gluons**, although the derivation that follows applies to any gauge group. For this reason, we call "gluon" the generic SU(N) gauge boson, and we search its propagator in the covariant gauges. What we will find is a unique feature: unphysical virtual states called Faddeev-Popov ghosts. The covariant gauge tries to force Lorentz invariance on these states, but, as it is known, unitarity for massless spin-1 particles and manifest Lorentz invariance generate a conflict, shown these artificial ghosts. In some non-covariant gauges, such as the axial gauge, ghosts are absent, but the covariant ones are simpler to compute, despite ghosts.

11.4.3 Faddeev-Popov Procedure

When fixing the gauge in a theory, it is often convenient to impose a covariant gauge condition, meaning that the resulting gauge-fixing term in the Lagrangian preserves manifest Lorentz covariance. For instance, this is commonly achieved in QED by adding a Lagrange multiplier $-\frac{1}{2\xi}(\partial_\mu A^\mu)^2$, which is a Lorentz scalar and goes to 0 when $\xi \to 0$, provided the Lorenz gauge condition is satisfied.

In a non-abelian theory with gauge group SU(N), we need $N^2 - 1$ gauge-fixing fields π^a, so that the infinitesimal transformation for the gauge field is parametrized by these as

$$A^a_\mu \to A^a_\mu + \frac{1}{g}\partial_\mu \pi^a + f_{abc} A^b_\mu \pi^c. \tag{11.158}$$

Since π^a is in the adjoint representation, this can be written more concisely as

$$A^a_\mu \quad \to \quad A^a_\mu + \frac{1}{g} D_\mu \pi^a, \tag{11.159}$$

where $D_\mu \pi^a = \partial_\mu \pi^a + g f^a_{bc} A^b_\mu \pi^c$ is the way a covariant derivative acts on a field transforming in the adjoint representation.

A more systematic way to derive the propagator is by using path integrals. Our path integral $f[A]$ relative to π fields is then given by the $-\frac{1}{2\xi}(\partial_\mu A^\mu)^2$ term with the Lorenz condition $\partial_\mu A^\mu = 0$, as

$$f[A] = \int \mathcal{D}\pi \exp\left[-i\int d^4x \frac{1}{2\xi}(\partial^\mu D_\mu \pi^a)^2\right]. \tag{11.160}$$

Unlike in the abelian case, f is not just a number, but is now a functional of A_μ. Nevertheless, we can still define $\alpha^a[A]$ as the gauge transformation parameters that take a given A^a_μ configuration to Lorenz gauge. That is, $\partial^\mu A^a_\mu = \frac{1}{g}\partial^\mu D_\mu \alpha^a[A]$ has a solution. Shifting π^a by $\frac{1}{g}\alpha^a[A]$ then gives

$$f[A] = \int \mathcal{D}\pi \exp\left[-i\int d^4x \frac{1}{2\xi}\left(\partial^\mu A^a_\mu - \partial^\mu D_\mu \pi^a\right)^2\right], \tag{11.161}$$

so that the full path integral is

$$\begin{aligned} Z[J=0] &= \int \mathcal{D}A_\mu \mathcal{D}\phi_i \, e^{i\int d^4x \mathcal{L}[A,\phi_i]} \\ &= \int \mathcal{D}\pi \mathcal{D}A_\mu \mathcal{D}\phi_i \frac{1}{f[A]} \exp\left[i\int d^4x \left(\mathcal{L}[A,\phi_i] - \frac{1}{2\xi}\left(\partial^\mu A^a_\mu - \partial^\mu D_\mu \pi^a\right)^2\right)\right] \\ &= \left(\int \mathcal{D}\pi\right)\int \mathcal{D}A_\mu \mathcal{D}\phi_i \frac{1}{f[A]} \exp\left[i\int d^4x \left(\mathcal{L}[A,\phi_i] - \frac{1}{2\xi}\left(\partial^\mu A^a_\mu\right)^2\right)\right], \end{aligned} \tag{11.162}$$

where we have redefined $A^a_\mu \to A^a_\mu + D_\mu \pi^a$ in the last step, which leaves $\mathcal{L}[A, \phi_i]$ unaffected. This redefinition removes π from the Lagrangian, thus the π integral returns only an unphysical constant.

This indicates that the additional π^a auxiliary modes are artifacts arising from the non-abelianity of the gauge terms, but they do not have any physical contribution as shown in the path integral, so they pose no real threat.

To simplify $f[A]$, we observe that it's quadratic in π, thus we can perform the Gaussian integral as a functional of A_μ, finding

$$f[A] = \sqrt{\frac{1}{\det(\partial_\mu D^\mu)^2}} \cdot \text{const.}, \tag{11.163}$$

so that

$$Z[J=0] = \text{const.} \cdot \int \mathcal{D}A_\mu \mathcal{D}\phi_i [\det(\partial_\mu D^\mu)] \exp\left[i\int d^4x \left(\mathcal{L}[A,\phi_i] - \frac{1}{2\xi}(\partial^\mu A^a_\mu)^2\right)\right]. \tag{11.164}$$

We know that a determinant can be written as a path integral over Grassmann numbers, introducing a fermionic field c, as

$$\det(\partial_\mu D^\mu) = \int \mathcal{D}\bar{c}\mathcal{D}c \exp\left(-i \int d^4x \bar{c}(\partial_\mu D^\mu)c\right). \tag{11.165}$$

Therefore, we have the gauge-fixed path integral for a non-abelian gauge theory:

$$Z[J=0] = \text{const.} \cdot \int \mathcal{D}A_\mu \mathcal{D}\phi_i \mathcal{D}\bar{c}\mathcal{D}c \exp\left[i \int d^4x \left(\mathcal{L}[A,\phi_i] - \frac{1}{2\xi}(\partial^\mu A^a_\mu)^2 - \bar{c}^a \partial^\mu D_\mu c^a\right)\right]. \tag{11.166}$$

The term for the field c lacks a spinor matrix. In fact, in order to preserve Lorentz invariance, it must transform as $c(x) \to c(\Lambda^{-1}x)$, thus with a trivial Lorentz matrix, that is, it transforms as a scalar, despite being anticommuting. This violates the spin-statistics theorem, since c^a must be spin-0 scalars to retain Lorentz invariance, but they anticommute, like fermion fields. These fields c^a are therefore unphysical, and are called the **Faddeev-Popov ghosts**.

Despite being unphysical, nothing prevents ghosts from appearing in the path integral. As with physical fields, one can expand the path integral in perturbation theory, generating Feynman diagrams involving these ghosts. For S-matrix elements of physical states, the ghosts will appear in internal lines. One way to understand why ghosts have to be fermionic is that they can cancel unphysical degrees of freedom of the gluons in loops. When we take the gluons off-shell, we are no longer guaranteed to have the right number of physical degrees of freedom. The ghosts are then fermionic because they need a –1 in loops to cancel the +1 from the unphysical polarizations. Yet, they have spin 0 and transform accordingly like scalars. We will now see how these ghosts actually come with an advantage.

11.4.4 BRST Invariance

In QED there are no gauge self-interactions, thus the gauge invariance breaking term $\frac{1}{2\xi}(\partial^\mu A_\mu)^2$ still allows to fix a gauge choice, since A_μ couples to a conserved current. In QCD, the gauge fields self-interact and do not couple to any conserved current. So, the $\frac{1}{2\xi}(\partial^\mu A^a_\mu)^2$ term, along with its associated Faddeev-Popov ghosts, is not so clearly an innocuous deformation. Remarkably, it turns out that ghosts are necessary for the QCD Lagrangian to retain an exact global symmetry, called **BRST invariance** (named after Becchi, Rouet, Stara and Tyutin). This is crucial in proving renormalizability of non-abelian gauge theories.
The non-abelian Faddeev-Popov Lagrangian is

$$\mathcal{L}_{\text{FP}} = \mathcal{L}[A^a_\mu, \phi_i] - \frac{1}{2\xi}(\partial^\mu A^a_\mu)^2 + (\partial_\mu \bar{c}^a)(D^\mu c_a), \tag{11.167}$$

where $D_\mu c^a = \partial_\mu c^a + g f^a_{bc} A^b_\mu c^c$. We could further fix the gauge with fields $\alpha^a(x)$, since the Lagrangian retains further symmetry under which $A^a_\mu \to A^a_\mu + \frac{1}{g}\partial_\mu \alpha^a$.

However, we notice that the equations of motion for c and $\bar{c}$ are $\partial_\mu D^\mu c^a = 0$ and $\partial_\mu D^\mu \bar{c}^a = 0$, thus we gauge transform with a term $\alpha^a = \theta c^a(x)$ instead. The transformations, that leave $\mathcal{L}[A^a_\mu, \phi_i]$ invariant, are

$$\phi_i \to \phi_i + i\theta\, c_a T^a_{ij}\phi^j, \tag{11.168}$$

$$A^a_\mu \to A^a_\mu + \frac{1}{g}\,\theta D_\mu c^a, \tag{11.169}$$

$$\bar{c}^a \to \bar{c}^a - \frac{1}{g}\,\theta \frac{1}{\xi}\,\partial^\mu A^a_\mu. \tag{11.170}$$

The transformation of $(\partial_\mu \bar{c}^a)$ is designed to exactly cancel the transformation of $-\frac{1}{2\xi}\,(\partial^\mu A^a_\mu)^2$. However, the $D_\mu c^a$ term is not invariant, because of the A^a_μ hidden in $D_\mu c^a$:

$$D_\mu c^a \;\to\; D_\mu c^a + \theta\, f^a_{bc}(D_\mu c^b)\, c^c. \tag{11.171}$$

To make this covariant derivative invariant, we will need to transform c^a as well. This can be done by defining the BRST transformation for c^a as

$$c^a \;\to\; c^a - \frac{1}{2}\,\theta\, f^a_{bc} c^b c^c. \tag{11.172}$$

Note that nowhere did we use that c^a and $\bar{c}^a$ were related in any way (as ψ and $\bar{\psi}$ are related by charge-conjugation invariance), thus, we are free to give them different transformation laws. In fact, it could be checked that

$$D_\mu c^a \;\to\; D_\mu c^a + \theta\, f^a_{bc}(D_\mu c^b)\, c^c - \theta\, f^a_{bc}\Big[(\partial_\mu c^b)\, c^c + g f^b_{ed} A^e_\mu c^d c^c\Big] = D_\mu c^a, \tag{11.173}$$

and that $\mathcal{L}_{\mathrm{FP}}$ is invariant.

We conclude that the Faddeev-Popov Lagrangian is BRST invariant. BRST invariance is then a global symmetry, parametrized by a Grassmann number θ.

One implication of such an invariance is related to renormalization. Since BRST invariance is an exact symmetry of the Lagrangian, it will be preserved in loops. Thus, one will not need any counterterms that violate BRST invariance. The Faddeev-Popov ghosts and anti-ghosts are critical in establishing BRST invariance of the gauge-fixed non-abelian Lagrangian, thus strongly constraining how they can appear at higher orders. The proof of renormalizability for non-abelian theories shows that all the infinities are canceled with the finite number of counterterms corresponding to terms in the most general BRST invariant Lagrangian.

Although many of the quantization tools introduced here were historically developed in the context of QCD, they are not specific to the strong interaction. Rather, they are generic structural ingredients of any non-Abelian gauge theory. What we saw, namely, gauge fixing, the Faddeev-Popov procedure, ghost fields, and BRST symmetry are required whenever the gauge group is non-Abelian, independently of the physical interpretation of the gauge bosons.

These have been therefore here introduced in order to give a consistent quantum formulation of the electroweak theory and, in particular, of the weak interactions *beyond the tree level*, cfr Sect. 11.5. Their role here is therefore conceptual and structural: they explain why the weak interaction, formulated as a non-Abelian gauge theory, can be consistently quantized and renormalized.

The same machinery acquires a direct and more concrete physical relevance in the theory of strong interactions. In QCD, discussed in the next chapter, the self-interaction of gluons, the structure of the gauge-fixed Lagrangian, and the presence of ghost fields are not only formal necessities but are essential for the perturbative description of hadronic processes, for the computation of loop corrections, and for the understanding of phenomena such as asymptotic freedom and the running of the strong coupling.

11.4.5 Breaking Symmetry with a Non-Abelian Gauge Theory

Let us examine the issue of symmetry breaking in a non-Abelian gauge theory. This is useful because the electroweak theory, examined later, exactly relies on this physics. We will examine the case of $SO(3)$ gauge theory with a ϕ^4-type interaction and a symmetry-breaking positive mass term, with a Lagrangian given by:

$$\mathcal{L} = \frac{1}{2} D^\mu \mathbf{\Phi} \cdot D_\mu \mathbf{\Phi} + \frac{1}{2} m^2 \mathbf{\Phi} \cdot \mathbf{\Phi} - \lambda (\mathbf{\Phi} \cdot \mathbf{\Phi})^2 - \frac{1}{4} G_{\mu\nu} \cdot G^{\mu\nu}, \tag{11.174}$$

where, in this case, $D_\mu \mathbf{\Phi} = \partial_\mu \mathbf{\Phi} - g \mathbf{\Phi} \times \mathbf{W}_\mu(x)$. Expanding the covariant derivative explicitly, the Lagrangian becomes

$$\mathcal{L} = \frac{1}{2} (\partial^\mu \mathbf{\Phi} - g \mathbf{\Phi} \times \mathbf{W}^\mu(x)) \cdot (\partial_\mu \mathbf{\Phi} - g \mathbf{\Phi} \times \mathbf{W}_\mu(x)) + \frac{1}{2} m^2 \mathbf{\Phi} \cdot \mathbf{\Phi} - \lambda (\mathbf{\Phi} \cdot \mathbf{\Phi})^2 - \frac{1}{4} G_{\mu\nu} \cdot G^{\mu\nu}. \tag{11.175}$$

This Lagrangian is symmetric under transformation

$$\mathbf{\Phi} \to \mathbf{\Phi} - \boldsymbol{\theta}(x) \times \mathbf{\Phi}, \quad \mathbf{W}_\mu \to \mathbf{W}_\mu - \frac{1}{g} \partial_\mu \boldsymbol{\theta}(x) - (\boldsymbol{\theta}(x) \times \mathbf{W}_\mu). \tag{11.176}$$

Driving exercise
To study spontaneous symmetry breaking, we now focus on the scalar potential associated with the Lagrangian, which determines the vacuum of the theory. The potential is

$$U(\phi_1, \phi_2, \phi_3) = -\frac{m^2}{2}(\phi_1^2 + \phi_2^2 + \phi_3^2) + \lambda(\phi_1^2 + \phi_2^2 + \phi_3^2)^2 \qquad (11.177)$$

and the critical points are obtained from $\nabla U = 0$. Therefore, we have:

$$\frac{\partial U}{\partial \phi_1} = -2\frac{m^2}{2}\phi_1 + 4\lambda(\phi_1^2 + \phi_2^2 + \phi_3^2)\phi_1 = 0 \Rightarrow m^2 - 4\lambda(\phi_1^2 + \phi_2^2 + \phi_3^2) = 0, \qquad (11.178)$$

$$\frac{\partial U}{\partial \phi_2} = -2\frac{m^2}{2}\phi_2 + 4\lambda(\phi_1^2 + \phi_2^2 + \phi_3^2)\phi_2 = 0 \Rightarrow m^2 - 4\lambda(\phi_1^2 + \phi_2^2 + \phi_3^2) = 0, \qquad (11.179)$$

$$\frac{\partial U}{\partial \phi_3} = -2\frac{m^2}{2}\phi_3 + 4\lambda(\phi_1^2 + \phi_2^2 + \phi_3^2)\phi_3 = 0 \Rightarrow m^2 - 4\lambda(\phi_1^2 + \phi_2^2 + \phi_3^2) = 0. \qquad (11.180)$$

Therefore, the potential has an infinite number of minima, which lie on the surface of a sphere

$$\phi_1^2 + \phi_2^2 + \phi_3^2 = \frac{m^2}{4\lambda}, \qquad (11.181)$$

with radius equal to

$$\phi_0 = |\mathbf{\Phi}_0| = \left(\frac{m^2}{4\lambda}\right)^{\frac{1}{2}}. \qquad (11.182)$$

Symmetry breaking involves picking one of the infinite number of equivalent vacua, which breaks global (and local) symmetry. For simplicity, we choose the one that points along the 3-direction in isospin space, that is:

$$\mathbf{\Phi}_0 = \begin{pmatrix} 0 \\ 0 \\ \phi_0 \end{pmatrix}. \qquad (11.183)$$

For the excitation above this ground state, we write:

$$\mathbf{\Phi} = \begin{pmatrix} \phi_1 \\ \phi_2 \\ \phi_0 + \chi(x) \end{pmatrix}, \qquad (11.184)$$

where $\chi(x)$ describes deviations in the 3-direction.
We now insert the expression obtained above into the Lagrangian given in Eq. (11.175).
For clarity, we analyse the result term by term.

– $\mathbf{\Phi} \cdot \mathbf{\Phi}$:

$$\begin{aligned}\mathbf{\Phi} \cdot \mathbf{\Phi} &= \phi_1^2 + \phi_2^2 + (\phi_3')^2 = \phi_1^2 + \phi_2^2 + (\phi_0 + \chi)^2 \\ &= \phi_1^2 + \phi_2^2 + \phi_0^2 + 2\phi_0\chi + \chi^2. \end{aligned} \tag{11.185}$$

Placing $\delta \equiv \phi_1^2 + \phi_2^2 + 2\phi_0\chi + \chi^2$ we get:

$$\mathbf{\Phi} \cdot \mathbf{\Phi} = \phi_0^2 + \delta. \tag{11.186}$$

– $(\mathbf{\Phi} \cdot \mathbf{\Phi})^2$:

$$(\mathbf{\Phi} \cdot \mathbf{\Phi})^2 = (\phi_0^2 + \delta)^2 = \phi_0^4 + 2\phi_0^2\delta + \delta^2, \tag{11.187}$$

but

$$\begin{aligned}\delta^2 &= (\phi_1^2 + \phi_2^2 + 2\phi_0\chi + \chi^2)^2 \\ &= \phi_1^4 + \phi_2^4 + 4\phi_0^2\chi^2 + \chi^4 + 2\phi_1^2\phi_2^2 + 2\chi^2\phi_1^2 + 2\phi_2^2\chi^2 + 4\phi_0\chi\phi_1^2 \\ &\quad + 4\phi_0\chi\phi_2^2 + 4\phi_0\chi^3, \end{aligned} \tag{11.188}$$

therefore

$$\begin{aligned}\phi_0^4 + 2\phi_0^2\delta + \delta^2 =& \phi_0^4 + 2\phi_0^2(\phi_1^2 + \phi_2^2 + 2\phi_0\chi + \chi^2) + \phi_1^4 + \phi_2^4 \\ &+ 4\phi_0^2\chi^2 + \chi^4 + 2\phi_1^2\phi_2^2 + 2\chi^2\phi_1^2 + 2\phi_2^2\chi^2 \\ &+ 4\phi_0\chi\phi_1^2 + 4\phi_0\chi\phi_2^2 + 4\phi_0\chi^3, \end{aligned} \tag{11.189}$$

which, approximated to the second order in the fields, becomes:

$$(\mathbf{\Phi} \cdot \mathbf{\Phi})^2 \simeq \phi_0^4 + 4\phi_0^3\chi + 2\phi_0^2(\phi_1^2 + \phi_2^2) + 6\phi_0^2\chi^2. \tag{11.190}$$

– $\frac{1}{2}D^\mu\mathbf{\Phi} \cdot D_\mu\mathbf{\Phi} = \frac{1}{2}(\partial^\mu\mathbf{\Phi} - g\mathbf{\Phi} \times \mathbf{W}^\mu(x)) \cdot (\partial_\mu\mathbf{\Phi} - g\mathbf{\Phi} \times \mathbf{W}_\mu(x))$.
We need to compute

$$D^\mu\mathbf{\Phi} = \partial^\mu\mathbf{\Phi} - g\mathbf{\Phi} \times \mathbf{W}^\mu(x),$$

recalling that $\mathbf{W}^\mu(x) = (W_1^\mu, W_2^\mu, W_3^\mu)$ and that $\mathbf{\Phi} = (\phi_1, \phi_2, \phi_0 + \chi)$. Hence,

$$\mathbf{\Phi} \times \mathbf{W}^\mu(x) \rightarrow \begin{cases}(\mathbf{\Phi} \times \mathbf{W}^\mu(x))_1 = (\phi_2 W_3^\mu - (\phi_0 + \chi)W_2^\mu) \\ (\mathbf{\Phi} \times \mathbf{W}^\mu(x))_2 = -(\phi_1 W_3^\mu - (\phi_0 + \chi)W_1^\mu) \\ (\mathbf{\Phi} \times \mathbf{W}^\mu(x))_3 = (\phi_1 W_2^\mu - \phi_2 W_1^\mu)\end{cases}, \tag{11.191}$$

which implies that:

$$
\begin{aligned}
(D^\mu \mathbf{\Phi}_1) &= \partial^\mu \phi_1 - g(\phi_2 W_3^\mu - (\phi_0 + \chi) W_2^\mu) \\
(D^\mu \mathbf{\Phi}_2) &= \partial^\mu \phi_2 + (\phi_1 W_3^\mu - (\phi_0 + \chi) W_1^\mu) \\
(D^\mu \mathbf{\Phi}_3) &= \partial^\mu \phi_3 - g(\phi_1 W_2^\mu - \phi_2 W_1^\mu) \\
&= \partial^\mu (\phi_0 + \chi(x)) - g(\phi_1 W_2^\mu - \phi_2 W_1^\mu) \\
&= \partial^\mu (\chi(x)) - g(\phi_1 W_2^\mu - \phi_2 W_1^\mu).
\end{aligned} \tag{11.192}
$$

We now evaluate the product $D^\mu \mathbf{\Phi} D_\mu \mathbf{\Phi} = \frac{1}{2} \Sigma_{i=1}^3 D^\mu \mathbf{\Phi}_i D_\mu \mathbf{\Phi}_i$.

1. $(D^\mu \mathbf{\Phi}_1)^2$:

$$
\begin{aligned}
(D^\mu \mathbf{\Phi}_1)^2 &= [\partial^\mu \phi_1 - g(\phi_2 W_3^\mu - (\phi_0 + \chi) W_2^\mu)]^2 \\
&= (\partial^\mu \phi_1)^2 - 2g \partial^\mu \phi_1 (\phi_2 W_3^\mu - (\phi_0 + \chi) W_2^\mu) \\
&\quad + g^2 (\phi_2 W_3^\mu - (\phi_0 + \chi) W_2^\mu)^2 \\
&= (\partial^\mu \phi_1)^2 - 2g(\partial^\mu \phi_1)\phi_2 W_3^\mu + 2g(\partial^\mu \phi_1) W_2^\mu \phi_0 \\
&\quad + 2g(\partial^\mu \phi_1) W_2^\mu \chi + g^2 (W_3^\mu)^2 \phi_2^2 \\
&\quad - 2g^2 W_3^\mu \phi_2 (\phi_0 + \chi) W_2^\mu + g^2 (W_2^\mu)^2 \phi_0^2 \\
&\quad + g^2 (W_2^\mu)^2 \chi^2 + 2g^2 (W_2^\mu)^2 \phi_0 \chi.
\end{aligned} \tag{11.193}
$$

Keeping terms up to second order in the fields, we obtain:

$$
(D^\mu \mathbf{\Phi}_1)^2 \simeq (\partial^\mu \phi_1)^2 + 2g\phi_0 (\partial^\mu \phi_1) W_2^\mu + g^2 \phi_0^2 (W_2^\mu)^2. \tag{11.194}
$$

2. The same procedure applies to $(D^\mu \mathbf{\Phi}_2)^2$:

$$
(D^\mu \mathbf{\Phi}_2)^2 \simeq (\partial^\mu \phi_2)^2 - 2g\phi_0 (\partial^\mu \phi_2) W_1^\mu + g^2 \phi_0^2 (W_1^\mu)^2. \tag{11.195}
$$

3. The same procedure applies to $(D^\mu \mathbf{\Phi}_3)^2$:

$$
(D^\mu \mathbf{\Phi}_3)^2 \simeq (\partial^\mu \chi)^2. \tag{11.196}
$$

We now continue the analysis by focusing on potential terms

$$
\frac{m^2}{2} \mathbf{\Phi} \cdot \mathbf{\Phi} - \lambda (\mathbf{\Phi} \cdot \mathbf{\Phi})^2, \tag{11.197}
$$

which, on the basis of the previous calculations, becomes:

$$
\frac{m^2}{2} (\phi_1^2 + \phi_2^2 + \phi_0^2 + 2\phi_0 \chi + \chi^2) - \lambda (\phi_0^4 + 4\phi_0^3 \chi + 2\phi_0^2 (\phi_1^2 + \phi_2^2) + 6\phi_0^2 \chi^2). \tag{11.198}
$$

In order to evaluate this expression, we collect the common terms.

– In χ:

$$\frac{m^2}{2}2\phi_0 - 4\lambda\phi_0^3 = \phi_0(m^2 - 4\lambda\phi_0^2) = \phi_0(m^2 - 4\lambda\frac{m^2}{4\lambda}) = 0. \quad (11.199)$$

– In χ^2:

$$\frac{m^2}{2} - 6\lambda\phi_0^2 = \frac{4\lambda\phi_0^2}{2} - 6\lambda\phi_0^2 = -4\lambda\phi_0^2. \quad (11.200)$$

The field is massive, with $m_{\chi^2} = 8\lambda\phi_0^2$.

– In ϕ_1^2, ϕ_2^2:

$$\frac{m^2}{2} - 2\lambda\phi_0^2 = \frac{4\lambda\phi_0^2}{2} - 2\lambda\phi_0^2 = 0. \quad (11.201)$$

The field is massless.

– Constant:

$$\lambda\phi_0^4. \quad (11.202)$$

Collecting all the results obtained for both the kinetic and potential terms, we obtain the following Lagrangian:

$$\begin{aligned}\mathcal{L} =& \frac{1}{2}[(\partial^\mu\phi_1)^2 + (\partial^\mu\phi_2)^2 + (\partial^\mu\chi)^2] - 4\phi_0^2\lambda\chi^2 \\ &+ g\phi_0[(\partial^\mu\phi_1)W_\mu^2 - (\partial^\mu\phi_2)W_\mu^1] + \frac{g^2\phi_0^2}{2}[(W_1^\mu)^2 + (W_2^\mu)^2] - \frac{1}{4}G^{\mu\nu}G_{\mu\nu} + \lambda\phi_0^4,\end{aligned} \quad (11.203)$$

where we ignore higher order terms. The first line describes the dynamics of the $\chi(x)$ scalar field, while the last line describes the dynamics of the $\boldsymbol{W}_\mu(x)$ vector field, but the terms that are mixed in the components of $\boldsymbol{W}_\mu$ and $\boldsymbol{\Phi}$ are not easy to interpret.

To solve it, we choose a gauge such that the excited-state field $\boldsymbol{\Phi}$ is aligned along the 3-direction in isospin space, a choice commonly referred to as the **Unitary Gauge**. This removes all mention of the components ϕ_1 and ϕ_2 and means that the excited state field becomes simply

$$\boldsymbol{\Phi} = \begin{pmatrix} 0 \\ 0 \\ \phi_0 + \chi(x) \end{pmatrix}. \quad (11.204)$$

The final symmetry-broken Lagrangian is given by

$$\begin{aligned}\mathcal{L} =& \frac{1}{2}(\partial^\mu\chi)^2 - 4\phi_0^2\lambda\chi^2 \\ &+ \frac{g^2\phi_0^2}{2}[(W_1^\mu)^2 + (W_2^\mu)^2] - \frac{1}{4}G^{\mu\nu}G_{\mu\nu} + \lambda\phi_0^4,\end{aligned} \quad (11.205)$$

which is the sum of Lagrangians for a massive, single-component scalar field χ and a massive vector field with two components W_μ^1 and W_μ^2. Symmetry breaking has caused the disappearance of the two massive scalar fields (ϕ_1 and ϕ_2) from our original Lagrangian and we are left with $\chi(x)$. The vector fields W_μ^1 and W_μ^2 have grown massive, taking on a mass $g\phi_0$. The field W_μ^3 remains massless.

11.4.6 The Weinberg-Salam Model

The Yang-Mills theory represented a fundamental step in our understanding of fundamental interactions, introducing the concept of non-Abelian gauge symmetry as a unifying principle of the forces. However, in its original form, it describes massless gauge bosons, in contrast with the experimental observations of the weak interaction. Based on these observations, Abdus Salam and Steven Weinberg developed in the 1960s a unified model that combines the electromagnetic and weak forces into a single electroweak interaction. At the core of this theory lies the idea that the universe is the outcome of a symmetry breaking phase transition, through which elementary particles acquire mass via the Higgs mechanism. The inclusion of this mechanism in a Yang-Mills type gauge structure, based on the group $SU(2) \times U(1)$, allows to explain the origin of the masses of the weak bosons $W^\pm$ and Z^0 and of the electrons, while preserving gauge invariance and the masslessness of the photon.

The Lagrangian for the Weinberg-Salam model is given by:

$$\mathcal{L} = \bar{L}i\gamma^\mu D_\mu L + \bar{R}i\gamma^\mu D_\mu R + (D^\mu\phi)^\dagger(D_\mu\phi) + \frac{m_{\mathrm{h}}^2}{2}\phi^\dagger\phi - \frac{\lambda}{4}(\phi^\dagger\phi)^2$$
$$- G_e(\bar{L}\phi R + \bar{R}\phi^\dagger L) - \frac{1}{4}\boldsymbol{G}_{\mu\nu}^{(W)}\boldsymbol{G}^{(W)\mu\nu} - \frac{1}{4}F_{\mu\nu}^{(B)}F^{(B)\mu\nu}. \qquad (11.206)$$

Let us now examine the meaning of the various terms in the Lagrangian.

11.4.6.1 Neutrinos and Electrons Massless Lagrangian

Before symmetry breaking neutrinos and electrons are both massless particles. Electrons are possible in both left- and right-handed forms, but neutrinos only exist in left-handed form. They are all fermions, which implies that the reference Lagrangian is the Dirac one, without mass terms:

$$\mathcal{L}_f = \bar{\nu}_e(x)i\partial\!\!\!/\nu_e(x) + \bar{e}_L(x)i\partial\!\!\!/ e_L(x) + \bar{e}_R(x)i\partial\!\!\!/ e_R(x). \qquad (11.207)$$

The equation can be rewritten in a more compact form by introducing the column vector

$$\Psi(x) = \begin{pmatrix} \nu_e(x) \\ e_L(x) \\ e_R(x) \end{pmatrix}. \tag{11.208}$$

We then observe that the above Lagrangian is symmetric under both $U(1)$ and $SU(2)$ local transformations. In particular, the $U(1)$ transformation introduces a phase factor $e^{i\beta(x)}$. In order to ensure invariance, a gauge field $B_\mu(x)$ is introduced, which transforms as $B_\mu \to B_\mu + \frac{1}{g'Y}\partial_\mu\beta(x)$, where Y is the weak hypercharge of the theory and the parameter g' tells us how strongly particles will couple to the hypercharge. For future convenience, we change the position of the hypercharge and we write:

$$e^{i\frac{Y}{2}\beta(x)}, \quad B_\mu \to B_\mu + \frac{1}{g'}\partial_\mu\beta(x). \tag{11.209}$$

The known values for the charge are given in the table below.

Weak hypercharge Y		
ν_e	e_L	e_R
-1	-1	-2

Therefore, Ψ transform as

$$\begin{pmatrix} \nu_e \\ e_L \\ e_R \end{pmatrix} \to \begin{pmatrix} e^{-i\frac{\beta(x)}{2}} & 0 & 0 \\ 0 & e^{-i\frac{\beta(x)}{2}} & 0 \\ 0 & 0 & e^{-i\beta(x)} \end{pmatrix} \begin{pmatrix} \nu_e \\ e_L \\ e_R \end{pmatrix}. \tag{11.210}$$

Finally, to ensure the invariance of the Lagrangian, the covariant derivative corresponding to the $U(1)$ transformation must be defined:

$$D_\mu\Psi = \partial_\mu\Psi - \frac{i}{2}g'YB_\mu(x)\Psi. \tag{11.211}$$

Concerning the $SU(2)$ transformation, the discussion proceeds in a similar manner. This transformation introduces a phase that depends on I, also known as weak isospin:

$$e^{iI\boldsymbol{\tau}\cdot\boldsymbol{\alpha}(x)}, \quad \boldsymbol{\tau}\cdot\mathbf{W}_\mu \to \boldsymbol{\tau}\cdot\mathbf{W}_\mu + \frac{1}{g}\boldsymbol{\tau}(\partial_\mu\boldsymbol{\alpha}(x)) - \boldsymbol{\tau}\cdot(\boldsymbol{\alpha}\times\mathbf{W}_\mu), \tag{11.212}$$

where, in this case, g is the coupling to the weak isospin.

The known values for the isospin are given in the table below.

Weak Isospin I			
	ν_e	e_L	e_R
I	$\frac{1}{2}$	$\frac{1}{2}$	0
I_3	$\frac{1}{2}$	$-\frac{1}{2}$	0

Therefore, Ψ transform as

$$\begin{pmatrix} \nu_e \\ e_L \end{pmatrix} \to e^{\frac{i}{2}\boldsymbol{\tau}\cdot\boldsymbol{\alpha}(x)} \begin{pmatrix} \nu_e \\ e_L \end{pmatrix} \; ; \quad e_R \to e_R. \tag{11.213}$$

Again, to ensure the invariance of the Lagrangian, the covariant derivative corresponding to the $SU(2)$ transformation must be defined:

$$D_\mu \Psi = \partial_\mu \Psi - igI\boldsymbol{\tau} \cdot \boldsymbol{W}_\mu \Psi. \tag{11.214}$$

In conclusion, the original column vector (11.208) can be decomposed into a left-handed component, which has the same values of Y and I, and a right-handed component, which differs from the others. Hence, we can express:

$$\Psi = \begin{pmatrix} L \\ R \end{pmatrix} \text{ with } L = \begin{pmatrix} \nu_e \\ e_L \end{pmatrix}, \quad R = e_R. \tag{11.215}$$

In terms of these new fields, the covariant derivatives are written:

$$D_\mu L = \partial_\mu L - \frac{i}{2} g\boldsymbol{\tau} \cdot \boldsymbol{W}_\mu L + \frac{i}{2} g' B_\mu(x) L, \tag{11.216}$$

$$D_\mu R = \partial_\mu R + ig' B_\mu(x) R, \tag{11.217}$$

and our resulting locally symmetric Lagrangian is given by:

$$\begin{aligned} \mathcal{L}_f =& \bar{L} i\gamma^\mu D_\mu L + \bar{R} i\gamma^\mu D_\mu R \\ & - \frac{1}{4}\boldsymbol{G}_{\mu\nu}^{(W)} \boldsymbol{G}^{(W)\mu\nu} - \frac{1}{4} F_{\mu\nu}^{(B)} F^{(B)\mu\nu}, \end{aligned} \tag{11.218}$$

where $\boldsymbol{G}_{\mu\nu}^{(W)} = \partial_\mu \boldsymbol{W}_\nu - \partial_\nu \boldsymbol{W}_\mu - g\boldsymbol{W}_\mu \times \boldsymbol{W}_\nu$ and $F_{\mu\nu}^{(B)} = \partial_\mu B_\nu - \partial_\nu B_\mu$, which are the correct contributions from the respective gauge fields for $SU(2)$ and $U(1)$ theories.

▶ **Driving exercise**

Invariance respect to the global SU(2) transformation of the Dirac Lagrangian.

The Dirac Lagrangian for two sort of fermions, each with mass m, is:

$$\mathcal{L} = \bar{f}(i\gamma^\mu \partial_\mu - m) f + \bar{g}(i\gamma^\mu \partial_\mu - m) g = \bar{\Psi}(i\gamma^\mu \partial_\mu - m)\Psi, \tag{11.219}$$

where

$$\Psi = \begin{pmatrix} f \\ g \end{pmatrix}, \quad \bar{\Psi} = \left(\bar{f}\ \bar{g}\right). \tag{11.220}$$

Consider now the global $SU(2)$ transformation:

$$\Psi \to e^{\frac{i}{2}\boldsymbol{\tau}\cdot\boldsymbol{\alpha}}\Psi, \tag{11.221}$$

$$\bar{\Psi} \to \bar{\Psi}e^{-\frac{i}{2}\boldsymbol{\tau}\cdot\boldsymbol{\alpha}}, \tag{11.222}$$

where $\boldsymbol{\tau} = (\tau_1, \tau_2, \tau_3)$ are the Pauli matrices.
Substituting these into the Lagrangian (11.219) we get:

$$\begin{aligned}
\mathcal{L} &\to \bar{\Psi}e^{-\frac{i}{2}\boldsymbol{\tau}\cdot\boldsymbol{\alpha}}(i\gamma^\mu\partial_\mu - m)e^{\frac{i}{2}\boldsymbol{\tau}\cdot\boldsymbol{\alpha}}\Psi \\
&= \bar{\Psi}e^{-\frac{i}{2}\boldsymbol{\tau}\cdot\boldsymbol{\alpha}}(i\gamma^\mu\partial_\mu)e^{\frac{i}{2}\boldsymbol{\tau}\cdot\boldsymbol{\alpha}}\Psi - \bar{\Psi}e^{-\frac{i}{2}\boldsymbol{\tau}\cdot\boldsymbol{\alpha}}me^{\frac{i}{2}\boldsymbol{\tau}\cdot\boldsymbol{\alpha}}\Psi \\
&= e^{-\frac{i}{2}\boldsymbol{\tau}\cdot\boldsymbol{\alpha}}e^{\frac{i}{2}\boldsymbol{\tau}\cdot\boldsymbol{\alpha}}\bar{\Psi}(i\gamma^\mu\partial_\mu)\Psi - e^{-\frac{i}{2}\boldsymbol{\tau}\cdot\boldsymbol{\alpha}}e^{\frac{i}{2}\boldsymbol{\tau}\cdot\boldsymbol{\alpha}}\bar{\Psi}m\Psi \\
&= \bar{\Psi}(i\gamma^\mu\partial_\mu)\Psi - \bar{\Psi}m\Psi \\
&= \bar{\Psi}(i\gamma^\mu\partial_\mu - m)\Psi,
\end{aligned} \tag{11.223}$$

which shows the invariance of the Lagrangian for this type of transformation.
If we instead consider the associated infinitesimal transformation:

$$\Psi \to (1 + \frac{i}{2}\boldsymbol{\tau}\cdot\boldsymbol{\alpha})\Psi, \tag{11.224}$$

$$\bar{\Psi} \to \bar{\Psi}(1 - \frac{i}{2}\boldsymbol{\tau}\cdot\boldsymbol{\alpha}), \tag{11.225}$$

we obtain:

$$\begin{aligned}
\mathcal{L} &\to \bar{\Psi}(1 - \frac{i}{2}\boldsymbol{\tau}\cdot\boldsymbol{\alpha})(i\gamma^\mu\partial_\mu - m)(1 + \frac{i}{2}\boldsymbol{\tau}\cdot\boldsymbol{\alpha})\Psi \\
&= \bar{\Psi}(i\gamma^\mu\partial_\mu - m)\Psi + \bar{\Psi}(i\gamma^\mu\partial_\mu - m)\frac{i}{2}\boldsymbol{\tau}\cdot\boldsymbol{\alpha}\Psi \\
&\quad - \bar{\Psi}\frac{i}{2}\boldsymbol{\tau}\cdot\boldsymbol{\alpha}(i\gamma^\mu\partial_\mu - m)\Psi - \cancel{\bar{\Psi}\frac{i}{2}\boldsymbol{\tau}\cdot\boldsymbol{\alpha}(i\gamma^\mu\partial_\mu - m)\frac{i}{2}\boldsymbol{\tau}\cdot\boldsymbol{\alpha}\Psi} \\
&= \bar{\Psi}(i\gamma^\mu\partial_\mu - m)\Psi
\end{aligned} \tag{11.226}$$

(where the last term has been simplified since it is of higher order), which turns out to be invariant in any case.

Note that the electric charge of the particle can be derived from the hypercharge Y and the isospin I through the Gell-Mann-Nishijima relation:

$$Q = I_3 + \frac{Y}{2}. \tag{11.227}$$

11.4.6.2 Higgs Lagrangian

We now introduce a massive, complex scalar field called the Higgs field. It has four components, which are conveniently arranged into a two-component vector as follows:

$$\phi = \begin{pmatrix} \phi^+ \\ \phi^0 \end{pmatrix} = \frac{1}{\sqrt{2}} \begin{pmatrix} \phi_3 + i\phi_4 \\ \phi_1 + i\phi_2 \end{pmatrix}. \tag{11.228}$$

In this way we have that:

$$\phi^\dagger \phi = (\phi^+)^* \phi^+ + (\phi^0)^* \phi^0 = \frac{1}{2}(\phi_1^2 + \phi_2^2 + \phi_3^2 + \phi_4^2), \tag{11.229}$$

similar to a magnitude.

The Higgs field is defined to have weak hypercharge $Y = +1$ and weak isospin $I = \frac{1}{2}$, consequently its transformations under $U(1)$ and $SU(2)$ take the form:

$$\begin{pmatrix} \phi^+ \\ \phi^0 \end{pmatrix} \to \begin{pmatrix} e^{i\frac{\beta}{2}} & 0 \\ 0 & e^{i\frac{\beta}{2}} \end{pmatrix} \begin{pmatrix} \phi^+ \\ \phi^0 \end{pmatrix}, \quad \begin{pmatrix} \phi^+ \\ \phi^0 \end{pmatrix} \to e^{\frac{i}{2}\boldsymbol{\tau}\cdot\boldsymbol{\alpha}} \begin{pmatrix} \phi^+ \\ \phi^0 \end{pmatrix}. \tag{11.230}$$

This choice of charges requires a covariant derivative for the Higgs field of the form:

$$D_\mu = \partial_\mu \phi - \frac{i}{2} g \boldsymbol{\tau} \cdot \mathbf{W}_\mu \phi - \frac{i}{2} g' B_\mu \phi. \tag{11.231}$$

Finally, the Lagrangian of the Higgs field is:

$$\mathcal{L}_\phi = (D^\mu \phi)^\dagger (D_\mu \phi) + \frac{m_{\mathrm{h}}^2}{2} \phi^\dagger \phi - \frac{\lambda}{4} (\phi^\dagger \phi)^2. \tag{11.232}$$

11.4.6.3 Interaction Lagrangian

Regarding the interaction, we know that it is the coupling between the Higgs field and the left- and right-handed components of the electron, e_L and e_R, that gives it its mass. This is given by:

$$\mathcal{L}_{\mathrm{I}} = -G_e (\bar{L} \phi R + \bar{R} \phi^\dagger L), \tag{11.233}$$

where G_e is the coupling strength.

11.4.7 Higgs Broken Symmetry and Origin of Electron Mass

The Higgs field is unstable to symmetry breaking. After about $10^{-12} s$ from the Big Bang, the Universe cooled and its ground state spontaneously broke the $SU(2) \otimes U(1)$ symmetry.

The potential for the Higgs field, described in Eq. (11.232), is very similar to that of the Lagrangian given in in Eq. (11.174). The process of symmetry breaking, and consequently the corresponding calculations, are therefore quite analogous. In the present case, the minimum of the potential is obtained for a value

$$(\phi^\dagger \phi)_0 = \frac{m_{\text{h}}^2}{\lambda} \equiv v^2. \tag{11.234}$$

We will choose to break the symmetry with a new ground state at

$$(\phi_1)_0^2 = \frac{2m_{\text{h}}^2}{\lambda}, \ (\phi_2)_0^2 = 0, \ (\phi_3)_0^2 = 0, \ (\phi_4)_0^2 = 0. \tag{11.235}$$

Writing $(\phi_1)_0 = \left(\frac{2m_{\text{h}}^2}{\lambda}\right)^{\frac{1}{2}} = \sqrt{2}v$ we then have a ground state field:

$$(\phi)_0 = \begin{pmatrix} \phi^+ \\ \phi^0 \end{pmatrix} = \begin{pmatrix} 0 \\ v \end{pmatrix}. \tag{11.236}$$

Choosing to work in the **Unitary Gauge**, the excitations of the field around the vacuum state are given by:

$$\phi = \begin{pmatrix} 0 \\ v + \frac{h(x)}{\sqrt{2}} \end{pmatrix}. \tag{11.237}$$

Symmetry breaking has profound consequences: it predicts the mass of the electron, the masslessness of both the neutrino and the photon, and the nonzero masses of the $W^{\pm}$ and Z^0 bosons.

▶ **Driving exercise**

Let us now give mass to the electron!

As a first step, we identify the term in the Lagrangian responsible for the fermion mass. The Dirac Lagrangian is:

$$\mathcal{L} = \bar{\psi}(\not{p} - m)\psi, \tag{11.238}$$

with $\psi = \psi_L + \psi_R$. Remembering that

$$\begin{cases} \psi_L = P_L \psi = \\ \psi_R = P_R \psi \end{cases} \Rightarrow \begin{cases} \bar{\psi}_L = \bar{\psi} P_R \\ \bar{\psi}_R = \bar{\psi} P_L \end{cases}, \tag{11.239}$$

with $P_L = \left(\frac{1-\gamma^5}{2}\right)$ and $P_R = \left(\frac{1+\gamma^5}{2}\right)$, the Lagrangian becomes:

$$\begin{aligned} \mathcal{L} &= (\bar{\psi}_L + \bar{\psi}_R)(\not{p} - m)(\psi_L + \psi_R) \\ &= \bar{\psi}_L \not{p} \psi_L + \bar{\psi}_L \not{p} \psi_R - \bar{\psi}_L m \psi_L - \bar{\psi}_L m \psi_R + \bar{\psi}_R \not{p} \psi_L \\ &\quad + \bar{\psi}_R \not{p} \psi_R - \bar{\psi}_R m \psi_L - \bar{\psi}_R m \psi_R \\ &= \bar{\psi}_L \not{p} \psi_L + \bar{\psi}_R \not{p} \psi_R - m(\bar{\psi}_L \psi_R + \bar{\psi}_R \psi_L), \end{aligned} \tag{11.240}$$

where in the last step we use the fact that:

$$\bar{\psi}_L \not{\partial} \psi_R = \bar{\psi} P_R \gamma^\mu \partial_\mu (P_R \psi) = \bar{\psi} \gamma^\mu P_L P_R \psi = 0 \tag{11.241}$$

and

$$\bar{\psi}_L m \psi_L = m \bar{\psi}_L \psi_L = m \bar{\psi} P_R P_L \psi = 0 = \bar{\psi}_R m \psi_R. \tag{11.242}$$

Therefore

$$\mathcal{L} = \bar{\psi}_L \not{\partial} \psi_L + \bar{\psi}_R \not{\partial} \psi_R - m(\bar{\psi}_L \psi_R + \bar{\psi}_R \psi_L). \tag{11.243}$$

This highlights the fact that the mass arises from a term involving the combination $(\bar{\psi}_L \psi_R + \bar{\psi}_R \psi_L)$.
In the Weinberg-Salam Lagrangian, this term is expected to appear in the interaction Lagrangian, expressed in the Eq.(11.233), $\mathcal{L}_I = -G_e(\bar{L}\phi R + \bar{R}\phi^\dagger L)$. We can now insert our broken symmetry excited state, Eq.(11.237), into this expression, analyzing the two terms separately:

$$\bar{L}\phi R = (\bar{\nu}_e \; \bar{e}_L) \begin{pmatrix} 0 \\ v + \frac{h(x)}{\sqrt{2}} \end{pmatrix} e_R = \bar{e}_L \left(v + \frac{h(x)}{\sqrt{2}} \right) e_R \tag{11.244}$$

and

$$\bar{R}\phi^\dagger L = \bar{e}_R \left(0 \; v + \frac{h(x)}{\sqrt{2}} \right) \begin{pmatrix} \nu_e \\ e_L \end{pmatrix} = \bar{e}_R \left(v + \frac{h(x)}{\sqrt{2}} \right) e_L. \tag{11.245}$$

Putting the two terms together we obtain:

$$\mathcal{L}_{int} = -G_e(\bar{e}_L v e_R + \bar{e}_R v e_L) - G_e \left(\bar{e}_L \frac{h(x)}{\sqrt{2}} e_R + \bar{e}_R \frac{h(x)}{\sqrt{2}} e_L \right). \tag{11.246}$$

The first term represents the mass term generated by the interaction with the vacuum expectation value of the Higgs field and it provides mass to the electron. It predicts the electron mass is equal to $m_e = G_e v$.
The second term corresponds to the interaction between the electron and the excitation of the Higgs field $h(x)$.

11.4.8 The Photon and the Gauge Bosons

Next, we demonstrate that the theory predicts a massless photon. To this end, we consider the portion of the Lagrangian in which the gauge fields couple to the Higgs field. According to the principle of minimal coupling, this corresponds to the term containing the covariant derivative of the Higgs field: $(D_\mu\phi)^\dagger(D^\mu\phi)$.

The covariant derivative for the Higgs field, defined in Eq. (11.231), can be equivalently written as:

$$D_\mu\phi = \partial_\mu\phi - \left(\frac{i}{2}g\boldsymbol{\tau}\cdot\boldsymbol{W}_\mu + \frac{i}{2}g'B_\mu\right)\phi. \tag{11.247}$$

Displaying the Pauli matrices:

$$\tau^1 = \begin{pmatrix} 0 & 1 \\ 1 & 0 \end{pmatrix}; \quad \tau^2 = \begin{pmatrix} 0 & -i \\ i & 0 \end{pmatrix}; \quad \tau^3 = \begin{pmatrix} 1 & 0 \\ 0 & -1 \end{pmatrix}, \tag{11.248}$$

we have that:

$$\begin{aligned}\boldsymbol{\tau}\cdot\boldsymbol{W}_\mu &= \tau^1 W_\mu^1 + \tau^2 W_\mu^2 + \tau^3 W_\mu^3 = \begin{pmatrix} 0 & 1 \\ 1 & 0 \end{pmatrix} W_\mu^1 + \begin{pmatrix} 0 & -i \\ i & 0 \end{pmatrix} W_\mu^2 + \begin{pmatrix} 1 & 0 \\ 0 & -1 \end{pmatrix} W_\mu^3 \\ &= \begin{pmatrix} W_\mu^3 & W_\mu^1 - iW_\mu^2 \\ W_\mu^1 + iW_\mu^2 & -W_\mu^3 \end{pmatrix}.\end{aligned} \tag{11.249}$$

Therefore, inserting the broken symmetry version of the Higgs field ϕ, the covariant derivative becomes:

$$\partial_\mu\begin{pmatrix} 0 \\ v + \frac{h(x)}{\sqrt{2}} \end{pmatrix} - \left[\frac{i}{2}g\begin{pmatrix} W_\mu^3 & W_\mu^1 - iW_\mu^2 \\ W_\mu^1 + iW_\mu^2 & -W_\mu^3 \end{pmatrix} + \frac{i}{2}g'B_\mu\right]\begin{pmatrix} 0 \\ v + \frac{h(x)}{\sqrt{2}} \end{pmatrix}. \tag{11.250}$$

We now expand the product of the two matrices:

$$\begin{aligned}&\begin{pmatrix} W_\mu^3 & W_\mu^1 - iW_\mu^2 \\ W_\mu^1 + iW_\mu^2 & -W_\mu^3 \end{pmatrix}\begin{pmatrix} 0 \\ v + \frac{h(x)}{\sqrt{2}} \end{pmatrix} \\ &= \begin{pmatrix} (W_\mu^1 - iW_\mu^2)(v + \frac{h(x)}{\sqrt{2}}) \\ -W_\mu^3(v + \frac{h(x)}{\sqrt{2}}) \end{pmatrix} = \begin{pmatrix} (W_\mu^1 - iW_\mu^2)v + (W_\mu^1 - iW_\mu^2)\frac{h(x)}{\sqrt{2}} \\ -W_\mu^3 v - W_\mu^3\frac{h(x)}{\sqrt{2}} \end{pmatrix}\end{aligned} \tag{11.251}$$

$$
\begin{aligned}
\Rightarrow D_\mu \phi = &\begin{pmatrix} 0 \\ \partial_\mu (v + \frac{h(x)}{\sqrt{2}}) \end{pmatrix} - \left[\frac{i}{2} g \begin{pmatrix} (W^1_\mu - i W^2_\mu) v + (W^1_\mu - i W^2_\mu) \frac{h(x)}{\sqrt{2}} \\ -W^3_\mu v - W^3_\mu \frac{h(x)}{\sqrt{2}} \end{pmatrix} \right. \\
&\left. + \frac{i}{2} g' B_\mu \begin{pmatrix} 0 \\ v + \frac{h(x)}{\sqrt{2}} \end{pmatrix} \right] \\
= &\frac{i}{2} \begin{pmatrix} g(W^1_\mu - i W^2_\mu) v + g(W^1_\mu - i W^2_\mu) \frac{h(x)}{\sqrt{2}} \\ i\sqrt{2}(\partial_\mu h(x)) + v(g' B_\mu - g W^3_\mu) + \frac{h(x)}{\sqrt{2}} (g' B_\mu - g W^3_\mu) \end{pmatrix}.
\end{aligned}
\tag{11.252}
$$

For simplicity, we introduce:

$$
D_\mu \phi = -\frac{i}{2} \begin{pmatrix} A_\mu \\ B_\mu \end{pmatrix}, \tag{11.253}
$$

with $A_\mu \equiv g(W^1_\mu - i W^2_\mu)(v + \frac{h(x)}{\sqrt{2}})$, $B_\mu \equiv i\sqrt{2}(\partial_\mu h(x)) + C_\mu$ and $C_\mu \equiv (g' B_\mu - g W^3_\mu)(v + \frac{h(x)}{\sqrt{2}})$. It follows that

$$
(D_\mu \phi)^\dagger = \frac{i}{2} (A^*_\mu \; B^*_\mu), \tag{11.254}
$$

and consequently

$$
(D_\mu \phi)^\dagger (D^\mu \phi) = \frac{1}{4} (A^*_\mu A^\mu + B^*_\mu B^\mu). \tag{11.255}
$$

We analyse the result term by term.

- Consider $A^*_\mu A^\mu$:

$$
\begin{aligned}
A^*_\mu A^\mu = &g^2 \left(v + \frac{h(x)}{\sqrt{2}} \right)^2 (W^1_\mu - i W^2_\mu)(W^1_\mu + i W^2_\mu) \\
= &g^2 \left(v + \frac{h(x)}{\sqrt{2}} \right)^2 \left[(W^1_\mu)^2 + (W^2_\mu)^2 \right].
\end{aligned}
\tag{11.256}
$$

- Consider $B^*_\mu B^\mu$:

$$
\begin{aligned}
B^*_\mu B^\mu = &\left(C^*_\mu - i\sqrt{2}(\partial_\mu h(x)) \right) \left(C^\mu + i\sqrt{2}(\partial^\mu h(x)) \right) \\
= &C^*_\mu C^\mu + 2 \partial_\mu h(x) \partial^\mu h(x) \\
= &(g' B_\mu - g W^3_\mu)^2 \left(v + \frac{h(x)}{\sqrt{2}} \right)^2 + 2 \partial_\mu h(x) \partial^\mu h(x).
\end{aligned}
\tag{11.257}
$$

Putting the two halves together we obtain:

$$
\begin{aligned}
(D_\mu\phi)^\dagger(D^\mu\phi) =& \frac{1}{4}\Bigg[g^2\left(v+\frac{h(x)}{\sqrt{2}}\right)^2[(W_\mu^1)^2+(W_\mu^2)^2] \\
&+(g'B_\mu-gW_\mu^3)^2\left(v+\frac{h(x)}{\sqrt{2}}\right)^2+2\partial_\mu h(x)\partial^\mu h(x)\Bigg] \\
=& \frac{1}{2}\partial_\mu h\partial^\mu h+\frac{g^2v^2}{4}(W_\mu^1)^2+\frac{g^2v^2}{4}(W_\mu^2)^2+\frac{v^2}{4}(gW_\mu^3-g'B_\mu)^2,
\end{aligned}
\tag{11.258}
$$

where, in the last step, only the terms of order zero in $h(x)$ were taken into account.

It is possible to note that a boson mass always appears in the Lagrangian in the form $1/2(mass)^2 \times (field)^2$: after breaking the symmetry of the Higgs field, two massive vector bosons, W_μ^1 and W_μ^2, are generated, each with mass $M_W^2 = g^2v^2/2$. Finally, the linear combination of gauge fields, $gW_\mu^3 - g'B_\mu$, also becomes massive, with a mass that depends on the relative values of the coupling constants g and g'.

We have thus obtained three massive fields, from which the Z^0 and $W^\pm$ bosons will later emerge.

It remains to determine where the photon comes from. It can be observed that the term $(g'W_\mu^3 + gB_\mu)$ is absent from Eq. (11.258). We therefore conclude that A_μ corresponds to this combination.

Since the particles depend on the ratio g'/g we introduce the right-angled triangle shown in Fig. 11.6, with sides g and g', and define the Weinberg angle as $\tan\theta_W = g'/g$.

We then define two new fields Z_μ and A_μ as:

$$
\begin{aligned}
\begin{pmatrix} Z_\mu \\ A_\mu \end{pmatrix} &= \begin{pmatrix} \cos\theta_\omega & -\sin\theta_\omega \\ \sin\theta_\omega & \cos\theta_\omega \end{pmatrix}\begin{pmatrix} W_\mu^3 \\ B_\mu \end{pmatrix} \\
&= \begin{pmatrix} W_\mu^3\cos\theta_\omega & -B_\mu\sin\theta_\omega \\ W_\mu^3\sin\theta_\omega & B_\mu\cos\theta_\omega \end{pmatrix}.
\end{aligned}
\tag{11.259}
$$

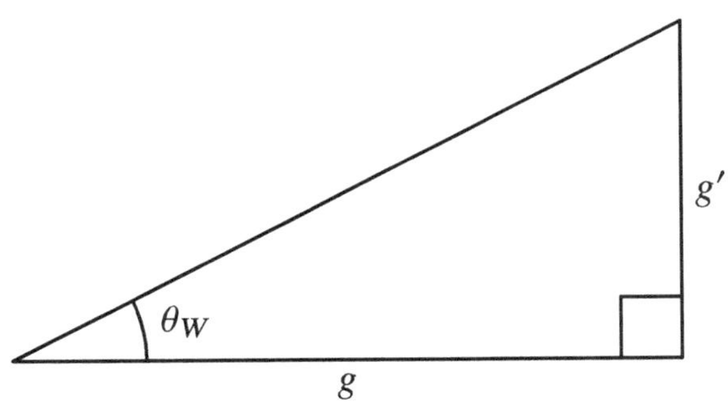

Fig. 11.6 Definition of the Weinberg angle θ_W in terms of g and g'

It follows that:

$$\begin{cases} Z_\mu = W^3_\mu \cos\theta_\omega - B_\mu \sin\theta_\omega \\ A_\mu = W^3_\mu \sin\theta_\omega + B_\mu \cos\theta_\omega \end{cases} \Rightarrow \begin{cases} B_\mu = A_\mu \cos\theta_\omega - Z_\mu \sin\theta_\omega \\ W^3_\mu = Z_\mu \cos\theta_\omega + A_\mu \sin\theta_\omega \end{cases} . \tag{11.260}$$

Starting from Eq. (11.258) and noting also that

$$g W^3_\mu - g' B_\mu = Z_\mu (g'^2 + g^2)^{\frac{1}{2}} \Rightarrow (g W^3_\mu - g' B_\mu)^2 = Z^2_\mu (g'^2 + g^2) \tag{11.261}$$

and that

$$\begin{aligned} \tan\theta_\omega &= \frac{g'}{g} \Rightarrow \cos\theta_\omega = \frac{g}{(g'^2 + g^2)^{\frac{1}{2}}} \\ \Rightarrow \cos^2\theta_\omega &= \frac{g^2}{(g'^2 + g^2)} \Rightarrow (g'^2 + g^2) = \frac{g^2}{\cos^2\theta_\omega}, \end{aligned} \tag{11.262}$$

the previous expression can be rewritten as:

$$(g W^3_\mu - g' B_\mu)^2 = Z^2_\mu (g'^2 + g^2) = Z^2_\mu \frac{g^2}{\cos^2\theta_\omega}. \tag{11.263}$$

Equation (11.258) therefore becomes:

$$(D_\mu \phi)^\dagger (D^\mu \phi) = \frac{1}{2}[\partial h(x)]^2 + \frac{g^2 v^2}{4}(W^1_\mu)^2 + \frac{g^2 v^2}{4}(W^2_\mu)^2 + \frac{v^2}{4}\frac{g^2}{\cos^2\theta_\omega} Z^2_\mu, \tag{11.264}$$

and from this equation we can read off that the mass of the Z^0 is $M_W / \cos\theta_\omega$.

The next step is to determine how the gauge fields couple to the fermions. For this purpose, we consider the covariant derivatives of the Eq. (11.218):

$$i\bar{R}\gamma^\mu(\partial_\mu + ig' B_\mu)R + i\bar{L}\gamma^\mu(\partial_\mu - \frac{i}{2} g \boldsymbol{\tau} \cdot \mathbf{W}_\mu + \frac{i}{2} g' B_\mu)L, \tag{11.265}$$

remembering that

$$\begin{aligned} R &= e_R \to \bar{R} = \bar{e}_R \\ L &= \begin{pmatrix} \nu_e \\ e_L \end{pmatrix} \to \bar{L} = \begin{pmatrix} \bar{\nu}_e & \bar{e}_L \end{pmatrix}. \end{aligned}$$

Let us first consider only the terms involving L.

– We consider $i\bar{L}\gamma^\mu \partial_\mu L$:

$$i\bar{L}\gamma^\mu \partial_\mu L = i \begin{pmatrix} \bar{\nu}_e & \bar{e}_L \end{pmatrix} \gamma^\mu \partial_\mu \begin{pmatrix} \nu_e \\ e_L \end{pmatrix} = i\bar{\nu}_e \gamma^\mu \partial_\mu \nu_e + i\bar{e}_L \gamma^\mu \partial_\mu e_L. \tag{11.266}$$

– We consider $i\bar{L}\gamma^\mu \frac{i}{2} g' B_\mu L$:

$$i\bar{L}\gamma^\mu \frac{i}{2} g' B_\mu L = i \begin{pmatrix} \bar{\nu}_e & \bar{e}_L \end{pmatrix} \gamma^\mu \frac{i}{2} g' B_\mu \begin{pmatrix} \nu_e \\ e_L \end{pmatrix}$$
$$= -\frac{g'}{2}\bar{\nu}_e\gamma^\mu B_\mu \nu_e - \frac{g'}{2}\bar{e}_L\gamma^\mu B_\mu e_L. \tag{11.267}$$

– $i\bar{L}\gamma^\mu \frac{i}{2} g \boldsymbol{\tau} \cdot \boldsymbol{W}_\mu L$:

$$i\bar{L}\gamma^\mu \frac{i}{2} g \boldsymbol{\tau} \cdot \boldsymbol{W}_\mu L = \frac{g}{2} \begin{pmatrix} \bar{\nu}_e & \bar{e}_L \end{pmatrix} \gamma^\mu \begin{pmatrix} W^3_\mu & W^1_\mu - iW^2_\mu \\ W^1_\mu + iW^2_\mu & -W^3_\mu \end{pmatrix} \begin{pmatrix} \nu_e \\ e_L \end{pmatrix} =$$
$$= \frac{g}{2}\bar{\nu}_e\gamma^\mu (W^3_\mu \nu_e + (W^1_\mu - iW^2_\mu) e_L)$$
$$+ \frac{g}{2}\bar{e}_L\gamma^\mu ((W^1_\mu + iW^2_\mu)\nu_e - W^3_\mu e_L)$$
$$= \frac{g}{2}\bar{\nu}_e\gamma^\mu W^3_\mu \nu_e + \frac{g}{2}\bar{\nu}_e\gamma^\mu (W^1_\mu - iW^2_\mu) e_L$$
$$+ \frac{g}{2}\bar{e}_L\gamma^\mu (W^1_\mu + iW^2_\mu)\nu_e - \frac{g}{2}\bar{e}_L\gamma^\mu W^3_\mu e_L. \tag{11.268}$$

Defining

$$W_\mu = \frac{W^1_\mu + iW^2_\mu}{\sqrt{2}} \rightarrow W^\dagger_\mu = \frac{W^1_\mu - iW^2_\mu}{\sqrt{2}}, \tag{11.269}$$

the expression becomes:

$$\frac{g}{2}(\bar{\nu}_e\gamma^\mu \nu_e - \bar{e}_L\gamma^\mu e_L) W^3_\mu + \frac{g}{\sqrt{2}}(\bar{\nu}_e\gamma^\mu W^\dagger_\mu e_L + \bar{e}_L\gamma^\mu W_\mu \nu_e). \tag{11.270}$$

Now let us consider the terms involving R.

– We consider $i\bar{R}\gamma^\mu \partial_\mu R$:

$$i\bar{R}\gamma^\mu \partial_\mu R = i\bar{e}_R\gamma^\mu \partial_\mu e_R. \tag{11.271}$$

– We consider $i\bar{R}\gamma^\mu g' B_\mu R$:

$$\bar{R}\gamma^\mu g' B_\mu R = g'\bar{e}_R\gamma^\mu B_\mu e_R. \tag{11.272}$$

Collecting all the terms obtained, we get:

$$i\bar{e}_R\gamma^\mu\partial_\mu e_R - g'\bar{e}_R\gamma^\mu e_R B_\mu + i\bar{\nu}_e\gamma^\mu\partial_\mu\nu_e + i\bar{e}_L\gamma^\mu\partial_\mu e_L - \frac{g'}{2}\bar{\nu}_e\gamma^\mu B_\mu\nu_e$$
$$-\frac{g'}{2}\bar{e}_L\gamma^\mu B_\mu e_L + \frac{g}{2}(\bar{\nu}_e\gamma^\mu\nu_e - \bar{e}_L\gamma^\mu e_L)W^3_\mu + \frac{g}{\sqrt{2}}(\bar{\nu}_e\gamma^\mu e_L W^\dagger_\mu + \bar{e}_L\gamma^\mu \nu_e W_\mu). \tag{11.273}$$

In order to simplify the calculations, we consider the various terms separately.

– We consider $i\bar{e}_R\gamma^\mu\partial_\mu e_R + i\bar{e}_L\gamma^\mu\partial_\mu e_L$:

$$i\bar{e}_R\gamma^\mu\partial_\mu e_R + i\bar{e}_L\gamma^\mu\partial_\mu e_L = i(\bar{e}_R\gamma^\mu\partial_\mu e_R + \bar{e}_L\gamma^\mu\partial_\mu e_L) = ie\gamma^\mu\partial_\mu e. \tag{11.274}$$

– We consider $-\frac{g'}{2}\bar{\nu}_e\gamma^\mu\nu_e B_\mu - \frac{g'}{2}\bar{e}_L\gamma^\mu e_L B_\mu + \frac{g}{2}(\bar{\nu}_e\gamma^\mu\nu_e - \bar{e}_L\gamma^\mu e_L)W^3_\mu$:

$$\begin{aligned}&-\frac{g'}{2}\bar{\nu}_e\gamma^\mu\nu_e B_\mu - \frac{g'}{2}\bar{e}_L\gamma^\mu e_L B_\mu + \frac{g}{2}\bar{\nu}_e\gamma^\mu\nu_e W^3_\mu - \frac{g}{2}\bar{e}_L\gamma^\mu e_L W^3_\mu\\ &= \frac{1}{2}\bar{\nu}_e\gamma^\mu\nu_e(-g'B_\mu + gW^3_\mu) - \frac{1}{2}\bar{e}_L\gamma^\mu e_L(g'B_\mu + gW^3_\mu),\end{aligned} \tag{11.275}$$

which, for Eq. (11.263), becomes:

$$\frac{1}{2}\bar{\nu}_e\gamma^\mu\nu_e Z_\mu\frac{g}{\cos\theta_\omega} - \frac{1}{2}\bar{e}_L\gamma^\mu e_L(g'B_\mu + gW^3_\mu). \tag{11.276}$$

By substituting the result back into Eq. (11.273), we obtain:

$$\begin{aligned}&i\bar{e}\gamma^\mu\partial_\mu e - g'\bar{e}_R\gamma^\mu e_R B_\mu + i\bar{\nu}_e\gamma^\mu\partial_\mu\nu_e + \frac{1}{2}\bar{\nu}_e\gamma^\mu\nu_e Z_\mu\frac{g}{\cos\theta_\omega}\\ &-\frac{1}{2}\bar{e}_L\gamma^\mu e_L(g'B_\mu + gW^3_\mu) + \frac{g}{\sqrt{2}}(\bar{\nu}_e\gamma^\mu e_L W^\dagger_\mu + \bar{e}_L\gamma^\mu\nu_e W_\mu).\end{aligned} \tag{11.277}$$

Taking into account the term

$$\frac{1}{2}\bar{\nu}_e\gamma^\mu\nu_e Z_\mu\frac{g}{\cos\theta_\omega} - g'\bar{e}_R\gamma^\mu e_R B_\mu - \frac{1}{2}\bar{e}_L\gamma^\mu e_L(g'B_\mu + gW^3_\mu)$$

and Eq. (11.260), we have:

$$\begin{aligned}&\frac{1}{2}\bar{\nu}_e\gamma^\mu\nu_e Z_\mu\frac{g}{\cos\theta_\omega} - g'\bar{e}_R\gamma^\mu e_R B_\mu - \frac{1}{2}\bar{e}_L\gamma^\mu e_L(g'B_\mu + gW^3_\mu)\\ =&\frac{1}{2}\bar{\nu}_e\gamma^\mu\nu_e Z_\mu\frac{g}{\cos\theta_\omega} + \bar{e}_R\gamma^\mu e_R(-g'A_\mu\cos\theta_\omega + g'Z_\mu\sin\theta_\omega)\\ &-\frac{1}{2}\bar{e}_L\gamma^\mu e_L(g'A_\mu\cos\theta_\omega - g'Z_\mu\sin\theta_\omega + gZ_\mu\cos\theta_\omega + gA_\mu\sin\theta_\omega).\end{aligned} \tag{11.278}$$

Let us once again consider the terms separately.

– We consider $-\bar{e}_R\gamma^\mu e_R(g'\cos\theta_\omega)A_\mu - \frac{1}{2}\bar{e}_L\gamma^\mu e_L(g'\cos\theta_\omega + g\sin\theta_\omega)A_\mu$. Observing that $\tan\theta_\omega = \frac{g'}{g} \Rightarrow g\sin\theta_\omega = g'\cos\theta_\omega$ we get:

$$\begin{aligned}&-[\bar{e}_R\gamma^\mu e_R(g\sin\theta_\omega)A_\mu + \frac{1}{2}\bar{e}_L\gamma^\mu e_L(g\sin\theta_\omega + g\sin\theta_\omega)A_\mu]\\ =&-[\bar{e}_R\gamma^\mu e_R + \bar{e}_L\gamma^\mu e_L]g\sin\theta_\omega A_\mu = -g\sin\theta_\omega\bar{e}\gamma^\mu e A_\mu.\end{aligned} \tag{11.279}$$

– We consider $\frac{1}{2}\frac{g}{\cos\theta_\omega}\bar{\nu}_e\gamma^\mu\nu_e Z_\mu + \bar{e}_R\gamma^\mu e_R(g'\sin\theta_\omega)Z_\mu - \frac{1}{2}\bar{e}_L\gamma^\mu e_L$ $(-g'\sin\theta_\omega Z_\mu + g\cos\theta_\omega Z_\mu)$.
Hence,

$$\begin{aligned}
&\frac{g}{\cos\theta_\omega}\left[\frac{1}{2}\bar{\nu}_e\gamma^\mu\nu_e + \frac{g'\sin\theta_\omega\cos\theta_\omega}{g}\bar{e}_R\gamma^\mu e_R\right.\\
&\left.-\frac{1}{2}\bar{e}_L\gamma^\mu e_L\frac{-g'\sin\theta_\omega + g\cos\theta_\omega}{g}\cos\theta_\omega\right]Z_\mu\\
&=\frac{g}{\cos\theta_\omega}\left[\frac{1}{2}\bar{\nu}_e\gamma^\mu\nu_e + \sin^2\theta_\omega\bar{e}_R\gamma^\mu e_R - \frac{1}{2}\bar{e}_L\gamma^\mu e_L\cos 2\theta_\omega\right]Z_\mu, \qquad (11.280)
\end{aligned}$$

because

$$\frac{g'\sin\theta_\omega\cos\theta_\omega}{g} = \frac{g\sin^2\theta_\omega}{g} = \sin^2\theta_\omega$$

and

$$\begin{aligned}
\frac{-g'\sin\theta_\omega + g\cos\theta_\omega}{g}\cos\theta_\omega &= \frac{-g'\sin\theta_\omega\cos\theta_\omega + g\cos^2\theta_\omega}{g}\\
&= \frac{-g\sin^2\theta_\omega + g\cos^2\theta_\omega}{g}\\
&= -\sin^2\theta_\omega + \cos^2\theta_\omega = \cos 2\theta_\omega. \qquad (11.281)
\end{aligned}$$

Collecting all the previous terms, we find that Eq. (11.265) can be written as:

$$\begin{aligned}
&i\bar{e}\gamma^\mu\partial_\mu e + i\bar{\nu}_e\gamma^\mu\partial_\mu\nu_e - g\sin\theta_\omega\bar{e}\gamma^\mu e A_\mu\\
&+\frac{g}{\cos\theta_\omega}\left[\sin^2\theta_\omega\bar{e}_R\gamma^\mu e_R - \frac{1}{2}\bar{e}_L\gamma^\mu e_L\cos 2\theta_\omega + \frac{1}{2}\bar{\nu}_e\gamma^\mu\nu_e\right]Z_\mu\\
&+\frac{g}{\sqrt{2}}(\bar{\nu}_e\gamma^\mu e_L W_\mu^\dagger + \bar{e}_L\gamma^\mu\nu_e W_\mu), \qquad (11.282)
\end{aligned}$$

where $e = e_L + e_R$. The charged fields W_μ and $W_\mu^\dagger$ correspond to the physical $W^\pm$ bosons.

In conclusion, we can observe that the field A_μ does not contain any mass term; therefore, the photon remains massless. In contrast, the Z^0 and $W^\pm$ bosons acquire mass ($W^\pm$ as a combination of the massive gauge fields W_μ^1 and W_μ^2).

In theories with a global continuous symmetry, spontaneous symmetry breaking leads to the appearance of massless scalar excitations, known as Goldstone bosons. These modes correspond to fluctuations of the field along the flat directions of the potential, reflecting the degeneracy of the vacuum manifold. Goldstone's theorem ensures that such massless excitations are an unavoidable consequence of breaking a continuous global symmetry.

Gauge theories, however, exhibit a fundamentally different behavior. When a local symmetry is spontaneously broken, the would-be Goldstone modes do not appear as physical particles in the spectrum. Instead, they are absorbed by the gauge fields, providing the additional longitudinal degrees of freedom required for a massive vector boson. This phenomenon is often summarized by the statement that the gauge bosons "eat" the Goldstone bosons.

From a more precise point of view, no symmetry is truly lost. The underlying Lagrangian remains exactly invariant under the local gauge transformations. What changes is the realization of the symmetry in the vacuum state and the choice of gauge. The disappearance of the Goldstone modes reflects the redundancy inherent in the gauge description rather than the elimination of physical degrees of freedom.

The Higgs mechanism can therefore be understood as a rearrangement of degrees of freedom. Before symmetry breaking, the theory contains massless gauge bosons with two transverse polarizations and a set of scalar fields. After symmetry breaking, some of the scalar degrees of freedom are converted into longitudinal polarizations of massive gauge bosons, while the remaining scalar excitation manifests itself as a physical Higgs particle.

This mechanism resolves the apparent contradiction between gauge invariance and massive vector bosons. Explicit mass terms for gauge fields are forbidden by local gauge symmetry. However, masses generated dynamically through interaction with a symmetry-breaking vacuum preserve gauge invariance at the fundamental level and ensure the renormalizability of the theory.

In the electroweak theory, the Higgs mechanism plays a dual role. It not only generates masses for the weak gauge bosons but also provides a consistent origin for fermion masses through Yukawa interactions. The observed pattern of fermion masses is encoded in the corresponding Yukawa couplings, while the gauge structure fixes the form of the interactions.

It is important to emphasize that the Higgs field does not merely "add mass" to particles in a mechanical sense. Rather, the masses emerge as a consequence of the interaction between quantum fields and a nontrivial vacuum structure. In this sense, mass is not a fundamental parameter inserted by hand but a dynamical property arising from symmetry breaking.

The experimental discovery of the Higgs boson confirms this picture and provides direct evidence that the electroweak vacuum is indeed nontrivial. Understanding the Higgs mechanism is therefore essential not only for particle physics but also for cosmology, where the electroweak phase transition plays a key role in the thermal history of the Universe.

11.5 Beyond the Tree Level

The SM, formulated as a gauge theory based on $SU(3)_c \otimes SU(2)_L \otimes U(1)_Y$ with spontaneous symmetry breaking, is *renormalizable*. In practical terms, renormalizability means that once a finite set of input parameters is fixed by experiment, loop-corrected predictions for infrared observables (cross sections, decay widths, asymmetries, and so on) are finite and unambiguous order by order in perturbation theory. This statement is highly nontrivial: individual Feynman graphs are typically ultraviolet (UV) divergent, and only the sum of virtual corrections and counterterms (together with real-emission contributions, when needed, to cure infrared singularities) defines a meaningful observable.

The strategy is standard:

- compute the relevant amplitudes at a fixed loop order (here, one loop), including self-energies, vertices, and box diagrams;
- regulate UV divergences (dimensional regularization is the canonical choice in gauge theories);
- introduce renormalized parameters and fields, together with counterterms that absorb the UV poles;
- impose a renormalization scheme (e.g. on-shell or $\overline{\text{MS}}$), which fixes the finite parts of counterterms and thus specifies how the renormalized parameters are defined.

Once this is done, the perturbative expansion yields finite S-matrix elements.

What "beyond tree level" really means. Tree-level (Born) amplitudes are the leading terms in the loop expansion of the S-matrix. Going beyond tree level is not merely adding "small corrections": it is the quantum completion required by unitarity and gauge invariance. Ultraviolet divergences arise in intermediate steps, but they are consistently reabsorbed into a redefinition of the parameters and fields, leaving finite predictions for physical observables.

In the electroweak sector, it is convenient to organize the discussion in terms of the following parameters: the gauge couplings, g and g', and the Higgs vacuum expectation value v. At tree level, one introduces the weak mixing angle θ_W via

$$s_W^2 \equiv \sin^2\theta_W = \frac{g'^2}{g^2+g'^2}, \qquad c_W^2 \equiv \cos^2\theta_W = \frac{g^2}{g^2+g'^2}. \tag{11.283}$$

Tree-level relations between (g, g', v) and the most precisely known electroweak inputs are then

$$\alpha_0(0) = \frac{g^2 s_W^2}{4\pi}, \qquad G_{F0} = \frac{1}{\sqrt{2}\, v^2}, \qquad M_{Z0}^2 = \frac{g^2 v^2}{4 c_W^2}, \tag{11.284}$$

where the subscript 0 emphasizes that Eq. (11.284) holds at tree level.

Beyond tree level, the measurable quantities α, G_F and M_Z receive radiative corrections. A conceptually clean way to proceed is to choose three inputs and express every other observable in terms of them. A natural choice is

- the fine-structure constant $\alpha(0)$ extracted at very low momentum transfer;
- the Fermi constant G_F extracted from the muon lifetime;
- the Z-boson mass M_Z determined at the resonance peak.

At the quantum level one may write, schematically,

$$\alpha = \alpha_0 + \delta\alpha, \qquad G_F = G_{F0} + \delta G_F, \qquad M_Z^2 = M_{Z0}^2 + \delta M_Z^2, \tag{11.285}$$

where the counterterms (and loop contributions) encoded in $\delta\alpha$, δG_F, δM_Z^2 depend on the renormalization scheme and on the full SM particle content. The crucial point is that once the three renormalized inputs are fixed, the parameters (g, g', v) are determined (to the same loop order), and the SM yields predictions for all other electroweak observables.

In practice, heavy masses such as the top-quark mass m_t and the Higgs mass M_H enter many precision observables only through loops; accordingly, they appear as *radiative parameters*. This is the origin of the classic statement that electroweak precision fits constrain m_t and M_H indirectly, long before (or even without) direct production.

11.5.1 One-Loop Electroweak Building Blocks

At one loop, the dominant universal structures can be organized in terms of gauge-boson self-energies (vacuum polarizations), vertex form factors, and box diagrams. For the electroweak gauge bosons, it is convenient to define the transverse vacuum polarization amplitudes by

$$\Pi_{ab}^{\mu\nu}(q^2) \equiv g^{\mu\nu}\, \Pi_{ab}(q^2) + \text{terms proportional to } q^\mu q^\nu, \qquad a, b = \gamma, Z, W, \tag{11.286}$$

where only the transverse part $\Pi_{ab}(q^2)$ contributes to physical amplitudes when the gauge bosons couple to conserved external currents.

Two sets of one-loop quantities are particularly important for the renormalization of the chosen electroweak inputs.

– **The effective $Zf\bar{f}$ couplings at $q^2 = M_Z^2$.** Including vertex and external-leg corrections, the Z-fermion vertex can be written as

$$V^{\mu}_{Zf\bar{f}} = -i\,\frac{g}{2c_W}\,\bar{\psi}_f\gamma^{\mu}\Big(g_V^f - g_A^f\gamma_5\Big)\psi_f, \tag{11.287}$$

with loop-corrected vector and axial couplings parametrized as

$$g_V^f = \sqrt{\rho}\left(T_3^f - 2\kappa_f\,Q_f\,s_W^2\right), \qquad g_A^f = \sqrt{\rho}\;T_3^f. \tag{11.288}$$

Here T_3^f is the weak isospin of the left-handed fermion, Q_f its electric charge, and the form factors ρ and κ_f encode the leading universal and flavor-dependent electroweak corrections; at tree level $\rho = \kappa_f = 1$.

– **The radiative correction to muon decay at $q^2 \simeq 0$.** At low momentum transfer the muon-decay amplitude can be expressed in terms of the charged-current four-fermion operator, corrected by vertex and box diagrams:

$$\mathcal{M}(\mu) = -i\,\delta G_{(B,V)}\Big[\bar{\psi}_e\gamma^{\mu}(1-\gamma_5)\psi_{\nu_e}\Big]\Big[\bar{\psi}_{\nu_\mu}\gamma_\mu(1-\gamma_5)\psi_\mu\Big], \tag{11.289}$$

where $\delta G_{(B,V)}$ collects the finite one-loop contributions from box and vertex topologies (and the external-leg renormalization consistently included).

With these ingredients, the radiative corrections to the *input parameters* in Eq. (11.285) can be expressed in terms of vacuum polarizations and $\delta G_{(B,V)}$. In the on-shell organization, one writes

$$\frac{\delta\alpha}{\alpha} = -\Pi_{\gamma\gamma}(0) - 2\,\frac{s_W}{c_W}\,\frac{\Pi_{\gamma Z}(0)}{M_Z^2}, \tag{11.290}$$

$$\frac{\delta M_Z^2}{M_Z^2} = -\frac{\Pi_{ZZ}(M_Z^2)}{M_Z^2}, \tag{11.291}$$

$$\frac{\delta G_F}{G_F} = \frac{\Pi_{WW}(0)}{M_W^2} + \frac{\delta G_{(B,V)}}{G_F}. \tag{11.292}$$

Equation (11.290) makes explicit a key conceptual point: the renormalization of the *measured* low-energy parameters is controlled by the gauge-boson two-point functions evaluated at specific kinematic points ($q^2 = 0$ for α and G_F, and $q^2 = M_Z^2$ for M_Z).

Why the vacuum polarization matters. Gauge-boson self-energies are not a technical nuisance: they encode the quantum response of the vacuum to gauge fields. The shift from $\alpha(0)$ to an effective electromagnetic coupling at $q^2 = M_Z^2$ is governed by $\Pi_{\gamma\gamma}(q^2)$ and $\Pi_{\gamma Z}(q^2)$; similarly, the relation between G_F and M_W is controlled by $\Pi_{WW}(0)$ together with genuinely non-factorizable contributions (box and vertex terms) summarized by $\delta G_{(B,V)}$ in Eq. (11.290).

11.5.2 From Renormalized Inputs to Predictions: Derived Observables and the Case of M_W

Once the renormalized input set $\mathcal{I}^i = \{\alpha, G_F, M_Z\}$ is fixed, any derived observable O (or any properly defined S-matrix element) can be expanded consistently at one loop. Denote by O_0 the tree-level prediction expressed in terms of tree-level inputs $\mathcal{I}_0^i$, and by δO the genuine one-loop correction at fixed bare parameters. The renormalization of the inputs induces additional shifts. One can write

$$O\big[\mathcal{I}_0^i(\mathcal{I}^i)\big] = O_0(\mathcal{I}^i) - \sum_i \frac{\partial O_0}{\partial \mathcal{I}^i}\,\delta\mathcal{I}^i + \delta O(\mathcal{I}^i) \equiv O_0(\mathcal{I}^i) + \Delta O(\mathcal{I}^i). \quad (11.293)$$

A canonical example is the prediction for the W mass. At tree level one has

$$M_{W0}^2 = c_{W0}^2\, M_{Z0}^2, \quad (11.294)$$

and, using the tree-level relation between s_W^2 and the inputs,

$$s_W^2 c_W^2 = \frac{\pi\alpha}{\sqrt{2}\, G_F\, M_Z^2}, \quad (11.295)$$

one may solve for c_W^2 and obtain the familiar tree-level prediction $M_{W0}^2(\mathcal{I}^i)$ as a closed function of (α, G_F, M_Z). Differentiating $M_{W0}^2(\mathcal{I}^i)$ with respect to the inputs and inserting Eq. (11.290) into Eq. (11.293), one finds the one-loop structure

$$M_W^2 = c_W^2 M_Z^2 - \frac{c_W^2 M_Z^2}{s_W^2 - c_W^2}\left(s_W^2\,\frac{\delta\alpha}{\alpha} - s_W^2\,\frac{\delta G_F}{G_F} - c_W^2\,\frac{\delta M_Z^2}{M_Z^2}\right) + \delta M_W^2, \quad (11.296)$$

with the genuine one-loop W self-energy contribution

$$\delta M_W^2 = -\,\Pi_{WW}(M_W^2). \quad (11.297)$$

Equations (11.296) and (11.297) illustrate the general architecture of electroweak precision physics: input renormalization plus process-dependent finite corrections.

Beyond one loop, the same logic persists, but additional parameters that enter at one loop (notably m_t and M_H) must themselves be renormalized consistently. This is the technical reason why two-loop electroweak precision physics is a substantial enterprise rather than a routine extension.

11.5.3 Z-Pole Physics and Precision Observables

The cleanest experimental arena for testing electroweak radiative corrections is the Z pole, where the process

$$e^+e^- \to Z \to f\bar{f} \tag{11.298}$$

is measured with extremely high statistics. The Z resonance profile is characterized in terms of the peak position M_Z, the total width Γ_Z, and the peak cross section. In the narrow-width approximation, one writes the peak cross section as

$$\sigma^0_{f\bar{f}} = \frac{12\pi}{M_Z^2}\frac{\Gamma_e \Gamma_f}{\Gamma_Z^2}. \tag{11.299}$$

A central theoretical object here is the Z partial width into a given fermion species. At tree level, expressing the result in terms of G_F and the effective vector/axial couplings, one has

$$\Gamma(Z \to f\bar{f}) = C_f \frac{G_F M_Z^3}{6\sqrt{2}\pi}\left[\left(g_A^f\right)^2 + \left(g_V^f\right)^2\right], \tag{11.300}$$

where C_f is the color/QCD factor ($C_f = 1$ for leptons, while for quarks it includes perturbative QCD corrections). Beyond tree level, Eq. (11.300) remains valid provided the effective couplings $g^f_{V,A}$ include the loop-induced form factors of Eq. (11.288), together with the appropriate QED final-state radiation corrections.

One particularly useful observable is the effective weak mixing angle extracted at the Z pole,

$$\sin^2\theta_{\text{eff}} \equiv \frac{1}{4}\left(1 - \frac{g_V^\ell}{g_A^\ell}\right). \tag{11.301}$$

The key point is that $\sin^2\theta_{\text{eff}}$ is not simply s_W^2 evaluated at tree level: it is a renormalized, process-dependent quantity sensitive to electroweak loops through κ_ℓ in Eq. (11.288). This is one of the cleanest ways in which genuine weak radiative corrections become experimentally visible.

The Z pole also provides a classic determination of the number of light (active) neutrino species. Defining the invisible width Γ_{inv} and using

$$\Gamma_{\text{inv}} = N_\nu \, \Gamma_\nu, \tag{11.302}$$

one obtains N_ν from the ratio $\Gamma_{\text{inv}}/\Gamma_\ell$ once the SM prediction for Γ_ν/Γ_ℓ is supplied.

Precision tests as loop-level consistency checks. At the Z pole, the experimental sensitivity is such that one can resolve loop-level deviations from Born predictions in quantities like g_A^ℓ and $\sin^2\theta_{\text{eff}}$. The agreement between (i) the measured Z line shape, (ii) the extracted effective couplings in Eq. (11.288), and (iii) the M_W-M_Z-G_F relation at one loop, is not merely a fit: it is a stringent quantum-level consistency test of the electroweak gauge structure.

Finally, it is important to stress that QED effects must be handled with care at the Z pole. The running of α from $q^2 \simeq 0$ to $q^2 = M_Z^2$ is conventionally written as

$$\alpha(M_Z^2) = \frac{\alpha(0)}{1-\Delta\alpha}, \tag{11.303}$$

with $\Delta\alpha$ receiving leptonic, hadronic, and top-quark contributions; the hadronic term dominates the uncertainty budget because it is controlled by low-energy, nonperturbative QCD. Initial-state radiation is accounted for by a convolution with a radiator function,

$$\sigma(s) = \int_0^{k_{\max}} dk\; H(k)\, \sigma_0\big(s(1-k)\big), \tag{11.304}$$

and final-state radiation induces multiplicative QED factors of order $1 + O(\alpha Q_f^2/\pi)$.

The conceptual endpoint of this discussion is the following. Tree-level formulae such as Eqs. (11.284), (11.299) and (11.300) provide an excellent first organization of electroweak phenomena. Precision data, however, probe the SM at the level where loop effects are numerically mandatory and the dependence on heavy scales (notably m_t and M_H) becomes visible through virtual corrections. The agreement between the loop-improved SM predictions and the Z-pole data is therefore best understood as a quantum-level verification of the full gauge structure of the theory, rather than as a mere refinement of Born-level phenomenology.

11.5.4 Standard Model Parameters and Precision Observables at the Z Pole

A decisive validation of the electroweak sector of the SM came from precision measurements at the Z resonance, performed at CERN (LEP) and at SLAC (SLD).

The main conceptual point is that these experiments did not merely confirm tree-level (Born) relations: they resolved genuinely weak radiative corrections and constrained virtual effects of heavy degrees of freedom. In practice, the comparison is made by expressing observables in terms of a small set of precisely measured inputs (typically α, G_F, M_Z) and computing the loop-improved predictions for widths, asymmetries and effective couplings.

Throughout this section we parametrize the loop-corrected $Zf\bar{f}$ vertex as in Eq. (11.287), with effective vector and axial couplings written in the ρ-κ_f form of Eq. (11.288). This organization isolates universal electroweak corrections (encoded in ρ) from the process-dependent running of the effective weak mixing angle (encoded in κ_f).

Z **partial widths.** At tree level, the partial width for the decay $Z \to f\bar{f}$ can be written in a compact form in terms of G_F and the effective couplings:

$$\Gamma(Z \to f\bar{f}) = C_f \, \frac{G_F M_Z^3}{6\sqrt{2}\pi} \left[\left(g_A^f\right)^2 + \left(g_V^f\right)^2 \right], \tag{11.305}$$

where C_f encodes color and QCD effects. For leptons one has $C_f = 1$, whereas for quarks QCD radiative corrections enhance the width, so that

$$C_f = 3\left[1 + \frac{\alpha_s(M_Z^2)}{\pi} + 1.409\,\frac{\alpha_s^2(M_Z^2)}{\pi^2} + \cdots\right]. \tag{11.306}$$

Beyond tree level, Eq. (11.305) remains structurally correct provided:

- the couplings g_V^f and g_A^f are understood as loop-improved quantities of Eq. (11.288);
- QED final-state radiation and electroweak vertex corrections are included consistently in the definition of the effective couplings and/or as explicit multiplicative factors, depending on the adopted scheme.

For orientation, the SM partial widths at the Z pole (representative values) are commonly quoted as shown below.

Channel $Z \to f\bar{f}$	Γ_f [MeV]
$\nu\bar{\nu}$	167.25
e^+e^-	84.01
$u\bar{u}$	300.30
$d\bar{d}$	383.10
$b\bar{b}$	376.00

From the experimental side, the quantities most directly constrained by the line-shape analysis are the charged-lepton partial width Γ_ℓ, the hadronic width $\Gamma_{\rm had}$, the total width Γ_Z, and the invisible width $\Gamma_{\rm inv}$. Denoting

$$\Gamma_{\rm inv} \equiv \Gamma_Z - 3\,\Gamma_\ell - \Gamma_{\rm had}, \tag{11.307}$$

one interprets $\Gamma_{\rm inv}$ as the sum of all kinematically accessible neutral weak decays not seen in the detector (in the SM, dominantly neutrinos).

What the Z widths actually test. The widths probe the *normalization* of the neutral current, i.e., the overall strength of the Z coupling to fermions, and they are therefore particularly sensitive to the universal form factor ρ in Eq. (11.288). Since the data are precise at the per-mille level, small shifts in g_A^f away from the Born value are experimentally resolvable.

Number of light neutrino species. Assuming that the invisible width is saturated by decays into light neutrinos, one writes

$$\Gamma_{\rm inv} = N_\nu \, \Gamma_\nu, \tag{11.308}$$

where Γ_ν is the partial width for a single neutrino species and N_ν counts the number of neutrino flavors with

$$m_\nu < \frac{M_Z}{2}. \tag{11.309}$$

Experimentally one often uses the ratio $\Gamma_{\rm inv}/\Gamma_\ell$, because many systematic uncertainties cancel, and compares it to the SM prediction for Γ_ν/Γ_ℓ. Taking the ratio of Eq. (11.308) to Γ_ℓ yields

$$N_\nu = \frac{\Gamma_{\rm inv}/\Gamma_\ell}{(\Gamma_\nu/\Gamma_\ell)_{\rm SM}}. \tag{11.310}$$

The result is that N_ν is extremely close to 3, implying that, under the assumption that families replicate the observed pattern, only three generations contain neutrinos light enough to be produced in Z decay.

Radiative corrections beyond pure QED. A key conceptual question is whether one can disentangle *genuine weak* radiative corrections from the more elementary QED effects (in particular, the running of α). A clean way to see this is to consider pairs of observables whose correlation is different in "QED-only" and in full electroweak theory. A standard example is the $(\sin^2\theta_{\rm eff}, \Gamma_\ell)$ plane. One defines the effective weak mixing angle through the leptonic couplings as

$$\sin^2\theta_{\rm eff} \equiv \frac{1}{4}\left(1 - \frac{g_V^\ell}{g_A^\ell}\right), \tag{11.311}$$

which is directly sensitive to the loop-induced factor κ_ℓ in Eq. (11.288). The important point is that including only the photon vacuum polarization (i.e., the running of α) shifts the prediction in a direction that is *incompatible* with the measured correlation once Γ_ℓ is simultaneously constrained. Full electroweak corrections, including W and Z loops and heavy-particle virtual effects, move the theory into the experimentally allowed region.

A second classic diagnostic is the radiative correction Δr_W, defined through the loop-improved relation among M_W, M_Z, $\alpha(M_Z^2)$ and G_F:

$$\left(1 - \frac{M_W^2}{M_Z^2}\right)\frac{M_W^2}{M_Z^2} = \frac{\pi\,\alpha(M_Z^2)}{\sqrt{2}\,G_F\,M_Z^2\,(1-\Delta r_W)}. \tag{11.312}$$

In Eq. (11.312), the effect of the running of α is treated explicitly via $\alpha(M_Z^2)$, so Δr_W isolates electroweak contributions not attributable to QED vacuum polarization alone. The phenomenological message is that Δr_W is decisively different from zero, implying that purely weak loop effects are experimentally measurable.

Two complementary probes of weak loops. The quantity $\sin^2\theta_{\text{eff}}$ in Eq. (11.311) is dominated by vertex and γ-Z mixing effects encoded in κ_ℓ, whereas Δr_W in Eq. (11.312) is controlled by gauge-boson self-energies and box/vertex corrections in muon decay. Their simultaneous agreement with the SM is therefore a nontrivial cross-check of the whole renormalized electroweak structure.

g_V^ℓ, g_A^ℓ and lepton universality. Lepton universality can be tested by extracting (g_V^ℓ, g_A^ℓ) separately for $\ell = e, \mu, \tau$ using leptonic widths and asymmetries. A global fit in the (g_V^ℓ, g_A^ℓ) plane typically shows that the three determinations are consistent within uncertainties, supporting the hypothesis that the neutral current treats the three charged leptons identically (up to their masses, which are negligible at the Z pole). An instructive detail is that the measured axial coupling deviates from the Born value $g_A^\ell = -1/2$ by several standard deviations, while remaining fully consistent with the loop-improved SM prediction. This is precisely the kind of statement that cannot be made without genuine electroweak radiative corrections.

Asymmetries and parity violation. Parity violation in Z exchange arises from the mismatch between left- and right-handed couplings, equivalently from the coexistence of vector and axial currents. A convenient combination is

$$A_f \equiv \frac{2g_V^f g_A^f}{(g_V^f)^2 + (g_A^f)^2}, \tag{11.313}$$

which measures the degree of parity violation in the $Zf\bar{f}$ coupling.

The forward-backward asymmetry for $e^+e^- \to f\bar{f}$ is defined as

$$A_{\mathrm{FB}} \equiv \frac{\sigma_F - \sigma_B}{\sigma_F + \sigma_B}, \tag{11.314}$$

where F and B refer to the direction of the outgoing fermion relative to the electron beam. At the Z pole, neglecting non-resonant contributions, one finds the well-known factorized expression

$$A_{\mathrm{FB}}^{0,f} = \frac{3}{4} A_e A_f. \tag{11.315}$$

Equation (11.315) shows explicitly that $A_{\mathrm{FB}}^{0,f}$ constrains only the product $A_e A_f$; disentangling A_e and A_f requires additional observables.

A particularly powerful handle is provided by τ polarization. Define the longitudinal polarization as

$$\mathcal{P}_\tau \equiv \frac{\sigma_R - \sigma_L}{\sigma_R + \sigma_L}, \tag{11.316}$$

with $\sigma_{R(L)}$ the cross section for producing a right(left)-handed τ^-. At the Z pole, the polarization as a function of the scattering angle θ can be written as

$$\mathcal{P}_\tau(\theta) = -\frac{A_\tau\,(1+\cos^2\theta) + 2A_e\cos\theta}{1+\cos^2\theta + 2A_e A_\tau\cos\theta}. \tag{11.317}$$

Equation (11.317) allows an independent extraction of A_e and A_τ, thereby testing lepton universality at the level of parity-violating couplings.

Finally, SLD measured the left-right asymmetry with polarized electrons,

$$A_{\mathrm{LR}} \equiv \frac{\sigma_L - \sigma_R}{\sigma_L + \sigma_R}. \tag{11.318}$$

At the Z pole one has the direct identification

$$A_{\mathrm{LR}} = A_e, \tag{11.319}$$

which provides one of the cleanest determinations of the electron neutral-current couplings, and thus of $\sin^2\theta_{\mathrm{eff}}$ through Eq. (11.311). Combining leptonic and heavy-flavor asymmetries then yields information on A_c and A_b, enabling stringent tests of the quark sector of the neutral current.

Higgs-mass sensitivity and global electroweak fits. Electroweak precision observables depend on heavy masses through loop corrections. The dependence on the Higgs mass M_H is typically logarithmic, arising from terms of the form $\log(M_H^2/M_Z^2)$ in the one-loop amplitudes. As a result, individual observables exhibit only a mild variation across a wide range of M_H, whereas a *global* fit to many observables can accumulate sensitivity. A convenient statistical diagnostic is the function

$$\Delta\chi^2(M_H) \equiv \chi^2(M_H) - \chi^2_{\min}, \tag{11.320}$$

whose minimum selects the preferred Higgs-mass region once all experimental inputs and theoretical uncertainties are included. The width of the corresponding $\Delta\chi^2$ band reflects both experimental errors and residual theoretical uncertainties from missing higher-order corrections.

Precision electroweak data as an indirect microscope. While direct production fixes masses by kinematics, precision observables probe heavy degrees of freedom virtually. The consistent description of Z widths, $\sin^2\theta_{\rm eff}$, M_W through Eq. (11.312), and a broad set of asymmetries through Eqs. (11.313) and (11.315) is best interpreted as a quantum-level confirmation of the renormalized electroweak gauge structure, not merely as an agreement of Born formulae with data.

As a final global comparison, one typically summarizes the status of the SM by listing several electroweak observables and reporting their "pull",

$$\text{Pull} \equiv \frac{O_{\rm meas} - O_{\rm fit}}{\sigma_{\rm meas}}, \tag{11.321}$$

which measures the deviation in units of the experimental standard deviation. The overall pattern of pulls at the Z pole is remarkably consistent with the SM at the loop level, and the few long-discussed tensions are quantitatively small compared to the scale at which electroweak radiative corrections are now resolved.

11.6 Neutrinos

Neutrinos are electrically neutral spin-$\frac{1}{2}$ fermions that participate in weak interactions and, unlike charged leptons, may admit several logically distinct realizations of the relativistic fermion concept. Their special role in particle physics stems from three intertwined facts:

- they carry no electric charge,
- they interact only through the weak force and gravity, and
- their masses are extremely small compared to those of all other known fermions.

From a purely kinematical point of view, neutrinos are described by spinorial fields transforming under representations of the Lorentz group. As for any spin-$\frac{1}{2}$ particle, the most general relativistic description is in terms of a four-component Dirac spinor transforming as

$$\left(\tfrac{1}{2}, 0\right) \oplus \left(0, \tfrac{1}{2}\right). \tag{11.322}$$

However, this representation is reducible, so it admits a decomposition into two irreducible chiral sectors through the projectors

$$P_{L,R} \equiv \frac{1 \mp \gamma^5}{2}, \qquad \psi_{L,R} = P_{L,R}\psi, \qquad \psi = \psi_L + \psi_R. \tag{11.323}$$

This observation already indicates that several inequivalent realizations of a "neutrino field" are possible, depending on whether one keeps both chiral components or only one of them, and on how the mass term (if any) couples these components.

Historically, neutrinos were first introduced as massless particles in order to account for the continuous energy spectrum of β decay. In this original picture, and for many decades thereafter, neutrinos were assumed to be strictly massless.

In such a case, the equations of motion decouple into two independent Weyl equations, and a neutrino can be consistently described by a single two-component chiral field. This led to the concept of the *Weyl neutrino*, which naturally accommodates the maximal parity violation observed in weak interactions, since only one chirality (the left-handed one) couples to the charged weak current.

The discovery of neutrino oscillations has demonstrated that neutrinos are not exactly massless. This empirical fact forces one to enlarge the theoretical framework and to address the question of how neutrino masses are generated and what is their fundamental (fermionic) nature.

Once a mass term is present, the two chiral sectors can no longer remain dynamically independent, and different logical possibilities arise:

- neutrinos may be *Dirac fermions*, in which case left- and right-chiral fields are independent degrees of freedom, coupled by a Dirac mass term and distinguished by a conserved lepton number;
- neutrinos may be *Majorana fermions*, in which case the neutrino field coincides with its charge conjugate and no conserved additive lepton number exists;
- in the strict massless limit, neutrinos reduce to *Weyl fermions*, each described by a single chiral two-component field.

Last but not least, the extreme smallness of neutrino masses, together with their neutrality, makes it plausible that the mechanism responsible for their mass generation is qualitatively different from that of charged fermions.

In the following, we first analyze the concept of *Weyl neutrinos*, which provides the natural starting point from the viewpoint of Lorentz representations and of the historical development of weak interactions. We then introduce *Dirac neutrinos*, emphasizing how the presence of a right-chiral field and of a conserved lepton number distinguishes them from the Weyl case. Finally, we shall turn to the possibility of *Majorana neutrinos*, which represent a special realization of the more general concept of Majorana fermions and lead to qualitatively new physical consequences, such as the violation of lepton number.

11.6.1 Weyl Neutrinos

In the minimal (pre-oscillation) picture, neutrinos were modeled as strictly massless and therefore described as irreducible two-component fermions. As discussed above, the four-component Dirac field may be decomposed into two irreducible chiral sectors by means of appropriate projectors.

A *Weyl neutrino* is, by definition, a field that resides entirely in one of these irreducible subspaces; in particular, in the Standard Model charged current only the left-chiral neutrino field appears.

Starting from the massless Dirac equation,

$$i\,\partial\!\!\!/\,\psi = 0, \tag{11.324}$$

and adopting the chiral (Weyl) basis in which

$$\gamma^\mu = \begin{pmatrix} 0 & \sigma^\mu \\ \bar{\sigma}^\mu & 0 \end{pmatrix}, \qquad \sigma^\mu = (\mathbb{I}, \boldsymbol{\sigma}), \qquad \bar{\sigma}^\mu = (\mathbb{I}, -\boldsymbol{\sigma}), \tag{11.325}$$

Equation (11.324) splits into two *decoupled* equations for the two-component spinors ψ_L and ψ_R,

$$i\,\bar{\sigma}^\mu \partial_\mu \psi_L = 0, \tag{11.326a}$$

$$i\,\sigma^\mu \partial_\mu \psi_R = 0. \tag{11.326b}$$

Equations (11.326) are the Weyl equations: they show that the massless Dirac field contains two independent propagating degrees of freedom, each described by a two-component fermion transforming irreducibly under the Lorentz group. In particular, a *left-handed Weyl neutrino* is described by ν_L satisfying

$$i\,\bar{\sigma}^\mu \partial_\mu \nu_L = 0. \tag{11.327}$$

The physical content of Eq. (11.327) becomes transparent on plane-wave solutions. Let

$$\nu_L(x) = u_L(\mathbf{p})\, e^{-ip\cdot x}, \qquad p^\mu = (E_\mathbf{p}, \mathbf{p}), \qquad E_\mathbf{p} = |\mathbf{p}|. \tag{11.328}$$

Then, Eq. (11.327) gives

$$\bar{\sigma}^\mu p_\mu\, u_L(\mathbf{p}) = 0 \iff \big(E_\mathbf{p} - \boldsymbol{\sigma}\cdot\mathbf{p}\big) u_L(\mathbf{p}) = 0 \iff \frac{\boldsymbol{\sigma}\cdot\mathbf{p}}{|\mathbf{p}|}\, u_L(\mathbf{p}) = u_L(\mathbf{p}). \tag{11.329}$$

Thus the two-component spinor is an eigenstate of the helicity operator $\boldsymbol{\sigma}\cdot\hat{\mathbf{p}}$ with eigenvalue $+1$ in the two-component notation appropriate to ν_L. Converting this statement to the four-component language, one recovers the standard result: for massless particles, chirality and helicity coincide, and a purely left-chiral field propagates only one helicity for the particle and the opposite helicity for the antiparticle.

Weyl neutrino as an irreducible relativistic degree of freedom. A Weyl neutrino is a two-component field transforming in either $(\frac{1}{2}, 0)$ (left-chiral) or $(0, \frac{1}{2})$ (right-chiral). For strictly massless propagation, Eqs. (11.326) decouple, and the two chiral sectors do not communicate. In this limit, chirality is Lorentz-invariant and coincides with helicity, so the neutrino can be consistently described by a single chiral field, e.g. ν_L obeying Eq. (11.327).

At the quantum level the two-component field admits the usual mode expansion. For a left-chiral Weyl neutrino one may write schematically

$$\nu_L(x) = \int \frac{d^3p}{(2\pi)^3} \frac{1}{\sqrt{2E_\mathbf{p}}} \left[a(\mathbf{p})\, u_L(\mathbf{p})\, e^{-ip\cdot x} + b^\dagger(\mathbf{p})\, v_L(\mathbf{p})\, e^{ip\cdot x}\right], \tag{11.330}$$

where a annihilates neutrinos and $b^\dagger$ creates antineutrinos. The key structural point is that the field is irreducible: there is no independent right-chiral neutrino degree of freedom in the Weyl description.

Historically, this framework matched the observed maximal parity violation of charged-current weak interactions. Conceptually, however, it is incompatible with nonzero neutrino masses: as soon as neutrinos are massive, the two chiral sectors can no longer remain dynamically isolated, and the Weyl description ceases to be a complete model of the neutrino.

11.6.2 Dirac Neutrinos

A *Dirac neutrino* is a massive neutral fermion described by a four-component spinor field ν whose left- and right-chiral components are *independent* degrees of freedom. Kinematically, the starting point is the massive Dirac equation

$$\big(i\,\partial\!\!\!/ - m_D\big)\nu = 0, \tag{11.331}$$

which necessarily couples the two chiralities. Indeed, applying the projectors to Eq. (11.331) yields the coupled system

$$i\,\partial\!\!\!/\,\nu_L - m_D\,\nu_R = 0, \tag{11.332a}$$
$$i\,\partial\!\!\!/\,\nu_R - m_D\,\nu_L = 0, \tag{11.332b}$$

so that the mass term provides the unique Lorentz-invariant mechanism that *communicates* between the irreducible Weyl sectors. Equivalently, the Dirac mass term in the Lagrangian reads

$$\mathcal{L}_{m_D} = -m_D\,\bar{\nu}\nu = -m_D\big(\bar{\nu}_L\nu_R + \bar{\nu}_R\nu_L\big), \tag{11.333}$$

which is meaningful only if *both* ν_L and ν_R exist as independent fields.

The neutrino produced in charged-current processes is left-chiral. Therefore, a Dirac neutrino hypothesis requires the introduction of a right-chiral neutral fermion ν_R which does not participate in the $SU(2)_L$ charged current. In the Standard Model gauge language, this is expressed by taking ν_L to belong to a lepton doublet, while ν_R is an electroweak singlet:

$$L = \begin{pmatrix}\nu_L\\ \ell_L\end{pmatrix}, \qquad \nu_R : \text{ singlet under } SU(2)_L. \tag{11.334}$$

A gauge-invariant origin of the Dirac mass term in Eq. (11.333) is then provided by a Yukawa interaction with the Higgs doublet Φ,

$$\mathcal{L}_Y = -y_\nu\,\bar{L}\,\tilde{\Phi}\,\nu_R + \text{h.c.}, \qquad \tilde{\Phi} \equiv i\sigma_2\,\Phi^\star, \tag{11.335}$$

which, after electroweak symmetry breaking $\langle\Phi\rangle = (0, v/\sqrt{2})^{\mathsf{T}}$, generates

$$m_D = \frac{y_\nu\,v}{\sqrt{2}}, \qquad \mathcal{L}_Y \longrightarrow -m_D\big(\bar{\nu}_L\nu_R + \bar{\nu}_R\nu_L\big), \tag{11.336}$$

reproducing Eq. (11.333).

What makes a neutrino "Dirac". The Dirac hypothesis is not merely the statement "the neutrino is massive". It is the structural claim that the neutrino carries a conserved global charge (lepton number) associated with the vector current

$$J_L^\mu \equiv \bar{\nu}\gamma^\mu \nu, \qquad \partial_\mu J_L^\mu = 0, \tag{11.337}$$

and that the right-chiral field ν_R is an independent degree of freedom, introduced so that the mass term Eq. (11.333) can be written without identifying the neutrino with its charge conjugate.

The mode expansion makes the particle-antiparticle distinction explicit. A quantized Dirac neutrino field can be written as

$$\nu(x) = \sum_r \int \frac{d^3p}{(2\pi)^3} \frac{1}{\sqrt{2E_{\mathbf{p}}}} \left[a_r(\mathbf{p})\, u_r(\mathbf{p})\, e^{-ip\cdot x} + b_r^\dagger(\mathbf{p})\, v_r(\mathbf{p})\, e^{ip\cdot x} \right], \tag{11.338}$$

with a_r and b_r annihilating neutrinos and antineutrinos, respectively. Unlike the Weyl case in Eq. (11.330), the Dirac field contains both chiralities and therefore supports two helicity states for the particle once $m_D \neq 0$. The relation between chirality and helicity is then only approximate: for $E \gg m_D$ the "wrong-helicity" component is suppressed by m_D/E.

Operationally, the Dirac and Weyl descriptions coincide in the strict massless limit: setting $m_D \to 0$ in Eq. (11.331) yields the decoupled Weyl equations in Eqs. (11.326). For $m_D \neq 0$, however, the two notions diverge sharply. A purely Weyl neutrino is intrinsically massless and irreducible; a Dirac neutrino is massive, requires the existence of ν_R, and admits a conserved lepton number current as in Eq. (11.337).

For later use, it is convenient to summarize the distinguishing features.

- **Degrees of freedom**: a Weyl neutrino carries two real on-shell degrees of freedom (one helicity for the particle and one for the antiparticle); a Dirac neutrino carries four real on-shell degrees of freedom (two helicities for the particle and two for the antiparticle).
- **Mass term**: the Weyl theory has no Lorentz-invariant Dirac mass term; the Dirac theory admits Eq. (11.333) and the coupled equations in Eqs. (11.332).
- **Global charge:** the Dirac theory supports a conserved $U(1)$ current Eq. (11.337); the Weyl description, by itself, does not require such a conserved additive charge.

11.6.3 Majorana Neutrinos

In Chap. 8, we discussed the Majorana fermions, focusing on their physical meaning as a reduction of degrees of freedom: the corresponding spinor is real, instead of complex. Here we specialize to neutrinos, emphasizing that:

- neutrinos sit in chiral electroweak multiplets,
- right-handed neutrinos, if they exist, are sterile under $SU(2)_L \otimes U(1)_Y$,
- a Majorana mass term is incompatible with the conservation of lepton number and therefore explicitly breaks any putative lepton-number symmetry.

Majorana fermion versus Majorana neutrino. A *Majorana fermion* is any neutral spin-$\frac{1}{2}$ field satisfying the constraint

$$\Psi = \Psi^C. \tag{11.339}$$

A *Majorana neutrino* is the specific realization of that possibility within the electroweak theory: the field content and gauge quantum numbers of the Standard Model restrict which Majorana mass terms can be written, and therefore how ν_L and a sterile ν_R (if present) can combine into physical mass eigenstates.

Neutrino charge conjugation in the chiral language. In the notation used in Sect. 8.10, charge conjugation acts antilinearly as

$$\Psi^C(x) \equiv C_0\, \Psi^\star(x), \tag{11.340}$$

where C_0 is representation-dependent and implements $C_0\gamma^\mu C_0^{-1} = -(\gamma^\mu)^\star$. The Majorana condition is then given by Eq. (11.339) At the level of chiral components, this constraint identifies a Weyl field with the charge conjugate of the opposite chirality.

For neutrinos, the relevant point is conceptual rather than representational: once Eq. (11.339) is imposed, "neutrino" and "antineutrino" are no longer independent quanta, hence any additive lepton number cannot be an exact symmetry.

Dirac versus Majorana mass: what changes for neutrinos. A Dirac neutrino requires two *independent* chiral fields ν_L and ν_R. Writing $\nu = \nu_L + \nu_R$ and using the usual projectors, the Dirac mass term reduces to the chirality-flipping bilinear

$$\mathcal{L}_{m,D} = -m_D\, \bar{\nu}\nu = -m_D\, (\bar{\nu}_L\nu_R + \bar{\nu}_R\nu_L)\,, \tag{11.341}$$

which preserves a global $U(1)$ rephasing symmetry $\nu \to e^{i\alpha}\nu$ and therefore admits a conserved lepton number current.

A Majorana mass term instead couples a chiral field to its own charge conjugate. Defining $\nu_{L,C} \equiv (\nu_L)^C$ and $\nu_{R,C} \equiv (\nu_R)^C$, one may write

$$\mathcal{L}_{m,L} = -\frac{1}{2}\left(\bar{\nu}_{L,C}\, m_L\, \nu_L + \bar{\nu}_L\, m_L\, \nu_{L,C}\right), \tag{11.342a}$$

$$\mathcal{L}_{m,R} = -\frac{1}{2}\left(\bar{\nu}_{R,C}\, m_R\, \nu_R + \bar{\nu}_R\, m_R\, \nu_{R,C}\right), \tag{11.342b}$$

which are Hermitian by construction. Crucially, under a global rephasing $\nu_L \to e^{i\alpha}\nu_L$ one has $\nu_{L,C} \to e^{-i\alpha}\nu_{L,C}$, hence $\mathcal{L}_{m,L}$ acquires phases $e^{\pm 2i\alpha}$ and is not invariant unless $\alpha = 0$. Therefore, whenever Eq. (11.342) is present, there is no exactly conserved lepton number: the mass term itself violates it by two units.

Gauge-theory viewpoint (why neutrinos are special). In the electroweak theory, ν_L belongs to the lepton doublet $L = (\nu_L, \ell_L)^T$ and is therefore charged under $SU(2)_L \otimes U(1)_Y$, whereas a putative ν_R is a gauge singlet. This asymmetry strongly constrains how m_L, m_D, and m_R can arise from gauge-invariant operators: m_D is tied to electroweak symmetry breaking (Yukawa coupling to the Higgs), while m_R is allowed already at the renormalizable level for a sterile singlet. The term m_L is not generated by a renormalizable gauge-invariant operator built solely from Standard Model fields, and it is naturally interpreted as an effective contribution from higher-dimensional operators (see also the discussion of the Weinberg operator later in the chapter).

Unified parametrization and diagonalization. Keeping all possibilities, the most general quadratic mass Lagrangian in the one-flavor case can be arranged in a symmetric form by working with the left-handed multiplet

$$N_L \equiv \begin{pmatrix} \nu_L \\ \nu_{R,C} \end{pmatrix}, \qquad N_{L,C} \equiv (N_L)^C = \begin{pmatrix} \nu_{L,C} \\ \nu_R \end{pmatrix}. \tag{11.343}$$

Then the mass sector can be written compactly as

$$\mathcal{L}_m = -\frac{1}{2}\left(\bar{N}_{L,C}\, M\, N_L + \bar{N}_L\, M\, N_{L,C}\right), \qquad M = \begin{pmatrix} m_L & m_D \\ m_D & m_R \end{pmatrix}. \tag{11.344}$$

The entries of M are *couplings*; the physical masses are obtained by diagonalizing the symmetric matrix in Eq. (11.344). In the one-flavor case, the eigenvalues are

$$m_\pm = \frac{1}{2}\left[(m_L + m_R) \pm \sqrt{(m_R - m_L)^2 + 4m_D^2}\right], \tag{11.345}$$

and the corresponding eigenfields satisfy a Majorana condition whenever at least one of m_L or m_R is nonzero.

Limiting cases and physical interpretation.

- *Pure Dirac limit.* If $m_L = m_R = 0$, Eq. (11.344) reduces to $M = \begin{pmatrix} 0 & m_D \\ m_D & 0 \end{pmatrix}$ and Eq. (11.345) gives $m_\pm = \pm m_D$. In this case the theory admits a conserved $U(1)$ and one recovers a Dirac neutrino with independent neutrino/antineutrino quanta.
- *Majorana regime.* If $m_L \neq 0$ and/or $m_R \neq 0$, the mass eigenfields are Majorana in the sense of Eq. (11.339), and lepton number is not an exact symmetry of the quadratic Lagrangian.
- *Sterile-singlet dominance.* If $m_R \gg m_D$ and $m_L \simeq 0$, Eq. (11.345) yields one heavy and one light eigenvalue, $m_+ \simeq m_R$ and $m_- \simeq m_D^2/m_R$, anticipating the see-saw scaling discussed later.

11.7 Neutrino Oscillations

Pause and think
Knowing neutrino physics and the fact that their mass is not a mirage, why would you expect to have neutrino oscillations?

The first experiments that allowed neutrinos to be observed and identified date back to the late 1950s and early 1960s. In 1956, Cowan and Reines detected the electron antineutrino through inverse β decay, associated with the standard process in Eq. (11.10).

In their experiment, antineutrinos were produced by a β-decaying source, which at that time consisted of a large water target placed near a nuclear reactor. The antineutrinos interacted with the protons contained in the water via

$$\bar{\nu}_e + p \longrightarrow e^+ + n. \tag{11.346}$$

The emitted positron subsequently annihilates with an electron in the medium, producing two γ rays of energy 0.5 MeV each. In order to reduce background and improve the detection signature, the water was doped with cadmium chloride. Cadmium efficiently captures the neutron, leading to the emission of an additional γ ray,

$$n + {}^{108}\mathrm{Cd} \longrightarrow {}^{109m}\mathrm{Cd} \longrightarrow {}^{109}\mathrm{Cd} + \gamma. \tag{11.347}$$

The coincidence between the prompt γ rays from positron annihilation and the delayed γ ray from neutron capture provided a clear experimental signature,

$$e^+ + e^- \longrightarrow 2\gamma. \tag{11.348}$$

In 1962, Lederman, Schwartz, and Steinberger demonstrated that the muon is associated with a neutrino distinct from the electron neutrino, the muon neutrino ν_μ, produced for instance in the decays

$$\pi^+ \longrightarrow \mu^+ + \nu_\mu, \quad K^+ \longrightarrow \mu^+ + \nu_\mu. \tag{11.349}$$

The interpretation of these results is that there exist distinct conserved lepton numbers:

1. the *electron lepton number*, associated with electrons and electron neutrinos;
2. the *muon lepton number*, associated with muons and muon neutrinos.

Let $|\nu_e\rangle$ and $|\nu_\mu\rangle$ denote neutrino states carrying electron and muon lepton number, respectively. We consider two limiting cases.

1. *Massless neutrinos*, $m_\nu = 0$ with $E_\nu = |\mathbf{p}|$. In this case, the mass matrix is diagonal and its eigenvalues vanish. The simultaneous eigenstates are $|\nu_e\rangle$ and $|\nu_\mu\rangle$, described by Dirac fields. The eigenstates of the mass matrix coincide with the flavor eigenstates.
2. *Massive neutrinos*, $m_\nu \neq 0$. In general, the mass matrix is not diagonal in the flavor basis $\{|\nu_e\rangle, |\nu_\mu\rangle\}$, but is diagonal in a different basis $\{|\nu'\rangle, |\nu''\rangle\}$, whose elements are mass eigenstates.

Oscillation occurs because flavor eigenstates are not mass eigenstates.

From lepton families, we can thus distinguish two- and three-flavor oscillations.

11.7.1 Two-Flavor Oscillation

The flavor states can be expressed as superpositions of mass eigenstates,

$$|\nu_e\rangle = \cos\theta\,|\nu_1\rangle + \sin\theta\,|\nu_2\rangle, \tag{11.350a}$$

$$|\nu_\mu\rangle = -\sin\theta\,|\nu_1\rangle + \cos\theta\,|\nu_2\rangle, \tag{11.350b}$$

where $|\nu_1\rangle$ and $|\nu_2\rangle$ are mass eigenstates, while $|\nu_e\rangle$ and $|\nu_\mu\rangle$ are flavor eigenstates. We assume that the initial neutrino state $|i\rangle$ is produced with a definite flavor.

▶ **Driving exercise**

The probability amplitude that the neutrino is detected with flavor j at time t is

$$\langle j|e^{-iHt}|i\rangle \equiv A(i \to j), \tag{11.351}$$

which encodes the time evolution $|i\rangle \to e^{-iHt}|i\rangle$. Denoting by $|\nu_a(0)\rangle$ a mass eigenstate, its time evolution is described by a plane wave, $|\nu_a(t)\rangle = e^{-iHt}|\nu_a(0)\rangle$. In the ultrarelativistic limit relevant for neutrinos, $|\mathbf{p}_a| \gg m_a$, the energy of the mass eigenstate is approximated as

$$H_a = \sqrt{p_a^2 + m_a^2} \simeq |\mathbf{p}_a| \left(1 + \frac{m_a^2}{2|\mathbf{p}_a|^2}\right) = |\mathbf{p}_a| + \frac{m_a^2}{2|\mathbf{p}_a|}. \tag{11.352}$$

Moreover, in this limit one may identify $t \simeq L$, where L is the distance traveled by the neutrino. The transition amplitude then becomes

$$A(i \to j) = \sum_{a,b} \langle j|\nu_a\rangle\langle\nu_a|e^{-i\hat{H}L}|\nu_b\rangle\langle\nu_b|i\rangle. \tag{11.353}$$

Since $\langle\nu_a|e^{-i\hat{H}L}|\nu_b\rangle = \delta_{ab}\, e^{-i\left(E_\nu + \frac{m_a^2}{2E_\nu}\right)L}$, the amplitude in Eq. (11.353) reduces to

$$A(i \to j) = e^{-iE_\nu L} \sum_a \langle j|\nu_a\rangle\langle\nu_a|i\rangle\, e^{-i\frac{m_a^2 L}{2E_\nu}}. \tag{11.354}$$

Let us now compute explicitly the probability for flavor conversion $\nu_e \to \nu_\mu$,

$$\begin{aligned} |A(\nu_e \to \nu_\mu)|^2 &= \left| \langle\nu_\mu|\nu_1\rangle\langle\nu_1|\nu_e\rangle e^{-i\frac{(m_1)^2 L}{2E_\nu}} + \langle\nu_\mu|\nu_2\rangle\langle\nu_2|\nu_e\rangle e^{-i\frac{(m_2)^2 L}{2E_\nu}} \right|^2 \\ &= \sin^2(2\theta)\, \sin^2\left(\frac{\Delta m^2 L}{4E_\nu}\right), \end{aligned} \tag{11.355}$$

where $\Delta m^2 = m_{\nu_2}^2 - m_{\nu_1}^2$. Thus, the probability of detecting a neutrino with a different flavor is

$$\Pr(\nu_e \to \nu_\mu) = \sin^2(2\theta)\, \sin^2\left(\frac{\Delta m^2 L}{4E_\nu}\right), \tag{11.356}$$

which depends on the neutrino energy E_ν and on the squared-mass difference Δm^2.

Two limiting cases are particularly relevant:

1. if $\Delta m^2 L \ll 4E_\nu$, the probability tends to zero and the detected neutrino coincides with the one produced;
2. if $\Delta m^2 L \sim 4E_\nu$, the probability is nonvanishing and neutrino *flavor oscillations* occur.

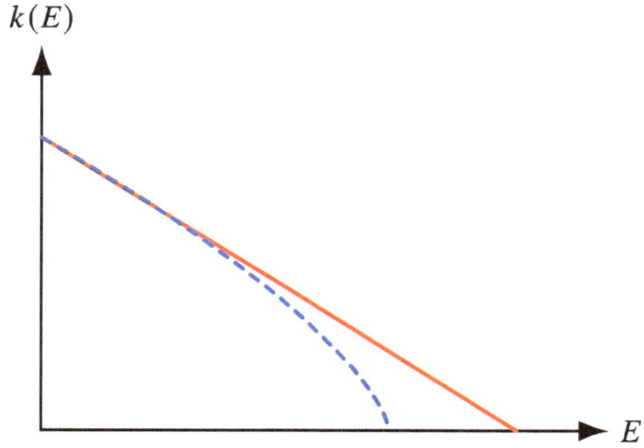

Fig. 11.7 Example of Kurie plot. The straight line represents the theoretical prediction, while the dashed one represents the experimental data

Reactor experiments probe mainly electron antineutrino disappearance, since their energies are too low for muon production in charged-current interactions. More generally, the observation of oscillations provides direct evidence that neutrinos have nonzero masses: neutrinos produced in fusion processes oscillate between electron and muon flavors during propagation. A neutrino produced initially as an electron neutrino propagates as a superposition of mass eigenstates, each accumulating a phase at a different rate. This relative phase difference, controlled by $\Delta m^2 \neq 0$, leads to periodic flavor conversion. At sufficiently large distances, the electron neutrino may no longer have a definite electron flavor, but rather appear as a superposition of flavors, eventually converting into a muon neutrino. Deviations between theoretical predictions and experimental observations can be analyzed, for instance, using the *Kurie plot* (see Fig. 11.7), traditionally employed in the study of β decay.

11.7.2 Three-Flavor Oscillation

In Nature, neutrinos come in three flavors, $\alpha = e, \mu, \tau$, and the most general relation between flavor eigenstates and mass eigenstates $|\nu_i\rangle$ $(i = 1, 2, 3)$ is written as

$$|\nu_\alpha\rangle = \sum_{i=1}^{3} U_{\alpha i} \, |\nu_i\rangle, \tag{11.357}$$

where U is a unitary 3×3 matrix, known as Pontecorvo-Maki-Nakagawa-Sakata (PMNS) matrix. Unitarity implies

$$\sum_i U_{\alpha i} U^*_{\beta i} = \delta_{\alpha\beta}, \qquad \sum_\alpha U^*_{\alpha i} U_{\alpha j} = \delta_{ij}. \tag{11.358}$$

A convenient parameterization of U is

$$U = \begin{pmatrix} 1 & 0 & 0 \\ 0 & c_{23} & s_{23} \\ 0 & -s_{23} & c_{23} \end{pmatrix} \begin{pmatrix} c_{13} & 0 & s_{13}e^{-i\delta} \\ 0 & 1 & 0 \\ -s_{13}e^{i\delta} & 0 & c_{13} \end{pmatrix} \begin{pmatrix} c_{12} & s_{12} & 0 \\ -s_{12} & c_{12} & 0 \\ 0 & 0 & 1 \end{pmatrix} \cdot \begin{pmatrix} e^{i\alpha_1/2} & 0 & 0 \\ 0 & e^{i\alpha_2/2} & 0 \\ 0 & 0 & 1 \end{pmatrix}, \tag{11.359}$$

where $c_{ij} = \cos\theta_{ij}$, $s_{ij} = \sin\theta_{ij}$, δ is the Dirac CP-violating phase, and $\alpha_{1,2}$ are Majorana phases, irrelevant for oscillation phenomena.

▶ **Driving exercise**
A neutrino produced with flavor α at $x = 0$ evolves as

$$|\nu_\alpha(L)\rangle = \sum_i U_{\alpha i} e^{-iE_i L} |\nu_i\rangle, \tag{11.360}$$

and the amplitude to detect it as flavor β at distance L is

$$A(\nu_\alpha \to \nu_\beta) = \sum_i U^*_{\beta i} U_{\alpha i}\, e^{-iE_i L}. \tag{11.361}$$

In the ultrarelativistic limit $E_i \simeq E_\nu + m_i^2/(2E_\nu)$, Eq. (11.361) becomes

$$A(\nu_\alpha \to \nu_\beta) = e^{-iE_\nu L} \sum_i U^*_{\beta i} U_{\alpha i} \exp\left(-i\frac{m_i^2 L}{2E_\nu}\right). \tag{11.362}$$

The transition probability is therefore

$$\begin{aligned} P_{\alpha\beta} = \delta_{\alpha\beta} &- 4\sum_{i>j} \Re\left(U_{\alpha i} U^*_{\beta i} U^*_{\alpha j} U_{\beta j}\right) \sin^2\left(\frac{\Delta m_{ij}^2 L}{4E_\nu}\right) \\ &+ 2\sum_{i>j} \Im\left(U_{\alpha i} U^*_{\beta i} U^*_{\alpha j} U_{\beta j}\right) \sin\left(\frac{\Delta m_{ij}^2 L}{2E_\nu}\right), \end{aligned} \tag{11.363}$$

with $\Delta m_{ij}^2 = m_i^2 - m_j^2$. The second line of Eq. (11.363) is odd under CP and vanishes if the phase δ in Eq. (11.359) is zero.

Three-flavor oscillations therefore involve:

- three mixing angles $\theta_{12}, \theta_{23}, \theta_{13}$,
- two independent mass splittings, usually chosen as Δm_{21}^2 and Δm_{31}^2,
- one Dirac CP phase δ.

Physical regimes. Depending on the ratio L/E_ν, different oscillation scales dominate: for solar neutrinos, the relevant phase is controlled by Δm^2_{21}, whereas for atmospheric and accelerator neutrinos the leading frequency is set by $\Delta m^2_{31} \simeq \Delta m^2_{32}$. In many experimental setups the three-flavor probability in Eq. (11.363) effectively reduces to a two-flavor form, but with parameters that are functions of the full PMNS matrix.

The three-flavor framework is essential for describing CP violation in the leptonic sector, encoded in the imaginary part in Eq. (11.363), and for addressing the problem of the neutrino mass ordering, namely whether m_3 is heavier (normal ordering) or lighter (inverted ordering) than the pair (m_1, m_2).

11.7.3 Oscillation in Action: Solving the Solar Neutrino Problem

The Sun derives its luminosity from thermonuclear fusion reactions occurring in its core, dominantly through the proton-proton (pp) chain and, at a subleading level, through the carbon-nitrogen-oxygen (CNO) cycle. Each fusion branch produces electron-flavor neutrinos with characteristic energy spectra. Given the solar density and temperature profiles, the equations of stellar structure and nuclear reaction rates lead to a quantitative prediction for the neutrino production rate per unit volume, $\Phi_\nu(r, E)$, from which the total flux at Earth is obtained as

$$\Phi^{\oplus}_{\nu_e}(E) = \frac{1}{4\pi D^2} \int d^3r \; \Phi_{\nu_e}(r, E), \tag{11.364}$$

where D is the Sun-Earth distance. The standard solar model (SSM) yields, after integrating over energy, a total electron-neutrino flux at Earth of order

$$\Phi^{\oplus}_{\nu_e} \sim 6.5 \times 10^{10}\,\mathrm{cm}^{-2}\,\mathrm{s}^{-1}, \tag{11.365}$$

dominated by low energy neutrinos from the pp chain.

The first direct test of Eq. (11.365) was performed by the radiochemical experiment of Davis and collaborators in the Homestake mine. Using the reaction

$$\nu_e + {}^{37}\mathrm{Cl} \longrightarrow e^- + {}^{37}\mathrm{Ar}, \tag{11.366}$$

the experiment measured a capture rate corresponding to roughly one third of the SSM prediction. This discrepancy could not be attributed to experimental error alone, since the reaction in Eq. (11.366) is well understood and sensitive primarily to the higher energy tail of the solar spectrum.

Subsequent experiments employing different detection principles confirmed the anomaly:

- gallium radiochemical detectors (SAGE, GALLEX), sensitive to low energy pp neutrinos through reactions such as $\nu_e + {}^{71}\text{Ga} \to e^- + {}^{71}\text{Ge}$;
- water Cerenkov detectors (Kamiokande, Super-Kamiokande), observing elastic scattering $\nu_x + e^- \to \nu_x + e^-$, with enhanced sensitivity to ν_e.

All of them consistently found that the observed ν_e flux was smaller than the SSM prediction, over a wide range of neutrino energies.

The *solar neutrino problem* is the persistent and systematic deficit between the electron-neutrino flux predicted by the SSM and the flux measured by experiments on Earth, with observed rates typically in the range 30–60% of the theoretical expectation.

The conceptual tension was immediate. Either the SSM was seriously flawed in its description of solar fusion reactions, or neutrinos themselves possessed properties not included in the minimal SM, where they are assumed to be exactly massless and non-mixing. For nearly two decades, improvements in solar modeling, nuclear cross sections and helioseismology consistently supported the reliability of the SSM, leaving little room to accommodate the observed deficit within astrophysical uncertainties.

The resolution became possible once neutrinos were allowed to have masses and mixing. If flavor eigenstates differ from mass eigenstates, an electron neutrino produced in the solar core propagates as a coherent superposition of mass eigenstates. As shown by the oscillation formalism, the probability that it is still detected as ν_e at Earth is

$$P_{ee}(E, L) = 1 - \sum_{i>j} 4|U_{ei}|^2|U_{ej}|^2 \sin^2\left(\frac{\Delta m_{ij}^2 L}{4E}\right), \tag{11.367}$$

so that flavor conversion into ν_μ or ν_τ naturally reduces the ν_e flux without changing the total neutrino emission of the Sun. Since early detectors were sensitive almost exclusively to ν_e, oscillations would manifest themselves precisely as a deficit of detected events.

A decisive experimental step was achieved by the Sudbury Neutrino Observatory (SNO), using heavy water as target. In this medium neutrinos can be detected through:

- charged-current reactions, sensitive only to ν_e,

$$\nu_e + d \to p + p + e^-;$$

- neutral-current reactions, equally sensitive to all active flavors,

$$\nu_x + d \to p + n + \nu_x.$$

By comparing the rates of these two channels, SNO could measure separately the ν_e flux and the total flux of all active neutrinos. The result was that the ν_e component is suppressed relative to the SSM prediction, while the total flux agrees with Eq. (11.365). This demonstrated, in a model-independent way, that the "missing" neutrinos had not disappeared but had changed flavor.

Resolution of the solar neutrino problem. The solar neutrino problem is resolved by recognizing that neutrino flavor eigenstates differ from mass eigenstates. Electron neutrinos produced in the Sun propagate as superpositions of mass eigenstates and undergo flavor conversion. Early detectors, being sensitive to ν_e mainly, observed only the surviving electron component, leading to an apparent deficit that is, in fact, a direct manifestation of neutrino mixing.

This was not an isolated anomaly: the same oscillation parameters that explain solar neutrino data also account for reactor antineutrino measurements at long baselines, notably in KamLAND. The consistency between astrophysical and terrestrial experiments leaves little doubt that the solar neutrino problem was the first clear window into neutrino mass and mixing.

11.7.4 Mikheyev-Smirnov-Wolfenstein Effect

Neutrino oscillations in vacuum are governed by the mismatch between flavor and mass eigenstates. In a medium, however, coherent forward scattering off the background modifies neutrino propagation by adding an *index-of-refraction* term to the effective Hamiltonian. The Mikheyev-Smirnov-Wolfenstein (MSW) effect is the resulting phenomenon whereby matter can *enhance* flavor conversion, even when the vacuum mixing angle is small. The essential point is that forward scattering changes the dispersion relation without absorption: it contributes a real, flavor-dependent potential, hence altering the oscillation frequency and the effective mixing in an energy- and density-dependent way.

For pedagogical clarity, we develop the effect first in two flavors, (ν_e, ν_x), with ν_x a linear combination of ν_μ, ν_τ, and then generalize the mapping to the three-flavor case relevant in solar and supernova contexts.

Vacuum mixing and Hamiltonian form. Let flavor and mass eigenstates be related by a single mixing angle θ,

$$\begin{pmatrix} \nu_e \\ \nu_x \end{pmatrix} = \begin{pmatrix} \cos\theta & \sin\theta \\ -\sin\theta & \cos\theta \end{pmatrix} \begin{pmatrix} \nu_1 \\ \nu_2 \end{pmatrix}, \tag{11.368}$$

with masses $m_{1,2}$ and $\Delta m^2 \equiv m_2^2 - m_1^2$. In the ultra-relativistic limit, the Schrödinger-like evolution equation in coordinate space reads

$$i\frac{d}{dx}\begin{pmatrix} \nu_e \\ \nu_x \end{pmatrix} = H_{\text{vac}} \begin{pmatrix} \nu_e \\ \nu_x \end{pmatrix}, \tag{11.369}$$

where, up to an irrelevant multiple of the identity, the vacuum Hamiltonian in the flavor basis is

$$H_{\text{vac}} = \frac{\Delta m^2}{4E} \begin{pmatrix} -\cos 2\theta & \sin 2\theta \\ \sin 2\theta & \cos 2\theta \end{pmatrix}. \tag{11.370}$$

Equation (11.370) encodes both the vacuum oscillation length $L_{\text{osc}} = 4\pi E/\Delta m^2$ and the vacuum mixing angle θ.

Origin of the matter potential. In ordinary matter composed of electrons, protons and neutrons, neutrinos undergo coherent forward scattering through weak interactions:

- all active flavors scatter via neutral currents (NC) on e, p, n,
- ν_e additionally scatters via charged currents (CC) on electrons.

For oscillations among *active* flavors, the NC contribution is common to all flavors and therefore drops out of flavor conversion probabilities: it amounts to adding $c\,\mathbb{I}$ to the Hamiltonian, which produces only a global phase. The CC contribution is flavor-selective and survives.

The forward CC scattering amplitude produces an effective potential $V_{\text{CC}} = \sqrt{2}\,G_F\,N_e$, where N_e is the electron number density along the neutrino trajectory. For antineutrinos the potential changes sign, $V_{\text{CC}} \to -V_{\text{CC}}$. Including matter effects, Eq. (11.369) becomes

$$i\frac{d}{dx}\begin{pmatrix} \nu_e \\ \nu_x \end{pmatrix} = [H_{\text{vac}} + H_{\text{mat}}] \begin{pmatrix} \nu_e \\ \nu_x \end{pmatrix}, \qquad H_{\text{mat}} = \begin{pmatrix} V_{\text{CC}} & 0 \\ 0 & 0 \end{pmatrix}. \tag{11.371}$$

Diagonalization in matter: effective mixing angle and splitting. It is convenient to rewrite the total Hamiltonian of Eq. (11.371) again up to a multiple of the identity. Subtracting $(V_{\mathrm{CC}}/2)\mathbb{I}$, one obtains

$$H_{\mathrm{eff}} = \frac{1}{2}\begin{pmatrix} -\Delta\cos 2\theta + V_{\mathrm{CC}} & \Delta\sin 2\theta \\ \Delta\sin 2\theta & \Delta\cos 2\theta - V_{\mathrm{CC}} \end{pmatrix}, \qquad \Delta \equiv \frac{\Delta m^2}{2E}. \tag{11.372}$$

The Hamiltonian in Eq. (11.372) is diagonalized by an *effective* mixing angle in matter, θ_m, defined by requiring that the off-diagonal terms vanish in the rotated basis. One finds

$$\tan 2\theta_m = \frac{\Delta\sin 2\theta}{\Delta\cos 2\theta - V_{\mathrm{CC}}}. \tag{11.373}$$

From Eq. (11.373) it follows that the mixing can be *enhanced* in matter. The effective eigenvalue splitting in matter is

$$\Delta_m \equiv \lambda_2 - \lambda_1 = \sqrt{(\Delta\cos 2\theta - V_{\mathrm{CC}})^2 + (\Delta\sin 2\theta)^2}. \tag{11.374}$$

Equations (11.373) and (11.374) are the basic analytic results: matter modifies both the mixing angle and the oscillation length,

$$L_{\mathrm{osc}}^{(m)} = \frac{2\pi}{\Delta_m}. \tag{11.375}$$

MSW resonance condition. Maximal mixing in matter, $\theta_m = \pi/4$, occurs when the denominator in Eq. (11.373) vanishes:

$$V_{\mathrm{CC}} = \Delta\cos 2\theta \qquad \Longleftrightarrow \qquad 2\sqrt{2}\,G_F N_e E = \Delta m^2 \cos 2\theta. \tag{11.376}$$

At resonance, matter "cancels" the vacuum diagonal term, and the eigenstates in matter become maximally mixed even if θ is small.

Equation (11.376) is an *energy-dependent* condition: for fixed density N_e, only neutrinos in a certain energy band experience resonant enhancement; conversely, for a given energy, resonance occurs only at a specific density. In the Sun and in supernova envelopes, where the density decreases with radius, neutrinos produced at high density may cross the resonant layer during propagation.

Constant-density matter: oscillation probability. If N_e is constant along the baseline, the evolution is identical to vacuum oscillations but with parameters replaced by their matter values. In particular, the conversion probability is

$$P(\nu_e \to \nu_x) = \sin^2(2\theta_m)\,\sin^2\!\left(\frac{\Delta_m L}{2}\right), \tag{11.377}$$

where θ_m and Δ_m are given by Eqs. (11.373) and (11.374). The survival probability is $P_{ee} = 1 - P(\nu_e \to \nu_x)$. Equation (11.377) makes explicit how matter can increase the oscillation amplitude through $\sin^2(2\theta_m)$.

Slowly-varying density: adiabatic evolution. In realistic media, $N_e = N_e(x)$ varies with position. The instantaneous matter eigenstates $|\nu_{1m}(x)\rangle, |\nu_{2m}(x)\rangle$ diagonalize the local Hamiltonian, with local mixing angle $\theta_m(x)$ determined by Eq. (11.373) upon replacing $V_{CC} \to V_{CC}(x)$. The evolution is *adiabatic* if the neutrino remains on the same instantaneous eigenstate as it propagates, i.e. if transitions between matter eigenstates are suppressed. A convenient adiabaticity criterion is

$$\gamma(x) \equiv \frac{\Delta_m(x)}{2\,|\theta'_m(x)|} \gg 1, \tag{11.378}$$

where a prime denotes d/dx. The most stringent condition is at resonance, where θ_m changes rapidly. Evaluating Eq. (11.378) at the resonant point yields the standard adiabatic parameter

$$\gamma_{\text{res}} = \frac{\Delta m^2}{2E}\frac{\sin^2 2\theta}{\cos 2\theta}\left|\frac{d\ln N_e}{dx}\right|^{-1}\Bigg|_{\text{res}}. \tag{11.379}$$

When $\gamma_{\text{res}} \gg 1$, the evolution through the resonance is adiabatic and flavor conversion can be very efficient.

In the adiabatic regime, the physical interpretation is particularly clear. Assume neutrinos are produced in the solar core at very high density, so that $V_{CC} \gg \Delta$ and Eq. (11.373) implies $\theta_m \simeq \pi/2$: the electron neutrino is then approximately aligned with one matter eigenstate, say $|\nu_{2m}\rangle$. As the neutrino propagates outward and the density decreases, the matter eigenstate continuously deforms into a vacuum mass eigenstate. Far outside the Sun, where $V_{CC} \to 0$, the eigenstate $|\nu_{2m}\rangle$ becomes $|\nu_2\rangle$. Thus, adiabatic propagation maps

$$|\nu_e\rangle_{\text{prod}} \simeq |\nu_{2m}\rangle_{\text{core}} \longrightarrow |\nu_2\rangle_{\text{vac}}. \tag{11.380}$$

Detection on Earth projects $|\nu_2\rangle$ back onto the flavor basis, yielding an electron-neutrino survival probability

$$P_{ee}^{\text{ad}} \simeq |\langle\nu_e|\nu_2\rangle|^2 = |U_{e2}|^2 \simeq \sin^2\theta, \tag{11.381}$$

where the last step refers to the two-flavor limit. Equation (11.381) shows that in the adiabatic MSW regime the survival probability becomes *energy independent* once the neutrino exits the medium, because phase information is washed out by production-region averaging and finite energy resolution.

Physical picture. Vacuum oscillations are driven by phase accumulation between mass eigenstates. The MSW effect is different: it is a *dynamical re-identification* of the propagating eigenstates induced by the medium. In an adiabatically varying density profile, a neutrino can emerge as a different vacuum mass eigenstate than the one dominantly present at production, leading to large flavor conversion without requiring large vacuum mixing.

Non-adiabatic crossing and the level-crossing probability. If the density changes too rapidly near resonance, transitions between matter eigenstates occur. One can parametrize the breakdown of adiabaticity by a "jump" probability P_c (often called the level-crossing probability), defined as the probability that the neutrino hops from one instantaneous matter eigenstate to the other at the resonance layer. The survival probability then interpolates between the adiabatic and non-adiabatic limits. In the two-flavor case, for a neutrino produced as ν_e at high density one finds

$$P_{ee} \simeq P_c \cos^2\theta + (1 - P_c)\sin^2\theta, \tag{11.382}$$

where $P_c \to 0$ reproduces the adiabatic result in Eq. (11.381), while $P_c \to 1$ corresponds to a complete non-adiabatic jump. In many practical applications P_c can be estimated by a Landau-Zener form,

$$P_c \simeq \exp\!\left(-\frac{\pi}{2}\gamma_{\text{res}}\right), \tag{11.383}$$

which shows explicitly how adiabaticity improves exponentially with γ_{res}.

Antineutrinos and the role of the mass ordering. For antineutrinos, $V_{\text{CC}} \to -V_{\text{CC}}$ in Eq. (11.373). The resonance condition, Eq. (11.376), then depends on the sign of $\Delta m^2 \cos 2\theta$. Thus, for a given channel, resonance can occur either for neutrinos or for antineutrinos, but not both, depending on the ordering of the masses. This sensitivity is central in supernova physics and in long-baseline oscillation experiments.

Embedding into the three-flavor framework. In solar applications, the dominant matter effect involves the (1, 2) sector of the PMNS matrix, because Δm^2_{21} sets the relevant oscillation scale. To excellent approximation one can map the two-flavor expressions above to the three-flavor case by the replacements

$$\theta \to \theta_{12}, \qquad \Delta m^2 \to \Delta m^2_{21}, \qquad V_{\text{CC}} \to V_{\text{CC}}\cos^2\theta_{13}, \tag{11.384}$$

while θ_{13} mainly introduces a small renormalization of the matter potential and an overall projection factor in the electron flavor content. The resulting survival probability smoothly connects the low-energy vacuum regime to the high-energy adiabatic MSW regime observed in solar neutrino data.

11.7.5 Are We Sure About the Physics of Neutrino Oscillation?

The standard derivation of oscillation formulas, such as Eq. (11.363), is based on quantum-mechanical superpositions of plane waves. While phenomenologically successful, this procedure raises conceptual questions when viewed from the perspective of relativistic quantum field theory.

▶ **Pause and think**
Before going on, try to single out difficulties and conceptual issues with neutrino oscillation in QFT.

The first difficulty is the definition of a "flavor state". In QFT, particle states are associated with irreducible representations of the Poincaré group and are labeled by mass and spin. Flavor states, defined by Eq. (11.357), are superpositions of states with different masses. Such superpositions are not eigenstates of the free Hamiltonian, nor do they transform irreducibly under the Poincaré group. This already signals that $|\nu_\alpha\rangle$ is not a standard one-particle state in the strict QFT sense.

A second issue concerns energy and momentum conservation. In the plane-wave derivation, one usually assumes either equal momenta or equal energies for the different mass eigenstates. Neither choice is exact: in a microscopic production process, energy and momentum are fixed by the external particles. Strict conservation would forbid the coherent production of different mass eigenstates unless they are exactly degenerate. Oscillations become possible only because neutrinos are produced and detected in processes with intrinsic uncertainties, allowing slightly different energies and momenta to contribute coherently.

This leads naturally to a wave-packet description. Each mass eigenstate is represented as a localized packet,

$$|\nu_i\rangle = \int d^3p \; f_i(\mathbf{p} - \mathbf{p}_0)\, |\nu_i(\mathbf{p})\rangle, \tag{11.385}$$

with f_i sharply peaked around $\mathbf{p}_0$. Different mass components then propagate with slightly different group velocities,

$$v_i = \frac{\partial E_i}{\partial p} \simeq 1 - \frac{m_i^2}{2E_\nu^2}. \tag{11.386}$$

After a coherence length, $L_{\text{coh}} \sim \frac{4\sqrt{2}E_\nu^2}{|\Delta m_{ij}^2|}\sigma_x$, where σ_x is the spatial width of the packet, different mass components separate and no longer interfere. Oscillations are therefore meaningful only for $L \ll L_{\text{coh}}$.

A deeper conceptual problem is that, within QFT, flavor is not associated with a conserved charge in the presence of mixing. One can define flavor fields $\nu_\alpha(x) = \sum_i U_{\alpha i}\nu_i(x)$, but the corresponding flavor number operators do not commute with the Hamiltonian. As a result, "flavor" is not a sharp quantum number at all times, but only an approximate label tied to the production and detection processes.

Oscillation as an interference phenomenon. Neutrino oscillation is not the time evolution of a single particle with a definite identity. It is an interference effect between different propagation amplitudes connecting the same initial and final macroscopic processes. The neutrino is never observed as an isolated system: it is always created in a weak interaction and destroyed in another.

From this viewpoint, the oscillation probability is not a property of a "free-flavor state", but of the full process

$$\text{source} \;\rightarrow\; \text{(neutrino propagation)} \;\rightarrow\; \text{detector},$$

computed in QFT as a single amplitude. In this approach, the standard formula Eq. (11.363) emerges as an excellent approximation when:

- neutrinos are ultra-relativistic,
- production and detection regions are localized,
- coherence is maintained over the baseline.

Thus, are we sure about neutrino oscillation? Experimentally, yes: the observed energy- and distance-dependent deficits and flavor conversions are unambiguous.

Conceptually, however, oscillation is not a fundamental notion in QFT, but rather an effective description of interference between amplitudes associated with different mass eigenstates. Its success relies on the fact that, in all realistic experiments, the conditions for coherence and localization are satisfied with remarkable accuracy.

11.7.6 See-Saw Mechanism

The see-saw mechanism provides a natural explanation for the extreme smallness of neutrino masses by exploiting the fact that neutrinos, being electrically neutral, can possess both Dirac and Majorana mass terms. In the most general case, one introduces for each flavor a left-handed neutrino field ν_L and a right-handed singlet field ν_R, and writes the mass Lagrangian as

$$\mathcal{L}_{\text{mass}} = -\frac{1}{2}\left(\bar{\nu}_L \; \overline{\nu_R^c}\right)\begin{pmatrix} m_L & m_D \\ m_D & m_R \end{pmatrix}\begin{pmatrix} \nu_L^c \\ \nu_R \end{pmatrix} + \text{h.c.}, \tag{11.387}$$

where m_D is a Dirac mass term generated after electroweak symmetry breaking, m_L and m_R are Majorana masses for left- and right-handed fields, and $\psi^c = C\bar{\psi}^T$ denotes charge conjugation. Gauge invariance of the Standard Model forbids m_L at the renormalizable level, while m_R is allowed because ν_R is a gauge singlet.

The physical masses are obtained by diagonalizing the symmetric mass matrix in Eq. (11.387). The eigenvalues are

$$m_{\pm} = \frac{1}{2}\left(m_L + m_R \pm \sqrt{(m_R - m_L)^2 + 4m_D^2}\right). \tag{11.388}$$

In the phenomenologically relevant limit

$$m_L \simeq 0, \qquad m_R \gg m_D, \tag{11.389}$$

one finds

$$m_1 \simeq \frac{m_D^2}{m_R}, \qquad m_2 \simeq m_R, \tag{11.390}$$

so that one eigenstate is extremely light and the other is extremely heavy. The corresponding eigenvectors show that the light state is predominantly left-handed, while the heavy state is predominantly right-handed:

$$\nu_1 \simeq \nu_L + \theta\, \nu_R^c, \quad \nu_2 \simeq \nu_R + \theta\, \nu_L^c, \tag{11.391}$$

with a small mixing angle

$$\theta \simeq \frac{m_D}{m_R} \ll 1. \tag{11.392}$$

Geometric meaning of the see-saw. As m_R increases, the heavy eigenvalue grows linearly with m_R, while the light eigenvalue decreases inversely with m_R. Pushing one mass up forces the other down: this is the algebraic origin of the "see-saw" behavior encoded in Eq. (11.390).

The Dirac mass m_D typically arises from a Yukawa interaction of the form

$$\mathcal{L}_Y = -y_\nu \, \bar{L} \, \tilde{\Phi} \, \nu_R + \text{h.c.}, \tag{11.393}$$

where L is the lepton doublet, $\tilde{\Phi} = i\sigma_2 \Phi^*$ is the conjugate Higgs doublet, and y_ν is a dimensionless Yukawa coupling. After symmetry breaking, $\langle \Phi \rangle = v/\sqrt{2}$, one has $m_D = y_\nu \frac{v}{\sqrt{2}}$. If m_R lies near a grand-unified or intermediate scale, say $m_R \sim 10^{10}$–10^{15} GeV, and m_D is of the order of charged-lepton or quark masses, then Eq. (11.390) yields light neutrino masses in the sub-eV range, in striking agreement with oscillation data.

In this framework, neutrinos are necessarily Majorana particles: both mass eigenstates satisfy a reality condition of the form $\nu_i = \nu_i^c$, so lepton number is not an exact symmetry. At low energies, however, lepton number violation is suppressed by the small ratio m_D/m_R, explaining why neutrino interactions appear approximately lepton-number conserving.

It is useful to contrast the see-saw picture with the possible realizations of neutrino fields.

- **Weyl neutrinos**. If neutrinos are exactly massless, they can be described by two-component Weyl spinors of definite chirality. In this case lepton number is an exact global symmetry and no mixing between left- and right-handed components occurs.
- **Dirac neutrinos**. If both ν_L and ν_R exist but $m_L = m_R = 0$, the only mass term is m_D. The neutrino is then a Dirac particle, distinct from its antiparticle, and lepton number is exactly conserved.
- **Majorana neutrinos**. If either m_L or m_R (or both) is nonzero, the neutrino mass eigenstates are Majorana fermions, coinciding with their own antiparticles. In the see-saw limit Eq. (11.389), both the light and heavy states are Majorana, but only the light one is accessible at low energies.

From the low-energy point of view, the see-saw mechanism can be encoded in an effective operator. Integrating out the heavy state ν_2 generates the dimension-five Weinberg operator

$$\mathcal{L}_{\text{eff}} = \frac{1}{\Lambda} (\bar{L} \, \tilde{\Phi})(\tilde{\Phi}^T L^c) + \text{h.c.}, \tag{11.394}$$

with $\Lambda \sim m_R$. After electroweak symmetry breaking, Eq. (11.394) yields $m_\nu \sim \frac{v^2}{\Lambda}$, which reproduces the see-saw scaling in Eq. (11.390).

11.8 Proposed Exercises

Exercise 1

In the four-fermion limit, charged-current weak processes are described by an effective interaction of the schematic form

$$\mathcal{L}_{\text{eff}} = -\frac{G_F}{\sqrt{2}} \left(\bar{\ell}\gamma^{\mu}(1-\gamma_5)\nu_{\ell}\right)\left(\bar{\nu}_{\ell'}\gamma_{\mu}(1-\gamma_5)\ell'\right) + \text{h.c.} \tag{11.395}$$

(for purely leptonic channels) and by analogous current-current operators for semileptonic decays.

(a) Starting from Eq. (11.395), write the tree-level invariant amplitude for $\mu^- \to e^- \bar{\nu}_e \nu_\mu$ and show that the squared amplitude, summed over final spins and averaged over the initial muon spin, can be expressed in terms of Lorentz invariants built from the external momenta. State clearly which masses you neglect and why the approximation is consistent.
(b) Using the standard three-body phase space, derive the scaling $\Gamma(\mu \to e\,\bar{\nu}_e\,\nu_\mu) \propto G_F^2\, m_\mu^5$, and compute the proportionality coefficient in the limit $m_e \to 0$. Comment on the physical origin of the fifth power of m_μ in $\Gamma(\mu \to e\,\bar{\nu}_e\,\nu_\mu)$.
(c) Assume that at energies well below the mediator mass the same process can be described by the exchange of a massive charged vector boson W with coupling g to the left-handed current. By matching the low-energy limit of the tree-level W-exchange amplitude to Eq. (11.395), obtain the relation

$$\frac{G_F}{\sqrt{2}} = \frac{g^2}{8M_W^2}. \tag{11.396}$$

Explain, in one concise paragraph, in which precise sense Eq. (11.396) is an infrared statement and what kinds of effects can correct it beyond tree level.

Exercise 2

Consider two-flavor neutrino mixing in vacuum and in matter. Let the flavor states be related to the mass eigenstates by a mixing angle θ.

(a) Starting from the decomposition of flavor states in terms of mass eigenstates, rederive the two-flavor vacuum transition probability

$$P(\nu_\alpha \to \nu_\beta; L) = \sin^2(2\theta)\,\sin^2\!\left(\frac{\Delta m^2 L}{4E}\right), \qquad \alpha \neq \beta. \tag{11.397}$$

(b) In matter, assume that only the charged-current forward scattering of ν_e contributes an effective potential V_{CC}. Write the effective Hamiltonian in the flavor basis and show that it can be written in the same form as the vacuum Hamiltonian but with an effective mixing angle θ_m and an effective splitting Δm_m^2. Derive the resonance condition and state it as a simple relation among V_{CC}, Δm^2, θ, and E.

(c) Define an adiabaticity parameter γ_{res} at resonance (you may take as definition any standard equivalent form based on comparing the local level splitting to the rate of change of θ_m). Explain why the limit $\gamma_{\text{res}} \gg 1$ leads to an energy-independent survival probability after leaving the medium.

(d) Consider a non-adiabatic crossing modeled by a level-crossing probability P_c. Show that the electron-neutrino survival probability can be written as

$$P_{ee} \simeq P_c \cos^2\theta + (1 - P_c)\sin^2\theta, \tag{11.398}$$

and discuss the limits $P_c \to 0$ and $P_c \to 1$.

Exercise 3

Weak charged currents couple left-handed quark fields through a unitary mixing matrix V. Work at leading order in the four-fermion description and focus on symmetry arguments and invariant measures.

(a) Consider a generic charged-current amplitude in which an up-type quark propagates virtually between two down-type external legs. Show that, when summed over internal up-type flavors, the result involves combinations of the form

$$\sum_{i=u,c,t} V_{is}^* V_{id}\, F(m_i^2), \tag{11.399}$$

where F is a loop function or an effective coefficient depending on the internal mass. Using only the unitarity of V, show that if $F(m_i^2)$ were independent of i the sum in Eq. (11.399) would vanish exactly. Explain why this observation captures the core logic behind the suppression of flavor-changing neutral currents in the Standard Model.

(b) Assume now that $F(m_i^2)$ is slowly varying and expand around a reference mass scale m_0^2:

$$F(m_i^2) = F(m_0^2) + F'(m_0^2)\,(m_i^2 - m_0^2) + \cdots . \tag{11.400}$$

Use Eq. (11.400) inside Eq. (11.399) and show that the leading nonzero contribution is controlled by mass splittings among the internal up-type quarks. State clearly which term cancels by unitarity and which term survives.

(c) In the three-generation case, CP violation requires a rephasing-invariant measure. Define the Jarlskog invariant

$$J \equiv \text{Im}\Big(V_{ij} V_{kl} V_{il}^* V_{kj}^*\Big), \tag{11.401}$$

with (i, k) and (j, l) chosen distinct. Explain why J is invariant under independent rephasings of quark fields. Then state the minimal physical conditions for $J \neq 0$ in terms of mixing angles, a complex phase, and non-degenerate quark masses.

Essential Bibliography

- Aitchison IJR, Hey AJG (2012) Gauge theories in particle physics, vols. I–II, 4th ed. CRC Press.
  A comprehensive and phenomenologically oriented treatment of non-abelian gauge theories and the electroweak model.
- Weinberg S (1996) The quantum theory of fields: modern applications, vol. II. Cambridge University Press (1996).
  A field-theoretic and symmetry-based derivation of the electroweak theory from first principles.
- Altarelli G (1999) The standard model and beyond. Nucl Phys B Proc Suppl 75:1–2.
  A phenomenology-oriented account of the Standard Model with emphasis on precision tests, useful for connecting the theoretical structure of electroweak dynamics to measurable observables and possible extensions beyond the Standard Model.

Strong Interactions 12

Symmetry is a guiding principle, but nature often hides it.

–Yoichiro Nambu

Abstract

This Chapter provides an introduction to the strong interaction as a fundamental force of Nature, emphasizing the interplay between experimental phenomenology, symmetry principles and the microscopic formulation of quantum chromodynamics (QCD). Starting from the empirical features of hadronic spectra, scattering processes, and resonance systematics, the discussion highlights how approximate flavor symmetries, Regge trajectories, and effective hadronic descriptions anticipated key aspects of the underlying theory. The emergence of quarks and color as unavoidable degrees of freedom is motivated through pre-gauge models and deep-inelastic scattering (DIS), ultimately leading to the non-Abelian gauge structure of QCD. The formulation of the QCD Lagrangian, its symmetries, and its distinctive properties, such as confinement and asymptotic freedom, are analyzed together with their phenomenological consequences. Quantization of the gluon field, the role of gauge fixing and ghosts, and the contrast with QED are discussed to clarify the conceptual structure of the theory. The Chapter concludes with heavy-ion collisions and the search for the quark-gluon plasma, illustrating how QCD connects microscopic dynamics to collective and thermodynamic behavior.

O. Luongo, *Fundamental Interactions*,
https://doi.org/10.1007/978-3-032-27413-7_12

Fig. 12.1 Portrait of Murray Gell-Mann, 1965. Reprinted from Wikimedia. Public domain

A key character: *Murray Gell-Mann (1929–2019), Fig.* 12.1.
Murray Gell-Mann was born in New York City into a family that strongly valued intellectual achievement and cultural breadth. From a very early age, he demonstrated exceptional intellectual ability, mastering advanced mathematical concepts well ahead of his peers and developing a broad curiosity for languages, history, and natural sciences. His precocious talent allowed him to enter Yale University at the age of fifteen, where he completed his undergraduate studies in physics with remarkable speed and distinction. After Yale, Gell-Mann pursued doctoral studies at the Massachusetts Institute of Technology, earning his PhD in physics in his early twenties. Following short appointments at the Institute for Advanced Study in Princeton and at the University of Chicago, he joined the California Institute of Technology, which became his primary academic home and the environment in which many of his most influential ideas were developed. The period of the late 1950s and early 1960s marked the decisive phase of Gell-Mann's scientific life. Faced with an ever-growing number of strongly interacting particles discovered in cosmic-ray and accelerator experiments, he became increasingly concerned with the absence of a unifying framework. In parallel with these developments, Gell-Mann introduced the concept of quarks (also George Zweig introduced such a concept, calling quarks with the name of *aces*, albeit he remained unpleasantly ignored). The notion of fractional charges, together with the insistence that such constituents might never appear as free asymptotic states, reflected Gell-Mann's

careful balance between bold theoretical innovation and empirical restraint. This intellectual posture proved crucial as the field moved toward the formulation of QCD, in which confinement and non-Abelian gauge dynamics became central elements. Gell-Mann was awarded the Nobel Prize in Physics in 1969 for his contributions to the classification of elementary particles and their interactions. In later years, he increasingly devoted his attention to interdisciplinary research, particularly the study of complex systems, while remaining intellectually engaged with developments in high-energy physics. Murray Gell-Mann passed away in 2019 and is remembered as a scientist who reshaped the conceptual foundations of the strong interaction, not only by proposing new entities or symmetries, but by demonstrating how deep theoretical order can emerge from apparent experimental chaos.

12.1 Phenomenology

The strong interaction is experimentally identified through three intertwined phenomenological pillars:

- a rich spectrum of colorless bound states (hadrons), organized according to approximate flavor symmetries and characteristic spectral regularities;
- large hadronic cross sections at low and intermediate energies, dominated by resonances and non-perturbative dynamics;
- the emergence, at high momentum transfer, of quasi-pointlike constituents (partons) and logarithmic scaling violations, consistent with a short-distance description in terms of QCD.

In the present section, *phenomenology* is used as an efficient "pre-QCD" language: it summarizes what experiments force us to accept about strong interactions, and it motivates the theoretical structures introduced later (confinement, asymptotic freedom, effective degrees of freedom, and factorization). The emphasis is explanatory rather than formal, with only the minimum amount of equations required to make the key scales and systematics transparent.

12.1.1 Scales and Observables

A useful phenomenological classification of strong interaction dynamics is in terms of the typical invariant momentum transfer Q, which fixes the spatial resolution $\Delta x \sim 1/Q$. At $Q \lesssim O(1\text{ GeV})$ the effective coupling becomes large, the relevant degrees of freedom are hadrons rather than quarks and gluons, and the description is intrinsically non-perturbative. In this regime, the most efficient organising principles are symmetries and their (approximate) realisations: isospin and $SU(3)_F$ flavor, chiral symmetry and its breaking, and selection rules for strong decays. At $Q \gg 1$ GeV, the interaction becomes sufficiently weak for a partonic interpretation, and controlled expansions become possible; however, the connection to measurements still involves long-distance physics through hadronisation, jet formation, and the universality of non-perturbative distributions.

The practical boundary between these regimes is not sharp and depends on the observable. A useful "rule of thumb" is that inclusive observables characterized by a hard scale Q admit a short-distance treatment already for Q of a few GeV, whereas exclusive observables and spectroscopy remain dominated by long-distance dynamics deep into the multi-GeV region. This is not a paradox: QCD admits *different expansions* in different corners of phase space, and phenomenology teaches us which degrees of freedom and which approximations are appropriate.

Experimentally, strong interactions manifest themselves through:

- *hadron spectroscopy*: masses, quantum numbers, decay modes and widths of mesons and baryons, including excited states and exotics; the existence of approximate multiplets under $SU(3)_F$ and isospin; and systematics such as Regge trajectories;
- *hadronic scattering*: large total cross sections at low energies, prominent resonance structures in partial waves, and a smoother high-energy behavior describable by Regge exchanges;
- *bound-state and confinement indicators*: the absence of isolated color-charged asymptotic states, and the success of potential models in capturing short-distance attraction together with long-distance confinement.

At low resolution, hadrons behave as composite, strongly correlated bound states with a dense excitation spectrum. At high resolution, the same hadrons reveal quasi-pointlike constituents whose distributions evolve slowly with the scale. Phenomenology provides the bridge between these two faces: it tells us what must be reproduced by any microscopic theory, and it anticipates why QCD needs both a weak-coupling ultraviolet description and a non-perturbative infrared sector.

12.1.2 Hadron Classification and the Observed Spectrum

Hadrons are color-singlet states classified by spin J, parity P, and (when applicable) charge conjugation C, together with flavor quantum numbers such as isospin I, hypercharge Y, strangeness S, charm, and bottomness, which will be defined later on. In a purely phenomenological language, the long-standing success of organising hadrons into approximate $SU(3)_F$ multiplets (octets, decuplets, nonets) anticipates QCD and reflects a simple hierarchy of scales: the light-quark mass splittings are small compared with the characteristic hadronic scale, so flavor symmetry is only mildly broken in many channels.

In this classification, mesons are grouped by J^{PC} and flavor, while baryons are organised by J^P and by their placement within flavor multiplets. Beyond the ground-state multiplets, one observes a robust phenomenological separation between *ground states* and *excitations*. Excited hadrons appear as a dense set of resonances whose properties (mass splittings, decay patterns, and selection rules) are naturally described by constituent-quark and flux-tube intuition, while their very existence indicates confining dynamics at large distances.

A striking aspect of strong interaction bound states is their genuinely relativistic character. In atoms, relativistic corrections are controlled by a small parameter; in hadrons there is no analogous expansion parameter tied to constituent velocities. This is reflected in the fact that spectroscopy often captures the gross features with "few-body" labels (valence assignments), while hard scattering reveals a much richer structure: hadrons contain an infinite tower of Fock components with quarks, antiquarks and gluons. The coexistence of these two descriptions is not contradictory; it is an expression of scale separation: spectroscopy is dominated by long-distance confining dynamics, while high-Q probes are sensitive to short-distance fluctuations.

What spectroscopy teaches before QCD is invoked.

- Strong dynamics produces many more bound states than electromagnetic binding, including a dense spectrum of excitations.
- Approximate flavor multiplets and selection rules imply underlying symmetries that are only weakly broken.
- The systematic appearance of excitations suggests a confining mechanism with a characteristic long-range behavior, rather than a finite set of discrete molecular-type bonds.

In scattering experiments, excited hadrons are identified as resonances in the energy dependence of cross sections or in partial-wave amplitudes. For an isolated and narrow resonance, the Breit-Wigner form provides a useful phenomenological parameterisation. In a given partial wave ℓ, one may write schematically

$$T_\ell(s) \simeq \frac{\sqrt{s}\,\Gamma_{\text{in}}(s)\,\Gamma_{\text{out}}(s)}{m_R^2 - s - i\sqrt{s}\,\Gamma_{\text{tot}}(s)}, \tag{12.1}$$

where s is the Mandelstam variable, m_R the resonance mass, and Γ_{tot} the total width. The energy dependence of the partial width reflects phase space and angular momentum barriers, while the numerator makes explicit that the observed signal depends on both production and decay couplings.

For pedagogical purposes, it is essential to separate two levels of interpretation. First, Eq. (12.1) is a convenient *parameterisation* of a peak. Second, the deeper field-theory statement is that resonances correspond to poles of the analytically continued S-matrix on unphysical Riemann sheets. This pole viewpoint becomes indispensable once channels are strongly coupled, thresholds are nearby, or several resonances overlap: in such situations, the assignment of a single "mass" and "width" is model-dependent, while pole positions provide a more robust characterization.

Two phenomenological observations shape the textbook narrative.

- *Density with energy*: as energy increases, resonances become more numerous and overlap more strongly. This implies that a purely hadronic description must be organised systematically (by symmetries and by controlled expansions), rather than by a finite list of degrees of freedom.
- *Correlated families*: resonances appear in patterns of correlated quantum numbers and decay modes, strongly suggesting an underlying organising principle. Regge trajectories provide one of the most economical summaries of this structure.

Masses and widths quoted in tables are not fundamental constants in the same sense as particle masses in a weakly coupled theory. They are effective parameters describing unstable, strongly interacting states whose very definition may depend on the nearby thresholds and on the analysis strategy. This is one reason why phenomenology benefits from symmetry-based classifications and from analyticity constraints.

12.1.3 Regge Trajectories and Spectral Systematics

A particularly economical way to organize a large fraction of the observed excited-hadron spectrum is the Chew-Frautschi plot, where one displays the spin J of resonances against their squared mass M^2 within families that share the same internal quantum numbers (flavor, parity, and, for mesons, charge conjugation when defined). Empirically, many light-quark mesons and several baryonic families populate near-linear trajectories,

$$J \simeq \alpha(M^2) \approx \alpha_0 + \alpha' M^2, \tag{12.2}$$

with an approximately universal slope α' across broad classes of trajectories. Equation (12.2) provides a remarkably compact summary of the data: it correlates the location of resonances over a wide range of spins and masses with just two parameters, an intercept α_0 and a slope α'.

The qualitative physical message of Eq. (12.2) is that confinement leaves a geometric imprint on the spectrum. In semiclassical language, highly excited hadrons can be pictured as extended objects whose angular momentum grows with their spatial extension. A useful intuition is that of a rotating color flux tube: if the energy stored in the tube is approximately proportional to its length, then long strings correspond to highly excited states, and J grows roughly linearly with M^2. In this sense, the empirical near-linearity in Eq. (12.2) is one of the clearest phenomenological hints that the long-distance dynamics is "string-like".

The slope α' can be related, at least parametrically, to a tension scale characteristic of the confining flux tube. In a simple rotating-string estimate, one expects a relation of the form

$$\alpha' \sim \frac{1}{2\pi\sigma}, \tag{12.3}$$

where σ denotes a string tension (energy per unit length). The precise coefficient and the interpretation of σ depend on the modeling (open string versus effective flux-tube descriptions, endpoint masses, spin effects), and therefore Eq. (12.3) should be read as a scale-setting statement rather than as a precision formula. Nevertheless, it encapsulates an important point: the approximate universality of α' across light-quark trajectories suggests a comparatively universal confining scale in the infrared, even though the short-distance dynamics and flavor content may differ.

What the Regge slope "measures". Equation (12.2) is a statement about excited states, hence about large distances. The parameter α' is therefore not a perturbative quantity. It encodes an infrared scale, often summarized in terms of an effective string tension via Eq. (12.3). This provides a practical bridge between spectroscopy and confinement: a straight line in a Chew-Frautschi plot is, in effect, a one-parameter fingerprint of long-range QCD dynamics.

Regge phenomenology is also valuable because it links spectroscopy to high-energy scattering in a unified analytic framework. In the language of the analytic S-matrix, the scattering amplitude may be decomposed into partial waves of angular momentum ℓ and then analytically continued to complex ℓ. A family of resonances appearing in the direct channel can be interpreted as the manifestation of a Regge pole in the complex angular-momentum plane. The same singularity, viewed in a crossed channel, acts as an effective exchange controlling the energy dependence of scattering amplitudes. Concretely, in appropriate kinematic regimes (large s at fixed t), Regge-pole dominance leads to a characteristic power-law behavior of amplitudes,

$$A(s,t) \sim \beta(t)\, s^{\alpha(t)}, \tag{12.4}$$

where $\alpha(t)$ is the Regge trajectory analytically continued away from the physical resonance region, and $\beta(t)$ is a residue encoding coupling information. Equation (12.4) makes explicit the dual role already implicit in Eq. (12.2): the same function α organizes both the resonance spectrum (time-like region) and the high-energy scattering systematics (space-like momentum transfer).

This dual interpretation provides a compact bridge between three otherwise disparate pieces of phenomenology:

- confinement and quasi-string excitations (suggested by the near-linearity in Eq. (12.2));
- the correlated pattern of excited hadrons (families classified by fixed internal quantum numbers);
- the analytic structure of scattering amplitudes (Regge poles controlling the asymptotic behavior in Eq. (12.4)).

It is important to appreciate both the power and the limitations of Regge systematics. The approximate linearity in Eq. (12.2) is not exact: thresholds distort trajectories, nearby channels induce mixing, and finite widths complicate the assignment of a resonance to a unique family. These effects become increasingly relevant for precision spectroscopy and for heavier flavors, where endpoint masses are not negligible and where the spectrum is influenced by additional dynamical scales. Moreover, for broad states the very notions of "mass" and "spin assignment" can be analysis-dependent, so the empirical scatter around Eq. (12.2) is not merely experimental noise but partly reflects genuine dynamical ambiguities.

Historically, Regge phenomenology is a pre-QCD language: it emerged from the attempt to organize strong interaction data using general principles such as analyticity, crossing symmetry, and unitarity, without assuming a microscopic Lagrangian. Its lasting value in a modern presentation is precisely that it remains an effective description of long-distance dynamics in a regime where straight perturbation theory is not applicable. In other words, Regge theory is not an alternative to QCD, but a robust set of infrared constraints and regularities that QCD must reproduce when projected onto hadronic degrees of freedom. This is why Regge ideas still appear, explicitly or implicitly, in modern approaches ranging from flux-tube models to string-inspired representations of confining dynamics.

12.1.4 A New Look at the Yukawa Picture

Before QCD was established as the microscopic theory of the strong interaction, the short-range character of the nuclear force suggested a field-theoretic origin in terms of the exchange of a massive mediator. In its minimal realisation, one introduces a nucleon field N coupled to a meson field ϕ through a Yukawa interaction,

$$\mathcal{L}_{\text{Yukawa}} = \bar{N}\,(i\partial\!\!\!/ - m_N)\,N + \frac{1}{2}\,(\partial_\mu\phi)(\partial^\mu\phi) - \frac{1}{2}\,m_\phi^2\,\phi^2 - g\,\bar{N}N\,\phi\,, \qquad (12.5)$$

or, for pions, an isovector generalisation with $\phi \to \boldsymbol{\pi}$ and $g\,\bar{N}N\phi \to g\,\bar{N}i\gamma_5\boldsymbol{\tau}\cdot\boldsymbol{\pi}N$ (or equivalently a pseudovector form). At tree level, single-meson exchange generates a momentum-space potential,

$$\widetilde{V}(\boldsymbol{q}) = -\frac{g^2}{\boldsymbol{q}^2 + m_\phi^2}, \tag{12.6}$$

which in coordinate space yields the Yukawa potential,

$$V(r) = -\frac{g^2}{4\pi}\frac{e^{-m_\phi r}}{r}, \qquad r \equiv |\boldsymbol{r}|. \tag{12.7}$$

Equation (12.7) makes the key phenomenological point transparent: the range of the interaction is set by the inverse mass of the exchanged meson, $r_{\text{range}} \sim 1/m_\phi$. The nuclear force is therefore short-ranged not because the strong interaction is intrinsically short-ranged, but because the relevant low-energy *hadronic* mediators are massive. In particular, pion exchange produces the longest-range part of the nuclear force because the pion is the lightest hadron.

The Yukawa picture is pedagogically valuable even in a modern QCD-based presentation, because it illustrates how low-energy strong dynamics can be organised efficiently in terms of hadronic degrees of freedom. At the same time, it hints at deeper microscopic origins: pions are now understood as pseudo-Goldstone bosons associated with spontaneously broken chiral symmetry, which explains both their small mass and the characteristic derivative structure of their interactions.

For realistic nucleon-nucleon scattering and nuclear structure, one-meson exchange must be supplemented by additional contributions: short-distance contact terms (encoding unresolved dynamics), multi-meson exchange, and form factors associated with hadron compositeness. In modern language, these ingredients are systematically organised by an effective field theory expansion constrained by chiral symmetry. From the phenomenological viewpoint, this transition can be motivated as follows.

- Single-meson exchange captures the leading long-range tail.
- Additional short-distance physics is unavoidable once probes resolve distances comparable to hadron sizes, and is most efficiently parameterised by local operators.
- Symmetries (isospin, approximate $SU(3)_F$, and chiral symmetry) restrict the allowed operator structures and correlate channels.

The Yukawa potential in Eq. (12.7) is a controlled low-energy statement: it follows from locality, Lorentz invariance, and the existence of a light mediator, and it explains the observed range of the nuclear force with minimal assumptions. Its limitations are equally instructive: once the probe energy increases, nucleons and mesons reveal internal structure, and the interaction must be organised systematically (today: by chiral EFT) rather than by adding a finite number of exchanges by hand.

12.1.5 From Phenomenology to QCD

The phenomenological ingredients reviewed above anticipate, in a remarkably direct way, the structural elements that reappear in the microscopic theory. The near-universality of Regge slopes and the emergence of approximately linear confining potentials point to an infrared dynamics that cannot be captured by a straight perturbative expansion, but must be encoded in the bound-state sector. Conversely, the empirical success of organising hadron properties by approximate flavor symmetries and by exchange mechanisms suggests that, at low energies, a controlled description must proceed via effective degrees of freedom and symmetry-based expansions.

At a practical level, the lesson is that strong interaction physics is intrinsically multi-scale. A modern treatment must therefore combine:

1. microscopic short-distance dynamics (where quarks and gluons are the appropriate variables),
2. long-distance hadronic degrees of freedom (where symmetry arguments are most efficient), and
3. matching relations ensuring that both descriptions are consistent in the overlap region.

These themes will recur in the next sections, where QCD will be formulated as an $SU(3)_c$ gauge theory and confronted with both high-energy and low-energy data.

12.2 Multiplets in Particle Physics

A striking empirical feature of strong interaction phenomenology is the appearance of *charged multiplets* of nearly degenerate hadrons. This quasi-degeneracy is not accidental: it signals that, to a good approximation, the strong dynamics is invariant under internal transformations that rotate one hadronic state into another without changing its strong interaction energy. From the modern viewpoint this is the first, most direct experimental hint that the strong interaction possesses approximate *flavor symmetries*, whose origin lies in the fact that the light-quark masses are small compared with the characteristic hadronic scale. In this subsection, we revisit the simplest realisation of this idea (isospin), explain how strangeness emerges from production systematics, and illustrate how internal symmetries generate quantitative relations among scattering amplitudes.

Isospin as an approximate internal symmetry

Historically, the paradigm example is the near equality of proton and neutron masses,

$$\frac{m_n - m_p}{m_p} \simeq 0.15 \times 10^{-2}, \tag{12.8}$$

which motivated Heisenberg to interpret (p, n) as two charge states of a single object, the *nucleon*. Phenomenologically, the underlying hypothesis is that the strong

Hamiltonian is approximately invariant under unitary transformations mixing proton and neutron states,

$$\begin{pmatrix} p \\ n \end{pmatrix} \longrightarrow U \begin{pmatrix} p \\ n \end{pmatrix}, \qquad U^\dagger U = \mathbb{I}. \tag{12.9}$$

Any such U may be factorized as $U = e^{i\phi} U_{\text{spec}}$ with $\det(U_{\text{spec}}) = 1$.

▶ **Driving Exercise**

The overall phase $e^{i\phi}$ corresponds to a global $U(1)$ invariance and is associated, via Noether's theorem, with conservation of baryon number. It is convenient to display this point explicitly by writing a schematic baryon current,

$$J_B^\mu = \bar{p}\gamma^\mu p + \bar{n}\gamma^\mu n + \cdots, \qquad \partial_\mu J_B^\mu = 0, \tag{12.10}$$

and the corresponding conserved charge,

$$N_B = \int d^3x\, J_B^0 = N_p - N_{\bar{p}} + N_n - N_{\bar{n}} + \cdots, \qquad \frac{dN_B}{dt} = 0. \tag{12.11}$$

Since baryon number conservation can be treated independently, the genuinely new phenomenological content of Eq. (12.9) resides in the *special* transformations, $\det(U) = 1$, forming the group $SU(2)$.

One therefore introduces the nucleon isodoublet,

$$N = \begin{pmatrix} \alpha \\ \beta \end{pmatrix}, \qquad p = \begin{pmatrix} 1 \\ 0 \end{pmatrix}, \qquad n = \begin{pmatrix} 0 \\ 1 \end{pmatrix}, \tag{12.12}$$

and assigns to it total isospin $I = \frac{1}{2}$. Proton and neutron are distinguished by the third component I_3,

$$I_3(p) = +\frac{1}{2}, \qquad I_3(n) = -\frac{1}{2}. \tag{12.13}$$

In complete analogy with angular momentum, irreducible representations of $SU(2)$ are labelled by (I, I_3) and have dimension $2I + 1$. Hence a doublet has two states, while a triplet has three. The canonical example of a triplet is the pion family (π^+, π^0, π^-), whose near-degeneracy makes isospin one of the most robust organising principles in low-energy hadron physics.

Why is isospin only approximate? Even if the strong interaction were exactly invariant under $SU(2)$ rotations in (u, d) flavor space, Nature would still break isospin through two effects: (i) the electromagnetic interaction distinguishes charges, and (ii) the quark masses satisfy $m_u \neq m_d$. Equation (12.8) is therefore not a prediction of isospin, but rather a phenomenological *diagnostic* showing that isospin breaking is small compared with the typical hadronic scale.

From a modern viewpoint, it is conceptually useful to interpret internal symmetries as acting on the fundamental degrees of freedom of the strong interaction. If $g \in G$ denotes a symmetry transformation of the underlying dynamics, then its action on nucleon states is represented by a matrix $U(g)$ of the form Eq. (12.9),

$$\begin{pmatrix} p \\ n \end{pmatrix} \longrightarrow U(g) \begin{pmatrix} p \\ n \end{pmatrix}, \tag{12.14}$$

and the composition law $g = g_2 g_1$ implies the representation property,

$$U(g_2)\, U(g_1) = U(g_2 g_1). \tag{12.15}$$

For group elements close to the identity, one may expand the representation matrix as

$$U(g) = \mathbb{I} + i \sum_{k=1}^{3} \alpha_k I_k, \tag{12.16}$$

with Hermitian generators I_k obeying the $SU(2)$ commutation relations

$$[I_i, I_j] = i\, \epsilon_{ijk} I_k. \tag{12.17}$$

The Lie algebra in Eq. (12.17) is the same as that of ordinary spatial rotations, but $SU(2)$ admits half-integer representations. This is precisely what is needed to accommodate doublets such as the nucleon.

12.2.1 Strangeness and the Gell-Mann-Nishijima Relation

In the early 1950s, the neutral hyperon Λ^0 was discovered via decays with lifetimes characteristic of weak processes, for instance

$$\Lambda^0 \longrightarrow p + \pi^-. \tag{12.18}$$

The surprise was not the decay itself, but the *production* of Λ in strong interactions. If Λ carried no new conserved quantum number, one might expect reactions such as

$$n + n \longrightarrow \Lambda + \Lambda \tag{12.19}$$

to occur copiously. Experimentally, however, Λ particles were observed to be produced in *associated* channels, i.e., together with other strange hadrons, rather than through channels described by Eq. (12.19). This phenomenological pattern led to the introduction of *strangeness* S, a quantum number conserved by the strong interaction (but not necessarily by the weak interaction).

Hadron quantum numbers are compactly encoded in the Gell-Mann-Nishijima relation,

$$Q = I_3 + \frac{B+S}{2}, \tag{12.20}$$

which relates electric charge Q to baryon number B, isospin projection I_3, and strangeness S. Phenomenologically, Eq. (12.20) achieves two goals. First, it correlates the charges of particles within a multiplet once B, S and I_3 are fixed. Second, together with the empirical strong conservation of S, it explains selection rules in production: processes that would change S cannot proceed via strong interactions, being therefore absent or strongly suppressed.

The appearance of strange hadrons in pairs is a typical signature of a quantum number conserved in the production process, but violated in subsequent decays. In modern language, strangeness conservation in the strong interaction reflects the fact that QCD interactions do not change quark flavor: the production of a strange hadron requires the creation of an $s\bar{s}$ pair, which naturally leads to associated channels.

Driving Exercise

Fields and Noether currents in an isospin-invariant hadronic Lagrangian. At energies not too high (typically $E \lesssim 1\,\mathrm{GeV}$), a useful effective description of hadronic physics can be written in terms of nucleon and pion fields. An $SU(2)$-invariant Lagrangian of the Yukawa type reads

$$\mathcal{L} = \bar{N}(i\partial\!\!\!/ - M)N + \frac{1}{2}\left(\partial_\mu \pi^i\, \partial^\mu \pi^i - m^2 \pi^i \pi^i\right) + g_{\pi N}\, \bar{N}\, i\gamma_5 \tau^i N\, \pi^i + \cdots\,, \tag{12.21}$$

where N is the nucleon isodoublet, π^i ($i = 1, 2, 3$) are real pion fields, τ^i are Pauli matrices in isospin space, and the factor γ_5 accounts for the negative intrinsic parity of the pion. The ellipsis denotes additional interactions (contact terms, higher-derivative operators, couplings to heavier mesons) whose role is to parameterise unresolved short-distance physics in a controlled way.

The infinitesimal $SU(2)$ transformations are

$$\delta N = -i\, \alpha_i\, \frac{\tau^i}{2}\, N, \tag{12.22}$$

$$\delta \pi^k = \alpha_i\, \epsilon_{kij}\, \pi^j, \qquad (i, j, k = 1, 2, 3), \tag{12.23}$$

and the relation to charged pion fields is

$$\pi^0 = \pi^3, \qquad \pi \equiv \frac{\pi^1 + i\pi^2}{\sqrt{2}}. \tag{12.24}$$

Noether's theorem yields three conserved isospin currents. In general form,

$$J^k_\mu = -i \sum_{\text{fields}} \frac{\partial \mathcal{L}}{\partial(\partial^\mu \psi_\alpha)} (I^k)_{\alpha\beta}\, \psi_\beta, \tag{12.25}$$

and for Eq. (12.21) one obtains

$$J^k_\mu = \bar{N}\gamma_\mu \frac{\tau^k}{2} N + (\partial_\mu \pi)^l\, \epsilon_{lks}\, \pi^s + \cdots . \tag{12.26}$$

The corresponding charges are

$$I^k = \int d^3x\, J^k_0 = \int d^3x \Big[N^\dagger \frac{\tau^k}{2} N + (\partial_0 \pi)^l\, \epsilon_{lks}\, \pi^s \Big] + \cdots , \tag{12.27}$$

and they satisfy the algebra in Eq. (12.17). The practical message for phenomenology is that approximate symmetry implies approximate conservation laws, and therefore robust relations among amplitudes, branching ratios, and cross sections.

Pion-nucleon scattering and isospin relations

Isospin symmetry constrains not only spectra but also scattering amplitudes. Let us focus on pion-nucleon scattering. One may label pion states by $I = 1$ and nucleon states by $I = \frac{1}{2}$, forming tensor-product states as

$$|\pi, m\rangle\, |N, l\rangle \equiv |\pi, m; N, l\rangle, \tag{12.28}$$

where $m = \{+1, 0, -1\}$ and $l = \{\pm\frac{1}{2}\}$ are the third components of isospin. The scattering amplitudes are S-matrix elements,

$$\langle \pi, m'; N, l' | S | \pi, m; N, l\rangle. \tag{12.29}$$

Coupling $1 \otimes \frac{1}{2}$ one obtains total isospin eigenstates with $I \in \{\frac{1}{2}, \frac{3}{2}\}$ and $I_3 = m + l$,

$$|\pi, m; N, l\rangle = \sum_{I=1/2,3/2} C(1, m; 1/2, l \,|\, I, m + l)\, |\pi, N; I, m + l\rangle, \tag{12.30}$$

with Clebsch-Gordan coefficients. Isospin invariance implies that S is diagonal in (I, I_3),

$$\langle \pi, N; I', I_3' | S | \pi, N; I, I_3 \rangle = \delta_{II'} \delta_{I_3 I_3'} S(I, I_3), \tag{12.31}$$

and Schur's lemma implies that within each irreducible representation the amplitude depends only on I,

$$S(I, I_3) = S(I). \tag{12.32}$$

Hence pion-nucleon scattering is governed by only two invariant amplitudes, $S(\frac{1}{2})$ and $S(\frac{3}{2})$.

The experimentally accessible channels include

$$\pi^+ + p \to \pi^+ + p, \tag{12.33}$$

$$\pi^- + p \to \pi^- + p, \tag{12.34}$$

$$\pi^- + p \to \pi^0 + n \qquad \text{(charge exchange)}. \tag{12.35}$$

Using the explicit Clebsch-Gordan coefficients, one finds the state decompositions

$$|\pi^+ p\rangle = |3/2, +3/2\rangle, \tag{12.36}$$

$$|\pi^- p\rangle = \sqrt{\frac{1}{3}}\, |3/2, -1/2\rangle - \sqrt{\frac{2}{3}}\, |1/2, -1/2\rangle, \tag{12.37}$$

$$|\pi^0 n\rangle = \sqrt{\frac{2}{3}}\, |3/2, -1/2\rangle + \sqrt{\frac{1}{3}}\, |1/2, -1/2\rangle, \tag{12.38}$$

and consequently the physical amplitudes become

$$A(\pi^+ p \to \pi^+ p) = S(3/2), \tag{12.39}$$

$$A(\pi^- p \to \pi^- p) = \frac{1}{3} S(3/2) + \frac{2}{3} S(1/2), \tag{12.40}$$

$$A(\pi^- p \to \pi^0 n) = \frac{\sqrt{2}}{3} \big[S(3/2) - S(1/2) \big]. \tag{12.41}$$

The phenomenological content of Eq. (12.41) is immediate: once the two isospin amplitudes are known (e.g. from two independent channels), all other channels are predicted at the level of isospin symmetry. This is a typical virtue of internal symmetries in hadron physics: they reduce the number of independent hadronic parameters and generate non-trivial relations among observables.

At pion momenta above ~200 MeV, the cross section for $\pi^+ p$ scattering is dominated by a strong resonance. Since $|\pi^+ p\rangle = |3/2, +3/2\rangle$ is a pure $I = \frac{3}{2}$ state, this resonance must carry $I = \frac{3}{2}$. Near the resonance peak, it is often a useful approximation to neglect $S(1/2)$ relative to $S(3/2)$. In that limit, squaring the amplitudes in Eq. (12.41) yields the characteristic ratios

$$\sigma(\pi^+ p) : \sigma(\pi^- p)_{\text{el}} : \sigma(\pi^- p)_{\text{ch.ex}} = 9 : 1 : 2, \tag{12.42}$$

which provides a compact illustration of how a symmetry argument translates directly into quantitative phenomenology.

Isospin relations such as Eq. (12.41) and the ratio in Eq. (12.42) are not exact in Nature. Electromagnetic effects, threshold differences, and $m_u \neq m_d$ induce corrections. Nevertheless, the relations remain extremely useful because they identify the dominant structures and provide controlled benchmarks against which symmetry breaking can be measured and modelled.

12.2.2 From Strangeness to Flavor $SU(3)$

The empirical taxonomy of low-energy hadron physics is organized by a small set of quantum numbers that remain meaningful across a wide range of reactions and spectroscopy data. Besides baryon number B and isospin (I, I_3), hadrons may carry an additional label, *strangeness* S, whose introduction was forced by a very specific experimental puzzle: strange particles are copiously *produced* in strong interactions, yet the lightest strange hadrons are *long-lived* on hadronic time scales. The resolution is that S is conserved by strong and electromagnetic processes, while it is violated by weak interactions (see Chap. 11). Consequently, the lightest hadrons with $S \neq 0$ (kaons and the Λ, Σ hyperons) cannot decay strongly into non-strange final states and therefore decay only weakly, with characteristic lifetimes in the range 10^{-12} - 10^{-8} s. In contrast, hadronic resonances decay strongly with widths $\Gamma \sim$ 10-100 MeV, i.e., on time scales $\tau \sim 10^{-23}$ s.

Lifetimes, selection rules, and the hierarchy of interactions

Before introducing the symmetry group, it is instructive to keep explicit the experimental hierarchy of time scales and selection rules. Tables 12.1 and 12.3 provide a compact summary for pseudoscalar mesons, stable baryons (against strong decay), and the first resonances. They collectively illustrate three robust facts.

- Strong decays are extremely rapid and conserve S (Table 12.3).
- Electromagnetic decays are slower than strong decays but still very fast on macroscopic scales (Table 12.2 includes $\Sigma^0 \to \Lambda\gamma$).
- Weak decays can change S (typically $|\Delta S| = 1$ for the lightest strange hadrons) and dominate whenever strong and electromagnetic selection rules forbid a decay (Tables 12.1 and 12.2).

Table 12.1 Representative decays of light pseudoscalar mesons

Particle	S	Dominant decay	ΔI	$\|\Delta S\|$
$\pi^{\pm}$	0	$\mu\nu_{\mu}$	1	0
π^0	0	$\gamma\gamma$	1	0
$K^{\pm}$	± 1	$\pi^{\pm}\pi^0,\ \mu\nu_{\mu},\ \pi\ell\nu_{\ell}$	$\frac{3}{2}, \frac{1}{2}$	1
K_L	± 1	$3\pi,\ \pi\ell\nu_{\ell}$	$\frac{1}{2}, \frac{3}{2}$	1
K_S	± 1	2π	$\frac{1}{2}$	1
η	0	$3\pi,\ \gamma\gamma$	≥ 1	0

Table 12.2 Representative weak/em decays of baryons stable under strong interactions. Branching fractions are indicated when given

Baryon	S	Dominant decay modes	Interaction
n	0	$p e^- \bar{\nu}_e$	Weak
Σ^+	-1	$p\pi^0$ (0.52), $n\pi^+$ (0.48), $\Lambda e^+\nu_e$ (2.0×10^{-5})	Weak
Σ^0	-1	$\Lambda\gamma$ (1.00)	Electromagnetic
Σ^-	-1	$n\pi^-$ (0.998), $\Lambda e^-\bar{\nu}_e$ (5.73×10^{-3}), $ne^+\nu_e$ (1.017×10^{-3})	Weak
Λ	-1	$p\pi^-$ (0.639), $n\pi^0$ (0.358), $pe^-\bar{\nu}_e$ (8.32×10^{-4})	Weak
Ξ^0	-1	$\Lambda\pi^0$ (0.99522), $\Sigma^+e^-\bar{\nu}_e$ (2.7×10^{-4})	Weak
Ξ^-	-1	$\Lambda\pi^-$ (0.99887), $\Sigma^0e^-\bar{\nu}_e$ (8.7×10^{-5}), $\Lambda e^-\bar{\nu}_e$ (5.63×10^{-4})	Weak

Table 12.3 Representative decays of the first hadronic resonances

Resonance	Dominant decay	Width/Lifetime	Interaction
ρ	2π	$\Gamma \simeq 149$ MeV	Strong
K^*	$K\pi$	$\Gamma \simeq 51$ MeV	Strong
ω	3π	$\Gamma \simeq 8.4$ MeV	Strong
ϕ	$K\bar{K}$	$\Gamma \simeq 4.3$ MeV	Strong
Δ	$N\pi$	$\Gamma \simeq 120$ MeV	Strong
Y^*	$\Lambda\pi,\ \Sigma\pi$	$\Gamma \simeq 3.6$ MeV	Strong
Ξ^*	$\Xi\pi$	$\Gamma \simeq 9$ MeV	Strong
Ω	$\Lambda K,\ \Xi\pi$	$\tau \sim 0.8 \times 10^{-10}$ s	Weak

Tables 12.1 and 12.3 encode, in a single glance, the logic that leads from spectroscopy to symmetry: the strong interaction does *not* distinguish strangeness, yet the spectrum contains families with different S. This is precisely the situation where an enlarged internal symmetry becomes the natural organizing principle.

So, how to pass from isospin to the Eightfold Way?
Isospin $SU(2)$ explains the near degeneracy of (p, n) and of the pion triplet by treating them as members of irreducible representations of a non-Abelian internal symmetry. The same strategy extends to strangeness: one enlarges the flavor space so that u, d, s are treated as approximately equivalent under the *strong* interaction.
Unlike isospin, flavor $SU(3)$ is only an approximate symmetry. The breaking is large enough to be visible already at the level of masses: the strange sector lies systematically higher than the non-strange sector, reflecting the fact that the s quark is substantially heavier than u and d. One therefore expects $SU(3)_F$ relations to be semi-quantitative, yet the symmetry remains powerful because it predicts the correct *multiplet structure* and correlates quantum numbers across families. This is the essence of the Eightfold Way.

12.2.3 Some Remarks on $SU(3)$ and Its Lie Algebra

The $SU(3)$ group enters strong interaction physics in three conceptually distinct roles:

1. it is the gauge group of QCD (*color* $SU(3)_c$);
2. it is approximately realized as a global symmetry of the light-hadron spectrum (*flavor* $SU(3)_F$), enabling the Eightfold Way classification;
3. in the chiral limit $m_u, m_d, m_s \to 0$, the QCD Lagrangian enjoys an approximate $SU(3)_L \times SU(3)_R$ symmetry, central to low-energy effective theories.

In the present phenomenological discussion, we focus on the minimal group-theory input needed to understand why $SU(3)_F$ is the natural extension of isospin and how its rank-two structure accommodates the commuting labels (I_3, Y).

Definition and exponential parameterization

The group $SU(3)$ is the set of unitary, unimodular 3×3 matrices U,

$$U^\dagger U = \mathbb{I}, \qquad \det(U) = 1, \tag{12.43}$$

namely a compact, simply connected Lie group of dimension 8. Any element can be written in exponential form

$$U(\Theta) = \exp\left(-i\sum_{a=1}^{8}\Theta_a\frac{\lambda_a}{2}\right), \tag{12.44}$$

with real parameters Θ_a and Hermitian, traceless generators λ_a. The generators satisfy

$$\lambda_a = \lambda_a^\dagger, \qquad \mathrm{Tr}(\lambda_a) = 0, \qquad \mathrm{Tr}(\lambda_a\lambda_b) = 2\delta_{ab}, \tag{12.45}$$

and an explicit representation is

$$\lambda_1 = \begin{pmatrix}0&1&0\\1&0&0\\0&0&0\end{pmatrix}, \quad \lambda_2 = \begin{pmatrix}0&-i&0\\i&0&0\\0&0&0\end{pmatrix}, \quad \lambda_3 = \begin{pmatrix}1&0&0\\0&-1&0\\0&0&0\end{pmatrix},$$
$$\lambda_4 = \begin{pmatrix}0&0&1\\0&0&0\\1&0&0\end{pmatrix}, \quad \lambda_5 = \begin{pmatrix}0&0&-i\\0&0&0\\i&0&0\end{pmatrix}, \quad \lambda_6 = \begin{pmatrix}0&0&0\\0&0&1\\0&1&0\end{pmatrix},$$
$$\lambda_7 = \begin{pmatrix}0&0&0\\0&0&-i\\0&i&0\end{pmatrix}, \quad \lambda_8 = \sqrt{\frac{1}{3}}\begin{pmatrix}1&0&0\\0&1&0\\0&0&-2\end{pmatrix}. \tag{12.46}$$

The set $\{i\lambda_a\}$ spans the Lie algebra $\mathfrak{su}(3)$: traceless, skew-Hermitian 3×3 matrices. The Lie product is the commutator, which automatically satisfies antisymmetry and the Jacobi identity.

▶ **Pause and think**
Try to define commutators, structure constants, and the rank-two Cartan subalgebra using the recipe of Chap. 1.

The commutation relations define the structure constants f_{abc},

$$\left[\frac{\lambda_a}{2}, \frac{\lambda_b}{2}\right] = if_{abc}\frac{\lambda_c}{2}, \tag{12.47}$$

with

$$f_{abc} = \frac{1}{4i}\mathrm{Tr}([\lambda_a, \lambda_b]\lambda_c). \tag{12.48}$$

The independent, non-vanishing f_{abc} are summarized in Table 12.4. In practice, f_{abc} quantify the non-Abelian nature of $SU(3)$ and control how generators mix under commutation, a feature that will reappear later in the QCD gauge-field self-interactions.

Table 12.4 Totally antisymmetric non-vanishing structure constants of $SU(3)$

abc	123	147	156	246	257	345	367	458	678
f_{abc}	1	$\frac{1}{2}$	$-\frac{1}{2}$	$\frac{1}{2}$	$\frac{1}{2}$	$\frac{1}{2}$	$-\frac{1}{2}$	$\frac{1}{2}\sqrt{3}$	$\frac{1}{2}\sqrt{3}$

Table 12.5 Totally symmetric non-vanishing d symbols of $SU(3)$

abc	118	146	157	228	247	256	338	344
d_{abc}	$\frac{1}{\sqrt{3}}$	$\frac{1}{2}$	$\frac{1}{2}$	$\frac{1}{\sqrt{3}}$	$-\frac{1}{2}$	$\frac{1}{2}$	$\frac{1}{\sqrt{3}}$	$\frac{1}{2}$
abc	355	366	377	448	558	668	778	888
d_{abc}	$\frac{1}{2}$	$-\frac{1}{2}$	$-\frac{1}{2}$	$-\frac{1}{2\sqrt{3}}$	$-\frac{1}{2\sqrt{3}}$	$-\frac{1}{2\sqrt{3}}$	$-\frac{1}{2\sqrt{3}}$	$-\frac{1}{\sqrt{3}}$

A crucial structural point for the Eightfold Way is that $SU(3)$ has rank two: there exist two independent commuting generators. In the Gell-Mann basis they are λ_3 and λ_8,

$$[\lambda_3, \lambda_8] = 0, \tag{12.49}$$

so one may simultaneously diagonalize them and label states by two commuting quantum numbers. It is natural to identify

$$I_3 = \frac{1}{2}\lambda_3, \qquad Y = \frac{1}{\sqrt{3}}\lambda_8. \tag{12.50}$$

The explicit matrix form anticipates the assignment of charges in the (u, d, s) triplet and provides the group-theoretic reason why the (I_3, Y) plane is the natural arena for the Eightfold Way weight diagrams.

Beyond commutators, the Gell-Mann matrices satisfy anticommutation relations

$$\{\lambda_a, \lambda_b\} = \frac{4}{3}\delta_{ab} + 2d_{abc}\lambda_c, \tag{12.51}$$

where the totally symmetric coefficients d_{abc} are given by

$$d_{abc} = \frac{1}{4}\mathrm{Tr}(\{\lambda_a, \lambda_b\}\lambda_c)\,. \tag{12.52}$$

Their independent non-zero values are summarized in Table 12.5. In effective hadronic Lagrangians, d_{abc} and f_{abc} control the most general $SU(3)$-invariant couplings, with the antisymmetric and symmetric tensors selecting different invariant structures.

It is sometimes convenient to adjoin a ninth matrix

$$\lambda_0 = \sqrt{\frac{2}{3}}\,\mathrm{diag}(1, 1, 1), \tag{12.53}$$

which extends the traceless basis of $\mathfrak{su}(3)$ to a basis of $\mathfrak{u}(3)$. In particular, any 3×3 matrix M can be expanded as

$$M = \sum_{a=0}^{8} \lambda_a M_a, \tag{12.54}$$

with coefficients extracted by orthogonality,

$$M_a = \frac{1}{2} \operatorname{Tr}(\lambda_a M).$$

This decomposition is more than a formal identity: in QCD it underlies the splitting of a generic matrix-valued field into singlet and octet components, and in phenomenology it clarifies why flavor multiplets naturally appear as octets plus singlets.

The Eightfold Way is the *phenomenological* statement that light hadrons fall into approximate $SU(3)_F$ multiplets, labeled by weights in the (I_3, Y) plane. QCD will later provide the *dynamical* explanation: the strong interaction is a local $SU(3)_c$ gauge theory of quarks and gluons, while the approximate global $SU(3)_F$ emerges because $m_u \simeq m_d \ll m_s$ is small compared with the hadronic scale. The present section isolates the symmetry logic directly from data, so that the transition to QCD reads as a completion, not as a leap.

12.3 Pre-gauge Scenarios

The emergence of the quark model did not occur as an abstract group-theory exercise. It was the endpoint of a sequence of increasingly constrained attempts to account for (i) the rapidly growing hadron spectrum, (ii) the near-degeneracy patterns captured by isospin and strangeness, and (iii) the selection rules distinguishing strong and electromagnetic processes from weak decays (see Chap. 11). In this pre-QCD era, the guiding principle was economy: one postulates a small set of "fundamental" constituents and demands that observed mesons and baryons be built as composites, while respecting the empirical quantum numbers (B, S, I, I_3, Q) and the classification logic behind the Eightfold Way.

This section reconstructs that logic in a historically faithful and technically transparent way. We start with the Sakata model, which is best understood as a baryon-based analogue of modern flavor symmetry. We then show why Sakata construction is simultaneously suggestive and ultimately insufficient, and how its shortcomings point unambiguously to quarks and finally to a three-valued internal degree of freedom, *color*. Throughout we separate: (i) *global flavor* $SU(3)_F$, an approximate symmetry of the strong interaction in the light sector, from (ii) *local color* $SU(3)_c$, the exact gauge symmetry of QCD introduced later.

- The data suggest approximate multiplets labeled by (I_3, Y), hence a rank-two internal symmetry; $SU(3)_F$ is the minimal simple choice.
- A baryon-based fundamental triplet (Sakata) is a natural first step, because baryons were the best established "elementary" hadrons experimentally.
- A consistent and economical dynamical picture requires truly fundamental triplet degrees of freedom with baryon number 1/3, i.e., quarks.
- Fermi statistics for qqq ground states forces an additional triplet label: color, with a totally antisymmetric singlet .

In the mid 1950s and early 1960s, before the quark model was proposed, it was reasonable to regard the lightest baryons as "fundamental" and attempt to build the rest of the hadron spectrum from them. The Sakata model assumes the existence of a fundamental baryon triplet (Sakata's "sakatons")

$$\Phi \equiv \begin{pmatrix} p \\ n \\ \Lambda \end{pmatrix}, \quad B(\Phi) = 1, \quad Q(p) = +1, \quad Q(n) = 0, \quad Q(\Lambda) = 0, \tag{12.55}$$

$$S(p) = S(n) = 0, \quad S(\Lambda) = -1, \tag{12.56}$$

and postulates that the strong interaction is approximately invariant under global transformations

$$\Phi \longrightarrow U\,\Phi, \qquad U \in SU(3)_F, \tag{12.57}$$

which generalize isospin rotations of the nucleon doublet (p, n). The two commuting labels of the rank-two group are identified with

$$I_3 = \frac{1}{2}\lambda_3, \qquad Y = \frac{1}{\sqrt{3}}\lambda_8, \qquad Q = I_3 + \frac{Y}{2}, \tag{12.58}$$

so that the empirical Gell-Mann-Nishijima structure becomes the group-theoretic statement that the electric charge operator lies in the Cartan subalgebra of $SU(3)_F$.

The key conceptual point is that Eq. (12.57) is *not* meant as an exact statement: the strange sector is heavier, and the symmetry is broken. However, the symmetry is expected to remain predictive at the level of multiplet structure and approximate relations among strong amplitudes, in direct analogy with the success of isospin.

12.3.1 Mesons and Baryons in the Sakata Model

Given the triplet in Eq. (12.56), Sakata constructs mesons as bilinears of a baryon and an antibaryon, i.e., as operators

$$M^i{}_j \sim \bar{\Phi}_j \, \Phi^i, \tag{12.59}$$

transforming in $\mathbf{3} \otimes \bar{\mathbf{3}}$ under $SU(3)_F$. The representation content follows from the general decomposition

$$\mathbf{3} \otimes \bar{\mathbf{3}} = \mathbf{8} \oplus \mathbf{1}. \tag{12.60}$$

This is seen explicitly by splitting $M^i{}_j$ into traceless and trace parts,

$$M^i{}_j = \left(M^i{}_j - \frac{1}{3}\delta^i{}_j M^k{}_k \right) + \frac{1}{3}\delta^i{}_j M^k{}_k. \tag{12.61}$$

The octet in Eq. (12.60) is then identified with the pseudoscalar octet (π, K, η_8) and, separately, with the vector octet (ρ, K^*, ω_8) at the level of classification. The singlet is an additional $SU(3)_F$-invariant meson (η_1 or ω_1 depending on the spin-parity channel).

In Sakata language, the eightfold pattern emerges because the bilinear $\bar{\Phi}_j \Phi^i$ carries two indices and therefore naturally separates into an adjoint (traceless) component plus a singlet trace. This already anticipates the quark-model formula $q\bar{q}$; the difference is that the "fundamental" fields are baryons rather than quarks.

If (p, n, Λ) are taken as fundamental, additional baryons must be constructed as composites. The minimal possibility is to form baryons as "trilinears" of the type $\Phi\Phi\bar{\Phi}$ (schematically), so that baryon number remains $B = 1$:

$$B \sim \Phi \, \Phi \, \bar{\Phi}, \qquad B(B) = 1. \tag{12.62}$$

At the level of $SU(3)_F$ representation theory, this lives in $\mathbf{3} \otimes \mathbf{3} \otimes \bar{\mathbf{3}}$, which contains octets and decuplets. Therefore, the Sakata model can *classify* the observed octet and decuplet structures, but it faces immediate dynamical and kinematical tensions.

- **Mass scale problem**. If mesons are baryon-antibaryon composites, a naive binding scale would be comparable to $2m_B$, while observed mesons are much lighter. Deep binding is conceivable, but then one must explain why an entire light octet exists with the observed quantum numbers and why it is not overwhelmed by broad annihilation channels.
- **Proliferation of states**. Composite constructions such as Eq. (12.62) generically produce many more multiplets than are observed at low mass, unless additional ad hoc dynamical selection rules are imposed.
- **Spin-statistics**. The spin and parity assignments of low-lying mesons and baryons become opaque if all of them are built from spin-1/2 baryons and antibaryons, because orbital excitations and strong annihilation channels are hard to control.

The lesson is that $SU(3)_F$ as a *classification symmetry* is sound, but the choice of (p, n, Λ) as *fundamental constituents* is not.

Sakata's construction captures the correct symmetry-theoretic backbone:

- a fundamental triplet transforming under $SU(3)_F$;
- mesons in $\mathbf{3} \otimes \bar{\mathbf{3}} = \mathbf{8} \oplus \mathbf{1}$;
- the physical relevance of the Cartan labels (I_3, Y) and the charge operator in Eq. (12.58).

The subsequent quark picture keeps *exactly* this structure, but replaces the baryon triplet by truly fundamental fields of baryon number $1/3$.

Pre-QCD mesons and baryons in the quark picture

The modern resolution is to postulate a triplet of light quark fields

$$q \equiv \begin{pmatrix} u \\ d \\ s \end{pmatrix}, \qquad Q_u = \frac{2}{3}, \qquad Q_d = Q_s = -\frac{1}{3}, \qquad B = \frac{1}{3}, \tag{12.63}$$

on which $SU(3)_F$ acts as a global approximate symmetry. The empirical Gell-Mann-Nishijima formula is then reproduced by the operator identity Eq. (12.58), with $Y = B + S$ and with $S(s) = -1$ at the level of flavor assignment.

Mesons as $q\bar{q}$ and the octet-singlet structure

Mesons are described as flavor bilinears

$$M^i{}_j \sim \bar{q}_j q^i, \tag{12.64}$$

transforming exactly as in Eq. (12.59). Therefore, the decomposition in Eq. (12.60) and the split in Eq. (12.61) apply unchanged. What changes relative to Sakata's hypothesis is the mass scale: $q\bar{q}$ bound states are naturally light because the constituents themselves are not baryons. In addition, the quantum numbers of meson nonets become transparent: the pseudoscalars are the lightest $L = 0$ states with $J^{PC} = 0^{-+}$, vectors are the $J^{PC} = 1^{--}$ partners, and higher multiplets correspond to orbital excitations.

At the phenomenological level, one distinguishes the $SU(3)_F$ octet state η_8 from the singlet η_1. Physical η and η' are mixtures because $SU(3)_F$ is broken and because the singlet channel is sensitive to effects not captured by naive symmetry arguments. The symmetry classification, however, remains the correct starting point.

▶ **Driving Exercise**

Baryons arise as three-quark composites. The flavor decomposition proceeds exactly as a tensor-product computation in $SU(3)_F$:

$$\mathbf{3} \otimes \mathbf{3} = \mathbf{6} \oplus \bar{\mathbf{3}}, \tag{12.65}$$

where $\mathbf{6}$ is symmetric and $\bar{\mathbf{3}}$ antisymmetric under exchange of the two factors. Multiplying by a third $\mathbf{3}$ yields

$$\begin{aligned}(\mathbf{6} \oplus \bar{\mathbf{3}}) \otimes \mathbf{3} &= (\mathbf{6} \otimes \mathbf{3}) \oplus (\bar{\mathbf{3}} \otimes \mathbf{3}) \\ &= (\mathbf{10} \oplus \mathbf{8}) \oplus (\mathbf{8} \oplus \mathbf{1}),\end{aligned} \tag{12.66}$$

that is, $\mathbf{3} \otimes \mathbf{3} \otimes \mathbf{3} = \mathbf{10} \oplus \mathbf{8} \oplus \mathbf{8} \oplus \mathbf{1}$. The observed light baryons populate an octet with $J = \frac{1}{2}$ and a decuplet with $J = \frac{3}{2}$ in their ground-state configurations.

The decomposition in Eq. (12.66) is purely a statement about flavor. To connect it with the observed spin assignments, one uses the standard constituent picture: for ground states $L = 0$ and the spatial wavefunction is symmetric. The decuplet corresponds to fully symmetric flavor and spin wavefunctions (aligned spins, $J = \frac{3}{2}$), while the octet corresponds to mixed symmetry, yielding $J = \frac{1}{2}$.

12.3.2 Young Tableaux for $SU(3)$ Multiplets

For $SU(3)$, Young tableaux encode the symmetry type of tensor indices and make the decomposition logic in Eqs. (12.60) and (12.66) essentially visual. The fundamental $\mathbf{3}$ is a single box □. The product $\mathbf{3} \otimes \mathbf{3}$ follows by adding a box:

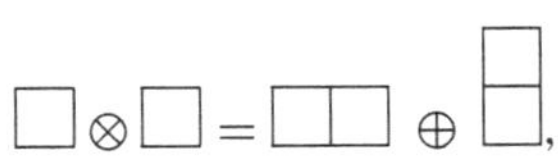

corresponding to $\mathbf{6} \oplus \bar{\mathbf{3}}$ in Eq. (12.65). Multiplying by one more □ gives

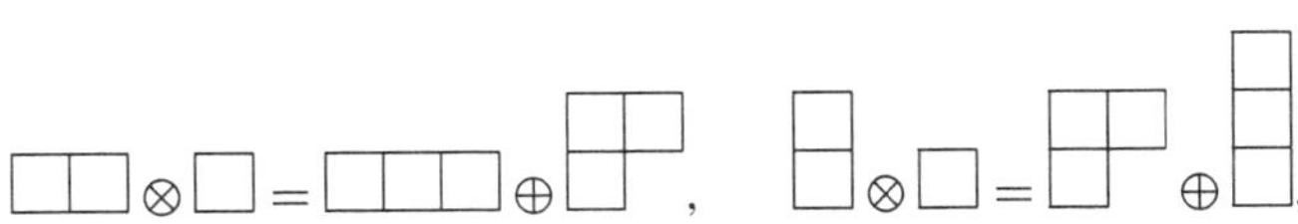

corresponding to $\mathbf{10} \oplus \mathbf{8}$ and $\mathbf{8} \oplus \mathbf{1}$, reproducing Eq. (12.66). The tableau 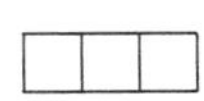 is the fully symmetric decuplet;  is the fully antisymmetric singlet.

12.3.3 Color is Unavoidable!

Quarks are spin-$\frac{1}{2}$ fermions, hence the *total* wavefunction of a system of identical quarks must be antisymmetric under exchange. Consider the Δ^{++} resonance, experimentally observed with $J = \frac{3}{2}$ and valence content uuu in the quark model. In the ground state one takes $L = 0$, so

$$\Psi_{\text{space}} \ \text{is symmetric.} \tag{12.67}$$

The $J = \frac{3}{2}$ state corresponds to aligned spins, hence

$$\Psi_{\text{spin}} \ \text{is fully symmetric.} \tag{12.68}$$

Since all three flavors are the same (u), the flavor part is also symmetric,

$$\Psi_{\text{flavour}} \ \text{is symmetric.} \tag{12.69}$$

Therefore, the product $\Psi_{\text{space}}\Psi_{\text{spin}}\Psi_{\text{flavour}}$ is symmetric, contradicting Fermi statistics. The only consistent resolution is an additional internal degree of freedom whose baryon wavefunction provides a totally antisymmetric factor.

Introduce color indices $a = 1, 2, 3$ and postulate that quarks transform in the fundamental of an exact $SU(3)_c$ symmetry. A color-singlet baryon is then

$$\Psi^{(B)}_{\text{color}} = \frac{1}{\sqrt{6}}\, \varepsilon_{abc}\, |a\, b\, c\rangle, \tag{12.70}$$

which is totally antisymmetric and restores the overall fermionic antisymmetry. Similarly, a color-singlet meson is

$$\Psi^{(M)}_{\text{color}} = \frac{1}{\sqrt{3}}\, \delta^{a}{}_{b}\, |a\, \bar{b}\rangle. \tag{12.71}$$

Color is therefore not a decorative label: it is forced by the coexistence of (i) spin-$\frac{1}{2}$ constituents, (ii) observed symmetric qqq flavor-spin states (notably the decuplet), and (iii) the empirical fact that baryons and mesons exist as colorless asymptotic states.

▶ **Pause and think**
How do you imagine to go beyond the Sakata model by including color?

The strong interaction puzzle must therefore reconcile three layers of description:

- the fundamental theory must be gauge invariant,
- we do expect a plausible perturbative regime,
- the emergence of hadrons as relativistic bound states in the IR.

Phrasing it differently, as central theme we do conclude that "strong coupling" in the IR is not synonymous with "loss of theoretical control". A variety of evidence suggests that effective couplings may remain moderate at low scales and that well-defined expansions can still be constructed, provided the correct degrees of freedom and boundary conditions are used.

12.3.4 The Parton Model

The parton model is the historically first framework that renders hard lepton-hadron scattering predictive without assuming an underlying gauge Lagrangian. Its central claim is a separation of time scales: when a probe transfers a large invariant momentum Q, the interaction time is so short that the target is effectively frozen into an incoherent ensemble of quasi-free constituents. Inclusive observables then admit a factorized description in terms of a pointlike lepton-parton subprocess and universal, non-perturbative parton distributions encoding the longitudinal momentum flow inside the hadron. In modern language, the parton model is the kinematic and conceptual template of QCD factorization, while remaining agnostic about the microscopic dynamics responsible for radiation, scaling violations, and the renormalization-group evolution of parton densities.

We focus on inclusive DIS,

$$\ell(k) + N(P) \to \ell(k') + X,$$

mediated by a virtual photon $\gamma^*(q)$ with

$$q^\mu \equiv k^\mu - k'^\mu, \qquad Q^2 \equiv -q^2 > 0. \tag{12.72}$$

Besides Q^2, two standard invariants are

$$\nu \equiv \frac{P \cdot q}{M}, \qquad W^2 \equiv (P+q)^2 = M^2 + 2P \cdot q - Q^2, \tag{12.73}$$

with M the nucleon mass. The Bjorken limit is defined by

$$Q^2 \to \infty, \qquad P \cdot q \to \infty, \qquad x_B \equiv \frac{Q^2}{2P \cdot q} \quad \text{(fixed)}, \tag{12.74}$$

so that the dimensionless scaling variable x_B remains finite. A convenient second variable is the inelasticity,

$$y \equiv \frac{P \cdot q}{P \cdot k}, \qquad (12.75)$$

which parameterizes the lepton-level phase space in a Lorentz-invariant way.

The object that encodes the strong interaction dynamics in inclusive DIS is the hadronic tensor,

$$W_{\mu\nu}(P, q) \equiv \frac{1}{4\pi} \int d^4 z \, e^{iq\cdot z} \langle P|[J_\mu(z), J_\nu(0)]|P\rangle. \qquad (12.76)$$

Lorentz covariance, parity, and current conservation restrict $W_{\mu\nu}$ to two scalar structure functions,

$$W_{\mu\nu} = \left(-g_{\mu\nu} + \frac{q_\mu q_\nu}{q^2}\right) F_1(x_B, Q^2) + \left(P_\mu - \frac{P \cdot q}{q^2} q_\mu\right)\left(P_\nu - \frac{P \cdot q}{q^2} q_\nu\right) \frac{F_2(x_B, Q^2)}{P \cdot q}. \qquad (12.77)$$

It is often useful to introduce the longitudinal structure function

$$F_L(x_B, Q^2) \equiv F_2(x_B, Q^2) - 2x_B F_1(x_B, Q^2), \qquad (12.78)$$

since F_L cleanly diagnoses physics beyond an idealized picture of strictly collinear, free spin-$1/2$ constituents.

The parton model becomes most transparent in a frame where the target carries a large light-cone momentum P^+, with $v^\pm \equiv (v^0 \pm v^3)/\sqrt{2}$. The hard probe resolves distances $\Delta z \sim 1/Q$ and interacts over a proper time $\Delta t \sim 1/Q$, which is much shorter than the characteristic time scale of binding dynamics in the target. This motivates the impulse approximation: the inclusive cross section is approximated as an incoherent sum of scatterings off individual constituents carrying a longitudinal momentum fraction x.

In its simplest form, one assumes an almost on-shell, collinear parton,

$$p^\mu \simeq x\, P^\mu, \qquad 0 < x < 1, \qquad (12.79)$$

and imposes the partonic "mass-shell" condition for the struck constituent after absorbing the photon,

$$(p + q)^2 \simeq 0, \qquad (12.80)$$

appropriate to massless quarks in the hard regime. Using Eqs. (12.79) in (12.80) and neglecting $P^2 = M^2$ in the Bjorken limit yields

$$0 \simeq (xP + q)^2 \simeq 2x\, P \cdot q - Q^2 \quad \Longrightarrow \quad x \simeq \frac{Q^2}{2P \cdot q} \equiv x_B. \qquad (12.81)$$

Equation (12.81) gives the operational meaning of Bjorken-x: the inclusive kinematics selects constituents carrying a momentum fraction x close to the measured x_B.

What is the physical meaning of Bjorken-x and why is it non-trivial? Equation (12.81) is not merely a kinematic identity. It is the quantitative expression of scale separation: the hard probe couples locally to a single constituent, while the soft binding dynamics enters only through universal probability densities for finding that constituent inside the hadron. This is why x_B becomes a genuine dynamical label of the target structure in inclusive measurements.

Let $f_i(x)$ denote the number density of partons of species i (quark flavors and antiquarks) with longitudinal momentum fraction x. In the parton model, inclusive DIS can be written as an incoherent sum of lepton-parton scatterings weighted by these densities. At leading order in the electromagnetic interaction, the current couples to quarks with charges e_i, and comparison of Eq. (12.76) with elastic scattering off a free spin-1/2 parton yields

$$
\begin{aligned}
F_2(x_B, Q^2) &= x_B \sum_i e_i^2 \Big[f_i(x_B) + \bar{f}_i(x_B) \Big], \\
F_1(x_B, Q^2) &= \frac{1}{2} \sum_i e_i^2 \Big[f_i(x_B) + \bar{f}_i(x_B) \Big].
\end{aligned}
\tag{12.82}
$$

Consequently, the Callan-Gross relation follows,

$$
F_2(x_B, Q^2) = 2x_B F_1(x_B, Q^2) \qquad \Longleftrightarrow \qquad F_L(x_B, Q^2) = 0. \tag{12.83}
$$

Equation (12.83) is the cleanest pre-QCD signature of spin-1/2 constituents: strictly collinear scattering off free fermions produces no longitudinal response. Experimentally, F_L is nonzero once finite transverse momentum, intrinsic motion, and radiative effects are included, so deviations from Eq. (12.83) are direct evidence that the microscopic dynamics involves field-theoretic radiation beyond the naive impulse picture.

Parton densities satisfy sum rules that encode conserved quantum numbers. For a proton target, baryon number conservation implies the valence sum rules

$$
\int_0^1 dx \, [u(x) - \bar{u}(x)] = 2, \qquad \int_0^1 dx \, [d(x) - \bar{d}(x)] = 1, \tag{12.84}
$$

while momentum conservation yields

$$
\sum_i \int_0^1 dx \, x \Big[f_i(x) + \bar{f}_i(x) \Big] + \int_0^1 dx \, x \, g(x) = 1, \tag{12.85}
$$

where $g(x)$ denotes the gluon momentum distribution. In the parton model, $g(x)$ may be introduced phenomenologically as the carrier of a missing momentum fraction.

In QCD, gluons are unavoidable dynamical degrees of freedom and Eq. (12.85) becomes a non-trivial quantitative constraint linking quarks and gluons.

A striking prediction of the parton model is Bjorken scaling: in the Bjorken limit, the structure functions depend predominantly on x_B rather than on Q^2,

$$F_{1,2}(x_B, Q^2) \simeq F_{1,2}(x_B). \tag{12.86}$$

This is not an exact symmetry of Nature. It is an emergent regularity reflecting the dominance of short-distance physics at large Q and the universality of the long-distance input encoded in $f_i(x)$. The model also implies a practical universality principle: the same parton distributions govern a variety of sufficiently hard and inclusive processes involving the same hadron. In modern language, Eq. (12.86) is the leading term in a more refined factorization statement that separates hard and soft physics at the level of amplitudes and systematically organizes corrections.

A useful phenomenological organization is:

- **short-distance part (hard scattering)**: computable from a pointlike lepton-parton interaction at a scale set by Q;
- **long-distance part (structure)**: encoded in universal functions $f_i(x)$, insensitive to the details of the hard probe;
- **power corrections**: finite-mass effects, target-mass contributions, and higher-twist correlations suppressed by powers of Λ/Q.

In particular, target-mass kinematics can be incorporated by replacing x_B with the Nachtmann variable

$$\xi_N \equiv \frac{2x_B}{1+\sqrt{1+\frac{4M^2x_B^2}{Q^2}}}, \tag{12.87}$$

and one expects, at large but finite Q^2, calculable expansions in M^2/Q^2 and additional genuinely non-perturbative contributions.

The parton model is "pre-gauge". Indeed, the parton model identifies the correct degrees of freedom for hard inclusive scattering and captures the leading scaling laws. However, it cannot predict the Q^2 evolution of $f_i(x)$, nor compute F_L beyond the vanishing result in Eq. (12.83), because both effects originate from radiation and virtual corrections governed by the underlying gauge dynamics. What experiments observe as logarithmic scaling violations and transverse-momentum broadening therefore demands a microscopic theory with gauge-boson emission, running couplings, and a systematic renormalization-group organization.

The empirical success of the parton model is thus inseparable from its limitations. Treating partons as free during the hard collision is justified only because the probe is short-lived; it does not explain why partons are confined, why gluons carry a substantial fraction of the hadron momentum, or why scaling is violated in a controlled logarithmic way. Once QCD is introduced, the following refinements become unavoidable.

- **Scaling violations**: Bjorken scaling in Eq. (12.86) is broken by logarithmic Q^2 dependence induced by gluon emission and loop effects, which reorganize into evolution equations for parton distributions.
- **Transverse momentum and jets**: collinearity in Eq. (12.79) is an idealization; QCD radiation generates transverse momentum and jet structure, motivating k_T-dependent refinements when observables are sensitive to transverse scales.
- **Gluon dominance at small** x: the gluon distribution in Eq. (12.85) becomes the driver of the dynamics at small x, and resummations beyond fixed-order perturbation theory may be required.
- **Higher twist and correlations**: multi-parton correlations enter as higher-twist operators, producing power-suppressed corrections that become relevant at moderate Q^2.

12.4 The QCD Lagrangian

QCD is the microscopic theory of strong interactions. Its defining structural statement is that *color* is an exact local $SU(3)_c$ gauge symmetry, with eight gauge bosons (gluons) in the adjoint representation and quark matter fields in the fundamental representation. Historically, color first entered hadron physics as the minimal internal degree of freedom required by Fermi statistics in baryons; in QCD it becomes a *dynamical* gauge charge, whose long-distance behavior explains why colored objects are not observed as asymptotic states.

At the hadronic scale, quarks are never directly measured as free particles. Consequently, quantities such as "the quark mass" are meaningful only after specifying a renormalization prescription (scheme) and a scale. In practice, light-quark masses are extracted indirectly from hadron properties (often using chiral constraints), while heavy-quark masses can be related more directly to quarkonium and heavy-meson spectra. A convenient summary of flavours, charges, and representative masses is given in Table 12.6.

▶ **Driving Exercise**
The QCD Lagrangian is obtained from free quark fields by imposing local $SU(3)_c$ invariance. For each flavour f, the quark field carries a color index

Table 12.6 Quark flavours with electric charges and indicative masses. The numerical values depend on the renormalization prescription and scale. For light quarks, one often quotes $\overline{\text{MS}}$ running masses at a low scale; heavy-quark masses may be given in $\overline{\text{MS}}$ or in short-distance threshold schemes. The top mass is extracted from collider observables and requires particular care when translated into a Lagrangian parameter

Flavour	u	d	s
Charge [e]	2/3	$-1/3$	$-1/3$
Mass [MeV]	5.1 ± 0.9	9.3 ± 1.4	175 ± 25
Flavour	c	b	t
Charge [e]	2/3	$-1/3$	2/3
Mass [GeV]	$1.15 - 1.35$	$4.0 - 4.4$	$174.3 \pm 3.2 \pm 4.0$

and forms a triplet,

$$q_f \equiv \begin{pmatrix} q_{f,r} \\ q_{f,g} \\ q_{f,b} \end{pmatrix}, \tag{12.88}$$

which transforms under a local gauge transformation $g(x) \in SU(3)_c$ as

$$q_f(x) \mapsto q'_f(x) = U[g(x)]\, q_f(x), \qquad U[g(x)] \equiv \exp\left[-i \sum_{a=1}^{8} \Theta_a(x)\, \frac{\lambda_a^C}{2}\right]. \tag{12.89}$$

Here λ_a^C are the Gell-Mann matrices acting in *color* space and $\Theta_a(x)$ are eight real gauge parameters.
Local invariance requires replacing ∂_μ by the covariant derivative

$$D_\mu \equiv \partial_\mu - ig\, \mathcal{A}_\mu(x), \qquad \mathcal{A}_\mu(x) \equiv \sum_{a=1}^{8} \frac{\lambda_a^C}{2}\, \mathcal{A}_{\mu,a}(x), \tag{12.90}$$

so that, in components,

$$D_\mu q_f = \partial_\mu q_f - ig \sum_{a=1}^{8} \frac{\lambda_a^C}{2}\, \mathcal{A}_{\mu,a}\, q_f. \tag{12.91}$$

A crucial point for strong interaction phenomenology is that the quark-gluon coupling is *flavour blind*: the same gauge coupling g appears for all flavours, reflecting the universality of color interactions.
With these definitions, the QCD Lagrangian reads

$$\mathcal{L}_{\text{QCD}} = \sum_{f=u,d,s,c,b,t} \bar{q}_f \big(i \not{D} - m_f\big) q_f - \frac{1}{4}\, \mathcal{G}^a_{\mu\nu}\, \mathcal{G}_a^{\mu\nu}, \tag{12.92}$$

where $\mathcal{G}^a_{\mu\nu}$ is the non-Abelian field strength tensor, discussed in the next subsection.

Gauge invariance fixes essentially *all* structural terms in Eq. (12.92):

- the quark kinetic term becomes $\bar{q}\, i \not{D} q$, introducing quark-gluon vertices proportional to g;
- the gluon kinetic term is built from a field strength $\mathcal{G}_{\mu\nu}$ that necessarily contains commutators, producing gluon self-interactions of strength g and g^2;
- the only renormalizable explicit mass parameters are the quark masses m_f, which break chiral symmetries but preserve color gauge symmetry.

12.4.1 Gauge-Field Transformation and Non-abelian Field Strength

Demanding invariance of the matter sector in Eq. (12.92) under Eq. (12.89) fixes the gauge-field transformation law:

$$\mathcal{A}_\mu(x) \mapsto \mathcal{A}'_\mu(x) = U[g(x)]\, \mathcal{A}_\mu(x)\, U^\dagger[g(x)] - \frac{i}{g} \left(\partial_\mu U[g(x)]\right) U^\dagger[g(x)]. \tag{12.93}$$

For global (space-time independent) transformations, the derivative term in Eq. (12.93) vanishes and $\mathcal{A}_\mu$ transforms in the adjoint representation. For local transformations, the inhomogeneous term is essential: it is precisely what makes D_μ covariant.

The non-Abelian field strength is defined by

$$\mathcal{G}_{\mu\nu} \equiv \frac{i}{g} [D_\mu, D_\nu] = \partial_\mu \mathcal{A}_\nu - \partial_\nu \mathcal{A}_\mu - ig\, [\mathcal{A}_\mu, \mathcal{A}_\nu], \tag{12.94}$$

or, in components,

$$\mathcal{G}^a_{\mu\nu} = \partial_\mu \mathcal{A}^a_\nu - \partial_\nu \mathcal{A}^a_\mu + g\, f_{abc}\, \mathcal{A}^b_\mu\, \mathcal{A}^c_\nu, \tag{12.95}$$

where f_{abc} are the totally antisymmetric $SU(3)$ structure constants (for which a representative list has been given in Table 12.4). The transformation law of $\mathcal{G}_{\mu\nu}$ follows directly from Eq. (12.94) and the covariance of D_μ:

$$\mathcal{G}_{\mu\nu}(x) \mapsto \mathcal{G}'_{\mu\nu}(x) = U[g(x)]\, \mathcal{G}_{\mu\nu}(x)\, U^\dagger[g(x)]. \tag{12.96}$$

Using Eq. (12.96), the purely gluonic part of Eq. (12.92) can be written in the compact form

$$-\frac{1}{2}\, \mathrm{Tr}_C\!\left(\mathcal{G}_{\mu\nu}\mathcal{G}^{\mu\nu}\right), \tag{12.97}$$

which is gauge invariant by cyclicity of the trace. In contrast with QED, the commutator term in Eq. (12.94) generates *three- and four-gluon vertices*. This is not a detail: gluon self-interactions are the seed of asymptotic freedom at short distances and of confinement at long distances.

12.4.2 The Strong CP Problem and Axions

Besides the gauge-invariant terms displayed in Eq. (12.92), QCD admits a further local operator of mass dimension four built from the dual field strength. Defining

$$\widetilde{\mathcal{G}}^{a\,\mu\nu} \equiv \frac{1}{2}\,\epsilon^{\mu\nu\rho\sigma}\,\mathcal{G}^{a}_{\rho\sigma}, \qquad \epsilon^{0123} = +1, \tag{12.98}$$

one may add to the Lagrangian the term

$$\mathcal{L}_\theta = \frac{g^2\bar{\theta}}{64\pi^2}\,\epsilon^{\mu\nu\rho\sigma}\sum_{a=1}^{8}\mathcal{G}^a_{\mu\nu}\mathcal{G}^a_{\rho\sigma} = \frac{g^2\bar{\theta}}{32\pi^2}\sum_{a=1}^{8}\mathcal{G}^a_{\mu\nu}\widetilde{\mathcal{G}}^{a\,\mu\nu}. \tag{12.99}$$

The density $\mathcal{G}^a_{\mu\nu}\widetilde{\mathcal{G}}^{a\,\mu\nu}$ is odd under parity and time reversal and therefore violates P and CP. In a purely classical treatment, Eq. (12.99) is a total derivative:

$$\mathcal{G}^a_{\mu\nu}\widetilde{\mathcal{G}}^{a\,\mu\nu} = \partial_\mu K^\mu, \tag{12.100}$$

where K^μ is the Chern-Simons current. Hence, for gauge fields that fall off sufficiently fast at infinity, $\mathcal{L}_\theta$ does not affect the Euler-Lagrange equations. Quantum mechanically, however, the vacuum structure of a non-Abelian gauge theory is richer: finite-action Euclidean configurations (interpolating between distinct gauge vacua) carry non-trivial topology, and the integral of Eq. (12.100) becomes quantized. The corresponding topological charge is

$$Q \equiv \frac{g^2}{32\pi^2}\int d^4x\,\sum_{a=1}^{8}\mathcal{G}^a_{\mu\nu}\widetilde{\mathcal{G}}^{a\,\mu\nu} \in \mathbb{Z}, \tag{12.101}$$

and the path integral receives contributions from sectors labeled by Q. The effect of $\mathcal{L}_\theta$ is then to weight each sector by a phase $e^{i\bar{\theta}Q}$, so $\bar{\theta}$ becomes a physically meaningful parameter.

The coefficient $\bar{\theta}$ is not merely the "bare" parameter appearing in Eq. (12.99). In the presence of quarks, chiral transformations can shift θ by phases in the quark mass matrix. The physically relevant combination is

$$\bar{\theta} = \theta + \arg\det M_q, \tag{12.102}$$

where M_q is the complex quark mass matrix in the basis appearing in Eq. (12.92). Equation (12.102) captures the key point: even if the fundamental θ term is set to zero at some scale, complex phases in the quark masses (ultimately related to electroweak CP violation) reinduce a non-vanishing $\bar{\theta}$ unless a further mechanism enforces its smallness.

The quantity $\bar{\theta}$ is physical. Indeed, classically, Eq. (12.99) looks innocuous because it is a total derivative, as shown by Eq. (12.100). Quantum mechanically, the space of gauge configurations splits into topological sectors labeled by the integer Q in Eq. (12.101). The parameter $\bar{\theta}$ controls the relative phase between these sectors and therefore affects observable amplitudes, despite not modifying the classical equations of motion.

The strong CP problem. If $\bar{\theta} \neq 0$, QCD violates CP already in the strong sector. The most sensitive low-energy diagnostic is the neutron electric dipole moment (EDM), since an EDM is a P- and T-odd quantity. Dimensional analysis and hadronic estimates imply that a generic value $\bar{\theta} \sim O(1)$ would induce a neutron EDM far above experimental bounds; phenomenology therefore requires $|\bar{\theta}| \ll 1$, in practice an extremely small number. The strong CP problem is the question: *why* is the CP-violating parameter $\bar{\theta}$ so tiny, given that

- Equation (12.99) is allowed by all local symmetries of QCD, hence its absence is not enforced by the gauge principle;
- CP violation exists in the Standard Model (e.g. in the CKM sector), so there is no fundamental reason for Nature to set all phases to zero;
- radiative stability does not protect $\bar{\theta} = 0$ unless an additional symmetry or dynamical relaxation mechanism is present.

In this sense, the problem is a naturalness puzzle: $\bar{\theta} \simeq 0$ is an empirical input that appears highly non-generic from the viewpoint of the effective QCD Lagrangian.

The conceptual tension. The Standard Model contains CP-violating phases that are measured in weak processes (see Chap. 11). QCD, however, allows an independent and much more efficient CP-violating operator, Eq. (12.99), whose effects would be visible in hadronic observables unless $|\bar{\theta}|$ is enormously suppressed. Explaining this suppression is the strong CP problem.

The Peccei-Quinn mechanism and the axion. A compelling resolution is to promote $\bar{\theta}$ from a fixed parameter to a dynamical variable. The Peccei-Quinn (PQ) mechanism postulates a global chiral $U(1)_{\text{PQ}}$ symmetry that is spontaneously broken at a scale f_a. The associated pseudo-Nambu-Goldstone boson is the *axion* field $a(x)$. The crucial requirement is that $U(1)_{\text{PQ}}$ be anomalous with respect to $SU(3)_c$, so that the axion couples to the topological density exactly like θ:

$$\mathcal{L}_a \supset \frac{g^2}{32\pi^2}\left(\bar{\theta} + \frac{a(x)}{f_a}\right)\sum_{a=1}^{8} \mathcal{G}^a_{\mu\nu}\widetilde{\mathcal{G}}^{a\,\mu\nu}. \tag{12.103}$$

The structure in Eq. (12.103) can be derived as follows. Under a $U(1)_{\text{PQ}}$ rotation, the fermion measure in the path integral is not invariant due to the Adler-Bell-Jackiw anomaly; the variation produces a term proportional to $\mathcal{G}\widetilde{\mathcal{G}}$. When the PQ symmetry is spontaneously broken, the axion is the phase field parameterizing the vacuum manifold, and its shift $a \to a + \alpha f_a$ reproduces the anomalous change of the effective action. The axion therefore inherits a shift symmetry at the classical level (ensuring technical naturalness) but couples anomalously to the QCD topological density as in Eq. (12.103).

The essential physics is that QCD generates an effective potential for the combination $\bar{\theta} + a/f_a$. At low energies, non-perturbative effects produce a vacuum energy that depends periodically on the effective angle,

$$V_{\text{eff}}\left(\bar{\theta} + \frac{a}{f_a}\right) = \chi\left[1 - \cos\left(\bar{\theta} + \frac{a}{f_a}\right)\right] + \cdots, \tag{12.104}$$

where χ is the topological susceptibility of QCD and the ellipsis denotes higher harmonics and small corrections. Minimizing Eq. (12.104) gives

$$\left\langle \bar{\theta} + \frac{a}{f_a} \right\rangle = 0 \pmod{2\pi}, \tag{12.105}$$

so the vacuum dynamically relaxes to a CP-conserving configuration, even if the microscopic theory starts with $\bar{\theta} \neq 0$. Expanding around the minimum, $a(x) = \langle a \rangle + \delta a(x)$, one finds that the axion acquires a mass from QCD dynamics,

$$m_a^2 f_a^2 = \chi, \tag{12.106}$$

so the axion is naturally light when the PQ scale f_a is high.

The PQ mechanism replaces the fine tuning $|\bar{\theta}| \ll 1$ by dynamics: QCD generates a potential for the effective angle $\bar{\theta} + a/f_a$, whose minimum occurs at CP conservation, Eq. (12.105). The axion is the corresponding pseudo-Goldstone boson; its mass is set by the QCD topological susceptibility through Eq. (12.106).

Pause and think
Ask yourself whether axions are mandatory or not.

Practical implications for a QCD-first presentation. For the purposes of introducing the gauge theory in Eq. (12.92), it is common to set $\bar{\theta} = 0$ and omit $\mathcal{L}_\theta$ in Eq. (12.99) as a working assumption. The discussion above clarifies why this is not a harmless simplification: it corresponds to taking as input either an unexplained empirical fact ($|\bar{\theta}| \ll 1$) or an additional dynamical ingredient such as the PQ symmetry and its axion. In what follows we work with Eq. (12.92), keeping in mind that the status of Eq. (12.99) is a distinct physical question rather than a matter of notation.

12.4.3 Axions as Dark Matter

The Peccei-Quinn mechanism solves the strong CP problem by dynamically relaxing the effective angle $\bar{\theta} + a/f_a$ to zero, as encoded in Eq. (12.105). Remarkably, the same dynamics implies that the axion field generically contributes to the energy density of the Universe in the form of a cold, non-relativistic condensate. In this sense, the axion is not merely an elegant theoretical completion of QCD, but also a well-motivated dark-matter candidate whose existence is tied to the topology of $SU(3)_c$ gauge fields and to the non-perturbative generation of the axion potential in Eq. (12.104).

Driving Exercise
From a dynamical θ angle to a coherent field mode. At energies well below the PQ scale, the axion is described by an effective field theory of a compact pseudo-Goldstone boson with an approximate shift symmetry, broken explicitly only by the QCD anomaly. Keeping only the leading terms relevant for the present discussion, the low-energy effective Lagrangian may be written as

$$\mathcal{L}_{\text{eff}}(a) = \frac{1}{2}\,\partial_\mu a\,\partial^\mu a - V_{\text{eff}}\left(\bar{\theta} + \frac{a}{f_a}\right) + \mathcal{L}_{\text{int}}(a), \tag{12.107}$$

where V_{eff} is the non-perturbative QCD potential introduced in Eq. (12.104), and $\mathcal{L}_{\text{int}}$ collects the model-dependent couplings of the axion to Standard Model fields (photons, fermions) that are relevant for detection but not essential for the basic dark-matter mechanism. Expanding Eq. (12.104) around the minimum in Eq. (12.105), writing $a(x) = \langle a\rangle + \delta a(x)$, one obtains to quadratic order

$$V_{\text{eff}}\left(\bar{\theta} + \frac{a}{f_a}\right) \simeq \frac{1}{2}\,m_a^2\,(\delta a)^2 \qquad \text{with} \qquad m_a^2 f_a^2 = \chi, \tag{12.108}$$

in agreement with Eq. (12.106). In this regime the axion behaves as a very weakly coupled, light scalar degree of freedom whose dominant self-interaction is fixed by the periodic completion of the potential.
The crucial point for dark matter is that, once the axion acquires a mass, any homogeneous component of the field (or, more generally, long-wavelength modes) evolves as a nearly free harmonic oscillator. Even without going into full cosmological details, one can capture the essence by considering the equation of motion implied by Eq. (12.107) in the presence of a friction term $3H\dot{a}$ generated by the expansion,

$$\ddot{a} + 3H\,\dot{a} + \frac{\partial V_{\text{eff}}}{\partial a} = 0. \tag{12.109}$$

When the effective friction dominates, the field is overdamped and essentially frozen. When the mass term dominates, the field undergoes coherent oscillations about the minimum. In the harmonic approximation of Eqs. (12.108) and (12.109) becomes

$$\ddot{\delta a} + 3H\,\dot{\delta a} + m_a^2\,\delta a = 0, \tag{12.110}$$

whose solution at late times is an oscillatory mode with slowly varying amplitude. The corresponding energy density,

$$\rho_a = \frac{1}{2}\,\dot{\delta a}^{\,2} + \frac{1}{2}\,m_a^2\,(\delta a)^2, \tag{12.111}$$

averages over many oscillations to $\langle\rho_a\rangle \simeq m_a^2\langle(\delta a)^2\rangle$. Since the comoving number of axion quanta in this coherent mode is conserved to excellent approximation once interactions are negligible, the energy density redshifts as that of non-relativistic matter. This is precisely the defining behavior of cold dark matter.

Pause and think
What does "cold" mean physically?

Why are axions naturally cold? After the onset of oscillations in Eq. (12.110), the axion field behaves as a coherent condensate of zero-momentum quanta. Because the axion is extremely weakly coupled, this condensate does not thermalize and its energy density redshifts approximately as a^{-3}, as expected for pressureless matter. The same non-perturbative QCD input that yields $m_a^2 f_a^2 = \chi$ in Eq. (12.108) therefore also sets the scale for the relic abundance.

Misalignment mechanism and parametric dependence. A robust production channel is the *misalignment mechanism*. The axion is an angular variable, so at high temperature the effective potential is suppressed and the field can take an initial value displaced from its late-time minimum. Denoting by $\theta_a \equiv a/f_a$ the dimensionless axion angle, the initial condition may be written as

$$\theta_a(t_\mathrm{i}) = \theta_\mathrm{i}, \qquad \dot{\theta}_a(t_\mathrm{i}) \simeq 0, \tag{12.112}$$

with θ_i a model-dependent angle typically expected to be of order unity unless selected by environmental or dynamical arguments. When the axion mass turns on, the field rolls and then oscillates around the minimum, producing a relic energy density governed by the initial displacement. Using Eq. (12.111) at the onset of oscillations one obtains, up to factors of order unity,

$$\rho_a(t_\mathrm{osc}) \sim \frac{1}{2}\, m_a^2(T_\mathrm{osc})\, f_a^2\, \theta_\mathrm{i}^2, \tag{12.113}$$

where t_osc is defined implicitly by the condition that the axion starts oscillating when the Hubble friction becomes subdominant,

$$3H(t_\mathrm{osc}) \sim m_a(T_\mathrm{osc}). \tag{12.114}$$

The subsequent redshifting as non-relativistic matter then fixes the present abundance in terms of f_a, θ_i, and the temperature dependence of the QCD topological susceptibility entering Eq. (12.106). The important message for a QCD-first presentation is the qualitative scaling: larger f_a corresponds to weaker couplings and typically to a lighter axion, but also to a larger field excursion in physical units and therefore to a larger potential energy stored in Eq. (12.113). This interplay is the origin of the classic axion dark-matter window.

Axion couplings and experimental handles. Although the dark-matter abundance is controlled primarily by the QCD potential, detection relies on the feeble interactions collected in $\mathcal{L}_\mathrm{int}$ of Eq. (12.107). In broad classes of models, the axion couples derivatively to fermions and, through the electromagnetic anomaly, to photons. A standard parametrization is

$$\mathcal{L}_{a\gamma\gamma} = -\frac{1}{4}\, g_{a\gamma\gamma}\, a\, F_{\mu\nu}\widetilde{F}^{\mu\nu}, \qquad g_{a\gamma\gamma} \sim \frac{\alpha}{2\pi f_a}\, C_{a\gamma}, \tag{12.115}$$

where $C_{a\gamma}$ is a model-dependent coefficient. If axions constitute (a significant fraction of) dark matter, the local field can be treated as a classical oscillation with frequency set by the axion mass,

$$a(t) \simeq a_0 \cos(m_a t), \qquad \rho_\mathrm{DM} \simeq \frac{1}{2} m_a^2 a_0^2, \tag{12.116}$$

so that the signal in resonant detectors is essentially monochromatic, with a small linewidth reflecting the virialized velocity dispersion of galactic dark matter. Equations (12.115) and (12.116) clarify why axion searches exploit coherent conversion mechanisms (haloscopes, helioscopes, resonant cavities) rather than hard scattering: the field is cold, weakly coupled, and best probed through macroscopic coherence.

Two complementary questions. The strong interaction viewpoint isolates the axion coupling $(\bar{\theta} + a/f_a)\, \mathcal{G}\tilde{\mathcal{G}}$ in Eq. (12.103) as the defining theoretical input. From there one can ask:

- does the PQ mechanism operate in Nature, dynamically enforcing Eq. (12.105) and solving the strong CP problem?
- If an axion exists, is its relic abundance set by misalignment as in Eqs. (12.113)–(12.114), so that it constitutes a significant component of dark matter?

These questions are logically distinct but experimentally intertwined, since the same scale f_a controls the axion mass through Eq. (12.106) and its observable couplings through Eq. (12.115).

12.4.4 Confinement

The defining infrared phenomenon of QCD is confinement: colored objects do not appear as asymptotic states. In the hadronic spectrum, observed particles are color singlets. In the simplest valence language this is encoded as

$$\text{mesons: } q\bar{q}, \qquad \text{baryons: } qqq, \tag{12.117}$$

but the field-theoretic statement is deeper: the physical Hilbert space consists of gauge-invariant states built from quark and gluon fields.

A useful gauge-invariant diagnostic is the static quark-antiquark potential extracted from a rectangular Wilson loop $W(C)$,

$$W(C) \equiv \frac{1}{N_c} \mathrm{Tr}_C\, \mathcal{P} \exp\left(ig \oint_C dx^\mu\, \mathcal{A}_\mu(x) \right), \tag{12.118}$$

with $N_c = 3$ and $\mathcal{P}$ denoting path ordering. For a loop of spatial length R and Euclidean time extent T, one defines

$$V(R) = - \lim_{T\to\infty} \frac{1}{T} \ln\langle W(R \times T)\rangle. \tag{12.119}$$

In a confining theory, $\langle W \rangle$ exhibits an area law, $\langle W \rangle \sim e^{-\sigma RT}$ at large R, T, implying

$$V(R) \simeq \sigma R \qquad (R \gg \Lambda_{\text{QCD}}^{-1}), \tag{12.120}$$

where σ is the string tension. The physical picture is that the color electric flux between static sources is squeezed into a tube, so that the energy grows linearly with separation until it becomes favourable to create a quark-antiquark pair from the vacuum, leading to hadronization rather than free color.

At short distances, one expects a Coulombic behavior from single-gluon exchange,

$$V(R) \simeq -\frac{C_F\,\alpha_s(R)}{R}, \qquad C_F = \frac{N_c^2 - 1}{2N_c} = \frac{4}{3}, \tag{12.121}$$

so the qualitative interpolation between Eqs. (12.120) and (12.121) captures a central empirical fact: QCD is weakly coupled at short distances and strongly coupled at large distances.

Confinement should be understood as a statement about the *spectrum and state space* of a gauge theory:

- physical asymptotic states are color singlets (gauge invariant);
- colored correlators are not directly observable and depend on gauge fixing;
- the long-distance dynamics is characterized by the nonperturbative scale Λ_{QCD} and a string tension σ.

This viewpoint prevents a common misconception: confinement is not simply "a large coupling", but a global property of the infrared theory.

12.4.5 Asymptotic Freedom

The ultraviolet behavior of QCD is controlled by the renormalization group. The running of the strong coupling $\alpha_s \equiv g^2/(4\pi)$ is governed by the beta function,

$$\beta(g) \equiv \mu \frac{dg}{d\mu}. \tag{12.122}$$

A hallmark of non-Abelian gauge theories is that gauge-boson self-interactions contribute with a sign opposite to matter screening. At one loop, one finds

$$\beta(g) = -\frac{\beta_0}{16\pi^2}\, g^3 + O(g^5), \qquad \beta_0 = \frac{11}{3} N_c - \frac{2}{3} n_f, \tag{12.123}$$

where n_f is the number of active quark flavours at scale μ. For QCD, $N_c = 3$, hence $\beta_0 > 0$ provided $n_f < 16.5$, which is satisfied in Nature. Therefore,

$$\beta(g) < 0, \tag{12.124}$$

and the coupling decreases at high energies: quarks and gluons behave as weakly interacting quasi-particles at short distances.

Integrating Eq. (12.123) yields the canonical logarithmic running,

$$\alpha_s(\mu) \simeq \frac{4\pi}{\beta_0 \ln(\mu^2/\Lambda_{\text{QCD}}^2)}, \tag{12.125}$$

where Λ_{QCD} is the nonperturbative integration constant that sets the scale for the transition between perturbative and nonperturbative regimes. Equation (12.125) is the field-theoretic backbone behind the parton model: hard processes probe short distances ($\mu \sim Q$ large) where $\alpha_s(Q)$ is small, enabling systematic expansions, while soft physics is sensitive to the hadronic scale Λ_{QCD}.

A useful physical separation is between:

- *hard* dynamics at scale Q, computable in perturbation theory via expansions in $\alpha_s(Q)$;
- *soft* dynamics at scales $\lesssim \Lambda_{\text{QCD}}$, encoded in hadronic matrix elements and in universal nonperturbative inputs (parton distributions, fragmentation functions, effective theories).

Asymptotic freedom is the reason this separation is meaningful in QCD.

12.4.6 Hadron Spectroscopy and Flavour Multiplets

The microscopic degrees of freedom in Eq. (12.92) are quark and gluon fields, whereas the physical spectrum consists of colorless hadrons. The map between the two is intrinsically nonperturbative; nevertheless, a remarkably large fraction of hadron spectroscopy can be understood already at the level of symmetry, representation theory, and controlled symmetry breaking. This is the precise sense in which the Eightfold Way is both historically and conceptually a gateway to QCD: it is a *phenomenological* classification principle that anticipates quarks and their quantum numbers, while remaining logically distinct from the *dynamical* gauge symmetry $SU(3)_c$.

Global flavour symmetry as a limiting pattern

Restrict attention to the three light flavours (u, d, s) and consider the idealized limit

$$m_u = m_d = m_s \equiv m_q, \tag{12.126}$$

in which the QCD Lagrangian in Eq. (12.92) is invariant under global $SU(3)_F$ rotations in flavour space,

$$q(x) \equiv \begin{pmatrix} u \\ d \\ s \end{pmatrix}, \qquad q(x) \mapsto U_F\, q(x), \qquad U_F \in SU(3)_F. \tag{12.127}$$

In this limit, hadrons built from light quarks are organized into irreducible representations of $SU(3)_F$. The appropriate commuting labels are the eigenvalues of I_3 and Y (or equivalently I_3 and S for fixed baryon number), reflecting the rank-two structure of $SU(3)$ already discussed in the Eightfold Way material. Empirically, the observed low-lying mesons and baryons indeed populate multiplets that match the dimensions and weight patterns of $SU(3)$ irreducible representations (octets, decuplets, singlets).

In the real world, the condition in Eq. (12.126) is only approximate, mainly because $m_s \gg m_{u,d}$. It is therefore useful to think of $SU(3)_F$ not as an exact symmetry, but as an organizing principle that becomes accurate when spectroscopy is examined *within* a given multiplet and when splittings are treated as controlled symmetry breaking.

Flavour versus color. The flavour group $SU(3)_F$ is an *approximate global* symmetry of the light-hadron spectrum (exact only in the limit of Eq. (12.126)). By contrast, $SU(3)_c$ is an *exact local* gauge symmetry of the QCD Lagrangian in Eq. (12.92). Confusing the two leads to conceptual errors: flavour explains multiplet patterns; color governs dynamics, enforces singlet physical states, and controls the running coupling.

▶ **Driving Exercise**

Spectroscopy is sensitive to color in a subtle way. Because physical states must be gauge invariant, observed hadrons are color singlets. In a valence description, this is encoded as

$$\text{mesons:}\, q\bar{q}, \qquad \text{baryons:}\, qqq, \tag{12.128}$$

but Eq. (12.128) should be read as the lowest Fock content consistent with gauge invariance, not as an exact statement of hadron structure. In QCD, gluons and sea quarks contribute at all scales; what is special about Eq. (12.128) is that the associated color tensors are the simplest singlet projectors.

For mesons, the unique singlet in $\mathbf{3} \otimes \bar{\mathbf{3}}$ is

$$\Psi^{(M)}_{\text{color}} = \frac{1}{\sqrt{3}} \, \delta^a{}_b \, |a \, \bar{b}\rangle, \tag{12.129}$$

while for baryons the singlet in $\mathbf{3} \otimes \mathbf{3} \otimes \mathbf{3}$ is the totally antisymmetric tensor,

$$\Psi^{(B)}_{\text{color}} = \frac{1}{\sqrt{6}} \, \varepsilon_{abc} \, |a \, b \, c\rangle. \tag{12.130}$$

Equation (12.130) is not merely a group-theory curiosity: it is the mechanism by which Fermi statistics is reconciled with the existence of baryons whose spin-flavour wavefunction is symmetric (e.g. the Δ-resonances). Color is thus *invisible* in hadron quantum numbers (because states are singlets) but *indispensable* for the internal consistency of the quark model and, in QCD, for the very definition of the physical state space.

From multiplets to mass relations: controlled $SU(3)_F$ breaking.
Once $SU(3)_F$ is explicitly broken by quark masses, the degeneracy within a multiplet is lifted. A standard and technically clean way to organize the splittings is to separate the QCD Hamiltonian into an $SU(3)_F$ symmetric part plus a small symmetry-breaking perturbation,

$$H = H_0 + H_{\text{br}}, \qquad [H_0, \, SU(3)_F] = 0, \tag{12.131}$$

where, at the level of quark masses, the leading flavour breaking is driven by

$$H_{\text{br}} \propto (m_s - m_q) \, \bar{q} \, \mathcal{P}_s \, q, \qquad m_q \equiv \frac{m_u + m_d}{2}, \tag{12.132}$$

with $\mathcal{P}_s$ a projector onto strange flavour. In the language of $SU(3)$, Eq. (12.132) contains a singlet component plus an octet component. The singlet shifts all masses equally, while the octet controls the *pattern* of splittings within a multiplet. To first order in perturbation theory, the mass of a hadron state $|R, \alpha\rangle$ belonging to an irreducible representation R is

$$M_{R,\alpha} = M^{(0)}_R + \langle R, \alpha | H_{\text{br}} | R, \alpha \rangle + \cdots, \tag{12.133}$$

where α labels weights (e.g. (I, Y)) inside the multiplet. Because an octet operator has restricted reduced matrix elements, Eq. (12.133) implies linear relations among the physical masses. For the baryon octet, this logic leads to the celebrated Gell-Mann-Okubo relation, commonly written as

$$\frac{1}{2}\big(M_N + M_\Xi\big) = \frac{1}{4}\big(3M_\Lambda + M_\Sigma\big), \tag{12.134}$$

and, for the pseudoscalar meson octet, to the squared-mass version

$$3m_{\eta_8}^2 + m_\pi^2 = 4m_K^2, \tag{12.135}$$

where η_8 denotes the octet component prior to singlet mixing. Here the important point is not how precisely Eqs. (12.134) and (12.135) reproduce data, since higher-order breaking and dynamical effects limit their accuracy, but rather that the *pattern* of splittings is controlled by symmetry and can be systematically organized.

Hadron multiplets and their splittings implement, at the observable level, the separation between:

- approximate *global* flavour symmetry $SU(3)_F$ (multiplet classification and mass relations such as Eq. (12.134));
- exact *local* color symmetry $SU(3)_c$ (singlet physical states and the internal antisymmetry of Eq. (12.130)).

This separation is a prerequisite for understanding why QCD is a gauge theory of color rather than a global symmetry model of flavours.

12.4.7 Infrared Behavior and Possible Freezing of $\alpha_s(Q)$

Strong interaction dynamics is organized around the emergence of the scale Λ_{QCD}, typically of order a few 10^2 MeV, corresponding to a characteristic hadronic length scale $\Lambda_{\text{QCD}}^{-1} \sim 1$ fm. In the ultraviolet, the running coupling is governed by asymptotic freedom, so that inclusive short-distance observables admit expansions in $\alpha_s(Q)$ evaluated at a hard scale Q, supplemented by resummations of large logarithms when multiple disparate scales are present.

The infrared regime is often summarized by the slogan "infrared slavery", suggesting that $\alpha_s(Q)$ diverges as $Q \to \Lambda_{\text{QCD}}$. Taken literally, this is not an observable statement. The coupling $\alpha_s(Q)$ is scheme-dependent, and its naive perturbative continuation towards the IR develops a Landau singularity. What is meaningful is the behavior of *effective charges* defined from infrared-safe observables, together with the consistency of data with models of the coupling at low scales.

A practical way to parametrize soft physics is to introduce an averaged coupling over the infrared region,

$$\alpha_0(\mu_I) \equiv \frac{1}{\mu_I}\int_0^{\mu_I} dQ\, \alpha_s(Q), \tag{12.136}$$

where μ_I is an infrared matching scale chosen above the genuinely nonperturbative domain but below typical hard scales (often μ_I of order a few GeV). Phenomenological fits to event-shape moments and related observables frequently prefer values $\alpha_0(\mu_I) \sim 0.5$ for such μ_I. Interpreting this number requires care: depending on the observable and scheme, it may represent a genuinely IR-finite effective charge ("freezing") or, alternatively, a compact way of parameterizing universal power corrections.

Two conceptual consequences are particularly important for the logic of this chapter.

First, *perturbative calculability* and *existence of colored asymptotic states* are logically distinct notions. Even if an effective coupling saturates in the IR, confinement may still hold, because confinement is not simply a statement about the numerical size of α_s but about the realization of color in the physical spectrum and the global structure of the gauge-field configurations.

Second, IR modelling is naturally tied to the potentials used in hadron phenomenology. A standard parametrization of the static interquark potential is

$$V(r) = kr - \frac{4}{3}\frac{\alpha_s}{r}, \tag{12.137}$$

where the color-Coulomb coefficient $4/3$ is the quadratic Casimir C_F of the fundamental representation in a color-singlet $q\bar{q}$ channel. The linear term encodes confinement at large distances and is often introduced as a manifestly nonperturbative ingredient. It is, however, technically illuminating to emphasize that in a gauge theory the temporal component of the gauge field is determined by Gauss' law and therefore depends on boundary data at spatial infinity. Different admissible boundary conditions correspond to different infrared realizations of the theory and can be viewed as a field-theoretic origin of an energy-density scale associated with confinement.

In the IR, statements should be formulated in terms of: (i) effective charges defined from observables, (ii) averaged couplings such as Eq. (12.136), (iii) a systematic separation between perturbative contributions and power corrections. The phrase "α_s becomes large" must always be tied to a concrete definition and scheme.

Throughout this chapter, whenever infrared modelling is required, we will make explicit which statements are scheme-dependent (e.g. the precise functional form of $\alpha_s(Q)$ below a few GeV) and which are structural (e.g. the emergence of Λ_{QCD}, confinement, and chiral symmetry breaking).

12.4.8 The Bound-State Problem in Strong Interactions

The characteristic objects of the strong interactions are hadrons, i.e., bound states of QCD. This immediately exposes the central technical obstruction: bound states do not appear at any finite order in ordinary perturbation theory. In a relativistic QFT, a bound state manifests as a pole of a Green function or of a scattering amplitude at a discrete invariant mass. Such a pole requires the resummation of an infinite class of diagrams; no fixed-order Feynman graph contains a genuine bound-state singularity.

A systematic formulation of relativistic bound states is provided by the Bethe-Salpeter equation for the four-point function. In momentum space, a schematic form reads

$$\Gamma = K + K\, G_0\, \Gamma, \tag{12.138}$$

where Γ is the Bethe-Salpeter vertex, K is the two-particle irreducible kernel, and G_0 is the product of constituent propagators. Equation (12.138) makes explicit how a bound-state pole emerges: it arises from the nonperturbative solution of an integral equation, even when K is approximated.

For pedagogy, it is useful to reinterpret "perturbation theory" not only as an expansion in the coupling, but also as a loop expansion. Loops are associated with powers of $\hbar$, so that one may attempt to define a Born-level bound state (no loops) and then correct it systematically by loop effects. In nonrelativistic QED, this program is clean: iterating the leading Coulomb kernel produces the Schrödinger equation, and the hydrogenic spectrum is nonanalytic in the coupling ($E \sim \alpha^2$), reflecting the necessity of resummation already at leading order.

In QCD, light hadrons are relativistic and the coupling becomes large at low scales, so the naive analogy fails. Nonetheless, several structural facts justify a refined viewpoint:

- multiple dynamical scales coexist, and controlled approximations may exist in sectors where a hierarchy is present (e.g. heavy quarks, high-energy factorization);
- confinement and the long-distance potential are tied to the solution of gauge constraints, rather than to a single numerical value of α_s;
- the physical Hilbert space consists of gauge-invariant states, and any bound-state formalism must implement color singletness at the level of states or operators.

A conceptually sharp entry point is the role of the temporal gauge field A^0. In canonical language, A^0 is not a propagating degree of freedom: instead, it is determined at each time by Gauss' law. In the Abelian case, one has

$$\nabla^2 A^0(\mathbf{x}, t) = -\rho(\mathbf{x}, t), \tag{12.139}$$

while in the non-Abelian case the constraint is nonlinear and involves covariant derivatives. The crucial point is that solving Eq. (12.139) (and its non-Abelian generalization) requires boundary data at spatial infinity. The standard Coulomb potential corresponds to selecting boundary conditions such that $A^0 \to 0$ as $|\mathbf{x}| \to \infty$. More general boundary data, consistent with global symmetries, can introduce a nontrivial energy-density scale and hence long-range contributions that are naturally associated with confinement. In this sense, the phenomenological form in Eq. (12.137) can be viewed as the simplest interpolation between Coulombic exchange at short distances and a confining long-distance component tied to the infrared structure of the gauge theory.

Pause and think
Why are bound states qualitatively different from scattering?

In scattering, the perturbative series is organized by powers of $\alpha_s(Q)$ at fixed kinematics. In bound-state physics, the relevant information is encoded in pole locations, which are nonanalytic in the coupling and require resummation already at "leading order". A meaningful systematic program therefore requires:

- identifying a lowest-order bound-state equation compatible with gauge invariance and the correct infrared boundary conditions;
- correcting its kernel by higher-loop effects without double-counting and while preserving symmetries order by order.

This viewpoint naturally brings into focus foundational issues that any relativistic bound-state formalism must confront:

- the transformation properties of equal-time wave functions under boosts, and the limits of naive Lorentz-contraction arguments for composite relativistic states;
- the role of higher Fock components and pair creation in an equal-time description, and how these are encoded in a gauge-invariant way;
- the consistent emergence of an effective potential description (when it exists) from QCD constraints and from scale separation, rather than from phenomenological insertion.

In later sections, these conceptual elements will be connected to concrete non-perturbative tools (lattice gauge theory), to low-energy physics (chiral dynamics), and to controlled expansions in heavy-quark systems, providing a coherent bridge between the QCD Lagrangian in Eq. (12.92) and the observed hadron spectrum.

12.4.9 Light and Heavy Quarks: Physical Scales and Chirality

Although the QCD Lagrangian in Eq. (12.92) contains six quark flavours, their role in low-energy strong interactions is highly hierarchical. A useful operational division is into the three *light* flavours u, d, s and the three *heavy* flavours c, b, t, in the sense that

$$\begin{pmatrix} m_u = 0.005\,\text{GeV} \\ m_d = 0.009\,\text{GeV} \\ m_s = 0.175\,\text{GeV} \end{pmatrix} \ll 1\,\text{GeV} \leq \begin{pmatrix} m_c = (1.15 - 1.35)\,\text{GeV} \\ m_b = (4.0 - 4.4)\,\text{GeV} \\ m_t = 174\,\text{GeV} \end{pmatrix}. \tag{12.140}$$

The reference scale 1 GeV is not arbitrary: it is set by typical masses of the lightest non-Goldstone hadrons, e.g. $m_\rho \simeq 770\,\text{MeV}$, and by the characteristic scale of spontaneous chiral symmetry breaking,

$$4\pi F_\pi \approx 1170\,\text{MeV}, \tag{12.141}$$

which governs the onset of strong-coupling, nonperturbative dynamics in the light sector.

A concrete way to appreciate the separation of regimes is provided by the lightest hadrons containing charm. The masses of $D^+ = c\bar{d}$ and $\Lambda_c^+ = udc$ are $(1869.4 \pm 0.5)\,\text{MeV}$ and $(2284.9 \pm 0.6)\,\text{MeV}$, respectively, so the threshold to produce open charm in e^+e^- annihilation satisfies

$$\sqrt{s}_{\text{thr}}(D^+D^-) \simeq 2m_{D^+} \approx 3.74\,\text{GeV}. \tag{12.142}$$

This is well beyond the genuinely low-energy regime in which the relevant degrees of freedom are hadrons and the dynamics is organized by chiral symmetry and its breaking. In practice, for many low-energy questions it is therefore consistent to approximate the full QCD Lagrangian by its three-flavour light-quark version and neglect effects due to virtual heavy pairs $h\bar{h}$.

The underlying principle is *decoupling*: for characteristic external momenta $Q \ll m_h$, heavy flavours do not propagate as active degrees of freedom, and their effects can be absorbed into renormalizations of couplings and higher-dimensional operators suppressed by powers of Q/m_h. In a renormalization-group language, when the scale μ is lowered below m_h one matches the six-flavour theory onto an effective theory with fewer active flavours, so that at sufficiently low μ only $n_f = 3$ remains dynamical. Accordingly, in the following we treat the low-energy theory as a light-flavour QCD effective description, while keeping in mind that the precise numerical values of quark masses depend on the renormalization scheme and scale.

A second, deeper point concerns the meaning of *light* quark masses. Comparing the proton mass, $m_p = 938\,\text{MeV}$, with the sum of current-quark masses,

$$m_p \gg 2m_u + m_d, \tag{12.143}$$

demonstrates that most of the nucleon mass is not a simple sum of the quark mass parameters appearing in Eq. (12.92). Unlike the hydrogen atom, whose total mass is essentially the sum of proton and electron masses corrected by a small binding energy, a hadron's mass is dominantly generated by nonperturbative gluon dynamics and by the strong interaction energy stored in fields and virtual pairs. This also clarifies why one must distinguish:

- *current* quark masses m_u, m_d, m_s (renormalized parameters of the QCD Lagrangian, defined in a scheme such as $\overline{\text{MS}}$ and entering, for example, the divergences of flavour currents), and
- *constituent* quark masses (model-dependent effective parameters of order 300-400 MeV used in nonrelativistic quark models to mimic dynamical mass generation).

Equation (12.143) therefore motivates the chiral-limit Lagrangian as a first conceptual starting point for low-energy QCD. Formally one defines the three-flavour chiral limit by

$$m_u, m_d, m_s \to 0, \tag{12.144}$$

and introduces the corresponding Lagrangian

$$\mathcal{L}^0_{\text{QCD}} = \sum_{l=u,d,s} \bar{q}_l \, i \not{D} \, q_l - \frac{1}{4} \mathcal{G}_{\mu\nu,a} \mathcal{G}^{\mu\nu}_a . \tag{12.145}$$

We emphasize that the covariant derivative in Eq. (12.145) acts on Dirac and color indices, but is flavour blind: the strong interactions couple universally to all quark flavours through the same gauge coupling.

To exhibit the global symmetries of Eq. (12.145), it is convenient to decompose each Dirac field into its chiral components. Introduce the chirality matrix

$$\gamma_5 = \gamma^5 = i\gamma^0\gamma^1\gamma^2\gamma^3, \qquad \gamma_5^\dagger = \gamma_5, \qquad \{\gamma^\mu, \gamma_5\} = 0, \qquad \gamma_5^2 = 1, \tag{12.146}$$

and recall the projectors

$$P_R = \frac{1}{2}(1 + \gamma_5) = P_R^\dagger, \qquad P_L = \frac{1}{2}(1 - \gamma_5) = P_L^\dagger. \tag{12.147}$$

The matrices $P_{R,L}$ satisfy the completeness relation

$$P_R + P_L = 1, \tag{12.148}$$

are idempotent,

$$P_R^2 = P_R, \qquad P_L^2 = P_L, \tag{12.149}$$

and are mutually orthogonal,

$$P_R P_L = P_L P_R = 0. \tag{12.150}$$

The chiral components of a quark field are then

$$q_R = P_R q, \qquad q_L = P_L q, \tag{12.151}$$

and their Dirac conjugates satisfy

$$\bar{q}_R = \bar{q}\, P_L, \qquad \bar{q}_L = \bar{q}\, P_R, \tag{12.152}$$

because $\bar{q} = q^\dagger \gamma^0$ and $\gamma^0 P_{R,L} = P_{L,R}\gamma^0$.

Chirality is a Lorentz-covariant notion defined at the level of the Dirac representation. Its relation to parity is immediate: under parity, the Dirac field transforms as

$$P : q(t, \mathbf{x}) \mapsto \gamma^0 q(t, -\mathbf{x}), \tag{12.153}$$

so that the chiral components are exchanged,

$$q_R(t, \mathbf{x}) \mapsto \gamma^0 q_L(t, -\mathbf{x}), \qquad q_L(t, \mathbf{x}) \mapsto \gamma^0 q_R(t, -\mathbf{x}), \tag{12.154}$$

which makes precise the statement that q_R and q_L are neither parity-even nor parity-odd objects.

In the standard Dirac representation, one may write the projectors explicitly as

$$P_R = \frac{1}{2}\begin{pmatrix} \mathbf{1}_{2\times2} & \mathbf{1}_{2\times2} \\ \mathbf{1}_{2\times2} & \mathbf{1}_{2\times2} \end{pmatrix}, \qquad P_L = \frac{1}{2}\begin{pmatrix} \mathbf{1}_{2\times2} & -\mathbf{1}_{2\times2} \\ -\mathbf{1}_{2\times2} & \mathbf{1}_{2\times2} \end{pmatrix}, \tag{12.155}$$

and one verifies that in the massless limit the chiral projections coincide with helicity eigenstates,

$$P_R u_+ = u_+, \qquad P_L u_+ = 0, \qquad P_R u_- = 0, \qquad P_L u_- = u_-. \tag{12.156}$$

Thus, chirality equals helicity only for $m = 0$ (or, more precisely, in the ultrarelativistic regime). For massive quarks chirality and helicity differ, and chiral symmetry is explicitly broken by mass terms.

The chiral decomposition also provides a systematic classification of fermion bilinears. Using the completeness of the projectors in Eq. (12.148) and the (anti)commutation of Dirac matrices with γ_5, one finds the general identities

$$\bar{q}\,\Gamma_i\, q = \begin{cases} \bar{q}_R\, \Gamma_1\, q_R + \bar{q}_L\, \Gamma_1\, q_L \text{ for } \Gamma_1 \in \{\gamma^\mu, \gamma^\mu\gamma_5\}, \\ \bar{q}_R\, \Gamma_2\, q_L + \bar{q}_L\, \Gamma_2\, q_R \text{ for } \Gamma_2 \in \{1, \gamma_5, \sigma^{\mu\nu}\}, \end{cases} \tag{12.157}$$

with $\sigma^{\mu\nu} = i[\gamma^\mu, \gamma^\nu]/2$. Equation (12.157) is purely algebraic and does not assume massless quarks; it simply states that vector and axial-vector bilinears preserve chirality, whereas scalar, pseudoscalar, and tensor bilinears flip chirality.

Applying Eq. (12.157) to the kinetic term in Eq. (12.145), one finds that the chiral-limit Lagrangian can be written as

$$\mathcal{L}^0_{\text{QCD}} = \sum_{l=u,d,s} \left(\bar{q}_{R,l}\, i\not{D}\, q_{R,l} + \bar{q}_{L,l}\, i\not{D}\, q_{L,l} \right) - \frac{1}{4}\mathcal{G}_{\mu\nu,a}\mathcal{G}^{\mu\nu}_a . \tag{12.158}$$

Because the covariant derivative is flavour independent, the light-quark sector possesses an invariance under independent global unitary transformations of left- and right-handed flavour triplets,

$$\begin{pmatrix} u_L \\ d_L \\ s_L \end{pmatrix} \mapsto U_L \begin{pmatrix} u_L \\ d_L \\ s_L \end{pmatrix} = \exp\left(-i \sum_{a=1}^{8} \Theta^L_a \frac{\lambda_a}{2} \right) e^{-i\Theta^L} \begin{pmatrix} u_L \\ d_L \\ s_L \end{pmatrix},$$
$$\begin{pmatrix} u_R \\ d_R \\ s_R \end{pmatrix} \mapsto U_R \begin{pmatrix} u_R \\ d_R \\ s_R \end{pmatrix} = \exp\left(-i \sum_{a=1}^{8} \Theta^R_a \frac{\lambda_a}{2} \right) e^{-i\Theta^R} \begin{pmatrix} u_R \\ d_R \\ s_R \end{pmatrix}, \tag{12.159}$$

where the Gell-Mann matrices λ_a act in *flavour* space and the overall phases $e^{-i\Theta^{L,R}}$ have been factored out for later convenience. Classically, Eq. (12.158) therefore has a global $\text{U}(3)_L \times \text{U}(3)_R$ symmetry, and by Noether's theorem one expects $2 \times (8 + 1) = 18$ conserved currents.

At this stage two subtleties are essential.

Two crucial refinements of the classical symmetry counting.

- The decomposition $\text{U}(3) = SU(3) \times \text{U}(1)$ implies $\text{U}(3)_L \times \text{U}(3)_R = SU(3)_L \times SU(3)_R \times \text{U}(1)_V \times \text{U}(1)_A$. The $\text{U}(1)_V$ corresponds to baryon number and remains an exact symmetry of QCD.
- The axial factor $\text{U}(1)_A$ is not an exact symmetry of the quantum theory: it is broken by the axial anomaly, so the corresponding Noether current is not conserved once quantization and renormalization are taken into account. In contrast, the non-Abelian chiral symmetry $SU(3)_L \times SU(3)_R$ is a good symmetry of Eq. (12.158) at the classical level, being broken by quark masses and by spontaneous symmetry breaking in the QCD vacuum.

The subsequent phenomenology of low-energy QCD is largely the physics of how the classical symmetry structure of Eq. (12.158) is realized in the quantum theory:

explicit breaking by the light quark masses, spontaneous breaking of $SU(3)_L \times SU(3)_R$ to $SU(3)_V$, and anomalous breaking of $\mathrm{U}(1)_A$. These mechanisms together explain why the pseudoscalar octet appears as pseudo-Goldstone bosons, why their masses are parametrically small compared to hadronic scales, and why there is no ninth light Goldstone boson associated with $\mathrm{U}(1)_A$.

12.4.10 Global Chiral Symmetries of Three-Flavour QCD

The light-quark Lagrangian in the chiral limit, Eq. (12.158), is invariant under independent global flavour rotations of the left- and right-handed triplets, Eq. (12.159). The associated Noether currents are most efficiently derived by promoting the global parameters to slowly varying local functions and extracting the coefficients of $\partial_\mu \Theta$. This is the Gell-Mann-Levý procedure: one computes the infinitesimal variation under the local form of Eq. (12.159) and reads off the conserved currents from the terms proportional to derivatives of the symmetry parameters.

To streamline the notation, assemble the light fields into flavour triplets

$$q_L = \begin{pmatrix} u_L \\ d_L \\ s_L \end{pmatrix}, \qquad q_R = \begin{pmatrix} u_R \\ d_R \\ s_R \end{pmatrix}, \tag{12.160}$$

where $q_{L,R} = P_{L,R} q$ as in Eq. (12.151). Under infinitesimal local transformations $U_{L,R}(x)$ with parameters $\Theta_a^{L,R}(x)$ and $\Theta^{L,R}(x)$, the variation of $\mathcal{L}^0_{\text{QCD}}$ takes the form

$$\delta\mathcal{L}^0_{\text{QCD}} = \bar{q}_R \left(\sum_{a=1}^{8} \partial_\mu \Theta_a^R \frac{\lambda_a}{2} + \partial_\mu \Theta^R \right) \gamma^\mu q_R + \bar{q}_L \left(\sum_{a=1}^{8} \partial_\mu \Theta_a^L \frac{\lambda_a}{2} + \partial_\mu \Theta^L \right) \gamma^\mu q_L, \tag{12.161}$$

so that Noether's theorem yields the chiral currents

$$\begin{aligned}
L^{\mu,a} &= \frac{\partial\, \delta\mathcal{L}^0_{\text{QCD}}}{\partial(\partial_\mu \Theta_a^L)} = \bar{q}_L \gamma^\mu \frac{\lambda^a}{2} q_L, \qquad \partial_\mu L^{\mu,a} = 0, \\
R^{\mu,a} &= \frac{\partial\, \delta\mathcal{L}^0_{\text{QCD}}}{\partial(\partial_\mu \Theta_a^R)} = \bar{q}_R \gamma^\mu \frac{\lambda^a}{2} q_R, \qquad \partial_\mu R^{\mu,a} = 0, \\
L^{\mu} &= \frac{\partial\, \delta\mathcal{L}^0_{\text{QCD}}}{\partial(\partial_\mu \Theta^L)} = \bar{q}_L \gamma^\mu q_L, \qquad \partial_\mu L^{\mu} = 0, \\
R^{\mu} &= \frac{\partial\, \delta\mathcal{L}^0_{\text{QCD}}}{\partial(\partial_\mu \Theta^R)} = \bar{q}_R \gamma^\mu q_R, \qquad \partial_\mu R^{\mu} = 0.
\end{aligned} \tag{12.162}$$

The octet currents $L^{\mu,a}$ transform as an (8, 1) multiplet and $R^{\mu,a}$ as a (1, 8) multiplet of $SU(3)_L \times SU(3)_R$. For many applications it is preferable to work with the parity eigen-combinations (vector and axial-vector currents)

$$V^{\mu,a} = R^{\mu,a} + L^{\mu,a} = \bar{q}\,\gamma^{\mu}\frac{\lambda^a}{2}q, \qquad A^{\mu,a} = R^{\mu,a} - L^{\mu,a} = \bar{q}\,\gamma^{\mu}\gamma_5\frac{\lambda^a}{2}q, \tag{12.163}$$

which satisfy the standard parity transformation laws

$$P:\ V^{\mu,a}(t,\mathbf{x}) \mapsto V^a_{\mu}(t,-\mathbf{x}), \qquad P:\ A^{\mu,a}(t,\mathbf{x}) \mapsto -A^a_{\mu}(t,-\mathbf{x}). \tag{12.164}$$

The singlet combinations define the $U(1)_V$ and $U(1)_A$ currents,

$$V^{\mu} = R^{\mu} + L^{\mu} = \bar{q}\gamma^{\mu}q, \qquad \partial_{\mu}V^{\mu} = 0, \tag{12.165}$$

and

$$A^{\mu} = R^{\mu} - L^{\mu} = \bar{q}\gamma^{\mu}\gamma_5 q. \tag{12.166}$$

The baryon-number current V^{μ} is conserved in the full quantum theory. By contrast, the singlet axial current A^{μ} is conserved only at the classical level: quantization and renormalization generate the axial anomaly, so that

$$\partial_{\mu}A^{\mu} = \frac{3g^2}{32\pi^2}\,\epsilon_{\mu\nu\rho\sigma}\,\mathcal{G}^{\mu\nu}_a\mathcal{G}^{\rho\sigma}_a, \qquad \epsilon_{0123} = 1, \tag{12.167}$$

where the factor of 3 reflects the number of light flavours in the present three-flavour discussion. This anomalous divergence is a structural feature of non-Abelian gauge theories: it is tied to the topology of gauge-field configurations and cannot be removed by counterterms without spoiling gauge invariance.

▶ **Pause and think**
Try to give your interpretation.

The classical symmetry group of Eq. (12.158) is $U(3)_L \times U(3)_R$, but the physically relevant quantum symmetries are more accurately summarized as:

- $U(1)_V$ (baryon number): exact and non-anomalous;
- $SU(3)_L \times SU(3)_R$: exact in the chiral limit at the level of the Lagrangian, explicitly broken by quark masses, and spontaneously broken by the QCD vacuum;
- $U(1)_A$: broken already in the chiral limit by the anomaly in Eq. (12.167).

12.4.11 Charge Operators and the Chiral Algebra

The global symmetry of $\mathcal{L}^0_{\text{QCD}}$ implies the corresponding symmetry of the chiral-limit Hamiltonian H^0_{QCD}: the generators are the space integrals of the time components of the conserved currents. Introducing the equal-time charge operators

$$Q^a_L(t) = \int d^3x\, q^\dagger_L(t, \mathbf{x}) \frac{\lambda^a}{2} q_L(t, \mathbf{x}), \qquad a = 1, \ldots, 8, \tag{12.168}$$

$$Q^a_R(t) = \int d^3x\, q^\dagger_R(t, \mathbf{x}) \frac{\lambda^a}{2} q_R(t, \mathbf{x}), \qquad a = 1, \ldots, 8, \tag{12.169}$$

$$Q_V(t) = \int d^3x \left[q^\dagger_L(t, \mathbf{x}) q_L(t, \mathbf{x}) + q^\dagger_R(t, \mathbf{x}) q_R(t, \mathbf{x})\right], \tag{12.170}$$

current conservation implies time independence, equivalently

$$[Q^a_L, H^0_{\text{QCD}}] = [Q^a_R, H^0_{\text{QCD}}] = [Q_V, H^0_{\text{QCD}}] = 0. \tag{12.171}$$

The Lie algebra of the symmetry group is encoded in the equal-time commutators of the charges. These are obtained by applying the canonical equal-time (anti)commutation relations of the quark fields to the commutator of color singlet quadratic forms. A convenient operator identity (for equal times) is

$$\begin{aligned}&\left[q^\dagger(t, \mathbf{x}) \Gamma_1 F_1 q(t, \mathbf{x}),\ q^\dagger(t, \mathbf{y}) \Gamma_2 F_2 q(t, \mathbf{y})\right] \\ &= \delta^3(\mathbf{x} - \mathbf{y}) \left[q^\dagger(t, \mathbf{x}) \Gamma_1 \Gamma_2 F_1 F_2 q(t, \mathbf{y}) - q^\dagger(t, \mathbf{y}) \Gamma_2 \Gamma_1 F_2 F_1 q(t, \mathbf{x})\right],\end{aligned} \tag{12.172}$$

where Γ_i are Dirac matrices and F_i are flavour matrices. Color indices are implicitly summed in each bilinear; since all operators considered here are color singlets, color plays no explicit role in the algebra.

▶ **Driving Exercise**
With appropriate insertions of $P_{L,R}$ and the flavour commutator $[\lambda_a/2, \lambda_b/2] = i f_{abc} \lambda_c/2$, one obtains the chiral algebra

$$\begin{aligned}&[Q^a_L, Q^b_L] = i f_{abc} Q^c_L, \quad [Q^a_R, Q^b_R] = i f_{abc} Q^c_R, \\ &[Q^a_L, Q^b_R] = 0, \quad [Q^a_L, Q_V] = [Q^a_R, Q_V] = 0,\end{aligned} \tag{12.173}$$

which reproduces the Lie algebra of $SU(3)_L \times SU(3)_R \times U(1)_V$. It is often useful to introduce vector and axial charges,

$$Q^a_V \equiv Q^a_R + Q^a_L, \qquad Q^a_A \equiv Q^a_R - Q^a_L, \tag{12.174}$$

for which Eq. (12.173) implies the characteristic relations

$$[Q^a_V, Q^b_V] = i f_{abc} Q^c_V, \qquad [Q^a_V, Q^b_A] = i f_{abc} Q^c_A, \qquad [Q^a_A, Q^b_A] = i f_{abc} Q^c_V. \tag{12.175}$$

Even without solving the QCD equations of motion explicitly, the symmetry of the Lagrangian plus the equal-time algebra guarantee that $Q^a_{L,R}$ generate the infinitesimal flavour rotations on the Hilbert space of H^0_{QCD}. Moreover, the commutator of these charges with any operator defines its transformation property under $SU(3)_L \times SU(3)_R \times U(1)_V$, which is the starting point for deriving Ward identities and soft-meson theorems.

For any operator $O(t, \mathbf{x})$ built from quark and gluon fields, its infinitesimal transformation under a symmetry generated by a charge Q is given by

$$\delta O(t, \mathbf{x}) = i\,[Q(t), O(t, \mathbf{x})]. \tag{12.176}$$

This algebraic relation remains meaningful even when the symmetry is only approximate, because it controls how explicit breaking enters Ward identities through current divergences.

12.4.12 Explicit Chiral Symmetry Breaking by Quark Masses and Current Divergences

In real QCD, the light quarks are not massless. The quark mass term

$$\mathcal{L}_M = -\bar{q}\, M\, q, \tag{12.177}$$

with the three-flavour mass matrix

$$M = \begin{pmatrix} m_u & 0 & 0 \\ 0 & m_d & 0 \\ 0 & 0 & m_s \end{pmatrix}, \tag{12.178}$$

breaks $SU(3)_L \times SU(3)_R$ explicitly because it couples left- and right-handed fields.

Driving Exercise

Using Eq. (12.157), the mass term can be written as

$$\mathcal{L}_M = -(\bar{q}_R M q_L + \bar{q}_L M q_R), \tag{12.179}$$

which makes the breaking pattern manifest: $\mathcal{L}_M$ transforms as $(3, 3^*) + (3^*, 3)$ under $SU(3)_L \times SU(3)_R$.

For later use, it is convenient to decompose the mass matrix in the basis of the nine λ matrices (including λ_0). In particular, since M is diagonal, only the diagonal generators contribute and one may always expand M as

$$M = \alpha\, \lambda_0 + \beta\, \lambda_3 + \gamma\, \lambda_8, \tag{12.180}$$

with coefficients fixed by m_u, m_d, m_s. This decomposition isolates the isospin-breaking piece (λ_3) from the $SU(3)$-breaking piece (λ_8), and is the microscopic origin of the hierarchy of symmetry violations: $m_u \simeq m_d$ implies approximate isospin, while $m_s \gg m_{u,d}$ implies only approximate $SU(3)$ flavour.

To compute the divergences of the currents in the presence of M, one evaluates the variation of $\mathcal{L}_M$ under the infinitesimal chiral transformations of Eq. (12.159). A direct calculation gives

$$\begin{aligned}\delta\mathcal{L}_M = -i\Big[&\bar{q}_R\Big(\sum_{a=1}^{8}\Theta_a^R\frac{\lambda_a}{2}+\Theta^R\Big)Mq_L - \bar{q}_R M\Big(\sum_{a=1}^{8}\Theta_a^L\frac{\lambda_a}{2}+\Theta^L\Big)q_L\\ &+\bar{q}_L\Big(\sum_{a=1}^{8}\Theta_a^L\frac{\lambda_a}{2}+\Theta^L\Big)Mq_R - \bar{q}_L M\Big(\sum_{a=1}^{8}\Theta_a^R\frac{\lambda_a}{2}+\Theta^R\Big)q_R\Big],\end{aligned} \tag{12.181}$$

and hence the current divergences

$$\begin{aligned}\partial_\mu L^{\mu,a} &= \frac{\partial\,\delta\mathcal{L}_M}{\partial\Theta_a^L} = -i\Big(\bar{q}_L\frac{\lambda_a}{2}Mq_R - \bar{q}_R M\frac{\lambda_a}{2}q_L\Big),\\ \partial_\mu R^{\mu,a} &= \frac{\partial\,\delta\mathcal{L}_M}{\partial\Theta_a^R} = -i\Big(\bar{q}_R\frac{\lambda_a}{2}Mq_L - \bar{q}_L M\frac{\lambda_a}{2}q_R\Big),\\ \partial_\mu L^{\mu} &= \frac{\partial\,\delta\mathcal{L}_M}{\partial\Theta^L} = -i(\bar{q}_L Mq_R - \bar{q}_R Mq_L),\\ \partial_\mu R^{\mu} &= \frac{\partial\,\delta\mathcal{L}_M}{\partial\Theta^R} = -i(\bar{q}_R Mq_L - \bar{q}_L Mq_R).\end{aligned} \tag{12.182}$$

Passing to vector and axial-vector currents and applying Eq. (12.157) to simplify the chiral structure, one obtains

$$\begin{aligned}\partial_\mu V^{\mu,a} &= -i\bar{q}_R\Big[\frac{\lambda_a}{2},M\Big]q_L - i\bar{q}_L\Big[\frac{\lambda_a}{2},M\Big]q_R = i\,\bar{q}\Big[M,\frac{\lambda_a}{2}\Big]q,\\ \partial_\mu A^{\mu,a} &= i\Big(\bar{q}_L\Big\{\frac{\lambda_a}{2},M\Big\}q_R - \bar{q}_R\Big\{\frac{\lambda_a}{2},M\Big\}q_L\Big) = i\,\bar{q}\Big\{\frac{\lambda_a}{2},M\Big\}\gamma_5 q,\end{aligned} \tag{12.183}$$

together with the singlet relations

$$\partial_\mu V^\mu = 0, \qquad \partial_\mu A^\mu = 2i\,\bar{q}\,M\,\gamma_5 q + \frac{3g^2}{32\pi^2}\epsilon_{\mu\nu\rho\sigma}\mathcal{G}_a^{\mu\nu}\mathcal{G}_a^{\rho\sigma}, \qquad \epsilon_{0123} = 1. \tag{12.184}$$

Equation (12.184) combines the explicit symmetry breaking due to quark masses with the anomalous contribution already present in the chiral limit.

A number of physically important corollaries follow directly from Eqs. (12.183) and (12.184).

- For *any* values of m_u, m_d, m_s, the individual flavour number currents $\bar{u}\gamma^\mu u$, $\bar{d}\gamma^\mu d$, and $\bar{s}\gamma^\mu s$ remain conserved in the strong interactions. This reflects the flavour blindness of the strong coupling and the diagonality of M in Eq. (12.178). The singlet vector current V^μ in Eq. (12.165) is their sum and is therefore always conserved.
- If $m_u = m_d = m_s$, then M is proportional to the identity in flavour space and Eq. (12.183) implies

$$[M, \lambda_a] = 0 \quad \Rightarrow \quad \partial_\mu V^{\mu,a} = 0, \tag{12.185}$$

  so the eight octet vector currents are conserved. This is the microscopic origin of the $SU(3)$ flavour symmetry underlying the Eightfold Way.
- Even when $m_u = m_d = m_s$, the octet axial currents are *not* conserved: their divergences are proportional to pseudoscalar bilinears via Eq. (12.183). This is the field-theoretic seed of PCAC: the axial current is "partially conserved" in the sense that its divergence is controlled by explicit symmetry breaking and couples strongly to the lowest pseudoscalar octet.
- The singlet axial current is special: in addition to explicit breaking, it has the anomalous divergence in Eq. (12.184). This is why the three-flavour chiral symmetry does not produce a ninth light Goldstone boson: $U(1)_A$ is not a true symmetry of the quantum theory.

12.5 Quantization of the Free Gluon Field

Once QCD has been specified at the level of the gauge-invariant Lagrangian, the next conceptual step is the construction of its quantum theory. The gluon is a massless spin-1 gauge boson and therefore carries redundant field components: the four-vector potential contains gauge degrees of freedom that are not physical. Any covariant quantization scheme must address two intertwined issues.

- **Gauge fixing**: one must add a gauge-fixing term to make the quadratic operator invertible and define propagators.
- **Physical subspace**: the covariant description introduces unphysical polarizations and an indefinite metric; physical states must be selected by a subsidiary condition (or, in modern language, by BRST cohomology).

The non-Abelian character of QCD enters through gluon self-interactions already at the classical level. Nevertheless, the quantization of the *free* field is controlled entirely by the quadratic part of the gauge-fixed action; interaction terms are then treated perturbatively. The purpose of this section is to make the free-field structure explicit, derive the propagator and the vertex factors that follow from the Yang-Mills action, and clarify why ghosts become unavoidable once one insists on a covariant gauge beyond tree level.

The adjective "free" refers to the quadratic dynamics obtained by retaining only terms bilinear in G^a_μ after gauge fixing. This defines the notion of asymptotic gluon modes and the propagator. The genuinely non-Abelian content then appears in the interaction vertices and, beyond tree level, in the ghost sector required by gauge fixing.

12.5.1 Gauge Fixing and Canonical Structure

Let $G^a_\mu(x)$ denote the gluon potential in the adjoint representation of $SU(3)_c$, with $a = 1, \ldots, 8$, and take the Minkowski metric $g_{\mu\nu} = \mathrm{diag}(1, -1, -1, -1)$. The Yang-Mills field strength is

$$G^a_{\mu\nu} = \partial_\mu G^a_\nu - \partial_\nu G^a_\mu - g_s f^{abc} G^b_\mu G^c_\nu, \tag{12.186}$$

and the gauge-invariant Lagrangian density reads

$$\mathcal{L}_{\mathrm{YM}} = -\frac{1}{4} G^a_{\mu\nu} G^{a\,\mu\nu}. \tag{12.187}$$

As in QED, covariant quantization requires gauge fixing. A standard choice is the covariant R_ξ gauge, implemented by adding

$$\mathcal{L}_{\mathrm{GF}} = -\frac{1}{2\xi} \left(\partial_\mu G^{a\,\mu}\right)\left(\partial_\nu G^{a\,\nu}\right), \tag{12.188}$$

where ξ is the gauge parameter. The gauge-fixed Lagrangian for the gluon field is therefore

$$\mathcal{L}_t(G) = \mathcal{L}_{\mathrm{YM}} + \mathcal{L}_{\mathrm{GF}}. \tag{12.189}$$

Quadratic dynamics and canonical momenta. In order to identify the free theory, one isolates the quadratic piece of Eq. (12.189) by dropping the g_s-dependent terms in Eq. (12.186). The resulting bilinear Lagrangian is

$$\mathcal{L}_0(G) = -\frac{1}{4}\left(\partial_\mu G^a_\nu - \partial_\nu G^a_\mu\right)\left(\partial^\mu G^{a\,\nu} - \partial^\nu G^{a\,\mu}\right) - \frac{1}{2\xi}\left(\partial_\mu G^{a\,\mu}\right)^2. \tag{12.190}$$

From Eq. (12.189), the canonical momenta are

$$\pi^{a\,\mu}(x) \equiv \frac{\partial \mathcal{L}_t}{\partial(\partial_0 G^a_\mu)} = G^{a\,\mu 0}(x) - \frac{1}{\xi}\, g^{\mu 0}\, \partial_\alpha G^{a\,\alpha}(x). \tag{12.191}$$

The second term in Eq. (12.191) is precisely the covariant analog of the QED remedy: it renders $\pi^{a\,0}$ non-vanishing, so that one can impose canonical commutators in a manifestly covariant scheme without running into the primary constraint $\pi^{a\,0} = 0$.

Equal-time canonical commutation relations are postulated as

$$\left[\, G^a_\mu(\mathbf{x}, t),\, \pi^{b\,\nu}(\mathbf{y}, t) \,\right] = i\, \delta^{ab}\, \delta_\mu^{\ \nu}\, \delta^{(3)}(\mathbf{x} - \mathbf{y}), \tag{12.192}$$

with all other equal-time commutators vanishing. These commutators quantize all four polarizations of the gluon field and thus enlarge the state space beyond the two physical helicities of a massless spin-1 particle.

Indefinite metric and subsidiary condition. The oscillator algebra extracted from Eq. (12.192) carries an intrinsic minus sign for the timelike component, reflecting an indefinite metric. In covariant gauges, this is not a flaw but an intermediate device: unitarity is recovered only after restricting to a physical subspace. The canonical analog of Gupta-Bleuler is to impose the gauge condition weakly on matrix elements, i.e., to require that physical vectors $|\Psi\rangle$ satisfy

$$\langle \Phi |\, \partial_\mu G^{a\,\mu}(x)\, | \Psi \rangle = 0 \qquad \text{for all physical } |\Phi\rangle, |\Psi\rangle. \tag{12.193}$$

Equation (12.193) ensures that negative-norm excitations decouple from gauge-invariant observables.

Covariant quantization introduces longitudinal and timelike modes as intermediate states, together with an indefinite metric. The subsidiary condition in Eq. (12.193) removes these unphysical modes from physical matrix elements. As a result, only the two transverse polarizations contribute to gauge-invariant observables, even though the field operator G^a_μ retains a manifestly Lorentz-covariant form.

12.5.2 From Mode Expansion to Ghosts

► **Driving Exercise**

For the free theory defined by Eq. (12.190), it is convenient to expand the field in plane waves. For a massless momentum $k^\mu = (k^0, \mathbf{k})$ with $k^0 = |\mathbf{k}|$, one introduces polarization four-vectors $\epsilon^{(\sigma)}_\mu(k)$, $\sigma = 0, 1, 2, 3$, such that the two transverse vectors satisfy

$$\epsilon^{(i)}_0(k) = 0, \qquad \epsilon^{(i)}(k) \cdot k = 0, \qquad i = 1, 2, \tag{12.194}$$

and a convenient covariant tetrad may be chosen with

$$\epsilon^{(i)}_\mu(k)\, \epsilon^{(j)\mu}(k) = -\delta^{ij}, \qquad \epsilon^{(0)}_\mu(k) = \delta_{\mu 0}, \qquad \epsilon^{(3)}_\mu(k) = \frac{k_\mu}{k^0} - \delta_{\mu 0}. \tag{12.195}$$

In this basis, $\epsilon^{(1)}$ and $\epsilon^{(2)}$ represent the two physical transverse polarizations, while $\epsilon^{(3)}$ and $\epsilon^{(0)}$ span the unphysical longitudinal and timelike directions. The free gluon field operator admits the expansion

$$G^b_\mu(x) = \frac{1}{(2\pi)^{3/2}} \int \frac{d^3k}{\sqrt{2k^0}} \sum_{\sigma=0}^{3} \left[\epsilon^{(\sigma)}_\mu(k)\, a_\sigma(b, \mathbf{k})\, e^{-ik\cdot x} + \epsilon^{(\sigma)*}_\mu(k)\, a^\dagger_\sigma(b, \mathbf{k})\, e^{ik\cdot x}\right], \tag{12.196}$$

and Eq. (12.192) implies the oscillator algebra. Writing it compactly by packaging polarizations into Lorentz indices, one obtains

$$\left[\, a_\mu(b, \mathbf{k}),\ a^\dagger_\nu(b', \mathbf{k}') \,\right] = -\delta_{bb'}\, g_{\mu\nu}\, \delta^{(3)}(\mathbf{k} - \mathbf{k}'), \tag{12.197}$$

which displays explicitly the negative sign associated with the scalar mode.

The propagator is defined as the inverse of the quadratic operator encoded in Eq. (12.190). In momentum space, one finds the standard covariant-gauge result

$$D^{ab}_{\mu\nu}(k) = -i\,\delta^{ab}\,\frac{1}{k^2 + i0}\left[g_{\mu\nu} - (1 - \xi)\frac{k_\mu k_\nu}{k^2 + i0}\right]. \tag{12.198}$$

In particular, Eq. (12.198) reduces to the Feynman-gauge form $D^{ab}_{\mu\nu}(k) = -i\delta^{ab} g_{\mu\nu}/(k^2 + i0)$ for $\xi = 1$, and to the Landau-gauge projector in the limit $\xi \to 0$.

The propagator in Eq. (12.198) depends on ξ, reflecting the fact that off-shell Green functions are gauge dependent. Physical observables are gauge independent: in perturbation theory, this statement is implemented by Slavnov-Taylor identities, the non-Abelian generalization of Ward identities.

Separation into quadratic and interaction terms. Retaining the full non-linear field strength in Eq. (12.186), the gauge-fixed Lagrangian in Eq. (12.189) may be split as

$$\mathcal{L}_t(G) = \mathcal{L}_0(G) + \mathcal{L}_{\text{self}}(G, g_s), \tag{12.199}$$

where $\mathcal{L}_0$ is given in Eq. (12.190) and the self-interaction terms are generated by the $g_s f^{abc} G^b_\mu G^c_\nu$ part of Eq. (12.186). They can be written as the sum of cubic and quartic contributions,

$$\mathcal{L}_{\text{self}}(G, g_s) = \mathcal{L}^{(1)}_{\text{self}} + \mathcal{L}^{(2)}_{\text{self}}, \tag{12.200}$$

with

$$\mathcal{L}^{(1)}_{\text{self}} = -\frac{1}{2}\, g_s\, f^{abc}\left(\partial_\mu G^a_\nu - \partial_\nu G^a_\mu\right) G^{b\,\mu} G^{c\,\nu}, \tag{12.201}$$

and

$$\mathcal{L}_{\text{self}}^{(2)} = -\frac{1}{4}\, g_s^2 f^{abc} f^{ade}\, G_\mu^b G_\nu^c G^{d\,\mu} G^{e\,\nu}. \tag{12.202}$$

Equations (12.201)–(12.202) exhibit the qualitative difference from QED: gluons carry the very charge associated with the gauge symmetry and therefore self-interact.

Momentum-space vertices. Treating $\mathcal{L}_{\text{self}}$ as the interaction in Eq. (12.199), the perturbative expansion produces the characteristic Yang-Mills vertex factors. With all momenta taken incoming and satisfying $p+q+k=0$, the triple-gluon vertex is

$$V_{\mu\nu\lambda}^{abc}(p,q,k) = g_s f^{abc}\Big[g_{\mu\nu}(p-q)_\lambda + g_{\nu\lambda}(q-k)_\mu + g_{\lambda\mu}(k-p)_\nu\Big], \tag{12.203}$$

while the quartic vertex takes the compact form

$$V_{\mu\nu\rho\sigma}^{abcd} = -ig_s^2\Big[f^{abe} f^{cde}(g_{\mu\rho}g_{\nu\sigma} - g_{\mu\sigma}g_{\nu\rho}) + \\ + f^{ace} f^{bde}(g_{\mu\nu}g_{\rho\sigma} - g_{\mu\sigma}g_{\nu\rho}) + f^{ade} f^{bce}(g_{\mu\nu}g_{\rho\sigma} - g_{\mu\rho}g_{\nu\sigma})\Big]. \tag{12.204}$$

Together with the propagator in Eq. (12.198), these are the building blocks of perturbative QCD.

▶ **Pause and think**
Why are ghosts unavoidable in a covariant non-Abelian gauge?

The free-field canonical discussion identifies the indefinite metric and the need for a physical subspace, Eq. (12.193). In the path-integral formulation, the same physics appears as a non-trivial Jacobian in the gauge-fixed functional measure: gauge fixing introduces the Faddeev-Popov determinant, which must be exponentiated by anticommuting ghost fields c^a, $\bar{c}^a$. In covariant gauges their Lagrangian is

$$\mathcal{L}_{\text{FP}} = \bar{c}^a\, \partial^\mu \big(D_\mu c\big)^a, \qquad (D_\mu c)^a = \partial_\mu c^a + g_s f^{abc} G_\mu^b c^c. \tag{12.205}$$

Although ghosts do not correspond to asymptotic particles, they contribute in loops and are essential for preserving gauge invariance and unitarity order by order. Equivalently, they ensure the correct cancellation of unphysical gluon polarizations in non-Abelian radiative corrections, in harmony with the subsidiary condition in Eq. (12.193).

The propagator in Eq. (12.198) follows from the quadratic Lagrangian in Eq. (12.190). Non-Abelian dynamics appears through the self-interactions in Eqs. (12.201)–(12.202), which generate the vertices in Eqs. (12.203) and (12.204). Beyond tree level, covariant quantization requires the ghost sector in Eq. (12.205) to maintain gauge invariance and to guarantee that ξ-dependence cancels from physical observables.

12.6 QED Versus QCD

QED and QCD are both local gauge field theories, yet their physical manifestations are profoundly different. The origin of this difference can be traced back to two structural facts.

- QED is based on an Abelian gauge symmetry $U(1)$; its gauge boson (the photon) carries no electric charge.
- QCD is based on a non-Abelian gauge symmetry $SU(3)$ in color space; its gauge bosons (gluons) carry color charge and therefore self-interact.

The interplay between locality, gauge invariance, and self-interactions determines the ultraviolet and infrared behavior of the couplings, the spectrum of asymptotic states, and the organization of observable matter.

12.6.1 Gauge Structure in Lagrangian Form

In a compact covariant notation, the QED Lagrangian reads

$$\mathcal{L}_{\text{QED}} = -\frac{1}{4}F_{\mu\nu}F^{\mu\nu} + \bar{\psi}\left(i\gamma^{\mu}D_{\mu} - m\right)\psi, \quad D_{\mu} = \partial_{\mu} + ieA_{\mu}, \quad F_{\mu\nu} = \partial_{\mu}A_{\nu} - \partial_{\nu}A_{\mu}, \tag{12.206}$$

and is invariant under local $U(1)$ transformations

$$\psi(x) \to e^{i\alpha(x)}\psi(x), \qquad A_{\mu}(x) \to A_{\mu}(x) - \frac{1}{e}\partial_{\mu}\alpha(x). \tag{12.207}$$

The non-Abelian generalization underlying QCD replaces the single gauge field A_{μ} by a matrix-valued gauge field $A^{a}_{\mu}T^{a}$ (with generators T^{a} of $SU(3)$), and the covariant derivative becomes

$$D_{\mu} = \partial_{\mu} - ig_s\, A^{a}_{\mu}T^{a}, \tag{12.208}$$

with g_s the strong coupling. The QCD field strength contains an additional nonlinear term:

$$G^{a}_{\mu\nu} = \partial_{\mu}A^{a}_{\nu} - \partial_{\nu}A^{a}_{\mu} + g_s f^{abc}A^{b}_{\mu}A^{c}_{\nu}, \tag{12.209}$$

where f^{abc} are the structure constants of $SU(3)$. The QCD Lagrangian is

$$\mathcal{L}_{\text{QCD}} = -\frac{1}{4}G^{a}_{\mu\nu}G^{a\,\mu\nu} + \sum_{f}\bar{q}_f\left(i\gamma^{\mu}D_{\mu} - m_f\right)q_f. \tag{12.210}$$

Equations (12.209) and (12.210) display the essential qualitative difference: gluon self-interactions follow directly from the non-Abelian gauge principle.

Abelian versus non-Abelian in one line. In QED, $F_{\mu\nu}$ is linear in A_μ and the photon is neutral; in QCD, $G^a_{\mu\nu}$ contains the quadratic term $g_s f^{abc} A^b_\mu A^c_\nu$, implying gluon-gluon interactions and nontrivial color dynamics.

12.6.2 The Many-Body Character of Hadrons

Opposite electric charges attract and can form neutral bound states (atoms, positronium) whose spectra are described with extreme accuracy by QED. QCD similarly admits quark-antiquark binding into mesons and three-quark binding into baryons. However, the interpretation of the *nuclear force* differs: it is naturally viewed as a residual interaction among color singlets, analogous in spirit to residual electromagnetic interactions among neutral atoms that bind into molecules.

A crucial refinement, emphasized by the relativistic quantum nature of gauge theories, is that hadrons are not fixed-particle-number systems. Instead, a physical hadron $|H\rangle$ admits an expansion in Fock components with variable numbers of quarks, antiquarks, and gluons:

$$|H\rangle = \psi_{q\cdots} \, |qqq\rangle + \psi_{q\cdots g} \, |qqq\, g\rangle + \psi_{q\cdots q\bar{q}} \, |qqq\, q\bar{q}\rangle + \cdots , \qquad (12.211)$$

where the minimal component is fixed by the hadron's global quantum numbers, but higher components are unavoidably generated by strong interactions. Equation (12.211) is the conceptual bridge between static quark-model classification and the dynamical, fluctuation-rich picture required by QCD.

12.6.3 Spectroscopy in Heavy-Quark QCD

QED accounts for atomic and positronium spectra with extraordinary precision because the effective coupling is small and the bound-state problem admits systematic expansions. In QCD, comparable quantitative control is achieved most cleanly in systems containing heavy quarks (charmonium and bottomonium), where the typical relative velocity is moderate and short-distance dynamics can be treated perturbatively, supplemented by confining contributions at larger separations. In addition, purely gluonic color-singlet bound states (glueballs) are allowed by Eq. (12.210) and are a robust theoretical prediction, even though their unambiguous experimental identification is challenging due to mixing with ordinary mesons.

12.7 Perturbation Theory

Perturbation theory provides the most direct route from the microscopic QCD Lagrangian to quantitative predictions at high energy. The operative idea is to expand around the free-field limit in powers of the strong coupling g_s or, equivalently, $\alpha_s \equiv g_s^2/(4\pi)$, and to identify classes of observables for which the resulting series is predictive. The physical reason this is possible in QCD is *asymptotic freedom*: for sufficiently hard momentum transfer $Q \gg \Lambda_{\mathrm{QCD}}$ the effective coupling becomes moderate, and short-distance dynamics can be organized systematically.

The central limitation is equally important. Many hallmark properties of the strong interactions are intrinsically non-perturbative: hadron masses, confinement, and matrix elements dominated by long-distance dynamics cannot be computed by a weak-coupling series. In practice, modern phenomenology is built from a *hybrid* architecture in which perturbative ingredients (hard functions, anomalous dimensions, splitting kernels, resummed logarithms) are combined with universal non-perturbative inputs (Parton Distribution Functions (PDFs), fragmentation functions, hadronic matrix elements), with the separation enforced by factorization.

Perturbative QCD is not the statement that hadrons are weakly coupled objects. It is the statement that for suitably inclusive or infrared-safe observables there exists a controlled separation of scales:

- the hard scattering at a scale Q is computable as an expansion in $\alpha_s(Q)$,
- long-distance structure is encoded in universal non-perturbative functions,
- residual sensitivity to momenta of order Λ_{QCD} appears as power corrections suppressed by $(\Lambda_{\mathrm{QCD}}/Q)^p$.

12.7.1 Factorization, Universality and Evolution

The canonical setting for perturbative QCD is a hadronic collision, in which a hard scale Q is generated (large invariant mass, large transverse momentum, or a large virtuality). At leading power in Λ_{QCD}/Q, the hadronic cross section for a generic $2 \to n$ process can be written in the factorized form

$$\sigma = \sum_{a,b} \int_0^1 \mathrm{d}x_a \int_0^1 \mathrm{d}x_b \; f_{a/h_1}(x_a, \mu_\mathrm{F})\, f_{b/h_2}(x_b, \mu_\mathrm{F})\, \mathrm{d}\hat{\sigma}_{ab\to n}(x_a, x_b, \mu_\mathrm{F}, \mu_\mathrm{R}) + \mathcal{O}\left(\left(\frac{\Lambda_{\mathrm{QCD}}}{Q}\right)^p\right), \tag{12.212}$$

where $d\hat{\sigma}_{ab\to n}$ is the short-distance partonic cross section, μ_F is the factorization scale, and μ_R the renormalization scale.

The functions $f_{i/h}(x, \mu_F)$ are PDFs. At leading power and in the collinear approximation, $f_{a/h}(x, \mu_F)$ admits the intuitive interpretation as the number density of partons of type a carrying longitudinal momentum fraction x of the parent hadron. More fundamentally, PDFs are gauge-invariant light-cone correlation functions. For a quark PDF, one may write schematically (omitting Wilson-line details and spin projectors)

$$f_{q/h}(x, \mu_F) \sim \int \frac{dz^-}{2\pi} e^{ixP^+z^-} \langle h(P)|\,\bar{q}(0)\,\Gamma\,\mathcal{W}(0, z^-)\,q(z^-)\,|h(P)\rangle\Big|_{\mu_F}, \tag{12.213}$$

where $\mathcal{W}$ is a light-like Wilson line ensuring gauge invariance and the subscript reminds that the operator requires renormalization at the scale μ_F. The key phenomenological statement is *universality*: assuming the factorization theorem appropriate to the observable holds, the same PDFs appear in different hard processes.

The factorized structure in Eq. (12.212) holds up to power corrections of order $(\Lambda_{QCD}/Q)^p$, with p depending on the process and on the measurement. Importantly, while the PDFs themselves are non-perturbative, their dependence on the factorization scale is perturbatively computable. The evolution is governed by the DGLAP equations

$$\frac{\partial f_{i/h}(x, \mu_F)}{\partial \ln \mu_F^2} = \sum_j \int_x^1 \frac{dz}{z} P_{ij}(z, \alpha_s(\mu_F))\, f_{j/h}\left(\frac{x}{z}, \mu_F\right), \tag{12.214}$$

where the splitting kernels P_{ij} encode collinear branching, e.g. $q \to qg$, $g \to gg$, and $g \to q\bar{q}$. Operationally, Eq. (12.214) resums potentially large collinear logarithms by transporting PDFs from a reference scale to the hard scale relevant for the process.

While Eq. (12.212) is the leading-power starting point for a large class of collider observables, its domain of validity is subtle and actively studied. Typical issues include enhanced power corrections, the treatment of increasingly exclusive measurements, and soft exchanges (such as Glauber modes) that can obstruct naive separations of scales. These effects do not invalidate the factorized picture, but they determine its precise scope and motivate effective-field-theory refinements.

The remainder $O\big(\Lambda_{QCD}/Q\big)^p$ in Eq. (12.212) quantifies sensitivity to non-perturbative physics beyond leading power. In precision applications, controlling this term can be as important as computing higher perturbative orders, because both effects can be numerically comparable when Q is large but not asymptotically infinite.

12.7.2 Partonic Cross Sections and the Diagrammatic Expansion

At the microscopic level, the hard scattering is described by a partonic cross section obtained from squared S-matrix elements. A convenient definition is

$$\hat{\sigma}_{ab\to n} = \frac{1}{2\hat{s}} \int \mathrm{d}\Phi_n \; |\mathcal{M}_{ab\to n}(p_1, \ldots, p_n)|^2 , \tag{12.215}$$

where $\frac{1}{2\hat{s}}$ is the flux factor, $\mathcal{M}_{ab\to n}$ the renormalized scattering amplitude, and $\mathrm{d}\Phi_n$ the n-particle Lorentz-invariant phase space (LIPS),

$$\mathrm{d}\Phi_n = (2\pi)^4 \, \delta^4\left(q_a + q_b - \sum_i p_i\right) \prod_{i=1}^{n} \frac{\mathrm{d}^4 p_i}{(2\pi)^4} \; 2\pi \, \delta(p_i^2 - m_i^2) \; \Theta(p_i^{(0)}). \tag{12.216}$$

The δ^4 distribution enforces momentum conservation, the on-shell $\delta(p_i^2 - m_i^2)$ imposes final-state kinematics, and $\Theta(p_i^{(0)})$ selects positive-energy solutions. The partonic centre-of-mass energy is $\hat{s} = x_a x_b s$, linking partonic and hadronic kinematics.

The S-matrix normalization is

$$\langle f \, | S - \mathbb{1} | \, i \rangle = i \, (2\pi)^4 \, \delta^4\left(q_a + q_b - \sum\nolimits_i p_i\right) \; \mathcal{M}_{i\to f}(p_1, \ldots, p_n). \tag{12.217}$$

In QCD, $\mathcal{M}$ depends on g_s and admits a perturbative expansion. Adopting the convention

$$\mathcal{M} = \sum_{k=0}^{\infty} g_s^{2k} \, \mathcal{M}_k , \tag{12.218}$$

the first non-vanishing term is the Leading-Order (LO) amplitude, the next the Next-to-Leading-Order (NLO) correction, and so on. Correspondingly, the partonic cross section expands as

$$\hat{\sigma} = \hat{\sigma}_{\mathrm{LO}} + \alpha_s \, \hat{\sigma}_{\mathrm{NLO}} + \alpha_s^2 \, \hat{\sigma}_{\mathrm{NNLO}} + \cdots . \tag{12.219}$$

Truncating Eq. (12.219) induces a residual dependence on the unphysical scales $(\mu_{\mathrm{F}}, \mu_{\mathrm{R}})$,

$$\begin{aligned} \mathcal{M}(p_1, \ldots, p_n) &\to \mathcal{M}(p_1, \ldots, p_n; \mu_{\mathrm{F}}, \mu_{\mathrm{R}}), \\ \hat{\sigma} &\to \hat{\sigma}(\mu_{\mathrm{F}}, \mu_{\mathrm{R}}), \end{aligned}$$

because the cancellation between explicit logarithms in loop corrections and implicit scale dependence of α_s and PDFs is incomplete at any finite order. The standard diagnostic is that the $(\mu_\mathrm{F}, \mu_\mathrm{R})$ sensitivity decreases as higher perturbative orders are included.

Driving Exercise

To make the abstract structure of Eqs. (12.215)–(12.219) concrete, consider the LO partonic process $q\bar{q} \to gg$. The LO amplitude receives three tree-level topologies: t-channel and u-channel quark exchange, and an s-channel contribution mediated by an intermediate gluon (which exists only because QCD is non-Abelian and has a three-gluon vertex). Applying the Feynman rules yields

$$i\mathcal{M}_{q\bar{q}\to gg} = -ig_s^2\, \epsilon^{\mu}_{1,\lambda_1}(k_1)\, \epsilon^{\nu}_{2,\lambda_2}(k_2)\, M_{\mu\nu}, \tag{12.220}$$

$$M_{\mu\nu} = (t^a t^b)_{ij} M^{(t)}_{\mu\nu} + (t^b t^a)_{ij} M^{(u)}_{\mu\nu} + i\, f^{abc} t^c\, M^{(s)}_{\mu\nu}, \tag{12.221}$$

with

$$M^{(t)}_{\mu\nu} = \bar{v}^{s_2}(p_2)\, \gamma_\nu \frac{\not{p}_1 - \not{k}_1}{(p_1-k_1)^2} \gamma_\mu\, u^{s_1}(p_1), \quad M^{(u)}_{\mu\nu} = \bar{v}^{s_2}(p_2)\, \gamma_\mu \frac{\not{p}_1 - \not{k}_2}{(p_1-k_2)^2} \gamma_\nu\, u^{s_1}(p_1), \tag{12.222}$$

$$M^{(s)}_{\mu\nu} = \bar{v}^{s_2}(p_2)\, \left[g^{\mu\nu}(\not{k}_2 - \not{k}_1) - \gamma^\nu(k_1+2k_2)^\mu + \gamma^\mu(2k_1+k_2)^\nu\right] u^{s_1}(p_1), \tag{12.223}$$

where s_1 and s_2 are the incoming spin labels. The separation between color and kinematics is explicit in Eq. (12.220): the generators t^a are those of $\mathrm{SU}(3)_c$ in the fundamental representation and f^{abc} are the structure constants.

The genuinely non-Abelian character is already visible at LO. In an Abelian theory (e.g. a hypothetical "strong QED" with neutral gauge bosons), the analogue of the s-channel contribution would be absent. Using the commutation relation $[t^a, t^b] = if^{abc}t^c$, the color decomposition can be rearranged as

$$M_{\mu\nu} = (t^a t^b)_{ij}\left[M^{(t)}_{\mu\nu} + M^{(u)}_{\mu\nu}\right] + if^{abc}t^c\left[M^{(s)}_{\mu\nu} - M^{(u)}_{\mu\nu}\right], \tag{12.224}$$

which isolates a "QED-like" bracket plus a purely non-Abelian contribution. To obtain the partonic cross section, one computes the squared amplitude $|\mathcal{M}|^2$ and performs the appropriate sums and averages. If polarizations are not analysed, one sums over gluon polarizations $\lambda_{1,2}$; since color is not measured, one sums over final-state color indices and averages over initial-state color and spin. The relevant quantity is therefore

$$|\mathcal{M}|^2 \to \overline{\sum}|\mathcal{M}|^2 = \prod_{\text{initial}} \frac{1}{N_{\text{spins}} N_{\text{cols}}} \sum_{\text{cols}} \sum_{\substack{s_1,s_2,\\ \lambda_1,\lambda_2}} |\mathcal{M}|^2. \tag{12.225}$$

The spin sums follow from completeness relations, $\sum_{s_1} u^{s_1}(p_1)\bar{u}^{s_1}(p_1) = \not{p}_1 + m, \sum_{s_2} v^{s_2}(p_2)\bar{v}^{s_2}(p_2) = \not{p}_2 - m,$ and the gluon polarization sum depends on the gauge choice. A convenient form that makes explicit the projection onto physical transverse polarizations is

$$\sum_{\lambda} \epsilon_{\lambda}^{\mu}(k)\,\epsilon_{\lambda}^{\star\nu}(k) = -g^{\mu\nu} + \frac{k^{\mu}n^{\nu} + k^{\nu}n^{\mu}}{k \cdot n}, \qquad n^2 = 0, \tag{12.226}$$

with n^{μ} a light-like reference vector satisfying $k \cdot n \neq 0$.

Equations (12.215)–(12.225) summarize the computational pipeline: diagrams → amplitudes → squared matrix elements → phase-space integrals → partonic cross sections → hadronic predictions via factorization in Eq. (12.212). Beyond LO, the principal conceptual complication is not the definition of the expansion in Eq. (12.219), but the consistent treatment of ultraviolet renormalization and infrared singularities, so that infrared-safe cross sections remain finite and predictive.

12.7.3 Renormalization and Infrared Structure

Perturbative predictions are constructed as a series around the Born-level dynamics. For amplitudes, one may write schematically

$$\mathcal{M} = \mathcal{M}^{(0)} + \alpha_s \mathcal{M}^{(1)} + \alpha_s^2 \mathcal{M}^{(2)} + \cdots, \tag{12.227}$$

where $\mathcal{M}^{(0)}$ is the first non-vanishing contribution and $\mathcal{M}^{(L)}$ denotes the L-loop correction together with the associated real-emission pieces required by unitarity when constructing infrared-safe observables. Squaring and integrating yields

$$\hat{\sigma} = \hat{\sigma}_{\mathrm{LO}} + \alpha_s \hat{\sigma}_{\mathrm{NLO}} + \alpha_s^2 \hat{\sigma}_{\mathrm{NNLO}} + \cdots. \tag{12.228}$$

The need for higher orders is dictated by physics. First, truncation leaves a dependence on the unphysical renormalization scale μ_R, because the running of $\alpha_s(\mu_R)$ is only partially compensated by explicit logarithms $\ln(\mu_R/Q)$ in loop corrections. Second, higher orders open new channels and kinematic configurations: real radiation allows recoil and non-trivial jet substructure that are absent at LO. Third, experimental precision is often at the few-percent level, whereas LO predictions typically carry uncertainties at the 30–50% level.

Born level and quantum corrections. The Born approximation captures the basic topology of a process, but quantum fluctuations modify both normalisations and shapes. Loop diagrams describe virtual fluctuations, while real-emission diagrams account for the fact that radiation is unavoidable in a gauge theory. Both are required for a physically meaningful prediction.

Dimensional regularization and renormalization. Loop and phase-space integrals are typically divergent. A regulator must control these divergences without spoiling gauge invariance and Lorentz symmetry. Dimensional regularization continues space-time to

$$d = 4 - 2\epsilon, \tag{12.229}$$

so that divergences appear as poles in ϵ. Consider the prototypical one-loop integral

$$I(\Delta) = \mu^{2\epsilon} \int \frac{\mathrm{d}^d \ell}{(2\pi)^d} \frac{1}{(\ell^2 - \Delta + i0)^2}. \tag{12.230}$$

Evaluating Eq. (12.230) yields

$$I(\Delta) = \frac{i}{(4\pi)^2} \left(\frac{1}{\epsilon} - \gamma_E + \ln 4\pi - \ln \frac{\Delta}{\mu^2} + O(\epsilon) \right). \tag{12.231}$$

Renormalization removes ultraviolet poles by redefining fields and parameters. For the coupling, one writes

$$g_{s,0} = \mu^{\epsilon} Z_g g_s(\mu), \tag{12.232}$$

with $Z_g = 1 + \sum_{n\geq 1} a_n/\epsilon^n$. In the $\overline{\text{MS}}$ scheme, only the pole terms (and universal constants) are subtracted. In practice, a perturbative calculation in the $\overline{\text{MS}}$ scheme proceeds through the following standard steps:

- regularize all integrals in $d = 4 - 2\epsilon$,
- renormalize by absorbing UV poles into parameter redefinitions,
- combine real and virtual contributions,
- take the limit $\epsilon \to 0$ for infrared-safe observables.

Infrared singularities and cancellation. In massless gauge theories, divergences arise not only from large loop momenta (UV) but also from soft and collinear configurations (IR). At NLO, a generic observable receives virtual and real contributions. Symbolically,

$$\sigma_{\rm NLO} = \int \mathrm{d}\Phi_n\, \mathrm{d}\sigma_n^{\rm V}(\epsilon) + \int \mathrm{d}\Phi_{n+1}\, \mathrm{d}\sigma_{n+1}^{\rm R}(\epsilon) + \int \mathrm{d}\Phi_n\, \mathrm{d}\sigma_n^{\rm C}(\epsilon), \qquad (12.233)$$

where $\mathrm{d}\sigma_n^{\rm C}$ subtracts initial-state collinear singularities to be absorbed into PDFs (mass factorization). Each term in Eq. (12.233) is separately divergent, but their sum is finite for infrared-safe observables.

An observable is infrared safe if adding a soft particle or splitting a particle collinearly does not change its value. Denoting by $\mathcal{F}_n$ the measurement function on n-particle phase space, infrared safety requires

$$\mathcal{F}_{n+1}(p_1, \ldots, p_n, k \to 0) = \mathcal{F}_n(p_1, \ldots, p_n), \qquad (12.234)$$

and analogously for collinear limits. This condasition formalises the physical statement that soft and collinear radiation is unresolvable in inclusive measurements, and therefore must not affect physical predictions.

Initial-state singularities, PDFs, and DGLAP. Collinear radiation off incoming partons produces singularities that do not cancel against virtual corrections; instead they are absorbed into the renormalized PDFs. The leading-power hadronic cross section reads

$$\sigma = \sum_{a,b} \int_0^1 \mathrm{d}x_a \mathrm{d}x_b\, f_{a/h_1}(x_a, \mu_F)\, f_{b/h_2}(x_b, \mu_F)\, \hat{\sigma}_{ab}(x_a, x_b, \mu_F, \mu_R) + \mathcal{O}\left(\left(\frac{\Lambda_{\rm QCD}}{Q}\right)^p\right), \qquad (12.235)$$

and the PDFs evolve with μ_F according to

$$\frac{\partial f_i(x, \mu_F)}{\partial \ln \mu_F^2} = \sum_j \int_x^1 \frac{\mathrm{d}z}{z} P_{ij}(z, \alpha_s(\mu_F))\, f_j\left(\frac{x}{z}, \mu_F\right). \qquad (12.236)$$

Equation (12.236) resums logarithms generated by multiple collinear emissions and is indispensable whenever the hard scale differs significantly from the scale at which PDFs are determined.

The same PDFs appear in DIS, Drell-Yan production, jet production and many other processes. This universality is one of the most powerful consequences of factorization and is repeatedly tested by global PDF fits.

12.7.4 Collider Tests, Jets and Deep-Inelastic Scattering

Perturbative QCD becomes empirically meaningful only once its predictions are tied to infrared-safe observables. High-energy experiments act as microscopes: the resolving power is set by the hard momentum transfer Q, corresponding to a transverse spatial scale

$$\Delta r \sim \frac{\hbar c}{Q}. \tag{12.237}$$

Event rates are controlled by luminosity and cross section,

$$N = \mathcal{L}\sigma, \tag{12.238}$$

which highlights the dual role of accelerator performance and physics channel selection.

> The experimental strategy is to (i) select an initial state with clean quantum numbers (e.g. e^+e^-, ep, pp), (ii) enforce a hard scale Q or a large transverse momentum p_T that legitimizes a partonic description, and (iii) use infrared-safe observables (jets, event shapes, inclusive rates) for which perturbation theory is meaningful.

Empirical determination of N_c: the R ratio and $\pi^0 \to \gamma\gamma$. At energies $\sqrt{s}$ above hadronic resonances and well below the Z pole, e^+e^- annihilation into hadrons is dominated by a virtual photon. The inclusive ratio

$$R(s) \equiv \frac{\sigma(e^+e^- \to \text{hadrons})}{\sigma(e^+e^- \to \mu^+\mu^-)} \tag{12.239}$$

cancels many normalisation factors and isolates the short-distance degrees of freedom. At tree level,

$$\sigma(e^+e^- \to \mu^+\mu^-) = \frac{4\pi\alpha^2}{3s}, \tag{12.240}$$

and for a quark flavour f of charge Q_f,

$$\sigma(e^+e^- \to q_f\bar{q}_f) = N_c\, Q_f^2\, \frac{4\pi\alpha^2}{3s}. \tag{12.241}$$

Summing over active flavours with $2m_f < \sqrt{s}$ gives

$$R(s) = N_c \sum_{f\ \text{active}} Q_f^2 \qquad \text{(leading order)}. \tag{12.242}$$

Including QCD corrections one finds

$$R(s) = N_c \sum_{f \text{ active}} Q_f^2 \left[1 + \frac{\alpha_s(s)}{\pi} + O(\alpha_s^2)\right], \tag{12.243}$$

where the α_s/π term encodes real and virtual gluon emission. The step-like behavior of $R(s)$ across quark thresholds and the overall normalisation are consistent with $N_c = 3$.

A complementary probe is provided by the chiral anomaly. The amplitude scales as

$$\mathcal{A}(\pi^0 \to \gamma\gamma) \propto N_c \left(Q_u^2 - Q_d^2\right) = N_c \left(\frac{4}{9} - \frac{1}{9}\right) = \frac{N_c}{3}, \tag{12.244}$$

and with standard anomaly normalisation the observed decay rate again favours $N_c = 3$.

Jets and event shapes: from partons to hadrons. Experiments observe hadrons and energy deposition, not quarks and gluons. To compare with perturbative predictions one constructs infrared-safe observables. Jet algorithms cluster final-state particles into jets approximating underlying partons. Sequential recombination algorithms define distances

$$d_{ij} = \min(p_{T,i}^{2p}, p_{T,j}^{2p}) \frac{\Delta R_{ij}^2}{R^2}, \qquad d_{iB} = p_{T,i}^{2p}, \tag{12.245}$$

with $\Delta R_{ij}^2 = (y_i - y_j)^2 + (\phi_i - \phi_j)^2$. Infrared and collinear safety ensures stability under soft emissions and collinear splittings.

Event shapes provide global measures of radiation. In e^+e^- annihilation the thrust is

$$T = \max_{\hat{n}} \frac{\sum_i |\mathbf{p}_i \cdot \hat{n}|}{\sum_i |\mathbf{p}_i|}, \qquad \tau = 1 - T. \tag{12.246}$$

The two-jet limit corresponds to $\tau \to 0$. In this region fixed-order perturbation theory produces large logarithms $\ln \tau$ that must be resummed to all orders for quantitative predictions.

> **Jets are not partons!** Jets are algorithm-dependent composite objects built from hadrons. Their usefulness lies in the fact that, for infrared-safe definitions, their properties can be computed perturbatively up to controllable power corrections.

A decisive qualitative test of QCD is the observation of hard gluon radiation in e^+e^- annihilation. Beyond the two-jet topology from $e^+e^- \to q\bar{q}$, QCD predicts

$$e^+e^- \to q\bar{q}g, \tag{12.247}$$

leading to three-jet events. The characteristic soft/collinear enhancement can be expressed schematically as

$$d\sigma_{q\bar{q}g} \sim \sigma_{q\bar{q}} \frac{\alpha_s}{2\pi} C_F \frac{d\omega}{\omega} \frac{d\theta^2}{\theta^2}, \tag{12.248}$$

where ω is the gluon energy and θ its angle with respect to the parent quark. Equation (12.248) explains why most additional activity is concentrated near jet axes unless the gluon is hard and well separated, in which case a third jet is resolved.

DIS and scaling violations. DIS provides a clean microscope of hadron structure. Consider inclusive lepton-nucleon scattering

$$\ell(k) + N(p) \to \ell(k') + X, \tag{12.249}$$

with $q \equiv k - k'$ and $Q^2 \equiv -q^2 > 0$. The invariants are

$$\nu \equiv \frac{p \cdot q}{M}, \qquad x_B \equiv \frac{Q^2}{2p \cdot q}, \qquad W^2 \equiv (p+q)^2 = M^2 + 2p \cdot q - Q^2, \tag{12.250}$$

where M is the nucleon mass. The inclusive cross section can be expressed in terms of leptonic and hadronic tensors,

$$\frac{d^2\sigma}{d\Omega\, dE'} = \frac{\alpha^2}{Q^4} \frac{E'}{E} L_{\mu\nu} W^{\mu\nu}. \tag{12.251}$$

Lorentz invariance, parity (for electromagnetic DIS), and current conservation imply the standard decomposition

$$W_{\mu\nu} = \left(-g_{\mu\nu} + \frac{q_\mu q_\nu}{q^2}\right) F_1(x, Q^2) + \left(p_\mu - \frac{p \cdot q}{q^2} q_\mu\right)\left(p_\nu - \frac{p \cdot q}{q^2} q_\nu\right) \frac{F_2(x, Q^2)}{p \cdot q}. \tag{12.252}$$

In the naive parton model, $F_{1,2}$ depend predominantly on x (Bjorken scaling). Moreover, for spin-$\frac{1}{2}$ partons one finds the Callan-Gross relation

$$F_2(x) = 2x\, F_1(x), \tag{12.253}$$

reflecting dominance of transverse photons for massless fermionic partons.

Exact scaling is broken by QCD radiative corrections. Gluon emission and splitting generate a logarithmic Q^2 dependence governed by DGLAP evolution. For example, a quark distribution obeys schematically

$$\frac{\partial q(x, Q^2)}{\partial \ln Q^2} = \frac{\alpha_s(Q^2)}{2\pi} \left[P_{qq} \otimes q + P_{qg} \otimes g\right], \tag{12.254}$$

where $(P \otimes f)(x) = \int_x^1 \frac{dz}{z} P(z) f(x/z)$. The observed scaling violations are a quantitative determination of α_s and a non-trivial validation of the non-Abelian gauge structure of QCD.

Theoretical uncertainties. Perturbative predictions are incomplete because the series is truncated. A standard estimate of missing higher orders is obtained by varying the unphysical scales μ_R and μ_F around a central choice $\mu_0 \sim Q$,

$$(\mu_R, \mu_F) = (c_R\mu_0, c_F\mu_0), \qquad c_{R,F} \in \left\{\frac{1}{2}, 1, 2\right\}, \tag{12.255}$$

excluding extreme ratios. Beyond scale variation, realistic uncertainty budgets include:

- uncertainties on PDFs and on the input value of $\alpha_s(M_Z)$,
- hadronization and underlying-event effects,
- uncertainties from resummation and matching procedures.

Scale variation is not a rigorous bound on the true error: it probes sensitivity to logarithmic terms controlled by the renormalization group, but it cannot capture all possible higher-order effects. Its credibility increases when successive perturbative orders show convergence and reduced scale dependence.

12.8 Lattice Quantum Chromodynamics

The thermodynamics of QCD in equilibrium can be determined from Euclidean lattice simulations with controlled systematics, whereas *transport* and *real-time emission rates* probe genuinely dynamical properties of the plasma and are substantially harder to access. The difficulty is structural: lattice QCD is formulated in Euclidean time, so one computes correlators at a discrete set of Euclidean time separations, while the quantities of phenomenological interest (conductivities, diffusion constants, viscosities, photon and dilepton rates) are encoded in Minkowski-space spectral functions in the low-frequency and low-momentum regime.

In this section, we develop a coherent lattice-based framework for transport and electromagnetic (EM) probes, emphasizing (i) the spectral representation of Euclidean correlators, (ii) Kubo relations, (iii) the inverse problem and its regularization, and (iv) selected state-of-the-art results and their interpretation near the QCD crossover.

12.8.1 Euclidean Correlators and Spectral Representations

On the lattice one evaluates thermal two-point functions of conserved currents and of the energy-momentum tensor. For a generic bosonic operator $\mathcal{J}$ (e.g. a vector current, a color-electric field insertion, or a component of $T_{\mu\nu}$), the Euclidean correlator at finite temperature is defined as

$$G(\tau, \mathbf{p}) = \int d^3x \, e^{-i\mathbf{p}\cdot\mathbf{x}} \, \langle \mathcal{J}(\tau, \mathbf{x}) \, \mathcal{J}(0, \mathbf{0}) \rangle_T \,, \qquad 0 \le \tau \le \frac{1}{T}. \tag{12.256}$$

At finite temperature, $G(\tau, \mathbf{p})$ admits a spectral representation in terms of the corresponding Minkowski spectral density $\rho(\omega, \mathbf{p}, T)$,

$$G(\tau, \mathbf{p}) = \int_0^\infty \frac{d\omega}{2\pi} \, \rho(\omega, \mathbf{p}, T) \, K(\omega, \tau, T), \qquad K(\omega, \tau, T) = \frac{\cosh\left(\omega\left(\tau - \frac{1}{2T}\right)\right)}{\sinh\left(\frac{\omega}{2T}\right)}. \tag{12.257}$$

Equation (12.257) is exact and expresses the Euclidean data as a Laplace-type transform of ρ. The kernel K is smooth and strongly suppresses detailed low-ω features when G is known only at a finite number of Euclidean points. Consequently, extracting transport coefficients (which depend precisely on ρ at small ω and small $|\mathbf{p}|$) requires either sophisticated inversion techniques or controlled modeling of the infrared structure of ρ.

▶ **Pause and think**
Before going on, imagine why transport is hard on the lattice.

The lattice provides $G(\tau, \mathbf{p})$ on N_τ time slices with statistical errors and correlated systematics. The map $\rho \mapsto G$ in Eq. (12.257) is a smoothing integral transform, hence its inversion is ill-posed: many distinct spectral functions reproduce the same discrete data within errors. Any extraction of low-frequency information must therefore incorporate additional physical constraints (sum rules, positivity, asymptotics, hydrodynamic expectations) in a controlled manner.

12.8.2 Kubo Relations and the Hydrodynamic Limit

Transport coefficients are defined as linear-response parameters governing the relaxation of conserved densities and currents. In thermal field theory they can be expressed as slopes or limits of spectral functions via Kubo formulas. For a conserved $U(1)$ current J_μ (e.g. the electromagnetic current), the electrical conductivity

σ is determined by the spatial components of the retarded correlator at vanishing momentum,

$$\sigma = \frac{1}{6} \lim_{\omega \to 0^+} \frac{1}{\omega} \sum_{i=1}^{3} \rho_{ii}(\omega, \mathbf{p} = \mathbf{0}, T), \tag{12.258}$$

where ρ_{ii} is the spectral function associated with J_i. In the same channel, the diffusion constant D is related to σ through an Einstein relation involving the corresponding static susceptibility χ,

$$D = \frac{\sigma}{\chi}, \qquad \chi = \left. \frac{\partial n}{\partial \mu} \right|_{\mu=0}. \tag{12.259}$$

Equations (12.258) and (12.259) clarify the role of *static* lattice inputs: susceptibilities are Euclidean thermodynamic derivatives, accessible with far better control than real-time slopes, and they can be used to form more robust combinations such as D.

For momentum transport, the shear viscosity η is obtained from the spectral function of the off-diagonal components of the energy-momentum tensor,

$$\eta = \lim_{\omega \to 0^+} \frac{1}{\omega} \rho_{T_{12}T_{12}}(\omega, \mathbf{0}, T), \tag{12.260}$$

and similarly the bulk viscosity ζ is extracted from appropriate projections of $T_{\mu\nu}$ correlators (in particular from the trace channel), again through a low-frequency limit analogous to Eq. (12.260). The practical lattice difficulty is twofold: the required correlators are noisy, and the inversion amplifies uncertainties precisely in the hydrodynamic regime $\omega \ll T$.

12.8.3 Inverse Problem: Reconstruction Strategies and Controlled Priors

Given $G(\tau, \mathbf{p})$ measured at discrete $\tau_n = n a_\tau$ with $n = 0, \ldots, N_\tau - 1$, Eq. (12.257) becomes a discretized integral equation. A direct χ^2 minimization over functions $\rho(\omega)$ is underdetermined because the number of degrees of freedom in ρ vastly exceeds the number of data points.

A standard resolution is to restrict the admissible spectral functions by introducing a regulator functional (a prior) and performing a Bayesian inference problem. Two families of methods are widely used.

- **Maximum Entropy Method (MEM)**: favors smooth spectra by maximizing an entropy functional relative to a default model, while fitting the Euclidean data.
- **Bayesian Reconstruction (BR) and related schemes**: impose positivity and asymptotic constraints and optimize a regulator that can be less smoothing than MEM, at the price of potential ringing artifacts when the data are sparse.

The key point is conceptual: reconstruction is meaningful only if the prior information is explicit and if the stability of the inferred infrared behavior is tested under variations of the default model, the basis representation, and the frequency window. In well-controlled analyses, the dominant uncertainty in transport coefficients is systematic and originates from the limited temporal interval and resolution of the lattice.

> A lattice extraction of a transport coefficient is credible only if it incorporates: (i) positivity of ρ in appropriate channels, (ii) correct large-ω behavior from perturbation theory, (iii) sum rules or Ward identities where available, (iv) hydrodynamic constraints at small ω, and (v) a sensitivity analysis demonstrating robustness of the low-ω limit under controlled deformations of the prior.

12.8.4 Electrical Conductivity and Charge Diffusion

The electrical conductivity σ quantifies the linear response of the medium to a weak external electromagnetic field. On the lattice, one computes Euclidean vector correlators and reconstructs the corresponding spectral function in the light-quark vector channel, then extracts σ using Eq. (12.258). In practice, results have been obtained in the quenched approximation and in dynamical simulations with $N_f = 2$ and $N_f = 2 + 1$, although continuum extrapolations and physical-mass simulations remain challenging in this context.

A particularly informative observable is the charge diffusion constant D defined in Eq. (12.259), where χ is the second-order fluctuation of the conserved charge. Empirically, the dimensionless combination $2\pi T D$ often exhibits a minimum near the crossover temperature, consistent with the expectation that the system is most strongly coupled in the vicinity of the transition.

12.8.5 Thermal Photons and Dileptons from Vector Spectral Functions

Electromagnetic radiation emitted by the quark-gluon plasma (QGP) provides penetrating probes: photons and dileptons escape the medium with minimal final-state interactions. The differential emission rates can be expressed in terms of the in-medium vector spectral function. In particular, the photon rate at momentum k is controlled by the lightlike limit of the vector correlator, while the dilepton rate at invariant mass M is controlled by the timelike regime of the same spectral function.

The lattice approach aims to determine $\rho_V(\omega, \mathbf{p}, T)$ with sufficient fidelity to obtain these rates in a model-independent way. A successful strategy is to combine:

- lattice Euclidean correlators at several **p**,
- perturbative inputs in regimes where weak-coupling expansions are reliable,
- flexible parameterizations (e.g. low-order polynomials or physically motivated ansätze) in the intermediate region where the lattice deviates from perturbative expectations.

This hybrid program yields spectral functions from which photon production rates can be estimated just above the transition. Typically, agreement with perturbative predictions is best at hard momenta $k \gtrsim 3T$, while at softer momenta perturbation theory tends to overestimate the rate, consistent with the emergence of strong-coupling effects.

For dileptons, low-mass emissivity is directly proportional to the vector spectral function and is sensitive to chiral symmetry restoration. In the hadronic regime, chiral symmetry breaking manifests as a splitting between chiral partners (e.g. vector and axialvector excitations). In-medium modifications of the ρ and a_1 channels, together with sum-rule constraints, can be confronted with lattice determinations of the chiral order parameter. The core idea is that the approach to chiral restoration implies progressive degeneracy of vector and axialvector spectral functions, and that this degeneracy is correlated with the temperature evolution of the quark condensate.

12.8.6 Heavy-Flavor Diffusion from Color-Electric Correlators

Heavy quarks serve as calibrated probes because their mass is parametrically larger than the temperature scale in the plasma. In the nonrelativistic regime, the long-wavelength motion of a heavy quark in a thermal medium is characterized by a momentum diffusion coefficient κ, which controls the stochastic broadening of the heavy-quark momentum distribution. On the lattice, κ can be related to the infrared limit of a spectral function associated with a color-electric field correlator evaluated along a temporal Wilson line.

Once κ is known, the spatial diffusion coefficient D follows from the nonrelativistic relation

$$D = \frac{2T^2}{\kappa}. \tag{12.261}$$

Moreover, κ sets the kinetic equilibration time scale for heavy quarks. A parametric estimate can be obtained by dimensional analysis and by matching to Langevin dynamics, yielding a characteristic relaxation time τ_{kin} that decreases with temperature and increases with the heavy-quark mass. The phenomenological relevance is immediate: if τ_{kin} is short compared to the lifetime of the plasma, heavy quarks approach local equilibrium; otherwise they retain memory of the early-time dynamics.

12.8.7 Shear Viscosity and the Nearly Ideal Fluid of the Quark-Gluon Plasma

One of the defining empirical discoveries in heavy-ion collisions is that the QGP behaves as a nearly ideal fluid. In hydrodynamic modeling this is encoded in a small shear viscosity-to-entropy ratio η/s. Gauge-string duality motivates a conjectured lower bound,

$$\frac{\eta}{s} \geq \frac{1}{4\pi}, \tag{12.262}$$

which provides a useful reference scale, even though QCD is neither exactly conformal nor exactly in the strong-coupling limit captured by simple holographic models.

Lattice determinations of η are substantially more difficult than those of σ because (i) the Euclidean correlators of $T_{\mu\nu}$ are extremely noisy and (ii) the corresponding spectral function is constrained weakly by the available Euclidean data. In the pure-gauge theory, multilevel algorithms can dramatically improve the signal-to-noise ratio by exploiting locality of the action; extending comparable noise-reduction techniques to full QCD with dynamical fermions is an active area of algorithmic development.

An important conceptual point is that lattice results and phenomenological extractions address complementary aspects: lattice computations attempt a first-principles determination of Eq. (12.260), while Bayesian analyses infer effective transport coefficients within a modeling framework calibrated to data. Consistency between the two approaches, within systematic uncertainties, is a nontrivial cross-check of our understanding of strongly coupled QCD matter.

12.8.8 Quarkonia, Spectral Functions and in-Medium Potentials

The in-medium behavior of heavy quarkonia (charmonium and bottomonium) provides another bridge between lattice QCD and experiment. Three complementary strategies are commonly used:

- compute a temperature-dependent static $q\bar{q}$ potential (or free energy) on the lattice and solve a Schrödinger-type equation for the bound-state correlator;
- compute Euclidean temporal correlators for quarkonium operators and reconstruct spectral functions via Bayesian methods;
- analyze spatial screening correlators, which probe in-medium screening masses without being limited by the finite Euclidean time interval.

The potential-based approach is naturally embedded in effective field theories for heavy quarks. In particular, one can integrate out the heavy mass scale and organize corrections in powers of $1/m$, leading to nonrelativistic QCD (NRQCD) and its potential version (pNRQCD). A key physical refinement is that the effective in-medium potential can acquire an *imaginary part*, reflecting thermal widths due to gluo-dissociation and inelastic scattering. The real part encodes Debye screening.

These features explain why melting is not purely a static screening phenomenon but also depends on dynamical broadening.

The direct reconstruction of quarkonium spectral functions from Euclidean correlators is again an ill-posed inverse problem. MEM-based reconstructions can oversmooth, while BR-type reconstructions can produce ringing; therefore, disappearance temperatures for specific states should be interpreted as intervals bracketed by controlled systematics rather than as sharp thresholds. Improving temporal resolution (larger N_τ) and statistics is essential for approaching a Bayesian continuum limit in which method-dependent artifacts diminish.

12.8.9 From Well-Defined Euclidean Observables to Phenomenological Parameters

A recurrent challenge is the translation between phenomenological parameters and lattice-accessible observables. Heavy-quark diffusion can be related to κ through Eq. (12.261); similarly, jet quenching parameters such as $\hat{q}$ and $\hat{q}_L$ can be defined within effective theories that resum soft momentum scales and, in favorable cases, become amenable to lattice simulation. Another important set of quantities includes a temperature-dependent Debye screening mass $m_D(T)$ and effective couplings $\alpha_s(r, T)$ extracted from static correlators and in-medium potentials, which can be used as inputs for transport models.

Finally, the open-quantum-systems viewpoint provides a systematic language for quarkonium in a thermal environment: the heavy $Q\bar{Q}$ pair is treated as a subsystem interacting with a bath, leading to stochastic evolution equations in which the real and imaginary parts of the in-medium potential govern coherent and dissipative dynamics, respectively. A fully consistent implementation must incorporate dissipation and decoherence in a way compatible with thermalization.

▶ **Pause and think**
Do you think there are possible problems?

The long-term goal is a quantitatively controlled, multi-channel determination of real-time properties of QCD from Euclidean lattice data. Progress hinges on: (i) improved algorithms for noisy correlators (especially with dynamical fermions), (ii) combined lattice-perturbation matching with explicit uncertainty budgets, (iii) systematic use of sum rules and Ward identities, and (iv) extended temporal resolution to stabilize the $\omega \to 0$ limits in Eqs. (12.258) and (12.260).

12.9 Quark-Gluon Condensates

The problem of determining the properties of nuclear matter at extremely large densities and temperatures had already been posed before QCD was established as the fundamental theory of strong interactions. Once confinement and asymptotic freedom were understood, it became clear that hadronic matter subjected to extreme compression and/or heating may enter a regime in which the relevant degrees of freedom are no longer hadrons, but their constituents, namely quarks and gluons. The very term *quark-gluon plasma* was argued since such a state may form in heavy-ion collisions once the energy density of the colliding system reaches values of order

$$\varepsilon \sim 1 \text{ GeV}/\text{fm}^3. \tag{12.263}$$

It is well known that, with a few exceptions, ordinary matter is encountered in three aggregate states: solid, liquid, and gas. A *plasma*, i.e., a partially or completely ionized gas, is commonly called the *fourth aggregate state of a substance*. A plasma can be produced either by increasing the temperature (thermal ionization) or by irradiating matter with electromagnetic waves (photoionization) and/or by bombarding it with charged particles. In a classical (electromagnetic) plasma, molecules are dissociated into atoms, which then become ions; at sufficiently high temperature the electronic and nuclear components effectively decouple due to ionization.

The QCD analogue of this ionization picture emerges when the temperature and/or density of hadronic matter become so large that the typical kinetic energy in the system overcomes the relevant binding mechanisms. For atomic nuclei, the reference scale is set by the Coulomb barrier (of order 10–20 MeV), and by the onset of short-range nuclear forces. As the temperature increases and the mean interparticle distance decreases toward the femtometer range,

$$\ell \sim 1 \text{ fm}, \tag{12.264}$$

the electromagnetic component becomes relatively unimportant, while strong interaction effects dominate. In this regime, nuclei dissociate into nucleons; concomitantly, light hadrons are produced, mainly light mesons ($\pi, \rho, \omega, \ldots$) and, to a lesser extent, baryons. Once all nuclei are dissociated one obtains, to a good approximation, a hadronic gas representing a system of essentially colorless mesons and baryons. Its properties are determined by QCD. (If, hypothetically, the strong interaction were switched off, a gas of charged hadrons would reduce to an ordinary electromagnetic plasma.)

A useful conceptual bridge between electromagnetic and strong interaction plasmas is the notion of *liberated constituents*: in an electromagnetic plasma the carriers are ions and electrons, while in a QGP the carriers are quarks and gluons. In both cases, macroscopic thermodynamic control parameters (temperature and chemical potentials) determine whether bound composites or quasi-free constituents dominate the dynamics.

For a statistical description, it is natural to use the grand-canonical ensemble, where the thermodynamic state is controlled by the temperature T and by chemical potentials associated with conserved charges. For strongly interacting matter, baryon number plays a distinguished role. Accordingly, one introduces a *baryon chemical potential* μ_B to control the mean baryon density.[1] The grand potential Ω is related to the pressure P as

$$\Omega(T, \mu_B, V) = -P(T, \mu_B)\, V, \tag{12.265}$$

and baryon number density is obtained via

$$n_B(T, \mu_B) = \frac{1}{V}\langle N_B\rangle = \left(\frac{\partial P}{\partial \mu_B}\right)_T. \tag{12.266}$$

Increasing either T or μ_B (hence n_B) enhances the overlap among hadrons. When the typical hadron separation becomes comparable with the confinement radius, quarks in neighboring hadrons begin to interact. In an intermediate regime, hadronic clusters can appear; beyond it, a unified quark-gluon medium may form.

The resulting state is termed the QGP: a strongly interacting state of matter characterized by the (effective) absence of confinement. In the QGP, color-charged quarks and gluons are no longer trapped inside individual hadrons; rather, they propagate as quasi-free excitations throughout the volume. In this sense, the QGP resembles an electron-ion plasma, with the crucial difference that color, not electric charge, is the relevant gauge charge. In natural conditions, such a phase is expected to have existed in the early Universe during the first

$$t \sim 10^{-5}\ \text{s} \tag{12.267}$$

after the Big Bang. It is also not excluded that QGP-like conditions may occur in the cores of the most massive neutron stars, where densities greatly exceed nuclear saturation values.

[1] At the quark level, one often uses μ_q, with $\mu_B = 3\mu_q$ when all light quark flavors share the same chemical potential.

A convenient reference scale for baryon density is the nuclear saturation density,

$$\rho_0 \simeq 0.15 \text{ baryon/fm}^3, \tag{12.268}$$

which corresponds to ordinary nuclear matter. A possible phase diagram of hadronic matter can be typically organized into four broad regions:

- *nuclear matter* (moderate T and $\rho \sim \rho_0$),
- *hadronic gas* (small ρ and relatively low T),
- a *mixed phase* where hadronic and quark-gluon degrees of freedom coexist,
- *QGP* (high T and/or high ρ).

The precise location and nature of the boundaries depend on the underlying equation of state and on the relevant microscopic dynamics. Nevertheless, this schematic partition captures the physical content of the transition from confined hadrons to deconfined quarks and gluons.

At low densities (in particular $\rho \ll \rho_0$) and temperatures below a characteristic hadronic scale, the matter is well described as a hadronic gas. The hadronic-gas regime can be typically associated with small baryon density and $T \lesssim 70\,\text{MeV}$. In this domain, mesons dominate, and their bosonic statistics makes the thermodynamics qualitatively different from that of nucleons in nuclear matter. In particular, mesons do not form compact nuclear-like aggregates in the same way nucleons do, and quantum collective effects and nonperturbative interactions remain important, especially as one approaches $\rho \sim \rho_0$ where a naive ideal-gas picture breaks down.

The region of high temperatures ($T > T_c$) or large baryon densities ($\rho \gtrsim \rho_0$ and beyond) is, instead, associated with the QGP. At large baryon density the typical interparticle separation is small, and the QCD property of asymptotic freedom becomes operative: at sufficiently high momentum transfer the effective strong coupling decreases. A convenient way to express this is to note that, at high temperature, typical momentum transfers are of order

$$q \sim T, \tag{12.269}$$

so that increasing T drives the running coupling $\alpha_s(q)$ toward smaller values. In this sense, for $T \gg T_c$ one expects $\alpha_s \to 0$ and an increasingly weakly coupled quark-gluon medium, although the approach can be slow and is not uniform across all scales.

- Deconfinement is not an "absolute" switch in a strict mechanical sense. Rather, it is a change in the effective degrees of freedom: from colorless hadrons to color-charged quarks and gluons.

– The control parameters are thermodynamic: raising T and/or μ_B increases phase-space overlap and modifies screening properties, enabling deconfinement.

A particularly transparent signature of confinement versus deconfinement is provided by the static quark-antiquark potential $V(r)$, which may be viewed as the free energy cost of separating a heavy $Q\bar{Q}$ pair by a distance r.

We recall what we presented in Eq. (12.137), in Sect. 12.4.7, this time by different temperature domains.

In the confining regime ($T < T_c$) the potential rises approximately linearly at large distances,

$$V(r) \propto \sigma r \qquad (T < T_c), \tag{12.270}$$

where σ is the string tension, reflecting the formation of a color flux tube. By contrast, lattice calculations indicate that at deconfinement the potential ceases to grow linearly and instead saturates, consistent with color screening in the medium:

$$V(r) \to \text{const.} \qquad (T > T_c), \tag{12.271}$$

or, more explicitly, one may expect a screened (Yukawa-like) behavior at intermediate distances,

$$V(r) \sim -\frac{\alpha}{r} e^{-m_D r}, \qquad (T > T_c), \tag{12.272}$$

where m_D is a (Debye) screening mass and α is an effective coupling. The flattening of $V(r)$ encodes the fact that the medium can screen color charge, preventing the indefinite energy growth associated with separating a color pair in vacuum.

Finally, the confinement-deconfinement transition is intertwined with the restoration of chiral symmetry: in QCD with (light) quarks, chiral symmetry is spontaneously broken in the hadronic phase and tends to be restored in the high-T and/or high-density regime.

12.10 Looking for Quark-Gluon Plasma

Theoretical arguments supporting the existence of a QGP motivate a central experimental question: *how can one infer, from the observed hadronic final state, that an intermediate deconfined phase was formed?* The most direct strategy is to create, in the laboratory, matter at extreme temperature and/or baryon density by colliding heavy nuclei at ultrarelativistic energies. In a central nucleus-nucleus collision, a macroscopic number of nucleons participates and deposits energy in a small space-time volume. A rapidly evolving, strongly interacting medium is formed, which expands and cools, eventually hadronizing into the particles measured by detectors.

A crucial conceptual obstacle is that the QGP (if formed) exists only transiently, while experiments access mainly long-lived hadrons after freeze-out. Therefore, a reliable QGP search relies on *indirect* signatures: observables whose interpretation requires a dynamical understanding of (i) pre-equilibrium and equilibration, (ii) hydrodynamic expansion, and (iii) hadronization and hadronic rescattering. The best-case scenario is not a single "smoking gun", but a coherent pattern of mutually consistent signals.

- *Control knobs*: beam energy, system size, and centrality tune the initial energy density and baryon stopping.
- *Early-time probes*: hard partons, heavy quarks, quarkonia, and penetrating electromagnetic radiation (photons, dileptons) are produced early and carry information about the hottest stages.
- *Bulk probes*: hadron yields, spectra, and collective flow constrain the equation of state and transport properties of the medium.
- *Consistency requirement*: different probes must be described within a single dynamical scenario.

12.10.1 From Heavy-Ion Collisions to a Thermodynamic Medium

When two heavy nuclei collide at high energy, of order 10^2 (or more) nucleons from each nucleus can participate in the interaction. In a sufficiently central collision, the overlap region is compressed and heated, producing a dense fireball. A key working hypothesis is that, after a short equilibration time, the produced matter can be described thermodynamically by a temperature T and (possibly) a baryon chemical potential μ_B. This hypothesis is non-trivial: it requires that interactions are strong enough to establish local (approximate) equilibrium before the system dilutes through expansion.

A practical criterion for thermodynamic applicability is the separation of timescales:

$$\tau_{\text{eq}} \ll \tau_{\text{exp}}, \tag{12.273}$$

where τ_{eq} is an equilibration (or isotropization) time and τ_{exp} is an expansion time. If Eq. (12.273) holds, the subsequent evolution is well approximated by relativistic hydrodynamics, with an equation of state $P(\varepsilon, n_B, \ldots)$ and transport coefficients (shear and bulk viscosity, conductivity).

The *mean interaction time* must be short compared with the time over which macroscopic parameters vary appreciably. When this fails, one is in a pre-equilibrium regime where a kinetic description (Boltzmann-type) is more appropriate.

12.10.2 Space-Time Size of the Emitting Source

One of the earliest experimental handles on the produced matter is its spacetime "extension". Correlation measurements of identical particles, such as pion interferometry (Hanbury Brown-Twiss, HBT), provide access to an effective source size.

A simple estimate highlights that the region of secondary particle emission in e^+e^- or pp interactions has a characteristic size of order 1 fm, while in heavy-ion collisions it can reach

$$R \sim 7 - 8 \text{ fm}, \tag{12.274}$$

exceeding the confinement radius and enabling, at least geometrically, a deconfined region to form.

Physically, the enhancement of R in nucleus-nucleus collisions reflects the superposition of many elementary interactions and the collective expansion of an extended medium. While HBT alone does not prove deconfinement, it provides essential baseline information on the macroscopic scale of the fireball.

12.10.3 Thermal Spectra and the Role of Rapid Expansion

A widely used diagnostic is the approximate exponential behavior of single-particle momentum spectra. For a thermalized source, the distribution of transverse momentum or transverse mass often exhibits an approximately Boltzmann-like form, schematically,

$$\frac{dN}{d^3p} \propto \exp\left(-\frac{E}{T}\right), \tag{12.275}$$

or, for midrapidity spectra in a simplified treatment,

$$\frac{dN}{dE} \sim \exp\left(-\frac{E}{T}\right). \tag{12.276}$$

This motivates the extraction of an effective slope parameter, commonly interpreted as a temperature.

However, it is remarkable to stress an important caveat: *heavy-ion collisions are characterized by rapid collective expansion, so that the observed slope is not a pure thermometer.* Flow "blue-shifts" spectra and generates mass-dependent hardening.

A more faithful interpretation requires a combined analysis of spectra and flow observables, within a hydrodynamic or blast-wave framework. Hence, exponential spectra are *consistent* with thermalization but not *unique evidence* for it.

12.10.4 Penetrating Probes and Electromagnetic Windows

Photons and dileptons are emitted throughout the evolution and interact only electromagnetically, thus they can escape with relatively little final-state interaction. Prompt photons are produced in elementary quark-gluon scatterings such as

$$qg \to q\gamma, \qquad q\bar{q} \to g\gamma, \tag{12.277}$$

and therefore constitute an early-time diagnostic of the partonic phase. In practice, disentangling these contributions from late hadronic sources and decay photons is experimentally challenging, but conceptually they provide a direct connection to deconfined degrees of freedom.

12.10.5 Quarkonium Suppression and Color Screening

One of the historically most discussed QGP signatures is the suppression of quarkonium states (notably J/ψ) relative to expectations from proton-nucleus and proton-proton baselines. The physical mechanism is Debye color screening in a deconfined medium: if the screening length becomes smaller than the quarkonium Bohr radius, the bound state cannot form or is dissolved. This is the QCD analogue of plasma screening in electrodynamics.

Let us recall that J/ψ suppression was observed at the CERN SPS (1989–1990) in oxygen-uranium and sulphur-uranium collisions at 200 GeV per nucleon, in a way incompatible with an extrapolation from proton-nucleus data. A compact theoretical representation shows J/ψ production versus the product of nucleon numbers $A_1 A_2$, where proton-nucleus data extrapolate smoothly while nucleus-nucleus collisions deviate, indicating an additional suppression mechanism in the dense medium.

From a modeling standpoint, it is useful to define a nuclear modification factor

$$R_{AA} = \frac{1}{\langle N_{\rm coll}\rangle} \frac{dN_{AA}/dy}{dN_{pp}/dy}, \tag{12.278}$$

where $\langle N_{\rm coll}\rangle$ is the mean number of binary nucleon-nucleon collisions. Values $R_{AA} < 1$ for quarkonia suggest suppression beyond trivial scaling. A complete interpretation must also incorporate cold-nuclear-matter effects and regeneration mechanisms at high energies; nevertheless, screening in a deconfined medium provides a natural microscopic origin.

A qualitative dissolution condition is

$$m_D^{-1}(T) \lesssim r_{\text{bind}}, \tag{12.279}$$

where m_D is a Debye screening mass and r_{bind} is the typical bound-state radius. Equation (12.279) encapsulates why quarkonium suppression is sensitive to the hottest stages of the collision.

12.10.6 Energy Density Estimates and the Bjorken Argument

A necessary (though not sufficient) condition for QGP formation is that the initial energy density exceeds the critical scale inferred from QCD thermodynamics.

The use of the Bjorken estimate for the energy density is based on measured transverse energy per unit rapidity in a boost-invariant picture, namely

$$\varepsilon(\tau_{\text{form}}) \simeq \frac{1}{A\,\tau_{\text{form}}} \left.\frac{dE_T}{dy}\right|_{y\simeq 0}, \tag{12.280}$$

where A is the transverse overlap area of the colliding nuclei and τ_{form} is a characteristic formation time.

In our notation, the overlap area is represented by A_{form} and one writes

$$\varepsilon(\tau_{\text{form}}) = \frac{dE_T/dy}{A_{\text{form}}\,\tau_{\text{form}}}. \tag{12.281}$$

A common estimate for A_{form} is based on the nuclear radius, $R \simeq r_0 A^{1/3}$ with $r_0 \simeq 1.2$ fm, so that $A_{\text{form}} \sim \pi R^2$ for central events.

The formation time τ_{form} is not directly measurable and is therefore the largest theoretical uncertainty in Eq. (12.281). An estimate for a typical transverse mass m_T is:

$$\tau_{\text{form}} = \frac{2\pi}{m_T}. \tag{12.282}$$

Using Eq. (12.282) with representative values of m_T yields formation times shorter at higher energies, hence larger initial energy densities for fixed dE_T/dy and geometry.

Plugging in experimental numbers, one obtains illustrative values for ε:

- SPS: $\varepsilon(\tau_{\text{form}}) \sim 1.5\ \text{GeV/fm}^3$ (for AGS energy) and $\sim 2.9\ \text{GeV/fm}^3$ (for SPS),
- RHIC: $\varepsilon(\tau_{\text{form}}) \sim 5.4\ \text{GeV/fm}^3$,

with the important remark that these are upper estimates because Eq. (12.281) is evaluated at early times, before full equilibration. It is also possible to report a thermalized estimate

$$\varepsilon(\tau_{\text{therm}}) \sim 5.4\ \text{GeV/fm}^3, \tag{12.283}$$

indicating that the energy density remains well above the canonical QGP threshold even after allowing for hydrodynamic expansion.

12.10.7 Baryon Stopping Power and Net-Baryon Rapidity Loss

To chart the QCD phase diagram, one must control not only energy density but also baryon density (or, equivalently, μ_B). Interestingly, the concept of *stopping power* may be used to quantify the longitudinal momentum loss of incoming baryons and thereby controls how much baryon number is transported to midrapidity.

Operationally, stopping is defined in terms of the average rapidity shift between the beam rapidity y_b and the mean rapidity of net baryons. Denoting the rapidity distribution of net baryons by dN/dy, one defines

$$\delta y = y_b - \langle y \rangle, \qquad \langle y \rangle = \frac{\int_0^{y_b} y \frac{dN}{dy} dy}{\int_0^{y_b} \frac{dN}{dy} dy}. \tag{12.284}$$

Equation (12.284) captures the idea that larger stopping corresponds to a larger baryon density around midrapidity, hence to larger μ_B. This becomes particularly relevant for dedicated programs aiming at high baryon density and moderate temperature.

12.10.8 Polyakov Loop and Critical Behavior

Beyond phenomenology, a central theoretical task is to characterize the confinement-deconfinement transition and to determine its nature (first order, second order, or crossover) as a function of baryon chemical potential. We can discuss lattice QCD and the Polyakov loop as an order parameter in the pure gauge theory.

The Polyakov loop is defined as the trace of the thermal Wilson line winding around Euclidean time:

$$P(\mathbf{x}) = \text{Sp} \prod_{n=1}^{N_\tau} U(\mathbf{x}, n + \hat{t}), \tag{12.285}$$

where $U(\mathbf{x}, n + \hat{t})$ is the link variable connecting a site to its neighbor in Euclidean time. In the continuum limit, the link variable is represented as

$$U(\mathbf{x}, n + \hat{t}) = \exp\left\{iag_0\frac{\lambda^a}{2}\,G_0^a(\mathbf{x}, n)\right\}, \qquad (12.286)$$

with lattice spacing a, bare coupling g_0, and gauge field G_0^a.

In the confining phase, the expectation value $\langle P\rangle$ vanishes in the infinite-volume limit (center symmetry unbroken), while in the deconfined phase it becomes nonzero. This can be linked to the free energy F_Q of a static quark:

$$\langle P\rangle \sim \exp\bigl(-\beta F_Q\bigr), \qquad (12.287)$$

so that $\langle P\rangle = 0$ corresponds to $F_Q \to \infty$ (confinement), whereas $\langle P\rangle \neq 0$ implies finite F_Q (deconfinement). The correlator of two Polyakov loops encodes the free energy $F(R)$ of a static quark-antiquark pair separated by a distance R:

$$\langle P(0)\,P^\dagger(R)\rangle \sim \exp(-\beta F(R)). \qquad (12.288)$$

In the confining phase, $F(R)$ grows with R, while in the deconfined phase it saturates due to screening.

It is also possible to introduce a lattice susceptibility associated with Polyakov loop fluctuations,

$$\chi_P = \langle \bar{P}^2\rangle - \langle \bar{P}\rangle^2, \qquad (12.289)$$

and notes that in a finite volume χ_P exhibits a peak near the transition. The scaling of the peak height with lattice size diagnoses the order of the transition. In particular, for a second-order transition the peak height grows with system size, while for a crossover the behavior is smoother.

The phase structure in the (T, μ_B) plane can be used as a theoretical framework for experimental scans in collision energy aimed at detecting critical behavior via fluctuations and correlations.

▶ **Pause and think**
Think why order parameters matter for experiments.

Even if the Polyakov loop is not directly measurable, it organizes the theoretical understanding of the transition. Experimental observables such as fluctuations of conserved charges (baryon number, strangeness, electric charge) are sensitive to the vicinity of a critical point through growing correlation lengths and non-Gaussian cumulants, thereby providing an indirect bridge between lattice QCD thermodynamics and heavy-ion phenomenology.

12.11 End of Our Journey

In this concluding section we revisit the architecture of fundamental interactions as codified in Table 12.7. The four sectors of Nature that have stood the test of increasingly precise experimental scrutiny, namely gravity, electromagnetism, the weak interaction, and the strong interaction, are each characterized by a unique gauge structure, a definable mediator spectrum, and observable consequences across disparate energy and length scales.

At the core of theoretical physics is the interplay between symmetries and dynamics: the long-range sectors arise from massless gauge quanta and exact (or nearly exact) symmetries, whereas short-range behavior is tied to spontaneous symmetry breaking or confinement. Table 12.7 collects these principles into a comparative framework that highlights both the universality and specificity of each interaction.

When viewed through the lens of modern effective field theory, each interaction can be associated with a leading effective Lagrangian density,

$$\mathcal{L}_{\text{eff}} = \mathcal{L}_{\text{gauge}} + \mathcal{L}_{\text{matter}} + \sum_{d>4} \frac{c^{(d)}}{\Lambda^{d-4}} \mathcal{O}^{(d)}, \tag{12.290}$$

where $\mathcal{L}_{\text{gauge}}$ encodes the renormalizable gauge dynamics, $\mathcal{L}_{\text{matter}}$ the interaction with matter fields, and the series of higher-dimensional operators $\mathcal{O}^{(d)}$ encapsulates subleading effects suppressed by an ultraviolet scale Λ. For each interaction in Table 12.7, Eq. (12.290) acts as the unifying scaffold that connects symmetry principles to observables.

12.11.1 Is There a Fifth Force?

The hypothesis of a fifth fundamental interaction is defined operationally as any long- or intermediate-range force that is not contained within the standard model gauge group or the geometric description of gravity in general relativity, and that survives all known renormalization and screening effects to impart an observable influence on matter. Because the existing four interactions already span a wide phenomenological landscape, any viable fifth force must either be extremely feeble, effectively short-ranged, or hidden via screening mechanisms.

A common theoretical template for such additional interactions is provided by a Yukawa-like modification of Newtonian gravity, which takes the form

$$V(r) = -\frac{G\, m_1 m_2}{r} \left[1 + \alpha\, e^{-r/\lambda}\right], \tag{12.291}$$

where α denotes a dimensionless coupling constant relative to standard gravity, and λ is the characteristic range. The Yukawa form in Eq. (12.291) arises generically when a new boson of mass $m_\phi = \hbar/(\lambda c)$ mediates an additional interaction between two point-like sources. Deviations from the pure Newtonian inverse-square law are then parameterized by α and λ.

A sensitive diagnostic for departures from universal free fall is the Eötvös parameter, defined for two different test bodies A and B as

$$\eta_{AB} \equiv 2\,\frac{a_A - a_B}{a_A + a_B}, \tag{12.292}$$

where $a_{A,B}$ are the measured accelerations of the bodies in the same external field. A nonzero value of η_{AB} would signal a violation of the Weak Equivalence Principle and, by extension, the presence of a fifth force acting with differential coupling to the composition of the test bodies.

- **Universality of free fall tests**. Precision torsion-balance experiments constrain η_{AB} to parts in 10^{-15} for carefully chosen materials, imposing stringent upper limits on any composition-dependent fifth force that couples to baryon number or other charges distinct from inertial mass.
- **Inverse-square law at short distances**. Laboratory probes that measure forces down to micron scales place bounds on $|\alpha|$ at the level of 10^{-2} âŁ“ 10^{-4} for ranges $\lambda \sim 10^{-6}$ mâŁ“10^{-3} m. These experiments effectively exclude large-coupling, short-range fifth-force components in this window.
- **Atomic precision measurements**. Techniques based on atomic spectroscopy or atom interferometry have emerged as complementary approaches. Frequency shifts and phase anomalies in such systems can place competitive constraints on light mediators that couple to electrons, nuclei, or combinations thereof.

Modern theoretical models accommodate fifth-force candidates in a variety of contexts: light scalar or vector fields with extremely small couplings to Standard Model states; "dark sector" mediators that communicate with ordinary matter via portal interactions; and screened scalar fields whose effective couplings become environment-dependent. In each case, the relevant phenomenology is captured either by a parameterized potential such as Eq. (12.291) or by effective operators of the form

$$\mathcal{L}_{\text{contact}} \supset \frac{g^2}{m_X^2}\,(\bar{\psi}\Gamma\psi)(\bar{\psi}\Gamma\psi), \tag{12.293}$$

where m_X is the mass of the mediator and g its coupling to matter fields ψ.

Table 12.7 Fundamental interactions and representative properties at (approximately) hadronic scales

Interaction	Gauge structure	Mediator	Spin	Range [m]	Rel. strength[†]	Distinguishing features
Gravity	(effective) diffeomorphism invariance	graviton (hyp.)	2	∞	$\sim 10^{-40}$	Universal coupling to the stress-energy tensor. Classical limit described by general relativity. Non-renormalizable in perturbation theory but predictive as an effective field theory below the Planck scale. Long-range due to a massless mediator
Electromag.	$U(1)_{\text{EM}}$	photon γ	1	∞	1/137	Abelian gauge theory. Long-range and unscreened in vacuum. Weakly coupled at laboratory energies. Sources are conserved electric currents. Mild running of α at hadronic scales
Weak	$SU(2)_L \times U(1)_Y$ (spontaneously broken)	$W^{\pm}$, Z^0	1	$\sim 10^{-17}$	$\sim 10^{-5}$	Short-range interaction set by M_W^{-1} and M_Z^{-1}. Chiral couplings to left-handed doublets. Low-energy limit described by four-fermion operators with Fermi constant G_F. Maximal parity violation in charged currents
Strong	$SU(3)_C$	gluons g^a	1	$\sim 10^{-15}$ (residual nuclear)	~ 1	Non-Abelian gauge theory with boson self-interactions. Asymptotic freedom at large momentum transfer and confinement at long distances. Intrinsically non-perturbative for $\mu \sim 1$ GeV with $\alpha_s = O(1)$. Long-distance physics governed by color-singlet hadrons and residual nuclear forces

[†] The "relative strength" is an order-of-magnitude comparison at hadronic scales and is only indicative. Gravity and electromagnetism are long-range because their mediators are effectively massless. The weak range is set by M_W and M_Z. The strong interaction length refers to the characteristic hadronic/nuclear scale relevant for residual interactions among color-singlet states

Though no definitive evidence for a fifth fundamental interaction has yet been obtained, the multi-pronged strategy delineated above remains a cornerstone of precision frontier physics. Prospective discoveries would not only expand Table 12.7 but also compel a structural rethinking of the symmetries that underlie our understanding of Nature.

Pause and think
Try to understand why cosmology has today become a concrete arena for discovering new physics and, in particular, possible evidence for a fifth force.

12.12 Proposed Exercises

Exercise 1

- Starting from the one-loop renormalization-group equation for QCD,

$$\mu \frac{d\alpha_s}{d\mu} = -\beta_0 \frac{\alpha_s^2}{2\pi}, \qquad \beta_0 = 11 - \frac{2}{3}n_f,$$

  derive the standard one-loop running coupling $\alpha_s(\mu)$ in terms of a reference scale Λ_{QCD}.
- Using your result, show that $\alpha_s(\mu) \to 0$ as $\mu \to \infty$ for $\beta_0 > 0$, and explain why this condition is satisfied in QCD for phenomenologically relevant n_f.
- Estimate $\alpha_s(\mu)$ at $\mu = 10$ GeV for $n_f = 5$ and $\Lambda_{\mathrm{QCD}} = 200$ MeV.

Exercise 2

- Consider a rectangular Wilson loop $W(R, T)$ of spatial length R and Euclidean time amplitude T in a confining gauge theory. Show that the large-T behavior can be parametrized as

$$\langle W(R, T)\rangle \sim \exp[-V(R)\, T], \qquad (T \to \infty),$$

  and interpret $V(R)$ as a static interquark potential.
- Assume an area law,

$$\langle W(R, T)\rangle \sim \exp[-\sigma\, R\, T],$$

  with string tension $\sigma > 0$. Use the above equations to infer the asymptotic form of $V(R)$ at large R.

– The phenomenological Cornell form is reported in Eq. (12.137). Explain which term is associated with confinement and which term encodes short-distance physics. Discuss briefly why lattice simulations are essential to determine σ from first principles.

Exercise 3

– In boost-invariant longitudinal expansion, the Bjorken estimate for the initial energy density at proper time τ_0 can be written as

$$\epsilon(\tau_0) \simeq \frac{1}{A_T\,\tau_0}\,\frac{dE_T}{dy},$$

where A_T is the effective transverse overlap area and dE_T/dy is the transverse energy per unit rapidity around midrapidity. Derive the above equation from energy conservation in a slab of width $d\eta$ at fixed proper time.
– Take $A_T = 150$ fm^2, $\tau_0 = 1$ fm, and $dE_T/dy = 600$ GeV. Estimate $\epsilon(\tau_0)$ in GeV/fm^3.
– Assuming, for an order-of-magnitude estimate, an ultrarelativistic equation of state $\epsilon = a\,T^4$ with $a = (\pi^2/30)\,g_\star$, discuss how lattice-QCD deviations from the Stefan-Boltzmann limit modify the inference of T from ϵ in the experimentally relevant range near a few times the crossover temperature.

Essential Bibliography

- Ellis RK, Stirling WJ, Webber BR (1996) QCD and collider physics. Cambridge University Press
  The standard reference for perturbative QCD. Provides a systematic treatment of factorization, parton distributions, renormalisation, infrared safety, and jet physics. Essential for understanding how asymptotic freedom translates into precision collider predictions.
- Muta T (2010) Foundations of quantum chromodynamics, 3rd edn. World Scientific
  A detailed field-theoretic exposition of QCD, including renormalisation, running coupling, anomalies, and the non-Abelian gauge structure. Particularly valuable for clarifying the conceptual differences between QED and QCD and for understanding color dynamics from first principles.
- Manohar AV, Wise MB (2000) Heavy quark physics. Cambridge University Press
  A focused account of heavy-quark symmetry, effective field theories, and the separation of scales in QCD. Useful for connecting spectroscopy, flavor physics, and systematic expansions in inverse heavy-quark mass.

Index

O. Luongo, *Fundamental Interactions*,
https://doi.org/10.1007/978-3-032-27413-7

FSC
www.fsc.org
MIX
Papier | Fördert
gute Waldnutzung
FSC® C083411

Kolibri 360 GmbH
Ferdinand-Jühlke-Straße 7
99095 Erfurt, Deutschland
produktsicherheit@kolibri360.de